Scaffolding in Tissue Engineering

Scaffolding in Tissue Engineering

Edited by

Peter X. Ma
Jennifer Elisseeff

CRC Taylor & Francis
Taylor & Francis Group
Boca Raton London New York

A CRC title, part of the Taylor & Francis imprint, a member of the Taylor & Francis Group, the academic division of T&F Informa plc.

Published in 2006 by
CRC Press
Taylor & Francis Group
6000 Broken Sound Parkway NW, Suite 300
Boca Raton, FL 33487-2742

Printed in the United States of America on acid-free paper
10 9 8 7 6 5 4 3 2 1

International Standard Book Number-10: 1-57444-521-9 (Hardcover)
International Standard Book Number-13: 978-1-57444-521-3 (Hardcover)
Library of Congress Card Number 2005041906

Library of Congress Cataloging-in-Publication Data

Scaffolding in tissue engineering / edited by Peter X. Ma and Jennifer H. Elisseeff.
p. cm.
Includes bibliographical references and index.
ISBN 1-57444-521-9
1. Tissue engineering. 2. Polymers in medicine. 3. Scaffolding. I. Ma, Peter X. II. Elisseeff, Jennifer H.

R857.T55S366 2005
610'28--dc22 2005041906

T&F informa

Taylor & Francis Group
is the Academic Division of T&F Informa plc.

Visit the Taylor & Francis Web site at
http://www.taylorandfrancis.com

and the CRC Press Web site at
http://www.crcpress.com

Preface

Transplantation provides hope for many patients with tissue loss or organ failure, but inherent limitations such as donor shortage and the required immunosuppression therapy seriously hinder the potential benefits that such an approach can provide for many patients. The problems of current transplantation therapy have stimulated research for alternative solutions. An emerging interdisciplinary and multidisciplinary field aims to recreate biologically functional tissues and organs. This field is either called tissue engineering or regenerative medicine, depending on who the researchers are and what their focus is. Man-made materials, not necessarily originally designed for body part replacements, are being used as implants and prostheses. These foreign materials' lack of biological function imposes serious health risks in patients.

Scaffolds play a critical role in tissue engineering by acting as temporary artificial extracellular matrices for cell accommodation, proliferation, and differentiated function as well as serving as three dimensional templates for neotissue/organ formation. The number of academic and industrial groups studying tissue engineering is rapidly increasing, leading to more journal articles on scaffolding and tissue engineering. While there are books addressing aspects of tissue engineering, there are no current, comprehensive texts focusing on scaffolding for tissue engineering. As the knowledge in scaffolding design increases with the development of more tissue engineering research programs, there is increased need for a comprehensive book focusing on scaffolding for tissue engineering.

This book, *Scaffolding in Tissue Engineering*, is intended to address the principles, methods, and applications of a broad range of tissue engineering scaffolds. We review the general principles of tissue engineering and provide an in-depth exploration of traditional and novel materials, three-dimensional scaffold design and fabrication technologies, important structural and functional scaffold modification, and various tissue engineering applications of scaffolds. While providing a comprehensive summary of the current knowledge and technologies in scaffolding design, we hope to provide insight into new trends and directions for scaffold development and an ever expanding range of tissue engineering applications. Therefore, this book is not only intended to be a textbook for advanced undergraduate and graduate students, but also a review and reference book for researchers in materials science, biomaterials, tissue engineering, biotechnology, biomedical engineering, biomedical sciences, and clinical laboratories.

Because of the multidisciplinary nature of the subject matter involved in this book, a large number of experts in engineering, life sciences, and clinical medicine with various tissue engineering expertise were invited to contribute. Without the participation of such a diverse group of people, we would not have been able to accomplish our goal of developing a comprehensive book on scaffolding in tissue engineering.

Peter X. Ma, Ph.D.
University of Michigan
Ann Arbor, Michigan, USA

Jennifer H. Elisseeff, Ph.D.
The Johns Hopkins University
Baltimore, Maryland, USA

Contributors

Adel Alhadlaq
Tissue Engineering Laboratory
University of Illinois
Chicago, Illinois

Kristi S. Anseth
Department of Chemical Engineering
University of Colorado at Boulder
Boulder, Colorado

Orasa Anusaksathien
Department of Periodontics/
Prevention/Geriatrics and Center
for Craniofacial Regeneration
School of Dentistry
University of Michigan
Ann Arbor, Michigan

Anthony Atala
Laboratory for Tissue Engineering
and Cellular Therapeutics
Department of Urology
Children's Hospital
Harvard Medical School
Boston, Massachusetts

John Barr
A.P. Pharma
Redwood City, California

Brian C. Baxter
A.P. Pharma
Redwood City, California

Sangeeta N. Bhatia
Department of Bioengineering
University of California at San Diego
La Jolla, California

Stephanie J. Bryant
University of Colorado at Boulder
Boulder, Colorado

David J. Carlsson
University of Ottawa Eye Institute
Ottawa, Ontario, Canada

Thomas Ming Swi Chang
Artificial Cells & Organs Research Center
MSSS-FRSQ Research Group on Blood
Substitutes in Transfusion Medicine
McGill University
Montreal, Quebec, Canada

Victor J. Chen
Department of Biomedical Engineering
University of Michigan
Ann Arbor, Michigan

Tien-Min G. Chu
Biomedical Engineering Program
School of Engineering and Technology
Department of Orthopaedic Surgery
School of Medicine
Indiana University, Purdue University
Indianapolis, Indiana

Clark K. Colton
Department of Chemical Engineering
Massachusetts of Technology
Boston, Massachusetts

Fu-Zhai Cui
Department of Materials Science
and Engineering
Tsinghua University
P.R. China

Klaas de Groot
Biomaterials Research Group
Leiden University
The Netherlands

Abraham J. Domb
Hebrew University of Jerusalem
School of Pharmacy
Jerusalem, Israel

Chang Du
Biomaterials Research Group
Leiden University
The Netherlands

Tirtsa Ehrenfreund-Kleinman
Hebrew University of Jerusalem
Jerusalem, Israel

Jennifer H. Elisseeff
Department of Biomedical Engineering
Whitaker Biomedical Engineering Institute
The Johns Hopkins University
Baltimore, Maryland

William V. Giannobile
Department of Periodontics/
Prevention/Geriatrics and
Center for Craniofacial Regeneration
School of Dentistry
University of Michigan
Ann Arbor, Michigan

Thomas J. Gill
Division of Sports Medicine
Department of Orthopaedic Surgery
Massachusetts General Hospital
Boston, Massachusetts

Jacob Golenser
Hebrew University of Jerusalem
Jerusalem, Israel

Erin Grassl
University of Minnesota
Minneapolis, Minnesota

May Griffith
University of Ottawa Eye Institute
Ottawa, Ontario, Canada

Changming Guo
Department of Orthopaedic Surgery
Singapore General Hospital
Singapore

Jorge Heller
A.P. Pharma
Redwood City, California

Theresa A. Holland
Rice University
Houston, Texas

Jeffrey O. Hollinger
Biomedical Engineering and Biological Sciences
Bone Tissue Engineering Center
Carnegie-Mellon University
Pittsburgh, Pennsylvania

Yen-Chen Huang
Department of Biomedical Engineering
University of Michigan
Ann Arbor, Michigan

Johnny Huard
Department of Molecular Genetics and Biochemistry
University of Pittsburgh
Pittsburgh, Pennsylvania

Xingyu Jiang
Harvard University
Cambridge, Massachusetts

Qi-ming Jin
Department of Periodontics/Prevention/
Geriatrics and Center for Craniofacial Regeneration
School of Dentistry
University of Michigan
Ann Arbor, Michigan

Yusuf Khan
School of Biomedical Engineering, Science and Health Systems
Drexel University
Philadelphia, Pennsylvania

Robert Langer
Chemical and Biomedical Engineering
Massachusetts Institute of Technology
Cambridge, Massachusetts

Cato T. Laurencin
Center for Advanced Biomaterials and Tissue Engineering
Department of Chemical Engineering
Drexel University
Philadelphia, Pennsylvania

Erin Lavik
Chemical and Biomedical Engineering
Massachusetts Institute of Technology
Cambridge, Massachusetts

Sang Cheon Lee
Departments of Pharmaceutics and Biomedical Engineering
Purdue University
West Lafayette, Indiana

Jessamine Ng Lee
Harvard University
Cambridge, Massachusetts

Kam W. Leong
Department of Biomedical Engineering
The Johns Hopkins University School of Medicine
Baltimore, Maryland

Amy S. Lewis
Department of Chemical Engineering
Massachusetts of Technology
Boston, Massachusetts

Fengfu Li
University of Ottawa Eye Institute
Ottawa, Ontario, Canada

Suming Li
Research Center on Artificial Biopolymers
CNRS
Faculty of Pharmacy
University of Montpellier
Montpellier, France

Chris Lohmann
Universitäts-Augenklinik
University of Regensburg
Regensburg, Germany

Peter X. Ma
Department of Biologic & Materials Science
Department of Biomedical Engineering
Macromolecular Science and Engineering Center
University of Michigan
Ann Arbor, Michigan

Jeremy J. Mao
Tissue Engineering Laboratory
University of Illinois
Chicago, Illinois

Kacey G. Marra
Departments of Surgery and Bioengineering
University of Pittsburgh
Pittsburgh, Pennsylvania

Antonios G. Mikos
Institute of Biosciences and Bioengineering
Rice University
Houston, Texas

David J. Mooney
Division of Engineering and Applied Sciences
Harvard University
Cambridge, Massachusetts

Mark P. Mooney
Departments of Oral Medicine and Pathology, Anthropology, Surgery, and Orthodontics
University of Pittsburgh
Pittsburgh, Pennsylvania

Priyabrata Mukherjee
Department of Biochemistry and Molecular Biology
Mayo Clinic Cancer Center
Rochester, Minnesota

Robert M. Nerem
Parker H. Petit Institute for Bioengineering and Bioscience
Georgia Institute of Technology
Atlanta, Georgia

Steve Y. Ng
A.P. Pharma
Redwood City, California

Bojana Obradovic
Department of Chemical Engineering
Faculty of Technology and Metallurgy
University of Belgrade
Belgrade, Yugoslavia

Kinam Park
Departments of Pharmaceutics and Biomedical Engineering
School of Pharmacy
Purdue University
West Lafayette, Indiana

Parul Natvar Patel
Laboratory of Reparative Biology & Bioengineering
Department of Plastic Surgery
The University of Texas M.D. Anderson Cancer Center
Houston, Texas

Charles W. Patrick, Jr.
Laboratory of Reparative Biology & Bioengineering
Department of Plastic Surgery
The University of Texas M.D. Anderson Cancer Center
Houston, Texas

Victor Prisk
Growth and Development Laboratory
Department of Orthopaedic Surgery
Children's Hospital of Pittsburgh
University of Pittsburgh
Pittsburgh, Pennsylvania

Milica Radisic
Massachusetts Institute of Technology
Harvard-MIT Division of Health Sciences and Technology
Cambridge, Massachusetts

Mark A. Randolph
Division of Plastic Surgery
Departments of Surgery and Orthopaedic Surgery
Massachusetts General Hospital
Boston, Massachusetts

Shaneen L. Rowe
Department of Biomedical Engineering
Rensselaer Polytechnic Institute
Troy, New York

Aliasger K. Salem
Department of Biomedical Engineering
The Johns Hopkins University School of Medicine
Baltimore, Maryland

Devang T. Shah
A.P. Pharama
Redwood City, California

Heather D. Sheardown
Department of Chemical Engineering
McMaster University
Hamilton, Ontario, Canada

Hui Rong Shen
A.P. Pharma
Redwood City, California

Shigeto Shimmura
Cornea Centre
Tokyo Dental College
Ichikawa General Hospital
Ichikawa, Chiba, Japan

Lisa Spirio
3-DMatrix, Inc.
Cambridge, Massachusetts

Myron Spector
Orthopaedic Research Laboratory
Brigham and Women's Hospital
Harvard Medical School
Boston, Massachusetts

Jan P. Stegemann
Department of Biomedical Engineering
Rensselaer Polytechnic Institute
Troy, New York

Yasuhiko Tabata
Department of Biomaterials, Field of Tissue Engineering
Institute for Frontier Medical Sciences
Kyoto University
Kyoto, Japan

Joerg K.V. Tessmar
University of Regensburg
Regensburg, Germany

Robert T. Tranquillo
Department of Chemical Engineering and Materials Science
University of Minnesota
Minneapolis, Minnesota

Valerie Liu Tsang
Department of Bioengineering
University of California, San Diego
La Jolla, California

Gordana Vunjak-Novakovic
Division of Health Sciences and Technology
Massachusetts of Technology
Cambridge, Massachusetts

Guobao Wei
Department of Biomedical Engineering
University of Michigan
Ann Arbor, Michigan

Jennifer L. West
Department of Bioengineering
Rice University
Houston, Texas

George M. Whitesides
Department of Chemistry and Chemical Biology
Harvard University
Cambridge, Massachusetts

Christopher G. Williams
Division of Plastic and Reconstructive Surgery
The Johns Hopkins University
Baltimore, Maryland

Ioannis V. Yannas
Fibers and Polymers Laboratory
Department of Mechanical Engineering
Massachusetts Institute of Technology
Cambridge, Massachusetts

Yoon Yeo
Departments of Pharmaceutics and Biomedical Engineering
Purdue University
West Lafayette, Indiana

Shuguang Zhang
Center for Biomedical Engineering
Massachusetts Institute of Technology
Cambridge, Massachusetts

Xiaojun Zhao
Center for Biomedical Engineering
Massachusetts Institute of Technology
Cambridge, Massachusetts

Beth Anne Zielinski
Bio Med Molecular Pharmacology, Physiology, and Biotechnology
Brown University
Providence, Rhode Island

Table of Contents

Part I

Scaffolding Materials

1 Biologically Active Scaffolds Based on Collagen–GAG Copolymers

Ioannis V. Yannas

CONTENTS

I. INTRODUCTION

Biologically active scaffolds are highly porous materials based on simple analogs of the extracellular matrix (ECM) that have induced synthesis of tissues and organs. Their activity has consisted so far in induction of complex synthetic processes, typically in anatomical sites where a mass of a tissue or organ has been recently deleted due to accidental trauma or surgical intervention. The results obtained so far strongly suggest that induced regeneration is the strategy of choice in several pathological situations that are currently treated with organ transplantation.

Scaffolds based on ECM analogs are characterized by a very high pore volume fraction, typically 0.90 to 0.95 or even higher, and an average pore diameter in the range of 5 to 500 μm. Their chemical composition has so far consisted of type I and II collagens as well as graft copolymers of collagen and a glycosaminoglycan (GAG). They are synthesized as highly porous graft polymers or copolymers using a process in which an aqueous suspension of the polymer, or copolymer, is freeze-dried under conditions that lead to the sublimation of ice crystals and the consequent formation of the desired pore structure, prior to chemical grafting of one of the polymers on the other. They can be sterilized and used as surgical devices for implantation. The majority of ECM analogs synthesized to date have been graft copolymers of collagen and a GAG; they will be referred to below as collagen–GAG (CG) scaffolds.

Although a very large number of scaffolds with structural characteristics similar to those described above have been prepared, only a few have shown dramatic biological activity. The few scaffolds that have displayed biological activity have induced regeneration of tissues and organs that do not regenerate spontaneously and have been referred to as *regeneration templates*. Since regeneration is an instance of induced *in vivo* synthesis, the activity of scaffolds has been described in terms of the quality of products induced to regenerate in the presence of these scaffolds. Their unique biological activity appears to lie primarily in their ability to regulate cell function through specific cell–matrix interactions involving integrins, mostly those of fibroblasts and myofibroblasts, and ligands on the matrix surface. Activity is conferred by chemical composition, which determines the *identity of ligands*; the specific surface of the porous network, which determines the *ligand density*; the orientation of pore channels which determines the *spatial configuration of ligands*; the degree of crystallinity of collagen which determines the down-regulation *in cytokine synthesis*; and the scaffold degradation rate which depends on the chemical composition as well as the cross-link density of the macromolecular network and determines the *duration of the active surface*.

The first biologically active scaffold was synthesized in 1974; its degradation behavior and exceptionally low antigenicity *in vivo*, as well as its thromboresistant behavior *in vitro*, were described soon thereafter.[1,2] The initial patent describing these scaffolds was granted in 1977.[3] Principles for synthesizing a biologically active scaffold, including the critical importance of the degradation rate,[4,5] were described in detail in 1980.[6] The first reports of induced regeneration of tissue in an adult (dermis) by a scaffold in animals [7–10] and humans[11] appeared from 1981 to1982; peripheral nerve regeneration across a gap of unprecedented length was reported in 1985;[12] and regeneration of the conjunctiva was reported in 2000.[13] Two scaffolds with regenerative activity (regeneration templates) based on ECM analogs have been approved by the FDA. They have been extensively used in treatment of burn patients and applications of plastic and reconstructive surgery (Integra® Dermis Regeneration Template); and for treatment of paralysis of arms, legs and facial nerves (Neuragen®).

We review briefly below the methodology for synthesis of these scaffolds, observations of tissue and organ regeneration induced following their use, as well as the structural basis of their biological activity.

II. PHYSICOCHEMICAL PROCESSES LEADING TO SYNTHESIS OF ECM ANALOGS[14]

Since water is the primary liquid medium for most ECM macromolecules in their native state, the processing of most ECM analogs is naturally conducted in this "universal" solvent. That is not to say that all of the macromolecular components of the ECM are decidedly hydrophilic. The primary structure of elastin comprises largely nonpolar amino acids; these macromolecular chains adopt the randomly coiled configuration in water, a sign of insolubility, and participate in formation of hydrophobic bonding that partly accounts for the rubber-elastic nature of their deformation. Each of the collagen chains that make up the triple helical molecule resembles a block copolymer: primarily polar amino acid sequences are interdigitated with primarily nonpolar sequences, resulting in a polymer that is grudgingly soluble in water.[15] In sharp contrast, each proteoglycan chain is crowded with side chains of the highly hydrophilic GAGs, with their highly water-soluble polysaccharide structure.[16] GAGs, rather than proteoglycans, are preferred in scaffold synthesis because they are much more easily extracted from tissues in relatively intact form and in large enough mass.

Since most ECM analogs that have been synthesized to date have been based on collagen and GAGs (CG scaffolds), it is not surprising that water (usually in the form of a dilute solution of acetic acid, an excellent collagen solvent) has been the main solvent used in the processing of these materials.

The acetic acid–water solvent medium is well-suited for highly swelling collagen fibers extracted from tissues. When swelling of the quaternary structure of collagen fibers is conducted below a pH level of 4.25 ± 0.30 in 0.05 *M* acetic acid, the essentially single-crystalline character of these fibers is partly lost: collagen fibers lose their characteristic banding periodicity (about 67 nm) without losing the triple helical structure of the constituent macromolecules.[17] This melting transformation is reversible: the banding periodicity is recovered when the fibers are transferred from the acidic medium to pH 7.[17] However, the structure of nonbanded collagen can be "locked" by covalent cross-linking at the acidic pH so that the fibers persist in the nonbanded state, even when they are exposed to a neutral medium. The persistence of the nonbanded structure appears to be critical to the biological activity of collagen associated with the synthesis of tissues and organs: banded collagen induces vigorous platelet aggregation while nonbanded collagen does not.[18] The consequences of this structural modification on biological activity are discussed below.

Grafting of collagen with GAG confers further control of the degradation rate of the scaffold, probably due to complexation with GAG of collagenase-sensitive sites on the substrate. A similar mechanism accounts for lack of solubility of collagen in the presence of GAG.[1] Addition of a dilute aqueous solution of GAG to the acidic medium causes de-swelling of collagen fibers and coprecipitation with GAG molecules. The interaction between the two macromolecules is apparently an ionic complexation, primarily between positively charged groups in collagen (e.g., primary amino groups) and negatively charged GAG groups (e.g., sulfate groups). Complexation by GAG probably results in partial blocking of electrostatically charged sites on collagen chains and in consequent reduction in solubility, leading to the de-swelling of collagen. As expected, the ionic complex is highly labile in neutral media and dissociates readily when the pH rises close to 7. The complex dissociates in media with high ionic strength, even when the pH is acidic.[14]

Construction of a highly porous solid from an aqueous suspension of CG particles requires freezing the solvent and the sublimation of the resulting ice crystals. This process, a modification of the well-known lyophilization (freeze-drying) process used to produce protein crystals from aqueous solution without denaturation, differs from the latter in two important respects. First, the freezing step is treated as a crystal nucleation process, carefully controlled to yield ice crystals that are sublimated, eventually leading to formation of pores of the desired size and axial orientation. Second, the container in which this process takes place becomes a piece of molding equipment in which the initially liquid CG suspension acquires its desired macroscopic shape in the solid state. The shape is dictated by the parameters of the surgical implantation protocol. For example, the final shape of the scaffold is preferably that of a two-dimensional sheet, about 1 mm in thickness, when used as a graft for a full-thickness skin wound, or as a conjunctiva graft; it is a cylinder, about 1 mm in diameter, for use as a nerve chamber in which the two stumps of a transected nerve trunk are inserted. During freeze-drying the CG scaffold acquires final values of its pore volume fraction, average pore diameter and average orientation of pore channels (random in the skin application and uniaxial in the nerve application).[19]

The partly dehydrated solid, with moisture content in the 5 to 10 wt% range, comprises the ionic CG complex that continues to be vulnerable to neutral pH exposure. If implanted without further treatment, the complex dissociates. Stability of the complex is provided by chemical grafting of the GAG macromolecules on the collagen fibers using a simple physicochemical process that does not require a chemical cross-linking agent.[15,17,20] This step amounts to drastic dehydration of the complex, a process that reduces the moisture content considerably below 1 wt%, thereby proceeding significantly beyond the dehydration level achieved by freeze-drying. Severe removal of water from the system apparently shifts the equilibrium of the condensation reaction between groups in adjacent chains (such as primary amino and carboxylic groups) toward increased concentration of the products (condensed groups). The result of this process is the formation of amide cross-links between neighboring chains that is enhanced by increasing the temperature and time of the treatment. The triple helical structure is not affected by temperatures as high as 120°C, provided that the water content at the start of the treatment is low enough to elevate the melting point to a high enough level.[15] The proposed chemical mechanism of this reaction is supported by substantial evidence.[14,20] The cross-linking process ties together not only GAG chains to collagen chains but also collagen chains to each other (and possibly a few GAG chains to each other).[14] This process has been referred to as dehydrothermal (DHT) cross-linking.[21]

Additional cross-linking may be required in order to control the degradation rate to appropriately low levels in different anatomical sites (see below). Degradation of collagen-based scaffolds *in vivo* is mediated by several metalloproteinases.[22] Physicochemical cross-linking by the DHT process clearly leads to insolubility of the macromolecular network, but the levels of resistance to enzymatic degradation gained are relatively modest. A number of methods for independently conferring increased resistance to degradation via cross-linking are available. One of these is grafting of collagen with GAG.[1] Use of aldehydes[14] and of 1-ethyl-3-(3-dimethylaminopropyl) carbodiimide (EDAC)[23–25] as cross-linking agents has also been made to prepare collagen-based networks with higher levels of resistance to degradation.

Detailed protocols for fabrication of devices for implantation based on scaffolds have been presented elsewhere.[26]

III. SYNTHESIS OF TISSUES AND ORGANS USING SCAFFOLDS

Healing of severe injuries occurs quite differently in the fetus and the adult. Unlike the fetus,[27] the adult mammal does not spontaneously regenerate organs that have been lost or removed either due to accident or deliberate excision. Based on data reported in the literature from several laboratories it has been concluded on several occasions that, in the adult, severe injuries in most organs (not only skin), in which a substantial amount of mass was lost, close spontaneously by a combination of two processes, contraction and scar formation. For the purpose of this review we will consider an anatomical structure to be an "organ" rather than "tissue" when it contains the members of the tissue triad, consisting of epithelial tissue-basement membrane-stroma.

The implantation of a variety of devices that incorporated CG scaffolds has induced the adult mammal to regenerate, at least in part, three of its organs. These are:

1. Skin, complete with a physiological dermis and epidermis but with no appendages, as confirmed by histological, immunohistochemical, ultrastructural and functional studies,[7–11,28–31] was regenerated in full-thickness skin wounds prepared by completely excising the epidermis and dermis, in the adult guinea pig, adult swine, and adult human (Figure 1.1).
2. The conjunctiva, in wounds prepared by fullthickness excision of the conjunctiva, including the stroma, in the adult rabbit, confirmed by histological and polarized microscopy studies.[13]
3. The peripheral nerve, confirmed primarily by morphological and electrophysiological studies,[12,32,33] in the fully transected rat sciatic nerve, with stumps initially separated by 15 mm (later separated by 22 mm and recently by as long as 30 mm) (Figure 1.2).

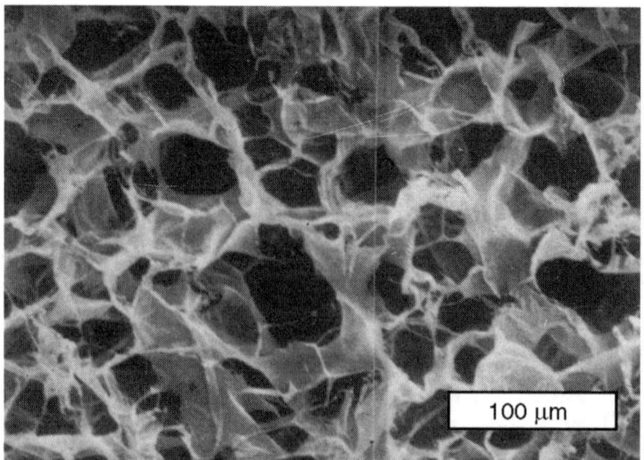

FIGURE 1.1 The porous structure of dermis regeneration template with a scaffold based on an ECM analog. This scaffold has induced partial regeneration of skin when used to graft a full-thickness skin wound in the human, swine or guinea pig. Scanning electron microscopy. (Courtesy of MIT.)

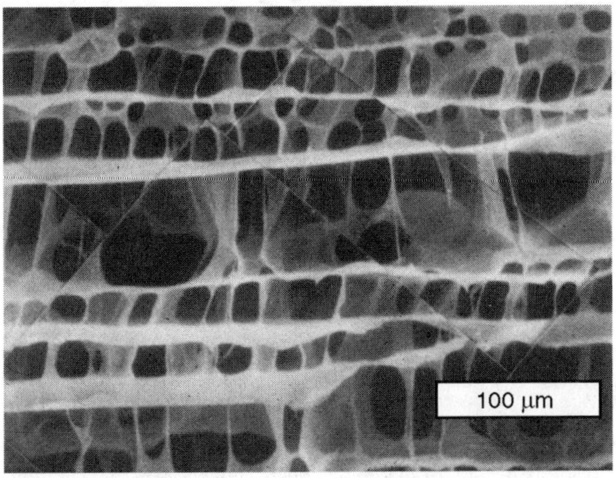

FIGURE 1.2 Nerve regeneration template. Used as a tube filling, this scaffold, also based on an ECM analog, has facilitated significant regeneration of a peripheral nerve across a gap in the rat's sciatic nerve. Scanning electron microscopy. (Courtesy of MIT.)

In these studies, the scaffolds were implanted either in the cell-free or in the cell-seeded state; cell-seeding appeared to affect the kinetics rather than the outcome. The morphological and functional studies leave no doubt that, in the presence of the scaffolds, either in a cell-free state or seeded prior to implantation, healing of these severely injured anatomical sites occurred by regeneration rather than by contraction and scar formation. However, it is emphasized that the induced regeneration has so far been partial, that is, perfectly physiological organs have not yet been regenerated in adults using these scaffolds. For example, regenerated skin was histologically and functionally different from scar and identical to physiological skin in almost all respects, including a physiological epidermis, well-formed basement membrane, well-formed capillary loops at the rete ridges of the dermal–epidermal junction, nerve endings with confirmed tactile and heat-cold feeling, and a physiological dermis; however, regenerated skin lacked organelles (hair follicles, sweat glands, etc.). Likewise, peripheral nerves were regenerated over unprecedented distances; however, regenerated nerves had conduction velocities that were about 70% of normal nerves and amplitudes that were only 30% of normal nerves; with ongoing studies, these outcomes are being continuously revised upward. It is clear that the field of induced organ regeneration in adults is still in its infancy. Nevertheless, we note that such dramatic biological activity, as has been displayed by these active scaffolds, has not been matched by cell suspensions (used by themselves), by solutions of growth factors or even by scaffolds based on synthetic polymers.

Clinical studies of skin regeneration with massively burned patients have been reported by several groups who used an FDA-approved device (Integra Dermis Regeneration Template).[34–38] FDA approval of this device has been recently extended to certain applications in plastic and reconstructive surgery; applications in this area[39] as well as the treatment of chronic wounds resulting from diabetes or venous stasis[40] have been reviewed. Results of studies leading to regeneration of a neomucosa,[41] and prevention of abdominal adhesions following surgery[42] have been also reported. A clinical study of a peripheral nerve regeneration device based on an ECM analog (Neuragen) has led to FDA approval for its use in the treatment of paralysis.

IV. STRUCTURAL BASIS OF BIOLOGICAL ACTIVITY

Substantial evidence supports the hypothesis that regeneration is induced by a scaffold-following selective down-regulation of contraction, the main engine for closure of severe wounds in the

TABLE 1.1
Structural Features of Two ECM Analogs with Significant Regenerative Activity

Structural Parameter of ECM Analog	ECM Analog for SKIN Regeneration[b]	ECM Analog for NERVE Regeneration[c]
Type I collagen/GAG (w/w)	98/2	98/2
Loss of banding in collagen fibers	Almost complete	Almost complete
Average molecular weight between cross-links,[a] M_c (kDa)	5–15	60
Average pore diameter (μm)	70–120	5–20
Pore channel orientation	Random	Axial

[a] Inverse measure of cross-link density of macromolecular network. The degradation rate increases with an increase in M_c.

[b] Yannas et al., 1989. Also used in conjunctiva regeneration (Hsu et al., 2000).

[c] Chang and Yannas, 1992.

majority of organs.[43] The term "selective down-regulation" suggests a specific inhibition of contractile cell function, with no apparent extension to several other processes involved in the complex healing process that leads to closure of the injured site. The kinetic evidence obtained during synthesis of skin and peripheral nerves is consistent with a two-step regeneration process that is induced in the presence of these templates. The first step is down-regulation of contraction; the second is the synthetic process that leads to a nearly physiological organ rather than to a scar.

The structural characteristics of templates that are required for down-regulation of contraction have been studied extensively in skin regeneration[28] and in peripheral nerve regeneration.[44] They are summarized in Table 1.1.

Two major observations that characterize down-regulation of contraction in skin regeneration models are a drastic reduction in the number of cells in the wound bed that stain with an antibody against α-smooth muscle actin (contractile cells or myofibroblasts), and a significant loss of planar orientation of contractile cell axes as well as loss of contractile cell connectivity. It has been hypothesized that the observed reduction in the number of contractile cells is primarily due to the significant inhibition of platelet degranulation. Such inhibition is a characteristic of collagen fibers that lack banding.[45] In the absence of platelet degranulation, there is a substantially reduced concentration of cytokines associated with wound healing, including transforming growth factor-beta (TGF-β) and platelet derived growth factor (PDGF). In particular, TGF-β has been clearly implicated in the differentiation of fibroblasts to contractile fibroblasts (myofibroblasts).[46] Furthermore, it has been observed that contractile cells bind extensively on the scaffold surface.[43] It has been hypothesized that such binding induces disorientation of axes of contractile cells as well as loss of contractile cell connectivity. These modifications of myofibroblast organization in the injured site are thought to be sufficient to lead to the observed substantial cancellation of the macroscopic contractile forces that normally close the wound.[28,43] It appears, therefore, that the regeneration template down-regulates contraction by a twofold action: by reducing the numbers of contractile cells available and by partially canceling out their ability to scale-up their individual contractile force to the macroscopic level that is required for wound closure.

V. SCAFFOLDS AS SOLID-STATE ENZYMES

The ECM analogs that have so far been shown to possess regenerative activity were highly porous polymers of type I collagen or graft copolymers of type I collagen and chondroitin 6-sulfate with the highly specific structure of the macromolecular network. The structure of ECM analogs that conferred regenerative activity in each case was determined empirically by conducting extensive animal studies with standardized models of organ loss, either of skin, the conjunctiva or of peripheral nerves. Spontaneous regeneration of these organs is not observed in the animal models described here; therefore, the observed regeneration, even partial, must have been induced in the presence of these scaffolds (occasionally seeded with autologous cells).

The activity of the ECM analog that induced regeneration of skin depended on choice of the appropriate collagen/GAG ratio, choice of an appropriately high ligand density (itself a function of a maximal average pore diameter) and choice of degradation rate (a function of the network cross-link density and of the collagen/GAG ratio). Significant structural differences were reported between the ECM analog that induces skin regeneration (dermis regeneration template) and that which induced nerve regeneration (nerve regeneration template; Table 1.1). However, the dermis regeneration template sufficed to regenerate the conjunctiva. An interesting finding was that ECM analogs that possessed structures with characteristics that fell outside the observed range had either seriously diminished regenerative activity or no discernible activity in the respective animal model (Figures 1.2–1.4).

The critical dependence of an ECM analog's regenerative activity on its structure suggests that the swollen and highly porous macromolecular network behaves as if it were the semisolid analog of an enzyme (Figure 1.4). Enzymes are protein catalysts that accelerate chemical reactions while remaining unaltered during the reaction. The activity of enzymes depends critically on the specificity of their three-dimensional configuration. Even slight variations in structure, such as those due to a chemical reaction or a mild physicochemical transition, usually lead to significant loss of enzymatic activity (denaturation).

Regeneration templates induce tissue and organ synthesis in two steps. First, the template blocks contraction of the wounded anatomical site; second, new tissue is synthesized while the template is undergoing degradation. The process of blocking contraction in a wounded anatomical site is an early event that appears to occur during the first week following implantation. At the end of this early process each of the two regeneration templates appear to have remained largely or entirely unchanged, while cells from the wounded site have migrated into the porous structure and have been bound to ligands on its extensive surface. The evidence is consistent with the view that the template becomes degraded during the synthetic step, the second in the sequence that hypothetically leads to regeneration. During the earlier, contraction-blocking step, the template appears to act as if it were a negative catalyst, a substance that inhibits contraction without

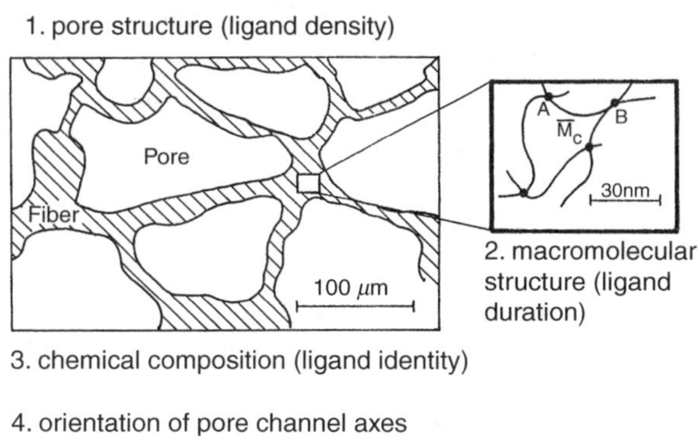

FIGURE 1.3 Schematic representation of several structural features of the nerve regeneration template critically involved in the process of nerve regeneration. These features are described either at the scale of the porous solid (scale of image: 100 μm) or that of the macromolecular network (scale: 30 nm). (Courtesy of MIT.)

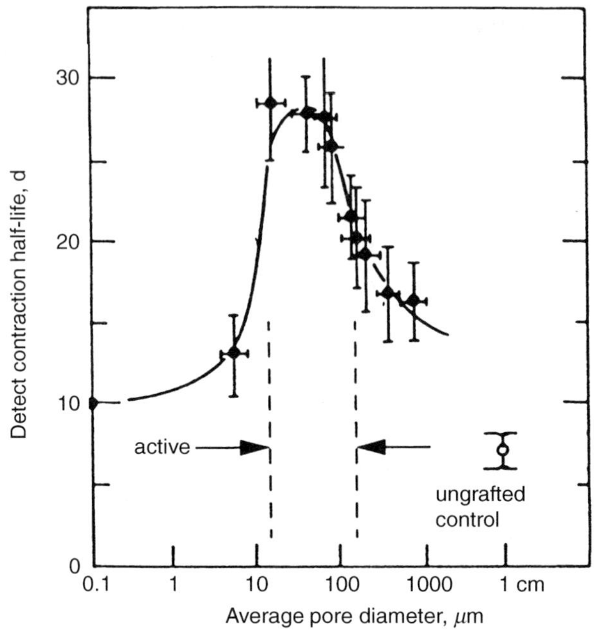

FIGURE 1.4 Data illustrating the dependence of biological activity of scaffolds on the average pore diameter of scaffold. All scaffolds were copolymers of type I collagen and chondroitin 6-sulphate that differed from each other only in average pore diameter (note logarithmic horizontal axis). Activity of grafted scaffolds was measured as the delay in half-life life of contraction of full-thickness skin wounds ("defect") in the guinea pig compared to ungrafted control. Activity increased with length of delay for onset of contraction of wound grafted with the scaffold (vertical axis). The maximum inhibition of contraction was observed with scaffolds that had average pore diameter between 20 and 120 μm, corresponding to the structure of the dermis regeneration template. (Courtesy of MIT.)

undergoing an apparent alteration during the process. In this paradigm the analog of the paradigm of an enzyme–substrate complex is the template–cell complex, formed by binding of cell integrins on matrix ligands. The active sites of this hypothetical enzyme are cell–scaffold binding sites. Fibroblast–scaffold binding sites have been partly identified during recent studies.[47]

Increased understanding of the mechanism by which certain ECM analogs induce regeneration of organs comprises, together with their effectiveness, a powerful argument in favor of pursuing future studies of synthesis of tissues and organs *in vivo* rather than *in vitro*.[48]

REFERENCES

1. Yannas, I. V., Burke, J. F., Huang, C., and Gordon, P. L., Suppression of in vivo degradability and of immunogenicity of collagen by reaction with glycosaminoglycans, *Polym. Prepr. Proc. Am. Chem. Soc.*, 16, 209–214, 1975.
2. Yannas, I. V., and Silver, F., Thromboresistant analogs of vascular tissue, *Polym. Prepr. Proc. Am. Chem. Soc.*, 16, 529–534, 1975.
3. Yannas, I. V., Burke, J. F., Gordon, P. L., and Huang, C., Multilayer membrane useful as synthetic skin, U.S. Patent 4, 060, 081, 1977 (November 29).
4. Yannas, I. V., Burke, J. F., Huang, C., and Gordon, P. L., Correlation of in vivo collagen degradation rate with in vitro measurements, *J. Biomed. Mater. Res.*, 9, 623–628, 1975.

5. Yannas, I. V., Burke, J. F., Umbreit, M., and Stasikelis, P., Progress in design of an artificial skin, *Fed. Proc.*, 38, 988, 1979.
6. Yannas, I. V., and Burke, J. F., Design of an artificial skin. I. Basic design principles, *J. Biomed. Mater. Res.*, 14, 65–81, 1980.
7. Yannas, I. V., Use of artificial skin in wound management, In *The Surgical Wound*, Dineen, P., ed., Lea and Febiger, Philadelphia, PA, pp. 171–190, 1981.
8. Yannas, I. V., Burke, J. F., Warpehoski, M., Stasikelis, P., Skrabut, E. M., Orgill, D., and Giard, D. J., Prompt, long-term functional replacement of skin, *Trans. Am. Soc. Artif. Intern. Org.*, 27, 19–22, 1981.
9. Yannas, I. V., Burke, J. F., Orgill, D. P., and Skrabut, E. M., Regeneration of skin following closure of deep wounds with a biodegradable template, *Trans. Soc. Biomater.*, 5, 24–29, 1982.
10. Yannas, I. V., Burke, J. F., Orgill, D. P., and Skrabut, E. M., Wound tissue can utilize a polymeric template to synthesize a functional extension of skin, *Science*, 215, 174–176, 1982.
11. Burke, J. F., Yannas, I. V., Quinby, W. C. Jr., Bondoc, C. C., and Jung, W. K., Successful use of a physiologically acceptable artificial skin in the treatment of extensive burn injury, *Ann. Surg.*, 194, 413–428, 1981.
12. Yannas, I. V., Orgill, D. P., Silver, J., Norregaard, T. V., Zervas, N. T., and Schoene, W. C., Polymeric template facilitates regeneration of sciatic nerve across 15-mm gap, *Trans. Soc. Biomater.*, 8, 1946, 1985; *Proc. Am. Chem. Soc. Div. Polym. Mater.*, 53, 216–218, 1985.
13. Hsu, W.-C., Spilker, M. H., Yannas, I. V., and Rubin, P. A. D., Inhibition of conjunctival scarring and contraction by a porous collagen-GAG implant, *Invest. Ophthalmol. Vis. Sci.*, 41, 2404–2411, 2000.
14. Yannas, I. V., Burke, J. F., Gordon, P. L., Huang, C., and Rubinstein, R. H., Design of an artificial skin. II. Control of chemical composition, *J. Biomed. Mater. Res.*, 14, 107–131, 1980.
15. Yannas, I. V., Collagen and gelatin in the solid state, *Rev. Macromol. Chem.*, C7, 49–104B, 1972.
16. Scott, J. E., Structure and function in extracellular matrices depend on interactions between anionic glycosaminoglycans, *Pathol. Biol. (Paris)*, 49, 284–289, 2001.
17. Yannas, I. V., Biologically active analogs of the extracellular matrix: artificial skin and nerves, *Angew. Chem.*, 29, 20–35, 1990.
18. Sylvester, M. F., Yannas, I. V., Salzman, E. W., and Forbes, M. J., Collagen banded fibril structure and the collagen-platelet reaction, *Thromb. Res.*, 55, 135–148, 1989.
19. Dagalakis, N., Flink, J., Stasikelis, P., Burke, J. F., and Yannas, I. V., Design of an artificial skin. Part III. Control of pore structure, *J. Biomed. Mater. Res.*, 14, 511–528, 1980.
20. Yannas, I. V., and Tobolsky, A. V., Crosslinking of gelatin by dehydration, *Nature*, 215, 509–510, 1967.
21. Silver, F. H., Yannas, I. V., and Salzman, E. W., In vitro blood compatibility of glycosaminoglycan-precipitated collagens, *J. Biomed. Mater. Res.*, 13, 701–716, 1979.
22. Visse, R., and Nagase, H., Matrix metalloproteinases and tissue inhibitors of metalloproteinases: structure, function, and biochemistry, *Circ. Res.*, 92, 827–839, 2003.
23. Osborne, C. S., Barbenel, J. C., Smith, D., Savakis, M., and Grant, M. H., Investigation into the tensile properties of collagen/chondroitin-6-sulphate gels: the effect of crosslinking agents and diamines, *Med. Biol. Eng. Comput.*, 36, 129–134, 1998.
24. Osborne, C. S., Reid, W. H., and Grant, M. H., Investigation into the biological stability of collagen/chondroitin-6-sulphate gels and their contraction by fibroblasts and keratinocytes: the effect of crosslinking agents and diamines, *Biomaterials*, 20, 283–290, 1999.
25. Lee, C. R., Grodzinsky, A. J., and Spector, M., The effects of cross-linking of collagen-glycosaminoglycan scaffolds on compressive stiffness, chondrocyte-mediated contraction, proliferation and biosynthesis, *Biomaterials*, 22, 3145–3154, 2001.
26. Chamberlain, L. J., and Yannas, I. V., Preparation of collagen-glycosaminoglycan copolymers for tissue regeneration, In *Tissue Engineering*, Morgan, J. R., and Yarmush, M. L., eds., Humana Press, Tolowa, NJ, pp. 3–17, 1998.
27. Dang, C., Ting, K., Soo, C., Longaker, M. T., and Lorenz, H. P., Fetal wound healing: current perspectives, *Clin. Plast. Surg.*, 30, 13–23, 2003.
28. Yannas, I. V., Lee, E., Orgill, D. O., Skrabut, E. M., and Murphy, G. F., Synthesis and characterization of a model extracellular matrix that induces partial regeneration of adult mammalian skin, *Proc. Natl Acad. Sci. USA*, 86, 933–937, 1989.

29. Murphy, G. F., Orgill, D. P., and Yannas, I. V., Dermal regeneration is induced by biodegradable collagen-glycosaminoglycan grafts, *Lab. Invest.*, 63, 305–313, 1990.
30. Compton, C. C., Butler, C. E., Yannas, I. V., Warland, G., and Orgill, D. P., Organized skin structure is regenerated in vivo from collagen-GAG matrices seeded with autologous keratinocytes, *J. Invest. Dermatol.*, 110, 908–916, 1998.
31. Butler, C. E., Yannas, I. V., Compton, C. C., and Orgill, D. P., Comparison of cultured and uncultured keratinocytes seeded into a collagen-GAG matrix for skin replacement, *Br. J. Plast. Surg.*, 52, 127–132, 1999.
32. Chamberlain, L. J., Yannas, I. V., Hsu, H.-P., Strichartz, G., and Spector, M., Collagen-GAG substrate enhances the quality of nerve regeneration through collagen tubes up to level of autograft, *Exp. Neurol.*, 154, 315–329, 1998.
33. Chamberlain, L. J., Yannas, I. V., Hsu, H.-P., Strichartz, G., and Spector, M., Near-terminus axonal structure and function following rat sciatic nerve regeneration through a collagen-GAG matrix in a 10-mm gap, *J. Neurosci. Res.*, 60, 666–677, 2000.
34. Heimbach, D., Luterman, A., Burke, J., Cram, A., Herndon, D., Hunt, J., Jordan, M., McManus, W., Solem, L., Warden, G., and Zawacki, B., Artificial dermis for major burns, *Ann. Surg.*, 208, 313–320, 1988.
35. Stern, R., McPherson, M., and Longaker, M. T., Histologic study of artificial skin used in the treatment of full-thickness thermal injury, *J. Burn Care Rehabil.*, 11, 7–13, 1990.
36. Burke, J. F., Observations on the development and clinical use of artificial skin — An attempt to employ regeneration rather than scar formation in wound healing, *Jpn. J. Surg.*, 17, 431–438, 1987.
37. Tompkins, R. G., Hilton, J. F., Burke, J. F., Schoenfeld, D. A., Hegarty, M. T., Bondoc, C. C., Quimby, W. C. Jr., Behringer, G. E., and Ackroyd, F. W., Increased survival after massive thermal injuries in adults: preliminary report using artificial skin, *Child Care Med.*, 17, 734–740, 1989.
38. Sheridan, R. L., Hegarty, M., Tompkins, R. G., and Burke, J. F., Artificial skin in massive burns — results to ten years, *Eur. J. Plast. Surg.*, 17, 91–93, 1994.
39. Orgill, D. P., Straus, F. H. 2nd, and Lee, R. C., The use of collagen-GAG membranes in reconstructive surgery, *Ann. NY Acad. Sci.*, 888, 233–248, 1999.
40. Boyce, S. T., Skin substitutes from cultured cells and collagen-GAG polymers, *Med. Biol. Eng. Comput.*, 36, 791–800, 1998.
41. Butler, C. E., Navarro, F. A., Park, C. S., and Orgill, D. P., Regeneration of neomucosa using cell-seeded collagen-GAG matrices in athymic mice, *Ann. Plast. Surg.*, 48, 298–304, 2002.
42. Butler, C. E., Navarro, R. A., and Orgill, D. P., Reduction of abdominal adhesions using composite collagen-GAG implants for ventral hernia repair, *J. Biomed. Mater. Res.*, 58, 75–80, 2001.
43. Yannas, I. V., *Tissue and Organ Regeneration in Adults*, Springer, New York, 2001.
44. Chang, A. S., and Yannas, I. V., Peripheral nerve regeneration, In *Neuroscience Year (Supplement 2 to Encyclopedia of Neuroscience)*, Smith, B., and Adelman, G., eds., Birkhaüser, Boston, pp. 125–126, 1992.
45. Sylvester, M. F., Yannas, I. V., Salzman, E. W., and Forbes, M. J., Collagen banded fibril structure and the collagen-platelet reaction, *Thromb. Res.*, 55, 135–148, 1989.
46. Desmoulière, A., Genioz, A., Gabbiani, F., and Gabbiani, G., Transforming growth factor-b1 induces a-smooth muscle actin expression in granulation tissue, myofibroblasts and in quiescent and growing cultured fibroblasts, *J. Cell Biol.*, 122, 103–111, 1993.
47. Sethi, K. K., Yannas, I. V., Mudera, V., Eastwood, M., McFarland, C., and Brown, R. A., Evidence for sequential utilization of fibronectin, vitronectin, and collagen during fibroblast-mediated collagen contraction, *Wound Repair Regen.*, 10, 307–408, 2002.
48. Yannas, I. V., Synthesis of organs: in vitro or in vivo?, *Proc. Natl Acad. Sci. USA*, 97, 9354–9356, 2000.

2 Alginate for Tissue Engineering

Peter X. Ma

CONTENTS

I. INTRODUCTION

Hydrogels are cross-linked hydrophilic polymers that contain large amounts of water without dissolution. They are attractive candidates for certain tissue engineering applications because of their capacity to fill irregularly shaped tissue defects, the allowance of minimally invasive procedures such as arthroscopic surgeries, and the ease of incorporation of cells or bioactive agents.[1–3] Hydrogels can be made from either synthetic or natural polymers as long as the polymers are hydrophilic and can be cross-linked in some way (either chemically or physically) to prevent the dissolution of the polymers. The most frequently studied synthetic polymers for tissue engineering are poly(ethylene glycol) (PEG), poly(vinyl alcohol) (PVA), acrylic polymers, and their derivatives.[1,2,4–8] The naturally derived polymers (macromolecules) most frequently used as hydrogel scaffolds in tissue engineering are alginate, hyaluronate, collagen, and their derivatives.[9–13] This chapter briefly reviews alginate biomaterials, their usage in tissue engineering, and other closely related biomedical applications.

II. SOURCES, STRUCTURES, AND PROPERTIES

Alginate can be extracted from certain seaweeds or produced by some bacteria. Alginate from brown algae was first reported by a British chemist (E.C.C. Stanford) as early as 1881, while the microbial alginate was first discovered more than 80 years later by Linker and Jones.[14,15] Owing to the significantly higher production cost of alginate from bacterial sources, the primary sources of current commercial alginate materials are various species of the brown algae. The major species used for alginate production are *Ascophylla*, *Laminaria*, and *Macrocystus* as reviewed by Sutherland.[16] The global alginate market is about 30,000 tons.[17] About 50% of the alginate is used in the food industry as thickening and emulsifying agents. For example, alginate is used in ice creams, frozen custards, cream and cake mixtures, beer, fruit drinks, and so forth. Alginate is also used in textile and paper industries to improve surface properties, in water treatment processes to

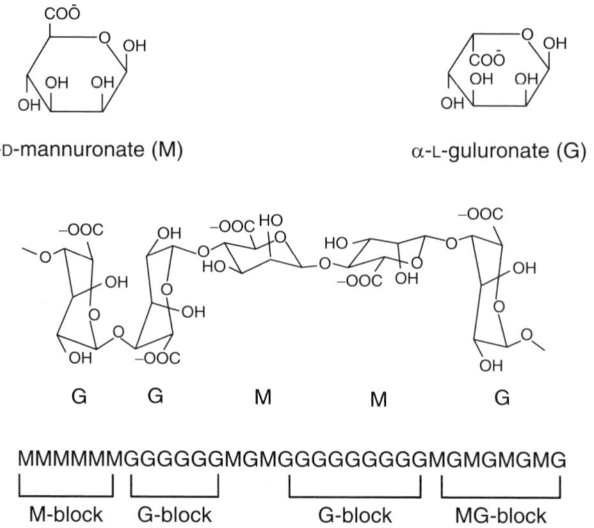

FIGURE 2.1 The chemical structures of mannuronate (M), guluronate (G), alginate, and its sequential structures.

increase aggregate size, in dentistry as impression materials, and in various formulations in cosmetic and pharmaceutical industries.

Alginic acid is a family of linear copolymers of 1,4-linked β-D-mannuronic acid (M) and α-L-guluronic acid (G) of varying compositional and sequential structures (Figure 2.1). The gelation capability, mechanical properties, and other physical characteristics of this natural polysaccharide are strongly dependent on the compositional and sequential structures.[10,18–20] The G and M units are C5 epimers, which are only different in their conformations. It is believed that the polymers are initially synthesized as poly(mannuronic acid) and some of the M units are subsequently converted into G units by C5-epimerases.[21] The alginates obtained from different algal species have different compositions. For example, alginate from *L. hyperborea* can contain a G-residue content as high as 75% (75% in outer cortex, 68% in the stipe, and 55% in the leaf), alginate from *M. pyrifera* contains a G-residue content of 39%, and alginate from *A. nodosum* can contain a G-residue content as low as 10% (10% in the fruiting body, and 36% in the old tissue).[22]

Gelation of alginate is based on the affinity of alginic acid towards certain ions and the ability to bind these ions selectively and co-operatively.[18] The G content and its sequential structure significantly affect mechanical properties of gel beads.[23] The consecutive G residues permit co-ordinated cavity organization at the molecular level (often called "egg-box" structure) to allow stable ionic binding formation. Therefore, longer G-block length increases binding strength and hence structural integrity and mechanical properties of the alginate gels.[10,18] It is worth mentioning that several investigators believed that M- and MG-blocks of alginates disadvantageously induced inflammatory response and fibrotic overgrowth of alginate capsules *in vivo*. However, recent studies seem to disagree and indicate that the response against contaminants of the alginate was mistakenly interpreted as a response against the M residues.[24–27]

For ionically cross-linked alginate gels, in addition to the composition and sequential structures of the alginate, the type of cross-linking cation is also critically important. The affinity for alginate is different for different cations. For example, the affinity of divalent cations decreases in the following order: $Pb^{2+} > Cu^{2+} > Cd^{2+} > Ba^{2+} > Sr^{2+} > Ca^{2+} > Co^{2+} = Ni^{2+} = Zn^{2+} > Mn^{2+}$. The rigidity of alginate gel beads made using these cations generally follows the same order with a few exceptions such as Cd^{2+} and Ni^{2+}.[22] However, Ca^{2+} remains the most frequently used cation for alginate formulations used in biomedical applications because it exists in large quantities in the human body and is considered biocompatible. The calcium alginate gel is also highly sensitive to chelating agents such as phosphate, citrate, EDTA and lactate. They are also weakened by antigelling cations such as Na^{+} or Mg^{2+}. In addition to ionic gels, alginic acid can also form acid gels under low pH values,[28] although they are not suitable for cell incorporation. More stable alginate gels may be achieved by covalent cross-linking, utilizing the carboxyl functionality. However, the cross-linking chemicals are usually toxic to cells and are not suitable for cell incorporation.

Since alginates have been used in a wide variety of biomedical applications including tissue engineering, ASTM has established standard guide for the characterization and testing of alginate materials intended for such applications (ASTM Standard F 2064).

III. VARIOUS BIOMEDICAL APPLICATIONS

A. Drug Delivery

Alginate has been widely used for drug delivery. Tablets and capsules are the most frequently used oral dosage forms. Sodium alginate has been used as a tablet binding agent, while alginic acid is used as a tablet disintegrant in compressed tablets designed for immediate drug release.[28] Alginate has also been used as a coating to achieve sustained release (slow release) since the alginate layer can serve as a barrier to reduce the diffusion rate of the drug compounds.

Controlled release systems for drug delivery have advanced substantially in recent years. These systems are designed to achieve predetermined and kinetically reproducible release profiles of drug compounds. For such applications, alginate has mainly been used in diffusion controlled delivery systems. These systems can be categorized as membrane systems and matrix systems. In the alginate membrane systems, the drug formulation (in the form of a solid, in a suspension, or in a solution) is encapsulated within a drug reservoir compartment. The release profiles of the drugs are controlled by the permeability of the alginate membrane. For such systems (e.g., alginate capsules), the thickness of the membrane can be used as a variable to control the release rate, and zero-order kinetics is often achievable.[28]

In the matrix system, the drug is homogeneously dispersed in an alginate material, either as microspheres, nanospheres, or conventional tablets.[28,29] When these systems are exposed to a medium or body fluid, the drug is released by diffusion mechanism. However, it is often more complicated than a simple diffusion mechanism because alginate swells in an aqueous fluid. Under certain conditions, dissolution and erosion may also occur, which complicate the release kinetics. Here, cross-linking density plays an important role and is often utilized to control the release characteristics. It should be noted that methodologies used in controlled release and gene therapies are increasingly used in advanced tissue engineering applications.

B. Microencapsulation

Immunoisolation is another important area in which various alginates have been the primary materials of interest. Immunoisolation is the enclosure of allogeneic (from the same species but not the recipient) or xenogeneic (from a species different from the recipient) cells or tissues in a semipermeable membrane or matrix in order to protect them from immune rejection. The applications can be in the form of either implants or extracorporeal devices. The core of this technology is semipermeable membranes or matrices that have well-defined molecular weight or size cutoff.[30–32] The materials should allow nutrients and metabolic wastes to diffuse through so that the encapsulated cells or tissues remain living and functional.[33,34] They should also allow the therapeutic molecules produced by the encapsulated cells or tissues to diffuse out to function as biohybrid artificial organs. However, they have to exclude host complement, immune effector cells, and antibodies from entering the implants.

Lim and Sun[35] pioneered the use of alginate as a microencapsulation material for islet transplantation in diabetic rodents, and successfully maintained glucose homeostasis using the encapsulated allogeneic islets. Since then, many articles have been published by various investigators involving the use of cross-linked alginate for cell encapsulation. Although there are many types of variations of the encapsulation process, the basic process is to extrude a cell containing sodium or potassium alginate solution dropwise into a solution that contains divalent or multivalent ions, which can react with the carboxyl groups of the alginate (G blocks are the domains for effective cross-linking) to quickly form gel beads. In the original process by Lim and Sun, the formed calcium-alginate gel beads were coated with poly(L-lysine) to stabilize the structure, and the stabilized gel beads were subsequently treated using sodium citrate to liquefy the alginate core (citrate is a strong chelating agent for calcium ions) to optimize cell survival. By treating these microcapsules with additional alginate and poly(L-lysine) solutions, "double-walled"[36] or even multilayered structures can be achieved to stabilize the beads and modulate the permeability. The gel beads are typically in the size range of between 300 and 800 μm. However, recent studies using larger beads (2 to 6 mm in diameter) without poly(L-lysine) coating showed surprisingly long-term islet xenograft survival,[37,38] although the mechanisms are not understood.[39]

In addition to the controlled release and microencapsulation of cells, alginate has been used in a wide range of other biomedical applications, such as gene therapy,[40,41] oral vaccination,[42] and wound dressings.[43–45]

IV. TISSUE ENGINEERING SCAFFOLDS

Microencapsulation of cells as discussed above can be considered to be a special case of tissue engineering. In such a case, the cells and the encapsulating alginate materials together serve as an artificial tissue/organ for a therapeutic purpose although they are not regenerated tissues/organs possessing the natural tissue composition and organization.

In the more generally accepted concept of tissue engineering, the artificial matrix (scaffold) only serves as a temporary guide (template) for cells to adhere, grow, function, synthesize their natural extracellular matrix (ECM), and eventually to generate new tissue. After the formation of the natural tissues/organs, the scaffold should in some way fade away to allow an entirely natural tissue replacement.

Although alginate is primarily used as hydrogel scaffolds in tissue engineering applications, it has also been processed into porous foams as scaffolds in the solid state by Cohen and others.[46–48] These alginate foams are fabricated using a freeze-drying technique. Aqueous sodium alginate solution is mixed with calcium gluconate solution and freeze-dried to form cross-linked alginate foams.[47] These porous alginate materials have been explored as scaffolds for the engineering of cardiac and liver tissues.[47,49] Alginate membranes have also been fabricated using sodium alginate solution and calcium chloride solution *in situ* as a means of guided tissue regeneration to repair bone defects.[50]

As a natural hydrophilic polysaccharide, alginate has been more frequently studied as a hydrogel scaffold than as solid-state porous scaffolds for tissue engineering applications. As a scaffold for cartilage tissue engineering, alginate hydrogel has been well discussed in the literature. The initial success of islet encapsulation in alginate gel beads[35] as a means of immunoisolation lead to very active research on chondrocyte culture studies in alginate gel beads.[51–54] Cultured chondrocytes in alginate gel beads express differentiated phenotype.[55–57] Recently, the gel beads have been used as a matrix to induce redifferentiation of cultured chondrocytes for tissue engineering applications.[58,59] Although the cell containing alginate gel bead cultures have demonstrated the advantages of alginate as a 3D matrix for differentiated chondrocyte function, these gel beads are not truly 3D tissue/organ constructs. A major limitation of the fast reaction chemistry used in gel bead formation is the inability to transfer this method to building 3D tissue equivalents or body parts.[10] The extremely fast reaction between calcium ions (usually from $CaCl_2$ solution) and the carboxyl groups in the polymer does not allow adequate mixing to uniformly distribute calcium ions in the alginate matrix.[10]

Paige and colleagues[60] incorporated chondrocytes into a sodium alginate solution in a mold and then immerse the cell–alginate containing mold into a bath of calcium chloride solution to cross-link. In some other cases, a membrane material is used as a cover or part of the mold to allow calcium ions to diffuse into the alginate solution.[22,61] This method (either with or without a membrane) allows the formation of 3D cell–gel constructs under mild conditions for living cell incorporation, but gradients of cross-linking density and alginate concentration exist in the gels formed,[61,62] resulting in structurally nonuniform constructs, which is not ideal for controlled uniform tissue formation.

A different approach has been used to fabricate cell–alginate gel constructs using calcium sulfate as the calcium source instead of calcium chloride. Calcium sulfate is less soluble than calcium chloride in water, leading to a slightly slower cross-linking reaction for alginate gel formation. This technique has been used by the Vacanti group as an injectable formulation for cartilage tissue engineering.[63,64] Chondrocytes are typically isolated enzymatically from articular cartilage of an animal joint. The cells are then suspended in an alginate solution and kept at low temperature on ice. Immediately before use, calcium sulfate powder is mixed into the cell containing alginate suspension and injected into defects. *In vivo* cartilage formation has been reported after the injection of cell–alginate gel constructs for 6 weeks or longer.[63,64] Alginate has also been combined with other biomolecules or synthetic materials for cartilage tissue engineering using either chondrocytes or bone marrow derived cells.[65–68]

Atala and colleagues used chondrocyte–alginate gel formulations as a treatment for vesicoureteral reflux in either an athymic mouse model using articular chondrocytes of calf shoulders[69] or a minipig model using cultured autologous auricular chondrocytes.[9] The alginate gel–cell suspensions were injected in the subureteral region of the refluxing ureter. They demonstrated cartilage formation *in vivo* and the success of correcting the reflux compared to the control site.[9]

In addition to engineering cartilage and repairing soft tissue defects, alginate gel has also been explored for bone tissue engineering.[70,71] Mooney group used the well-known RGD adhesion peptide to modify alginate gels.[72] The RGD containing peptide was reported to promote osteoblast adhesion and spreading on alginate gel surface. Cao and colleagues used autologous bone marrow stromal cells (MSCs) incorporated into a calcium alginate gel to repair cranial bone defects. New bone tissues were observed at the defects of experimental group as early as six weeks post repairing, but not in control groups. Complete repair of the defect of experimental group was reported to occur at 18 weeks.[71]

V. CONTROLLED GELATION AND UNIFORM TISSUE ENGINEERING CONSTRUCTS

Although alginate gels using calcium chloride or calcium sulfate as a cross-linking agent have been explored for tissue engineering applications, a disadvantage of using these fast dissolving calcium compounds is that the gelation kinetics is difficult to control. The fast cross-linking reaction does not allow uniform gel structure formation, which is important in tissue engineering scaffolds. Structural uniformity is not only important for uniform cell distribution, but also for well-controlled materials properties, including controlled construct geometry, mechanical properties, and diffusion characteristics.

The Messersmith group designed thermally triggerable liposomes constructed of 90% dipalmitoylphosphatidylcholine and 10% dimyristoylphosphatidylcholine to entrap $CaCl_2$ for controlled gelation of alginate solution. These liposomes released greater than 90% of entrapped Ca^{2+} when heated to 37°C. A precursor fluid containing liposomes suspended in aqueous sodium alginate remained fluid for several days at room temperature but gelled rapidly when heated to 37°C, as a result of Ca^{2+} release and formation of cross-linked Ca-alginate. This new approach may be useful for developing rapidly gelling injectable alginate solution that can be stored at room temperature and injected in a minimally invasive manner into a body tissue or cavity, upon which rapid solidification would occur.[73]

In our laboratory, we have developed techniques to control gelation rate for uniform cell–alginate gel construct formation using calcium carbonate, or a mixture of calcium carbonate and calcium sulfate as sources of calcium ions.[10] As a means of immunoisolation of cells (discussed earlier in this chapter), alginate gel beads are commonly prepared by dripping sodium or potassium alginate solution into an aqueous solution of calcium ions typically made from calcium chloride ($CaCl_2$), as schematically shown in Figure 2.2(a). The fast gelation rate with $CaCl_2$ results in a variation of cross-linking density and a polymer concentration gradient within the gel beads, as schematically illustrated in the inset of Figure 2.2(a).[22] In contrast, $CaCO_3$ has very low solubility in H_2O, allowing its uniform distribution in a cell incorporated alginate solution before gelation

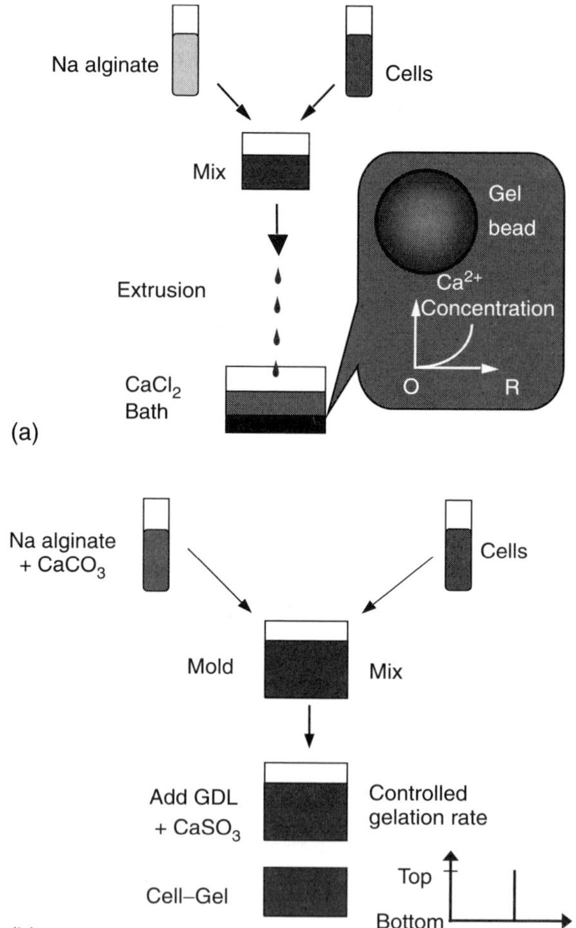

FIGURE 2.2 Schematic illustration of alginate gel formation: (a) Cell–gel bead preparation using fast dissolving $CaCl_2$, (b) Fabrication of homogeneous 3D cell–gel constructs using the controlled gelation technique (calcium ions from both $CaCO_3$/GDL and $CaSO_3$).

occurs. We then add D-glucono-δ-lactone (GDL) solution into the suspension. GDL gradually hydrolyzes to become an acid to locally lower pH value, which promotes dissolution of $CaCO_3$ to release calcium ions inside the alginate–cell suspension. The released calcium ions can immediately cross-link the surrounding alginic acid molecules and neutralize the acid, leading to structurally uniform, mechanically strong and neutral alginate gel–cell construct formation in a predesigned 3D shape using a mold.

We have demonstrated that the formation of a gel of uniform structure and consistent material properties requires working time for uniform dispersion of calcium throughout the gel before the cross-linking occurs.[10] While a slower gelation rate is advantageous in achieving uniform structure and improved mechanical integrity, a faster gelation rate is critical for certain applications. For example, the gels (with or without cells) can be potentially used as repairing materials to block a leakage in a blood vessel or other organs. A fast gelation rate is essential to ensure gel formation before being flushed away. In another application, the gel with incorporated cells can be constructed into a complex shaped tissue structure. The surgeon needs a certain working time to shape the material before it gels. We further developed a method of combining $CaCO_3$ and $CaSO_4$ into one system to control gelation rate and maintain structural homogeneity (Figure 2.2(b)). Gelation rate increased with increasing total calcium content and the $CaSO_4$ portion (Figure 2.3). This technique is an effective way to control gelation time, covering a broad range of gelation time from a few minutes to a few days (Figure 2.3; please note that the time axis is on a logarithm scale). The slow and controlled gelation systems allow for uniform cell–gel construct fabrication for tissue engineering applications. Three-dimensional gel constructs of various molded shapes have been fabricated (Figure 2.4(a)) and cells have been uniformly incorporated into the alginate gel and cultured *in vitro* (Figure 2.4(b)).

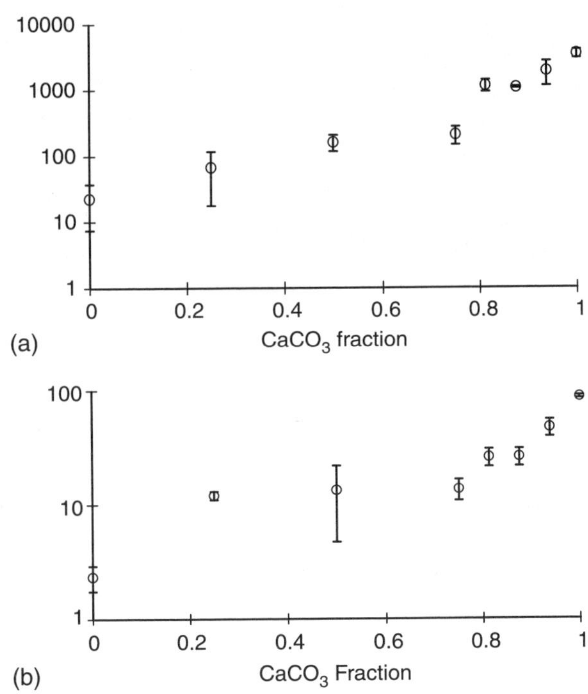

FIGURE 2.3 Gelation times for alginate gels (from *L. hyperborea*) made with varying ratios of $CaCO_3$:$CaSO_4$: (a) Total calcium content of 1X; (b) Total calcium content of 3X, where the basic calcium ion to carboxyl molar ratio of 0.18 was designated as 1X.[10] (From Kuo, C. K., and Ma, P. X., Ionically cross-linked alginate hydrogels as scaffolds for tissue engineering: 1. Structure, gelation rate and mechanical properties, *Biomaterials* 22, 511–521, 2001. With permission from Elsevier.)

(a)
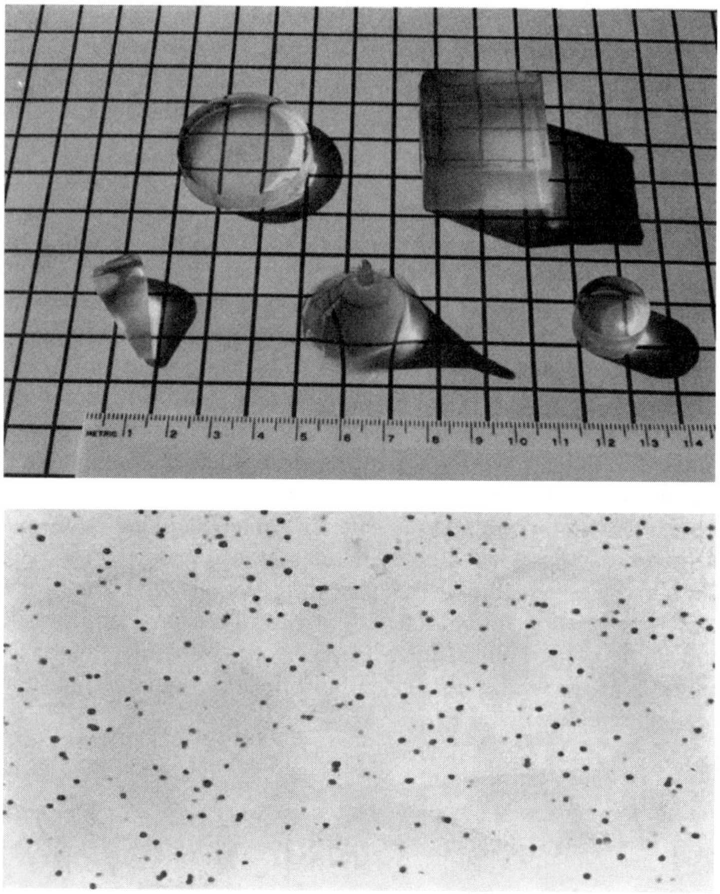
(b)

FIGURE 2.4 (a) Alginate gels of various molded shapes made from 1.5% alginate solution and 1.5X $CaCO_3$; (b) MC3T3-E1 osteoblasts incorporated into 3.2% LH alginate gels with 1.5X $CaCO_3$, cultured *in vitro* for 3 weeks. (From Kuo, C. K., and Ma, P. X., Ionically crosslinked alginate hydrogels as scaffolds for tissue engineering: 1. Structure, gelation rate and mechanical properties, *Biomaterials*, 22, 511–521, 2001. With permission from Elsevier.)

After successfully achieving controlled gelation rate and structural uniformity of the alginate–cell constructs, the physical properties of the ionically cross-linked alginate gels were investigated. It has been demonstrated that both compressive modulus and compressive strength increase with cross-linking density. Homogeneous gels prepared using slow gelation techniques always have higher modulus and strength than less homogeneous gels prepared using a fast gelation procedure.[10] Both molecular weight and the chemical structure affect the gel mechanical properties significantly. Higher molecular weight and higher guluronic acid content (G content) contribute to higher modulus and strength of the 3D alginate gel, which is consistent with results reported for gel beads.[22] In Figure 2.5, LH alginate has higher G-content but lower molecular weight than MP alginate. Our data show that at low cross-linking densities, the molecular weight dominates the mechanical properties, while at high cross-linking densities, the G-content dominates the mechanical properties (Figure 2.5). These understandings of structure–property relationships may help tissue engineers in tailoring the mechanical properties of alginate gel–cell constructs by varying formulations. Diffusivity of alginate gels may also be tailored similarly.[74]

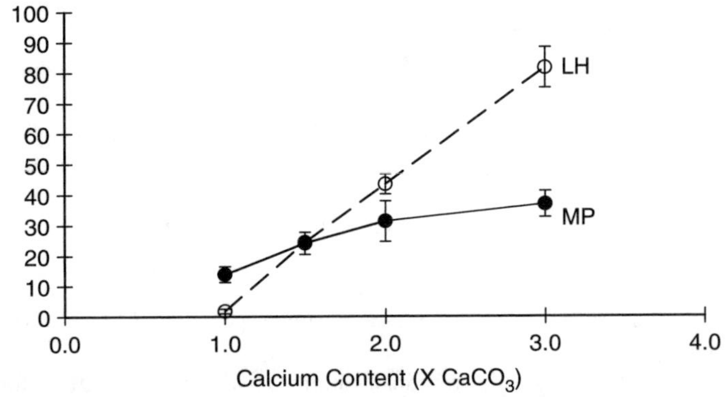

FIGURE 2.5 Comparison of compressive modulus of alginate gels made with 1.5% LH alginate (high G and low molecular weight) vs. those made with 1.5% MP alginate (low G and high molecular weight) and varying amounts of $CaCO_3$–GDL. (From Kuo, C. K., and Ma, P. X., Ionically crosslinked alginate hydrogels as scaffolds for tissue engineering: 1. Structure, gelation rate and mechanical properties, *Biomaterials*, 22, 511–521, 2001. With permission from Elsevier.)

VI. CONCLUSIONS

Alginate is biocompatible, and has been used in a variety of biomedical applications including as tissue engineering scaffolds. Alginate based hydrogels are attractive scaffolding materials because of their capacity to fill irregular tissue defects, suitability for minimally invasive procedures (such as injection and arthroscopic surgeries), and the ease of incorporation of cells and bioactive agents. In addition, there is extensive research history and literature on structure–property relationships of these materials. Recent advances in methodologies on gelation rate control, improvement of structural uniformity, chemical modifications, controlled diffusion properties, *in vitro* cell cultures, and animal model studies have further demonstrated the potential of alginate as a scaffolding material for tissue engineering. However, the issue of how to use alginate for load bearing tissue engineering requires further investigation. The incorporation and controlled release of growth and differentiation factors in various scaffolds are new and exciting developments in the tissue engineering field. Potentially important research areas include how to maintain the sustained bioactivities of such molecules in alginate hydrogels, how to manipulate the dissolution and degradation of alginate in tissue engineering environments, and how alginate interacts with host systems (such as the immune system, biological molecules, and various cells) in the long term. In summary, significant progress has been made in using alginate as tissue engineering scaffolds, although more research is needed to develop ideal alginate scaffolds for tissue engineering.

REFERENCES

1. Lutolf, M. P., Lauer-Fields, J. L., Schmoekel, H. G., Metters, A. T., Weber, F. E., Fields, G. B., and Hubbell, J. A., Synthetic matrix metalloproteinase-sensitive hydrogels for the conduction of tissue regeneration: engineering cell-invasion characteristics, *Proc. Natl Acad. Sci. USA*, 100, 5413–5418, 2003.
2. Hoffman, A. S., Hydrogels for biomedical applications, *Adv. Drug Deliv. Rev.*, 54, 3–12, 2002.
3. Ma, P. X., Tissue Engineering, In *Encyclopedia of Polymer Science and Technology*, Kroschwitz, J. I., ed., Wiley, Hoboken, NJ, 2003 (URL: www.mrw.interscience.wiley.com/epst).
4. Elisseeff, J., Anseth, K., Sims, D., McIntosh, W., Randolph, M., Yaremchuk, M., and Langer, R., Transdermal photopolymerization of poly(ethylene oxide)-based injectable hydrogels for tissue-engineered cartilage, *Plast. Reconstr. Surg.*, 104, 1014–1022, 1999.

5. Nguyen, K. T., and West, J. L., Photopolymerizable hydrogels for tissue engineering applications, *Biomaterials*, 23, 4307–4314, 2002.
6. Burdick, J. A., and Anseth, K. S., Photoencapsulation of osteoblasts in injectable RGD-modified PEG hydrogels for bone tissue engineering, *Biomaterials*, 23, 4315–4323, 2002.
7. Schmedlen, R. H., Masters, K. S., and West, J. L., Photocrosslinkable polyvinyl alcohol hydrogels that can be modified with cell adhesion peptides for use in tissue engineering, *Biomaterials*, 23, 4325–4332, 2002.
8. Stile, R. A., and Healy, K. E., Thermo-responsive peptide-modified hydrogels for tissue regeneration, *Biomacromolecules*, 2, 185–194, 2001.
9. Atala, A., Kim, W., Paige, K., Vacanti, C., and Retik, A., Endoscopic treatment of vesicoureteral reflux with a chondrocyte–alginate suspension, *J. Urol.*, 152, 641–643, 1994, discussion 644.
10. Kuo, C. K., and Ma, P. X., Ionically crosslinked alginate hydrogels as scaffolds for tissue engineering: 1. Structure, gelation rate and mechanical properties, *Biomaterials*, 22, 511–521, 2001.
11. Baier Leach, J., Bivens, K. A., Patrick, C. W. Jr., and Schmidt, C. E., Photocrosslinked hyaluronic acid hydrogels: natural, biodegradable tissue engineering scaffolds, *Biotechnol. Bioeng.*, 82, 578–589, 2003.
12. Wakitani, S., Goto, T., Pineda, S. J., Young, R. G., Mansour, J. M., Caplan, A. I., and Goldberg, V. M., Mesenchymal cell-based repair of large, full-thickness defects of articular cartilage, *J. Bone Joint Surg.*, 76, 579–592, 1994.
13. Ceballos, D., Navarro, X., Dubey, N., Wendelschafer-Crabb, G., Kennedy, W. R., and Tranquillo, R. T., Magnetically aligned collagen gel filling a collagen nerve guide improves peripheral nerve regeneration, *Exp. Neurol.*, 158, 290–300, 1999.
14. Linker, A., and Jones, R. S., A polysaccharide resembling alginic acid from a *Pseudomonas* microorganism, *Nature*, 204, 187–188, 1964.
15. Linker, A., and Jones, R. S., A new polysaccharide resembling alginic acid isolated from pseudomonads, *J. Biol. Chem.*, 241, 3845–3851, 1966.
16. Sutherland, I. W., Alginate, In *Biomaterials: Novel Materials from Biological Sources*, Byrom, D., ed., Stockton Press, New York, pp. 307–331, 1991.
17. Sabra, W., Zeng, A. P., and Deckwer, W. D., Bacterial alginate: physiology, product quality and process aspects, *Appl. Microbiol. Biotechnol.*, 56, 315–325, 2001.
18. Draget, K. I., Skjak-Braek, G., and Smidsrod, O., Alginate based new materials, *Int. J. Biol. Macromol.*, 21, 47–55, 1997.
19. Draget, K. I., Ostgaard, K., and Smidsrod, O., Homogeneous alginate gels: a technical approach, *Carbohydr. Polym.*, 14, 159–178, 1991.
20. LeRoux, M. A., Guilak, F., and Setton, L. A., Compressive and shear properties of alginate gel: effects of sodium ions and alginate concentration, *J. Biomed. Mater. Res.*, 47, 46–53, 1999.
21. Valla, S., Li, J., Ertesvag, H., Barbeyron, T., and Lindahl, U., Hexuronyl C5-epimerases in alginate and glycosaminoglycan biosynthesis, *Biochimie*, 83, 819–830, 2001.
22. Smidsrod, O., and Skjak-Braek, G., Alginate as immobilization matrix for cells, *Trends Biotechnol.*, 8, 71–78, 1990.
23. Thu, B., Smidsrod, O., and Skjak-Braek, G., Alginate gels — some structure–function correlations relevant to their use as immobilization matrix for cells, In *Immobilized Cells: Basics and Applications*, Wijffels, R. H, Buitelaar, R. M., Bucke, C. and Tramper, J., eds., Elsevier Science B.V., Amsterdam, pp. 19–30, 1996.
24. Zimmermann, U., Mimietz, S., Zimmermann, H., Hillgartner, M., Schneider, H., Ludwig, J., Hasse, C., Haase, A., Rothmund, M., and Fuhr, G., Hydrogel-based non-autologous cell and tissue therapy, *Biotechniques*, 29, 564–572, 2000, see also p. 574 and 576 passim.
25. De Vos, P., De Haan, B., and Van Schilfgaarde, R., Effect of the alginate composition on the biocompatibility of alginate–polylysine microcapsules, *Biomaterials*, 18, 273–278, 1997.
26. De Vos, P., Van Straaten, J. F., Nieuwenhuizen, A. G., de Groot, M., Ploeg, R. J., De Haan, B. J., and Van Schilfgaarde, R., Why do microencapsulated islet grafts fail in the absence of fibrotic overgrowth?, *Diabetes*, 48, 1381–1388, 1999.
27. Zimmermann, U., Klock, G., Federlin, K., Hannig, K., Kowalski, M., Bretzel, R. G., Horcher, A., Entenmann, H., Sieber, U., and Zekorn, T., Production of mitogen-contamination free alginates with

variable ratios of mannuronic acid to guluronic acid by free flow electrophoresis, *Electrophoresis*, 13, 269–274, 1992.
28. Tonnesen, H. H., and Karlsen, J., Alginate in drug delivery systems, *Drug Dev. Ind. Pharm.*, 28, 621–630, 2002.
29. Rajaonarivony, M., Vauthier, C., Couarraze, G., Puisieux, F., and Couvreur, P., Development of a new drug carrier made from alginate, *J. Pharm. Sci.*, 82, 912–917, 1993.
30. Humes, H. D., Fissell, W. H., Weitzel, W. F., Buffington, D. A., Westover, A. J., MacKay, S. M., and Gutierrez, J. M., Metabolic replacement of kidney function in uremic animals with a bioartificial kidney containing human cells, *Am. J. Kidney Dis.*, 39, 1078–1087, 2002.
31. Del Guerra, S., Bracci, C., Nilsson, K., Belcourt, A., Kessler, L., Lupi, R., Marselli, L., De Vos, P., and Marchetti, P., Entrapment of dispersed pancreatic islet cells in CultiSpher-S macroporous gelatin microcarriers: preparation, in vitro characterization, and microencapsulation, *Biotechnol. Bioeng.*, 75, 741–744, 2001.
32. Colton, C. K., Engineering a bioartificial kidney, *Nat. Biotechnol.*, 17, 421–422, 1999.
33. Lysaght, M. J., and Aebischer, P., Encapsulated cells as therapy, *Sci. Am.*, 280, 76–82, 1999.
34. Uludag, H., De Vos, P., and Tresco, P. A., Technology of mammalian cell encapsulation, *Adv. Drug Deliv. Rev.*, 42, 29–64, 2000.
35. Lim, F., and Sun, A. M., Microencapsulated islets as bioartificial endocrine pancreas, *Science*, 210, 908–910, 1980.
36. Weber, C. J., Zabinski, S., Koschitzky, T., Wicker, L., Rajotte, R., D'Agati, V., Peterson, L., Norton, J., and Reemtsma, K., The role of CD4 + helper T cells in the destruction of microencapsulated islet xenografts in nod mice, *Transplantation*, 49, 396–404, 1990.
37. Lanza, R. P., Kuhtreiber, W. M., Ecker, D., Staruk, J. E., and Chick, W. L., Xenotransplantation of porcine and bovine islets without immunosuppression using uncoated alginate microspheres, *Transplantation*, 59, 1377–1384, 1995.
38. Yang, H., O'Hali, W., Kearns, H., and Wright, J. R. Jr., Long-term function of fish islet xenografts in mice by alginate encapsulation, *Transplantation*, 64, 28–32, 1997.
39. Yang, H., and Wright, J. R., Microencapsulation Methods: Alginate (Ca^{2+}-Induced Gelation), In *Methods of Tissue Engineering,* Atala, A. and Lanza, R. P., eds., Academic Press, San Diego, CA, pp. 787–801, 2002.
40. Hortelano, G., Xu, N., Vandenberg, A., Solera, J., Chang, P. L., and Ofosu, F. A., Persistent delivery of factor IX in mice: gene therapy for hemophilia using implantable microcapsules, *Hum. Gene Ther.*, 10, 1281–1288, 1999.
41. Chang, P. L., Encapsulation for somatic gene therapy, *Ann. NY Acad. Sci.*, 875, 146–158, 1999.
42. Burkoth, A. K., Burdick, J., and Anseth, K. S., Surface and bulk modifications to photocrosslinked polyanhydrides to control degradation behavior, *J. Biomed. Mater. Res.*, 51, 352–359, 2000.
43. Wilson, P. R., Dressed to heal: new options for graft site dressing, *Australas. J. Dermatol.*, 37, 157–158, 1996.
44. Lin, S. S., Ueng, S. W., Lee, S. S., Chan, E. C., Chen, K. T., Yang, C. Y., Chen, C. Y., and Chan, Y. S., In vitro elution of antibiotic from antibiotic-impregnated biodegradable calcium alginate wound dressing, *J. Trauma*, 47, 136–141, 1999.
45. Edwards, J. V., Bopp, A. F., Batiste, S. L., and Goynes, W. R., Human neutrophil elastase inhibition with a novel cotton–alginate wound dressing formulation, *J. Biomed. Mater. Res.*, 66A, 433–440, 2003.
46. Shapiro, L., and Cohen, S., Novel alginate sponges for cell culture and transplantation, *Biomaterials*, 18, 583–590, 1997.
47. Zmora, S., Glicklis, R., and Cohen, S., Tailoring the pore architecture in 3-D alginate scaffolds by controlling the freezing regime during fabrication, *Biomaterials*, 23, 4087–4094, 2002.
48. Miralles, G., Baudoin, R., Dumas, D., Baptiste, D., Hubert, P., Stoltz, J. F., Dellacherie, E., Mainard, D., Netter, P., and Payan, E., Sodium alginate sponges with or without sodium hyaluronate: in vitro engineering of cartilage, *J. Biomed. Mater. Res.*, 57, 268–278, 2001.
49. Dar, A., Shachar, M., Leor, J., and Cohen, S., Optimization of cardiac cell seeding and distribution in 3D porous alginate scaffolds, *Biotechnol. Bioeng.*, 80, 305–312, 2002.
50. Ishikawa, K., Ueyama, Y., Mano, T., Koyama, T., Suzuki, K., and Matsumura, T., Self-setting barrier membrane for guided tissue regeneration method: initial evaluation of alginate membrane made with sodium alginate and calcium chloride aqueous solutions, *J. Biomed. Mater. Res.*, 47, 111–115, 1999.

51. Hauselmann, H., Fernandes, R., Mok, S., Schmid, T., Block, J., Aydelotte, M., Kuettner, K., and Thonar, E., Phenotypic stability of bovine articular chondrocytes after long-term culture in alginate beads, *J. Cell Sci.*, 107(Pt 1), 17–27, 1994.
52. Hauselmann, H., Masuda, K., Hunziker, E., Neidhart, M., Mok, S., Michel, B., and Thonar, E., Adult human chondrocytes cultured in alginate form a matrix similar to native human articular cartilage, *Am. J. Physiol.*, 271, C742–C752, 1996.
53. Loredo, G. A., Koolpe, M., and Benton, H. P., Influence of alginate polysaccharide composition and culture conditions on chondrocytes in three-dimensional culture, *Tissue Eng.*, 2, 115–125, 1996.
54. Mok, S., Masuda, K., Hauselmann, H., Aydelotte, M., and Thonar, E., Aggrecan synthesized by mature bovine chondrocytes suspended in alginate. Identification of two distinct metabolic matrix pools, *J. Biol. Chem.*, 269, 33021–33027, 1994.
55. Bonaventure, J., Kadhom, N., Cohen-Solal, L., Ng, K. H., Bourguignon, J., Lasselin, C., and Freisinger, P., Reexpression of cartilage-specific genes by dedifferentiated human articular chondrocytes cultured in alginate beads, *Exp. Cell Res.*, 212, 97–104, 1994.
56. Murphy, C. L., and Sambanis, A., Effect of oxygen tension and alginate encapsulation on restoration of the differentiated phenotype of passaged chondrocytes, *Tissue Eng.*, 7, 791–803, 2001.
57. Homicz, M. R., Chia, S. H., Schumacher, B. L., Masuda, K., Thonar, E. J., Sah, R. L., and Watson, D., Human septal chondrocyte redifferentiation in alginate, polyglycolic acid scaffold, and monolayer culture, *Laryngoscope*, 113, 25–32, 2003.
58. Masuda, K., Sah, R. L., Hejna, M. J., and Thonar, E. J., A novel two-step method for the formation of tissue-engineered cartilage by mature bovine chondrocytes: the alginate-recovered-chondrocyte (ARC) method, *J. Orthop. Res.*, 21, 139–148, 2003.
59. Lee, D. A., Reisler, T., and Bader, D. L., Expansion of chondrocytes for tissue engineering in alginate beads enhances chondrocytic phenotype compared to conventional monolayer techniques, *Acta Orthop. Scand.*, 74, 6–15, 2003.
60. Paige, K., Cima, L., Yaremchuk, M., Schloo, B., Vacanti, J., and Vacanti, C., De novo cartilage generation using calcium alginate–chondrocyte constructs, *Plast. Reconstr. Surg.*, 97, 168–178, 1996, discussion 179–180.
61. Skjåk-Bræk, G., Grasdalen, H., and Smidsrød, O., Inhomogeneous polysaccharide ionic gels, *Carbohydr. Polym.*, 10, 31–54, 1989.
62. Thu, B., Gaserod, O., Paus, D., Mikkelsen, A., Skjak-Braek, G., Toffanin, R., Vittur, F., and Rizzo, R., Inhomogeneous alginate gel spheres: an assessment of the polymer gradients by synchrotron radiation-induced X-ray emission, magnetic resonance microimaging, and mathematical modeling, *Biopolymers*, 53, 60–71, 2000.
63. Paige, K., Cima, L., Yaremchuk, M., Vacanti, J., and Vacanti, C., Injectable cartilage, *Plast. Reconstr. Surg.*, 96, 1390–1398, 1995, discussion 1399–1400.
64. Cao, Y., Rodriguez, A., Vacanti, M., Ibarra, C., Arevalo, C., and Vacanti, C. A., Comparative study of the use of poly(glycolic acid), calcium alginate and pluronics in the engineering of autologous porcine cartilage, *J. Biomater. Sci., Polym. Ed.*, 9, 475–487, 1998.
65. Perka, C., Spitzer, R. S., Lindenhayn, K., Sittinger, M., and Schultz, O., Matrix-mixed culture: new methodology for chondrocyte culture and preparation of cartilage transplants, *J. Biomed. Mater. Res.*, 49, 305–311, 2000.
66. de Chalain, T., Phillips, J. H., and Hinek, A., Bioengineering of elastic cartilage with aggregated porcine and human auricular chondrocytes and hydrogels containing alginate, collagen, and kappa-elastin, *J. Biomed. Mater. Res.*, 44, 280–288, 1999.
67. Caterson, E. J., Nesti, L. J., Li, W. J., Danielson, K. G., Albert, T. J., Vaccaro, A. R., and Tuan, R. S., Three-dimensional cartilage formation by bone marrow-derived cells seeded in polylactide/alginate amalgam, *J. Biomed. Mater. Res.*, 57, 394–403, 2001.
68. Elisseeff, J. H., Lee, A., Kleinman, H. K., and Yamada, Y., Biological response of chondrocytes to hydrogels, *Ann. NY Acad. Sci.*, 961, 118–122, 2002.
69. Atala, A., Cima, L., Kim, W., Paige, K., Vacanti, J., Retik, A., and Vacanti, C., Injectable alginate seeded with chondrocytes as a potential treatment for vesicoureteral reflux, *J. Urol.*, 150, 745–747, 1993.
70. Alsberg, E., Anderson, K. W., Albeiruti, A., Franceschi, R. T., and Mooney, D. J., Cell-interactive alginate hydrogels for bone tissue engineering, *J. Dent. Res.*, 80, 2025–2029, 2001.

71. Shang, Q., Wang, Z., Liu, W., Shi, Y., Cui, L., and Cao, Y., Tissue-engineered bone repair of sheep cranial defects with autologous bone marrow stromal cells, *J. Craniofac. Surg.*, 12, 586–593, 2001.
72. Rowley, J. A., Madlambayan, G., and Mooney, D. J., Alginate hydrogels as synthetic extracellular matrix materials, *Biomaterials*, 20, 45–53, 1999.
73. Westhaus, E., and Messersmith, P. B., Triggered release of calcium from lipid vesicles: a bioinspired strategy for rapid gelation of polysaccharide and protein hydrogels, *Biomaterials*, 22, 453–462, 2001.
74. Kuo, C. K., and Ma, P. X., Diffusivity of three-dimensional, ionically crosslinked alginate hydrogels, *Polym. Prepr.*, 41, 1661–1662, 2000.

3 Polysaccharide Scaffolds for Tissue Engineering

Tirtsa Ehrenfreund-Kleinman, Jacob Golenser, and Abraham J. Domb

CONTENTS

I. INTRODUCTION

Tissue engineering is an interdisciplinary field that applies the principles of engineering and the life sciences to the development of biological substitutes that restore, maintain, or improve tissue function. From both a therapeutic and an economic standpoint, the potential impact of tissue engineering is enormous.[1]

One of the earliest demonstrations that engineering tissue was possible was by Bisceglie in the 1930s, when he encased mouse tumor cells in a polymer membrane and inserted them into a pig's abdominal cavity. These studies showed that cells could survive and would not be destroyed by the immune system. This approach was an early example of cell encapsulation, which allows nutrients and wastes to diffuse through membranes, yet prevents immune cells or large molecules such as antibodies from entering. In the 1970s and 1980s, cells on sheets of collagen, or collagen glycosaminoglycan composites, were used in tissue regeneration in an attempt to create new skin.[2,3] These sheets were two-dimensional (2D) systems, and the materials composing them were naturally occurring. The next critical step was creating three-dimensional (3D) structures that enabled a large number of cells to be housed, required for creating tissues in three dimensions. Liver cells were used in early studies of this approach,[4] since then, more than 20 tissue types have been studied.

The strategy of tissue engineering generally involves the following steps:

1. Identify, isolate, and produce an appropriate cell source in sufficient amount.
2. Synthesize and manufacture into the desired shape and dimensions of an appropriate biocompatible carrier that can be used as a cell carrier.

3. Uniformly seed the cells onto or into the carrier and incubate for a predetermined period in a bioreactor.
4. Implant the seeded carrier in the proper animal model. Depending on the site and the structure, vascularization may be necessary.[5]

The field of polysaccharides as biomaterials is poised to experience rapid growth. Three factors have specifically contributed to this growing recognition of polysaccharide potential. The first is the long-standing recognition that the oligosaccharide components of most glycoproteins are essential for normal function of theses molecules. This has been coupled with a large and growing body of information pointing to the critical role of saccharide moieties in numerous cell-signaling schemes and in the area of immune modulation in particular. The second factor has been the realization that the so-called extracellular matrix (ECM) polysaccharides play critical roles in modulating the activities of signaling molecules as well as mediating certain intercellular signaling directly. The third factor is the tissue engineering research and the associated need for new materials with specific, controllable biological activity, and biodegradability.

In general, polysaccharides share a number of features that make them, as a class, particularly desirable starting materials for a number of biomaterial applications. The numerous hydroxyl groups present on the typical molecule provide numerous sites for the attachment of side groups. Such groups can provide specific functionality and biological recognition features, or may serve simply to modulate mechanical or biological properties of the parent molecule. Polysaccharides are also hydrophilic, and have the potential to be processed as hydrogels of various densities. The charged members of the polysaccharide family possess additional features that may facilitate the development of useful materials. Natural polysaccharides and their constituent monomers and degradation products can generally be considered to be either nontoxic or of low toxicity. Furthermore, polysaccharides are considered biodegradable.[6,7]

This overview will discuss the main polysaccharides used as scaffolds for tissue engineering, their characterization, and their use as cell carriers.

II. CHITOSAN

Chitosans are partially or fully deacetylated derivatives of chitin, the primary structural polymer in arthropod exoskeletons. It is a linear polysaccharide, consisting of β(1–4) linked D-glucosamine residues (Table 3.1) that are readily available for chemical reaction and salt formation with acids, and a variable number of randomly located *N*-acetyl-glucosamine groups, depending on the degree of deacetylation. Important characteristics of chitosan are its molecular weight, viscosity, degree of deacetylation, crystallinity index, number of monomeric units, water retention value, pKa, and hydration. Chitosan has a high charge density, and adheres to negatively charged surfaces and chelates metal ions. Chitosan is insoluble at an alkaline and neutral pH. The solubility of chitosan in inorganic acids is limited when compared with its solubility in common organic acids. Upon dissolution, amine groups of the polymer become protonated, with a resultant positively charged soluble polysaccharide (RNH_3^+). However, chitosan salts (glutamate, chloride, etc.) are soluble in water, their solubility being dependent on the degree of deacetylation. Solubility is also greatly influenced by the addition of salt to the solution. The higher the ionic strength, the lower the solubility.[6–10]

Chitosan biocompatibility was investigated in a mice model by implantation of porous chitosan scaffolds and their evaluation. Histological assessment indicated neutrophile accumulation within the implant, which resolved with time, and there were minimal signs of any inflammatory reaction to the material itself. Cellular immune responses indicated a very low incidence of chitosan-specific reactions. Collagen presence indicated that a connective tissue matrix was deposited within the chitosan implant.[11]

TABLE 3.1
Schematic Chemical Structures of Polysaccharides

Polysaccharide	Schematic Chemical Structure
Chitosan	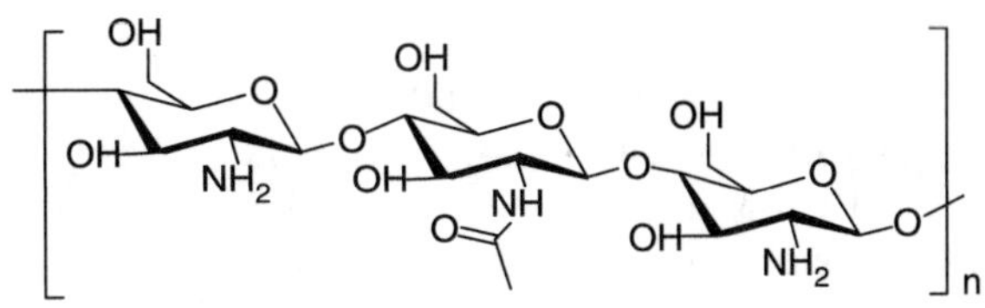
Alginate	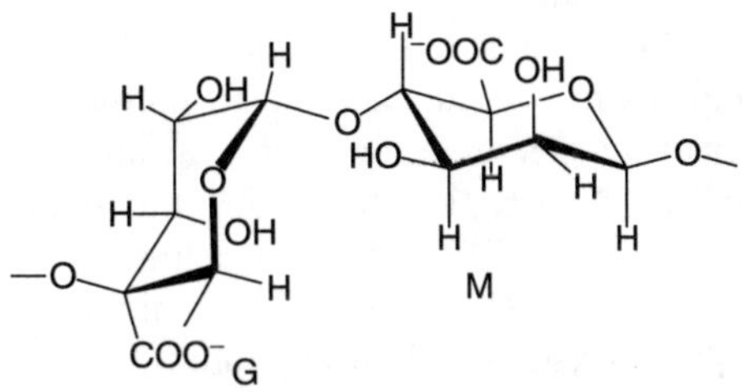
Hyaluronic acid	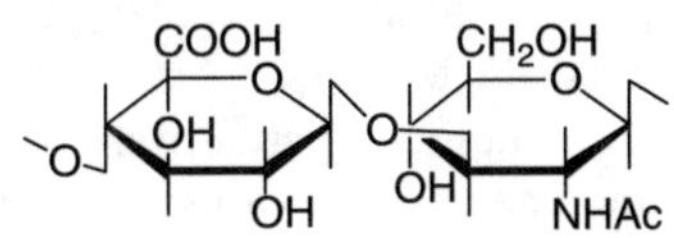
Heparin	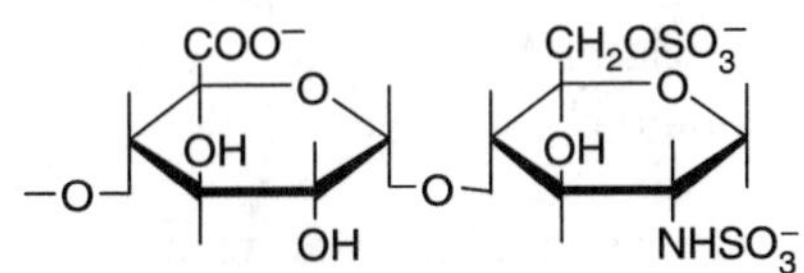
Chondroitin sulfate	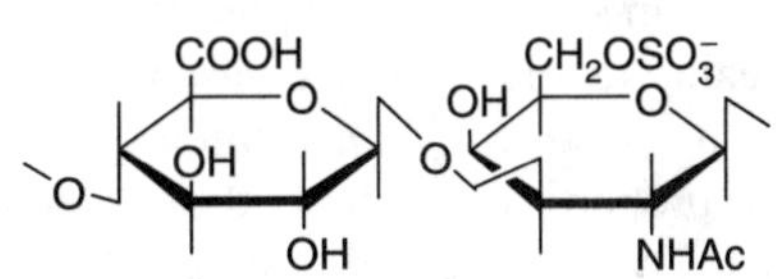
Dextran	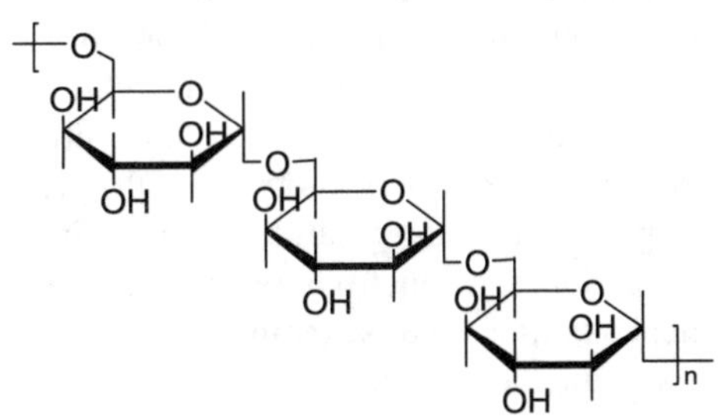
Arabinogalactan	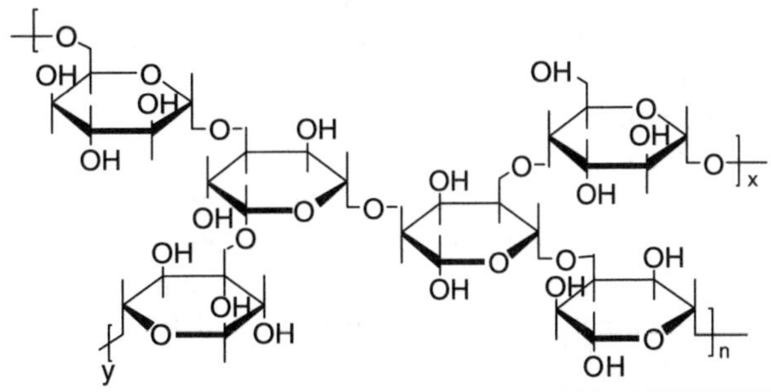

The most tried method to prepare chitin or chitosan scaffolds is by various lyophilization strategies in suitable molds. During the freezing process, ice crystals nucleate from solution and grow along the lines of thermal gradients. Ice removal by lyophilization generates a porous material, where the mean pore size can be controlled by varying the freezing rate. Pore orientation can be directed by controlling the geometry of the temperature gradients during freezing. Madihally et al.[12] demonstrated the effects of freezing temperature and chitosan concentration on the mean pore diameter of cylindrical chitosan scaffolds. The mean pore diameter could be controlled within the range of 40 to 250 μm, by varying the freezing temperature and hence the cooling rate. A lesser effect of chitosan solution concentration was also observed, with smaller pores being formed at higher concentrations. Since ice crystal growth and, hence, pore diameter are functions of the temperature gradient, pore diameters can be expected to vary with radial position in cylindrical scaffolds. In fact, pore diameter increased significantly moving from the edge towards the center of the sample. Chow et al.[13] produced porous chitin matrices with pore sizes ranging from 100 to 500 μm using the same technique. In an attempt to extend the upper pore size limit of chitin matrices obtained by lyophilization, they developed a novel method designated as an "internal bubbling process" (IBP). This was achieved by adding $CaCO_3$ to the chitin solution, casting into a mold to give a $CaCO_3$–chitin gel, which was next submerged in 1*N* HCl solution to generate gaseous CO_2 creating IBP throughout the gel.[13,14] The mechanical properties of chitosan scaffolds formed by the lyophilization technique are mainly dependent on the pore sizes and pore orientations. Tensile testing of hydrated samples shows that porous membranes have greatly reduced elastic moduli compared to nonporous chitosan membranes. The extensibility (maximum strain) of porous membranes varied from values similar to nonporous chitosan ($\sim$30%) to greater than 100% as a function of both pore size and orientation.[6]

Chitosan scaffolds supported the attachment, morphology, and proliferation of various kinds of cells, including: chondrocytes,[15] dermal fibroblasts,[16] hepatocytes,[17] and adrenal chromaffin cells.[18]

The combination of chitosan with other polymers appears to be common in various reports. Chitosan-gelatin scaffolds were prepared by Mao et al. as artificial skin, and presented smaller pore size as freezing temperature decreased. Coculturing keratinocytes with fibroblasts on the scaffolds resulted in a construct of an artificial bilayer skin *in vitro*, which was flexible and had good mechanical properties.[19,20] Risbud et al. have studied chitosan–gelatin hydrogels and their cell interactions. The hydrogel did not exert a cytotoxic effect on macrophages, and primary human respiratory epithelial cells cultured on the hydrogel showed proper attachment, and normal morphology and growth.[21] Chitosan–gelatin scaffolds were also evaluated as a substrate for inducing chondrocyte growth and differentiation. Chondrocytes were observed to attach, and exhibited differentiated phenotype with proper cell–cell contact.[22] Zhao et al. prepared a hydroxyapetite/chitosan–gelatin scaffold with 90% porosity. The scaffolds were used to examine the proliferation and functions of neonatal rat caldaria osteoblasts. Histological and immunohistochemical staining and scanning electron microscopy (SEM) observation indicated that the osteoblasts attached to and proliferated on the scaffolds.[23] Chitosan collagen scaffolds were examined for their ability to regulate cellular activity. SEM analysis indicated that the addition of chitosan to the collagen matrix reinforced the structure and increased pore size. In addition, cell function based on cytokine release was augmented.[24] Macroporous chitosan/calcium phosphate (β-tricalcium phosphate and calcium phosphate inverted glass) scaffolds have also been prepared. Here, the role of chitosan was to provide a scaffold form, while the calcium phosphates' bioactivity presumably encouraged osteoblast attachment and strengthened the scaffold. The composite scaffold was found to be stronger, bioactive, and biodegradable, the effect being dependent on the ratio of chitosan to the two types of calcium phosphates. Human osteoblast-like MG63 cells were cultured on the scaffolds and enhanced the phenotype expression of the cells, compared with chitosan scaffolds.[25,26] Lee et al.[27] seeded osteoblastic cells on chitosan–tricalcium phosphate scaffolds, and found them to support the proliferation and differentiation of the cells in 3D structure.

Chitosan–polyvinyl pyrrolidone hydrogels have been evaluated and found not to have significant interactions with endothelial cells.[28] Zhu et al. utilized the reaction between the amino group on chitosan and the carboxylic acid group on amino acids to attach various amino acids (lysine, arginine, aspartic acid, phenylalanine) onto chitosan. These amino acid functionalized chitosan moieties were subsequently entrapped onto poly-L-lactic acid (PLA) surfaces. The amino acid–chitosan–PLA membranes demonstrated good cyto-compatibility to chondrocytes, behaving much like glycosaminoglycans (GAGs) found in tissue.[29] Cai et al.[30] reported similar results using a carbodiimide process to link chitosan onto the PLA surface. To improve endothelial cell adhesion and growth on chitosan, cell adhesive peptide Gly-Arg–Gly-Asp was photochemically grafted to its surface, and was found to support the proliferation of human endothelial cells compared to chitosan.[31] Sodium alginate and chitosan sponges were prepared via freeze drying process in order to assess the utility of mixed sponges as matrices for tissue engineering. The sponges had a flexible yet strong texture, as assessed macroscopically. Measurement of the resistance to compression ("hardness") indicated that the chitosan sponges were the "hardest," while the alginate sponges showed the least resistance to compression, with all sponges showing a high degree of recovery. SEM studies indicated that the mixed systems had a less defined microstructure than the single component sponges. This was ascribed to the two polysaccharides interacting in aqueous solution via coulombic forces, leading to a more randomly ordered network being formed on freezing.[32] Yang et al. prepared porous scaffolds of alginate/galactosylated chitosan (ALG/GC) by lyophilization for liver tissue engineering. Primary hepatocytes in ALG/GC sponges showed higher cell attachment and viability than in alginate alone owing to the specific interaction of the asialoglycoprotein receptors on hepatocyte with the galactose residues on ALG/GC sponges. Improvements in spheroid formation and long-term liver-specific functions of the immobilized hepatocyte were also observed in ALG/GC sponge.[33] Chitosan–calcium phosphate composites were investigated as injectable resorbable scaffolds for bone tissue regeneration. In response to pH, chitosan solution gel changes from slightly acidic to physiological. At pH lower than 6.5, the chitosan-calcium phosphate suspension is a paste-like moldable/injectable system, and at physiological pH, the polymer undergoes a phase transition, resulting in entrapment of calcium phosphate component within the reversible gel matrix.[34]

III. ALGINATE

Alginic acids, or alginates, are isolated from several species of brown algae. Alginates are composed of two monomers, α-L-guluronic acid (G) and β-D-mannuronic acid (M). The polymer is structured as a block copolymer with alternating blocks of G and M. Within the blocks, residues are linked by α(1–4) and β(1–4) glycosidic bonds, respectively (Table 3.1). In some regions, single M and G units may alternate, giving rise to MG blocks. Sizes of the three blocks can vary over a wide range, giving rise to alginates of different properties. Most notably, gels of alginate richer in the G blocks have a higher elastic modulus and also show higher solute diffusity. These observations suggest a more open and more ordered polymer network for G-rich alginate gels. The average molecular weights range from 200 to 500 kDa.[8,10]

The biocompatibility of alginate materials is dependent on the composition of the alginates. *In vivo* studies with empty alginate-poly-L-lysine capsules have shown that the cellular overgrowth on capsules implanted intraperitoneally was composed of fibroblasts in cases of alginate with intermediate G content. But the overgrowth was mainly composed of macrophages when high G alginate was used. There were no evidence of other immune cell elements such as B or T lymphocytes.[35–37] In other studies, purified alginates with a high M content (68%) showed no mitogenic activity towards murine lymphocytes *in vitro*, and absence of a mitogen induced foreign body reaction after implantation in rats.[38]

Shapiro et al.[39] compared ionically cross-linked alginate sponges prepared from three different cross-linkers: calcium gluconate, calcium chloride, and strontium chloride. They showed that the type of cross-linker affects the sponge properties such as pore size, wall thickness, and mechanical properties. They showed also that the type of alginate (i.e., the comonomer ratio of G to M residues) had a significant effect on the sponge microstructure. The effect of G content in alginate on sponge morphology is in accordance with the theory of alginate gelation by ionic cross-linking, which correlates the gel forming capabilities and gel pore size with the poly-G content of the polymer. According to the "egg box" model of Grant et al.[40] the bivalent cations bridge the negatively charged guluronic acid residues on the alginate, and the mannuronic residues play only a subordinate role in the gel framework. The pore size and architecture of alginate sponges can be controlled by the freezing regimen during fabrication,[41] as described above for chitosan scaffolds. Alginate beads and scaffolds were investigated as a platform for various cell types. Leor et al.[42] showed that tissue engineered cardiac graft, constructed from cardiomyocytes seeded within an alginate scaffold, is capable of preventing the deterioration in cardiac function after myocardial infarction in rats. These scaffolds proved the ability to achieve 3D, high density cardiac constructs with a uniform cell distribution due to the hydrophilic nature of the alginate scaffold, its greater than 90% porosity, and interconnected pore structure. The highly dense cardiac constructs maintained high metabolic activity in culture, and some of the aggregates were contracting spontaneously within the matrix pores.[43] Alginate hydrogel supported Schwann cells viability and function *in vitro*, and may be useful in peripheral nerve tissue engineering.[44] Park et al.[45] investigated the use of calcium alginate as a matrix for cartilage generation with autogenous chondrocytes, and showed findings characteristic of natural cartilage after implantation of calcium alginate seeded chondrocytes in rabbits. Chondrocytes growth on alginate was shown also by Marijnissen et al. and Loty et al.[46,47] Kuo et al.[48] reported that osteoblastic cells were uniformly incorporated in the alginate gels. Hepatocytes' behavior within 3D porous alginate scaffolds was investigated by Glicklis et al. Due to the hydrophilic nature of alginate, hepatocytes' seeding was efficient. The cells maintained viability, and appeared to synthesize fibronectin and secrete albumin.[49] Alginate beads supplemented with hyaluronan or fibrin showed increased chondrocyte proliferation compared to controls.[50]

Macroporous alginate beads with an interconnected pore structure have been developed by incorporating gas pockets within alginate beads, stabilizing the gas bubbles with surfactants, and subsequently removing the gas. The beads maintained their porous structure and allowed cell invasion *in vivo*. Invasion of macrophages and fibroblasts was noted throughout the matrices at 2 weeks post implantation. Additionally, blood vessels containing erythrocytes were observed in the polymer interior thus demonstrating that the pores were large enough to allow vascularization of the alginate bead.[51] Blends composed of alginate and PLA in which alginate was used to improve cell loading and retention, and PLA provided appropriate mechanical support and stability were prepared. SEM revealed abundant chondrocytes with a rounded morphology in the PLA/alginate blend, and histological and immunohistochemical analysis showed development of a cartilaginous phenotype.[52,53] In order to improve cell interaction, alginate was covalently modified with arginine–glycine–aspartate (RGD) (cell adhesion tripeptide) containing cell adhesion ligands utilizing aqueous carbodiimide chemistry. Mouse skeletal myoblasts adhered to RGD modified alginate surfaces, proliferated, fused into multinucleated myofibrils, and expressed heavy chain myosin which is a differentiation marker for skeletal muscle.[54] RGD modified alginates were used to promote osteoblast adhesion and spreading. A better adhesion of cells *in vitro*, and an increase of *in vivo* bone formation compared with unmodified alginate was reported.[55] Alginate has been also modified with lectin, a carbohydrate specific binding protein, to enhance ligand-specific binding properties.[56] A novel bone morphogenetic protein (BMP-2) derived oligopeptide was covalently bound to alginate. After implantation of the BMP-2 oligopeptide to alginate hydrogel into the calf muscle of rats, ectopic bone formation was observed in the hydrogel, suggesting that this modified alginate might provide an alternative system for topical delivery of the morphogenetic signal of BMP-2.[57] Covalent cross-linking of alginate with various types of molecules and different cross-linking densities has been attempted to precisely control the mechanical and swelling properties of alginate gels. Lee et al.[58] cross-linked alginate hydrogels with molecules including adipic dihydrazide, lysine, and poly(ethylene glycol)-diamines. Bouhadir et al. isolated poly(guluronate), the portion of the alginate that is responsible for its gelling behavior. Poly(guluronate) was oxidized with periodate, and cross-linked with adipic dihydrazide to yield hydrogels with a wide range of mechanical properties, and with cell adhesion peptides coupled to their backbones.[59–61] Alginate oxidized to a low extent ($\sim 5\%$), and degraded with a rate depending on the pH and temperature of the solution. This polymer was still capable of being ionically cross-linked with calcium ions to form gels, which degraded faster than alginate gels in solution. In addition, the use of these degradable alginate-derived hydrogels greatly improved cartilage-like tissue formation *in vivo*, as compared with alginate hydrogels.[62] Slowly polymerizable calcium alginate gels as injectable delivery vehicles of isolated chondrocytes were prepared by mixing chondrocytes with alginate solution and injected subcutaneously in mice. After implantation, cartilage formation was observed. The chondrocytes exhibited the ability to recreate hyaline cartilage at the microscopic level, and immunohistochemistry showed the presence of sulfated mucopolysaccharides and collagen type II in the matrix.[34,63]

IV. GLYCOSAMINOGLYCANS (GAGs)

GAGs are polysaccharides which occur ubiquitously within the ECM of most animals. They are unbranched heteropolysaccharides consisting of repeated disaccharide units, with the general structure [uronic acid–amino sugar]$_n$. GAGs play a major role in organizing and determining the properties and functionality of the ECM. There are six different types of GAGs: chondroitin sulfate (CS), dermatan sulfate, keratan sulfate, heparan sulfate, heparin, and hyaluronic acid (HA). GAGs properties include biological activities like interactions with enzymes, growth factors, and ECM proteins. These biological activities can be exploited by binding, complexing, or covalently linking GAG moieties to other polymers with superior structural or mechanical characteristics.[6]

A. Hyaluronic Acid (HA)

HA is a natural mucopolysaccharide which consists of alternating residues of D-glucuronic acid, and *N*-acetyl-D-glucosamine (Table 3.1). HA functions as the backbone of the proteoglycan aggregates necessary for the functional integrity of articulate cartilage. HA is unique in its dual functionality; the high molecular weight HA is chondroinductive while low molecular weight HA oligomers are angiogenic. Thus, the high molecular weight form of HA fabricated as a delivery vehicle (with appropriately seeded progenitor cells) could be used to provide chondrogenic tissue at a particular repair site, and the breakdown products (small oligomers of hyaluronic acid) can cause an *in vivo* angiogenic response.[8,64]

The tolerability and safety of hyaluronan-based 3D scaffolds as a cell culture vehicle for mesenchymal progenitor cells was investigated. Rabbit autologous cells were cultured in hyaluronan-based scaffold and implanted in a full thickness osteochondral lesion. *In vitro* studies showed that the cells adhere and proliferate onto the hyaluronan-derived scaffold. Human stem cells have shown to produce the main ECM molecules, accompanied by an occasional synthesis of mature type II collagen. *In vivo* data demonstrated that, with or without mesenchymal progenitors, the biomaterial did not elicit any inflammatory response and was completely degraded within 4 months after implantation. With regard to the efficacy of this cell therapy, there was histological evidence that lesions filled with the biomaterial, either seeded or unseeded with cells, achieved a faster and better healing compared to empty controls. These data suggest that the hyaluronan-based scaffolds are well tolerated and safe and may be a valuable delivery vehicle for tissue engineering in the repair of articular cartilage defects.[65] Hyaluronan-based biomaterials present a suitable substrate also for human umbilical vein endothelial cells adhesion, proliferation, and reorganization.[66] Insoluble sponges composed of gelatin and sodium hyaluronate were achieved by dipping the soluble sponge into 90% (w/v) acetone/water mixture containing water soluble carbodiimide. The sponges exhibited a cross-linking degree of 10 to 35%, a mean pore size of 40 to 160 μm, porosity of 35 to 67%, and a tensile strength of 10 to 30 gf cm^{-2}. The porosity showed a tendency to increase with HA content, resulting in an increased water uptake.[67] The potential of a matrix containing esterified hyaluronic acid and gelatin, for the delivery of bone marrow-derived mesenchymal progenitor cells for the repair of chondral and osseous defects has been proven.[68] Leach et al. prepared a range of glycidyl methacrylate–HA (GMHA) conjugates, which were subsequently photopolymerized to form cross-linked hydrogels. A range of hydrogel degradation rates was achieved as well as a corresponding, modest range of material properties (e.g., swelling, mesh size). Increased amounts of conjugated methacrylate groups corresponded with increased cross-link densities and decreased degradation rates and yet had an insignificant effect on human aortic endothelial cell cytocompatibility and proliferation. Rat subcutaneous implants of the GMHA hydrogels showed good biocompatibility, little inflammatory response, and similar levels of vascularization at the implant edge compared with those of fibrin positive controls.[69] Nam et al. incorporated photodimerizable groups to hyaluronic acid, and prepared degradable hydrogels by UV radiation of the polymer. The pore size and distribution were controlled by changing the production conditions. These scaffolds are applicable for tissue regeneration.[70] *In situ* photopolymerization was applied to hyaluronan by Smeds et al.[71] first, methacrylated-polysaccharide was prepared and then exposed to argon ion laser to transform the pink viscous liquid into a clear, soft, flexible hydrogel. Cross-linked HA and collagen matrices were prepared by using water soluble carbodiimide. The swelling of the scaffolds obtained depended on the cross-linking, and SEM analysis showed multipore structures.[72] Scaffolds of type I collagen and hyaluronan were seeded with bovine nucleus pulposus or anulus fibrosus cells and maintained in culture in the presence of growth factors to try to generate a tissue whose properties could mimic those of the nucleus pulposus with respect to proteoglycan content. The results demonstrated that although it is possible to maintain functional disc cells in a biomatrix, it will be necessary to optimize proteoglycan synthesis and retention if any resulting tissue is to be of value in the biological repair of the degenerate disc.[73] The effect of the addition of HA to collagen scaffold on chondrocyte function *in vitro* was investigated by Allemann et al.[74] A small amount of HA in collagen scaffolds enhanced chondrogenesis, but a greater amount was inhibitory.

Composites of hyaluronic acid and the electrically conducting polymer polypyrrole were prepared in order to combine inherent biological properties which can specifically trigger desired cellular responses with electrical properties that have been shown to improve the regeneration of several tissues including bone and nerve. Polypyrrole–hyaluronic acid bilayer films were compatible *in vitro*, after seeding PC-12 cells. After subcutaneous implantation in rats, no significant long-term inflammation was noticeable, and enhanced vascularization in the vicinity of the implant was observed.[75,76] Covalently substituted HA with alkyl chains have been evaluated in a rat model of cartilage defect in comparison to alginate sponges alone, or combined with HA. Best macroscopic and histological scores were obtained with HA-alkyl chains in comparison to alginates.[77] Thermoresponsive hyaluronans were prepared by graft polymerization of *N*-isopropylacrylamide (NIPAM) on HA using dithiocarbamate. The PNIPAM-grafted HA were water-soluble at room temperature, while they precipitated at temperatures above approximately 34°C in water. A markedly reduced adhesion of endothelial cells to the film was observed, indicating that the PNIPAM–HA film may serve as a noncell adhesive matrix.[78] Wollina et al. presents a case report with bullosa dystropicans of the Hallopeau–Siemens type (HS-EBD, an autosomal-recessive blistering disease) treated with autologous keratinocytes on an esterified hyaluronic acid membrane. Two out of three wounds treated showed a complete integration of the graft. They improved markedly with a stable result over 12 months.[79] Hyaluronan benzylic esters scaffolds have been investigated for their physicochemical properties and degradability. The sponges were thermally stable up to 220°C, they swelled in aqueous media, and the structure appeared with open interconnective pores.[80,81] The ability of a nonwoven mesh of the hyaluronan benzyl ester-HYAFF 11, derived from the total esterification of sodium hyaluronate with benzyl alcohol on the free carboxyl groups of the glucuronic acid units, to support the growth of human chondrocytes and to maintain their original phenotype was investigated by Grigolo et al. Their data provided an *in vitro* demonstration for the therapeutic potential of HYAFF 11 as a delivery vehicle in a tissue engineered approach towards the repair of articular cartilage defects.[82] HYAFF 11 was also utilized to evaluate the growth of keratinocytes, fibroblasts, and chondrocytes by Brun et al. After *in vitro* culturing, keratinocytes gave rise to a well-differentiated epithelial layer, while fibroblasts were able to synthesize all the main extracellular molecules. When these two tissues were cocultured, a skin equivalent was formed with a dermal–epidermal junction. Chondrocytes were able to organize themselves into nodules, in which type II collagen was present.[83] HYAFF 11 was tested as osteogenic or chondrogenic delivery vehicle for rabbit mesenchymal progenitor cells, and compared with a well characterized porous calcium phosphate. HYAFF 11 sponge bind 90% more cells than ceramic, and retained its integrity after incubation in animals. The HA-based delivery vehicle was superior to calcium phosphate with respect to the number of cells loaded, and with regard to the amount of bone and cartilage formed. Additionally, HA-based vehicles have the advantage of degradation/resorption characteristics that allow complete replacement of the implant with newly formed tissue.[84] Benzyl ester of hyaluronan was also used by Hollander et al. for the reconstruction of extensive traumatic soft tissue defects. In injured patients, cultured "neodermis" consisting of cultured autologous fibroblasts grown on biocompatible 3D scaffolds composed of benzyl ester of hyaluronan was grafted on conditioned defect areas. The results showed grafting with cultured autologous fibroblasts revealed a suitable dermal tissue replacement. Epithelialization was evident after transplantation of keratinocytes. Final closure of the defects with "normoelastic" tissue properties was achieved after thin mesh-grafting.[85] Polypeptide resurfacing was applied to enhance the attachment of cells to HA. HA strands were cross-linked by glutaraldehyde and then resurfaced with poly-D-lysine, poly-L-lysine, glycine, or glutamine. After *in vitro* incubation with fibroblasts, and *in vivo* implantation, the results showed that both polylysines enhanced fibroblast adhesion to cross-linked HA strands, and modified HA had good biocompatibility, both *in vitro* and *in vivo*.[86]

B. Heparin

Heparin is a large polyanionic molecule expressed throughout the body (Table 3.1). Heparin is synthesized and stored exclusively in mast cells, and is coreleased with histamine, whereas the closely related molecule heparin sulfate is expressed, as part of a proteoglycan, on cell surfaces and throughout tissue matrices. Heparin binds highly specifically to many different growth factors, which become deactivated and stabilized. Heparin stimulates proliferation in some cell types,[87,88] and inhibits growth in others.[89,90] It also modulates several phases of wound healing.[91]

For wound healing and regeneration applications, heparin is generally complexed with, or covalently linked to, another biopolymer that serves as the main structural component. Under these circumstances, heparin can exert a number of influences dictated by its biological activities. Growth factors released in the vicinity of the implant by neutrophils and macrophages may be sequestered and stabilized within the implant structure by immobilized heparin, and subsequently released and utilized by ingrowing tissue cells. The bound heparin may also facilitate the binding and organization of deposited ECM components to the implant and consequently enhance the integration of the implant with neotissue and existing tissue. Alternatively, heparin released from an implant may activate and protect growth factors from degradation in the vicinity of the implant, which may lead to accelerated healing or enhanced angiogenesis.[92]

Heparin–chitosan ionically linked complexes were prepared and investigated by Kratz et al. The effect of heparin–chitosan on the stimulation of re-epithelialization of full thickness wounds in human skin was investigated in an *in vitro* model. After incubation, heparin–chitosan gel stimulated nine tenths of the full thickness wounds to re-epithelialize compared with only three tenths of the wounds covered with chitosan alone, and both dermal and epidermal cells were viable after incubation time. Furthermore, the stimulatory effect of the heparin–chitosan complexes depended on the concentration of heparin in the complex. The effect of heparin–chitosan complexes on wound healing in human skin was also investigated. The chitosan–heparin membrane stimulated healing of the donor sites when judged macroscopically. The authors hypothesize that the beneficial effects of the chitosan–heparin membrane result from slow release of heparin into the wound area which protects locally produced growth factors.[92,93] In order to improve endothelial cell retention, a poly(carbonate-urea) urethane graft (MyoLink™) was covalently bonded with RGD or heparin. RGD/heparin-bound graft had a significantly better cell retention than native MyoLink, and the cells exhibited higher metabolic activity.[94] Heparin–alginate matrix covalently cross-linked with ethylenediamine implanted subcutaneously in rats showed cellular infiltration and angiogenesis, which suggests that this matrix may be useful for not only the construction of transplantable blood vessels of small diameter, but also the induction of angiogenesis in regenerated skin constructed by tissue engineering.[95] Pieper et al. showed the attachment of glycosaminoglycan heparin sulfate to collagenous matrices modulates the tissue response *in vivo*. In their study, porous type I collagen matrices were cross-linked to chondroitin sulfate or heparan sulfate using water soluble carbodiimide, and implanted in rats. The cross-linked matrices retained their scaffold integrity during implantation, and supported the interstitial deposition and organization of ECM. Heparin sulfate promoted angiogenesis and caused a reduced foreign body reaction.[96]

C. Chondroitin Sulfate (CS)

Chondroitins are linear polysaccharides made up mainly of repeating disaccharide units of 2-acetamido-2-deoxy-β-D-galactose (GalNAc) and β-D-glucuronic acid (GlcA) linked 1–4 and 1–3, respectively (Table 3.1), and are involved in the structure of cartilage, tendons, and cornea. Chondroitin can be found in bovine cornea or can be obtained from CS by chemical treatment, whereas chondroitins 4- and 6-sulfate are most widely distributed in animal tissues. The difference between chondroitin, chondroitin 4-sulfate, and chondroitin 6-sulfate is the presence of an esterified sulfate group on the C-4 and C-6 position of GalNAc.

Cross-linked collagen–CS matrices were synthesized by using water soluble carbodiimide. Immobilized CS diminished the susceptibility of the matrices towards degradation by proteolytic enzymes, increased the water binding capacity, and decreased the denaturation temperature and tensile strength of the matrix. Immobilized CS bound anti-CS antibodies and was susceptible to chondroitinase digestion, demonstrating its bioavailability. The modified matrices were not cytotoxic, as was established using human myoblast and fibroblast culture systems. CS in the scaffolds influenced the bioactivity of the seeded chondrocytes; cell proliferation and the total amount of proteoglycans retained in the matrix were significantly higher, and a good preservation of the chondrocytic phenotype of the seeded cells was observed.[97–100] CS and chitosan were ionically cross-linked to develop a biomaterial to support chondrogenesis, and improve bone formation. Chondrocytes seeded on CS–chitosan maintained many characteristics of the differentiated chondrocytic phenotype, including round morphology, limited mitosis, collagen type II, and proteoglycan production. CS–chitosan also induced increased osteoblast migration and proliferation as compared with chitosan sponge alone.[101,102]

V. DEXTRAN

Dextran is a linear polysaccharide composed of glucose units linked via β(1–6) bonds (Table 3.1). Ionically cross-linked dextran sulfate–chitosan matrices were prepared, and their potential use for controlling the proliferation of vascular endothelial and smooth muscle cells was investigated.

Dextran sulfate–chitosan supported proliferation of both cell types *in vitro*. *In vivo*, dextran sulfate–chitosan scaffolds stimulated cell proliferation and the formation of a thick layer of dense granulation tissue.[103] Degradable porous scaffolds were developed by blending polylactide (PLA) with dextran, using solvent casting and particle leaching technique. The attachment of mouse 3T3 fibroblasts on PLA–dextran scaffolds was improved as compared to PLA alone, and their morphology was maintained.[104]

VI. ARABINOGALACTAN

Arabinogalactan (AG) is a highly branched natural polysaccharide with high water solubility (70%). It is extracted from the *Larix* tree and it is available in a 99.9% pure form with reproducible molecular weight and physicochemical properties. The basic building units of AG are arabinose and galactose in a ratio of approximately 1:6, linked via β(1–3), β(1–4) and β(1–6) bonds (Table 3.1).[105] The high solubility in water, biocompatibility, biodegradability, and the ease of chemical modification in aqueous media makes it an attractive polymer for the synthesis of scaffolds for possible use in tissue engineering.

AG-based sponges were synthesized by cross-linking the oxidized AG with oligoamines or polyamines such as alkanamines, chitosan, or peptides (Figure 3.1).

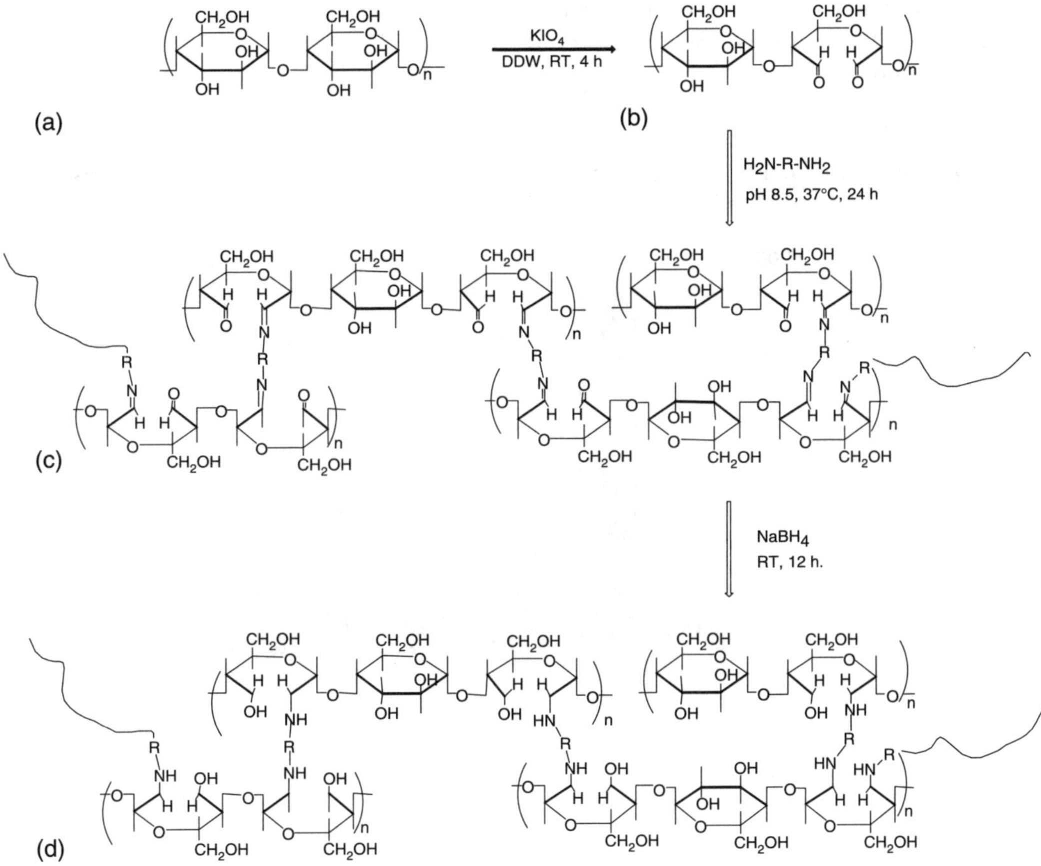

FIGURE 3.1 Schematic representation of the synthesis of AG-based sponges. (a) Natural polysaccharide, (b) oxidized polysaccharide, (c) imine polysaccharide-based sponge, (d) amine polysaccharide-based sponge. (From Ref. 107, with permission.)

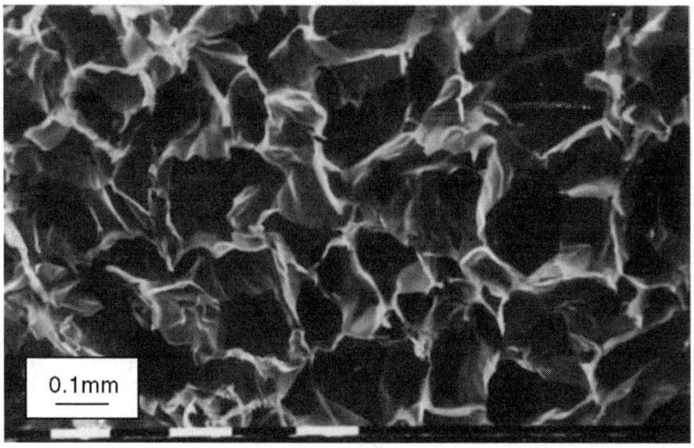

FIGURE 3.2 Representative scanning electron micrographs of 20% AG-chitosan sponge. (From Ref. 107, with permission.)

Arabinogalactan-based sponges cross-linked with the oligoamines: spermine, spermidine, or butanediamine at different ratios absorbed water, about 15 to 20 times of their initial weight during the first hour of incubation in aqueous media. The mean pore size of the sponges as measured from SEM images was about 80 μm, and their 3D structure was maintained for 10 weeks in solution. Arabinogalactan-based sponges cross-linked with chitosan, also absorbed a large amount of water, about 30 times of their initial weight during incubation in aqueous media. Sponges containing different amounts of AG (5 to 50% w/w) exhibited increased swelling as the amount of AG decreased. The decrease in swelling with the increase in AG amount is caused by a

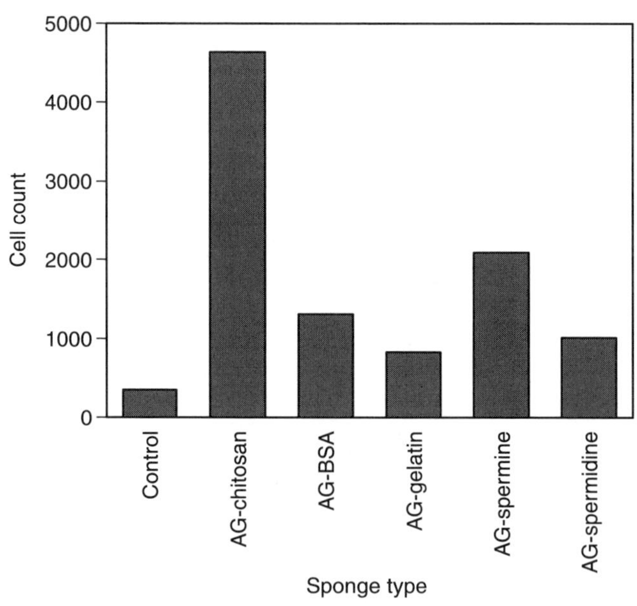

FIGURE 3.3 bEnd2 cell growth on arabinogalactan (AG) based sponges. bEnd2 cells were distributed the sponges, incubated for 1 h allowing cell attachment following addition of medium and 48 h incubation at 37°C. The number of viable cells was determined using MTT assay. Controls consisted of sponges without cells treated in a similar way. (From Ref. 107, with permission.)

more rigid and dense arrangement of the cross-linked polymer chains, which reduced the water penetration inside the sponge. The mean pore size of the sponges was about 200 μm, which is suitable for cell growth on biomaterial scaffolds (Figure 3.2). The degradation rate of arabinogalactan-based sponges was modified by several chemical strategies in order to tailor the degradation rate of sponges to different tissues requirements. Oxidation of the sponges with sodium-periodate or sodium-chlorite were the most efficient in increasing the sponges degradation rate, and the sponges could be programmed to be degraded within 1 to 12 weeks, as needed. The biocompatibility of AG-chitosan sponges was evaluated after subcutaneous implantation in BALB/c mice. The histological results showed a transient and self-limited inflammatory reaction in the area bordering the implanted sponge, and appearance of blood vessels inside and around the sponge. In addition, noninvasive MRI studies showed vascularization inside and around the sponge, supporting the histological results.

Arabinogalactan was cross-linked also with the proteins bovine serumalbumin (BSA) and gelatin to obtain 3D scaffolds. These sponges swelled in aqueous media, and exhibited mean pore size of about 100 μm.

Cell growth on AG-based scaffolds was determined using bEnd2 cells, a mouse brain endothelioma cell line. The cells were seeded on the sponges for 48 h, and the numbers of viable cells after incubation, as determined using the MTT assay, were 4640, 1320, 830, 2090, and 1010 cells/sponge for AG–chitosan, AG–BSA, AG–gelatin, AG–spermine, and AG–spermidine, respectively (Figure 3.3). The better cell growth on polysaccharide–chitosan sponges can be explained by the known ability of chitosan to facilitate cell adhesion and proliferation.[106,107]

VII. SUMMARY

The progress in the investigation of polysaccharides as scaffolds for tissue engineering has been reviewed. Polysaccharides are attractive biomaterials due to their biocompatibility, biodegradability, and targeted biological activity, and the research results indicate their potential to be used as cell carriers.

REFERENCES

1. Vacanti, J. P., and Langer, R., Tissue engineering: the design and fabrication of living replacement devices for surgical reconstruction and transplantation, *Lancet*, 354, SI32–SI34, 1999.
2. Bell, E., Ehrlich, H. P., Buttle, D. J., and Nakatsuji, T., Living tissue formed in vitro and accepted as skin-equivalent tissue of full thickness, *Science*, 211, 1052–1054, 1981.
3. Burke, J. F., Yannas, I. V., Quinby, W. C., Bondoc, C. C., and Jung, W. K., Successful use of a physiologically acceptable artificial skin in the treatment of extensive burn injury, *Ann. Surg.*, 194, 413–428, 1981.
4. Vacanti, J. P., Morse, M. A., Saltzman, W. M., Domb, A. J., Perezatayde, A., and Langer, R., Selective cell transplantation using bioabsorbable artificial polymers as matrices, *J. Pediatr. Surg.*, 23, 3–9, 1988.
5. Langer, R., Tissue engineering, *Mol. Ther.*, 1, 12–15, 2000.
6. Matthew, H. W. T., *Polymers For Tissue Engineering Scaffolds*, 2000.
7. Suh, J. K. F., and Matthew, H. W. T., Application of chitosan-based polysaccharide biomaterials in cartilage tissue engineering: a review, *Biomaterials*, 21, 2589–2598, 2000.
8. Lee, K. Y., and Mooney, D. J., Hydrogels for tissue engineering, *Chem. Rev.*, 101, 1869–1879, 2001.
9. Singla, A. K., and Chawla, M., Chitosan: some pharmaceutical and biological aspects — an update, *J. Pharm. Pharmacol.*, 53, 1047–1067, 2001.

10. Dumitriu, S., *Polysaccharides as Biomaterials*, 2000.
11. VandeVord, P. J., Matthew, H. W. T., DeSilva, S. P., Mayton, L., Wu, B., and Wooley, P. H., Evaluation of the biocompatibility of a chitosan scaffold in mice, *J. Biomed. Mater. Res.*, 59, 585–590, 2002.
12. Madihally, S. V., and Matthew, H. W. T., Porous chitosan scaffolds for tissue engineering, *Biomaterials*, 20, 1133–1142, 1999.
13. Chow, K. S., Khor, E., and Wan, A. C. A., Porous chitin matrices for tissue engineering: fabrication and in vitro cytotoxic assessment, *J. Polym. Res. — Taiwan*, 8, 27–35, 2001.
14. Khor, E., and Lim, L. Y., Implantable applications of chitin and chitosan, *Biomaterials*, 24, 2339–2349, 2003.
15. Nettles, D. L., Elder, S. H., and Gilbert, J. A., Potential use of chitosan as a cell scaffold material for cartilage tissue engineering, *Tissue Eng.*, 8, 1009–1016, 2002.
16. Ma, J. B., Wang, H. J., He, B. L., and Chen, J. T., A preliminary in vitro study on the fabrication and tissue engineering applications of a novel chitosan bilayer material as a scaffold of human neofetal dermal fibroblasts, *Biomaterials*, 22, 331–336, 2001.
17. Elcin, Y. M., Dixit, V., and Gitnick, G., Hepatocyte attachment on biodegradable modified chitosan membranes: in vitro evaluation for the development of liver organoids, *Artif. Organs*, 22, 837–846, 1998.
18. Elcin, A. E., Elcin, Y. M., and Pappas, G. D., Neural tissue engineering: adrenal chromaffin cell attachment and viability on chitosan scaffolds, *Neurol. Res.*, 20, 648–654, 1998.
19. Mao, J. S., Zhao, L. G., de Yao, K., Shang, Q. X., Yang, G. H., and Cao, Y. L., Study of novel chitosan-gelatin artificial skin in vitro, *J. Biomed. Mater. Res. Part A*, 64A, 301–308, 2003.
20. Mao, J. S., Zhao, L. G., Yin, Y. J., and Yao, K. D., Structure and properties of bilayer chitosan-gelatin scaffolds, *Biomaterials*, 24, 1067–1074, 2003.
21. Risbud, M., Endres, M., Ringe, J., Bhonde, R., and Sittinger, M., Biocompatible hydrogel supports the growth of respiratory epithelial cells: possibilities in tracheal tissue engineering, *J. Biomed. Mater. Res.*, 56, 120–127, 2001.
22. Risbud, M., Ringe, J., Bhonde, R., and Sittinger, M., In vitro expression of cartilage-specific markers by chondrocytes on a biocompatible hydrogel: implications for engineering cartilage tissue, *Cell Transplant.*, 10, 755–763, 2001.
23. Zhao, F., Yin, Y. J., Lu, W. W., Leong, J. C., Zhang, W. J., Zhang, J. Y., Zhang, M. F., and Yao, K. D., Preparation and histological evaluation of biomimetic three-dimensional hydroxyapatite/chitosan–gelatin network composite scaffolds, *Biomaterials*, 23, 3227–3234, 2002.
24. Tan, W., Krishnaraj, R., and Desai, T. A., Evaluation of nanostructured composite collagen-chitosan matrices for tissue engineering, *Tissue Eng.*, 7, 203–210, 2001.
25. Zhang, Y., and Zhang, M. Q., Synthesis and characterization of macroporous chitosan/calcium phosphate composite scaffolds for tissue engineering, *J. Biomed. Mater. Res.*, 55, 304–312, 2001.
26. Zhang, Y., Ni, M., Zhang, M. Q., and Ratner, B., Calcium phosphate–chitosan composite scaffolds for bone tissue engineering, *Tissue Eng.*, 9, 337–345, 2003.
27. Lee, Y. M., Park, Y. J., Lee, S. J., Ku, Y., Han, S. B., Choi, S. M., Klokkevold, P. R., and Chung, C. P., Tissue engineered bone formation using chitosan/tricalcium phosphate sponges, *J. Periodont.*, 71, 410–417, 2000.
28. Risbud, M. V., Bhonde, M. R., and Bhonde, R. R., Effect of chitosan-polyvinyl pyrrolidone hydrogel on proliferation and cytokine expression of endothelial cells: implications in islet immunoisolation, *J. Biomed. Mater. Res.*, 57, 300–305, 2001.
29. Zhu, H. U., Ji, J., Lin, R. G., Gao, C. G., Feng, L. X., and Shen, J. C., Surface engineering of poly(D′L-lactic acid) by entrapment of chitosan-based derivatives for the promotion of chondrogenesis, *J. Biomed. Mater. Res.*, 62, 532–539, 2002.
30. Cai, K. Y., Yao, K. D., Cui, Y. L., Lin, S. B., Yang, Z. M., Li, X. Q., Xie, H. Q., Qing, T. W., and Luo, J., Surface modification of poly (D,L-lactic acid) with chitosan and its effects on the culture of osteoblasts in vitro, *J. Biomed. Mater. Res.*, 60, 398–404, 2002.
31. Chung, T. W., Lu, Y. F., Wang, S. S., Lin, Y. S., and Chu, S. H., Growth of human endothelial cells on photochemically grafted Gly-Arg–Gly-Asp (GRGD) chitosans, *Biomaterials*, 23, 4803–4809, 2002.

32. Lai, H. L., Abu'Khalil, A., and Craig, D. Q. M., The preparation and characterisation of drug-loaded alginate and chitosan sponges, *Int. J. Pharm.*, 251, 175–181, 2003.
33. Yang, J., Chung, T. W., Nagaoka, M., Goto, M., Cho, C. S., and Akaike, T., Hepatocyte-specific porous polymer-scaffolds of alginate/galactosylated chitosan sponge for liver-tissue engineering, *Biotechnol. Lett.*, 23, 1385–1389, 2001.
34. Gutowska, A., Jeong, B., and Jasionowski, M., Injectable gels for tissue engineering, *Anat. Rec.*, 263, 342–349, 2001.
35. de Vos, P., De Haan, B., and Van Schilfgaarde, R., Effect of the alginate composition on the biocompatibility of alginate-polylysine microcapsules, *Biomaterials*, 18, 273–278, 1997.
36. de Vos, P., De Haan, B., Wolters, G. H. J., Strubbe, J. H., and Van Schilfgaarde, R., Improved biocompatibility but limited graft survival after purification of alginate for microencapsulation of pancreatic islets, *Diabetologia*, 40, 262–270, 1997.
37. de Vos, P., Hoogmoed, C. G., and Busscher, H. J., Chemistry and biocompatibility of alginate-PLL capsules for immunoprotection of mammalian cells, *J. Biomed. Mater. Res.*, 60, 252–259, 2002.
38. Klock, G., Pfeffermann, A., Ryser, C., Grohn, P., Kuttler, B., Hahn, H. J., and Zimmermann, U., Biocompatibility of mannuronic acid-rich alginates, *Biomaterials*, 18, 707–713, 1997.
39. Shapiro, L., and Cohen, S., Novel alginate sponges for cell culture and transplantation, *Biomaterials*, 18, 583–590, 1997.
40. Grant, G. T., Morris, E. R., Rees, D. A., Smith, P. J. C., and Thom, D., Biological interactions between polysaccharides and divalent cations — egg-box model, *FEBS Lett.*, 32, 195–198, 1973.
41. Zmora, S., Glicklis, R., and Cohen, S., Tailoring the pore architecture in 3-D alginate scaffolds by controlling the freezing regime during fabrication, *Biomaterials*, 23, 4087–4094, 2002.
42. Leor, J., Aboulafia-Etzion, S., Dar, A., Shapiro, L., Barbash, I. M., Battler, A., Granot, Y., and Cohen, S., Bioengineered cardiac grafts — a new approach to repair the infarcted myocardium?, *Circulation*, 102, 56–61, 2000.
43. Dar, A., Shachar, M., Leor, J., and Cohen, S., Cardiac tissue engineering — optimization of cardiac cell seeding and distribution in 3D porous alginate scaffolds, *Biotechnol. Bioeng*, 80, 305–312, 2002.
44. Mosahebi, A., Simon, M., Wiberg, M., and Terenghi, G., A novel use of alginate hydrogel as schwann cell matrix, *Tissue Eng.*, 7, 525–534, 2001.
45. Park, D. J., Bong, J. P., Park, S. Y., and Hong, K. S., Cartilage generation using alginate-encapsulated autogenous chondrocytes in rabbits, *Ann. Otol. Rhinol. Laryngol.*, 109, 1157–1161, 2000.
46. Loty, S., Sautier, J. M., Loty, C., Boulekbache, H., Kokubo, T., and Forest, N., Cartilage formation by fetal rat chondrocytes cultured in alginate beads: a proposed model for investigating tissue–biomaterial interactions, *J. Biomed. Mater. Res.*, 42, 213–222, 1998.
47. Marijnissen, W., van Osch, G., Aigner, J., van der Veen, S. W., Hollander, A. P., Verwoerd-Verhoef, H. L., and Verhaar, J. A. N., Alginate as a chondrocyte-delivery substance in combination with a non-woven scaffold for cartilage tissue engineering, *Biomaterials*, 23, 1511–1517, 2002.
48. Kuo, C. K., and Ma, P. X., Ionically crosslinked alginate hydrogels as scaffolds for tissue engineering: Part 1. Structure, gelation rate and mechanical properties, *Biomaterials*, 22, 511–521, 2001.
49. Glicklis, R., Shapiro, L., Agbaria, R., Merchuk, J. C., and Cohen, S., Hepatocyte behavior within three-dimensional porous alginate scaffolds, *Biotechnol. Bioeng.*, 67, 344–353, 2000.
50. Lindenhayn, K., Perka, C., Spitzer, R. S., Heilmann, H. H., Pommerening, K., Mennicke, J., and Sittinger, M., Retention of hyaluronic acid in alginate beads: aspects for in vitro cartilage engineering, *J. Biomed. Mater. Res.*, 44, 149–155, 1999.
51. Eiselt, P., Yeh, J., Latvala, R. K., Shea, L. D., and Mooney, D. J., Porous carriers for biomedical applications based on alginate hydrogels, *Biomaterials*, 21, 1921–1927, 2000.
52. Caterson, E. J., Nesti, L. J., Li, W. J., Danielson, K. G., Albert, T. J., Vaccaro, A. R., and Tuan, R. S., Three-dimensional cartilage formation by bone marrow-derived cells seeded ion polylactide/alginate amalgam, *J. Biomed. Mater. Res.*, 57, 394–403, 2001.
53. Caterson, E. J., Li, W. J., Nesti, L. J., Albert, T., Danielson, K., and Tuan, R. S., Polymer/alginate amalgam for cartilage-tissue engineering, *Reparative Medicine: Growing Tissues and Organs*, 134–138, 2002.

54. Rowley, J. A., Madlambayan, G., and Mooney, D. J., Alginate hydrogels as synthetic extracellular matrix materials, *Biomaterials*, 20, 45–53, 1999.
55. Alsberg, E., Anderson, K. W., Albeiruti, A., Franceschi, R. T., and Mooney, D. J., Cell-interactive alginate hydrogels for bone tissue engineering, *J. Dent. Res.*, 80, 2025–2029, 2001.
56. Sultzbaugh, K. J., and Speaker, T. J., A method to attach lectins to the surface of spermine alginate microcapsules based on the avidin biotin interaction, *J. Microencapsul.*, 13, 363–376, 1996.
57. Suzuki, Y., Tanihara, M., Suzuki, K., Saitou, A., Sufan, W., and Nishimura, Y., Alginate hydrogel linked with synthetic oligopeptide derived from bmp-2 allows ectopic osteoinduction in vivo, *J. Biomed. Mater. Res.*, 50, 405–409, 2000.
58. Lee, K. Y., Rowley, J. A., Eiselt, P., Moy, E. M., Bouhadir, K. H., and Mooney, D. J., Controlling mechanical and swelling properties of alginate hydrogels independently by cross-linker type and cross-linking density, *Macromolecules*, 33, 4291–4294, 2000.
59. Bouhadir, K. H., Yue, I., Chen, L., Hausman, D., and Mooney, D. J., Synthesis and cross-linking of partially oxidized alginate for tissue engineering applications, *Abstr. Pap. Am. Chem. Soc.*, 216, 078-BTEC, 1998.
60. Bouhadir, K. H., Yue, I., Hausman, D., and Mooney, D. J., Cross-link limit-oxidized alginate for tissue engineering applications, *J. Dent. Res.*, 77, 1169, 1998.
61. Bouhadir, K. H., Hausman, D. S., and Mooney, D. J., Synthesis of cross-linked poly(aldehyde guluronate) hydrogels, *Polymer*, 40, 3575–3584, 1999.
62. Bouhadir, K. H., Lee, K. Y., Alsberg, E., Damm, K. L., Anderson, K. W., and Mooney, D. J., Degradation of partially oxidized alginate and its potential application for tissue engineering, *Biotechnol. Prog.*, 17, 945–950, 2001.
63. Paige, K. T., Cima, L. G., Yaremchuk, M. J., Vacanti, J. P., and Vacanti, C. A., Injectable cartilage, *Plast. Reconstr. Surg.*, 96, 1390–1398, 1995.
64. Caplan, A. I., Tissue engineering designs for the future: new logics, old molecules, *Tissue Eng.*, 6, 1–8, 2000.
65. Radice, M., Brun, P., Cortivo, R., Scapinelli, R., Battaliard, C., and Abatangelo, G., Hyaluronan-based biopolymers as delivery vehicles for bone-marrow-derived mesenchymal progenitors, *J. Biomed. Mater. Res.*, 50, 101–109, 2000.
66. Tonello, C., Zavan, B., Cortivo, R., Brun, P., Panfilo, S., and Abatangelo, G., In vitro reconstruction of human dermal equivalent enriched with endothelial cells, *Biomaterials*, 24, 1205–1211, 2003.
67. Choi, Y. S., Hong, S. R., Lee, Y. M., Song, K. W., Park, M. H., and Nam, Y. S., Studies on gelatin-containing artificial skin: II. Preparation and characterization of cross-linked gelatin-hyaluronate sponge, *J. Biomed. Mater. Res.*, 48, 631–639, 1999.
68. Angele, P., Kujat, R., Nerlich, M., Yoo, J., Goldberg, V., and Johnstone, B., Engineering of osteochondral tissue with bone marrow mesenchymal progenitor: cells in a derivatized hyaluronan–gelatin composite sponge, *Tissue Eng.*, 5, 545–553, 1999.
69. Leach, J. B., Bivens, K. A., Patrick, C. W., and Schmidt, C. E., Photocrosslinked hyaluronic acid hydrogels: natural, biodegradable tissue engineering scaffolds, *Biotechnol. Bioeng.*, 82, 578–589, 2003.
70. Nam, H. S., Kim, J. H., An, J. H., and Chung, D. J., Synthesis of hyaluronic acid scaffold for tissue engineering and evaluation of its drug release behaviors, *Polymer — Korea*, 25, 476–485, 2001.
71. Smeds, K. A., and Grinstaff, M. W., Photocrosslinkable polysaccharides for in situ hydrogel formation, *J. Biomed. Mater. Res.*, 54, 115–121, 2001.
72. Taguchi, T., Ikoma, T., and Tanaka, J., An improved method to prepare hyaluronic acid and type II collagen composite matrices, *J. Biomed. Mater. Res.*, 61, 330–336, 2002.
73. Alini, M., Li, W., Markovic, P., Aebi, M., Spiro, R. C., and Roughley, P. J., The potential and limitations of a cell-seeded collagen/hyaluronan scaffold to engineer an intervertebral disc-like matrix, *Spine*, 28, 446–453, 2003.
74. Allemann, M., Mizuno, S., Eid, K., Yates, K. E., Zaleske, D., and Glowacki, J., Effects of hyaluronan on engineered articular cartilage extracellular matrix gene expression in 3-dimensional collagen scaffolds, *J. Biomed. Mater. Res.*, 55, 13–19, 2001.

75. Schmidt, C., Rivers, T., Hudson, T., and Collier, J., Modification of electroactive biomaterials for neural engineering applications, *Conducting Polymers and Polymer Electrolytes: From Biology to Photovoltaics*, 154–165, 2003.
76. Collier, J. H., Camp, J. P., Hudson, T. W., and Schmidt, C. E., Synthesis and characterization of polypyrrole–hyaluronic acid composite biomaterials for tissue engineering applications, *J. Biomed. Mater. Res.*, 50, 574–584, 2000.
77. Dausse, Y., Grossin, L., Miralles, G., Pelletier, S., Mainard, D., Hubert, P., Baptiste, D., Gillet, P., Dellacherie, E., Netter, P., and Payan, E., Cartilage repair using new polysaccharidic biomaterials: macroscopic, histological and biochemical approaches in a rat model of cartilage defect, *Osteoarthritis Cartilage*, 11, 16–28, 2003.
78. Ohya, S., Nakayama, Y., and Matsuda, T., Thermoresponsive artificial extracellular matrix for tissue engineering: hyaluronic acid bioconjugated with poly(*N*-isopropylacrylamide) grafts, *Biomacromolecules*, 2, 856–863, 2001.
79. Wollina, U., Konrad, H., and Fischer, T., Recessive epidermolysis bullosa dystrophicans (hallopeau-siemens) — improvement of wound healing by autologous epidermal grafts on an esterified hyaluronic acid membrane, *J. Dermatol.*, 28, 217–220, 2001.
80. Milella, E., Brescia, E., Massaro, C., and Ramires, P. A., Chemico-physical properties of hyaluronan-based sponges, *J. Biomed. Mater. Res.*, 52, 695–700, 2000.
81. Milella, E., Brescia, E., Massaro, C., Ramires, P. A., Miglietta, M. R., Fiori, V., and Aversa, P., Physico-chemical properties and degradability of non-woven hyaluronan benzylic esters as tissue engineering scaffolds, *Biomaterials*, 23, 1053–1063, 2002.
82. Grigolo, B., Lisignoli, G., Piacentini, A., Fiorini, M., Gobbi, P., Mazzotti, G., Duca, M., Pavesio, A., and Facchini, A., Evidence for redifferentiation of human chondrocytes grown on a hyaluronan-based biomaterial (HYAFF® 11): Molecular, immunohistochemical and ultrastructural analysis, *Biomaterials*, 23, 1187–1195, 2002.
83. Brun, P., Cortivo, R., Zavan, B., Vecchiato, N., and Abatangelo, G., In vitro reconstructed tissues on hyaluronan-based temporary scaffolding, *J. Mater. Sci.-Mater. Med.*, 10, 683–688, 1999.
84. Solchaga, L. A., Dennis, J. E., Goldberg, V. M., and Caplan, A. I., Hyaluronic acid-based polymers as cell carriers for tissue-engineered repair of bone and cartilage, *J. Orthop. Res.*, 17, 205–213, 1999.
85. Hollander, D. A., Soranzo, C., Falk, S., and Windolf, J., Extensive traumatic soft tissue loss: reconstruction in severely injured patients using cultured hyaluronan-based three-dimensional dermal and epidermal autografts, *J. Trauma-Inj. Infect. Crit. Care*, 50, 1125–1136, 2001.
86. Hu, M., Sabelman, E. E., Lai, S., Timek, E. K., Zhang, F., Hentz, V. R., and Lineaweaver, W. C., Polypeptide resurfacing method improves fibroblast's adhesion to hyaluronan strands, *J. Biomed. Mater. Res.*, 47, 79–84, 1999.
87. Maurer, A. M., Han, Z. C., Dhermy, D., and Briere, J., Glycosaminoglycans enhance human leukemic-cell growth in-vitro, *Leuk. Res.*, 18, 837–842, 1994.
88. Flint, N., Cove, F. L., and Evans, G. S., Heparin stimulates the proliferation of intestinal epithelial-cells in primary culture, *J. Cell Sci.*, 107, 401–411, 1994.
89. Fager, G., Hansson, G. K., Ottosson, P., Dahllof, B., and Bondjers, G., Human arterial smooth-muscle cells in culture — effects of platelet-derived growth-factor and heparin on growth-in vitro, *Exp. Cell Res.*, 176, 319–335, 1988.
90. Cavari, S., Ruggiero, M., and Vannucchi, S., Antiproliferative effects of heparin on normal and transformed NIH 3T3 fibroblasts, *Cell Biol. Int.*, 17, 781–786, 1993.
91. McPherson, J. M., Ledger, P. W., Ksander, G., Sawamura, S. J., Conti, A., Kincaid, S., Michaeli, D., and Clark, R. A. F., The influence of heparin on the wound-healing response to collagen implants in vivo, *Collagen Rel. Res.*, 8, 83–100, 1988.
92. Kratz, G., Back, M., Arnander, C., and Larm, O., Immobilised heparin accelerates the healing of human wounds in vivo, *Scand. J. Plast. Reconstr. Surg. Hand Surg.*, 32, 381–385, 1998.
93. Kratz, G., Arnander, C., Swedenborg, J., Back, M., Falk, C., Gouda, I., and Larm, O., Heparin–chitosan complexes stimulate wound healing in human skin, *Scand. J. Plast. Reconstr. Surg. Hand Surg.*, 31, 119–123, 1997.
94. Tiwari, A., Salacinski, H. J., Punshon, G., Hamilton, G., and Seifalian, A. M., Development of a hybrid cardiovascular graft using a tissue engineering approach, *FASEB J.*, 16, 791–796, 2002.

95. Tanihara, M., Suzuki, Y., Yamamoto, E., Noguchi, A., and Mizushima, Y., Sustained release of basic fibroblast growth factor and angiogenesis in a novel covalently crosslinked gel of heparin and alginate, *J. Biomed. Mater. Res.*, 56, 216–221, 2001.
96. Pieper, J. S., van Wachem, P. B., van Luyn, M. J. A., Brouwer, L. A., Hafmans, T., Veerkamp, J. H., and van Kuppevelt, T. H., Attachment of glycosaminoglycans to collagenous matrices modulates the tissue response in rats, *Biomaterials*, 21, 1689–1699, 2000.
97. Yang, H. Y., and Zhang, Q. Q., Preparation and characterization of collagen-GAGs bioactive matrices for tissue engineering, *J. Mater. Sci. Technol.*, 17, 495–500, 2001.
98. van Susante, J. L. C., Pieper, J., Buma, P., van Kuppevelt, T. H., van Beuningen, H., van der Kraan, P. M., Veerkamp, J. H., van den Berg, W. B., and Veth, R. P. H., Linkage of chondroitin-sulfate to type I collagen scaffolds stimulates the bioactivity of seeded chondrocytes in vitro, *Biomaterials*, 22, 2359–2369, 2001.
99. Pieper, J. S., Oosterhof, A., Dijkstra, P. J., Veerkamp, J. H., and van Kuppevelt, T. H., Preparation and characterization of porous crosslinked collagenous matrices containing bioavailable chondroitin sulphate, *Biomaterials*, 20, 847–858, 1999.
100. Pieper, J. S., van der Kraan, P. M., Hafmans, T., Kamp, J., Buma, P., van Susante, J. L. C., van den Berg, W. B., Veerkamp, J. H., and van Kuppevelt, T. H., Crosslinked type II collagen matrices: preparation, characterization, and potential for cartilage engineering, *Biomaterials*, 23, 3183–3192, 2002.
101. Park, Y. J., Lee, Y. M., Lee, J. Y., Seol, Y. J., Chung, C. P., and Lee, S. J., Controlled release of platelet-derived growth factor-BB from chondroitin sulfate–chitosan sponge for guided bone regeneration, *J. Control. Release*, 67, 385–394, 2000.
102. Sechriest, V. F., Miao, Y. J., Niyibizi, C., Westerhausen-Larson, A., Matthew, H. W., Evans, C. H., Fu, F. H., and Suh, J. K., GAG-augmented polysaccharide hydrogel: a novel biocompatible and biodegradable material to support chondrogenesis, *J. Biomed. Mater. Res.*, 49, 534–541, 1999.
103. Chupa, J. M., Foster, A. M., Sumner, S. R., Madihally, S. V., and Matthew, H. W. T., Vascular cell responses to polysaccharide materials: in vitro and in vivo evaluations, *Biomaterials*, 21, 2315–2322, 2000.
104. Cai, Q., Yang, J. A., Bei, J. Z., and Wang, S. G., A novel porous cells scaffold made of polylactide-dextran blend by combining phase-separation and particle-leaching techniques, *Biomaterials*, 23, 4483–4492, 2002.
105. Groman, E. V., Enriquez, P. M., Jung, C., and Josephson, L., Arabinogalactan for hepatic drug delivery, *Bioconjug. Chem.*, 5, 547–556, 1994.
106. Ehrenfreund-Kleinman, T., Gazit, Z., Gazit, D., Azzam, T., Golenser, J., and Domb, A. J., Synthesis and biodegradation of arabinogalactan sponges prepared by reductive amination, *Biomaterials*, 23, 4621–4631, 2002.
107. Ehrenfreund-Kleinman, T., Golenser, J., and Domb, A. J., Polysaccharide scaffolds prepared by cross-linking of polysaccharides with chitosan or proteins for cell growth. *J. Bioact. Compat. Polym.*, 18 (5), 323–338, 2003.

4 Role of Gelatin in the Release Carrier of Growth Factor for Tissue Engineering

Yasuhiko Tabata

CONTENTS

I. WHAT IS TISSUE ENGINEERING?

A. Factors Necessary for Tissue Engineering

When the body tissue or organ is severely injured, largely lost or dysfunctional, it is clinically treated with either reconstructive surgery or organ transplantation. There is no doubt that reconstructive surgery has saved or improved countless patients' lives. However, the biomedical materials or devices used can neither prevent progressive patient deterioration nor substitute all organ functions. One of the largest problems for organ transplantation is the shortage of donor tissues or organs. Additionally, the permanent medication of immunosuppressive agents causes various side-effects, such as viral infections and carcinogenesis. Because of these issues, a therapeutic trial to induce the regeneration of patients' tissues and organs by making use of their own self-healing potential has been initiated. Tissue engineering, a research and development field of biomedical engineering and technology, has been used in this new therapeutic trial. The objective of tissue engineering is to manipulate cells to induce tissue regeneration, therapeutically substituting the biological functions of defective or lost tissues and damaged organs.

It is undoubtedly necessary for tissue regeneration to make use of key tissue cells. However, by providing the cells to a target site only, tissue regeneration is not always successful. In addition to cells, it is necessary to build an appropriate environment which enables the cells to induce regeneration at the site. It has been recognized that cells generally interact with the extracellular matrix (ECM) to live and function in the body. Considering this *in vivo* system, it is easy to understand the necessity of creating a suitable environment for the induction of tissue regeneration. For example, one of the possible ways to create a suitable environment is to provide scaffolds for cell attachment and the subsequent proliferation and differentiation from the target site. The scaffold is the second factor in regeneration induction. The scaffold should have a porous structure which facilitates either cell infiltration into the matrix or oxygen and nutrition supply to cells. The scaffold material must be biologically compatible with the cells. Since the remaining scaffold often physically impairs the tissue regeneration process, it is crucial to control the biodegradability of matrix material.

If the tissue to be repaired has a high regeneration potential, new tissue will form in a scaffold matrix implanted with active cells taken from surrounding healthy tissue. However, if the regeneration potential of the tissue is very low because of, for instance, low concentration of the cells and growth factors responsible for new tissue generation, additional means are required. In this case, it will be a practically promising way to make use of cells and growth factor. The cells are used alone or with the scaffold to induce tissue regeneration. However, the direct injection of growth factor (in the solution form) into the regeneration site is often ineffective, as the growth factor injected rapidly deactivates or diffuses out of the site. To enable the growth factor to efficiently exert the biological effects, the technology and methodology of drug delivery systems (DDS) are practical. One promising technique is the controlled release of growth factor at the site of action over an extended time period by inserting the growth factor into an appropriate carrier. It is expected that the growth factor is protected from proteolysis for at least as long as it is incorporated in the release carrier, for prolonged retention of the activity *in vivo*. The release carrier should degrade in the body since it is not needed after the growth factor release is complete. Thus, to successfully induce tissue regeneration using scaffolds and DDS technologies it is vital to create a suitable environment. Without the efficient creation of a regeneration environment, the possibilities of regenerative medicine for patients cannot be realized.

B. Biodegradable Materials for Tissue Engineering

Considering the applications of scaffolds and DDSs described above, it is undoubtedly preferable to take advantage of biodegradable materials for tissue engineering. Thus, here the materials are briefly explained. The word "biodegradability" is defined as the ability of a material to enzymatically or nonenzymatically degrade and disappear from the original body site. Many materials have been investigated for their medical and pharmaceutical applications.[1] Table 4.1 summarizes synthetic and natural polymers of a biodegradable nature. Among them, homopolymer of lactide and its copolymers with glycolide and ε-caprolactone are the most extensively investigated synthetic biodegradable polymers used clinically. This is because their biodegradable profile can be readily controlled by changing the molecular weight and copolymer compositions. In addition, it is recognized that the constituting monomers are nontoxic because they are all metabolites in the body. Other synthetic polymers have been investigated for their biomedical and pharmaceutical applications. In comparison with the synthetic polymers, natural polymers have longer histories of medical and pharmaceutical usage. Proteins (collagen, gelatin, fibrin, and albumin), polysaccharides (chitin, hyaluronic acid, cellulose, and dextran) and their chemical derivatives have been clinically employed. Generally, the degradation of polymers is driven by oxidative, hydrolytic, and enzymatic cleavage of their main chain. Most synthetic polymers are fundamentally degraded by hydrolysis although polyamino acids show enzymatic degradation. In comparison, natural polymers degrade enzymatically. There is also a form of degradation in which a polymer becomes water-soluble with the chemical elimination of the side chain, and disappears from the original site. The natural polymers are being used the hydrogel prepared through their cross-linking. No metal materials in the body exist,

TABLE 4.1
Synthetic and Natural Polymers of Biodegradable Nature Based on the Chemical Structure

Synthetic Polymer			Natural Polymer (Animals, Plants, Microbes)		
The Name and Structure of Chemical Bond		**Example**	**The Name and Structure of Chemical Bond**		**Example**
Ester	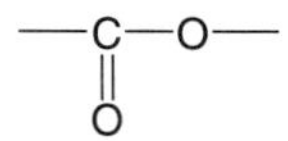	Polylactide, polyglycolide, lactideglycolide copolymer, poly (ε-caprolactone), poly(p-dioxane), poly(β-malic acid)	Ester	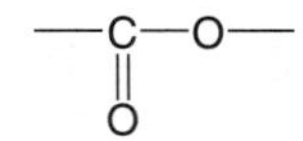	Poly(β-hydroxybutyrate), poly(malic acid)
Anhydride	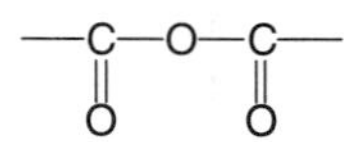	Poly(anhydrides)			
Ortho ester	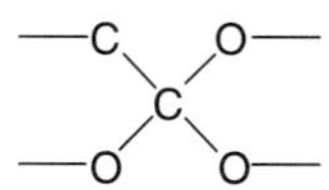	Poly(ortho esters)			
Carbonate	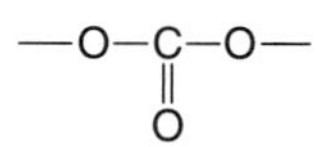	Polycarbonates	Glycoside (Polysaccharide)	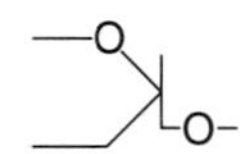	Chitin, chitosan, hyaluronic acid, pectin, pectic acid, galactan, starch, dextran, pullulan, agarose, heparin, alginate, chondroitin-6-sulfate
Phosphazene	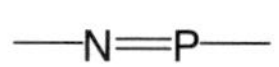	poly(phosphazenes)			
Peptide	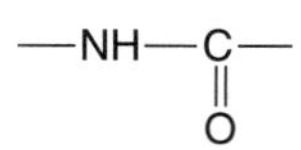	Poly(amino acids)	Peptide (Protein)	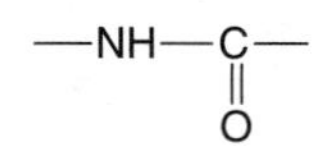	Collagen, gelatin, fibrin, albumin, gluten, polypeptides, elastin, fibroin, enzyme
Phosphoric ester	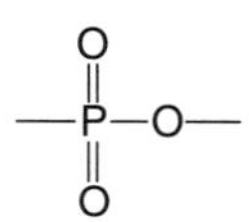	Poly(phosphoric ester-urethanes)	Phosphoric ester (Nucleic acid)	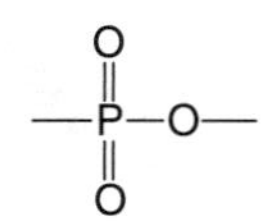	Deoxyribonucleic acid (DNA), ribonucleic acid (RNA)
Carbon–carbon	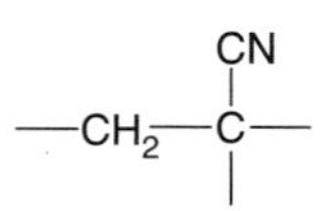	Poly(cyanoacrylates)			

being corroded. Among ceramics, only the tricalcium phosphate and calcium carbonate are biodegradable.

II. TISSUE ENGINEERING TECHNOLOGY TO INDUCE TISSUE REGENERATION

A. Role of Biomaterials in Tissue Engineering

Tissue engineering approaches include two types of tissue construction and regeneration: *in vitro* and *in vivo* tissue engineering. *In vitro* tissue engineering involves tissue regeneration and organ substitution (bioartificial organs). If a tissue can be reconstructed *in vitro* in factories or laboratories on a large scale, we can supply the tissue construct to patients when it is needed. This is an ideal therapy with commercial potential. However, with the cell culture technology currently available, it is quite difficult to arrange the environment required for tissue construction. Although there have been attempts to reconstruct various tissues *in vitro*, only skin dermis reconstruction has been successful. Another application of the *in vitro* tissue engineering is using allogeneic or xenogeneic cells to substitute organ functions. Such engineered organs are called bioartificial organs. For example, liver and pancreatic Langerhans cells are combined with man-made immunoisolation membranes in a bid to maintain cell functions.

However, it is usually anticipated that most of the biological substances necessary for tissue regeneration are automatically supplied by the host living body. Therefore, *in vivo* tissue engineering is becoming more and more popular. If the healthy ECM is still available in the body, no artificial scaffold is required. However, for the regeneration of a large defect it is necessary to use cells, the scaffold, and growth factor, or a combination of the three.

There are some cases in which growth factor is required for the *in vivo* tissue engineering. The type of growth factors depends on the tissue and the site where the regeneration is expected. As described above, it is undoubtedly indispensable for successful tissue regeneration with growth factors to make advantage of DDS technology. Research on *in vivo* tissue regeneration through a combination of growth factors and various carriers has demonstrated that good growth factor-carrier combination induces tissue regeneration. However, despite the necessity of growth factor release, few basic researches have been performed to evaluate the release effect on tissue regeneration. There are some reports that using growth factor in solution form was effective in enhancing tissue regeneration. However, the dosage is too high to clinically accept the therapeutic trials. Recently the gene encoding growth factor has been applied for tissue engineering.[2–4] The idea is that transfected cells secrete a growth factor for a certain time period, promoting tissue regeneration. If similar tissue regeneration can be achieved with either a growth factor or the gene, it would be preferable, from a clinical perspective, to use growth factor, provided the controlled release of protein is possible. A combination of growth factors, with or without preseeded cells, will accelerate the tissue regeneration.

A body defect is often filled with the fibrous tissue produced by the fibroblasts ubiquitous to the body and can rapidly proliferate. Once such fibrosis takes place, any tissue repair or regeneration at the defect can no longer be expected. To prevent generation of the fibrous tissue, we need the barrier membrane of biomaterials which will protect the defect space from the undesired tissue growth to make space for tissue regeneration and repair. Tissue regeneration at a defect protected by a barrier membrane will also be accelerated by combining cells and growth factor. For every role, biodegradable biomaterials will be useful since they can be chemically designed to gradually degrade, harmonizing the process of tissue regeneration, and finally disappearing in the body after their task is accomplished. Gelatin is one of the most suitable candidates because it has been extensively used for industrial, pharmaceutical, and medical purposes. The biosafety of gelatin has been proved through its long clinical usage. Biodegradability can also be controlled by cross-linking conditions which have been experimentally and practically proven to function efficiently.

The hydrogel of gelatin and the tissue engineering applications, especially the potential as DDS materials, are overviewed in the following section.

III. CONTROLLED RELEASE OF GROWTH FACTOR FROM BIODEGRADABLE HYDROGEL

A. Concept of Growth Factor Release

The drugs of interest for tissue regeneration are growth factor and the related gene. One of the largest problems in protein release technology is the loss of biological activity in a protein released from a formulation of protein and release carrier. It has been demonstrated that this activity loss often results from denaturation and deactivation of protein during the formulation process for protein release. Therefore, new methods to prepare the protein–carrier formulation should be exploited to minimize protein denaturation. Various materials of polymers and ceramics have been investigated for protein formulation. However, strong interactions of protein with the materials often cause denaturation. When exposed to harsh environmental changes, such as heating and exposure to sonication or organic solutions, protein is generally denatured, losing its biological activity.[5] From this viewpoint, polymer hydrogel may be a preferable candidate as a protein release carrier because of its biosafety and high inertness toward protein drugs. However, it will be impossible to achieve the controlled release of protein over a long time period from the hydrogel since the rate of protein release is generally diffusion-controlled through aqueous channels in the inside of hydrogels. Thus, one possible means of protein release from hydrogel is to immobilize a growth factor in a biodegradable hydrogel which, as a result of hydrogel water-solubilization with the biodegradation, will ensure protein release.

Chemical and physical methods are available to immobilize the growth factor in the hydrogel. Chemical methods have been investigated, but often result in the denaturation and activity lowering of protein because the functional groups of protein are chemically modified by immobilization reaction. Therefore, physical methods are preferable for growth factor integrity. In fact, such a physical immobilization is often observed in growth factors existing in the body.[6] Some growth factors are naturally stored in the body; they are stored ionically with the acidic polysaccharides of ECM (such as heparin sulfate and heparin) because most of them have a positively charged site on the molecular surface. It is recognized that, when needed, the complexed growth factor is water-solubilized by the enzymes secreted from the surrounding cells released from the ECM complex during ECM degradation. This complexation also protects growth factors from their denaturation and general enzymatic degradation *in vivo*.

We have designed and artificially created the release system of growth factors which mimics that in the living body. Figure 4.1 shows the simulated controlled release of growth factor from a biodegradable polymer hydrogel based on physicochemical interaction forces between the growth factor and polymer molecules. For example, a hydrogel is prepared from a biodegradable polymer with negative charges. The growth factor, with a positively charged site, is electrostatically interacted with the polymer chain for physical immobilization of the growth factor in the hydrogel carrier. If an environmental change, such as increased ionic strength, occurs, the immobilized growth factor will be released from the factor–carrier formulation. Even if such an environmental change does not take place, degradation of the carrier itself will also lead to growth factor release. Because degradation is more likely to happen *in vivo* than an environmental change, it is preferable to prepare the release carrier from biodegradable polymers. To release a negatively charged protein and plasmid DNA, a biodegradable polymer with positive charges is used for hydrogel preparation. In addition to the electrostatic interaction, other interaction forces, such as hydrogen bonding, hydrophobic interaction, and co-ordination bonding forces, will be available to design this hydrogel release system.[7,8]

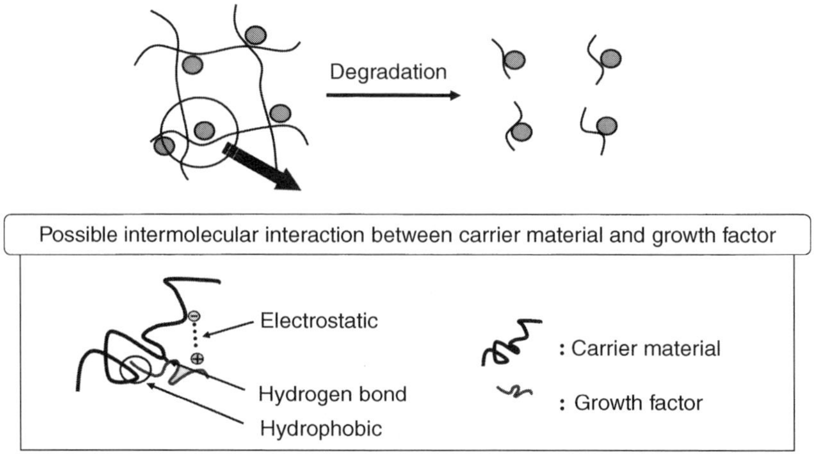

FIGURE 4.1 Conceptive illustration of growth factor release from biodegradable hydrogel based on physicochemical interaction forces.

B. Characterization of Gelatin and Gelatin Hydrogel

For growth factor release on the basis of physicochemical interaction forces, it is imperative to employ a bio-safe polymer with groups able to interact with the factor as the carrier material. In addition, if short-term biodegradability is required, the material used will be limited to natural polymers with charged groups. Gelatin is a biodegradable material which has been confirmed as bio-safe through extensive clinical applications. Another unique advantage of gelatin is its electrical nature which can be readily modified by the collagen processing method and conditions. For example, the alkaline process denatures collagen and hydrolyses the amide bond of glutamine and asparagine side chains to yield a high density of carboxyl groups, which makes the gelatin negatively charged. An acid process denatures collagen without hydrolysis to yield a positive charged gelatin of "basic" type. It was found that as expected, positively charged growth factors, such as basic fibroblast growth factor (bFGF), transforming growth factor β1 (TGF-β1), hepatocyte growth factor (HGF) or platelet-derived growth factor (PDGF) were physicochemically immobilized into the hydrogel of acidic gelatin, mainly due to the electrostatic interaction.[9] Additionally, it is easy to chemically derivatize gelatin because it composes of amino acids with different functional groups and the derivatization has been extensively investigated and performed. For example, cationized gelatin with different extents of aminization can be prepared through adding the reaction of amino groups to carboxyl groups. The cationized gelatin is cross-linked to prepare hydrogels for the controlled release of plasmid DNA because it forms a polyion complex with the plasmid DNA.[10,11]

Animal experiments revealed that the hydrogels prepared through glutaraldehyde cross-linking of the acidic gelatin degraded with time in the body.[12] The degradation period of hydrogels depends on their water content which is a measure of their cross-linking extent: the higher the water content of the hydrogels, the faster their *in vivo* degradation. The cross-linking extent of hydrogels could be changed by the conditions of hydrogel preparation. When traced in the back subcutis of mice by use of gelatin hydrogels incorporating ^{125}I-labeled bFGF or ^{125}I-labeled gelatin hydrogels incorporating bFGF, the residual radioactivities decreased with implantation time and the decrement rate increased with the increased water content of hydrogels. The time profile of bFGF radioactivity remaining in the hydrogel depended on the hydrogel degradability and was in accordance with that of remaining hydrogel radioactivity[12] (Figure 4.2). These findings strongly indicate that the growth factor release is governed mainly by hydrogel degradation as described in Figure 4.1. If the release

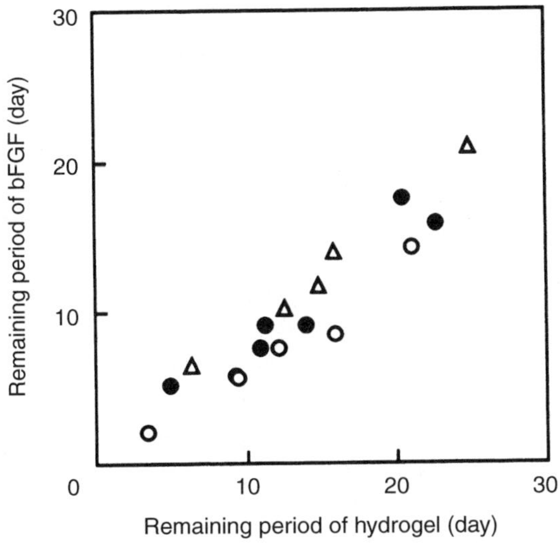

FIGURE 4.2 Relationship between the radioactivity remaining periods of bFGF and gelatin hydrogel at which 80% (clear circle), 50% (filled circle), and 20% of initial radioactivity (clear triangle) remain, after implantation of ^{125}I-labeled bFGF-incorporating acidic gelatin hydrogels and ^{125}I-labeled gelatin hydrogels incorporating bFGF with different water contents into the back subcutis of mice.

is regulated by the simple diffusion of growth factor from the hydrogel carrier, it is likely that the release rate became higher with a decrease in the carrier size because of higher surface area/carrier volume ratio. On the contrary, for this release system, since the time period of bFGF release can be controlled only by changing that of hydrogel degradation,[9,12] the release profile is not influenced by the shape of the hydrogel carrier (for example, disks, tubes, sheets, granules, and microspheres). It should be noted that gelatin hydrogel microspheres of injectable size achieved the controlled release of bFGF to induce angiogenesis.[13] For the controlled release of growth factors, the mechanical property of gelatin hydrogel is not always a problem. However, depending on the objective of hydrogel use, modification and improvement of hydrogel's mechanical property is required for future applications using gelatin.[14]

IV. TISSUE REGENERATION BY GELATIN HYDROGELS INCORPORATING GROWTH FACTOR

As described in the previous section, the gelatin hydrogel was found to be a good carrier for the controlled release of growth factor. Successful tissue regeneration and organ substitution by cell transplantation has been achieved by using gelatin hydrogel for growth factor release (Table 4.2). In this section, several results of tissue regeneration achieved by this release system are introduced to emphasize the significance of gelatin hydrogel-based DDS in tissue engineering.

A. Angiogenesis

bFGF was originally characterized as an *in vitro* growth factor for fibroblasts and capillary endothelial cells and as a potent *in vivo* mitogen and chemoattractant for a wide range of cells. In addition, it is reported that bFGF has a variety of biological activities[15] and effectively enhances wound healing through inducing of angiogenesis. Whcn a gelatin hydrogel incorporating bFGF was subcutaneously implanted into a mouse back, an angiogenic effect was observed around the implanted site, in marked contrast to the sites implanted with a bFGF-free empty gelatin

TABLE 4.2
Approaches to Tissue and Organ Regeneration Based on the Controlled Release of Growth Factors from Biodegradable Hydrogels

Materials	Growth Factor	Animals	Effect	Objective
Acidic gelatin (isoelectric point, 5.0)	bFGF	Mice, rats, and dogs	Angiogenesis	Transplantation of Langerhans' islands for diabetes therapy
		Rats	Angiogenesis	Transplantation of hepatocytes for therapy of enzyme deficiency disease
		Rats	Angiogenesis	Transplantation of renal epithelial cells
		Rats and dogs	Angiogenesis	Transplantation of cardiomyocytes
		Rats and guinea pigs	Angiogenesis	Repairing the dermal skin layer
		Rats and pigs	Angiogenesis	Treatment of cardiac infarction
		Rabbits	Angiogenesis	Treatment of lower limb ischemia
		Rats, dogs, and monkeys	Osteogenesis and angiogenesis	Repairing the sternum and connective tissue
		Rats, rabbits, and monkeys	Osteogenesis	Repairing the skull
		Rats, rabbits, dogs, and monkeys	Osteogenesis	Repairing long bones
		Mice	Adipogenesis	Repairing breast tissue and soft tissue reconstruction
		Mice	Angiogenesis and activation of hair follicle tissue	Promoting hair growth
		Dogs	Periodontium repair	Repairing periodontium
		Dogs	Peripheral nerve repair	Nerve repairing
		Dogs	Osteogenesis	Repairing the mandible
	TGF-β1	Rabbits and monkeys	Osteogenesis	Repairing the skull, long bones, and the mandible
		Goats	Chondrogenesis	Repairing tracheal cartilages

	HGF	Mice	Angiogenesis and activation of hair follicle tissue	Promoting hair growth
		Rats and pigs	Angiogenesis and inhibition of apoptosis	Treatment of dilated cardiomyopathy
	bFGF/ TGF-β1	Rabbits	Osteogenesis	Repairing the skull
	CTGF	Rabbits	Chondrogenesis	Repairing articular cartilage
Basic gelatin (isoelectric point, 9.0)	BMP-2	Rats, dogs, and monkeys	Osteogenesis	Repairing the skull and the mandible
		Dogs	Chondrogenesis	Repairing tracheal cartilages
collagen	TGF-β1	Rabbits	Osteogenesis	Repairing the skull
	VEGF	Pigs	Angiogenesis	Treatment of myocardiac infarction
		Rabbits	Angiogenesis	Promoting engraftment of soft tissue grafts
		Rabbits	Osteogenesis	Osteogenesis for spinal fusion
		Mice	Angiogenesis and activation of hair follicle tissue	Promoting hair growth

bFGF = fibroblast growth factot; TGF = transforming growth factor; HGF = hepatocyte growth factor; CTGF = connective tissue growth factor; VEGF = vascular endothelial growth factor; BMP = bone morphogenetic protein.

hydrogel or injected with an aqueous solution of bFGF.[9,12] No angiogenesis was induced by the injection of bFGF solution even when the dose was increased to 1 mg per site, because of a rapid excretion of bFGF from the injected site.[16] On the contrary, the gelatin hydrogel incorporating bFGF induced significant angiogenesis at a dose as low as 30 μg per site. The maintenance period of hydrogel-induced angiogenic effect could be regulated by the hydrogel water content and prolonged as the water content became lower.[12,16] It is likely that the hydrogels with lower water contents are degraded and consequently release bFGF of biological activity *in vivo* more slowly than those with higher water contents, leading to a prolonged angiogenic effect. A similarly enhanced and prolonged angiogenic effect was also observed upon using a gelatin hydrogel incorporating bFGF of microsphere type.[13] The hydrogel of cationized gelatin also functioned well to release the plasmid DNA of FGF4 and induce therapeutically acceptable angiogenesis in the hind-limb ischemia, in contrast to that of free plasmid DNA.[17]

The technology to induce *in vivo* angiogenesis is very important for tissue engineering, and the objective is the therapy of ischemic disease by inducing angiogenesis which is necessary for cell transplantation. Here, as in the former example, gelatin hydrogels incorporating bFGF were shown to have a good therapeutic effect on the hind-limb ischemia. When injected into the muscle infarction which was prepared by ligating the femoral artery and its branches of streptozotocin-induced diabetic rats, the gelatin microspheres incorporating bFGF increased the number of collateral vessels in the ischemic muscle (Figure 4.3). No angiogenic effect was observed from the injection of bFGF solution at the same dose.

Sufficient supply of nutrients and oxygen to cells transplanted in the body is indispensable for cell survival and the maintenance of biological functions. Without sufficient supply, the cells preseeded in a scaffold for tissue regeneration would hardly survive following implantation of the scaffold into the body. This is the issue when allogeneic or xenogeneic cells are transplanted into the body for organ substitution. For successful cell transplantation, the problem of nutrient and oxygen supply is as large as the problem of immuno-isolation. For this purpose, it is promising to induce, in advance, angiogenesis at the site of cells transplanted. Here is one example to show the necessity of prior angiogenesis. Gelatin microspheres incorporating bFGF were injected into the infarction site of rat heart to induce angiogenesis. Upon intramuscularly injecting into the ischemic site one week after bFGF-induced angiogenesis, autologous rat fetal cardiomyocytes remained alive in the myocardium even four weeks later. The thickness of myocardial muscles and the cardiac functions recovered to a significant extent compared with the injection of either microspheres incorporating bFGF or cardiomyocytes alone[18] (Figure 4.4). Such an angiogenic effect on the prolonged cell survival and improved cell function has been observed for transplantation of pancreatic islets, hepatocytes, and gene-engineered cells.[8,19,20] Recently, cultured skin composed of epidermal and dermal skin layers has been constructed *in vitro*.[21] However, even

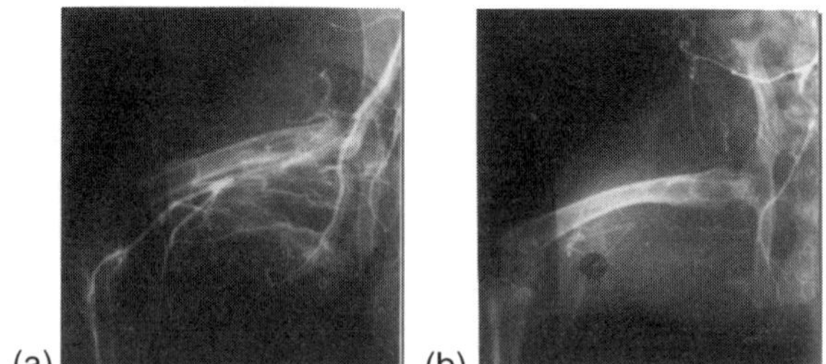

FIGURE 4.3 Angiograms of hind-limb ischemia in rats four weeks after ligation of the femoral artery and its branches of streptozotocin-induced diabetic rats. Gelatin microspheres incorporating bFGF was injected into the ischemic muscle one week after ligation (a) and free bFGF (b). The bFGF dose is 100 μg per muscle.

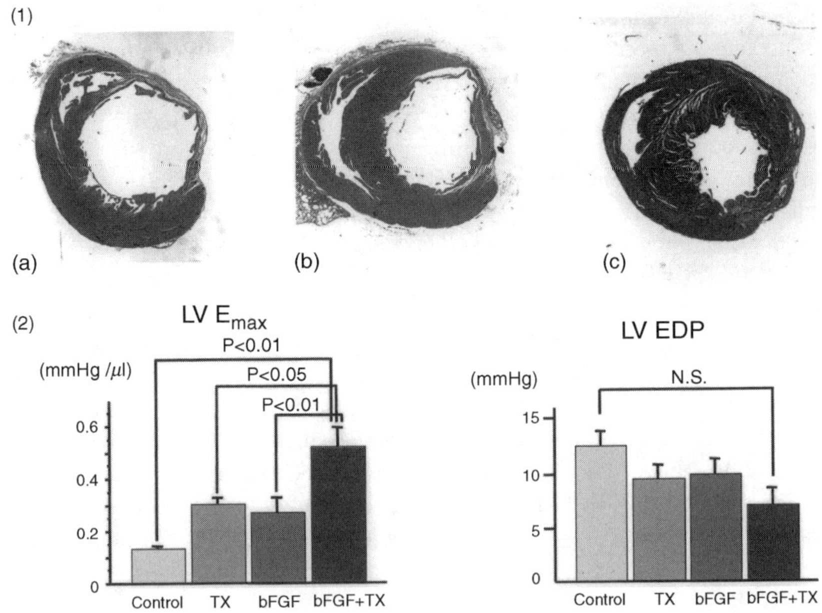

FIGURE 4.4 (1) Histrologic cross-sections of rat hearts four weeks after different applications: (a) control group; (b) cardiomyocyte transplantation (TX) group; (c) cardiomyocyte transplantation one week after bFGF-induced in advance angiogenesis (FGF-TX) group. (Hematoxylin and Eosin staining, original magnification ×1). (2) Indices of left ventricular maximum time-varying elastance (LVEmax) and left ventricular end-diastolic pressure (LVEDP) in the heart of rats from the control, TX, FGF, and FGF-TX groups four weeks after each application. Error bars show standard error of the mean.

if the skin construct is applied to a skin defect, the success rate of engraftment is very low because of a poor supply of oxygen and nutrients to the skin construct grafted. When gelatin microspheres incorporating bFGF were combined with the skin construct and applied the engraftment rate improved remarkably. Fast induction of angiogenesis was observed in the dermal layer of graft and both the epidermis and dermis were healthy.[22] These findings clearly indicate that prior induction of angiogenesis by the gelatin system of bFGF release is effective in achieving successful cell transplantation.

B. Bone Regeneration

Recent studies have demonstrated that gelatin hydrogels also enable bFGF to release and augment activity for bone regeneration and repairing bone defects of rabbits and monkeys.[23,24] For example, when a gelatin hydrogel incorporating bFGF was implanted into a monkey skull defect, it promoted bone regeneration at the defect and, 21 weeks after implantation, completely closed the defect. In comparison, neither bFGF-free, empty gelatin hydrogel, or the same dose of bFGF in solution form was effective, and remarkable growth of soft connective tissues was observed in the bone defect. The bone mineral density (BMD) at the skull defect treated with gelatin hydrogels incorporating bFGF was significantly higher than that of gelatin hydrogels that were bFGF free, irrespective of the hydrogel water content. The BMD of empty gelatin hydrogels was similar to that of the untreated group, indicating that the hydrogel presence did not disturb bone healing at the defect. For the bFGF-incorporated hydrogel implantation, the number of osteoblasts in the active form increased near the edge of newly formed bone tissue and there were increased cell numbers that were retained for long time periods. Prolonging the period of bFGF release by decreasing the hydrogel water content extends the retention period. It is likely that controlled release from the gelatin hydrogel enabled bFGF to activate osteoblasts for an extended time period, resulting in induced regeneration of skull bone. We have succeeded in accelerating the bone repairing at the sternum of normal and streptozotocin-induced diabetic rats by use of gelatin hydrogel that incorporates bFGF.[25,26] This hydrogel also promoted the wound healing of soft tissues around the sternum. In addition to bFGF, TGF-β1 and bone morphogenetic protein (BMP) also promote bone regeneration.[27–30] The controlled release of TGF-β1 from the gelatin hydrogel repaired the skull defects of rabbits and monkeys though bone regeneration, in marked contrast to free TGF-β1 even at the higher doses.[28] The hydrogels of a gelatin type enabled BMP-2 to control the *in vivo* release and induce formation of bone tissue ectopically or orthotopically; at low doses the BMP-2 solution was not effective. The time period of BMP-2 release can also be regulated by changing the hydrogel biodegradation.[31]

There are some cases in which the controlled release of growth factor does not repair bone in a large defect. In one trial, we have utilized cells with osteogeneic potentials and combined them with the release system of growth factor. Among them are mesenchymal stem cells (MSC) isolated from the bone marrow. We have demonstrated that application of the combined MSC and gelatin microspheres incorporating TGF-β1 is capable of completely closing a large-size defect of rabbit skulls with newly formed bone tissue, in marked contrast to that of either material.[32]

C. Adipogenesis

When gelatin microspheres incorporating bFGF were mixed with a basement membrane extract (Matrigel) and subcutaneously implanted into the mouse back, *de novo* formation of adipose tissue was observed at the implanted site.[33,34] A histological study revealed that newly formed matured adipocytes were observed in the tissue mass following the implantation of the mixed microspheres and Matrigel, in contrast to that of other material. Recently, we succeeded in regenerating human

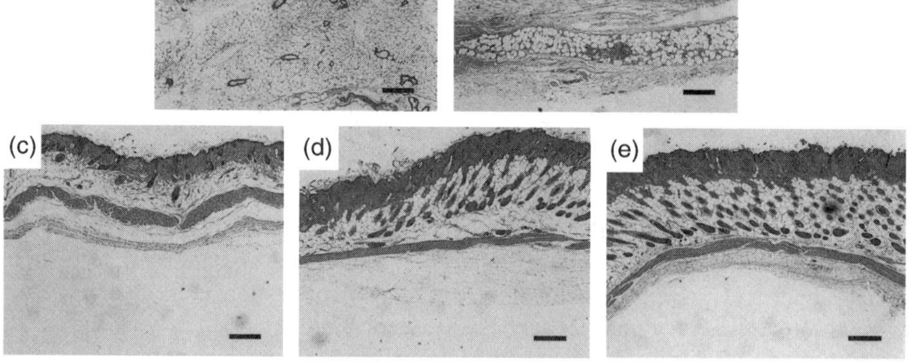

FIGURE 4.5 *De novo* formation of adipose tissue in the mouse subcutis six weeks after implantation of a collagen sponge including the mixture of preadipocytes and gelatin microspheres incorporating bFGF: (a) a collagen sponge including the mixture of preadipocytes and gelatin microspheres incorporating bFGF, (b) a collagen sponge including the mixture of preadipocytes and free bFGF, (c) a collagen sponge including preadipocytes, (d) a mixture of preadipocytes and gelatin microspheres incorporating bFGF, and (e) a collagen sponge including gelatin microspheres incorporating bFGF. (Magnification ×100, Sudan III staining) The bFGF dose was 10 μg per site and the hydrogel water content was 95.0 wt%. The bar length corresponds to 300 μm.

adipose tissue in the back subcutis of nude mice through the combination of preadipocytes isolated from human fat tissues, gelatin microspheres incorporating bFGF, and a collagen sponge (Figure 4.5).[35] The human preadipocytes were proliferated, mixed with the gelatin microspheres incorporating bFGF, and incorporated into a collagen sponge. After implantation into the back subcutis of mice, *de novo* formation of human adipose tissue was observed at the implanted site of the sponge. A combination of all three materials was required for adipogenesis. It is likely that the released bFGF increases the number of preadipocytes, and the rate of differentiation, in matured adipocytes in the collagen sponge (which functions as a scaffold), resulting in achievement of *de novo* adipogenesis.

V. CONCLUDING REMARKS

For regeneration of body tissues, a variety of growth factors act on cells forming a complex network; the timing, site, and concentration of action are delicately regulated in the body. It is likely that the mechanism of living systems will be clarified with the current rapid progress in cell biology, molecular biology, and embryology. Even so, it is impossible to artificially imitate the mechanism with the limited technology currently available. However, clarification will enable us to determine which growth factor best achieves the regeneration of target tissue. If the key growth factor is supplied to the necessary site at the appropriate time in a suitable concentration, we believe that the living body system will be naturally directed toward the process of tissue regeneration. There is no doubt that the controlled release of growth factors *in vivo* is an essential technology. However, the present technology of controlled release is still incapable of completely regulating the amount of growth factor released or regulating the release time period. Additionally, it is not always enough to control the amount and period of growth factor released. Therefore, at the present time, one possibility is to release a certain growth factor necessary for an increase in the number of precursor, blastic or stem cells *in vivo*. It should be noted that it is practically impossible to control cell differentiation with the currently available technology for growth factor release.

As described above, tissue engineering will be the third choice of therapeutic medicine, equal to reconstruction surgery and organ transplantation. To apply tissue engineering to clinical medicine, substantial collaborative research among materials, pharmaceutical, biological, medical sciences, and clinical medicine is needed to reach academic and technical maturity in tissue engineering. However, little research has been reported on DDS research aiming at tissue regeneration and organ substitution. One possible reason is the limited availability of growth factors together with their high cost. There are technologies of DDS other than the controlled release: prolongation of drug life-time, improvement of drug absorption, and drug targeting. For example, a promising way to promote tissue regeneration is by making use of the technology to target a growth factor to the required site of regeneration and prolonging the *in vivo* life-time. Another possibility is using the DDS technology to create nonviral vectors for gene transfection to prepare gene-engineered cells for tissue engineering.

Tissue engineering is still in its infancy as a field of research and it will be some time before the full potential of the field is recognized. We are sure that the readers of this chapter recognize the increasing significance of DDS in developing tissue engineering. We hope this chapter provokes interest and further research in this field.

REFERENCES

1. Ueda, H., and Tabata, Y., Polyhydroxyalkanonate derivatives in current clinical applications and trials, *Adv. Drug Deliv. Rev.*, 55, 501–518, 2003.

2. Bonadio, J., Goldstein, S. A., and Lecy, R. J., Gene therapy for tissue repairing and regeneration, *Adv. Drug. Deliv. Rev.*, 33, 53–69, 1998.
3. Lee, J. S., and Feldman, A. M., Gene therapy for therapeutic myocardial angiogenesis: a promising synthesis of two emerging technologies, *Nat. Med.*, 4, 739–742, 1998.
4. Bonadio, J., Smiley, E., Patil, P., and Goldstein, S., Localized, directed plasmid gene delivery in vivo: prolonged therapy results in reproducible tissue engineering, *Nat. Med.*, 5, 753–759, 1999.
5. Tabata, Y., The importance of drug delivery systems in tissue engineering, *Pharm. Sci. & Tech. Today*, 3, 80–89, 2000.
6. Taipale, J., and Keski-Oja, J., Growth factors in the extracellular matrix, *FASEB J.*, 11, 51–59, 1997.
7. Tabata, Y., Matsui, Y., and Ikada, Y., Growth factor release from amylopectin hydrogel based on copper coordination, *J. Control. Release.*, 56, 135–148, 1998.
8. Tabata, Y., and Ikada, Y., Protein release from gelatin matrices, *Adv. Drug Deliv. Rev.*, 31, 287–301, 1998.
9. Fukunaka, Y., Iwanaga, K., Morimoto, K., Kakemi, M., and Tabata, Y., Controlled release of plasmid DNA from cationized gelatin hydrogels based on hydrogel degradation, *J. Control. Release*, 80, 333–343, 2002.
10. Kushibiki, T., Tomoshige, R., Fukunaka, Y., Kakemi, M., and Tabata, Y., In vivo release and gene expression of plasmid DNA by hydrogels of gelatin with different cationization extents, *J. Control. Release.*, 2005, (in press).
11. Tabata, Y., Nagano, A., and Ikada, Y., Biodegradation of hydrogel carrier incorporating fibroblast growth factor, *Tissue Eng.*, 5, 127–138, 1999.
12. Tabata, Y., Hijikata, S., Munirzzaman, Md., and Ikada, Y., Neovascularization through biodegradable gelatin microspheres incorporating basic fibroblast growth factor, *J. Biomater. Sci., Polym. Ed.*, 10, 79–94, 1999.
13. Rifkin, D. B., and Moscatlli, D., Structural characterization and biological functions of basic fibroblast growth factor, *J. Cell Biol.*, 109, 1–6, 1989.
14. Kuijpers, A. J., Engbers, G. H., Krijgsveld, J., Zaat, S. A., Dankert, J., and Feijen, J., Cross-linking and characterization of gelatin matrices for biomedical application, *J. Biomater. Sci. Polym. Ed.*, 11, 225–243, 2000.
15. Tabata, Y., and Ikada, Y., Vascularization effect of basic fibroblast growth factor released from gelatin hydrogels with different biodegradabilities, *Biomaterials.*, 20, 2169–2175, 1999.
16. Kasahara, H., Tanaka, E., Fukuyama, N., Sato, E., Sakamoto, H., Tabata, Y., Ando, K., Iseki, H., Shinozaki, Y., Kimura, K., Kuwabara, E., Koide, S., Nakazawa, H., and Mori, H., Biodegradable gelatin hydrogel potentiates the angiogenic effect of fibroblast growth factor 4 plasmid in rabbit hindlimb ischemia, *J. Am. Coll. Cardiol.*, 41, 1056–1062, 2003.
17. Sakakibara, Y., Nishimura, K., Tambara, K., Yamamoto, M., Lu, F., Tabata, Y., and Komeda, M., Prevascularization with gelatin microspheres containing basic fibroblast growth factor enhances the benefits of cardiomyocyte transplantation, *J. Thorac. Cardiovasc. Surg.*, 124(1), 50–56, 2002.
18. Balamurugan, A. N., Gu, Y., Wang, W., Miyamoto, M., Hori, H., Inoue, K., and Tabata, Y., Effect of hepatocyte growth factor (HGF) is a mitogen and an insulinotropic agent for fetal islet cells in vitro, *Pancreas*, 26(1), 103–104, 2003.
19. Ogawa, K., Asonuma, K., Inomata, Y., Kim, I., Ikada, Y., and Tanaka, K., The efficacy of prevascularization by basic FGF for hepatocyte transplantation using polymer devices in rats, *Cell Transplant.*, 10(8), 723–729, 2001.
20. Saito, A., Kazuma, J. J., Iino, N., Cho, K., Yamazaki, H., Oyama, Y., Takeda, T., Orlando, R. A., Shimizu, F., Tabata, Y., and Gejyo, F., Bioengineered implantation of megalin-expressing cells: a potential intracorporeal therapeutic model for uremic toxin protein clearance in renal failure, *J. Am. Soc. Nephrol.*, 2003, (in press).
21. Kim, B. M., Suzuki, S., Nishimura, Y., Um, S. C., Morota, K., Maruguchi, T., and Ikada, Y., Cellular artificial skin substitute produced by short period simultaneous culture of fibroblasts and keratinocytes, *Br. J. Plast. Surg.*, 52(7), 573–578, 1999.
22. Saso, Y., Kawazoe, T., Goji, M., Suzuki, S., Tabata, Y., Tomihata, K., and Morita, S., The effect of basic fibroblast growth factor (bFGF) impregnated gelatin microspheres on proliferation of fibroblasts and formation of epithelial layer after grafting cell-preconfluent cultured skin, *Jpn. J. Burn Injuries*, 29, 24–30, 2003.

23. Tabata, Y., Yamada, K., Miyamoto, S., Nagata, I., Kikuchi, H., Aoyama, I., Tamura, M., and Ikada, Y., Bone regeneration by basic fibroblast growth factor complexed with biodegradable hydrogel, *Biomaterials*, 19, 807–815, 1998.
24. Tabata, Y., Yamada, K., Hong, L., Miyamoto, S., Yamada, K., Hashimoto, N., and Ikada, Y., Skull bone regeneration in primates in response to basic fibroblast growth factor, *J. Neurosurgery*, 91, 851–856, 1999.
25. Iwakura, A., Tabata, Y., Miyao, M., Ozeki, M., Tamura, N., Ikai, A., Nishimura, K., Nakamura, T., Shimizu, Y., Fujita, M., and Komeda, M., Novel method to enhance sternal healing after harvesting bilateral internal thoracic arteries with use of basic fibroblast growth factor, *Circulation*, 102(19 Suppl 3), III307–III311, 2000.
26. Iwakura, A., Tabata, Y., Tamura, N., Doi, K., Nishimura, K., Nakamura, T., Shimizu, Y., Fujita, M., and Komeda, M., Gelatin sheet incorporating basic fibroblast growth factor enhances healing of devascularized sternum in diabetic rats, *Circulation*, 104(12 Suppl 1), I325–I329, 2001.
27. Hong, L., Tabata, Y., Yamamoto, M., Miyamoto, S., Yamada, K., Hashimoto, N., and Ikada, Y., Comparison of bone regeneration in a rabbit skull defect by recombinant human BMP-2 incorporated in biodegradable hydrogel and in solution, *J. Biomater. Sci., Polym. Ed.*, 9, 1001–1014, 1998.
28. Hong, L., Tabata, Y., Miyamoto, S., Yamamoto, M., Yamada, K., Hashimoto, N., and Ikada, Y., Bone regeneration at rabbit skull defects treated with transforming growth factor-β1 incorporated into hydrogels with different levels of biodegradability, *J. Neurosurgery*, 92, 315–325, 2000.
29. Yamamoto, M., Tabata, Y., Hong, L. et al., Bone regeneration by transforming growth factor-β1 released from a biodegradable hydrogel, *J. Control. Release*, 64, 133–142, 2000.
30. Hong, L., Tabata, Y., Miyamoto, S., Miyamoto, S., Hashimoto, N., and Ikada, Y., Promoted bone healing at rabbit skull gap between autologous bone fragment and the surrounding intact bone with biodegradabale microspheres containing transforming growth factor-β1, *Tissue Eng.*, 6, 331–340, 2000.
31. Yamamoto, M., Takahashi, Y., and Tabata, Y., Controlled release by biodegradable hydrogels enhances the ectopic bone formation of bone morphogenetic protein, *Biomaterials*, 2003, (in press).
32. Tabata, Y., Hong, L., Miyamoto, S., Miyao, M., Hashimoto, N., and Ikada, Y., Bone formation at a rabbit skull defect by autologous bone marrow combined with gelatin microspheres containing TGF-β1, *J. Biomater. Sci., Polym. Ed.*, 11, 891–901, 2000.
33. Tabata, Y., Miyao, M., Inamoto, T., Ishii, T., Hirano, Y., Yamaoki, Y., and Ikada, Y., De novo formation of adipose tissue by controlled release of basic fibroblast growth factor, *Tissue Eng.*, 6, 279–289, 2000.
34. Kimura, Y., Ozeki, M., Inamoto, T., and Tabata, Y., Time course of de novo adipogenesis in Matrigel by gelatin microspheres incorporating basic fibroblast growth factor, *Tissue Eng.*, 8, 603–613, 2002.
35. Kimura, Y., Ozeki, M., Inamoto, T., and Tabata, Y., Adipose tissue engineering based on human preadipocytes combined with gelatin microspheres containing basic fibroblast growth factor, *Biomaterials*, 24, 2513–2521, 2003.

5 Fibrillar Fibrin Gels

Erin Grassl and Robert T. Tranquillo

CONTENTS

I. INTRODUCTION

Collagen, which has seen widespread use in tissue engineering, has several advantages as a scaffold, including the ability to entrap cells directly as it is reconstituted into a gel. However, there are some drawbacks to its use, particularly the suppression of cell proliferation and protein synthesis.[1,2] An alternative biopolymer scaffold that has many of the same features as collagen is fibrin, a protein involved in clotting. It also forms a fibrillar network which can directly entrap the cells and exhibits similar rheology to collagen allowing for cell-mediated compaction and consequent alignment of the fibrils and cells.[3] In addition, it promotes cell proliferation and ECM synthesis and remodeling,[4–6] since its purpose is a temporary scaffold to be remodeled and replaced by new tissue during wound healing.

Fibrin has been, and continues to be widely used in surgical applications as a tissue sealant,[7] although this requires much higher concentrations than the 3.5 mg ml^{-1} of a clot which cells invade and remodel. More recently, it has been examined as a scaffold for tissue engineering. Fibrin possesses several qualities in addition to those already mentioned that make it ideal for use in tissue engineering. It is biocompatible, biodegradable, and can be produced from human serum, making it possible to use autologous sources. In the following sections we will provide background on the structure and biochemistry of fibrin, as well as an overview of its interactions with cells. We will then finish with a discussion of the tissue engineering applications currently being pursued by researchers.

II. STRUCTURE, BIOCHEMISTRY, AND RHEOLOGICAL PROPERTIES OF FIBRIN

Fibrinogen, the monomeric form of fibrin, is a 340 kDa protein made up of three pairs of non-identical polypeptide chains.[8] It consists of three main domains; a central domain containing fibrinopeptide E and two each of fibrinopeptides A and B, and two terminal domains containing fibrinopeptide D and sites that participate in cross-linking, as shown in Figure 5.1.

The enzyme thrombin cleaves the fibrinopeptides A and B, allowing the fibrinogen to undergo spontaneous fibrillogenesis, forming a linear fibril. The degree of lateral association of these fibrils depends on the conditions under which the fibrin gels, including pH, ionic strength, and concentration of fibrinogen and thrombin, resulting in fibers with diameters ranging from 10 to 200 nm.[9] At lower pH and ionic strength there is more lateral association, whereas at higher pH and ionic strength there is little lateral association. More lateral association results in a gel comprising thicker fibers which, compared to one with fine fibrils, has a higher modulus, creeps more at short times, and creeps less at long times.[10]

The degradation process of fibrin is termed fibrinolysis, and involves a complex cascade of enzymes. The cascade of enzymes leads to the activation of plasminogen, which becomes the proteolytic enzyme plasmin. Plasmin is a serine protease that attacks not only fibrin, but other plasma proteins as well. *In vivo*, this activity is regulated through the presence of inhibitors of both plasmin and plasminogen activators. The use of these inhibitors *in vitro* will be discussed later.

Greater resistance to fibrinolysis can be achieved through fibrin cross-linking. *In vivo*, enzymes termed transglutaminases participate in this cross-linking through the formation of a bond between primary amines at the γ-carboxamide group of glutamine residues. This forms either ε-(γ-glutamyl) lysine or (γ-glutamyl) polyamine bonds which are covalent, stable, and resistant to proteolysis.[11]

Transglutaminases are found in a variety of tissues. Factor XIII is the enzyme responsible for much of the cross-linking *in vivo*. It is found in both platelets and plasma and participates in cross-linking by mediating the formation of isopeptide bonds between D domains.[12] Factor XIII is composed of two a-chains and two b-chains. It is a proenzyme activated by thrombin to Factor XIIIa. In addition, it facilitates the cross-linking of other proteins, such as collagen and fibronectin, to fibrin.[12] Another transglutaminase that has been suggested to stabilize fibrin to some extent is

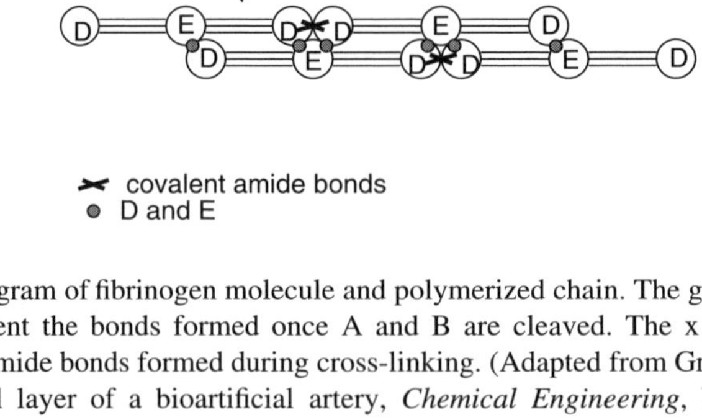

FIGURE 5.1 Simple diagram of fibrinogen molecule and polymerized chain. The gray circles between D and E fibrinopeptides represent the bonds formed once A and B are cleaved. The x between two D peptides represents the covalent amide bonds formed during cross-linking. (Adapted from Grassl, E. D., Enhancing the properties of the medial layer of a bioartificial artery, *Chemical Engineering*, University of Minnesota, Minneapolis, pp. 130, 2002. With permission.)

guinea pig liver transglutaminase.[11,13] Liver transglutaminase has similar binding characteristics, but with a lower affinity for fibrin than factor XIIIa. The binding of liver transglutaminase to fibrin is time- and temperature-dependent and involves binding sites similar to those involved in Factor XIII cross-linking.[13] Liver transglutaminase may be a suitable alternative to Factor XIII for use in the fabrication of cross-linked fibrin tissue equivalents. In addition to these mammalian sources of transglutaminase, several bacterial and fungal transglutaminases have been identified, though only one has been extensively purified and characterized.[14]

For use in tissue engineering applications, fibrin gels are typically prepared by combining fibrinogen and thrombin solutions containing calcium ions. The fibrinogen and thrombin are usually derived from blood plasma including human,[5,15] porcine,[16] or bovine,[17] and several commercial sources exist. Fibrinogen can also be synthesized in a recombinant form by Chinese hamster ovary (CHO) cells.[18] The resulting gel consists of two component phases: a fibrillar network and an interstitial fluid, often tissue culture medium. The interaction of these two phases determines the response of the gel to an applied force. Fibrin gels respond similarly to collagen gels, which also consist of a fibrillar network and interstitial fluid, exhibiting viscoelastic fluid behavior in shear and compression.[3] The mechanical properties depend on concentration, fibril size, and degree of interaction (cross-linking), among other things. For example, in tensile tests of high and low concentration fibrin sealants, the high concentration sealant exhibited higher values for ultimate tensile strength and elastic modulus and less strain hardening than the lower concentration sealant.[19]

III. INTERACTION OF CELLS WITH FIBRIN

A. Fibroblasts in Fibrin

The interactions between the cells and matrix, as well as the proliferation and collagen production of fibroblasts, have been explored by several researchers. Compaction of the fibrin matrix by fibroblasts has been shown by the Tranquillo group to generate predictable alignment, as has also been seen in collagen.[20,21] In addition to matrix compaction and alignment, Tuan et al.[6] have demonstrated that fibroblasts proliferate significantly and are very active synthetically, replacing the fibrin matrix with collagen and other unidentified ECM. As early as the second day of culture, collagen was detected, with highly cross-linked collagen detected after 6 days in culture. Coustry et al.[4] also found a significant increase in the production of both collagen and total protein when fibroblasts were cultured in fibrin gels instead of collagen gels. Clark et al.[22] also observed an improvement in collagen production in the presence of the growth factor TGF-β. While the collagen-synthetic response of fibroblasts to TGF-β was attenuated when seeded in collagen, the response in fibrin was similar to that seen on tissue culture plastic. The addition of 100 pM TGF-β to the culture medium of fibroblasts in fibrin resulted in a 4.4- and 3.4-fold increase in collagen and non-collagen synthesis, respectively, when compared to fibroblasts in fibrin without the growth factor. More recently, Neidert et al.[23] obtained similar results in a study of the effect of TGF-β and insulin on fibroblasts cultured in either fibrin or collagen. This study, and others looking at the use of fibrin for heart valves, is described in more detail in Section V.E.

B. Smooth Muscle Cells in Fibrin

Several studies have examined the effect of fibrin and fibrinogen on the adhesion, migration, and proliferation of cultured smooth muscle cells (SMC). Similar to the effect seen with fibroblasts, fibrin promotes the proliferation of SMC in culture. SMC attach to a fibrin clot at 3 h, begin to proliferate at 6 h, and show marked proliferation at 24 h.[5] Naito et al. found that cultured SMC attach to and migrate on fibrinogen- and fibrin-coated dishes, as well as fibrin gels.[5,17,24,25] SMC from explanted tissue also migrate into fibrin gel after an initial lag period.[26]

While there have been many studies examining migration and proliferation of SMC in fibrin, few studies have examined the effect of fibrin on protein synthesis by SMC. Some evidence for the production of ECM can be seen in studies of atherosclerotic fibrous plaques in vessel walls, where SMC appear to proliferate and produce collagen to replace fibrin as they break it down.[27] More recent *in vitro* studies of SMC in fibrin, within the context of a bioartificial artery, have also shown that SMC break down fibrin and replace it with collagen.[28–30] SMC cultured in fibrin gels with complete medium (M199 with 10% fetal bovine serum, 1% penicillin/streptomycin, 1% L-glutamine, and 50 μg ml^{-1} ascorbic acid) produced 3.2 to 4.9 times the amount of collagen as SMC in collagen gels. They also produced a detectable amount of elastin in fibrin, but not in collagen. This will be discussed in more detail in Section V.

C. Other Cell Types

In addition to fibroblasts and SMC, many other cell types have been grown on and in fibrin. Endothelial cells were found to attach and reach confluence. On gels consisting of thicker fibers (and larger pores, presumably) the cells invaded the gel and aligned with the fibrin fibers, whereas on thin fibers they were more randomly distributed on the surface.[31] Other cells that have been cultured in fibrin with some degree of success include chondrocytes, which tend to de-differentiate into a fibroblast-like morphology, periosteal-derived cells, and nucleus pulposus cells.[32] Additional cell types that have been used in fibrin for tissue engineering applications will be discussed in a later section.

IV. CHALLENGES USING FIBRIN

A. Controlling Fibrinolysis during Formation of New ECM

One of the challenges in using fibrin as a scaffold for engineered tissues is the potentially rapid degradation of the fibrin matrix. The rate and extent of degradation varies with cell type and culture conditions, but in some cases it is too rapid compared to the rate at which ECM is produced by the cells present to replace it. While fibroblasts slowly degrade the fibrin while replacing it with new ECM, smooth muscle cells rapidly degrade the fibrin, resulting in almost complete degradation before significant ECM is produced to replace it.

Rapid degradation of the fibrin matrix can be controlled with several inhibitors. One of these is the serine protease inhibitor aprotinin. Tuan et al.[33] used aprotinin in concentrations of 500 KIU/ml to successfully combat the degradation of fibrin by fibroblasts in the absence of serum. Others have also used aprotinin to prevent degradation during culture of other cell types in fibrin.[7,32,34] Another inhibitor, a lysine analog that competitively inhibits attachment of plasmin and plasminogen to fibrin, is ε-aminocaproic acid (ACA). Herbert et al.[35] used ACA successfully to regulate fibrinolysis in studies of neurite growth in fibrin. Our own studies[29] have shown that ACA can be as effective as aprotinin at appropriate concentrations. Figure 5.2 compares the inhibition of fibrinolysis by aprotinin and ACA in fibrin gels containing SMC. The fibrin gels contained 5% fluorescently labeled fibrinogen. The fluorescence in the medium above the samples was measured at various time points and used as an indication of fibrin degradation.[29]

B. Removing (or Masking) Residual Fibrin

It is not clear whether residual fibrin would be prothrombogenic. If so, this may be another consideration in using fibrin as a scaffold, particularly in applications that will have contact with blood. One possibility for removing residual fibrin is to add plasmin or upregulate the activation of plasmin, which should selectively degrade the fibrin without any significant effect on collagen or elastin. However, initial studies with constructs prepared from SMC in fibrin in our lab did not show extensive fibrinolysis. Another possibility is to remove any inhibitor present and allow the cells to degrade it naturally. Our initial studies show that this may be a promising strategy, but needs to be examined further. However, as already mentioned, residual fibrin may not be a problem, particularly in applications where there is little contact with blood, or where the residual fibrin is masked with an

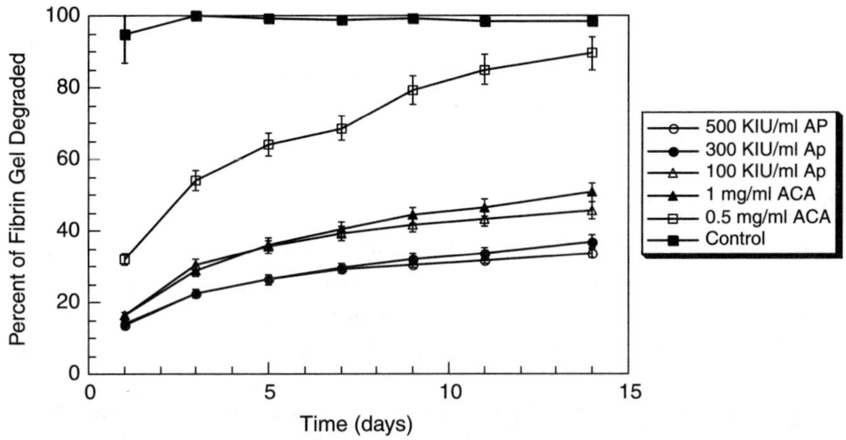

FIGURE 5.2 Degradation of fibrin by neonatal SMC at 4 million cells ml^{-1}, cultured in medium supplemented with various concentrations of aprotinin and ACA. (Taken from Grassl, E. D., Enhancing the properties of the medial layer of a bioartificial artery, *Chemical Engineering*, University of Minnesota, Minneapolis, pp. 130, 2002. With permission.)

endothelium as in the case of blood vessels. In addition, the cells may continue to break down (and remodel) the fibrin once implanted. This is an issue that needs further examination.

C. Effects of Fibrin Degradation Products (FDPs)

Studies have examined the effect of various fibrin (and fibrinogen) degradation products on several different cell lines under different culture conditions. Most of the work has been done on endothelial cells (EC) and suggests that some FDPs could be damaging to endothelial cells, but results are somewhat conflicting. Dang et al.[36] found that the fibrinogen fragment D, but not E, caused EC to retract from each other and round up. The effect did not appear to be cytotoxic and was reversible since the cells could be replated. However, other studies[16,37] suggest that fragment D does not affect EC, but that certain fractions of low molecular weight FDPs damage EC, causing them to detach from their substrate.

Other cell types have also shown responses to FDPs. Fragment B, which is released during formation of fibrin and may be trapped within the matrix, has been shown to cause direct cell migration of neutrophils and fibroblasts, as well as changes in neutrophil morphology and secretion of enzymes.[38] Fragment D induced proliferation of human hemopoietic cells,[39] while fragment E resulted in IL-6 production by rat peritoneal macrophages[40] and fibrinogen synthesis by cultured rat hepatocytes.[41]

It is not clear whether these results accurately reflect what would be seen in a tissue-engineered construct, since it is a more complex system than the cultured cells in these studies. The potential damage to ECs is of particular concern and warrants a more careful examination using the conditions appropriate for tissue engineering applications.

V. TISSUE ENGINEERING APPLICATIONS

A. Fibrin as a Cell Delivery Vehicle

Simple tissue engineering applications of fibrin have stemmed from its use as a sealant in surgery. The use of fibrin glue has been examined for the delivery of cells to a specific site or attachment of cells to another tissue or matrix. An example of this is its use as a delivery vehicle for urothelial cells in urethral reconstruction.[42] Bach et al. suspended cells in fibrin glue and applied the mixture to a connective tissue capsule tube formed *in vivo*. The urothelial cells spread and formed an adherent and confluent cell layer 2 weeks after implantation, at which time the fibrin clot had already been replaced with other connective tissue. This type of technique has also been used with skin grafts[7] and a composite neotrachea.[43]

B. Artificial Ocular Surface

Han et al.[44] examined the use of fibrin as a matrix for corneal epithelial stem cells to produce a bioengineered ocular surface. This work was based on the same idea that fibrin provides a favorable matrix environment for epithelial cell growth and differentiation during wound healing. The fibrinogen and thrombin was isolated from blood plasma in the lab and contained a number of other plasma proteins. This demonstrated the possibility of using autologous sources for the fibrin gel. Human corneal stem cells suspended in the fibrin proliferated and exhibited markers of differentiation, expressing keratin 3, as well as keratin 19 in some cases. Keratin 3 is a marker of corneal-type epithelial differentiation, while keratin 19 is proposed as a marker for corneal stem cells. No functional tests were performed, but the tissue was found to be soft, pliable, and elastic.

C. Skin Grafts

As already mentioned, fibrin has been used in skin graft applications. In addition to its use as a delivery vehicle for cells, it has been used for a more structural approach. In some cases, acellular fibrin was used to induce migration of surrounding cells into the wounded area.[7] More complicated skin grafts have involved fibrin in a composite structure. Meana et al.[34] seeded keratinocytes on a fibrin gel containing fibroblasts. After 15 days of culture a confluent bilayer of keratinocytes had formed, and staining for collagen IV and laminin suggested the formation of a basal membrane between the "dermal" and "epidermal" layers. The fibrin used for these studies was obtained from a cryoprecipitate of human plasma and therefore contained additional components that would not be found in purified commercially available fibrinogen. It was not determined whether the cryoprecipitate contained growth factors necessary for the growth of keratinocytes, and therefore it is not clear whether the commercially available fibrinogen would yield similar results.

D. Bio-Artificial Arteries

The previous examples have involved fairly simple geometries. However, fibrin gel can be molded into various shapes, and has been examined for the use in more complicated geometries, such as a bio-artificial artery. As in the case of skin grafts, previous work on bioartificial arteries has focused on using collagen as a natural biopolymer.[45–50] However, these constructs typically lack the necessary strength and elasticity (i.e., cell-produced elastic fibers) found in the native artery. Therefore, fibrin was examined as an alternative natural biopolymer because of the tendency of cells to remodel it and replace it with new ECM. In addition, the ability to align the fibrin may be used to provide an aligned template for the newly synthesized matrix, providing the circumferential alignment seen in the native vessel.

Initial studies of 4×10^6 SMC ml^{-1} in 3 mg ml^{-1} fibrin gels prepared from commercially available fibrinogen and thrombin, showed that SMC produced 3.2 to 4.9 times more collagen when seeded in fibrin than in collagen.[29] The amount of ECM produced was further increased by adding growth factors such as TGF-β and insulin. After a 3 week incubation with 1 mg ml^{-1} ACA (fibrinolysis inhibitor), 5 ng ml^{-1} TGF-β and 2 μg ml^{-1} insulin, the collagen content of the SMC-seeded fibrin gels was 6 times the content of SMC-seeded fibrin gels without TGF-β or insulin. This resulted in a 15- to 20-fold improvement in the mechanical properties (ultimate tensile strength and tangent modulus). After 6 weeks, the collagen content and mechanical properties more than doubled what was seen at 3 weeks.[30] In addition, elastin was found in these

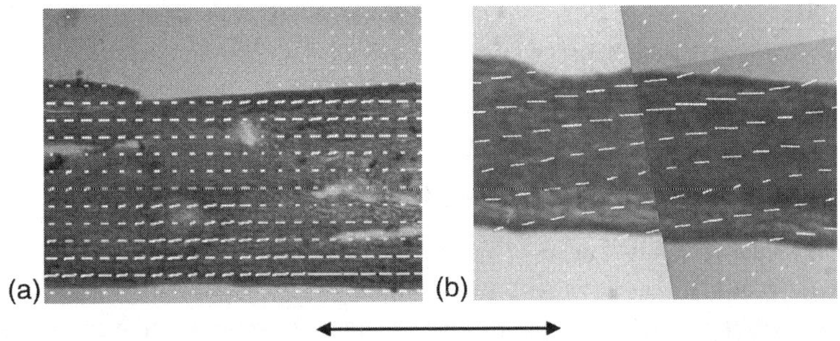

FIGURE 5.3 Alignment maps of a fibrin ME incubated for (a) 1 week and (b) 6 weeks with 1 mg/ml ACA, 1 ng/ml TGF-β, and 2 mg/ml insulin. The vectors indicate the direction of alignment, with the length indicating the relative strength of alignment. The arrow indicates the circumferential direction. (Adapted from Bromberek, B. A., Enever, P. A., Shreiber, D. I., Caldwell, M. D., and Tranquillo, R. T., Macrophages influence a competition of contact guidance and chemotaxis for fibroblast alignment in a fibrin gel coculture assay, *Exp. Cell Res.*, 275(2), 230–242, 2002. With permission.)

samples, though it made up only 2 to 17% of the protein in the samples.[51] The ultimate tensile strength and modulus of these constructs were on the order of rat aorta, but the burst strength was still quite low, around 120 to 140 mm Hg, suggesting that further work is needed to produce an adequate artificial artery. More recent work aimed at optimization of the culture conditions has suggested that even greater ECM synthesis can be achieved. When incubated in DMEM supplemented with TGF-β, insulin and 10% FBS, more than 50% of the original fibrin was replaced by elastin and collagen after 4 weeks.[52] With these optimized conditions, burst pressures up to 1100 mm Hg have been achieved (unpublished data). Another exciting result from this work was verification of the hypothesis that an aligned fibrin template would lead to deposition of new ECM with the same alignment. Figure 5.3 demonstrates the alignment at 1 week when the matrix is mostly fibrin, and alignment in the same direction at 6 weeks when there is substantial collagen produced.

E. Heart Valves

Another cardiovascular application of fibrin is in the fabrication of a tissue-engineered heart valve or venous valve. Ye et al.[53] and Neidert et al.[23] examined the behavior of fibroblasts in fibrin for use in cardiovascular tissue engineering applications such as the heart valve. Ye et al. combined myofibroblasts with commercially available fibrinogen and thrombin to form a fibrin gel with 750,000 cells ml^{-1} and 3.5 mg ml^{-1} fibrin. They found that the cells would completely degrade the fibrin with 5 μg ml^{-1} or less aprotinin. There was some fibrinolysis with 15 μg ml^{-1}, but no visible fibrinolysis with 20 μg ml^{-1}. They also observed uniform cell distribution and a 20% reduction in thickness of the gel after 1 month. A later study from the same group,[54] examined the effect of different methods of fixing the fibrin to a mold to prevent compaction, and used the results to design a molding technique for forming an aortic valve conduit out of myofibroblasts and fibrin gel. However, the mechanical properties of the myofibroblast-seeded fibrin gel were not examined in either study.

Neidert et al.[23] examined ways to enhance the remodeling and assessed the effect on the mechanical properties. Unlike what was observed with SMC and myofibroblasts, the fibroblasts did not degrade the fibrin rapidly, therefore no inhibitor was necessary. However, the effect of TGF-β and insulin was similar to what was seen with SMC. After 51 days with 5 ng ml^{-1} TGF-β and 2 μg ml^{-1} insulin added to the medium, the fibrin constructs contained nearly 8 times the amount of collagen as those without the additives. In addition, the ultimate tensile strength (UTS) and modulus were also improved by the addition of these growth factors. Under the optimum conditions, the final constructs consisted of more than 30% collagen and had a modulus and UTS within an order of magnitude of a human heart valve leaflet.

VI. SUMMARY

Fibrin has several characteristics that makes it desirable for tissue engineering. It is biocompatible (able to be made from autologous sources), biodegradable, and capable of entrapping cells directly. In addition, it has adequate mechanical properties and promotes cell growth and remodeling. Studies of cells cultured in fibrin have shown that several different cell types can be successfully grown in fibrin. While there are some challenges to be addressed, fibrin has shown promise as a biopolymer scaffold in several tissue engineering applications including ocular implants, skin, blood vessels, and heart valves. More work is necessary to further develop these applications and examine new ones.

REFERENCES

1. Clark, R. A. F., Nielsen, L. D., Welch, M. P., and McPherson, J. M., Collagen matrices attenuate the collagen-synthetic response of cultured fibroblasts to TGF-beta, *J. Cell Sci.*, 108, 1251–1261, 1995.
2. Thie, M., Schlumberger, W., Semich, R., Rauterberg, J., and Robenek, H., Aortic smooth muscle cells in collagen lattice culture: effects on ultrastructure, proliferation and collagen synthesis, *Eur. J. Cell Biol.*, 55, 295–304, 1991.
3. Tranquillo, R. T., Self-organization of tissue-equivalents: the nature and tole of contact guidance, *Biochem. Soc. Symp.*, 65, 27–42, 1999.
4. Coustry, F., Gillery, P., Maquart, F. X., and Borel, J. P., Effect of transforming growth factor beta on fibroblasts in three-dimensional lattice cultures, *FEBS Lett.*, 262(2), 339–341, 1990.
5. Ishida, T., and Tanaka, K., Effects of fibrin and fibrinogen-degradation products on the growth of rabbit aortic smooth muscle cells in culture, *Atherosclerosis*, 44, 161–174, 1982.
6. Tuan, T., Song, A., Chang, S., Younai, S., and Nimni, M., In vitro fibroplasia: matrix contraction, cell growth, and collagen production of fibroblasts cultured in fibrin gels, *Exp. Cell Res.*, 223, 127–134, 1996.
7. Currie, L. J., Sharpe, J. R., and Martin, R., The use of fibrin glue in skin grafts and tissue-engineered skin replacements: a review, *Plast. Reconstr. Surg.*, 108(6), 1713–1726, 2001.
8. Doolittle, R. F., Fibrinogen and fibrin, *Annu. Rev. Biochem.*, 53, 195–229, 1984.
9. Kaibara, M., Fukada, E., and Sakaoku, K., Rheological study on network structure of fibrin clots under various conditions, *Biorheology*, 18, 23–35, 1981.
10. Kramer, O., ed., *Biological and Synthetic Polymer Networks*, Elsevier Applied Science, London, 1988.
11. Greenberg, C. S., Birckbichler, P. J., and Rice, R. H., Transglutaminases: multifunctional cross-linking enzymes that stabilize tissues, *FASEB J.*, 5(15), 3071–3077, 1991.
12. Ariens, R. A. S., Lai, T.-S., Weisel, J. W., Greenberg, C. S., and Grant, P. J., Role of factor XIII in fibrin clot formation and effects of genetic polymorphisms, *Blood*, 100, 743–754, 2002.
13. Achyuthan, K. E., Mary, A., and Greenberg, C. S., The binding sites on fibrin(ogen) for guinea pig liver transglutaminase are similar to those of blood coagulation factor XIII. Characterization of the binding of liver transglutaminase to fibrin, *J. Biol. Chem.*, 263(28), 14296–14301, 1988.
14. Griffin, M., Casadio, R., and Bergamini, C. M., Transglutaminases: nature's biological glues, *Biochem. J.*, 368, 377–396, 2002.
15. Blomback, B., Carlsson, K., Hessel, B., Liljeborg, A., Procyk, R., and Aslund, N., Native fibrin gel networks observed by 3D microscopy, permeation and turbidity, *Biochim. Biophys. Acta*, 997, 96–110, 1989.
16. Watanabe, K., and Tanaka, K., Influence of fibrin, fibrinogen, and fibrinogen degradation products on cultured endothelial cells, *Atherosclerosis*, 48, 57–70, 1983.

17. Naito, M., Nomura, H., and Iguchi, A., Migration of cultured vascular smooth muscle cells into non-crosslinked fibrin gels, *Thromb. Res.*, 84(2), 129–136, 1996.
18. Gorkun, O. V., Veklish, Y. I., Weisel, J. W., and Lord, S. T., The conversion of fibrinogen to fibrin: recombinant fibrinogen typifies plasma fibrinogen, *Blood*, 89(12), 4407–4414, 1997.
19. Sierra, D. H., Eberhardt, A. W., and Lemons, J. E., Failure characteristics of multiple-component fibrin-based adhesives, *J. Biomed. Mater. Res.*, 59, 1–11, 2002.
20. Barocas, V. H., Girton, T. S., and Tranquillo, R. T., Engineered alignment in media-equivalents: magnetic prealignment and mandrel compaction, *J. Biomech. Eng.*, 120, 660–666, 1998.
21. Bromberek, B. A., Enever, P. A., Shreiber, D. I., Caldwell, M. D., and Tranquillo, R. T., Macrophages influence a competition of contact guidance and chemotaxis for fibroblast alignment in a fibrin gel coculture assay, *Exp. Cell Res.*, 275(2), 230–242, 2002.
22. Clark, R. A. F., McCoy, G. A., Folkvord, J. M., and McPherson, J. M., TGF-beta1 stimulates cultured human fibroblasts to proliferate and produce tissue-like fibroplasia: a fibronectin matrix-dependent event, *J. Cell. Physiol.*, 170, 69–80, 1997.
23. Neidert, M. R., Lee, E. S., Oegema, T. R., and Tranquillo, R. T., Enhanced fibrin remodeling in vitro with TGF-beta1, insulin and plasmin for improved tissue-equivalents, *Biomaterials*, 23(17), 3717–3731, 2002.
24. Naito, M., Hayashi, T., Kuzuya, M., Funaki, C., Asai, K., and Kuzuya, F., Effects of fibrinogen and fibrin on the migration of vascular smooth muscle cells in vitro, *Atherosclerosis*, 83(1), 9–14, 1990.
25. Naito, M., Funaki, C., Hayashi, T., Yamada, K., Asai, K., Yoshimine, N., and Kuzuya, F., Substrate-bound fibrinogen, fibrin and other cell attachment-promoting proteins as a scaffold for cultured vascular smooth muscle cells, *Atherosclerosis*, 96(2-3), 227–234, 1992.
26. Nomura, H., Naito, M., Iguchi, A., Thompson, W. D., and Smith, E. B., Fibrin gel induces the migration of smooth muscle cells from rabbit aortic explants, *Thromb. Haemost.*, 82, 1347–1352, 1999.
27. Smith, E. B., Fibrinogen, fibrin, and the arterial wall, *Eur. Heart J.*, 16(Suppl. A), 11–15, 1995.
28. Grassl, E. D., Enhancing the properties of the medial layer of a bioartificial artery, *Chemical Engineering*, University of Minnesota, Minneapolis, pp. 130, 2002.
29. Grassl, E. D., Oegema, T. R., and Tranquillo, R. T., Fibrin as an alternative biopolymer to type I collagen for fabrication of a media-equivalent, *J. Biomed. Mater. Res.*, 60(4), 607–612, 2002.
30. Grassl, E. D., Oegema, T. R., and Tranquillo, R. T., A fibrin-based arterial media-equivalent, *J. Biomed. Mat. Res.*, 66A, 550–561, 2003.
31. Shats, E. A., Nair, C. H., and Dhall, D. P., Interaction of endothelial cells and fibroblasts with modified fibrin networks: role in atherosclerosis, *Atherosclerosis*, 129, 9–15, 1997.
32. Perka, C., Arnold, U., Spitzer, R.-S., and Lindenhayn, K., The use of fibrin beads for tissue engineering and subsequential transplantation, *Tissue Eng.*, 7(3), 359–361, 2001.
33. Tuan, T. L., and Grinnell, F., Fibronectin and fibrinolysis are not required for fibrin gel contraction by human skin fibroblasts, *J. Cell. Physiol.*, 140(3), 577–583, 1989.
34. Meana, A., Iglesias, J., Del Rio, M., Larcher, F., Madrigal, B., Fresno, M. F., Martin, C., San Roman, F., and Tevar, F., Large surface of cultured human epithelium obtained on a dermal matrix based on live fibroblast-containing fibrin gels, *Burns*, 24, 621–630, 1998.
35. Herbert, C. B., Bittner, G. D., and Hubbell, J. A., Effects of fibrinolysis on neurite growth from dorsal root ganglia cultured in two- and three-dimensional fibrin gels, *J. Comp. Neurol.*, 365, 380–391, 1996.
36. Dang, C. V., Bell, W. R., Kaiser, D., and Wong, A., Disorganization of cultured vascular endothelial cell monolayers by fibrinogen fragment D, *Science*, 227, 1487–1490, 1984.
37. Lorenzet, R., Sobel, J. H., Bini, A., and Witte, L. D., Low molecular weight fibrinogen degradation products stimulate the release of growth factors from endothelial cells, *Thromb. Haemost.*, 68(3), 357–363, 1992.
38. Senior, R. M., Skogen, W. F., Griffin, G. L., and Wilner, G. D., Effects of fibrinogen derivatives upon the inflammatory response, *J. Clin. Invest.*, 77, 1014–1019, 1986.
39. Hatzfeld, J. A., Hatzfeld, A., and Maigne, J., Fibrinogen and its fragment D stimulate proliferation of human hemopoietic cells in vitro, *Proc. of the Natl Acad. of Sci.*, 79, 6280–6284, 1982.
40. Lee, M. E., Rhee, K. J., and Nham, S. U., Fragment E derived from both fibrin and fibrinogen stimulates interleukin-6 production in rat peritoneal macrophages, *Mol. Cells*, 9(1), 7–13, 1999.

41. Qureshi, G. D., Guzelian, P. S., Vennart, R. M., and Evans, H. J., Stimulation of fibrinogen synthesis in cultured rat hepatocytes by fibrinogen fragment E, *Biochim. Biophys. Acta*, 844, 288–295, 1985.
42. Bach, A. D., Bannasch, H., Galla, T. J., Bittner, K. M., and Stark, G. B., Fibrin glue as a matrix for cultured autologous urothelial cells in urethral reconstruction, *Tissue Eng.*, 7(1), 45–53, 2001.
43. Doolin, E. J., Strande, L. F., Sheng, X., Hewitt, C. W., Engineering a composite neotrachea with surgical adhesives, *J. Pediatr. Surg.*, 37(7), 1034–1037, 2002.
44. Han, B., Schwab, I. R., Madsen, T. K., and Isseroff, R. R., A fibrin-based bioengineered ocular surface with human corneal epithelial stem cells, *Cornea*, 21(5), 505–510, 2002.
45. Girton, T. S., Oegema, T. R., Grassl, E. D., Isenberg, B. C., and Tranquillo, R. T., Mechanisms of stiffening and strengthening in media-equivalents fabricated using glycation, *J. Biomech. Eng.*, 122, 216–223, 2000.
46. Girton, T. S., Oegema, T. R., and Tranquillo, R. T., Exploiting glycation to stiffen and strengthen tissue-equivalents for tissue engineering, *J. Biomed. Mater. Res.*, 46, 1999.
47. Hirai, J., Kanda, K., Oka, T., and Matsuda, T., Highly oriented, tubular hybrid vascular tissue for a low pressure circulatory system, *ASAIO J.*, 40(3), M383–M388, 1994.
48. L'Heureux, N., Germain, L., Labbe, R., and Auger, F. A., In vitro construction of a human blood vessel from cultured vascular cells: a morphological study, *J. Vasc. Surg.*, 17, 499–509, 1993.
49. Seliktar, D., Black, R. A., Vito, R. P., and Nerem, R. M., Dynamic mechanical conditioning of collagen-gel blood vessel constructs induces remodeling in vitro, *Ann. Biomed. Eng.*, 28, 351–362, 2000.
50. Weinberg, C. B., and Bell, E., A blood vessel model constructed from collagen and cultured vascular cells, *Science*, 231, 397–400, 1986.
51. Long, J. L., and Tranquillo, R. T., Elastic fiber production in cardiovascular tissue-equivalents, *Matrix Biol.*, 22(4), 339–350, 2003.
52. Ross, J. J., and Tranquillo, R. T., ECM gene expression correlates with in vitro tissue growth and development in fibrin gel remodeled by neonatal smooth muscle cells, *Matrix Biol.*, 22(6), 477–490, 2003.
53. Ye, Q., Zund, G., Benedikt, P., Jockenhoevel, S., Hoerstrup, S. P., Sakyama, S., Hubbell, J. A., and Turina, M., Fibrin gel as a three dimensional matrix in cardiovascular tissue engineering, *Eur. J. Cardiothorac. Surg.*, 17, 587–591, 2000.
54. Jockenhoevel, S., Zund, G., Hoerstrup, S. P., Chalabi, K., Sachweh, J. S., Demircan, L., Messmer, B. J., and Turina, M., Fibrin gel — advantages of a new scaffold in cardiovascular tissue engineering, *Eur. J. Cardiothorac. Surg.*, 19, 424–430, 2001.

6 Photopolymerization of Hydrogel Scaffolds

Stephanie J. Bryant and Kristi S. Anseth

CONTENTS

I. INTRODUCTION

Photopolymerization reactions have been explored since the times of the ancient Egyptians who used sunlight to cross-link linens during mummification. Since then, photopolymerizations have been utilized in a diverse array of applications ranging from high-tech integrated circuits to printing plates for newspapers to polymeric dental fillings. The versatility of photopolymerizations affords material processing with high reaction rates at room temperature, low energy input, and spatial and temporal control. Owing to the mild reaction conditions, photopolymerization reactions have become a rapidly explored new approach for creating biomaterials.

In particular, photopolymerized hydrogels are very promising materials for tissue engineering due to their high water content and tissue-like elastic properties. Hydrogels are hydrophilic cross-linked materials that are insoluble, but capable of swelling in aqueous solutions. The high degree of swelling facilitates diffusion of oxygen and nutrients into the gel, making them quite suitable for tissue engineering scaffolds. Hydrogels can be formed via a photoinitiated chain polymerization to form highly cross-linked networks with controlled macroscopic properties.

Photopolymerization utilizes light to convert a liquid monomer or macromer (i.e., macromolecular monomer) solution to a solid three-dimensional network via a chain polymerization mechanism. Photopolymerization reactions that form cross-linked networks follow the characteristic steps of chain polymerizations with photoinitiation, propagation, and termination. Photopolymerizations often proceed via radical or cationic initiation mechanisms; however, the cationic initiation mechanism is inhibited by water, which limits its usefulness in creating cell scaffolds. This chapter focuses on the synthesis of cell scaffolds for tissue engineering using photoinitiated *radical* chain polymerizations. It is important to note that an emerging area of research is in controlled radical photopolymerizations, such as, photoiniferters and step growth photopolymerizations (e.g., thiolene); however, these topics will not be reviewed here.

Photopolymerization reactions can occur at physiological temperature and pH, resulting in materials that can be gelled directly in the presence of cells or tissues (e.g., *in vivo* formation). This process allows the invasiveness of many surgical procedures to be minimized as liquid solutions are easily introduced via a syringe and photocured arthroscopically with fiber optic cables. Nonetheless, the choice of monomer (or macromer) and the reaction conditions must be *carefully* chosen to yield strong and suitable scaffolds that will meet the design requirements of a specific tissue without leading to adverse conditions for cells in the scaffold and the surrounding tissue. Additional attention to the reaction conditions may be required if cells are to be present during the polymerization reaction. Suitable reaction conditions for cell photo-encapsulation have been found that work for a number of different cell types including chondrocytes,[1,2] osteoblasts,[3] human dermal fibroblasts,[4] and smooth muscle cells.[5] However, in designing a cell scaffold, a number of factors above and beyond cell viability must be considered for its ultimate success. In particular, a scaffold must be mechanically strong such that it does not fall apart when placed *in vivo*, where physiological strains can be high; a scaffold must exhibit adequate transport properties for nutrients and signaling molecules to diffuse into the gel; and the degradation rate of the scaffold must be controlled and tailored to match tissue regeneration.

The goal of this chapter is to give the reader a knowledge base from which to design systematically cell scaffolds derived from photopolymerized hydrogels. In this chapter, the photopolymerization reaction mechanisms and their implications in the final network structure, and on cell survival, are described. Key hydrogel properties important to tissue engineering (e.g., water content, mechanical properties, diffusion properties, degradation) and their relation to the network structure via the cross-linking density are also described. Finally, a short review of the literature that has made use of photopolymerized hydrogels as cell scaffolds is presented.

II. PHOTOPOLYMERIZATION MECHANISMS

A. PHOTOINITIATION

The first step in the photopolymerization process is photoinitiation, which requires an initiator molecule that absorbs light at the wavelength of interest. Photoinitiator molecules (In) classically absorb light in either the ultraviolet (200–400 nm) or visible light (400–800 nm) range. Upon light absorption, the excited photoinitiator molecules (In^*) typically follow one of two pathways: (1) dissociate into primary radicals ($R\cdot$) (reaction 6.1), or (2) react with a second species ($A-H$) via hydrogen abstraction to form secondary radicals ($R\cdot$) (reaction 6.2).

$$\mathrm{In} \xrightarrow{h\nu} \mathrm{In}^* \longrightarrow 2R\cdot \tag{6.1}$$

$$\mathrm{In} \xrightarrow{h\nu} \mathrm{In}^* \xrightarrow{A-H} 2R\cdot \tag{6.2}$$

The radicals readily react with vinyl groups (C=C) to initiate the chain polymerization reaction.

1. Types of Initiators

The largest class of photoinitiators is the aromatic carbonyl-containing ketones. For example, 2,2-dimethoxy-2-phenylacetophenone (Irgacure 651) is a common UV photoinitiator that is widely used in many different applications. 4-(2-Hydroxyethoxy) phenyl-(2-propyl) ketone (Irgacure 2959) is a UV photoinitiator that has been used more recently in developing photopolymerized cell scaffolds due to its increased water solubility compared to many of the other UV photoinitiators.[2,4] (*Irgacure* is a trademarked name by Ciba Specialty Chemicals). The chemical structures of these UV photoinitiators are shown in Figure 6.1(a) and (b).

Camphorquinone is one of the most common visible light photoinitiators used in dentistry and is often used in conjunction with a hydrogen donating amine. More recently, eosin Y has been used as a visible light initiator in conjunction with triethanolamine for encapsulating islets.[6] The chemical structures for these visible light photoinitiators are shown in Figure 6.1(c) and (d). Much of the research involving hydrogel fabrication for cell scaffolds uses UV photoinitiators because of their superior initiating efficiency compared to visible light initiators. However, photosensitizers are available which can increase the rate of initiation or shift the wavelength where initiation occurs. For example, isopropyl thioxanthone shifts the initiating wavelength to the visible light region and can be used in conjunction with any UV photoinitiator.[7] 1-Vinyl 2-pyrrolidinone has been used along with eosin Y to increase the rate of initiation and decrease the fraction of extractable soluble polymer (i.e., not cross-linked into the gel).[6] For a more in-depth discussion on different types of photoinitiators, see Rabek.[8]

A large number of photoinitiators are commercially available and several have been examined to determine cytotoxicity levels.[9–11] Table 6.1 lists several UV and visible light initiating systems that have been used successfully in biology related applications.

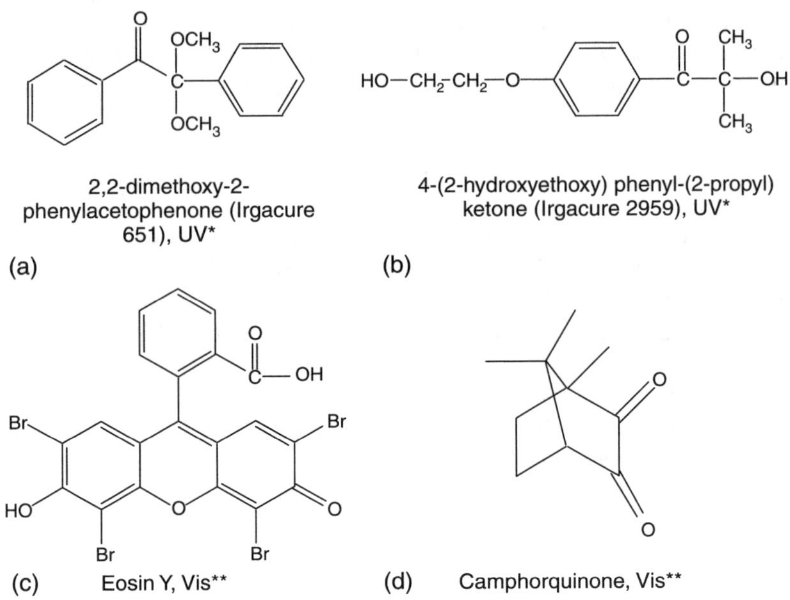

FIGURE 6.1 Photoinitiator structures (* follow reaction scheme 1 (In $\xrightarrow{h\nu}$ In* $\longrightarrow$ 2*R*·) and ** follow reaction scheme 2 (In $\xrightarrow{h\nu}$ In* $\xrightarrow{A-H}$ 2*R*·)). UV = ultraviolet photoinitiators; Vis = visible light photoinitiators.

TABLE 6.1
Photoinitiating Systems and Their Applications

Chemical Name	Light	Gel Chemistry	Applications
2,2-Dimethoxy-2-phenyl-acetophenone (Irgacure 651[a])	UV	PEG	*In vivo* blood vessel glue[49]
			Smooth muscle cell encapsulation[5]
1-Hydroxycyclohexyl phenyl ketone (Irgacure 184[a])	UV	PEG	Chondrocyte encapsulation[1,41]
4-(2-Hydroxyethoxy) phenyl-(2-propyl) ketone (Irgacure 2959[a])	UV	PEG	Chondrocyte encapsulation[38,39]
		PEG	Osteoblast encapsulation[3]
		PVA	Human dermal fibroblasts encapsulation[4]
		HA	Subcutaneous implants in rats[19]
Eosin Y and triethanolamine[b]	Vis	PEG	Encapsulation of islets of langerhans[6]

[a] Trademarked by Ciba Geigy

[b] Initiator follows reaction scheme (6.2) where an amine provides readily abstractable hydrogens; PEG = poly(ethylene glycol); PVA = poly(vinyl alcohol); HA = hyaluronic acid

2. Rate of Initiation

The rate of photoinitiation, R_i, is given by Ref. 12

$$R_{\mathrm{i}} = \frac{2\phi f(2.303\varepsilon)I_0\lambda[\mathrm{In}]}{N_{\mathrm{AV}}hc} \tag{6.3}$$

where ϕ is the quantum yield (the number of radicals produced per light photon absorbed); f is the photoinitiator efficiency (the fraction of radicals that actually initiate polymerization); ε is the molar extinction coefficient at the initiating wavelength, λ; I_0 is the incident light intensity (power/area); [In] is the initiator concentration; N_{AV} is Avogadro's number; h is Planck's constant; c is the velocity of light; and 2 implies two radicals are formed per dissociated initiator molecule. This form of the equation is valid when the light intensity does not vary appreciably with thickness. In general, light intensity (I) is a function of the molar extinction coefficient, initiator concentration and thickness (b) according to Ref. 12

$$I = I_0 \mathrm{e}^{-2.303\varepsilon[\mathrm{In}]b} \tag{6.4}$$

For most tissue engineering applications, uniform light intensity is an appropriate assumption where low initiator concentrations (<0.1 wt%) are employed; however, care must be taken to ensure this assumption is valid.

The choice of initiator and the initiating conditions are critical because the initiation step controls the polymerization rate and impacts the final gel structure. The initiator chemistry will set the initiator efficiency, the molar extinction coefficient, and limit the range of acceptable initiating wavelengths. Typical ranges for initiator efficiency are 0.3 to 0.8.[13] The molar extinction coefficient can range over several orders of magnitude from 10 to 10^5 l/mol-cm depending on the initiating wavelength and the initiator chemistry. At 365 nm, the molar extinction coefficient for Irgacure 651 is 15 times higher than that of Irgacure 2959.[11] For a given initiator chemistry, the rate of initiation is easily controlled through the initiator concentration and the initiating light intensity.

The initiating light intensity depends on the wavelength of interest and the light source. Monochromatic, longwave UV light sources (365 nm) are inexpensive and provide intensities up to 20 mW/cm^2. However, mercury and xenon arc lamps provide a much higher intensity

($>$100 mW/cm^2), but the light is polychromatic and band pass filters must be used to eliminate short-wave UV radiation. Commercially available dental curing units provide 100 to 1000 mW/cm^2 of 470 to 490 nm light, and visible lasers can provide intensities up to several W/cm^2.

B. Propagation and Termination

The rate of polymerization, R_p is given by Ref. 12

$$R_p = k_p[\mathrm{DB}]\left(\frac{R_i}{2k_t}\right)^{1/2} \tag{6.5}$$

where k_p is the rate constant for propagation; [DB] is the double bond concentration; and k_t is the rate constant for bimolecular termination. The rate of polymerization can be easily determined using differential scanning calorimetry where the heat flow is proportional to R_p. A typical plot of R_p as a function of polymerization time is shown in Figure 6.2 for the formation of a cross-linked hydrogel. In cross-linking polymerizations, the kinetic constants are not constant, but rather complex functions that depend on mass transfer limitations of the reacting specie.[14] As the macromer solution is converted to a growing, polymer network, the solution viscosity increases dramatically, resulting in a decrease in the radical mobility. This decrease leads to a decrease in k_t and a build up in radical concentration. The higher radical concentration leads to an increasing R_p with time, despite the fact that the [DB] is decreasing. This phenomenon is termed autoacceleration. Near the peak maximum in R_p, the mobility of the double bonds also becomes diffusion controlled as the network cross-linking density continues to increase with reaction time. When this happens, k_p decreases, leading to an overall decrease in R_p, and this behavior is termed autodeceleration.

The overall rate of polymerization can be modulated by changes in the initiating conditions or the double bond concentration. For example, an increase in the double bond concentration leads to higher rates of polymerization, and subsequently, shorter polymerization times for a given initiating system.[15] The total bond conversion can be calculated from the R_p data by integrating the curve over time. In most hydrogel systems, 100% conversion of the double bonds is reached, but in glassy, highly cross-linked networks, such as dental materials the maximum double bond

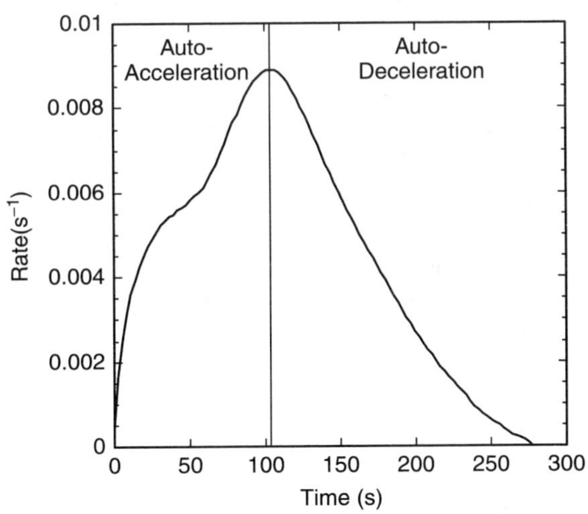

FIGURE 6.2 A typical polymerization rate curve as a function of exposure time (acquired during the formation of a PEG hydrogel).

conversion is often less than 100%. In many clinical applications, the polymerization time is an important factor, and therefore, optimizing R_p with time while maintaining cytocompatible conditions is essential, particularly when cells are present during the polymerization reaction.

C. Advantages/Disadvantages

The advantages of radical photopolymerizations are many. Photopolymerization reactions often happen in a matter of seconds to several minutes which are acceptable clinical timescales. The highly energy efficient reaction yields cost effectiveness. One of the unique facets of photopolymerization is the temporal control and spatial resolution of the initiation process. Temporal control gives the operator the ability to turn on and off the reaction by simply shuttering the light source. Spatial resolution allows polymerization to occur only in designated areas (i.e., exposed to the initiating light).

Oxygen inhibition is one of the major disadvantages of photopolymerization. Molecular oxygen is a radical scavenger and a radical quencher that can readily terminate initiating and propagating radicals. This inhibition can easily be overcome by purging the system with an inert gas, for example, nitrogen, or by simply covering the free surface, with a glass slide for instance, to minimize oxygen diffusion into the sample during polymerization. However, when cells are present, oxygen must be available, and therefore, a small amount of oxygen inhibition, particularly at the air–macromer interface, will inevitably occur. The fast initiating rates and reaction times help to minimize this adverse effect. Of further note, radical photopolymerization reactions of cross-linked networks generally result in volume shrinkage, incomplete double bond conversion, and an inability to cure optically thick samples. However, for highly swollen photopolymerized hydrogels, the former two disadvantages are negligible and the latter can be overcome by the appropriate choice of initiator chemistry and initiator concentration.

III. MULTIFUNCTIONAL MACROMERS

A. Types of Materials

Macromers are typically used as precursors to form hydrogel cell scaffolds because of their low toxicity compared to monomers. For example, poly(2-hydroxyethyl methacrylate) gels are used in contact lenses and many biomaterials applications. These gels can be formed through a photopolymerization of the monomer, 2-hydroxyethyl methacrylate, with a small amount of a dimethacrylate cross-linker to form gels with readily tailored properties. However, despite the biocompatibility of these polymer gels, the low molecular weight monomers are cytotoxic and the polymerizations cannot be conducted in the presence of cells or *in vivo*.

Macromers are macromolecular monomers or polymers that contain two or more vinyl groups, acrylates and methacrylates being the most common. Their chemical structures are given in Figure 6.3. The methyl group on the vinyl functionality of the methacrylate serves to stabilize the radical leading to a less reactive radical, which decreases k_p and leads to slower polymerization rates, but is less toxic to cells.

The modification of polymers with vinyl groups to form photoreactive macromers often occurs through an available hydroxyl group. Both synthetic and natural polymers have been modified with acrylates or methacrylates. Synthetic polymers include poly(ethylene glycol) (PEG)[16] and poly(vinyl alcohol) (PVA),[4,15,17] while natural polymers have primarily been polysaccharide based. The polysaccharides include hyaluronic acid,[18–21] alginate,[21] agarose,[22] and chondroitin sulfate.[23] Chemical structures of several degradable macromers are given in Figure 6.4. PEG has two hydroxyl groups at each end of the polymer that can be modified with a vinyl group to form a divinyl macromer. On the other hand, PVA and polysaccharides have an abundance of

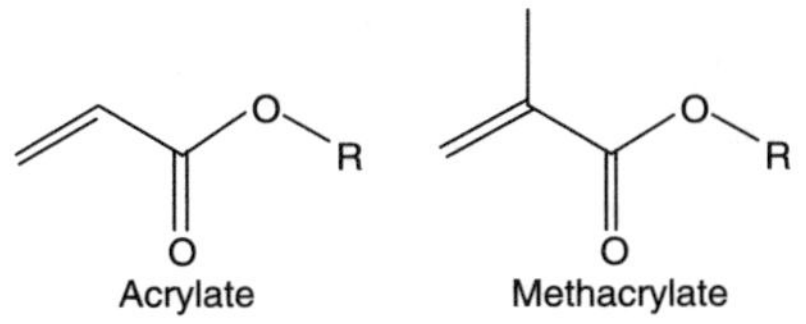

FIGURE 6.3 Typical vinyl groups used to create multifunctional macromolecular monomers, in other words, macromers.

hydroxyl groups available along the polymer chain that can be modified to form multivinyl macromers.

The ease with which molecules can be modified with acrylate or methacrylate groups results in a plethora of chemistries that can be photopolymerizable. For example, proteins or oligomeric amino acids can easily be incorporated into the scaffold by fabricating macromers that are terminated with a vinyl group on one end and a protein covalently linked to the other end.[24]

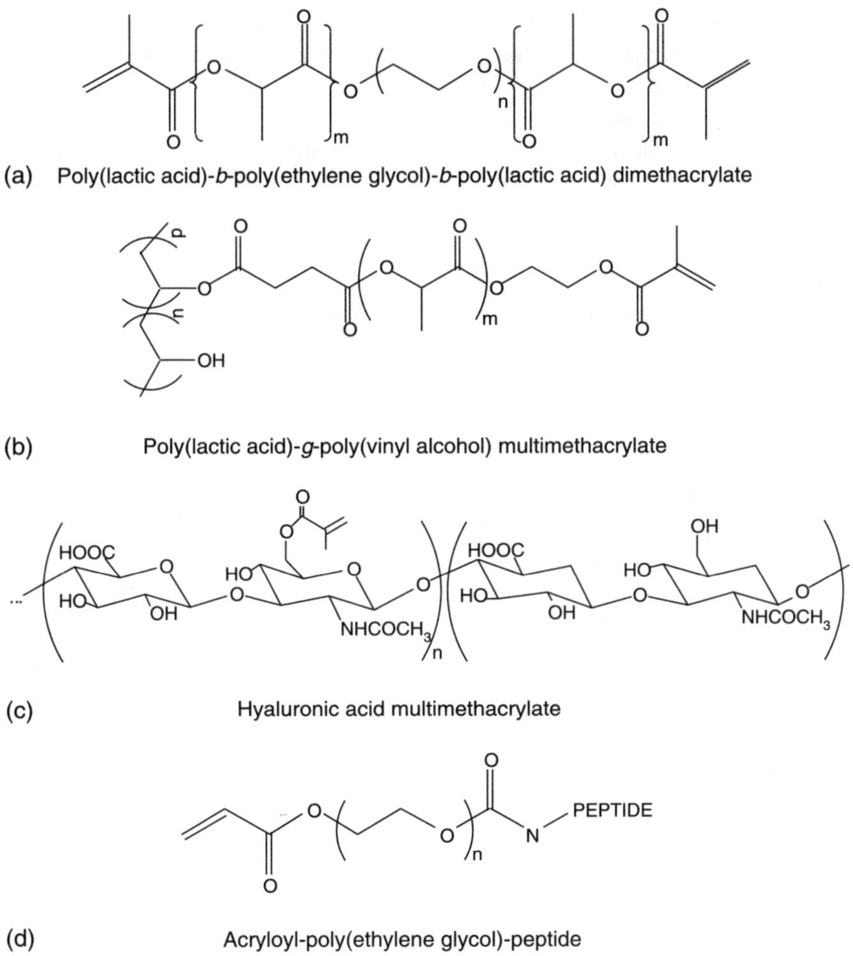

FIGURE 6.4 Chemical structures of biodegradable, multifunctional and photo-crosslinkable macromers: (a) poly(lactic acid)-*b*-poly(ethylene glycol)-*b*-poly(lactic acid) dimethacrylate,[16] (b) poly(lactic acid)-*g*-poly(vinyl alcohol) multimethacrylate,[48] (c) hyaluronic acid multimethacrylate,[21] and (d) acryloyl-poly(ethylene glycol)-peptide.[24]

Figure 6.4(d) shows the structure of an acrylated PEG covalently linked to a peptide. The peptide can be short oligomeric amino acid sequences, such as RGD,[24] or large proteins such as growth factors.[25] Since many of the synthetic and natural gel chemistries do not promote cell adhesion and cell spreading, incorporation of adhesion peptides, or other proteins that promote cell–gel interactions, may be essential to the scaffold design. The choice of peptide(s) will be highly dependent on the cell type(s).

B. Cross-Linking Mechanism

To form a cross-linked network, a minimum of two vinyl groups on a macromer molecule is required. Figure 6.5 illustrates "ideal" network formation from divinyl (Figure 6.5(a)) and multivinyl (Figure 6.5(b)) macromers. The initiating radicals react with the carbon–carbon double bonds on a macromer. The radical propagates through the double bond to react with another double bond in solution to form growing kinetic chains, which can be several hundred repeat units long (depending on R_i and other polymerization parameters). When acrylates are employed, the kinetic chains are polyacrylate, and when methacrylates are employed the kinetic chains are polymethacrylate. In the case where divinyl macromers are employed, as in the case of PEG, the core PEG molecule acts as the cross-link between the kinetic chains. When multivinyl macromers (>2 functional groups) are used, the cross-links become the degradable blocks linking the core polymer (e.g., PVA) and the kinetic chains together.

The polymerization reactions typically occur in an aqueous solvent such as deionized water, buffer or cell culture medium. When a high solvent concentration is used, nonidealities in network formation result, such as cyclization. Cyclization occurs when a propagating radical reacts with a pendant double bond on its own kinetic chain, resulting in a cycle instead of a cross-link (as depicted in Figure 6.5(a)). As a result, a decrease in cross-linking efficiency occurs. For example, when a 10% PEG macromer concentration is used to form a gel, the cross-linking density is 3 times lower than a gel formed from a 20% PEG macromer concentration under identical photoinitiation conditions.[26]

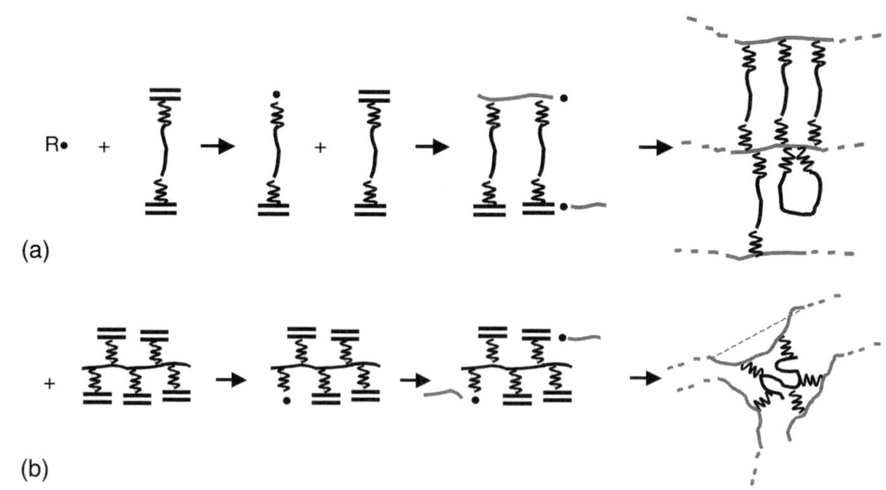

FIGURE 6.5 Network formation by divinyl macromers (e.g., poly(lactic acid-*b*-poly(ethylene glycol)-*b*-poly(lactic acid) dimethacrylate). (a) multivinyl macromers (e.g., poly(lactic acid)-*g*-poly(vinyl alcohol) multimethacrylate), and (b) The gray lines represent the kinetic chains that are formed during polymerization. When (meth)acrylates are employed as the vinyl group, the kinetic chains are poly(meth)acrylate. The [core symbol] is the core molecule (e.g., PEG, PVA), [degradable symbol] is the degradable unit and = is the polymerizable double bond group.

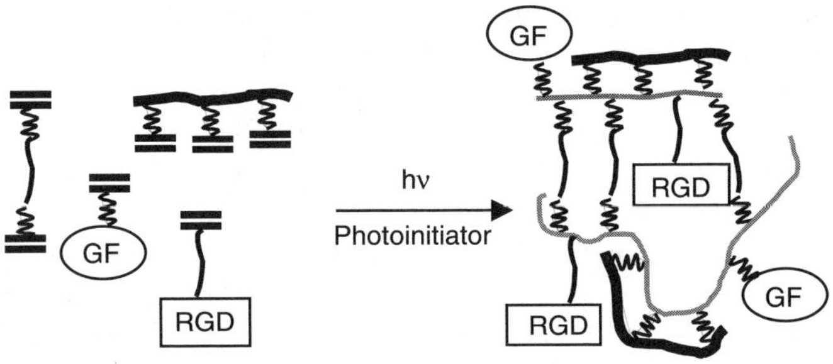

FIGURE 6.6 Multifunctional, multichemical, and biomimetic hydrogels are easily fabricated by simply mixing different macromers in solution prior to polymerization. GF = growth factor; RGD = adhesive peptide sequence.

Finally, through derivatization of the macromers, and the ease with which copolymerizations of multiple macromers can be performed, it is possible to incorporate a plethora of different components into a single hydrogel with great fidelity. By simply mixing different components in controlled ratios prior to polymerization, a variety of functionalities may be incorporated into the network. For example, pendant adhesive peptide functionalities, growth factors tethered through degradable blocks, natural polymers (e.g., hyaluronic acid), and synthetic polymers can be incorporated into one gel to fabricate a biomimetic cell scaffold that may be used to enhance tissue regeneration (as depicted in Figure 6.6).

C. Macromer Properties

Selecting a macromer entails the choice of chemistry, molecular weight, and the degree of functionality (i.e., the number of double bonds per polymer chain), all of which will influence the gel structure and gel properties. The macromer chemistry will influence the hydrophilicity and solvent–polymer interaction parameters of the hydrogel, and therefore, the overall water content. The molecular weight of the macromer is important in both the role it plays in the gel structure and in its role during degradation. An increase in the macromer molecular weight typically decreases the cross-linking density for a given degree of functionality.[17] The highest degree of functionality for PEG is two. However, for PVA and the polysaccharides, the degree of functionality can be varied giving additional control over the final gel properties. For example, the cross-linking density can be increased by simply increasing the number of functional groups on the polymer chains for a given molecular weight and macromer concentration.

D. Degradation Mechanisms

An important characteristic of networks formed from chain polymerizations of multivinyl macromers is that the carbon–carbon-based kinetic chains (i.e., polyacrylate or polymethacrylate) are nondegradable, so strategies to synthesize degradable gels focus on creating materials with degradable cross-links. The two main degradation mechanisms that have been investigated for photopolymerized hydrogels are hydrolytic degradation and enzymatic degradation. In synthetic gels, the degradable units are incorporated into the macromer backbone between the polymer and the vinyl group, as indicated in Figures 6.4 and 6.5.

Poly(α-hydroxy esters) have been the most commonly used polymers to make hydrolytically labile gel scaffolds. Sawhney et al.[16] described the synthesis of a triblock copolymer poly(lactic acid)-*b*-PEG-*b*-poly(lactic acid) end capped with acrylate functional groups to form a degradable photo-crosslinked hydrogel. The degradation profile and rate are controlled through the length of the degradable block (e.g., poly(lactic acid)) and the chemistry. For example, an increase in the degradable block molecular weight increases the probability that an ester bond will be cleaved and the cross-link broken. The main chemistries that have been investigated as hydrolytically degradable segments in hydrogel scaffolds are poly(glycolic acid), poly(lactic acid) and poly(ϵ-caprolactone), which all hydrolyze at different rates.

In enzymatic degradation, the cells dictate the degradation behavior. For the biological polysaccharides, enzymes that cleave the polymers naturally exist *in vivo*. For example, hyaluronic acid is cleaved by hyaluronidase, and chondroitin sulfate is cleaved by chondroitinase. To develop synthetic hydrogels that degrade enzymatically, degradable linkers have been replaced with short proteolytically labile peptide sequences. Degradable gels have been fabricated that respond to cell migration.[27] In this case, the peptide sequence was derived from fibrinogen, which is recognized and cleaved by plasmin, a protein released by migrating cells. Entrapped fibroblasts were shown to migrate into a cross-linked gel with a similar peptide sequence where otherwise no migration occurred.[28] This peptide sequence can be fine tuned to incorporate degradation sensitive to a number of different cell responses, such as extracellular matrix rebuilding.

IV. HYDROGEL PROPERTIES AND CHARACTERIZATION

Desirable hydrogel properties are dictated, in part, by the tissue application and, in part, by the cell types. From a purely *materials* perspective, some of the most important gel properties are the gel water content to maintain cell viability, the gel mechanics to withstand the physiological strains *in vivo* or mechanical conditioning *in vitro*, diffusion of nutrients (and waste) through the gel, and the degradation profiles to match tissue regeneration. All of these properties are closely related through the cross-linking density of the hydrogel.

A. Equilibrium Swelling Ratio

The amount of water present in a hydrogel is a function of the chemical potential, which drives water into the gel, and the elastic retractive forces of the cross-links, which restrict water absorption. At equilibrium, these forces are equal. The equilibrium degree of swelling is a measure of the amount of water the hydrogel imbibes, and is expressed as the equilibrium volume swelling ratio (Q) or equilibrium mass swelling ratio (q).

The equilibrium swelling ratios are determined experimentally by submerging the hydrogels in an aqueous solution, typically a buffered solution at physiologic pH, and allowing the system to equilibrate. Q can be determined by measuring the equilibrium swollen volume (V_s) and the dry polymer volume (V_d) using Archimedes' buoyancy principle or q can be determined by measuring the equilibrium swollen mass (M_s) and the dry polymer mass (M_d).[29]

$$Q = \frac{V_s}{V_d} = 1 + \frac{\rho_p}{\rho_s}(q - 1) \tag{6.6}$$

Here ρ_p is the density of the polymer and ρ_s is the density of the solvent and

$$q = \frac{M_s}{M_d} \tag{6.7}$$

The equilibrium swelling ratio is a function of the cross-linking density and the polymer–solvent interaction parameter. For a given chemistry and solvent, a more tightly cross-linked gel will exhibit greater retractive forces, resulting in less water absorption into the gel. Figure 6.7 illustrates the decrease in Q as a function of cross-linking density for a PEG hydrogel.

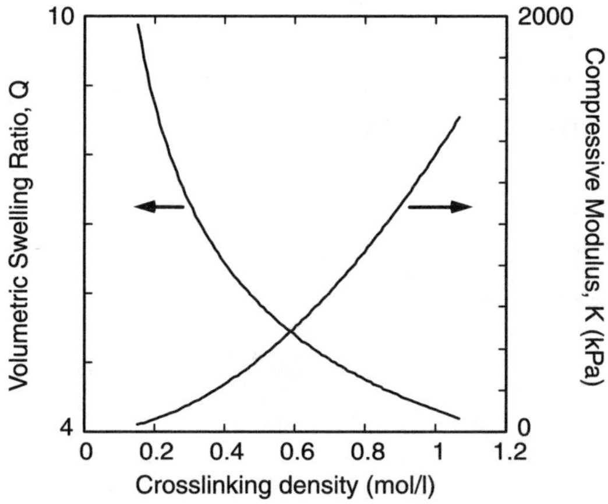

FIGURE 6.7 Hydrogel properties for poly(ethylene glycol) photo-crosslinked hydrogels as a function of cross-linking density. Data adapted from Bryant et al.[38]

B. Gel Mechanics

The mechanical properties of the hydrogel are dependent on two factors: the cross-linking density and the equilibrium volumetric swelling ratio. An increase in the cross-linking density results in a decrease in the equilibrium water content and an increase in the mechanical properties like compressive modulus. Figure 6.7 shows the increase in compressive modulus as a function of cross-linking density for PEG hydrogels. In many tissue applications, a high water content and good mechanics are equally important properties for a cell scaffold. Thus, a delicate balance must be sought between gels by minimizing the amount of water to maintain cell viability while maximizing the gel mechanics to provide adequate function.

C. Cross-Linking Density

From thermodynamics, Flory and Rehner[30] derived the following equation which relates the average molecular weight between cross-links, $\bar{M}_c$, to the equilibrium polymer fraction, $\nu_{2,s}$.

$$\frac{1}{\bar{M}_c} = \frac{2}{\bar{M}_n} - \frac{(\bar{\nu}/V_1)[\ln(1-\nu_{2,s}) + \nu_{2,s} + \chi_{12}\nu_{2,s}^2]}{\left[\nu_{2,s}^{1/3} - \frac{\nu_{2,s}}{2}\right]} \tag{6.8}$$

Here, $\bar{M}_n$ is the number average molecular weight in the absence of any cross-linking, $\bar{\nu}$ is the specific volume of the polymer $(1/\rho_P)$, V_1 is the molar volume of the solvent; $\nu_{2,s}$ is the equilibrium polymer volume fraction $(1/Q)$, and χ_{12} is the solvent-polymer interaction parameter. When network imperfections resulting from chain ends are negligible ($\bar{M}_n \gg \bar{M}_c$), then the term $2/\bar{M}_n$ reduces to zero.

The interaction between the polymer and the solvent, represented by χ, is an important parameter that influences the degree of swelling. In general, a good solvent has a $\chi < 0.5$, while a poor solvent has a $\chi > 0.5$. Selected values of χ are given in Table 6.2. For example, in water PVA has a higher χ value compared to PEG, and therefore, PEG gels will tend to swell to a greater degree for a given cross-linking density and temperature.

Rubber elasticity theory can be used to explain the relationship between gel mechanical properties and the molecular structure. Here, the relationship between the molecular weight

TABLE 6.2
Selected Values for Parameters Used to Determine Gel Structure

Polymer-solvent interaction parameter (χ)	
PEG in water	0.426[50]
PVA in water	0.49[51]
Poisson's ratio (ν)	
PEG	0.5[a]
PVA	0.426–0.447[52]
Characteristic ratio (C_n)	
PEG	4.0[50]
PVA	8.3[53]
Bond lengths (l)	
C–C	1.54 Å
C–O	1.43 Å

[a] Reasonable approximation for elastic materials[31]

between cross-links and compressive modulus (K) is given by Refs. 31,32

$$\frac{1}{\bar{M}_c} = \frac{2}{\bar{M}_n} + \frac{3(1-2\nu)}{2(1+\nu)\rho_p RT}\frac{K}{(\nu_{2,s})^{1/3}} \tag{6.9}$$

where ν is Poisson's ratio for the hydrogel (selected values are given in Table 6.2); ρ_p is the density of the dry polymer; R is the gas constant; and T is the temperature in Kelvin.

The gel cross-linking density, ρ_x, can then be calculated by Ref. 29

$$\rho_x = \frac{1}{\bar{v}\bar{M}_c} \tag{6.10}$$

For high degrees of swelling, $Q > 10$, where chain ends are negligible, the above Equation 6.8 and Equation 6.9 can be reduced to the following relationships, respectively,

$$Q \propto \rho_x^{-3/5} \tag{6.11}$$

$$K \propto \rho_x^{6/5} \tag{6.12}$$

D. Transport Properties

To maintain cell viability and function, a number of different molecules from glucose to large proteins, such as growth factors and other signaling factors, must diffuse readily through the gel. The mesh size, ξ, is the distance in length scale (usually reported in angstroms) between two adjacent cross-links and relates to the maximum size of a solute that can diffuse through the cross-linked gel. The average mesh size for swollen hydrogels can be estimated from Ref. 33

$$\xi = \nu_{2,s}^{-1/3} C_n^{1/2} \ln^{1/2} \tag{6.13}$$

where $\nu_{2,s}$ is the polymer volume fraction in the swollen gel (as defined previously); l is the bond length; C_n is the characteristic ratio of the polymer; and n is the number of bonds between the cross-links (e.g., for PVA $n = 2\bar{M}_c/M_r$ because there are two bonds within each repeat unit (M_r is the molecular weight of the repeating unit)). Selected values for C_n and l are given in Table 6.2.

With the knowledge of the gel mesh size, the diffusivity of any given solute within a hydrogel (D_g) can be estimated using a free-volume approach where in Ref. 34

$$\frac{D_g}{D_0} = \left(1 - \frac{r_s}{\xi}\right)\exp\left(-Y\frac{\nu_{2,s}}{1-\nu_{2,s}}\right) \tag{6.14}$$

where D_0 is the solute diffusivity in the pure solvent (e.g., water); r_s is the Stokes–Einstein hydrodynamic radius of the solute; and Y is the ratio of the critical volume required for a successful translational movement of the solute molecule to the average free volume per molecule of the liquid (usually assumed to be unity). D_0 can be estimated from the Stokes–Einstein equation:

$$D_0 = \frac{k_B T}{6\pi\eta r_s} \tag{6.15}$$

where k_B is Boltzmann's constant; T is temperature; and η is viscosity of water at T. Combining D_g with standard transport equations, concentration profiles, i.e., release rates, can be predicted as a function of time.[35]

E. Mass Loss and Degradation

The most common type of gel degradation mechanism that has been quantitatively characterized is hydrolytic degradation.[7,36,37] Since hydrogels are highly swollen and hydrophilic, degradation occurs through the bulk of the gel. During degradation the ester bonds are cleaved. However, for the scaffold to lose mass, an ester bond within the degradable block must be cleaved on each end of the cross-link for the case where a gel is made from divinyl macromers (Figure 6.8(a)). In the case where multivinyl macromers are employed, a higher fraction of ester bonds must be cleaved along the polymer chain before any mass is lost (Figure 6.8(b)). In both cases, the kinetic chains are attached to the network by numerous cross-links, and therefore are not released until late in

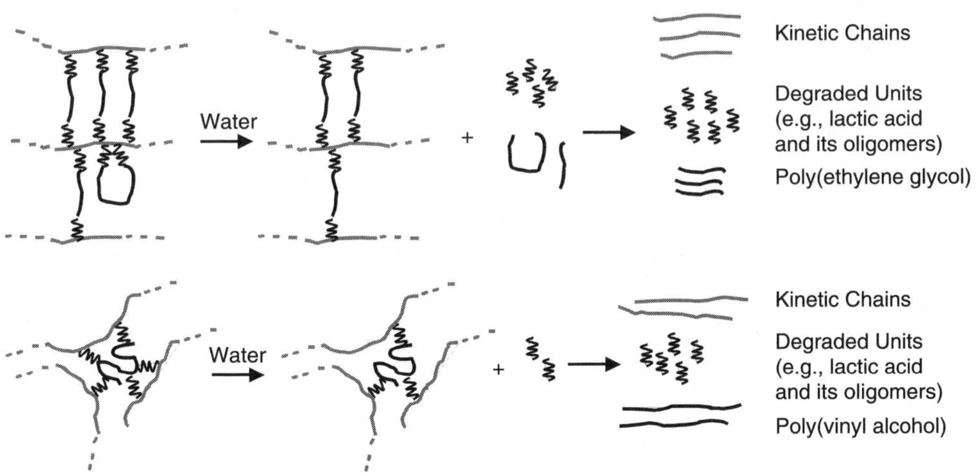

FIGURE 6.8 Degradation scheme for hydrogels formed from (a) divinyl, and (b) multivinyl macromers. The degradation products include poly(meth)acrylate kinetic chains, degraded ester groups (e.g., monomeric or oligomeric lactic acids) and the starting polymer (e.g., PEG in the divinyl case and PVA in the multivinyl case).

the degradation. During degradation, the polymer (e.g., PEG, PVA), polymer kinetic chains, and low molecular weight species are released from the network resulting in an overall mass loss.

Degradation is often measured by the mass loss as a function of degradation time and is defined by

$$\% \text{ Mass Loss} = \frac{M_{\mathrm{id}} - M_{\mathrm{fd}}}{M_{\mathrm{id}}} \times 100 \tag{6.16}$$

where M_{id} is the initial dry polymer mass calculated by multiplying the initial wet weight of the gel by weight fraction of macromer in solution; M_{fd} is the dry polymer mass at a given time in the degradation.

Typical curves for mass loss as a function of degradation time for cross-linked hydrogels are shown in Figure 6.9. As the cross-links are hydrolyzed and eroded from the gel, mass loss increases with degradation time until there are no longer a sufficient number of cross-links to maintain a three-dimensional network (fewer than two cross-links per kinetic chain), at which time the gel dissolves. At this point, a sharp increase in mass loss is observed, and this phenomenon is referred to as reverse gelation. The degradation profile can be controlled in a variety of ways. In Figure 6.9(a), the structure of the gel was kept constant while the degradation rate was decreased leading to slower degradation times. The degradation rate can be changed through the chemistry of the degradable segment (e.g., lactide vs. glycolide). The kinetic chain length also influences the mass loss profiles as shown in Figure 6.9(b). Longer kinetic chains increase the overall degradation time *and* increase the amount of mass that must be lost before reverse gelation occurs. With longer kinetic chains, more ester bonds must be cleaved before the chains are released from the gel. Changes in the kinetic chain length are accomplished by varying the rate of initiation.

For highly swollen gels with $Q > 10$, the rate of hydrolysis will often follow pseudo first-order rate kinetics where[7]

$$\frac{\mathrm{d}n_t}{\mathrm{d}t} = -k' n_t \tag{6.17}$$

and n_t represents the number of moles of degradable blocks and is directly related to the cross-linking density; k' is the pseudo first order reaction rate constant for the hydrolysis of the degradable

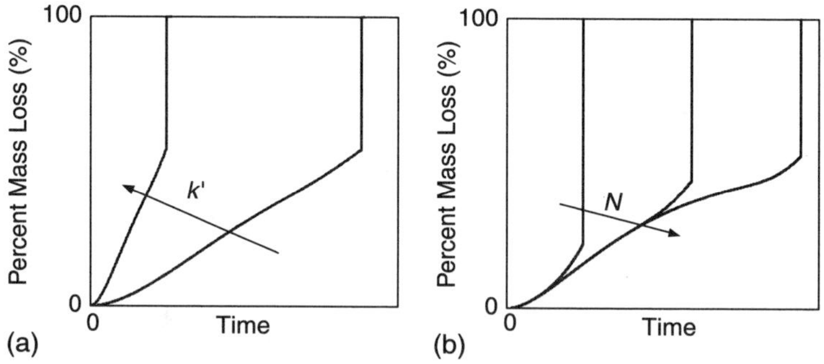

FIGURE 6.9 Typical mass loss profiles for degrading, cross-linked hydrogels. (a) For a given macromer chemistry and functionality, the mass loss curve is shifted to a shorter degradation time when the hydrolysis kinetic constant (k') increases, (b) For a given macromer chemistry, when the number of cross-links per kinetic chain (N, analogous to kinetic chain length) increases, the overall degradation time increases and the % mass loss required to reach reverse gelation increases. Reverse gelation refers to the phenomenon when there are fewer than two cross-links per kinetic chain and dissolution of the gel occurs resulting in a sharp increase in mass loss.

block; and t is the degradation time. As the cross-links are hydrolyzed during degradation, the cross-linking density will then decrease exponentially with degradation time.

$$\rho_x \propto e^{-k't} \tag{6.18}$$

Then, Equation 6.11 and Equation 6.18 can be combined to obtain the following relationship for highly swollen gels:

$$Q \propto \mathrm{e}^{(3/5)\,k't} \tag{6.19}$$

Combining Equation 6.12 and Equation 6.18 gives

$$K \propto \mathrm{e}^{(-6/5)\,k't}$$

and combining equations Equation 6.11, Equation 6.13 and Equation 6.18 give the following relationship for mesh size as a function of degradation time

$$\xi \propto \rho_x^{-7/10} \propto \mathrm{e}^{(7/5)k't} \tag{6.20}$$

Finally, combining Equation 6.14 and Equation 6.20 gives

$$D_{\mathrm{g}} \propto 1 - \mathrm{e}^{(-7/5)\,k't} \tag{6.21}$$

Through these relationships, a great deal of information can be elucidated for cross-linked hydrogels that, through their cross-links, degrade from the experimentally determined equilibrium swelling ratio data. Information on how molecules diffuse through the gel is important in understanding both nutrient diffusion to the cells within the scaffold, as well as diffusion of extracellular matrix molecules through the gel.[38,39] Plotting volumetric swelling ratio, Q, as a function of degradation time allows the calculation of the hydrolysis kinetic constant, which can then be used to predict changes in cross-linking density, and subsequently, compressive modulus, mesh size and diffusion coefficients as a function of degradation time. This information is easily obtained and highly valuable to designing successful cell scaffolds that will meet specific tissue requirements.

V. CELL ENCAPSULATION

Photopolymerization is an attractive method to encapsulate cells in three-dimensional hydrogels because of its mild reaction conditions and ability to be polymerized *in vivo*. Cells can be encapsulated uniformly, at varying densities and in spatially regulated patterns, through photoinitiated polymerizations. However, photopolymerization adds complexity to the gelling process with the presence of photoinitiator molecules, initiating light, and free radicals associated with initiator molecules and propagating polymer macromolecules. Therefore, careful attention to cytocompatibility is required. Previously, we have shown that photoinitiator chemistry and initiator concentration influence cell viability.[11] Low intensity ultraviolet light, ≤ 6 mW/cm^2 at 365 nm, and low intensity visible light, ≤ 80 mW/cm^2 at 470 to 490 nm, do not adversely effect cells (e.g., fibroblasts) based on a cell metabolism assay. When cells, specifically chondrocytes, were encapsulated in a PVA hydrogel with two different photoinitiators but identical photoinitiating conditions, a high cell survival was observed when Irgacure 2959 was used whereas very little cell survival was observed when Irgacure 651 was used. Figure 6.10 illustrates the high cell viability after 5 weeks of culture for chondrocytes encapsulated in a nondegrading PVA hydrogel. In a different study, human dermal fibroblasts

FIGURE 6.10 Chondrocytes photoencapsulated in a nondegradable PVA hydrogel (10 wt% methacrylate poly(vinyl alcohol) (13–23 K molecular weight), 0.5 mol% methacrylated, photoinitiation conditions: 0.05 wt% Irgacure 2959, $I_o \approx 8$ mW/cm^2, 10 min exposure) remain viable in culture for up to 5 weeks. A fluorescent cell viability assay was used to visualize simultaneously live cells (shown in green) and dead cells (shown in red). Original magnification $\times 40$.

were found to remain greater than 80% viable up to 2 weeks in PVA hydrogels modified with adhesive peptides.[4]

VI. APPLICATIONS

A. Tissue Engineering

Photopolymerized hydrogels have been used to encapsulate several different cell types and serve as cell delivery vehicles. Islets of Langerhans were the first cells encapsulated in photopolymerized hydrogel that served as an immunoprotective barrier for the development of an artificial pancreas.[40] For tissue engineering applications, there have been several studies investigating chondrocytes photoencapsulated in PEG hydrogels. Elisseeff et al.[41] conducted the first successful investigation of transdermally photoencapsulating chondrocytes in PEG hydrogels formed subcutaneously in athymic mice. Figure 6.11 demonstrates that chondrocytes photoencapsulated in degrading PEG hydrogels are metabolically active and produce a cartilaginous tissue rich in glycosaminoglycans and collagen, the two main components of cartilage. The histological micrograph illustrates the distribution of glycosaminoglycans (shown in red) that form a tissue-engineered cartilage. PEG hydrogels have been fabricated to exhibit properties similar to native cartilage. When chondrocytes were encapsulated in these gels, they remained viable and functional; however, degradation of the scaffold was required to attain a uniform distribution of extracellular matrix molecules throughout the gel scaffold.[38,39] In addition, chondrocytes have been encapsulated in gels up to 8 mm thick, and the cells remained viable and functional as a function of thickness.[2]

Incorporation of adhesive peptide sequences such as RGD into PEG hydrogel scaffolds have been shown to increase mineralization deposition by encapsulated osteoblasts.[3] The incorporation

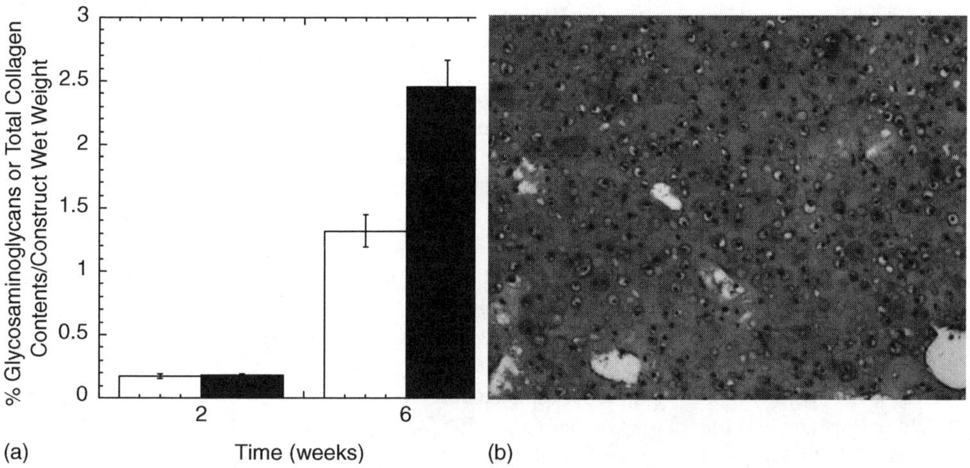

FIGURE 6.11 Tissue engineered cartilage produced by photoencapsulating chondrocytes in a degradable PEG hydrogel. (a) The cartilaginous tissue is comprised of glycosaminoglycans (open bars) and collagen (black bars), the two main components of cartilage. (b) A histological micrograph of the cartilaginous tissue produced after 4 weeks *in vitro* where the glycosaminoglycans are stained red and the cell nuclei are stained black. Original magnification × 100.

of adhesive peptides *and* proteolytic peptides promoted cell migration through PEG hydrogels by aortic smooth muscle cells.[42] Endothelial cells were photoencapsulated and remained proliferative in a biodegradable poly(propylene funarate-coethylene glycol) hydrogel when implanted subcutaneously in rats.[43]

B. Drug Delivery

Photopolymerized hydrogels may also be useful as drug delivery vesicles for tissue engineering applications. A number of model compounds, drugs, and proteins have been incorporated in degrading and nondegrading cross-linked hydrogels and their subsequent release behavior investigated.[35,44,45] Mason et al.[35] provided a first principles analysis to predict release profiles from degrading gels by combining the approaches in Section 6.4.4 with similar transport models. Most importantly, proteins have been shown to remain active after photoencapsulation. For example, tissue plasminogen activator and urokinase plasminogen activator were released from PEG hydrogels and found to reduce postsurgical adhesion in rat models.[46] In another study, osteoinductive growth factors were photoentrapped in a PEG hydrogel, and the release of the growth factor significantly enhanced mineralization by osteoblasts during early culture times.[47]

REFERENCES

1. Elisseeff, J., McIntosh, W., Anseth, K., Riley, S., Ragan, P., and Langer, R., Photoencapsulation of chondrocytes in poly(ethylene oxide)- based semi-interpenetrating networks, *J. Biomed. Mater. Res.*, 51, 164–171, 2000.
2. Bryant, S. J., and Anseth, K. S., The effects of scaffold thickness on tissue engineered cartilage in photocrosslinked poly(ethylene oxide) hydrogels, *Biomaterials*, 22, 619–626, 2001.
3. Burdick, J. A., and Anseth, K. S., Photoencapsulation of osteoblasts in injectable RGD-modified PEG hydrogels for bone tissue engineering, *Biomaterials*, 23, 4315–4323, 2002.

4. Schmedlen, K. H., Masters, K. S., and West, J. L., Photocrosslinkable polyvinyl alcohol hydrogels that can be modified with cell adhesion peptides for use in tissue engineering, *Biomaterials*, 23, 4325–4332, 2002.
5. Mann, B. K., and West, J. L., Cell adhesion peptides alter smooth muscle cell adhesion, proliferation, migration, and matrix protein synthesis on modified surfaces and in polymer scaffolds, *J. Biomed. Mater. Res.*, 60, 86–93, 2002.
6. Cruise, G. M., Hegre, O. D., Scharp, D. S., and Hubbell, J. A., A sensitivity study of the key parameters in the interfacial photopolymerization of poly(ethylene glycol) diacrylate upon porcine islets, *Biotechnol. Bioeng.*, 57, 655–665, 1998.
7. Metters, A. T., Anseth, K. S., and Bowman, C. N., Fundamental studies of a novel, biodegradable PEG-b-PLA hydrogel, *Polymer*, 41, 3993–4004, 2000.
8. Rabek, J. F., *Mechanisms of Photophysical Processes and Photochemical Reactions in Polymers.* Wiley, New York, 1987.
9. Hanks, C. T., Strawn, S. E., Wataha, J. C., and Craig, R. G., Cytotoxic effects of resin components on cultured mammalian fibroblasts, *J. Dent. Res.*, 70, 1450–1455, 1991.
10. Atsumi, T., Murata, J., Kamiyanagi, I., Fujisawa, S., and Ueha, T., Cytotoxicity of photosensitizers camphorquinone and 9-fluorenone with visible light irradiation on a human submandibular-duct cell line in vitro, *Arch. Oral Biol.*, 43, 73–81, 1998.
11. Bryant, S. J., Nuttelman, C. R., and Anseth, K. S., Cytocompatibility of ultraviolet and visible light photoinitiating systems on cultured NIH/3T3 fibroblasts in vitro, *J. Biomater. Sci. Polym. Ed.*, 11, 439–457, 2000.
12. Odian, G., *Principles of Polymerization*, Wiley, New York, 1991.
13. Kurdikar, D. L., and Peppas, N. A., Method of determination of initiator efficiency — application to uv polymerizations using 2,2-dimethoxy-2-phenylacetophenone, *Macromolecules*, 27, 733–738, 1994.
14. Anseth, K. S., and Bowman, C. N., Reaction diffusion enhanced termination in polymerizations of multifunctional monomers, *Polym. React. Eng.*, 1, 499 1993.
15. Martens, P., and Anseth, K. S., Characterization of hydrogels formed from acrylate modified poly(vinyl alcohol) macromers, *Polymer*, 41, 7715–7722, 2000.
16. Sawhney, A. S., Pathak, C. P., and Hubbell, J. A., Bioerodible hydrogels based on photopolymerized poly(ethylene glycol)-co-poly(alpha-hydroxy acid) diacrylate macromers, *Macromolecules*, 26, 581–587, 1993.
17. Martens, P., Holland, T., and Anseth, K. S., Synthesis and characterization of degradable hydrogels formed from acrylate modified poly(vinyl alcohol) macromers, *Polymer*, 43, 6093–6100, 2002.
18. Adolphe, M., and Demignot, S., Modulation of differentiated functions of cultured chondrocytes. Advantages for therapy, *Bull. Acad. Natl Med.*, 184, 593–604, 2000.
19. Leach, J. B., Bivens, K. A., Patrick, C. W., and Schmidt, C. E., Photocrosslinked hyaluronic acid hydrogels: natural, biodegradable tissue engineering scaffolds, *Biotechnol. Bioeng.*, 82, 578–589, 2003.
20. Park, Y. D., Tirelli, N., and Hubbell, J. A., Photopolymerized hyaluronic acid-based hydrogels and interpenetrating networks, *Biomaterials*, 24, 893–900, 2003.
21. Smeds, K. A., and Grinstaff, M. W., Photocrosslinkable polysaccharides for in situ hydrogel formation, *J. Biomed. Mater. Res.*, 54, 115–121, 2001.
22. De Paepe, I., Declercq, H., Cornelissen, M., and Schacht, E., Novel hydrogels based on methacrylate-modified agarose, *Polym. Int.*, 51, 867–870, 2002.
23. Luo, N., Bryant, S., and Anseth, K., Photopolymerizable PVA and chondroitin sulfate hydrogels for cartilage tissue engineering, pp. 327. *Proceedings of the 27th Annual Meeting of the Society for Biomaterials*, St. Paul, Minnesota, 2001.
24. Hern, D. L., and Hubbell, J. A., Incorporation of adhesion peptides into nonadhesive hydrogels useful for tissue resurfacing, *J. Biomed. Mater. Res.*, 39, 266–276, 1998.
25. Mann, B. K., Schmedlen, R. H., and West, J. L., Tethered-TGF-beta increases extracellular matrix production of vascular smooth muscle cells, *Biomaterials*, 22, 439–444, 2001.
26. Bryant, S. J., Photocrosslinkable hydrogels as cell-scaffolds for tissue engineering cartilage: a study examining gel properties, degradation, mechanical loading, and clinical relevance, *Chem. Eng.*, 379 2002.

27. West, J. L., and Hubbell, J. A., Polymeric biomaterials with degradation sites for proteases involved in cell migration, *Macromolecules*, 32, 241–244, 1999.
28. Halstenberg, S., Panitch, A., Rizzi, S., Hall, H., and Hubbell, J. A., Biologically engineered protein-graft-poly(ethylene glycol) hydrogels: a cell adhesive and plasm in-degradable biosynthetic material for tissue repair, *Biomacromolecules*, 3, 710–723, 2002.
29. Peppas, N. A., ed., *Hydrogels in Medicine and Pharmacy*, CRC Press, Boca Raton, FL, 1986.
30. Flory, P. J., and Rehner, R. Jr., Statistical mechanics of crosslinked polymer networks. I. Rubberlike elasticity, *J. Chem. Phys.*, 11, 521 1943.
31. Flory, P. J., ed., *Principles of Polymer Chemistry*, Cornell University Press, Ithaca, NY, 1953.
32. Anseth, K. S., Bowman, C. N., and BrannonPeppas, L., Mechanical properties of hydrogels and their experimental determination, *Biomaterials*, 17, 1647–1657, 1996.
33. Canal, T., and Peppas, N. A., Correlation between mesh size and equilibrium degree of swelling of polymeric networks, *J. Biomed. Mater. Res.*, 23, 1183–1193, 1989.
34. Lustig, S. R., and Peppas, N. A., Solute diffusion in swollen membranes. 9. Scaling laws for solute diffusion in gels, *J. Appl. Polym. Sci.*, 36, 735–747, 1988.
35. Mason, M. N., Metters, A. T., Bowman, C. N., and Anseth, K. S., Predicting controlled-release behavior of degradable PLA-b-PEG- b-PLA hydrogels, *Macromolecules*, 34, 4630–4635, 2001.
36. Metters, A. T., Anseth, K. S., and Bowman, C. N., A statistical kinetic model for the bulk degradation of PLA-b- PEG-b-PLA hydrogel networks: incorporating network non- idealities, *J. Phys. Chem. B*, 105, 8069–8076, 2001.
37. Martens, P., Metters, A. T., Anseth, K. S., and Bowman, C. N., A generalized bulk-degradation model for hydrogel networks formed from multivinyl cross-linking molecules, *J. Phys. Chem. B*, 105, 5131–5138, 2001.
38. Bryant, S. J., and Anseth, K. S., Hydrogel properties influence ECM production by chondrocyte photoencapsulated in poly(ethylene glycol) hydrogels, *J. Biomed. Mater. Res.*, 59, 63–72, 2001.
39. Bryant, S. J., and Anseth, K. S., Controlling the spatial distribution of ECM components in degradable PEG hydrogels for tissue engineering cartilage, *J. Biomed. Mater. Res.*, 64A, 70–79, 2003.
40. Sawhney, A. S., Pathak, C. P., and Hubbell, J. A., Modification of islet of langerhans surfaces with immunoprotective poly(ethylene glycol) coatings via interfacial photopolymerization, *Biotechnol. Bioeng.*, 44, 383–386, 1994.
41. Elisseeff, J., Anseth, K., Sims, D., McIntosh, W., Randolph, M., and Langer, R., Transdermal photopolymerization for minimally invasive implantation, *Proc. Natl Acad. Sci. USA*, 96, 3104–3107, 1999.
42. Mann, B. K., Gobin, A. S., Tsai, A. T., Schmedlen, R. H., and West, J. L., Smooth muscle cell growth in photopolymerized hydrogels with cell adhesive and proteolytically degradable domains: synthetic ECM analogs for tissue engineering, *Biomaterials*, 22, 3045–3051, 2001.
43. Suggs, L. J., and Mikos, A. G., Development of poly(propylene fumarate-co-ethylene glycol) as an injectable carrier for endothelial cells, *Cell Transplant.*, 8, 345–350, 1999.
44. West, J. L., and Hubbell, J. A., Photopolymerized hydrogel materials for drug-delivery applications, *React. Polym.*, 25, 139–147, 1995.
45. Lu, S. X., and Anseth, K. S., Release behavior of high molecular weight solutes from poly(ethylene glycol)-based degradable networks, *Macromolecules*, 33, 2509–2515, 2000.
46. Hill-West, J. L., Dunn, R. C., and Hubbell, J. A., Local release of fibrinolytic agents for adhesion prevention, *J. Surg. Res.*, 59, 759–763, 1995.
47. Burdick, J. A., Mason, M. N., Hinman, A. D., Thorne, K., and Anseth, K. S., Delivery of osteoinductive growth factors from degradable PEG hydrogels influences osteoblast differentiation and mineralization, *J. Controlled Release*, 83, 53–63, 2002.
48. Nuttelman, C. R., Henry, S. M., and Anseth, K. S., Synthesis and characterization of photocrosslinkable, degradable poly(vinyl alcohol)-based tissue engineering scaffolds, *Biomaterials*, 23, 3617–3626, 2002.
49. Dumanian, G. A., Dascombe, W., Hong, C., Labadie, K., Garrett, K., Sawhney, A. S., Pathak, C. P., Hubbell, J. A., and Johnson, P. C., A new photopolymerizable blood vessel glue that seals human vessel anastomoses without augmenting thrombogenicity, *Plast. Reconstr. Surg.*, 95, 901–907, 1995.
50. Merrill, E. W., Dennison, K. A., and Sung, C., Partitioning and diffusion of solutes in hydrogels of poly(ethylene oxide), *Biomaterials*, 14, 1117–1126, 1993.

51. Orwoll, R.A., and Arnold, P.A., Polymer-solvent interaction parameter, In *Physical Properties of Polymers, Handbook*, Marh, J. E., Ed., American Institute of Physics Press, Woodbury, NY, 1996.
52. Urayama, K., Takigawa, T., and Masuda, T., Poisson ratio of poly(vinyl alcohol) gels, *Macromolecules*, 26, 3092–3096, 1993.
53. Kurata, M., and Tsunashima, Y., Viscosity-molecular weight relationships and unperturbed dimensions of linear chain molecules, In *Polymer Handbook*, Brandrup, J., and Immergut, E. H., Eds., John Wiley and Sons, New York, 1989.

FURTHER READING

For additional reading and detailed analysis, and derivation of the equations summarized in this chapter, we suggest three textbooks: *Hydrogels in Medicine and Pharmacy* edited by Nikolaos A. Peppas,[29] *Principles of Polymerization* by George Odian,[12] and *Principles of Polymer Chemistry* by Paul J. Flory.[31]

7 Poly(ortho Esters)

Jorge Heller, John Barr, Devang T. Shah, Steve Y. Ng, Hui Rong Shen, and Brian C. Baxter

CONTENTS

I. INTRODUCTION

Poly(ortho esters) have been under development since 1970, and during that time four polymer families have been developed. These have recently been comprehensively reviewed[1–3], therefore, no attempt to comprehensively review poly(ortho esters) again will be made here. We will present the most important features of the latest family, poly(ortho ester) IV, with particular emphasis and details of polymer fabrication by extrusion.

The four families of poly(ortho esters) are shown in Scheme 7.1.

II. SYNTHESIS

As is evident from Scheme 7.1, POE IV is a variation of POE II in that a latent acid catalyst has been incorporated into the polymer backbone. The synthesis of POE IV is shown in Scheme 7.2.[4]

The synthesis involves the use of two compounds that are not commercially available, the diketene acetal 3,9-diethylidene-2,4,8,10-tetraoxaspiro[5.5] undecane and a diol containing a short segment of poly(glycolic acid), the latent catalyst.

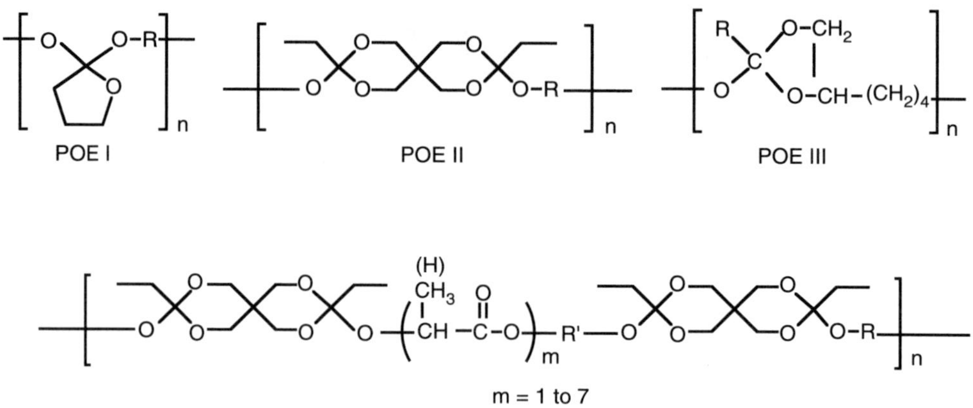

POE IV

SCHEME 7.1

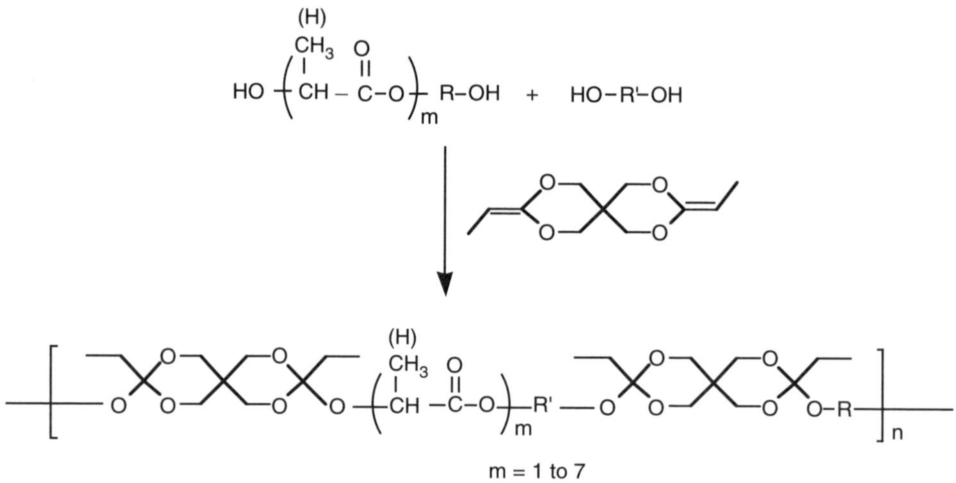

SCHEME 7.2

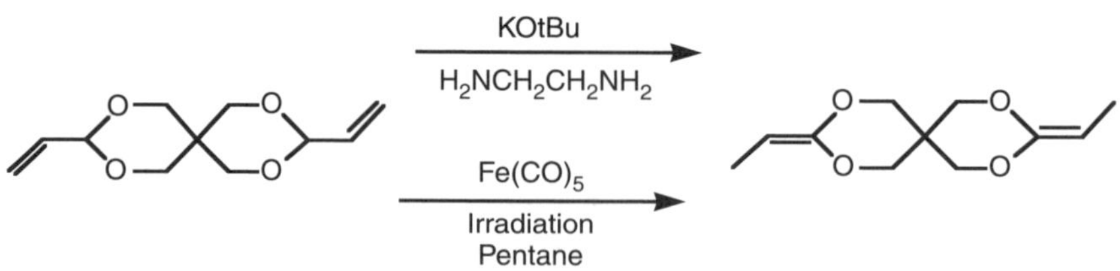

SCHEME 7.3 POE, poly(ortho) ester.

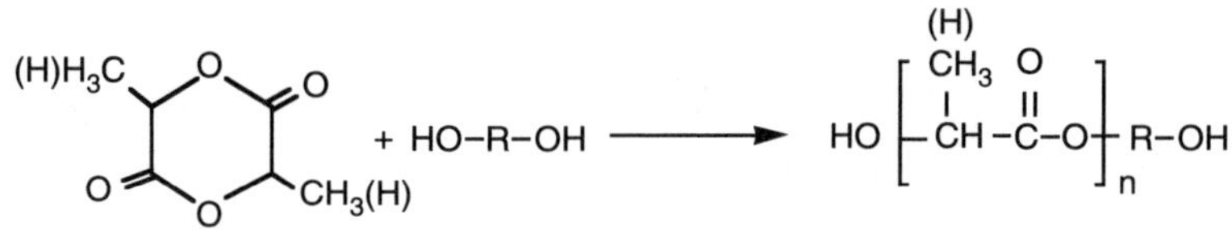

SCHEME 7.4

The diketene acetal is prepared as shown in Scheme 7.3. The photochemical rearrangement is a recent development and a significant improvement. The latent acid diol is prepared as shown in Scheme 7.4.

Both syntheses are now well worked out and kilogram quantities of these compounds prepared under good manufacturing practice (GMP) have been completed. The reaction between glycolide and a diol produces a range of products in varying concentrations with n-values varying between 1 and 7, as shown by the gel permeation chromatogram trace in Figure 7.1. To control erosion rates, the exact structure of the latent acid diol is not important, and it is the total concentration of the α-hydroxy acid segments in the polymer that determines erosion rates.

III. HYDROLYSIS

The hydrolysis proceeds in three consecutive steps.[5] In the first step, the glycolic acid segment in the polymer backbone hydrolyzes to generate a polymer fragment containing a carboxylic acid end-group that will catalyze ortho ester hydrolysis. A second cleavage produces free glycolic acid that also catalyzes hydrolysis of the ortho ester links. Further hydrolysis of the polymer then proceeds to first generate the diol, or mixture of diols used in the synthesis and pentaerythritol dipropionate, followed by ester hydrolysis to produce pentaerythritol and propionic acid.

Scheme 7.5 shows details of polymer hydrolysis. In this particular scheme we have depicted the latent acid as a dimer of lactic acid since at the time this study was carried out, we believed that this was the correct structure resulting from a reaction between lactide and a diol. We now know that this is not correct, and further, we now use glycolic acid instead of lactic acid. However, the conclusions reached in that study are in full agreement with all of our subsequent studies.

The most significant finding of the hydrolysis study is the linearity of weight loss and the concomitant release of lactic and propionic acid as shown in Figure 7.2. While linear rate of weight loss alone does not necessarily indicate surface erosion,[6] the concomitant linear weight loss and

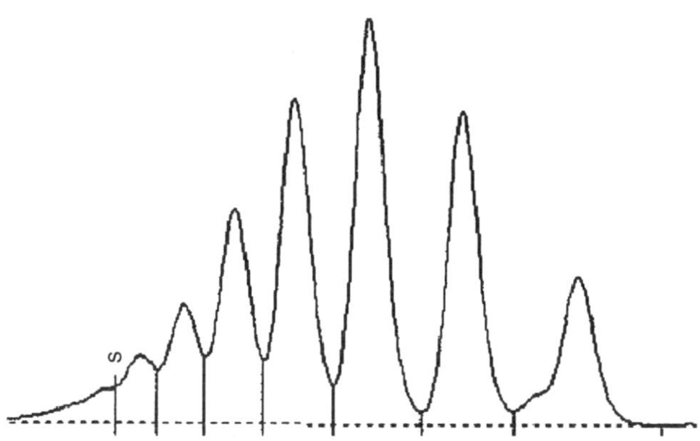

FIGURE 7.1 Gel permeation chromatograph of reaction products between lactide and triethylene glycol. (From Heller, J., Barr, J., Ng, S. Y., Schwach-Abdellaoui, K., and Gurny, R., Poly(ortho esters): synthesis, characterization, properties and uses, *Adv. Drug Deliv. Rev.*, 54, 1015–1039, 2002. With permission.)

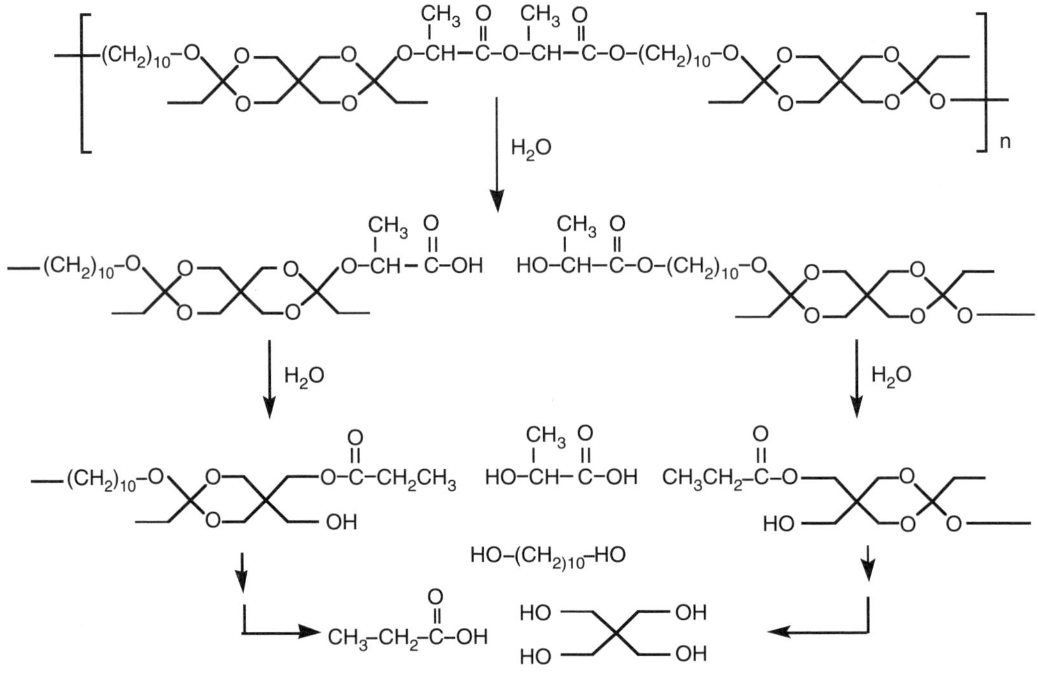

SCHEME 7.5

release of lactic acid argues convincingly for a process confined predominantly to the surface layers of the polymer matrix.

The erosion process is shown schematically in Figure 7.3. Surface erosion demands a much higher rate of hydrolysis in the surface layers of a solid device as compared to the interior of the device, and pure surface erosion can only take place if no hydrolysis occurs in the interior of the device. Clearly, this can only occur if no water penetrates the polymer, and since no polymer is so

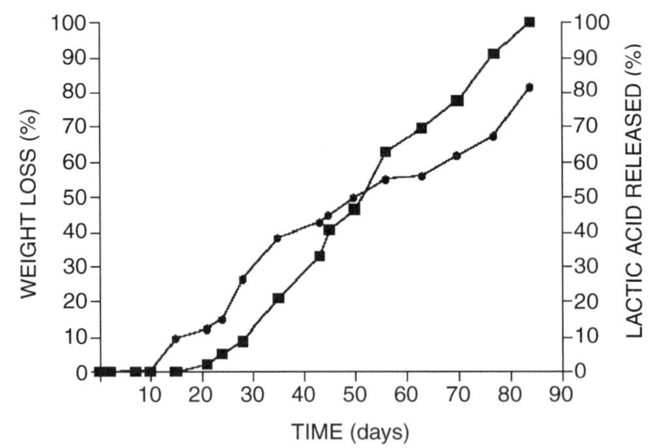

FIGURE 7.2 The relationship between lactic acid release (●) and weight loss (■) for a poly(ortho ester) prepared from 3,9 diethylidene-2,4,8,10-tetraoxaspiro [5.5] undecane, 1,10-decanediol and 1,10-decanediol lactide (100/70/30). 0.13 M, pH 7.4 sodium phosphate buffer at 37°C. (From Schwach-Abdellaoui, K., Heller, J., and Gurny, R., Hydrolysis and erosion studies of autocatalyzed poly(ortho esters) containing lactoyl-lactic acid dimers, *Macromolecules*, 32, 301–307, 1999. With permission.)

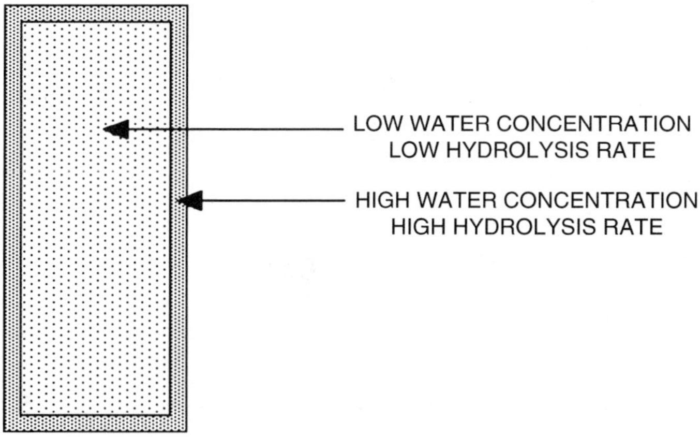

FIGURE 7.3 Schematic representation of proposed erosion mechanism.

hydrophobic that no water can penetrate the matrix, some hydrolysis does take place in the interior of the matrix.

An erosion process that is confined predominantly to the surface layers has a number of important benefits. First, if the drug is well immobilized in the matrix, its release is controlled by polymer erosion so that an ability to control polymer erosion translates into an ability to control rate of drug release. Second, because drug release is controlled by polymer erosion, drug release and polymer erosion take place concomitantly and when drug release has been completed, no polymer remains. Third, because most of the hydrolysis occurs in the outer layers of the device, acidic hydrolysis products can diffuse away from the device and do not accumulate in the bulk material. Thus, the interior of the matrix does not become highly acidic, as is the case of poly(lactide-*co*-glycolide) copolymers, or poly(lactic acid), and acid-sensitive drugs can be released without loss of activity.

An ongoing study is designed to provide a direct measurement of the pH in the matrix as a function of depth.[7] To do so, pH-sensitive nitroxide radicals with different pKa values are incorporated into the matrix and the EPR signal analyzed. Such data provide a direct measure of the pH in different layers within the device and the mobility of the nitroxide radicals provides a measure of water penetration into the matrix. Preliminary data are consistent with a surface erosion process and indicate that the pH in the outer eroding layer is about 5.4, and the pH in the matrix interior is about 6.4. The exact values may require a slight adjustment pending a determination of the pKa of the nitroxide radicals in water and in the interior of the hydrophobic polymer environment.

IV. CONTROL OF MECHANICAL AND THERMAL PROPERTIES

One of the important attributes of POE IV is the ability to vary thermal and mechanical properties by choosing the appropriate R-group in the diol and in the latent acid. The use of rigid diols produces materials with high glass transition temperatures while the use of flexible diols produces materials with low glass transition temperatures. The use of very flexible diols produces materials that are semisolids at room temperature. Thus, poly(ortho esters) can be used as solid, or semisolid devices.

A. Solid Devices

An example of the effect of diol structure on the glass transition temperature is shown in Figure 7.4 where the ratio of a rigid diol, *trans*-cyclohexanedimethanol and a flexible diol, 1,6-hexanediol in the polymer has been varied.[8] Clearly, by an appropriate choice of diol ratios, any glass transition

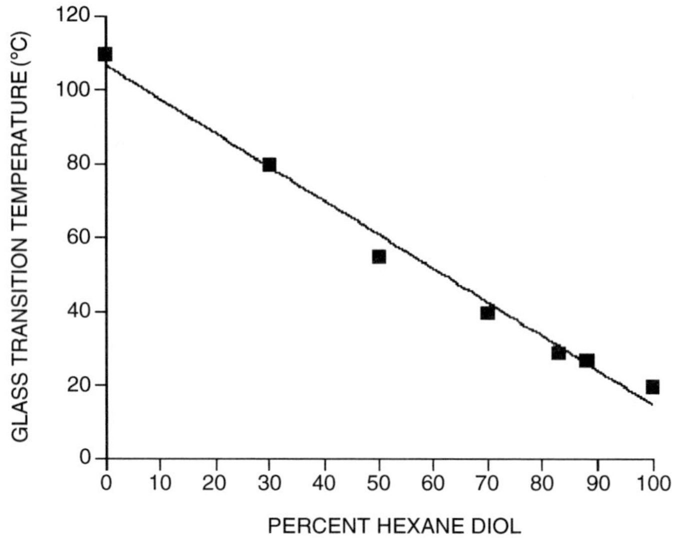

FIGURE 7.4 Glass transition temperature of 3,9-diethylidene-2,4,8,10-tetraoxaspiro [5.5] undecane, *trans*-cyclohexanedimethanol, 1,6-hexanediol polymer as a function of mole% 1,6-hexanediol. (From Heller, J., Penhale, D. W. H., Fritzinger, B. K., Rose, J. E., and Helwing, R. F., Controlled release of contraceptive steroids from biodegradable poly(ortho esters), *Contracept. Deliv. Syst.*, 4, 43–53, 1983. With permission.)

temperature between 110 and 20°C can be selected. By using even more flexible diols, materials with a glass transition temperature as low as −20°C have been prepared.[9]

Figure 7.4 has been generated with polymers with no latent acid. However, the latent acid diol does have a significant effect on the glass transition temperature, as shown in Figure 7.5.[10] Thus, both diol structure, latent acid diol structure, and their ratios must be considered when designing polymers with the desired thermal and mechanical characteristics.

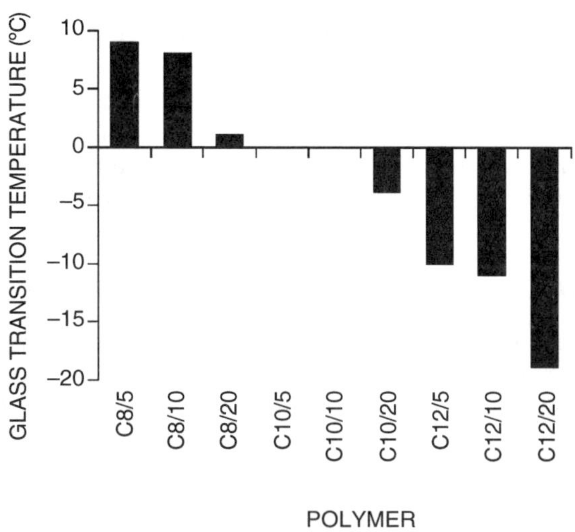

FIGURE 7.5 Glass transition temperatures for poly(ortho ester) prepared from 3,9-diethylidene-2,4,8,10-tetraoxaspiro [5.5] undecane and *n*-octanediol, *n*-decanediol and *n*-dodecanediol, each with 5, 10 and 20 mol% of the corresponding dilactide. (From Heller, J., Barr, J., Ng, S. Y., Schwach-Abdellaoui, K., and Gurny, R., Poly(ortho esters): synthesis, characterization, properties and uses, *Adv. Drug Deliv. Rev.*, 54, 1015–1039, 2002. With permission.)

1. Polymer Storage Stability

As shown in Figure 7.6, poly(ortho esters) have excellent stability, and when stored under anhydrous conditions, are stable at room temperature.[3] The particular polymer used in this stability study is a hydrophilic polymer containing 40 mol% latent acid that has been ground to produce microparticles, thus greatly increasing surface area. This is a very rapidly eroding polymer that will completely erode in a matter of a few days if placed in an aqueous buffer. Despite this high reactivity, when stored under anhydrous conditions, it is stable for a number of months.

2. Polymer Fabrication

A successful drug delivery system requires the development of suitable polymers and fabrication methods that in concert can produce devices able to achieve the desired drug release profiles. While the need to develop polymers is well recognized, the need to develop proper fabrication methods is not always appreciated. Desired release profiles that are free from drug burst and are reasonably linear can best be achieved with devices that are fabricated to minimize internal porosity, and where the drug is uniformly dispersed in the matrix with minimal particle-to-particle contact.

While there are many different types of devices used in controlled drug delivery, the two most frequently used devices are microspheres and strands prepared by an extrusion process. Of these, strands prepared by extrusion have a number of significant advantages. Dominant among these is the ability to fabricate devices without the use of solvents, and the ability to prepare dense devices with drugs that are uniformly dispersed along the length of the strand. However, extrusion requires temperatures that are typically 30 to 40°C above the glass transition temperature, and such temperatures could compromise the thermal stability of some drugs. For this reason, it is imperative that the polymers used have sufficient design flexibility to tailor glass transition temperatures to the desired values.

Since extrusion requires the use of elevated temperatures and a typical small scale extrusion as used in these studies requires about 20 to 30 min, we have investigated potential changes in

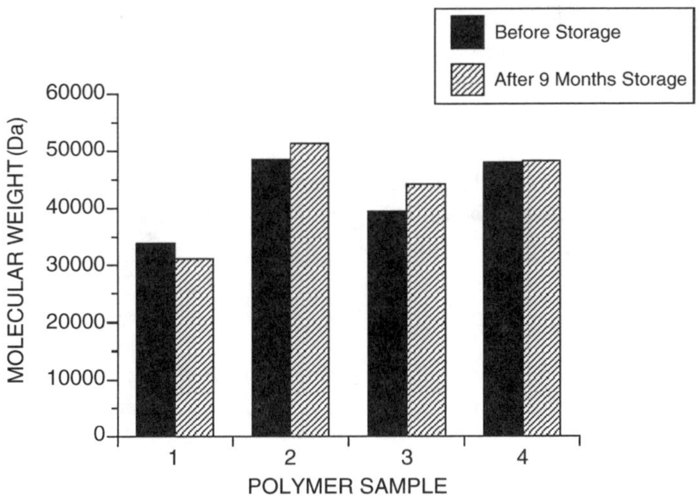

FIGURE 7.6 Stability of a polymer prepared from 3,9 diethylidene-2,4,8,10-tetraoxaspiro [5.5] undecane, *cis/trans*-cyclohexanedimethanol, triethylene glycol and triethylene glycol glycolide (100/35/25/40) stored at room temperature and under anhydrous conditions. (From Heller, J., Barr, J., Ng, S. Y., Schwach-Abdellaoui, K., and Gurny, R., Poly(ortho esters): synthesis, characterization, properties and uses, *Adv. Drug Deliv. Rev.*, 54, 1015–1039, 2002. With permission.)

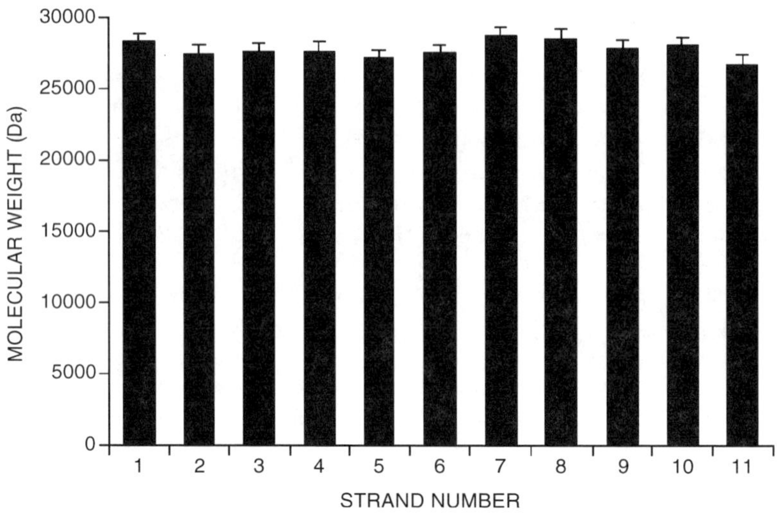

FIGURE 7.7 Molecular weight of each segment of an entire extruded strand cut into 10–1 cm sections along the entire length of the strand. Polymer prepared from 3,9 diethylidene-2,4,8,10-tetraoxaspiro [5.5] undecane, *cis/trans*-cyclohexanedimethanol, triethylene glycol, 1,10-decanediol and triethylene glycol glycolide (100/40/10/49.9/0.1).

molecular weight as a function of time by sectioning the entire extruded strand into 10 mm pieces and determining the molecular weight of selected pieces. Results are shown in Figure 7.7.

In this particular case, there is no significant change in molecular weight along the length of the extruded strand despite a long exposure to 95°C, the extrusion temperature.

Equally important is a determination of the uniformity of drug along the entire length of the extruded strand. To optimize the extrusion process, three experiments were carried out. In one experiment, the entire drug was added to the total amount of polymer and the mixture then blended in the extruder at 95°C for 5 min. As shown in Figure 7.8, drug content was well below the theoretical 20 wt% loading. In a second experiment, one half of the polymer was added to the extruder followed by the drug, and this was then followed by the second half of the polymer. The mixture was then blended for 5 min. Using this procedure, the theoretical drug loading was achieved. In a third experiment the blending time was increased to 8 min. Within experimental error, there was no change.

We have also investigated possible changes in molecular weight of the strands following γ-ray irradiation at 2.4 Mrad. Results are shown in Figure 7.9. While we do observe a decrease, for this particular molecular weight range, it is minimal.

3. Drug Release

Drug release from poly(ortho esters) has been extensively investigated and in this chapter we will only choose a few examples that best illustrate the potential of this polymer system.

a. Release of Bovine Serum Albumin from Extruded Strands

Since it is well known that most proteins that are exposed to an organic solvent–water interface become denatured, fabrication methods that avoid the use of solvents are of interest. One such procedure is an extrusion method where finely ground polymer and a micronized protein are intimately mixed and then extruded into thin strands at temperatures low enough so that protein activity is not compromised.

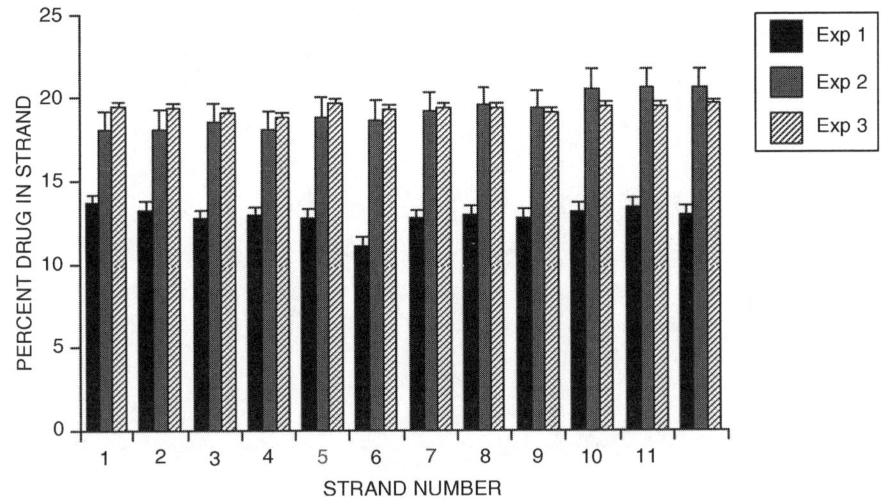

FIGURE 7.8 Uniformity of a model drug loading as a function of three experimental procedures described in the text. Drug content shown for each segment of an entire extruded strand cut into 10–1 cm sections along the entire length of the strand. Polymer prepared from 3,9 diethylidene-2,4,8,10-tetraoxaspiro [5.5] undecane, *cis/trans*-cyclohexanedimethanol, triethylene glycol, 1,10-decanediol and triethylene glycol glycolide (100/40/10/49.9/0.1).

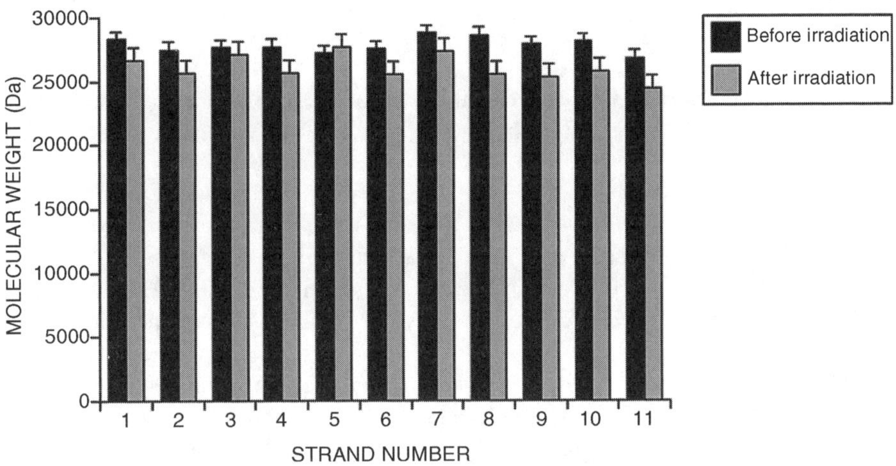

FIGURE 7.9 Polymer molecular weight before and after irradiation at 2.4 Mrad for each segment of an entire extruded strand cut into 10–1 cm sections along the entire length of the strand. Polymer prepared from 3,9 diethylidene-2,4,8,10-tetraoxaspiro [5.5] undecane, *cis/trans*-cyclohexanedimethanol, triethylene glycol, 1,10-decanediol and triethylene glycol glycolide (100/40/10/49.9/0.1).

Figure 7.10 shows release of FITC-BSA from extruded strands and the weight loss of the strands as a function of time.[11] Three features are notable. First, there is only a minimal burst despite the fact that 15 wt% of a water-soluble material has been incorporated, second, there is a significant lag before release of FITC-BSA begins and third, release is linear and concomitant with weight loss.

While a long induction period may be desirable in some applications, for example, in vaccine delivery, for general protein delivery it is not desirable and we have investigated means of

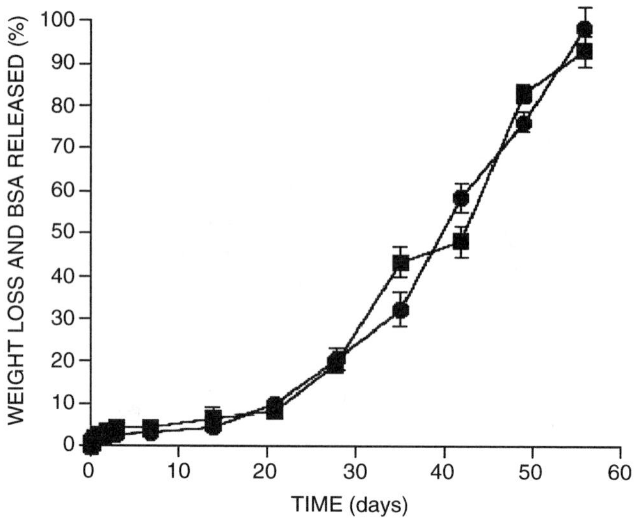

FIGURE 7.10 Release of FITC-BSA (●) and weight loss (■) from a poly(ortho ester) prepared from 3,9-diethylidene-2,4,8,10-tetraoxaspiro [5.5] undecane, 1,4-pentanediol and 1,6-hexanediol glycolide (100/95/5). Strands, 1 × 10 mm, extruded at 70°C. 0.01 M phosphate buffered saline, pH 7.4, 37°C. FITC-BSA loading 15 wt%. (From Rothen-Weinhold, A., Schwach-Abdelaoui, K., Barr, J., Ng, S. Y., Shen, H.-R., Gurny, R., and Heller, J., Release of BSA from poly(ortho ester) extruded thin strands, *J. Control. Release*, 71, 31–37, 2001. With permission.)

eliminating the induction period. One successful means of accomplishing this was to use a block copolymer of poly(ortho ester) and poly(ethylene glycol), and this will be discussed in Section V.

b. Delivery of DNA

The delivery of DNA from poly(ortho esters) is of particular interest, because we have previously demonstrated an erosion-controlled release as well as an essentially neutral pH in the interior of the matrix so that incorporation of DNA into the polymer and its subsequent release should not adversely affect DNA stability.

Release of DNA from 5 μm microspheres prepared by a double emulsion method is shown in Figure 7.11.[12] When the microspheres were placed in a pH 7.2 buffer and release of DNA followed using the pico green method, release was slow, corresponding to slow polymer erosion. However, when the pH of the buffer was changed to 5.0, the pH of endosomes, an immediate and significant acceleration of DNA release was noted. Since a lowered pH results in an increased erosion rate, this plot provides unequivocal evidence of an erosion-controlled DNA release.

We have also investigated the release of DNA from nanospheres with a diameter slightly less than 1 μm.[13] The release is shown in Figure 7.12. Preliminary evidence indicates that DNA expression was observed after injection into Zebra fish, again indicating that DNA released from poly(ortho esters) retains is functionality.

c. Delivery of 5-Fluorouracil

Unlike BSA or DNA that are high in molecular weight, water-soluble molecules, 5-fluorouracil (5-FU) is a small water-soluble molecule. Thus, during drug release studies using 5-FU, there is the potential for a significant diffusional component to the release mechanism. However, as shown in Figure 7.13, when 5-FU release from thin wafers and weight loss of the wafers was determined,

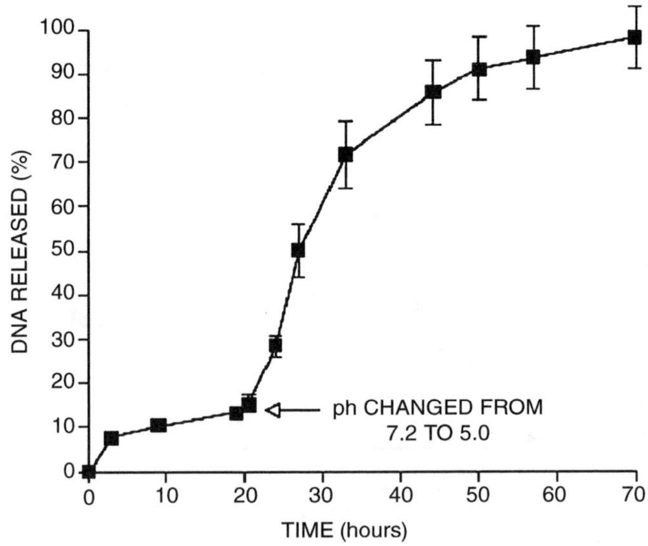

FIGURE 7.11 Release of DNA from a poly(ortho ester) prepared from 3,9-diethylidene-2,4,8,10-tetraoxaspiro [5.5] undecane, triethylene glycol, 1,2-propane diol and triethylene glycol glycolide (100/35/15/45/5). Microspheres, 5 μm, phosphate buffer (pH 7.4) or sodium acetate buffer (pH 5.0) at 37°C.

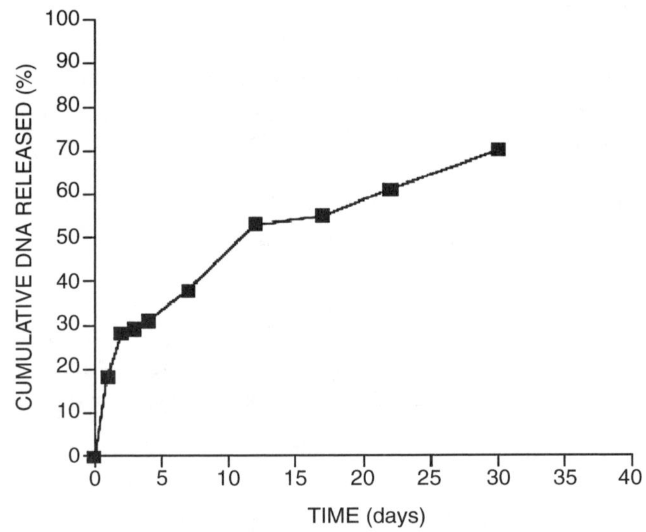

FIGURE 7.12 Release of DNA from nanospheres (1 μm) from a poly(ortho ester) prepared from 3,9-diethylidene-2,4,8,10-tetraoxaspiro [5.5] undecane, cyclohexanedimethanol, triethylene glycol and triethylene glycol lactide (100/89/10/1). DNA loading 0.75 μg/mg.

within experimental error, both processes occurred concomitantly suggesting that the dominant drug release mechanism was polymer erosion.[14]

This is an encouraging result and indicates that a wide range of therapeutic agents can be delivered from poly(ortho esters), as has been validated in numerous studies.

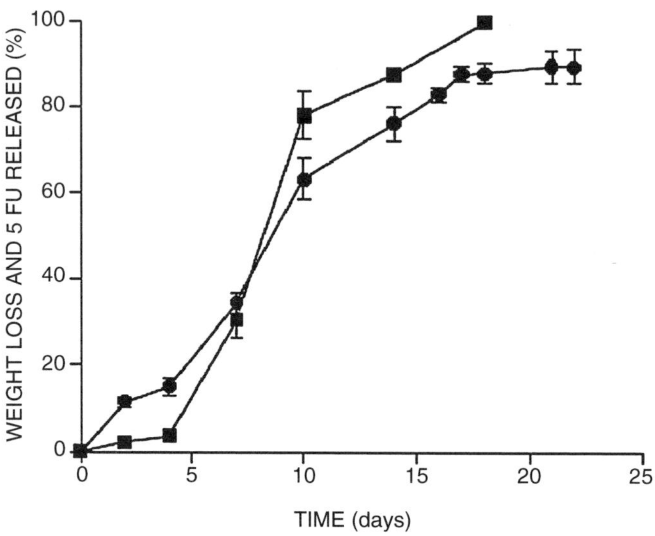

FIGURE 7.13 Polymer weight loss (●) and 5-fluorouracil (5-FU) release (■) from a polymer prepared from 3,9-diethylidene-2,4,8,10-tetraoxaspiro [5.5] undecane, 1,3-propanediol and triethylene glycol diglycolide (90/10). Drug loading 20 wt%. 0.05 *M* phosphate buffer, pH 7.4, 37°C. (From Heller, J., Barr, J., Ng, S. Y., Shen, H.-R., Schwach-Abdellaoui, K., Einmahl, S., Rothen-Weinhold, A., and Gurny, R., Poly(ortho esters) — their development and some recent applications, *Eur. J. Pharm. Biopharm.*, 50, 121–128, 2000. With permission.)

B. Semisolid Devices

To prepare semisolid materials it is necessary to use highly flexible diols, and in order to allow direct injection, their viscosity must be limited by preparing polymers with molecular weights no higher than about 6 kDa. However, even such low molecular weight materials cannot be injected unless an excipient is used to further reduce viscosity. Our work with semisolid materials has concentrated on those using triethylene glycol and those using 1,10-decanediol. Semisolids based on triethylene glycol produce somewhat hydrophilic materials while semisolids based on 1,10-decanediol produce hydrophobic materials.

The most significant advantage of semisolid materials is that therapeutic agents can be readily incorporated by mixing at ambient temperature and without the use of solvents. Such conditions are mild enough to incorporate even the most sensitive therapeutic agents. Mixing can be accomplished on a small scale by using a mortar and pestle, but on a larger scale is carried out using a three-roll mill, or mechanical mixers.

1. Polymer Molecular Weight Control

Polymer molecular weight control can be achieved by using an excess of diol relative to the diketene acetal, but can also be accomplished by using a chain-stopper. In this approach a calculated amount of a monofunctional alcohol is used.[15] As shown in Scheme 7.6, when *n*-decanol is used as a chain-stopper in combination with 1,10-decanediol, both polymer ends have *n*-decanol residues so that this chain-stopped material is somewhat more hydrophobic relative to a stoichiometry-controlled material that has terminal hydroxyl groups.

The use of chain-stoppers allows excellent and reproducible molecular weight control by varying the ratios of 1,10-decanediol to *n*-decanol as shown in Figure 7.14.[15] The existence of terminal methyl groups has been established by ^{1}H NMR studies.[15]

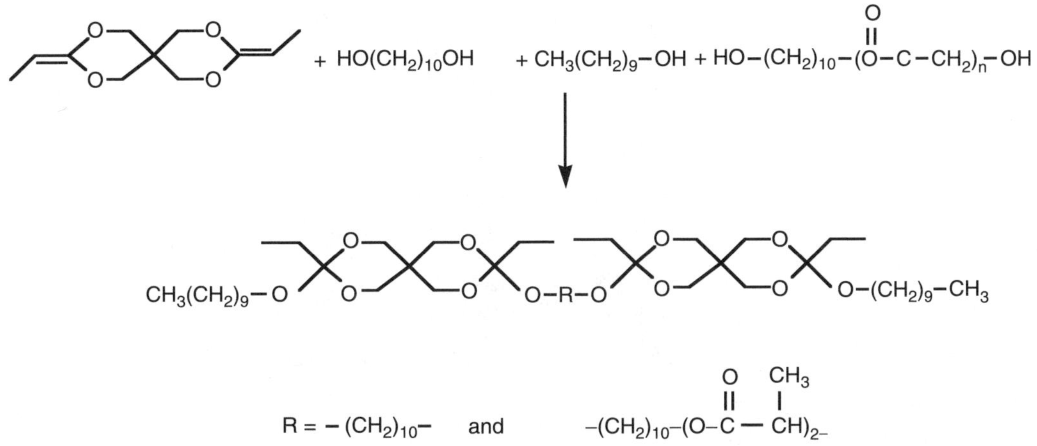

SCHEME 7.6

2. Polymer Stability

As already demonstrated for solid poly(ortho esters), semisolid materials also show excellent stability when stored under anhydrous conditions. Results of one such study is shown in Figure 7.15.[16]

3. Polymer Erosion and Drug Release

Semisolid materials based on triethylene glycol are somewhat hydrophilic and are designed for relatively short lifetimes, while semisolid materials based on decanediol are more hydrophobic and are designed for longer term lifetimes. *In vivo* lifetimes of these two materials implanted in rats and followed by magnetic resonance imaging (MRI) is shown in Figure 7.16. Clearly, the more hydrophobic material has a significantly longer lifetime.

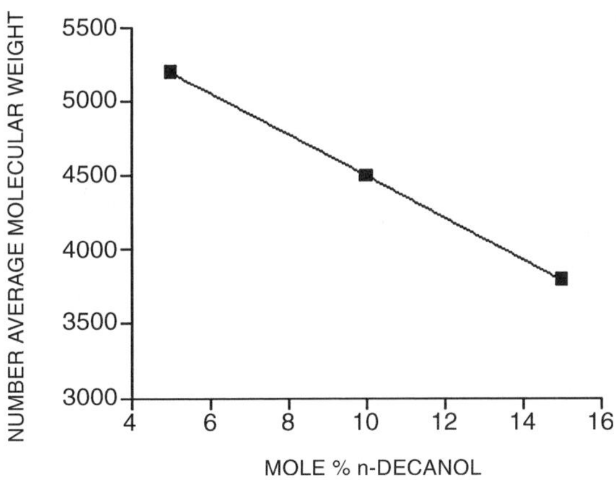

FIGURE 7.14 Effect of n-decanol on the molecular weight of a poly(ortho ester) prepared from 3,9-diethylidene-2,4,8,10-tetraoxaspiro[5.5] undecane, 1,10-decanediol and 1,10-decanediol lactide (100/70/30). (From Schwach-Abdellaoui, K., Heller, J., Barr, J., and Gurny, R., Control of molecular weight for autocatalyzed poly(ortho esters) obtained by polycondensation reaction, *Int. J. Polym. Anal. Charact.*, 7, 145–161, 2002. With permission.)

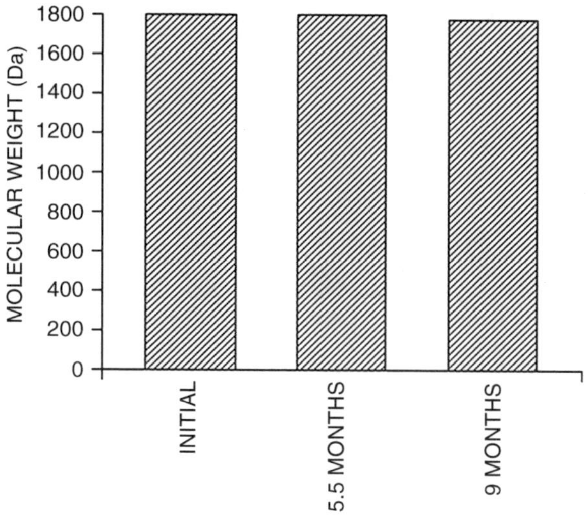

FIGURE 7.15 Room temperature stability for a polymer prepared from 3,9-diethylidene-2,4,8,10-tetraoxaspiro[5.5] undecane, triethylene glycol and triethylene glycol glycolide (60/50/50). Material stored under anhydrous conditions. (From Heller, J., Barr, J., Ng, S. Y., Shen, H.-R., Gurny, R., Schwach-Abdelaoui, K., Rothen-Weinhold, A., and van der Weert, M., Development of poly(ortho esters) and their application for bovine serum albumin and bupivacaine delivery, *J. Control. Release*, 78, 133–141, 2002. With permission.)

Drug release from semisolid materials is complicated by their low molecular weights which makes immobilization of drugs in the matrix difficult. Consequently, unless the molecular weight of the incorporated drug is high enough to impede diffusion, or the therapeutic agent is highly insoluble in water and in the matrix, the dominant drug release mechanism is diffusion.

An example of the ability of semisolid materials to deliver proteins is shown in Figure 7.17, where FITC-BSA has been incorporated and rate of release followed.[17] While no weight loss determinations have been carried out, visually, drug depletion and compete polymer erosion occurred simultaneously.

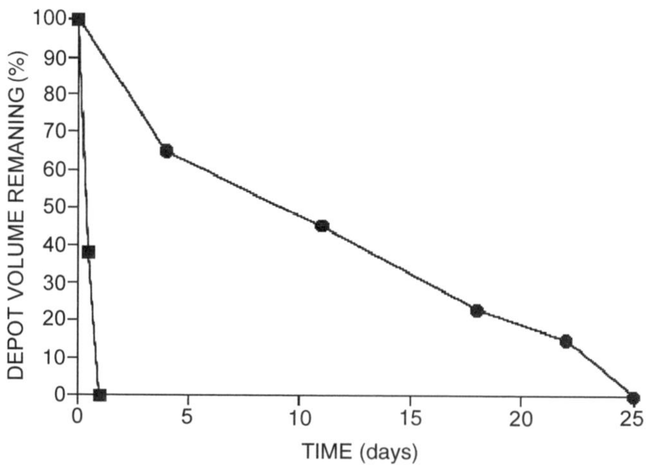

FIGURE 7.16 *In vivo* polymer erosion measurements using magnetic resonance imaging (MRI). Rats were injected subcutaneously with (■) polymer prepared from 3,9-diethylidene-2,4,8,10-tetraoxaspiro [5.5] undecane triethylene glycol and triethylene glycol glycolide, (●) polymer prepared from 3,9-diethylidene-2,4,8,10-tetraoxaspiro [5.5] undecane, decanediol and triethylene glycol glycolide.

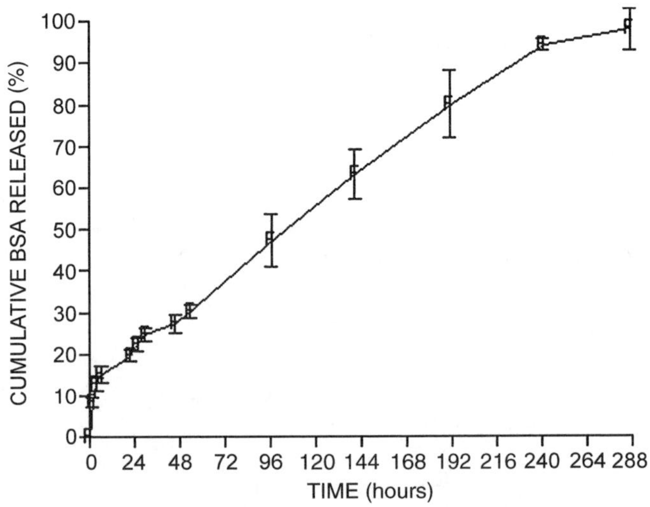

FIGURE 7.17 Release of BSA from a semisolid poly(ortho ester) prepared from 3,9-diethylidene-2,4,8,10-tetraoxaspiro [5.5] undecane, triethylene glycol and triethylene glycol glycolide (100/95/5). BSA loading 4 wt%. 0.1 *M* phosphate buffer, pH 7.3, 37°C. (From Heller, J., Barr, J., Ng, S. Y., Shen, H.-R., Gurny, R., Schwach-Abdelaoui, K., Rothen-Weinhold, A., and van der Weert, M., Development of poly(ortho esters) and their application for bovine serum albumin and bupivacaine delivery, *J. Control. Release*, 78, 133–141, 2002. With permission.)

V. BLOCK COPOLYMERS

The synthesis of poly(ortho esters) has been extended by the synthesis of block copolymers of poly(ortho esters) and poly(ethylene glycol). The synthesis of AB blocks is shown in Scheme 7.7[11] and the synthesis of ABA block copolymers in Scheme 7.8.[3]

Such materials are useful as matrices for protein delivery and as building blocks for micelles.

A. Matrices for Protein Delivery

Block copolymers prepared from hydrophilic and hydrophobic segments may represent a desirable matrix for protein delivery, due to the incompatibility of hydrophilic proteins with hydrophobic polymers that may lead to protein aggregation. Thus, the introduction of hydrophilic domains into hydrophobic polymers may prevent the aggregation problem and result in better release kinetics and better retention of protein activity.

While we have not yet investigated retention of protein activity, we have demonstrated improved release kinetics, as shown by FITC-BSA release from an AB Block copolymer, shown in Figure 7.18.[11]

These release kinetics represent a significant improvement over those achieved with a hydrophobic poly(ortho ester) as already shown in Figure 7.10.

B. Micelles

When amphiphilic polymers are dispersed in water, they spontaneously self-assemble into micellar structures.[18] Such structures have a number of important applications, but perhaps the most important application is tumor targeting using the enhanced permeability and retention effect.[19] The basis of this phenomenon is the finding that the vasculature in tumors and also that of inflamed tissues is hyperpermeable, so that macromolecules, such as polymer–drug conjugates, can only undergo extravasation at the hyperpermeable sites. Additionally, it has also been shown that there is an inherent lack of lymphatic drainage in solid tumors resulting in passive accumulation of

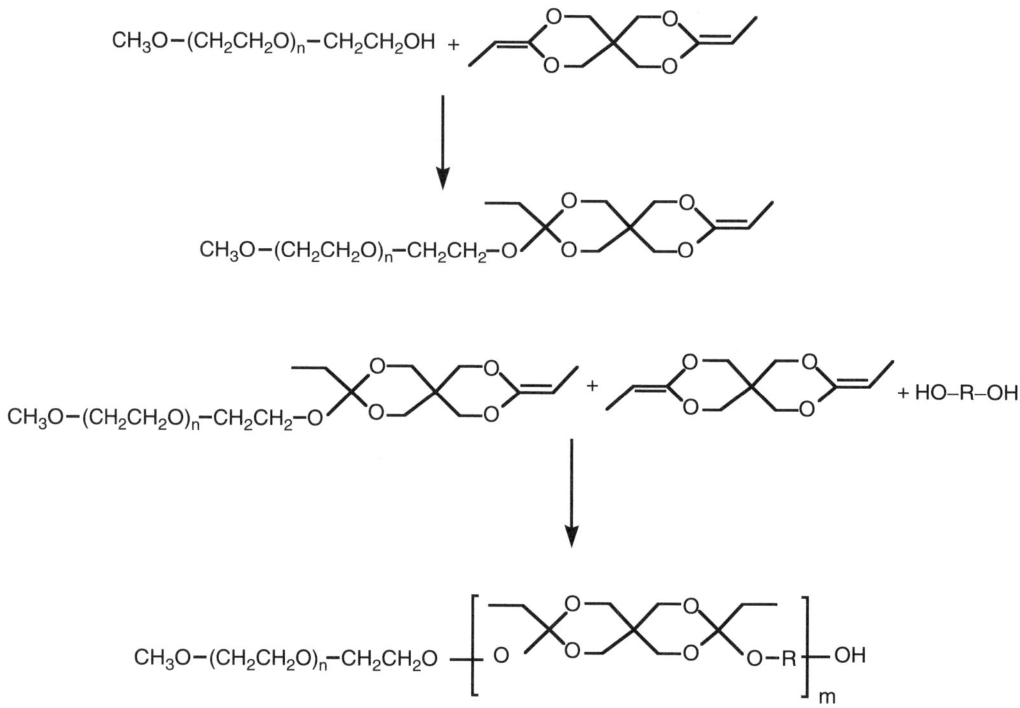

SCHEME 7.7

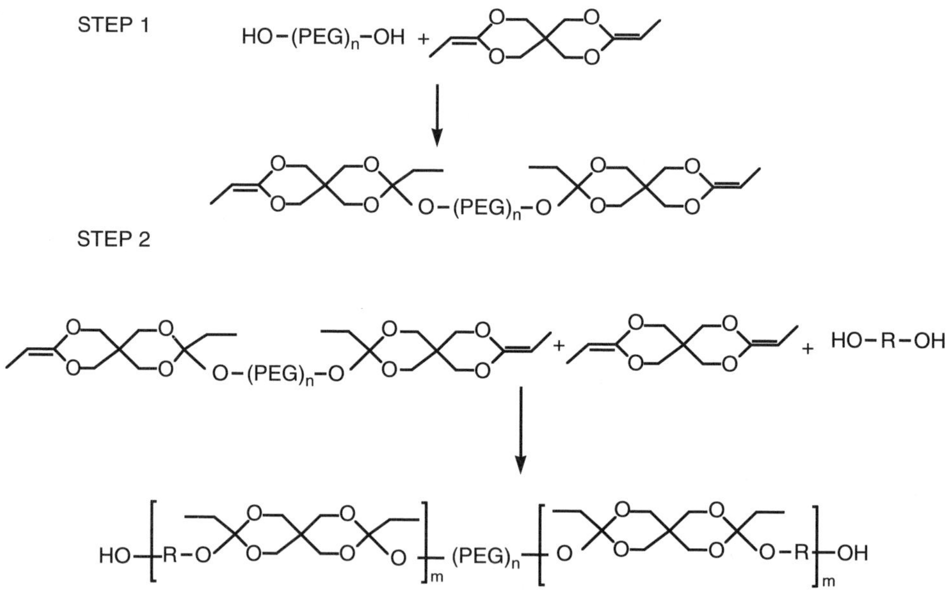

SCHEME 7.8 PEG, polyethylene glycol.

macromolecules with time. This passive accumulation will then alter the biodistribution of a chemotherapeutic agent and allow high tumor drug concentration without a concomitant high systemic drug concentration.

In the actual application of micelles in tumor targeting, a hydrophobic anticancer agent is physically entrapped in the hydrophobic core of the micelle and the formulation injected

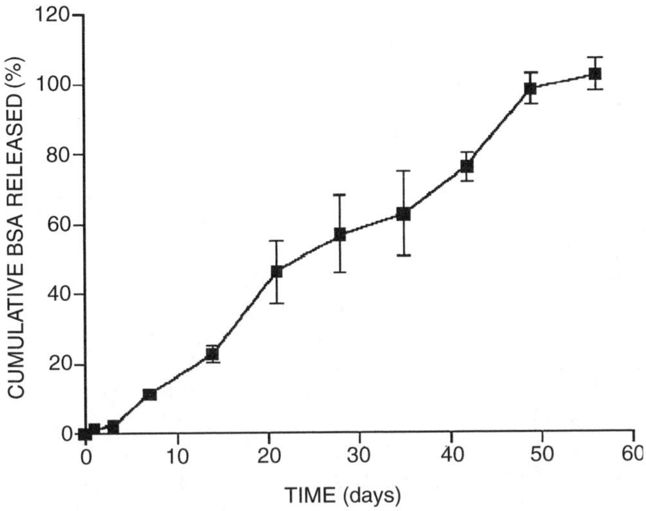

FIGURE 7.18 Release of FITC-BSA from an AB-block copolymer containing 6 wt% 2 kDa poly(ethylene glycol). Poly(ortho ester) prepared from 3,9-diethylidene-2,4,8,10-tetraoxaspiro [5.5] undecane, 1,3-propanediol and triethylene glycol glycolide (100/85/15). Strands, 1 × 10 mm, extruded at 70°C. 0.01 M phosphate buffered saline, pH 7.4, 37°C. FITC-BSA loading 15 wt%. (From Rothen-Weinhold, A., Schwach-Abdelaoui, K., Barr, J., Ng, S. Y., Shen, H.-R., Gurny, R., and Heller, J., Release of BSA from poly(ortho ester) extruded thin strands, *J. Control. Release*, 71, 31–37, 2001. With permission.)

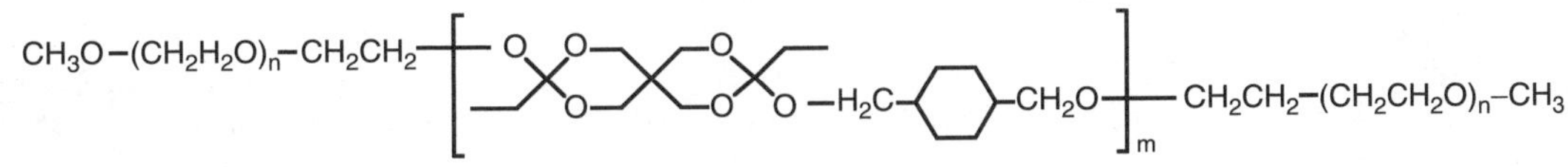

SCHEME 7.9

intravenously. The physical entrapment of drugs is preferable to one where the drug is chemically attached to the hydrophobic portion of the micelle, or a water-soluble polymer because the drug has not been chemically modified thus making regulatory approval easier.

Micelles based on poly(ethylene glycol) and poly(ortho esters) are of particular interest because poly(ortho esters) are highly hydrophobic materials so that the entrapment of highly hydrophobic drugs such as doxorubicin, or taxol, should be very efficient. Additionally, poly(ortho esters) are pH-sensitive, and because the pH in the interior of tumors is about 5.5, once internalized, the micelles should rapidly degrade and deliver their payload.

Micelles with a structure shown in Scheme 7.9 are currently under investigation.[20] In this particular composition, a poly(ortho ester) segment based on *trans*-cyclohexanedimethanol was chosen because this diol will produce the most hydrophobic structure.

This ongoing work has shown that these amphiphilic block copolymers form micellar structures in water and buffers with a CMC in the range of 3×10^{-4} to 5×10^{-4} g/l which is low enough to assure retention of micelle integrity upon intravenous injection. The size as determined by dynamic light scattering was in the 40 to 70 nm range. We have found that the micelles can be stored in lyophilized form for at least 8 months and easily reconstituted to the original properties. The micelles are stable in PBS at pH 7.4 and 37°C for 5 days and in citrate buffer at pH 5.5 and 37°C for 2 h. Stability in the presence of bovine serum albumin depends on the structure of the block copolymer and especially the length of the POE block.

Thus, these micelles appear to be a promising system and are currently under active development as a carrier for paclitaxel.

REFERENCES

1. Heller, J., Poly(ortho esters), *Adv. Polym. Sci.*, 107, 41–92, 1993.
2. Heller, J., and Gurny, R., Poly(ortho esters), In *Encyclopedia of Controlled Drug Delivery*, Mathiowitz, E., ed., Wiley, New York, pp. 852–874, 1999.
3. Heller, J., Barr, J., Ng, S. Y., Schwach-Abdellaoui, K., and Gurny, R., Poly(ortho esters): synthesis, characterization, properties and uses, *Adv. Drug Deliv. Rev.*, 54, 1015–1039, 2002.
4. Ng, S. Y., Vandamme, T., Taylor, M. S., and Heller, J., Synthesis and erosion studies of self-catalyzed poly(ortho esters), *Macromolecules*, 30, 770–772, 1997.
5. Schwach-Abdellaoui, K., Heller, J., and Gurny, R., Hydrolysis and erosion studies of autocatalyzed poly(ortho esters) containing lactoyl-lactic acid dimers, *Macromolecules*, 32, 301–307, 1999.
6. Shaw, S. S., Cha, Y., and Pitt, C. G., Poly(glycolic acid-*co*-lactic acid): diffusion or degradation controlled drug delivery ?, *J. Control. Release*, 18, 261–270, 1992.
7. Carpancioni, S., Schwach-Abdellaoui, K., Herrmann, W., Brosig, H., Borchet, H.-B., Heller, J., and Gurny, R., In vitro monitoring of poly(ortho esters) degradation by electron paramagnetic resonance imaging, *Macromolecules*, 36, 6135–6141, 2003.
8. Heller, J., Penhale, D. W. H., Fritzinger, B. K., Rose, J. E., and Helwing, R. F., Controlled release of contraceptive steroids from biodegradable poly(ortho esters), *Contracept. Deliv. Syst.*, 4, 43–53, 1983.
9. Heller, J., Rime, A.-F., Rao, S. S., Fritzinger, B. K., and Ng, S. Y., Poly(ortho esters) for the continuous and pulsatile delivery of proteins, In *Trends and Future Perspectives in Peptide and Protein Delivery*, Lee, V. H. L., Hashida, M., and Mizushima, Y., eds., Harwood Academic Publishers, Switzerland, pp. 39–56, 1995.
10. Sintzel, M. B., Heller, J., Ng, S. Y., Taylor, M. S., Tabatabay, C., and Gurny, R., Synthesis and characterization of self-catalyzed poly(ortho esters), *Biomaterials*, 19, 791–800, 1998.
11. Rothen-Weinhold, A., Schwach-Abdelaoui, K., Barr, J., Ng, S. Y., Shen, H.-R., Gurny, R., and Heller, J., Release of BSA from poly(ortho ester) extruded thin strands, *J. Control. Release*, 71, 31–37, 2001.
12. Wang, C., Ge, Q., Ting, D., Nguyen, D., Shen, H.-R., Chen, J., Eisen, H. N., Heller, J., Langer, R., and Putnam, D., Molecularly engineered poly(ortho ester) microspheres for enhanced delivery of DNA vaccines, *Nat. Mater.*
13. Yang, Y.-Y., unpublished work carried out at the University of Singapore.
14. Heller, J., Barr, J., Ng, S. Y., Shen, H.-R., Schwach-Abdellaoui, K., Einmahl, S., Rothen-Weinhold, A., and Gurny, R., Poly(ortho esters) — their development and some recent applications, *Eur. J. Pharm. Biopharm.*, 50, 121–128, 2000.
15. Schwach-Abdellaoui, K., Heller, J., Barr, J., and Gurny, R., Control of molecular weight for autocatalyzed poly(ortho esters) obtained by polycondensation reaction, *Int. J. Polym. Anal. Charact.*, 7, 145–161, 2002.
16. Heller, J., Barr, J., Ng, S. Y., Shen, H.-R., Schwach-Abdellaoui, K., Gurny, R., Vivien-Castioni, N., Loup, P. J., Baehni, P., and Mombelli, A., Development and applications of injectable poly(ortho esters) for pain control and peridontal treatment, *Biomaterials*, 23, 4397–4404, 2002.
17. Heller, J., Barr, J., Ng, S. Y., Shen, H.-R., Gurny, R., Schwach-Abdelaoui, K., Rothen-Weinhold, A., and van der Weert, M., Development of poly(ortho esters) and their application for bovine serum albumin and bupivacaine delivery, *J. Control. Release*, 78, 133–141, 2002.
18. Kataoka, K., Kwon, G. S., Yokoyama, M., Okano, T., and Sakurai, Y., Block copolymer micelles as vehicles for drug delivery, *J. Control. Release*, 24, 119–132, 1993.
19. Matsumira, Y., and Maeda, H., A new concept for macromolecular therapeutics in cancer chemotherapy: mechanism of tumoritropic accumulation of proteins and antitumor agent SMANCS, *Cancer Res.*, 46, 6387–6392, 1986.
20. Toncheva, V., Schacht, E., Ng, S. Y., Barr, J., and Heller, J., Use of block copolymers of poly(ortho esters) and poly(ethylene glycol) micellar carriers as potential tumour targeting systems, *J. Drug Target*, 11, 345–353, 2003.

Part II

Scaffold Fabrication Technologies

8 Salt Leaching for Polymer Scaffolds: Laboratory-Scale Manufacture of Cell Carriers

Joerg K.V. Tessmar, Theresa A. Holland, and Antonios G. Mikos

CONTENTS

I. INTRODUCTION

Despite recent progress and advances in processing techniques for tissue engineering scaffolds, further improvements are still necessary for laboratory processing of polymers into porous cell carriers with controllable porosity, pore size, outer shape, and polymer crystallinity. For example, the machinery to produce woven scaffolds for cell culture experiments in tissue engineering is often unavailable and too expensive for many academic research labs or smaller research facilities. In addition, the variety of polymers capable of achieving application tailored characteristics and specific properties are greatly limited.

Accordingly, to produce small, polymeric scaffolds with adequate porosity for use in cell culture experiments, the technique of porogen leaching is often undertaken. A suitable porogen is first combined with a solution of polymer in an appropriate mold for the manufacture of solid polymer–porogen constructs. The porogen is then leached out to form highly porous sponges for the cultivation of cells. This procedure, known as solvent-casting and particulate-leaching, takes advantage of the fact that inorganic salts (sodium chloride[1,2] and/or ammonium carbonate[3]) and

TABLE 8.1
Examples of Used Polymers and Porogens and Applied Techniques

Polymer	Porogen	Processing Technique	Reference
PCL	NaCl	Freeze-drying, salt leaching	15,16
PHA	NaCl	Thermo welding around mold, salt leaching	33
PLA, PCL	NaCl	Coagulation, compression molding, salt leaching	34
PLGA	NaCl and NH_4HCO_3	Gas foaming by elevated temperature	3
PLGA	Glucose	Solvent casting, leaching	4
PLGA	NaCl	Reinforcement with hydroxyapatite, salt leaching	5
PLGA	NaCl	Solvent merging of polymer and salt particles	18
PLGA	NaCl	NaCl salt fusion with elevated humidity, gas foaming	27
PLGA	NaCl	Gas Foaming, Salt Leaching	14,35
PLGA, PLLA	NaCl	Melt Extrusion of Polymer Salt Premixture	6
PLGA, PLLA	NH_4HCO_3	Gas Foaming in Acidic Solution	36,37
PLLA	NaCl	Solvent Casting, Salt Leaching	1
PLLA	NaCl	Solid Free-Form Fabrication and Salt Leaching	11
PLLA, PLGA	NaCl	Solvent Casting, Salt Leaching, Lamination	2
PPF	NaCl	Photo Cross-Linking, Leaching with Water	24

other recently used porogens (e.g., sugars[4]) are not soluble in the organic solvents typically used to dissolve biodegradable polymers. After casting the polymer–porogen mixture, the porogen can be removed later with water, leaving the water-insoluble polymer behind (Table 8.1).

The main advantage of this processing technique is the fact that only small quantities of polymer are necessary to obtain the final polymer scaffolds, since this method circumvents complex machinery that must be filled with large amounts of polymer. For example, the use of large extrusion or spinning machines to manufacture woven poly(glycolic acid) meshes requires kilograms of polymer to fill machines reservoirs, always leading to some polymer remaining unused in these chambers. Therefore, this salt leaching technique is especially useful for biomaterials in the development stage, where only small polymer amounts are available, and the amount of wasted polymer should be minimized as much as possible. Although the technique of salt leaching is especially suitable for small-scale applications, bulk expansion is easily accomplished. However, size limitations do exist on the achievable height of the resulting scaffolds, necessitating more sophisticated methods to achieve homogenous pore size distribution in these larger scaffolds.

II. DEVELOPMENT OF THE SOLVENT-CASTING AND PARTICULATE-LEACHING PROCEDURE

The method of salt leaching of scaffolds was first described in 1993 and 1994[1,2] to demonstrate that biodegradable polymers could be processed into various porous cell foams by using grinded sodium chloride as a porogen. Originating as a technique to produce single layered membranes, the method of solvent-casting and particulate-leaching was further developed to obtain more complex, laminated, three-dimensional constructs for use as cell carriers.[2] More recent steps in the development of these carriers include the introduction of reinforcing and tissue-guiding fibers in polymer scaffolds to improve tissue regeneration.[5]

To make these porous cell carriers suitable for additional applications, like peripheral nerve growth or the development of artificial blood vessels, precise three-dimensional shapes were required and led to the development of sophisticated extrusion technologies[6,7] and methods of adhering porous membranes to the desirable shapes.[8,9] Additional processing procedures have further improved the properties of these cell carriers. As this chapter will later describe, modifications to the sodium chloride crystal size and methods of linking sodium chloride crystals to bigger crystal structures has allowed for successful control of the porosity and pore size of the resulting carriers.

In addition to the manufacturing process, shape, and porosity of the polymeric scaffolds, the degradation rate and the overall fate of these biomaterials in the human body are also crucial scaffold properties that will ultimately determine the extent of tissue in-growth. Factors, which determine the degradation time, include the polymer's chemical structure, molecular weight, and crystallinity within the resulting scaffold. The scaffold fabrication process can dramatically impact these properties. For example, heat or chemical changes resulting from sterilization techniques, the leaching process, or other chemical or physical stresses during scaffold manufacturing may alter polymer properties. Therefore, a detailed evaluation of the degradation kinetics of the final scaffolds is required to predict and interpret the outcome of tissue culture experiments.[10] Furthermore, degradation studies allow for additional improvements to processing techniques to achieve scaffolds with more suitable degradation profiles.

Porogen leaching can be coupled with many of the other processing methods described in this book to develop scaffolds for tissue engineering with bioactive properties, precise shapes, and interconnected pore networks. For example, current methods involve integration of growth factors into the polymeric phase and onto the surface of salt leached cell carriers to enhance and facilitate tissue growth. Various shapes and sizes of scaffolds have been obtained by using solid free form fabrication methods.[11,12] Here, three-dimensional printing is combined with the salt leaching procedure to achieve the desired shape and increased porosity. The obtained scaffolds have variable outer shapes, and during manufacture, different polymers can be used to produce laminated structures for improved adhesion and cellular organization. Salt leaching can also be combined with gas foaming methods to produce highly porous scaffolds. The incorporated salt assures the interconnectivity of the pores generated by the foaming process and allows for successful cell in-growth into the scaffolds.[13,14] Additionally, the salt can also be incorporated in a manner which induces phase separation of the polymer and solvent, leading to even higher porosity and pore interconnectivity.[15,16]

III. SCAFFOLD MANUFACTURING

Numerous salt leaching procedures are reported in literature for the production of porous scaffolds for tissue engineering. These various methods attempt to address several different problems, including the size of the scaffolds, their overall porosity, and achievable pore sizes. Common steps in scaffold manufacturing techniques, which involve salt leaching, are given in Figure 8.1 and are described in detail in Section III.A–III.F.

A. Salt and Polymer Preparation

To obtain cell carriers, raw materials must be initially prepared for the scaffold assembly. First the desired porogen is ground and sieved to obtain the required pore size of the final scaffolds. One of the easiest methods to obtain small sodium chloride particulates is the grinding with analytical mills or air attrition mills. However, the resulting ground salt particles often have a wide size distribution, and therefore, analytical sieves must be used to obtain fractions with narrow size distributions. The salt should then be stored in tightly sealed containers to prevent recrystallization and particulate clumping due to the hygroscopic nature of these small salt particles.

In addition, polymers must often be prepared for the scaffold fabrication process. For example, if large polymer particles or sheets are present after the polymer synthesis, methods must be undertaken to facilitate homogenous mixing with the porogen. The predominant method of polymer grinding involves the use of liquid nitrogen during milling. Liquid nitrogen serves a dual role,

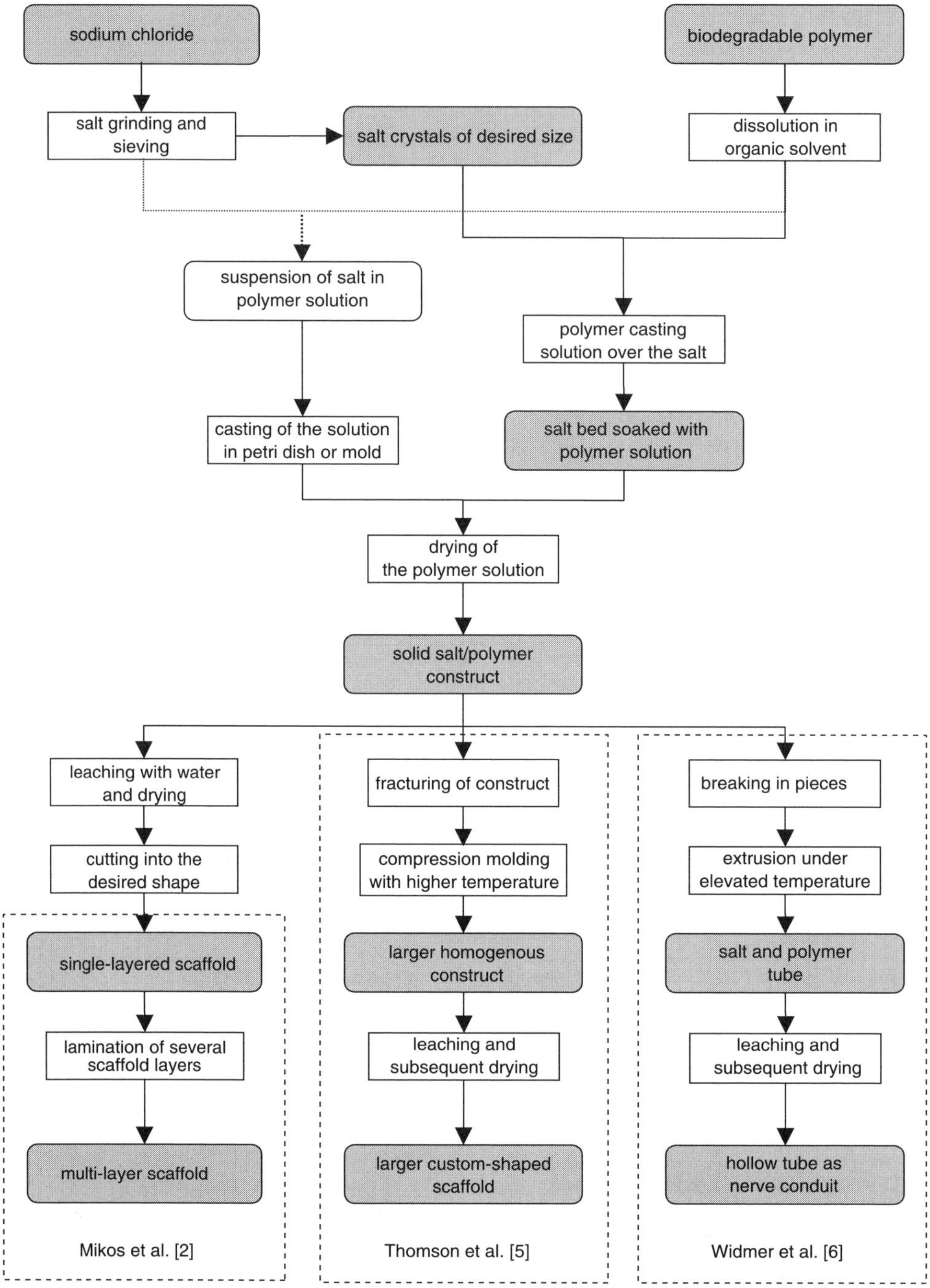

FIGURE 8.1 Flow chart of the manufacturing process for different salt leached scaffolds described.

making the polymers more brittle and preventing thermal degradation of the polymer due to heat that is developed during the grinding process. The resulting polymer particulates are then dissolved in a volatile organic solvent. The extent of polymer solubility and porogen insolubility is crucial in the selection of this third component. Commonly used organic solvents include methylene chloride and chloroform.

B. Salt and Polymer Mixing

A highly concentrated and viscous organic solution of the polymer is then mixed with the sieved salt particles. Often, the salt particles are added directly to the polymer solution to create a salt suspension, which is then cast into Teflon® molds or antiadhesive Petri dishes. Alternatively, the polymer solution can be cast over a salt bed and the resulting suspension dried in a fume hood. In both cases, the aim is to produce porous scaffolds with an evenly distributed pore size throughout the entire scaffold height. However, salt sedimentation in the polymer solution and the formation of thin films on the scaffold surface often prevent uniform scaffold porosity. Figure 8.2 displays scaffold imperfections resulting from rapid surface drying. These thin film layers can decrease cellular accessibility to the porous carrier from the top surface.

To prevent these imperfections, the quantity and viscosity of polymer used in the scaffold fabrication process must be optimized for every process. Alternatively, inhomogeneous scaffolds can be avoided by using predried salt polymer mixtures. These dry mixtures are broken into pieces to fill molds and then compacted under high pressure and temperature to produce more homogenous and mechanically stable scaffolds.[5,17] This method also allows for creation of larger constructs with irregular shapes, which are especially needed for orthopedic tissue engineering applications. A final variation of procedures, which involve organic solvents, attempts to prevent inhomogeneities by compacting a powder mixture of similar sized polymer and salt particles in a Teflon mold.[18] Afterwards, the polymer particles are linked together by applying the organic solvent to the scaffold

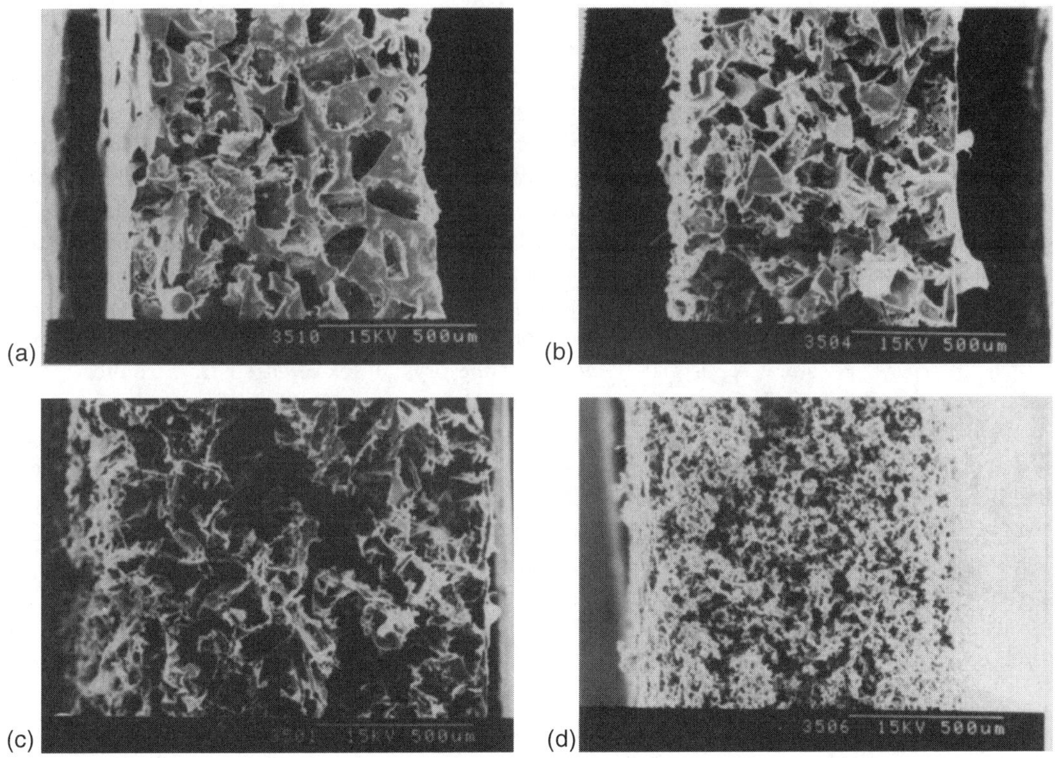

FIGURE 8.2 Scanning electron microscopy (SEM) photomicrographs of cross-sections of semi crystalline PLLA foams prepared with 70 wt% (a), 80 wt% (b), and 90 wt% (c,d) NaCl particles in the size range from 106 to 150 μm (a, b, c) and from 0 to 53 μm (d). (Reproduced from Mikos, A. G., Thorsen, A. J., Czerwonka, L. A., Bao, Y., Langer, R., Winslow, D. N., and Vacanti, J. P., Preparation and characterization of poly(L-lactic acid) foams, *Polymer*, 35, 1068–1077, 1994. With permission.)

mold to dissolve a small outer layer of the polymer particulates. Then the remaining solvent is removed by a vacuum pump as the partially dissolved polymer particles coagulate to form the stable polymer network of the final scaffold.

C. Polymer/Salt Construct Drying

After using one of the many methods for combining polymer and salt particles to give three-dimensional structures, the resulting polymer/salt constructs are then dried extensively using high vacuum (approximately 100 μm Hg) to remove any residual organic solvent that would be cytotoxic to cells seeded on the scaffolds. Typical drying times for this procedure are between 24 and 72 h under high vacuum. Extended drying times should be used for larger and less porous scaffolds to reduce the amount of residual solvent to acceptable ranges for application to the human body. For example, the USP mandates that chloroform levels must be below 60 μg per g for pharmaceutic pills, tablets, and capsules.[19] Possible treatments to further reduce the amount of residual solvents in scaffolds included the use of elevated temperatures and supercritical carbon dioxide.

D. Construct Modifications

An issue of critical importance during this drying stage involves the polymer crystallinity in the final scaffolds. Accurate control over crystallinity is necessary to achieve a defined degradation behavior and degradation time, since polymer degradation is significantly influenced by the different rates of chain cleavage in crystalline and amorphous regions of polymer scaffolds.[21,22] If semicrystalline polymers (L-PLA) are used, the drying time and the conditions used during polymer hardening can be altered to achieve varying degrees of crystallinity in the dried constructs. To achieve scaffolds with a reproducible amount of crystallinity or no crystallinity, it may be desirable to heat-treat the resulting constructs or use polymer-quenching techniques. These methods use defined heating and cooling regimes to control crystallization and the amount of crystalline polymer chains.[1] The heat of crystallization of PLLA/NaCl composites could be reduced from over -40 J/g to almost 0 J/g, indicating the formation of an amorphous polymer scaffold. Research with semicrystalline polymers has shown that crystalline polymers degrade much slower than amorphous polymers, and thus, result in much longer, stable scaffolds with improved mechanical properties. However, some reports also show that the crystallites released from semicrystalline polymers may result in inflammatory reactions and adverse tissue-responses.[23]

Additional procedures after the drying stage can be undertaken to create more complicated structures for certain applications. For example, hollow cylinders of porous poly(lactic acid) scaffolds can be extruded to produce conduits for nerve growth.[6] A special extrusion chamber is used to obtain tubes with an outer and inner diameter of 3.2 and 1.6 mm, respectively (Figure 8.3). These scaffolds can then be used for peripheral nerve regeneration by injecting Schwann cells embedded in a viscous collagen solution into the inner channel of these porous structures.[7]

E. Porogen Leaching

After drying and modifying the polymer/salt constructs, the incorporated salt is finally leached out using double distilled water to leave behind a water-insoluble polymer foam. Constructs are placed into an excess of water to dissolve the incorporated salt. Since the polymer/salt constructs often contain 70 to 95% salt by weight, the water is exchanged every 6 h for 48 h to assure complete leaching of the incorporated salt. During this step, the exchanged water must be removed very gently to prevent the breaking and destruction of the polymer scaffolds, which may be fragile due to their sometimes extensive swelling in the water.

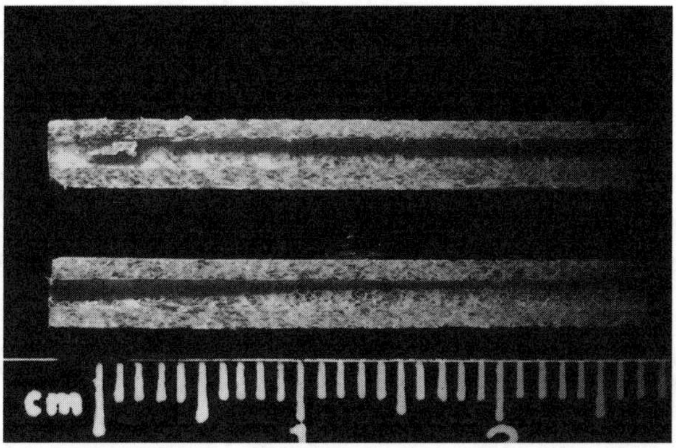

FIGURE 8.3 Optical micrograph of a conduit fabricated from PLGA with a salt weight fraction of 90% and a salt crystal size between 150 and 300 μm. The 300 mm-long conduit was injected with 0.3 ml of a suspension of red microspheres of 6 μm and cut along its axis. (Reproduced from Widmer, M. S., Gupta, P. K., Lu, L., Meszlenyi, R. K., Evans, G. R., Brandt, K., Savel, T., Gurlek, A., Patrick, C. W. Jr., and Mikos, A. G., Manufacture of porous biodegradable polymer conduits by an extrusion process for guided tissue regeneration, *Biomaterials*, 19, 1945–1955, 1998. With permission.)

F. Scaffold Drying and Shaping

After the leaching process, the scaffolds are dried extensively to remove water from the swollen polymer foam to prevent a further decrease in the polymer's molecular weight. Then the dry polymer foams are ready for cell culture use or for further processing refinements. To obtain the desired shapes, dry scaffolds are often cut with cork borers, razor blades, or scissors. The shape of the final scaffold is mainly determined by the desired application. If the scaffolds are used for a cell culture study, the outer shape is generally kept as simple and as small as possible to reduce the necessary amount of culture media and cells. However, if the scaffolds are used to treat a certain defect model, their shape must often be adjusted to match the defect or the desired cell type.

To achieve larger scaffolds, lamination of several layers has been proposed to achieve the necessary dimensions while maintaining scaffold porosity. To adhere layers to each other, dry scaffolds are briefly dipped into chloroform or methylene chloride and then laminated (Figure 8.4). If only a very thin layer is wetted with solvent, porosity imperfections can be avoided, and the resulting laminated structure can be seeded with cells and supplied with culture media.[2] This adhesive technique has also been used to obtain cylindrical shapes for engineered blood vessels by rolling a membrane around Teflon-coated cylinders.[8]

IV. IMPROVEMENTS IN SCAFFOLD PREPARATION AND PROCESSING

As previously mentioned, the method of salt-leaching can be used to fabricate porous scaffolds based on numerous polymers. Of course, the most widely used polymers for the fabrication of tissue engineering scaffolds include various kinds of poly(lactic acid) and different copolymers of poly(lactic acid) and poly(glycolic acid). However, other liquid cross-linkable polymers, like poly(propylene fumarate) (PPF) have been successfully processed into porous scaffolds with this technique.[24] Depending on the intrinsic properties of the polymer used in scaffold fabrication, additional modifications may be required to ensure the resulting scaffolds possess properties suitable for use *in vivo* or *in vitro*. Accordingly, researchers are currently developing numerous methods of altering the mechanical properties, porosity, and bioactivity of scaffolds.

FIGURE 8.4 Scanning electron microscopy (SEM) images of cross-sections of a laminated PLLA foam (a) and one of its constituent layers (b) before lamination (lamination areas are indicated with triangles). (Reproduced from Mikos, A. G., Sarakinos, G., Leite, S. M., Vacanti, J. P., and Langer, R., Laminated three-dimensional biodegradable foams for use in tissue engineering, *Biomaterials*, 14, 323–330, 1993. With permission.)

A. Mechanical Stability

The mechanical properties of cell scaffolds are important to ensure proper support at the defect site while preventing stress shielding. In an effort to improve the mechanical stability of scaffolds, researchers have investigated the use of hydroxyapatite fibers to reinforce the scaffold material. These fibers were also able to help guide tissue growth due to their osteoconductivity.[5] However, as Figure 8.5 demonstrates, it is difficult to incorporate these fibers into scaffolds without sacrificing pore size and structure. Scaffolds with a high content of fibers provide enhanced mechanical stability, indicated by their increased compressive strength and compressive modulus. However, these scaffolds lacked the necessary porosity to allow for sufficient cell in-growth.[5]

Another method to increase mechanical strength of polymeric scaffolds centers on the choice of polymer. For example, polymers, like poly(propylene fumarate), possess unsaturated double bonds, which can be cross-linked to form strong networks. By cross-linking the polymer around embedded salt particles, one can obtain much stronger scaffolds (Figure 8.6). Possible mechanisms to initiate polymer cross-linking include thermal treatment of radical initiators or UV light treatment of photo initiators. The resulting, solid salt/polymer constructs can then be processed with the methods described above, using porogen leaching to produce a highly porous structure for cell seeding or implantation.[24]

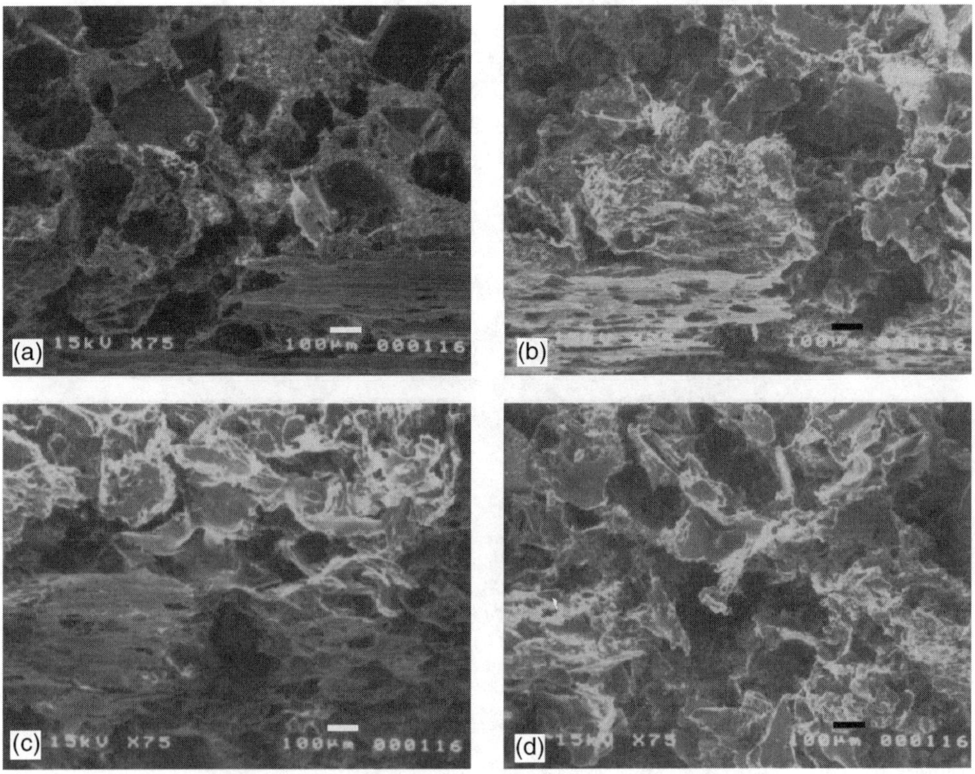

FIGURE 8.5 Hydroxyapatite fiber reinforced scaffolds with different salt weight fractions: (a) 60%, (b) 70%, (c) 80%, and (d) 90%. (Reproduced from Thomson, R. C., Yaszemski, M. J., Powers, J. M., and Mikos, A. G., Hydroxyapatite fiber reinforced poly(alpha-hydroxy ester) foams for bone regeneration, *Biomaterials*, 19, 1935–1943, 1998. With permission.)

Recent investigations also demonstrated the usefulness of nanomaterials for the reinforcement of scaffolds for tissue engineering applications; incorporation of inorganic nanoplatelets into PLLA scaffolds changed the crystallization of the polymer and increased the tensile modulus of the obtained scaffolds.[25]

B. Pore Interconnectivity

Another important property of scaffolds is the interconnectivity of their pore network. Studies demonstrate that the porosity and mechanical stability of scaffolds can be modified by changing the initial salt content (50 to 90% by weight) and salt size (53 to 150 μm; Figure 8.2).[1] These scaffolds had a porosity suitable for the in-growth of cells and sufficient diffusion of the nutrients necessary for their survival.[17,26]

Recent studies have also shown that crystal fusion techniques can be used to increase the amount and size of connections between the pores formed by the salt crystals. High humidity (95%) was used to associate the sodium chloride particles to form large sodium chloride aggregates.[27] The resulting scaffolds exhibited increased pore connectivity, which is favorable for cellular in-growth. These scaffolds also exhibited increased compressive modulus, presumably due to the formation of stiffer structures around these salt aggregates (Figure 8.7).

FIGURE 8.6 Scanning electron microscopy (SEM) images of photo cross-linked PPF scaffolds. Scaffolds were constructed with 70 wt% (a), 80 wt% (b), and 90 wt% (c) NaCl and PPF. Increase salt content leads to more interconnected pores with a content of 80 wt% and greater. (Reproduced from Fisher, J. P., Holland, T. A., Dean, D., Engel, P. S., and Mikos, A. G., Synthesis and properties of photocross-linked poly(propylene fumarate) scaffolds, *J. Biomater. Sci. Polym. Ed.*, 12, 673–687, 2001. With permission.)

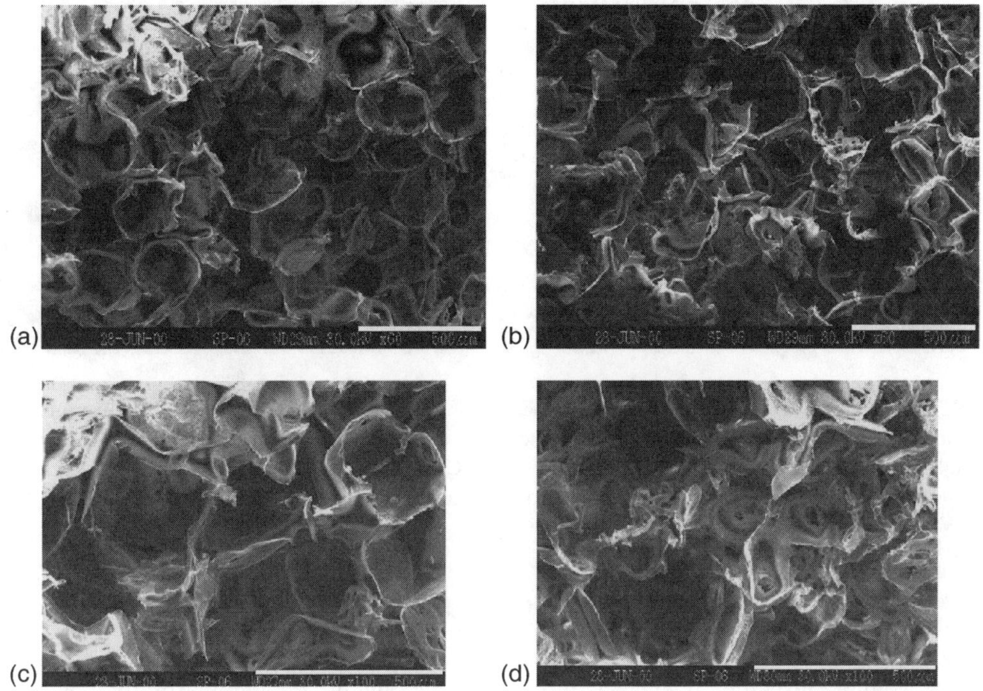

FIGURE 8.7 Scanning electron micrographs of the cross-sections of solvent cast polymer scaffolds prepared using the salf fusion process. One hour of salt fusion results in the presence of a low density of holes (31 ± 10 μm) in pore walls (a,c), while a 24-h fusion treatment results in a high density of larger (78 ± 21 μm) holes in the pore walls (b,d). (Reproduced from Murphy, W. L., Dennis, R. G., Kileny, J. L., and Mooney, D. J., Salt fusion: an approach to improve pore interconnectivity within tissue engineering scaffolds, *Tissue Eng.*, 8, 43–52, 2002. With permission.)

C. Scaffold Bioactivity

Another desirable modification to these scaffolds is the incorporation of bioactive substances, like growth factors. However, once these scaffolds undergo salt leaching and become porous, any incorporated factors are often quickly released due to the high surface-to-volume ratio of these porous scaffolds. For example, a recently published study compared the release kinetics of basic fibroblastic growth factor (bFGF) from salt leached scaffolds with bFGF release from scaffolds fabricated from a technique involving supercritical carbon dioxide. Much higher bFGF burst release was observed from the salt leached scaffolds when compared to the other scaffolds.[28] However, significantly more active bFGF was shown to be released from the salt leached scaffolds than the scaffolds prepared with supercritical carbon dioxide, indicating a degradation of the growth factor due to the higher temperatures during the manufacturing process with CO_2. Therefore, it seems essential to develop more sophisticated techniques to load salt-leached scaffolds with growth factors and also to retain an appropriate amount of the active factors in the scaffolds for a later release.

To immobilize the proteins in scaffolds, growth factors are often complexed or incorporated into prefabricated microparticles. For example, alginate microparticles were used to slow the release of vascular endothelial growth factor from tissue-engineering scaffolds.[29] Another technique involves coating preleached scaffolds with growth factors (Figure 8.8). While this method eliminates protein loss during scaffold processing, the growth factor is only incorporated into the top layer of the scaffold, which may also lead to fast initial release rates.[30]

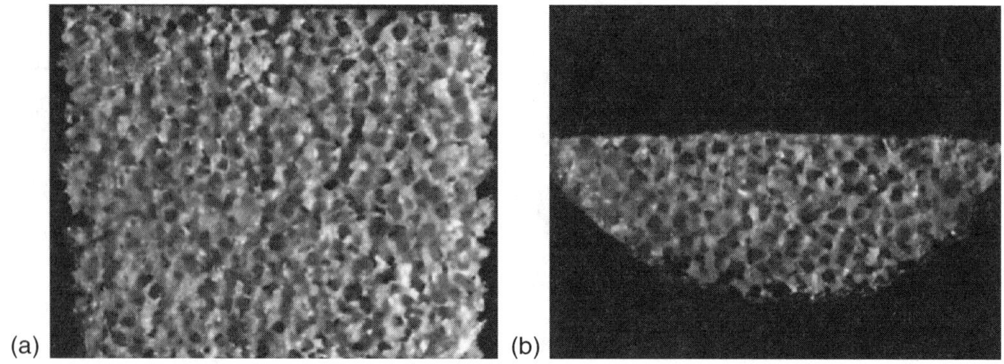

FIGURE 8.8 Coating of salt leached scaffolds with red model drug to demonstrate the uniformity of the emulsion coating of scaffolds for controlled release of growth factors [longitudinal (a) and lateral (b) cross-sectional]. (Reproduced from Sohier, J., Haan, R. E., de Groot, K., and Bezemer, J. M., A novel method to obtain protein release from porous polymer scaffolds: emulsion coating, *J. Controlled Release*, 87, 57–68, 2003. With permission.)

V. DIRECTIONS FOR FUTURE RESEARCH

Despite these recent improvements to the mechanical properties, porosity, and bioactivity of scaffolds, future research is needed to overcome many remaining limitations in the manufacturing process. First of all there are a limited number of polymers that are compatible with the salt leaching process, since a polymer must be soluble in an organic solvent that is a poor solvent for the porogen. Additionally, few solvents exist for these techniques since a solvent must be biocompatible or easy to remove from the scaffold. For example, the only available solvents for poly(glycolic acid) are hexafluoroacetone and hexafluoroisopropanol, which are both highly toxic to cells. In some cases, residual solvents in scaffolds may be removed with supercritical carbon dioxide, allowing for the retention of pore architecture.[20] However, future research is necessary to broaden the choice of polymer and solvent so that materials can be tailored to a particular application.

Furthermore, studies should be undertaken to examine the effect which water may have on scaffold properties during the leaching procedure. Polymers exposed to water for long time-periods may begin to degrade. Additionally, the water involved in the leaching process poses a problem to bioactive substances incorporated into scaffolds. Further methods of protecting these molecules during scaffold fabrication and leaching are required to formulate scaffolds with the necessary biological properties. Here, the application of newly developed biomaterials with covalently immobilized growth factors on the scaffold's surface might help to overcome these challenges.[31,32]

VI. CONCLUSIONS

The procedure of salt leaching allows for the easy fabrication of scaffolds with controllable porosity and pore size using various biodegradable polymers. This technique allows cell carriers for tissue engineering applications to be formed with minimal lab equipment and polymer amounts. Several recent modifications to the common method of producing salt leached scaffolds demonstrate that tremendous room for improvement still exists to manufacture scaffolds with precise chemical, physical, and biological properties. However, the ease in implementing and refining the methods for producing salt leached scaffolds, ensures that this technique will remain one of the most preferred methods for the manufacture of small-scale tissue engineering scaffolds.

ACKNOWLEDGMENTS

Research for the development of biodegradable polymer scaffolds for tissue engineering applications has been funded by the National Institutes of Health (R01 AR42639; R01 AR48756) (AGM). JKVT acknowledges financial support (TE366-1/1) by the German Research Foundation (DFG). TAH was supported by a Whitaker Foundation Graduate Fellowship.

REFERENCES

1. Mikos, A. G., Thorsen, A. J., Czerwonka, L. A., Bao, Y., Langer, R., Winslow, D. N., and Vacanti, J. P., Preparation and characterization of poly(L-lactic acid) foams, *Polymer*, 35, 1068–1077, 1994.
2. Mikos, A. G., Sarakinos, G., Leite, S. M., Vacanti, J. P., and Langer, R., Laminated three-dimensional biodegradable foams for use in tissue engineering, *Biomaterials*, 14, 323–330, 1993.
3. Lin, H., Kuo, C., Yang, C. Y., Shaw, S., and Wu, Y., Preparation of macroporous biodegradable PLGA scaffolds for cell attachment with the use of mixed salts as porogen additives, *J. Biomed. Mater. Res.*, 63, 271–279, 2002.
4. Holy, C. E., Dang, S. M., Davies, J. E., and Shoichet, M. S., In vitro degradation of a novel poly(lactide-*co*-glycolide) 75/25 foam, *Biomaterials*, 20, 1177–1185, 1999.
5. Thomson, R. C., Yaszemski, M. J., Powers, J. M., and Mikos, A. G., Hydroxyapatite fiber reinforced poly(alpha-hydroxy ester) foams for bone regeneration, *Biomaterials*, 19, 1935–1943, 1998.
6. Widmer, M. S., Gupta, P. K., Lu, L., Meszlenyi, R. K., Evans, G. R., Brandt, K., Savel, T., Gurlek, A., Patrick, C. W. Jr., and Mikos, A. G., Manufacture of porous biodegradable polymer conduits by an extrusion process for guided tissue regeneration, *Biomaterials*, 19, 1945–1955, 1998.
7. Evans, G. R. D., Brandt, K., Katz, S., Chauvin, P., Otto, L., Bogle, M., Wang, B., Meszlenyi, R. K., Lu, L., Mikos, A. G., and Patrick, C. W., Bioactive poly(L-lactic acid) conduits seeded with Schwann cells for peripheral nerve regeneration, *Biomaterials*, 23, 841–848, 2002.
8. Mooney, D. J., Breuer, M. D., McNamara, K., Vacanti, J. P., and Langer, R., Fabricating tubular devices from polymers of lactic and glycolic acid for tissue engineering, *Tissue Eng.*, 1, 107–118, 1995.
9. Mooney, D. J., Organ, G., Vacanti, J. P., and Langer, R., Design and fabrication of biodegradable polymer devices to engineer tubular tissues, *Cell Transplant.*, 3, 203–210, 1994.
10. Lu, L., Peter, S. J., Lyman, M. D., Lai, H. L., Leite, S. M., Tamada, J. A., Uyama, S., Vacanti, J. P., Langer, R., and Mikos, A. G., In vitro and in vivo degradation of porous poly(D,L-lactic-*co*-glycolic acid) foams, *Biomaterials*, 21, 1837–1845, 2000.
11. Zeltinger, J., Sherwood, J. K., Graham, D. A., Mueller, R., and Griffith, L. G., Effect of pore size and void fraction on cellular adhesion, proliferation, and matrix deposition, *Tissue Eng.*, 7, 557–572, 2001.
12. Griffith, L. G., and Lopina, S., Microdistribution of substratum-bound ligands affects cell function: hepatocyte spreading on PEO-tethered galactose, *Biomaterials*, 19, 979–986, 1998.
13. Mooney, D. J., Baldwin, D. F., Suh, N. P., Vacanti, J. P., and Langer, R., Novel approach to fabricate porous sponges of poly(D,L-lactic-*co*-glycolic acid) without the use of organic solvents, *Biomaterials*, 17, 1417–1422, 1996.
14. Harris, L. D., Kim, B. S., and Mooney, D. J., Open pore biodegradable matrices formed with gas foaming, *J. Biomed. Mater. Res.*, 42, 396–402, 1998.
15. Groot de, J. H., Nijenhuis, A. J., Bruin, P., Pennings, A. J., Veth, R. P., Klompmaker, J., and Jansen, H. W., Use of porous biodegradable polymer implants in meniscus reconstruction. 1) Preparation of porous biodegradable polyurethans for the reconstruction of meniscus lesions, *Colloid Polym. Sci.*, 268, 1073–1081, 1990.
16. Elema, H., Groot de, J. H., Nijenhuis, A. J., Pennings, A. J., Veth, R. P., Klompmaker, J., and Jansen, H. W., Use of porous biodegradable polymer implants in meniscus reconstruction. 2) Biological evaluation of porous biodegradable polymer implants in menisci, *Colloid Polym. Sci.*, 268, 1082–1088, 1990.
17. Goldstein, A. S., Juarez, T. M., Helmke, C. D., Gustin, M. C., and Mikos, A. G., Effect of convection on osteoblastic cell growth and function in biodegradable polymer foam scaffolds, *Biomaterials*, 22, 1279–1288, 2001.

18. Liao, C., Chen, C., Chen, J., Chiang, S., Lin, Y., and Chang, K., Fabrication of porous biodegradable polymer scaffolds using a solvent merging/particulate leaching method, *J. Biomed. Mater. Res.*, 59, 676–681, 2002.
19. USP. *Chemical Tests: Organic Volatile Impurities. USP 24*, The United States Pharmacopeial Convention, Inc, Rockville, MD, 2004, pp. 1877–1878.
20. Koegler, W. S., Patrick, C., Cima, M. J., and Griffith, L. G., Carbon dioxide extraction of residual chloroform from biodegradable polymers, *J. Biomed. Mater. Res.*, 63, 567–576, 2002.
21. Middleton, J. C., and Tipton, A. J., Synthetic biodegradable polymers as orthopedic devices, *Biomaterials*, 21, 2335–2346, 2000.
22. von Burkersroda, F., Schedl, L., and Gopferich, A., Why degradable polymers undergo surface erosion or bulk erosion, *Biomaterials*, 23, 4221–4231, 2002.
23. Bergsma, J. E., Rozema, F. R., Bos, R. R. M., Boering, G., de Bruijn, W. C., and Pennings, A. J., In vivo degradation and biocompatibility study of in vitro pre-degraded as-polymerized polylactide particles, *Biomaterials*, 16, 267–274, 1995.
24. Fisher, J. P., Holland, T. A., Dean, D., Engel, P. S., and Mikos, A. G., Synthesis and properties of photocross-linked poly(propylene fumarate) scaffolds, *J. Biomater. Sci. Polym. Ed.*, 12, 673–687, 2001.
25. Lee, J. H., Park, T. G., Park, H. S., Lee, D. S., Lee, Y. K., Yoon, S. C., and Nam, J. D., Thermal and mechanical characteristics of poly(-lactic acid) nanocomposite scaffold, *Biomaterials*, 24, 2773–2778, 2003.
26. Goldstein, A. S., Zhu, G., Morris, G. E., Meszlenyi, R. K., and Mikos, A. G., Effect of osteoblastic culture conditions on the structure of Poly(D,L-lactic-*co*-glycolic acid) foam scaffolds, *Tissue Eng.*, 5, 421–433, 1999.
27. Murphy, W. L., Dennis, R. G., Kileny, J. L., and Mooney, D. J., Salt fusion: an approach to improve pore interconnectivity within tissue engineering scaffolds, *Tissue Eng.*, 8, 43–52, 2002.
28. Hile, D. D., Amirpour, M. L., Akgerman, A., and Pishko, M. V., Active growth factor delivery from poly(D,L-lactide-*co*-glycolide) foams prepared in supercritical CO_2, *J. Controlled Release*, 66, 177–185, 2000.
29. Murphy, W. L., Peters, M. C., Kohn, D. H., and Mooney, D. J., Sustained release of vascular endothelial growth factor from mineralized poly(lactide-*co*-glycolide) scaffolds for tissue engineering, *Biomaterials*, 21, 2521–2527, 2000.
30. Sohier, J., Haan, R. E., de Groot, K., and Bezemer, J. M., A novel method to obtain protein release from porous polymer scaffolds: emulsion coating, *J. Controlled Release*, 87, 57–68, 2003.
31. Otsuka, H., Nagasaki, Y., and Kataoka, K., Surface characterization of functionalized polylactide through the coating with heterobifunctional poly(ethylene glycol)/polylactide block copolymers, *Biomacromolecules*, 1, 39–48, 2000.
32. Tessmar, J. K., Mikos, A. G., and Goepferich, A., Amine-reactive biodegradable diblock copolymers, *Biomacromolecules*, 3, 194–200, 2002.
33. Sodian, R., Sperling, J. S., Martin, D. P., Egozy, A., Stock, U., Mayer, J. E. Jr., and Vacanti, J. P., Fabrication of a trileaflet heart valve scaffold from a polyhydroxyalkanoate biopolyester for use in tissue engineering, *Tissue Eng.*, 6, 183–187, 2000.
34. Hou, Q., Grijpma, D. W., and Feijen, J., Porous polymeric structures for tissue engineering prepared by a coagulation, compression moulding and salt leaching technique, *Biomaterials*, 24, 1937–1947, 2003.
35. Sheridan, M. H., Shea, L. D., Peters, M. C., and Mooney, D. J., Bioabsorbable polymer scaffolds for tissue engineering capable of sustained growth factor delivery, *J. Controlled Release*, 64, 91–102, 2000.
36. Yoon, J. J., and Park, T. G., Degradation behaviors of biodegradable macroporous scaffolds prepared by gas foaming of effervescent salts, *J. Biomed. Mater. Res.*, 55, 401–408, 2001.
37. Nam, Y. S., Yoon, J. J., and Park, T. G., A novel fabrication method of macroporous biodegradable polymer scaffolds using gas foaming salt as a porogen additive, *J. Biomed. Mater. Res.*, 53, 1–7, 2000.

9 Polymer Phase Separation

Victor J. Chen and Peter X. Ma

CONTENTS

I. INTRODUCTION

Synthetic biodegradable polymers such as poly(L-lactic acid) (PLLA), poly(glycolic acid) (PGA), and poly(D,L-lactic-*co*-glycolic acid) (PLGA) have been widely used as scaffolding materials in tissue engineering. Because these materials are biocompatible, degradable by hydrolysis, and are among the few synthetic polymers approved by the Food and Drug Administration (FDA) for certain clinical applications, these polyesters remain popular for use as scaffolding polymers. When preparing these polymers for scaffolding purposes, it is important that the scaffold perform a number of critical functions. The scaffold should have an open porous structure for uniform cell seeding and for mass transport of nutrients and metabolic waste removal, it should have a suitable surface for cell attachment, proliferation, and differentiation, and it should provide a three-dimensional template that guides tissue growth.[1–5] Several techniques have been utilized to fabricate porous scaffolds from these synthetic polyesters possessing these characteristics, including particulate leaching,[6,7] textile technologies,[8–10] membrane lamination,[11] melt molding,[12] emulsion freeze-drying,[13] gas foaming,[14] and three-dimensional printing.[15–18]

Thermally induced phase separation (TIPS) of polymer solutions has been previously used to fabricate synthetic membranes for nonmedical applications, and has also been used in the field of drug delivery to fabricate microspheres for the incorporation of biological and pharmaceutical agents.[19] Recently, the TIPS process has become a popular technique for fabricating porous scaffolds for tissue engineering.[3,5,20–27] In this process, the temperature of a polymer solution is decreased to induce a phase separation into two phases, one having a high polymer concentration (polymer-rich phase) and one having a low polymer concentration (polymer-lean phase). The solvent occupying the polymer-lean phase is later removed by extraction, evaporation, or sublimation which leaves behind

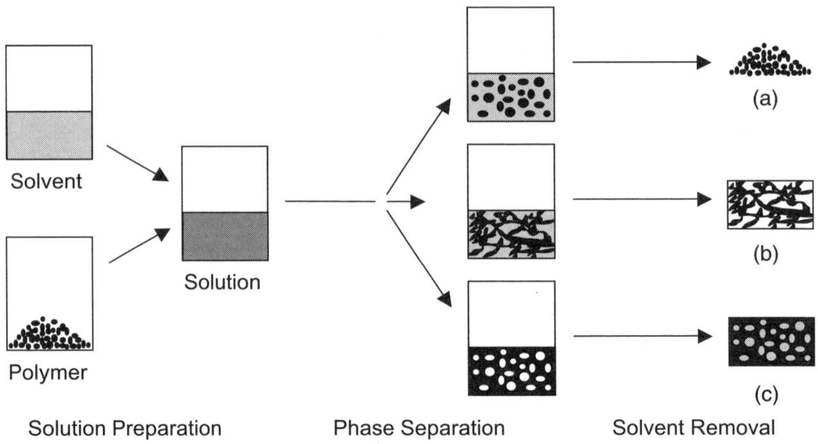

FIGURE 9.1 Schematic illustration of scaffold fabrication with phase separation processes: (a) powder, (b) scaffolds with continuous network, and (c) foam with closed pores. (From Zhang, R., and Ma, P. X., Processing of polymer scaffolds: phase separation, In *Methods of Tissue Engineering*, Atala, A., and Lanza R. P., eds., Academic Press, San Diego, pp. 715–724, 2002. With permission.)

open pores, and the polymer in the polymer-rich phase solidifies into the skeleton of the polymer foam. By varying the types of polymer and solvent, polymer concentration, and phase-separation temperature, the micro and macrostructure of the porous polymer foam can be altered. In this chapter, we will address advances in TIPS techniques for porous polymer scaffold fabrication, and how the TIPS process has been utilized to control pore morphology on a micrometer and nanometer level (Figure 9.1). We will also address how the TIPS process can be combined with other popular fabrication techniques such as particulate leaching and three-dimensional printing, and how the properties of these scaffolds can be tailored for tissue engineering applications. Protocols developed in our laboratory will be used as examples of these processes.

II. SOLID–LIQUID PHASE SEPARATION

TIPS of polymer solutions can be a complicated process, depending on the thermodynamic and kinetic behavior of the polymer solution under certain conditions.[28,29] For a system where the solvent crystallization temperature (freezing point) is higher than the liquid–liquid phase separation temperature, the system can separate by lowering the temperature, a process we will call solid–liquid phase separation. In this process, the solvent is crystallized and the polymer is expelled from the solvent crystallization front. After the solvent is removed, the pores that remain have morphologies similar to the solvent crystallite geometries. Thus, the pore architecture of the scaffold will depend greatly on the solvent used, the polymer concentration, the phase-separation temperature, and the temperature gradient applied to the polymer solution.

A. Preparation of Polymer Matrices

Foams of PLLA and PLGA can be prepared with the solid–liquid phase separation process by decreasing the temperature of the polymer solutions to induce solvent crystallization, and subsequently sublimating the solvent.[5,23,24] Typically, the chosen polymer is stirred in a solvent (such as dioxane or benzene) to form a solution. Depending on the solvent choice and phase separating conditions, the foams can be controlled to form either random or oriented pore architectures.

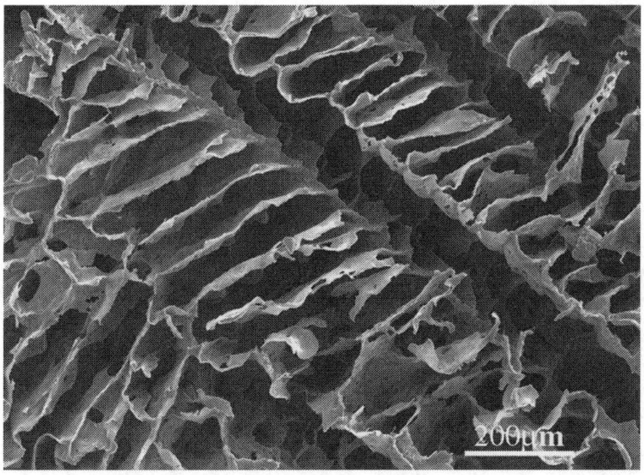

FIGURE 9.2 SEM micrograph of a PLLA scaffold prepared from PLLA-dioxane solution with a thermally induced solid–liquid phase separation process. (From Zhang, R. Y., and Ma, P. X., Poly(alpha-hydroxyl acids) hydroxyapatite porous composites for bone-tissue engineering. I. Preparation and morphology, *J. Biomed. Mater. Res.*, 44, 446–455, 1999. With permission.)

For foams with randomly oriented pore architectures, the polymer solution is quickly transferred to a refrigerator or freezer at a preset temperature to crystallize the solvent and induce solid–liquid phase separation. After freeze-drying the solvent, the resulting samples have a highly anisotropic morphology. In the case of a PLLA–dioxane solution undergoing solid–liquid phase separation, the porous foam contains an internal ladder-like structure (Figure 9.2).[23] Although a temperature gradient is not intentionally applied to this phase-separating system, a local gradient from the surface to the center of the sample may have led to the anisotropic pore structure, leading to channels that are parallel to the temperature gradient and have repeating partitions with nearly uniform spacing perpendicular to the solidification direction.

For foams with oriented pore architectures, the phase separation is performed under a uniaxial temperature gradient which is created by insulating the side walls of a mold containing the polymer solution, and placing the mold on a metal block in the freezer to induce heat conduction in the longitudinal direction. When a solution of PLLA–benzene undergoes solid–liquid phase separation under an intentionally applied uniaxial temperature gradient, the scaffold contains well-defined

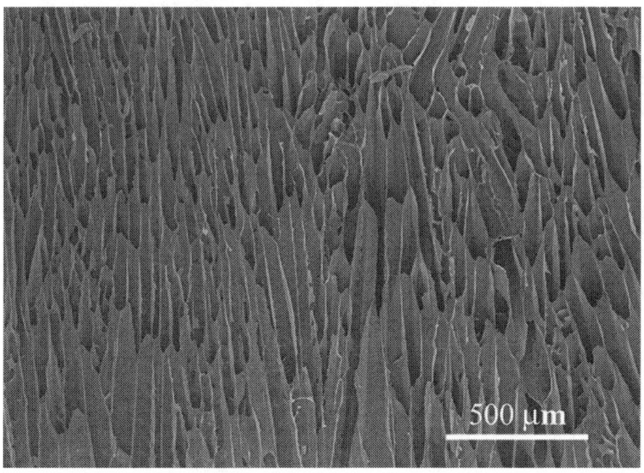

FIGURE 9.3 SEM longitudinal-section micrograph of PLLA scaffold prepared from PLLA-benzene solution with a thermally induced solid–liquid phase separation process involving a uniaxial temperature gradient. (From Ma, P. X., and Zhang, R. Y., Microtubular architecture of biodegradable polymer scaffolds, *J. Biomed. Mater. Res.*, 56, 469–477, 2001. With permission.)

channels parallel to the longitudinal directions of solidification, and the ladder-like partitions can be avoided when polymer concentration is sufficiently low (Figure 9.3).[5] By using a solvent with different crystallization properties (freezing point ~6°C for benzene, ~12°C for dioxane), one can affect the nature of the solid–liquid phase separation and create polymer foams with various pore morphologies.

In addition, the solid–liquid phase separation technique can be used to create composite scaffolds. For instance, by mixing the osteoconductive mineral hydroxyapatite (HAP) with the polymer solution and subsequently phase separating the solution, HAP particles can be integrated into the polymer matrix to increase the osteoconductivity of a scaffold for bone tissue engineering.[23] In studies where osteoblasts were cultured in both composite and polymer foams, the polymer–HAP composite scaffolds displayed more promising results for bone tissue engineering.[30]

III. LIQUID–LIQUID PHASE SEPARATION

For the previous sections that discussed solid–liquid phase separation processes, the crystallization temperature of the solvent in the polymer solution was higher than the liquid–liquid phase separation temperature. However, when the solvent crystallization temperature is much lower than the phase separation temperature, a liquid–liquid phase separation takes place if the temperature of the polymer solution is decreased. Figure 9.4 shows a schematic phase diagram for a binary polymer–solvent system. For a solution where a polymer is in a solvent of lower molecular mass, a strongly asymmetric phase diagram is typical. At high temperatures, the solution is in the one-phase region and is homogeneous. When a homogeneous polymer solution is cooled to a temperature-composition point below the binodal envelope, a liquid–liquid phase separation to a polymer-lean and a polymer-rich phase can take place. These systems are characterized by an upper critical solution temperature, and although a variety of different forms of behavior can occur, phase separation usually occurs by one of two generic forms, namely, nucleation and growth (NG) or spinodal decomposition (SD).[28,31]

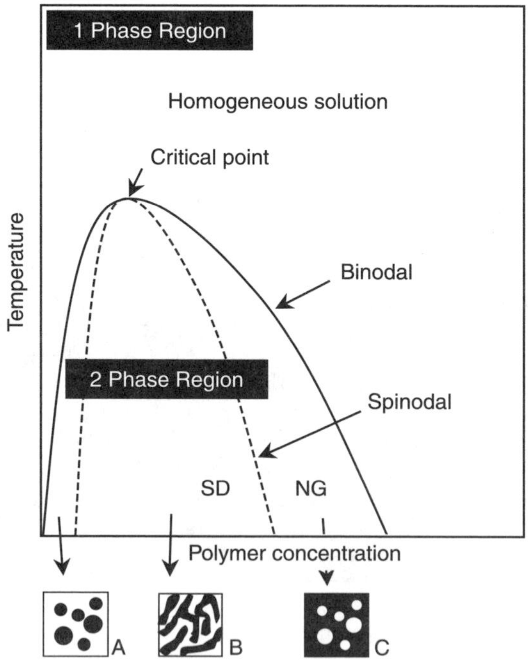

FIGURE 9.4 Schematic equilibrium temperature–composition phase diagram for a polymer solution system.

Inside the binodal envelope, these two phase separating mechanisms take place in two thermodynamically characteristic regions that are separated by the spinodal envelope. The region between the binodal and spinodal envelopes is thermodynamically metastable, meaning that solutions are stable with respect to small fluctuations in the composition, and the liquid–liquid phase separation occurs by nucleation and growth. In the metastable region of low polymer concentration (Figure 9.4, Inset A), droplets of polymer-rich phase are formed and dispersed in a matrix of polymer-lean phase, and the resulting structure after removing the solvent is a powder like polymer solid (Figure 9.1). In the metastable region of high polymer concentration (Figure 9.4, Inset C), droplets of polymer-lean phase are formed in a matrix of polymer-rich phase, resulting in a foam with a closed-pore structure (Figure 9.1). Under the spinodal envelope, the region is thermodynamically unstable and any fluctuations in composition result in a decrease in free energy and trigger a wave of fluctuations throughout the solution; this phase separation occurs via spinodal decomposition.[28,32] In this unstable region (Figure 9.4, Inset B), a bicontinuous pattern in which both the polymer-rich and polymer-lean phases are completely interconnected is formed. The resulting scaffold is a foam with a continuous pore network (Figure 1).

When a polymer solution involves a semicrystalline polymer, the phase separation becomes significantly more complicated because of the potential for the polymer to crystallize. If the temperature of the polymer solution is low enough, the solution will experience driving forces for both liquid–liquid phase separation and polymer crystallization. In this situation, kinetic phenomena are important. Generally, liquid–liquid phase separation is faster than polymer crystallization. The solution usually undergoes liquid–liquid phase separation first, and when the solution is cooled enough, the polymer-rich phase will be able to crystallize which stabilizes the structure of the phase separated solution. In the end, the morphology of the phase separated solution is largely dependent on the liquid–liquid phase separation.

In some solutions, liquid–liquid phase separation can be combined with gelation during the cooling process.[28,33] In this process, the whole polymer solution solidifies into a gel; a network of physically cross-linked polymer chains with solvent trapped within the network. In a semicrystalline polymer solution, the interlocking of small crystal agglomerates may play a key role for the formation of the gel. The highly porous structure is obtained by removing the solvent from the gel.[20]

When the polymer crystallization temperature is higher than the spinodal temperature, the polymer can crystallize in the solution during the cooling process before the spinodal phase separation occurs. If the polymer solution is held long enough at a temperature higher than the spinodal temperature, the solution will experience a crystallization-induced phase separation, another type of solid–liquid phase separation. The polymer-rich phase forms by nucleation and growth of polymer crystals, and the polymer platelets formed are suspended in a matrix of the polymer-lean phase or precipitated from the solution. When the polymer concentration is high enough, gelation can also take place in this phase separation process, whereupon it can be used to fabricate polymer foams.

A. Preparation of Continuous Network Polymer Matrices

To fabricate matrices having pore structures controlled by liquid–liquid phase separation, one must first choose a solvent with a freezing point lower than the phase separation temperature of the polymer solution. For example, a mixture of poly(α-hydroxyl acids) in a dioxane-H_2O system can be used to attain liquid–liquid phase separation.[25,26] In this case, a clear polymer solution is quenched to a preset temperature to induce liquid–liquid phase separation. The frozen solution is then freeze-dried.

Depending on the phase separation conditions, it was mentioned in a previous section that either a powdery material or a foam with an isotropic continuous pore network (Figure 9.5) can be obtained by means of liquid–liquid phase separation. The final structure of the phase separated

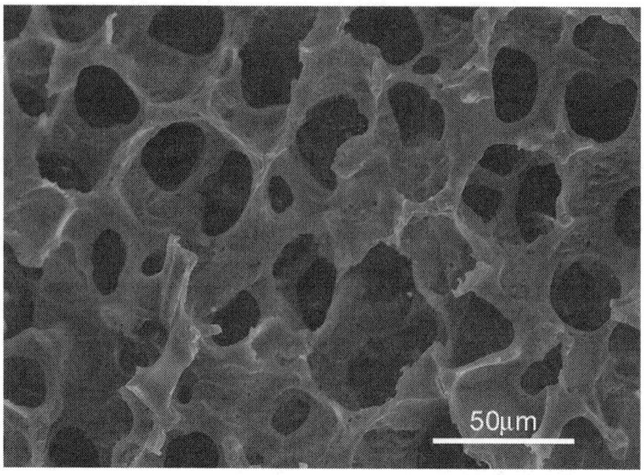

FIGURE 9.5 SEM micrograph of a PLGA scaffold prepared with a thermally induced liquid–liquid phase separation. (From Ma, P. X., and Zhang, R. Y., Synthetic nano-scale fibrous extracellular matrix, *J. Biomed. Mater. Res.*, 46, 60–72, 1999. With permission.)

polymer matrix is dependent on the concentration of the polymer solution, the quenching temperature, and the molecular weight of the polymer. Generally, a powder-like structure is obtained from a polymer solution with a very low concentration, whereas a foam with an isotropic pore structure is obtained from a solution of a relatively higher concentration. At the same concentration, a polymer with higher molecular weight favors the formation of a uniform isotropic pore structure. In a certain polymer concentration range, different quenching temperatures also have an effect on the pore structures. For instance, a lower quenching temperature (higher cooling rate) generally results in an interconnected pore network with uniform pore size, whereas a higher quenching temperature (lower cooling rate) results in larger pores with a wider pore size distribution.

Pore sizes in foams obtained from liquid–liquid phase separation usually range from several microns to tens of microns, and this fine-scaled "early-stage" structure can be "frozen-in" if the polymer solution gels or if the solvent freezes. In these polymer solutions, the dynamics of early-stage structure development are expected to be very fast, leaving little room for experimental control. Another process that affects the system phase morphology, coarsening, occurs during the late stages of phase separation and may be of more technological interest.[34,35] When a sample is held at the phase separation temperature, the droplets that are well separated and have well-defined interfaces tend to increase in average size (coarsen), while the smaller droplets which have a higher solubility tend to dissolve. The driving force for such an increase in average droplet size is the tendency of the system to minimize its interfacial free energy by minimizing the interfacial area between the polymer-rich and polymer-lean phases. By utilizing this phenomenon, one could control the pore sizes of matrices prepared by liquid–liquid phase separation, especially for porous scaffolds in tissue engineering where ideal pore sizes vary for different cells and tissues.[34]

B. Preparation of Polymer Matrices with a Nano-Fibrous Network

To mimic the fibrous three-dimensional structure of natural type I collagen, synthetic nano-fibrous matrices can be fabricated by thermally inducing the gelation of a PLLA solution. For this process, a solution of PLLA is dissolved in a chosen solvent system (e.g., THF, DMF, pyridine, THF–methanol, dioxane–methanol, dioxane–H_2O, dioxane–acetone, dioxane–pyridine) to create a homogeneous solution. The example given here will be for a solution of PLLA dissolved in THF. A homogeneous PLLA–THF solution is placed into a refrigerator or freezer at the desired

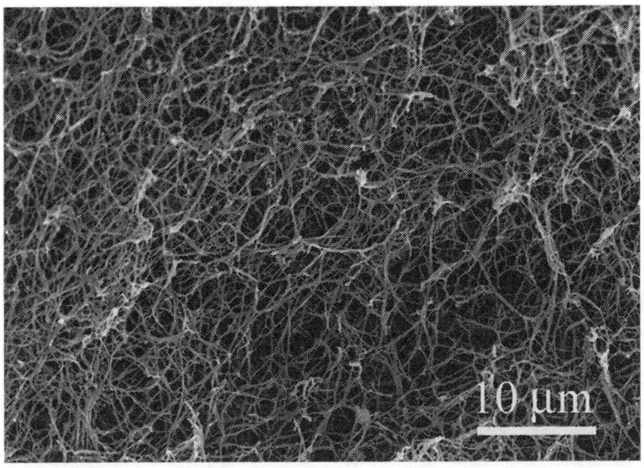

FIGURE 9.6 SEM micrograph of a PLLA matrix with a nano-fibrous network prepared with a thermally induced gelation process (gelled at −18°C). (From Ma, P. X., and Zhang, R. Y., Synthetic nano-scale fibrous extracellular matrix, *J. Biomed. Mater. Res.*, 46, 60–72, 1999. With permission.)

temperature (lower than 15°C for PLLA–THF solution). Gelation of the solution depends on the temperature, the solvent, and the PLLA concentration of the solution.[25] After gelation, the gel is immersed in distilled water for solvent exchange and subsequently, the gel is removed from the water and freeze-dried to yield a three-dimensional nano-fibrous network (Figure 9.6).

This phase-separation/gelation process of PLLA solutions offers a unique and convenient method of fabricating fibrous matrices. Unlike the fabrication of nonwoven meshes with textile technologies or the electrospinning of polymer nano-fibers, almost no equipment is required for the gelation process. Additionally, the process is highly reproducible, is very simple to perform, and by adjusting processing parameters such as the gelation temperature or polymer concentration, one can control the fiber density, porosity, and mechanical properties of the nano-fibrous matrices.[25] Furthermore, as we will see in a later section, this phase separation procedure can be combined with three-dimensional molds to form nano-fibrous scaffolds of complex geometries.

The synthetic fibrous network possesses fiber diameters ranging from 50 to 500 nm, resulting in a matrix morphologically similar to type I collagen. In addition, the fibrous matrix possesses a very high surface area-to-volume ratio which has been suggested to improve cell adhesion, which consequently affects cell migration, proliferation, and differentiation.[25,36–40] In initial studies, nano-fibrous matrices have been shown to provide a more favorable environment for cell attachment than nonfibrous scaffolds.[41] As a result, synthetic nano-fibrous extracellular matrices might also provide a more suitable environment for cell proliferation, differentiated function, and three-dimensional tissue formation.

C. Preparation of Polymer Matrices with Platelet-Like Structures

Highly porous PLLA matrices having a platelet-like structure are prepared from PLLA solutions undergoing thermally induced gelation at relatively high temperatures. For this process, the PLLA solution is prepared in the same manner as described for nano-fibrous matrices. However, the gelation temperature should be higher than the temperature used to induce nano-fiber formation. For a PLLA–THF solution, foams with platelet-like structures form at a gelation temperature higher than 20°C (Figure 9.7). A possible mechanism for the formation of the platelet-like structure is the nucleation and growth of PLLA crystals at a temperature higher than that of the spinodal decomposition temperature.

FIGURE 9.7 SEM micrograph of a PLLA scaffold with a platelet-like structure prepared with a thermally induced gelation process (gelled at 23°C). (From Ma, P. X., and Zhang, R. Y., Synthetic nano-scale fibrous extracellular matrix, *J. Biomed. Mater. Res.*, 46, 60–72, 1999. With permission.)

D. Preparation of Nano-Fibrous Matrices with Macroporous Structures

In scaffolding for tissue engineering, the three-dimensional nature of the pore structure plays an important role in determining the quality of the tissue being developed. Macroscopic pores (>100 μm) in a scaffold play a significant role in cell seeding distribution, cell migration throughout the three-dimensional space, neo-vascularization after implantation of the scaffold *in vivo*, and perfusion of soluble signaling molecules, nutrients, and metabolic waste removal. To create a nano-fibrous matrix with a macroscopic pore network, a combination of gelation and porogen leaching can be used. By casting a polymer solution into a preformed mold containing a porogen, followed by subsequent phase separation and porogen leaching, scaffolds with various pore geometries can be created. Furthermore, the molds holding the porogen can be designed into various anatomical forms since the simplicity of the casting and gelation processes allow for the fabrication of scaffolds with complex geometries.

1. Creating Macroporous Structures with Salt or Sugar Particles

Salt or sugar particles can be used as a porogen to generate macroporous networks in scaffolds with nano-fibrous pore walls. In one example, PLLA–THF polymer solution is dripped slowly onto sugar particles in a mold and then cooled to a preset gelation temperature. After gelation, the gel–sugar composite is immersed in distilled water to simultaneously extract the solvent and leach the sugar from the composite. The gel is freeze-dried, resulting in a three-dimensional nano-fibrous matrix with macropores left behind from the leached porogen (Figure 9.8).

2. Creating Macroporous Structures with Sugar Fibers

Another porogen leaching technique that can be used to create predesigned macroporous structures is the use of sugar fibers. The fibers are drawn from molten sugar and are organized into the desired architecture. For example, a network of perpendicular fibers can be built layer by layer, with each layer subjected to a high humidity environment to adhere the sugar fibers, until the assembly is complete. Afterwards, the PLLA solution is dripped onto the porogen frame and cooled to the desired gelation temperature. Then, solvent extraction, porogen leaching, and freeze-drying are performed in the same manner as with sugar particles. Figure 9.9 shows a schematic diagram of the perpendicular sugar fiber assembly and the resulting nano-fibrous network with perpendicular channels.

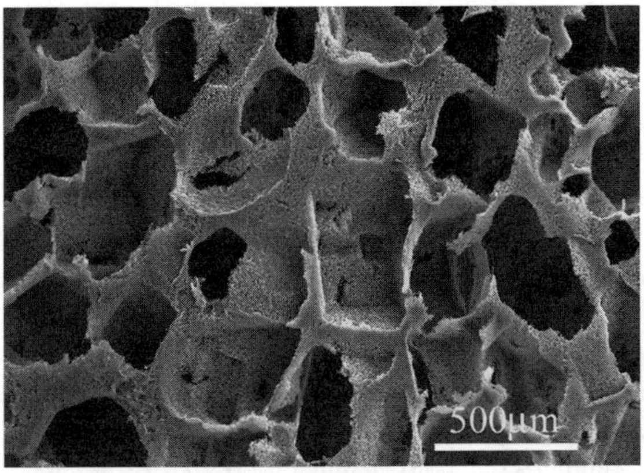

FIGURE 9.8 SEM micrograph of a nano-fibrous PLLA scaffold prepared by combining a sugar particle-leaching technique and a thermally induced gelation process. (From Zhang, R. Y., and Ma, P. X., Synthetic nano-fibrillar extracellular matrices with predesigned macroporous architectures, *J. Biomed. Mater. Res.*, 52, 430–438, 2000. With permission.)

3. Creating Macroporous Structures with Paraffin Microspheres

To improve the interconnectivity of pores throughout a three-dimensional matrix, our laboratory developed an emulsification technique to create paraffin microspheres that can be thermally bonded prior to polymer solution casting. The microspheres are created by emulsifying paraffin

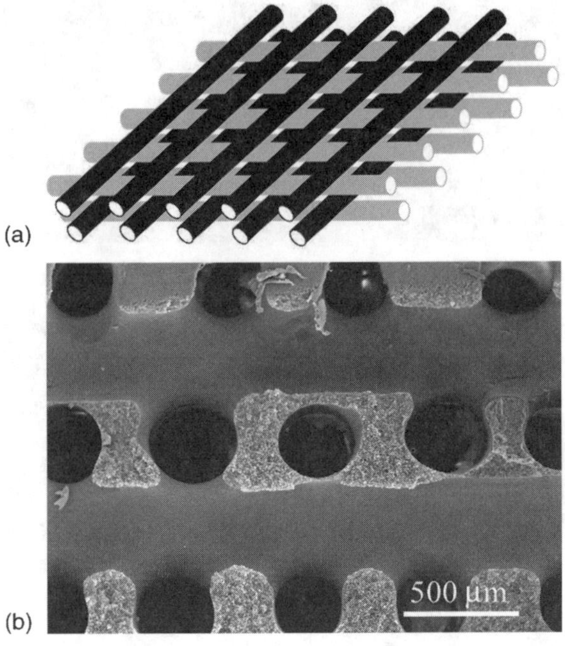

FIGURE 9.9 Nano-fibrous PLLA scaffold prepared by combining a sugar fiber-leaching technique and a thermally induced gelation process. (a) Schematic diagram of perpendicular sugar fiber assembly. (b) SEM micrograph of nano-fibrous matrix prepared with a perpendicular tubular macropore network. (From Zhang, R. Y., and Ma, P. X., Synthetic nano-fibrillar extracellular matrices with predesigned macroporous architectures, *J. Biomed. Mater. Res.*, 52, 430–438, 2000. With permission.)

in a solution of poly(vinyl alcohol) and water. After drying, the microspheres are separated into the proper sizes ranging from 150 to 500 μm.[4] In this example, the paraffin mold is formed by placing the spheres (diameter = 250–420 μm) into a mold and thermally bonding them. By varying the size range and heat treatment of the microspheres used in the paraffin assembly, one can control the average macropore size and the interconnectivity of the pores throughout the scaffold.

Since paraffin dissolves in THF, the PLLA–THF solutions used in the above sections cannot be utilized with paraffin porogens. Instead, PLLA is dissolved in solvent systems such as dioxane–pyidine, dioxane–methanol, or dioxane–H_2O. For this example, PLLA is dissolved in a 1:1 (v:v) dioxane–pyridine system. The solution is cast onto the paraffin mold dropwise, and quickly vacuum-treated in a heated vacuum oven to remove air trapped in the paraffin sphere assembly. Then, the polymer solution is induced to phase separate at −70°C. After phase separation, the gel/paraffin composite is submersed in cold hexane (−20°C) for solvent exchange. The gel can then be cut to the desired size, and placed in room temperature hexane to extract the remaining solvent and to leach out the paraffin. Hexane is exchanged with cyclohexane, the samples are frozen, and the frozen gel is lyophilized. The result is a scaffold with interconnected spherical macropores and nano-fibrous pore walls (Figure 9.10).

(a)
(b)

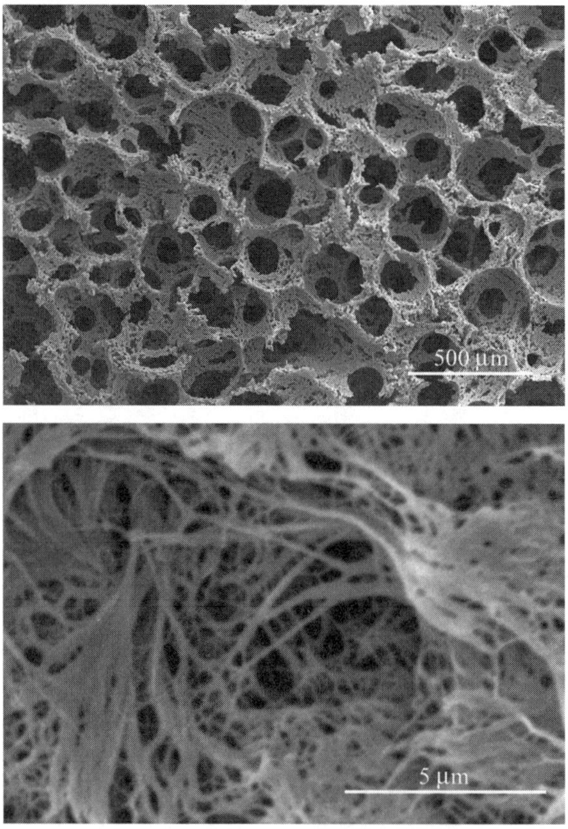

FIGURE 9.10 SEM micrographs of a nano-fibrous PLLA scaffold with interconnected spherical macropores prepared by combining a paraffin sphere-leaching technique and a thermally induced gelation process. (From Chen, V. J., and Ma, P. X., Nano-fibrous poly(L-lactic acid) scaffolds with interconnected spherical macropores, *Biomaterials*, (in press). With permission.)

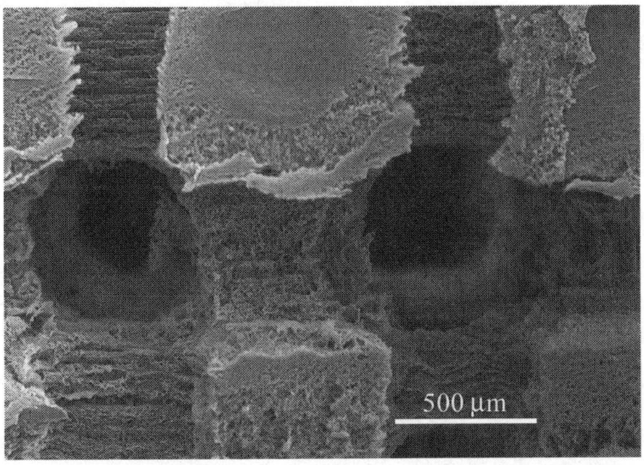

FIGURE 9.11 SEM micrograph of a nano-fibrous PLLA scaffold with interconnected cylindrical channels prepared by combining solid free form fabrication and thermally induced gelation.

4. Creating Macroporous Structures with Solid Freeform Fabrication

A scaffold design technique that is rapidly growing in popularity is rapid prototyping or solid freeform (SFF) fabrication. In this process, a computer-aided design (CAD) program can be used to design a three-dimensional negative mold of the desired scaffold. Since the printing process is completely automated, the scaffold designer is allowed to create scaffolds with very controlled and reproducible scaffold morphologies, pore geometries, and pore interconnectivities. The mold is then created layer-by-layer on a three-dimensional printer. While SFF fabrication gives the scaffold designer the ability to control several geometric parameters, this method does have certain limitations with resolution and feature size. For instance, the sizes and shapes of the pores are limited by the print nozzle diameter and materials involved in the printing process, sometimes making it difficult to form pores of complex geometries or smaller pores where support material used during the build process is difficult to remove.[27,42] By combining phase separation with SFF fabrication, nano-scaled features, such as the nano-fibers demonstrated with porogens mentioned earlier, can be integrated into scaffolds with macroscopic features created from the SFF fabrication.

To combine phase separation with SFF fabrication, a polymer solution can be cast into a wax mold created from the three-dimensional printer. The solution is phase separated, the mold is dissolved in a solvent, and the scaffold is freeze-dried to complete the fabrication. Figure 9.11 shows a nano-fibrous scaffold with interconnected cylindrical channels created from a PLLA–THF solution. The solution is cast into the wax mold and phase separated at -70°C. Afterwards, the THF is exchanged with water, the wax is leached with cyclohexane and ethanol, the ethanol is exchanged with water, and the sample is then frozen and lyophilized.

IV. CONCLUSIONS

Phase separation of polymer solutions can be a complex process. However, if utilized properly, preparations of polymer scaffolds using phase separation can be relatively simple procedures that yield highly organized structures. While solid–liquid phase separation can be used to create various channels and partitions at the micrometer level, liquid–liquid phase separation can yield structures at the micrometer and nanometer level, including platelet and nanometer-scaled fibrous structures. By combining polymer phase separation with various porogen leaching techniques, these nanometer-scaled features can be easily incorporated into three-dimensional constructs ultimately designed to aid neo-tissue formation. These advantages associated with polymer phase separation continue to make this technique an effective method of fabricating three-dimensional scaffolds for tissue engineering.

REFERENCES

1. Langer, R., and Vacanti, J. P., Tissue engineering, *Science*, 260, 920–926, 1993.
2. Ma, P. X., Tissue engineering, In *Encyclopedia of Polymer Science and Technology*, Kroschwitz, J. I., ed., Wiley, Hoboken, NJ, 2004.
3. Zhang, R. Y., and Ma, P. X., Synthetic nano-fibrillar extracellular matrices with predesigned macroporous architectures, *J. Biomed. Mater. Res.*, 52, 430–438, 2000.
4. Ma, P. X., and Choi, J. W., Biodegradable polymer scaffolds with well-defined interconnected spherical pore network, *Tissue Eng.*, 7, 23–33, 2001.
5. Ma, P. X., and Zhang, R. Y., Microtubular architecture of biodegradable polymer scaffolds, *J. Biomed. Mater. Res.*, 56, 469–477, 2001.
6. Mikos, A. G., Thorsen, A. J., Czerwonka, L. A., Bao, Y., Langer, R., Winslow, D. N., and Vacanti, J. P., Preparation and characterization of poly(L-lactic acid) foams, *Polymer*, 35, 1068–1077, 1994.
7. Ma, P. X., and Langer, R., Fabrication of biodegradable polymer foams for cell transplantation and tissue engineering, In *Tissue Engineering Methods and Protocols*, Morgan, J. and Yarmush, M., eds., Humana Press Inc., Totowa, NJ, pp. 47–56, 1999.
8. Cima, L. G., Vacanti, J. P., Vacanti, C., Ingber, D., Mooney, D., and Langer, R., Tissue engineering by cell transplantation using degradable polymer substrates, *J. Biomech. Eng. Trans. ASME*, 113, 143–151, 1991.
9. Freed, L. E., Vunjaknovakovic, G., Biron, R. J., Eagles, D. B., Lesnoy, D. C., Barlow, S. K., and Langer, R., Biodegradable polymer scaffolds for tissue engineering, *Bio/Technology*, 12, 689–693, 1994.
10. Ma, P. X., and Langer, R., Degradation, structure and properties of fibrous nonwoven poly(glycolic acid) scaffolds for tissue engineering, In *Polymers in Medicine and Pharmacy*, Mikos, A. G., Leong, K. W., Radomsky, M. L., Tamada, J. A. and Yaszemski, M. J., eds., MRS, Pittsburgh, pp. 99–104, 1995.
11. Mikos, A. G., Sarakinos, G., Leite, S. M., Vacanti, J. P., and Langer, R., Laminated 3-dimensional biodegradable foams for use in tissue engineering, *Biomaterials*, 14, 323–330, 1993.
12. Thomson, R. C., Yaszemski, M. J., Powers, J. M., and Mikos, A. G., Fabrication of Biodegradable Polymer Scaffolds to Engineer Trabecular Bone, *J. Biomater. Sci. Polym. Ed.*, 7, 23–38, 1995.
13. Whang, K., Thomas, C. H., Healy, K. E., and Nuber, G., A novel method to fabricate bioabsorbable scaffolds, *Polymer*, 36, 837–842, 1995.
14. Mooney, D. J., Baldwin, D. F., Suh, N. P., Vacanti, L. P., and Langer, R., Novel approach to fabricate porous sponges of poly(D,L-lactic-co-glycolic acid) without the use of organic solvents, *Biomaterials*, 17, 1417–1422, 1996.
15. Sachs, E., Cima, M., Williams, P., Brancazio, D., and Cornie, J., 3-dimensional printing — rapid tooling and prototypes directly from a CAD model, *J. Eng. Ind. Trans. ASME*, 114, 481–488, 1992.
16. Park, A., Wu, B., and Griffith, L. G., Integration of surface modification and 3D fabrication techniques to prepare patterned poly(L-lactide) substrates allowing regionally selective cell adhesion, *J. Biomater. Sci. Polym. Ed.*, 9, 89–110, 1998.
17. Hutmacher, D. W., Scaffold design and fabrication technologies for engineering tissues — state of the art and future perspectives, *J. Biomater. Sci. Polym. Ed.*, 12, 107–124, 2001.
18. Taboas, J. M., Maddox, R. D., Krebsbach, P. H., and Hollister, S. J., Indirect solid free form fabrication of local and global porous, biomimetic and composite 3D polymer-ceramic scaffolds, *Biomaterials*, 24, 181–194, 2003.
19. Hutmacher, D. W., Scaffolds in tissue engineering bone and cartilage, *Biomaterials*, 21, 2529–2543, 2000.
20. Zhang, R., and Ma, P. X., Processing of polymer scaffolds: phase separation, In *Methods of Tissue Engineering*, Atala, A., and Lanza, R. P., eds., Academic Press, San Diego, pp. 715–724, 2002.

21. Lo, H., Ponticiello, M. S., and Leong, K. W., Fabrication of controlled release biodegradable foams by phase separation, *Tissue Eng.*, 1, 15–28, 1995.
22. Schugens, C., Maquet, V., Grandfils, C., Jerome, R., and Teyssie, P., Polylactide macroporous biodegradable implants for cell transplantation 2. Preparation of polylactide foams by liquid–liquid phase separation, *J. Biomed. Mater. Res.*, 30, 449–461, 1996.
23. Zhang, R. Y., and Ma, P. X., Poly(alpha-hydroxyl acids) hydroxyapatite porous composites for bone-tissue engineering. I. Preparation and morphology, *J. Biomed. Mater. Res.*, 44, 446–455, 1999.
24. Zhang, R. Y., and Ma, P. X., Porous poly(L-lactic acid)/apatite composites created by biomimetic process, *J. Biomed. Mater. Res.*, 45, 285–293, 1999.
25. Ma, P. X., and Zhang, R. Y., Synthetic nano-scale fibrous extracellular matrix, *J. Biomed. Mater. Res.*, 46, 60–72, 1999.
26. Nam, Y. S., and Park, T. G., Porous biodegradable polymeric scaffolds prepared by thermally induced phase separation, *J. Biomed. Mater. Res.*, 47, 8–17, 1999.
27. Chen, V. J., and Ma, P. X., Nano-fibrous poly(L-lactic acid) scaffolds with interconnected spherical macropores, *Biomaterials*, 25, 2065–2073, 2004.
28. van de Witte, P., Dijkstra, P. J., van den Berg, J. W. A., and Feijen, J., Phase separation processes in polymer solutions in relation to membrane formation, *J. Membr. Sci.*, 117, 1–31, 1996.
29. Zhang, R., and Ma, P. X., Processing of polymer scaffolds: phase separation, In *Methods of Tissue Engineering*, Atala, A. and Lanza, R., eds., Academic Press, San Diego, CA, pp. 715–724, 2001.
30. Ma, P. X., Zhang, R. Y., Xiao, G. Z., and Franceschi, R., Engineering new bone tissue in vitro on highly porous poly(alpha-hydroxyl acids)/hydroxyapatite composite scaffolds, *J. Biomed. Mater. Res.*, 54, 284–293, 2001.
31. Larson, R. G., *The Structure and Rheology of Complex Fluids*, Oxford University Press, New York, 1999, pp. 388–398.
32. Cahn, J. W., Phase separation by spinodal decomposition in isotropic systems, *J. Chem. Phys.*, 42, 93–99, 1965.
33. Guenet, J. M., *Thermoreversible Gelation of Polymers and Biopolymers*, Academic Press, London, 1992.
34. Wei, G., and Ma, P. X., Structure and properties of nano-hydroxyapatite/polymer composite scaffolds for bone tissue engineering, *Biomaterials*, 25, 4749–4757, 2004.
35. Aubert, J. H., Structural coarsening of demixed polymer-solutions, *Macromolecules*, 23, 1446–1452, 1990.
36. Ma, P. X., and Langer, R., Fabrication of biodegradable polymer foams for cell transplantation and tissue engineering, In *Tissue Engineering Methods and Protocols*, Yarmush, M. and Morgan, J., eds., Humana Press Inc., Totowa, NJ, pp. 47–56, 1998.
37. Goodman, S. L., Sims, P. A., and Albrecht, R. M., Three-dimensional extracellular matrix textured biomaterials, *Biomaterials*, 17, 2087–2095, 1996.
38. Palecek, S. P., Integrin-ligand binding properties govern cell migration speed through cell-substratum adhesiveness, *Nature*, 385, 537–540, 1997.
39. Folkman, J., and Moscona, A., Role of cell-shape in growth-control, *Nature*, 273, 345–349, 1978.
40. Benya, P. D., and Shaffer, J. D., Dedifferentiated chondrocytes reexpress the differentiated collagen phenotype when cultured in agarose gels, *Cell*, 30, 215–224, 1982.
41. Woo, K. M., Chen, V. J., and Ma, P. X., Nano-fibrous scaffolding architecture selectively enhances protein adsorption contributing to cell attachment, *J. Biomed. Mater. Res.*, 67A, 531–537, 2003.
42. Yang, S. F., Leong, K. F., Du, Z. H., and Chua, C. K., The design of scaffolds for use in tissue engineering. Part II. Rapid prototyping techniques, *Tissue Eng.*, 8, 1–11, 2002.

10 Solid Freeform Fabrication of Tissue Engineering Scaffolds

Tien-Min G. Chu

CONTENTS

I. INTRODUCTION

Internal architecture of scaffolds, including pore size, pore shape, and connectivity, are critical to the *in vivo* and mechanical performance of the scaffolds. It affects the degree of bone regeneration,[1–5] influences the path of bone regeneration,[6] and determines the mechanical properties of the scaffolds.[7]

However, in many scaffold manufacturing techniques, the control over the internal architecture and interconnectivity is limited. Current scaffold processing techniques, including solvent casting, freeze-drying, phase separation, and gas forming, can now produce interconnected microporosity with a certain degree of oriented pore structure.[8] However, the connectivity created by these techniques is a result of many processing variables, including the rate of solvent evaporation and the three-dimensional contact between the porogen particles, rather than the demonstration of a predetermined engineering design.[9] Precise control over the internal architecture and interconnectivity with these scaffolds manufacturing methods is extremely difficult, if not impossible.

For this reason, tissue engineers have been exploring a new manufacturing technique that would allow for a higher degree of architectural control. A new manufacturing technique that has drawn significant interest from the tissue engineering community is solid freeform fabrication (SFF).[10–12] SFF is a collective term for a group of technologies that can manufacture objects in a layer-by-layer fashion from the three-dimensional computer design of the object.[13] This layer-by-layer fashion allows the manufacturing of objects with complex internal architecture, not possible with traditional manufacturing methods. In this chapter, the background of SFF will first be introduced. The SFF technologies developed for scaffold fabrication will be discussed.

A. Background of SFF

SFF was initially developed by the engineering community for fabricating prototype engineering parts, thus the name "rapid prototyping" (RP) was also used widely. More than 20 SFF technologies have been developed since 1987 when the first commercial SFF machine was introduced. These technologies differentiate themselves mainly by how the layers are laid down, solidified, and attached to previous layers. Among these SFF technologies, several have been modified or developed toward the manufacturing of tissue engineering scaffolds. Given the scope of this chapter, only these SFF technologies will be reviewed. The list includes: three-dimensional printing (3DP), fused deposition modeling (FDM), ink-jet printing and indirect casting (IC), stereolithography (SL), selective laser sintering (SLS), and a few other extrusion-based technologies. Also reviewed in this chapter are some later developments in this area that involve the incorporation of cells and proteins in the scaffold fabricating process. These techniques include protein printing and cell and organ printer.

Though every SFF differs significantly in the way they build models, they do share several common features. These features will be introduced first in the following section.

B. Common Features of SFF

All SFF technologies employ the following three steps in their process: (1) data input, (2) data file preparation, and (3) object building. Some SFF technologies require an additional step of postprocessing to either remove the excessive material trapped inside the void space in the built object, or to increase the strength of the built object. Since SFF technologies are based on computer controlled layer-by-layer manufacturing, a data file describing the three-dimensional geometrical information of the target object provides the basis for the computer controlled building process. A common source for this data file is the three-dimensional computer model of the object created by computer aided design (CAD) software. Regardless of the type of CAD software used, the CAD file typically needs to be converted to a particular file format, STL format,[14] before it can be accepted by the particular controlling software that comes with each SFF machine. After receiving the STL file, the SFF machine software then further prepares the file by "slicing" the three-dimensional file into many two-dimensional layer files and then generates accompanying hardware control file from the two-dimensional slices. The control files contain the parameters to direct the operation of specific machine hardware.

A second type of data source is the images generated by computer tomography (CT) or magnetic resonance imaging (MRI). The image files generated by CT or MRI generally require an intermediate processing step to identify the object of interest on each two-dimensional image slice. The two-dimensional slice file can then be converted into files that SFF machine software can manipulate. Commercial software is available to process and convert files obtained from a large selection of CT or MRI scanners.[15]

II. SCAFFOLDS BY SFF

In this section, the operation principles of tissue engineering related SFF technology will be described first, followed by a description of the type of scaffolds and the research associated with each technology.

A. Three-Dimensional Printing

1. The Technology

3DP was first developed at Massachusetts Institute of Technology. Therics Incorporated (Princeton, NJ) later developed commercial machines based on the 3DP technology for producing products in the health care market. The layout of the system consists of a build platform set on top of an elevator that can travel in the up–down or *Z*-direction, a powder dispensing roller on one side of the platform, and a printhead mounted on *X*, *Y* rails similar to conventional graphic plotter (Figure 10.1). Polymer or ceramic powders, typically PLA, PGA, and poly(D,L-lactide-*co*-glycolide) (PLGA), are used as building materials. In object building, the computer directs the printhead to dispense liquid binder in a pattern conforming to the shape of the first layer image in the series of two-dimensional images generated in the file preparation process. A binder typically used for the aforementioned polymer is chloroform. The surface of the powder in contact with the solvent is partially dissolved, allowing the particles in the print path to attach to one another. After one layer is built, the platform moves downward to a specified distance to allow the powder dispensing roller to travel across the platform and coat the entire layer with fresh powder. The computer then directs the printhead to print binder on the fresh powder to build the next layer. The process is repeated until the entire object is made. In building scaffold-like structures where multiple internal channels are present, there will be loose powder trapped inside the channels and will require a postprocessing step to remove the trapped powders.

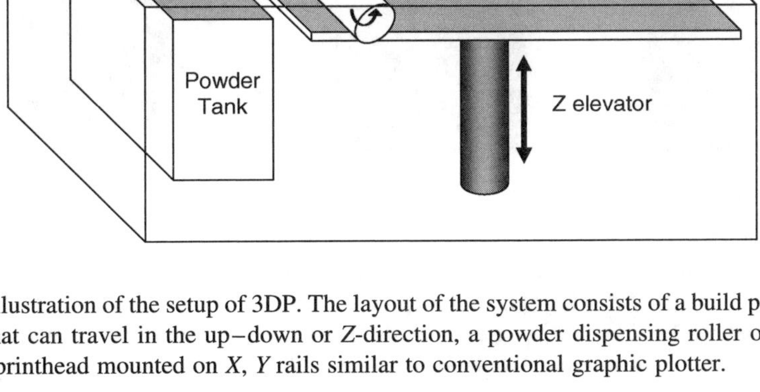

FIGURE 10.1 Illustration of the setup of 3DP. The layout of the system consists of a build platform set on top of an elevator that can travel in the up–down or *Z*-direction, a powder dispensing roller on one side of the platform, and a printhead mounted on *X*, *Y* rails similar to conventional graphic plotter.

2. Application in Tissue Engineering Scaffolds

This technology is the one of the most documented technologies so far in SFF related scaffold research. Using different binder powder pairs, researchers have fabricated scaffolds from a variety of materials, including poly(L-lactide) (PLLA),[16] PLGA,[17–19] hydroxyapatite (HA),[20,21] tricalcium phosphate (TCP),[22] PLGA/TCP,[19,23,24] and poly(desaminotyrosyl-tyrosine ethyl ester carbonate) (poly-(DTE carbonate).[24] Kim et al.[17] fabricated scaffolds from PLGA that contained interconnected channels of 800 μm in diameters. Salt particles were incorporated in the powder bed to increase the final porosity. After salt leaching, a total porosity of 60% was achieved. Roy et al. [23] created PLGA/TCP scaffolds that contained 1.6×1.6 mm and 1.6×1.0 mm square channels. Salt particles of 80 and 87.5 wt% in the size range of 150 μm were incorporated into the powder bed to create porosity gradients in the scaffolds. Scaffolds with the same architectural design were also made from PLGA/TCP and poly-(DTE carbonate).[24] The most sophisticated scaffold was created by Sherwood et al.[19] A composite osteochondral scaffold with three distinct regions was fabricated. The bone region, shaped as a clove, was fabricated with PLGA/TCP powder mixed with 55 wt% salt particles, the cartilage region was fabricated with PLGA/PLA mixed with 90 wt% salt particles, and the transition region consisting of three sections was fabricated with PLGA/PLA powder mixed with 30, 15, and 5 wt% of salt particles. The transition region was designed and incorporated in the scaffold to prevent delamination from occurring between the 90% porous PLGA/PLA cartilage region and the 50% porous PLGA/TCP bone region (Figure 10.2).

In mechanical property evaluation, the tensile strength and modulus of solid specimens fabricated with 3DP were found to be comparable between specimens with and without further densification by isostatic pressing.[16] However, the tensile strength (4 to 14 MPa) of these specimens fabricated from 3DP is lower than the typical strength value from PLLA of similar molecular weight (mol wt = 50,000, 28 MPa).[25] In a separate study, Sherwood[19] found that tensile and compressive strengths of 25% porous PLGA/TCP scaffolds (5.7 and 13.5 MPa) were comparable to that of human cancellous bone. The tensile and compressive strength decreased to 1.6 and 2.5 MPa when the porosity increased to 55%.

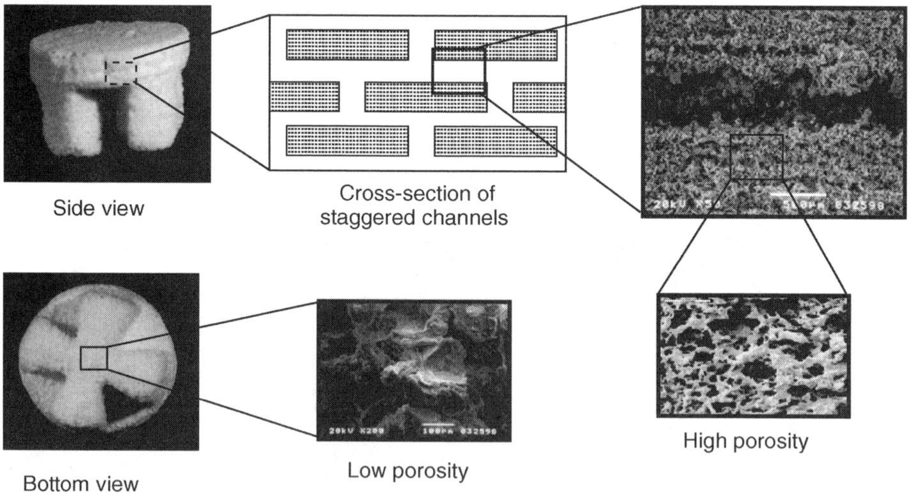

FIGURE 10.2 Figure showing the construction of osteochondral scaffold fabricated with TheriForm™ process. (From Sherwood, J. K., Riley, S. L., Palazzolo, R., Brown, S. C., Monkhouse, D. C., Coates, M., Griffith, L. G., Landeen, L. K., and Ratcliffe, A., A three-dimensional osteochondral composite scaffold for articular cartilage repair, *Biomaterials*, 23(24), 4739–4751, 2002. With permission.)

In the *in vitro* evaluation, Park et al.[18] made a patterned film of PLLA with 0.5 to 1 mm poly(ethylene-oxide; PEO)-poly(propylene-oxide; PPO) lines spaced at 4 mm by printing a PEO–PPO–PEO copolymer onto PLA substrates. Using 3T3 mouse fibroblast, the authors showed that the cell attachments on the printed lines were reduced to 20% or less compared to the cell attachments on the control. The PEO–PPO–PEO copolymer was further modified with 1-amino-1-deoxy-β, D-galactose, a ligand for the asialoglycoprotein receptor in hepatocytes. By culturing primary rat hepatocytes on the PEO–PPO–PEO–galactose modified PLLA substrates, the authors demonstrated a longer and more stable attachment of hepatocytes compared to cells cultured on unmodified PLLA surfaces. The results demonstrated that 3DP can be used as a cell patterning tool for hepatocytes. Kim et al.[17] studied the use of bioreactor for rat hepatocytes seeded PLGA scaffolds. A significantly higher oxygen partial pressure, glucose level, and a more physiologic pH were found in the scaffolds cultured under flow condition. The hepatocytes were found to attach, spread, and infiltrate about 200 μm into the three-dimensional PLGA scaffolds after seeding for 2 days in the flow condition. In the area of bone and cartilage tissue engineering, using the composite osteochondral scaffold, Sherwood et al.[19] successfully achieved a preferential seeding and adhesion of ovine articular condrocytes to the 90% porous PLGA/PLA cartilage part of the scaffold and prevented the cell from attaching to the PLGA/TCP bone part of the scaffold. The top seeding method was found to result in higher cell, S-GAG, and collagen than the rotational seeding method.

Using a rabbit cranium model, Roy et al.[23] and Simon et al.[24] evaluated the *in vivo* performance of 3DP scaffolds. After implanting the PLGA/TCP scaffolds in rabbit cranium for 8 weeks, Roy et al.[23] found that the bone regeneration in PLGA/TCP scaffolds that contained designed channels was significantly higher than bone regeneration in scaffolds without designed channels. Simon et al.[24] found that the printed PLGA/TCP collapsed to 45% of their original cross-section by Week 8 and completely collapsed at Week 16. Scaffolds made from poly-(DTE carbonate), however, retain their geometry for the same experimental duration.

The removal of chloroform from the 3DP scaffolds has been a concern.[10] Giordano[16] showed that 0.5 wt% of chloroform still remained after dry the parts in vacuum for 1 week. In scaffolds used in a later study,[19] the use of liquid CO_2 extraction was found to improve solvent removal; a residual chloroform of lower than 100 ppm was achieved (Sherwood, J.K., personal communication, 2003). The use of solvent, however, precludes the incorporation of biological factors and cells in 3DP technique.

B. Fused Deposition Modeling

1. The Technology

FDM was developed and commercialized by Stratasys Incorporated (Eden Prairie, MN). In this technology, filaments of thermal plastics are used as the build material. The machine consists of a build platform set on top of an elevator with a computer controlled heated nozzle mounted on *X*, *Y* rails. The filaments are extruded through the heated nozzle and laid down as strings on the build platform. After the nozzle lays down strings of material in the defined pattern, the build platform is lowered and the second layer is built on top of the first layer. The process repeats itself until the object is completely built. Polycaprolactone (PCL) is typically used on FDM for tissue engineering scaffold fabrication.[26] Polycaprolactone was chosen for its low glass transition temperature and high decomposition temperature. When building solid models, the filaments are laid down side-by-side to build up the volume. When building scaffold-like structures, the spacing between the filaments is increased to provide porosity in the scaffold. The diameter of the extruded filament, distance between the filaments, and the change in lay down directions between successive layers are used to control the porosity of the scaffold. Figure 10.3 shows examples of scaffolds with 0/60/120° and 0/90° lay down patterns. By optimizing the liquefier temperature and the roller speed, Zein et al.[27] were able to produce porous scaffold from two different nozzle sizes, 0.010

FIGURE 10.3 Figure showing PCL scaffold manufactured by FDM. (a) and (b): 0/60/120° lay down pattern. (c) and (d): 0/90° lay down pattern. Porosity levels: (a) – 68%, (b) – 80%, (c) – 50%, (d) – 75%. The micron bar shown in the photograph is 500 μm (courtesy of Dr. Hutmacher).

and 0.016 in. By changing the distance between the filaments, the orientation of the filaments, and the layer thickness, scaffolds with 48 to 77% porosity were made from filaments in the diameter of 260 to 370 μm and spaced at between 160 and 700 μm.

Unlike 3DP, there is no unbound loose powder and there is no solvent removal required in FDM. However, this technology requires the process to take place at an elevated temperature to melt the thermal plastic filament. The use of elevated temperature precludes the incorporation of biological factors or cells in the process.

2. Application in Tissue Engineering Scaffolds

In mechanical property evaluation, when scaffolds made from a lay down pattern of 0/60/120° were tested in saline, the compressive stiffness and 1% yield strength was found to be 29.4 and 2.3 MPa; a value significantly higher than that of the scaffolds printed with a lay down pattern of 0/72/144/36/108°.[26] The stress–strain curve demonstrated the deformation of a typical honeycomb structure; linear elastic at the beginning, followed by a plateau, then by a densification region of sharp increase in stress. A feature common to all SFF scaffolds — the anisotropy of their mechanical properties — was demonstrated by Zein et al.[27] SFF builds scaffold in a layer-by-layer fashion. Therefore, the mechanical property of the scaffold normal to the build plane can vary significantly from the in plane property. Zein et al.[27] showed that the PCL scaffold fabricated with a 0/90° lay down pattern had a compressive yield strength of 3.32 MPa when measured in the plane of layers and 2.58 MPa when measured at a direction normal to the build plane. The potential application of this feature in tissue regeneration, however, was not discussed.

An early *in vitro* study using human fibroblasts and osteoprogenitor cells showed that the cells can attach and spread on the PCL scaffolds.[26] However, the attachment rate of human chondrocytes on the PCL scaffolds were not satisfactory.[28] In another study, mesenchymal progenitor cells and calvarial osteoblasts from rabbits were found to populate throughout the PCL scaffolds 3 weeks after seeding.[29] Finally, a prototype osteochondral construct was made by coculturing human marrow stromal cells and chondrocytes on two separate regions of PCL scaffolds. A mixture of cell types was observed at the interface after 11 weeks.[30]

In an *in vivo* study, chondrocytes harvested from the ears of Yorkshire pigs were seeded to the PCL scaffolds, encapsulated with fibrin glue, and cultured for 1 week before the scaffolds were transplanted subcutaneously.[28] Cartilage-like tissue was found in the scaffold after 18 weeks of implantation. Type I, II, III, and IX collagen were found in the retrieved samples. In the same paper, chondrocytes harvested from the ears of New Zealand white rabbits were seeded with and without fibrin glue encapsulation onto PCL scaffold for 26 days prior to implantation. The PCL scaffolds were implanted subcutaneously in the back of rabbits and were evaluated.[28] Scaffolds with allogeneic fibrin glues encapsulation showed 80% occupancy of the cells in the pore channels after 8 weeks of implantation. Both elastic cartilage-like tissue and islands of mineralization were observed in the retrieved samples. In another study, bone marrow coated scaffolds were used to evaluate healing in bony orbital wall defects.[31] PCL scaffolds coated with heparinized bone marrow freshly harvested from the iliac crest of Yorkshire pigs were implanted in defects created in the medial orbital walls of the same animal from which the marrow was extracted. After 3 months of implantation, a bone growth of 14.1% was found in the coated scaffold, compared to 4.5% in the uncoated scaffolds. Finally, bone marrow progenitor cells and calvarial osteoblasts seeded PCL scaffolds were evaluated using a critical sized rabbit calvaria model.[32] After 3 months of implantation, the area of mineralization in the defects repaired with cell seeded scaffolds was 60% higher compared to that in the unrepaired defects. The yield strength from the push-out test of the cell seeded scaffold reached 85 to 90% of that from the normal calvaria bone.

C. Ink-Jet Printing and Indirect Casting

1. The Technology

Ink-jet printing was first developed by Sanders Design International and was later commercialized by SolidScape Incorporated (Merrimack, NH). The layout of the system consists of a build platform set on top of an elevator with a rolling cutter blade on one side of the platform and two print jets mounted on *X*, *Y* rails. Two print materials are used. The build material consists of polysulfonate and the support material is wax. The build jet first lays down the design pattern by printing droplets of polysulfonates onto the platform. The support jet then prints support material around the printed pattern. After printing, the cutter blade comes over and cuts the build layer to a predetermined layer thickness, thereby controlling the accuracy in the *Z*-direction. The build jet then prints build material for the next layer. The process repeats itself until the entire object is completed.

2. Application in Tissue Engineering Scaffolds

The technology was not used for direct manufacturing of scaffolds, but rather, was used to build casting molds for use in the indirect casting (IC) technique developed by Chu et al.[33] The technique was first developed to fabricate bioceramic scaffolds.[34] Highly loaded ceramic suspension, formulated from bioceramic powder and acrylic monomers, were cast into molds made with stereolithography or ink-jet printing. Thermal initiator was used in the ceramic–acrylic suspension to initiate the polymerization reaction in the acrylic monomers, which turned the ceramic filled acrylic suspension into a solid ceramic–polymer composite. The ceramic–polymer composite and the mold were subjected to heat treatment to pyrolyze the mold and the polymer binder and, at a higher temperature, sinter the ceramic powder. The designed channels for the scaffolds were revealed when the mold is removed by pyrolysis.

In an early study, sintered hydroxyapatite scaffolds with channel size from 366 to 968 μm and porosity level from 26 to 52% were made.[34] With various combination of ceramic powder and monomers, scaffolds from a variety of materials were fabricated, including HA/TCP biphasic bioceramic,[35] calcium phosphate cements,[36] poly(propylene fumarate)/TCP composites,[37]

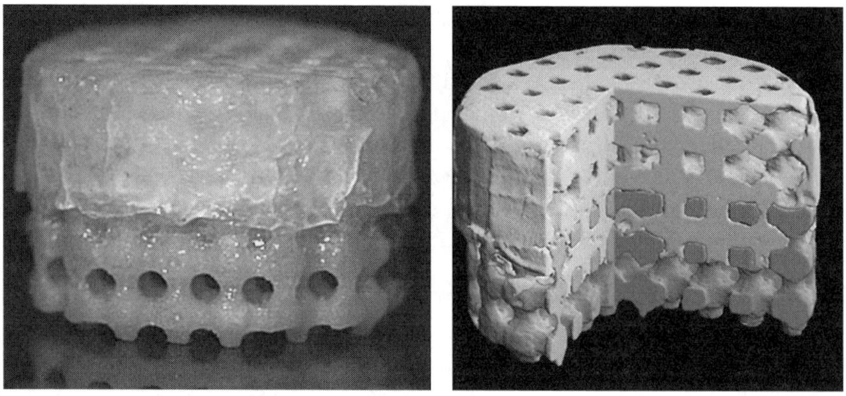

FIGURE 10.4 Figure showing composite scaffold of PLA/HA by indirect casting technique. Top half of the scaffold was PLA and the bottom half was HA. On the right is the colored CT scan of the scaffold showing the interdigitation of two materials at the interface. (From Taboas, J. M., Maddox, R. D., Krebsbach, P. H., and Hollister, S. J., Indirect solid free form fabrication of local and global porous, biomimetic and composite 3D polymer-ceramic scaffolds, *Biomaterials*, 24(1), 181–194, 2003. With permission.)

and TCP.[38] By using the sintered HA from IC as ceramic casting mold, Taboas et al.[9] further used the technique to fabricate osteochondral scaffolds with PLA on one end and HA on the other end, both contain designed pore channels (Figure 10.4).

In the mechanical property evaluation, the compressive strength and compressive modulus of a 40% porous HA scaffold was found to be 30 MPa and 1.4 GPa.[39] In the PPF/TCP scaffolds, the compressive strength reduces from of 18.9 to 1.6 MPa when the porosity was increased from 40 to 80%.[37]

In the *in vivo* study, HA scaffolds with two different architectural designs were used: one with a large 3 mm center channel and smaller 380 μm channels radiating from the center of the cylinder toward outer surface, one with 450 μm orthogonal channels in X–Z-directions.[40] After 9 weeks of implantation in the mandible of Yucatan minipigs, bone was found to penetrate the scaffolds to 1.4 mm underneath the surface of the scaffolds and occupy 30 to 59% of the channel space in this 1.4 mm penetration zone. In the space enclosed by this penetration zone very limited bone formation was found (2%). Though higher amount of bone formation was found in the orthogonal design scaffolds, the difference was not statistically significant, possibly due to the small number of animals tested ($N = 6$). When PPF/TCP scaffolds with the same dimension to the HA scaffolds were implanted in the same animal model, a higher bone growth (61%) and a more even distribution of bone formation was found.[37]

D. Stereolithography

1. The Technology

SL was developed by 3D Systems (Valencia, CA) and is the first commercially available SFF technology.[41] The build material for SL is a photosensitive liquid resin that can be cured by ultraviolet (UV) laser. The layout of the machine consists of a build platform held by an elevator and immersed just below the surface of a tank of liquid resin with a UV laser source on top of the machine. The computer directs the UV laser to raster on the resin surface to solidify the liquid resin into the shape of the first layer image via a photopolymerization reaction. The elevator then moves downward at a specified distance, bringing along the cured first layer, which is attached to the build platform and allows for the uncured fresh resin to flow over the first cured layer. The computer then directs the UV to raster on the fresh resin surface again to build the second layer. The process is repeated until the entire object is made. The finished part is then removed from the platform and rinsed in solvent to wash off the uncured resin on the surface. After washing, the part is again exposed to UV light in a postcuring unit to further increase the degree of polymerization in the resin.

2. Application in Tissue Engineering Scaffolds

The use of SL for bioceramic scaffolds fabrication was first explored by Chu et al.[42–44] Hydroxyapatite powder was dispersed in low viscosity arcrylate with photoinitiators to form a UV curable HA slurry. The formulated HA slurry was used in place of the regular SL resin on the SL machine to build scaffolds. A prototype HA orbital floor model was successfully built on the SL machine.[44] HA scaffold of 50 vol.% macroporosity was made by firing the cured HA slurry at 1350°C.[45] With a similar approach, porous calcium phosphate implants were fabricated by Porter et al.[46] Amorphous calcium phosphate powder was dispersed in commercial SL resin to formulate UV curable slurry. The slurry was used as the build material on SL machine to build scaffolds. The cured slurry was then sinter at 585 and 600°C to produce calcium phosphate implants with 27.7 and 22.9% microporosity. Using a UV curable monomer solution based on diethyl fumarate (DEF) and poly(propylene fumarate) (PPF), Cooke et al.[47] fabricated DEF-*co*-PPF biodegradable scaffolds. Solid test sample of 50 mm diameter and 4 mm thick was successfully made.

E. Selective Laser Sintering

1. The Technology

SLS was originally developed at the University of Texas at Austin and was commercialized by DTM Corporation in 1986,[48] which was later acquired by 3D Systems (Valencia, CA). The build material for the system is either polymer or polymer coated ceramic powder. The layout of the system consists of a build platform set on top of an elevator with a powder dispensing roller on one side of the platform and a CO_2 laser on the top of the machine. The powder bed on the platform is preheated to a temperature just below the glass transition temperature of the polymer to minimize the energy required in the subsequent fusing processing. In object building, the computer directs the CO_2 laser to raster on the polymer or polymer-coated ceramic powder bed causing the powder particles to fuse. After one layer is built, the platform moves downward to a specified distance to allow the powder dispensing roller to travel across the platform and coat the entire layer with fresh powder. The computer then directs the laser to raster on the fresh powder to build the next layer. The process is repeated until the entire object is made. In building scaffold-like structures where multiple internal channels are present, there are loose, unsintered powder trapped inside the channels and will require a postprocessing step to remove the trapped powders.

2. Application in Tissue Engineering Scaffolds

Lee[49] fabricated calcium phosphate implants by sintering poly(methyl methacrylate) coated hydroxyapatite powders on SLS. To demonstrate the potential of using SLS for manufacturing drug delivery device, Low et al.[50] manufactured cylinders that have low porosity on the surface and high porosity inside the cylinder from a polyamide. By optimizing the powder temperature and the laser powder, Tan et al.[51] were able to build composite scaffolds from polyetheretherketon (PEEK) and hydroxyapatite powder. Using porcine marrow stromal cells, Das et al.[52] studied the *in vitro* response of Nylon-6 scaffolds made with SLS. When cultured with Nylon-6 conditioned media, the cells showed viability similar to cells cultured on tissue culture polystyrene plates. The cell viability reduces to below 70% when cells were cultured directly on the Nylon-6 disks. In the *in vivo* study,[52] specimens with 1200 μm beams intersecting orthogonally at a spacing of 800 μm between beams were implanted in the mandible of Yucatan minipigs for 6 weeks. In the retrieved samples, 43% of the void space was filled with newly formed bones.

The polymers currently studied on SLS for tissue engineering application are nondegradable, for example, Nylon-6,[52] PEEK,[51] and Polyamide.[50] An addition of biodegradable polymer to the available material list on SLS will make the technique more attractive to tissue engineers. Due to heat dissipation, particles outside the laser scanning path are sometime also fused and cause an irregular boundary and roughness on the part surface.[10] Furthermore, for building scaffold-like structures where multiple internal channels are present, the loose, unsintered powder trapped inside the channels needs to be removed to reveal the channels. The partial fusion of the particles can make the cleaning of the loose powder difficult. The trapped loose particles were suspected to have caused a reduction in cell viability in the *in vitro* study using Nylon-6.[52]

F. Other Extrusion-Based Technologies

There were several other extrusion-based technologies similar to FDM developed for scaffold fabrication. These technologies all use pressurized nozzles to extrude the build material into the form of filaments which solidify on the platform through either solvent evaporation or temperature induced phase transformation. These technologies are reviewed together in this section.

1. Robocast

Robocast was first developed at the Sandia National Laboratory[53] for fabricating prototypes from ceramics. The technology has a setup similar to FDM with a build platform set on top of an elevator with an extrusion nozzle mounted on X, Y rails. Instead of using spools of thermal plastic polymer in the extrusion head, highly concentrated ceramic paste was used as building material. The paste is extruded through the extrusion head in the shape of thin filaments. After extrusion, water in the ceramic paste evaporates and induces a pseudoplastic to dilatent transition in the paste, which stiffens the filament and provides structural stability to the scaffold. The scaffold is then subject to firing to remove the organic binders in the paste and to sinter the ceramic powder. Stromal and mesenchymal cells were used to evaluate the fabricated hydroxyapatite scaffolds.[54] HA scaffolds with channels of 270 μm were seeded with second and third passage cells from bone marrow aspirates from rat femurs and tibias. The cells were found to attach and uniformly distributed on the scaffolds after 1 day. The cells were found to form networks after 1 week.

A similar system was developed by de Sousa and Evans.[55] Instead of the aqueous-based paste system used in Robocast, a nonaqueous-based paste was used as build material. The extruded filaments from the paste were stiffened on the build platform through the evaporation of organic solvents. Using this technology, HA scaffolds were built and human osteoblast sarcoma (HOS) cells were found to attach to the scaffolds after culturing for 48 h.

2. Bioplotter

Bioplotter (Envision Tec, Germany) was developed by Landers et al.[56] to fabricate scaffolds from hydrogel for soft tissue engineering. Hydrogel plotting material is loaded in an extruder head and extruded through a nozzle, which defines the diameter of the filaments. Two types of plotting materials with different gelation mechanisms are used: (1) thermoreversible gels like agar, which solidifies through temperature induced phase change, and (2) alginate, which solidifies through chemically-induced cross-linking. Since hydrogel filament can deform under its own weight, the hydrogel is dispensed into a liquid of matching density that can compensate the gravity by buoyancy force. Agar scaffolds were made by printing agar solution into gelatin solution.[57] The strands were 500 μm spaced at 500 μm, with a total porosity between 35 and 40%. The surface of the scaffolds was further modified with hyaluronic acid and alginic acid to improve cell attachments. A seeding efficiency of 20 to 35% was found when the scaffolds were seeded with human osteosarcoma cell line (CAL-72) or mouse fibroblasts.

G. Protein and Cell Printing

Although the above SFF technologies have provided the means to control the internal architecture of scaffolds, the use of SFF to control the spatial distribution of cells and protein inside the three-dimensional scaffolds remains a challenge. Several researchers have investigated and documented the progress in this area of cell and protein printing.

1. Protein Printing

Early in 1988 and in 1994, Klebe[58,59] used a modified Hewlett Packard ink-jet printhead to print fibronectin onto thin polystyrene sheets. Cells from a BALB-3T3 transformed cell line were cultured on the fibronectin imprinted sheet to show the printing pattern. Wilson and Boland[60] later developed a protein printing machine based on commercial printers. Bovine serum albumin (BSA), biotinylate BSA, streptavidin, and Lucifer yellow labeled biotin were successfully printed in the designed pattern on glass slides. A similar technology, Biological Architectural Tool (BAT) was also developed with the goal of depositing extracellular matrix proteins.[61] Four printheads were used to enable printing multiples materials in each layer. Preliminary data in the report showed that the machine is capable of printing Type I collagen, fibronectin, and Type I collagen + laminin.

2. Cell and Organ Printer

Klebe et al.[59] documented one of the earliest works on cell printing. CHO Chinese hamster ovary cells loaded in a fluorescence activated cell sorter (FACS) were printed onto a computer controlled *X-Y* translation table in 500 μm-wide lines (about 10 cell diameters). Based on an HP 660C printer body, Wilson and Boland[60] designed a more sophisticated cell printer capable of printing large mammalian cells. The printer has nine piezoelectric pumps, each with an inlet connected to the cell suspensions. However, the nozzles currently have to work one at a time due to software limitations. Bovine aortal endothelial cells and a smooth muscle cell line (ATTC) passage ten were successfully printed onto a reconstituted basement membrane gel and a collagen gel. Damage to the cells did occur during and/or after the printing process. Up to 25% cell death was observed in the printed pattern. Cell–cell attachment is another critical issue in cell printing. Layers of cells show very limited attachment when cell layers were simply stacked and pressed together.[58] Instead of using monolayer cells, Boland et al.[62] cultured aggregates of bovine aortal endothelial cells on collagen gel printed from the cell and organ printer. Fusion and cell spreading was observed in the cell aggregates.

III. CONCLUDING REMARKS

SFF is an exciting new tool for tissue engineering. Scaffolds with controlled internal architecture and interconnectivity now can be made with SFF from many biomaterials used in traditional scaffold manufacturing process, including PLA, PGA, PLGA, PCL, and bioceramics. The many *in vivo* and *in vitro* results have demonstrated the feasibility of the SFF manufacturing technique for tissue engineering scaffolds. Some unique features of SFF scaffolds are further summarized here.

A. Connectivity at Wide Range of Porosity Levels

High porosity level (typically $>90\%$) is used in traditional scaffold processing technique to ensure sufficient pore connectivity for cell attachment and nutrient diffusion. The high porosity also leads to very low mechanical properties in these scaffolds. When the porosity is decreased to increase the mechanical property, the interconnectivity is usually compromised. This strength-porosity-interconnectivity relation of traditional scaffolds is a dilemma to their development toward load bearing conditions. Connectivity in three-dimensional at a lower porosity level is a potential strategy to fulfill the requirements for load bearing scaffolds. With SFF, the three-dimensional interconnection of the scaffold can be maintained at a wide porosity levels. SFF scaffolds can provide connectivity for cell migration and nutrient diffusion at a lower porosity level that supports the required scaffold strength. However, direct evidence to support this potential advantage for load bearing application of SFF scaffold over traditional scaffolds is needed.

B. Versatility in Internal Architecture and External Geometry Control

The versatility of SFF in internal architecture control has been the main focus of this chapter. Of equally importance is the versatility of SFF in controlling the external geometry of the scaffolds. Using CT or MRI as the data source, surgical models made with SFF have been used for surgical planning and as templates for implant placement.[63–67] In the same way, scaffolds can be made with external geometry conforming to the patients' anatomic structure. The external geometry of the scaffolds can also be designed and customized to fulfill the engineering requirement of the scaffold.[19]

C. Ability to Create Discrete Material and Design Domains

Several researchers have demonstrated the use of SFF to manufacture scaffolds with distinct material and design domains. Examples are the osteochondral scaffolds from 3DP,[19] FDM,[30] and IC.[9] Gradients of material can even be incorporated into the scaffold in the manufacturing process. The initial results for the osteochondral scaffolds were extremely encouraging. However, a long-term functional analysis for the osteochondral construct is yet to be demonstrated.

Although still at the initial stage of their development, protein and cell printings are drawing the attention of the tissue engineering community for their potential in tissue engineering complex tissues. Although preliminary cell fusions from the printed cells were demonstrated, a functional analysis of cell agglomeration and differentiation in three-dimensional printed scaffold is yet to be explored. The challenges to overcome the angiogenesis and nutrient diffusion to the cells after they are printed into the scaffold remain.

ACKNOWLEDGMENTS

The author would like to thank Dr. Sherwood, Dr. Hutmacher, and Dr. Taboas for providing the original photographs used in this chapter.

REFERENCES

1. Jin, Q. M., Takita, H., Kohgo, T., Atsumi, K., Itoh, H., and Kuboki, Y., Effects of geometry of hydroxyapatite as a cell substratum in BMP-induced ectopic bone formation, *J. Biomed. Mater. Res.*, 52(4), 491–499, 2000.
2. Pineda, L. M., Busing, M., Meinig, R. P., and Gogolewski, S., Bone regeneration with resorbable polymeric membranes. III. Effect of poly(L-lactide) membrane pore size on the bone healing process in large defects, *J. Biomed. Mater. Res.*, 31(3), 385–394, 1996.
3. Robinson, B. P., Hollinger, J. O., Szachowicz, E. H., and Brekke, J., Calvarial bone repair with porous D,L-polylactide, *Otolaryngol. Head Neck Surg.*, 112(6), 707–713, 1995.
4. Whang, K., Healy, K. E., Elenz, D. R., Nam, E. K., Tsai, D. C., Thomas, C. H., Nuber, G. W., Glorieux, F. H., Travers, R., and Sprague, S. M., Engineering bone regeneration with bioabsorbable scaffolds with novel microarchitecture, *Tissue Eng.*, 5(1), 35–51, 1999.
5. Ishaug-Riley, S. L., Crane-Kruger, G. M., Yaszemski, M. J., and Mikos, A. G., Three-dimensional culture of rat calvarial osteoblasts in porous biodegradable polymers, *Biomaterials*, 19(15), 1405–1412, 1998.
6. Chang, B. S., Lee, C. K., Hong, K. S., Youn, H. J., Ryu, H. S., Chung, S. S., and Park, K. W., Osteoconduction at porous hydroxyapatite with various pore configurations, *Biomaterials*, 21(12), 1291–1298, 2000.

7. Liu, D., Control of pore geometry on influencing the mechanical property of porous hydroxyapatite ceramic, *J. Mater. Sci. Lett.*, 15, 419–421, 1996.
8. Schugens, C., Grandfils, C., Jerome, R., Teyssie, P., Delree, P., Martin, D., Malgrange, B., and Moonen, G., Preparation of a macroporous biodegradable polylactide implant for neuronal transplantation, *J. Biomed. Mater. Res.*, 29(11), 1349–1362, 1995.
9. Taboas, J. M., Maddox, R. D., Krebsbach, P. H., and Hollister, S. J., Indirect solid free form fabrication of local and global porous, biomimetic and composite 3D polymer-ceramic scaffolds, *Biomaterials*, 24(1), 181–194, 2003.
10. Yang, S., Leong, K. F., Du, Z., and Chua, C. K., The design of scaffolds for use in tissue engineering. Part II. Rapid prototyping techniques, *Tissue Eng.*, 8(1), 1–11, 2002.
11. Hutmacher, D. W., Scaffolds in tissue engineering bone and cartilage, *Biomaterials*, 21(24), 2529–2543, 2000.
12. Hutmacher, D. W., Scaffold design and fabrication technologies for engineering tissues — state of the art and future perspectives, *J. Biomater. Sci. Polym. Ed.*, 12(1), 107–124, 2001.
13. Chua, C. K., Leong, K. F., and Lim, C. S., Rapid prototyping process chain, In *Rapid Prototyping — Principles and Applications*, World Scientific Publishing Co., Singapore, pp. 25–33, 2003.
14. Fedchenko, R. P., and Jacobs, P. F., Introduction, In *Stereolithography and other RP&M Technologies*, Jacobs, P. F., ed., Society of Manufacturing Engineers, Dearborn, MI, pp. 1–26, 1996.
15. Jacobs, P. F., Special applications of RP&M, In *Stereolithography and other RP&M Technologies*, Jacobs, P. F., ed., Society of Manufacturing Engineers, Dearborn, MI, pp. 317–366, 1996.
16. Giordano, R. A., Wu, B. M., Borland, S. W., Cima, L. G., Sachs, E. M., and Cima, M. J., Mechanical properties of dense polylactic acid structures fabricated by three dimensional printing, *J. Biomater. Sci. Polym. Ed.*, 8(1), 63–75, 1996.
17. Kim, S. S., Utsunomiya, H., Koski, J. A., Wu, B. M., Cima, M. J., Sohn, J., Mukai, K., Griffith, L. G., and Vacanti, J. P., Survival and function of hepatocytes on a novel three-dimensional synthetic biodegradable polymer scaffold with an intrinsic network of channels, *Ann. Surg.*, 228(1), 8–13, 1998.
18. Park, A., Wu, B., and Griffith, L. G., Integration of surface modification and 3D fabrication techniques to prepare patterned poly(L-lactide) substrate allowing regionally selective cell adhesion, *J. Biomater. Sci. Polym. Ed.*, 9, 89–110, 1998.
19. Sherwood, J. K., Riley, S. L., Palazzolo, R., Brown, S. C., Monkhouse, D. C., Coates, M., Griffith, L. G., Landeen, L. K., and Ratcliffe, A., A three-dimensional osteochondral composite scaffold for articular cartilage repair, *Biomaterials*, 23(24), 4739–4751, 2002.
20. Yoo, J., Zhou, M., West, T., Liu, Q., Wang, C.-C., and Saini, S., Fabrication of biomaterial structures using a three dimensional printing technique, 29th Annual Meeting of Society for Biomaterials, Society for Biomaterials, Reno, NV, p. 36, 2003.
21. Dutta Roy, T., Simon, J., Beam, H., McGlohorn, J., Rekow, E., Ricci, J., Thompson, V., and Parsons, J., Bone formation in 3D fabricated hydroxyapatite scaffolds with pores less than 100 microns, 29th Annual Meeting of Society for Biomaterials, Society for Biomaterials, Reno, NV, p. 351, 2003.
22. Resnell, A., Shah, R., McGlohorn, J., Saini, S., and Monkhouse, D. C., *In vitro characterization of a bone void filler fabricated using TheriForm 3D Printing Technology, 29th Annual Meeting of Society for Biomaterials*, Society for Biomaterials, Reno, NV, p. 352, 2003.
23. Roy, T. D., Simon, J. L., Ricci, J. L., Rekow, E. D., Thompson, V. P., and Parsons, J. R., Performance of degradable composite bone repair products made via three-dimensional fabrication techniques, *J. Biomed. Mater. Res.*, 66A(2), 283–291, 2003.
24. Simon, J. L., Roy, T. D., Parsons, J. R., Rekow, E. D., Thompson, V. P., Kemnitzer, J., and Ricci, J. L., Engineered cellular response to scaffold architecture in a rabbit trephine defect, *J. Biomed. Mater. Res.*, 66A(2), 275–282, 2003.
25. Engelberg, I., and Kohn, J., Physico-mechanical properties of degradable polymers used in medical applications: a comparative study, *Biomaterials*, 12(3), 292–304, 1991.
26. Hutmacher, D. W., Schantz, T., Zein, I., Ng, K. W., Teoh, S. H., and Tan, K. C., Mechanical properties and cell cultural response of polycaprolactone scaffolds designed and fabricated via fused deposition modeling, *J. Biomed. Mater. Res.*, 55(2), 203–216, 2001.
27. Zein, I., Hutmacher, D. W., Tan, K. C., and Teoh, S. H., Fused deposition modeling of novel scaffold architectures for tissue engineering applications, *Biomaterials*, 23(4), 1169–1185, 2002.

28. Hutmacher, D. W., Ng, K. W., Kaps, C., Sittinger, M., and Klaring, S., Elastic cartilage engineering using novel scaffold architectures in combination with a biomimetic cell carrier, *Biomaterials*, 24, 4445–4458, 2003.
29. Schantz, J. T., Teoh, S. H., Lim, T. C., Endres, M., Lam, C. X., and Hutmacher, D. W., Repair of calvarial defects with customized tissue-engineered bone grafts I. Evaluation of osteogenesis in a three-dimensional culture system, *Tissue Eng.*, 9(Suppl. 1), S113–S126, 2003.
30. Cao, T., Ho, K. H., and Teoh, S. H., Scaffold design and in vitro study of osteochondral coculture in a three-dimensional porous polycaprolactone scaffold fabricated by fused deposition modeling, *Tissue Eng.*, 9(Suppl. 1), S103–S112, 2003.
31. Rohner, D., Hutmacher, D. W., Cheng, T. K., Oberholzer, M., and Hammer, B., In vivo efficacy of bone-marrow-coated polycaprolactone scaffolds for the reconstruction of orbital defects in the pig, *J. Biomed. Mater. Res.*, 66B(2), 574–580, 2003.
32. Schantz, J. T., Hutmacher, D. W., Lam, C. X., Brinkmann, M., Wong, K. M., Lim, T. C., Chou, N., Guldberg, R. E., and Teoh, S. H., Repair of calvarial defects with customised tissue-engineered bone grafts II. Evaluation of cellular efficiency and efficacy in vivo, *Tissue Eng.*, 9(Suppl. 1), S127–S139, 2003.
33. Chu, T. M., Brady, G. A., Miao, W., Halloran, J. W., Hollister, S., and Brei, D., Ceramic SFF by direct and indirect stereolithography, Materials Research Society Symposium, Materials Research Society, Pittsburgh, PA, pp. 119–123, 1999.
34. Chu, T. M., Halloran, J. W., Hollister, S., and Feinberg, S. E., Hydroxyapatite implants with designed internal architecture, *J. Mater. Sci.: Mater. Med.*, 12, 471–478, 2001.
35. Diggs, A., Hollister, S. J., and Chu, T. M., Fabrication and characterization of 3-D biphasic ceramic scaffolds composed of hydroxyapatite and β-tricalcium phosphate, Society for Biomaterials 29th Annual Meeting, Society for Biomaterials, Reno, NV, p. 353, 2003.
36. Jongpaiboonkit, L., Chu, T. M., and Halloran, J. W., Characterization and rheology behavior of DCPD cement slips for casting complex scaffold, 29th Annual Meeting of Society for Biomaterials, Society for Biomaterials, Reno, NV, p. 646, 2003.
37. Chu, T. M., Flanagan, C. L., Hollister, S., Feinberg, S. E., Fisher, J. P., and Mikos, A. G., The mechanical and in vivo performance of 3-D poly(propylene fumarate)/tricalcium phosphate scaffolds, 29th Annual Meeting of Society for Biomaterials, Society for Biomaterials, Reno, NV, p. 660, 2003.
38. Limpanuphap, S., and Derby, B., Manufacture of biomaterials by a novel printing process, *J. Mater. Sci.: Mater. Med.*, 13, 1163–1166, 2002.
39. Chu, T. M., Orton, D. G., Hollister, S. J., Feinberg, S. E., and Halloran, J. W., Mechanical and in vivo performance of hydroxyapatite implants with controlled architectures, *Biomaterials*, 23(5), 1283–1293, 2002.
40. Chu, T. M., Hollister, S. J., Halloran, J. W., Feinberg, S. E., and Orton, D. G., Manufacturing and characterization of 3-D hydroxyapatite bone tissue engineering scaffolds, *Ann. NY Acad. Sci.*, 961, 114–117, 2002.
41. Chua, C. K., Leong, K. F., and Lim, C. S., Liquid-based rapid prototyping systems, In *Rapid Prototyping — Principles and Applications*, World Scientific Publishing Co., Singapore, pp. 35–110, 2003.
42. Chu, T. M., Halloran, J. W., and Wagner, W. C., Ultraviolet curing of highly loaded hydroxyapatite suspension, In *Bioceramics: Materials and Applications II*, Rusin, R. P. and Fischman, G. S., eds., American Ceramic Society, Westerville, OH, pp. 57–66, 1996.
43. Chu, T. M., Halloran, J. W., and Wagner, W. C., Hydroxyapatite suspension for implant fabrication by stereolithography, In *Case Studies in Ceramic Product Development*, Ghosh, A., Barks, R. E. and Hiremath, B., eds., American Ceramic Society, Westerville, OH, pp. 119–125, 1997.
44. Levy, R. A., Chu, T. M., Halloran, J. W., Feinberg, S. E., and Hollister, S., CT-generated porous hydroxyapatite orbital floor prosthesis as a prototype bioimplant, *AJNR Am. J. Neuroradiol.*, 18(8), 1522–1525, 1997.
45. Chu, T. M., Hollister, S., Feinberg, S. E., and Halloran, J. W., Manufacturing of biomaterial scaffolds using Image Centered Engineering methods, Keystone Symposium on Tissue Engineering, Copper Mountain, CO, 1998.
46. Porter, N. L., Pilliar, R. M., and Grynpas, M. D., Fabrication of porous calcium polyphosphate implants by solid freeform fabrication: a study of processing parameters and in vitro degradation characteristics, *J. Biomed. Mater. Res.*, 56(4), 504–515, 2001.

47. Cooke, M. N., Fisher, J. P., Dean, D., Rimnac, C., and Mikos, A. G., Use of stereolithography to manufacture critical-sized 3D biodegradable scaffolds for bone ingrowth, *J. Biomed. Mater. Res.*, 64B(2), 65–69, 2003.
48. Chua, C. K., Leong, K. F., and Lim, C. S., Powder-based rapid prototyping systems, In *Rapid Prototyping — Principles and Applications*, World Scientific Publishing Co., Singapore, pp. 173–236, 2003.
49. Lee, G., and Barlow, J. W., Selective laser sintering of bioceramic materials for implants, *Solid Freeform Fabrication Symposium*, Austin, TX, pp. 376–380, 1993.
50. Low, K. H., Leong, K. F., Chua, C. K., Du, Z., and Cheah, C. M., Characterization of SLS parts for drug delivery devices, *Rapid Prototyping J.*, 7(5), 262–267, 2001.
51. Tan, K. H., Chua, C. K., Leong, K. F., Cheah, C. M., Cheang, P., Abu Bakar, M. S., and Cha, S. W., Scaffold development using selective laser sintering of polyetheretherketone-hydroxyapatite biocomposite blends, *Biomaterials*, 24(18), 3115–3123, 2003.
52. Das, S., Hollister, S. J., Flanagan, C., Adewunmi, A., Bark, K., Chen, C., Ramaswamy, K., Rose, D., and Widjaja, E., Freeform fabrication of Nylon-6 tissue engineering scaffolds, *Rapid Prototyping J.*, 9(1), 43–49, 2003.
53. Smay, J. E., Cesarano, J. III, and Lewis, J. A., Colloidal inks for directed assembly of 3-D periodic structures, *Langmuir*, 18, 5429–5437, 2002.
54. Dellinger, J., Cesarano, J. III, Smay, J. E., and Jamison, R., 3-D periodic hydroxyapatite scaffolds: cell responses to tailored porosities, 29th Annual Meeting of Society for Biomaterials, Society for Biomaterials, Reno, NV, p. 31, 2003.
55. de Sousa, F., and Evans, J., Sintered hydroxyapatite latticework for bone substitute, *J. Am. Ceram. Soc.*, 86, 517–519, 2003.
56. Landers, R., Pfister, A., Hubner, U., John, H., Schmelzeisen, R., and Hulhaupt, R., Fabrication of soft tissue engineering scaffolds by means of rapid prototyping techniques, *J. Mater. Sci.*, 37, 3107–3116, 2002.
57. Landers, R., Hubner, U., Schmelzeisen, R., and Mulhaupt, R., Rapid prototyping of scaffolds derived from thermoreversible hydrogels and tailored for applications in tissue engineering, *Biomaterials*, 23(23), 4437–4447, 2002.
58. Klebe, R. J., Cytoscribing: amethod for micropositioning cells and the construction of two- and three-dimensional synthetic tissues, *Exp. Cell Res.*, 179, 362–373, 1988.
59. Klebe, R. J., Thomas, C. A., Grant, G. M., Grant, A., and Gosh, R., Cytoscription: computer controlled micropositioning of cell adhesion proteins and cells, *J. Tissue Culture Methods*, 16, 189–192, 1994.
60. Wilson, W. C. Jr., and Boland, T., Cell and organ printing 1: protein and cell printers, *Anat. Rec.*, 272A(2), 491–496, 2003.
61. Stone, A. L., Smith, C. M., Kachurin, A. M., Stewart, R. L., Parkhill, R. L., Simpkins, M., Warren, W. L., and Williams, S. K., Angiogenic potentials of polymers coated with extracellular matrix proteins using a three dimensional printing system, 29th Annual Meeting of Society for Biomaterials, Society for Biomaterials, Reno, NV, p. 31, 2003.
62. Boland, T., Mironov, V., Gutowska, A., Roth, E. A., and Markwald, R. R., Cell and organ printing 2: fusion of cell aggregates in three-dimensional gels, *Anat. Rec.*, 272A(2), 497–502, 2003.
63. Vrielinck, L., Politis, C., Schepers, S., Pauwels, M., and Naert, I., Image-based planning and clinical validation of zygoma and pterygoid implant placement in patients with severe bone atrophy using customized drill guides. Preliminary results from a prospective clinical follow-up study, *Int. J. Oral Maxillofac Surg.*, 32(1), 7–14, 2003.
64. Chang, P. S., Parker, T. H., Patrick, C. W. Jr., and Miller, M. J., The accuracy of stereolithography in planning craniofacial bone replacement, *J. Craniofac. Surg.*, 14(2), 164–170, 2003.
65. Perez-Arjona, E., Dujovny, M., Park, H., Kulyanov, D., Galaniuk, A., Agner, C., Michael, D., and Diaz, F. G., Stereolithography: neurosurgical and medical implications, *Neurol. Res.*, 25(3), 227–236, 2003.
66. Tardieu, P. B., Vrielinck, L., and Escolano, E., Computer-assisted implant placement. A case report: treatment of the mandible, *Int. J. Oral Maxillofac. Implants*, 18(4), 599–604, 2003.
67. Sarment, D. P., Al-Shammari, K., and Kazor, C. E., Stereolithographic surgical templates for placement of dental implants in complex cases, *Int. J. Periodontics Restorative Dent.*, 23(3), 287–295, 2003.

11 Gas Foaming to Fabricate Polymer Scaffolds in Tissue Engineering

Yen-Chen Huang and David J. Mooney

CONTENTS

I. INTRODUCTION

Tissue engineering aims to regenerate or replace the structural and functional characteristics of lost or damaged tissues. Strategies to accomplish this goal generally include one or a combination of the following components depending on the desired outcome: biocompatible scaffolds, bio-inductive factors and transplanted cells. For either the delivery of growth factors or cells in tissue engineering, it is often desirable to confine them within a space that can be controllably localized, as opposed to systemic administration. In this respect, the use of scaffolds as delivery vehicles is critical. Thus, a major emphasis in biomaterials research for tissue engineering applications is the fabrication of three-dimensional scaffolds that mimic aspects of the nature extracellular matrix in an attempt to replace or restore the tissue of interest.

A variety of natural and synthetic materials have been utilized to fabricate tissue engineering scaffolds in order to meet the design criteria important to a specific application.[1] Naturally derived polymer scaffolds, such as those composed of collagen, have been widely employed due to the potential advantage of specific cell interactions.[2] However, these materials potentially present risks of antigenicity or recipient exposure to viruses and other infectious diseases due to their natural origins.[3] In addition, these materials are isolated from human or animal tissues, which have limited quantity and can exhibit batch-to-batch variations. Alternatively, synthetic biodegradable polyesters, for example, poly(D,L-lactic-coglycolic acid) (PLGA), have also been widely investigated and applied to fabricate tissue engineering scaffolds.[2] Synthetic materials typically exhibit controllable and consistent properties, can be reliably produced, and these polymers can display appropriate biocompatibility for many applications.[4]

PLGA (Figure 11.1) can be fabricated into porous structures using a variety of processes. The majority of processes for fabricating scaffolds involve the principle of phase transition, including

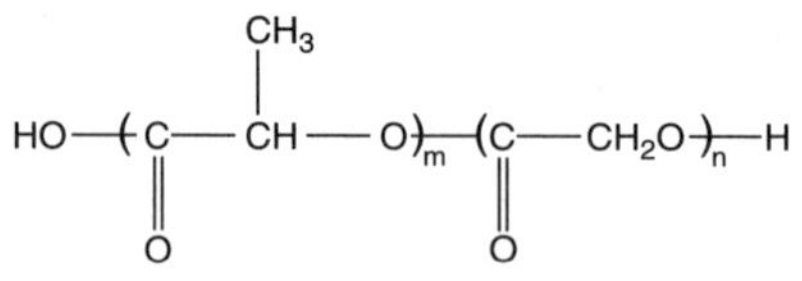

FIGURE 11.1 Chemical structure of poly(lactic-coglycolic acid).

phase separation,[5] solvent-casting particulate-leaching (SCPL),[6] supercritical fluid,[7,8] fiber extrusion and fabric foaming,[9] and gas foaming[10] processes. Many of these processes utilize organic solvents, whose residues may remain after processing and damage cells and nearby tissues, and denature any biologically active factors incorporated in the scaffolds. Fiber extrusion and fabric foaming require high temperatures, above the transition temperature of the polymer, so are often not suitable for amorphous polymers like PLGA. High temperatures also have the potential to inactivate any biologically active species present.[10] Gas foaming, in contrast, does not require organic solvents or high temperatures in the fabrication process. The gas foaming and the particulate leaching processes have been combined to allow for the fabrication of open pore matrices without the use of organic solvents.[10]

The choice of biomaterials for a scaffold and the method for processing and fabrication are critical for the success of tissue formation. This will allow the development of polymer scaffolds with desired mechanical integrity, retention of the activity of growth factors, and the control of release kinetics from the polymer that could potentially follow the natural course of tissue development. PLGA has been found to be an appropriate conductive material, and processing this polymer with the gas foaming method has proven to be feasible in various tissue engineering applications. This chapter will provide a review of PLGA polymers, the gas foaming process, and the delivery of bio-inductive growth factors and cells using highly porous PLGA three-dimensional biodegradable scaffolds fabricated using the gas foaming process. Specific examples in angiogenesis and bone regeneration will be discussed.

II. PLGA AS A TISSUE ENGINEERING MATERIAL

Polymer materials with appropriate properties for utilization as three-dimensional scaffolds are essential for tissue development in many applications. Synthetic polymers are attractive for forming tissue engineering scaffolds due to their reproducible, large-scale production, and readily manipulated range of mechanical properties and degradation rates. Polymers based on naturally occurring α-hydroxy acids have found extensive use in tissue engineering, and synthetic scaffolds are commonly produced from poly(glycolic acid) (PGA), poly(D,L-lactic acid) (PLA), and PLGA.[4,11] PGA, PLA, and PLGA have been approved for human use since the 1970's in the USA, when they were introduced as suture material. These polymers have also been applied to other areas of medicine, including drug delivery, orthopedic reconstruction, pulmonary surgery, and tissue engineering of skin, cartilage, nerve, and trabecular bone.[1,12]

PLGA scaffolds have an excellent ability to be customized to meet a particular absorption time requirement.[3] Degradation of PLGA *in vivo* proceeds by nonenzymatic simple hydrolysis into lactic and glycolic acid products,[13] which are eliminated by the respiratory system.[2] Additionally, neither PLGA nor its degradation products are typically associated with severe inflammation, local tissue damage, or local or systemic toxicity *in vivo*, demonstrating its biocompatibility. The rate of PLGA degradation is dependent on its crystallinty and hydrophobicity, which is affected by several factors: (1) the ratio of lactic to glycolic acid,[14] (2) the stereo-regularity of the monomer units, which affects chain packing, (3) the randomness of lactic and glycolic acid residues along the backbone, (4) molecular weight,[1] and (5) the inherent viscosity.[15] PLA is more hydrophobic than PGA, because of the extra methyl group, and the ester bond is less prone to hydrolysis due to steric hindrance. Therefore, degradation rates are typically faster with higher glycolide content.[1]

Numerous investigators have effectively used PLGA for tissue engineering purposes, and demonstrated its compatibility *in vitro* and *in vivo*.[10,13,15] The limitations associated with the use of PLGA include its acidic degradation products that can affect the local pH and its lack of functional groups for covalent modification.[2]

III. THE GAS FOAMING PROCESS

The gas foaming process is extremely useful for the fabrication of highly porous PLGA scaffolds for the delivery of growth factors and cells in tissue engineering applications. The gas foaming process involves exposing PLGA to high-pressure (e.g., 800 psi) carbon dioxide (CO_2) to saturate the polymer with gas. The CO_2 dissolves to an appreciable degree in the PLGA under these conditions. A thermodynamic instability is then created by decreasing the gas pressure to ambient pressure. The dissolved CO_2 becomes unstable and will phase separate from the PLGA. The CO_2 molecules cluster to minimize the free energy, and this causes pore nucleation. Growth of pores occurs as the CO_2 molecules from the surrounding polymer diffuse into the pore nuclei.[16] This introduction of pores to the PLGA causes a significant expansion of the polymer volume, and a concomitant decrease in polymer density.[10] The resulting structure, upon completion of the foaming process, is a highly porous (e.g., 95%) three-dimensional scaffold, with a predominantly closed pore structure due to the depletion of CO_2 in the areas between the nucleating pores.[16] Early experiments investigating the effects of processing parameters on the foaming process demonstrated that the use of CO_2 produced more porous matrices compared to those obtained using N_2 and He. This may be due to intermolecular interactions between CO_2 molecules and the carbonyl groups of PLGA.[17] Scaffolds fabricated with copolymers of PLGA also foamed much better than either the homopolymer PGA or PLA, likely due to their amorphous nature and resultant ability to dissolve CO_2. Polymers with a low molecular weight (low intrinsic viscosity) resulted in scaffolds with higher porosity than polymers of the same composition but a higher molecular weight.[18]

The porosity of scaffolds produced with this process can be significantly altered by incorporation of a porogen (pore forming agent). Scaffolds have been fabricated using sodium chloride or sucrose as the porogen. In this situation, as originally dispersed PLGA particles expand during the foaming process, they fuse together around the porogen to create a continuous polymer matrix surrounding the porogen, and also entrap any other molecules originally present in the mixture.[10] In this approach, the mixed polymer and porogen is exposed to the high-pressure environment until saturation with CO_2, which is typically completed in approximately 24 h.[15] Subsequent leaching of the scaffolds in either deionized water or phosphate buffered saline (PBS) following the foaming process removes the porogen and creates an interconnected pore structure throughout the sponge[10] (Figure 11.2). All processing steps are typically performed at ambient

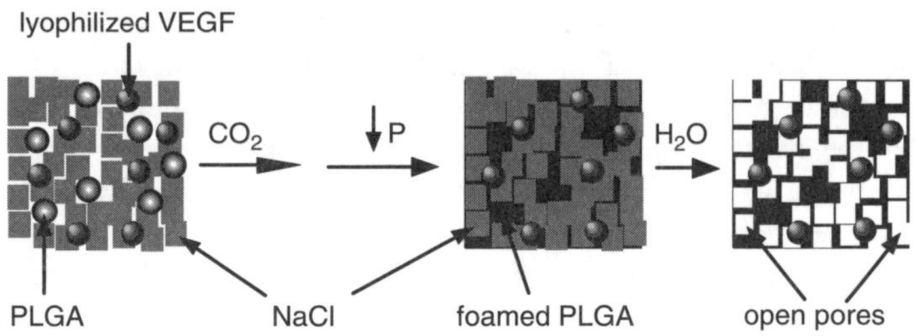

FIGURE 11.2 Schematic of the gas foaming process. Lyophilized growth factors (e.g., VEGF) mixed with PLGA particles and porogen (e.g., NaCl) are subject to high pressure CO_2 incubation, followed by rapid decrease of pressure to ambient. The open pore structure of the foamed scaffold can be created by leaching the scaffold in PBS, yielding a highly porous three-dimensional scaffold.

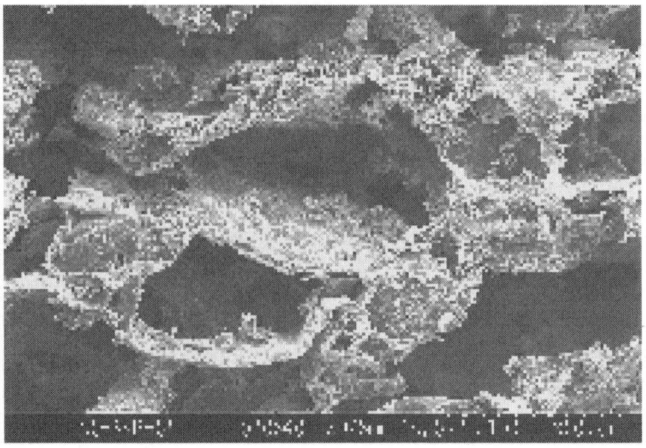

FIGURE 11.3 Scanning electron photomicrograph of cross-section of a 85:15 PLGA scaffold fabricated with the gas foaming method. The scaffold was prepared at 91 wt% (Sucrose: PLGA). Original magnification of photomicrograph was 150 ×.

temperature. The technical details of the gas foaming process protocol (e.g., processing parameters and characterization of fabricated scaffolds) have been described in detail previously.[19] Scanning electron microscopy (SEM) analysis of PLGA scaffolds fabricated with this method illustrates the highly porous nature of these scaffolds (Figure 11.3). The fabrication process has also been recently modified using a salt fusion method to further increase the interconnectivity of the pores.[20]

IV. PROTEIN DELIVERY FROM GAS-FOAMED PLGA SCAFFOLDS

A major motivation for the development of the gas foaming process was the anticipated benefits for the incorporation and release of biologically active macromolecules. Conventional delivery methods for these types of drugs, such as intravenous or bolus injection, demonstrated limited therapeutic effect especially in long-term treatment.[21–25] This is partially related to the wide-spread circulation of factors throughout the body and insufficient local concentration with these delivery approaches.[26,27] A major issue for protein delivery is the short half-lives of proteins *in vivo*, typically in the range of minutes to hours.[25,28] Although a larger dose of protein may be administered to offset degradation, the large dose may result in toxic effects and exert effects at non-targeted sites.

Polymeric systems with controlled release properties are attractive and can be successfully utilized to deliver defined doses of protein or growth factors at desired release rates directly to targeted sites. The release profiles of molecules from such carriers can be controlled by both diffusion and polymer degradation.[1,2] Diffusion is typically regulated by the amount of factor incorporated within the system, while the polymer degradation rate can be controlled by choice of the composition of the polymer. Delivery of proteins from polymer scaffolds allows for localized controlled and sustained delivery of growth factor, and thereby focuses the effect at the targeted tissues.[29] The gas foaming method has shown significant utility in fabricating controlled release system in applications such as angiogenesis and bone regeneration.

A. Angiogenesis

New blood vessel formation is critical in many areas of medicine, including the treatment of ischemic diseases that involve the blockage of existing vessels, and in tissue engineering. An important issue for the survival of transplanted cells in tissue engineering is the rapid formation of an efficient vasculature for nutrient supply and waste exchange. This is especially important for larger implants, as it is has been shown that cells more than 200 μm away from capillaries suffer from limited oxygen supply that can hamper or prevent tissue formation.[30] Strategies to promote new blood vessel networks will be essential in virtually all engineered tissues.[31] A number of growth factors have been identified that are involved in the formation of new capillary vessels via their branching from preexisting vessels, a process called angiogenesis. Many of these factors have been cloned, and recombinant variants are available.[27,32] A focus in this field has been on the use of vascular endothelial growth factor (VEGF), as it is known to be more specific to endothelial cells than many other factors.[33,34] Scaffolds for the controlled and localized release of VEGF have been fabricated using the gas foaming method, to assess their applicability in promoting angiogenesis. These scaffolds typically displayed a small initial burst release of VEGF, followed by a sustained release for several weeks while maintaining a high percentage of the bioactivity of the released factor.[18]

The incorporation efficiency of growth factors within the scaffold is an important parameter that is essential for effective localized growth factor delivery. In order to enhance the incorporation efficiency and to further control the release kinetics, growth factors were first encapsulated within alginate beads prior to the mixing with PLG polymer and salt for foaming. This approach was motivated by the finding that growth factors incorporated within alginate beads demonstrate favorable release kinetics, and the bioactivity of factors is not only maintained but actually enhanced.[35] The release kinetics of PLGA scaffolds encapsulating alginate-incorporated VEGF were very similar to the release obtained from alginate beads alone (not incorporated into PLGA matrices) initially, but were much slower after the initial period. This approach was recently modified by simply using alginate as an excipient.[29]

The molecular level mechanisms underlying angiogenesis involve a complex cascade of events, and distinct growth factors are involved at different stages of vascular development.[36,37] In general, these types of complex biological events are usually triggered by a number of biomolecules interacting simultaneously, sequentially, or a combination of both types of interactions. The development of polymers capable of delivering multiple growth factors with distinct kinetics would allow for better control and development of engineered tissues such as blood vessels, and potentially improve the treatment of diseases. In the process of angiogenesis, although VEGF is a well-known activator, its presence alone is not sufficient to form a mature vascular network. Platelet derived growth factor (PDGF) is another growth factor that promotes the maturation of newly formed blood vessels.[38–40] Due to the competing effects of VEGF and PDGF during angiogenesis, the time frame of availability for each factor needs to be tightly controlled to avoid antagonistic actions. VEGF and PDGF were delivered together to determine if gas-foamed scaffolds were capable of dual factor delivery and formation of mature blood vessels. PDGF was preencapsulated in PLG microspheres before combination with VEGF and particulate PLGA for the gas foaming process, in order to confer distinct release kinetics for each factor.[41] The sequential delivery of these two factors led to significant increase not only in blood vessel density, but also greatly increased the extent of blood vessel maturity (Figure 11.4). This approach may be useful when promoting regeneration of many other tissues, once the sequence of growth factors involved in the development of these tissues are delineated.

B. Bone Regeneration

PLGA scaffolds fabricated using the gas foaming process can provide an osteoconductive physical structure and osteoinductive environment by the delivery of osteogenic growth factors and cells for bone regeneration. The basic considerations for scaffolds to be used for bone tissue engineering include biocompatibility, degradability, mechanical integrity, osteoconductivity, and their ability to

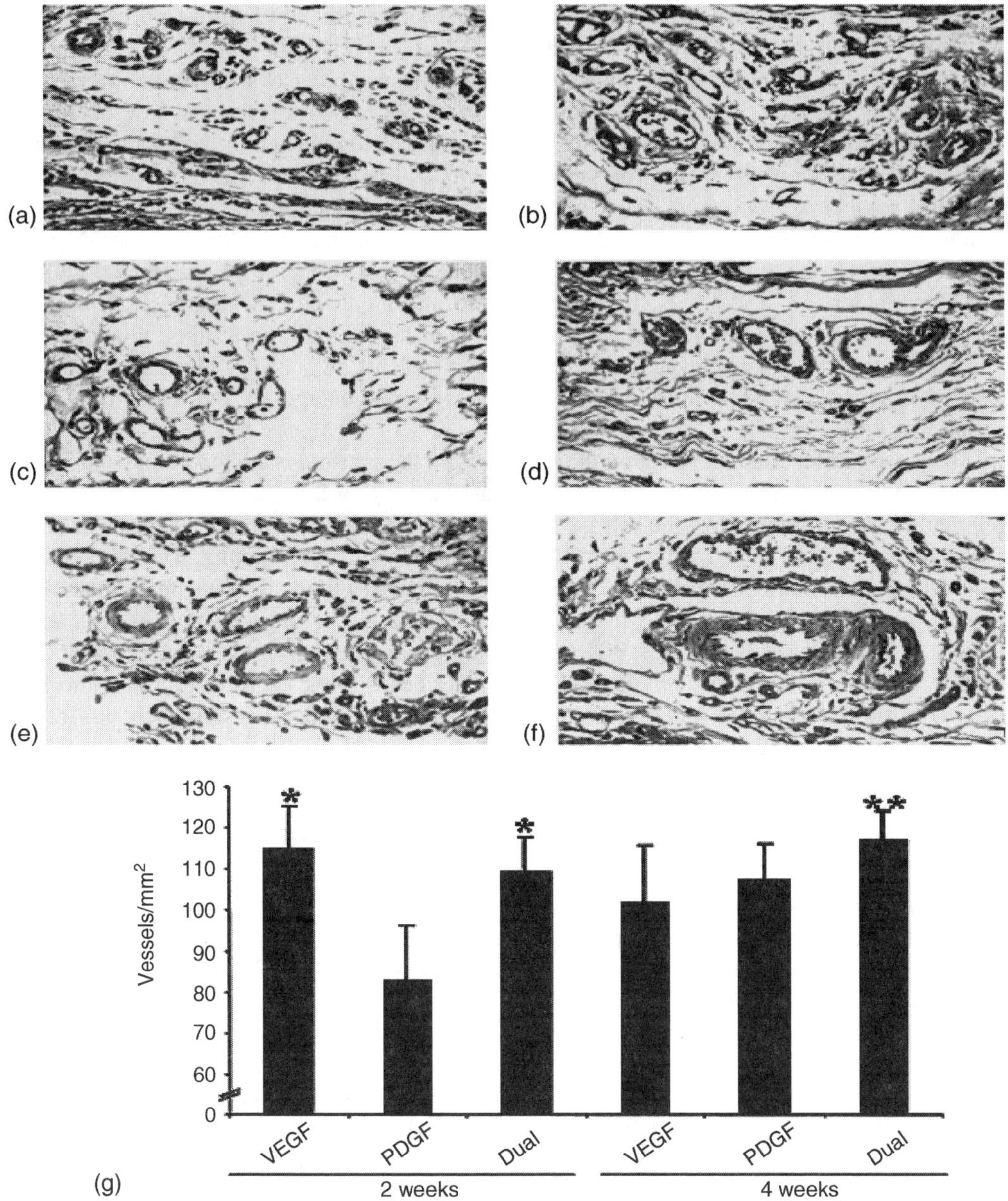

FIGURE 11.4 Sequential delivery of VEGF and PDGF led to significant increase not only in blood vessel density, but also greatly increased the extent of blood vessel maturity. Scaffolds that rapidly release VEGF followed by a slower release of PDGF were used in these experiments. Tissue sections from the following conditions were stained with hematoxylin and eosin: (1) scaffolds containing VEGF only, at 2 weeks (a) and 4 weeks (b); (2) scaffolds containing PDGF only, at 2 weeks (c) and 4 weeks (d); and (3) scaffolds releasing both VEGF and PDGF, at 2 weeks (e) and 4 weeks (f). Quantification of vascular density within tissue sections in each condition (g). * indicates statistical significance relative to blank ($p < .05$); ** indicates statistical significance relative to VEGF and PDGF ($p < .05$). (From Richardson, T. P., Peters, M. C., and Mooney, D. J., Processing of polymer scaffolds: gas foaming process, In *Methods of Tissue Engineering*, Atala, A. and Lanza, R., eds., Academic Press, New York, pp. 733–740, 2002. With permission.)

deliver inductive factors and cells. A number of studies demonstrated that orthopedic or dental implant materials coated with bonelike mineral (BLM) display increased osteoconductivity. However, little work has been performed to determine if the formation of a continuous bone-like mineral layer within three-dimensional, porous, degradable scaffolds would improve their ability to promote bone regeneration. To improve the osteoconductivity of gas-foamed biodegradable PLGA scaffolds, a novel approach to enhance the formation of a BLM film on PLGA was adapted to allow a continuous BLM layer to be formed on these scaffolds.[42] An advantage of this approach is that the nucleation and growth of a continuous carbonated apatite mineral layer on the interior surface of scaffolds can be achieved through a one-step, room temperature incubation process. The compressive moduli of SBF-coated polymer scaffolds also increased significantly as compared to uncoated control scaffolds[43] (Figure 11.5). The BLM film increased the amount of bone regeneration when these scaffolds were placed in critical sized cranial defects.[44]

In addition to regenerating the mineralized tissue component of bone, nerves, and blood vessels it must also be integrated within this dynamic tissue for full functionality. Vascularization is an essential component of de novo bone formation and a high degree of vascularity is present in natural bone. It has also been recently shown that bone formation can be enhanced in response to the presence of VEGF.[45] Biomineralized PLGA scaffolds incorporating VEGF were fabricated to determine if they could be used to enhance bone regeneration. Sustained release of bioactive VEGF was achieved in the presence of the bone-like mineral while maintaining the bioactivity of VEGF, and no compromise in total scaffold porosity was shown[46] (Figure 11.6). This strategy of combining a conductive substrate and inductive bioactive factors was shown to significantly enhance bone regeneration *in vivo*.[44] These scaffolds can also be seeded with preosteoblasts, and demonstrate bone growth *in vitro*.[47] The engineered bone tissue displays characteristics and developmental stages similar to that of native bone.

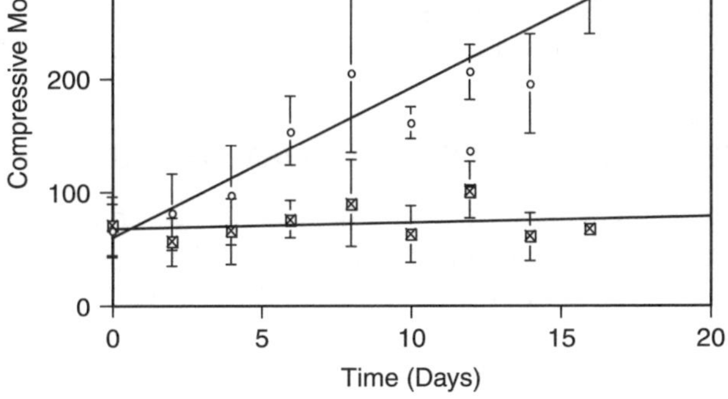

FIGURE 11.5 The compressive modulus vs. incubation time for mineralized and control scaffolds. Scaffolds were incubated in SBF (○), or in Tris buffer (pH 7.4) (⊠). The compressive moduli of SBF-coated polymer scaffolds increased significantly over time as compared to uncoated control scaffolds. (From Murphy, W. L., Kohn, D. H., and Mooney, D. J., Growth of continuous bonelike mineral within porous poly(lactide-co-glycolide) scaffolds in vitro, *J. Biomed. Mater. Res.*, 50(1), 50–58, 2000. With permission.)

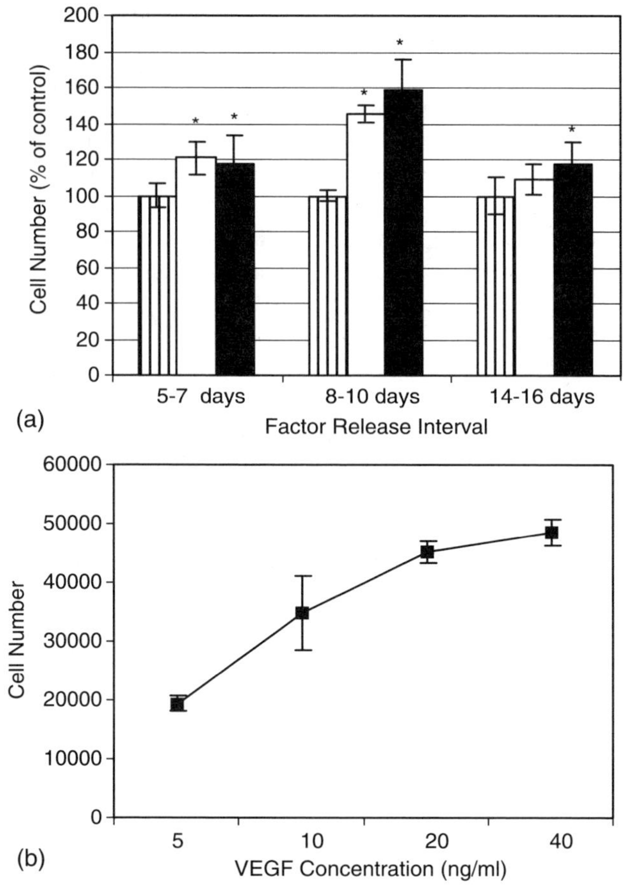

FIGURE 11.6 (a) The effect of VEGF release from mineralized (■) and nonmineralized (□) scaffolds on human dermal microvascular endothelial cells. Quantification of cell numbers for each release time period are displayed as percentage of the control value (scaffolds without VEGF, striped column) during that period. *'s indicate statistical significant relative to controls. (b) The mitogenic effect of VEGF on human dermal microvascular endothelial cells was demonstrated by the dose-response curve. Sustained release of bioactive VEGF was achieved in the presence of the bone-like mineral while maintaining the bioactivity of VEGF. (From Murphy, W. L. et al. Sustained release of vascular endothelial growth factor from mineralized poly(lactide-co-glycolide) scaffolds for tissue engineering, *Biomaterials*, 21(24), 2521–2527, 2000. With permission.)

V. GENE DELIVERY FROM GAS-FOAMED PLGA SCAFFOLDS

Gene therapy utilizing scaffolds fabricated with the gas-foamed process may allow for localized gene delivery in a sustained and controlled manner, in order to achieve therapeutic effects at specific tissue sites within a desired time frame. Gene delivery has been widely employed in the study of systemic DNA transfer to mediate genetic related physiological malfunction or disease. Recently, it has been applied to tissue engineering by either delivery of microspheres encapsulating DNA[48,49] or localized release of DNA from polymer scaffolds.[15,50] The advantages for a gene therapy approach to tissue regeneration, as opposed to delivering recombinant proteins, include that gene therapy promises to be cost effective and that DNA is inherently more stable than protein.[51] DNA has the potential to provide long-term delivery due to long-term expression by the targeted cells. In addition, since most proteins are highly modified posttranslationally, problems can develop with efficient cell receptor targeting when using recombinant proteins. TGF-β for example requires a very complex biosynthetic processing for proper synthesis and function. Normally, TGF-β is delivered as a latent extracellular complex that needs additional modification before targeting cell surface receptors.[52] As delivered genes are expressed by the target cells in the body, the protein production process can utilize the cells' own ability to perform these posttranslational modifications. The relative utility of gene therapy vs. protein delivery depends on many factors, including the type or duration of treatment of the target tissue and disease, and both approaches will likely find significant use in medicine.

The efficiency of gene transfer can be enhanced by delivering DNA with vector vehicles. There are currently two categories of vectors, viral and nonviral. The most common viral vector systems include retroviral, adeno-associated virus (AAV), and herpes simplex virus (HSV). Cationic liposomes, polylysine complexes, and other polymers comprise the nonviral vectors.[53] Although viral vectors generally display high efficiency, their application has been hampered by a variety of issues, including a limitation on the size of inserted DNA for certain viruses, uncontrollable expression of inserted genes, high immunogenicity that could result in patient death, and the potential for the insertion of the new gene into the host cell's genome in a manner that leads to mutagenesis.[54] Delivery of plasmid DNA has been pursued as an alternative approach to circumvent these issues. Plasmid DNA is economical and relatively simple to manufacture, and has a stable, flexible chemistry that lends itself for polymer-based drug delivery systems. Also, plasmid DNA has limited safety issues associated with toxicity and immunogenicity.[55] Direct DNA delivery through a polymeric vehicle may circumvent the issue of viral vector immunogenicity, and the use of plasmid DNA addresses the issue of adverse effects resulting from integration with host DNA.

Polymeric gene delivery vehicles have the potential to effectively and safely deliver plasmid to target cells. The gas foaming process has been utilized to deliver plasmid encoding for platelet-derived growth factors, and this resulted in increased blood vessel and granulation tissue formation[15] (Figure 11.7). The transfection accomplished from a single plasmid copy using nonviral vectors is transient, but the sustained release of the DNA in this approach provides high levels of expression over a time frame that is controllable when engineering a tissue. This approach enables control over temporal gene expression as the new tissues form, and undesired activity is presumably lost after the polymeric vehicle is exhausted of plasmid and tissue formation is complete.

Despite potential advantages of plasmid delivery approaches to tissue regeneration, plasmid DNA typically demonstrates low transfection efficiency *in vivo*. This is a result of its chemical, enzymatic, and colloidal instability, macrophage uptake, and nonspecific targeting.[56]

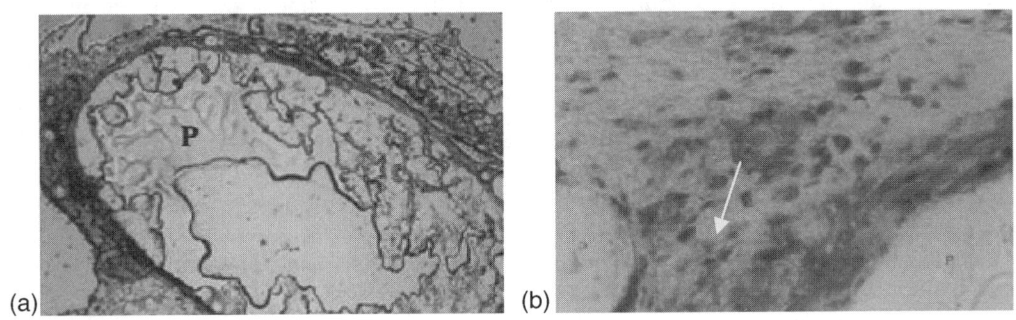

FIGURE 11.7 Tissue section of a gas-foamed scaffold incorporating plasmid DNA encoding for nt β-gal retrieved 4 weeks post implantation. Tissue sections were stained with X-gal (a). At higher magnification, cell with stained nuclei (arrow) demonstrates *in vivo* transfection of the nt β-gal plasmid (b). P indicates polymer, G indicates granulation tissue. (From Shea, L. D. et al. DNA delivery from polymer matrices for tissue engineering, *Nat. Biotechnol.*, 17(6), 551–554, 1999. With permission.)

Therefore, gene therapy success is largely dependent on the development of vectors that allow efficient delivery to targeted sites with a high expression level.[57] To enhance the efficiency of plasmid delivery, cationic polymer condensing agents such as poly(ethylenimine) (PEI), have been utilized as the delivery vehicle for the plasmid. It may be ideal to combine the localized and sustained DNA delivery properties of gas-foamed scaffolds with the transfection efficiency of PEI-condensed DNA. However, the presence of NaCl as the porogen in the gas foaming process may dissociate the DNA condensates, resulting in fragmented, inactive DNA.[58] Therefore, an alternative porogen, sucrose, was used during the fabrication process to bypass this issue. Gas-foamed scaffolds produced using sucrose as the porogen demonstrated high retention efficiency of condensed DNA, while at the same time maintaining the porosity and mechanical integrity of the scaffolds, as compared to using NaCl as the porogen. Release of uncondensed DNA from sponges was rapid, while release was significantly slower when condensed DNA was incorporated. The local delivery of condensed DNA from these scaffolds enhanced transfection efficiency compared to that of uncondensed DNA. *In vitro* transfection was observed only in sponges incorporating condensed DNA, while no transfection was observed for sponges incorporating uncondensed DNA.[59] This last study demonstrates the feasibility of incorporating freeze-dried PEI–DNA condensates within PLGA sponges to efficiently transfect cells. Recent studies demonstrated that gene expression may be observed up to 15 weeks *in vivo* using β-galactosidase as a marker gene (data not shown). Further studies are being performed to test the ability of this system to promote bone formation via the delivery of plasmids encoding for therapeutic proteins.

VI. CONCLUSIONS AND FUTURE DIRECTIONS

The gas foaming process has proven to be an extremely useful approach to fabricate highly porous three-dimensional scaffolds capable of sustained and sequential release of growth factors and DNA, and the transplantation of specific cell populations in localized regions. This process for fabricating scaffolds is advantageous in that the delivered biofactors maintain their intact bioactive state. This may allow much lower doses of these molecules to be utilized while still achieving therapeutic results, avoiding systemic exposure and the related side effects that are commonly observed with traditional approaches to deliver these factors. The utility of this system in tissue engineering has been demonstrated in various applications, including angiogenesis and bone regeneration. This delivery system can be designed to allow the delivery of combinations of various cell types or growth factors implicated in the regeneration of the desired engineered tissue, and can also be used as a model system to gain insight into the complex and dynamic process of tissue formation.

The utility of scaffolds fabricated with the gas foaming method can be further modified or enhanced by the addition of conductive coatings or cell responsive signals for specific applications. For example, bone regeneration may be greatly improved by implanting scaffolds combining the biomineralization process, and the delivery of plasmid DNA and growth factors implicated in the regeneration process. It may also be useful for the polymers used in scaffold fabrication to possess the capability of adapting or responding to the local environment so that the delivery could be controlled at that level. The interaction between cells and scaffold could be further controlled by the coupling or attachment of specific cell adhesion molecules on the polymer surface.[60,61] The temporal and spatial guidance of tissue growth using micro or nanopatterning could provide more precise manipulation for tissue development. Recently, it has also been shown that the extent of cell infiltration into the scaffolds fabricated by the gas foaming process can be regulated by the size of PLGA particle size.[62] This scaffold system may find great utility in the future in the delivery of stem cells, coupled with the appropriate delivery of multiple growth factors to stimulate the cells to differentiate to the desired cell types in the target tissue.

REFERENCES

1. Wong, W. H., and Mooney, D., Synthesis and properties of biodegradable polymers used as synthetic matrices for tissue engineering, In *Synthetic Biodegradable Polymer Scaffolds*, Atala, A., and Mooney, D., eds., Birkhauser, Boston, pp. 51–82, 1997.
2. Kim, B. S., and Mooney, D. J., Development of biocompatible synthetic extracellular matrices for tissue engineering, *Trends Biotechnol.*, 16(5), 224–230, 1998.
3. Burg, K. J., Porter, S., and Kellam, J. F., Biomaterial developments for bone tissue engineering, *Biomaterials*, 21(23), 2347–2359, 2000.
4. Huang, S. J., Biodegradable polymers, In *Polymers-Biomaterials and Medical Application*, Kroshwitz, I. J., ed., Wiley, New York, 1989.
5. Lo, H., Ponticiello, M. S., and Leong, K. W., Fabrication of controlled release biodegradable foams by phase seperation, *Tissue Eng.*, 1, 15–28, 1995.
6. Mikos, A. G., Thorsen, A. J., Czerwonka, L. A., Bao, Y., and Langer, R., Preparation and characterization of poly(L-lactic acid) foams, *Polymer*, 35, 1068–1077, 1994.
7. Hile, D. D., Amirpour, M. L., Akgerman, A., and Pishko, M. V., Active growth factor delivery from poly(D,L-lactide-co-glycolide) foams prepared in supercritical CO(2), *J Controlled Release*, 66(2–3), 177–185, 2000.
8. Howdle, S. M., Michael, S. W., Whitaker, M. J., Popov, V. K., Davies, M. C., Mandel, F. S., Wang, J. D., and Shakesheff, K. M., Supercritical fluid mixing: preparation of thermally sensitive polymer composites containing bioactive materials, *Chem. Commun.*, 109–110, 2001.
9. Freed, L. E., Vunjak-Novakovic, G., Biron, R. J., Eagles, D. B., Lesnoy, D. C., Barlow, S. K., and Langer, R., Biodegradable polymer scaffolds for tissue engineering, *Biotechnology (NY)*, 12(7), 689–693, 1994.
10. Harris, L. D., Kim, B. S., and Mooney, D. J., Open pore biodegradable matrices formed with gas foaming, *J. Biomed. Mater. Res.*, 42(3), 396–402, 1998.
11. Kim, B. S., and Mooney, D. J., Development of biocompatible synthetic extracellular matrices for tissue engineering, *Trends Biotechnol.*, 16(5), 224–230, 1998.
12. Thomson, R. C., Yaszemski, M. J., Powers, J. M., and Mikos, A. G., Fabrication of biodegradable polymer scaffolds to engineer trabecular bone, *J. Biomater. Sci. Polym. Ed.*, 7(1), 23–38, 1995.
13. Thomson, R. C., Mikos, A. G., Beahm, E., Lemon, J. C., Satterfield, W. C., Aufdemorte, T. B., and Miller, M. J., Guided tissue fabrication from periosteum using preformed biodegradable polymer scaffolds, *Biomaterials*, 20(21), 2007–2018, 1999.
14. Gilding, D. K., Biodegradable polymers, In *Biocompatibility of Clinical Implant Materials*, Williams, D. F., ed., CRC Press, Boca Raton, FL, 1981.
15. Shea, L. D., Smiley, E., Bonadio, J., and Mooney, D. J., DNA delivery from polymer matrices for tissue engineering, *Nat. Biotechnol.*, 17(6), 551–554, 1999.
16. Mooney, D. J., Baldwin, D. F., Suh, N. P., Vacanti, J. P., and Langer, R., Novel approach to fabricate porous sponges of poly(D,L-lactic-co-glycolic acid) without the use of organic solvents, *Biomaterials*, 17, 1417–1422, 1996.
17. Kazarian, S. G., Vincent, M. F., Bright, F. V., Liotta, C. L., and Eckert, C. A., Specific intermolecular interaction of carbon dioxide with polymers, *J. Am. Chem. Soc.*, 118, 1729–1736, 1996.
18. Sheridan, M. H., Shea, L. D., Peters, M. C., and Mooney, D. J., Bioadsorbable polymer scaffolds for tissue engineering capable of sustained growth factor delivery, *J. Controlled Release*, 64(1–3), 91–102, 2000.
19. Richardson, T. P., Peters, M. C., and Mooney, D. J., Processing of polymer scaffolds: gas foaming process, In *Methods of Tissue Engineering*, Atala, A., and Lanza, R., eds., Academic Press, New York, pp. 733–740, 2002.
20. Murphy, W. L., Dennis, R. G., Kileny, J. L., and Mooney, D. J., Salt fusion: an approach to improve pore interconnectivity within tissue engineering scaffolds, *Tissue Eng.*, 8(1), 43–52, 2002.
21. Khan, T. A., Sellke, F. W., and Laham, R. J., *Ther. Angiogenesis Coron. Artery Dis.*, 4(1), 65–74, 2002.
22. Laham, R., Angiogenesis (clinical trials), *Can J Cardiol*, 17(Suppl A), 29A–32A, 2001.
23. Post, M. J., Laham, R., Selke, F. W., and Simons, M., Therapeutic angiogenesis in cardiology using protein formulations, *Cardiovasc. Res.*, 49(3), 522–531, 2001.

24. Freedman, S. B., and Isner, J. M., Therapeutic angiogenesis for coronary artery disease, *Ann. Intern. Med.*, 136(1), 54–71, 2002.
25. Edelman, E. R., Nugent, M. A., and Karnovsky, M. J., Perivascular and intravenous administration of basic fibroblast growth factor: vascular and solid organ deposition, *Proc. Natl Acad. Sci. USA*, 90(4), 1513–1517, 1993.
26. Lazarous, D. F., Shou, M., Scheinowitz, M., Hodge, E., Thirumurti, V., Kitsiou, A. N., Stiber, J. A., Lobo, A. D., Hunsberger, S., Guetta, E., Epstein, S. E., and Unger, E. F., Comparative effects of basic fibroblast growth factor and vascular endothelial growth factor on coronary collateral development and the arterial response to injury, *Circulation*, 94(5), 1074–1082, 1996.
27. Yancopoulos, G. D., Davis, S., Gale, N. W., Rudge, J. S., Wiegand, S. J., and Holash, J., Vascular-specific growth factors and blood vessel formation, *Nature*, 407(6801), 242–248, 2000.
28. Bowen-Pope, D. F. Malpass, T. W., Foster, D. M., and Ross, R., Platelet-derived growth factor in vivo: levels, activity, and rate of clearance, *Blood*, 64(2), 458–469, 1984.
29. Peters, M. C., Polverini, P. J., and Mooney, D. J., Engineering vascular networks in porous polymer matrices, *J. Biomed. Mater. Res.*, 60(4), 668–678, 2002.
30. Colton, C. K., Implantable biohybrid artificial organs, *Cell Transplant.*, 4(4), 415–436, 1995.
31. Hirschi, K. K., Skalak, T. C., Peirce, S. M., and Little, C. D., Vascular assembly in natural and engineered tissues, *Ann. N. Y. Acad. Sci.*, 961, 223–242, 2002.
32. Conway, E. M., Collen, D., and Carmeliet, P., Molecular mechanisms of blood vessel growth, *Cardiovasc. Res.*, 49(3), 507–521, 2001.
33. Polverini, P. J., The pathophysiology of angiogenesis, *Crit. Rev. Oral. Biol. Med.*, 6(3), 230–247, 1995.
34. Nor, J. E., Christensen, J., Mooney, D. J., and Polverini, P. J., Vascular endothelial growth factor (VEGF)-mediated angiogenesis is associated with enhanced endothelial cell survival and induction of Bcl-2 expression, *Am. J. Pathol.*, 154(2), 375–384, 1999.
35. Peters, M. C., Isenberg, B. C., Rowley, J. A., and Mooney, D. J., Release from alginate enhances the biological activity of vascular endothelial growth factor, *J. Biomater. Sci.-Polym. Ed.*, 9(12), 1267–1278, 1998.
36. Benjamin, L. E., Hemo, I., and Keshet, E., A plasticity window for blood vessel remodelling is defined by pericyte coverage of the preformed endothelial network and is regulated by PDGF-B and VEGF, *Development*, 125(9), 1591–1598, 1998.
37. Darland, D. C., and D'Amore, P. A., Blood vessel maturation: vascular development comes of age, *J. Clin. Invest.*, 103(2), 157–158, 1999.
38. Cao, R., Brakenhielm, E., Li, X., Pietras, K., Widenfalk, J., Ostman, A., Eriksson, U., and Cao, Y., Angiogenesis stimulated by PDGF-CC, a novel member in the PDGF family, involves activation of PDGFR-alphaalpha and -alphabeta receptors, *Faseb J*, 16(12), 1575–1583, 2002.
39. Dunn, I. F., Heese, O., and Black, P. M., Growth factors in glioma angiogenesis: FGFs, PDGF, EGF, and TGFs, *J. Neurooncol.*, 50(1–2), 121–137, 2000.
40. Thommen, R., Humar, R., Misevic, G., Pepper, M. S., Hahn, A. W., John, M., and Battegay, E. J., PDGF-BB increases endothelial migration on cord movements during angiogenesis in vitro, *J. Cell. Biochem.*, 64(3), 403–413, 1997.
41. Richardson, T. P., Peters, M. C., Ennett, A. B., and Mooney, D. J., Polymeric system for dual growth factor delivery, *Nat. Biotechnol.*, 19(11), 1029–1034, 2001.
42. Murphy, W. L., and Mooney, D. J., Bioinspired growth of crystalline carbonate apatite on biodegradable polymer substrata, *J. Am. Chem. Soc.*, 124(9), 1910–1917, 2002.
43. Murphy, W. L., Kohn, D. H., and Mooney, D. J., Growth of continuous bonelike mineral within porous poly(lactide-co-glycolide) scaffolds in vitro, *J. Biomed. Mater. Res.*, 50(1), 50–58, 2000.
44. Murphy, W. L., Simmons, C. A., Kaigler, D., and Mooney, D. J., Bone regeneration via a mineral substrate and induced angiogenesis, *J. Dent. Res.*, 83(3), 204–210, 2004.
45. Street, J., Bao, M., deGuzman, L., Bunting, S., Peale, F. V., Jr., Ferrara, N., Steinmetz, H., Hoeffel, J., Cleland, J. L., Daugherty, A., van Bruggen, N., Redmond, H. P., Carano, R. A., and Filvaroff, E. H., Vascular endothelial growth factor stimulates bone repair by promoting angiogenesis and bone turnover, *Proc. Natl Acad. Sci. USA*, 99(15), 9656–9661, 2002.

46. Murphy, W. L., Peters, M. C., Kohn, D. H., and Mooney, D. J., Sustained release of vascular endothelial growth factor from mineralized poly(lactide-co-glycolide) scaffolds for tissue engineering, *Biomaterials*, 21(24), 2521–2527, 2000.
47. Shea, L. D., Wang, D., Franceschi, R. T., and Mooney, D. J., Engineered bone development from a pre-osteoblast cell line on three-dimensional scaffolds, *Tissue Eng.*, 6(6), 605–617, 2000.
48. Cohen, H. et al., Sustained delivery and expression of DNA encapsulated in polymeric nanoparticles, *Gene Ther.*, 7(22), 1896–1905, 2000.
49. Hedley, M. L., Curley, J., and Urban, R., Microspheres containing plasmid-encoded antigens elicit cytotoxic T-cell responses, *Nat. Med.*, 4(3), 365–368, 1998.
50. Bonadio, J., Smiley, E., Patil, P., and Goldstein, S., Localized, direct plasmid gene delivery in vivo: prolonged therapy results in reproducible tissue regeneration, *Nat. Med.*, 5(7), 753–759, 1999.
51. Fang, J. et al., Stimulation of new bone formation by direct transfer of osteogenic plasmid genes, *Proc. Natl Acad. Sci. USA*, 93(12), 5753–5758, 1996.
52. Yin, W., Smiley, E., Germiller, J., Mecham, R. P., Florer, J. B., Wenstrup, R. J., and Bonadio, J., Isolation of a novel latent transforming growth factor-beta binding protein gene (LTBP-3), *J. Biol. Chem.*, 270(17), 10147–10160, 1995.
53. Romano, G., Pacilio, C., and Giordano, A., Gene transfer technology in therapy: current applications and future goals, *Stem Cells*, 17, 191–202, 1999.
54. Bonadio, J., Goldstein, S. A., and Levy, R. J., Gene therapy for tissue repair and regeneration, *Adv. Drug Deliv. Rev.*, 33, 53–69, 1998.
55. Winn, S. R., Hu, Y., Sfeir, C., and Hollinger, J. O., Gene therapy approaches for modulating bone regeneration, *Adv. Drug Deliv. Rev.*, 42(1–2), 121–138, 2000.
56. Mahato, R. I., Non-viral peptide-based approaches to gene delivery, *J. Drug Target*, 7(4), 249–268, 1999.
57. Li, S., and Huang, L., Nonviral gene therapy: promises and challenges, *Gene Ther.*, 7, 31–34, 2000.
58. Adami, R. C., Collard, W. T., Gupta, S. A., Kwok, K. Y., Bonadio, J., and Rice, K. G., Stability of peptide-condensed plasmid DNA formulations, *J. Pharm. Sci.*, 87(6), 678–683, 1998.
59. Huang, Y. C., Connell, M., Park, Y., Mooney, D. J., and Rice, K. G., Fabrication and in vitro testing of polymeric delivery system for condensed DNA, *J. Biomed. Mater. Res. A*, 67(4), 1384–1392, 2003.
60. Park, A., Wu, B., and Griffith, L. G., Integration of surface modification and 3D fabrication techniques to prepare patterned poly(L-lactide) substrates allowing regionally selective cell adhesion, *J Biomater. Sci. Polym. Ed.*, 9(2), 89–110, 1998.
61. Eid, K. et al., Effect of RGD coating on osteocompatibility of PLGA-polymer disks in a rat tibial wound, *J. Biomed. Mater. Res.*, 57(2), 224–231, 2001.
62. Riddle, K. and Mooney, D. J., Role of PLG Particle Size on Gas Foamed Scaffolds. Unpublished Results.

12 Injectable Systems for Cartilage Tissue Engineering

Christopher G. Williams and Jennifer H. Elisseeff

CONTENTS

I. INTRODUCTION

The goal of this chapter is to review prior work on the development of injectable hydrogel tissue engineering systems with a focus on those systems used in conjunction with cells to facilitate the repair or creation of cartilage. A detailed review of the physical properties of hydrogels is not the aim of the chapter; however, important considerations will be briefly discussed. More expansive reviews on the chemistry and physical characteristics of hydrogels already exist in the literature (for review see Refs. 1,2). The authors will also describe their current research in the area of injectable tissue engineering systems utilizing photosensitive polymer solutions to create three-dimensional, cell-laden hydrogel constructs.

II. CLINICAL NEED FOR INJECTABLE TISSUE ENGINEERING SYSTEMS

Current medical and surgical therapies can produce remarkable results for many diseases and pathologies related to cartilage and bone tissues. The treatments for complex fractures, end-stage arthritic joint disease, limb deformities, and craniofacial pathologies have all recently seen dramatic improvements. Patients with disorders of the musculoskeletal and craniofacial systems have more therapeutic options today than ever before, however, there are still vast improvements in technologies and therapies that need to be realized, especially in the areas of repairing articular cartilage and severe bone defects. According to one study there are approximately 250,000 total knee and 250,000 total hips prostheses being placed per year for the treatment of end-stage degenerative joint diseases. Some of the current therapies used to correct articular cartilage defects in joints include penetration of subchondral bone,[3–5] mosaiplasty or autograph osteochondral transplant,[6,7] osteochondral allograft placement,[8] partial and total joint arthroplasty, and recently, autologous chondrocyte transplantation[9,10] (for an extensive review see[11]). The need for cartilage in the craniofacial skeleton is also high. Cartilage tissue is often harvested from distant sites and used in nasal and ear reconstructions in plastic and reconstructive surgery. The temporomandibular joint of the jaw bone is a complex articulation that can be involved in numerous pathologies leading to cartilage failure and wearing that require invasive surgical correction.[12]

While providing some benefit, current surgical procedures have important shortcomings such as suboptimal long-term outcome, implantation and long-term presence of alloplastic material, increased donor site and joint morbidity, invasive surgical approach, risk of infection, and structural failure. Allografting and autografting strategies have other shortcomings, such as the possibility of disease transmission, rejection of allograft tissue, insufficient autologous resources, contour irregularities, major histoincompatibility, graft-vs.-host disease, and potential need for immunosuppression.[11–14]

III. TISSUE ENGINEERING STRATEGIES

The emerging field of tissue engineering has widened the search for better and less invasive treatments for many disease processes. Tissue engineering is a broad, continually evolving, multidisciplinary field that holds great possibilities for future medical therapies in almost all areas of medicine to restore, maintain, and enhance tissues.[15,16] In the broadest sense, tissue engineering aims to regenerate tissues using cells, biomaterial scaffolds, and biological signals.[16] Tissue engineering is a $600 million per year industry with approximately 70 companies attempting to develop new products. However, as of 2001, the Food and Drug Administration had approved fewer than five engineered tissues for clinical use.[15]

When one considers the field of tissue engineering, there are three classical components one usually bears in mind when designing an application: the cell population(s), the matrix or scaffold, and bioactive factors. There are an infinite number of combinations using one, two, or all three components, depending on the ultimate application and the theoretical and practical considerations of the investigators. Certain applications may be amenable to using cells alone such as autologous chondrocyte transplantation.[9,10] Other applications may require only a scaffold or a bioactive factor, such as placing devitalized bone graphs and osteoinductive scaffolds or injecting bone morphogenic proteins in fracture sites to improve healing.[12] Still other applications will best be solved using a combination of all three, providing a critical mass of living cells to repair or enhance the local tissue, a physical scaffold to protect and localize the cells, and biological signals to direct the cells in their anabolic pathways. This chapter focuses on injectable cell-laden systems for cartilage tissue engineering.

IV. THE DRIVE FOR INJECTABLE SYSTEMS

A major focus in today's medical environment is to limit the trauma created by physicians while treating patients' illnesses. From large incisions to laparoscopy and from 2-week hospital stays to outpatient surgery, the goal is the same — to reduce the cost and invasiveness of medical care. The same is true in tissue engineering efforts. Early tissue engineering attempts often used devices and polymer scaffolds that had to be surgically implanted, such as preformed meshes or rigid scaffolds.[17–19] Surgical implantation implies incisions, anesthesia, and the complications of the two. Ultimately, some applications may require this invasive approach. However, this chapter focuses on new techniques and technologies that will allow tissue engineers and surgeons to accomplish many goals in skeletal tissue engineering in a less invasive manner, perhaps removing the need for surgical implantation in many applications.

One attractive method to reduce the invasiveness of a tissue engineering system is to make it injectable, and thus eliminate the need for, or reduce the size of, any incisions required to implant the material. For a system to be injectable, it has to be able to flow through a small-bore needle or trocar. Therefore, by necessity, most of these systems exist, at least preliminarily, in a liquid state. Since this chapter emphasizes systems that can be used in conjunction with the living cells, these systems must be cytocompatible to be able to encapsulate cells. To design a system that encapsulates cells directly, polymer scaffolds have to be water soluble, able to produce cell–polymer suspensions, and able to undergo physical or chemical changes inducing polymerization under physiologic conditions. Control of the polymerization reaction is also desirable. The biocompatibility of the polymer and its breakdown products with the host tissue is also an important factor in designing these systems.[20]

An injectable scaffold offers many advantages over the preformed scaffolds used in many tissue engineering approaches. An injectable or flowable material can fill any defect shape and has the potential to polymerize *in situ* once it has been injected. It may incorporate various therapeutic agents (e.g., growth factors) by simple mixing, and chemistries are readily available to incorporate bioactive and biodegradable linkages into these systems. Injectable systems can predictably and uniformly encapsulate cells, delivering them into a specific therapeutic site.[2] These liquid systems are designed without the residual toxic solvents that may be present in preformed solid scaffolds. Finally, these systems do not require invasive surgical procedures for placement. Instead of large, debilitating incisions and postoperative pain, many scaffold systems and cellular therapies for tissue repair and augmentation will be injectable through needles or small ports similar to those used in joint arthroscopy (or smaller), limiting unwanted scarring and hastening postprocedural recovery.[2]

Hydrogels are a class of polymer scaffolds which meet the criteria for injectable systems that could be used in tissue engineering applications.[21] Hydrogel scaffolds often provide the researcher with a powerful ability to deliver synthetic or natural scaffolds and cells by injection because of the liquid state of many of these systems. A gel is defined as a three-dimensional network swollen by a solvent. The solvent, water in this case, is usually the major component of the gel system.[2,22] Hydrogels are formed by cross-linking water-soluble polymers to form insoluble, hydrophilic polymer networks. The networks are capable of absorbing water from a small fraction of their dry weight up to thousands of times their dry weight in water.[1] (For more complete reviews see Refs. 21,23,24).

The aqueous environment in hydrogels is advantageous because it can protect cells, fragile drugs, and factors such as proteins and nucleic acids. Swollen hydrogels facilitate diffusion of nutrients and waste into and out of the gels. Many are biocompatible.[1] Another "significant advantage of hydrogels as matrices for tissue engineering vs. more hydrophobic alternatives such as PLGA is the ease in which one may covalently incorporate cell membrane receptor peptide ligands (and bioactive factors) in order to stimulate adhesion, spreading, and growth of cells within the hydrogel matrix."[1,2]

A significant potential disadvantage of hydrogel systems, particularly those used to encapsulate cells, is that they typically lack significant mechanical strength which can pose difficulties in handling or limit the applications if they are to bear a high degree of mechanical loading. Certain hydrogels may be difficult to sterilize.[1]

Hydrogels can be designed with an almost limitless combination of characteristics. Hydrogels can be chemically stable, partially degradable or fully degradable depending on the mechanism by which the polymer network cross-links are formed and the monomers used. Also, they can be designed to be "reversible" or "irreversible," depending on the physical and chemical forces that induce polymerization. "Reversible" or "physical gels" are polymer networks that are held together by weaker forces such as molecular entanglement, hydrogen bonding, or hydrophilic–hydrophobic interactions. These reversible gels can often be changed from liquid to solid simply by changing the environmental conditions, such as pH, ionic concentrations, or temperature.[2] "Irreversible" or "chemical hydrogels" are typically formed by more stable covalent bonds between monomers and, once polymerized they remain polymerized.[1]

The biologic uses of hydrogels have been of interest to researchers since the 1960s with a report of HEMA-based hydrogels.[25] Important early work by Lim and Sun demonstrated the promises of cell-laden hydrogels by successfully encapsulating pancreatic islet cells in calcium alginate microspheres, a classic "reversible" hydrogel system.[26] The islet cells not only survived encapsulation, but they maintained their endocrine responsiveness and functions. Interest continued to mount as more and more applications were reported.[21,27–31]

V. *IN SITU* POLYMERIZATION

Recently, injectable approaches to tissue engineering using *in situ* gel-forming systems have become more prominent. *In situ* gelling or polymerizing systems are particularly attractive because they enable the investigator to control the polymerization of the hydrogel system. That is, they are usually in a liquid or injectable state prior to use. However, when placed in the body or defect these systems are meant to treat, they polymerize in a controllable fashion forming its final three-dimensional structure. *In situ* polymerization strategies are diverse and limited only to the imaginations of researchers. Some of the mechanisms that have been utilized in *in situ* polymerizing formulations take advantage of environmental changes like temperature, pH, or ionic concentrations, similar to those used to make "reversible hydrogels." Other examples, like hyaluronic acid solutions, are called high-viscosity, shear-thinning gels because they can be injected through small bore needles, but once the injection shearing force has been removed, a thick gel is formed *in situ*.[2] Other mechanisms take advantage of reactions driven by specific controllable initiators such as photopolymerization or enzymatic cleavage of a propeptide (like fibrinogen). The rest of the chapter will be dedicated in reviewing recent works using many of these systems for cartilage tissue engineering.

VI. THERMORESPONSIVE HYDROGELS

Thermoresponsive gels are one class of *in situ* polymerizing hydrogels that have been used successfully. These gels polymerize as the temperature of the solutions increases or decreases. For clinical use in the body, the temperature at which they polymerize, the "lower critical solution temperature," is often around body temperature and triggers the gel to polymerize after it has been delivered into the patient.[32] Agarose is an example of a thermoresponsive polymer that gels as its temperature decreases. Conversely, Pluronic is a copolymer of polyethylene oxide (PEO) and polypropylene oxide (PPO) that exists in a liquid state when cold and polymerizes into a thick gel at warmer, more physiologic temperatures. Both have been used to encapsulate cells.

Agarose is a polysaccharide containing L- and D-galactose residues and is derived from certain seaweeds.[11] This hydrogel has a significant history in tissue engineering, dating from the 1980s, due to its ability to encapsulate cells and deserves mention despite the fact that this system has not been used as an injectable scaffold. Benya and Shaffer[28] performed important early work with agarose gels and chondrocytes. They showed that it was possible to expand chondrocytes *in vitro*; however, the cells de-differentiated, undergoing dramatic phenotypic changes when cultivated in monolayer. Importantly, they also demonstrated that the de-differentiated chondrocytes redifferentiated when cultured after encapsulation in agarose gels. The chondrocytes also produced an extracellular matrix of collagen type II and glycosaminoglycans in this three dimensional hydrogel culture, hinting perhaps at a therapeutic strategy for cartilage repair.[28]

Despite its frequent use in *in vitro* studies on chondrocyte behavior, few *in vivo studies* using agarose have been performed because of its poor biodegradability and its induction of extensive foreign body cell reactions *in vivo*.[11] No studies were found that injected this hydrogel with cells *in vivo*. However, Rahfoth and colleagues[33] attempted to repair articular cartilage defects in rabbits by using allogenic chondrocytes in surgically implanted agarose gels. These gels were precultured *in vitro* for 2 weeks prior to implantation. They showed that there was some inhibitory response to healing when agarose gels were implanted alone, which was partially overcome by adding the allogenic chondrocytes. Other limitations of agarose preparations are that they often have to be boiled at high temperatures to initiate the polymerization which may damage cells, and there is minimal to no control over the polymerization of the polymer once the temperature begins to drop or after it is injected into the body.

Pluronic has been used in various tissue engineering applications for making cartilaginous tissues. Cao et al.[34] used Pluronic F-127 to encapsulate and inject autologous chondrocytes in a porcine model. They investigated its potential application for nipple reconstruction, which is a common difficulty of breast reconstruction after mastectomy for breast cancer extirpations.[34] They demonstrated that after 10 weeks of *in vivo* culture in swine, the sites injected with Pluronic F-127 hydrogel and chondrocytes closely resembled the human female nipple-areola complex with respect to size, shape, and texture. Histology revealed nodules characteristic of elastic cartilage. Similarly, Saim et al.[35] also used Pluronic F-127 to encapsulate and inject autologous chondrocytes in a swine model but were interested in ear reconstruction. The authors revealed favorable results with elastic cartilage being produced in the subcutaneous injection sites after 10 weeks.[35]

PEO/PPO copolymers have been used for articular cartilage applications using various techniques.[36–38] Weng et al.[38] "painted" pluronic/chondrocyte suspensions onto surfaces of PGA/PLA scaffolds previously seeded with osteoblast/alginate suspensions to create osteochondral complexes resembling a temporomandibular joint. Liu et al.[36] used PEO/PPO/chondrocytes suspensions to seed PGA scaffolds in 8 mm articular cartilage defects in pigs and showed healing after 24 weeks with the hyaline cartilage with biomechanical and biological contents improving over time. Neither of these applications were performed in a minimally invasive manner, but the injectable nature of these copolymers provides engineers with the potential to modify procedures and perform them in a less invasive manner.

Other injectable thermoresponsive polymers used in cartilage tissue engineering are poly(*n*-isopropylacrylamide)-grafted gelatin[39] and poly(DL-lactic acid-coglycolic acid)/(poly(ethylene glycol).[40] The initial *in vitro* work performed by Ibusuki et al.[39] demonstrated chondrocyte survivability and extracelluar matrix production over a 12 week period. However, further work is needed to demonstrate a minimally invasive application of this new polymer for cartilage tissue repair. Jeong et al.[40] demonstrated improved repair of cartilage defects after injecting the defect with chondrocytes encapsulated in their thermoresponsive polymer.

VII. IONIC GELS

Another strategy used to create injectable *in situ* polymerizing polymers takes advantage of ionic interactions in the polymer solutions. Alginate is a polysaccharide also derived from seaweed and is composed of mannuronic and guluronic acid. It is a classic example of an anionic polymer that binds divalent cations (e.g., Ca^{2+}) to form a water insoluble polymer network. The rate of polymerization of alginates can be partially controlled by changing the type and concentration of cation and the concentration of alginate. When calcium chloride is used, alginates tend to polymerize very quickly. However, to develop a slower gelling solution Paige et al.[41] mixed alginate solutions with calcium sulfate which made it injectable within a clinically acceptable time frame.

There have been a large number of *in vitro* studies using alginate gels for chondrocyte culture.[27,29,30,42–45] Some of the earliest studies have been cited; however, it is not within the scope of this chapter to review them all. Instead, we will briefly discuss a selection of studies using *in vivo* models with alginate and cells for various cartilage tissue engineering applications. Again, some applications described were not injected; however, several authors have shown that with simple modifications that the alginate system is indeed injectable.[41]

Paige et al.[46] demonstrated that bovine chondrocytes encapsulated in 1.5% alginate solutions and then surgically implanted in the subcutaneous space of the nude mouse survived and produced matrix for up to 12 weeks. The same authors modified this system and were able to produce a truly injectable hydrogel/cell system.[41] Paige et al.[41] injected their "slowly polymerizing" alginate–chondrocyte solutions under the panniculus carnosus of nude mice and demonstrated that the chondrocytes survived and produced an extracellular matrix characteristic of cartilage, opening the door for a less invasive strategy for cartilage repair or augmentation.

Importantly, injectable alginate and chondrocytes have been used successfully in human patients.[47] Tissue engineered cartilaginous tissue was used to treat women with urinary incontinence to bulk up their incontinent external bladder sphincter. The extra tissue around the sphincter reduced the symptoms of urinary incontinence in 28 out of 32 women.[47]

An important study by Fragonas et al.[48] demonstrated that alginate gels containing chondrocytes could be polymerized *in situ* into full thickness osteochondral defects in rabbits. Another notable feature of this study was that allogenic, not autologous, cells were also used to repair the cartilage defects. The defects were examined after 1, 2, 4, and 6 months. These authors concluded their article with the statement: " it appears to be a relevant fact that the regenerative process induced by the implanted chondrocytes culminates in the formation of a tissue which presents the characteristics of a normal articular cartilage with a regular distribution of cells of various sizes and shapes in the different layers and a matrix which stains regularly for glycosaminoglycans."[48]

Another notable paper was written by Diduch et al.[49] who used bone marrow derived stem cells (instead of chondrocytes) in alginate beads to fill full-thickness osteochondral defects in rabbits. There are other *in vivo* studies that used alginate gels in conjunction with chondrocytes[50–52] as well as periosteal cells[53] for the creation of tissue engineered cartilaginous tissues.

Alginate gels can also be chemically modified to improve cell adhesion and cellular differentiation.[54,55] One study showed enhanced cellular proliferation and osteogenesis both *in vitro* and *in vivo* when an RGD-based amino acid sequence was covalently grafted into the alginate hydrogel.[54] A second study investigated the use of RGD-based polypetides and its role in myogenic phenotype in alginate cultures.[55]

VIII. FIBRIN GLUE

Fibrin glue is another hydrogel system that has been used for a wide rage of applications in tissue engineering. Fibrin is a naturally occurring peptide that is the product of propeptide fibrinogen after it has been enzymatically cleaved by thrombin. Its natural physiologic role is important in the formation of blood clots after damage to blood vessels as it traps platelets and red blood cells to form a thrombus. Because of its importance in wound repair, it has been studied as a scaffold for cartilage repair. As early as 1983, Albrecht[56] used fibrin adhesive and cartilage chips to fill experimentally made osteochondral articular defects in adult rabbits and reported complete closure of the defects with hyaline cartilage after 40 weeks. Homminga and colleagues[57] tested fibrin glue as a vehicle in chondrocyte transplantation in the early 1990s. Their *in vitro* experiments showed that chondrocytes encapsulated in fibrin glue hydrogels multiplied, retained their morphology, and produced extracellular matrix in the fibrin glue as long as the cells were surrounded by the glue. However, they noted that the fibrin glue began to disintegrate after only 3 days and was accelerated by higher cell concentrations.[57] Hendrickson et al.[58,59] also studied a fibrin glue system *in vitro* and *in vivo* using an equine model.

Despite these several studies, the fibrin glue systems used were not injected in a minimally invasive manner into animals. Improving Sims' and Ting's non-injectable fibrin glue systems, Silverman et al.[60] varied the concentrations of fibrinogen and thrombin to enhance the injectable nature of the fibrin glue and chondrocyte suspension. They injected fibrin glue and ovine chondrocytes into the subcutaneous space of nude mice and found that 40×10^6 cells/ml gave the best overall neocartilage in this system. The same paper also demonstrated that the degradation of the fibrin constructs could be inhibited by adding a fibrinolysis inhibitor, aprotinin, or by simply increasing the original concentration of fibrinogen. The same year, Meinhart, Fussenegger, and Hobling[61] similarly showed that they could stabilize or inhibit the degradation of fibrin glue and chondrocyte constructs *in vitro* by using aprotinin or tranexamic acid.

Potential limitations for the clinical application of naturally occurring hydrogels such as collagens and fibrin glues include the gels' purported lack of mechanical integrity and the lack of control over polymerization reactions. There can also be difficulties in controlling the consistency, the stability, and the degradation of these naturally occurring compounds.[62] A case in point; van Susante and colleagues[63] performed a large animal study with goats, using fibrin glue and heterologous (xenogenic) articular chondrocytes from rabbits to fill osteochondral defects. They found that the fibrin glue showed noticeable mechanical and physical degradation after only 3 weeks in the weight bearing area of the joint. The authors comment that the large size of the defect (1 cm diameter) may have been too big for the fibrin glue to withstand the mechanical forces: "the active disintegration of the glue and non-sufficient potential to resist mechanical forces from the joint are the main reason why in this experiment" the fibrin glue preparation that was used "failed to offer persistent support to the grafted cells."[63]

Other possible applications of fibrin glue and chondrocyte therapy include ear reconstruction or soft tissue filler. In 1998, Sims et al.[64] encapsulated bovine chondrocytes in fibrin glue and showed that neocartilage could be produced *in vivo* after surgical implantation of their fibrin/chondrocyte constructs into the subcutaneous pockets of nude mice. Contemporaneously with Sims and his colleagues, Ting et al.[65] utilized fibrin glue hydrogels to encapsulate human costal chondrocytes prefabricated in shape of human nasal cartilages and cultured *in vitro* for 4 weeks then surgically implanted subcutaneously in nude mice for 4 weeks. The authors demonstrated that the constructs kept their basic shape and produced near normal amounts GAG and collagen.[65] More recent work has focused on encapsulating chondrocytes isolated from elastic cartilage (ears) with fibrin glue.[66–68] Potential advantages for the use of fibrin glue in this instance are that the ear cartilage does not have to bear weight; however, it must maintain its mechanical integrity while staying flexible and elastic. Neovius and Kratz, as well as Xu et al.[67,68] demonstrated that the elastic chondrocytes, similar to articular chondrocytes, maintained their own phenotypes and continued to express extracellular matrix specific for elastic cartilage. Xu et al.[68] cleverly sandwiched the fibrin glue and chondrocyte matrix in between lyophilized perichondrium demonstrating that the ear cartilages maintained their gross shape and, importantly, an impressive degree of flexibility and strength.

IX. COLLAGEN

Collagen is another natural substrate that has a long history in the literature. Collagen preparations come in many forms such as dry sponges, fibrillar meshes, and gels. It is usually used as a scaffold for transplanted cells. However, the authors could not find any paper that described a one-step, cell-laden, injectable, *in situ* polymerizing method using this natural substrate. Instead, several groups have transplanted chondrocyte-laden collagen gels that have been either preformed *ex vivo* and "precultured" *in vitro* for a period of time, or immediately surgically implanted after *ex vivo* polymerization.[69–73] Im et al.[74] used collagen gels in conjunction with bone marrow-derived mesenchymal stem cells in full-thickness osteochondral lesions in rabbits, while Wakitani et al.[75] performed a similar study in human subjects. The plethora of *in vivo* studies using collagen would benefit from a less invasive delivery system.

X. POLYESTERS

Our discussion of the classic aliphatic polyesters polylactic (PLA) and polyglycolic (PGA) acids will be strictly limited to their inclusion into photopolymerizable block polymers with PVA and

PEG.[40,62] In this form they are included into hydrogels, as opposed to the solid scaffolds in which they are traditionally used.

XI. POLY(ETHYLENE GLYCOL)

Poly(ethylene oxide) or poly(ethylene glycol) has been approved by the FDA for several medical applications due to its biocompatibility and low toxicity.[32] Early work with injectable cartilage includes a paper that describes the use of a viscous poly(ethylene oxide) solution as a cell carrier.[76] This study demonstrated that 20% poly(ethylene oxide) solutions, which are biocompatible and biodegradable synthetic polymers, can be utilized for the encapsulation of isolated chondrocytes and maintenance of three-dimensional spatial support for neocartilage tissue development. The authors injected poly(ethylene oxide) hydrogel/chondrocyte suspensions subcutaneously in nude mice and then incubated for 6 and 12 weeks *in vivo*. Histological and biochemical findings demonstrated neocartilage generation. The group envisioned using this type of system as a soft tissue filler or to help sculpt areas of the craniofacial skeleton.[76]

Dr. Hubbell's group has taken a dramatic step in designing a new generation of PEG-based hydrogels for tissue engineering by adding adhesion peptide motifs (RGDSP) and substrates for matrix metalloproteinases (MMPs).[77] The RGD peptides provide an adhesive scaffold for the encapsulated cells to attach onto in the otherwise non-adhesive gel, while the MMP sensitive substrates allow the cells to degrade, and remodel the synthetic scaffold with natural extracellular proteases. They also added biological signals into the gels by adding bone morphogenic protein-2. They presented *in vitro* data suggesting that the gel must first be degraded by MMPs to release the majority of the BMP. Thus, they provided specific and local delivery of the bioactive factor as it is physiologically needed. Finally, in an *in vivo* rat skull model, they showed complete filling of a critical sized defect using their adhesive (RGD), MMP sensitive hydrogels when 5 μg of BMP-2 was added per 100 μl defect after 5 weeks. Their results were similar to, if not slightly better than, collagen sponges containing the same amount of BMP-2.[77] This system was specifically tested for bone tissue engineering; however, it could easily be adapted for numerous tissues including cartilage. The one drawback of the system was that the gel was preprepared and surgically implanted, thus, it loses the minimal invasiveness of the system. Modification of the delivery system would be beneficial.

One other method for implanting scaffold material deserves mentioning not because it is a hydrogel, but because of its sheer novelty. One team of investigators have created "open porous scaffold structures" in a minimally invasive manner by injecting a thread-like material through a small cannula.[78] A stream of water is injected through the cannula in order to drag the implant material off a pivoting spool and into the injection site. The result is an injected "tangle" of thread-like material that forms an open-porous structure into which cells can secrete extracellular matrix.[78] This system is yet to be tested for injectable cartilage.

XII. PHOTOPOLYMERIZATION

Several laboratories, including our own, have focused on a different technique for creating injectable hydrogels suitable for tissue engineering applications. We have studied photopolymerization reactions, which are chemical reactions driven by light sensitive "initiators." This type of technology is used in applications ranging from printing to dentistry to optical materials.[79] These systems are being used for an increasing number of biomedical applications because of their ability to rapidly convert liquid monomer or macromer solutions to a gel under physiologic conditions.[80] Other important advantages of these systems include powerful spatial and temporal control of the reaction's kinetics, minimal heat production, and adaptability for *in situ* polymerization by adapting light sources to fit clinical scenarios.[20,62,81–83] A photon from a light source excites or dissociates the light sensitive initiator, called photoinitiators, into a high-energy radical state. This radical then drives a chemical reaction that induces the polymerization of a macromer solution. There are a variety of photoinitiators, each with its own light wavelength excitation spectrum. Visible light initiators and UV sensitive initiators exist. Nguyen and West[83] have written a concise and very informative review on photopolymerizing hydrogels and their applications in tissue engineering.

XIII. TOXICITY STUDY OF PHOTOINITIATORS

One concern about photopolymerization reactions is that the free radicals formed by the photoinitiators might react adversely with cell membranes, proteins, and DNA. However, several groups have adapted this technology and have made it cytocompatible and applicable to cell-based therapies in tissue engineering of bone and cartilage.[20,80,84–88] Addressing the concerns of cellular cytocompatibility, Bryant, Nuttleman, and Anseth tested the *in vitro* cytotoxicity of a variety of UV and visible light initiating systems on dermal fibroblasts. They found that Igracure 2959 was the least toxic UV sensitive initiator and triethanolamine was the least toxic visible light sensitive initiator. They confirmed that Irgacure 2959 was cytocompatible with bovine chondrocytes when photoencapsulated into PEG-based hydrogels for their applications in cartilage tissue engineering[80] (Figure 12.1).

In our and other's laboratories, there existed a belief that different cell types had varying sensitivities to the same photoinitiator. To investigate this more closely, we explored the issue of several UV sensitive photoinitiators' cytotoxicity using six cell lines employed in a variety of tissue engineering applications. We studied the effects of Irgacure 2959 (2-hydroxy-1-[4-(hydroxyethoxy)phenyl]-2-methyl-1-propanone), Irgacure 651 (2,2-dimethoxy-2-phenylacetophenone), and Irgacure 184 (1-hydroxycyclohexyl-1-phenyl ketone) on bovine chondrocytes, goat bone marrow-derived mesenchymal stem cells, human bone marrow-derived mesenchymal stem cells (Clonetics), rabbit corneal epithelial cells (ATCC), human fetal osteoblasts[89] (ATCC) and human embryonic germ cells (LVEC cell line).[90] Confirming Bryant's[80] results, we similarly found that I2959 was the least cytotoxic in all concentrations (0.03, 0.05, and 0.1 w/v%) for all cell types tested. I184 and I651 showed significant toxicity in almost all situations (Figure 12.2).

FIGURE 12.1 Live-dead cell assay staining live bovine chondrocytes green and dead bovine chondrocytes red. This was taken after photoencapsulation of the cells in a PEG-based hydrogel and 3 days of *in vitro* culture. (The original photomicrograph was taken at 200×. The meter bar represents 100 μm).

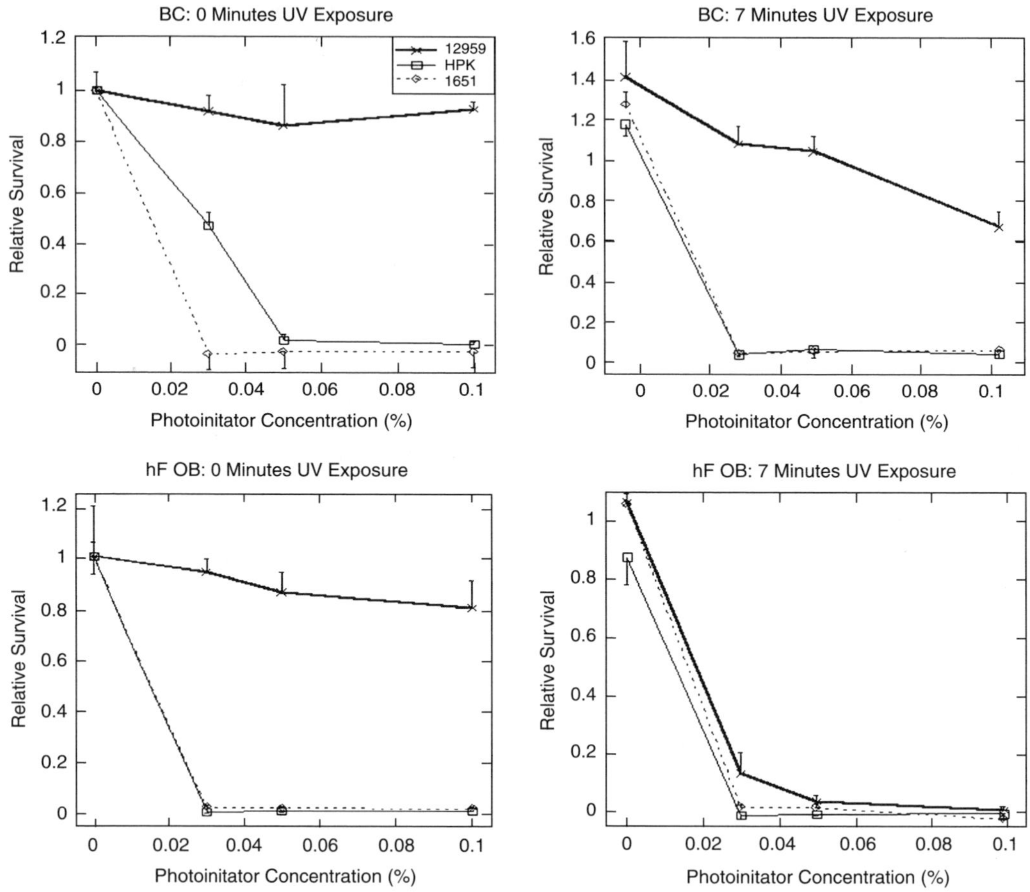

FIGURE 12.2 Toxicity results of bovine chondrocytes (BC) (top right and left) and human fetal osteoblasts (hFOBs) (bottom right and left). Three photoinitiators were used: I2959 (thick black line), HPK (thin black line), and I651 (dotted line). Two UV-exposure times are shown, 0 min (no UV exposure, left sided grafts) and 7 min of UV exposure (right sided grafts). Error bars represent the standard deviations of the mean values ($n = 3$). (From Williams, C. G. et al., Variable cytocompatibility of six cell lines with photinitiators used for polymerizing hydrogels and cell encapsulation, *Biomaterials*, 26(11), 1211–1218, 2005. With permission.)

Interestingly, we found that different cell types had different cytocompatibility profiles for the same conditions using I2959. Bovine chondrocytes, goat mesenchymal stem cells, and human mesenchymal stems cells were the least affected, while rabbit corneal epithelial cells, human embryonic germ cells (LVECs), and fetal human osteoblasts were the most adversely affected by I2959. The toxicity profiles seemed to follow the population doubling time. That is, it appeared that the slowest dividing cells were affected least while the fastest dividing cells were affected most[91] (Figures 12.3 and 12.4).

XIV. CARTILAGE TISSUE ENGINEERING WITH PHOTOPOLYMERIZING HYDROGELS

We began exploring the use of photopolymerizing PEO-based hydrogels for tissue engineering of cartilage in the late 1990s.[82,86] We first demonstrated that bovine chondrocytes survived photoencapsulation in three-dimensional hydrogels *in vitro* and *in vivo* for at least 2 weeks. The cells produced and secreted chondrogenic extracellular matrix proteins such as collagen type II

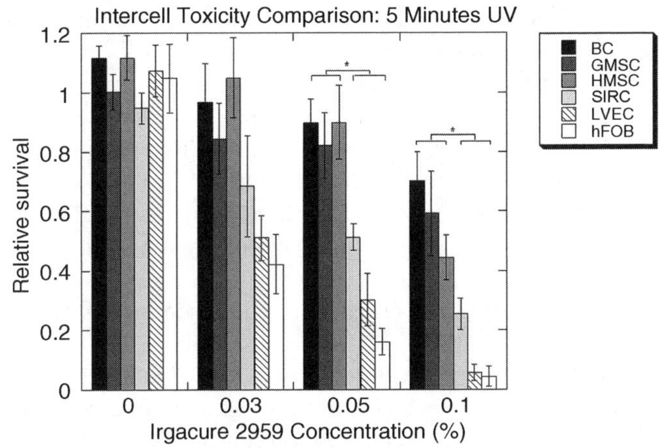

FIGURE 12.3 Bar graph of six cell types, from left to right, BC, GMSC, HMSC, SIRC, LVEC, hFOB at 5 min of UV exposure with four concentrations of Irgacure 2959 (0, 0.03, 0.05, and 0.1% w/v). Note the general trend of decreasing cell line survival as the concentration of the photoinitiator increases. Also note the differences in cell types as the concentration increases. (From Williams, C. G. et al., Variable cytocompatibility of six cell lines with photoinitiators used for polymerizing hydrogels and cell encapsulation, *Biomaterials*, 26(11), 1211–1218, 2005. With permission.)

and glycosaminoglycans into the hydrogels. We also showed that the mechanical properties of the chondrocyte–hydrogel constructs improved towards normal articular cartilage over a culture period of 6 weeks.[20] Potential clinical applicability for this system was also demonstrated by transdermally photopolymerizing the cell-laden hydrogels after subcutaneous injection of the polymer solutions.[82,86] That is, the cell/polymer solutions were first injected under the skin of nude mice. Then, they were polymerized by shining a low-intensity 365 nm UV lamp onto the skin for several minutes. These polymerized cell/polymer matrices survived *in vivo* culture and elaborated a cartilaginous matrix. However, these first systems for chondrocyte encapsulation were not biodegradable.

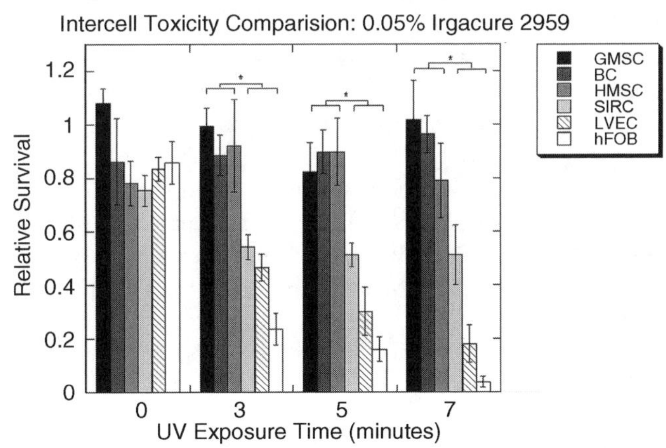

FIGURE 12.4 Bar graph of six cell types, from left to right, BC, GMSC, HMSC, SIRC, LVEC, hFOB at 0.05% w/v Irgacure 2959 over four UV exposure times (0, 3, 5, and 7 min). Note the general trend of decreasing cell line survival of the cell lines as the UV exposure times increase. Also note the different survival rates for the various cell types. (From Williams, C. G. et al., Variable cytocompatibility of six cell lines with photinitiators used for polymerizing hydrogels and cell encapsulation, *Biomaterials*, 26(11), 1211–1218, 2005. With permission.)

Bryant and Anseth[62,92–94] have written a series of papers investigating the use of biodegradable and non-biodegradable PEO-based photopolymerizing hydrogels for cartilage tissue engineering. Their series of papers on *in vitro* experiments explored the effects of polymer cross-linking density, scaffold thickness, mechanical properties, and biodegradability on cell survival, extracellular matrix production, and the special deposition of extracellular matrix. They showed that chondrocytes survive *in vitro* culture in PEO-based hydrogels up to 8 mm thick. There was no detectable adverse affect of hydrogel thickness on chondrocyte survival or matrix production in different layers of the 10% w/w poly(ethylene oxide) dimethacrylate hydrogels, indicating that the diffusion of nutrients and waste was appropriate even in the deepest portions. This suggests the possibility that chondrocyte–PEO constructs could be made *in vivo* to the depth of even the thickest human cartilage.

In another paper Bryant and Anseth[94] explored how intrinsic hydrogel characteristics like cross-linking density, equilibrium water content, and compressive modulus affect chondrocyte behavior *in vitro*. They showed that non-degrading gels with varying cross-linking densities have various physical distributions of the glycosaminoglycans produced, but do not change the total amount produced. Gels with higher water content (thus lower cross-linking density) allowed the GAG to diffuse throughout the gels. Gels with a lower water content (and thus higher cross-linking density) prohibited GAG diffusion, keeping the matrix protein immediately pericellular. In contrast, total collagen production was affected by the compressive modulus of the gels. That is, in their culture system, gels with a moderate compressive modulus of 360 kPa showed an increased production in type II collagen, in comparison to gels with higher or lower compressive moduli.

A subsequent paper studied how changing the biodegradation rate of PEG-based photopolymerizing hydrogels, specifically, poly(lactic acid)-*b*-poly(ethylene glycol)-*b*-poly(lactic acid), affected chondrocyte behavior and productive capacity.[62] The authors controlled the number of biodegradable linkages from 50 to 85% by creating copolymer hydrogels of PEG and PEG–LA–DA and photoencapsulated bovine articular chondrocytes. By controlling the number of biodegradable linkages, they controlled the rate of polymer mass loss in the hydrogels. The authors showed that as the polymer degradation rate increased, the DNA rose early in the culture period, suggesting that the cells were able to divide better in the first 2 weeks of *in vitro* culture. Collagen content in the hydrogels was statistically higher in the faster degrading hydrogels at all points during culture. At the end of the culture period the collagen content was greater by almost an order of magnitude in the gels with 85% degradable linkages vs. those with 50% degradable linkages. In contrast, GAG content was significantly lower in the faster degrading hydrogels, perhaps because the matrix protein diffused out of the more degraded hydrogels.[62]

Anseth's[92,95] group has various other papers demonstrating the ability to control degradation using other photoinitiating polymers such as poly(vinyl alcohols) (PVA) and poly anhydrides.[81,85,96–98] Unfortunately, we cannot discuss them all.

Recently, Park, Tirelli, and Hubbell[99] published a paper describing the creation of a methacrylated hyaluronic acid and PEODA-based polymers to create a photopolymerizable system for creation of biodegradable hydrogels. However, the group only cultured fibroblasts on top of hydrogels to demonstrate cell adhesion and did not fully encapsulate cells inside the hydrogel. This would be an interesting potential system for cartilage repair if further work is presented.

XV. CHONDROGENESIS OF MESENCHYMAL STEM CELLS IN A PHOTOPOLYMERIZING HYDROGEL

Recently, our lab has been interested in investigating the possibility of utilizing multipotential stem cells in conjunction with our photopolymerizing PEG-based polymers for cartilage tissue engineering. There are two potential methods to do this. First, one could isolate mesenchymal stem cells, differentiate them into chondrocytes, and subsequently encapsulate them. Alternatively, one could encapsulate undifferentiated mesenchymal stem cells and attempt to induce chondrogenic differentiation in the hydrogels. We chose the second method because we thought it would ultimately be more cost effective should a clinical therapeutic technology be developed.

We encapsulated 20×10^6 cells/ml undifferentiated bone marrow-derived mesenchymal stem cells (MSCs) isolated from the femurs of three to three and an half year old goats in 10% w/v PEGDA hydrogels using 0.05% Irgacure 2959 and 5 min UV exposure (365 nm, ~4 mW/cm^2, Glowmark). The molds held 75 μl of solution, were 6 mm in diameter, and approximately 3 to 4 mm in thickness. We then cultured the cell–polymer constructs in control and chondrogenic media with TGF-β1 (10 ng/ml) for up to 6 weeks *in vitro* on an orbital shaker at 75 rpm in standard culture conditions. We performed biochemical assays for DNA, GAG, and collagen, histological analyses, and RT–PCR for various chondrogenic markers at 0 days, 3 weeks, and 6 weeks.

Our results demonstrated that previously undifferentiated MSCs were able to differentiate *in vitro* into a chondrogenic lineage after encapsulation in our PEG-based hydrogels when cultured in the presence of TGF-β1 (TGF-β3 has also worked well in subsequent unpublished studies).

Histologic study of fixed slides after culture showed that the experimental groups cultured with TGF-β1 were strongly positive for GAG by Safranin-O/Fast Green staining compared to the other groups (Figure 12.5(b,c)). As the culture time extended, the amount of GAG increased and staining was seen pericellularly, as well as in the matrix between the cells, indicating that the GAG had diffused throughout the PEODA gels. However, constructs cultured without TGF-β1 (Figure 12.5(d)), consistently showed only a few scattered cells producing GAG by spontaneous differentiation.

Immunohistochemical staining for aggrecan and link protein showed strong positive staining in the 6 wk + TGF group but revealed negative or sporadic, weakly positive cells in the 6 wk − TGF

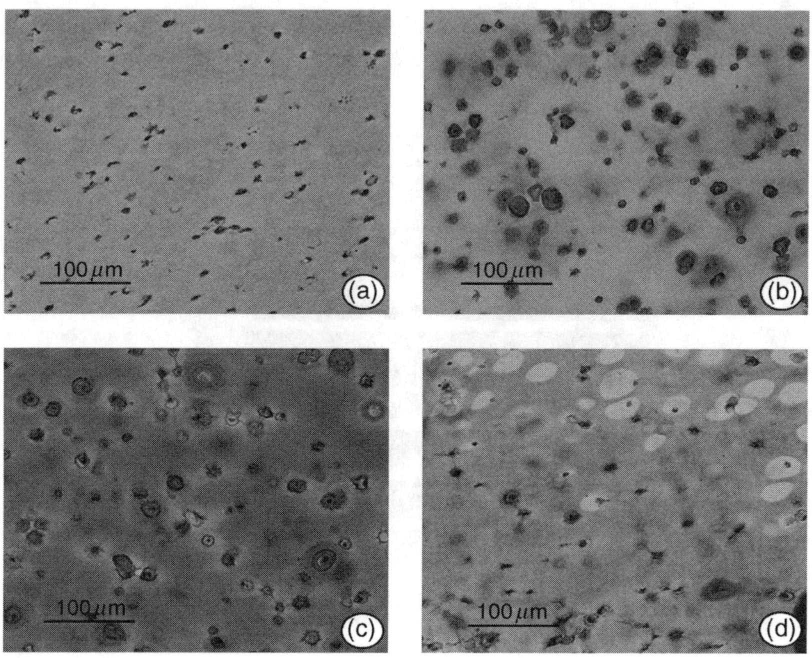

FIGURE 12.5 Paraffin embedded histological sections of PEGDA–MSC hydrogels for Day 0 (a), 3 wk + TGF (b), 6 wk + TGF (c), and 6 wk − TGF (d) constructs stained with Safranin-O/Fast Green. Originally acquired at 200×. This dye combination stains GAG red and nuclei green. The scale bars are 100 μm. (From Williams, C. G., Kim, T. K., Taboas, A., Malik, A., Manson, P., Elisseeff, J. In vitro chondrogenesis of bone marrow derived stem cells in a photopolymerizable hydrogel, *Tissue Eng.*, 9(4), 679–688, 2003. With permission.)

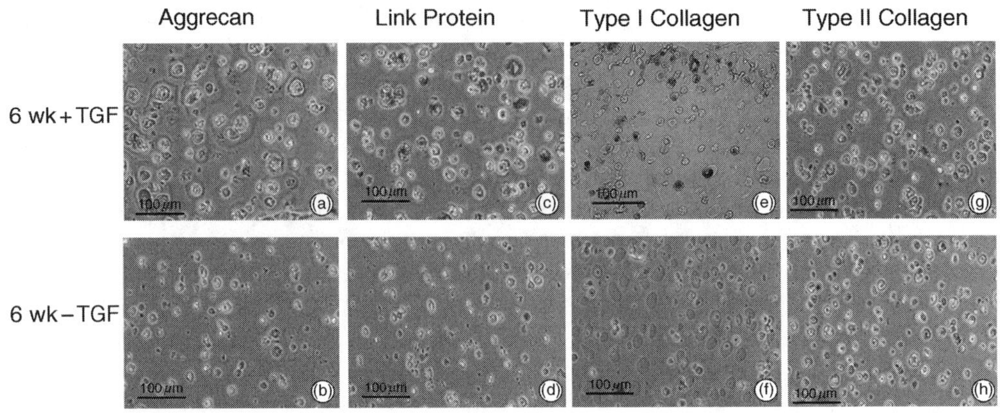

FIGURE 12.6 Immunohistochemical staining of paraffin embedded sections for 6 wk + TGF (a, c, e, g) and 6 wk − TGF (b, d, f, h)) constructs. Antibodies for aggrecan (a, b), link protein (c, d), type I collagen (e, f), and type II collagen (g, h) were used. Controls were negative (not shown). Images were originally acquired at 100×, and the scale bar represents 100 μm. Dark red staining is positive for the antigen. (From Williams, C. G., Kim, T. K., Taboas, A., Malik, A., Manson, P., Elisseeff, J. In vitro chondrogenesis of bone marrow derived stem cells in a photopolymerizable hydrogel, *Tissue Eng*., 9(4), 679–688, 2003. With permission.)

group (Figure 12.6(a–d)). Staining for type I collagen was positive on both the 6 wk +/− TGF sections (Figure 12.6(e,f)). Interestingly, type II collagen staining was also noted on sections from both 6 wk +/− TGF; the 6 wk + TGF section, however, demonstrated more intense staining (Figure 12.6(g,h)).

RT–PCR demonstrated that the previously undifferentiated MSCs shifted their genetic expression during the culture period (Figure 12.7). Type I collagen was absent in monolayer

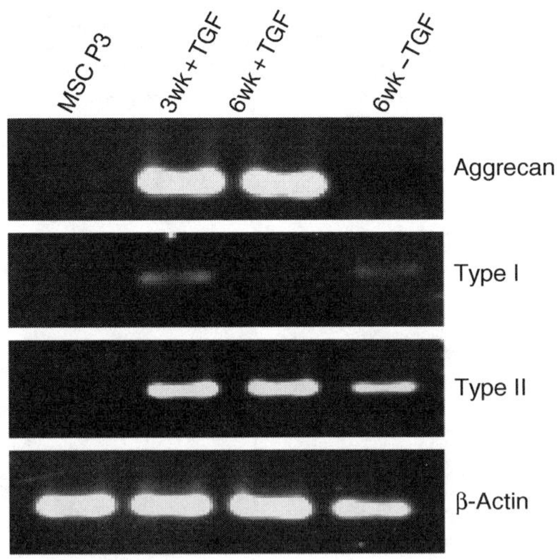

FIGURE 12.7 RT–PCR products for MSC passage 3 monolayer culture and 6 wk − TGF as controls *flank the two sides of the gel.* 3 wk + TGF and 6 wk + TGF *RT–PCR products are in the center two lanes of the gel.* Primers used include aggrecan, type I collagen, type II collagen, and β-actin (from top to bottom of the gels) (From Williams, C. G., Kim, T. K., Taboas, A., Malik, A., Manson, P., Elisseeff, J. In vitro chondrogenesis of bone marrow derived stem cells in a photopolymerizable hydrogel, *Tissue Eng*., 9(4), 679–688, 2003. With permission.)

culture, almost absent in 6 wk + TGF, but weakly present in 3 wk + TGF and 6 wk − TGF. Type II collagen was not expressed in monolayer culture of the MSCs, present in low quantities in 6 wk − TGF, but strongly present in the 3 wk + TGF and 6 wk + TGF constructs. Aggrecan gene expression was almost entirely absent in monolayer controls, absent in the 6 wk − TGF, but strongly present in the 3 and 6 wk + TGF constructs.

The DNA content (Figure 12.8(a)) of the MSC constructs revealed a statistically significant increase to 1101 ng (± 170 ng) of DNA/mg dw in the 6 wk + TGF group from an initial (0 day) value of 882 ng (± 94 ng)/mg dw ($p = .036$) and a significant decrease to 681 ng (± 43 ng)/mg dw in the 6 wk − TGF group ($p = .028$). These results are supported by an observation that cells seemed to divide and formed small aggregates in the 6 wk + TGF group (Figure 12.8(c)). Correlating with the RT–PCR and histological findings, the hydrogels showed a significant increase in total collagen and GAG produced (% dry weight) by detection of chondroitin sulfate and hydroxyproline when the constructs were cultured with TGF-β1 (Figure 12.8(b,c)). Collagen production significantly increased over the culture period when compared to the Day 0 group: from 0 to 5.0% dw in 6 wk + TGF ($p = .001$); and to 1.4% dw in 6 wk − TGF ($p = .029$).

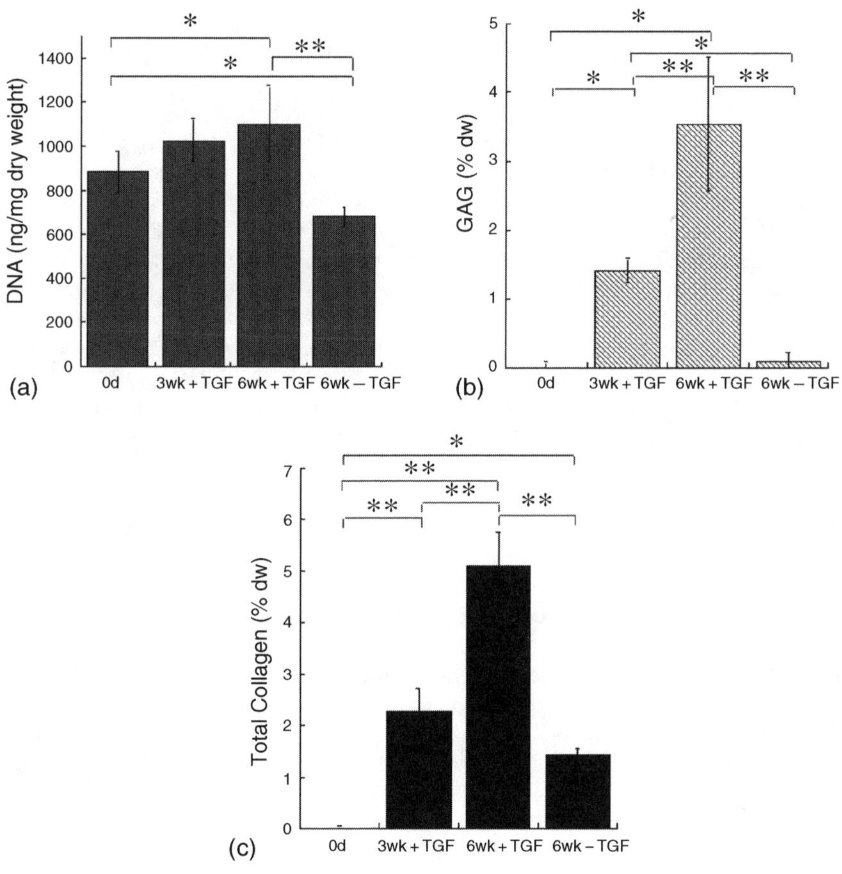

FIGURE 12.8 Results of biochemical assays for Day 0, 3wk + TGF, 6wk + TGF, and 6wk − TGF constructs are depicted: DNA content (ng of DNA/mg dry weight) (a), GAG content (% dry weight of construct) (b), total collagen content (% dry weight of construct), (c), $*p < .05$, $**p < .001$. (From Williams, C. G., Kim, T. K., Taboas, A., Malik, A., Manson, P., Elisseeff, J. In vitro chondrogenesis of bone marrow derived stem cells in a photopolymerizable hydrogel, *Tissue Eng.*, 9(4), 679–688, 2003. With permission.)

GAG synthesis increased from 0% dw in Day 0 constructs to 3.5%dw in the 6 wk + TGF constructs ($p = .001$). 6 wk − TGF control hydrogels showed no difference when compared to the Day 0 time point ($p > .05$).

XVI. CONCLUSION

The field of cartilage tissue engineering is an exciting one. Many groups are attempting to design new and successful strategies to reconstruct or repair articular and non-weight bearing cartilage tissues. As medicine advances and the drive towards minimally invasive therapies continues to grow, engineers will continue developing novel methods for injectable therapies to treat damaged or absent cartilage. This chapter has been a brief review of a vast number of studies in this field. We have strived to emphasize papers that demonstrated an injectable, cell-based approach for cartilage tissue engineering with notable exceptions throughout the chapter. *In vitro* and *in vivo* studies were mentioned where considered appropriate. Our lab focuses on one method to create an *in situ* photopolymerizing hydrogel for cartilage tissue repair and augmentation.

REFERENCES

1. Hoffman, A. S., Hydrogels for biomedical applications, *Adv. Drug Deliv. Appl.*, 43, 3–12, 2001.
2. Gutowska, A., Jeong, B., and Jasionowski, M., Injectable gels for tissue engineering, *Anat. Rec.*, 263(4), 342–349, 2001.
3. Friedman, M. J., Berasi, C. C., Fox, J. M., Del Pizzo, W., Snyder, S. J., and Ferkel, R. D., Preliminary results with abrasion arthroplasty in the osteoarthritic knee, *Clin. Orthop.*, 182, 200–205, 1984.
4. Johnson, L. L., Arthroscopic abrasion arthroplasty historical and pathologic perspective: present status, *Arthroscopy*, 2(1), 54–69, 1986.
5. Buckwalter, J. A., and Lohmander, S., Operative treatment of osteoarthrosis. Current practice and future development, *J. Bone Joint Surg. Am.*, 76(9), 1405–1418, 1994.
6. Matsusue, Y., Yamamuro, T., and Hama, H., Arthroscopic multiple osteochondral transplantation to the chondral defect in the knee associated with anterior cruciate ligament disruption, *Arthroscopy*, 9(3), 318–321, 1993.
7. Bobic, V., Arthroscopic osteochondral autograft transplantation in anterior cruciate ligament reconstruction: a preliminary clinical study, *Knee Surg. Sports Traumatol. Arthrosc.*, 3(4), 262–264, 1996.
8. McDermott, A. G., Langer, F., Pritzker, K. P., and Gross, A. E., Fresh small-fragment osteochondral allografts. Long-term follow-up study on first 100 cases, *Clin. Orthop.*, 197, 96–102, 1985.
9. Grande, D. A., Pitman, M. I., Peterson, L., Menche, D., and Klein, M. The repair of experimentally produced defects in rabbit articular cartilage by autologous chondrocyte transplantation, *J. Orthop. Res.*, 7(2), 208–218, 1989.
10. Brittberg, M., Lindahl, A., Nilsson, A., Ohlsson, C., Isaksson, O., and Peterson, L., Treatment of deep cartilage defects in the knee with autologous chondrocyte transplantation, *N. Engl. J. Med.*, 331(14), 889–895, 1994.
11. Hunziker, E. B., Articular cartilage repair: basic science and clinical progress. A review of the current status and prospects, *Osteoarthritis Cartilage*, 10(6), 432–463, 2002.
12. Warren, S. M., Fong, K. D., Chen, C. M., Loboa, E. G., Cowan, C. M., Lorenz, H. P., and Longaker, M. T., Tools and techniques for craniofacial tissue engineering, *Tissue Eng.*, 9(2), 187–200, 2003.
13. Mulliken, J. B., and Glowacki, J., Induced osteogenesis for repair and construction in the craniofacial region, *Plast. Reconstr. Surg.*, 65(5), 553–560, 1980.
14. Bostrom, R., and Mikos, A. G., Synthetic biodegradable polymer scaffolds, In *Tissue Engineering of Bone*, Atala, A. et al., ed., Birkhauser, Boston, pp. 215–234, 1997.
15. Lysaght, M. J., and Reyes, J., The growth of tissue engineering, *Tissue Eng.*, 7(5), 485–493, 2001.

16. Lanza, R. P., Langer, R., Vacanti, J., eds., *Principles of Tissue Engineering*, 2nd edn., Academic Press, San Diego, 2000.
17. Vacanti, C. A., Langer, R., Schloo, B., and Vacanti, J. P., Synthetic polymers seeded with chondrocytes provide a template for new cartilage formation, *Plast. Reconstr. Surg.*, 88(5), 753–759, 1991.
18. Freed, L. E., Marquis, J. C., Nohria, A., Emmanual, J., Mikos, A. G., and Langer, R., Neocartilage formation in vitro and in vivo using cells cultured on synthetic biodegradable polymers, *J. Biomed. Mater. Res.*, 27(1), 11–23, 1993.
19. Cao, Y., Vacanti, J. P., Paige, K. T., Upton, J., and Vacanti, C. A., Transplantation of chondrocytes utilizing a polymer–cell construct to produce tissue-engineered cartilage in the shape of a human ear, *Plast. Reconstr. Surg.*, 100(2), 297–302, 1997; discussion 303–304.
20. Elisseeff, J., McIntosh, W., Anseth, K., Riley, S., Ragan, P., and Langer, R., Photoencapsulation of chondrocytes in poly(ethylene oxide)-based semi-interpenetrating networks, *J. Biomed. Mater. Res.*, 51(2), 164–171, 2000.
21. Oxley, H. R., Corkhill, P. H., Fitton, J. H., and Tighe, B. J., Macroporous hydrogels for biomedical applications: methodology and morphology, *Biomaterials*, 14(14), 1064–1072, 1993.
22. Higgs, P. G., and Ball, R. C., Polymers and gels, *Physical Networks*, Elsevier, New York, pp. 185–194, 1990.
23. Peppas, N., *Hydrogels in Medicine and Pharmacy*, CRC Press, Boca Raton, FL, 1987.
24. Jen, A., Wake, M., and Mikos, A. G., Review: hydrogels for cell immobilization, *Biotechnol. Bioeng.*, 50(4), 357–364, 1996.
25. Wichterle, O., and Lim, D., Hydrophylic gels in biologic use, *Nature*, 185, 117, 1960.
26. Lim, F., and Sum, A. M., Microencapsulated islets as bioartificial pancreas, *Science*, 210, 908–910, 1980.
27. Guo, J. F., Jourdian, G. W., and MacCallum, D. K., Culture and growth characteristics of chondrocytes encapsulated in alginate beads, *Connect. Tissue Res.*, 19(2–4), 277–297, 1989.
28. Benya, P. D., and Shaffer, J. D., Dedifferentiated chondrocytes reexpress the differentiated collagen phenotype when cultured in agarose gels, *Cell*, 30(1), 215–224, 1982.
29. Hauselmann, H. J. Aydelotte, M. B., Schumacher, B. L., Kuettner, K. E., Gitelis, S. H., and Thonar, E. J., Synthesis and turnover of proteoglycans by human and bovine adult articular chondrocytes cultured in alginate beads, *Matrix*, 12(2), 116–129, 1992.
30. Tamponnet, C., Ramdi, H., Guyot, J. B., and Lievremont, M., Rabbit articular chondrocytes in alginate gel: characterisation of immobilized preparations and potential applications, *Appl. Microbiol. Biotechnol.*, 37(3), 311–315, 1992.
31. Corkhill, P. H., Fitton, J. H., and Tighe, B. J., Towards a synthetic articular cartilage, *J. Biomater. Sci. Polym. Ed.*, 4(6), 615–630, 1993.
32. Lee, K. Y., and Mooney, D. J., Hydrogels for tissue engineering, *Chem. Rev.*, 101(7), 1869–1879, 2001.
33. Rahfoth, B., Weisser, J., Sternkopf, F., Aigner, T., von der Mark, K., and Brauer, R., Transplantation of allograft chondrocytes embedded in agarose gel into cartilage defects of rabbits, *Osteoarthritis Cartilage*, 6(1), 50–65, 1998.
34. Cao, Y. L., Lach, E., Kim, T. H., Rodriguez, A., Arevalo, C. A., and Vacanti, C. A., Tissue-engineered nipple reconstruction, *Plast. Reconstr. Surg.*, 102(7), 2293–2298, 1998.
35. Saim, A. B., Cao, Y. L., Weng, Y. L., Chang, C. N., Vacanti, M. A., Vacanti, C. A., and Eavey, R. D., Engineering autogenous cartilage in the shape of a helix using an injectable hydrogel scaffold, *Laryngoscope*, 110(10), 1694–1697, 2000.
36. Liu, Y., Chen, F., Liu, W., Cui, L., Shang, Q., Xia, W., Wang, J., Cui, Y., Yang, G., Liu, D., Wu, J., Xu, R., Buonocore, S. D., and Cao, Y., Repairing large porcine full-thickness defects of articular cartilage using autologous chondrocyte-engineered cartilage, *Tissue Eng.*, 8(4), 709–721, 2002.
37. Xia, W., Cao, Y., and Shang, Q., [An experimental study of tissue engineered autologous cartilage by using an injectable polymer], *Zhonghua Zheng Xing Wai Ke Za Zhi*, 17(5), 302–305, 2001.
38. Weng, Y., Cao, Y., Silva, C. A., Vacanti, M. P., and Vacanti, C. A., Tissue-engineered composites of bone and cartilage for mandible condylar reconstruction, *J. Oral Maxillofac. Surg.*, 59(2), 185–190, 2001.

39. Ibusuki, S., Fujii, Y., Iwamoto, Y., and Matsuda, T., Tissue-engineered cartilage using an injectable and in situ gelable thermoresponsive gelatin: fabrication and in vitro performance, *Tissue Eng.*, 9(2), 371–384, 2003.
40. Jeong, B., Lee, K. M., Gutowska, A., and An, Y. H., Thermogelling biodegradable copolymer aqueous solutions for injectable protein delivery and tissue engineering, *Biomacromolecules*, 3(4), 865–868, 2002.
41. Paige, K. T., Cima, L. G., Yaremchuk, M. J., Vacanti, J. P., and Vacanti, C. A., Injectable cartilage, *Plast. Reconstr. Surg.*, 96(6), 1390–1398, 1995; discussion 1399–1400.
42. Atala, A., Cima, L. G., Kim, W., Paige, K. T., Vacanti, J. P., Retik, A. B., and Vacanti, C. A., Injectable alginate seeded with chondrocytes as a potential treatment for vesicoureteral reflux, *J. Urol.*, 150(2 Pt 2), 745–747, 1993.
43. Grandolfo, M., D'Andrea, P., Paoletti, S., Martina, M., Silvestrini, G., Bonucci, E., and Vittur, F., Culture and differentiation of chondrocytes entrapped in alginate gels, *Calcif. Tissue Int.*, 52(1), 42–48, 1993.
44. Ramdi, H., Tahri Jouti, M. A., and Lievremont, M., Immobilized articular chondrocytes: in vitro production of extracellular matrix compounds, *Biomater. Artif. Cells Immobil. Biotechnol.*, 21(3), 335–341, 1993.
45. Bonaventure, J., Kadhom, N., Cohen-Solal, L., Ng, K. H., Bourguignon, J., Lasselin, C., and Freisinger, P., Reexpression of cartilage-specific genes by dedifferentiated human articular chondrocytes cultured in alginate beads, *Exp. Cell Res.*, 212(1), 97–104, 1994.
46. Paige, K. T., Cima, L. G., Yaremchuk, M. J., Schloo, B. L., Vacanti, J. P., and Vacanti, C. A., De novo cartilage generation using calcium alginate–chondrocyte constructs, *Plast. Reconstr. Surg.*, 97(1), 168–178, 1996; discussion 179–180.
47. Bent, A. E., Tutrone, R. T., McLennan, M. T., Lloyd, L. K., Kennelly, M. J., and Badlani, G., Treatment of intrinsic sphincter deficiency using autologous ear chondrocytes as a bulking agent, *Neurourol. Urodyn.*, 20(2), 157–165, 2001.
48. Fragonas, E., Valente, M., Pozzi-Mucelli, M., Toffanin, R., Rizzo, R., Silvestri, F., and Vittur, F., Articular cartilage repair in rabbits by using suspensions of allogenic chondrocytes in alginate, *Biomaterials*, 21(8), 795–801, 2000.
49. Diduch, D. R., Jordan, L. C., Mierisch, C. M., and Balian, G., Marrow stromal cells embedded in alginate for repair of osteochondral defects, *Arthroscopy*, 16(6), 571–577, 2000.
50. Park, D. J., Bong, J. P., Park, S. Y., and Hong, K. S., Cartilage generation using alginate-encapsulated autogenous chondrocytes in rabbits, *Ann. Otol. Rhinol. Laryngol.*, 109(12 Pt 1), 1157–1161, 2000.
51. Yang, W. D., Chen, S. J., Mao, T. Q., Chen, F. L., Lei, D. L., Tao, K., Tang, L. H., and Xiao, M. G., A study of injectable tissue-engineered autologous cartilage, *Chin. J. Dent. Res.*, 3(4), 10–15, 2000.
52. Chang, S. C., Rowley, J. A., Tobias, G., Genes, N. G., Roy, A. K., Mooney, D. J., Vacanti, C. A., and Bonassar, L. J., Injection molding of chondrocyte/alginate constructs in the shape of facial implants, *J. Biomed. Mater. Res.*, 55(4), 503–511, 2001.
53. Perka, C., Schultz, O., Spitzer, R. S., and Lindenhayn, K., The influence of transforming growth factor beta 1 on mesenchymal cell repair of full-thickness cartilage defects, *J. Biomed. Mater. Res.*, 52(3), 543–552, 2000.
54. Alsberg, E., Anderson, K. W., Albeiruti, A., Franceschi, R. T., and Mooney, D. J., Cell-interactive alginate hydrogels for bone tissue engineering, *J. Dent. Res.*, 80(11), 2025–2029, 2001.
55. Rowley, J. A., and Mooney, D. J., Alginate type and RGD density control myoblast phenotype, *J. Biomed. Mater. Res.*, 60(2), 217–223, 2002.
56. Albrecht, F. H., [Closure of joint cartilage defects using cartilage fragments and fibrin glue], *Fortschr. Med.*, 101(37), 1650–1652, 1983.
57. Homminga, G. N., Buma, P., Koot, H. W., van der Kraan, P. M., and van den Berg, W. B., Chondrocyte behavior in fibrin glue in vitro, *Acta Orthop. Scand.*, 64(4), 441–445, 1993.
58. Hendrickson, D. A., Nixon, A. J., Grande, D. A., Todhunter, R. J., Minor, R. M., Erb, H., and Lust, G., Chondrocyte-fibrin matrix transplants for resurfacing extensive articular cartilage defects, *J. Orthop. Res.*, 12(4), 485–497, 1994.
59. Hendrickson, D. A., Nixon, A. J., Erb, H. N., and Lust, G., Phenotype and biological activity of neonatal equine chondrocytes cultured in a three-dimensional fibrin matrix, *Am. J. Vet. Res.*, 55(3), 410–414, 1994.

60. Silverman, R. P., Passaretti, D., Huang, W., Randolph, M. A., and Yaremchuk, M. J., Injectable tissue-engineered cartilage using a fibrin glue polymer, *Plast. Reconstr. Surg.*, 103(7), 1809–1818, 1999.
61. Meinhart, J., Fussenegger, M., and Hobling, W., Stabilization of fibrin–chondrocyte constructs for cartilage reconstruction, *Ann. Plast. Surg.*, 42(6), 673–678, 1999.
62. Bryant, S. J., and Anseth, K. S., Controlling the spatial distribution of ECM components in degradable PEG hydrogels for tissue engineering cartilage, *J. Biomed. Mater. Res.*, 64(1), 70–79, 2003.
63. van Susante, J. L., Buma, P., Schuman, L., Homminga, G. N., van den Berg, W. B., and Veth, R. P., Resurfacing potential of heterologous chondrocytes suspended in fibrin glue in large full-thickness defects of femoral articular cartilage: an experimental study in the goat, *Biomaterials*, 20(13), 1167–1175, 1999.
64. Sims, C. D., Butler, P. E., Cao, Y. L., Casanova, R., Randolph, M. A., Black, A., Vacanti, C. A., and Yaremchuk, M. J., Tissue engineered neocartilage using plasma derived polymer substrates and chondrocytes, *Plast. Reconstr. Surg.*, 101(6), 1580–1585, 1998.
65. Ting, V., Sims, C. D., Brecht, L. E., McCarthy, J. G., Kasabian, A. K., Connelly, P. R., Elisseeff, J., Gittes, G. K., and Longaker, M. T., In vitro prefabrication of human cartilage shapes using fibrin glue and human chondrocytes, *Ann. Plast. Surg.*, 40(4), 413–420, 1998; discussion 420–421.
66. Xu, J. W., Nazzal, J., Peretti, G. M., Kirchhoff, C. H., Randolph, M. A., and Yaremchuk, M. J. ,Tissue-engineered cartilage composite with expanded polytetrafluoroethylene membrane, *Ann. Plast. Surg.*, 46(5), 527–532, 2001.
67. Neovius, E. B., and Kratz, G., Tissue engineering by cocultivating human elastic chondrocytes and keratinocytes, *Tissue Eng.*, 9(2), 365–369, 2003.
68. Xu J. W., Johnson, T. S., Hettiaratchy, S., Motarjem, P. M., Weinand, C., Randolph, M. A., and Yaremchuk, M. J., Tissue engineered flexible ear-shaped cartilage. *48th Annual Meeting of the Plastic Surgery Research Council*. Las Vegas, NV, 2003.
69. Wakitani, S., Goto, T., Young, R. G., Mansour, J. M., Goldberg, V. M., and Caplan, A. I., Repair of large full-thickness articular cartilage defects with allograft articular chondrocytes embedded in a collagen gel, *Tissue Eng.*, 4(4), 429–444, 1998.
70. Kawamura, S., Wakitani, S., Kimura, T., Maeda, A., Caplan, A. I., Shino, K., and Ochi, T., Articular cartilage repair. Rabbit experiments with a collagen gel-biomatrix and chondrocytes cultured in it, *Acta Orthop. Scand.*, 69(1), 56–62, 1998.
71. Katsube, K., Ochi, M., Uchio, Y., Maniwa, S., Matsusaki, M., Tobita, M., and Iwasa, J., Repair of articular cartilage defects with cultured chondrocytes in Atelocollagen gel. Comparison with cultured chondrocytes in suspension, *Arch. Orthop. Trauma Surg.*, 120(3–4), 121–127, 2000.
72. Uchio, Y., Ochi, M., Matsusaki, M., Kurioka, H., and Katsube, K., Human chondrocyte proliferation and matrix synthesis cultured in Atelocollagen gel, *J. Biomed. Mater. Res.*, 50(2), 138–143, 2000.
73. de Chalain, T., Phillips, J. H., and Hinek, A., Bioengineering of elastic cartilage with aggregated porcine and human auricular chondrocytes and hydrogels containing alginate, collagen, and kappa-elastin, *J. Biomed. Mater. Res.*, 44(3), 280–288, 1999.
74. Im, G. I., Kim, D. Y., Shin, J. H., Hyun, C. W., and Cho, W. H., Repair of cartilage defect in the rabbit with cultured mesenchymal stem cells from bone marrow, *J. Bone Joint Surg. Br.*, 83(2), 289–294, 2001.
75. Wakitani, S., Imoto, K., Yamamoto, T., Saito, M., Murata, N., and Yoneda, M., Human autologous culture expanded bone marrow mesenchymal cell transplantation for repair of cartilage defects in osteoarthritic knees, *Osteoarthritis Cartilage*, 10(3), 199–206, 2002.
76. Sims, C. D., Butler, P. E., Casanova, R., Lee, B. T., Randolph, M. A., Lee, W. P., Vacanti, C. A., and Yaremchuk, M. J., Injectable cartilage using polyethylene oxide polymer substrates, *Plast. Reconstr. Surg.*, 98(5), 843–850, 1996.
77. Lutolf, M. P., Weber, F. E., Schmoekel, H. G., Schense, J. C., Kohler, T., Muller, R., and Hubbell, J. A., Repair of bone defects using synthetic mimetics of collagenous extracellular matrices, *Nat. Biotechnol.*, 21, 513–518, 2003.
78. Wintermantel, E., Mayer, J., Blum, J., Eckert, K. L., Luscher, P., and Mathey, M., Tissue engineering scaffolds using superstructures, *Biomaterials*, 17(2), 83–91, 1996.
79. Scranton, A. B., and Bowman, C. N., *Photopolymerization Fundamentals and Applications*, ACS Publishers, New Orleans, 1996.

80. Bryant, S. J., Nuttelman, C. R., and Anseth, K. S., Cytocompatibility of UV and visible light photoinitiating systems on cultured NIH/3T3 fibroblasts in vitro, *J. Biomater. Sci. Polym. Ed.*, 11(5), 439–457, 2000.
81. Burkoth, A. K., and Anseth, K. S., A review of photocrosslinked polyanhydrides: in situ forming degradable networks, *Biomaterials*, 21(23), 2395–2404, 2000.
82. Elisseeff, J., Anseth, K., Sims, D., McIntosh, W., Randolph, M., Yaremchuk, M., and Langer, R., Transdermal photopolymerization of poly(ethylene oxide)-based injectable hydrogels for tissue-engineered cartilage, *Plast. Reconstr. Surg.*, 104(4), 1014–1022, 1999.
83. Nguyen, K. T., and West, J. L., Photopolymerizable hydrogels for tissue engineering applications, *Biomaterials*, 23(22), 4307–4314, 2002.
84. Sawhney, A., Pathak, C., and Hubbell, J., Bioerodible hydrogels based on photopolymerized poly(ethylene glycol)-co-poly(a-hydroxy acid) diacrylate macromers, *Macromolecules*, 26, 581–587, 1993.
85. Anseth, K. S., Shastri, V. R., and Langer, R., Photopolymerizable degradable polyanhydrides with osteocompatibility, *Nat. Biotechnol.*, 17(2), 156–159, 1999.
86. Elisseeff, J., Anseth, K., Sims, D., McIntosh, W., Randolph, M., and Langer, R., Transdermal photopolymerization for minimally invasive implantation, *Proc. Natl Acad. Sci. USA*, 96(6), 3104–3107, 1999.
87. Burdick, J. A., and Anseth, K. S., Photoencapsulation of osteoblasts in injectable RGD-modified PEG hydrogels for bone tissue engineering, *Biomaterials*, 23(22), 4315–4323, 2002.
88. Williams, C. G., Kim, T. K., Taboas, A., Malik, A., Manson, P., Elisseeff, J. In vitro chondrogenesis of bone marrow derived stem cells in a photopolymerizable hydrogel, *Tissue Eng.*, 9(4), 679–688, 2003.
89. Harris, S. A., Enger, R. J., Riggs, B. L., and Spelsberg, T. C., Development and characterization of a conditionally immortalized human fetal osteoblastic cell line, *J. Bone Miner. Res.*, 10(2), 178–186, 1995.
90. Shamblott, M. J., Axelman, J., Wang, S., Bugg, E. M., Littlefield, J. W., Donovan, P. J., Blumenthal, P. D., Huggins, G. R., and Gearhart, J. D., Derivation of pluripotent stem cells from cultured human primordial germ cells, *Proc. Natl Acad. Sci. USA*, 95(23), 13726–13731, 1998.
91. Williams, C. G., Malik, A., Kim, T. K., Manson, P., and Elisseeff, J. H., Variable cytocompatibility of six cell lines with photinitiators used for polymerizing hydrogels and cell encapsulation, *Biomaterials*, 26(11), 1211–1218, 2005.
92. Bryant, S. J., Nuttelman, C. R., and Anseth, K. S., The effects of crosslinking density on cartilage formation in photocrosslinkable hydrogels, *Biomed. Sci. Instrum.*, 35, 309–314, 1999.
93. Bryant, S. J., and Anseth, K. S., The effects of scaffold thickness on tissue engineered cartilage in photocrosslinked poly(ethylene oxide) hydrogels, *Biomaterials*, 22(6), 619–626, 2001.
94. Bryant, S. J., and Anseth, K. S., Hydrogel properties influence ECM production by chondrocytes photoencapsulated in poly(ethylene glycol) hydrogels, *J. Biomed. Mater. Res.*, 59(1), 63–72, 2002.
95. Martens, P. J., Bryant, S. J., and Anseth, K. S., Tailoring the degradation of hydrogels formed from multivinyl poly(ethylene glycol) and poly(vinyl alcohol) macromers for cartilage tissue engineering, *Biomacromolecules*, 4(2), 283–292, 2003.
96. Burkoth, A. K., Burdick, J., and Anseth, K. S., Surface and bulk modifications to photocrosslinked polyanhydrides to control degradation behavior, *J. Biomed. Mater. Res.*, 51(3), 352–359, 2000.
97. Muggli, D. S., Burkoth, A. K., and Anseth, K. S., Crosslinked polyanhydrides for use in orthopedic applications: degradation behavior and mechanics, *J. Biomed. Mater. Res.*, 46(2), 271–278, 1999.
98. Young, J. S., Gonzales, K. D., and Anseth, K. S., Photopolymers in orthopedics: characterization of novel crosslinked polyanhydrides, *Biomaterials*, 21(11), 1181–1188, 2000.
99. Park, Y. D., Tirelli, N., and Hubbell, J. A., Photopolymerized hyaluronic acid-based hydrogels and interpenetrating networks, *Biomaterials*, 24(6), 893–900, 2003.

13 Immunoisolation Techniques

Beth Anne Zielinski

CONTENTS

I. INTRODUCTION

Replacement of the vital functional physiology of irreparably damaged native organs has been the goal of both transplantation medicine and immunoisolatory medicine. Although transplants of allogeneic tissue carry the promise of complete metabolic restoration, they are beset with several problems. Limited donor supply of transplantable organs along with the immunosuppressive regimen that follows transplantation hinders sufficient organ supply and efficacy. Alternative treatments to transplantation such as administration of therapeutic proteins designed to replace metabolic deficiencies are also limited. Restrictions include the routes of administration, enzymatic degradation, inability to maintain therapeutic levels of drug, bioavailability, and ultimately, cost. This type of therapy requires stringent monitoring of delivered doses of therapeutic proteins since the system itself is not reactive to changes in the patient's physiological status. Immunoisolation and transplantation of protein-secreting cells is a viable option for the replacement of protein-secreting tissues and potentially whole organs.

Immunoisolation or encapsulated cell therapy is the process of encapsulating or sequestering metabolically active cells within a selective membrane. This membrane allows for bidirectional diffusion of nutrients and cellular secretogues while preventing the entry of host immune molecules and cells that could potentially kill the cellular implant. In addition to delivering potentially inexhaustible supplies of therapeutic protein, implants such as these may also be responsive to metabolic changes in the patient that result in modification of cellular secretions.

The first serious investigative efforts in the field of immunoisolation began with the implantation of encapsulated islets for the treatment of diabetes.[1,2] Chick et al. (1977) and Lim and Sun (1980) successfully maintained glucose homeostasis in chemically induced diabetic rats using encapsulated allogeneic islets.[1,2] Since that time, strides in the field of encapsulated cell therapy have been made not only in the area of diabetes, but also in the areas of chronic pain,[3–5] neurodegenerative diseases including Parkinson's disease,[6–9] amyotrophic lateral sclerosis (ALS),[10,11] dwarfism,[12] anemia,[13,14] hemophilia,[15,16] and cancer.[17] Human trials have been initiated for diabetes,[18,19] chronic pain,[20,21] and ALS.[22,23]

This review is intended to address the principles of immunoisolation and the technological developments underlying progress in this field. The theory of immunoisolation, capsule format criteria, cell sourcing, and the issues of immunological recognition and rejection will be discussed. Finally, modifications to the design of the immunoisolatory system will be proposed for future study and review.

II. THEORY AND CAPSULE FORMAT

Immunoisolation is based on the premise that allogeneic and xenogeneic cells, once sequestered within a selectively permeable membrane, are protected from host immune destruction and are able to deliver specific therapeutic proteins to the host over an extended period of time. In addition to protecting the transplanted cells from destruction, the semipermeable membrane also prevents outgrowth of the encapsulated cells into host parenchyma. This permits the use of mitotically active cells. Bidirectional diffusion of oxygen, carbon dioxide, soluble nutrients, and cellular secretogues including therapeutic proteins allows for sustained viability of transplanted cells and delivery of the molecules of interest. Membranes are fabricated from either natural or synthetic material and can be designed to have various pore sizes depending upon the intended application. The specific materials used will be discussed in a later section. Once manufactured, membranes are classified according to their particular molecular weight cutoffs (MWCO). The MWCO determines the types of molecules that are able to diffuse in to and out of the capsule according to size. The molecular specificities of the membranes are nominal since the capsules are not fabricated with absolute pore sizes and exhibit a pore size distribution. Although the MWCO restricts the movement of most molecules larger than the selective range, imperfections in the membrane can result in the diffusion of even larger molecules. This can lead to unintended sensitization of the host and immune destruction of transplanted cells. On the other hand, molecules within the selective range may be restricted from entering or leaving the capsule due to steric hindrance, charge, and hydrophobicity. This could also lead to immune destruction as well as the restricted delivery of therapeutic protein. Available barriers range from those considered semipermeable with a MWCO of 30 kD to those considered to be microporous having pore sizes up to 0.6 μm. Immunogenicity of the encapsulated cell types plays a key role in governing the selection of a membrane with the appropriate MWCO.[15] Semipermeable membranes are usually considered for those applications where immunogenic xenogeneic and allogeneic cells are used. Microporous membranes are usually considered for those applications where larger amounts of soluble proteins are to be delivered and the long-term viability of encapsulated cells is not a primary requirement. Such a strategy could be employed for tumor cell encapsulation where the transient release of tumor antigen from encapsulated tumor cells results in host sensitization, the generation of antitumor immune responses *in situ*, as well as encapsulated tumor cell destruction.[17] Specific immunological responses to encapsulated cells and shed molecules will be discussed in a later section.

The types of immunoisolatory systems used over the past 25 years can be categorized into two main groups; vascular perfusion devices that are implanted directly in contact with the host's circulatory system,[24,25] and nonvascular devices that are implanted subcutaneously, intramuscularly, or intraperitoneally. Vascular perfusion devices have waned in popularity as a result of their intrinsic complexity and the need for long-term anticoagulation in order to prevent thrombosis. Recently, however, resurgence in the development of the artificial liver has resulted in novel designs that may preclude hemoincompatibility.[26] These designs are intended as a bridge to transplant and not as long-term replacement therapy.

Nonvascular devices can be subdivided into two classifications; spherical microcapsules and larger polymer-based macrocapsules.[27] Macrocapsules can be designed as cylindrical hollow fibers, flat sheet, and planar devices.[22] Although all of these devices are very different in configuration, they are all engineered with one dimension below 1 mm in order to maximize the bidirectional diffusion of nutrients and cellular secretogues. Conformal coatings of cells have also been investigated as a way to reduce diffusion distances between the encapsulated cells and the host interstitum.[28]

Spherical microcapsules can be formed from organic polymers such as sodium alginate and poly-L-lysine,[2] agarose,[29,30] polyethylene glycol,[31] glycol chitosan, and multilayered glycol chitosan–alginate complexes.[32] The method most commonly used to form microcapsules from organic polymers is interfacial precipitation.[33] Interfacial precipitation involves the gelation of a polyanionic polymer–cell suspension, such as alginate and islets, in a bath containing a divalent cation such as calcium chloride. Once the cell-containing spheres are formed, they can be laminated with alternating coats of polylysine and alginate. The alginate core is then liquefied using sodium citrate. Liquefaction of the sphere's core allows for additional space within the capsule for cellular movement and growth. In the case of alginate microcapsules, permeability of the membrane and membrane strength are controlled by the percentage of alginate used, the molecular weight of the poly-L-lysine, and the number of additional polylysine–alginate layers applied.[34,35] Modifications to enhance membrane strength by reducing polylysine molecular weight or altering alginate concentration usually have secondary effects that negatively impact membrane permeability and diffusive properties. Attempts have been made to increase the strength of microcapsule membranes by combining cells with water-insoluble polyacrylates and precipitating the selective membrane in an aqueous bath.[36–39] Viability of the encapsulated cells appears to be marginal, however, due to contact between the encapsulated cells and organic solvent and inadequate diffusive properties of the encapsulating membrane.

Polyelectrolyte coacervation is another method that is used to construct hydrogel microcapsules using binary polymer blends. The hydrogel membrane is formed by the complexation of oppositely charged polymers resulting in the formation of an interpenetrating hydrogel network. Examples of binary polymer blends are alginate with protamine, and carboxymethyl cellulose and chitosan.[33] These complexes also exhibit an inverse relationship between permeability and molecular strength. A limiting factor in all of the methods discussed thus far is the inability to achieve independent control of both permeability and mechanical strength. This has lead investigators to explore the option of loading prefabricated polymer macrocapsules with either cells or cell–matrix combinations.

Macrocapsules are fabricated from preformed hollow-fiber or planar membranes prepared by the phase inversion of water-insoluble polymers quenched in a nonsolvent bath or in a saturated atmosphere. The capsules consist of an asymmetric porous structure containing a selective skin that determines MWCO and hence diffusive capacity and a more open structure responsible for mechanical support. Polymer precipitation time, polymer-solvent compatibility, and solvent concentration can influence phase separation resulting in the formation of a wide range of selective membranes which have vastly different MWCOs.[33] Membrane strength is directly related to wall thickness and is often coupled to decreased diffusive capacity and lower permeability.[33] Unlike cell-loaded microcapsules, prefabricated polymer macrocapsules can be analyzed and tested for specific characteristics such as MWCO prior to cell loading. Once characterized, macrocapsules are loaded and sealed appropriately. Due to their inherently larger size, macrocapsules are able to sequester larger cell volumes than microcapsules and can be scaled more easily to clinical applications. Furthermore, these larger capsules can accommodate the addition of a variety of luminal matrices such as alginate, chitosan, and cross-linked collagen in order to optimize cell viability and function.[40,41] Several Phase I and Phase II clinical trials have been conducted using cell-loaded prefabricated polymer hollow fibers.[19,20,23]

III. CELL SOURCING

Encapsulations of primary cell types such as Islets of Langerhans have dominated early literature.[1,2,25,42–44] Primary cells are isolated from the excised glands of donor animals. Most encapsulation systems designed to date have used allogeneic and xenogeneic cell sources. Following excision, the glands are mechanically digested, and enzymatically treated. Isolated cells are subsequently adapted to *in vitro* cell culture conditions and then finally encapsulated in the preferred system. Primary cells isolated in this manner offer advantages over other cell sources such as cell lines and modified cell lines. These benefits include the potential to provide regulated release of several cellular products at one time and the ability to afford insight into therapeutic efficacy due to their postmitotic nature. In order to achieve therapeutic efficacy, however, macrocapsules must be loaded to capacity with mitotically inactive primary cells. This can lead to the exhaustion of cell sources and an inability to supply necessary yields.

Alternatives to primary cell sources include mitotically active cells and genetically engineered cell lines. These cell sources are immortalized thus demonstrating the ability to proliferate indefinitely. Capsules can be seeded with suboptimal cell numbers and allowed to support continued growth until the cells have virtually filled the capsule. Within the encapsulated environment, however, cell proliferation may be constrained by contact inhibition and metabolic factors.[45] Dividing cell lines can also be engineered to secrete desired gene products.[9,11,14,46] This allows investigators to use easily accessible cell types such as fibroblasts and tailor these cells and the encapsulated product for a very specific application. Xenogeneic as well as allogeneic cells can be used. The disparity between the host and cell source determines the selectivity and thus the MWCO of the isolating membrane that is used. Although cell lines appear to be generally advantageous over primary cells for reasons just mentioned, genetically engineered cell lines usually only have the capability of secreting one bioactive cellular product at a time. Safety of the implanted host is another serious consideration, as these immortalized cells may be able to grow into host interstitum following escape through a defect in the isolating membrane. Risk of immune evasion may be even greater when these cells are implanted into so-called immunopriveleged sites such as the central nervous system.[47]

IV. HOST IMMUNE RESPONSES TO ENCAPSULATED CELLS

The fundamental theory of immunoisolation has focused on the physical separation of implanted cells and the cellular components of the host's immune system. Isolation of allogeneic or xenogeneic cells within a semipermeable membrane that prevents contact and interaction with host immune cells has been thought to be sufficient for implant metabolic function and survival. Although this premise remains as an extremely important consideration, it is now known that host immune systems are modulated by a myriad of soluble molecules including cytokines and chemotactic factors that not only affect host immune cell reactions, but also have the potential to affect the viability of the encapsulated cells directly. In order to ultimately achieve the goal of successful immunoisolation, issues such as capsule biocompatibility, the innate immune response, and direct and indirect pathways of antigen recognition need to be addressed.

The initial rate-limiting property of any implanted biomaterial that is to be used for cell encapsulation is biocompatibility. Materials must be immunologically inert and therefore must not cause the development of chronic inflammatory responses and foreign body reactions. Within 24 h of implantation, encapsulating polymer membranes become virtually encased in a layer of host protein. Adsorption of protein onto capsule surfaces is a dynamic process that leads to the accumulation and activation of local cell populations including resident macrophages and fibroblasts. Activation of this innate immune response can continue for approximately 7 days, at which time the response either diminishes and is replaced by fibrotic growth or continues to proliferate into chronic inflammation and the rejection of the capsule.[48,49] Either condition can result in the impedance of bidirectional diffusion resulting in chronic nutritional deprivation that eventually destroys the encapsulated cells irrespective of specific immune damage. The initial responses of protein adsorption and acute inflammation are due to both the presence of foreign biomaterials as well as the surgical procedure of implantation.

Although the preliminary response of the host is generated towards the biomaterial component of the implant, success or failure of the device may rest upon the type of cell encapsulated. Early models of encapsulation have been based upon the theory of direct antigen presentation. Graft-derived antigen presenting cells (APC) complex soluble antigen with major histocompatibility molecules (MHC II) and present this complex to host T cells. APC-T cell contact interactions along with costimulation leads to graft recognition and rejection. According to this theory, physical separation of the graft and the host by a semipermeable membrane is adequate for maintaining graft function and viability. Immune recognition of foreign tissue is much more complex, however, and involves capture of a soluble antigen and the production of chemotactic factors and cytotoxic molecules such as cytokines, antibodies, and complement proteins by the host.[54] If the membrane is permissible, graft-derived antigens can potentially diffuse across the semipermeable membrane and can be processed and presented by the host APCs. This leads to the generation of antibodies with specificity for the grafted cells. In conjunction with complement proteins and soluble immune factors, antibodies may be able to traverse the membrane and destroy encapsulated cells. The indirect pathway just described is responsible for the failures of many allogeneic and most xenogeneic implants.[50] Lastly, in addition to the generation of both nonspecific and specific immune responses, animal models themselves can differ in the immune reactions that they initiate towards implantations of the same encapsulated cells.[50] Larger animal models tend to be more sensitive to the presence of foreign molecules. Scale up from traditional rodent and other small animal models to preclinical and clinical trials has met with much failure and disappointment.[50]

V. CONCLUSION

Continuous delivery of therapeutic proteins by encapsulated cells is a promising alternative to traditional modes of treatment for many conditions such as neurodegenerative disorders, diabetes, chronic pain and cancer. Immunoisolation provides a method for achieving the sustained release of specific proteins that target host tissue directly. Issues of toxicity and unwanted side effects that plague traditional therapeutic approaches can be circumvented by using this unique delivery system. Immune recognition of encapsulated cells is evaded by incorporating a selectively permeable membrane that acts as an immunological sieve. Semipermeable membranes are designed to block the diffusion of soluble immune molecules as well as activated immune cells. Simultaneously, bioactive molecules produced by the encapsulated cells and essential nutrients are allowed access to the host and encapsulated cells respectively. Balance between these components results in a therapeutic system that can not only maintain delivery of bioactive molecules within the therapeutic window, but may also be reactive to metabolic changes in the host. This dynamic equilibrium is the goal of immunoisolatory technology. Genetic engineering of implanted cells has lead to more targeted and effective delivery. Optimization of the selective membrane remains an engineering challenge. Investigators have focused their efforts on designing hydrogel composite membranes, uniform nanoporous and micromachined capsules and vascularizing membranes to attain optimal cell viability and protein delivery.[51–53] With the advancement of genetic engineering and capsule design, immunoisolation offers the promise of biologically "smart" therapeutic systems that can be applied to virtually any physiological disorder.

REFERENCES

1. Chick, W. L., Perna, J. J., Lauris, V., Law, D., Galetti, P. M., Panol, G., Whittemore, A. D., Like, A. A., Colton, C. K., and Lysaght, M. J., Artificial pancreas using live beat cells; effects on glucose homeostasis in diabetic rats, *Science*, 197, 780–782, 1977.
2. Lim, F., and Sun, A. M., Microencapsulated islets as bioartificial endocrine pancreas, *Science*, 210, 908–910, 1980.
3. Sagen, J., Hama, A. T., Winn, S. R., Pain reduction by spinal implantation of xenogeneic chromaffin cells immunologically-isolated in polymer capsules, *Neurosci. Abstr.*, 19, 234 1993.

4. Joseph, J. M., Goddard, M. B., Mills, J., Padrun, V., Zurn, A., Zielinski, B., Favre, J., Gardaz, J. P., Mosimann, F., Sagen, J., Transplantation of encapsulated bovine chromaffin cells in the sheep subarachnoid space: a preclinical study for the treatment of cancer pain, *Cell Transplant.*, 3, 355–364, 1994.
5. Decostard, I., Buscher, E., Gilliard, N., Saydoff, J., Zurn, A., and Aebischer, P., Intrathecal implants of bovine chromaffin cells alleviate mechanical allodynia in a rat model of neuropathic pain, *Pain*, 76, 159–166, 1998.
6. Tresco, P. A., Winn, S. R., Tan, S., Jaeger, C. B., Greene, L. A., and Aebsicher, P., Polymer-encapsulated PC12 cells: long-term survival and associated reduction in lesion-induced rotational behaviour, *Cell Transplant.*, 1, 255–264, 1992.
7. Aebischer, P., Goddard, M., Signore, A. P., and Timpson, R. L., Functional recovery in hemiparkinsonian primates transplanted with polymer-encapsulated PC12 cells, *Exp. Neurol.*, 126, 151–158, 1994.
8. Tseng, J. L., Baetge, E. E., Zurn, A. D., and Aebsicher, P., GDNF reduces drug-induced rotational behavior after medial forebrain bundle transaction by a mechanism not involving striatal dopamine, *J. Neurosci.*, 17, 325–333, 1997.
9. Sautter, J., Tseng, J. L., Braguglia, D., Aebischer, P., Spenger, C., Seiler, R. W., Widmer, H. R., and Zurn, A. D., Implants of polymer-encapsulated genetically modified cells releasing glial cell line-derived neurotrophic factor improve survival, growth, and function of fetal dopaminergic grafts, *Exp. Neurol.*, 149, 230–236, 1998.
10. Sagot, Y., Tan, S. A., Baetge, E., Schmalbruch, H., Kato, A. C., and Aebischer, P., Polymer encapsulated cell lines genetically modified to release ciliary neurotrophic factor can slow down progressive motor neuropathy in the mouse, *Eur. J. Neurosci.*, 7, 1313–1322, 1995.
11. Tan, S. A., Deglon, N., Zurn, A. D., Baetge, E. E., Bamber, B., Kato, A. C., and Aebischer, P., Rescue of motor neurons from axotomy-induced cell death by polymer encapsulated cells genetically engineered to release CNTF, *Cell Transplant.*, 5, 577–587, 1996.
12. Chang, P. L., Shen, N., and Wescott, A. J., Delivery of recombinant gene products with microencapsulated cells in vivo, *Hum. Gene Ther.*, 4, 433–440, 1993.
13. Koo, J., and Chang, T. M. S., Secretion of erythropoietin from microencapsulated rat kidney cells: preliminary results, *Int. J. Artif. Organs*, 16, 557–560, 1993.
14. Rinsch, C., Regulier, E., Deglon, N., Dalle, B., Beuzard, Y., and Aebischer, P., A gene approach to regulated delivery of erythropoietin as a function of oxygen retention, *Hum. Gene Ther.*, 8, 1881–1889, 1997.
15. Colton, C. K., Engineering challenges in cell-encapsulation technology, *Trends Biotechnol.*, 14, 158–162, 1996.
16. Brauker, J. H., Martinson, L. A., Hill, R. S., Young, S. K., Carr-Brendel, V. E., and Johnson, R. C., Neovascularization of immunoisolation membranes: the effect of membrane architecture and encapsulated tissue, *Transplant. Proc.*, 24, 2924 1992.
17. Geller, R. L., Neuenfeldt, S., Levon, S. A., Maryanov, D. A., Thomas, T. J., and Brauker, J. H., Immunoisolation of tumor cells: generation of antitumor immunity through indirect presentation of antigen, *J. Immunother.*, 20, 131–137, 1997.
18. Soon-Shiong, P., Feldman, E., Nelson, R., Heintz, R., Yao, O., Zheng, T., Merideth, N., Skjak-Braek, G., Espevik, T., Smidsrod, O., and Sandford, P., Long-term reversal of diabetes by the injection of immunoprotected islets, *Proc. Natl Acad. Sci. USA*, 90, 5843–5847, 1994.
19. Scharp, D. W., Swanson, C. J., Olack, B. J., Latta, P. P., Hegre, O. D., Doherty, E. J., Gentile, F. T., Flavin, K. S., Ansara, M. F., and Lacy, P. E., Protection of encapsulated human islets implanted without immunosuppression in patients with type I or type II diabetes and in nondiabetic control subjects, *Diabetes*, 43, 1167–1170, 1994.
20. Aebischer, P., Buscher, E., Joseph, J. M., Favre, J., Lysaght, M., and Rudnick, S., Transplantation in humans of encapsulated xenogeneic cells without immunosuppression, *Transplantation*, 58, 1275–1277, 1994.
21. Buscher, E., Goddard, M., Heyd, B., Joseph, J. M., Favre, J., De Tribolet, N., Lysaght, M., and Aebischer, P., Immunoisolated xenogeneic chromaffin cell therapy for chronic pain, *Anesthesiology*, 85, 1005–1012, 1996.

22. Ezzell, C., Tissue engineering and the human body shop: encapsulated-cell transplants enter the clinic, *J. NIH Res.*, 7, 47–51, 1995.
23. Aebischer, P., Scluep, M., Deglon, N., Joseph, J. M., Hirt, L., Heyd, B., Goddard, M., Hammang, J. P., Zurn, A. D., Kato, A. C., Regli, R., and Baetge, E. E., Intrathecal delivery of CNTF using encapsulated genetically modified xenogeneic cells in amyotrophic lateral sclerosis patients, *Nat. Med.*, 2, 1041 1996.
24. Sullivan, S. J., Maki, T., Borland, K. M., Mahoney, M. D., Soloman, B. A., Muller, T. E., Monaco, A. P., and Chick, W. L., Biohybrid artificial pancreas: long-term implantation studies in diabetic, pancreatectomized dogs, *Science*, 252, 718–721, 1991.
25. Maki, T., Lodge, J. P. A., Carretta, M., Ohzato, H., Barland, K. M., Sullivan, S. J., Staruk, J., Muller, T. E., Soloman, B. A., Chick, W. L. et al., Treatment of severe diabetes mellitus for more than one year using a vascularized hybrid artificial pancreas, *Transplantation*, 55, 713–718, 1993.
26. Stange, J., Hassanein, T. I., Mehta, R., Mitzer, S. R., and Bartlett, R. H., The molecular adsorbents recycling system as a liver support system based on albumin dialysis: a summary of preclinical investigations, prospective, randomized, controlled clinical trial and clinical experience from 19 centers, *Artif. Organs*, 26, 103–110, 2002.
27. Geller, R. L., Loudovaris, T., Neunfeldt, S., Johnson, R. C., and Brauker, J. H., Use of an immunoisolation device for cell transplantation and tumor immunotherapy, *Ann. NY Acad. Sci.*, 831, 438–451, 1997.
28. Hill, R. S., Cruise, G. M., Lamberti, F. V., Yu, X., Garufis, C. L., Yu, Y., Mundwiler, K. E., Cole, J. F., Hubbell, J. A., Hegre, O. D., and Scharp, D. W., Immunoisolation of adult porcine islets for the treatment of diabetes mellitus: the use of photopolymerizable polyethylene glycol in the conformal coating of mass-isolated porcine islets, *Ann. NY Acad. Sci.*, 831, 332–343, 1997.
29. Iwata, H. Y., Murakami, Y., and Ikada, Y., Control of complement activities for immunoisolation, *Ann. NY Acad. Sci.*, 875, 7–23, 1999.
30. Kobayashi, T., Aomatsu, Y., Iwata, H., Kin, T., Kanehiro, H., Hisanaga, M., Ko, S., Nagao, M., and Nakajima, Y., Indefinite islet protection from autoimmune destruction in nonobese diabetic mice by agarose microencapsulation without immunosuppression, *Transplantation*, 75, 619–625, 2003.
31. Chen, J. P., Chu, I. M., and Shiao, M. Y., Microencapsulation of islets in PEG-amine modified alginate-poly (L-lysine)-alginate microcapules for constructing bioartificial pancreas, *J. Ferment. Bioeng.*, 86, 185–190, 1998.
32. Sakai, S., Ono, T., Ijima, H., and Kawakami, K., Control of molecular weight cut-off for immunoisolation by multilayering glycol chitosan–alginate polyion complex on alginate-based microcapsule, *J. Microencapsulation*, 17, 691–699, 2000.
33. Chaikof, E. L., Engineering and material considerations in islet cell transplantation, *Annu. Rev. Biomed. Eng.*, 1, 103–127, 1999.
34. Thu, B., Bruheim, P., Espevik, T., Smidsrod, O., Soon-Shiong, P., and Skjak-Braek, G., Alginate polycation microcapsules. I. Interaction between alginate and polycation, *Biomaterials*, 17, 1031–1040, 1996.
35. Thu, B., Bruheim, P., Espevik, T., Smidsrod, O., Soon-Shiong, P., and Skjak-Braek, G., Alginate polycation microcapsules. II. Some functional properties, *Biomaterials*, 17, 1069–1079, 1996.
36. Boag, A. H., and Sefton, M. V., Microencapsulation of human fibroblasts in a water-soluble polyacrylate, *Biotechnol. Bioeng.*, 30, 954–962, 1987.
37. Brauker, J. H., Martinson, L. A., Hill, R. S., Young, S. K., Carr-Brendel, V. E., and Johnson, R. C., Neovascularization of immunoisolation membranes: the effects of membrane architecture and encapsulated tissue, *Transplant. Proc.*, 24, 2924 1992.
38. Brauker, J. H., Frost, G. H., Dwarki, V., Nijjar, T., Chin, R., Carr-Brendel, V., Jasunas, C., Hodgett, D., Stone, W., Cohen, L. K., and Johnson, R. C., Sustained expression of high levels of human factor IX from human cells implanted within an immunoisolation device into athymic rodents, *Hum. Gene Ther.*, 9, 879–888, 1998.
39. Broughton, R. L., and Sefton, M. V., Effect of capsule permeability on growth of CHO cells in Eudragit RL microcapsules: use of FITC-dextran as a marker of capsule quality, *Biomaterials*, 10, 462–465, 1990.
40. Lanza, R. P., Butler, D. H., Borland, K. M., Harvey, J. M., Fanstman, D. L., Soloman, B. A., Muller, T. E., Rupp, R. G., Maki, T., Monaco, A. P., Successful xenotransplantation of a diffusion-based

biohybrid artificial pancreas: a study using canine, bovine, and porcine islets, *Transplant. Proc.*, 24, 669–671, 1992.
41. Zielinski, B. A., and Aebischer, P., Chitosan as a matrix for mammalian cell encapsulation, *Biomaterials*, 15, 1049–1056, 1994.
42. Lacy, P. E., Hegre, O. D., Gerasimidi-Vazeou, A., Gentile, F. T., and Dionne, K. E., Maintenance of normoglycemia in diabetic mice by subcutaneous xenografts of encapsulated islets, *Science*, 254, 1782–1784, 1991.
43. Lanza, R. P., Lodge, P., Borland, K. M., Carretta, M., Sullivan, S. J., Beyer, A. M., Muller, T. E., Soloman, B. A., Maki, T., Monaco, A. P. et al., Transplantation of islet allografts using a diffusion-based biohybrid artificial pancreas: long-term studies in diabetic, pancreatectomized dogs, *Transplant. Proc.*, 25, 978–980, 1993.
44. Soon-Shiong, P., Feldman, E., Nelson, R., Heintz, R., Yao, O., Zheng, T., Merideth, N., Skjak-Braik, G., Espevik, T., Smidsrod, O., and Sandford, P., Long-term reversal of diabetes by the injections of immunoprotected islets, *Proc. Natl Acad. Sci. USA*, 90, 5843–5847, 1993.
45. Lysaght, M. J., Frydel, B., Gentile, F., Emerich, D., and Winn, S., Recent progress in immunoisolated cell therapy, *J. Cell. Biochem.*, 56, 196–203, 1994.
46. Chang, P. L., Sheng, N., and Westcott, A. J., Delivery of recombinant gene products with microencapsulated cells in vivo, *Hum. Gene Ther.*, 4, 433–440, 1993.
47. Morris, P. J., Immunoprotection of therapeutic cell transplants by encapsulation, *Trends Biotechnol.*, 14, 163–167, 1996.
48. de Vos, P., van Hoogmoed, C. G., de Haan, B. J., and Busscher, H. J., Tissue responses against immunoisolating alginate–PLL capsules in the immediate post-transplant period, *J. Biomed. Mater. Res.*, 62, 430–437, 2002.
49. Grey, D. W. R., An overview of the immune system with specific reference to membrane encapsulation and islet transplantation, *Ann. NY Acad. Sci.*, 944, 226–239, 2001.
50. Gill, R. G., Use of small animal models for screening immunoisolation approaches to cellular transplantation, *Ann. NY Acad. Sci.*, 944, 35–46, 2001.
51. Risbud, M. V., Bhonde, M. R., and Bhonde, R. R., Effect of chitosan–polyvinyl pyrrolidone hydrogel on proliferation and cytokine expression of endothelial cells: implications in islet immunoisolation, *J. Biomed. Mater. Res.*, 57, 300–305, 2001.
52. Leoni, L., and Desai, T. A., Nanoporous biocapsules for the encapsulation of insulinoma cells: biotransport and biocompatibility considerations, *IEEE Trans. Biomed. Eng.*, 48, 1335–1341, 2001.
53. Tao, S. L., and Desai, T. A., Microfabricated drug delivery systems: from particles to pores, *Adv. Drug Deliv. Rev.*, 24, 315–328, 2003.
54. Gray, D. W. R., An overview of the immune system with specific reference to membrane encapsulation and islet transplantation, *Ann. NY Acad. Sci.*, 944, 226–239, 2001.

APPENDIX 1. IMMUNOISOLATION

Schematic depiction of encapsulated cells implanted into a nonautologous host (Scheme 13.1) Bidirectional diffusion of nutrients and cellular secretogues into and out of the capsule occurs along with movement of immune molecules that may have the potential to traverse the capsule wall. Movement into the capsule is influenced by structural integrity of the membrane, MWCO of the membrane and intensity of the host reaction that is elicited.

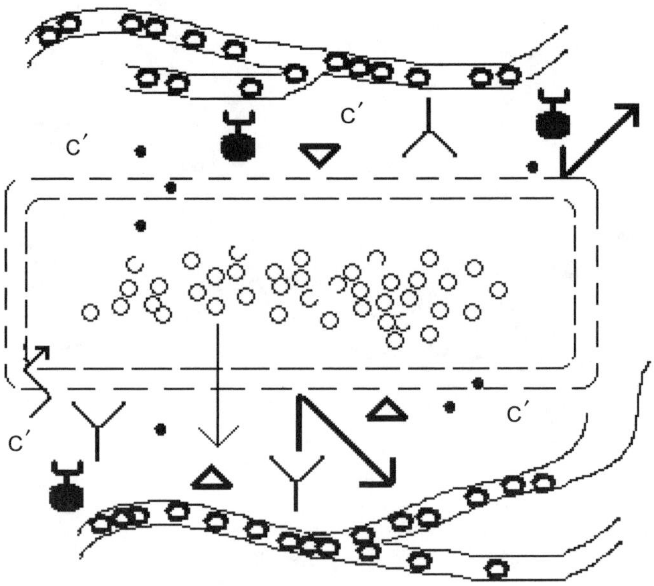

SCHEME 13.1 Bombardment of encapsulated cells by host immune components. = host immune effector cells; = host antibody molecules; ∇ = proinflammatory molecules; C′ = complement proteins; ○ = encapsulated cells; ● = cellular secretogues.

14 Self-Assembled Monolayers in Mammalian Cell Cultures

Xingyu Jiang, Jessamine Ng Lee, and George M. Whitesides

CONTENTS

I. INTRODUCTION

Self-assembled monolayers (SAMs) are materials comprising ordered monolayers of organic molecules supported on solid substrates.[1,2] The most extensively investigated types of SAMs form when molecules having the general formula $HS(CH_2)_nY$ self-assemble on films of gold, silver, or palladium.[1,3,4] In general, for SAMs formed from XRY, –X is a functional group on the molecule that has a high chemical affinity to a solid substrate, –Y represents a group whose functions can be tailored to generate a surface that has a particular property, and –R is an organic linker in between. Interactions between the –R groups, as well as those between the –Y groups, also contribute to the stability of the system. SAMs offer better control of surface chemistry than polymers, metals, and metal oxides. By making it possible to control the molecular level structure of surfaces, SAMs have allowed applications ranging from the fabrication of surfaces that are tailored to present biomolecular ligands, through to the development of supports for attached tissue cultures, to the testing of substrates for tissue engineering. SAMs are also widely used in a range of non-biological applications (micro and nanolithography,[5] preparation of substrates for studies of wetting,[6,7] studies of adhesion and tribology,[8,9] studies of nucleation sites for crystallization,[10] modification of surfaces of electrodes,[11] and studies of mechanisms of electron transport in organic molecules).[12,13] This chapter focuses on applications of SAMs in cell and tissue cultures.

II. PREPARATIONS OF SAMs AND TYPES OF SAMs

SAMs are formed by immersing gold-coated substrates in solutions of organic disulfides or alkanethiols.[14] Table 14.1 lists a number of different types of SAMs.

A. SAMs of Alkanethiols on Gold and Silver

Alkanethiolates with the general formula $HS(CH_2)_nY$ self-assemble into ordered monolayers on gold and silver. This type of SAM is the most widely used in studies related to biology,[15,16] Although silver forms more ordered SAMs than gold, silver has two characteristics that make it less attractive than gold for biological applications: it is toxic to cells, and it oxidizes to silver oxide

TABLE 14.1
Substrates and Ligands that Form SAMs

Substrate	Organic Precursor	Reference
Au	RSSR′ R = aliphatic or aromatic	103
—	RSH	103
—	RSR	103
Ag	RSH	103
Cu	RSH	103
Pd	RSH	4
Pt	RNC/RSH	103
SiO_2/glass	$RSiX_3$, X = Cl, OH, OCH_3, OCH_2CH_3	104
Si/SiH	$(RCOO)_2$	105
—	RCH = CHR′	106
—	RLi, RMgX	107
GaAs	RSH	108
InP	RSH	109
In_2O_3/SnO_2	RPO_3H_2	110
Al_2O_3/other metal oxides	$RCOOH/RCONHOH/ROPO_3^{2-}/RPO_3^{2-}$	103

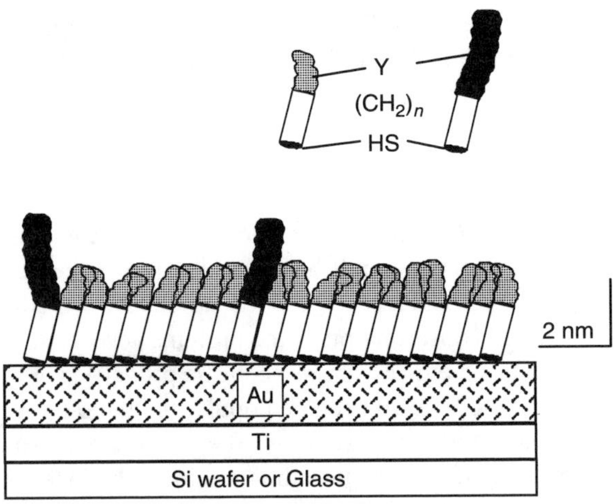

FIGURE 14.1 SAMs are formed by $HS(CH_2)_mY_1$ and $HS(CH_2)_nY_2$ on gold. A generic scenario of a mixed SAM formed by two different types of alkanethiols is shown.

upon contact with air. SAMs on gold are therefore the preferred materials for studies in biology. This chapter will focus on the SAMs formed by thiols on gold.

Mixed SAMs can be formed from a mixture of $HS(CH_2)_mY_1$ and $HS(CH_2)_nY_2$[17] (where $-Y_1$ and $-Y_2$ represent different types of chemical groups, and m and n can be the same or different; Figure 14.1). Mixed SAMs have important applications in cell and tissue engineering (see Sections VI–VII in this chapter).

B. SAMs of Alkylsiloxanes

SAMs of alkylsiloxanes on Si/SiO_2 are useful for functionalizing surfaces of glass and silicon wafers.[15] These SAMs have found many applications in controlling cell attachment,[18] patterning neurons,[19,20] and patterning DNA on chips.[21,22] This class of SAMs has been extensively reviewed elsewhere and will not be addressed in this chapter.[23]

C. Other Types of SAMs

Table 14.1 lists some of the other types of SAMs. These SAMs are not as widely used as the SAMs formed by thiols on gold, but we list them for completeness; some of them have been extensively characterized while others have not.[5]

III. CHARACTERIZATION OF SAMs

SAMs on gold are molecularly well-defined interfacial materials. They are thin films that self-assemble onto a solid substrate with semicrystalline or polycrystalline packing, and whose chemical and physical properties are largely controlled by the characteristics of the terminal groups.[24] Many techniques used for analyzing surfaces have been employed to characterize the structure and composition of SAMs and systems involving SAMs.

Contact angle goniometry gives information about the homogeneity of the SAM by examining the hydrophobicity (i.e., the wetting properties) of the surface; this property is largely controlled by the terminal groups presented on the SAMs.[1,14] Infrared (IR) spectroscopy is used to characterize the ordering of SAMs.[25] Ellipsometry measures the thickness of an organic film formed on a metallic surface, and it is particularly useful for analyzing the amount of the adsorbent on the surface.[26,27] Mass spectrometry (MS) identifies the types of organic functional groups that can be ejected in ionized form from the surface.[28,29] X-ray photoelectron spectroscopy (XPS) provides the elemental composition of the surface. It is useful for confirming the elemental composition of various types of SAMs and characterizing the adsorbed proteins or polymers on SAMs.[30–32] Scanning probe microscopy provides information about the composition of SAMs on the atomic scale.[33,34]

Surface plasmon resonance (SPR) spectroscopy allows *in situ* measurements of the thickness of adsorbed organic films on the surfaces of metals based on the shift in the resonance angle of surface plasmons.[35–37] Quartz crystal microbalance (QCM) is an acoustic method that gives real-time information about the amount and conformation of the proteins adsorbed,[38] based on the change in resonance frequency of a crystal as the mass and viscoelastic properties of the adsorbed material on the quartz are varied.[39] SPR and QCM have become popular methods for measuring the adsorption of proteins on surfaces for the following reasons: they provide thermodynamic and kinetic information about adsorption, they do not require the proteins to be labeled, and they require only modest amounts of proteins. Typically, SPR provides higher signal-to-noise ratios than QCM, although QCM provides additional information about the change in conformation of the proteins as they adsorb, and about the viscoelastic properties of the adsorbed protein film.

These methods generate qualitative and quantitative information — especially composition and coverage — about the properties of SAMs. Some of them are also used to study the interactions of biomolecules and cells with SAMs.

IV. INTRODUCTION OF FUNCTIONAL GROUPS TO THE SURFACES OF SAMs

SAMs have the capacity to accommodate a wide range of chemical groups on a surface. Thus, diverse biochemical functionalities and macroscopic properties can be incorporated. Because SAMs enable molecular level control of interfaces, they allow the study of structure–property relationships. This capability to fine-tune the surface-properties of SAMs is the major advantage of SAMs over organic polymers, metals, or metal oxides for fundamental studies.

There are two main methods of introducing chemical groups onto surfaces using SAMs. One method is to synthesize the alkanethiol or organic disulfide that already terminates in the appropriate functional group. The second method is to introduce functional groups onto the surface

TABLE 14.2
Biologically Important Ligands that Have Been Introduced onto the Termini of SAMs

Molecule on SAM	Molecule in Solution/Function	Reference
Benzensulfonamide	Carbonic andydrase	70
Nitrilotriacetic acid	Ni (II) and His-tag	71
Biotin	Avidin/strepavidin/antibiotin	52
RGD	Integrin	42
Ang-1	Integrin (α5)	111
Mannose	FimH adhesin	43
Globotriose	IgG/IgM	44
Acridine	DNA	112
D-Ala-D-Ala	Vancomycin	113
PEG and others	Resist nonspecific adsorption of proteins	26
Carbohydrates	Lectins	77
Kinase substrates	Kinase	75
Phosphonate	Cutinase	72

after the formation of a SAM. Using these reactions, small molecules, peptides, sugars, proteins, and DNA can be incorporated onto surfaces (Table 14.2).

A. The Synthesis of SAMs

Many types of thiols have been synthesized. For example, we synthesized thiols that resist the non-specific adsorption of proteins, such as polyethylene glycol (PEG)-terminated thiols.[40,41] We also made thiols that terminate in a peptide sequence (e.g., RGD) and promote adhesion of cells.[42] We, among others, synthesized thiols that terminate in carbohydrate groups to carry out fundamental studies of the interactions between proteins and surfaces.[43,44]

B. *In Situ* Transformation

An alternative strategy for introducing functional groups onto a surface is to carry out chemical reactions on the terminal groups of the SAM after its formation on the surface. The approach using activated esters and related groups is one of the most commonly used strategies to introduce functional groups onto the surfaces of SAMs. Conversion to functional moieties usually involve intermediates such as succinimidyl or pentafluorephenyl esters,[45,46] interchain anhydrides,[47] maleimide,[48] aldehydes,[49] or acid chlorides/fluorides.[50]

Mrksich and coworkers used electrochemistry to immobilize or release functional groups on SAMs. In one such system, a monolayer of hydroquinone was electrooxidized to quinone, and the quinone subsequently reacted with a solution of cyclopentadiene to give the Diels–Alder adduct on the surface.[51] In a second example, a monolayer of a quinone propionic ester was selectively reduced by applying a reductive potential to the gold substrate, and this reduction caused the release of biotin from the SAM.[52]

Other types of chemical reactions have been employed to introduce functional groups onto surfaces, for example, photochemical reactions that carry out changes on SAMs.[53] A more extensive review on chemical reactions on SAMs covers the details of all types of *in situ* transformations on SAMs.[54]

Each of the two strategies used for introducing functional groups onto the termini of SAMs has its advantages and disadvantages. The synthesis of desirable thiols is often laborious and cannot be used to introduce functional groups that are incompatible with sulfides or disulfides. However, the formation of the SAM is convenient (simple immersion of the gold coated substrate in solutions of desired thiols) once the thiols are made. *In situ* transformation of SAMs is often more straightforward in synthesis and allows several layers of transformations, but it is difficult to achieve 100% conversion and to control the exact densities of ligands in mixed SAMs.[54] *In situ* transformation, therefore, typically leads to heterogeneous surfaces. Many types of biologically relevant groups have been introduced onto the surfaces of SAMs; Table 14.2 lists some of them.

V. CHEMICAL TRANSFORMATIONS THAT MODIFY THE INTERACTIONS BETWEEN THE SAM AND THE SUBSTRATE

This section covers chemical reactions that change the sulfur–gold bond and alter the overall structures of SAMs. The ability to selectively change the properties of a surface using this class of chemical reactions has important applications in biology.

A. Using Electrochemistry to Desorb SAMs

When an electrical potential is applied across the monolayer, the thiols making up the SAM desorb and the integrity of the SAM is destroyed.[55] Electrochemical desorption of SAMs is used to release biological ligands from surfaces,[56,57] and this process has been extensively studied, characterized, and applied to various systems.[55,58,59] Electrochemists have focused on the desorption of SAMs on the dependence of the strength of the applied potential, the length of the thiol chains making up the SAM, and the conditions in solution (such as pH and temperature).[60,61] Desorption of PEG-terminated SAMs enables adhesion of proteins and attachment of cells in areas that were originally inert, and allows patterned cells to be released from areas of confinement.[59,62] This process is useful for studying the spreading of cells.

B. Oxidation of SAMs and Other Systems

Light and air can irreversibly oxidize thiols and cause them to desorb from surfaces of gold.[63] SAMs were used as photoresists for the patterning of micrometer sized features and can be used to study the adhesion and patterning of mammalian cells.[64,65]

Langer and coworkers[66] made a system of "switchable" SAMs that can reversibly modify the wettability of a surface by controlling electrochemical changes on the surface.

VI. INTERACTIONS OF PROTEINS WITH SAMs

By enabling molecular level control of interfaces, SAMs allow systematic studies of the interactions between proteins and surfaces. Because the interactions between cells and surfaces are mediated by proteins, these studies also provide the basis for studies of interactions between cells and surfaces. SPR, QCM, and ellipsometry are the main tools used to quantify the amounts of proteins adsorbed onto a given surface (see Section III).

A. Using SAMs as Model Surfaces to Study Adsorption of Proteins

Proteins adsorb irreversibly to many kinds of surfaces.[26] Adsorption of proteins on surfaces is a complex issue, dominated by hydrophobic, electrostatic, and van der Waals interactions. The adsorption is highly dependent on the structures and properties of both the proteins and the surfaces. We will only present a brief overview of this issue and point out related studies important for tissue engineering.[16]

Since SAMs allow the incorporation of different types and densities of chemical functional groups onto surfaces, macroscopic properties of the surface, such as wettability, can be systematically varied. Studies of the wettability of surfaces using SAMs have improved the understanding of the physical parameters that determine the adsorption of proteins. For instance, in general, the more hydrophobic a surface is, the more proteins adsorb (with some exceptions).[67]

B. "Inert" Surfaces

We refer to surfaces that resist the non-specific adsorption of protein as "inert" surfaces. PEG-terminated SAMs were initially identified as being inert.[26] We have screened many types of surfaces to identify additional types of inert surfaces using the anhydride chemistry approach (see Section IV.B).[68] There are certain common features associated with surfaces that are inert (such as overall electrical neutrality, polarity, H-bond donor/acceptor), but there are still no general principles for designing inert surfaces. PEG and related surfaces are the most inert surfaces that have been studied thus far.[68]

C. Non-Specific Adsorption of Proteins on Hydrophobic SAMs

SAMs allow systematic studies of interactions of proteins with surfaces presenting hydrophobic groups of well-defined shapes. We used mixed SAMs presenting hydrophobic headgroups in a background of PEG-terminated thiols as a model to study the adsorption of proteins using SPR. The extent to which adsorbed proteins undergo conformational rearrangements appears to depend on the density of the hydrophobic groups at the surface, and on the concentration of proteins in solution.[69]

D. Biospecific Adsorption of Proteins on SAMs

Using inert surfaces, it is possible to generate surfaces that present specific ligands that immobilize only desired proteins.[36,70] These surfaces allow studies of the thermodynamics and kinetics of the interactions between a type of ligand and its receptor on the surface.[71,72]

E. Electroactive Proteins on SAMs

It is also possible to probe the conformation of redox active proteins on surfaces through the use of electrochemistry and SAMs. The formal redox potential of cytochrome c (cyt c), when it is adsorbed onto a SAM gives information about the conformation and orientation of the protein. The conformation and orientation of the protein adsorbed, and thus redox potential, changes depending on the type of monolayer presented on the surface.[73] For instance, differences in formal redox potentials of cyt c shows that the conformation of cyt c adsorbed on SAMs terminating in methyl, amine, and aromatic groups is different than on SAMs terminating in trimethylammonium and carboxylic acid groups. The rates of electron transfer in these SAMs are also different.

F. Controlling the Orientation of Proteins on Surfaces

How is the conformation of an adsorbed protein determined by the properties of the surface? Some studies on the adsorption of proteins on SAMs have demonstrated how conformations of proteins can be determined by the charge and hydrophobicity of the surface. It is still difficult to determine, *a priori*, the conformation of any given protein on a surface with known physical and chemical properties, although in certain cases, controlling the orientation of proteins on surfaces with either electrostatic forces or biospecific binding is certainly possible.[72,74]

G. Other Biomolecules: DNA, Lipids, and Others

In addition to proteins, it is possible to immobilize other classes of biologically important molecules on SAMs, such as peptides,[75] DNA,[56] lipids,[76] and carbohydrates.[77] Some of these molecules have found uses in tissue engineering.[16]

SAM-based systems have dramatically improved the understanding of how proteins interact with surfaces, especially the influence of charge and hydrophobicity of the surface on the adsorption of proteins. The effect of pH and ionic strength on the adsorption of certain proteins yielded information about the mechanism of adsorption.[39] Principles derived from these studies have enabled the development of new types of inert surfaces, as well as the design of materials having properties that interact favorably with proteins.[16]

VII. CONTROLLING ADSORPTION OF PROTEINS AND PATTERNING OF CELLS; APPLICATION OF SAMs IN CELL BIOLOGY

Most mammalian cells need to attach to a surface and spread, in order to go through the cell cycle and maintain basal cellular functions.[78] The attachment of mammalian cells can be controlled using SAMs. There are several levels in controlling the attachment of cells: controlling the composition and density of ligands for cell-surface receptors, minimizing non-specific attachment, and spatially confining the patterns of cells.[42,79,80] These technologies provide the opportunity to study in detail the biology of attached cells.

A. Controlling the Composition and Density of Ligands on the Surface

It is possible to achieve biospecific attachment of mammalian cells using a SAM presenting the peptide sequence arginine–glycine–aspartate (RGD) in a background of PEG groups.[42] By controlling the relative amount of RGD-terminated thiols in the mixed SAM, we can control the adhesion of cells on the substrate. This system allows systematic studies of the fundamental mechanisms of adhesion on surfaces, as well as the spreading and detachment of cells.

B. Soft Lithography, μCP, and Patterning Proteins

"Soft lithography" refers to a set of technologies capable of generating micro- and nanofeatures rapidly using flexible polymers such as poly(dimethylsiloxane) (PDMS).[5] Micro-contact printing (μCP) is an important technique in soft-lithography, where a stamp carrying an ink is used to "print" thiols onto gold to form spatially confined SAMs (Figure 14.2).[81] This method has become a standard technology to generate patterns of proteins and cells, and is generally applicable to many types of proteins. Once the stamp is made, it is straightforward to form patterns of SAMs on a surface. This process does not require the use of a clean room and can be implemented under ambient conditions of most biology laboratories. Figure 14.2 shows how this process can be used to generate arrays of proteins.

To pattern islands of proteins, a thiol that terminates in a methyl group (such as $HS(CH_2)_{17}CH_3$) is first printed onto the gold using a PDMS stamp. The surface is then incubated with a second thiol

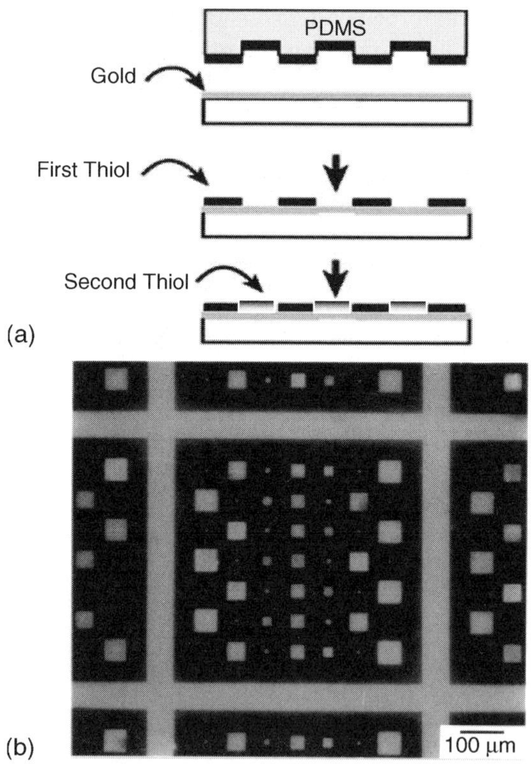

FIGURE 14.2 Generation of micropatterns of proteins by μCP of SAMs. (a) A schematic illustration of the procedure of μCP. (b) Incubation of a chip that has SAMs patterned by μCP with a solution of fibronectin, an extracellular matrix protein, results in arrays of proteins. The proteins were visualized by antibodies against fibronectin.

that terminates in an inert group (such as PEG, $-(CH_2CH_2O)_{n=3-6}OH$) in order to fill the uncovered surfaces of gold with PEG. Patterns of proteins are made by the adsorption of proteins from a solution containing them.

C. Patterning Cells

Most mammalian cells require adsorbed extracellular matrix (ECM) proteins or adhesion molecules to attach to surfaces.[78] Defined patterns of these proteins or molecules can restrict the spreading of cells.[79,82] The geometry (i.e., the sizes and shapes) of attached cells can be conveniently controlled by the geometry of the stamps used in μCP. Figure 14.3 illustrates the result of a single bovine endothelial (BCE) cell patterned using μCP of SAMs.

By varying the sizes and total attachment areas of a single cell using patterned arrays of ECM proteins, we determined that it is the total area of projection of a cell, instead of the total area of attachment, that is important in determining whether a cell undergoes growth or programmed death (i.e., apoptosis).[83]

D. Dynamic Patterning of Cells and the Patterning of Multiple Proteins and Cells

When cells are patterned onto SAMs, it is possible to release them controllably from their patterns of confinement; we call this process "dynamic patterning."[59,84] Cells can be released by electrochemical desorption of SAMs,[59] and we believe that this straightforward method will allow cell-based assays based on cell motility for the screening of drugs. Mrksich and coworkers[84] electrooxidized a mixed SAM of hydroquinone and PEG (that was overall inert to the spreading and attachment of cells) to immobilize RGD-tethered cylcopentadiene; the attachment of this peptide enabled the release of cells controllably from their patterns of confinement.

There have been other methods to pattern cells into defined geometries, for example, using agar,[85] siloxane-based chemistry on glass or silicone chips,[20,86] polymers,[87] and elastic membranes.[88] Micropatterning of cells with SAMs on gold is the only process to date that is

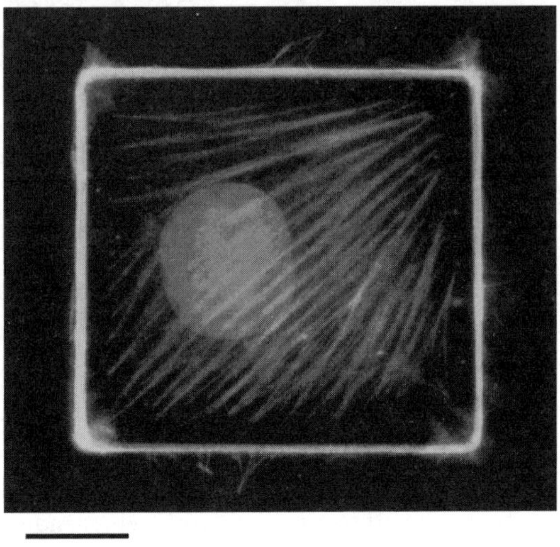

FIGURE 14.3 A single bovine capillary endothelial cell patterned on a square microisland. The cell is stained with a fluorescent probe that binds actin. Also shown is the nucleus.

simple to implement, allows control of cell adhesion, achieves high levels of fidelity, is able to maintain long-term stability, and is convenient for the execution of dynamic patterning.

E. Directional Control of Cell Migration

Defining the shape of a cell can direct cell migration.[89] When cells that are confined to square patterns are stimulated by growth factors, they extend their lamellipodia, filopodia, and microspikes preferentially to the corners of the squares. Their behaviors are similar on other shapes having corners, such as triangles, pentagons, and hexagons.[90]

Technologies based on SAMs can be used to control the attachment of cells that have defined sizes, shapes, and extent of spreading. These technologies yield information about the cell that is difficult to obtain through biochemical methods.

VIII. THE BRIDGE TO TISSUE ENGINEERING

A tissue is a complex structure made up of different types of cells with specific arrangements that carry out certain functions. The generation of tissues from a few undifferentiated cells is a complex, well-regulated process that is only partially understood.[91] Tissue engineers have not yet been able to "grow" vascularized organs *in vitro*. There have only been a few examples of successful demonstrations of simple man-made tissues generated *in vitro*.[92,93] Many complex problems limit our ability to generate vascularized organs *in vitro*. One major issue is the ability to control the interaction between cells and substrates. Using SAMs for the control of this interaction has opened up avenues to the *in vitro* production of several tissue-like biomaterials, although the area of culturing different types of cells with the correct structure and arrangements using SAMs, or any other technology, is still in its infancy. In some cases, patterned cells mimic certain structures found in tissues or organs. There are some initial demonstrations of tissue-like functions of cultured cells in SAM-based and non-SAM-based systems. This section outlines cultured cells that mimic the structures and functions of certain tissues.

A. Angiogenesis

Angiogenesis is the process by which new blood vessels form; it is an important physiological process in areas ranging from tissue generation to the growth of tumors.[94] By controlling the area of cell spreading, we can control switching between growth, differentiation, and apoptosis during angiogenesis.[95] When cells are allowed to spread fully (on islands $>1500\ \mu m^2$), they grow, and when spreading is restricted to small areas (on islands $<500\ \mu m^2$), they undergo apoptosis. When spreading of an endothelial cell is confined to intermediate areas (narrow lines of 10 μm), the cells neither divide nor undergo apoptosis. Rather, they differentiate into hollow tubes that resemble blood vessels, but when cells are confined to lines of 30 μm, they do not differentiate into tubular structures (Figure 14.4).

B. Patterning of Neurons

Patterning networks of neurons provide insight into the cellular basis of neuroscience. A future idea is to interface defined patterns of neurons with silicon-based computers. Neurons from both invertebrates (such as aplysia) and vertebrates (such as goldfish and rat) can be patterned into defined geometries. Some of these neurons produce functional synapses that mimic synapses in the brain. These primitive neuronal networks are useful in understanding how electrical signals propagate in the nervous system.[19,96]

One important aspect of the patterning of neurons and the understanding of neuronal development is to specify the polarity of neurons on a given substrate. Micro-fluidic gradients of chemoattractants or repellants can control the polarity of cultured rat hippocampal neurons.[97]

(a)
(b)
(c)

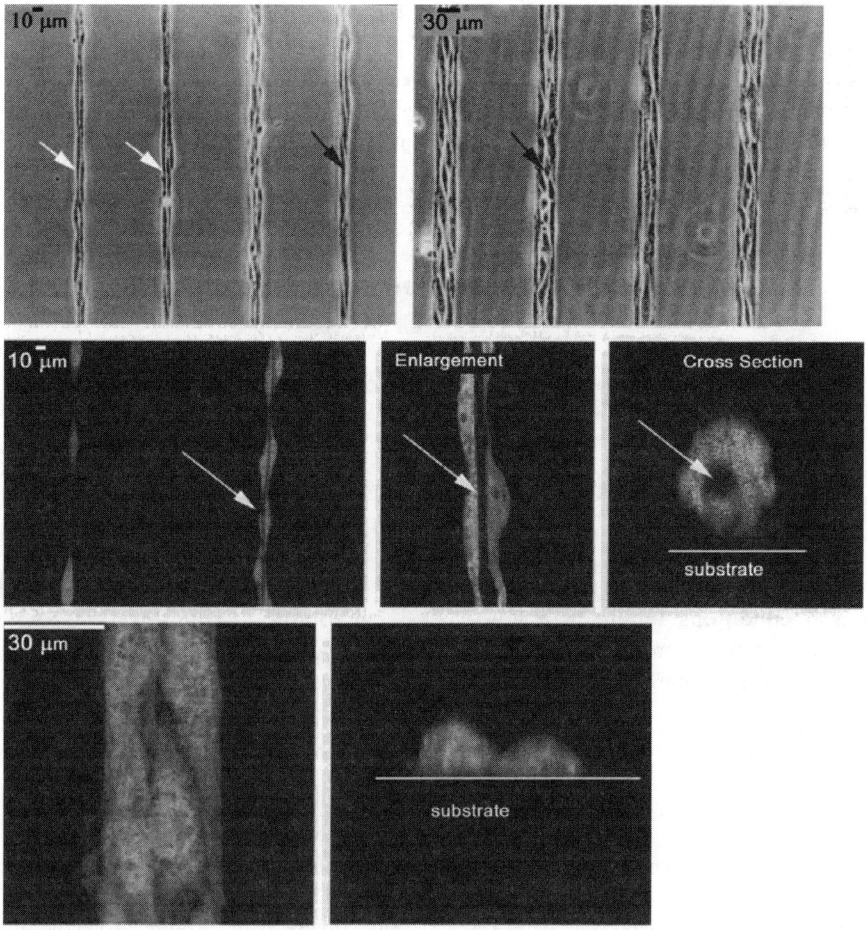

FIGURE 14.4 Formation of capillary tubes by endothelial cells on striped patterns. Cells cultured on 10 μm lines differentiate into capillaries, while those on 30 μm lines do not. (a) Phase-contrast images of endothelial cells cultured on 10 or 30 μm lines. White arrows indicate chords of cells that formed a tube with a central lumen, while black arrows indicate those that did not. (b) and (c) are fluorescent micrographs illustrating the tubes on 10 μm lines (b) vs. those on 30 μm lines (c).

C. Coculturing of Different Types of Cells

The ability to achieve controlled patterns of two or more types of cells is a key issue in tissue engineering because the development of tissues usually involves the mutual interactions of several different types of cells.[91] The ability to pattern two or more types of cells is one route to achieve artificial tissue *in vitro*. Using SAMs and related methods, it is possible to pattern several different types of cells on a solid substrate.

Using electrochemical methods on SAMs,[84] Mrksich and coworkers[98] demonstrated culturing of different populations of cells using SAMs. There have been other methods that utilize the printing from elastic stamps or membranes to pattern several different types of cells. For instance, Chen and coworkers[99] used multi-level stamps to pattern multiple types of proteins to control the attachment of different types of cells, and Toner and coworkers[100] used elastic membranes and stamps to make cocultures of hepatocytes and non-parenchymal cells. Their strategy can define the extent of the contacts and the ratio of population between hepatocytes and non-parenchymal cells, and provide information on how the interaction between the two types of cells determines the phenotype of hepatocytes.

Applications of SAMs in tissue engineering are limited by the fact that SAMs can only provide two-dimensional surfaces, while most tissues need to fold into three-dimensional structures to function properly. Methods using SAMs and related systems allowed the fabrication of pseudo three-dimensional structures.[101]

IX. OTHER APPLICATIONS OF SAMs IN BIOCHEMISTRY AND BIOTECHNOLOGY

SAMs are widely applicable in many fields of biochemistry and biotechnology.[102] Methods that render the surface inert, and methods that enable biospecific binding on surfaces, are both useful in various types of screening assays involving both proteins and cells.[15]

X. CONCLUSIONS AND SUMMARY

SAMs are a class of materials that can be molecularly controlled. They are robust under a variety of physiological conditions for up to a few weeks, but long-term robustness of SAMs in tissue implants have not been carefully evaluated. The key application of SAMs in tissue engineering is likely to stem from its capacity to tune the extent and geometry of attachment of various types of cells.

REFERENCES

1. Bain, C. D., and Whitesides, G. M., Molecular-level control over surface order in self-assembled monolayer films of thiols on gold, *Science*, 240, 62–63, 1988.
2. Nuzzo, R. G., and Allara, D. L., Adsorption of bifunctional organic disulfides on gold surfaces, *J. Am. Chem. Soc.*, 105, 4481–4483, 1983.
3. Laibinis, P. E., Bain, C. D., Nuzzo, R. G., and Whitesides, G. M., Structure and wetting properties of omega-alkoxy-*N*-alkanethiolate monolayers on gold and silver, *J. Phys. Chem. B*, 99, 7663–7676, 1995.
4. Love, J. C., Wolfe, D. B., Haasch, R., Chabinyc, M. L., Paul, K. E., Whitesides, G. M., and Nuzzo, R. G., Formation and structure of self-assembled monolayers of alkanethiolates on palladium, *J. Am. Chem. Soc.*, 125, 2597–2609, 2003.
5. Xia, Y., and Whitesides, G. M., Soft lithography, *Angew. Chem.-Int. Ed.*, 37, 550–575, 1998.
6. Bain, C. D., and Whitesides, G. M., Depth sensitivity of wetting — monolayers of omega-mercapto ethers on gold, *J. Am. Chem. Soc.*, 110, 5897–5898, 1988.
7. Abbott, N. L., Gorman, C. B., and Whitesides, G. M., Active control of wetting using applied electrical potentials and self-assembled monolayers, *Langmuir*, 11, 16–18, 1995.
8. Bliznyuk, V. N., Everson, M. P., and Tsukruk, V. V., Nanotribological properties of organic boundary lubricants: langmuir films versus self-assembled monolayers, *J. Tribol.-Trans. ASME*, 120, 489–495, 1998.
9. Mikulski, P. T., and Harrison, J. A., Packing-density effects on the friction of *n*-alkane monolayers, *J. Am. Chem. Soc.*, 123, 6873–6881, 2001.
10. Aizenberg, J., Black, A. J., and Whitesides, G. M., Control of crystal nucleation by patterned self-assembled monolayers, *Nature*, 398, 495–498, 1999.
11. Lamp, B. D., Hobara, D., Porter, M. D., Niki, K., and Cotton, T. M., Correlation of the structural decomposition and performance of pyridinethiolate surface modifiers at gold electrodes for the facilitation of cytochrome *c* heterogeneous electron-transfer reactions, *Langmuir*, 13, 736–741, 1997.

12. Holmlin, R. E., Chen, X., Chapman, R. G., Takayama, S., and Whitesides, G. M., Zwitterionic SAMs that resist nonspecific adsorption of protein from aqueous buffer, *Langmuir*, 17, 2841–2850, 2001.
13. Rampi, M. A., and Whitesides, G. M., A versatile experimental approach for understanding electron transport through organic materials, *Chem. Phys.*, 281, 373–391, 2002.
14. Bain, C. D., and Whitesides, G. M., Correlations between wettability and structure in monolayers of alkanethiols adsorbed on gold, *J. Am. Chem. Soc.*, 110, 3665–3666, 1988.
15. Mrksich, M., and Whitesides, G. M., Using self-assembled monolayers to understand the interactions of man-made surfaces with proteins and cells, *Annu. Rev. Biophys. Biomol. Struct.*, 25, 55–78, 1996.
16. Whitesides, G. M., Ostuni, E., Takayama, S., Jiang, X., and Ingber, D. E., Soft lithography in biology and biochemistry, *Annu. Rev. Biomed. Eng.*, 3, 335–373, 2001.
17. Bain, C. D., and Whitesides, G. M., Formation of 2-component surfaces by the spontaneous assembly of monolayers on gold from solutions containing mixtures of organic thiols, *J. Am. Chem. Soc.*, 110, 6560–6561, 1988.
18. Culp, L. A., and Sukenik, C. N., Cell type-specific modulation of fibronectin adhesion functions on chemically-derivatized self-assembled monolayers, *J. Biomater. Sci.-Polym. Ed.*, 9, 1161–1176, 1998.
19. Ravenscroft, M. S., Bateman, K. E., Shaffer, K. M., Schessler, H. M., Jung, D. R., Schneider, T. W., Montgomery, C. B., Custer, T. L., Schaffner, A. E., Liu, Q. Y., Li, Y. X., Barker, J. L., and Hickman, J. J., Developmental neurobiology implications from fabrication and analysis of hippocampal neuronal networks on patterned silane-modified surfaces, *J. Am. Chem. Soc.*, 120, 12169–12177, 1998.
20. Kleinfeld, D., Kahler, K. H., and Hockberger, P. E., Controlled outgrowth of dissociated neurons on patterned substrates, *J. Neurosci.*, 8, 4098–4120, 1988.
21. Chrisey, L. A., Lee, G. U., and O'ferrall, C. E., Covalent attachment of synthetic DNA to self-assembled monolayer films, *Nucleic Acids Res.*, 24, 3031–3039, 1996.
22. Guo, Z., Guilfoyle, R. A., Thiel, A. J., Wang, R., and Smith, L. M., Direct fluorescence analysis of genetic polymorphisms by hybridization with oligonucleotide arrays on glass supports, *Nucleic Acids Res.*, 22, 5456–5465, 1994.
23. Ulman, A., Self-assembled monolayers of alkyltrichlorosilanes: building blocks for future organic materials, *Adv. Mater.*, 2, 573–582, 1990.
24. Ostuni, E., Yan, L., and Whitesides, G. M., The interaction of proteins and cells with self-assembled monolayers of alkanethiols on gold and silver, *Colloids Surface B*, 15, 3–30, 1999.
25. Porter, M. D., Bright, T. B., Allara, D. L., and Chidsey, C. E. D., Spontaneously organized molecular assemblies. 4. Structural characterization of n-alkyl thiol monolayers on gold by optical ellipsometry, infrared spectroscopy, and electrochemistry, *J. Am. Chem. Soc.*, 109, 3559–3568, 1987.
26. Prime, K. L., and Whitesides, G. M., Self-assembled organic monolayers: model systems for studying adsorption of proteins at surfaces, *Science*, 252, 1164–1167, 1991.
27. Prime, K. L., and Whitesides, G. M., Adsorption of proteins onto surfaces containing end-attached oligo(ethylene oxide) — a model system using self-assembled monolayers, *J. Am. Chem. Soc.*, 115, 10714–10721, 1993.
28. Li, Y., Huang, J., Mciver, R. T. Jr., and Hemminger, J. C., Characterization of thiol self-assembled films by laser desorption fourier transform mass spectrometry, *J. Am. Chem. Soc.*, 114, 2428–2432, 1992.
29. Su, J., and Mrksich, M., Using mass spectrometry to characterize self-assembled monolayers presenting peptides, proteins, and carbohydrates, *Angew. Chem.-Int. Ed.*, 41, 4715–4718, 2002.
30. Bain, C. D., Troughton, E. B., Tao, Y. T., Evall, J., Whitesides, G. M., and Nuzzo, R. G., Formation of monolayer films by the spontaneous assembly of organic thiols from solution onto gold, *J. Am. Chem. Soc.*, 111, 321–335, 1989.
31. Laibinis, P. E., Hickman, J. J., Wrighton, M. S., and Whitesides, G. M., Orthogonal self-assembled monolayers: alkanethiols on gold and alkane carboxylic acids on alumina, *Science*, 245, 845–847, 1989.
32. Yan, L., Huck, W. T. S., Zhao, X.-M., and Whitesides, G. M., Patterning thin films of poly(ethylene imine) on a reactive SAM using microcontact printing, *Langmuir*, 15, 1208–1214, 1999.

33. Li, L., Chen, S., and Jiang, S., Protein adsorption on alkanethiolate self-assembled monolayers: nanoscale surface structural and chemical effects, *Langmuir*, 19, 2974–2982, 2003.
34. Vanderah, D. J., Gates, R. S., Silin, V., Zeiger, D. N., Woodward, J. T., Meuse, C. W., Valincius, G., and Nickel, B., Isostructural self-assembled monolayers. 1. Octadecyl 1-thiaoligo(ethylene oxides), *Langmuir*, 19, 2612–2620, 2003.
35. Mrksich, M., Sigal, G. B., and Whitesides, G. M., Surface-plasmon resonance permits in-situ measurement of protein adsorption on self-assembled monolayers of alkanethiolates on gold, *Langmuir*, 11, 4383–4385, 1995.
36. Lofas, S., and Johnsson, B., A novel hydrogel matrix on gold surfaces in surface plasmon resonance sensors for fast and efficient covalent immobilization of ligands, *J. Chem. Soc. Chem. Commun.*, 1526–1528, 1990.
37. Lofas, S., Malmqvist, M., Ronnberg, I., Stenberg, E., Liedber, B., and Lundstrom, I., Bioanalysis with surface plasmon resonance, *Sens. Actuator B—Chem.*, 5, 79–84, 1991.
38. Shen, D., Huang, M., Chow, L. M., and Yang, M., Kinetic profile of the adsorption and conformational change of lysozyme on self-assembled monolayers as revealed by quartz crystal resonator, *Sens. Actuator B—Chem.*, B77, 664–670, 2001.
39. Hook, F., Rodahl, M., Kasemo, B., and Brzezinski, P., Structural changes in hemoglobin during adsorption to solid surfaces: effects of pH, ionic strength, and ligand binding, *Proc. Natl Acad. Sci. USA*, 95, 12271–12276, 1998.
40. Pale-Grosdemange, C., Simon, E. S., Prime, K. L., and Whitesides, G. M., Formation of self-assembled monolayers by chemisorption of derivatives of oligo(ethylene glycol) of structure $HS(CH_2)11(OCH_2CH_2)_mOH$ on gold, *J. Am. Chem. Soc.*, 113, 12–20, 1991.
41. Ostuni, E., Chapman, R. G., Holmlin, R. E., Takayama, S., and Whitesides, G. M., A survey of structure–property relationships of surfaces that resist the adsorption of protein, *Langmuir*, 17, 5605–5620, 2001.
42. Roberts, C., Chen, C. S., Mrksich, M., Martichonok, V., Ingber, D. E., and Whitesides, G. M., Using mixed self-assembled monolayers presenting RGD and (EG)(3)OH groups to characterize long-term attachment of bovine capillary endothelial cells to surfaces, *J. Am. Chem. Soc.*, 120, 6548–6555, 1998.
43. Liang, M. N., Smith, S. P., Metallo, S. J., Choi, I. S., Prentiss, M., and Whitesides, G. M., Measuring the forces involved in polyvalent adhesion of uropathogenic *Escherichia coli* to mannose-presenting surfaces, *Proc. Natl Acad. Sci. USA*, 97, 13092–13096, 2000.
44. Svedhem, S., Oehberg, L., Borrelli, S., Valiokas, R., Andersson, M., Oscarson, S., Svensson, S. C. T., Liedberg, B., and Konradsson, P., Synthesis and self-assembly of globotriose derivatives: a model system for studies of carbohydrate–protein interactions, *Langmuir*, 18, 2848–2858, 2002.
45. Lahiri, J., Isaacs, L., Tien, J., and Whitesides, G. M., A strategy for the generation of surfaces presenting ligands for studies of binding based on an active ester as a common reactive intermediate: a surface plasmon resonance study, *Anal. Chem.*, 71, 777–790, 1999.
46. Yang, Z., Frey, W., Oliver, T., and Chilkoti, A., Light-activated affinity micropatterning of proteins on self-assembled monolayers on gold, *Langmuir*, 16, 1751–1758, 2000.
47. Yan, L., Marzolin, C., Terfort, A., and Whitesides, G. M., Formation and reaction of interchain carboxylic anhydride groups on self-assembled monolayers on gold, *Langmuir*, 13, 6704–6712, 1997.
48. Houseman, B. T., Gawalt, E. S., and Mrksich, M., Maleimide-functionalized self-assembled monolayers for the preparation of peptide and carbohydrate biochips, *Langmuir*, 19, 1522–1531, 2003.
49. Horton, R. C. Jr., Herne, T. M., and Myles, D. C., Aldehyde-terminated self-assembled monolayers on gold: immobilization of amines onto gold surfaces, *J. Am. Chem. Soc.*, 119, 12980–12981, 1997.
50. Niemz, A., Jeoung, E., Boal, A. K., Deans, R., and Rotello, V. M., Divergent surface functionalization using acid fluoride-functionalized self-assembled monolayers, *Langmuir*, 16, 1460–1462, 2000.
51. Yousaf, M. N., and Mrksich, M., Diels–Alder reaction for the selective immobilization of protein to electroactive self-assembled monolayers, *J. Am. Chem. Soc.*, 121, 4286–4287, 1999.
52. Hodneland, C. D., and Mrksich, M., Biomolecular surfaces that release ligands under electrochemical control, *J. Am. Chem. Soc.*, 122, 4235–4236, 2000.

53. Delamarche, E., Sundarababu, G., Biebuyck, H., Michel, B., Gerber, C., Sigrist, H., Wolf, H., Ringsdorf, H., Xanthopoulos, N., and Mathieu, H. J., Immobilization of antibodies on a photoactive self-assembled monolayer on gold, *Langmuir*, 12, 1997–2006, 1996.
54. Sullivan, T. P., and Huck, W. T. S., Reactions on monolayers: organic synthesis in two dimensions, *Eur. J. Org. Chem.*, 17–29, 2003.
55. Walczak, M. M., Popenoe, D. D., Deinhammer, R. S., Lamp, B. D., Chung, C. K., and Porter, M. D., Reductive desorption of alkanethiolate monolayers at gold — a measure of surface coverage, *Langmuir*, 7, 2687–2693, 1991.
56. Huang, E., Satjapipat, M., Han, S., and Zhou, F., Surface structure and coverage of an oligonucleotide probe tethered onto a gold substrate and its hybridization efficiency for a polynucleotide target, *Langmuir*, 17, 1215–1224, 2001.
57. Kim, K., Yang, H., Kim, E., Han, Y. B., Kim, Y. T., Kang, S. H., and Kwak, J., Electrochemical deprotection for site-selective immobilization of biomolecules, *Langmuir*, 18, 1460–1462, 2002.
58. Gorman, C. B., Biebuyck, H. A., and Whitesides, G. M., Control of the shape of liquid lenses on a modified gold surface using an applied electrical potential across a self-assembled monolayer, *Langmuir*, 11, 2242–2246, 1995.
59. Jiang, X., Ferrigno, R., Mrksich, M., and Whitesides, G. M., Electrochemical desorption of self-assembled monolayers noninvasively releases patterned cells from geometrical confinements, *J. Am. Chem. Soc.*, 125, 2366–2367, 2003.
60. Kawaguchi, T., Yasuda, H., Shimazu, K., and Porter, M. D., Electrochemical quartz crystal microbalance investigation of the reductive desorption of self-assembled monolayers of alkanethiols and mercaptoalkanoic acids on Au, *Langmuir*, 16, 9830–9840, 2000.
61. Yang, D. F., Almaznai, H., and Morin, M., Vibrational study of the fast reductive and the slow oxidative desorptions of a nonanethiol self-assembled monolayer from a Au(111) single crystal electrode, *J. Phys. Chem. B*, 101, 1158–1166, 1997.
62. Tender, L. M., Worley, R. L., Fan, H., and Lopez, G. P., Electrochemical patterning of self-assembled monolayers onto microscopic arrays of gold electrodes fabricated by laser ablation, *Langmuir*, 12, 5515–5518, 1996.
63. Huang, J., and Hemminger, J. C., Photooxidation of thiols in self-assembled monolayers on gold, *J. Am. Chem. Soc.*, 115, 3342–3343, 1993.
64. Huang, J., Dahlgren, D. A., and Hemminger, J. C., Photopatterning of self-assembled alkanethiolate monolayers on gold: a simple monolayer photoresist utilizing aqueous chemistry, *Langmuir*, 10, 626–628, 1994.
65. Brewer, N. J., Rawsterne, R. E., Kothari, S., and Leggett, G. J., Oxidation of self-assembled monolayers by UV light with a wavelength of 254 nm, *J. Am. Chem. Soc.*, 123, 4089–4090, 2001.
66. Lahann, J., Mitragotri, S., Tran, T.-N., Kaido, H., Sundaram, J., Choi, I. S., Hoffer, S., Somorjai, G. A., and Langer, R., A reversibly switching surface, *Science*, 299, 371–374, 2003.
67. Sigal, G. B., Mrksich, M., and Whitesides, G. M., Effect of surface wettability on the adsorption of proteins and detergents, *J. Am. Chem. Soc.*, 120, 3464–3473, 1998.
68. Chapman, R. G., Ostuni, E., Takayama, S., Holmlin, R. E., Yan, L., and Whitesides, G. M., Surveying for surfaces that resist the adsorption of proteins, *J. Am. Chem. Soc.*, 122, 8303–8304, 2000.
69. Ostuni, E., Grzybowski, B. A., Mrksich, M., Roberts, C. S., and Whitesides, G. M., Adsorption of proteins to hydrophobic sites on mixed self-assembled monolayers, *Langmuir*, 19, 1861–1872, 2003.
70. Mrksich, M., Grunwell, J. R., and Whitesides, G. M., Biospecific adsorption of carbonic anhydrase to self-assembled monolayers of alkanethiolates that present benzenesulfonamide groups on gold, *J. Am. Chem. Soc.*, 117, 12009–12010, 1995.
71. Sigal, G. B., Bamdad, C., Barberis, A., Strominger, J., and Whitesides, G. M., A self-assembled monolayer for the binding and study of histidine-tagged proteins by surface plasmon resonance, *Anal. Chem.*, 68, 490–497, 1996.
72. Hodneland, C. D., Lee, Y.-S., Min, D.-H., and Mrksich, M., Selective immobilization of proteins to self-assembled monolayers presenting active site-directed capture ligands, *Proc. Natl Acad. Sci. USA*, 99, 5048–5052, 2002.

73. Chen, X., Ferrigno, R., Yang, J., and Whitesides, G. M., Redox properties of cytochrome c adsorbed on self-assembled monolayers: a probe for protein conformation and orientation, *Langmuir*, 18, 7009–7015, 2002.
74. Chen, S., Liu, L., Zhou, J., and Jiang, S., Controlling antibody orientation on charged self-assembled monolayers, *Langmuir*, 19, 2859–2864, 2003.
75. Houseman, B. T., Huh, J. H., Kron, S. J., and Mrksich, M., Peptide chips for the quantitative evaluation of protein kinase activity, *Nat. Biotechnol.*, 20, 270–274, 2002.
76. Lahiri, J., Kalal, P., Frutos, A. G., Jonas, S. T., and Schaeffler, R., Method for fabricating supported bilayer lipid membranes on gold, *Langmuir*, 16, 7805–7810, 2000.
77. Houseman, B. T., and Mrksich, M., Carbohydrate arrays for the evaluation of protein binding and enzymatic modification, *Chem. Biol.*, 9, 443–454, 2002.
78. Ruoslahti, E., and Pierschbacher, M. D., New perspectives in cell adhesion: RGD and integrins, *Science*, 238, 491–497, 1987.
79. Singhvi, R., Kumar, A., Lopez, G. P., Stephanopoulos, G. N., Wang, D. I. C., Whitesides, G. M., and Ingber, D. E., Engineering cell-shape and function, *Science*, 264, 696–698, 1994.
80. Lopez, G. P., Albers, M. W., Schreiber, S. L., Carroll, R., Peralta, E., and Whitesides, G. M., Convenient methods for patterning the adhesion of mammalian-cells to surfaces using self-assembled monolayers of alkanethiolates on gold, *J. Am. Chem. Soc.*, 115(13), 5877–5878, 1993.
81. Kumar, A., and Whitesides, G. M., Features of gold having micrometer to centimeter dimensions can be formed through a combination of stamping with an elastomeric stamp and an alkanethiol ink followed by chemical etching, *Appl. Phys. Lett.*, 63, 2002–2004, 1993.
82. Lopez, G. P., Biebuyck, H. A., Harter, R., Kumar, A., and Whitesides, G. M., Fabrication and imaging of 2-dimensional patterns of proteins adsorbed on self-assembled monolayers by scanning electron-microscopy, *J. Am. Chem. Soc.*, 115, 10774–10781, 1993.
83. Chen, C. S., Mrksich, M., Huang, S., Whitesides, G. M., and Ingber, D. E., Geometric control of cell life and death, *Science*, 276, 1425–1428, 1997.
84. Yousaf, M. N., Houseman, B. T., and Mrksich, M., Turning on cell migration with electroactive substrates, *Angew. Chem.-Int. Ed.*, 40, 1093–1096, 2001.
85. Moriguchi, H., Wakamoto, Y., Sugio, Y., Takahashi, K., Inoue, I., and Yasuda, K., An agar-microchamber cell-cultivation system: flexible change of microchamber shapes during cultivation by photo-thermal etching, *Lab Chip*, 2, 125–132, 2002.
86. Stenger, D. A., Georger, J. H., Dulcey, C. S., Hickman, J. J., Rudolph, A. S., Nielsen, T. B., Mccort, S. M., and Calvert, J. M., Coplanar molecular assemblies of aminoalkylsilane and perfluorinated alkylsilane — characterization and geometric definition of mammalian-cell adhesion and growth, *J. Am. Chem. Soc.*, 114, 8435–8442, 1992.
87. Hyun, J., Ma, H., Zhang, Z., Beebe, T. P. Jr., and Chilkoti, A., Universal route to cell micropatterning using an amphiphilic comb polymer, *Adv. Mater.*, 15, 576–579, 2003.
88. Ostuni, E., Kane, R., Chen, C. S., Ingber, D. E., and Whitesides, G. M., Patterning mammalian cells using elastomeric membranes, *Langmuir*, 16, 7811–7819, 2000.
89. Parker, K. K., Brock, A. L., Brangwynne, C., Mannix, R. J., Wang, N., Ostuni, E., Geisse, N. A., Adams, J. C., Whitesides, G. M., and Ingber, D. E., Directional control of lamellipodia extension by constraining cell shape and orienting cell tractional forces, *FASEB J.*, 16, 1195–1204, 2002.
90. Brock, A., Chang, E., Ho, C.-C., Leduc, P., Jiang, X., Whitesides, G. M., and Ingber, D. E., Geometric determinants of directional cell motility revealed using microcontact printing, *Langmuir*, 19, 1611–1617, 2003.
91. Alberts, B., Johnson, A., Lewis, J., Raff, M., Keith, R., and Walter, P., *Molecular Biology of the Cell*, Garland Science, Taylor & Francis Group, New York, 2002.
92. Vacanti, JP, Vacanti, CA, and Langer, RS. In U.S. 17 pp., *Cont.-in-part of U.S.* 5,041,138. Massachusetts Institute of Technology, U.S.; Children's Medical Center Corporation, U.S., 1998.
93. Langer, R., and Vacanti, J. P., Tissue engineering, *Science*, 260, 920–926, 1993.
94. Folkman, J., Angiogenesis: what makes blood vessels grow?, *News Physiol. Sci.*, 1, 199–202, 1986.
95. Dike, L. E., Chen, C. S., Mrksich, M., Tien, J., Whitesides, G. M., and Ingber, D. E., Geometric control of switching between growth, apoptosis, and differentiation during angiogenesis using micropatterned substrates, *In Vitro Cell. Dev. Biol.-Anim.*, 35, 441–448, 1999.

96. Bi, G.-Q., and Poo, M.-M., Distributed synaptic modification in neural networks induced by patterned stimulation, *Nature*, 401, 792–796, 1999.
97. Dertinger, S. K. W., Jiang, X., Li, Z., Murthy, V. N., and Whitesides, G. M., Gradients of substrate-bound laminin orient axonal specification of neurons, *Proc. Natl Acad. Sci. USA*, 99, 12542–12547, 2002.
98. Yousaf, M. N., Houseman, B. T., and Mrksich, M., Using electroactive substrates to pattern the attachment of two different cell populations, *Proc. Natl Acad. Sci. USA*, 98, 5992–5996, 2001.
99. Tien, J., Nelson, C. M., and Chen, C. S., Fabrication of aligned microstructures with a single elastomeric stamp, *Proc. Natl Acad. Sci. USA*, 99, 1758–1762, 2002.
100. Bhatia, S. N., Balis, U. J., Yarmush, M. L., and Toner, M., Effect of cell–cell interactions in preservation of cellular phenotype: cocultivation of hepatocytes and nonparenchymal cells, *FASEB J.*, 13, 1883–1900, 1999.
101. Yamato, M., Konno, C., Utsumi, M., Kikuchi, A., and Okano, T., Thermally responsive polymer-grafted surfaces facilitate patterned cell seeding and co-culture, *Biomaterials*, 23, 561–567, 2001.
102. Mrksich, M., and Whitesides, G. M., Patterning self-assembled monolayers using microcontact printing — a new technology for biosensors, *Trends Biotechnol.*, 13, 228–235, 1995.
103. Bain, C. D., and Whitesides, G. M., Modeling organic-surfaces with self-assembled monolayers, *Angew. Chem.-Int. Ed.*, 28, 506–512, 1989.
104. Ulman, A., *Introduction to Thin Organic Films: from Langmuir-Blodgett to Self-Assembly*, Academic Press, Boston, 1991.
105. Linford, M. R., and Chidsey, C. E. D., Alkyl monolayers covalently bonded to silicon surfaces, *J. Am. Chem. Soc.*, 115, 12631–12632, 1993.
106. Linford, M. R., Fenter, P., Eisenberger, P. M., and Chidsey, C. E. D., Alkyl monolayers on silicon prepared from 1-alkenes and hydrogen-terminated silicon, *J. Am. Chem. Soc.*, 117, 3145–3155, 1995.
107. Bansal, A., Li, X., Lauermann, I., Lewis, N. S., Yi, S. I., and Weinberg, W. H., Alkylation of Si surfaces using a two-step halogenation/grignard route, *J. Am. Chem. Soc.*, 118, 7225–7226, 1996.
108. Bain, C. D., A new class of self-assembled monolayers: organic thiols on gallium arsenide, *Adv. Mater.*, 4, 591–594, 1992.
109. Gu, Y., Lin, Z., Butera, R. A., Smentkowski, V. S., and Waldeck, D. H., Preparation of self-assembled monolayers on InP, *Langmuir*, 11, 1849–1851, 1995.
110. Gardner, T. J., Frisbie, C. D., and Wrighton, M. S., Systems for orthogonal self-assembly of electroactive monolayers on Au and ITO: an approach to molecular electronics, *J. Am. Chem. Soc.*, 117, 6927–6933, 1995.
111. Carlson, T. R., Feng, Y., Maisonpierre, P. C., Mrksich, M., and Morla, A. O., Direct cell adhesion to the angiopoietins mediated by integrins, *J. Biol. Chem.*, 276, 26516–26525, 2001.
112. Higashi, N., Takahashi, M., and Niwa, M., Immoblization of DNA through intercalation at self-assembled monolayers on gold, *Langmuir*, 15, 111–115, 1999.
113. Rao, J. H., Yan, L., Lahiri, J., Whitesides, G. M., Weis, R. M., and Warren, H. S., Binding of a dimeric derivative of vancomycin to L-LYS-D-ALA-D- lactate in solution and at a surface, *Chem. Biol.*, 6, 353–359, 1999.

15 PuraMatrix: Self-Assembling Peptide Nanofiber Scaffolds

Shuguang Zhang, Xiaojun Zhao, and Lisa Spirio

CONTENTS

I. INTRODUCTION

The fields of tissue engineering and regenerative medicine require two key complementary components: (1) a suitable biological scaffold that creates a microenvironment niche for a given cell type, and (2) that the given cell type can rapidly integrate and coalesce into the needed tissue.

The three-dimensional assembly of cells in a biological microenvironment is designed to recapitulate normal tissue function.[1] Often, this involves the use of biomaterials. The task of the biomedical engineering is to design and to select optimal biomaterials with the properties most closely matched to those needed for a particular application.[2–6]

Both stem and progenitor cells hold promise for tissue engineering and for reparative and regenerative medicine. In most cases, these cells must be paired with microenvironments, which ensure proper expansion or differentiation through mimicry *in vivo* microenvironments, basement membranes with a critical constituent being the extracellular matrix (ECM).[7–9] This fibrous protein scaffold not only provides the appropriate three-dimensional architecture, but also promotes signaling pathways influencing critical cell functions such as proliferation, differentiation, and migration.[10,11] Moreover, researchers in the fields of tissue engineering, stem cell biology, and cancer biology are realizing that the ECM and a fine-tuned three-dimensional microenvironment are not only critical for fully understanding of cell biology, but also for developing successful clinically relevant therapies.

However, almost all cell biology research uses two-dimensional constructs to study single cell populations treated selectively with soluble factors in a homogeneous environment.[12] This greatly simplifies the research models, but also mitigates the powerful effects of ECM and three-dimensional culture conditions on the cells of interest and, in some cases, might be misleading. Recreating three-dimensional conditions and structures *in vitro*, rather than growing cells in two-dimensional Petri dishes, will provide tools for more accurate biological analyses of cells and tissues.[13,14]

A. Animal Derived ECM and Synthetic Scaffolds

The past two decades have witnessed the quest for synthetic biocompatible polymers, hydrogels, and or animal derived materials.[2–6] Until recently, three-dimensional cell culture has required either synthetic scaffolds, which fail to approximate the physical nanoscale size and chemical attributes of native ECM, or animal derived materials, which may confound cell culture with undefined or inconsistent variables.

Biomaterials such as PLLA and PGA biopolymers, calcium phosphate mesh, PEG gels, methylcellulose, alginates, and agarose have shown only limited success, partly due to either their large microfiber size relative to cells (about the same size as most tissue cells), acidic breakdown products, charge density, lower nutrient diffusion rates, or their inability to allow the creation of functional microenvironments by the cells.

Animal derived biomaterials including bovine collagen and gelatin, fibronectin, intestinal submucosa, cadaver tissue, fibrin, and Matrigel may help to create the right microenvironments, but complicated research and therapies, with their potential risk of other unknown material contaminations, such as viruses, prions, unknown proteins, and factors, raises issues about cell signaling, protein content, and reproducibility.

We have described a family of synthetic self-assembling peptide scaffolds (SAPS), now called *PuraMatrix* because of its synthetic nature and extreme purity of a single peptide component. Its basic constituents are the natural L-amino acids, thus the breakdown products can be reused by cells. PuraMatrix not only can be used as a two-dimensional coating or a three-dimensional encapsulation of cells similar to the ECM, but can also be tailor-made for particular cells, tissues, and cell-based therapies.

B. PuraMatrix Peptide Nanofiber Scaffolds

Over the past 10 years, a new class of biologically inspired peptide biomaterials has been discovered and developed in the context of cell culture, stem cell biology, and tissue engineering. These self-assembling peptide scaffolds have been used successfully as a synthetic *in vitro* and *in vivo* ECM, proving themselves as a critical component to successful three-dimensional cell growth.[15–22,35]

1. Nanofiber Scale Synthetic ECM

PuraMatrix is a 16 amino acid synthetic peptide that is resuspended in water to generate a range of solution concentrations. Upon the introduction of millimolar amounts of monovalent cations, either through the addition of salt solution, cell culture media (e.g., DMEM), or injection of the material *in vivo*, PuraMatrix undergoes self-assembly into nanofibers, ~10 nm in diameter, on a scale similar to the *in vivo* ECM. The physical size relative to cells and proteins, the amphiphilic peptides' charge density, and water structuring abilities mimic the *in vivo* ECM. These well-ordered nanofibers create three-dimensional porous scaffolds that are very difficult or impossible to synthetically produce by other fabrication techniques. The nanofiber density and average pore size (between ~5 and 200 nm) correlates with the concentration of peptide solution that is used to produce the material, which can be varied from 0.1 to 5% in water (1 to 50 mg/ml w/v) depending on the application.

2. Defined Three-Dimensional Microenvironments for Cell Biology

PuraMatrix mimics several important aspects of the *in vivo* environment — a synthetic ECM — enabling defined cell culture conditions while allowing cells to proliferate and differentiate within a three-dimensional context, easily migrate within that microenvironment, and create their own microenvironments quickly including production of their own ECM.

C. Three-Dimensional Cell Culture vs. Two-Dimensional

The advancement of biological study often requires the development of new materials, methods and tools. The introduction of the Petri dish over 100 years ago provided an indispensable tool for culturing cells *in vitro*, thus permitting the detailed dissection of seemingly intractable biology and physiology systems into manageable units and well-defined studies. This simple dish has had profound impact on our understanding of complex biology, especially cell biology and neurobiology.

However, the Petri dish culture system, including multiwell plates, glass cover slips, and so on, is less than ideal for several reasons:

1. It is a two-dimensional system that sharply contrasts to the three-dimensional environment of natural tissues both animal and plant.
2. The Petri dish surface without coating is rigid and impermeable, again, in sharp contrast to the *in vivo* environment where cells intimately interact with the ECM and with each other.
3. The tissue cell monolayers on a coated two-dimensional surface, such as poly-L-lysine, collagen gels, fibronectin, and laminin as well as other synthetic materials containing segments of adhesion motifs, have only part of the cell surface attached to the materials

and interact neighboring cells. The remaining parts are often directly exposed to the culture media, unlike the tissue environment where every cell intimately interact with its neighbors and the ECM.

4. The transport phenomena of two-dimensional and three-dimensional are drastically different. In two-dimensional culture systems, cytokines, chemokines, and growth factors quickly diffuse in the media across the culture dish. Again, this sharply contrasts to the *in vivo* environment where chemical and biological gradients play a vital role in signal transduction, chemotaxis, cell–cell communication and development. Cells respond to local concentrations of a variety of molecules, which in traditional cell culture, are distributed homogenously. Yet, most cells, by virtue of being embedded in a tissue, live in an environment that contains a dynamic gradient of nutrients and secreted factors — a "spatial heterogeneity" that cannot be emulated using conventional two-dimensional tissue culture. Since many cell types secrete their own ECM, a synthetic scaffold can be used even if not initially coated with the particular ECM proteins.
5. Cells cultured on a two-dimensional Petri dish are not readily transportable, and it is difficult to move cells from one environment to another without incurring changes in the cell–material and cell–cell interactions. For example, enzymatic cell harvesting using trypsin or mechanical disruption using rubber policeman both may have adverse effects on cell–environment interactions. Culturing to on two-dimensional surfaces to cell confluence requires frequent passages, including more reagent costs and cell manipulation. In contrast, cells cultured in three-dimensions are more readily transportable without significantly harming cell–material and cell–cell interactions, thus providing a significantly new way to study cell biology.

D. Nanoscale Fibers vs. Microscale

In the last two decades, several biopolymers, such as PLLA, PLGA, PLLA–PLGA copolymers, and other biomaterials including alginate, agarose, and collagen gels have been developed to culture cells in three-dimensions.[2–6] These culture systems have significantly advanced our understanding of cell–material interactions and fostered a new field of tissue engineering. However, these biomaterials are often made of microfibers with diameters of 10 to 100 μm — drastically different in size, surface interaction, porosity, and concentration relative to the native ECM and cells interacting with it. Therefore, cells attached on microfibers are in fact two-dimensional, despite the various curvatures associated with the large diameter microfibers. In order to culture cells in a truly three-dimensional microenvironment, the fibers must be significantly smaller than cells so that the cells are surrounded by the scaffold, similar to the extracellular environment and native ECM.

We believe that the development of new biological materials, particularly those biologically inspired nanoscale scaffolds mimicking an *in vivo* environment that serve as permissive substrates for cell growth, differentiation, and biological function, is a key area of study. These materials will be useful not only in furthering our understanding of cell biology in a three-dimensional environment, but also in advancing medical technology, tissue engineering, regenerative biology, and medicine.

E. Ideal Synthetic Biological Scaffolds

The ideal biological scaffold and its building blocks should meet several criteria:

1. They are derived from chemically defined, synthetic sources which are present in native tissue,
2. They are amenable to design and modification to customize specific bioactive and functional requirements,

3. They allow cell attachment, migration, cell–cell, cell–substrate interactions, and recovery of cells from the scaffold,
4. They exhibit no cytotoxicity or biocompatibility problems while chemically compatible with aqueous solutions, cell culture and physiological conditions,
5. They are compatible with microscopy, molecular biology analysis, and flow cytometry,
6. They are sterile and stable enough for shelf life, transportation, bioproduction, and closed system cell therapy culture,
7. They are economically viable and scaleable material production, purification, and processing,
8. They exhibit a controlled rate of material biodegradation *in vivo* with nondetectable immune responses and inflammation,
9. They foster cell migration and angiogenesis to rapidly integrate with tissues in the body,
10. They are injectable along with cells, compatible with cell delivery and surgical tools.

II. SELF-ASSEMBLING PEPTIDES

A. Discovery and Development of Self-Assembling Peptides

1. Simple Repeating Units of Amino Acids Assemble into Nanofiber Scaffolds

The first molecule of this class of self-assembling peptides, EAK16-II, a 16 amino acid peptide, was found as a segment in a yeast protein, zuotin which was originally characterized by binding to left-handed Z-DNA.[15] Zuotin is a 433-residue protein with a domain consisting of 34 amino acid residues with alternating alanines and alternating charges of glutamates and lysines of remarkable regularity, AGARAEAEAKAKAEAEAKAKAESEAKANASAKAD.[15] We subsequently reported a class of biological materials made from self-assembling peptides.[16–22] This biological scaffold consists of greater than 99% water content (peptide content 1–10 mg/ml). They form scaffolds when the peptide solution is exposed to physiological media or salt solution.[23–26]

The scaffolds consist of alternating amino acids that contain 50% charged residues.[16–20,23–27] These peptides are characterized by their periodic repeats of alternating ionic hydrophilic and hydrophobic amino acids. Thus, the β-sheets have distinct polar and nonpolar surfaces.[16–20] A number of additional self-assembling peptides including RAD16-I and RAD16-II, in which arginine and aspartate residues substitute lysine and glutamate have been designed and characterized for salt facilitated scaffold formation.[16–20] Stable macroscopic matrix structures have been fabricated through the spontaneous self-assembly of aqueous peptide solutions introduced into physiological salt containing solutions. Several peptide scaffolds have been shown to support cell attachment of a variety of mammalian primary and tissue culture cells.[16–22]

2. Amenable to Design Incorporating Functional Motifs

Since these peptides are synthetic and molecular engineered, they can be modified through the incorporation of desired functional motifs for specific ECM–cell interactions. We have tailor-made several self-assembling peptides containing functional sequences incorporated into the scaffold and shown some important implications for three-dimensional cell culture and tissue engineering [unpublished results].

B. Structural Properties of Self-Assembling Peptides

In general, these self-assembling peptides form very stable β-sheet structures in water. Although, sometimes, they may not form long nanofibers, their β-sheet structure remains largely unaffected.[16,17]

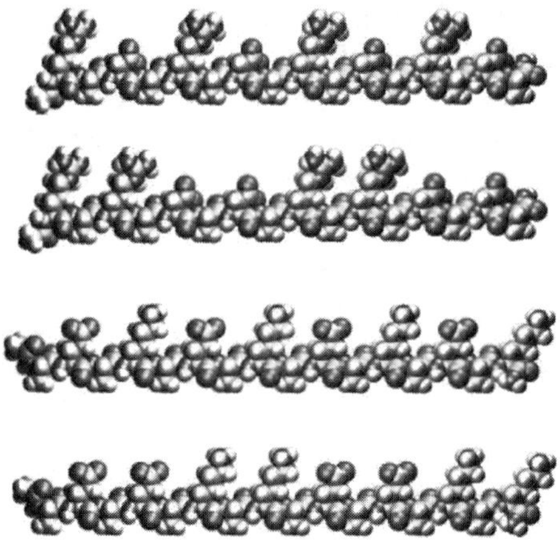

FIGURE 15.1 Molecular models of several self-assembling peptides, RAD16-I, RAD16-II, EAK16-I, and EAK16-II. Each molecule is ~5 nm in length with eight alanines on one side and four negative and four positive charge amino acids in an alternating arrangement on the other side.

One of the possible reasons is their unique structure. The alternating alanine residues in PuraMatrix are similar to silk fibroin such that the alanines can pack into interdigital hydrophobic interactions (Figure 15.1). The ionic complementary sides have been classified into several moduli (Modulus I, Modulus II, Modulus III, Modulus IV, etc., and mixtures thereof). This classification scheme is based on the hydrophilic surface of the molecules that have alternating positively and negatively charged amino acids, alternating by one residue, two residues, three residues, and so on. For example, charge arrangements for Modulus I, Modulus II, Modulus III, and Modulus IV are $-+-+-+-+$, $--++--++$, $---+++$, and $----++++$, respectively. These well-defined sequences allow the peptides to undergo ordered self-assembly, resembling some situations found in well-studied polymer assemblies.

III. PEPTIDE NANOFIBER SCAFFOLDS

A. EAK16-II

The EAK16-II, AEAEAKAKAEAEAKAK, is the first member in the self-assembling peptide family. EAK16-II was the first peptide to be characterized in detail[16–18,28] and has also been shown to retain β-sheet structure for extended periods of time. (One sample was stable for over 10 years; Zhang, unpublished results). An EAK membrane-like scaffold was first discovered in the tissue culture media where PC12 cells were used to test for EAK16-II cytotoxicity. The EAK scaffold showed no apparent toxicity; instead, the PC12 cells were found to attach onto the membranous materials where EAK16-II was added and not in the dishes where EAK8, a single unit of AEAEAKAK, was used.[16] Furthermore, the EAK16-II nanofibers were found to support neurite outgrowth in nerve growth factor (NGF) treated PC-12 cells and synapse formation in primary rat hippocampal neurons.[16] The membranous material was examined under scanning electron microscopy (SEM) to reveal a well-ordered nanofiber structure (Figure 15.2). Later, using atomic force microscopy (AFM), the nanofiber structure was confirmed.[29]

B. RADA16

We then designed several peptides altering the amino acid sequences containing RAD motif. RADA16-I, RADARADARADARADARADA, and RAD16-II, RARADADARARADADADA, were studied (Figure 15.3). These peptides have motif RAD that is similar to the ubiquitous integrin

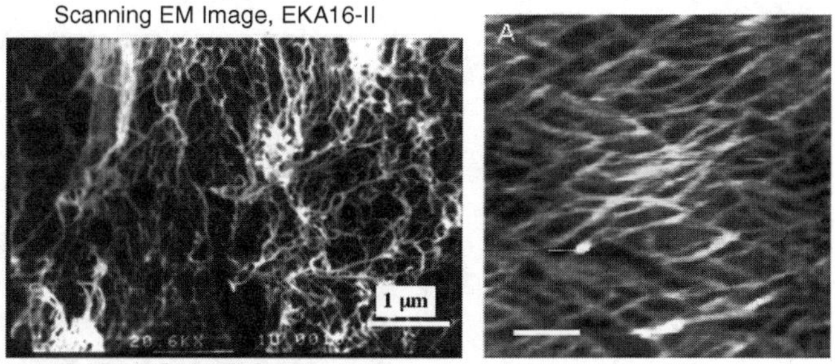

FIGURE 15.2 EAK16-II nanofibers. EAK16-II nanofiber formation was demonstrated using (SEM)[16] and AFM.[29] The nanofibers are ~10 to 20 in diameter with structural regularity due to the repeating peptide sequences and amphiphilic nature of the peptides. The scale bar in AFM is 200 nm.

receptor binding site RGD. Although it is not known if these RAD repeats in the PuraMatrix scaffold behave similarly to RGD motifs, they have been studied in the context of cell attachment across a number of cell lines.[18,19] These peptides form well-ordered nanofibers, similar to EAK16. Interestingly, RADARGDARADARGDA, and RADA8, RADARADA, did not form stable β-sheets or nanofibers when studied under the identical conditions as RAD16. These observations suggest that the formation of stable β-sheet is important not only for nanofiber, but also for scaffold formation.

Furthermore, increasing concentration of RADA8 added into RAD16-I inhibited the long and well-ordered RADA16 nanofiber formation [Zhang, unpublished results]. Other variations of RAD16 including RAD16-IV and DAR16-IV have also been studied.[30–33] DAR16 has an identical composition to RADA16-I but a different sequence arrangement. When it formed a β-sheet, the nanofibers are formed but upon heating, the DAR16 transformed into an α-helical structure and nanofibers were no longer observed [Zhao, et al., unpublished results]. These results indicate that simple properties, such as length of the peptide and order of amino acids within that peptide, determine its ability to form and maintain stable β-sheet nanofibers.

C. KFE8 AND KLD12

Self-assembling peptides less than 16 residues have also been studied including KFE8 and KLD12 (Figure 15.4).[20,23–27,34] Many self-assembling peptides that form scaffolds have

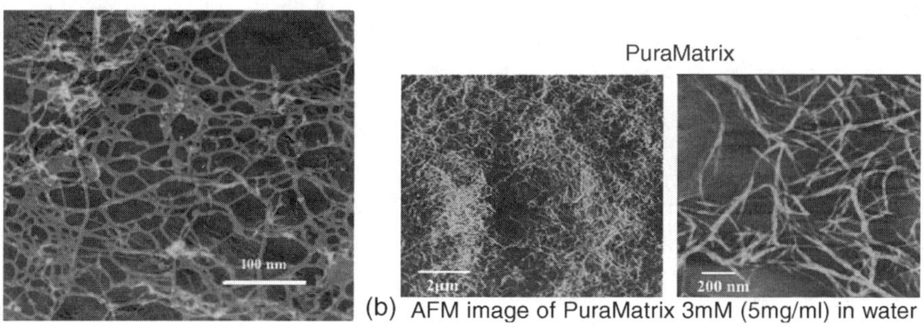

FIGURE 15.3 PuraMatrix. (a) Transmission electron microscopy (TEM) and (b) AFM. Note the nanofiber scale fiber with pores ranging from 5 to 200 nm, the right pore size for biomolecular diffusion. This is in sharp contrast to the microfibers of traditional polymer scaffolds, where the fiber diameter is ~1 to 50 μm and the pores range from 10 to 200 μm.[2–4]

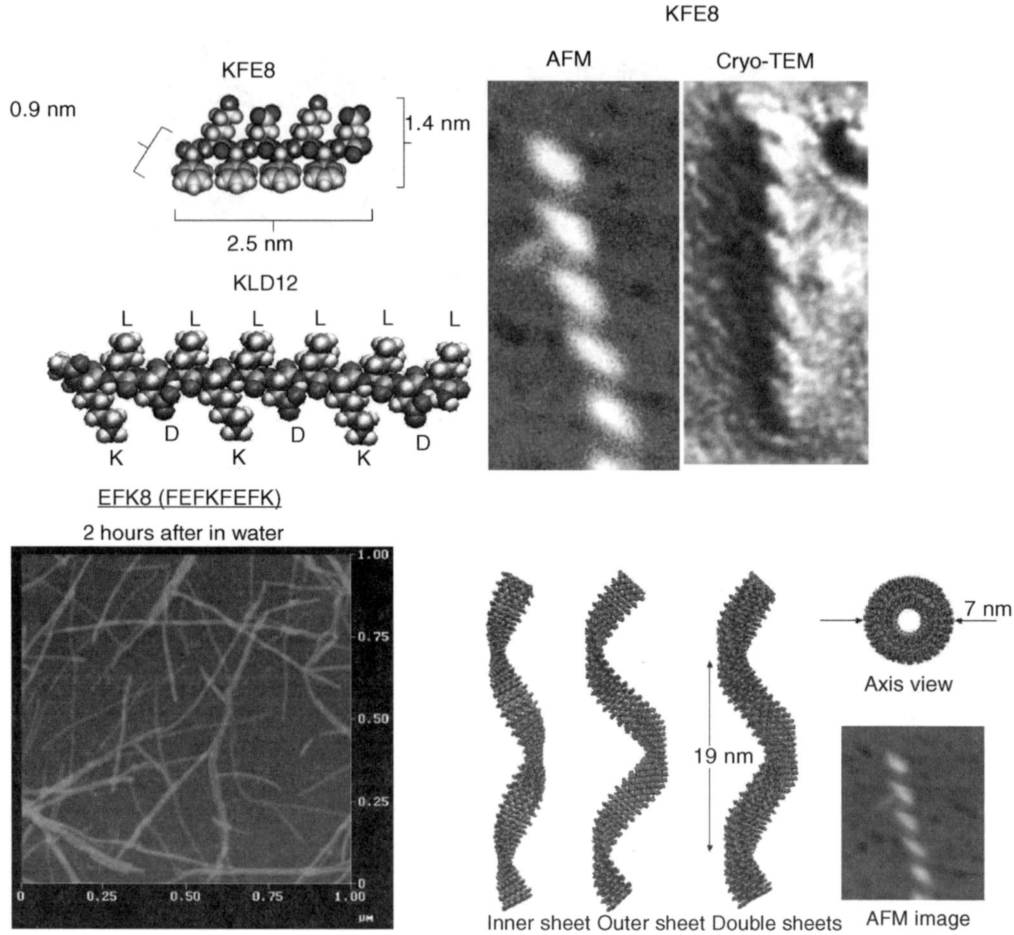

FIGURE 15.4 Molecular model of KFE8 and KLD12 and AFM/TEM image of KFE8. These peptides undergo self-assembly to form left-handed nanofibers with a diameter of approximately 7 nm as single fiber and thicker when bundle with other fibers. After 2 h, the nanofiber length increases. The left-handedness is unmistakable from the detail AFM (left) and Quick-freeze/deep-etch TEM (right) images.[27] Molecular simulation of KFE8 in left-handed double helix. The simulation was carried out using CHARMM program (updated 2002). There are an inner helix and an outer helix with hydrophobic Phe inside away from water and hydrophilic Lys and Glu exposed to water. The diameter is 7 nm and the pitch is 19 nm as reproducibly observed experimentally by AFM.[27]

been reported and the numbers are increasing.[31–33] We now understand that the formation of the scaffold and its mechanical properties are influenced by several factors, one of which is the level of hydrophobicity.[20,23–27,34] That is, in addition to the ionic complementary interactions, the extent of the hydrophobic residues, Ala, Val, Ile, Leu, Tyr, Phe, Trp (or single letter code, A, V, I, L, Y, P, W) can significantly influence the mechanical properties of the scaffolds and the speed of their self-assembly. The higher the content of hydrophobicity, the shorter the length required, the easier the scaffold formation is, and the better their mechanical strength and gelation properties.[20,23–27,34] Interestingly, different cells may behave differently in different peptide scaffolds depending on the culture conditions. This again demonstrates the importance of tailor-making scaffolds for different tissue cells microenvironment.

IV. PURAMATRIX *IN VITRO* CELL CULTURE EXAMPLES

PuraMatrix has been used to culture diverse types of tissue cells including stem and progenitor cells, as well as differentiated cell types and organotypic tissue slices.[15–22,35] A subset of cell types cultured in PuraMatrix is exemplified below (Table 15.1).

A. Hepatocytes

Hepatocytes have been cultured on PuraMatrix and Matrigel where they attach, proliferate, exhibited proper spheroid morphology, and created tight junctions. On the other hand, hepatocytes cultured on collagen coated Petri dish did not survive long-term, and did not attach when cultured on plastic dish alone.

Hepatocyte morphology was evaluated at Day 2 and Day 5 (Figure 15.5). Cytochrome p450 1A1 was evaluated as follows on Day 10. Substrate 8 μm 7-ethoxyresorufin metabolized by cells for 30 min at 37°C to resorufin conjugate. Fluorescence was measured at 530 nm excitation and 590 nm emission. Acceptance criteria requires hepatocytes cultured on the Matrigel substrate is twice the

TABLE 15.1
A Variety of Tissue Cells and Tissues Cultured on PuraMatrix

Mouse fibroblast	Bovine calf and adult chondrocytes
Chicken embryo fibroblast	Bovine endothelial cells
Chinese hamster ovary	Rat adult liver progenitor cells
Rat pheochromocytoma	Rat cardiac myocytes
Rat neural stem cells	Rat hippocampal neural tissue slice
Mouse embryonic stem (ES) cells	Mouse neural colony stem cells
Mouse cerebellum granule cells	Mouse and rat hippocampal cells
Bovine osteoblasts	Hamster pancreas cells
Human cervical carcinoma	Human osteosarcoma
Human hepatocellular carcinoma	Human neuroblastoma
Human embryonic kidney	Human foreskin fibroblast
Human epidermal keratinocytes	Human neural stem cells
Human hepatocytes	Human ES cells

These cells include stable cell lines, primary isolated cells from animals, progenitor, and stem cells.

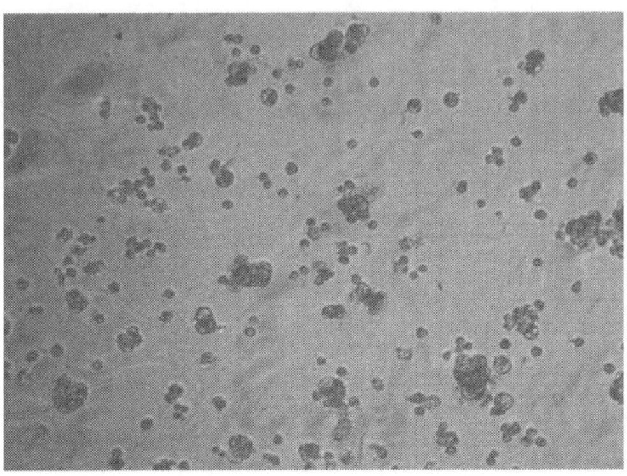

FIGURE 15.5 Primary rat hepatocytes cultured on PuraMatrix at Day 5. Cells begin to form clusters.

level of hepatocytes on the collagen I substrate. Hepatocytes on PuraMatrix grew much better than collagen I, thus suggesting PuraMatrix is a suitable scaffold for hepatocytes culture *in vitro.*

B. Adult Liver Progenitor Cells

A hepatocyte progenitor cell line has also been cultured using three-dimensional PuraMatrix. These cells exhibited nonexponential cell kinetics, acquired spheroidal morphology, and produces progeny cells with mature hepatocyte properties.[21] The differentiated progeny cells display increased expression of albumin and several other indicators of hepatocyte maturation, including binucleation, upregulation of transcription factor C/EBPα, and expression of cytochrome P450's CYP1A1, CYP1A2, and CYP2E1 (Figure 15.6). In contrast, markers of hepatocyte progenitors, α-fetoprotein and CK8, are unchanged. These relationships suggest production of transition spheroidal units comprised of asymmetrically cycling adult progenitor cells and their differentiating progeny. All three cytochrome p450 enzyme activities are 3-methylcholanthrene-inducible in such spheroids. These results demonstrated the ability of a designed biological scaffold to provide a microenvironment in which adult progenitor cells regain their intrinsic ability to continuously produce differentiating progeny cells. This three-dimensional nanofiber scaffold may provide a physiological approach for biomedical applications and pharmaceutic high throughput drug screening tools with more reliable content outcome.[21]

C. Chondrocytes Form Molded Cartilage in Cell Culture

In choosing a scaffold for cartilage repair, it is important to identify a material that can maintain a normal rate of proliferation of differentiated chondrocytes and a high rate of chondrocyte synthesis of specific ECM macromolecules, including Type II collagen and GAGs, until repair evolves into a steady state tissue maintenance. Kisiday et al. first used RADA16 scaffold, but it did not give the optimal results because it is mechanically rather weak. We then designed the KLD12 (****n-KLDLKLDLKLDL-c) peptide scaffold reasoning that because leucine is more hydrophobic than alanine, the leucines would probably pack more tightly in the nanofibers in aqueous conditions and thus provide a higher mechanical strength.[20,25,26]

In Alan Grodzinsky's laboratory, Kisiday et al. used the self-assembling peptide KLD12 scaffold as a model for cartilage repair and developed a method to encapsulate chondrocytes within the scaffold. During 4 weeks of culture *in vitro*, chondrocytes seeded within the peptide scaffold developed a cartilage-like ECM, rich in proteoglycans and Type II collagen, indicative of a stable chondrocyte phenotype (Figure 15.7). Time dependent accumulation of this ECM was paralleled by increases in material stiffness, indicative of deposition of mechanically functional newly formed tissue. The content of viable differentiated chondrocytes within the peptide scaffold increased at a rate that was fourfold higher than that of parallel, chondrocyte seeded agarose culture, a well-defined reference chondrocyte culture system. These results demonstrate the potential of a tailor-made peptide scaffold as a scaffold for the synthesis and accumulation of a true cartilage-like ECM in a three-dimensional cell culture for cartilage tissue repair. The peptide KLD12 used in this study represents a designed, self-assembling peptide made through molecular engineering, which can be modified to suit specific cell and tissue application interests.[20]

FIGURE 15.6 Promotion of hepatocyte differentiation by PuraMatrix. Lig-8 hepatic progenitor cells were cultured in two-dimensional adherent culture or in three-dimensional PuraMatrix. Spheroids were isolated and analyzed approximately 16 h after transfer to adherent culture. **a**, **e**, **i**, adherent cell colonies (phase contrast); **b**, **f**, **j**, respective *in situ* immunofluorescence for adherent colonies with anti-C/EBPα, antialbumin, and anti-CYP1A1/1A2 antibodies; **c**, **g**, **k**, isolated scaffold spheroids (phase contrast; note binucleated cells); **d**, **h**, **l**, respective *in situ* immunofluorescence for isolated scaffold spheroids with anti-C/EBPα, antialbumin, and anti-CYP1A1/1A2 antibodies.

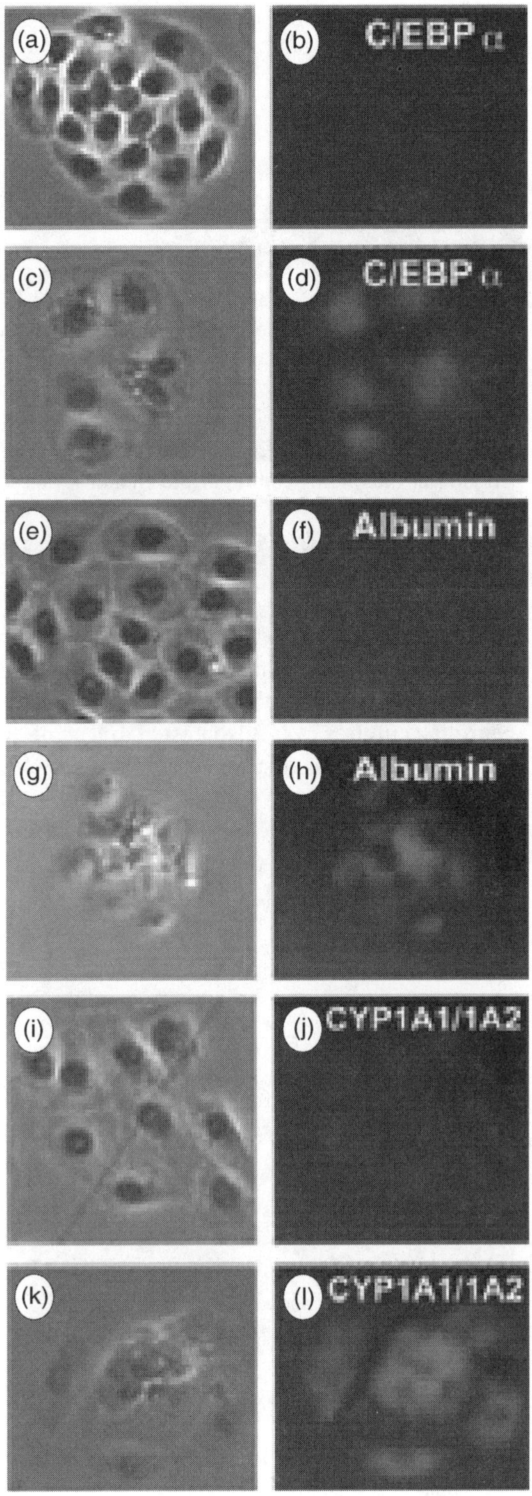
(a)
(b) C/EBP α
(c)
(d) C/EBP α
(e)
(f) Albumin
(g)
(h) Albumin
(i)
(j) CYP1A1/1A2
(k)
(l) CYP1A1/1A2

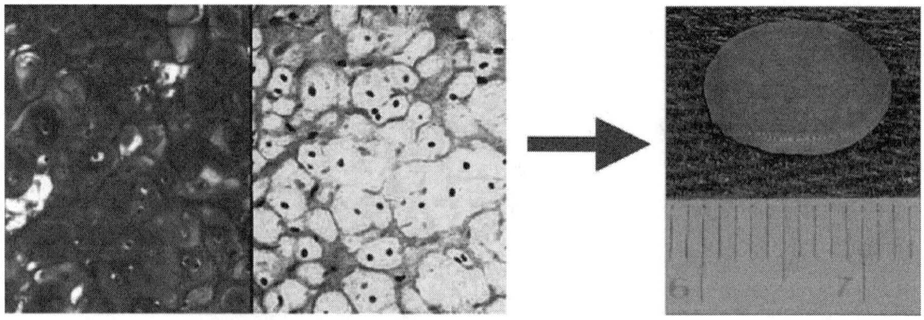

FIGURE 15.7 Peptide KLD12 (KLDLKLDLKLDL) chondrocytes in the peptide scaffold and cartilage. The chondrocytes stained with toluidine blue (TB) showing abundant glycosaminoglycans (GAG) production (left panel) and antibody to Type II collagen demonstrating abundant Type II collagen production (right panel). A piece of premolded cartilage with encapsulated chondrocytes in the peptide nanofiber scaffold. The cartilage formed over a 3 to 4 weeks period after the initial seeding of the chondrocytes.[20]

D. Extensive Neurite Outgrowth and Active Synapse Formation on PuraMatrix

PuraMatrix serves as a substrate and portable membrane media to support both neuronal attachment and differentiation in the form of extensive neurite outgrowth. Functional synapse formation also occurs between attached neurons when the cells are grown on PuraMatrix (Figure 15.8).[19] Neurite outgrowth from primary neuronal cultures was also tested by using several nerve cell types, including primary dissociated neurons from the mouse cerebellum and rat hippocampus. Cerebellar granule neurons undergo postnatal development and are morphologically distinguishable from other cerebellar cells. PuraMatrix scaffolds support extensive neurite out-growth from cerebellar granule neurons prepared from 7 days old mice and the neurites were readily visualized in two different focal planes, suggesting that the neurites closely follow the contours of the matrices. The primary cerebellar neuronal cultures on the PuraMatrix scaffolds were maintained for up to 4 weeks. Dissociated mouse and rat hippocampal neurons also attach and project neurites on the scaffolds (Table 15.2).

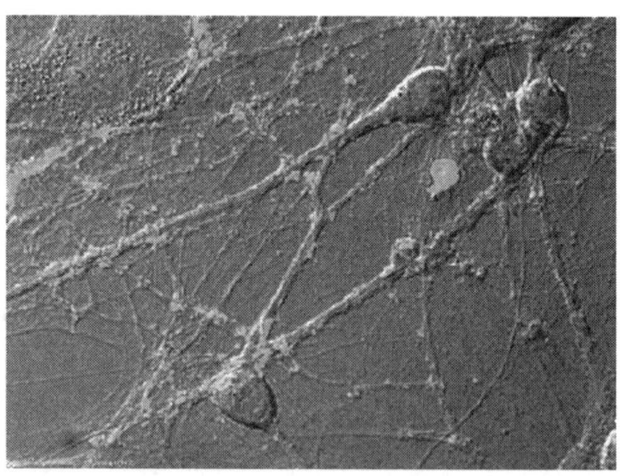

FIGURE 15.8 Active synapses on the peptide surface. Primary rat hippocampal neurons form active synapses on peptide scaffolds. The confocal images shown bright discrete green dot labeling indicative of synaptically active membranes after incubation of neurons with the fluorescent lipophilic probe FM-143. FM-143 can selectively trace synaptic vesicle turnover during the process of synaptic transmission. The active synapses on the peptide scaffold are fully functional, indicating that the PuraMatrix is a permissible substrate for neurite outgrowth and active synapse formation.[19]

TABLE 15.2
Neurtite Outgrowth on PuraMatrix Membrane Surface

Cell Type	Length of Process (μm)	Cell Sources
NGF treated rat PC12	400–500	Cultured cell line
NGF preprimed PC12	400–500	Cultured cell line
Human SY5Y neuroblastoma	400–500	Cultured cell line
Mouse cerebellar granule neurons	200–300	Primary cells[a]
Mouse hippocampal neurons	100–200	Primary cells[a]
Rat hippocampal neurons	200–300	Primary cells[b]

Cells were seeded onto PuraMatrix scaffold coating. The cell bearing PuraMatrix was transferred to dishes with fresh medium. Maximum neurite length was estimated visually with scale bars three to seven days after cell attachment for primary cultures and 10 to 14 days after matrix attachment for the cultured cell lines.

[a] Seven day old mouse.

[b] One day old rat.

E. Organotypic Hippocampal Tissue Culture in PuraMatrix

PuraMatrix has also been used to isolate and expand self-renewing neural cells *ex vivo*. Neurogenesis occurs in restricted areas of postnatal mammalian brain including the dentate gyrus and subvetricular zone (SVZ).[22] PuraMatrix has been used to entrap migrating neural cells (potential neuroprogenitors) from postnatal hippocampal organotypic cultures in three-dimensional PuraMatrix (Figure 15.9). In those experiments, brain tissues from the rat neonatal hippocampus were laid on top of preformed PuraMatrix and cultured for up to 2 weeks. Within a few hours, cell division activity was observed only at the interface zone.[22] The migrating neural cells with mitotic activity from the dentate gyrus, CA1, CA2, and CA3 regions of the postnatal rat hippocampus were greatly enriched in organotypic culture and they were entrapped in PuraMatrix. After a few hours, the cells exhibited high proliferating activity, as measured by incorporation of $BrdU^+$ cells at the "interface zone" between the tissue slice and the culture surface. Using PuraMatrix for neural progenitor cell isolation and expansion *in vitro* may thus have applications in developing new strategies and for cell-based therapies in regenerative medicine.

F. Osteoblasts

Osteoblasts (ATCC MT3T3) have been maintained with increase cell density successfully in PuraMatrix up to 1 month (Figure 15.10). These cells not only proliferated but also formed gap junctions connecting osteoblasts in a manner that is also found in native bone tissues.

V. STANDARD *IN VITRO* TOXICOLOGY AND BIOCOMPATIBILITY STUDIES

In addition to over 10 years of cell culture studies in our laboratory and elsewhere, standard *in vitro* toxicology studies have been completed, including EN/ISO tests for cytotoxicity and hemocompatibility. The tests below were completed at a Federal Drug Agency (FDA) certified toxicology testing company (Toxikon Corp, Bedford, MA) using established standards of measure on the commercially available PuraMatrix (RADI-16; Table 15.3).

A. Cytotoxicity

The agar diffusion test (ISO 10993-5) measures a material's effect on cell cultures, which are extremely sensitive to minute quantities of diffusible chemicals and readily

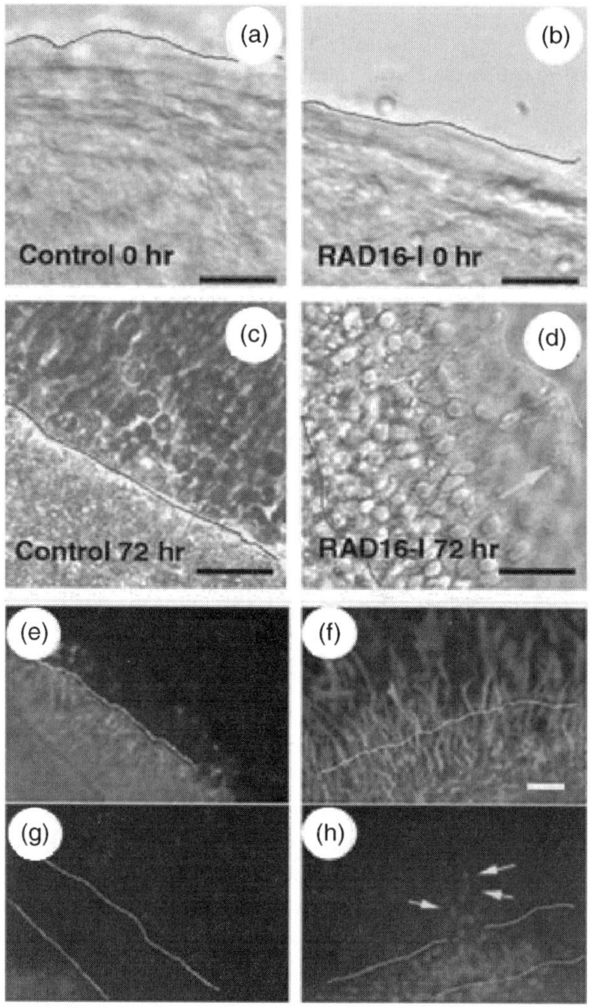

FIGURE 15.9 Hippocampal organotypic slice cultures cultured on peptide scaffolds develop extended tissue scaffolds.[22] Hippocampal slices were cultured organotypically either on control membrane or on RAD16-I peptide scaffolds layers ($\sim$500 μm thick). A time lapse experiment was carried out to follow up the tissue scaffold growth from the perimeter of the dentate gyrus region. (a) time 0 (0 H) of control slice culture; (b) time 0 (0 H) of scaffold slice culture; (c) 72 h of control slice culture; (d) 72 h of scaffold slice culture. The red line indicates the original border of the tissue slice and the yellow in d the extended tissue scaffold. The yellow arrow in d indicates the direction of tissue scaffold growth and extension. Black bars indicate 100 μm. (e) 72 h control slice culture immunostained for GFAP (glia cell marker); (f) 72 h scaffold slice culture immunostained for GFAP; (g) same optical layer as e immunostained for NeuN (neuron marker); h) same optical layer as f immunostained for NeuN (red). Red lines in e and f and yellow in g and h indicates the original perimeter of the tissue slice. The white line in e and g indicates the extended tissue scaffold in the control cultures. The white line in f and h is use to compare the over extension obtained on peptide scaffolds cultures. Yellow arrows in h indicate neuron cells staining with NeuN migrating into scaffold slice cultures. The white bar in f is 100 μm.

display characteristic signs of toxicity in the presence of potentially harmful diffusible substances. The biological reactivity of a mammalian monolayer, L929 mouse fibroblast cell culture, in response to the test article was determined as "no reactivity" in tests.

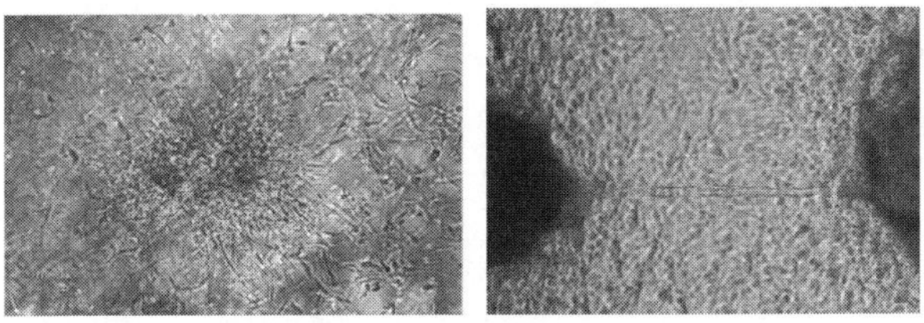

FIGURE 15.10 Osteoblasts in PuraMatrix. Cells form clusters (left panel) and junctions connected by cells (right panel) (image courtesy of 3DMatrix).

B. Hemolysis

The hemolysis test (direct contact, ISO-10993-4) measures the ability of a material to cause red blood cells to rupture. This test is derived from well-established National Institute of Health (NIH) protocols and is performed in triplicate. This test uses rabbit blood in direct contact with the test material and the degree of hemolysis is measured spectrophotometrically. PuraMatrix tested "nonhemolytic" in tests carried out by an independent FDA certified testing agency.

C. Coagulation Prothrombin Time

Prothrombin time (human plasma, ISO 10993-4) measures the effect of a test article extract on human blood coagulation time. This assay has become a suitable clinical means of determining the presence and functioning ability of prothrombin in the process of coagulation. PuraMatrix did not have an adverse effect on prothrombin coagulation time of human plasma in tests carried out by an independent FDA certified testing agency.

VI. *In Vivo* Biocompatibility and Toxicology Studies

Several kinds of biocompatibility and toxicity examinations of PuraMatrix have been carried out in numerous animal models, including a number of standard FDA/ISO toxicology and *in vivo* implant studies, to assess the safety of PuraMatrix. PuraMatrix has passed every one. These tests point to the biocompatibility of this particular peptide material, but more long-term studies are required for the commercialization of therapeutic applications (Table 15.4). These tests were outsourced and conducted under GLP conditions by Toxikon Corporation, an FDA certified facility in Bedford, Mass.

TABLE 15.3
***In Vivo* Biocompatibility and Toxicology Tests**

Test	Result[a]
Cytotoxicity:	
Agar diffusion ISO 10993-5	Noncytotoxic
Hemolysis:	
Direct contact (ISO 10993-4)	Nonhemolytic
Prothrombin time assay:	
Human plasma (ISO 10993-4)	No adverse effect on prothrombin coagulation

[a] These tests are contracted to the FDA certified commercial toxicology test facility (Toxikon Corp, Bedford, MA).

TABLE 15.4
PuraMatrix Used in *In Vivo* Animal Studies

Test	Result
ADME tox: 14 day implant	No problematic organ accumulation, excretion. 14C Radiolabeled PuraMatrix 10% excreted by day 14
Rabbit muscle implant: 14 day (ISO 10993-6)	Nontoxic score across 13 categories of reactions
Rabbit pyrogenicity: 24 h (ISO-10993-11)	No effect on animals
Intracutaneous implant: 72 h (ISO 10993-12)	Primary irritation index = 0.0 "Negligible" No febrile reaction from any animal over 24 h after intravenous injection of dilute solution
CNS lesions in hamsters: 2–60 days	Reduced scarring and reinnervation across the severed optic tract caudal to the lesion only in PuraMatrix treated animals[35]

Moreover, previously studies of animal reactions[19,35] described lack of immunogenicity and inflammation in rat, rabbit, goat, and hamster models.

A. ADME and Biodegradability

In order to measure the local and whole animal distribution of PuraMatrix, and given that previous experiments were unable to raise antibodies to it, we generated a carbon radioisotope labeled version of ^{14}C-PuraMatrix. The radioisotope ^{14}C was internally labeled at the third alanine site (Acetyl-(RADA)-(R-[^{14}C(U)-Ala]-D-A)-$(RADA)_2$-$CONH_2$) as opposed to a labeled acetyl group, which could be cleaved off, in order to better characterize the adsorption, degradation, metabolism, and excretion (ADME) of the PuraMatrix material *in vivo*.

The radioisotope labeled ^{14}C-PuraMatrix was then implanted into Sprague Dawley® rats in a femur defect model. Urine and stool samples were collected and the radioactivity was counted.

B. Rabbit Muscle Implant (2 Weeks)

This test assesses the local effects of material on contact with living tissue. Test article was implanted into paravertebral muscle of three New Zealand white rabbits, with negative control (GelFoam) implanted in the contralateral muscle of each animal. Healing was allowed for 2 weeks. Animals were sacrificed and implants excised. Excised implants were examined macroscopically with a magnifying lens and fixed in formalin. Histologic slides of hematoxylin and eosin (H&E) and Mason's Trichrome stained sections were prepared, studied microscopically by a board certified veterinary pathologist, and evaluated on a scale of 0 to 3.

PuraMatrix implants retained the initial implant volume, collagen fibers, and vascularization of the injection site (Figure 15.11). PuraMatrix was rated nontoxic (0.13) on a scale ranging from nontoxic (<1), slightly toxic (1 to <2), mildly toxic (2 to <3), moderately toxic (3 to >4), to severely toxic (<4), in tests carried out by an independent FDA certified testing agency.

C. Intracutaneous Reactivity

Irritation reactivity tests assess the localized reaction of tissues, including breached tissue and blood contact, to device materials or extracts. A 0.5% w/v peptide PuraMatrix solution was extracted in NaCl and CSO at a ratio of 200 mg/1.0 ml at 70 ± 2°C for 24 h. Control extracts were prepared, in a similar manner to the test article. Three rabbits were injected intracutaneously, using one side of the

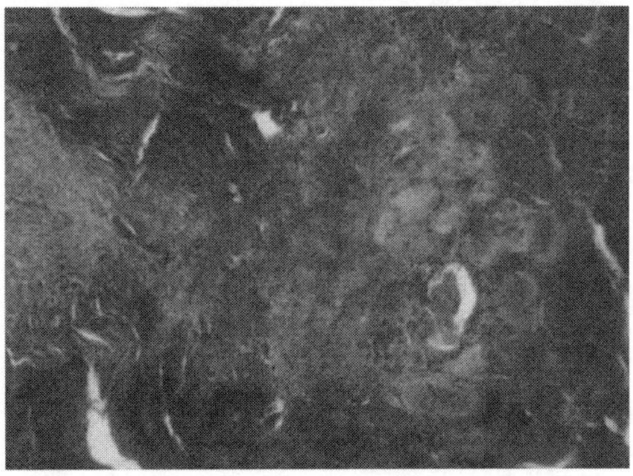

FIGURE 15.11 PuraMatrix creates a permissive environment for tissue bulking (2 weeks postinjection).

animal for the test article extracts and the other side for the control extracts, at 0.2 ml per site. The injected sites were examined at 24, 48, and 72 h post inoculation for gross evidence of tissue reaction such as erythema, edema, and necrosis. A primary irritation index of 0.5 or less will be considered a negligible irritant, where 0.5 to less than 2 are slight irritants, 2 to less than 5 are moderate irritants, and greater than 5 are severe irritants. The test sites injected with PuraMatrix did not exhibit any signs of erythema or edema over the 72 h observation point. The primary irritation index for the material is 0.0 in tests carried out by an independent FDA certified testing agency.

D. Rabbit Pyrogen

The purpose of the test was to detect the risk of a patient to a febrile reaction as a result of the administration of the test article extract. The test article was prepared by mixing 9 ml of the test article with 9 ml of 0.9% USP sodium chloride for injection (NaCl) and the resulting gel extracted at a ration of 0.2 gm/ml. The test article extract was administered by intravenous injection at 10 ml/kg. The rectal temperatures of the injected rabbits were compared with the temperature of a control rabbit similarly injected with 0.9% USP Sodium Chloride for Injection (NaCl). The baseline temperature of the rabbits, determined no more than 30 min prior to injection of the test article extract, were used to exclude rabbits whose body temperatures vary by more than 1°C from each other and whose temperatures are greater than 39.8°C. Body temperatures were recorded at 30 min intervals between one and 3 h subsequent to injection. If no rabbit exhibited a rise in temperature of 0.5°C or more above its baseline temperature, then the product met the requirements for the absence of pyrogens (Table 15.4). None of the animals injected with the test article extract exhibited signs of pyrogenic response, in tests carried out by an independent FDA certified testing agency.

VII. FUTURE PERSPECTIVES

A. Compatible with Bioproduction and Clinical Applications

PuraMatrix fulfills both traditional biological materials and a nanoporous/nanofiber hydrogel to enable proper three-dimensional cell growth and the creation of microenvironments surrounding cell colonies. For cells that prefer culture on a two-dimensional surface, better results were obtained by culturing cells on PuraMatrix floating sheets, allowing a three-dimensional nutrient bath within the two-dimensional growth context.

PuraMatrix mimics important aspects of the *in vivo* environment, while eliminating complicating variables traditionally experienced from animal derived materials. Unlike most

TABLE 15.5
PuraMatrix Attributes and Advantages Comparison Table

Attributes	PuraMatrix Synthetic ECM	Natural ECM	Synthetic Scaffolds	PuraMatrix Advantage
Scaffold Properties				**Defined, consistent bare ECM analog**
Composition	RAD_{16} 16 peptide monomer in 0.5 to 1.0% w/v patented worldwide	Collagen, matrigel, cadaver tissue, basement membrane	PLA, PLGA, carbon fiber, calcium phosphate	Animal free, reproducible cell culture and cell signaling
Fiber size & sequence	7 to 10 nm diameter Bare ECM free of unwanted signals — add components as needed	5 to 10 nm diameter complex undefined protein sequences	10,000 to 100,000 nm looks 2D relative to cell	Approximates *in vivo* ECM nanoscale without the complexity
Pore size	50 to 400 nm	50 to 400 nm	20,000 to 1×10^6 nm	Encapsulates like ECM
Water content	99.5 to 99.9%	80 to 97%	60 to 80%	Better hydration and nutrient diffusion
Mechanical strength	Low to medium, cells can migrate within it	Low to medium	Medium to high	More rapid cell migration, ingrowth
Physical loading	Able to simulate physiological loading and flow	Depends on material	Often brittle, shift load from cells	Create physiologically relevant microenvironment
Cell Encapsulation and Handling				**Stable, Injectable, Customizable**
Scaffold formation	Fibers and gel form around cells with simple addition of culture media or injection *in vivo*	Require refrigeration or complex processing	Preformed scaffolds tough to seed with cells	Superior encapsulation allows cells to create own microenvironments and surrounding ECM
Combination with bioactives	Bioactives, ECM proteins can be added for tailored reproducible 3D cultures	Inconsistent levels of proteins and GF	Yes, but not true microenvironments	Enables consistent, defined ECM microenvironments
Allows cell attachment, migration, angiogenesis	Yes, enables anchorage dependent cell culture	Yes	Some	Allows cell–cell interactions, migration and invasion assays
Sterile	Gamma irradiation or pregelled filter sterilization *in situ*	No, destroys material and proteins	No gamma, often limited to ethylene oxide	Able to use gold standard sterilization, and filter sterilize *in situ if* needed
Injectable	Yes, will not swell beyond injected volume. Compatible with cardiovascular catheters	Yes, when chilled	Not injectable	Injectable along with cells, yet gel forms upon introduction *in vivo*

Clinical cell culture "closed system"	Sterility (gamma irradiation) and injectability enables closed system culture for bioproduction and clinical cell expansion	Animal components discouraged by FDA	Often hard to sterilize via gamma irradiation	Gold standard sterility, combined with injectability and easy handling
Stable	Shelf stable at room temperature and across broad temperature range for at least 2 years.	Natural product requires refrigeration with short shelf life	Shelf stable only while dry	Ideal stability independent of water content
Molding, coating, sheeting	Amenable to casting, coating, layering, 3D printing	Often monolayer only or limited sheets	Fixed shapes can be cut to shape	Simple to configure as needed
Cell Recovery and Analysis				**Research and Clinical Compatibility**
Microscopy	Transparent	Often cloudy	Often opaque	Easy visualization
Cell passaging harvesting, separation	Spin cells out, wash, replate, or re-encapsulate	Trypsin, collagenase	Hard to harvest cells without disruption	Easy cell harvesting with high cell viability
Molecular biology	Protein monomer has simple fingerprint to distinguish	Variation and noisy background data	Not compatible	Straightforward westerns, southerns, northerns
Flow cytometry	Compatible without separation	Must separate	Not compatible	Compatible
Closed system cell harvesting	Compatible	Depends on material	Not compatible	Requirement for clinical and bioproduction apps
Biocompatibility				**Superior *in vivo*, rigorously tested**
Nonimmunogenic	No discernable antibodies, foreign body response, chronic inflammation	More immune response and inflammation	Foreign body response, scarring, acidic breakdown	Superior in direct comparison studies
Biodegradeable	Yes, rapid due to low material/water ratio	Yes, but heavy lymphcyte activity	Yes, but illicits foreign body reaction.	More rapid, but rate can be varied & controlled
Swelling/absorption	Nonswelling upon injection	Nonswelling	Often swell, causing irritation	Consistent control of injected volume
Ingrowth, angiogenesis	Weak gel & cell migration allow rapid ingrowth	Undefined cell signaling	Large structure inhibits ingrowth	More rapid ingrowth

other scaffolds, PuraMatrix can be sterilized through UV radiation or filtration, and has proven itself shelf stable at room temperature for over 10 years. PuraMatrix not only lowers the handling, processing, transportation, and inventory costs, but also meets the stringent bioproduction requirements through its synthetic composition, free of the undesired factors and adventitious agents from animal product. PuraMatrix not only provides the three-dimensional context for cell research, but they are forward compatible with bioproduction and clinical requirements necessary for eventual tissue engineering and stem cell-based therapies.

B. Synthetic Origin, Clinical-Grade Quality, Clinical Delivery

Because PuraMatrix is synthetic and sterile, it is thus suitable for bioproduction, and can be readily in this and clinical settings, unlike many animal derived materials (bovine, pig or others) that lack the consistent quality control, thus complicating clinical reproducibility and risking the introduction of undesirable contaminants (cell signaling factors, prions, adventitious agents, etc.). PuraMatrix is currently manufactured in large quantities, some under good manufacturing processes (GMP) grade processes and commercially available. Additionally, PuraMatrix can be used for closed, sterile system culture *in vitro* and can also be injected *in vivo* without eliciting immune responses, unlike collagens and alginates. Upon injection or pipetting into solution, the PuraMatrix volume does not swell, retaining a consistent volume since it already has more than 99% water content.

C. Tailor-Made PuraMatrix

In order to fabricate tailor-made PuraMatrix, it is crucial to understand the finest detail of peptide and protein structures, and their influence on the nanofiber structural formation and stability. Since there is a vast array of possibilities to form countless structures, a firm understanding of all available amino acids, their properties, and the peptide and protein secondary structures is a prerequisite for further progress in the fabrication of peptide and protein materials.[36,37] We are moving in that direction, which will further accelerate new scaffold development.[38–42]

To date, PuraMatrix has been used in diverse cell and tissue systems from a variety of sources. This demonstrates a promising prospect in further improvement for specific needs since tissues are known to reside in different microenvironments. The PuraMatrix used thus far are general peptide nanofiber scaffolds and not tailor-made for specific tissue environment. We are designing second generation PuraMatrix scaffolds and to show that these tailor-made PuraMatrix incorporating specific functional motifs will perform as superior scaffolds in specific applications. They may not only create a fine-tuned microenvironment for three-dimensional tissue cell cultures, but also may enhance cell–materials' interactions, cell proliferation, migration, differentiation, and the performance of their biological function. The ultimate goal is to produce tailor-made scaffolds for particular tissue engineering and for regenerative and reparative medical therapies.

ACKNOWLEDGMENTS

We would like to thank members of our laboratory, past and present, for making discoveries and conducting exciting research. We gratefully acknowledge the supports by grants from the US Army Research Office, Office of Naval Research, Defense Advanced Research Project Agency (BioComputing), DARPA/Naval Research Laboratories, DARPA/AFO, NSF-MIT BPEC and NSF CCR-0122419 to MIT Media Laboratory's Center for Bits and Atoms, the National Institute of Health, the Whitaker Foundation, Du Pont-MIT Alliance, and Menicon, Ltd, Japan. We also acknowledge the Intel Corp., educational donation of a computing cluster to the Center for Biomedical Engineering at MIT.

REFERENCES

1. Sipe, J., Tissue engineering and reparative medicine, *Ann. NY Acad. Sci.*, 961, 1–9, 2002.
2. Ratner, B., Hoffman, A., Schoen, F., Lemons, J., eds., *Biomaterials Science*, Adademic Press, New York, 1996.
3. Lanza, R., Langer, R., and Vacanti, J., *Principles of Tissue Engineering*, 2nd ed., Academic Press, San Diego, CA, 2000.
4. Atala, A., and Lanza, R., *Methods of Tissue Engineering*, Academic Press, San Diego, CA, 2001.
5. Hoffman, A. S., Hydrogels for biomedical applications, *Adv. Drug Deliv. Rev.*, 43, 3–12, 2002.
6. Palsson, B., Hubell, J., Plonsey, R., and Bronzino, J. D., *Tissue Engineering: Principles and Applications in Engineering*, CRC Press, Boca Raton, FL, 2003.
7. Ayad, S., Boot-Handford, R. P., Humphreise, M. J., Kadler, K. E., and Shuttleworth, C. A., *The Extracellulat Matrix: Facts Book*, 2nd ed., Academic Press, San Diego, CA, 1998.
8. Kreis, T., and Vale, R., *Guide Book to the Extracellular Matrix, Anchor, and Adhesion Proteins*, 2nd ed., Oxford University Press, Oxford, 1999.
9. Yannas, I., *Tissue and Organ Regeneration in Adults*, Springer, New York, 2001.
10. Mooney, D. J., Hansen, L. K., Langer, R., Vacanti, J. P., and Ingber, D. E., Extracellular matrix controls tubulin monomer levels in hepatocytes by regulating protein turnover, *Mol. Biol. Cell*, 5, 1281–1288, 1994.
11. Friedl, P., Zanker, K., and Brocker, E., Cell migration strategies in 3-D extracellular matrix: differences in morphology, cell matrix adhesions, and integrin function, *Microsc. Res. Tech.*, 43, 369–378, 1998.
12. Folch, A., and Toner, M., Microengineering of cellular interactions, *Annu. Rev. Biomed. Eng.*, 2, 227–256, 2000.
13. Cukierman, E., Pankov, R., Stevens, D. R., and Yamada, K. M., Taking cell-matrix adhesions to the third dimension, *Science*, 294, 1708–1712, 2001.
14. Abbott, A., and Cyranoski, D., Biology's new dimensions, *Nature*, 420, 870–872, 2003.
15. Zhang, S., Lockshin, C., Herbert, A., Winter, E., and Rich, A., Zuotin, a putative Z-DNA binding protein in saccharomyces cerevisiae, *EMBO. J.*, 11, 3787–3796, 1992.
16. Zhang, S., Holmes, T., Lockshin, C., and Rich, A., Spontaneous assembly of a self-complementary oligopeptide to form a stable macroscopic membrane, *Proc. Natl Acad. Sci. USA*, 90, 3334–3338, 1993.
17. Zhang, S., Lockshin, C., Cook, R., and Rich, A., Unusually stable beta-sheet formation of an ionic self-complementary oligopeptide, *Biopolymers*, 34, 663–672, 1994.
18. Zhang, S., Holmes, T., DiPersio, M., Hynes, R. O., Su, X., and Rich, A., Self-complementary oligopeptide matrices support mammalian cell attachment, *Biomaterials*, 16, 1385–1393, 1995.
19. Holmes, T. C., Delacalle, S., Su, X., Rich, A., and Zhang, S., Extensive neurite outgrowth and active neuronal synapses on peptide scaffolds, *Proc. Natl Acad. Sci. USA*, 97, 6728–6733, 2000.
20. Kisiday, J., Jin, M., Kurz, B., Hung, H., Semino, C. E., Zhang, S., and Grodzinsky, A. J., Self-assembling peptide hydrogel fosters chondrocyte extracellular matrix production and cell division: implications for cartilage tissue repair, *Proc. Natl Acad. Sci. USA*, 99, 9996–10001, 2002.
21. Semino, C. E., Merok, J. R., Crane, G., Panagiotakos, G., and Zhang, S., Functional differentiation of hepatocyte-like spheroid structures from putative liver progenitor cells in three-dimensional peptide scaffolds, *Differentiation*, 71, 262–270, 2003.
22. Semino, C. E., Kasahara, J., Hayashi, Y., and Zhang, S., Entrapment of hippocampal neural cells in self-assembling peptide scaffold, *Tissue Eng.*, 10, 643–655, 2004.
23. León, E. J., Verma, N., Zhang, S., Lauffenburger, D. A., and Kamm, R. D., Mechanical properties of a self-assembling oligopeptide matrix, *J. Biomater. Sci. Polym. Ed.*, 9, 297–312, 1998.
24. Caplan, M., Moore, P., Zhang, S., Kamm, R. D., and Lauffenburger, D. A., Self-assembly of a beta-sheet oligopeptide is governed by electrostatic repulsion, *Biomacromolecules*, 1, 627–631, 2000.
25. Caplan, M., Schwartzfarb, E. M., Zhang, S., Kamm, R. D., and Lauffenburger, D., Control of self-assembling oligopeptide matrix formation through systematic variation of amino acid sequence, *Biomaterials*, 23, 219–227, 2002.

26. Caplan, M. R., Schwartzfarb, E. M., Zhang, S., Kamm, R. D., and Lauffenburger, D. A., Effects of systematic variation of amino acid sequence on the mechanical properties of a self-assembling, oligopeptide biomaterial, *J. Biomater. Sci. Polym. Ed.*, 13, 225–236, 2002.
27. Marini, D., Hwang, W., Lauffenburger, D. A., Zhang, S., and Kamm, R. D., Left-handed helical ribbon intermediates in the self-assembly of a beta-sheet peptide, *Nano Lett.*, 2, 295–299, 2002.
28. Zhang, S., Design and exploitation of self-assembling ionic complementary peptide systems: a model for peptide biomaterial engineering, In *Perspective in Protein Engineering 1996 (CDROM Edition)*, Geisow, M., ed., Biodigm Ltd, UK, 1996, ISBN0-9529015-0-1.
29. Hong, Y., Legge, R. L., Zhang, S., and Chen, P., Effect of amino acid sequence and pH on nanofiber formation of self-assembling peptides EAK16-II and EAK16-IV, *Biomacromolecules*, 4, 1433–1442, 2003.
30. Altman, M., Lee, P., Rich, A., and Zhang, S., Conformational behavior of ionic self-complementary peptides, *Protein Sci.*, 9, 1095–1105, 2000.
31. Zhang, S., and Altman, M., Peptide self-assembly in functional polymer science and engineering, *React. Funct. Polym.*, 41, 91–102, 1999.
32. Zhang, S., and Altman, M., Self-assembling peptide systems in biology, engineering and medicine, In *Crete Meeting Proceedings*, Aggeli, A., Boden, N., and Zhang, S., eds., Kluwer Academic Publishers, Dordrent, The Netherlands, pp. 343–360, 2001.
33. Zhang, S., Emerging biological materials through molecular self-assembly, *Biotechnol. Adv.*, 20, 321–339, 2002.
34. Hwang, W., Marini, D., Kamm, R. D., and Zhang, S., Supramolecular structure of helical ribbons self-assembled from a beta-sheet peptide, *J. Chem. Phys.*, 118, 389–397, 2003.
35. Ellis-Behnke, R., Semino, C. E., Zhang, S., and Schneider, G. E., *Peptide Nanofiber Scaffold for Brain Lesion Repair.* (Submitted for publication), 2005.
36. Branden, C.-I., and Tooze, J., *Introduction to Protein Structure*, 2nd ed., Garland Publishing, New York, 1999.
37. Zhang, S., Fabrication of novel materials through molecular self-assembly, *Nat. Biotechnol.*, 21, 1171–1178, 2003.
38. Zhang, S., Marini, D., Hwang, W., and Santoso, S., Design nano biological materials through self-assembly of peptide & proteins, *Curr. Opin. Chem. Biol.*, 6, 865–871, 2002.
39. Zhang, S., Building from bottom-up, *Mater. Today*, 6, 20–27, 2003.
40. Zhang, S., Beyond the petri dish, *Nat. Biotechnol.*, 22, 151–152, 2004.
41. Zhang, S., Wet or let die, *Nat. Mater.*, 3, 7–8, 2004.
42. Santoso, S., and Zhang, S., Nanomaterials through molecular self-assembly, In *Encyclopedia of Nanoscience and Nanotechnology*, Nalwa, H. S., Ed., American Scientific Publisher, Stevenson Ranch, CA, USA, 2004.

Part III

Materials Modifications and Properties

16 Polymer/Ceramic Composite Scaffolds for Bone Tissue Engineering

Guobao Wei and Peter X. Ma

CONTENTS

I. INTRODUCTION

Tissue loss and organ failure resulting from injuries or diseases remain frequent and serious health problems despite great advances in medical technologies and life sciences. In the U.S. alone, bone fractures and damage result in more than 1.3 million surgical procedures each year.[1] Currently, bone grafting procedures are employed to promote the healing of fracture nonunion and the repairing of other bone defects.[2] Autologous bone grafting, although effective as a clinical gold standard for bone repair, is seriously challenged by the availability of sufficient donor tissue supply and problems associated with donor site morbidity.[3,4] The efforts to address these problems and limitations have led to the development of new biomaterials and alternative therapies, among which a tissue engineering approach holds great promise. In this approach, neotissues are constructed by seeding functional cells, originated from either donors or differentiated stem cells, within a biodegradable three-dimensional porous scaffold, followed by *in vitro* culture and *in vivo* implantation to the desired sites.[5–7] Since most primary organ cells, including bone cells, are believed to be anchorage-dependent and require specific environments for growth, the success of tissue engineering relies greatly on the development of suitable scaffolds for *in vitro* tissue culture and *in vivo* neotissue formation. The scaffold should be highly porous for a sufficient amount of cell ingrowth and survival, it should have sufficient mechanical strength for temporary physical support, and it should be conductive to guide tissue regeneration in three dimensions.

Being similar to the major inorganic component of natural bone, bioceramics, including hydroxyapatite (HAP) and tricalcium phosphate (TCP),[8–10] are widely used as scaffolding materials for bone repair. These materials are demonstrated to have good osteoconductivity and bone bonding ability.[11] However, the main limitation for the use of ceramics is their inherent brittleness and difficulty of processing.[12] Furthermore, in fabrication of ceramic scaffolds, it is not easy to achieve a high porosity and a high total surface area which are necessary for anchorage-dependent cells such as bone cells to survive and differentiate. Biodegradable polymers, particularly the family of poly(α-hydroxy esters) including poly(lactic acid) (PLA), poly(glycolic acid) (PGA), and their copolymers (PLGA) have also been fabricated into a porous three-dimensional scaffold to engineer osseous tissues with varying successes.[7,13,14] These polymers are demonstrated to be biocompatible and biodegradable. The physical, chemical, surface, and degradation properties can be achieved by varying the ratio of lactide to glycolide in PLGA copolymer. In addition, they also have great three-dimensional design flexibility and can be readily fabricated into highly porous scaffolds with different structures and sizes to meet the needs of a specific tissue-engineering application.[15–19] PLGA (75:25) was processed into porous foams using a solvent casting and particulate-leaching technique, and cultured with stromal osteoblastic cells *in vitro*.[7] Unfortunately, in such PLGA scaffolds, the osteoblasts penetration and tissue regeneration occurred only in the surface layer (maximum of 240 μm) and no neotissue regenerated in the central areas. Other problems associated with scaffolds fabricated using the particulate-leaching technique were poor mechanical properties and their rapid decline of mechanical strength with increasing porosity, which is disadvantageous for structural tissue engineering applications.[20]

Although poly(α-hydroxy esters) or bioceramics alone have been used as scaffolds showing feasibility and the potential of bone tissue engineering, individually they are not ideal scaffolds for bone regeneration because of their own limitations. Compact PLLA/HAP composites (with little porosity) as bone fillers or implant materials have shown some favorable properties over both pure PLLA and pure HAP.[21,22] Polymer/ceramic composite scaffolds have received increasing attention for bone tissue engineering due to the fact that the composite scaffold combines favorable properties of polymer and ceramic phases while maintaining highly porous three-dimensional structures.[23–27] This chapter will review some of the recent work on synthetic polymer/ceramic porous composite scaffolds, and discuss the improvement of bone regeneration in such composite scaffold systems where bone composition and structure are mimicked.

II. POLYMER/HYDROXYAPATITE COMPOSITE SCAFFOLDS

From a biomimetic view, natural bone is a composite of organic collagen and inorganic mineral in forms of hydroxyapatite. To mimic this composite characteristic of natural bone, poly(α-hydroxy acids)/HAP scaffolds have been created through a thermally induced phase separation technique (TIPS) in our lab for bone regeneration.[23,24] In this process, HAP particles are first dispersed uniformly in a polymer solution at a high temperature, the resulting mixture is then subjected to solid–liquid or liquid–liquid phase separation by lowering the solution temperature. Subsequent removal of the solidified solvent-rich phase by sublimation leaves a porous polymer/ceramic composite scaffold. Although the addition of HAP may perturb the phase separation process to some extent,[23] composite scaffolds with porosity as high as 95% and open pore structure can be achieved. The scaffold microarchitecture can be controlled by the polymer concentration, HAP content, phase-separation temperature, and the solvent used (Figure 16.1 and Figure 16.2). Polymer/HAP composite scaffolds showed significant improvement in their compressive modulus and compressive yield strength over pure polymer scaffolds. Similar results of improvement in mechanical properties were also observed in a study by Thomson et al.,[28] where HAP fibers were incorporated into PLGA foams using a traditional particulate leaching technique. However, the improvement in mechanical properties was achieved only on scaffolds with low-porosity (47%) whereas high-porosity scaffolds (81%, suitable for cell ingrowth) were not reinforced by the introduction of HAP short fibers. For the same high porosity (>90%), the compressive modulus of pure polymer scaffolds fabricated with TIPS is about 20 times higher than those of the same porosity made with the well-documented salt-leaching technique. These results suggested that phase separation may be a preferred method to create three-dimensional porous composite scaffolds with high porosity, good mechanical properties, and controlled microstructures.

In a study of Laurencin and co-workers,[26,27] primary osteoblast culture on PLGA (50:50)/HAP composite scaffolds showed some promising features in cell attachment and differentiation function. However, the matrix porosity was quite low (only 33% based on the mixture of

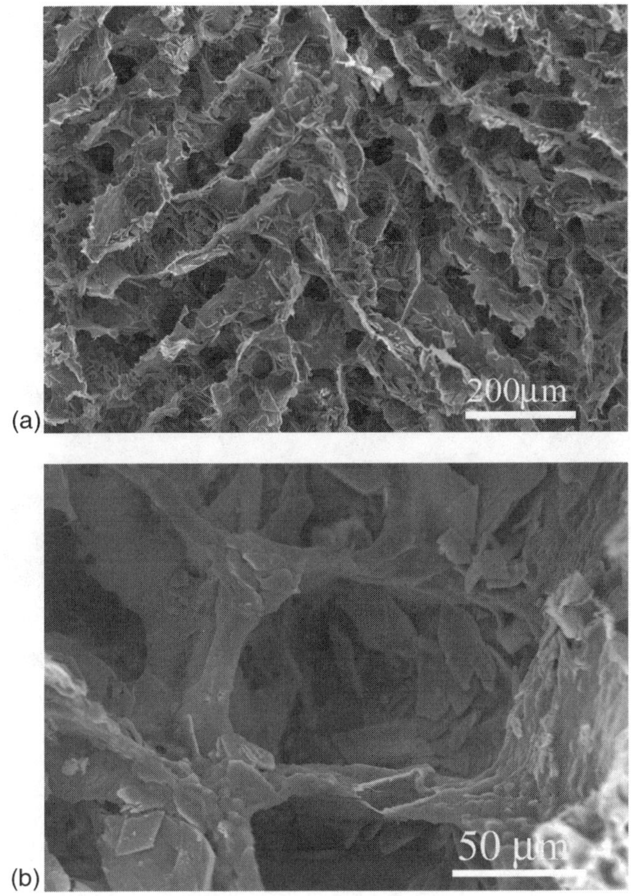

FIGURE 16.1 SEM micrograph of PLLA/mHAP (50:50) scaffold fabricated with TIPS using dioxane as solvent. Original magnification: (a) × 100; (b) × 500. From Zhang and Ma,[23] p. 450. © 1999 by John Wiley and Sons. Reprinted by permission of John Wiley and Sons.

PLGA:HAP:NaCl = 1:1:1 in salt-leaching technique), which might not be ideal for long-term cell survival, proliferation, and tissue formation. In a recent study of Ma et al.,[24] phase separated PLLA/HAP composite scaffolds were seeded with MC3T3-E1 osteoblasts and the cell activities in the scaffold were examined with histology, DNA assay, and Northern blot analysis. Compared to pure polymer scaffolds in which tissue matrix can only form at the top surface layer of the foam, the composite scaffolds supported uniform cell seeding, cell ingrowth, and tissue formation throughout,[24] including the very center of the scaffold (Figure 16.3). The polymer/HAP scaffolds have been shown to have higher osteoblast survival rate, more uniform cell distribution and growth, improved new tissue formation, and enhanced bone specific gene expression *in vitro*. In further studies,[29,30] the adsorption of serum proteins onto PLLA/HAP scaffolds was performed to examine the mechanism by which cell activities are regulated. It was found that PLLA/HAP composite scaffold adsorbed significantly greater amounts of serum proteins than the PLLA scaffolds and exhibited a different adsorption profile from plain PLLA scaffolds. The enhancement of protein adsorption including fibronectin and vitronectin from serum was thought to play an important role in suppressing apoptotic cell death in highly porous biodegradable PLLA/HAP scaffolds.[30] The composite material improved biocompatibility and hard tissue integration in a way that ceramic particles, which are embedded into the polymer matrix, allow for increased initial adsorption of serum proteins compared to the more hydrophobic plain polymer surface. The ability of selective

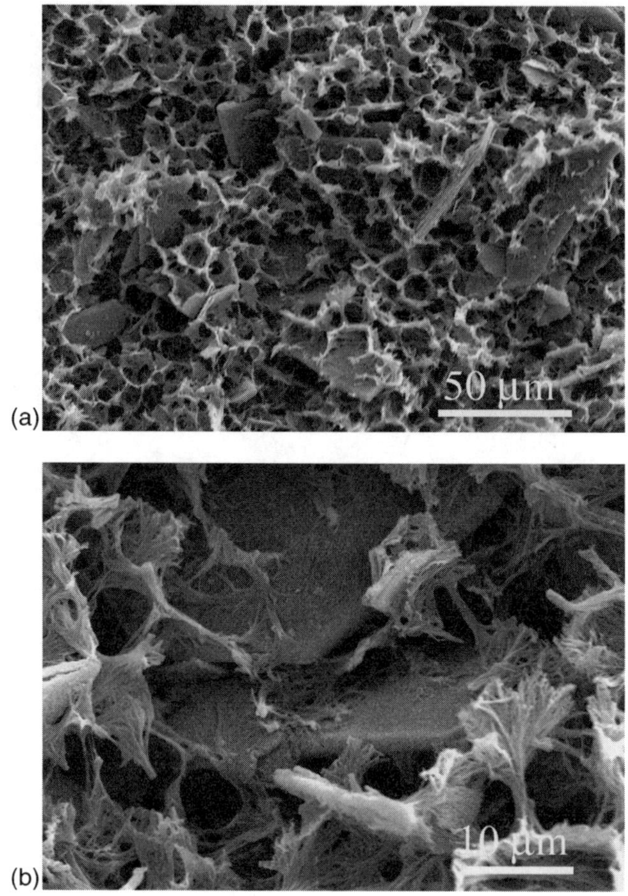

FIGURE 16.2 SEM micrograph of PLLA/mHAP (50:50) scaffold fabricated with TIPS using dioxane/water (90:10) mixture solvent. Original magnification: (a) ×500; (b) ×2000. From Zhang and Ma,[23] p. 453. © 1999 by John Wiley and Sons. Reprinted by permission of John Wiley and Sons.

adsorption of some proteins onto composite scaffolds is of interest and importance for the scaffolding system because it may also impart osteoinductivity to the scaffold when the adsorption of certain osteoinductive factors (e.g., BMPs) reaches their thresholds.[31]

Although aliphatic polyesters such as PLLA, PGA, and PLGA are biocompatible and have been approved by the U.S. Food and Drug Administration (FDA) for certain human clinical applications, there is a concern with the release of acidic degradation by-products. The reduced local pH may accelerate the polymer degradation rate and induce an inflammation.[32,33] The addition of HAP into the polymer scaffold improved the structural stability and mediated the degradation behavior.[34] The basic resorption products of HAP would buffer the acidic resorption by-products of the aliphatic polyester and may thereby help to avoid the formation of an unfavorable environment for the cells due to a decreased pH. The buffering effect of HAP on pH was significant at the late stage of the degradation when large amounts of degradation products were generated and retained within pores.[34]

III. POLYMER/HYDROXYAPATITE NANO COMPOSITE SCAFFOLD

The inorganic component (partially carbonated hydroxyapatite) in natural bone has a crystal size at nanometer scale, which is considered to be important for the mechanical properties of the bone.[35]

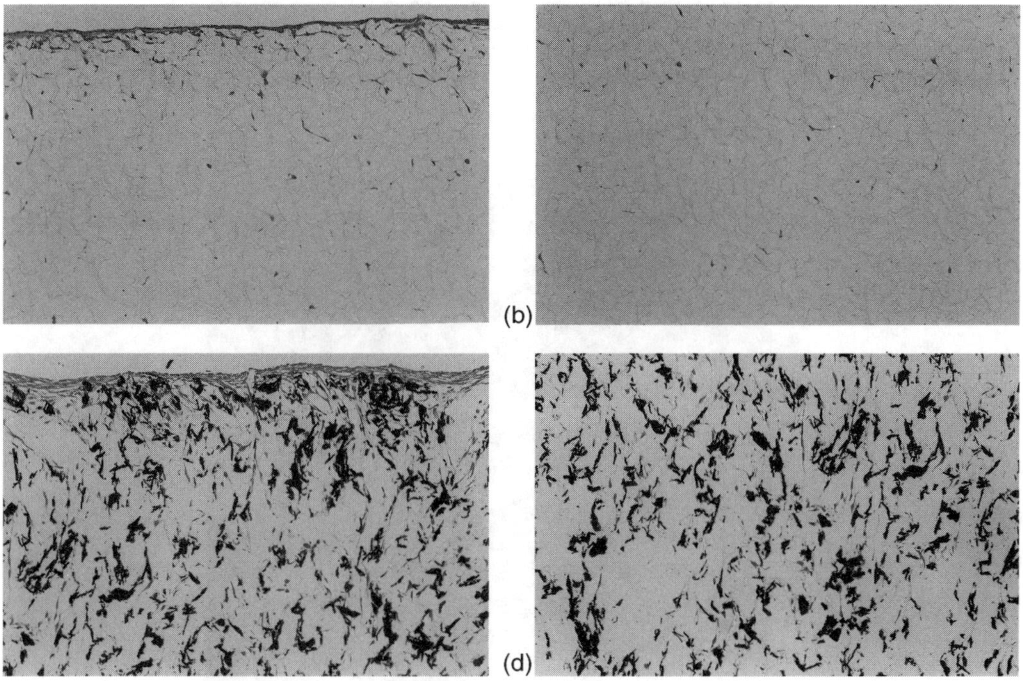

FIGURE 16.3 Comparison of osteoblastic cell distribution in highly porous PLLA and PLLA/HAP composite scaffolds 1 week after cell seeding (von Kossa's silver nitrate staining; original magnification × 100): (a) the surface area of an osteoblast–PLLA construct; (b) the center of an osteoblast–PLLA construct; (c) the surface area of an osteoblast–PLLA/HAP construct; (d) the center of an osteoblast–PLLA construct. From Zhang and Ma,[24] p. 288. © 2001 by John Wiley and Sons. Reprinted by permission of John Wiley and Sons.

In addition, Webster and colleagues[36,37] have shown significant increase in protein adsorption and osteoblasts adhesion on the nano-sized ceramic surface compared to traditional micron-sized ceramic surface, possibly due to increased surface area and hydrophilicity. Increase in osteoblast adhesion was observed with the decrease in the size of the particles, which was thought to be modulated by the absorbed serum proteins. In application of three-dimensional porous scaffold design, novel nano-hydroxyapatite/polymer composite scaffolds have been developed by Wei and Ma.[38] Regular anisotropic three-dimensional pore structure (Figure 16.4) similar to plain polymers or a fibrous morphology (Figure 16.5) could be obtained, using dioxane and dioxane/water solvent systems, respectively. As compared to PLLA/mHAP (micro-sized hydroxyapatite) scaffolds prepared with the same solvent systems (Figure 16.1 and Figure 16.2), PLLA/nHAP (nano-sized hydroxyapatite) scaffold exhibited more regular and controllable morphologies. When phase separation was performed with a uniaxial temperature gradient, oriented microtubular PLLA/nHAP composite scaffolds (Figure 16.6) can also be obtained, whereas such pore morphologies can not be obtained for micron-sized HAP. Nano HAP particles dispersed uniformly in and on the pore walls of the scaffolds, which was thought to be responsible for the further improvement of mechanical properties and protein adsorption capability on the composite scaffolds. Recent research in collagen/HAP nanocomposite scaffolds has demonstrated that better osteoconductivity would be achieved if synthetic HAP in a composite scaffold could more resemble bone minerals in composition, size, and morphology.[39,40] These properties suggest that the HAP/polymer nanocomposite scaffolds may serve as a better three-dimensional substrate for cell attachment and migration in the process of bone tissue engineering.

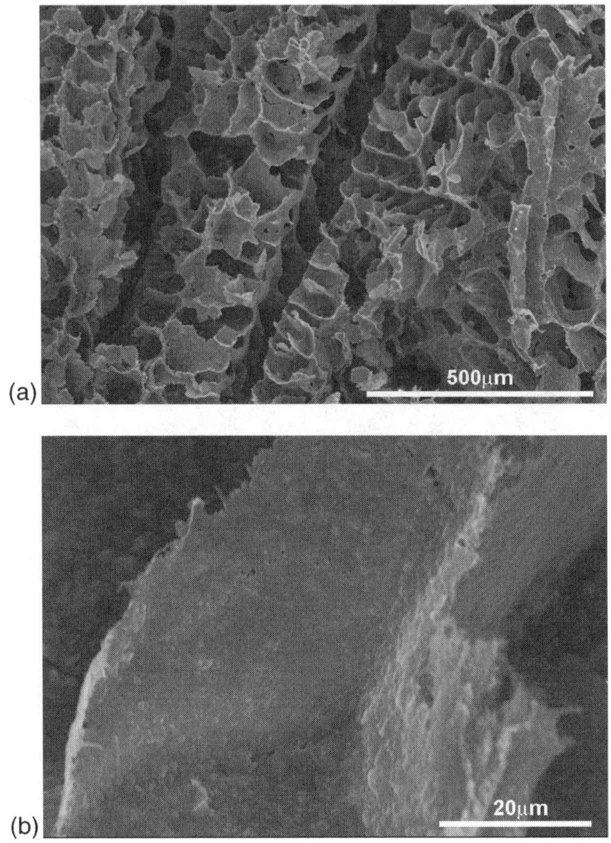

FIGURE 16.4 SEM micrograph of PLLA/nHAP (50:50) scaffold fabricated with dioxane. Original magnification: (a) ×100; (b) ×1000. From Wei and Ma.[38] © 2003 by Elsevier Science Ltd. Reprinted by permission of Elsevier Science Ltd.

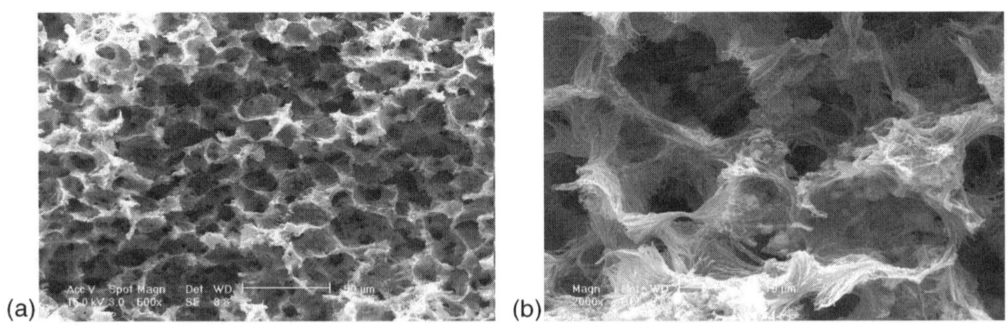

FIGURE 16.5 SEM micrograph of PLLA/nHAP (50:50) scaffold fabricated with dioxane/water (90:10). Original magnification: (a) ×500; (b) ×2000. From Wei and Ma.[38] © 2003 by Elsevier Science Ltd. Reprinted by permission of Elsevier Science Ltd.

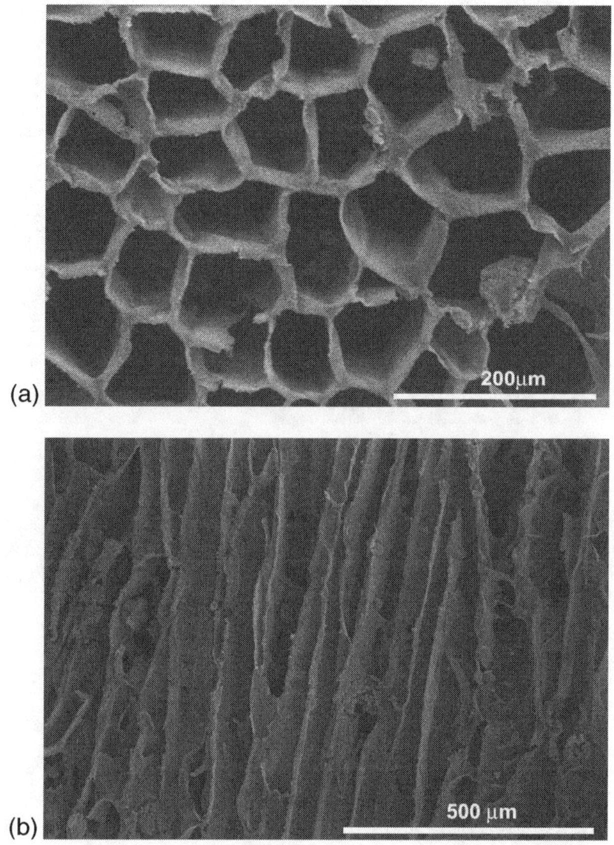

FIGURE 16.6 SEM micrograph of tubular PLLA/nHAP (70:30) scaffold fabricated with TIPS using benzene as solvent. (a) cross-section, × 200; and (b) longitudinal section, × 100. Note that longitudinal section means the section is made parallel to the heat transfer direction. From Wei and Ma.[38] © 2003 by Elsevier Science Ltd. Reprinted by permission of Elsevier Science Ltd.

IV. POLYMER/APATITE COMPOSITE SCAFFOLD BY BIOMIMETIC PROCESS

Bone mineral is predominated by nonstoichiometric hydroxyapatite (apatite) which contains other ions such as carbonate, and has lower crystallinity than synthetic stoichiometric HAP. It is believed that a bone-like apatite layer plays a key role for the integration of materials interface to host bone. The presence of sparingly soluble HA coatings lead to a tissue response in which bone grows along the coating and forms a mechanically strong interface.[41] Such a bioactive apatite layer could be created on the pore wall surface of a porous scaffold via a biomimetic process in which prefabricated scaffolds are incubated in a metaphysic simulated body fluid (SBF).[25] SBF solution utilized contains inorganic ion concentrations 1.5 times those of human blood plasma and has a pH of 7.4 at 37°C. After incubation in SBF for a period of time, apatite particles are successfully grown on the inner pore surfaces of the porous polymer scaffold (Figure 16.7). The apatite particles formed are similar to the bonelike apatite based on SEM, EDS, IR, and XRD analyses. The density and particle size of the apatite can be controlled by incubation time. A large number of carboxyl groups in the poly(α-hydroxy acids) backbone act as inducers for the formation of bone-like apatite.[42] However, in three-dimensional porous scaffold, the carboxylic acid formed via hydrolysis has dual effects in apatite nucleation and growth.[43] On one hand, the carboxylic acid groups may promote the accumulation of calcium ions and facilitate the nucleation of apatite crystals on pore wall

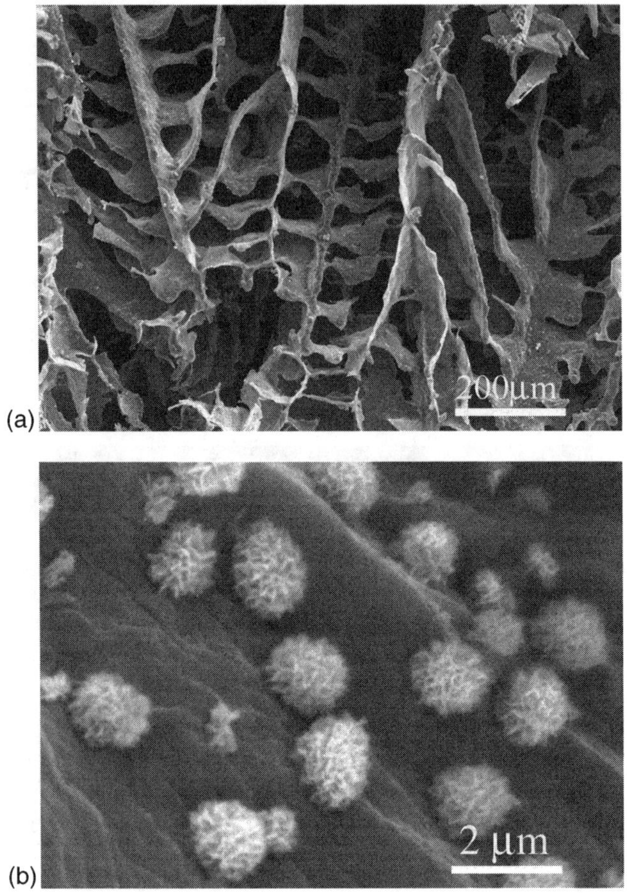

FIGURE 16.7 SEM micrograph of PLLA-apatite composite scaffold prepared by SBF incubation. (a) × 100 and (b) × 10,000. From Zhang and Ma,[25] p. 288. © 1999 by John Wiley and Sons. Reprinted by permission of John Wiley and Sons.

surface, which is similar to two-dimensional film. On the other hand, the carboxylic acid groups may effectively lower the local surface pH value in a three-dimensional porous scaffold which is unfavorable for the formation of apatite crystals. Our research has shown that the incorporation of nano-sized HAP can promote and improve the deposition of bone-like apatite crystals on the porous composite scaffolds. Both the increase of amount and the improvement of the uniformity of apatite particle deposition have been demonstrated on nHAP/polymer composite scaffolds, more so than on pure polymer scaffolds. Similar results were found for bone-like apatite growth on other polymeric sponges such as gelatin[44] and chitin.[45] Since these naturally derived polymers lack functional groups to initiate apatite deposition, only a small amount of apatite was observed on pure polymer sponges. Incorporation of HAP can significantly increase the deposition amount and rate. Biomimetic deposition of bone-like apatite may be of direct interest for the development of a bone tissue engineering composite scaffold or for assessing the calcification function of currently existing scaffolds.

V. CONCLUSIONS

Highly porous polymer/ceramic composite scaffolding appears to be a promising substrate for bone tissue engineering due to its excellent mechanical properties and osteoconductivity. Among current fabrication techniques, TIPS is an excellent method for the construction of such composite scaffolds due to its capability in controlling the scaffold characteristics such as porosity, pore size, pore morphologies as well as in improving mechanical properties. Incorporation of HAP into a highly porous biodegradable polymer scaffold may provide several advantages for bone tissue engineering: a better environment for cell seeding, survival, growth, and differentiated function because of the osteoconductive function imparted by HAP; improved mechanical properties which are necessary for load-bearing defects; and a buffered pH environment of local acidic molecules produced by degradation of polyesters. While mimicking the natural bone structure and composition presents new prospects with regard to biomimetic scaffold design, much work is needed to optimize the properties of the scaffold *in vitro,* and examine its effectiveness *in vivo.*

REFERENCES

1. Langer, R., and Vacanti, J. P., Tissue engineering, *Science*, 260, 920–926, 1993.
2. Goldberg, V. M., and Stevenson, S., Natural history of autografts and allogragts, *Clin. Orthop.*, 225, 7–16, 1987.
3. Goldberg, C. F., The current status of bone and joint transplants, *Clin. Orthop.*, 87, 165–166, 1972.
4. Goldberg, V. M., and Stevenson, S., Bone graft options: fact and fancy, *Othopedics*, 17, 809–810, 1994.
5. Ma, P. X., and Langer, R., Degradation, structure and properties of fibrous nonwoven poly(glycolic acid) scaffolds for tissue engineering, In *Polymers in Medicine and Pharmacy*, Mikos, A. G., Leong, K. W., Yaszemski, M. J., Tamada, J. A., and Radomsky, M. L., eds., MRS, Pittsburgh, PA, pp. 99–104, 1995.
6. Ma, P. X., and Langer, R., Fabrication of biodegradable polymer foams for cell transplantation and tissue engineering, In *Tissue Engineering Methods and Protocols*, Yarmush, M., and Morgan, J., eds., Humana Press, Totowa, NJ, pp. 47–56, 1999.
7. Ishaug, S. L., Crane, G. M., Miller, M. J., Yasko, A. W., Yaszemski, M. J., and Mikos, A. G., Bone formation by three-dimensional stromal osteoblast culture in biodegradable polymer scaffolds, *J. Biomed. Mater. Res.*, 36, 17–28, 1997.
8. Flautre, B., Descamps, M., Delecourt, C., Blary, M. C., and Hardouin, P., Porous HA ceramic for bone replacement: role of the pores and interconnections — experimental study in the rabbit, *J. Mater. Sci. Mater. Med.*, 12, 679–682, 2001.
9. Li, S. H., deGroot, K., and Layrolle, P., Bioceramic scaffold with controlled porous structure for bone tissue engineering, *Key Eng. Mater.*, 218–220, 25–30, 2002.
10. Sun, J. S., Liu, H. C., Chang, W. H. S., Li, J., Lin, F. H., and Tai, H. C., Influence of hydroxyapatite particle size on bone cell activities: an in vitro study, *J. Biomed. Mater. Res.*, 39, 390–397, 1998.
11. LeGeros, R. Z., Properties of osteoconductive biomaterials: calcium phosphates, *Clin. Orthop. Relat. Res.*, 395, 81–98, 2002.
12. Cooke, F. W., Ceramics in orthopedic surgery, *Clin. Orthop.*, 276, 135–146, 1992.
13. Schwartz, I., Robinson, B. P., Szachowicz, E. H., and Brekke, J., Calvarial bone repair with porous D,L-polylactide, *Otolaryngol. Head Neck Surg.*, 112, 707–713, 1995.
14. Ishaug-Riley, S. L., Crane, G. M., Curlek, A., Miller, M. J., Yasko, A. W., Yaszemski, M. J., and Mikos, A. G., Ectopic bone formation by marrow stromal osteoblast transplantation using poly(DL-lactic-*co*-glycolic acid) foams implanted into the rat mesentery, *J. Biomed. Mater. Res.*, 36, 1–8, 1997.
15. Mikos, A. G., Thorsen, A. J., Czerwonka, L. A., Bao, Y., Langer, R., Winslow, D. N., and Vacanti, J. P., Preparation and characterization of poly(L-lactic acid) foams, *Polymer*, 35, 1068–1077, 1994.
16. Ma, P. X., and Zhang, R. Y., Microtubular architecture of biodegradable polymer scaffolds, *J. Biomed. Mater. Res.*, 56, 469–477, 2001.
17. Whang, K., Thomas, G. H., and Healy, K. E., A novel method to fabricate bioabsorbable scaffolds, *Polymer*, 36, 837–842, 1995.
18. Harris, L. D., Kim, B. S., and Mooney, D. J., Open pore biodegradable matrices formed with gas foaming, *J. Biomed. Mater. Res.*, 42, 396–402, 1998.

19. Borden, M., Attawia, M., and Laurencin, C. T., The sintered microsphere matrix for bone tissue engineering: in vitro osteoconductivity studies, *J. Biomed. Mater. Res.*, 61, 421–429, 2002.
20. Thomson, R. C., Yaszemski, M. J., and Mikos, A. G., Fabrication of biodegradable polymer scaffolds to engineer trabecular bone, *J. Biomater. Sci. Polym. Ed.*, 7, 23–38, 1995.
21. Flahiff, C. M., Blackwell, A. S., Hollis, J. M., and Feldman, D. S., Analysis of a biodegradable composite for bone healing, *J. Biomed. Mater. Res.*, 32, 419–424, 1996.
22. Verheyen, C., Klein, C., Deblieckhogervorst, J. M. A., Wolke, J. G. C., Vanblitterswijn, C. A., and Degroot, K., Evaluation of hydroxylapatite poly(L-lactide) composites — physicochemical properties, *J. Mater. Sci.-Mater. Med.*, 4, 58–65, 1993.
23. Zhang, R. Y., and Ma, P. X., Poly(alpha-hydroxyl acids) hydroxyapatite porous composites for bone-tissue engineering. I. Preparation and morphology, *J. Biomed. Mater. Res.*, 44, 446–455, 1999.
24. Ma, P. X., Zhang, R. Y., Xiao, G. Z., and Franceschi, R., Engineering new bone tissue in vitro on highly porous poly(α-hydroxyl acids)/hydroxyapatite composite scaffolds, *J. Biomed. Mater. Res.*, 54, 284–293, 2001.
25. Zhang, R. Y., and Ma, P. X., Porous poly(L-lactic acid)/apatite composites created by biomimetic process, *J. Biomed. Mater. Res.*, 45, 285–293, 1999.
26. Laurencin, C. T., Attawia, M. A., Elgendy, H. E., and Herbert, K. M., Tissue engineered bone-regeneration using degradable polymers: the formation of mineralized matrices, *Bone*, 19, S93–S99, 1996.
27. Attawia, M. A., Herbert, K. M., and Laurencin, C. T., Osteoblast-like cell adherance and migration through 3-dimensional porous polymer matrices, *Biochem. Biophys. Res. Commun.*, 213, 639–644, 1995.
28. Thomson, R. C., Yaszemski, M. J., Power, J. M., and Mikos, A. G., Hydroxyapatite fiber reinforced poly(a-hydroxy ester) foams for bone regeneration, *Biomaterials*, 19, 1935–1943, 1998.
29. Woo, K. M., Zhang, R. Y., Deng, H. Y., and Ma, P. X., Protein-mediated osteoblast survival and migration on biodegradable polymer/hydroxyapatite scaffolds, *Trans. Soc. Biomater.*, 25, 605, 2002.
30. Woo, K. M., Wei, G., and Ma, P. X., Enhancement of fibronectin- and vitronectin-adsorption to polymer/hydroxyapatite scaffolds suppresses the apoptosis of osteoblasts (abstr), *J. Bone Miner. Res.*, 17(Suppl. 1), M49, 2002.
31. Ripamonti, U., Osteoinduction in porous hydroxyapatite implanted in heterotopic sites of different animal models, *Biomaterials*, 17, 31–35, 1996.
32. Vert, M., Mauduit, J., and Li, S., Biodegradation of PLA/GA polymers: increasing complexity, *Biomaterials*, 15, 1209–1213, 1994.
33. Mainil-Varlet, P., Rahn, B., and Gogolewski, S., Long term in vivo degradation and bone reaction to various polylactides. 1. One-year results, *Biomaterials*, 18, 257–266, 1997.
34. Zhang, R. Y., and Ma, P. X., Degradation behavior of porous poly(α-hydroxy acids)/hydroxyapatite composite scaffolds, *Polym. Prepr.*, 41, 1618–1619, 2000.
35. Rho, J. Y., Kuhn-Spearing, L., and Zioupos, P., Mechanical properties and the hierarchical structure of bone, *Med. Eng. Phys.*, 20, 92–102, 1998.
36. Webster, T. J., Ergun, C., Dpremus, R. H., Siegel, R. W., and Bizios, R., Specific proteins mediate enhanced osteoblast adhesion on nanophase ceramics, *J. Biomed. Mater. Res.*, 51, 475–483, 2000.
37. Webster, T. J., Siegel, R. W., and Bizios, R., Osteoblast adhesion on nanophase ceramics, *Biomaterials*, 20, 1221–1227, 1999.
38. Wei, G., and Ma, P. X., Structure and properties of nano-hydroxyapatite/polymer composite scaffolds for bone tissue engineering, *Biomaterials*, 25, 4749–4757, 2004.
39. Du, C., Cui, F. Z., Feng, Q. L., Zhu, X. D., and deGroot, K., Tissue response to nano-hydroxyapatite/collagen composite implants in marrow cavity, *J. Biomed. Mater. Res.*, 42, 540–548, 1998.
40. Du, C., Cui, F. Z., Feng, Q. L., Zhu, X. D., and deGroot, K., Three-dimensional nano-HAP/collagen matrix loading with osteogenic cells in organ culture, *J. Biomed. Mater. Res.*, 44, 407–415, 1999.
41. Cook, S. D., Thomas, K. A., Dalton, J. E., Volkman, T. K., Whitecloud, T. S. 3rd., and Kay, J. F., Hydroxylapatite coating of porous implants improves bone ingrowth and interface attachment strength, *J. Biomed. Mater. Res.*, 26, 989–1001, 1992.
42. Tanahashi, M., and Matsuda, T., Surface functional group dependence on apatite formation on self-assembled monolayers in a simulated body fluid, *J. Biomed. Mater. Res.*, 34, 305–315, 1997.

43. Zhang, R.Y., and Ma, P.X., Biomimetic polymer/apatite composite scaffolds for mineralized tissue engineering, *Macromol. Biosci.*, 4, 100–111, 2004.
44. Bigi, A., Boanini, E., Panzavolta, S., Roveri, N., Rubini, K., Bonelike apatite growth on hydroxyapatite-gelatin sponges from simulated body fluid, *J. Biomed. Mater. Res.*, 59, 709–715, 2002.
45. Zhang, Y., and Zhang, M. Q., Synthesis and characterization of macroporous chitosan/calcium phosphate composite scaffolds for tissue engineering, *J. Biomed. Mater. Res.*, 55, 304–312, 2001.

17 Polymer/Calcium Phosphate Scaffolds for Bone Tissue Engineering

Cato T. Laurencin and Yusuf Khan

CONTENTS

I. INTRODUCTION

It is projected that in the next 20 years, the population of Americans over the age of 60 will almost double.[1] As this population increases, the necessity of care mechanisms for the elderly increases as well. Nowhere else in healthcare is this more profoundly felt than in the orthopedic realm. In 1999 there were approximately 500,000 bone graft procedures performed,[2] and in 2002 the market share of these procedures was over \$7 billion in the U.S. alone. For these reasons, orthopedic treatments are, and will continue to be, a critical health care issue.

II. BONE GRAFTS AND BONE GRAFT SUBSTITUTES

Currently the preferred option for bone repair is the use of autografts. With a success rate of 80 to 90%, autografts are considered the gold standard of bone grafts.[3] The high rate of success can be attributed to the nature of autografts; tissue harvested for an injury from a remote site of the patient, often the iliac crest. By harvesting tissue from the patient, there is a guaranteed biocompatibility, and no risk of disease transmission. Further, the harvested tissue is biologically intact, possessing the cells, proteins, and factors necessary for complete healing. However, the process of harvesting the tissue can lead to complications independent of the initial injury since the tissue harvest can be as invasive as the initial injury, resulting in excessive pain, infection, and possibly mechanical weakening of the donor site.[4] These complications are collectively known as donor-site morbidity and are seen in approximately 20% of all procedures. There are also constraints on the amount of tissue that can be harvested from the site, presenting limitations in supply of harvested tissue.

In the event of these complications or as a supplement to autograft, surgeons turn to allograft. Tissue that is taken from cadavers and processed to minimize potential disease transmission and bio-incompatibility helps reduce the limitations on quantity associated with autografts and eliminates concerns related to donor-site morbidity. However, despite the processing undertaken to minimize the potential immunogenic response between donor and recipient, allografts still possess the risk of disease transmission, and as recently as November 2001, have been implicated in the transmission of disease in a patient undergoing reconstructive knee surgery.[5] Although the bone repair and restoration strategies presently available are good, they have their limitations and further options are necessary.

Currently available alternatives to bone grafts incorporate both synthetic and biological materials. Most bone graft substitutes are formed from a composite of one or more materials from an extensive list that includes allograft bone chips, demineralized bone matrix, natural and recombinant growth factors, degradable and nondegradable polymers, bioactive glass, calcium sulfates, and calcium phosphates. Generally each of these materials have both benefits and drawbacks as bone graft substitutes, but by combining them with other materials, it is possible to impart the benefits of each material to the implant site, while minimizing the drawbacks of each material. For instance, allograft bone chips provide a porous interconnected structure that is appropriate for new cell migration, but alone cannot maintain a shape. Therefore, they are mixed with polymers to allow for molding and sculpting into a defect site. Demineralized bone matrix has excellent bone restorative properties, but once again has no maintainable shape or structure and is, therefore, mixed with polymeric materials to impart this property that, on their own, would have limited bone healing capabilities. An approach to bone repair that moves one step beyond that of the presently available bone graft substitutes involves tissue engineering.

III. TISSUE ENGINEERING

Tissue engineering has been defined by Laurencin[6] as "the application of biological, chemical, and engineering principles toward the repair, restoration, or regeneration of living tissues using biomaterials, cells, and factors alone or in combination." So by developing a scaffold that is biocompatible and biodegradable and allows cells to attach, proliferate, and migrate throughout its structure, the damaged tissue can be replaced slowly over time and can completely replace the synthetic scaffold initially implanted. This approach holds great promise as a bone repair model provided that the nature and structure of bone is considered in the scaffold design. For bone tissue engineering, one may look to autografts to form a short list of requirements for a successful scaffold. An autograft finds its success through several attributes:

Biocompatibility: the lack of immunogenic response.

Osteoconductivity: the quality of a porous interconnected structure that permits new cells to attach, proliferate, and migrate through the structure, and also allows for nutrient/waste exchange and new vessel penetration.

Osteoinductivity: possessing the necessary proteins and growth factors that induce mesenchymal stem cells and other osteoprogenitor cells toward the osteoblast lineage. This is essential when the defect is of critical size, or of sufficient size to prevent spontaneous healing.

Osteogenicity: the osteoblasts that are at the site of new bone formation are able to produce mineral to calcify the collagen matrix that forms the substrate for the new bone.

Osteointegrity: the newly formed mineralized tissue must be able to form an intimate bond with the implant material.

Mechanical match: having similar mechanical properties to the neighboring tissue of the implant site to prevent mechanical mismatch which can lead to stress shielding and bone resorption.

TABLE 17.1
Parameters for a Successful Tissue Engineered Scaffold as Indicated by Autograft

Autograft Quality	Polymers	Ceramics	Composite
Biocompatibility	Yes	Yes	Yes
Osteoconductivity	Yes	Yes	Yes
Osteoinductivity	No (but can add factors)	No (but can add factors)	Yes (with delivered factors)
Osteogenicity	No	Yes	Yes
Osteointegrity	No	Yes	Yes
Mechanical Match	Yes	No	Yes

Neither polymers nor ceramics can accomplish all parameters alone, but this can be achieved when formed into a composite.

These characteristics are the keys to a successful bone tissue engineering implant and no one material to date can satisfy each requirement. Therefore, for an ideal bone tissue engineering scaffold, it is a reasonable strategy to combine two or more materials into a composite.

As indicated from Table 17.1, neither polymers nor ceramics alone can possess all the necessary components of an ideal bone tissue engineered scaffold, but by combining them into one material an ideal composite material can be developed (see Figure 17.1).

However, within both polymers and ceramics there are several choices. Some of the more extensively studied polymers for bone tissue engineering composite scaffolds include polyesters like polylactide, polyglycolide, their copolymer poly(lactide-*co*glycolide), collagen, chitosan, starch-based polymers, ethylene alcohol polymers, gelatins, and polymethylmethacrylates (PMMA). The ceramics used in composites include calcium sulfate, bioactive glass, and calcium phosphate. This manuscript, although not an exhaustive list, will present several of the composites currently under study as bone tissue engineering scaffolds, but will concentrate on the use of calcium phosphates in several forms in conjunction with different polymers, and will examine how these calcium phosphates have been used to form composite scaffolds suitable for bone tissue engineering, and what makes the different calcium phosphates used unique.

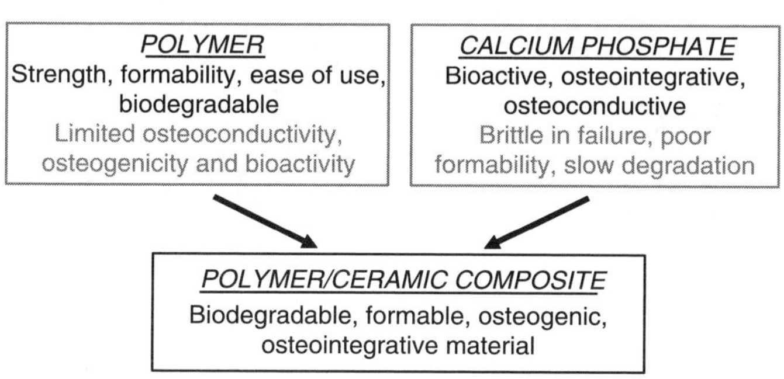

FIGURE 17.1 Neither the polymer nor the calcium phosphate alone possesses each of the necessary parameters for the successful scaffold, but a composite of a polymer and ceramic together can take advantage of the benefits of each material while minimizing the shortcomings.

IV. POLYMER/CALCIUM PHOSPHATE SCAFFOLDS

A. Why Calcium Phosphate

The use of calcium phosphate in a composite scaffold for bone tissue engineering is logical if one considers the structure of bone itself. Bone is formed from approximately 8% water, 22% proteins, and 70% mineral.[7] The mineral component of bone is a form of calcium phosphate known as calcium hydroxyapatite. Like other apatites it has a calcium and phosphate component in a ratio of 1.67 (Ca/P ratio) but is associated with a hydroxyl group. The stoichiometrically stable form of hydroxyapatite (HA) is that with a Ca/P ratio of 1.67 as can be seen by its molecular formula $Ca_{10}(PO_4)_6(OH)_2$. There are several other nonapatitic calcium phosphates that are distinguished from one another by their molecular formulae, Ca/P ratios (see Table 17.2), crystal structures as seen by x-ray diffraction patterns, and solubilities, with those Ca/P ratios that differ from 1.67 having higher dissolution rates than stoichiometric hydroxyapatite.

Several composites have been studied with varied calcium phosphates and an array of polymers including polylactides, poly(lactide-*co*-glycolides), PMMAs, chitosan, collagen, and others. Some have sought to design a structure suitable for bone attachment and ingrowth, others have added calcium phosphates in hopes of imparting certain mechanical properties, and others have examined the influence of the calcium phosphate on both *in vitro* cell growth and *in vivo* healing. Regardless of how the list of composite materials is organized, it is extensive.

B. Hydroxyapatite

Stoichiometric hydroxyapatite has been used in conjunction with different polymers for different purposes. Several groups have investigated the benefits of adding hydroxyapatite to a scaffold to increase mechanical properties. Deng et al. added nanocrystalline hydroxyapatite to a polylactide solution to form solvent-cast composite matrices, and found a steady increase in tensile modulus as hydroxyapatite loading increased from a low of 1.66 GPa for polymer without hydroxyapatite up to 2.47 GPa for a 10.5% hydroxyapatite content.[8] It was theorized that the increase in tensile modulus was due to: (1) an increase in rigidity over the polymer alone when the hydroxyapatite was added, and (2) the resulting strong adhesion between the two materials. Wang et al. combined polyamide, a bio-inert polymer, with both microcrystalline and nanocrystalline hydroxyapatite and compared the resulting bending strength and tensile strength. As the ceramic content of each composite increased, so did the bending strength. For both bending and tensile strength, the addition of nanocrystalline hydroxyapatite increased the properties over those with microcrystalline hydroxyapatite.[9] It was theorized that the smaller crystals of the nanocrystalline HA resulted in higher surface areas and thus greater surface energy, surface activity, and therefore bonding between the polymer and the HA. Finally, Abu Bakar et al. examined the effect of varying amounts of hydroxyapatite added to polyetheretherketone as an injection-molded composite by varying hydroxyapatite content between

TABLE 17.2
Different Forms of Calcium Phosphate, Their Molecular Formulae, and the Corresponding Calcium/Phosphorus Ratio

Calcium Phosphate Form	Molecular Formula	Calcium/Phosphorus Ratio
Tricalcium phosphate (α and β) (TCP)	$Ca_3(PO_4)_2$	1.5
Hydroxyapatite (HA)	$Ca_{10}(PO_4)_6(OH)_2$	1.67
Calcium-deficient hydroxyapatite (CDHA)	$Ca_9(HPO_4)(PO_4)_5(OH)$	1.5
Tetracalcium phosphate (TTCP)	$Ca_4(PO_4)_2O$	2.0
Octacalcium phosphate (OCP)	$Ca_8(HPO_4)_2(PO_4)_4{\cdot}5H_2O$	1.33

0 and 40% by volume. Results indicate that Young's modulus increased from approximately 3 to 15 GPa as hydroxyapatite content increased from 0 to 40%, but tensile strength decreased from 80 to 44 MPa along the same increase in HA content.[10] Examination of the load–displacement curves indicated that as the HA content decreased the brittle nature of the composite also decreased. This is in good agreement with the common complaint of ceramics failing in a brittle fashion in load-bearing situations, and also supports the theory behind forming composites between polymers and ceramics; the polymers become stronger with the addition of the ceramic, and the brittle nature of the ceramic is reduced with the addition of the polymer. Balac et al. attempted to understand the effect of hydroxyapatite particle shape and volume fraction in a polylactide/collagen/hydroxyapatite composite scaffold using finite element analysis, and found fewer stress concentrations throughout the matrix with an increased hydroxyapatite volume fraction, but a reduced dependence on this as the hydroxyapatite particles were modeled as spherical, suggesting yet another design consideration for the composite scaffold.[11]

The physical shape and structure of a composite scaffold is of utmost importance in bone tissue engineering. It is generally agreed that a porous structure with an interconnected network of pores is the preferred design for bone tissue engineering, allowing for cells to migrate throughout the structure and for cellular nutrients and waste to be freely exchanged. Several approaches using polymer/calcium phosphate composites have been undertaken toward this goal, either through direct pore formation, pore generation in a solid, or the agglomeration of particles to create porous void spaces, to name a few. Taboas et al.[12] has formed porous interconnected composite structures through two of these approaches. Utilizing a CT image of femoral bone, a template of trabecular bone was developed and used to form a porous interconnected structure via three-dimensional printing. From this, a negative template was made into which hydroxyapatite was added and then sintered to both form the structure and to burn out the organic material of the original negative template, from which the hydroxyapatite structure was formed with pores ranging from 500 to 600 μm. To form the composite, polylactide was melted into the porous structure to form an interlocking mesh between the two materials. This design approach allowed for precise control of pore size and interconnectivity. Marra et al.[13] formed three-dimensional composite structures with a pore system generated through particle leaching. Polycaprolactone and poly(lactide-*co*-glycolide) blend solutions were mixed with both hydroxyapatite particles and NaCl crystals and were allowed to harden through solvent evaporation. After the composite solidified the NaCl particles were dissolved with water, leaving a porous structure. Compressive modulus increased from 2.5 to 12.5 MPa with the addition of 10% hydroxyapatite particles. However, there have been some questions as to the interconnectivity of the pore system when using the particle leaching technique in which poor tissue ingrowth was seen and was attributed to poor nutrient transport.[14] Ma et al.[15] utilized a phase separation technique to impart porosity to a polylactide/hydroxyapatite composite scaffold by cooling the polymer/HA solution causing a solid–liquid phase separation between the polymer and solvent, and then vacuum dried the samples, removing the liquid phase and thus leaving a pore structure. The resulting structure had a cross-section that appeared honeycomb-like in appearance with a pore size range from approximately 10 to 500 μm in diameter. The addition of the hydroxyapatite increased the compressive modulus by about 40%. Laurencin et al. has used microspheres fused together to form a porous interconnected structure between the microspheres, and has combined hydroxyapatite particles with the microspheres to form a composite matrix. This matrix showed a mineralized surface in cell culture that only appeared on scaffolds containing cells, indicating that mineralization did not take place spontaneously but was deposited by cells.[16] In other studies, Borden et al. formed a composite of poly(lactide-*co*-glycolide) and hydroxyapatite using a modified microsphere-based scaffold in which gelatinous microspheres were mixed with hydroxyapatite. Mechanical properties ranged from 630 to 1650 MPa compressive modulus depending on the hydroxyapatite content, but the pore structure was deemed insufficiently interconnected.[17] More recently, Laurencin et al. examined the potential for a composite scaffold to deliver BMP-2. A poly(lactide-*co*glycolide)/hydroxyapatite composite scaffold delivered BMP-2 to a heterotopic bone site in SCID mice, suggesting that the composite scaffold was also functional as a delivery vehicle *in vivo*.[18] Still others have used polymers to assist in the generation of pores but have removed this material from the final scaffold. Li et al. has developed a porous structure using dual-phase mixing of PMMA and hydroxyapatite. The two components were added together to form a slurry, and allowed to harden. After hardening, the composite was fired at 1200°C for approximately 2.5 days, during which the PMMA was pyrolyzed, leaving a porous hydroxyapatite structure in which the HA was sintered to form a solid scaffold. This resulted in a 50% porous structure with some but not complete interconnectivity, so a foaming agent was added to the slurry prior to hardening to increase overall porosity to approximately 70%, and also to increase interconnectivity of the porous structure. Although technically not a composite in its final form, a composite scaffold was used in this method of scaffold formation.

C. Other Calcium Phosphates

Hydroxyapatite, in its crystalline form and with a Ca/P ratio of 1.67, is very slow to degrade. In fact, very often it is either nondegradable or incapable of being resorbed by osteoclasts during the remodeling phase.[19] This can lead to hydroxyapatite particles that are associated with composite implant materials remaining long after the implant has either degraded or been resorbed, or with nondegradable implants long after the surrounding bone has grown into the implant material. Although these particles may lay dormant for several years, the nondegradable hydroxyapatite particles do not lend themselves to tissue engineering applications as they may prevent bone remodeling over time.[20] As a result, alternative calcium phosphates to stoichiometric hydroxyapatite have been incorporated into composite scaffolds as well.

Tricalcium phosphate (TCP) differs from hydroxyapatite in both molecular formula and Ca/P ratio (see Table 17.2), as does calcium-deficient hydroxyapatite (CDHA) and carbonated hydroxyapatite (CHA) in which a carbonate ion (CO_3) is incorporated into the ceramic. All three are less stable than stoichiometric hydroxyapatite, which makes them attractive for certain applications. With the less stable forms of hydroxyapatite, osteoclasts can more readily resorb the ceramic, and these less stable forms dissolve more rapidly than stoichiometric HA, allowing for bone remodeling and complete scaffold replacement over time. Given these parameters and the other benefits of adding a ceramic to the polymeric scaffold, it becomes clear that the addition of these nonstoichiometric calcium phosphates may have benefits for bone tissue engineering. Toward this goal, Lin et al.[21] combined polylactide (PLA) with TCP to make composite rods for internal fixation by dissolving the polymer and creating a suspension of TCP particles within the polymer solution. The suspension was added to rod-shaped molds and heated to harden the polymer. It was determined that the elastic modulus of the rods increased with increasing TCP content from 21 GPa with no TCP to 47 GPa with 50% TCP, indicating that TCP shares the same reinforcing capabilities as hydroxyapatite. It was also found that the addition of TCP slowed the rate of degradation of the rods. It was theorized that this may be due to the TCP dissolving in solution and causing hydroxyapatite to precipitate on its surface, which was less susceptible to dissolution and thus protected the composite from a more rapid dissolution/degradation. Further, with the TCP protected from dissolution, it formed a protective seal between the aqueous environment and the PLA, thus reducing the rate of hydrolytic degradation that PLA undergoes in an aqueous environment. A similar study by Imai et al.[22] examined the effect of blending TCP with a polylactide/poly(ethylene; hexamethylene/sebacate) blend on degradation and found results similar to Lin et al. However, they attributed the delayed degradation of the polymer in the presence of TCP to the ability of the TCP to neutralize the acidic oligomers that result from the degradation of the polylactide. These oligomers, if left unchecked, will result in the autocatalytic degradation of the polymer. This theory of the calcium phosphate regulating the acidic degradation products is further supported by Schiller et al.[23] Although Schiller suggests that calcium carbonate is the most effective buffer around physiological pH and hydroxyapatite and β-TCP are better buffers at nonphysiological pH, closer to pH 4. It is reasonable to suppose that the presence of the resorbable TCP with a calcium carbonate content may not only improve mechanical properties, but alter degradation rates and possibly reduce acidic degradation products.

CDHA is appealing not only for the above-mentioned reasons, but also because careful analysis has determined that the mineral content of bone is actually carbonated CDHA.[24,25] With this in mind, Durucan et al.[24] have synthesized a CDHA through hydrolysis of TCP at 37°C and combined it with polylactide and poly(lactide-*co*-glycolide) to form composites. X-ray diffraction patterns of the resulting CDHA show an amorphous or nanocrystalline ceramic that strongly resembles HA with traces of α-TCP.[26] Composites formed with the calcium phosphate and poly(lactide-*co*-glycolide) showed a range of tensile and flexural strengths as the ceramic content was changed. An 80% ceramic content of the composite yielded a tensile strength of 13.3 MPa and a flexural strength of 24.8 MPa, while decreasing the ceramic content to 60% showed a slight decrease in tensile strength to 12 MPa but an increase in flexural strength to 36.1 MPa. Laurencin et al. has developed a composite microsphere-based scaffold in which composite microspheres are formed with a nanocrystalline hydroxyapatite within each microsphere.[27,28] This was accomplished by synthesizing the calcium phosphate within each microsphere as the microsphere was formed in a water/oil/water emulsion system. Through this approach, it was possible to form a composite scaffold with well-dispersed hydroxyapatite in a form that would be easily resorbed and lend itself to calcium phosphate reprecipitation *in vitro* and *in vivo*. Linhart et al.[29] formed a carbonated apatite in which hydroxyl groups were replaced with carbonate groups, reducing the crystallinity and the thermodynamic stability, and thus increasing the solubility of the ceramic. To test the biocompatibility of the composite, osteoblasts were seeded on the composite and found to attach, proliferate, and express alkaline phosphatase, collagen, and osteocalcin, indicating that osteoblasts maintained their phenotype and were functionally active. Further, after 4 weeks of culture, nanocrystalline hydroxyapatite was seen to be forming from the reprecipitation of the dissolution products of the carbonated apatite, namely calcium phosphate, and carbonate. This reprecipitation of dissolution products onto the scaffold surface has been seen elsewhere, and has stimulated studies where the precipitation of calcium phosphate is not a by-product, but the goal.

D. Formation of Calcium Phosphate Precipitates

Upon degradation, the calcium and phosphate that is liberated from the calcium phosphate will be available for new bone formation. This may happen because the dissolution of calcium phosphate results in an increased concentration of calcium and phosphate ions locally, resulting in apatite crystals precipitating out of solution, incorporating other locally available ions such as carbonate, and forming new bone mineral.[30] With this precipitation, calcium phosphate crystals form while present osteoblasts mineralize the collagen fibrils of the extracellular matrix, facilitating the attachment of the composite scaffold to neighboring bone at the implant site. The presence of calcium phosphate in an implant *in vivo* may also initiate further calcium phosphate development by preexisting osteoblasts, either through the liberation of calcium or through the stimulation of calcium phosphate crystal formation. Once new mineral crystals are formed, they may form seeds which can stimulate additional crystals.[31] Also, calcium phosphates have been shown to be osteoconductive when implanted next to preexisting bone, and in some cases osteoinductive as evidenced by the ectopic bone formation within a porous hydroxyapatite block that was implanted into abdominal muscles of baboons.[32] For all these reasons, it has become of interest to form an apatitic layer onto a material rather than rely on the preimplantation content of the composite. This notion of initiating apatite deposition may improve mechanical properties of a polymeric implant by providing a ceramic coating on a polymeric material. Zhang et al.[33] has shown that a calcium phosphate coating can be precipitated onto a poly(L-lactic acid) surface by incubating the polymer in an ion-rich simulated body fluid solution. A three-dimensional scaffold was incubated at 37°C for up to 30 days, at which time a continuous coating was seen throughout the scaffold that was determined to be bone-like apatite based on x-ray diffraction analysis, infrared spectroscopy, and Ca/P ratio. Further, the compressive modulus of the scaffold increased approximately 25% after 30 days of incubation as the apatite coating increased, from approximately 6.5 to 8 MPa. This was done with no modification to the polymer itself. A more recent analysis of particular polymer types has shown the poly(L-lactic acid) to support a more abundant apatite deposition than the copolymer poly(lactide-*co*-glycolide).[34] However, other groups have sought to modify the polymeric surface to encourage rapid deposition of a ceramic coating. Song et al.[35] modified a poly(2-hydroxyethyl methacrylate) (pHEMA) polymeric scaffold to form a carboxylate-rich outer layer that would encourage the attachment and deposition of a ceramic layer. It was felt that by exposing carboxyl groups for ceramic deposition, a better bond would be formed between the polymer and ceramic. Hydroxyapatite was dissolved in an acidic solution into which pHEMA hydrogels were added. The solution was slowly heated to 90–95°C and incubated overnight. The liberated ions from the dissolved hydroxyapatite precipitated on the surface of the treated pHEMA and x-ray diffraction confirmed a nanocrystalline layer of hydroxyapatite had formed on the surface of the polymer.

By examining how ceramics precipitate and deposit on a scaffold surface, one can design scaffolds accordingly and enhance the healing and ultimate bone repair of the defect site. This dissolution/reprecipitation process has been described in detail by Ducheyne et al.[36] Initially, ions are liberated from the ceramic to the surrounding fluid via dissolution. These dissolution rates are determined by: (1) the stoichiometric stability of the ceramic, with those differing from the thermodynamically stable 1.67 having higher dissolution rates, and (2) the crystallinity of the ceramic, with those of smaller crystal size and thus more surface area having higher dissolution rates. Then the ions undergo reprecipitation from the surrounding solution onto the surface of the material. As described above, the rate or degree of reprecipitation can be augmented by modifying the surface of the polymer. Studies examining the effect of dissolution rates on overall healing have shown that increased rates of dissolution result in enhanced healing and mineral formation.[37–39] With this, several researchers have embraced the notion of dissolution and reprecipitation but have bypassed the dissolution step. To mimic the ion rich milieu seen after ceramic dissolution, some or all ions commonly seen in the body (Na^+, Ca^{2+}, K^+, Mg^{2+}, Cl^-, HPO_4^{2-}, HCO_3^-, SO_4^{2-}) are mixed to form a simulated body solution. The ions present in this solution precipitate out and form a ceramic coating on the polymeric material much like it occurs from ion dissolution described above. Varma et al.[40] formed chitosan films with a phosphorylated surface and incubated the films in a calcium salt, inducing a calcium phosphate coating over the chitosan. This composite was incubated in a simulated body fluid for 20 days and formed a porous, poorly crystalline hydroxyapatite coating on top of the calcium phosphate layer. Lawson et al.[41] performed a similar experiment in which collagen sheets were incubated in reservoirs of calcium and phosphate ions. It was determined that the calcium phosphate phase that precipitated was dependent on the pH of the solution, with higher pH's producing a calcium phosphate similar to poorly crystalline hydroxyapatite, and as the concentration of ions in the solution increased, so did both the amount of ceramic that precipitated on the surface, and correspondingly, the compressive modulus of the composite. These poorly crystalline surfaces are generally characterized by a smaller grain size, usually submicron in size, suggesting a nanocrystalline phase. This smaller grain size has been shown to enhance both osteoblast adhesion and osteoclast functionality.[42,43] Webster et al.[42] examined the attachment of osteoblasts, fibroblasts, and endothelial cells to nanophase ceramics and found increased adhesion with osteoblasts after 4 h of attachment as compared to standard preparations of the same ceramics. Further study indicated enhanced osteoclast activity on similar nanophase surfaces as measured by the synthesis of tartrate resistant acid phosphatase and resorption pits after 13 days in culture.[43] Normal osteoclast activity on precipitated ceramic layers has also been reported elsewhere.[44]

V. POLYMER/CERAMIC COMPOSITES IN THE MARKETPLACE

The use of calcium phosphates in commercially available bone graft substitutes is widespread, with products consisting of crystalline hydroxyapatite, resorbable β-TCP, and biphasic calcium phosphates that combine the two to tune the degradability. These calcium phosphates have been formed into powders, particulates, and solid blocks that possess minimal pore structures and well structured porous networks that look very similar to trabecular bone. However, most suffer from the same shortcomings that have lead researchers to investigate the efficacy of combining these ceramics with polymers, although only a few products are currently used in the clinical setting that combine either a natural or synthetic polymer with a calcium phosphate. One of those products is Healos® from Orquest Incorporated in which natural collagen fibers are coated with hydroxyapatite and formed into highly porous thin sheets that are designed to mimic the natural healing process of bone by being completely resorbed by the body. Healos was designed to perform as an independent spinal fusion device but has also been investigated for applications in combination with Orquest's interbody fusion cage implants.[45] Zimmer Incorporated has Collagraft, which combines both TCP and hydroxyapatite with collagen. Of the synthetic polymers, Orthovita Incorporated has one composite available. Rhakoss®, a resin composite, comes as a solid product in various forms for spinal applications.

VI. CONCLUSIONS

The addition of calcium phosphate to a polymeric scaffold has several advantages. With the inorganic component of bone largely being calcium phosphate, it is logical to utilize it in a scaffold designed for bone repair. It can increase the strength of a polymeric scaffold while the presence of the polymer reduces the brittle nature of a purely ceramic scaffold. It can also encourage the deposition of newly forming mineral. If a soluble form of calcium phosphate is chosen it can supply necessary ions to form newly precipitated mineral layers while acting as a local buffer for the acidic degradation products seen in some commonly used degradable polymers. The polymeric portion of the composite can also be modified to enhance the deposition of new mineral as well. Much work has been done and much remains to be completed in the area of composite scaffolds for bone tissue engineering, and with the increase in both an aging population and overall bone repair incidents, this work is eagerly anticipated.

REFERENCES

1. Aging in the Americas into the XXII Century. Bureau of the Census, United States Census 2000.
2. Bostrom, M. P., Saleh, K. J., and Einhorn, T. A., Osteoinductive growth factors in preclinical fracture and long bone defects models, *Orthop. Clin. North Am.*, 30(4), 647–658, 1999.
3. Cook, S. D., Baffes, G. C., Wolfe, M. W., Sampath, T. K., Rueger, D. C., and Whitecloud, T. S., The effect of recombinant human osteogenic protein-1 on healing of large segmental bone defects, *J. Bone Joint Surg. Am.*, 76(6), 827–838, 1994.
4. Fleming, J. E., Cornell, C. N., and Muschler, G. F., Bone cells and matrices in orthopedic tissue engineering, *Orthop. Clin. North Am.*, 31(3), 357–374, 2000.
5. CDC second document Center for Disease Control: Update: Allograft-Associated Bacterial Infections — United States, *Morb. Mortal. Wkly Rep.*, 51(10), 207–210, 2002.
6. Laurencin, C. T., Ambrosio, A. A., Borden, M. D., and Cooper, J. A., Tissue engineering: orthopaedic applications, In *Annual Review of Biomedical Engineering*, Yarmush, M. L., ed., Annual Reviews Inc., Palo Alto, p. 19, 1999.
7. Einhorn, T., Biomechanics of bone, In *Principles of Bone Biology*, Bilezikian, J. P., Raisz, L. G. and Rodan, G. A., eds., Academic Press, San Diego, CA, p. 35, 1996.

8. Deng, X., Hao, J., and Wang, C., Preparation and mechanical properties of nanocomposites of poly(D,L-lactide) with Ca-deficient hydroxyapatite nanocrystals, *Biomaterials*, 22(21), 2867–2873, 2001.
9. Wang, X., Li, Y., Wei, J., and de Groot, K., Development of biomimetic nano-hydroxyapatite/poly(hexamethylene adipamide) composites, *Biomaterials*, 23, 4787–4791, 2002.
10. Abu Bakar, M. S., Cheng, M. H. W., Tang, S. M., Yu, S. C., Liao, K., Tan, C. T., Khor, K. A., and Cheang, P., Tensile properties, tension–tension fatigue and biological response of polyetherketone-hydroxyapatite composites for load-bearing orthopedic implants, *Biomaterials*, 24, 2245–2250, 2003.
11. Balac, I., Uskokovic, P. S., Aleksic, R., and Uskokovic, D., Predictive modeling of the mechanical properties of particulate hydroxyapatite reinforced polymer composites, *J. Biomed. Mater. Res.*, 63(6), 793–799, 2002.
12. Taboas, J. M., Maddox, R. D., Krebsbach, P. H., and Hollister, S. J., Indirect solid freeform fabrication of local and global porous, biomimetic and composite 3D polymer-ceramic scaffolds, *Biomaterials*, 24(1), 181–194, 2003.
13. Marra, K. G., Szem, J. W., Kumta, P. N., DiMilla, P. A., and Weiss, L. E., In vitro analysis of biodegradable polymer blend/hydroxyapatite composites for bone tissue engineering, *J. Biomed. Mater. Res.*, 47(3), 324–335, 1999.
14. Ishaug-Riley, S. L., Crane-Kruger, G. M., Yaszemski, M. J., and Mikos, A. G., Three-dimensional culture of rat calvarial osteoblasts in porous biodegradable polymers, *Biomaterials*, 19(15), 1405–1412, 1998.
15. Ma, P. X., Zhang, R., Xiao, G., and Franceschi, R., Engineering new bone tissue in vitro on highly porous poly(alpha-hydroxyl acids)/hydroxyapatite composite scaffolds, *J. Biomed. Mater. Res.*, 54(2), 284–293, 2001.
16. Laurencin, C. T., Attawia, M. A., Elgendy, H. E., and Herbert, K. M., Tissue engineered bone-regeneration using degradable polymers: the formation of mineralized matrices, *Bone*, 19(1S), 93S–99S, 1996.
17. Borden, M. D., Attawia, M. A., Khan, Y., and Laurencin, C. T., Tissue engineered microsphere-based matrices for bone repair: design and evaluation, *Biomaterials*, 23(2), 551–559, 2002.
18. Laurencin, C. T., Attawia, M. A., Lu, L. Q., Borden, M. D., Lu, H. H., Gorum, W. J., and Lieberman, J. R., Poly(lactide-*co*-glycolide)/hydroxyapatite delivery of BMP-2-producing cells: a regional gene therapy approach to bone regeneration, *Biomaterials*, 22(11), 1271–1277, 2001.
19. Van Landuyt, P., Li, F., Keustermans, J. P., Streydio, J. M., Delannay, F., and Munting, E., The influence of high sintering temperatures on the mechanical properties of hydroxylapatite, *J. Mater. Sci. Mater. Med.*, 6, 8–13, 1995.
20. Sarkar, M. R., Wachter, N., Patka, P., and Kinzl, L., First histological observations on the incorporation of a novel calcium phosphate bone substitute material in human cancellous bone, *J. Biomed. Mater. Res. (Appl. Biomater.)*, 58, 329–334, 2001.
21. Lin, F. H., Chen, T. M., Lin, C. P., and Lee, C. J., The merit of sintered PDLLA/TCP composites in management of bone fracture internal fixation, *Artif. Organs*, 23(2), 186–194, 1999.
22. Imai, Y., Fukuzawa, A., and Watanabe, M., Effect of blending tricalcium phosphate on hydrolytic degradation of a block polyester containing poly(L-lactic acid) segment, *J. Biomater. Sci. Polym. Ed.*, 10(7), 773–786, 1999.
23. Schiller, C., and Epple, M., Carbonated calcium phosphates are suitable pH-stabilizing fillers for biodegradable polyesters, *Biomaterials*, 24(12), 2037–2043, 2003.
24. Durucan, C., and Brown, P. W., Low temperature formation of calcium-deficient hydroxyapatite-PLA/PLGA composites, *J. Biomed. Mater. Res.*, 51(4), 717–725, 2000.
25. Posner, A. S., Crystal chemistry of bone mineral, *Physiol. Rev.*, 49, 760–792, 1969.
26. Durucan, C., and Brown, P. W., Calcium-deficient hydroxyapatite–PLGA composites: mechanical and microstructural investigation, *J. Biomed. Mater. Res.*, 51(4), 726–734, 2000.
27. Ambrosio, A. A., Sahota, J. S., Khan, Y., and Laurencin, C. T., A novel amorphous calcium phosphate polymer ceramic for bone repair: I. Synthesis and characterization, *J. Biomed. Mater. Res.*, 58(3), 295–301, 2001.
28. Khan, Y, Katti, DS, and Laurencin, CT, Novel polymer-synthesized ceramic composite-based system for bone repair: an in vitro evaluation, *J. Biomed. Mater. Res.* 69-A(4), 728–737, 2004.

29. Linhart, W., Peters, F., Lehmann, W., Schwarz, K., Schilling, A. F., Amling, M., Rueger, J. M., and Epple, M., Biologically and chemically optimized composites of carbonated apatite and polyglycolide as bone substitution materials, *J. Biomed. Mater. Res.*, 54(2), 162–171, 2001.
30. Daculsi, G., LeGeros, R. Z., Nery, E., Lynch, K., and Kerebel, B., Transformation of biphasic calcium phosphate ceramics in vivo: ultrastructural and physicochemical characterization, *J. Biomed. Mater. Res.*, 23, 883–894, 1989.
31. Daculsi, G., LeGeros, R. Z., Heughebaert, M., and Barbieux, I., Formation of carbonate-apatite crystals after implantation of calcium phosphate ceramics, *Calcif. Tissue Int.*, 46, 20–27, 1990.
32. Ripamonti, U., Osteoinduction in porous hydroxyapatite implanted in heterotopic sites of different animal models, *Biomaterials*, 17(1), 31–35, 1995.
33. Zhang, R., and Ma, P. X., Porous poly(L-lactic acid)/apatite composites created by biomimetic process, *J. Biomed. Mater. Res.*, 45(4), 285–293, 1999.
34. Zhang, R., and Ma, P. X., Biomimetic polymer/apatite composite scaffolds for mineralized tissue engineering, *Macromol. Biosci.*, 4(2), 100–111, 2004.
35. Song, J., Saiz, E., and Bertozzi, C. R., A new approach to mineralization of biocompatible hydrogel scaffolds: an efficient process toward 3-dimensional bonelike composites, *JACS*, 125, 1236–1243, 2003.
36. Ducheyne, P., Bianco, P., Radin, S., and Schepers, E., Bioactive materials: mechanisms and bioengineering considerations, In *Bone-Bonding*, Ducheyne, P., Kokubo, T. and van Blitterswijk, C. A., eds., Reed Healthcare Communications, pp. 1–12, 1992.
37. Ducheyne, P., Beight, J., Cuckler, J., Evans, B., and Radin, S., Effect of calcium phosphate coating characteristics on early post-operative bone tissue growth, *Biomaterials*, 11(8), 531–540, 1990.
38. de Bruijn, J. D., Bovell, Y. P., and van Blitterswijk, C. A., Structural arrangements at the interface between plasma sprayed calcium phosphates and bone, *Biomaterials*, 15(7), 543–550, 1994.
39. Morgan, J., Holtman, K. R., Keller, J. C., and Stanford, C. M., In vitro mineralization and implant calcium phosphate-hydroxyapatite crystallinity, *Implant Dent.*, 5(4), 264–271, 1996.
40. Varma, H. K., Yokogawa, Y., Espinosa, F. F., Kawamoto, Y., Nishizawa, K., Nagata, F., and Kameyama, T., Porous calcium phosphate coating over phosphorylated chitosan film by a biomimetic method, *Biomaterials*, 20(9), 879–884, 1999.
41. Lawson, A. C., and Czernuszka, J. T., Production and characterization of a collagen-calcium phosphate composite for use as a bone substitute, *Mater. Res. Soc. Symp. Proc.*, 550, 273–278, 1999.
42. Webster, T. J., Ergun, C., Doremus, R. H., Siegel, R. W., and Bizios, R., Specific proteins mediate enhanced osteoblast adhesion on nanophase ceramics, *J. Biomed. Mater. Res.*, 51(3), 475–483, 2000.
43. Webster, T. J., Ergun, C., Doremus, R. H., Siegel, R. W., and Bizios, R., Enhanced osteoclast-like cell functions on nanophase ceramics, *Biomaterials*, 22(11), 1327–1333, 2001.
44. Leeuwenburgh, S., Layrolle, P., Barrere, F., de Bruijn, J., Schoonman, J., van Blitterswijk, C. A., and de Groot, K., Osteoclastic resorption of biomimetic calcium phosphate coatings in vitro, *J. Biomed. Mater. Res.*, 56(2), 208–215, 2001.
45. Kroeber, MW, Ashford, F, Slucky, AV, and Lotz, JC, Cervical Interbody Fusion: Comparison of BAK Cages with Hydroxyapatite-Collagen Matrix Versus Standard Single Level Plate Fixation Transactions of the 45th Annual Meeting Orthopaedic Research Society, Anaheim, CA, 1999.

18 Hydroxyapatite/Collagen Scaffolds

Chang Du, Fu-Zhai Cui, and Klaas de Groot

CONTENTS

I. INTRODUCTION

Bone regeneration through tissue engineering is one of the important focuses of research in a wide variety of areas including orthopedic, craniofacial, maxillofacial research, and dentistry. It may provide a promising alternative for clinical practices to overcome the inherent limitations of the currently available solutions for bone reconstruction surgery. The challenging situations are those severe traumatic or degenerative skeletal problems where massive bone loss or complicated defect occurs.

The current golden standard for bone defect treatment is autogenous bone graft, usually harvested from the iliac crest. Compared to other options such as allograft or artificial bone substitute materials, autografts possess biological advantages with several crucial components. These include: (1) living osteogenic cells that may contribute to bone regeneration; (2) osteoinductive factors that may trigger and stimulate the differentiation and proliferation of osteoprogenitor and osteogenic cells; (3) extracellular matrix that serves as a reservoir for these cells and factors as well as a scaffold for bone ingrowth (osteoconduction); and (4) the autogenic origin of aforementioned cells, factors, and matrix that pose no risk of transmitting diseases and inducing immunological rejection. The major concerns for autografts are the shortage of sufficient amount of bone, donor site morbidity, and additional operations and pains to the patient.[1]

A basic strategy of bone tissue engineering[2,3] employs a three-dimensional porous scaffold to deliver bone-inducing factors or osteogenic cells to the defect site. Such engineering constructs of scaffolds with cells and/or factors are expected to substitute autogenous bone graft by providing similar crucial components. Various materials have been investigated as carriers for bone-related cells or bone morphogenetic protein (BMP).[4–8] The composite of hydroxyapatite (HA) and collagen is attracting interest primarily due to its compositional analogy to bone matrix.

II. HA/COLLAGEN COMPOSITE AS BONE SUBSTITUTE

The simplest way to construct a HA/collagen composite is mechanically mixing the two materials: HA ceramic powders and collagen suspension. In such a combination, collagen can serve as an effective fixation agent to confine particulate ceramics and reduce their migration after implantation.[9,10] The HA/collagen mixture can be further processed into various forms, such as film[11] or microspheres.[12,13] The ceramic particles can act as fillers to stiffen the organic matrix.[11] The microspheres obtained by emulsifying method can be used as carriers for osteoblast cell growth.[12] Furthermore, the spherical gel beads can be optimized for developing an injectable bone substitute, thus minimizing invasive operations for patients.[13]

Both HA ceramic and collagen have a good record in clinical application. Calcium phosphate ceramics in various forms, especially HA with bone–bonding properties, are probably the most effective artificial bone substitutes in orthopedic and dental surgery.[14,15] Collagen has long been used for the repair of lesions in ligaments, tendons, and other soft tissues, as well as for wound dressings or hemostatics. The usage of allogenic or xenogenic collagen-based materials for bone repair occasionally present an immune reaction, but no complications or adverse clinical outcomes were associated with such a response.[16] In a randomized multicenter clinical trial, Chapman et al. compared a composite material consisting of Type I bovine collagen, a biphasic calcium phosphate ceramic and autogenous marrow, with an autogenous iliac crest bone graft, for the treatment of acute long bone fractures. They observed similar rates of union and functional measures between the two groups and a similar prevalence of complications, except that the groups with autogenous grafts gave a higher rate of infection. A rare occurrence (12 out of 128 patients) of antibody titer to bovine collagen was observed without complications to fracture healing.[17]

One problem with the HA/collagen mixture is the far lower mechanical property than that of bone. Such materials cannot provide structural support for load-bearing applications. Another problem may come from the difference in the biodegradability of the two components. Collagen can be remodeled quickly, while the sintered HA crystals can rarely be degraded even after long-term implantation.[18] The fairly slow degradation of HA ceramics may retard or inhibit bone formation in the longer term, as has been suggested in several studies where HA ceramics/collagen were used as carriers for osteoinductive agent BMP.[19,20]

Current experiences with various bone substitute materials including HA/collagen composite may suggest some criteria for an ideal scaffold for bone tissue engineering. Several key requirements besides the basic necessity of biocompatibility should be fulfilled. First, bioactivity and bone-bonding property: the scaffold should provide a favorable microenvironment for the attachment, proliferation, and differentiation of osteoblasts or osteoprogenitor cells. The deposition of bone-like mineral and collagen onto the scaffold surface can establish a chemical continuity between bone and the material. The bone formation thus starts directly from the surface of the scaffold, which has been referred to as bonding osteogenesis.[21,22] Extensive work on bioactive ceramics, glasses, or glass-ceramics have indicated that *in vivo* formation of a Ca–P rich layer and surface bone-like apatite is the key step in bone-bonding behavior of these materials.[23–25] Second, biodegradability: the scaffold should serve as a temporary aid for the cell growth and bone formation. The material should be able to degrade into nontoxic products for being resorbed or cleaned by the body. The degradation rate should be comparable to the rate of bone formation. In addition, non- or less-inflammatory type degradation is desired when cell-mediated degradation activity is involved. It is worthwhile to note that bone as a living tissue is under constant metabolism of both its inorganic and organic components during a remodeling process. The resorption of bone is mediated by osteoclasts through direct chemical and enzymatic processes. Furthermore, the resorption and formation of bone are adapted to the strains and stresses on it, as represented by Wolff's law. Third, mechanical adaptability: the mechanical properties (strength, stiffness, toughness, etc.) of the scaffold should be comparable or slightly higher than that of bone. Initially, the scaffold should be able to provide a structural support when necessary. In the longer term, the scaffold should be able to adapt to the changing environment along with the new bone formation to avoid stress shielding to the bone. Bone has an excellent combination of strength and toughness that can be attributed not only to its inorganic and organic components, but also to its complex structure.[26,27] To serve as an ideal scaffold for bone tissue engineering, a HA/collagen composite should be a bone-like material not only in its composition but also in its structure and function.

III. BIOMIMETIC HA/COLLAGEN SCAFFOLD

In the pursuit of a not-yet-available ideal scaffold, a biomimetic strategy applying principles that nature has employed to make biological materials seems to be promising. As an important biomineralized material, bone has been a good model that inspires new designs and technologies for the synthesis of advanced materials.[28–30] With respect to the medical application for bone defect repair, biomimetic synthesis may involve the close imitation of bone from composition, structure, hierarchical organization and processing pathway, to property and function. Table 18.1 shows a simple comparison of bone, biomimetic HA/collagen composite, and conventional HA ceramic.

The fresh bone consists of inorganic mineral, organic matrix, and water. The precise composition varies with species, age, sex, health state, and the type of bone tissue. In general, the three components have a weight ratio 65:25:10 and approximately equal proportions by volume.[31,32] The major organic constituent is Type I collagen that represents about 90 to 95% in weight. Others are noncollagenous proteins, phosphoproteins, glycoproteins, lipid, and proteoglycans. The mineral component of bone is made of calcium phosphates, with the major phase as carbonate apatite crystal called dahllite. The bone apatite has a calcium deficiency ranging between 5 to 10%. This nonstoichiometric HA has carbonate substitutions and low crystallographic order as well as nanometer sized crystals.[33,34] Other phases like amorphous calcium phosphate (ACP) or octacalcium phosphate (OCP) probably also exist as a precursor for apatite.[35–37] The collagen molecules assemble into fibrils in a quarter staggered pattern.[38] The overlapping of adjacent molecules and the gaps left between the ends of molecules give a periodic structure of about 67 nm and provide structural scaffold for mineral deposition. The crystals are initially formed in the gap region and eventually grow into overlapping area. The crystal plates are arranged in parallel layers within a specific fibril. The crystallographic *c* axis of the apatite is oriented to the fibril axis.[39,40]

The unique ultrastructural characteristics of bone mineral make it different from the sintered ceramic in many properties, such as its dissolution behavior and its affinity with organic matrix. Daculsi et al.[41] investigated the dissolution of biological and ceramic apatite and indicated that lattice defects and surfaces of the crystals were the starting points of the process. Therefore, the low crystallinity, carbonate substitution, and nanometer size are important factors for the turnover of the minerals. With respect to the mechanical property, the interfacial bonding between minerals and organic matrix makes an important contribution to the overall outcome.[42,43] The parallel arrangement of collagen fibrils and crystal arrays may contribute to the anisotropic property of compact bone with a higher strength principally along its long axis. The microstructure of bone brings another level of complexity. The cancellous bone has a low apparent density and low mechanical properties with a highly interconnected porous structure. The compact bone is highly dense, strong, and tough for bearing load. In general, cancellous bone has a compressive strength in the range of 1 to 100 MPa and an elastic modulus of 1 to 2 GPa.[44] Compact bone has a tensile strength in the range of 50 to 170 MPa and compressive strength of 130 to 200 MPa. The modulus of elasticity is about 10 to 20 GPa.[45]

The biomimetic approaches generally apply a wet-chemistry-based low temperature process under environmentally benign conditions. They are advantageous to produce materials with nanoscale structures and therefore promise to mimic natural bone at the ultrastructural level. A key step in biomimetic HA/collagen composite synthesis is the growth of calcium phosphate minerals on/in a collagen matrix in aqueous media.[46–49] By combining the organic matrix regulation effect with inorganic solution chemistry, the size, shape, crystallinity, and phase selectivity of the mineral crystals and crystal organization may be tailored. On the one hand, the organic matrix such as collagen may regulate the nucleation and growth of the inorganic crystal.[50,51] Wang et al.[52] developed a coprecipitation method to produce HA/collagen composite, in which collagen was first dispersed in acidic calcium phosphate solution and the mineralization of collagen was induced by increasing solution pH. The resulting material proved to be a nanocomposite in which nanometer-sized bone-like apatite was induced and uniformly embedded in the collagen matrix. On the other

TABLE 18.1
A Comparison of Bone, Biomimetic Composite, and Conventional HA Ceramic

	Bone		Biomimetic Composite		Conventional Ceramic
	Mineral	**Matrix**	**Inorganic**	**Organic**	
Composition	Dahllite (ACP) (OCP)	Collagen glycoprotein proteoglycan lipid	Apatite(CO_3) (ACP) (OCP)	Collagen	Hydroxyapatite
Ultrastructure	Nanoscale crystals form periodical array in matrix with preferential orientation		Nanoscale crystals form random or oriented array in matrix		Microscale crystals fuse together
Microstructure	Cancellous: porous, cortical: dense		Porous or dense		Porous or dense
Temperature	37°C		<100°C		1000–1500°C
Hierarchical organization	From nanometer to centimeter		?		?
Mechanical property	Toughness, strength		Modulus and strength reach lower limit of bone, improving toughness		High modulus, high strength, brittle

hand, an inorganic solution condition may influence the electrochemical properties of the organic matrix[53–55] and in turn affect the mineralizing process.[56] Du et al.[57] developed a method, starting from a basic solution with high ionic concentration and high pH value, to produce several composites including poorly crystalline carbonate-apatite/collagen by controlling reaction kinetics. Besides aforementioned efforts using preformed collagen matrix, recent development in biomimetic composite synthesis has applied the idea of combining collagen matrix self-assembly and mineral formation in a one process step.[58,59] At specific conditions, self-organization of calcified collagen fibrils into fibers can be obtained where the *c*-axes of HA nanocrystals were aligned along collagen fibers.[59]

Some biological evaluations on such biomimetic composites have shown promising results for their potential application in bone tissue engineering.[57,59–61] In an implantation experiment in a rabbit marrow cavity,[60] nanophase HA/collagen composite showed biodegradation and bioactive properties. Solution-mediated dissolution and giant cell mediated resorption of the implant occurred at the interface as well as new bone formation. Figure 18.1 shows bone formation apparently following an active excavation into the dense implant made of this composite. An intimate contact between bone and the implant can be noted with the cracks being present away from the bone-implant interface. In a study of Kikuchi et al.[59] the involvement of osteoclast-like cells was observed in the resorption of self-organized bone-like HA/collagen nanocomposites and osteoblasts were observed to deposit new bone tissue around the composites. An *in vitro* study performed by Leeuwenburgh et al.[62] showed that biomimetic calcium phosphate coating, though without collagen, could be degraded through osteoclastic activity. Collectively, these studies suggest that biomimetic composites may possibly be incorporated into bone metabolism in a similar way to a bone remodeling process instead of being a permanent implant. Such a performance of biomimetic composites could likely result from their resemblance to bone in both composition and ultrastructure. Furthermore, a porous scaffold of biomimetic composites with complex configuration is readily produced because of the solution-based process. Figure 18.2a shows the porous structure of a nanophase HA/collagen composite made through the calcification of a

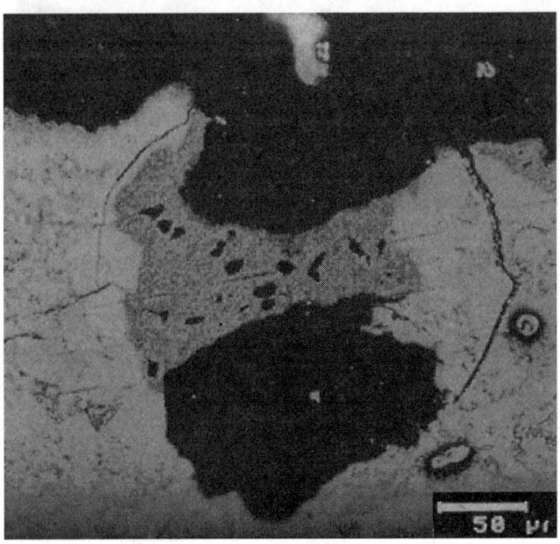

FIGURE 18.1 Backscattered electron image of scanning electron microscopy shows the excavation into the dense implant made from biomimetic HA/collagen composite and the new bone formation with direct apposition on the material. (Adapted from Du, C., Cui, F. Z., Feng, Q. L., Zhu, X. D., and de Groot, K., Tissue response to nano-hydroxyapatite/collagen composite implants in marrow cavity, Copyright © 1998 John Wiley & Sons, Inc. *J. Biomed. Mater. Res.*, 42, 540–548, 1998. With permission.)

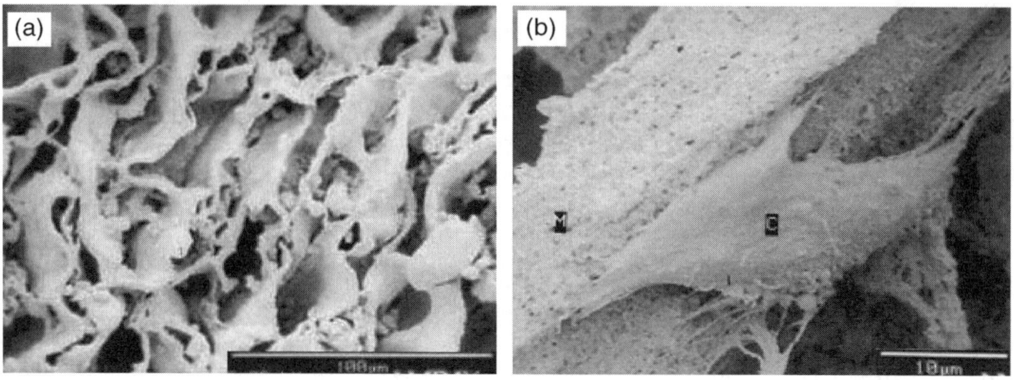

FIGURE 18.2 Scanning electron micrographs show (a) the porous structure of biomimetic HA/collagen composite, and (b) a bone-derived osteogenic cell (C) adhering to the mineralized collagen (M). (Adapted from Du, C., Cui, F. Z., Zhu, X. D., and de Groot, K., Three dimensional nano-HAp/collagen matrix loading with osteogenic cells in organ culture, Copyright © 1999 John Wiley & Sons, Inc. *J. Biomed. Mater. Res.*, 44, 407–415, 1999. With permission.)

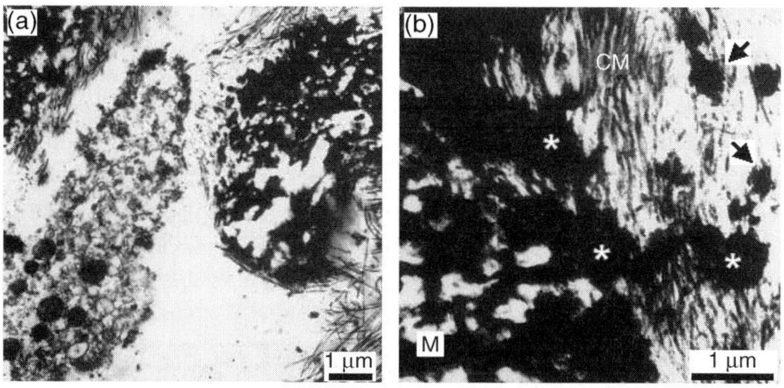

FIGURE 18.3 Transmission electron micrographs show (a) bone-derived cell depositing collagenous matrix on a biomimetic HA/collagen composite, and (b) the mineralization of the collagenous matrix (CM) on the composite (M). The mineralization process seemed to proceed as separate mineralized loci (arrows) and the loci expanded to fuse together as heavily mineralized area (asterisks). (Adapted from Du, C., Cui, F. Z., Zhang, W., Feng, Q. L., Zhu, X. D., and de Groot, K., Formation of calcium phosphate/collagen composites through mineralization of collagen matrix, Copyright © 2000 John Wiley & Sons, Inc. *J. Biomed. Mater. Res.*, 50, 518–527, 2000. With permission.)

commercial collagen hemostatic sponge. In an *in vitro* tissue culture experiment,[61] osteogenic cells were obtained as they migrated out of bone fragments. The cell can well adhere to the porous composite and produce fibrillar extracellular matrix (Figure 18.2b). In another study with a composite consisting of poorly crystalline carbonate-apatite and collagen,[57] the material seemed to be a promising scaffold for bone-derived osteogenic cells (Figure 18.3). The cell deposited collagenous matrix on the composite (Figure 18.3a) and the composite provided the initial mineralizing front for this matrix (Figure 18.3b).

IV. CHALLENGES AND PERSPECTIVES

The hierarchical organization is a challenging issue for both biomimetic and conventional scaffold processing. The structural hierarchy of bone is closely related to its mechanical property.

Weiner and Wagner[63] have managed to distinguish seven levels of organization in bone starting from three basic components (Type I collagen fibrils, apatite crystals, and water) to the macroscopic whole bone. The length scale spans over nanometer to centimeter. There is a different structural diversity and mechanical variation at each level and they collectively contribute to the overall property of bone. Using the mineralized collagen fibril as a basic building unit, bone can adapt to various mechanical functions by adopting various fibril array patterns. By the different arrangement of similar bone lamella, highly dense (cortical) and highly porous (cancellous) bone can be formed. The structural design and architecture of a scaffold, even without such complex hierarchy, may prove to be crucial for its successful application. The pore size, geometry, orientation, and interconnection, as well as porosity may determine the final outcome of a porous scaffold.

There is generally a lack of suitable techniques to make artificial materials with a multileveled structure. The conventional ceramic processing needs high temperature sintering to fabricate large bulk objects with sound mechanical properties, but such high temperature treatment could wipe off any deliberately introduced nanostructural features, not to mention the incorporation of temperature-sensitive materials like proteins. In the case of biomimetic approaches, a processing gap between the nanoscale building units (mineralized collagen fibrils) and the higher-order architectures does limit their potential. The mechanical property data of the biomimetic composites can only reach to the lower limit of those reported for bone.[48,57]

Concerning the manufacture of a large body with intricate three-dimensional architecture, an advanced material processing technique is known as "rapid prototyping" or "free-form manufacturing." This technique employs an incremental, layer-by-layer process to produce a three-dimensional body from a computer model.[28] Such layer-wise processing has shown some promise for overcoming the long-range diffusion problem associated with the solution-based mineralizing system.[64] With respect to the chemical construction of complex architectures, the idea of a templated morpho-synthesis seems promising. Recent progress has highlighted the assembly of preformed building blocks such as preformed nanoparticles as well as molecular precursors.[65,66] It is likely that the combination of a novel material processing technique with innovative biomimetic chemical synthesis strategies is a reliable pathway toward the replication or mimicry of bone hierarchy.

REFERENCES

1. Damien, J. C., and Parson, J. R., Bone graft and bone graft substitutes: a review of current technology and applications, *J. Appl. Biomater.*, 2, 187–208, 1991.
2. Langer, R., and Vacanti, J. P., Tissue engineering, *Science*, 260, 920–926, 1993.
3. Crane, G. M., Ishaug, S. L., and Mikos, A. G., Bone tissue engineering, *Nat. Med.*, 1, 1322–1324, 1995.
4. Ohgushi, H., Goldberg, V. M., and Caplan, A. I., Heterotopic osteogenesis in porous ceramics induced by marrow cells, *J. Orthop. Res.*, 7, 568–578, 1989.
5. Paralkar, V. M., Nandedkar, A. K. N., Pointer, R. H., Kleinman, H. K., and Reddi, A. H., Interaction of osteogenin, a heparin binding bone morphogenetic protein, with type IV collagen, *J. Biol. Chem.*, 265, 17281–17284, 1990.
6. Kawai, T., Mieki, A., Ohno, Y., Umemura, M., Kataoka, H., Kurita, S., Koie, M., Jinde, T., Hasegawa, J., and Urist, M. R., Osteoinductive activity of composites of bone morphogenetic protein and pure titanium, *Clin. Orthop. Rel Res.*, 290, 296–305, 1993.
7. Ishaug, S. L., Yaszemski, M. J., Bizios, R., and Mikos, A. G., Osteoblast function on synthetic biodegradable polymers, *J. Biomed. Mater. Res.*, 28, 1445–1453, 1994.
8. Laurencin, C. T., EI-Amin, S. F., Ibim, S. E., Willoughby, D. A., Attawia, M., Allcock, H. R., and Ambrosio, A. A., A highly porous 3-dimensional polyphosphazene polymer matrix for skeletal tissue regeneration, *J. Biomed. Mater. Res.*, 30, 133–138, 1996.

9. Richard, K., Gongloff, R. K., and Montogomery, C. K., Experimental study of the use of collagen tubes for implantation of particulate hydroxyapatite, *J. Oral. Maxillofac. Surg.*, 43, 845–849, 1985.
10. Bell, R., and Beirne, O. R., Effect of hydroxyapatite, tricalcium phosphate, and collagen on the healing of defects in the rat mandible, *J. Oral. Maxillofac. Surg.*, 46(7), 589–594, 1988.
11. Bigi, A., Panzavolta, S., and Roveri, N., Hydroxyapatite-gelatin films: a structural and mechanical characterization, *Biomaterials*, 19, 739–744, 1998.
12. Hsu, F. Y., Chueh, S. C., and Wang, Y. J., Microspheres of hydroxyapatite/reconstituted collagen as supports for osteoblast cell growth, *Biomaterials*, 20, 1931–1936, 1999.
13. Flautre, B., Pasquier, G., Blary, M. C., Anselme, K., and Hardouin, P., Evaluation of hydroxyapatite powder coated with collagen as an injectable bone substitute: microscopic study in rabbit, *J. Mater. Sci. Mater. Med.*, 7, 63–67, 1996.
14. Jarcho, M., Calcium phosphate ceramics as hard tissue prosthetics, *Clin. Orthop. Rel. Res.*, 157, 259–278, 1981.
15. de Groot, K., Bioceramics consisting of calcium phosphate salts, *Biomaterials*, 1, 47–50, 1980.
16. Delustro, F., Dasch, J., Keefe, J., and Ellingsworth, L., Immune responses to allogeneic and zenogeneic implants of collagen and collagen derivatives, *Clin. Orthop. Rel. Res.*, 260, 263–279, 1990.
17. Chapman, M. W., Bucholz, R., and Cornell, C., Treatment of acute fractures with a collagen–calcium phosphate graft material: a randomized clinical trial, *J. Bone Joint Surg. Am.*, 79-A, 495–502, 1997.
18. Hemmerle, J., Leize, M., and Voegel, J.-C., Long-term behaviour of a hydroxyapatite/collagen–glycosaminoglycan biomaterial used for oral surgery: a case report, *J. Mater. Sci. Mater. Med.*, 6, 360–366, 1995.
19. Takaoka, K., Nakahara, H., Yoshikawa, H., Masuhara, K., Tsuda, T., and Ono, K., Ectopic bone induction on and in porous hydroxyapatite combined with collagen and bone morphogenetic protein, *Clin. Orthop. Rel. Res.*, 234, 250–254, 1988.
20. Asahina, I., Watanabe, M., Sakurai, N., Mori, M., and Enomoto, S., Repair of bone defect in primate mandible using a bone morphogenetic protein (BMP) — hydroxyapatite — collagen composite, *J. Med. Dent. Sci.*, 44, 63–70, 1997.
21. Osborn, J. F., and Newesly, H., Bonding osteogenesis induced by calcium phosphate ceramic implants, In *Biomaterials*, Winter, G. D., Gibbons, D. F., and Plenk, H., eds., Wiley, New York, pp. 55–58, 1982.
22. Williams, D. F., Black, J., and Doherty, P. J., Second concensus conference on definitions in biomaterials, In *Advances in Biomaterials: Biomaterial-Tissue Interfaces*, Doherty, P. J., Williams, D. F., and Lee, A. J. C., eds., Elsevier Science Publishers, Amsterdam, pp. 525–533, 1991.
23. Tracy, B. M., and Doremus, R. H., Direct electron microscopy studies of the bone-hydroxyapatite interface, *J. Biomed. Mater. Res.*, 18, 719–726, 1984.
24. Gross, U., and Strunz, V., The interface of various glasses and glass-ceramics with a bony implantation bed, *J. Biomed. Mater. Res.*, 19, 251–271, 1985.
25. Neo, M., Nakamura, T., Ohtsuki, C., Kokubo, T., and Yamamuro, T., Apatite formation on three kinds of bioactive material at an early stage in vivo: a comparative study by transmission electron microscopy, *J. Biomed. Mater. Res.*, 27, 999–1006, 1993.
26. Bonfield, W., and Grynpas, M. D., Anisotropy of the Young's modulus of bone, *Nature (London)*, 270, 453–454, 1977.
27. Piekarski, K., Analysis of bone as a composite structure, *Int. J. Eng. Sci.*, 11, 557–565, 1973.
28. Heuer, A. H., Fink, D. J., Laraia, V. J., Arias, J. L., Calvert, P. D., Kendall, K., Messing, G. L., Blackwell, J., Rieke, P. C., Thompson, D. H., Wheeler, A. P., Veis, A., and Caplan, A. I., Innovative materials processing strategies: a biomimetic approach, *Science*, 255, 1098–1105, 1992.
29. Weiner, S., and Addadi, L., Design strategies in mineralized biological materials, *J. Mater. Chem.*, 7(5), 689–702, 1997.
30. Mann, S., Molecular recognition in biomineralization, *Nature*, 332, 119–124, 1988.
31. Carter, D. R., and Spengler, D. M., Mechanical properties and composition of cortical bone, *Clin. Orthop.*, 135, 193–217, 1978.
32. LeGeros, R. Z., *Calcium Phosphates in Oral Biology and Medicine*, Karger AG, 1991.
33. Lowenstam, H. A., and Weiner, S., *On Biomineralization*, Oxford University Press, New York, 1989.

34. Landis, W. J., Song, M. J., Leith, A., McEwen, L., and McEwen, B. F., Mineral and organic matrix interaction in normally calcifying tendon visualized in three dimensions by high-voltage electron microscopic tomography and graphic image reconstruction, *J. Struct. Biol.*, 110, 39–54, 1993.
35. Termine, J. D., and Eanes, E. D., Comparative chemistry of amorphous and apatitic calcium phosphate preparations, *Calcif. Tissue Res.*, 10, 171–197, 1972.
36. Madsen, H. E. L., and Thorvardarson, G., Precipitation of calcium phosphate from moderately acid solution, *J. Cryst. Growth*, 66, 369–376, 1984.
37. Brown, W. E., Crystal growth of bone mineral, *Clin. Orthop.*, 44, 205–220, 1966.
38. Hodge, A. J., and Petruska, J. A., Recent studies with the electron microscope on ordered aggregates of the tropocollagen molecule, In *Aspects of Protein Structure*, Ramachandran, G. N., ed., Academic, New York, pp. 289–300, 1963.
39. Landis, W. J., Song, M. J., Leith, A., McEwen, L., and McEwen, B. F., Mineral and organic matrix interaction in normally calcifying tendon visualized in three dimensions by high-voltage electron microscopic tomography and graphic image reconstruction, *J. Struct. Biol.*, 110, 39–54, 1993.
40. Matsuchima, N., Akiyama, M., and Terayama, Y., Quantitative analysis of the orientation of mineral in bone from small angle X-ray scattering patterns, *Jpn J. Appl. Phys.*, 21, 186–189, 1982.
41. Daculsi, G., LeGeros, R. Z., and Mitre, D., Crystal dissolution of biological and ceramic apatites, *Calcif. Tissue Int.*, 45, 95–103, 1989.
42. Walsh, W. R., Ohno, M., and Guzelsu, N., Bone composite behavior: effects of mineral organic bonding, *J. Mater. Sci. Mater. Med.*, 5, 72–79, 1994.
43. Bundy, K. J., Determination of mineral–organic bonding effectiveness in bone: theoretical considerations, *Ann. Biomed. Eng.*, 13, 119–135, 1985.
44. Gibson, L. J., The mechanical behaviour of cancellous bone, *J. Biomech.*, 18, 317–328, 1985.
45. Reilly, D. T., and Burstein, A. H., The elastic and ultimate properties of compact bone tissue, *J. Biomech.*, 8, 393–405, 1975.
46. Mathers, N. J., and Czernuszka, J. T., Growth of hydroxyapatite on type I collagen, *J. Mater. Sci. Lett.*, 10, 992–993, 1991.
47. Clarke, K. I., Graves, S. E., Wong, A. T. C., Triffitt, J. T., Francis, M. J. O., and Czernuszka, J. T., Investigation into the formation and mechanical properties of a bioactive material based on collagen and calcium phosphate, *J. Mater. Sci. Mater. Med.*, 4, 107–110, 1993.
48. Tenhuisen, K. S., Martin, R. I., Klimkiewicz, M., and Brown, P. W., Formation and properties of a synthetic bone composite: hydroxyapatite-collagen, *J. Biomed. Mater. Res.*, 29, 803–810, 1995.
49. Doi, Y., Horiguchi, T., Moriwaki, Y., Kitago, H., Kajimoto, T., and Iwayama, Y., Formation of apatite-collagen complexes, *J. Biomed. Mater. Res.*, 31, 43–49, 1996.
50. Glimcher, M. J., The role of collagen and phosphoproteins in the calcification of bone and other collagenous tissues, In *Calcium in Biological Systems*, Rubin, R. P., Weiss, G. B. and Putney, J. W., eds., Plenum, New York, pp. 607–616, 1985.
51. Christiansen, D. L., and Silver, F. H., Mineralization of an axially aligned collagenous matrix: a morphological study, *Cells Mater.*, 3, 177–188, 1993.
52. Wang, R. Z., Cui, F. Z., Lu, H. B., Wen, H. B., Ma, C. L., and Li, H. D., Synthesis of nanophase hydroxyapatite/collagen composite, *J. Mater. Sci. Lett.*, 14, 490–492, 1995.
53. Bensusan, H. B., and Hoyt, B. L., The effect of various parameters on the rate of formation of fibers from collagen solutions, *J. Am. Chem. Soc.*, 80, 719–724, 1958.
54. Eastoe, J. E., Long, J. E., and Willan, A. L. D., The amide nitrogen content of gelatins, *Biochem. J.*, 78, 51–56, 1961.
55. Weinstock, A., King, P. C., and Wuthier, R. E., The ion-binding characteristics of reconstituted collagen, *Biochem. J.*, 102, 983–988, 1967.
56. Iijima, M., Moriwaki, Y., Yamaguchi, R., and Kuboki, Y., Effect of solution pH on the calcium phosphates formation and ionic diffusion on and through the collagenous matrix, *Connect. Tissue Res.*, 36, 73–83, 1997.
57. Du, C., Cui, F. Z., Zhang, W., Feng, Q. L., Zhu, X. D., and de Groot, K., Formation of calcium phosphate/collagen composites through mineralization of collagen matrix, *J. Biomed. Mater. Res.*, 50, 518–527, 2000.
58. Bradt, J. H., Mertig, M., Teresiak, A., and Pompe, W., Biomimetic mineralization of collagen by combined fibril assembly and calcium phosphate formation, *Chem. Mater.*, 11, 2694–2701, 1999.

59. Kikuchi, M., Itoh, S., Ichinose, S., Shinomiya, K., and Tanaka, J., Self-organization mechanism in a bone-like hydroxyapatite/collagen nanocomposite synthesized in vitro and its biological reaction in vivo, *Biomaterials*, 22, 1705–1711, 2001.
60. Du, C., Cui, F. Z., Feng, Q. L., Zhu, X. D., and de Groot, K., Tissue response to nano-hydroxyapatite/collagen composite implants in marrow cavity, *J. Biomed. Mater. Res.*, 42, 540–548, 1998.
61. Du, C., Cui, F. Z., Zhu, X. D., and de Groot, K., Three dimensional nano-HAp/Collagen matrix loading with osteogenic cells in organ culture, *J. Biomed. Mater. Res.*, 44, 407–415, 1999.
62. Leeuwenburgh, S., Layrolle, P., Barrere, F., de Bruijn, J., Schoonman, J., van Blitterswijk, C. A., and de Groot, K., Osteoclastic resorption of biomimetic calcium phosphate coating in vitro, *J. Biomed. Mater. Res.*, 56, 208–215, 2001.
63. Weiner, S., and Wagner, H. D., The material bone: structure-mechanical function relations, *Ann. Rev. Mater. Sci.*, 28, 271–298, 1998.
64. Calvert, P., Frechette, J., and Souvignier, C., Mineralization of multilayer hydrogels as a model for mineralization of bone, pp.153–159. *Proceedings of Materials Research Society Symposium*. Warrendale, PA, USA, 1998.
65. Mann, S., The chemistry of form, *Angew. Chem. Int. Ed.*, 39, 3392–3406, 2000.
66. Davis, S. A., Breulmann, M., Rhodes, K. H., Zhang, B., and Mann, S., Template-directed assembly using nanoparticle building blocks: a nanotectonic approach to organized materials, *Chem. Mater.*, 13, 3218–3226, 2001.

19 Bioactive Hydrogels: Mimicking the ECM with Synthetic Materials

Jennifer L. West

CONTENTS

I. THE EXTRACELLULAR MATRIX — A PROTOTYPICAL BIOACTIVE HYDROGEL

In all soft tissues in our bodies, cells grow within or on a hydrogel scaffold consisting of a variety of cross-linked proteins and polysaccharides called the extracellular matrix (ECM). The ECM provides mechanical support for cells within the tissue and also has direct biological interactions with the tissue cells, influencing cell adhesion, growth, migration, gene expression, morphology, and differentiation. Ideally in applications such as tissue engineering, we would like to replace many of the functions of the ECM in order to control and optimize the tissue formation process, but biomaterials used as scaffolds in tissue engineering generally serve solely as mechanical support structures. Using natural polymers such as collagen or fibrin gels provides many of the biological activities and cues, but controlling these interactions is difficult, and frequently the mechanical properties of these protein-based materials are insufficient. One option may be to create biohybrid materials composed primarily of synthetic polymers but modified with bioactive moieties such as peptides, proteins or polysaccharides. This may allow one to provide the requisite mechanical support as well as provide signals to control tissue formation and differentiation. Some of the modifications that could mimic, at least to some extent, the biological functions of the ECM would include incorporation of cell adhesive ligands, immobilization of agents with growth factor activity, and inclusion of proteolytically degradable domains within the polymer structure to allow hydrogel degradation in response to cellular remodeling activity.

II. MODIFICATION OF HYDROGEL MATERIALS WITH CELL ADHESION LIGANDS

In the ECM, cell adhesion occurs due to interaction between cell surface receptors, such as the integrins, and specific ligands on ECM proteins. This activity has been replicated in several hydrogel systems through the attachment of certain peptides to the hydrogel network. Numerous

TABLE 19.1
Short peptides derived from ECM proteins can sometimes mediate robust, biospecific cell adhesion. Examples of several such peptides and the ECM protein in which they are found are listed

Peptide Sequence	ECM Protein
RGD	Fibronectin, vitronectin, laminin, collagen
YIGSR	Laminin
IKVAV	Laminin
REDV	Fibronectin
DGEA	Collagen
KQAGDV	Fibrinogen
VAPG	Elastin

short peptide sequences derived from cell adhesive ECM proteins have been identified that are able to bind to cell surface receptors and mediate cell adhesion with affinity and specificity similar to that observed with intact proteins. Peptides are generally preferable to intact proteins as they are not subject to denaturation and may be less susceptible to proteolysis. Some of these cell adhesive peptides, along with the ECM proteins from which they are derived, are shown in Table 19.1. When modifying materials with adhesive peptides, however, one must be aware that activation of cell signaling pathways upon receptor binding can alter many other cellular activities including proliferation, migration, and synthesis of new ECM.[1,2]

Perhaps the most extensively studied cell adhesion peptide is the sequence Arg–Gly–Asp (RGD). The RGD sequence is found in many cell adhesion proteins and binds to integrin receptors on a wide variety of cell types.[3] Modification of hydrogel materials with RGD peptides may enhance cell adhesion to the hydrogel and activate integrin signaling pathways. A number of studies have been performed with RGD-containing peptides grafted to polymers. In initial studies, RGD peptides were immobilized on a polymer-modified glass substrate, and cell adhesion parameters were correlated with peptide density.[4] A peptide density of 10 fmol/cm^2 was sufficient to support adhesion and spreading of fibroblast cells, clustering of integrin receptors, and organization of actin stress fibers. At 1 fmol/cm^2, cells were fully spread, but the morphology of the stress fibers was abnormal, and the cells did not form focal contacts.

RGD peptides have been used to modify a wide variety of hydrogel materials including PEG,[5–8] PVA,[9] *N*-(2-hydroxypropyl)methacrylamide (HPMA[10]), alginate,[11] and hyaluronic acid (HA[12]). In all of these studies, the covalently incorporated RGD was capable of mediating cell adhesion via integrin receptors. These types of RGD-modified hydrogels have shown promise in several animal models of wound healing. For example, porous hydrogels formed from HPMA–RGD supported ingrowth of axons in lesions formed in both the optic tract and the cerebral cortex of rats.[10] Proteolytically degradable PEG hydrogels modified with RGD peptides have been used in a rat critical size cranial defect model of bone healing. The RGD-containing hydrogels were shown to support cell infiltration with remodeling to bony tissue within 5 weeks.[7,8]

When designing peptide-modified materials, it is often desirable to start with intrinsically cell nonadhesive polymers, effectively providing a blank slate upon which one can build desired biological interactions. This provides a great deal of control over the activation of cellular signaling pathways and could potentially be used to control cellular differentiation. Many hydrogel materials are highly resistant to protein adsorption and cell adhesion, making them excellent base materials for bioactive modifications. For example, photo-crosslinked PEG[5,6] and PVA[9] hydrogels have been

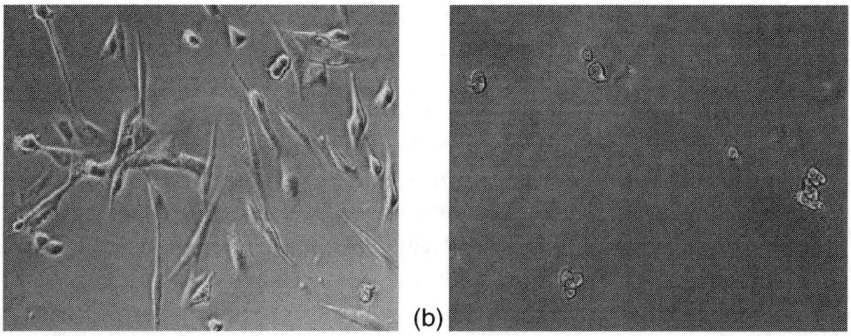

(a) (b)

FIGURE 19.1 PVA hydrogels were prepared with RGD peptides grafted to a fraction of the pendant groups. The PVA was also modified with methacrylamide groups, and hydrogels were formed via photopolymerization. Human dermal fibroblast cell adhesion occurred on hydrogels modified with RGD peptides (a) but not on those without cell adhesion ligands (b).

particularly interesting for this approach. Figure 19.1 shows interaction of fibroblasts with an unmodified PVA hydrogel and with a RGD-containing PVA hydrogel, demonstrating adhesion and spreading only with inclusion of the adhesion ligand. In addition to providing controlled and specific biological interactions, modification of cell nonadhesive materials, if adhesion ligands with appropriate selectivity are utilized, can allow one to generate cell-specific materials, that is, materials that are adhesive to certain cell types only. For example, the peptide REDV has been shown to be adhesive to endothelial cells but not smooth muscle cells, fibroblasts, or platelets,[13] and thus may be useful in development of endothelial cell linings on vascular grafts. Figure 19.2 shows a monolayer of human umbilical vein endothelial cells grown on a REDV-containing PEG hydrogel. The elastin-derived peptide VAPG has been shown to be a selective and robust adhesion ligand for smooth muscle cells, and PEG hydrogel materials have also been modified with this peptide to create a smooth muscle specific scaffold material for vascular tissue engineering.[14]

III. GROWTH FACTOR IMMOBILIZATION

Growth factors are polypeptides involved in the regulation of a variety of cellular activities such as growth and differentiation as well as many metabolic processes. Many growth factors

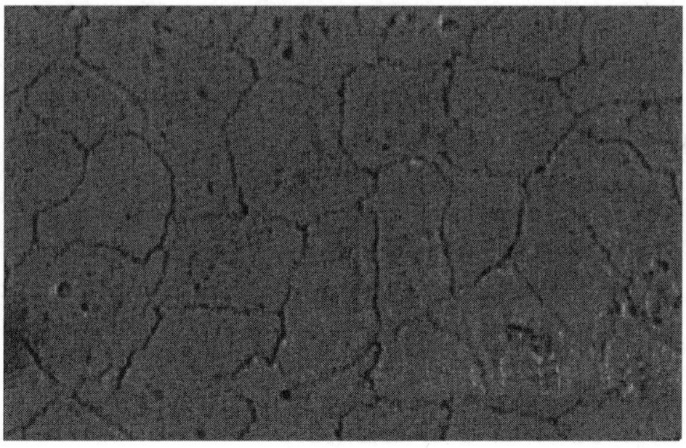

FIGURE 19.2 Human umbilical vein endothelial cells were seeded on the surface of a photopolymerized PEG hydrogel that contained REDV peptides. Cells were stained with silver nitrate to highlight the endothelial cell borders. Normal cell morphology is apparent. Smooth muscle cells, fibroblasts, and platelets are not able to adhere to the REDV peptide.

have been identified with a wide range of activities. Potential applications include acceleration of wound healing, tissue engineering, and control over differentiation of stem cells. Modification of hydrogel biomaterials with growth factors may target activity to the desired tissue and may maintain biological function for a prolonged period of time. Growth factors that have been covalently immobilized to hydrogel biomaterials include epidermal growth factor (EGF), basic fibroblast growth factor (bFGF), vascular endothelial cell growth factor (VEGF), nerve growth factor (NGF), and transforming growth factor-beta (TGF-β).

EGF is a 6 kDa polypeptide that is mitogenic and chemotactic for many of the cell types involved in wound healing. EGF was also the first growth factor that was shown to remain active after covalent immobilization. In this first study, EGF was immobilized to aminated glass via a star PEG spacer molecule and shown to retain mitogenic activity, whereas, physically adsorbed EGF showed no activity.[15] Prior to this work, it was believed that receptor dimerization and internalization were required for growth factor activity. However, that does not appear to be true. Recently, EGF has been immobilized within photo-crosslinked PEG hydrogels by attaching the EGF to an acrylate-terminated PEG spacer. Upon mixing the monoacrylate–PEG–EGF with a diacrylate PEG derivative and photo-crosslinking, a hydrogel is formed with EGF bound to the hydrogel network. Hydrogels with immobilized EGF and RGD were shown to support enhanced cell proliferation and migration as compared to those with only RGD.[16]

bFGF is a 16 kDa polypeptide that is mitogenic for many cell types and appears to play an important role in angiogenesis. bFGF strongly binds to heparin, and thus heparinized materials have been widely used for complexation and presentation of bFGF.[17–19] To create bioactive hydrogels with high affinity for bFGF, one can take advantage of its high affinity for heparin. By creating hydrogels modified with the heparin-binding peptide derived from anti-thrombin III and then exposed to heparin, one can immobilize bFGF in its most bioactive form, bound to heparin in precisely the same manner as it is in the ECM.[20] This format sequesters and protects the growth factor, but allows it to become bioavailable in response to cellular demands as occurs in the ECM.

VEGF has been shown to be a potent stimulator of angiogenesis. While a number of growth factors have been shown to stimulate angiogenesis, the VEGF family of growth factors is particularly interesting because their mitogenic activity is essentially limited to endothelial cells. PEG hydrogels have been prepared that were covalently modified with both RGD and VEGF.[21] These materials supported robust angiogenic activity in both the chick chorioallantoic membrane and after subcutaneous implantation in rats. NGF is similarly cell-specific, but for neuronal cells. NGF has been covalently immobilized to poly(2-hydroxyethylmethacrylate) hydrogels with good retention of bioactivity.[22]

While in many cell types it does not stimulate growth, TGF-β plays a very important role in regulating ECM metabolism. For example, upon exposure to TGF-β synthesis of collagen and elastin is greatly upregulated in vascular smooth muscle cells.[23] In tissue engineered vascular grafts, a potential strategy to improve mechanical properties is to increase ECM deposition. Thus, PEG-based hydrogel scaffolds have been synthesized with TGF-β tethered to the hydrogel network.[24] Relative to soluble TGF-β, tethering this growth factor to the polymeric biomaterial actually enhanced its ability to stimulate ECM protein production by entrapped vascular smooth muscle cells. This effect may be due to the ability of the cells to interact with TGF-β via the appropriate receptors but not internalize it. In addition, engineered tissues formed with tethered TGF-β in the scaffold formulation were shown to have improved tensile strength.[24]

IV. PROTEOLYTICALLY DEGRADABLE HYDROGELS

The majority of biomaterials currently used in tissue engineering applications are designed to degrade via hydrolysis. With many of these materials, such as poly(lactide-*co*-glycolide), the degradation rate can be tailored to some degree by altering factors such as the copolymer composition, degree of crystallinity, or initial molecular weight. However, in these types of applications, it can be very difficult to predict *a priori* what the rate of tissue formation will be and thus, what the rate of polymer degradation should be. In the ECM, during processes such as wound healing, the rate of scaffold (ECM) degradation is precisely matched to the rate of tissue formation. ECM is degraded by proteolytic enzymes, such as the matrix metalloproteases and plasmin, produced by cells during cell migration and tissue remodeling.[25] Many of the natural polymer scaffolds, such as collagen and fibrin gels, are degraded by the same cellular mechanisms. Recently, synthetic hydrogels have been designed that are similarly degraded by cellular proteolytic activity.[6,26] These materials are block copolymers of synthetic polymers, such as polyethylene glycol (PEG), and peptides that are substrates for proteolytic enzymes. When exposed to solutions containing the targeted protease, PEG–peptide copolymer hydrogels degraded, whereas they remained stable in the absence of protease or in the presence of a nontargeted protease. Furthermore, cells can migrate through PEG–peptide hydrogels that contain both proteolytically degradable peptide sequences and cell adhesive sequences.[6,7,27] In this case, degradable sequences are required so that cells can create pathways through the solid structure for migration, and cell adhesion ligands are required to permit cells to develop the traction force required for migration. Such materials can be formed by combining the proteolytically degradable PEG–diacrylate derivative with one or more PEG–monoacrylate derivatives coupled to adhesion peptides and/or growth factors. This system provides significant flexibility in material design since several polymers can be synthesized and characterized, and then a wide variety of bioactive hydrogels formed from different combinations. Using these types of bioactive hydrogels, the concentration of RGD in the hydrogel was shown to have a profound effect on cell migration, with a biphasic response where optimal migration is observed at intermediate densities of adhesion ligands.[7,27] Additionally, when hydrogels were used with varying susceptibility to degradation by matrix metalloproteases, highest migration occurred in the hydrogels that were most easily degraded by the cellular proteases.[7] Figure 19.3 demonstrates the typical morphology of smooth muscle cells migrating through a RGD-containing hydrogel that was targeted for degradation by matrix metalloproteases.

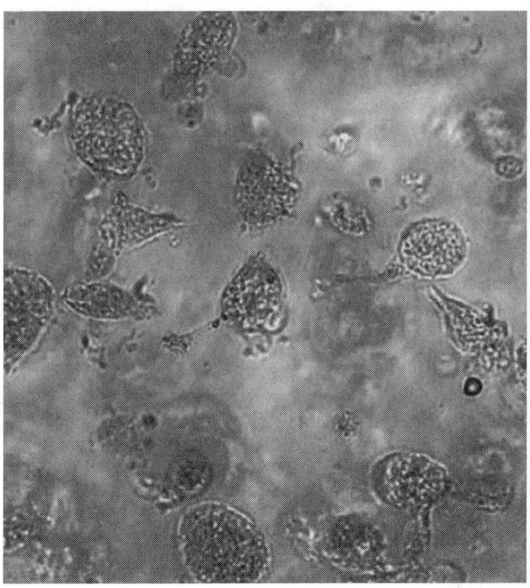

FIGURE 19.3 Smooth muscle cells were encapsulated in a photopolymerized PEG hydrogel that contained RGD peptides and sequences susceptible to degradation by matrix metalloproteases. Cells within this material actively migrate, sending processes into the hydrogel material.

An alternative approach to generate hydrogels containing both proteolytically degradable domains and cell adhesion ligands has been to generate biofunctional artificial proteins via recombinant techniques.[28] Small artificial proteins were generated that contained plasmin-sensitive degradation sites, RGD sequences for cell adhesion, and sites to allow cross-linking to PEG to form hydrogel materials. Hydrogel cross-linking was performed by Michael-type addition between thiol groups on cysteine residues in the artificial protein and PEG-diacrylate. Fibroblasts were able to adhere to and degrade these hydrogel materials. An advantage of this approach is the ability to generate artificial proteins with a variety of biological functions and thus form hydrogel materials that closely mimic the ECM.

V. CONCLUSIONS

The ability to synthesize bioactive hydrogels that replicate many of the normal functions of the ECM may substantially improve our ability to control cell growth, migration, differentiation, and protein expression for a wide variety of tissue engineering applications. Such materials will likely need to do much more than simply mediate cell adhesion, for example reconstituting many of the complex signals and interactions provided by the ECM environment. Ideally, such materials should support cell adhesion, degrade in response to cellular proteases so that scaffold degradation is coupled to the tissue formation process, and provide additional cues, such as those that can be provided by immobilized growth factors. Such multifunctional bioactive hydrogels have already been developed using the PEG–peptide copolymers and artificial protein cross-linked PEG systems discussed above. Furthermore, bioactive hydrogels have been shown to support cell adhesion, proliferation, and tissue formation. Many additional applications for bioactive hydrogels in areas such as wound healing and tissue engineering are likely in the near future. As additional advances are made in cellular and molecular biology that improve our understanding of the complex structure and functions of the ECM, the functionality that can be incorporated into these types of bioactive hydrogels should greatly expand, further enabling new and exciting applications.

REFERENCES

1. Mann, B. K., Tsai, A. T., Scott-Burden, T., and West, J. L., Modification of surfaces with cell adhesion peptides alters extracellular matrix deposition, *Biomaterials*, 20, 2281–2286, 1999.
2. Mann, B. K., and West, J. L., Cell adhesion peptides alter smooth muscle cell adhesion, proliferation, and matrix protein synthesis on modified surfaces and in polymer scaffolds, *J. Biomed. Mater. Res.*, 60, 86–93, 2002.
3. Humphries, M. J., The molecular basis and specificity of integrin–ligand interactions, *J. Cell Sci.*, 97, 585–592, 1990.
4. Massia, S. P., and Hubell, J. A., An RGD spacing of 440 nm is sufficient for integrin mediated fibroblast spreading and 140 nm for focal contact and stress fiber formation, *J. Cell Biol.*, 114, 1089–1100, 1991.
5. Hern, D. L., and Hubbell, J. A., Incorporation of adhesion peptides into nonadhesive hydrogels useful in tissue resurfacing, *J. Biomed. Mater. Res.*, 39, 266–276, 1998.
6. Mann, B. K., Gobin, A. S., Tsai, A. T., Schmedlen, R. H., and West, J. L., Smooth muscle cell growth in photopolymerized hydrogels with cell adhesive and proteolytically degradable domains: synthetic ECM analogs for tissue engineering, *Biomaterials*, 22, 3045–3051, 2001.
7. Lutolf, M. P., Lauer-Fields, J. L., Schmoekel, H. G., Metters, A. T., Weber, F. E., Fields, G. B., and Hubbell, J. A., Synthetic matrix metalloproteinase-sensitive hydrogels for the conduction of tissue regeneration: Engineering cell-invasion characteristics, *Proc. Natl Acad. Sci. USA*, 100, 5413–5418, 2003.

8. Lutolf, M. P., Weber, F. E., Schmoekel, H. G., Schense, J. C., Kohler, T., Muller, R., and Hubbell, J. A., Repair of bone defects using synthetic mimetics of collagenous extracellular matrices, *Nat. Biotech.*, 21, 513–519, 2003.
9. Schmedlen, R. H., Masters, K. S., and West, J. L., Photocrosslinkable PVA hydrogels that can be modified with cell adhesion peptides for use in tissue engineering, *Biomaterials*, 23, 4325–4332, 2002.
10. Plant, G. W., Woerly, S., and Harvey, A. R., Hydrogel containing peptide or aminosugar sequences implanted into the rat brain: Influences on cellular migration and axonal growth, *Exp. Neurol.*, 143, 287–299, 1997.
11. Rowley, J. A., Madlambayan, G., and Mooney, D. J., Alginate hydrogels as synthetic extracellular matrix materials, *Biomaterials*, 20, 45–53, 1999.
12. Park, Y. D., Tirelli, N., and Hubbell, J. A., Photopolymerized hyaluronic acid-based hydrogels and interpenetrating networks, *Biomaterials*, 24, 893–900, 2003.
13. Hubbell, J. A., Massia, S. P., and Drumheller, P. D., Endothelial cell-selective tissue engineering in the vascular graft via a new receptor, *Bio/Technology*, 9, 568–572, 1991.
14. Gobin, A. S., and West, J. L., Val–Ala–Pro–Gly, an elastin-derived non-integrin ligand: smooth muscle cell adhesion and specificity, *J. Biomed. Mater. Res.*, 67, 255–259, 2003.
15. Kuhl, P. R., and Griffith-Cima, L. G., Tethered epidermal growth factor as a paradigm for growth factor-induced stimulation from the solid phase, *Nat. Med.*, 2, 1022–1027, 1996.
16. Gobin, A. S., and West, J. L., Effects of epidermal growth factor on fibroblast migration through biomimetic hydrogels, *Biotech. Progr.*, 2003, 10.1021/bp0341390.
17. Doi, K., and Matsuda, T., Enhanced vascularization in a microporous polyurethane graft impregnated with basic fibroblast growth factor and heparin, *J. Biomed. Mater. Res.*, 34, 361–370, 1997.
18. Bos, G. W., Scharenborg, N. M., Poot, A. A., Engbers, G. H. M., Beugling, T., van Aken, W. G., and Feijen, J., Proliferation of endothelial cells on surface-immobilized albumin–heparin conjugate loaded with basic fibroblast growth factor, *J. Biomed. Mater. Res.*, 44, 330–340, 1999.
19. Wissink, M. J. B., Beernink, R., Poot, A. A., Engbers, G. H. M., Beugling, T., van Aken, W. G., and Feijen, J., Improved endothelialization of vascular grafts by local release of growth factor from heparinized collagen matrices, *J. Control. Release*, 64, 103–114, 2000.
20. Sakiyama-Elbert, S. E., and Hubbell, J. A., Development of fibrin derivatives for controlled release of heparin-binding growth factors, *J. Control. Release*, 65, 389–402, 2000.
21. Zisch, A. H., Lutolf, M. P., Ehrbar, M., Raeber, G. P., Rizzi, S. C., Davies, N., Bezuidenhout, D., Djonov, V., Zilla, P., and Hubbell, J. A., Cell-demanded release of VEGF from synthetic, biointeractive cell-ingrowth matrices for vascularized tissue growth, *FASEB J.*, 2003, 10.1096/fj.02-1041fje.
22. Kapur, T. A., and Shoichet, M. S., Chemically bound nerve growth factor for neural tissue engineering applications, *J. Biomater. Sci. Polym. Ed.*, 14, 383–394, 2003.
23. Amento, E. P., Ehsani, N., Plamer, H., and Libby, P., Cytokines and growth factors positively and negatively regulate interstitial collagen gene expression in vascular smooth muscle cells, *Arterioscler. Thromb.*, 11, 1223–1230, 1991.
24. Mann, B. K., Schmedlen, R. H., and West, J. L., Tethered TGF-beta increases extracellular matrix production of vascular smooth muscle cells, *Biomaterials*, 22, 439–444, 2001.
25. Rabbani, S. A., Metalloproteases and urokinase in angiogenesis and tumor progression, *In Vivo*, 12, 135–142, 1998.
26. West, J. L., and Hubbell, J. A., Polymeric biomaterials with degradation sites for proteases involved in cell migration, *Macromolecules*, 32, 241–244, 1999.
27. Gobin, A. S., and West, J. L., Cell migration through defined synthetic ECM analogs, *FASEB J.*, 16, 751–753, 2002.
28. Halstenberg, S., Panitch, A., Rizzi, S., Hall, H., and Hubbell, J. A., Biologically engineered protein-graft-poly(ethylene glycol) hydrogels: a cell adhesive and plasmin-degradable biosynthetic material for tissue repair, *Biomacromolecules*, 3, 710–723, 2002.

20 Albumin Modification

Sang Cheon Lee, Yoon Yeo, and Kinam Park

CONTENTS

I. INTRODUCTION

Albumin is the most abundant protein in plasma. Albumin concentration is about 40 mg/ml, while the total protein concentration in plasma is about 70 mg/ml. The structure and physicochemical properties of human serum albumin (HSA) have been well characterized.[1–5] HSA has 585 amino acid residues with the molecular weight of about 69,000 g/mol. The isoelectric point of HSA varies from 4.7 to 5.5 depending on the amount of bound fatty acid, and HSA has a net negative charge of 18 at pH 7.4. The albumin molecule is a prolate ellipsoid in shape with the hydrated dimension of $150 \times 40 \times 40$ Å^3. The three-dimensional structure of HAS is stabilized by 17 disulfide bridges and the protein is capable of withstanding a range of solution pHs and temperatures without being denatured.[3,6,7]

HSA plays an important role in the transport of numerous hydrophobic molecules, such as fatty acids, naphthoquinone derivatives, bilirubin, urobilin, steroid hormones, and bile acid salts, as well as many extraneous substances, such as penicillin, sulfonamides, and mercury. Albumin, due to its relatively high concentration and small size, is responsible for about 80% of the total osmotic effect of the plasma proteins. It is widely used in plasma-expanders for the treatment of shock and burns, and to compensate for blood loss as a result of surgery, accidents, or hemorrhages.[7]

Albumin is one of the most biocompatible macromolecules having low antigenicity and immunogenicity.[8] For this reason, albumin has been widely employed as one of the building components of coating materials of biomaterial surfaces, cross-linked gels, and nano/microspheres.[9–13] This chapter is focused on biomedical applications of albumin-modified materials, which have shown high potential in tissue engineering and controlled drug delivery.

II. ALBUMIN MODIFICATIONS WITH HEPARIN FOR SURFACE COATINGS

Albumin immobilization on biomaterial surfaces by physical adsorption or covalent grafting is known to reduce surface-induced platelet adhesion and activation.[14–19] For this reason, biomaterial surfaces have been grafted with albumin to make surfaces thromboresistant. Human and bovine serum albumins (BSAs) have been most frequently used in the surface modification of biomaterials.

Heparin is a polysaccharide with numerous sulfate groups.[20] It has been widely used as an anticoagulant. Excellent reviews are available on the preparation and properties of heparinized polymeric materials.[20–24] Heparin immobilization on the artificial surfaces has also been shown to reduce surface thrombogenicity.[25,26] Furthermore, heparin coated surfaces significantly reduced the adsorption of procoagulant and proinflammatory enzymes. Heparin has been used in the surface modification of biomaterials along with albumin to make thromboresistant surfaces.[27,28]

A. Immobilization of Albumin–Heparin Conjugates

Covalently bound conjugates of albumin and heparin were used to improve the blood-compatibility of biomaterials.[29] The conjugates were synthesized by coupling reaction at pH 5.1 to 5.2 using 1-ethyl-3-(dimethylaminopropyl)carbodiimide. The conjugates were isolated from the mixtures of unreacted albumin and heparin using diethylaminoethyl cellulose and Cibacron Blue Sepharose chromatography. The heparin content in the conjugates was in the range 10 to 16 wt%. Albumin–heparin conjugates were coated on surfaces either by physical adsorption or by covalent grafting.[29,30] Polyurethane catheters coated with albumin–heparin conjugates showed a fivefold reduction in platelet adhesion as compared with uncoated catheters.[31] Biomaterials coated with the conjugates showed a significant increase in the clotting time and the prolonged recalcification time as compared with the control surfaces.[27,30] The interaction of antithrombin III (ATIII) with polystyrene (PS) surfaces preadsorbed or covalently immobilized with albumin–heparin conjugates was studied.[28,32] An enzyme immunoassay and surface plasmon resonance study showed that the ATIII binding property of heparin in the albumin–heparin conjugate was maintained, and a large amount of ATIII was bound to the PS surface modified with albumin–heparin conjugate but not to the albumin-immobilized PS surface.[28,33] These results show that albumin–heparin conjugates combine the individual beneficial effects of albumin and heparin.

Surfaces coated with albumin–heparin conjugates were used as substrates for the seeding of endothelial cells (ECs).[34–36] Human umbilical vein endothelial cells were grown on the conjugate-grafted polystyrene in the presence of basic fibroblast growth factor (bFGF). As illustrated in Figure 20.1, bFGF bound to the conjugates accelerated adhesion of ECs leading to formation of a confluent monolayer. In addition, the seeding of ECs on the substrates significantly reduced the number of adherent platelets.

B. Surface Coating by Layer-by-Layer Assembly of Albumin and Heparin

Albumin–heparin multilayer coatings on solid substrates were prepared via layer-by-layer (LBL) assembly of albumin and heparin.[37–39] LBL assembly of polyelectrolytes is one of the most versatile and inexpensive methods for thin film depositions.[40,41] Multilayer films of albumin and heparin were prepared by alternative adsorption from solutions at pH 3.9.[38] The primary albumin monolayer was irreversibly adsorbed on the substrates mainly by hydrophobic interaction. Upon adsorption to the surface, albumin is known to expose its hydrophobic core to the surface, and such hydrophobic

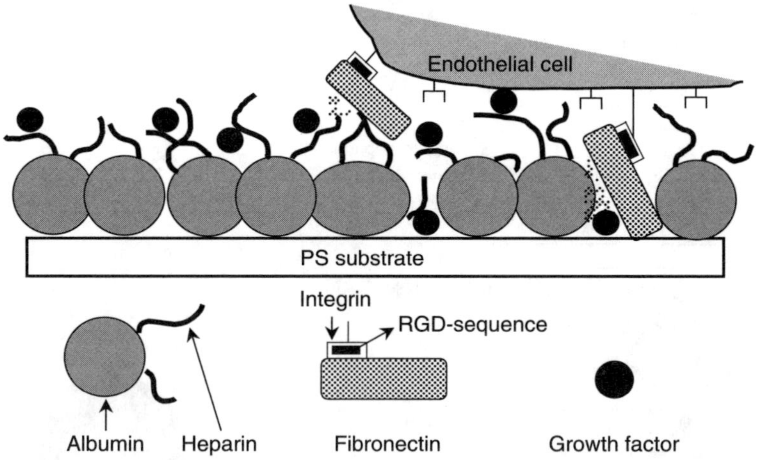

FIGURE 20.1 Schematic illustration of adhesion of an endothelial cell on surface-immobilized albumin–heparin conjugates loaded with bFGF. Integrin on endothelial cell membrane binds to cell adhesive proteins, such as fibronectin and growth factor, which interact with heparin. (From Bos, G. W., Scharenborg, N. M., Poot, A. A., Engbers, G. H. M., Beugeling, T., van Aken, W. G., and Feijen, J., Proliferation of endothelial cells on surface-immobilized albumin–heparin conjugate loaded with basic fibroblast growth factor, *J. Biomed. Mater. Res.*, 44, 330–340, 1999. With permission.)

interaction is so strong that the adsorption is almost irreversible. Electrostatic association of heparin with the adsorbed albumin layer exhibits a strong pH-dependence. At neutral pH where the overall charge of albumin is negative, heparin did not bind to the albumin-modified surface. At acidic pH's below the isoelectric point of albumin (ca. pH 4.9), however, the interaction became markedly pronounced reaching about 0.42 g of heparin bound per 1 g of albumin at pH 3.[39]

A multilayer consisting of five albumin–heparin bilayers, that is, ten alternating layers of albumin and heparin, showed the fixed stoichiometry of bilayers and approximately constant amount of albumin and heparin bound in each layer. This indicates that surfaces of any shape and size can be coated with a multilayer of a well-defined and uniform thickness. Stable multilayers in physiological solutions were obtained by the covalent cross-linking of albumin–heparin assembly in 0.5% glutaraldehyde. Cross-linked multilayered albumin–heparin films adhered to the substrate tightly through the first albumin layer on surfaces. Surfaces with multilayer films were tested *in vitro* with respect to fibrinogen and ATIII from citrated human plasma. Adsorption of fibrinogen on albumin–heparin multilayers was decreased. Adsorption of ATIII on surfaces, as expected, increased with increasing the content of heparin in the assembly. The long-term durability of the multilayered and the cross-linked albumin–heparin coating was supported by a 21-day implantation of the coated polyurethane plates in a goat heart.[37] This indicated that large plasma proteins with high affinity to surfaces, such as fibrinogen and kininogen, could not penetrate through the outer layers of the cross-linked assemblies to replace the surface-adsorbed albumin molecules.

C. Surface Coating by Polymerized Mixed Heparin–Albumin

Recently, a novel technique for the formation of polymerized mixed heparin–albumin surfaces layer was proposed.[42] The procedure comprised four steps as shown in Figure 20.2(a). The first reaction was photograft polymerization of acrylic acid (AA) on segmented polyurethane (SPU) film impregnated with camphorquinone (CQ; a visible light-induced radical producer). The second step was absorption of a mixed solution of styrenated albumin, styrenated heparin, and a water-soluble carboxylated CQ, followed by visible light irradiation as a third step. Styrenated albumin and styrenated heparin (Figure 20.2(b) were prepared by partial condensation of the amino or carboxylic group present in albumin and heparin with 4-vinylbenzoic acid for albumin and

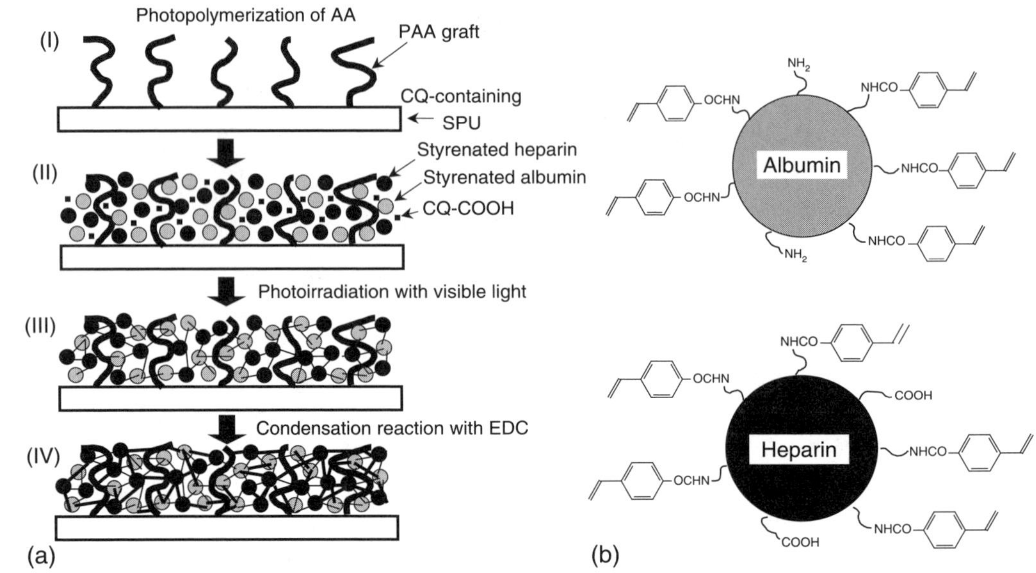

FIGURE 20.2 (a). Schematic representation for preparation of the copolymerized surface layer of styrenated albumin and styrenated heparin. (I) photograft polymerization to form a PAA-grafted surface; (II) absorption of a mixed solution of styrenated albumin, styrenated heparin, and CQ–COOH; (III) photo-polymerization by visible light; and (IV) coupling using a condensation reagent. (b). Schematic structures of styrenated albumin and styrenated heparin. (From Magoshi, T., and Matsuda, T., Formation of polymerized mixed heparin/albumin surface layer and cellular adhesional responses, *Biomacromolecules*, 3, 976–983, 2002. With permission.)

4-vinylaniline for heparin.[43] The last step was covalent bonding between poly(acrylic acid) (PAA) chains and polymerized biomacromolecules as well as between polymerized biomacromolecules to form stable and dense immobilized multilayers covalently fixed on the polymer surface.

The amount of albumin in the monolayer was reported to be 0.1 to 0.8 $\mu g/cm^2$, and that of heparin immobilized on a PAA-grafted surface, which was induced by plasma glow discharge treatment and subsequent thermal polymerization, was 1.23 $\mu g/cm^2$. The thickness of albumin- or heparin-adsorbed surfaces was estimated to be less than a few hundred angstroms. Visible light-induced photo-polymerization of styrenated albumin and styrenated heparin was effective in increasing the thickness of the protein- or polysaccharide-immobilized layers on surfaces. Using dye-complexation methods, the amounts of albumin and heparin immobilized on PAA-grafted surface were estimated to be 19.7 $\mu g/cm^2$ and 12.8 $\mu g/cm^2$, respectively. The polymerized layer was about 100 times thicker than the albumin layer or the heparin layer induced by conventional methods. Platelet adhesion was markedly reduced on polymerized mixed albumin–heparin layers, as compared with nontreated SPU and SPU–PAA surfaces. Bovine ECs were seeded and cultured on the polymerized surfaces. Adhesion and proliferative potentials of endothelial cells were comparable to those of commercial tissue culture dishes. This reflected that the increased hydrophobicity of the mixed albumin/heparin surface through the modification with the styryl group might have facilitated its interaction with the cell membrane.

III. ALBUMIN-CROSS-LINKED GELS

A. PEG Hydrogels Cross-Linked with Albumin

Biomedical applications of hydrogels are extending into areas such as implant materials,[44,45] contact lenses,[46,47] wound dressing,[48] and controlled release devices.[49–51] Recently, development of hybrid biomaterials based on blends of synthetic and natural polymers has attracted much attention. These hybrid polymers, referred to as "bioartificial polymeric materials" has been designed to combine good mechanical properties of synthetic polymers with good biocompatibility of natural polymers, such as proteins and polysaccharides.[52–54] Hydrogels composed of poly(ethylene glycol) (PEG) and BSA have been described as a bioartificial hydrogel.[52,55,56] Use of PEG in the synthesis of such bioartificial hydrogels was expected to lower the immunogenicity of materials and reduce the protein adsorption or cell adhesion. PEG–BSA hydrogel was prepared in a two-step synthetic route, which included activation of PEG (molecular weights: 2000 to 35,000) with *p*-nitrophenyl chloroformate at 60°C followed by polymerization between activated PEG and BSA in an aqueous buffered solution at pH 8.5 to 9.4 (Figure 20.3).[52]

A three-dimensional network of PEG and BSA was formed by the reaction between activated OH groups of PEG ends and free NH_2 groups of BSA. The pore size and reticulation was controlled by varying the length of the PEG chain. PEG–BSA hydrogels showed a very high swelling ability with equilibrium water content of around 95% in physiological solutions. The swelling rates of hydrogels increased as the molecular weight of PEG increased. PEG–BSA hydrogels, even in the fully swollen state, had good mechanical properties showing good elastic and compression properties.[57]

Hydrogels based on PEG and BSA were used as a matrix for immobilization of enzymes, such as L-asparaginase, acid phosphatase, arginase, apyrase, and glutaminase.[58–63] Enzymes could be immobilized into the hydrogel matrix by dissolving them in reaction solutions before cross-linking polymerization. At a physiological pH, immobilized L-asparaginase retained 90% of its activity as

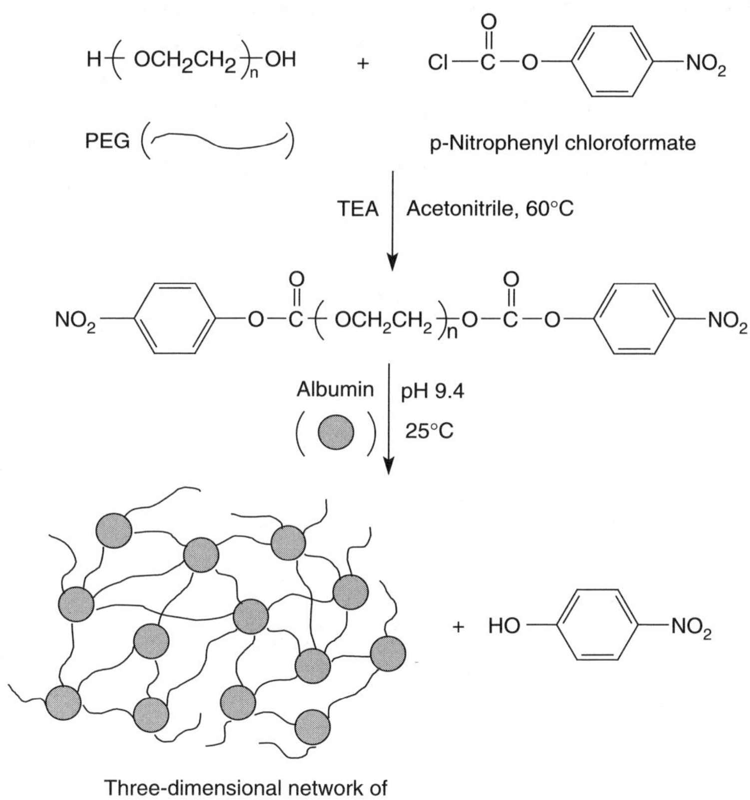

FIGURE 20.3 Schematic illustration of synthesis of PEG–BSA hydrogels. (From D'Urso, E. M., and Fortier, G., New bioartificial polymeric material: poly (ethylene glycol) crosslinked with albumin. I. Syntheis and swelling properties, *J. Bioact. Compat. Polym.*, 9, 367–387, 1994. With permission.)

compared with only 43% for the native form. Furthermore, the immobilized enzyme maintained its initial activity, more than 90% after 50 days of incubation at 37°C, whereas a half life of 2 days was observed with the native enzyme at the same incubation condition.[61] L-Asparaginase-immobilized hydrogels implanted in the peritoneal cavity of rats depleted the plasma level of asparagines for 7 days.[60] The *in vivo* evaluations showed that PEG–BSA matrix effectively protected the enzyme from rapid inactivation, and thus 75% of the enzyme was still active after 1 month of implantation. Although hydrogels, after 10 to 14 days *in vivo*, were surrounded by a fibrotic capsule with a few inflammatory sites, they showed high stability, and, even after 4 months of implantation, 12% of the initial L-asparaginase activity was present.

PEG–BSA hydrogels were also used as a controlled release device for hydrophilic and hydrophobic drugs such as theophylline, acetaminophen, hydrocortisone, and tetracycline as well as a small protein molecule, lysozyme (molecular weight ≅ 14,400 g/mol).[57,64] The release experiment demonstrated that the discharge of different types of drugs and an enzyme from the hydrogel matrix followed a Fickian diffusion-controlled mechanism. Release rates can be further controlled by adjusting the composition of hydrogels. In addition, diffusion coefficients of drugs were slightly increased with increasing the molecular mass of PEG, which controlled the porosity and the equilibrium water content of hydrogels.

B. Vinyl Polymer Hydrogels Cross-Linked with Albumin

Albumin modified with double bonds was used as a cross-linking agent in the polymerization of vinyl monomers.[65] For hydrogels cross-linked with albumin, albumin was first modified with glycidyl acrylate (GA) to introduce double bonds. The epoxide groups of glycidyl acrylate preferentially reacted with amine groups of albumin. The number of vinyl groups in modified albumin was controlled by adjusting the reaction time and/or the concentration of GA. The water-soluble vinyl monomers, such as acrylic acid, acrylamide, and vinylpyrrolidone were then polymerized in the presence of modified albumin as a cross-linker, as shown in Figure 20.4.[65–71]

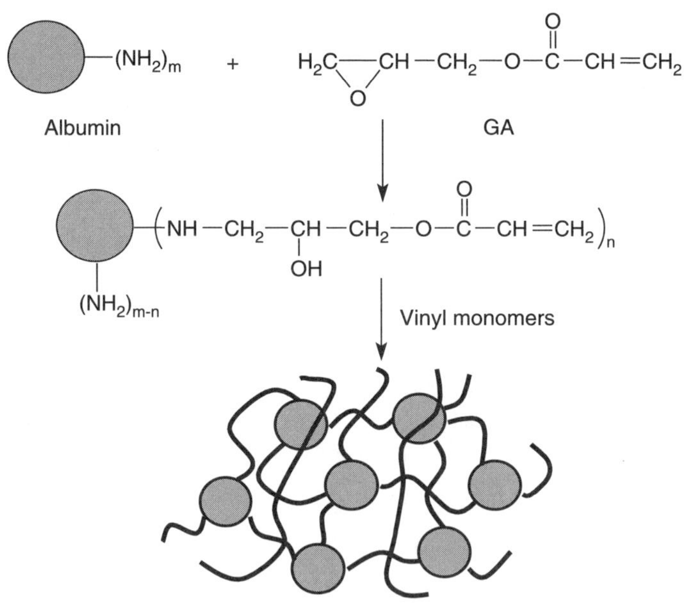

FIGURE 20.4 Synthesis of albumin-cross-linked hydrogels using vinyl monomers and modified albumin as a cross-linker.

The ability of modified albumin for cross-linking was dependent on the extent of albumin modification.[65,66] Albumin-cross-linked hydrogels were shown to be susceptible to degradation by enzymes, such as trypsin, pepsin, and chymotrypsin.[65,66] The presence of enzymes led to the disruption of the gel structure, thereby complete dissolution of the hydrogels. The degradation mode of hydrogels, that is, bulk degradation versus surface degradation, was largely dependent on the extent of albumin modification as well as the polymerization condition, such as the initiator concentration. Bulk degradation was found more pronounced when the degree of albumin incorporation was high. On the other hand, surface degradation was dominant when the degree of modification was low, for example, less than 30% of the available amine groups for modification.[70] These hydrogels were shown to remain in the stomach of dogs for up to 60 h. Such a long gastric retention resulted in the improved bioavailability of a loaded drug. The absorption of flavin mononucleotide released from the hydrogel dosage form was extended for more than 50 h.[72]

Semi-interpenetrating networks (IPN) based on albumin-cross-linked polyvinylpyrrolidone (PVP) hydrogels were prepared by introducing another vinyl monomer, acryloxyethyltrimethylammonium chloride (AETAC).[73] The IPN approach is useful for producing two-phase hydrogel systems possessing good swelling and mechanical properties. γ-Irradiation of monomers resulted in enzyme-digestible two-phase hydrogel systems, which consisted of albumin-cross-linked PVP as an inner phase and a polyelectrolyte homopolymer of AETAC as an outer phase. The outer phase enhanced swelling, whereas the inner phase maintained the mechanical integrity. This approach may offer an opportunity to combine two different monomers with different properties into one hydrogel system.

C. Cross-Linked Heparinized Albumin Gels

Heparinized albumin gels were prepared as a sealant of prosthetic vascular grafts.[12] Cross-linked albumin gels were prepared by using EDC and *N*-hydroxysuccinimide (NHS). Heparin was then immobilized by pipetting the heparin solution containing EDC and NHS on an air-dried albumin gel. Both albumin gels and heparinized albumin gels were tested for their *in vitro* stabilities, bFGF-binding properties, and cellular interactions. Human umbilical vein endothelial cells rapidly adhered and spread on albumin gels as well as heparinized albumin gels, but proliferation was only observed on the heparinized albumin gels. This result coincided with the observation that the binding of bFGF to the heparinized albumin gel was 35% higher than the binding to the nonheparinized albumin gel. Growth of endothelial cells was found only on the heparinized albumin gel loaded with bFGF and not on the bFGF-loaded albumin gel. Seeding of endothelial cells on the heparinized albumin gel significantly reduced the adsorption of platelets to this surface, and no spreading of platelets was observed on substrates seeded with endothelial cells.

IV. ALBUMIN MODIFICATIONS FOR NANO/MICROSPHERES

Due to the biodegradability, low toxicity and immunogenicity, and capability of encapsulating a wide variety of drugs in a relatively nonspecific fashion,[74] albumin has been widely used as drug carriers in various forms.[75] Numerous nano/microspheres made of albumin have been prepared during the past few decades.[9,10] Albumin microspheres have been used to deliver various drugs including steroids, antibacterial drugs, and cytotoxic agents such as adriamycin (ADR), 5-fluorouracil, and mitomycin C. Albumin microspheres can be used for drug targeting to different organs depending on the particle sizes.[75] Albumin microspheres can be made responsive to the magnetic field and can be used as site-specific drug delivery systems.[76] Radiolabeled albumin microspheres are used as diagnostic agents [77] and for tumor treatment [78] or organ imaging.[79] Modified albumins have been used to produce nano/microspheres in an attempt to improve the drug loading efficiency and prolong the *in vivo* release rate.

A. Microspheres of Albumin–Heparin Conjugates

Albumin–heparin conjugates were used for making microspheres in drug delivery applications.[80–85] Introduction of heparin in microspheres led to increase in hydrophilicity as well as enhanced blood and tissue compatibility. Furthermore, heparin immobilization in albumin microspheres improved the loading efficiency of ionic drugs by providing ion-exchange properties.[80,81] Albumin–heparin microspheres were prepared by emulsifying an aqueous solution of albumin–heparin conjugates in an oil continuous phase, followed by stabilization of the emulsified droplets with a chemical cross-linker, such as glutaraldehyde. Microspheres could be prepared with a diameter of 5 to 35 μm by adjusting the emulsification conditions. Positively charged ADR formed an ionic complex with heparin allowing a high loading efficiency up to 33% of the heparin content of the conjugate.[81] *In vitro* ADR release was influenced by the ionic strength of the release medium due to the ion-exchange property of microspheres. In a saline solution, 90% of ADR was released within 45 min, whereas in nonionic media, such as distilled water, only 30% of ADR was released at the same time period. Preclinical studies of ADR encapsulated within albumin–heparin microspheres showed that the peak plasma concentration of ADR and the level in nontarget tissues were significantly reduced, as compared with those of a free ADR.[85] Also, no acute toxicity appeared from the ADR-loaded microspheres as opposed to the same dose of free ADR.

B. Nanospheres of Surface-Modified Albumin with PEG

Surface-modified albumin nanospheres with a diameter of about 100 nm were prepared using PEG-modified human serum albumin (PEG–HSA). HSA was grafted with copolymers of poly(thioetheramido acid) and PEG (PTAAC–PEG–HSA) as well as poly(amidoamine) and PEG.[86] PTAAC and PAA provide an easy way of attaching acidic and basic groups, respectively. Nanoparticles with a hydrated PEG steric layer avoided uptake by the reticuloendothelial system (RES) to escape to the target sites. Nanoparticles of PEG–HSA had a surface surrounded by a hydrated PEG layer to offer favorable properties as a drug carrier by reducing the adsorption of plasma proteins and avoiding the nonspecific clearance by the RES uptake.[87] The nanospheres were readily produced using a pH-coacervation method,[13] followed by cross-linking with glutaraldehyde. The zeta potential values showed that the surface charge of surface-modified albumin nanospheres was much lower that that of the unmodified albumin nanospheres. Existence of a PEG hydrated barrier surrounding each nanosphere was confirmed by electrolyte- and pH-induced flocculation tests. The PEG-HSA nanospheres showed the reduced adsorption of plasma proteins on the particle surface as compared with unmodified HSA nanospheres. Rose Bengal was loaded as a model drug into nanospheres either during the particle formation or by adsorption after particle preparation.[88] The amount of RB incorporated in PTAAC-PEG-modified HSA nanospheres was up to 5% w/w. Release of Rose Bengal from the nanospheres was usually accelerated in the presence of trypsin. Such acceleration, however, was retarded when the surfaces of the nanospheres were modified with PEG. This indicated that the hydrophilic PEG barrier surrounding the nanospheres protected albumin nanospheres from enzymatic attack, thereby delaying the RB release significantly.

V. ALBUMIN MODIFICATIONS FOR CARRIER SYSTEMS

Macromolecular drug conjugates of albumin were used to increase the water-solubility of poorly soluble drugs, such as paclitaxel, and thus to improve their chemotherapeutic potentials.[89,90] Owing to the preferential albumin accumulation in tumor issues,[91–94] albumin–paclitaxel conjugates showed a good activity on neoplastic cell lines, reduction *in vivo* toxicity, and increased plasma concentration. Albumin has been chemically modified by glycosylation,[95,96] cationization,[97–99] complexation with gold,[100,101] and insulin [102] to improve transport across the blood–brain barrier.

Colloidal gold particles adsorbed with albumin were used as a macromolecular tracer in ultrastructural studies of vascular permeability.[100,101] The insulin-complexed albumin was assumed to enhance albumin adsorption on brain microvessels and facilitate the transport across them. These modifications allow albumin to become useful carriers for the delivery of bioactive agents into the brain parenchyma to treat tumor and infections of the central nervous system.[98,103] The negative charge-modified HSA with antiviral activity against HIV 1 was covalently coupled to conventional liposome carriers, consisting of phosphatidylcholine, cholesterol and maleimido-4-(p-phenylbutyryl)phosphatidylethanolamine[104] Modified HSA–liposome conjugates had a long circulation time in blood and proved to be attractive carriers for anti-HIV agents to T-lymphocytes.

VI. FABRICATION OF ALBUMIN WAFERS USING A COAXIAL ULTRASONIC ATOMIZER

In engineering artificial tissues, scaffolds are used as three-dimensional templates for initial cell attachment and subsequent tissue formation.[105] Various organic or inorganic materials of natural or synthetic origin have been investigated as scaffold materials. Examples include aliphatic polyesters, PEG, hydroxyapatite, glass ceramics, collagen, and hyaluronic acid. In general the scaffold materials should be biocompatible, and preferably bioerodible and resorbable at a controlled rate.[105] A number of techniques have been applied to process bioerodible and bioresorbable materials into three-dimensional scaffolds. Traditional methods include textile nonwovens,[106] solvent casting in combination with particulate leaching,[107] membrane lamination,[108] and melt molding. Given the biocompatibility, albumin is an excellent candidate for scaffold fabrication. Recently we have found a new way of making albumin wafers using ultrasonic atomizers. The method is relatively simple and easy, and the porosity of the albumin matrix is controllable based on the concentrations of albumin and the cross-linker.

A. Principles of Ultrasonic Atomization

When a liquid film is introduced onto a vibrating surface, such that the direction of vibration is perpendicular to the surface, the liquid film absorbs the vibration energy and creates unique capillary waves, which form regularly alternating crests and troughs in the liquid film.[109] Beyond the critical amplitude, the capillary waves cannot maintain their stability, and the waves collapse and tiny droplets of liquid emerge from the top of the waves.

An ultrasonic atomizer is a device used to generate such vibrations leading to atomization of a liquid. The atomizer body consists of three principal sections: front horn, the atomizing section; rear horn, the rear section; and a section consisting of a pair of disc-shaped piezoelectric transducers. Working in unison, these three elements provide a means for creating the vibration required to atomize liquids delivered to the atomizing surface. Liquid enters through a fitting on the rear, passes through the tube and then the central axis of the front horn. Finally, the liquid reaches the atomizing surface where atomization takes place. Piezoelectric transducers convert electrical energy provided by an external power source into high-frequency mechanical motion.[109] The vibration energy applied to the atomizing surface lets the liquid overcome the surface tension and spread on the surface forming a liquid film. The liquid film absorbs the underlying vibration energy and generates capillary waves. When the amplitude of the capillary waves exceeds a critical value, the waves collapse ejecting small droplets of the liquid.

Ultrasonic atomizers have been used in industrial and research applications related to electronics and biomedical areas, mainly for surface coating and liquid dispensing.[109] For example, the electronics industry utilized the ultrasonic spray technology in applying solder flux to circuit assembly, depositing film coatings on semiconductors, or coating the interior of illumination devices. In the biomedical industry, the ultrasonic atomizers have been used for coating the interior of blood-collecting tubes with anticoagulants, applying adhesives to sutures, or dispensing reagents into well-plates for diagnostic testing.[109] The popularity of the ultrasonic atomizer in such areas is mainly attributed to the ability to produce droplets of small size and low inertia. The velocity of the droplets produced from an ultrasonic atomizer is 1 to 10% that of a hydraulic or air-atomizing nozzle, which virtually eliminates overspray problems. The soft and low velocity spray provides a unique opportunity to make albumin wafers.

B. Fabrication of Albumin Wafers Using a Coaxial Ultrasonic Atomizer

Albumin wafers can be produced by depositing mists of solutions of albumin and a cross-linker such as glutaraldehyde on a water-immiscible platform such as Teflon plate or oil. In particular, a coaxial atomizer makes it possible to deliver the two reactants separately, which otherwise cannot be handled as a liquid mixture, and spray them simultaneously. Figure 20.5 illustrates the coaxial atomizer that has been used to generate albumin wafers. The coaxial atomizer consists of two coaxial cables that allow separate delivery of albumin and glutaraldehyde. The two aqueous solutions are fed into the atomizer using syringe pumps at controlled flow rates. Upon onset of the ultrasonic vibration on the atomizer, both liquids are fragmented into microdroplets and then are collected on a Teflon plate. The albumin droplets are mixed with the glutaraldehyde droplets midair and deposit on the collecting plate to form a thin wafer of cross-linked albumin. Figure 20.6 shows examples of the albumin wafer.

When no cross-linker was involved, the albumin droplets collected on the plate did not stay as a thin film but form multiple islets of liquid droplets due to the high interfacial tension between the aqueous solution and the Teflon platform. Figure 20.6(a) shows discontinuous solid deposits that formed after water evaporates. On the other hand, glutaraldehyde as a cross-linker stabilized the liquid film into a solid one before the liquid segregates (Figure 20.6(b) and (c)). The cross-linked wafer is a porous thin film and the porosity decreases as the concentration of glutaraldehyde solution increases. The thickness of the wafer and the surface area-to-volume ratio were readily controlled by the amount of the liquids applied to a fixed area as well as by concentrations of albumin and glutaraldehyde.

Two-dimensional precursors can be fabricated into three-dimensional scaffolds by laminating multiple membranes.[108] Recently, new techniques known as rapid prototyping technologies were introduced as a way to create three-dimensional objects.[105] Through repetitive deposition and processing of material layers using computer-controlled tools, three-dimensional object models could be built based on two-dimensional data obtained from a computer-aided design (CAD) model as well as magnetic resonance imaging (MRI) or computerized tomography (CT) scans.[105] The wafer fabrication process introduced above can be easily modified to build three-dimensional scaffolds. The high biocompatibility of albumin products and the simplicity of the process are advantages of this approach.

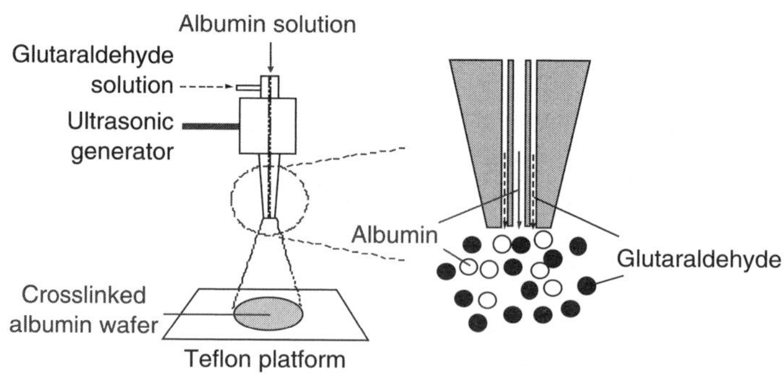

FIGURE 20.5 Schematic description of a coaxial ultrasonic atomizer for generation of albumin wafers.

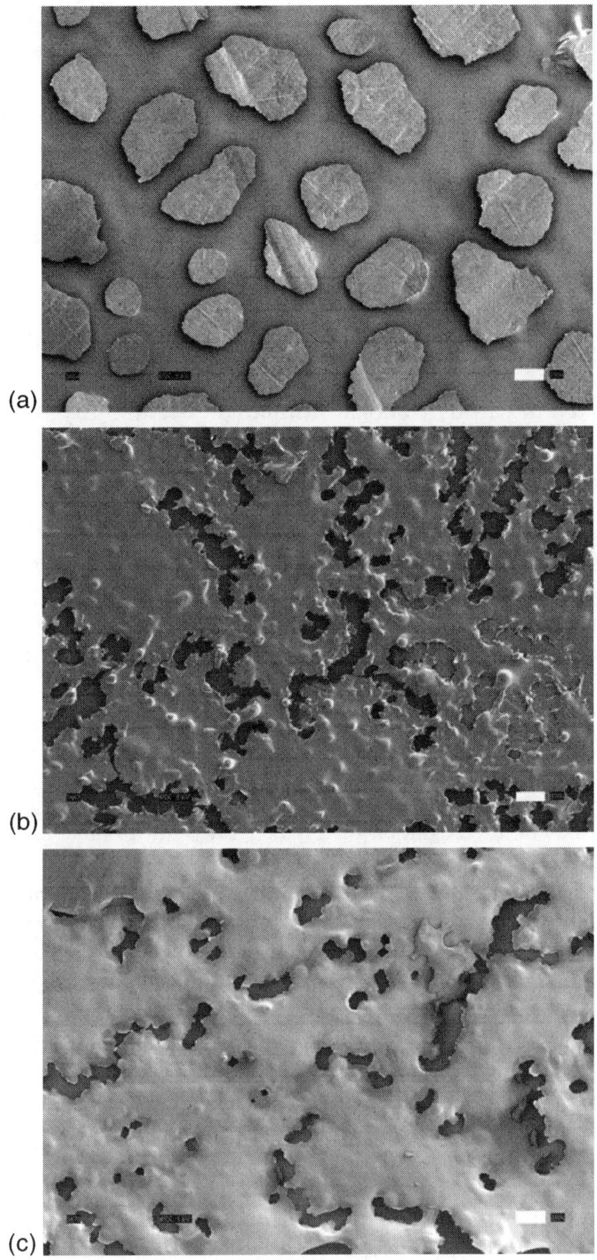

FIGURE 20.6 Scanning electron microscopic images of the albumin wafers: 10% albumin solution was atomized without a cross-linker (a), along with 1.25% glutaraldehyde solution (b), and along with 12.5% glutaraldehyde solution (c). The mixtures of droplets were collected on 150 cm^2 Teflon plate. The albumin solution was fed at 0.2 ml/min through the inner cable, and either water (a) or glutaraldehyde solution (b and c) was introduced at 0.2 ml/min through the outer cable for 10 min. Scale bar (white) = 100 μm.

VII. SUMMARY

Albumin has been modified with a variety of agents for applications in pharmaceuticals and biomedical engineering. Biocompatible and bioerodible properties of albumin without producing any undesirable molecules have made it highly attractive for applications in tissue engineering and controlled drug delivery. In tissue engineering applications, albumin can be mixed with other molecules having supplementary properties. Albumin modified with heparin through chemical conjugations as well as physical interactions could be used as a coating matrix on biomaterial surfaces. Such modifications can provide surfaces with highly favorable environment for proliferation of endothelial cells. Albumin can also be incorporated with bioerodible synthetic polymers or naturally occurring polymers to make scaffolds. High versatility of albumin for creating a broad range of formulations in different forms and sizes has made it one of the most widely used materials in pharmaceutical and biomedical applications.

ACKNOWLEDGMENTS

This study was supported in part by National Institute of Health through grant GM67044.

REFERENCES

1. Rosener, V. M., Oratz, M., and Rothschild, M. A., *Albumin Structure, Function, and Use*, Pergamon, New York, 1977.
2. Peters, T., Serum albumin, *Adv. Clin. Chem.*, 13, 37–111, 1970.
3. Peters, T., Serum albumin, In *The Plasma Proteins: Structure, Function, and Genetic Control*, Putnam, F. W., ed., Academic Press, New York, pp. 133–181, 1975.
4. Peters, T., Serum albumin, *Adv. Protein Chem.*, 37, 161–245, 1985.
5. Andrade, J. D., and Hlady, V., Plasma protein adsorption: the big twelve, *Ann. NY Acad. Sci.*, 516, 158–172, 1987.
6. Gilpin, R. K., Ehtesham, S. E., and Gregory, R. B., Liquid chromatographic studies of the effect of temperature on the chiral recognition of tryptophan by silica-immobilized bovine albumin, *Anal. Chem.*, 63, 2825–2828, 1991.
7. Fleer, R., Yeh, P., Amellal, N., Maury, I., Fournier, A., Bacchetta, F., Baduel, P., Jung, G., L'Hote, H., Fukuhara, H., and Mayaux, J. F., Stable multicopy vectors for high-level secretion of recombinant human serum albumin by *Kluyveromyces* yeasts, *Bio/Technology*, 9, 968–975, 1991.
8. Komatsu, T., Hamamatsu, K., Takeoka, S., Nishide, H., and Tsuchida, E., Human serum albumin-bound synthetic hemes as an oxygen carrier: determination of equilibrium constants for heme binding to host albumin, *Artif. Cells Blood Substit. Immobil. Biotechnol.*, 26, 519–527, 1998.
9. Arshady, R., Albumin microspheres and microcapsules: methodology of manufacturing techniques, *J. Control. Release*, 14, 111–131, 1990.
10. Arshady, R., Albumin microspheres and microcapsules: morphology of manufacturing techniques, *J. Control. Release*, 14, 111–131, 1990.
11. Amiji, M. M., Kamath, K. R., and Park, K., Albumin-modified biomaterial surfaces for reduced thrombogenicity, In *Encyclopedic Handbook of Biomaterials and Bioengineering. Part B. Applications*, Wise, D. L., Trantolo, D. J., Altobelli, D. E., Yaszemski, M. J., Gresser, J. D. and Schwartz, E. R., eds., Dekker, New York, pp. 1057–1070, 1995.
12. Bos, G. W., Scharenborg, N. M., Poot, A. A., Engbers, G. H. M., Beugeling, T., van Aken, W. G., and Feijen, J., Endothelialization of cross-linked albumin–heparin gels, *Thromb. Haemostasis*, 82, 1757–1763, 1999.
13. Lin, W., Coombes, A. G. A., Davies, M. C., Davis, S. S., and Illum, L., Preparation of sub 100 nm human serum albumin nanospheres using a pH-coacervation method, *J. Drug Target.*, 1, 237–243, 1993.
14. Mulvihill, J. N., Faradji, A., Oberling, F., and Cazenave, J.-P., Surface passivation by human albumin of plasmapheresis circuits reduces platelet accumulation and thrombus formation. Experimental and clinical studies, *J. Biomed. Mater. Res.*, 24, 155–163, 1990.
15. Kottke-Marchant, K., Anderson, J. M., Umemura, Y., and Marchant, R. E., Effect of albumin coating on the in vitro blood compatibility of Dacron arterial prostheses, *Biomaterials*, 10, 147–155, 1989.

16. Kim, S. W., and Feijen, J., Surface modification of polymers for improved blood compatibility, *CRC Crit. Rev. Biocompat.*, 1, 229–260, 1985.
17. Park, K., Mosher, D. F., and Cooper, S. L., Acute surface-induced thrombosis in the canine ex vivo model: importance of protein composition of the initial monolayer and platelet activation, *J. Biomed. Mater. Res.*, 20, 589–612, 1986.
18. Absolom, D. R., Zingg, W., and Neumann, A. W., Protein adsorption on polymer particles: role of surface properties, *J. Biomed. Mater. Res.*, 21, 161–171, 1987.
19. Amiji, M. M., Park, H., and Park, K., Study on the prevention of surface-induced platelet activation by albumin coating, *J. Biomater. Sci. Polym. Ed.*, 3, 375–388, 1992.
20. Mulloy, B., Mourao, P. A., and Gray, E., Structure/function studies of anticoagulant sulphated polysaccharides using NMR, *J. Biotechnol.*, 77, 123–135, 2000.
21. Ehrlich, J., Long term thromboresistance of heparinized materials, *Polym. Eng. Sci.*, 15, 281–285, 1975.
22. Leininger, R. I., Polymeric materials that don't clot blood, *Chem. Tech. (Leipzig)*, 5, 172–176, 1975.
23. Wilson, J. E., Heparinized polymers as thromboresistant biomaterials, *Polym.-Plast. Technol. Eng.*, 16, 119–220, 1981.
24. Bjork, I., and Lindahl, U., Mechanism of the anticoagulant action of heparin, *Mol. Cell. Biochem.*, 48, 161–182, 1982.
25. Gott, V. L., Whiiffen, J. D., and Datton, R. C., Heparin bonding on colloidal graphite surfaces, *Science*, 142, 1297 1963.
26. Crowther, M. A., Ginsberg, J. S., and Hirsh, J., Pratical aspects of anticoagulant therapy, In *Hemostasis and Thrombosis. Basic Principles and Clinical Practice*, Colman, R. W., Hirsh, J., Marder, V. J., Clowes, A. W. and Geroge, J. N., eds., Lippincott Williams and Wilkins, Philadelphia, PA, pp. 1497–1516, 2001.
27. Hennink, W. E., Kim, S. W., and Feijen, J., Inhibition of surface induced coagulation by preadsorption of albumin–heparin conjugates, *J. Biomed. Mater. Res.*, 18, 911–926, 1984.
28. van Delden, C. J., Lens, J. P., Kooyman, R. P., Engbers, G. H., and Feijen, J., Heparinization of gas plasma-modified polystyrene surfaces and the interactions of these surfaces with proteins studied with surface plasmon resonance, *Biomaterials*, 18, 845–852, 1997.
29. Hennink, W. E., Feijen, J., Ebert, C. D., and Kim, S. W., Covalently bound conjugates of albumin and heparin: synthesis, fractionation and characterization, *Thromb. Res.*, 29, 1–13, 1983.
30. Hennink, W. E., Dost, L., Feijen, J., and Kim, S. W., Interaction of albumin–heparin conjugate preadsorbed surfaces with blood, *Trans. Am. Soc. Artif. Intern. Organs*, 29, 200–205, 1983.
31. Engbers, G. H., Dost, L., Hennink, W. E., Aarts, P. A., Sixma, J. J., and Feijen, J., An in vitro study of the adhesion of blood platelets onto vascular catheters. Part I, *J. Biomed. Mater. Res.*, 21, 613–627, 1987.
32. van Delden, C. J., Engbers, G. H., and Feijen, J., Interaction of antithrombin III with surface-immobilized albumin–heparin conjugates, *J. Biomed. Mater. Res.*, 29, 1317–1329, 1995.
33. Hennink, W. E., Ebert, C. D., Kim, S. W., Breemhaar, W., Bantjes, A., and Feijen, J., Interaction of antithrombin III with preadsorbed albumin–heparin conjugates, *Biomaterials*, 5, 264–268, 1984.
34. Bos, G. W., Scharenborg, N. M., Poot, A. A., Engbers, G. H. M., Beugeling, T., van Aken, W. G., and Feijen, J., Adherence and proliferation of endothelial cells on surface-immobilized albumin–heparin conjugate, *Tissue Eng.*, 4, 267–279, 1998.
35. Bos, G. W., Scharenborg, N. M., Poot, A. A., Engbers, G. H. M., Beugeling, T., van Aken, W. G., and Feijen, J., Proliferation of endothelial cells on surface-immobilized albumin–heparin conjugate loaded with basic fibroblast growth factor, *J. Biomed. Mater. Res.*, 44, 330–340, 1999.
36. Bos, G. W., Schrenborg, N. M., Poot, A. A., Engbers, G. H. M., Beugeling, T., van Aken, W. G., and Feijen, J., Blood compatibility of surfaces with immobilized albumin–heparin conjugate and effect of endothelial cell seeding on platelet adhesion, *J. Biomed. Mater. Res.*, 47, 279–291, 1999.
37. Brynda, E., Houska, M., Jirouskova, M., and Dyr, J. E., Albumin and heparin multilayer coatings for blood-contacting medical devices, *J. Biomed. Mater. Res.*, 51, 249–257, 2000.
38. Brynda, E., and Houska, M., Multiple alternating molecular layers of albumin and heparin on solid surfaces, *J. Colloid Interface Sci.*, 183, 18–25, 1996.

39. Houska, M., and Brynda, E., Interactions of proteins with polyelectrolytes at solid/liquid interfaces: sequential adsorption of albumin and heparin, *J. Colloid Interface Sci.*, 188, 243–250, 1997.
40. Kotov, N. A., Layer-by-layer assembly of nanoparticles and nanocolloids: intermolecular interactions, structure and materials perspectives, In *Multilayer Thin Films. Sequential Assembly of Nanocomposite Materials*, Decher, G. and Schlenoff, J. B., eds., Wiley-VCH, Germany, pp. 207–243, 2003.
41. Decher, G., Polyelectrolyte multilayers, an overview, In *Multilayer Thin Films. Sequential Assembly of Nanocomposite Materials*, Decher, G. and Schlenoff, J. B., eds., Wiley-VCH, Germany, pp. 1–46, 2003.
42. Magoshi, T., and Matsuda, T., Formation of polymerized mixed heparin/albumin surface layer and cellular adhesional responses, *Biomacromolecules*, 3, 976–983, 2002.
43. Matsuda, T., and Magoshi, T., Preparation of vinylated polysaccharides and photofabrication of tubular scaffolds as potential use in tissue engineering, *Biomacromolecules*, 3, 942–950, 2002.
44. Sefc, L., Pradny, M., Vacik, J., Michalek, J., Povysil, C., Vitkova, I., Halaska, M., and Simon, V., Development of hydrogel implants for urinary incontinence treatment, *Biomaterials*, 23, 3711–3715, 2002.
45. Vijayasekaran, S., Fitton, J. H., Hicks, C. R., Chirila, T. V., Crawford, G. J., and Constable, I. J., Cell viability and inflammatory response in hydrogel sponges implanted in the rabbit cornea, *Biomaterials*, 19, 2255–2267, 1998.
46. Hiratani, H., and Alvarez-Lorenzo, C., Timolol uptake and release by imprinted soft contact lenses made of *N,N*-diethylacrylamide and methacrylic acid, *J. Control. Release*, 83, 223–230, 2002.
47. Karlgard, C. C. S., Wong, N. S., Jones, L. W., and Moresoli, C., In vitro uptake and release studies of ocular pharmaceutical agents by silicon-containing and p-HEMA hydrogel contact lens materials, *Int. J. Pharm.*, 257, 141–151, 2003.
48. Kirker, K. R., Luo, Y., Nielson, J. H., Shelby, J., and Prestwich, G. D., Glycosaminoglycan hydrogel films as bio-interactive dressings for wound healing, *Biomaterials*, 23, 3661–3671, 2002.
49. Colombo, P., Swelling-controlled release in hydrogel matrices for oral route, *Adv. Drug Deliv. Rev.*, 11, 37–57, 1993.
50. Jeyanthi, R., and Rao, K. P., Controlled release of anticancer drugs from collagen-poly (HEMA) matrices, *J. Control. Release*, 13, 91–98, 1990.
51. Changez, M., Burugapalli, K., Koul, V., and Choudhary, V., The effect of composition of poly (acrylic acid)-gelatin hydrogel on gentamicin sulphate release: in vitro, *Biomaterials*, 24, 527–536, 2003.
52. D'Urso, E. M., and Fortier, G., New hydrogel based on polyethylene glycol crosslinked with bovine serum albumin, *Biotechnol. Tech.*, 8, 71–76, 1994.
53. Katre, N. V., The conjugation of proteins with polyethylene glycol and other polymers: altering properties of proteins to enhance their therapeutic potential, *Adv. Drug Deliv. Rev.*, 10, 91–114, 1993.
54. Brooks, D. E., van Alstine, J. M., Sharp, K. A., and Stocks, S. J., PEG-derivatized ligands with hydrophobic and immunological specificity, In *Poly (ethylene glycol) Chemistry: Biotechnical and Biomedical Applications*, Harris, J. M., ed., Plenum Press, New York, pp. 57–71, 1992.
55. D'Urso, E. M., and Fortier, G., New bioartificial polymeric material: poly (ethylene glycol) crosslinked with albumin. I. Syntheis and swelling properties, *J. Bioact. Compat. Polym.*, 9, 367–387, 1994.
56. Gayet, J. C., He, P., and Fortier, G., Bioartifical polymeric material: poly (ethylene glycol) cross-linked with albumin. II. Mechanical and thermal properties, *J. Bioact. Compat. Polym.*, 13, 179–197, 1998.
57. Gayet, J. C., and Fortier, G., Drug release from new bioartificial hydrogel, *Artif. Cells Blood Substit. Immobil. Biotechnol.*, 23, 605–611, 1995.
58. Demers, N., Agostinelli, E., Averill-Bates, D. A., and Fortier, G., Immobilization of native and poly (ethylene glycol)-treated ("PEGylated") bovine serum amine oxidase into a biocompatible hydrogel, *Biotechnol. Appl. Biochem.*, 33, 201–207, 2001.
59. Belgoudi, J., and Fortier, G., Poly (ethylene glycol)-bovine serum albumin hydrogel as a matrix for enzyme immobilization. In vitro biochemical characterization, *J. Bioact. Compat. Polym.*, 14, 31–53, 1999.
60. Jean-Francois, J., D'Urso, E. M., and Fortier, G., Immobilization of L-asparaginase into a biocompatible poly (ethylene glycol)-albumin hydrogel: evaluation of performance in vivo, *Biotechnol. Appl. Biochem.*, 26, 203–212, 1997.

61. Jean-Francois, J., and Fortier, G., Immobilization of L-asparaginase into a biocompatible poly (ethylene glycol)-albumin hydrogel, *Biotechnol. Appl. Biochem.*, 23, 221–226, 1996.
62. D'Urso, E. M., and Fortier, G., Albumin-poly (ethylene glycol) hydrogel as matrix for enzyme immobilization: biochemical characterization of cross-linked acid phosphates, *Enzyme Microb. Technol.*, 18, 482–488, 1996.
63. D'Urso, E. M., Jean-Francois, J., Doillon, C. J., and Fortier, G., Poly (ethylene glycol)-serum albumin hydrogel as matrix for enzyme immobilization: biomedical applications, *Artif. Cells Blood Substit. Immobil. Biotechnol.*, 1995.
64. Gayet, J. C., and Fortier, G., High water content BSA–PEG hydrogel for controlled release device: evaluation of the drug release properties, *J. Control. Release*, 38, 177–184, 1996.
65. Park, K., Enzyme-digestible swelling hydrogels as platforms for long-term oral drug delivery: synthesis and characterization, *Biomaterials*, 9, 435–441, 1988.
66. Shalaby, W. S. W., and Park, K., Biochemical and mechanical characterization of enzyme-digestible hydrogels, *Pharm. Res.*, 7, 816–823, 1990.
67. Shalaby, W. S. W., Peck, G., and Park, K., Release of dextromethorphan hydrobromide from freeze-dried enzyme-degradable hydrogels, *J. Control. Release*, 16, 355–364, 1991.
68. Shalaby, W. S. W., Blevins, W. E., and Park, K., Enzyme-degradable hydrogels: properties associated with albumin-cross-linked polyvinylpyrrolidone hydrogels, In *Water-soluble Polymers, Synthesis, Solution Properties, and Application*, Shalaby, S. W., McCormick, C. L. and Butler, G. B., eds., American Chemical Society, Washington, DC, pp. 484–492, 1991.
69. Shalaby, W. S. W., Blevins, W. E., and Park, K., Gastric retention of enzyme-digestible hydrogels in the canine stomach under fasted and fed conditions: a preliminary analysis using new analytical techniques, In *Polymeric Drugs and Drug Delivery Systems*, Dunn, R. and Ottenbrite, R. M., eds., American Chemical Society, Washington, DC, pp. 237–248, 1991.
70. Shalaby, W. S. W., Chen, M., and Park, K., A mechanistic assessment of enzyme-induced degradation of albumin-cross-linked hydrogels, *J. Bioact. Compat. Polym.*, 7, 257–274, 1992.
71. Shalaby, W. S. W., Blevins, W. E., and Park, K., The use of ultrasound imaging and fluoroscopic imaging to study gastric retention of enzyme-digestible hydrogels, *Biomaterials*, 13, 289–296, 1992.
72. Shalaby, W. S. W., Blevins, W. E., and Park, K., In vitro and in vivo studies of enzyme-digestible hydrogels for oral drug delivery, *J. Control. Release*, 19, 131–144, 1992.
73. Shalaby, W. S. W., Jackson, R., Blevins, W. E., and Park, K., Synthesis of enzyme-digestible, interpenetrating hydrogel network by gamma-irradiation, *J. Bioact. Compat. Polym.*, 8, 3–23, 1993.
74. Katti, D., and Krishnamurti, N., Preparation of albumin microspheres by an improved process, *J. Microencapsulation*, 16, 231–242, 1999.
75. Tuan Giam Chuang, V., Kragh-Hansen, U., and Otagiri, M., Pharmaceutical strategies utilizing recombinant human serum albumin, *Pharm. Res.*, 19, 569–577, 2002.
76. Morimoto, Y., and Fujimoto, S., Albumin microspheres as drug carriers, *Crit. Rev. Ther. Drug Carrier Syst.*, 2, 19–63, 1985.
77. Caner, B., Ercan, M. T., and Erbas, B., Diagnostic use of radiolabeled particulates, *Microspheres Microcapsules Liposomes*, 3, 281–319, 2001.
78. Wunderlich, G., Pinkert, J., Andreeff, M., Stintz, M., Knapp, F. F. Jr., Kropp, J., and Franke, W.-G., Preparation and biodistribution of rhenium-188 labeled albumin microspheres B 20: a promising new agent for radiotherapy, *Appl. Radiat. Isotopes*, 52, 63–68, 2000.
79. Perkins, A. C., and Frier, M., Experimental biodistribution studies of 99m Tc-recombinant human serum albumin (rHSA): a new generation of radiopharmaceutical, *Eur. J. Nucl. Med.*, 21, 1231–1233, 1994.
80. Cremers, H. F. M., Feijen, J., Kwon, G., Bae, Y. H., Kim, S. W., Noteborn, H. P. J. M., and McVie, J. G., Albumin–heparin microspheres as carriers for cytostatic agents, *J. Control. Release*, 11, 167–179, 1990.
81. Cremers, H. F. M., Verrijk, R., Noteborn, H. P. J. M., Kwon, G., Bae, Y. H., Kim, S. W., and Feijen, J., Adriamycin loading and release characteristics of albumin–heparin conjugate microspheres, *J. Control. Release*, 29, 143–155, 1994.
82. Kwon, G. S., Bae, Y. H., Cremers, H., Feijen, J., and Kim, S. W., Release of macromolecules from albumin–heparin microspheres, *Int. J. Pharm.*, 79, 191–198, 1992.

83. Cremers, H. F. M., Kwon, G., Bae, Y. H., Kim, S. W., Verrijk, R., Noteborn, H. P. J. M., and Feijen, J., Preparation and characterization of albumin–heparin microspheres, *Biomaterials*, 15, 38–48, 1994.
84. Cremers, H. F. M., Wolf, R. F. E., Blaauw, E. H., Schakenraad, J. M., Lam, K. H., Nienuwenhuis, P., Verrijk, R., Kwon, G., Bae, Y. H., Kim, S. W., and Feijen, J., Degradation and intraheptic compatibility of albumin–heparin conjugate microspheres, *Biomaterials*, 15, 577–585, 1994.
85. Cremers, H. F. M., Verrijk, R., Bayon, L. G., Wesseling, M. M., Wondergem, J., Heuff, G., Meijer, S., Kwon, G. S., Bae, Y. H., Kim, S. W., and Feijen, J., Improved distribution and reduced toxicity of adriamycin bound to albumin–heparin microspheres, *Int. J. Pharm.*, 120, 51–61, 1995.
86. Lin, W., Garnett, M. C., Davies, M. C., Bignotti, F., Ferruti, P., Davis, S. S., and Illum, L., Preparation of surface-modified albumin nanospheres, *Biomaterials*, 17, 559–565, 1997.
87. Lin, W., Garnett, M. C., Schacht, E., Davis, S. S., and Illum, L., Preparation and in vitro characterization of HSA–mPEG nanoparticles, *Int. J. Pharm.*, 189, 161–170, 1999.
88. Lin, W., Garnett, M. C., Davis, S. S., Schacht, E., Ferruti, P., and Illum, L., Preparation and characterization of rose Bengal-loaded surface-modified albumin nanoparticles, *J. Control. Release*, 71, 117–126, 2001.
89. Dosio, F., Arpicco, S., Brusa, P., Stella, B., and Cattel, L., Poly (ethylene glycol)-human serum albumin–paclitaxel conjugates: preparation, characterization and pharmacokinetics, *J. Control. Release*, 76, 107–117, 2001.
90. Dosio, F., Brusa, P., Crosasso, P., Arpicco, S., and Cattel, L., Preparation, characterization and properties in vitro and in vivo of a paclitaxel–albumin conjugate, *J. Control. Release*, 47, 293–304, 1997.
91. Fiume, L., Busi, C., Mattioli, A., and Spinosa, G., Targeting of antiviral drugs bound to protein carriers, *CRC Crit. Rev. Ther. Drug Carrier Syst.*, 4, 265–284, 1988.
92. Matsumura, Y., and Maeda, H., A new concept for macromolecular therapeutics in cancer chemotherapy: mechanism of tumoritropic accumulation of proteins and the antitumor agent smancs, *Cancer Res.*, 46, 6387–6392, 1986.
93. Cerrottini, J. C., and Isliker, H., Transport of cytostatic agents by plasma proteins. Penetration of serum albumin into tumor cells, *Eur. J. Cancer*, 3, 111–124, 1967.
94. Chu, B. C. F., and Whiteley, J. M., High molecular weight derivatives of methotrexate as chemotherapic agents, *Mol. Pharmacol.*, 13, 80–88, 1977.
95. Williams, S. K., Devenny, J. J., and Bitensky, M. W., Micropinocytotic ingestion of nonenzymatically glucosylated proteins by capillary endothelium, *Microvasc. Res.*, 28, 311–321, 1981.
96. Predescu, D., Simionescu, M., Simionescu, N., and Palade, G. E., Binding and transcytosis of glycoalbumin by the microvascular endothelium of the murine myocardium: evidence that glycoalbumin behaves as a bifunctional ligand, *J. Cell Biol.*, 107, 1729–1738, 1988.
97. Smith, K. R., and Borchardt, R. T., Permeability and mechanism of albumin, cationized albumin, and glycosylated albumin transcellular transport across monolayers of cultured bovine brain capillary endothelial cells, *Pharm. Res.*, 6, 466–473, 1989.
98. Pardridge, W. M., Triguero, D., Buciak, J., and Yang, J., Evaluation of cationized rat albumin as a potential blood–brain barrier drug transport vector, *J. Pharmacol. Exp. Ther.*, 255, 893–899, 1990.
99. Shimon-Hophy, M., Wadhwani, K. C., Chandrasekaran, K., Larson, D., Smith, Q. R., and Rapoport, S. I., Regional blood–brain barrier transport of cationized bovine serum albumin in awake rats, *Am. J. Physiol.*, 261, R478–R483, 1991.
100. Handley, D. A., and Chien, S., Colloidal gold labelling studies related to vascular and endothelial function, hemostasis and receptor-mediated processing of plasma macromolecules, *Eur. J. Cell Biol.*, 43, 163–174, 1987.
101. Villashi, S., Preparation and application of albumin–gold complex, In *Colloid Gold*, Hayat, M. A., ed., Academic Press, Orlando, FL, pp. 163–174, 1989.
102. Vorbrodt, A. W., Dobrogowska, D. H., and Lossinsky, A. S., Ultrastructural study on the interaction of insulin–albumin–gold complex with mouse brain microvascular endothelial cells, *J. Neurocytol.*, 23, 201–208, 1994.

103. Pardridge, W. M., Kumagai, A. K., and Eisenberg, J. B., Chimeric peptides as a vehicle for peptide pharmaceutical delivery through the blood–brain barrier, *Biochem. Biophys. Res. Commun.*, 146, 307–313, 1987.
104. Kamps, J. A. A. A., Swart, P. J., Morselt, H. W. M., Pauwels, R., De Bethune, M.-P., Clercq, E. D., Meijer, A. K. F., and Scherphof, G. L., Preparation and characterization of conjugates of (modified) human serum albumin and liposomes: drug carriers with an intrinsic anti-HIV activity, *Biochimica et Biophysica Acta*, 1278, 183–190, 1996.
105. Hutmacher, D. W., Scaffold design and fabrication technologies for engineering tissues-state of the art and future perspectives, *J. Biomater. Sci. Polym. Ed.*, 12, 107–124, 2001.
106. Sittinger, M., Reitzel, D., Dauner, M., Hierlemann, H., Hammer, C., Kastenbauer, E., Planck, H., Burmester, G. R., and Bujia, J., Resorbable polyesters in cartilage engineering: affinity and biocompatibility of polymer fiber structures to chondrocytes, *J. Biomed. Mater. Res. Appl. Biomater.*, 33, 57–63, 1996.
107. Widmer, M. S., Gupta, P. K., Lu, L., Meszlenyi, R. K., Evans, G. R. D., Brandt, K., Savel, T., Gurlek, A., Patrick, C. W. Jr., and Mikos, A. G., Manufacture of porous biodegradable polymer conduits by an extrusion process for guided tissue regeneration, *Biomaterials*, 19, 1945–1955, 1998.
108. Mikos, A., Sarakinos, G., Leite, S., Vacanti, J., and Langer, R., Laminated three-dimensional biodegradable foams for use in tissue engineering, *Biomaterials*, 14, 323–330, 1993.
109. Berger, H. L., *Ultrasonic liquid atomization*, Partridge Hill Publishers, 1998.

21 Modified Alginates for Tissue Engineering

Yen-Chen Huang and David J. Mooney

CONTENTS

I. INTRODUCTION

The tremendous need to replace lost or injured tissues and organs has driven the development of the tissue engineering field. Successful tissue engineering strategies generally require the combination of physical scaffolds, bioactive signaling molecules, and cells to mimic the *in vivo* environment of the developing or regenerating tissues.[1,2] The scaffolds serve to deliver cells or growth factors to a desired site in the body and also serve as a synthetic extracellular matrix to define the space for tissue development and shape of the engineered tissues. In these ways, the scaffold allows both implanted and host cells to control tissue formation.[3,4] Scaffolds can potentially also control and regulate how the biological components, growth factors, and cells are delivered to take part in this process. The interaction of these three components may confer additive, synergistic, or antagonistic effects to dictate the course of tissue development. Polymeric materials are attractive for forming the scaffolds in these approaches, and a variety of synthetic and naturally derived polymers have been utilized.[5] These materials can be processed into different physical forms, including nonwoven mesh, porous scaffolds, or hydrogels. The type of polymeric materials to be applied may depend on the nature of the engineered tissues and their locations.

Hydrogels are an attractive form of biomaterial for use in tissue engineering since they can be used to encapsulate and deliver cells and biologically active macromolecules, can be readily processed into a variety of physical forms, and can be injected into the body in a minimally invasive manner. The high water content leads to structures similar to the natural extracellular matrix, and hydrogels typically possess good biocompatibility.[6] These features have made hydrogel forming materials attractive candidates for fabrication of delivery vehicle for cells and growth factors.[7] Various types of hydrogels, including polysaccharides (e.g., alginate, chitoson, agarose, hyaluronic acid), poly(vinyl alcohol) (PVA), and poly(ethylene glycol) (PEO) have been investigated in tissue engineering applications.[8] Chitosan has been explored in tissue engineering approaches due to its low toxicity, biocompatibility, and structural similarity to natural glycosaminoglycans,[9] and has been used to regenerate cartilage,[10] and bone.[11] Agarose is another form of algae polysaccharide, which forms hydrogels that are sensitive to temperature change.[5] The physical structure and cell interaction capability of agarose hydrogels can be manipulated by modifying their concentration, chemical structure, and gelling condition. Agarose has been combined with other synthetic polymers to regenerate nerve[12] and bone.[13] Hyaluronic acid is one of the proteoglycans in extracellular matrix, and has been utilized to engineer skin[14] and bone[15] due to its inherent cell interaction ability. However, hyaluronic acid requires a rigorous purification process and it may induce an immune response.[16] PVA typically forms hydrogels via the cross-linking action of agents such as glutaraldehyde.[17] To bypass the toxicity issue caused by these agents, physical cross-linking approaches such as repeated freezing and thawing have been developed.[18] PVA hydrogels have been used for the regeneration of cartilage[19] and bone[20] in tissue engineering. However, PVA is generally not degradable in a physiological environment and may be confined to long-term use. PEO has been widely investigated for the surface modification of biomaterials,[21] bioconjugation,[22] and drug delivery.[23] For example, PEO has been modified with galactose moieties to promote adhesiveness with liver cells.[24] Certain copolymers of PEO form hydrogels that exhibit thermally triggered gelation behavior.[25] Specific copolymers of PEO have been utilized to regenerate bone.[26]

Alginate is one class of hydrogel forming material that has been widely investigated in a variety of medical applications, including tissue engineering. Alginate is a natural polysaccharide typically extracted from brown algae, and is an anionic block copolymer composed of two uronic acids, α-D-mannuronic acid (M) and β-L-guluronic acid (G). These sugars can be organized into blocks of poly (L-guluronate), poly (D-mannuronate), and alternating sequences of both sugars.[27] Alginate hydrogels form in the presence of divalent cations (e.g., Ca^{2+}, Sr^{2+}, Ba^{2+}). Two guluronate residues on different polymer chains bind to a calcium ion, and formation of "junction zones" composed of more than 20 guluronate–calcium ion pairs leads to gelling.[28] The gelling of these hydrogels can be reversed by an exchange of divalent ions with alkali-metal cations or through sequestering of the cation responsible for gelling.[29] Hydrogels formed from alginate have been utilized for making dental impressions, wound dressings, and cell immobilization matrices due to their high biocompatibility, low toxicity, and gentle gelling properties.[30,31]

The utilization of alginate in tissue engineering applications, however, is limited by the narrow range of mechanical properties available from this material, uncontrollable degradation, and lack of cellular interactions. One would ideally form hydrogels with a controllable and wide range of mechanical properties, to allow for their use in a variety of applications involving different tissues. Ionically cross-linked alginate dissolves in an uncontrollable and unpredictable manner due to the loss of divalent cations into the surrounding tissue and subsequent dissolution. Alginate hydrogels present minimal protein adsorption due to their hydrophilic nature, which discourages cells to adhere to alginate hydrogels. Furthermore, mammalian cells do not interact with alginate because of the absence of receptors required for direct cellular binding. A wide range of physical properties in alginate gels can be achieved with the modification of the chemical structure of the sugar residues, and the use of various covalent and noncovalent cross-linking molecules. The interaction of cells with alginate can be regulated by conjugation of cell adhesion molecules to the alginate chains. This chapter will review approaches developed in our laboratories to modify the physical and biological properties of alginate hydrogels, in order to tailor gels for use in specific tissue engineering applications. Applications involving regulation of skeletal muscle cell phenotype, cartilage and bone regeneration, and drug delivery will be discussed.

II. ALGINATE MODIFICATION

A. Control of Mechanical Properties

The major motivation for controlling the mechanical properties of alginate hydrogels is to meet the physical property requirements of the desired engineered tissue at a specific site in the body. The mechanical properties of alginate are generally attributed to the concentration of G-blocks and their block length in the polymer because ionic cross-linking occurs between these blocks.[28,32] The mechanical properties of alginates can be varied in both external and internal gelation approaches by choosing the appropriate divalent ion type and concentration and the G-block lengths.[33] However, the range of mechanical properties achieved with these approaches is limited.

One approach to control and widen the range of mechanical properties exhibited by alginate hydrogels is to covalently cross-link the chains using various diamines, multiamines, or hydrazide containing cross-linking molecules. In this approach, a water soluble carbodiimide, 1-ethyl-3-(dimethylaminopropyl) carbodiimide (EDC) is commonly utilized to activate the carboxylate groups of alginate hydrogels, followed by addition of a bifunctional cross-linker to covalently cross-link these groups.[34] The methyl ester of L-lysine, poly (ethylene glycol)-diamines, and adipic dihydrazide have been used as cross-linkers.[34] The unreacted EDC and contaminants can be eliminated upon completion of the cross-linking by thorough purification. The use of carbodiimide chemistry is potentially favorable when forming preformed matrices, in contrast to *in situ* gel formation in the presence of cells. In the former, different molds can be fabricated to form gels into the desired size and shape, and the gels can then be frozen, lyophilized, and stored until actual application.[35]

Another approach to varying the mechanical properties of alginate is the partial oxidation of alginate to generate aldehydes, which can subsequently be used as covalent cross-linking sites (Figure 21.1). Sodium periodate has been utilized to generate aldehyde groups on alginate, and the oxidized alginates are then cross-linked with a functional group that is reactive in aqueous solutions.[36] One such group is the hydrazide group, which reacts with aldehyde groups to form hydrazone bonds. A suitable bifunctional cross-linker is the water soluble adipic dihydrazide molecule. The partial oxidation of alginates can also be used to precisely regulate the stiffness of the polyguluronate backbone, to further control the gel physical properties.[37] The mechanical and swelling properties of covalently cross-linked hydrogels formed with both the carbodiimide

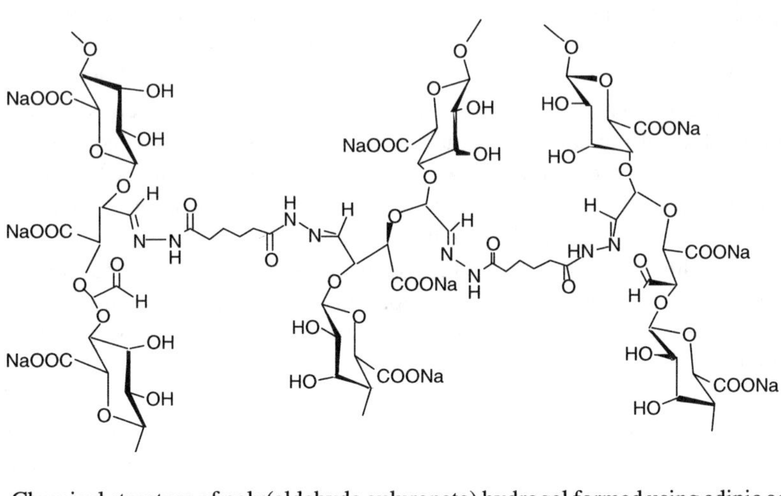

FIGURE 21.1 Chemical structure of poly(aldehyde guluronate) hydrogel formed using adipic acid dihydrazide as the cross-linker. (From Lee, K. Y., Bouhadir, K. H., and Mooney, D. J., Degradation behavior of covalently cross-linked poly(aldehyde guluronate) hydrogels, *Macromolecules*, 33(1), 97–101, 2000. With permission.)

chemistry and oxidation of alginate can be controlled by the concentration of the polymer and the type and density of cross-linker.[38]

To improve long-term stability of alginate hydrogels, the mechanical properties of alginate hydrogels can be increased by forming ionic complexes with polycations. In this approach, alginate hydrogels are frequently modified with an external coating of a cationic polymer such as poly(L-lysine), poly(ethyleneimine), or poly(allylamine).[39,40] An ionic complex is formed with the ammonium groups in these polymers and the carboxylate groups in alginate, enhancing the resistance of gels to ion exchange, thus increasing the stability of alginate hydrogels and decreasing their permeability.[40–43] This type of alginate hydrogels is typically utilized for applications involving drug delivery and immunoisolated cell transplantation.[44]

B. Controlling Degradability

The ability of alginate hydrogels to degrade over time and break down into low molecular weight oligomers that can be cleared from the body are important requirements for many tissue engineering applications. Commercially available alginates typically have high molecular weights, and mammalian cells do not possess enzymes to hydrolyze the β-glycoside bonds in alginate chains.[45] Gels formed from these alginates, even if they dissolve *in vivo*, will not lead to lower molecular weight species that can be readily cleared the body. Ionically cross-linked hydrogels typically dissolve via an ion exchange process involving the loss of divalent ions. This process is generally uncontrollable and unpredictable. Alginate hydrogels of lower molecular weight chains may dissolve more quickly due to more rapid loss of divalent cations. It is reasonable to break down alginate into low molecular weight polymers, usually smaller than the size that can be cleared by the kidney, prior to hydrogel formation to achieve more controllable properties. The generation of low molecular weight alginates can be achieved through several approaches involving the scission of high molecular weight alginate chains. Specific approaches include acid hydrolysis,[46] enzymatic degradation,[47] and irradiation.[48] Formation of hydrogels from low molecular weight alginate chains leads to gels with controllable degradability, and the breakdown products can be cleared from the body. For example, to form alginate hydrogels with controllable degradability while maintaining their gelation ability, alginates have been partially oxidized with sodium periodate to generate aldehyde groups, and then covalently cross-linked with adipic dihydrazide. The cross-linking bonds are susceptible to hydrolytic cleavage in aqueous conditions, and the degradation rate can be controlled by the percentage of oxidization and cross-linking density.[49,50] Following the dissolution of the hydrogel, low molecular weight sugar chains are released. The mechanical properties and degradation time can be decoupled by utilizing partially bound cross-linking molecules that are capable of reversibly cross-linking the polymer to form the hydrogel. In addition, alginates that are slightly oxidized were found to be susceptible to hydrolytic scission in aqueous media, depending on the pH and temperature of the solution.[35] Furthermore, porous alginate beads with interconnected pores have been formed by cross-linking low molecular weight alginate with calcium ions.[51] The porous beads were implanted subcutaneously and allowed cell infiltration *in vivo*. One simple and practical approach to breaking down alginate into low molecular weight chains involves exposure to gamma irradiation, which does not result in deleterious side products.[52] Recently, the hydrogels formed from these alginates were successfully applied in the engineering of bone tissue.[53]

Finally, the use of bimodal molecular weight distributions of alginates (i.e., a mixture of high molecular weight and irradiated low molecular weight alginates) to form gels has been investigated as a means to control independently the viscosity of the pregel solution, the mechanical properties of gels, and their subsequent degradation.[54] This approach results in a continuous increase in the elastic moduli of the gels with total solids concentration, and this can be achieved with minimal increases in the viscosity of the pregelled solution. Alginate hydrogels with a bimodal molecular weight distribution may have comparable mechanical properties to unary high molecular weight alginate gels, while the degradation rate can be adjusted by the ratio of high molecular weight alginate to low molecular weight alginate. The combination of a low degree of partial oxidization and bimodal molecular weight distribution may offer an even wider degree of control over gel degradation.

C. Control of Cell Interactions

Appropriate cell-scaffold interactions are essential for developing suitable tissue engineering scaffolds, as cellular processes such as proliferation, migration, and differentiation are mediated by cell adhesion.[55] Alginate lacks specific adhesion sites for mammalian cells. However, modification of alginates with cell adhesion peptide sequences allows precise control over cellular adhesion by varying the peptide sequence, distribution, and density.[56,57] One of the most commonly used cell adhesion peptides is arginine–glycine–aspartic acid (RGD), a sequence common to many proteins present in the extracellular matrix. Gels formed from alginate containing covalently coupled RGD peptides demonstrated improved cell adhesion, proliferation, and differentiation for a variety of cell types.[58,59] The cell adhesion ligands are coupled to the alginate backbone using EDC, a water soluble carbodiimide, to activate the carboxylate groups in the sugar residues. The amine terminal of the peptide then reacts with the activated carboxylate groups to form amide bonds (Figure 21.2). Cell adhesion ligands have also been incorporated into oxidized alginate via reductive amination chemistry.[49] In this situation, the amine group of the peptide reacts with the pendant aldehyde group of the oxidized alginate to form an imine bond. The imine bone is selectively reduced with sodium cyanoborohydride to form an amine bond, while maintaining the reactivity of the unreacted aldehyde groups for further cross-linking with adipic dihydrazide. Carbohydrates such as galactose derivatives have also been used to confer cellular interaction properties to alginate, by covalent grafting these molecules to alginate hydrogels. Adhesion and spheroid formation by hepatocytes were improved when encapsulated within galactosylated alginate hydrogels.[60]

One concern with the encapsulation of cells within alginate hydrogels is that the high shear stress exerted upon cells during the mixing of the high viscosity pregelled solutions may decrease cell viability. It may be desirable in many situations to decrease the viscosity of the pregelled solution to avoid the need for high shear mixing. One approach to decreasing the shear forces required for the mixing is to decrease the molecular weight of the polymer used to form the gel, while maintaining its gel forming ability. Decreasing the molecular weight of alginate using irradiation has been demonstrated to decrease the viscosity of pregelled solutions, while maintaining a high elastic moduli of the gels following cross-linking.[61] Immobilization of cells with these polymers greatly increases the cell viability, as compared to using standard high molecular weight polymer chains to form the gels. Furthermore, the solids concentration of the gels formed with the low molecular weight alginate could be increased to further enhance the moduli of gels without significantly deteriorating the viability of immobilized cells, probably due to the limited increase in the viscosity of these former solutions.

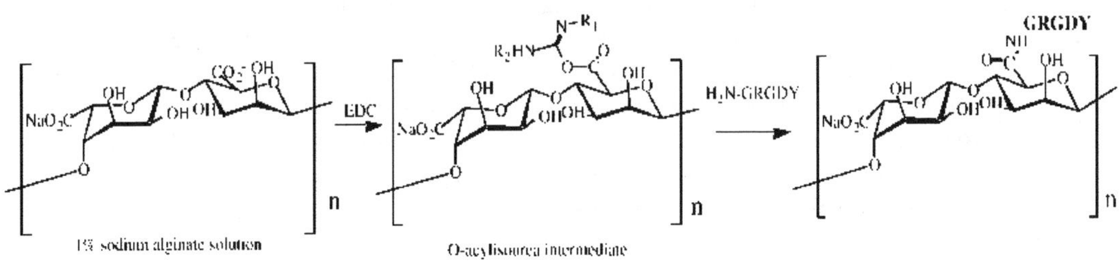

FIGURE 21.2 Reaction scheme for coupling cell adhesion peptides to alginate. The cell adhesion ligands are coupled to the alginate backbone using EDC to activate the carboxylate groups in the alginate chain. The amine terminal of the GRGDY pentapeptide then reacts with the activated carboxylate groups to form amide bonds. (From Rowley, J. A., Madlambayan, G., and Mooney, D. J., Alginate hydrogels as synthetic extracellular matrix materials, *Biomaterials*, 20(1), 45–53, 1999. With permission.)

III. APPLICATION OF MODIFIED ALGINATES IN TISSUE ENGINEERING

A. Regulation of Skeletal Muscle Cell Phenotype

The utility of cell adhesion, peptide modified, alginate hydrogels in regulating cell phenotype has been demonstrated with skeletal myoblasts. Mouse skeletal myoblasts seeded onto the surface of peptide modified hydrogels adhered, while cells did not attach in appreciable numbers to nonmodified alginate.[59] Myoblast adhesion and spreading on these RGD modified hydrogels was inhibited with soluble peptide added to the seeding medium, illustrating the specificity of the adhesion. The subsequent proliferation and differentiation of the myoblasts could be regulated by varying the alginate monomeric ratio and the type and density of adhesion ligands at the substrate surface. Specific combinations of alginate type and RGD density were required to obtain efficient myoblast differentiation on these materials.[62] Alginate hydrogels can also be modified to spatially regulate the function of skeletal muscle cells interacting with the polymers by incorporating two distinct cell fate directing signals (proliferation vs. differentiation) into the hydrogels.[63] In certain tissue engineering applications, subsets of cells with distinct phenotypes may be needed to precisely control tissue formation. For example, proliferation may be desirable if a larger cell population is needed to grow the tissue, prior to terminal cell differentiation and loss of growth capability. RGD sequences provide the cell adhesion functionality of human fibronectin, which is associated with the proliferative phase of myoblasts.[64] Calcium is an important first and second messenger signaling molecule associated with differentiation in many cell types, including skeletal muscle.[65] To manipulate calcium signaling, alginates of different chemistries were utilized. High mannuronic (M) acid alginates bind calcium less strongly than the high guluronic (G) acid alginates,[28] leading to its more rapid release from high M gels. To determine if alginate hydrogels could be designed to spatially regulate cell fate, RGD sequences were immobilized onto the alginates and the concentration of calcium released from the gels was controlled by the M/G ratio. Myoblast maintained the capability to adhere and multiply on both high M and G alginate hydrogels. However, myoblast fusion was greatly enhanced on the high G alginates when compared to high M alginates, most probably due to the decreased free calcium concentration (Figure 21.3). Strikingly, proliferation was observed in the cells directly adhering to the gels, while differentiation of cells into myoblasts was only observed in cells removed from the surface. Thus, two cell populations with distinct phenotypes could be simultaneously present.

B. Regeneration of Cartilage and Bone

To date, a major application of alginate hydrogels involves cartilage and bone regeneration. Alginate hydrogels modified with RGD peptides have been demonstrated to promote osteoblast adhesion and spreading, whereas minimal cell adhesion was observed on unmodified hydrogels. Furthermore, the modified alginate hydrogels led to increased osteoblast differentiation, and resulted in mature ectopic bone formation *in vivo*[66] (Figure 21.4). The quality of new bone formed by transplanted osteoblasts was greater when RGD ligands were coupled to the alginate. A remarkable observation was that cotransplantation of chondrocytes and osteoblasts on modified alginate hydrogels led to the formation of large amounts of new bone that expanded in size with time. Furthermore, the osteoblasts and chondrocytes organized themselves into structures that morphologically and functionally resembled growth plates[58] (Figure 21.5).

An important issue in this approach to tissue engineering is the appropriate degradation rate for the biomaterials. Ideally, it will degrade rapidly enough to provide continuous additional space for transplanted cells to deposit new extracellular matrix and form new tissue. Biodegradable alginate hydrogels have been developed to address this issue, and it has been shown that controlling the degradability of modified alginate hydrogels can affect the extent of bone formation. For example, the degradation time of gels formed of poly(aldehyde guluronate) (PAG), obtained from the acid hydrolysis and subsequent oxidation of alginate, has been modified by altering the extent of

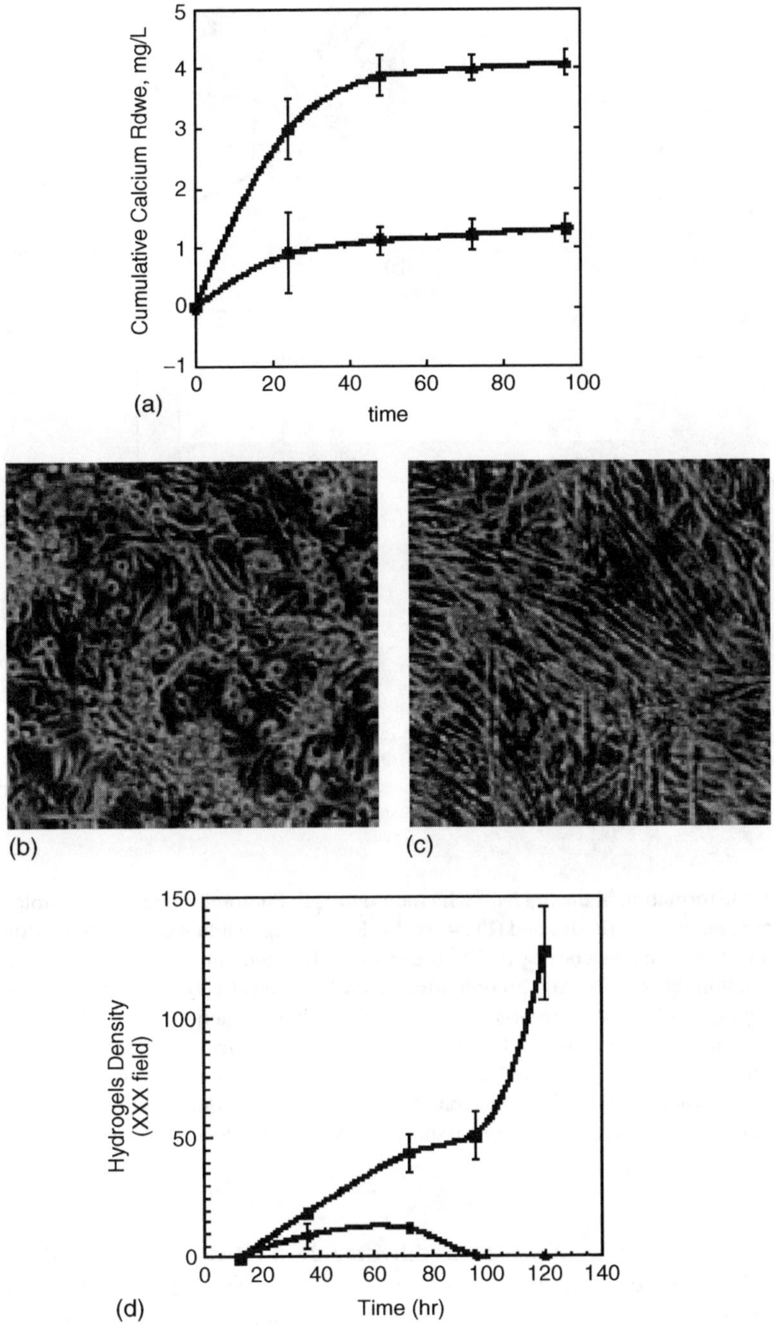

FIGURE 21.3 Hydrogels formed from high M alginate (▲) release calcium into the surrounding medium more rapidly than alginates composed of high G (●) content (a). Myoblast fusion observed on high M gels (b) was less significant when compared to that on high G gels (c) after 5 days. Quantification of the myotube density (d) on high M gels (●) and high G gels (■) corroborates this observation. Original magnification was 10 × (b, c). (From Rowley, J. A., Sun, Z. X., Goldman, D., and Mooney, D. J., Biomaterials to spatially regulate cell fate, *Adv. Mater.*, 14(12), 886–889, 2002. With permission.)

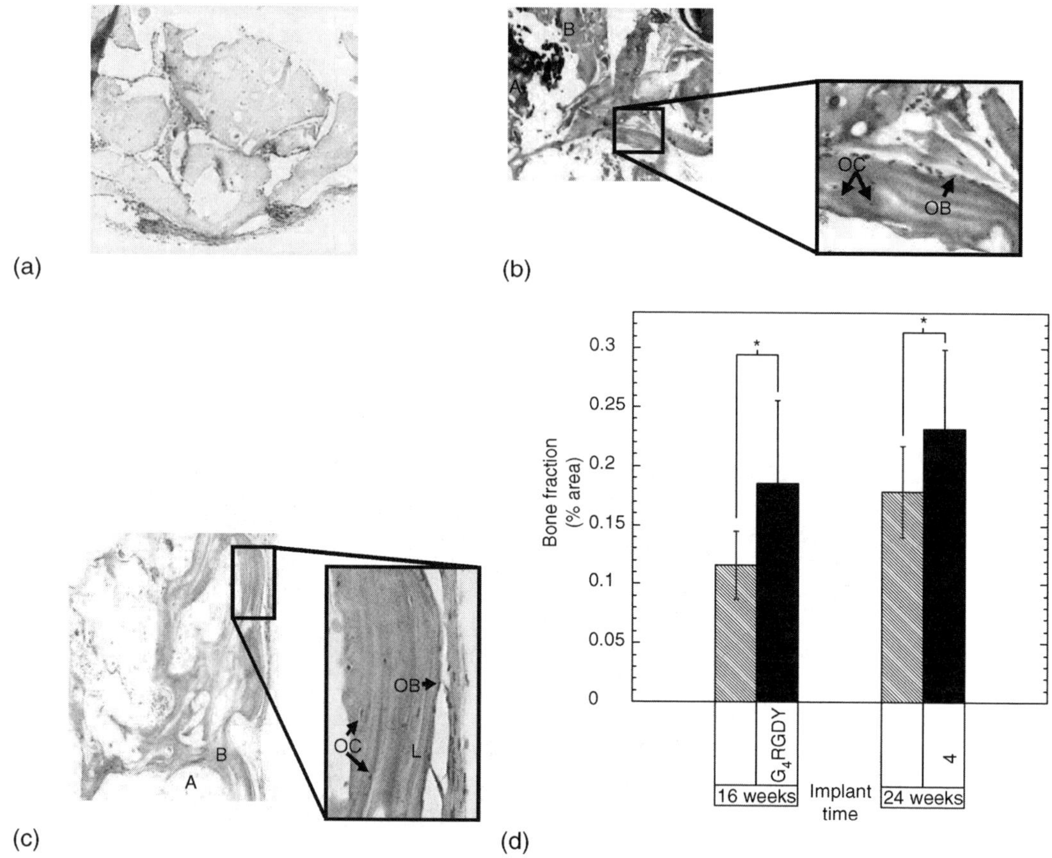

FIGURE 21.4 Bone formation in the *in vivo* SCID mouse model. Photomicrographs of histological sections of *in vivo* bone formation at (a) 4, (b) 16, and (c) 24 weeks. Remaining residual G_4RGDY-modified alginate (A), newly formed bone tissue (B), osteocytes (OC), osteoblasts (OB), and lamellae within trabecular bone (L) are displayed in the photomicrographs. All photomicrographs were taken at 100 × magnification or 400 × (higher magnification regions). (d) Histomorphometric analysis of bone fraction of cell/G_4RGDY-alginate and cell/unmodified alginate constructs at 6 and 24 weeks. The modified alginate hydrogels led to increased ectopic bone formation relative to the unmodified alginate hydrogels. (*) indicates statistical significance between conditions, $p < .05$. (From Alsberg, E., Anderson, K. W., Albeiruti, A.,Franceschi, R. T., and Mooney, D. J., Cell-interactive alginate hydrogels for bone tissue engineering, *J. Dent. Res.*, 80(11), 2025–2029, 2001. With permission.)

covalent cross-linking with adipic acid dihydrazide.[67] PAG cross-linked with a higher concentration of adipic acid dihydrazide formed hydrogels that degraded more slowly *in vitro*, and promoted more bone formation by transplanted osteoblasts *in vivo* than gels with a lower concentration of adipic acid dihydrazide. The latter gels degraded within a few days, and this time frame was probably insufficient to allow new tissue formation. Recently, γ-irradiation has also been utilized to produce alginate hydrogels with different degradation rates.[53] The quality of bone engineered using degradable alginate hydrogels was both structurally and mechanically superior to that using nondegradable alginate hydrogels. Importantly, the irradiated alginate constructs led to formation of bone throughout the bulk of the implants, and may provide an appropriate approach for dictating the degradation rate to promote optimal bone formation. Alginates containing coupled morphogenic protein-2 (BMP-2) derived oligopeptides have also been used to promote bone formation.[68]

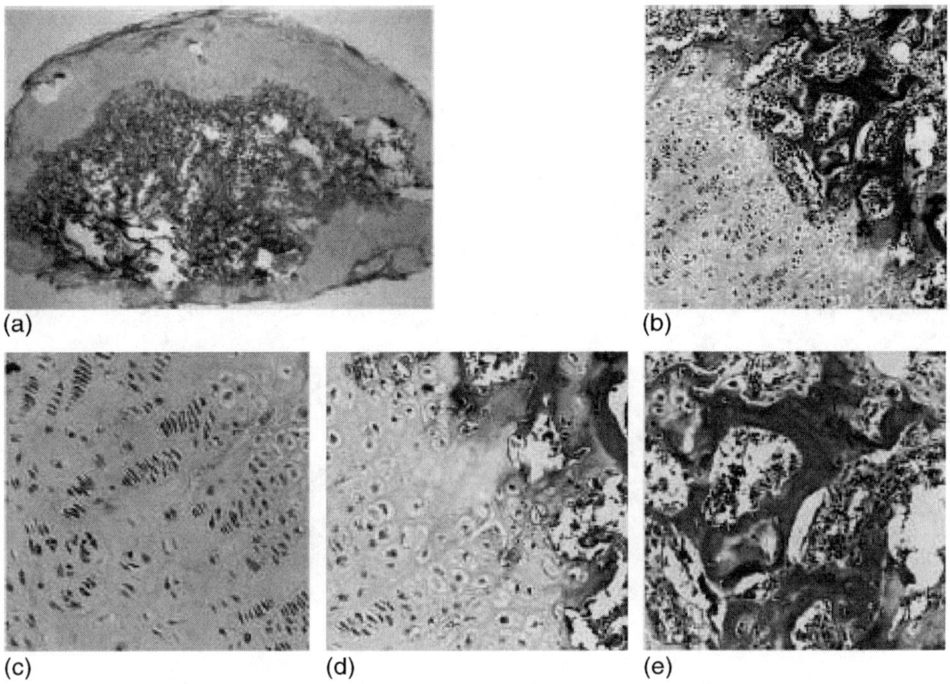

FIGURE 21.5 Cotransplantation of osteoblasts and chondrocytes via RGD-coupled alginate hydrogels resulted in the formation of growth plate-like structures *in vivo*. Growth plate-like structures can be observed at low magnification (a) (×20). Interface comprising of cartilage and bone tissue regions formed with cotransplantation displayed a structure similar to that seen in developing long bones (b) (×100). Higher magnification of the cartilage (c), transition (d), and bone and marrow space (e) regions exhibited cellular and tissue organization seen in growth plate (×200). (From Alsberg, E., Anderson, K. W., Albeiruti, A., Rowley, J. A., and Mooney, D. J., Engineering growing tissues, *Proc. Natl Acad. Sci.* USA, 99(19), 12025–12030, 2002. With permission.)

C. Growth Factor Delivery

Hydrogels formed from various alginate polymers, which have the capability for controlled release of growth factors, may find great utility in tissue engineering. The localized release of growth factors from these alginate hydrogels provides multiple advantages over the traditional delivery methods for proteins, which mainly consist of systematic delivery and bolus injection of protein solutions. These conventional delivery methods suffer from disadvantages, such as the short time frame during which the growth factor is present at the desired site, and repeated administration is often needed to overcome this problem. Modified alginates with controlled properties can potentially deliver and modulate the local concentration of growth factors for prolonged periods of time, and minimize the number of required doses. The applicability of these materials for growth factor delivery has been demonstrated in the context of angiogenesis, which involves the formation of new blood vessels from pre-existing blood vessels. The development of an appropriate vascular network is essential for engineering a variety tissue types, especially those with larger volumes, as oxygen, nutrient, and metabolite exchange limit survival of cells. Angiogenesis is generally driven by several growth factors, with vascular endothelial growth factor (VEGF) being extensively investigated.[69] The incorporation of VEGF in alginate beads not only led to sustained release of the factor, but also maintained its biological activity. Interestingly, the released VEGF was even more potent than the same mass of VEGF directly added to cell cultures.[70] Another important angiogenic factor, basic fibroblast growth factor (bFGF), has also been delivered via heparin–alginate gels in therapeutic angiogenesis applications.[71–73] Recently, alginate hydrogels have been utilized to compare the role of VEGF and bFGF in angiogenesis in SCID mice. Both factors demonstrated utility in promoting angiogenesis as expected from previous studies, but VEGF delivery led to a more dense neovasculature in this system.[74] Alginate delivery systems have also been utilized to investigate the role of the local mechanical environment in regulating growth factor release from polymer systems. Strikingly, local stresses and strains modulate the release of VEGF, and regulated the extent of vessel formation[75] (Figure 21.6). A critical feature in this effect is the reversible binding of the growth factor to the polymer scaffold.[76]

Alginate hydrogels also have potential utility for the treatment of cancer, as sequential or simultaneous localized delivery of multiple antineoplastic agents can be achieved with appropriately modified alginate. Localized delivery of these agents may be safer and more

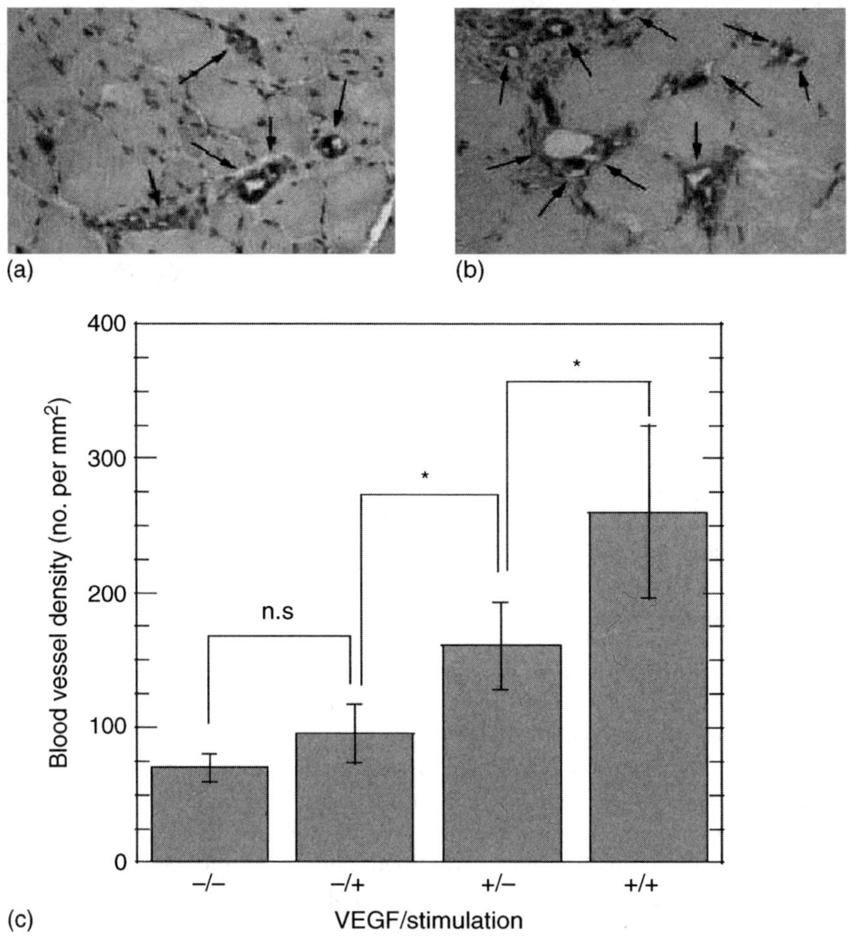

FIGURE 21.6 Local mechanical stimulation modulated the release of VEGF, and regulated the extent of vessel formation *in vivo*. VEGF incorporated in hydrogels with no mechanical stimulation (+/−) (a) exhibited a lower degree of blood vessel density when compared to (b) VEGF incorporated in hydrogels subjected to mechanical stimulation (+/+). (c) Quantification of blood vessel density revealed that (a) was greatly increased with the delivery of hydrogels encapsulating VEGF that were subjected to cyclic mechanical stimulation (+/+) when compared to other conditions. Gels without VEGF and no stimulation (−/−), or under mechanical stimulation (−/+) were used as controls. Original magnification 400 ×. Arrows indicate CD31 stained blood vessels. n.s. = no statistical difference. (∗) indicates statistical significance, $p < .05$. (From Lee, K. Y., Peters, M. C., Anderson, K. W., and Mooney, D. J., Controlled growth factor release from synthetic extracellular matrices, *Nature*, 408(6815), 998–1000, 2000. With permission.)

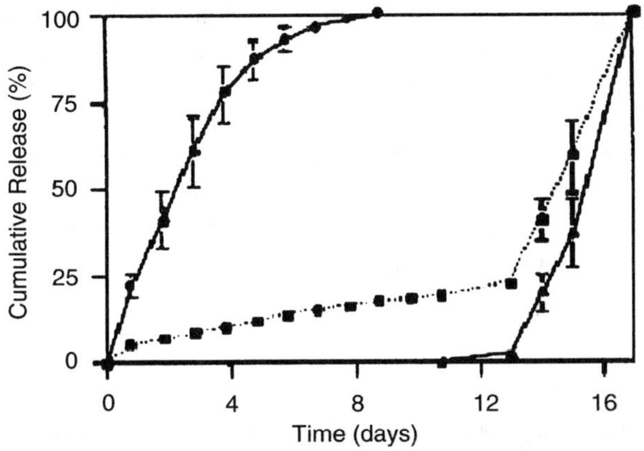

FIGURE 21.7 The release profile of (●) methotrexate, (■) doxorubicin, and (▲) mitoxantrone from covalently cross-linked oxidized alginate hydrogels incorporating all three drugs. (From Bouhadir, K. H., Alsberg, E., and Mooney, D. J., Hydrogels for combination delivery of antineoplastic agents, *Biomaterials*, 22(19), 2625–2633, 2001. With permission.)

effective, while bypassing the side effects associated with the systemic delivery of anticancer drugs.[77–80] Distinct release kinetics can be achieved with this system by exploiting different mechanisms of drug-alginate interactions. Drugs may be incorporated within the pores of the hydrogel and released by diffusion. In contrast, drugs may also be covalently attached to the hydrogel backbone by a hydrolytically labile linker and released following the chemical hydrolysis of the linker. Finally, drugs may be ionically complexed to the hydrogel and released after the dissociation of this complex (Figure 21.7).[81] Various combinations of antineoplastic agents could be delivered, with their release dependent on their chemical structure and alginate modification.

IV. CONCLUSION AND FUTURE PERSPECTIVES

Alginate has been modified in a variety of ways to enhance the physical and biological features of the gels formed from these materials, and these gels have been successfully utilized in various tissue engineering applications. Covalent cross-linking has been used to broaden the range of mechanical properties available in alginate gels. The degradability of these alginate hydrogels can be controlled to meet the specific needs of a particular application, and the gels can be designed to release only low molecular weight chains that can be cleared from the body. Moreover, alginate can be modified with cell adhesion ligands to promote the adhesion, proliferation, and differentiation of a variety cell types used to engineer tissues. The performances of these gels (e.g., growth factor release) can further be designed to respond to the external environment.

The potential of modified alginate hydrogels as scaffold and delivery systems can be further explored in various applications, and additional features can be engineered into these systems to make them respond dynamically to environmental cues. For example, hydrogels capable of responding to pH and temperature have been used for growth factor release.[82,83] Hydrogels that exhibit shape memory properties when rehydrated have been utilized for cartilage formation *in vivo*.[84] In the future, controlling the presentation of adhesion ligands by mechanical stimulation may allow one to control cell phenotype when engineering tissues in a mechanically dynamic environment. The pattern of adhesion ligands and degradability of the hydrogel may also be potentially coupled to guide tissue formation while maintaining mechanical strength. Recently, modified alginate has even been employed to develop cell cross-linkable hydrogels.[85] Gel formation in this system may be regulated by the concentration and molecular weight of polymer chains, density of adhesion ligands, and affinity of ligands for specific cell receptors. In the future, physically, chemically and biologically modified alginate hydrogels may be used for the simultaneous delivery of growth factors and cells, in order to obtain desired tissue formation in an optimal manner. For example, the differentiation of bone marrow stromal cells down specific lineages (e.g., osteogenic) may be controlled by the type and density of adhesion ligands in combination with the delivery of growth factors utilizing irradiated alginate hydrogels. The versatility of modified alginate hydrogels enhances their ability to promote the engineering and regeneration of numerous tissues, and future delineation of signals regulating these processes will allow these signals to be integrated into alginate hydrogels.

REFERENCES

1. Griffith, L. G., and Naughton, G., Tissue engineering — current challenges and expanding opportunities, *Science*, 295(5557), 1009–1014, 2002.
2. Langer, R., and Vacanti, J. P., Tissue engineering, *Science*, 260(5110), 920–926, 1993.
3. Putnam, A. J., and Mooney, D. J., Tissue engineering using synthetic extracellular matrices, *Nat. Med.*, 2(7), 824–826, 1996.
4. Vacanti, J. P., Langer, R., Upton, J., and Marler, J. J., Transplantation of cells in matrices for tissue regeneration, *Adv. Drug Deliv. Rev.*, 33(1–2), 165–182, 1998.
5. Wong, W. H., and Mooney, D., Synthesis and properties of biodegradable polymers used as synthetic matrices for tissue engineering, In *Synthetic Biodegradable Polymer Scaffolds*, Atala, A. and Mooney, D., eds., Birkhauser, Boston, MA, pp. 51–82, 1997.
6. Jhon, M. S., and Andrade, J. D., Water and hydrogels, *J. Biomed. Mater. Res.*, 7(6), 509–522, 1973.
7. Ratner, B. D., and Hoffman, A. S., In *Hydrogels for Medical and Related Application*, Andrade, J. D., ed., American Chemical Society, Washington, DC, p. 1, 1976.
8. Lee, K. Y., and Mooney, D. J., Hydrogels for tissue engineering, *Chem. Rev.*, 101(7), 1869–1879, 2001.
9. Suh, J. K., and Matthew, H. W., Application of chitosan-based polysaccharide biomaterials in cartilage tissue engineering: a review, *Biomaterials*, 21(24), 2589–2598, 2000.
10. Mattioli-Belmonte, M., Gigante, A., Muzzarelli, R. A., Politano, R., De Benedittis, A., Specchia, N., Buffa, A., Biagini, G., and Greco, F., *N,N*-dicarboxymethyl chitosan as delivery agent for bone morphogenetic protein in the repair of articular cartilage, *Med. Biol. Eng. Comput.*, 37(1), 130–134, 1999.
11. Park, Y. J., Lee, Y. M., Park, S. N., Sheen, S. Y., Chung, C. P., and Lee, S. J., Platelet derived growth factor releasing chitosan sponge for periodontal bone regeneration, *Biomaterials*, 21(2), 153–159, 2000.
12. Borkenhagen, M., Clemence, J. F., Sigrist, H., and Aebischer, P., Three-dimensional extracellular matrix engineering in the nervous system, *J. Biomed. Mater. Res.*, 40(3), 392–400, 1998.
13. Tabata, M., Shimoda, T., Sugihara, K., Ogomi, D., Serizawa, T., and Akashi, M., Osteoconductive and hemostatic properties of apatite formed on/in agarose gel as a bone-grafting material, *J. Biomed. Mater. Res.*, 67B(2), 680–688, 2003.
14. Choi, Y. S., Hong, S. R., Lee, Y. M., Song, K. W., Park, M. H., and Nam, Y. S., Studies on gelatin-containing artificial skin: II. Preparation and characterization of cross-linked gelatin-hyaluronate sponge, *J. Biomed. Mater. Res.*, 48(5), 631–639, 1999.
15. Bulpitt, P., and Aeschlimann, D., New strategy for chemical modification of hyaluronic acid: preparation of functionalized derivatives and their use in the formation of novel biocompatible hydrogels, *J. Biomed. Mater. Res.*, 47(2), 152–169, 1999.
16. Liu, L. S., Thompson, A. Y., Heidaran, M. A., Poser, J. W., and Spiro, R. C., An osteoconductive collagen/hyaluronate matrix for bone regeneration, *Biomaterials*, 20(12), 1097–1108, 1999.
17. Dai, W. S., and Barbari, T. A., Gel-impregnated pore membranes with mesh-size asymmetry for biohybrid artificial organs, *Biomaterials*, 21(13), 1363–1371, 2000.
18. Hassan, C. M., Stewart, J. E., and Peppas, N. A., Diffusional characteristics of freeze/thawed poly(vinyl alcohol) hydrogels: applications to protein controlled release from multilaminate devices, *Eur. J. Pharm. Biopharm.*, 49(2), 161–165, 2000.

19. Martens, P. J., Bryant, S. J., and Anseth, K. S., Tailoring the degradation of hydrogels formed from multivinyl poly(ethylene glycol) and poly(vinyl alcohol) macromers for cartilage tissue engineering, *Biomacromolecules*, 4(2), 283–292, 2003.
20. Taguchi, T., Kishida, A., and Akashi, M., Apatite formation on/in hydrogel matrices using an alternate soaking process (III): effect of physico-chemical factors on apatite formation on/in poly(vinyl alcohol) hydrogel matrices, *J. Biomater. Sci. Polym. Ed.*, 10(8), 795–804, 1999.
21. Sofia, S. J., Premnath, V. V., and Merrill, E. W., Poly(ethylene oxide) grafted to silicon surfaces: grafting density and protein adsorption, *Macromolecules*, 31(15), 5059–5070, 1998.
22. Zalipsky, S., Functionalized poly(ethylene glycol) for preparation of biologically relevant conjugates, *Bioconjug. Chem.*, 6(2), 150–165, 1995.
23. Coombes, A. G., Scholes, P. D., Davies, M. C., Illum, L., and Davis, S. S., Resorbable polymeric microspheres for drug delivery — production and simultaneous surface modification using PEO–PPO surfactants, *Biomaterials*, 15(9), 673–680, 1994.
24. Griffith, L. G., and Lopina, S., Microdistribution of substratum-bound ligands affects cell function: hepatocyte spreading on PEO-tethered galactose, *Biomaterials*, 19(11–12), 979–986, 1998.
25. Zhou, D., Alexandridis, P., and Khan, A., Self-assembly in a mixture of two poly(ethylene oxide)-*b*-poly(propylene oxide)-*b*-poly(ethylene oxide) copolymers in water, *J. Colloid Interface Sci.*, 183(2), 339–350, 1996.
26. Sakkers, R. J., Dalmeyer, R. A., de Wijn, J. R., and van Blitterswijk, C. A., Use of bone-bonding hydrogel copolymers in bone: an in vitro and in vivo study of expanding PEO–PBT copolymers in goat femora, *J. Biomed. Mater. Res.*, 49(3), 312–318, 2000.
27. Sutherland, I. W., Alginates, In *Biomaterials: Novel Materials from Biological Sources*, Byron, D., ed., Stockton Press, New York, pp. 309–331, 1991.
28. Smidsrod, O., and Skjak-Braek, G., Alginate as immobilization matrix for cells, *Trends Biotechnol.*, 8(3), 71–78, 1990.
29. Rees, D. A., Shapely polysaccharides. The eighth Colworth Medal Lecture, *Biochem. J.*, 126(2), 257–273, 1972.
30. Doyle, J. W., Roth, T. P., Smith, R. M., Li, Y. Q., and Dunn, R. M., Effects of calcium alginate on cellular wound healing processes modeled in vitro, *J. Biomed. Mater. Res.*, 32(4), 561–568, 1996.
31. Schmidt, R. J., Chung, L. Y., Andrews, A. M., Spyratou, O., and Turner, T. D., Biocompatibility of wound management products: a study of the effects of various polysaccharides on murine L929 fibroblast proliferation and macrophage respiratory burst, *J. Pharm. Pharmacol.*, 45(6), 508–513, 1993.
32. Draget, K. I., Skjak-Braek, G., and Smidsrod, O., Alginate based new materials, *Int. J. Biol. Macromol.*, 21(1–2), 47–55, 1997.
33. Moe, S., Skjak-Braek, G., Smidsrod, O., and Ichijo, H., Calcium alginate gel fibers-influence of alginate source and gel structure on fiber strength, *J. Appl. Polym. Sci.*, 51, 1771–1775, 1994.
34. Eiselt, P., Lee, K. Y., and Mooney, D. J., Rigidity of two-component hydrogels prepared from alginate and poly(ethylene glycol)-diamines, *Macromolecules*, 32(17), 5561–5566, 1999.
35. Bouhadir, K. H., Lee, K. Y., Alsberg, E., Damm, K. L., Anderson, K. W., and Mooney, D. J., Degradation of partially oxidized alginate and its potential application for tissue engineering, *Biotechnol. Progr.*, 17(5), 945–950, 2001.
36. Birnbaum, S., Pendleton, R., Larsson, P., and Mosbach, K., Covalent stabilization of alginate gels for the entrapment of living whole cells, *Biotechnol. Lett.*, 3, 393–400, 1981.
37. Lee, K. Y., Bouhadir, K. H., and Mooney, D. J., Evaluation of chain stiffness of partially oxidized polyguluronate, *Biomacromolecules*, 3(6), 1129–1134, 2002.
38. Lee, K. Y., Rowley, J. A., Eiselt, P., Moy, E. M., Bouhadir, K. H., and Mooney, D. J., Controlling mechanical and swelling properties of alginate hydrogels independently by cross-linker type and cross-linking density, *Macromolecules*, 33(11), 4291–4294, 2000.
39. Clayton, H. A., James, R. F., and London, N. J., Islet microencapsulation: a review, *Acta Diabetol.*, 30(4), 181–189, 1993.
40. Wang, F. F., Wu, C. R., and Wang, Y. J., Preparation and application of poly(vinylamine)/alginate microcapsules to culturing of a mouse erythroleukemia cell line, *Biotechnol. Bioeng.*, 40, 1115–1118, 1992.

41. Thu, B., Bruheim, P., Espevik, T., Smidsrod, O., Soon-Shiong, P., and Skjak-Braek, G., Alginate polycation microcapsules. II. Some functional properties, *Biomaterials*, 17(11), 1069–1079, 1996.
42. Thu, B., Bruheim, P., Espevik, T., Smidsrod, O., Soon-Shiong, P., and Skjak-Braek, G., Alginate polycation microcapsules. I. Interaction between alginate and polycation, *Biomaterials*, 17(10), 1031–1040, 1996.
43. Gaserod, O., Sannes, A., and Skjak-Braek, G., Microcapsules of alginate-chitosan. II. A study of capsule stability and permeability, *Biomaterials*, 20(8), 773–783, 1999.
44. Shoichet, M. S., Li, R. H., White, M. L., and Winns, S. R., Stability of hydrogels used in cell encapsulation: An in vivo comparison of alginate and agarose, *Biotechnol. Bioeng.*, 40, 374–381, 1996.
45. Al-Shamkhani, A., and Duncan, R., Radioiodination of alginate via covalently bound tyrosinamide allows for monitoring of its fate in vivo, *J. Bioact. Compat. Polym.*, 10, 4–13, 1995.
46. Bouhadir, K. H., Kruger, G. M., Lee, K. Y., and Mooney, D. J., Sustained and controlled release of daunomycin from cross-linked poly(aldehyde guluronate) hydrogels, *J. Pharm. Sci.*, 89(7), 910–919, 2000.
47. Nakada, H. I., and Sweeny, P. C., Alginic acid degradation by eliminases from abalone hepatopancreas, *J. Biol. Chem.*, 242(5), 845–851, 1967.
48. Nagasawa, N., Mitomo, H., Yoshii, F., and Tamikazu, K., Radiation-induced degradation of sodium alginate, *Polym. Degrad. Stab.*, 69, 275–285, 2000.
49. Bouhadir, K. H., Hausman, D. S., and Mooney, D. J., Synthesis of cross-linked poly(aldehyde guluronate) hydrogels, *Polymer*, 40(12), 3575–3584, 1999.
50. Lee, K. Y., Bouhadir, K. H., and Mooney, D. J., Degradation behavior of covalently cross-linked poly(aldehyde guluronate) hydrogels, *Macromolecules*, 33(1), 97–101, 2000.
51. Purwanto, Z. I., Broeck, L. A., Schols, H. A., Pilnik, W., and Voragen, A. G. J., Degradation of low molecular weight fragments of pectin and alginates by gamma-irradiation, *Acta Ailment*, 27, 29–42, 1998.
52. Luan, L. Q., Hien, N. Q., Nagasawa, N., Kume, T., Yoshii, F., and Nakanishi, T. M., Study on the biological effect of radiation-degraded alginate on flower plants in tissue culture, *Biotechnol. Appl. Biochem.*, 2003.
53. Alsberg, E., Kong, H. J., Hirano, Y., Smith, M. K., Albeiruti, A., and Mooney, D. J., Regulating bone formation via controlled scaffold degradation, *J. Dent. Res.*, 82(11), 903–908, 2003.
54. Kong, H. J., Lee, K. Y., and Mooney, D. J., Decoupling the dependence of rheological/mechanical properties of hydrogels from solids concentration, *Polymer*, 43(23), 6239–6246, 2002.
55. Lanza, R. P., Langer, R. S., and Vacanti, J. P., *Principles of Tissue Engineering*, Academic Press, San Diego, CA, p. 995, 2000.
56. Harrison, D., Johnson, R., Tucci, M., Puckett, A., Tsao, A., Hughes, J., and Benghuzzi, H., Interaction of cells with UHMWPE impregnated with the bioactive peptides RGD, RGE or Poly-L-lysine, *Biomed. Sci. Instrum.*, 34, 41–46, 1997.
57. Rezania, A., and Healy, K. E., The effect of peptide surface density on mineralization of a matrix deposited by osteogenic cells, *J. Biomed. Mater. Res.*, 52(4), 595–600, 2000.
58. Alsberg, E., Anderson, K. W., Albeiruti, A., Rowley, J. A., and Mooney, D. J., Engineering growing tissues, *Proc. Natl Acad. Sci. USA*, 99(19), 12025–12030, 2002.
59. Rowley, J. A., Madlambayan, G., and Mooney, D. J., Alginate hydrogels as synthetic extracellular matrix materials, *Biomaterials*, 20(1), 45–53, 1999.
60. Yang, J., Goto, M., Ise, H., Cho, C. S., and Akaike, T., Galactosylated alginate as a scaffold for hepatocytes entrapment, *Biomaterials*, 23(2), 471–479, 2002.
61. Kong, H. J., Smith, M. K., and Mooney, D. J., Designing alginate hydrogels to maintain viability of immobilized cells, *Biomaterials*, 24(22), 4023–4029, 2003.
62. Rowley, J. A., and Mooney, D. J., Alginate type and RGD density control myoblast phenotype, *J. Biomed. Mater. Res.*, 60(2), 217–223, 2002.
63. Rowley, J. A., Sun, Z. X., Goldman, D., and Mooney, D. J., Biomaterials to spatially regulate cell fate, *Adv. Mater.*, 14(12), 886–889, 2002.
64. Gullberg, D., and Ekblom, P., Extracellular matrix and its receptors during development, *Int. J. Dev. Biol.*, 39(5), 845–854, 1995.
65. Berchtold, M. W., Brinkmeier, H., and Muntener, M., Calcium ion in skeletal muscle: its crucial role for muscle function, plasticity, and disease, *Physiol. Rev.*, 80(3), 1215–1265, 2000.

66. Alsberg, E., Anderson, K. W., Albeiruti, A., Franceschi, R. T., and Mooney, D. J., Cell-interactive alginate hydrogels for bone tissue engineering, *J. Dent. Res.*, 80(11), 2025–2029, 2001.
67. Lee, K. Y., Alsberg, E., and Mooney, D. J., Degradable and injectable poly(aldehyde guluronate) hydrogels for bone tissue engineering, *J. Biomed. Mater. Res.*, 56(2), 228–233, 2001.
68. Suzuki, Y., Tanihara, M., Suzuki, K., Saitou, A., Sufan, W., and Nishimura, Y., Alginate hydrogel linked with synthetic oligopeptide derived from BMP-2 allows ectopic osteoinduction in vivo, *J. Biomed. Mater. Res.*, 50(3), 405–409, 2000.
69. Elcin, Y. M., Dixit, V., and Gitnick, G., Extensive in vivo angiogenesis following controlled release of human vascular endothelial cell growth factor: implications for tissue engineering and wound healing, *Artif. Organs*, 25(7), 558–565, 2001.
70. Peters, M. C., Isenberg, B. C., Rowley, J. A., and Mooney, D. J., Release from alginate enhances the biological activity of vascular endothelial growth factor, *J. Biomater. Sci.-Polym. Ed.*, 9(12), 1267–1278, 1998.
71. Edelman, E. R., Mathiowitz, E., Langer, R., and Klagsbrun, M., Controlled and modulated release of basic fibroblast growth factor, *Biomaterials*, 12(7), 619–626, 1991.
72. Edelman, E. R., Nugent, M. A., Smith, L. T., and Karnovsky, M. J., Basic fibroblast growth factor enhances the coupling of intimal hyperplasia and proliferation of vasa vasorum in injured rat arteries, *J. Clin. Invest.*, 89(2), 465–473, 1992.
73. Sellke, F. W., Laham, R. J., Edelman, E. R., Pearlman, J. D., and Simons, M., Therapeutic angiogenesis with basic fibroblast growth factor: technique and early results, *Ann. Thorac. Surg.*, 65(6), 1540–1544, 1998.
74. Lee, K. Y., Peters, M. C., and Mooney, D. J., Comparison of vascular endothelial growth factor and basic fibroblast growth factor on angiogenesis in SCID mice, *J. Control. Release*, 87(1–3), 49–56, 2003.
75. Lee, K. Y., Peters, M. C., Anderson, K. W., and Mooney, D. J., Controlled growth factor release from synthetic extracellular matrices, *Nature*, 408(6815), 998–1000, 2000.
76. Lee, K. Y., Peters, M. C., and Mooney, D. J., Controlled drug delivery from polymers by mechanical signals, *Adv. Mater.*, 13(11), 837–839, 2001.
77. Andersson, M., Madsen, E. L., Overgaard, M., Rose, C., Dombernowsky, P., and Mouridsen, H. T., Doxorubicin versus methotrexate both combined with cyclophosphamide, 5-fluorouracil and tamoxifen in postmenopausal patients with advanced breast cancer — a randomised study with more than 10 years follow-up from the Danish Breast Cancer Cooperative Group. Danish Breast Cancer Cooperative Group (DBCG), *Eur. J. Cancer*, 35(1), 39–46, 1999.
78. Ghielmini, M., Zappa, F., Menafoglio, A., Caoduro, L., Pampallona, S., and Gallino, A., The high-dose sequential (Milan) chemotherapy/PBSC transplantation regimen for patients with lymphoma is not cardiotoxic, *Ann. Oncol.*, 10(5), 533–537, 1999.
79. Ikeda, M., Okada, S., Ueno, H., Okusaka, T., Tanaka, N., Kuriyama, H., and Yoshimori, M., A phase II study of sequential methotrexate and 5-fluorouracil in metastatic pancreatic cancer, *Hepatogastroenterology*, 47(33), 862–865, 2000.
80. Miller, K. D., McCaskill-Stevens, W., Sisk, J., Loesch, D. M., Monaco, F., Seshadri, R., and Sledge, G. W. Jr., Combination versus sequential doxorubicin and docetaxel as primary chemotherapy for breast cancer: a randomized pilot trial of the Hoosier Oncology Group, *J. Clin. Oncol.*, 17(10), 3033–3037, 1999.
81. Bouhadir, K. H., Alsberg, E., and Mooney, D. J., Hydrogels for combination delivery of antineoplastic agents, *Biomaterials*, 22(19), 2625–2633, 2001.
82. Markland, P., Zhang, Y., Amidon, G. L., and Yang, V. C., A pH- and ionic strength-responsive polypeptide hydrogel: synthesis, characterization, and preliminary protein release studies, *J. Biomed. Mater. Res.*, 47(4), 595–602, 1999.
83. Sahoo, S. K., De, T. K., Ghosh, P. K., and Maitra, A., pH- and thermo-sensitive hydrogel nanoparticles, *J. Colloid Interface Sci.*, 206(2), 361–368, 1998.
84. Thornton, A. J., Alsberg, E., Albertelli, M., and Mooney, D. J., Shape defining scaffolds for minimally invasive tissue engineering, *Transplantation*, 77(12), 1798–1803, 2004.
85. Lee, K. Y., Kong, H. J., Larson, R. G., and Mooney, D. J., Hydrogel formation via cell cross-linking, *Adv. Mater.*, 15, 1828–1832, 2003.

22 Polymeric Scaffolds for Gene Delivery and Regenerative Medicine

Aliasger K. Salem and Kam W. Leong

CONTENTS

I. TISSUE ENGINEERING

Tissue engineering offers the possibility to create tissues *in vitro* and replace failing or malfunctioning organs *in vivo*.[1,2] There are inherent difficulties in current organ and tissue transplantations strategies because acute donor shortages have stranded a significant number of patients on the waiting list. This list has increased from 19,095 patients in 1989, to 74,800 by February 2001 in the U.S.A. alone.[3,4] Furthermore, those patients fortunate enough to receive transplantations may require immunosuppression therapies for the rest of their lives to defend against the associated risks of rejection. The lack of donor tissue availability and donor site morbidity further hampers transplantation of tissues. Replacement with mechanical devices is limited by an increased risk of inflammation and infection. Mechanical devices also lack the mechanism for self-repair, and such devices will not grow concurrently with the patient.[4–6]

The potential impact of tissue engineering from both a therapeutic and an economic standpoint is enormous. Organ failures and diseases are increasing with human life expectation.[7,8] The success of alternative donor sources from other species like the pig still remains in doubt because of potentially transferable diseases, such as the pig endogenous retrovirus.[3] One form of tissue engineering involves the creation of tissues by transplanting cells removed from the patient or a close relative and the subsequent seeding of the cells into an implant, which serves as both a substrate and a physical support for the isolated cells. Using cells from the same genotype should avoid many of the problems associated with immune rejection of foreign tissue. As these cells are capable of proliferation, a small number of harvested cells can be expanded to a sufficient cell mass to replace the organ function. Therefore, it is not required to sacrifice the entire organ of the donor.[9]

Tissue engineering can be classified into two main areas: *in vivo* and *in vitro*. Once a tissue can be generated on a large scale *in vitro*, it can then become a viable supply of new tissue for patients. *In vitro* tissue engineering thus requires a specifically designed environment for regeneration. This is in contrast to the *in vivo* approach, where the living body provides the microenvironment with the appropriate biochemical and biomechanical stimuli for tissue regeneration. To date, most effort has been concentrated on creating tissue with singular cell types *in vitro* and thus the formation of only simple avascular structures. Therefore, *in vitro* tissue engineering, while theoretically desirable, has concentrated on tissues such as dermis, epidermis, and articular cartilage.[10,11]

In vivo tissue regeneration attempts to achieve natural regeneration of tissues and organs by harnessing the natural healing process of the body. For large defects, it is necessary to use a scaffold as support for the tissue to grow. The scaffold may be used either with or without cell seeding prior to implantation. Scaffolds without cells serve to imitate the natural extracellular matrix (ECM) of the body. Tissue regeneration is, in this case, dependent on the ingrowth from the surrounding tissue in a process that is known as tissue induction. Infiltration of progenitor or stem cells from the site of implantation into the scaffold also plays an important role. Vascularization of a scaffold is a typical example of tissue induction.[12]

Tissues engineering using cell seeded scaffolds have been applied to tissues such as liver,[13,14] blood vessels,[15,16] nerve,[17] skin,[11] cartilage,[18] and bone.[19] Significant challenges to this approach include the design and fabrication of a suitable scaffold able to promote cell adhesion and to support cell growth, proliferation, and differentiation and induced formation of natural tissue. In many cases, biocompatible, biodegradable polymers are used to either induce surrounding tissue and cell ingrowth or to serve as a temporary scaffold for transplanted cells to attach, grow, and maintain differentiated functions.[20]

The hypothesis that cells on polymer scaffolds could give rise to organized tissue originates from the following biological observations:

1. Most tissues undergo constant remodeling.[21]
2. Dissociated mature cells can reorganize themselves into their native histological structures when placed in ideal cell culture conditions.[22]
3. While isolated cell populations are capable of histological reorganization, this is limited when they are delivered as a cell suspension because they lack a template to guide restructuring.[23]
4. The quantity of tissue for implantation is restricted by diffusion requirements for gas and nutrient exchange.[22]

The strategy of tissue engineering generally involves the following steps (Figure 22.1): Depending on the target organ, a suitable cell source is identified, isolated, and produced in sufficient numbers.[3,24] A biocompatible material that can be used as a cell substrate or cell encapsulation material is isolated or synthesized and processed into the required shape.[9] The material is seeded uniformly with cells, which can then be grown in a bioreactor [25] and, finally, the material cell construct is placed into the target *in vivo* site where, depending on the site and the structure, vascularization may be necessary.[23]

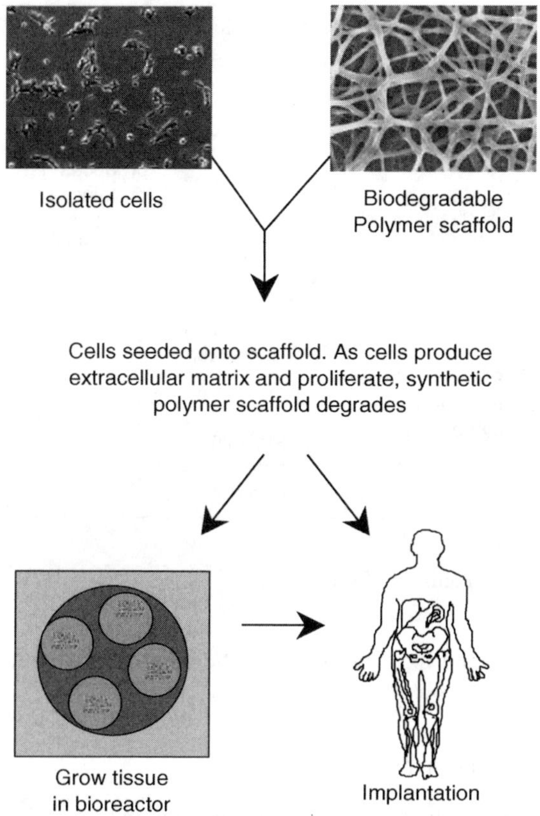

FIGURE 22.1 Schematic representation of a typical tissue engineering approach. Specific cell populations are harvested from the appropriate tissue and seeded onto a biodegradable polymer scaffold.

II. GENE DELIVERY

Gene therapy aims to treat diseases by delivering a foreign gene or DNA to the target cell or tissue for the expression of a desirable gene product, which is a protein.[26] It has been proposed for treating inherited single-order genetic disorders, cancer, cardiovascular diseases, and infectious diseases, among many others.[27] For example, the demonstration of efficient DNA delivery with emulsion coated stents on arterial walls in pig stent angioplasty studies highlights the potential for gene therapy in cardiovascular diseases.[28] Other examples include delivering DNA that encodes for normal clotting factors which could cure hemophilia,[29] delivering the p53 gene to colorectal cancer cells for enhanced responsiveness to antiangiogenic therapy,[30] or in a genetic immunization approach transferring HIV DNA plasmids to AIDS patients in order to stimulate a strong T-cell mediated immune response.[31]

Other than retroviral transfection, which can integrate the foreign gene into the genome of the host, other modes of transfection can only produce transient foreign gene expression. As such, many gene therapies are unsuitable for diseases that require prolonged expression of the gene product. Repeated doses of viral vectors cause immunologic and toxic side effects.[32] Sustained release of genetic vectors may prolong the transgene expression. However, like the delivery of other delicate bioactive agents, controlled release of bioactive genetic vectors is a significant challenge. Before describing gene delivery from tissue engineering-based scaffolds, it would be informative to discuss the mechanisms of gene delivery, introduce some of the materials commonly used in both

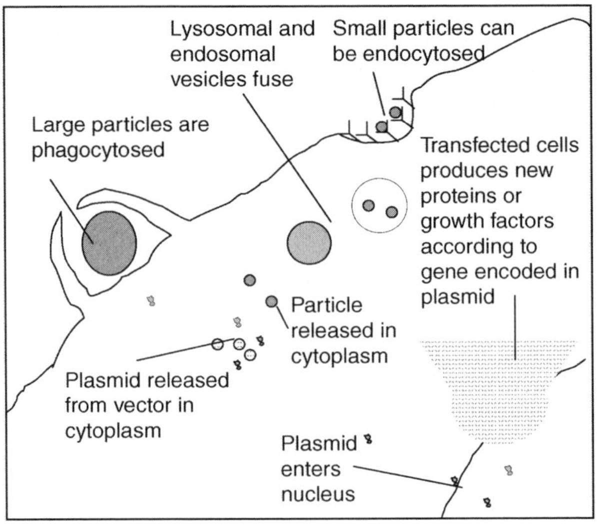

FIGURE 22.2 Schematic of mechanisms of transfection.

gene delivery and tissue engineering, and describe the ideal characteristics of a scaffold in tissue engineering applications.

A. Mechanism of Transfection

The mechanism with which cells take up DNA (as illustrated in Figure 22.2) is complicated and has been the subject of numerous reviews.[26,33–35] Typically negatively charged DNA is complexed with cationic vectors. With large particles, the route of uptake is predominantly phagocytic. If the particles are less than between 150 and 200 nm in diameter, then the complexes can be taken up by endocytosis.[36] This involves the particles being placed in pits that form endocytic vesicles by pinching away from the cell surface. These vesicles fuse with lysosomes followed by lysis. Receptor mediated endocytosis can be promoted by the use of cell specific ligands, such as transferrin, to target receptors on the surface of clathrin-coated pits that subsequently form the endosomes.[37,38] However, the low pH and harsh enzymatic environment of the endosomes and lysosomes would degrade the DNA if it is not shielded by a vector. Once escaped into the cytosol, it is expected that the DNA should be dissociated from the vector prior to nuclear entry, through pores that are typically 10 to 50 nm in diameter. However, there is evidence that the DNA/vector complex can enter the nucleus, probably through porous nuclear membranes as the cell undergoes mitosis. Although many of these steps have been identified as barriers to efficient transfection, the high efficiency with which naked DNA transfects muscle in comparison to vector-DNA conjugates/complexes highlights the immense complexities of gene delivery. We now describe some polymeric materials that have been investigated for scaffolds in tissue engineering and in gene delivery.

III. MATERIALS FOR GENE DELIVERY AND SCAFFOLDS IN TISSUE ENGINEERING

The first step in the design of a gene delivery vehicle or a scaffold for cell transplantation is the choice of a suitable material. The material must be biocompatible and preferably biodegradable to avoid the risk of complications that may be associated with the long-term presence of a foreign material in the body. Over the last century, materials such as metals, ceramics, and polymers have been extensively used for surgical implantations.[39] Metals and ceramics have contributed to major advances in the medical field, particularly in orthopedic tissue replacement.[40] However, in comparison to polymers, these materials are difficult to process and lack biodegradability. In the case of gene delivery, materials should ideally protect the DNA during transport through the cell before decomplexing, or unpackaging, to release the DNA for nuclear entry. The use of polymers can be subdivided into two categories: natural polymers and synthetic polymers.

A. Natural Derived Polymers

Proteins and sugars derived from natural extracellular matrices, such as collagen and glycosaminoglycan, have been used to repair nerve,[41] skin,[42] cartilage,[18] and bone.[43] Chemical cross-linking by glutaraldehyde has been proposed to control the stability and degradation rate of these matrices, whereas porosity has been controlled using both chemical and physical techniques.[44] However, collagenous scaffolds that have been cross-linked with glutaraldehyde can exhibit immunogenicity, calcification, and fibrous scarring during long-term implantation.[9] Chemical cross-linking in gene delivery applications could also bind DNA so strongly that decomplexation does not occur or damages the bioactivity of the DNA.[27] However, other collagen derivatives such as atelocollagen have shown great promise as gene delivery devices.[45]

Gelatin is a material that has shown potential in gene delivery applications. Our group has demonstrated the application of gelatin nanospheres for controlled gene delivery.[46–48] DNA release is controlled by the degree of cross-linking, and stimulation of receptor mediated endocytosis could be achieved by the covalent attachment of transferrin to the gelatin. This material has several advantages in gene delivery in that the gelatin nanosphere protects the DNA against degradation in serum, and can be used for codelivery of other biological agents such as chloroquine. Gelatin sponges seeded with adult mesenchymal stem cells and cultured in TGF-beta 3 supplemented media have produced cartilage-like ECM. *In vivo*, the scaffold was found to have good biocompatibility and immunological properties.[49] This suggests that gelatin has substantial potential as a gene delivery scaffold for tissue engineering.

Numerous other natural materials have found wide use in tissue engineering applications. Chitosan is a natural polysaccharide, whose structural characteristics are similar to glycosaminoglycans. Chitosan is highly soluble in an acidic environment and insoluble under neutral conditions. It has been used in a variety of biomedical applications, such as hemodialysis membranes, drug and DNA delivery systems, artificial skin, orthopedic, and dental coating materials.[50–54] Chitosan has been demonstrated to be neither toxic nor hemolytic with strong protection against nuclease degradation.[55,56] Administration of chitosan intravenously does not result in accumulation within the liver.[57] An example of its application has been with oral allergen gene immunization with chitosan-DNA nanoparticles. The nanoparticles were effective in modulating murine anaphylactic responses, indicating strong potential for prophylactic utility in treating food allergies.[58] The addition of cell targeting proteins such as transferrin to the chitosan-DNA complexes has also been shown to enhance reporter gene expression.[50] In specific cells, chitosan particles in the 50 to 100 nm size ranges have been shown to produce higher gene expression than polyethylenimine (PEI).[59] The degree of deacetylation can further be used to optimize transfection efficiency.[60] Chitosan is thus a strong candidate for effective gene delivery and tissue engineering.

Alginates are water soluble polysaccharides that have the ability to form cross-linked gels in the presence of multivalent ions, and thus potentially have a number of applications in gene delivery and tissue engineering.[61] Alginate is widely available (isolated from seaweed), readily forming a gel via calcium cross-linking, and exhibits reasonable biocompatibility. For example, alginate has been utilized in DNA vaccine-based therapies.[62] Mucosal immunization using LacZ encoding DNA entrapped within alginate was shown to produce significant immune responses. However, a disadvantage in its use in tissue engineering applications is that the calcium ions on which gelation is dependent can be lost in ionic exchange either in culture or *in vivo*.[23]

B. Synthetic Polymers

There are a wide variety of synthetic polymers that have been investigated for biomaterial and tissue engineering applications. For example, poly (vinyl alcohol) (PVA), poly (*N*-isopropyacrylamide) (PNIPAAm) and its derivatives have been shown to have great potential as delivery vehicles for cartilage and the pancreas. However, the nondegradable cross-links and the toxic cross-linking molecules used with polymers such as PNIPAAm diminish their appeal.[63–65]

The use of biodegradable materials has proven to be immensely important in medical applications over the last three decades. Polymers prepared from glycolic and lactic acid have found a multitude of uses in the medical industry, beginning with the biodegradable sutures first approved in the 1960s.[9] Since that time, diverse products based on lactic and glycolic acid and on other materials such as poly (ε-caprolactone) homopolymers and copolymers have been accepted for use as medical devices.[66,67] Poly(lactide-*co*-glycolide; PLGA) has subsequently become established as a material with significant potential for controlled release of plasmid DNA.[68,69] PLGA has also been used for controlled release of recombinant adenoviruses.[70,71] Release can be achieved over periods longer than 10 days with significantly reduced immunogenicity *in vivo*.[70,72] Sustained release of the adenovirus from the microspheres resulted in greater than 45-fold reductions in antiadenovirus titers in comparison with direct treatment of the adenovirus.

Other polymers that have been extensively investigated include polyphosphoesters,[73] polyorthoesters,[74] polyphosphazenes,[75] and other biodegradable polymers.[7] The degradation mechanism varies with each of these polymer types. For example, polyorthoesters can be surface eroding, while polyesters degrade in bulk. The degradation of polyphosphazenes can be controlled by changes in the structure of the side chain rather than the backbone as with the polyesters, while the properties of polyphosphoesters can be manipulated by adjusting either the backbone or the sidechain.[73] Polyanhydride derivatives such as copolymers of fumaric and sebacic acid (poly FA:SA 20:80) have found utility in both tissue engineering and gene delivery applications.[76] Other derivatives such as poly[α-(4 aminobutyl)-L-glycolic acid] (PAGA), a biodegradable analog of poly L lysine (PLL), have good cytotoxicity characteristics and PAGA has been demonstrated to produce significant cytokine gene expression of mRNA and proteins *in vitro* and *in vivo*.[77,78] Polyphosphoesters (PPE) and its derivatives have been utilized for biodegradable scaffolds in tissue engineering and controlled delivery of growth factors and DNA.[73] For example, PPE has been used to form nerve guide conduits and controlled release microspheres to provide prolonged site specific nerve growth factor (NGF).[79] A derivative of PPE, poly(2-aminoethyl propylene phosphate) (PPE-EA) has achieved efficient gene expression, with better tissue response than either polyethylenimine or poly-L-lysine in mouse muscle.[80]

The molecular properties of an increasing number of synthetic polymers such as molecular weight, molecular weight distribution, composition, and molecular architecture can be manipulated to modify their physicomechanical properties. Given the high processing ability of synthetic polymers, it is therefore possible to have porous materials with well-controlled microstructure and good mechanical properties.

IV. DESIGN CRITERIA FOR POLYMER SCAFFOLDS IN TISSUE ENGINEERING

There are a number of biological, physical, and chemical features desirable for the implementation of scaffolds for cell transplantation and tissue ingrowth. These are broadly split into five main categories:

1. Biocompatibility or cell interactive properties.
2. Porosity.
3. Biodegradability.

4. Mechanical compatibility.
5. Controlled release function in the scaffold.

A. Biocompatibility and Cell–Polymer Interactions

In vivo, or in culturing of cells in serum media, cells recognize synthetic materials through a complex protein layer, which forms immediately on the material upon contact with body fluids. This relation between material properties and cellular responses, mediated by the intervening protein layer, has complicated the development of biomaterials. Before the advent of biodegradable materials in the use of surgical procedures, polymer implants were intended to remain inert, thus unaffected by reactions with the surrounding tissues. As a result, cell–polymer interactions were first studied for the purpose of preventing or at least minimizing the interactions.[40] Recent thinking favors the incorporation of biologically active structures into the material to permit direct cellular interaction with the material.[81,82] These advances are based on the molecular biology of cell adhesion; specifically, the identification of small domains in the adhesion proteins of the ECM has been critical. The active peptide sequence responsible for the interaction with the cell-surface receptors can be synthesized and incorporated into the materials, either in bulk or immobilized on the surface. These approaches toward biofunctionalization provide a means to control biological interactions directly through material design.

Hubbell and coworkers have characterized the conditions under which the minimal cell binding oligopeptide sequences (such as arginyl-glycyl-aspartate (RGD), which are found in many cell adhesion proteins) is capable of supporting adhesion in cells such as the fibroblast.[83] It was observed that a surface density of 10 fmol/cm^2 of surface coupled RGD ligand would be enough to promote normal cell spreading, clustering of the cell-surface receptors, and organization of a normal cytoskeleton, corresponding to about 140 nm distance between peptide ligands.[84]

Numerous schemes have been developed for the incorporation of such adhesion ligands into both biostable and biodegradable polymer surfaces. These methods are generally based on adsorption, physical immobilization, or covalent binding. For example, Langer and coworkers have developed copolymers of polylactide and lysine to provide sites for facile grafting of such peptide-based adhesion ligands.[85–87] PLL is highly effective at condensing and delivering DNA efficiently. Therefore, the potential use of a scaffold with a PLL component for complexing DNA is also feasible. However, the cytotoxicity of PLL is a serious drawback.

B. Pore Size and Morphology

Three-dimensional porous polymer matrices possess several advantages over conventional cell culture dish, including an increased area for cell anchorage and an increased volume for cell growth, migration, and effective fluid phase transport of nutrients. In addition to the size of the pores of the scaffold, the morphology can critically determine the performance of an implanted scaffold, including the rate of tissue ingrowth. A high surface area favors cell attachment and growth, whereas a large pore volume is required to accommodate and subsequently deliver a cell mass sufficient for tissue repair.[88] This has been demonstrated by transplantation of hepatocytes for the engineering of new liver tissue.[89,90] When scaffolds are implanted *in vivo*, porous polymeric implants are often invaded by vascularized fibrous tissue. Predicting such behavior is important because ingrowth in this manner can improve the survival of cells such as the hepatocytes, but has also shown to dramatically decrease the porosity of the implant.[91,92] Implants show optimal vascular induction where pores are large enough for cell and tissue infiltration but not large enough to allow fibrous deposit.[93] Highly porous biomaterials are also desirable for the easy diffusion of waste products from the implant,[23,91] which is a major requirement for regeneration of highly metabolic tissues.

The optimum porosity varies between tissue types, from cartilage with minimal porosity constraints to liver, which requires pores with a minimum diameter of 60 μm.[9,90,94] A scaffold suitable for organ and tissue regeneration must also contribute towards the organization and direction of cell growth and ECM production. Porous matrices with well-defined networks of interconnected pores take on a significant role in this organization.[88,95] The degree of porosity also has a significant impact on the rate of degradation. In the case of scaffolds composed of polyesters, the higher porosity can reduce the accumulation of the acidic degradation products, thus diminishing the impact of autocatalytic degradation.[95] In terms of gene delivery, it has been reported that cells seeded on polyethylene terephtalate (PET) matrices with a lower porosity (ca. 87%) have higher gene expression levels than cells in matrices with a higher porosity (ca. 90%). Thus, porosity of a scaffold may also impact on transfection efficiencies.[96]

C. Biodegradability

When the ECM production is large enough to provide cells with a natural environment, the polymer scaffold degrades away at a controlled rate. Ideally, it should be completely resorbed, and natural physiological and metabolic pathways should eliminate the biodegradation products in order to avoid risks of unfavorable tissue reactions. The ideal lifetime of a biodegradable polymer scaffold depends on the application and particularly on the time required for the tissue or organ regeneration. Thus, the success of a biomaterial in tissue engineering depends largely on how its biodegradation rate can be controlled. Certain scaffolds are thought to exhibit accelerated degradation at times owing to autocatalysis, which may be a function of porosity. Poly(α-hydroxy acids) break down through a hydrolytic degradation pathway that leads to lactic acid/glycolic acid, which enter the tricarboxylic acid cycle and are eventually excreted. Thus, as these materials undergo hydrolytic breakdown, they release acidic by-products. An acidic environment can accelerate hydrolysis of these polymers.[97,98] If the implants are structurally able to allow sufficient fluid flow through the interior, then the by-products can be evacuated quickly. If, however, diffusion is restricted by a nonporous environment, then acidic by-products can accumulate within the implant, resulting in adverse reactions from the surrounding tissue.[99,100] One of the major advantages of degradation is that the scaffold can act as a controlled release device, delivering growth factors or plasmids over a sustained period of time.[101–108]

D. Scaffolds as Controlled Release Devices

A last factor that can influence the choice of polymer as biomaterial is the possibility of incorporating bioactive molecules, such as drugs, plasmids, or growth and differentiation factors.[109] Indeed, the controlled release of tissue specific growth factors from the polymer scaffold may, in some cases, considerably enhance the process of organ and tissue regeneration. Examples of growth factors include NGF for nervous tissue regeneration,[110] basic fibroblast growth factor (bFGF) for wound healing,[24] bone morphogenetic proteins (BMPs) for cartilage and bone remodeling,[111] and angiogenic growth factors for the control of vascularization.[112] In tissue-engineered devices, there are two potentially different delivery systems. Growth factors can be incorporated directly into the scaffold during or after fabrication.[108,113–116] In a biodegradable system, the growth factor would be released as the scaffold degrades to induce tissue regeneration. The growth factor, directly incorporated into a biodegradable polymer scaffold, is released by a diffusion-controlled mechanism that is regulated by the median pore size.[117] The protein can also be released by an erosion mechanism or through a combination with diffusion. Alternatively, the growth factor delivery device, in the form of microparticles, nanoparticles, or fibers, can be incorporated into the scaffold.[118] This method of delivery is also desirable because growth factors have short biological half-lives. For example, platelet-derived growth factor (PDGF) has a half-life of less than 2 min when injected intravenously.[113]

Specific growth factors released from the delivery device to influence cell migration, proliferation and differentiation, or to improve engraftment of seeded cells can lead to more efficient tissue regeneration.[113] There should be no interference of the growth factor delivery device and the tissue engineered device. The two components should function synergistically. Growth factors released from a device may interact with matrix proteins in the scaffold or in the surrounding tissue to enhance their local bioavailability or provide increased stability.[109,119] A novel approach to the use of growth factors is to immobilize them onto the surface of scaffolds, which, in conjunction with adhesive peptides, could mimic membrane anchored growth factors such as heparin-binding epidermal growth factor.[120]

V. GENE DELIVERY FROM SCAFFOLDS

As discussed above, many current tissue regeneration therapies are based on the controlled release of proteins and growth factors from scaffolds that promote tissue formation. However, a disadvantage of these systems is the decreased protein stability within the delivery system.[121,122] The fabrication process to encapsulate growth factors can damage the bioactivity. The harmful factors include sonication, organic solvents, high temperatures, and high concentration of surfactants.[27] Such conditions can promote therapeutic protein degradation, decreased potency, and increased risk of immune toxicity. As a result, maintaining bioactivity of agents such as growth factors or recombinant cytokines has been difficult. A high dose is often required to maintain the protein in the therapeutic range. High doses, however, tend to cause systemic toxicity.

A critical direction in transplantation therapies has been the genetic modification of cells to prevent rejection of allogenic and xenogenic tissues.[123] The cells can produce therapeutic proteins in a localized manner at physiological levels for prolonged periods of time, thus avoiding local toxicity. The transplantation of transfected cells on porous polymer matrices has resulted in enhanced cell survival and vascularization due to local expression of therapeutic proteins.[124] Transfected cells have also been utilized for the production of dopamine in the brain and BMP-2 growth factor in bone regeneration.[125,126] Cell transplantation, however, suffers from the need to harvest and grow the cells *in vitro*, which is typically followed by inefficient seeding and cell survival during transplantation.[122] Such genetically altered cells often rapidly lose their transgene expression and their efficacy.

Therefore, delivery of genes from scaffolds designed for tissue engineering is an attractive approach. Instead of cell transplantation, scaffolds providing controlled release of plasmids could transfect infiltrating cells during tissue induction, which could then promote healing or tissue formation (Figure 22.3). Concerning handling characteristics such as fabrication and storage, plasmid DNA can be more robust than the protein it will encode the cell to produce.[27] Delivering this DNA in a controlled release fashion ensures that the cells can provide a localized sustained expression of the bioactive protein. Furthermore, scaffold delivery of multiple genes can be

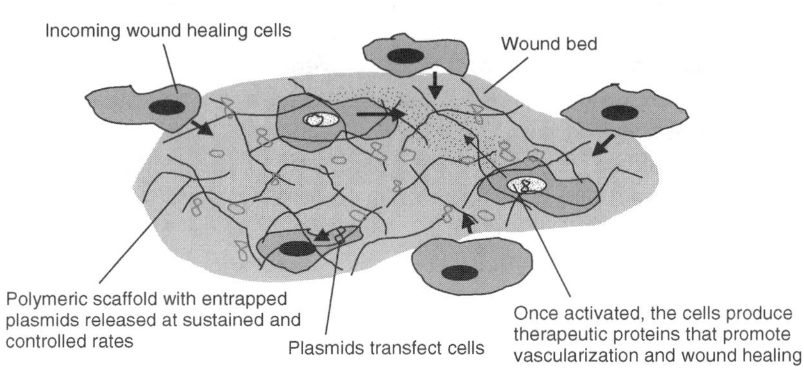

FIGURE 22.3 Schematic showing gene delivery from scaffolds approach to tissue engineering.

controlled to match the changing expression of genes that are required for optimal tissue induction or formation.[122] Additionally, plasmids that diffuse from the local site of delivery would not cause the toxicity that is associated with high doses of growth factors or proteins.[127] Finally, Xie and coworkers have shown that three-dimensional transfection may promote a higher gene expression level and longer expression time in comparison to two-dimensional transfection.[96]

Plasmid delivery from scaffolds for tissue regeneration has been demonstrated by a number of groups. Bonadio and coworkers pioneered this approach by entrapping plasmids encoding the gene for human parathyroid hormones in collagen matrices. This porous sponge was labeled "the gene activated matrix" (GAM).[127] When the GAM was placed in tibia or femur defects, dose dependent bone tissue growth was observed over a period of 6 weeks. In contrast, sham controls resulted in no bone growth. A direct correlation was found between osteoclast numbers and bone healing with DNA dose. Even the lowest quantities of plasmid resulted in some mRNA and protein expression. Mooney and coworkers used PLGA-based porous scaffolds to provide controlled release of DNA to the localized target.[122] In aqueous conditions, DNA could be released for up to 1 month with intact biological activity. *In vivo* studies using PDGF encoding plasmids in the porous scaffolds implanted subcutaneously resulted in efficient localized healing. Berry and coworkers[128] have shown that collagen scaffolds with entrapped plasmids placed within the proximal and distal ends of a severed rat optic nerve resulted in DNA delivery to the nerve cell body in the retina. An additional approach to gene delivery has been the tethering of plasmidpolylysine vectors to the surface of scaffolds. Shea and coworkers showed that transfection by surfaces presenting DNA with both HEK293 and 3T3 cells resulted in expression levels up to 100-fold greater than bulk delivery of the complexes.[129] Positively transfected cells were observed only where DNA complexes were tethered, indicating the potential for spatial control over gene transfection. This is consistent with the report by Luo and Saltzman[130] that physical concentration of DNA increases transfection efficiency. Using a similar biotin–avidin based interaction, viral vectors have also been successfully tethered to a surface for efficient gene transfection.[131]

Viral vectors have also been delivered from scaffolds. Viral vectors such as adenoviruses have been widely investigated because of the high efficiency of their transfection.[132,133] Transgene expression mediated by adenoviruses is transient. The use of polymeric scaffolds to provide a sustained release of viral vectors is attractive. For example, Siemens and coworkers tested a number of polymer matrices for the delivery of the canarypox virus to prostate cancer cells. In their study, gelatin sponges were found to be most effective for viral gene delivery both *in vivo* and *in vitro*.[134] Kalyanasundaram and coworkers[135] have shown that a combination of alginate and gelatin microspheres stabilized by calcium ions achieved a sustained release of adenovirus with a nominal loss of bioactivity. This may be a more viable combination in a scaffold delivering viral vectors.

In summary, gene delivery scaffolds for tissue engineering should meet several design criteria:

1. The surface should permit cell adhesion and growth,[82]
2. Neither the polymer nor its degradation products should provoke inflammation or toxicity when implanted *in vivo*,[136]
3. The material should be reproducibly processable into three-dimensional structures,
4. There should be controlled porosity in order to provide a high surface area for cell–polymer interactions, sufficient space for ECM regeneration, and minimal diffusional constraints during *in vitro* culture,[137]
5. The scaffold should resorb once it has served its purpose of providing a template for the regenerating tissue, since foreign materials carry a permanent risk of inflammation,[138]
6. The scaffold degradation rate should be adjustable to match the rate of tissue regeneration by the cell type of interest,[139]
7. Plasmids should be encapsulated and delivered without loss of bioactivity.[27,122]

Such scaffolds should allow attachment of isolated cells to a polymeric support structure that has suitable surface properties for guiding the reorganization and growth of cells. These should be designed so that the cells could survive by diffusion once the cell–polymer construct was implanted. Ideally, the cell–polymer construct would become vascularized in conjunction with expansion of the cell mass following implantation, and both these processes could be influenced if desired by release of plasmids that promote vascularization by encoding cells to produce therapeutic proteins.[27,122] The ability of scaffolds to conform to all these conditions is also highly dependent on the manufacturing process.

VI. FABRICATION METHODS

The conventional processing techniques to produce porous structures used in the polymer industry are unsuitable for producing tissue engineering scaffolds. Indeed, additives such as surfactants, plasticizers, stabilizers and lubricants are commonly used, which can be toxic to cells. The properties of the polymer, the components involved, and the shape of the scaffold required dictate the choice of processing technique for the manufacture of polymer scaffolds.[140]

The traditional methods for scaffold fabrication include fiber bonding, solvent casting, membrane lamination and melt molding.[141,142] Mooney et al. demonstrated that an effective method of stabilizing poly(glycolic acid; PGA) scaffolds was to spray solutions of poly(L-Lactic acid; PLLA) or PLGA in chloroform (1 to 15% w/v) over a PGA mesh using a nitrogen stream to atomize the polymer solution. Since PGA is poorly soluble in chloroform, the PGA fibers are effectively unchanged by the process.[143] Porosity in scaffolds is commonly formed by particulate leaching. This is achieved by evaporating chloroform from solutions of PLLA containing sodium chloride particles. These polymer films with entrapped salt particles are then leached in water to remove the particles resulting in highly porous films.[90] These same films can then also be formed into hollow tubes.[100] A similar system has been used by Shoichet and coworkers[97] where glucose crystals are dispersed within a PLGA solution in dimethyl sulfoxide (DMSO). Wan and coworkers have shown that coatings of various porosities can be obtained by immersing mandrels coated with a solution of PPE in chloroform into nonsolvent immersion baths, followed by freeze or vacuum drying. The porosity of the coatings decreased with an increase in polymer molecular weight, drying time before precipitation, and concentration of polymer solution. However, these methods appear to be most effective for only thin films or structures with thin walls.[144]

Novel methods of manufacturing scaffolds include a prototyping technology where a binder is expelled through a print head nozzle onto a powder bed. The scaffold is built by layers of powder and binder in a method that can create complex three-dimensional shapes. However, the highly toxic solvents prevent incorporation of biological agents such as cells, plasmids or growth factors during fabrication.[19] Wintermantel et al. have developed a novel method for scaffold formation using minimally invasive approaches. They introduced a thread-like material and delivered it through an injection canal (e.g., cannula). A fluid stream acted as a carrier for the lycra monfil™-based material, which was unreeled from a spool creating a porous scaffold in the form of a tangle.[145] Another minimally invasive technique has been demonstrated by Salem et al. in which biotinylated polymeric microparticle and cell slurries were self-assembled into porous scaffolds upon coinjection with the cross-linking protein, avidin.[146] Langer and coworkers have produced macroporous polymer foams by a hydrocarbon templating method in which a viscous polymer solution of PLA in chloroform and a particulate hydrocarbon porogen such as paraffin was compacted in a Teflon mold. The polymer/solvent/porogen phase was then extracted in a hydrocarbon solvent such as hexane, which is a nonsolvent for the polymer but miscible with the polymer solvent. This resulted in the porogen becoming extracted and rapid precipitation of a porous polymer phase.[147] Vacanti and coworkers have shown that electrospinning forms scaffolds that are suitable for bone regeneration. In this process, polymer fibers with nanometer dimensions are formed by subjecting a fluid jet to a high electric field.[148] This same process has also been used to produce PLGA and PLA–PEG based scaffolds for DNA delivery.[149]

Defined three-dimensional biodegradable foams have also been shaped by lamination of highly porous PLLA and PLGA membranes previously prepared by solvent casting and salt leaching. The membranes with the appropriate shape are solvent impregnated, then stacked up in a three-dimensional assembly with continuous pore structure. Computer assisted modeling can then help to design templates with the desired implant shape. The *in vivo* prevascularization of these laminated foams has been demonstrated by the injection of a sufficient mass of hepatocytes for liver regeneration.[91] Finally, Mooney et al. have demonstrated that the use of supercritical carbon dioxide (CO_2) is one of the promising approaches to creating polymer scaffolds.[150] This is achieved by exposing poly(α-hydroxy acid)s to CO_2 gas (5.5 MPa, 72 h). When the CO_2 gas pressure is decreased, the thermodynamic instability leads the dissolved CO_2 to nucleate and form pores within the polymer matrix. As the method avoids the use of organic solvents,[151] it is possible to then incorporate plasmids or DNA into the scaffold without loss of biological activity.[122]

VII. CONCLUSIONS

Many of the barriers in tissue engineering and the gene delivery fields can be overcome through the merging of the two disciplines. The pioneering work by Bonadio, Levy, Mooney, and many others has highlighted the potential of gene delivery in tissue engineering devices. Plasmids delivered by porous biodegradable scaffolds can fill diseased tissue or defects to endow infiltrating repair cells with therapeutic properties that can enhance the healing process. Controlled release of plasmids can provide sustained expression of the growth factors and proteins in the regenerating site. Present and future work must focus on combining the other factors that are critical in scaffold designs for tissue engineering, such as porosity and biocompatibility, with more sophisticated gene vector technology.

REFERENCES

1. Langer, R., and Vacanti, J. P., Tissue engineering, *Science*, 260, 920–926, 1993.
2. Mooney, D. J., and Mikos, A. G., Growing new organs, *Sci. Am.*, 280, 60–65, 1999.
3. Stock, U. A., and Vacanti, J. P., Tissue engineering: current state and prospects, *Ann. Rev. Med.*, 52, 443–451, 2001.
4. Vacanti, J. P., Looking back and looking ahead, *Tissue Eng.*, 7, 107–109, 2001.
5. Vacanti, C. A., and Vacanti, J. P., The science of tissue engineering, *Orthop. Clin. North Am.*, 31, 351 2000.
6. Vacanti, J. P., and Langer, R., Tissue engineering: the design and fabrication of living replacement devices for surgical reconstruction and transplantation, *Lancet*, 354, SI32–SI34, 1999.
7. Peppas, N. A., and Langer, R., New challenges in biomaterials, *Science*, 263, 1715–1720, 1994.
8. Davis, M., and Vacanti, J. P., Toward development of an implantable tissue engineered liver, *Biomaterials*, 17, 365–372, 1996.
9. Cima, L. G., Vacanti, J. P., Vacanti, C., Ingber, D., Mooney, D., and Langer, R., Tissue engineering by cell transplantation using degradable polymer substrates, *J. Biomech. Eng.*, 113, 143–149, 1991.
10. Elisseeff, J., Anseth, K., Sims, D., McIntosh, W., Randolph, M., and Langer, R., Transdermal photopolymerization for minimally invasive implantation, *Proc. Natl Acad. Sci. USA*, 96, 3104–3107, 1999.
11. Ma, L., Gao, C. Y., Mao, Z. W., Zhou, J., Shen, J. C., Hu, X. Q., and Han, C. M., Collagen/chitosan porous scaffolds with improved biostability for skin tissue engineering, *Biomaterials*, 24, 4833–4841, 2003.
12. Peters, M. C., Polverini, P. J., and Mooney, D. J., Engineering vascular networks in porous polymer matrices, *J. Biomed. Mater. Res.*, 60, 668–678, 2002.

13. Miyazawa, M., Torii, T., and Koyama, I., Liver reproduction by tissue engineering: control of hepatocyte–hepatocyte and hepatocyte-scaffold adhesion by shear stress and extracellular matrix, *Gastroenterology*, 124, A724, 2003.
14. Lu, H. F., Lim, W. S., Wang, J., Tang, Z. Q., Zhang, P. C., Leong, K. W., Chia, S. M., Yu, H., and Mao, H. Q., Galactosylated PVDF membrane promotes hepatocyte attachment and functional maintenance, *Biomaterials*, 24, 4893–4903, 2003.
15. Berglund, J. D., and Galis, Z. S., Designer blood vessels and therapeutic revascularization, *Br. J. Pharmacol.*, 140, 627–636, 2003.
16. Niklason, L. E., Gao, J., Abbott, W. M., Hirschi, K. K., Houser, S., Marini, R., and Langer, R., Functional arteries grown in vitro, *Science*, 284, 489–493, 1999.
17. Bellamkonda, R., and Aebischer, P., Review: tissue engineering in the nervous system, *Biotechnol. Bioeng.*, 43, 543–554, 1994.
18. LeBaron, R. G., and Athanasiou, K. A., Ex vivo synthesis of articular cartilage, *Biomaterials*, 21, 2575–2587, 2000.
19. Hutmacher, D. W., Scaffolds in tissue engineering bone and cartilage, *Biomaterials*, 21, 2529–2543, 2000.
20. Christenson, L., Mikos, A. G., Gibbons, D. F., and Picciolo, G. L., Biomaterials for tissue engineering: summary, *Tissue Eng.*, 3, 71–76, 1997.
21. Healy, K. E., Thomas, C. H., Rezania, A., Kim, J. E., McKeown, P. J., Lom, B., and Hockberger, P. E., Kinetics of bone cell organization and mineralization on materials with patterned surface chemistry, *Biomaterials*, 17, 195–208, 1996.
22. Marler, J. J., Upton, J., Langer, R., and Vacanti, J. P., Transplantation of cells in matrices for tissue regeneration, *Adv. Drug Deliv. Rev.*, 33, 165–182, 1998.
23. Kim, B. S., and Mooney, D. J., Development of biocompatible synthetic extracellular matrices for tissue engineering, *Trends Biotechnol.*, 16, 224–230, 1998.
24. Bonassar, L. J., and Vacanti, C. A., Tissue engineering: the first decade and beyond, *J. Cell. Biochem.*, 297–305, 1998.
25. Burg, K. J. L., Holder, W. D., Culberson, C. R., Beiler, R. J., Greene, K. G., Loebsack, A. B., Roland, W. D., Eiselt, P., Mooney, D. J., and Halberstadt, C. R., Comparative study of seeding methods for three-dimensional polymeric scaffolds, *J. Biomed. Mater. Res.*, 51, 642–649, 2000.
26. Luo, D., and Saltzman, W. M., Synthetic DNA delivery systems, *Nat. Biotechnol.*, 18, 33–37, 2000.
27. Bonadio, J., Tissue engineering via local gene delivery: update and future prospects for enhancing the technology, *Adv. Drug Deliv. Rev.*, 44, 185–194, 2000.
28. Klugherz, B. D., Jones, P. L., Cui, X. M., Chen, W. L., Meneveau, N. F., DeFelice, S., Connolly, J., Wilensky, R. L., and Levy, R. J., Gene delivery from a DNA controlled-release stent in porcine coronary arteries, *Nat. Biotechnol.*, 18, 1181–1184, 2000.
29. Connelly, S., and Kaleko, M., Gene therapy for hemophilia A, *Thromb. Haemost.*, 78, 31–36, 1997.
30. Bischoff, J. R., Kim, D. H., Williams, A., Heise, C., Horn, S., Muna, M., Ng, L., Nye, J. A., SampsonJohannes, A., Fattaey, A., and McCormick, F., An adenovirus mutant that replicates selectively in p53-deficient human tumor cells, *Science*, 274, 373–376, 1996.
31. Marasco, W. A., Intrabodies: turning the humoral immune system outside in for intracellular immunization, *Gene Ther.*, 4, 11–15, 1997.
32. Kaplan, J. M., Armentano, D., Sparer, T. E., Wynn, S. G., Peterson, P. A., Wadsworth, S. C., Couture, K. K., Pennington, S. E., StGeorge, J. A., Gooding, L. R., and Smith, A. E., Characterization of factors involved in modulating persistence of transgene expression from recombinant adenovirus in the mouse lung, *Hum. Gene Ther.*, 8, 45–56, 1997.
33. Godbey, W. T., and Mikos, A. G., Recent progress in gene delivery using non-viral transfer complexes, *J. Control. Release*, 72, 115–125, 2001.
34. Pouton, C. W., and Seymour, L. W., Key issues in non-viral gene delivery, *Adv. Drug Deliv. Rev.*, 46, 187–203, 2001.
35. Kataoka, K., and Harashima, H., Gene delivery systems: viral vs. non-viral vectors, *Adv. Drug Deliv. Rev.*, 52 2001, 151–151.
36. Conner, S. D., and Schmid, S. L., Regulated portals of entry into the cell, *Nature*, 422, 37–44, 2003.
37. Marsh, M., and McMahon, H. T., Cell biology — The structural era of endocytosis, *Science*, 285, 215–220, 1999.

38. Salem, A. K., Searson, P. C., and Leong, K. W., Multifunctional nanorods for gene delivery, *Nat. Mater.*, 2, 668–671, 2003.
39. Kohn, J., Tissue engineering: an overview, *MRS Bulletin*, 21, 18–19, 1996.
40. Agrawal, C. M., Reconstructing the human body using biomaterials, *J. Miner. Met. Mater. Soc.*, 50, 31–35, 1998.
41. Yoshii, S., Oka, M., Shima, M., Taniguchi, A., and Akagi, M., 30 mm regeneration of rat sciatic nerve along collagen filaments, *Brain Res.*, 949, 202–208, 2002.
42. Pachence, J. M., Collagen-based devices for soft tissue repair, *J. Biomed. Mater. Res.*, 33, 35–40, 1996.
43. Winn, S. R., Uludag, H., and Hollinger, J. O., Carrier systems for bone morphogenetic proteins, *Clin. Orthop. Rel. Res.*, S95–S106, 1999.
44. Cavallaro, J. F., and Kemp, P. D., Collagen fabrics as biomaterials, *Biotechnol. Bioeng.*, 43, 781–791, 1994.
45. Ochiya, T., Takahama, Y., Nagahara, S., Sumita, Y., Hisada, A., Itoh, H., Nagai, Y., and Terada, M., New delivery system for plasmid DNA in vivo using atelocollagen as a carrier material: the Minipellet, *Nat. Med.*, 5, 707–710, 1999.
46. Leong, K. W., Mao, H. Q., Truong-Le, V. L., Roy, K., Walsh, S. M., and August, J. T., DNA-polycation nanospheres as non-viral gene delivery vehicles, *J. Control. Release*, 53, 183–193, 1998.
47. Truong-Le, V. L., August, J. T., and Leong, K. W., Controlled gene delivery by DNA-gelatin nanospheres, *Hum. Gene Ther.*, 9, 1709–1717, 1998.
48. Truong-Le, V. L., Walsh, S. M., Schweibert, E., Mao, H. Q., Guggino, W. B., August, J. T., and Leong, K. W., Gene transfer by DNA-gelatin nanospheres, *Arch. Biochem. Biophys.*, 361, 47–56, 1999.
49. Ponticiello, M. S., Schinagl, R. M., Kadiyala, S., and Barry, F. P., Gelatin-based resorbable sponge as a carrier matrix for human mesenchymal stem cells in cartilage regeneration therapy, *J. Biomed. Mater. Res.*, 52, 246–255, 2000.
50. Mao, H. Q., Roy, K., Troung-Le, V. L., Janes, K. A., Lin, K. Y., Wang, Y., August, J. T., and Leong, K. W., Chitosan-DNA nanoparticles as gene carriers: synthesis, characterization and transfection efficiency, *J. Control. Release*, 70, 399–421, 2001.
51. Park, Y. J., Lee, Y. M., Park, S. N., Sheen, S. Y., Chung, C. P., and Lee, S. J., Platelet derived growth factor releasing chitosan sponge for periodontal bone regeneration, *Biomaterials*, 21, 153–159, 2000.
52. Chenite, A., Chaput, C., Wang, D., Combes, C., Buschmann, M. D., Hoemann, C. D., Leroux, J. C., Atkinson, B. L., Binette, F., and Selmani, A., Novel injectable neutral solutions of chitosan form biodegradable gels in situ, *Biomaterials*, 21, 2155–2161, 2000.
53. Gingras, M., Paradis, I., and Berthod, F., Nerve regeneration in a collagen-chitosan tissue-engineered skin transplanted on nude mice, *Biomaterials*, 24, 1653–1661, 2003.
54. Zhang, Y., Ni, M., Zhang, M. Q., and Ratner, B., Calcium phosphate–chitosan composite scaffolds for bone tissue engineering, *Tissue Eng.*, 9, 337–345, 2003.
55. Quong, D., and Neufeld, R. J., DNA protection from extracapsular nucleases, within chitosan- or poly-L-lysine-coated alginate beads, *Biotechnol. Bioeng.*, 60, 124–134, 1998.
56. Quong, D., Yeo, J. N., and Neufeld, R. J., Stability of chitosan and poly-L-lysine membranes coating DNA-alginate beads when exposed to hydrolytic enzymes, *J. Microencapsul.*, 16, 73–82, 1999.
57. Richardson, S. C. W., Kolbe, H. J. V., and Duncan, R., Potential of low molecular mass chitosan as a DNA delivery system: biocompatibility, body distribution and ability to complex and protect DNA, *Int. J. Pharm.*, 178, 231–243, 1999.
58. Roy, K., Mao, H. Q., Huang, S. K., and Leong, K. W., Oral gene delivery with chitosan-DNA nanoparticles generates immunologic protection in a murine model of peanut allergy, *Nat. Med.*, 5, 387–391, 1999.
59. Erbacher, P., Zou, S. M., Bettinger, T., Steffan, A. M., and Remy, J. S., Chitosan-based vector/DNA complexes for gene delivery: biophysical characteristics and transfection ability, *Pharm. Res.*, 15, 1332–1339, 1998.
60. Kiang, T., Wen, J., Lim, H., and Leong, K. W., The effect of the degree of deacetylation on the efficiency of gene transfection, *Biomaterials*, 25 (22), 5293–5301, 2004.
61. Kong, H. J., Smith, M. K., and Mooney, D. J., Designing alginate hydrogels to maintain viability of immobilized cells, *Biomaterials*, 24, 4023–4029, 2003.

62. Mittal, S. K., Aggarwal, N., Sailaja, G., van Olphen, A., HogenEsch, H., North, A., Hays, J., and Moffatt, S., Immunization with DNA, adenovirus or both in biodegradable alginate microspheres: effect of route of inoculation on immune response, *Vaccine*, 19, 253–263, 2000.
63. Peng, T., and Cheng, Y. L., PNIPAAm and PMAA co-grafted porous PE membranes: living radical co-grafting mechanism and multi-stimuli responsive permeability, *Polymer*, 42, 2091–2100, 2001.
64. Li, R. H., White, M., Williams, S., and Hazlett, T., Poly(vinyl alcohol) synthetic polymer foams as scaffolds for cell encapsulation, *J. Biomater. Sci.-Polym. Ed.*, 9, 239–258, 1998.
65. Lee, K. Y., and Mooney, D. J., Hydrogels for tissue engineering, *Chem. Rev.*, 101, 1869–1879, 2001.
66. Ural, E., Kesenci, K., Fambri, L., Migliaresi, C., and Piskin, E., Poly(D,L-lactide/epsilon-caprolactone)/hydroxyapatite composites, *Biomaterials*, 21, 2147–2154, 2000.
67. Yin, M., and Baker, G. L., Preparation and characterization of substituted polylactides, *Macromolecules*, 32, 7711–7718, 1999.
68. Jones, D. H., Corris, S., McDonald, S., Clegg, J. C. S., and Farrar, G. H., Poly(DL-lactide-*co*-glycolide)-encapsulated plasmid DNA elicits systemic and mucosal antibody responses to encoded protein after oral administration, *Vaccine*, 15, 814–817, 1997.
69. Wang, D. Q., Robinson, D. R., Kwon, G. S., and Samuel, J., Encapsulation of plasmid DNA in biodegradable poly(D,L-lactic-*co*-glycolic acid) microspheres as a novel approach for immunogene delivery, *J. Control. Release*, 57, 9–18, 1999.
70. Beer, S. J., Matthews, C. B., Stein, C. S., Ross, B. D., Hilfinger, J. M., and Davidson, B. L., Poly(lactic-glycolic) acid copolymer encapsulation of recombinant adenovirus reduces immunogenicity in vivo, *Gene Ther.*, 5, 740–746, 1998.
71. Beer, S. J., Hilfinger, J. M., and Davidson, B. L., Extended release of adenovirus from polymer microspheres: potential use in gene therapy for brain tumors, *Adv. Drug Deliv. Rev.*, 27, 59–66, 1997.
72. Matthews, C. B., Jenkins, G., Hilfinger, J. M., and Davidson, B. L., Poly-L-lysine improves gene transfer with adenovirus formulated in PLGA microspheres, *Gene Ther.*, 6, 1558–1564, 1999.
73. Zhao, Z., Wang, J., Mao, H. Q., and Leong, K. W., Polyphosphoesters in drug and gene delivery, *Adv. Drug Deliv. Rev.*, 55, 483–499, 2003.
74. Heller, J., Barr, J., Ng, S. Y., Abdellauoi, K. S., and Gurny, R., Poly(ortho esters): synthesis, characterization, properties and uses, *Adv. Drug Deliv. Rev.*, 54, 1015–1039, 2002.
75. Lakshmi, S., Katti, D. S., and Laurencin, C. T., Biodegradable polyphosphazenes for drug delivery applications, *Adv. Drug Deliv. Rev.*, 55, 467–482, 2003.
76. Mathiowitz, E., Jacob, J. S., Jong, Y. S., Carino, G. P., Chickering, D. E., Chaturvedi, P., Santos, C. A., Vijayaraghavan, K., Montgomery, S., Bassett, M., and Morrell, C., Biologically erodable microsphere as potential oral drug delivery system, *Nature*, 386, 410–414, 1997.
77. Lim, Y. B., Han, S. O., Kong, H. U., Lee, Y., Park, J. S., Jeong, B., and Kim, S. W., Biodegradable polyester, poly alpha-(4 aminobutyl)-L-glycolic acid, as a non-toxic gene carrier, *Pharm. Res.*, 17, 811–816, 2000.
78. Maheshwari, A., Mahato, R. I., McGregor, J., Han, S. O., Samlowski, W. E., Park, J. S., and Kim, S. W., Soluble biodegradable polymer-based cytokine gene delivery for cancer treatment, *Mol. Ther.*, 2, 121–130, 2000.
79. Wang, S., Wan, A. C. A., Xu, X. Y., Gao, S. J., Mao, H. Q., Leong, K. W., and Yu, H., A new nerve guide conduit material composed of a biodegradable poly(phosphoester), *Biomaterials*, 22, 1157–1169, 2001.
80. Wang, J., Mao, H. Q., and Leong, K. W., A novel biodegradable gene carrier based on polyphosphoester, *J. Am. Chem. Soc.*, 123, 9480–9481, 2001.
81. Drumheller, P. D., and Hubbell, J. A., Surface immobilization of adhesion ligands for investigation of cell-substrate interactions, *Tissue Eng.*, 1581–1583, 1995.
82. Shin, H., Jo, S., and Mikos, A. G., Biomimetic materials for tissue engineering, *Biomaterials*, 24, 4353–4364, 2003.
83. Massia, S. P., and Hubbell, J. A., Covalent surface immobilization of Arg-Gly-Asp and Tyr-Ile-Gly-Ser-Arg-containing peptides to obtain well-defined cell-adhesive substrates, *Anal. Biochem.*, 187, 292–301, 1990.
84. Massia, S. P., and Hubbell, J. A., An RGD spacing of 440 nm is sufficient for integrin $\alpha_v\beta_3$-mediated spreading and 140 nm for focal contact and stress fiber formation, *J. Cell Biol.*, 114, 1089–1100, 1991.

85. Hrkach, J. S., Ou, J., Lotan, N., and Langer, R., Synthesis of poly(L-lactic acid-*co*-L-lysine) graft-copolymers, *Macromolecules*, 28, 4736–4739, 1995.
86. Cook, A. D., Hrkach, J. S., Gao, N. N., Johnson, I. M., Pajvani, U. B., Cannizzaro, S. M., and Langer, R., Characterization and development of RGD-peptide-modified poly(lactic acid-*co*-lysine) as an interactive, resorbable biomaterial, *J. Biomed. Mater. Res.*, 35, 513–523, 1997.
87. Barrera, D. A., Zylstra, E., Lansbury, P. T. Jr., and Langer, R., Synthesis and RGD peptide modification of a new biodegradable copolymer: poly(lactic acid-*co*-lysine), *J. Am. Chem. Soc.*, 115, 11010–11011, 1993.
88. Freed, L. E., Vunjak-Novakovic, G., Biron, R. J., Eagles, D. B., Lesnoy, D. C., Barlow, S. K., and Langer, R., Biodegradable polymer scaffolds for tissue engineering, *Biotechnology*, 12, 689–693, 1994.
89. Cima, L. G., Ingber, D. E., Vacanti, J. P., and Langer, R., Hepatocyte culture on biodegradable polymeric substrates, *Biotechnol. Bioeng.*, 38, 145–158, 1991.
90. Kaufmann, P. M., Heimrath, S., Kim, B. S., and Mooney, D. J., Highly porous polymer matrices as a three-dimensional culture system for hepatocytes, *Cell Transplant.*, 6, 463–468, 1997.
91. Mikos, A. G., Sarakinos, G., Lyman, M. D., Ingber, D. E., Vacanti, J. P., and Langer, R., Prevascularization of porous biodegradable polymers, *Biotechnol. Bioeng.*, 42, 716–723, 1993.
92. Mooney, D. J., Sano, K., Kaufmann, P. M., Majahod, K., Schloo, B., Vacanti, J. P., and Langer, R., Long-term engraftment of hepatocytes transplanted on biodegradable polymer sponges, *J. Biomed. Mater. Res.*, 37, 413–420, 1997.
93. Sharkawy, A. A., Klitzman, B., Truskey, G. A., and Reichert, W. M., Engineering the tissue which encapsulates subcutaneous implants. II. Plasma-tissue exchange properties, *J. Biomed. Mater. Res.*, 40, 586–597, 1998.
94. Athanasiou, K. A., Schmitz, J. P., and Agrawal, C. M., The effects of porosity on in vitro degradation of polylactic acid polyglycolic acid implants used in repair of articular cartilage, *Tissue Eng.*, 4, 53–63, 1998.
95. Lu, L., Peter, S. J., Lyman, M. D., Lai, H. L., Leite, S. M., Tamada, J. A., Uyama, S., Vacanti, J. P., Langer, R., and Mikos, A. G., In vitro and in vivo degradation of porous poly(DL-lactic-*co* glycolic acid) foams, *Biomaterials*, 21, 1837–1845, 2000.
96. Xie, Y. B., Yang, S. T., and Kniss, D. A., Three-dimensional cell-scaffold constructs promote efficient gene transfection: implications for cell-based gene therapy, *Tissue Eng.*, 7, 585–598, 2001.
97. Holy, C. E., Dang, S. M., Davies, J. E., and Shoichet, M. S., In vitro degradation of a novel poly(lactide-*co*-glycolide) 75/25 foam, *Biomaterials*, 20, 1177–1185, 1999.
98. Lunt, J., Large-scale production, properties and commercial applications of polylactic acid polymers, *Polym. Degrad. Stab.*, 59, 145–152, 1998.
99. Agrawal, C. M., McKinney, J. S., Lanctot, D., and Athanasiou, K. A., Effects of fluid flow on the in vitro degradation kinetics of biodegradable scaffolds for tissue engineering, *Biomaterials*, 21, 2443–2452, 2000.
100. Mooney, D. J., Breuer, C., McNamara, K., Vacanti, J. P., and Langer, R., Fabricating tubular devices from polymers of lactic and glycolic acid for tissue engineering, *Tissue Eng.*, 1, 107–118, 1995.
101. Langer, R., Controlled release (of peptides and proteins) and tissue engineering, *Biopolymers*, 71, 284–284, 2003.
102. Jang, J. H., and Shea, L. D., Controllable delivery of non-viral DNA from porous scaffolds, *J. Control. Release*, 86, 157–168, 2003.
103. Chen, R. R., and Mooney, D. J., Polymeric growth factor delivery strategies for tissue engineering, *Pharm. Res.*, 20, 1103–1112, 2003.
104. Saltzman, W. M., and Olbricht, W. L., Building drug delivery into tissue engineering, *Nat. Rev. Drug Discovery*, 1, 177–186, 2002.
105. Richardson, T. P., Peters, M. C., Ennett, A. B., and Mooney, D. J., Polymeric system for dual growth factor delivery, *Nat. Biotechnol.*, 19, 1029–1034, 2001.
106. Lee, K. Y., Peters, M. C., and Mooney, D. J., Controlled drug delivery from polymers by mechanical signals, *Adv. Mater.*, 13, 837, 2001.
107. Richardson, T. P., Murphy, W. L., and Mooney, D. J., Polymeric delivery of proteins and plasmid DNA for tissue engineering and gene therapy, *Crit. Rev. Eukaryot. Gene Expr.*, 11, 47–58, 2001.

108. Whitaker, M. J., Quirk, R. A., Howdle, S. M., and Shakesheff, K. M., Growth factor release from tissue engineering scaffolds, *J. Pharm. Pharmacol.*, 53, 1427–1437, 2001.
109. Lee, K. Y., Peters, M. C., Anderson, K. W., and Mooney, D. J., Controlled growth factor release from synthetic extracellular matrices, *Nature*, 408, 998–1000, 2000.
110. Xu, X. Y., Yee, W. C., Hwang, P. Y. K., Yu, H., Wan, A. C. A., Gao, S. J., Boon, K. L., Mao, H. Q., Leong, K. W., and Wang, S., Peripheral nerve regeneration with sustained release of poly(phosphoester) microencapsulated nerve growth factor within nerve guide conduits, *Biomaterials*, 24, 2405–2412, 2003.
111. Saito, N., and Takaoka, K., New synthetic biodegradable polymers as BMP carriers for bone tissue engineering, *Biomaterials*, 24, 2287–2293, 2003.
112. Li, J., Zhang, Y. P., and Kirsner, R. S., Angiogenesis in wound repair: angiogenic growth factors and the extracellular matrix, *Microsc. Res. Tech.*, 60, 107–114, 2003.
113. Babensee, J. E., McIntire, L. V., and Mikos, A. G., Growth factor delivery for tissue engineering, *Pharm. Res.*, 17, 497–504, 2000.
114. Sheridan, M. H., Shea, L. D., Peters, M. C., and Mooney, D. J., Bioadsorbable polymer scaffolds for tissue engineering capable of sustained growth factor delivery, *J. Control. Release*, 64, 91–102, 2000.
115. Kim, S. E., Park, J. H., Cho, Y. W., Chung, H., Jeong, S. Y., Lee, E. B., and Kwon, I. C., Porous chitosan scaffold containing microspheres loaded with transforming growth factor-beta 1: implications for cartilage tissue engineering, *J. Control. Release*, 91, 365–374, 2003.
116. Perets, A., Baruch, Y., Weisbuch, F., Shoshany, G., Neufeld, G., and Cohen, S., Enhancing the vascularization of three-dimensional porous alginate scaffolds by incorporating controlled release basic fibroblast growth factor microspheres, *J. Biomed. Mater. Res. Part A*, 65A, 489–497, 2003.
117. Whang, K., Tsai, D. C., Nam, E. K., Aitken, M., Sprague, S. M., Patel, P. K., and Healy, K. E., Ectopic bone formation via rhBMP-2 delivery from porous bioabsorbable polymer scaffolds, *J. Biomed. Mater. Res.*, 42, 491–499, 1998.
118. Mooney, D. J., Kaufmann, P. M., Sano, K., Schwendeman, S. P., Majahod, K., Schloo, B., Vacanti, J. P., and Langer, R., Localised delivery of epidermal growth factor improves the survival of transplanted hepatocytes, *Biotechnol. Bioeng.*, 50, 422–429, 1996.
119. Babensee, J. E., Anderson, J. M., McIntire, L. V., and Mikos, A. G., Host response to tissue engineered devices, *Adv. Drug Deliv. Rev.*, 33, 111–139, 1998.
120. Ito, Y., Tissue engineering by immobilized growth factors, *Mater. Sci. Eng. C-Biomim. Mater. Sensors Syst.*, 6, 267–274, 1998.
121. Langer, R., Drug delivery and targeting, *Nature*, 392, 5–10, 1998.
122. Shea, L. D., Smiley, E., Bonadio, J., and Mooney, D. J., DNA delivery from polymer matrices for tissue engineering, *Nat. Biotechnol.*, 17, 551–554, 1999.
123. Giannoukakis, N., Thomson, A. W., and Robbins, P. D., Gene therapy in transplantation, *Gene Ther.*, 6, 1499–1511, 1999.
124. Nor, J. E., Christensen, J., Mooney, D. J., and Polverini, P. J., Vascular endothelial growth factor (VEGF)-mediated angiogenesis is associated with enhanced endothelial cell survival and induction of Bcl-2 expression, *Am. J. Pathol.*, 154, 375–384, 1999.
125. Laurencin, C. T., Attawia, M. A., Lu, L. Q., Borden, M. D., Lu, H. H., Gorum, W. J., and Lieberman, J. R., Poly(lactide-*co*-glycolide)/hydroxyapatite delivery of BMP-2-producing cells: a regional gene therapy approach to bone regeneration, *Biomaterials*, 22, 1271–1277, 2001.
126. Vallbacka, J. J., Nobrega, J. N., and Sefton, M. V., Tissue engineering as a platform for controlled release of therapeutic agents: implantation of microencapsulated dopamine producing cells in the brains of rats, *J. Control. Release*, 72, 93–100, 2001.
127. Bonadio, J., Smiley, E., Patil, P., and Goldstein, S., Localized, direct plasmid gene delivery in vivo: prolonged therapy results in reproducible tissue regeneration, *Nat. Med.*, 5, 753–759, 1999.
128. Berry, M., Gonzalez, A. M., Clarke, W., Greenlees, L., Barrett, L., Tsang, W., Seymour, L., Bonadio, J., Logan, A., and Baird, A., Sustained effects of gene-activated matrices after CNS injury, *Mol. Cell. Neurosci.*, 17, 706–716, 2001.
129. Segura, T., and Shea, L. D., Surface-tethered DNA complexes for enhanced gene delivery, *Bioconjugate Chem.*, 13, 621–629, 2002.

130. Luo, D., and Saltzman, W. M., Enhancement of transfection by physical concentration of DNA at the cell surface, *Nat. Biotechnol.*, 18, 893–895, 2000.
131. Hobson, D. A., Pandori, M. W., and Sano, T., In situ transduction of target cells on solid surfaces by immobilized viral vectors, *BMC Biotechnol.*, 1, 2003.
132. Ragot, T., Vincent, N., Chafey, P., Vigne, E., Gilgenkrantz, H., Couton, D., Cartaud, J., Briand, P., Kaplan, J. C., Perricaudet, M., and Kahn, A., Efficient adenovirus-mediated transfer of a human minidystrophin gene to skeletal-muscle of Mdx mice, *Nature*, 361, 647–650, 1993.
133. Lasalle, G. L., Robert, J. J., Berrard, S., Ridoux, V., Stratfordperricaudet, L. D., Perricaudet, M., and Mallet, J., An adenovirus vector for gene-transfer into neurons and glia in the brain, *Science*, 259, 988–990, 1993.
134. Siemens, D. R., Austin, J. C., Hedican, S. P., Tartaglia, J., and Ratliff, T. L., Viral vector delivery in solid-state vehicles: gene expression in a murine prostate cancer model, *J. Natl Cancer Inst.*, 92, 403–412, 2000.
135. Kalyanasundaram, S., Feinstein, S., Nicholson, J. P., Leong, K. W., and Garver, R. I., Coacervate microspheres as carriers of recombinant adenoviruses, *Cancer Gene Ther.*, 6, 107–112, 1999.
136. Peter, S. J., Miller, M. J., Yasko, A. W., Yaszemski, M. J., and Mikos, A. G., Polymer concepts in tissue engineering, *J. Biomed. Mater. Res.*, 43, 422–427, 1998.
137. Tjia, J. S., and Moghe, P. V., Analysis of 3-D microstructure of porous poly(lactide-glycolide) matrices using confocal microscopy, *J. Biomed. Mater. Res.*, 43, 291–299, 1998.
138. Ronneberger, B., Kissel, T., and Anderson, J. M., Biocompatibility of ABA triblock copolymer microparticles consisting of poly(L-lactic-*co*-glycolic-acid) A-blocks attached to central poly(oxyethylene) B-blocks in rats after intramuscular injection, *Eur. J. Pharm. Biopharm.*, 43, 19–28, 1997.
139. Alsberg, E., Kong, H. J., Hirano, Y., Smith, M. K., Albeiruti, A., and Mooney, D. J., Regulating bone formation via controlled scaffold degradation, *J. Dent. Res.*, 82, 903–908, 2003.
140. Thomson, R. C., Wake, M. C., Yaszemski, M. J., and Mikos, A. G., Biodegradable polymer scaffolds to regenerate organs, *Adv. Polym. Sci.*, 122, 245–274, 1995.
141. Thomson, R. C., Yaszemski, M. J., Powers, J. M., and Mikos, A. G., Fabrication of biodegradable polymer scaffolds to engineer trabecular bone, *J. Biomater. Sci.-Polym. Ed.*, 7, 23–38, 1995.
142. Lu, L. C., and Mikos, A. G., The importance of new processing techniques in tissue engineering, *MRS Bulletin*, 21, 28–32, 1996.
143. Mooney, D. T., Mazzoni, C. L., Breuer, C., McNamara, K., Hern, D., Vacanti, J. P., and Langer, R., Stabilized polyglycolic acid fibre based tubes for tissue engineering, *Biomaterials*, 17, 115–124, 1996.
144. Wan, A. C. A., Mao, H. Q., Wang, S., Leong, K. W., Ong, L., and Yu, H., Fabrication of poly(phosphoester) nerve guides by immersion precipitation and the control of porosity, *Biomaterials*, 22, 1147–1156, 2001.
145. Wintermantel, E., Mayer, J., Blum, J., Eckert, K. L., Luscher, P., and Mathey, M., Tissue engineering scaffolds using superstructures, *Biomaterials*, 17, 83–91, 1996.
146. Salem, A. K., Rose, F., Oreffo, R. O. C., Yang, X. B., Davies, M. C., Mitchell, J. R., Roberts, C. J., Stolnik-Trenkic, S., Tendler, S. J. B., Williams, P. M., and Shakesheff, K. M., Porous polymer and cell composites that self-assemble in situ, *Adv. Mater.*, 15, 210–213, 2003.
147. Shastri, V. P., Martin, I., and Langer, R., Macroporous polymer foams by hydrocarbon templating, *Proc. Natl Acad. Sci. USA*, 97, 1970–1975, 2000.
148. Yoshimoto, H., Shin, Y. M., Terai, H., and Vacanti, J. P., A biodegradable nanofiber scaffold by electrospinning and its potential for bone tissue engineering, *Biomaterials*, 24, 2077–2082, 2003.
149. Luu, Y. K., Kim, K., Hsiao, B. S., Chu, B., and Hadjiargyrou, M., Development of a nanostructured DNA delivery scaffold via electrospinning of PLGA and PLA-PEG block copolymers, *J. Control. Release*, 89, 341–353, 2003.
150. Mooney, D. J., Baldwin, D. F., Suh, N. P., Vacanti, L. P., and Langer, R., Novel approach to fabricate porous sponges of poly(D,L-lactic-*co*-glycolic acid) without the use of organic solvents, *Biomaterials*, 17, 1417–1422, 1996.
151. Howdle, S. M., Watson, M. S., Whitaker, M. J., Popov, V. K., Davies, M. C., Mandel, F. S., Wang, J. D., and Shakesheff, K. M., Supercritical fluid mixing: preparation of thermally sensitive polymer composites containing bioactive materials, *Chem. Commun.*, 109–110, 2001.

23 Degradation of Biodegradable Aliphatic Polyesters

Suming Li

CONTENTS

I. INTRODUCTION

During the past 50 years, synthetic polymers have changed the everyday life of our society. In the meantime, surgeons and pharmacists have tried to utilize these materials as biomaterials.[1,2] About 30 years ago, a distinction was made between permanent and temporary therapeutic applications. Permanent applications require biostable polymeric materials which are not subject to degradation in the body. In contrast, temporary applications require a material for the healing time only. In this regard, degradable polymers became of great interest in surgery, in pharmacology as well as in tissue engineering. The first degradable synthetic polymer was poly(glycolic acid) (PGA), which was synthesized in 1954.[3] This polymer was first discarded because of its poor thermal and hydrolytic stabilities which precluded any permanent application. It was later realized that one could take advantage of the hydrolytic sensitivity of PGA to make polymeric devices which can degrade in a human body. This led to the first bioresorbable suture material made of a synthetic polymer.[4,5] It is worth noting that terminology is one of the sources of confusion in the field. Nowadays, people tend to use the word "degradable" as a general term and reserve "biodegradable" for polymers which are biologically degraded by enzymes introduced *in vitro* or generated by surrounding living cells. A polymer able to degrade, and to have its degradation by-products assimilated or excreted by a living system, is then designated as "bioresorbable."[6]

Most of the degradable and biodegradable polymers contain hydrolysable linkages, namely, ester, orthoester, anhydride, carbonate, amide, urea, and urethane, in their backbones.[7,8] The ester bond-containing aliphatic polyesters are particularly interesting because of their outstanding biocompatibility and variable physical, chemical, and biological properties.[9–11] The main members of the aliphatic polyester family are listed in Table 23.1. In this family, poly(lactic acid) (PLA), poly(glycolic acid), and poly(ε-caprolactone) (PCL) are the most investigated. A number of

TABLE 23.1
Aliphatic Polyesters

Polymer	Acronym	Structure
Poly(glycolic acid)	PGA	$-[-O-CH_2-CO-]_n-$
Poly(lactic acid)	PLA	$-[-O-{}^{*}C(H)(CH_3)-CO-]_n-$
Poly(ε-caprolactone)	PCL	$-[-O-(CH_2)_5-CO-]_n-$
Poly(valerolactone)	PVL	$-[-O-(CH_2)_4-CO-]_n-$
Poly(ε-decalactone)	PDL	$-[-O-{}^{*}C(H)((CH_2)_3CH_3)-(CH_2)_4-CO-]_n-$
Poly(1,4-dioxane-2,3-one)		$-[-O-(CH_2)_2-O-CO-CO-]_n-$
Poly(1,3-dioxane-2-one)		$-[-O-(CH_2)_3-O-CO-]_n-$
Poly(*para*-dioxanone)	PDO	$-[-O-(CH_2)_2-O-CH_2-CO-]_n-$
Poly(hydroxy butyrate)	PHB	$-[-O-{}^{*}C(H)(CH_3)-CH_2-CO-]_n-$
Poly(hydroxy valerate)	PHV	$-[-O-{}^{*}C(H)(CH_2-CH_3)-CH_2-CO-]_n-$
Poly(β-malic acid)	PMLA	$-[-O-{}^{*}C(H)(COOH)-CH_2-CO-]_n-$

products have reached the stage of clinical usage such as Dexon®, Vicryl®, Maxon® and Monocryl® sutures, Lactomer® and Absolok® clips and staples, Biofix® and Phusiline® plates and screws, as well as Decapeptyl®, Lupron Depot®, Zoladex®, Adriamycin®, Capronor® drug delivery devices.[12]

High molecular weight (MW) PLA, PGA, PCL and their copolymers are obtained by ring opening polymerization of cyclic esters, that is, lactide, glycolide and ε-caprolactone, respectively.[13,14] The direct polycondensation of corresponding hydroxy acids leads to low MW oligomers only.[15–17] In the case of LA-containing polymer chains, the chirality of LA units provides a worthwhile means to adjust bioresorption rates as well as physical and mechanical characteristics.[18–20] The use of PGA homopolymer is limited to suture material because of its high crystallinity and the absence of practical solvents. In contrast, PLA, PCL and various copolymers have been used, largely to make implants, microparticles, nanoparticles, and scaffolds in tissue engineering.

For the sake of simplicity, PLA stereocopolymers and PLAGA copolymers are identified in this chapter by acronyms PLA_X or PLA_XGA_Y, where X is the percentage of L-LA units in the monomer feed, Y is that of GA units, and the rest is the percentage of D-LA units. Similarly, copolymers of lactides and ε-caprolactone can be denoted as PLA_XCL_Z where X and Z represent the percentages of L-LA and CL units, respectively. This nomenclature presents the advantage of clearly reflecting the chemical and configurational compositions of the polymers since the average polymer chain composition is generally close to that of the feed.[21]

The *in vivo* and *in vitro* degradations of aliphatic polyesters have been extensively investigated during the past two decades. It was established that degradation is catalyzed by carboxyl endgroups formed by chain cleavage, and that amorphous regions are preferentially degraded. Nevertheless, discrepancies or controversial data existed in the literature, especially concerning the degradation rates, degradation location (at the surface or in the bulk) and the involvement of enzymes *in vivo*. Our group is one of the pioneers to undertake long-term systematic investigations. Important advances have been accomplished in the understanding of the hydrolytic degradation characteristics, particularly faster internal degradation and degradation-induced morphological and compositional changes. In this chapter, we will describe in detail the degradation mechanism and the effects of various factors on the degradation of PLA, PGA, and PCL polymers. Discussion is largely based on recent advances in the domain.

II. DEGRADATION MECHANISM

The degradation mechanism of PLAGA polymers has been extensively investigated. In general, degradation is considered as a hydrolytic process, although enzymatic degradation can contribute during the later stages of *in vivo* degradation. The cleavage of an ester bond yields a carboxyl endgroup and a hydroxyl one. The thus formed carboxyl endgroups are capable of catalyzing the hydrolysis of other ester bonds, a phenomenon called autocatalysis as shown below[22,23]:

$$R \sim COO \sim R' + H_2O \xrightarrow{\sim COOH} R \sim COOH + HO \sim R'$$

The kinetics of the autocatalyzed hydrolytic degradation were derived according to the following equation:

$$d[COOH]/dt = -d[E]/dt = k[E]\cdot[H_2O]\cdot[COOH] \tag{23.1}$$

where [COOH], $[H_2O]$, and [E] represent carboxyl endgroup, water and ester concentrations in the polymer matrix, respectively. After a series of integrations and simplifications, one obtains the following relationship:

$$\mathrm{Ln}(\overline{DP}_n/\overline{DP}_{n_0}) = \mathrm{Ln}(\bar{M}_n/\bar{M}_{n_0}) = -k't \tag{23.2}$$

According to Equation 23.2, semilog plots of $\overline{DP}_n$ or of $\bar{M}_n$ vs. degradation time should be linear prior to the onset of weight loss.

The autocatalytic degradation kinetics were validated by experimental data. Figure 23.1 shows the plot of Ln(*Mn*) vs. degradation time of initially amorphous PLA_{100}. An almost linear relationship was obtained. The reaction constant, k', was found to be 0.0584 week^{-1}.

From the macroscopic viewpoint, degradation of aliphatic polyesters has been regarded as a homogeneous phenomenon,[24–31] although surface erosion was claimed occasionally.[32–34] The discovery of faster degradation inside large-size PLA specimens greatly changed the understanding of the hydrolytic degradation of PLAGA polymers.[35] The heterogeneous degradation was assigned to diffusion-reaction phenomena as summarized in the following.

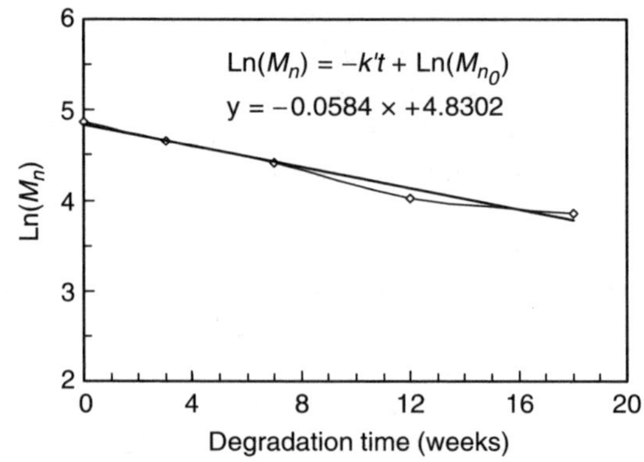

FIGURE 23.1 Ln(*Mn*) vs. time relationship of initially amorphous PLA_{100} during hydrolytic degradation in pH = 7.4 phosphate buffer at 37°C.

Typically, the polymer matrix is initially homogeneous in the sense that the average MW is the same throughout the matrix. Once placed in an aqueous medium, water penetrates into the specimen leading to hydrolytic cleavage of ester bonds. Each ester bond cleavage forms a carboxyl endgroup which, according to autocatalysis, accelerates the hydrolytic reaction of the remaining ester bonds.[36] However, no autocatalysis can occur at the surface because of two factors. First, as the aqueous medium is always buffered, *in vitro* and *in vivo*, the carboxyl endgroups present at the surface are neutralized and lose their catalytic potency. Second, when soluble oligomeric species are generated, those which are close to the surface can escape from the matrix before total degradation, while those located inside can hardly diffuse out of the matrix. Therefore, autocatalysis occurs only in the bulk, thus leading to a surface/interior differentiation. As the degradation proceeds, more and more carboxyl endgroups are formed inside to accelerate the internal degradation and enhance the surface/interior differentiation. Bimodal MW distributions are observable due to the presence of two populations of macromolecules degrading at different rates. Finally, hollow structures are formed when the internal material, which is totally transformed to soluble oligomers, dissolves in the aqueous medium (Figure 23.2(a)). This is the case of amorphous polymers like PLA_{50}, $PLA_{62.5}$, PLA_{75}, and $PLA_{37.5}GA_{25}$.[35,37,38] In contrast, in the case of crystallizable polyesters like $PLA_{87.5}$, PLA_{96}, PLA_{100}, $PLA_{75}GA_{25}$, and $PLA_{85}GA_{15}$, no hollow structures were obtained due to the crystallization of internal degradation by-products. Nevertheless, surface–interior differentiation was also observed with faster internal degradation.[37,39–42] For example, after as long as 50 weeks in a phosphate buffer, PLA_{96} exhibited a very pronounced surface/interior differentiation, the interior being composed of highly crystalline material (Figure 23.2(b)).

The heterogeneity of the autocatalytic degradation described above well explained the bimodal SEC chromatograms mentioned in literature and systematically assigned to semicrystallinity, even though, in some cases, the considered matrices were intrinsically amorphous.[18,43] Since its discovery, it was confirmed by many authors.[44–47] Schmitt et al.,[44] Ali et al.,[45] and Zhang et al.[46] detected visually a faster internal degradation of PLAGA and PLA_{50}. In a long-term *in vivo* study of PLA_{100} degradation of up to 116 weeks, Pistner et al. observed a well-defined surface/interior differentiation.[47] Gogolewski et al. first declared that no apparent difference was found between surface and center of various PLA stereocopolymer samples up to 6 months degradation,[48] although nearly all SEC chromatograms displayed a bimodal distribution. Afterwards, these authors observed heterogeneous cross-sections for degraded cylindrical bars made of PLA_{95} and $PLA_{92.5}$[49] and attributed them to chain orientation, although, in contrast to injection molded samples, no chain

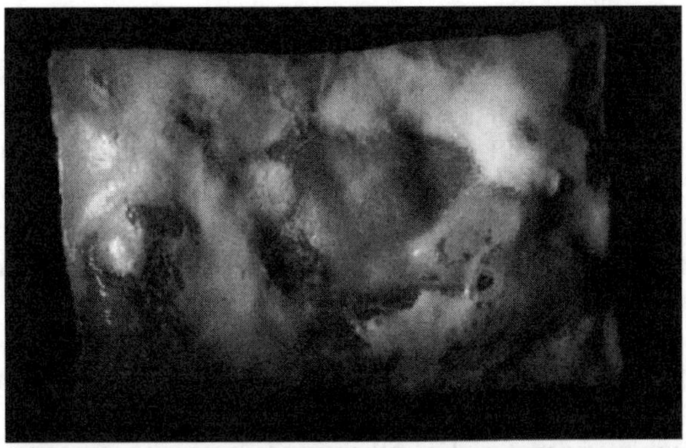

(a)

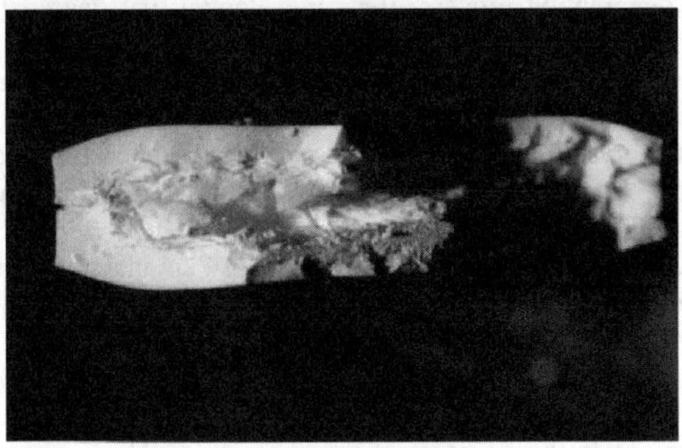

(b)

FIGURE 23.2 (a) Hollow structure of a PLA_{50} specimen after 12 week degradation in pH = 7.4 phosphate buffer at 37°C, and (b) cross-section of a PLA_{96} specimen after 50 week degradation in pH = 7.4 phosphate buffer at 37°C.

orientation occurs in the case of compression molded polymer samples. Grijpma and Pennings investigated the hydrolytic degradation of rectangular bars machined from as-polymerized copolymers of L-lactide with D-lactide, glycolide, ε-caprolactone, and so on.[50] The authors claimed that a hollowing out of the samples was not observed in any of the selected copolymers during degradation. However, as shown above, heterogeneous degradation does not necessarily yield hollow structures.

The faster internal degradation was also detected for lactide/caprolactone copolymers,[51] lactide/glycine copolymers,[27,28] lactide/1,5-dioxepan-2-one copolymers,[52] poly(trimethylene carbonate),[53] and poly(*para*-dioxanone) as well,[54] thus indicating that it is a general mechanism for hydrolytically degradable polymers, provided catalytic groups are formed during degradation and that the sample size is large enough.

Insofar as PCL is concerned, hydrolytic degradation is very slow. However, the degradation mechanism is the same as that of PLAGA polymers. Typically, degradation of PCL begins with random hydrolytic cleavage of ester linkages autocatalyzed by carboxyl endgroups. The MW of polymer decreases continuously, but there is no weight loss during this first stage of degradation. The second phase is characterized by the onset of weight loss due to the formation of low MW fragments small enough to diffuse out of the bulk.[22,51,55] No faster internal degradation has been reported for the PCL homopolymer thus far, probably because of the combination of high crystallinity, hydrophobicity and a low degradation rate.

III. DEGRADATION CHARACTERISTICS

The hydrolytic degradation of aliphatic polyesters is a complex process involving several phenomena, namely, water absorption, ester bond cleavage, neutralization of carboxyl endgroups at the surface, autocatalysis inside, and the diffusion and solubilization of soluble oligomers. These phenomena depend on many factors, such as matrix morphology, chemical composition and configurational structure, MW, size, distribution of chemically reactive compounds within the matrix, and the nature of the degradation media.[11] These factors' influence is discussed in the following.

A. Polymer Morphology

The morphology of a polymeric material, in other words, amorphousness or semicrystallinity, plays a critical role in the degradation process. It is now known that degradation of semicrystalline polyesters in aqueous media proceeds in two stages. The first stage consists of water diffusion into the amorphous regions with random hydrolytic scission of ester bonds. The second stage starts when most of the amorphous regions are degraded. The hydrolytic attack then progresses from the edge towards the center of the crystalline domains. This phenomenon was first observed by Fischer et al. who obtained a trimodal MW distribution after chemical degradation of $PLA_{92.5}$ single crystals, the three peaks corresponding, respectively, to onefold, twofold, and threefold the crystalline lamellae thickness.[56] Later on, many authors reported preferential degradation of amorphous areas in the cases of semicrystalline Vicryl® films, pellets, or sutures,[57,58] Dexon® sutures,[59,60] and PLA_{100} plates and rods.[61,62] An increase of crystallinity was detected in all cases.[57–62]

Chain orientation in both crystalline and amorphous regions could also play an important role in the degradation of polymers. In the case of melt-spun fibers, for example, alternative crystalline and amorphous regions arrange in the direction of the fiber axis.[59] The scission of tie-chain segments leads to rapid loss of tensile properties. Chain orientation along the fiber axis impeded water penetration and enhanced the resistance to hydrolytic attack.[60] It is also worthwhile to note the presence of imperfections and defective crystalline regions. When the spherulitic crystallization develops within a matrix containing impurities, monomers or oligomers, these noncrystallizable species are often concentrated at the interspherulitic boundaries.[57] These defects are generally preferentially degraded with the amorphous regions.

The systemic investigation of the *in vitro* degradation of PLAGA polymers agreed well with the preferential degradation in amorphous zones of semicrystalline PLA_{100}. Bimodal MW distributions were detected at the later stages of degradation, the two peaks corresponding to onefold and twofold the thickness of crystalline lamellae, respectively.[39] On the other hand, crystallization was observed in the case of initially amorphous PLA_{100} and PLA_{96}, obtained by quenching[39,40] as shown in Figure 23.3. Crystallization mainly resulted from low MW chains whose Tg was lower than that of longer chains. With the plasticizing effect of absorbed water, which further decreased the Tg, these short chains were mobile enough to crystallize under the degradation conditions. The resulting crystallites were more resistant to further degradation. As a consequence, a very narrow peak corresponding to low MW crystalline zones was detected in SEC chromatograms (Figure 23.4).

Among the various PLA_X stereocopolymers, the polymer is intrinsically amorphous if x is in the 10 to 90 range.[11] For PLA_XGA_Y copolymers, the polymer is intrinsically amorphous when y is in the 10 to 70 range (L-LA/GA) or in the 0 to 70 range (DL-LA/GA).[13] However, one must be careful with these divisions because of the variable extents of transesterification reactions, which tend to randomize the comonomer distribution along the chains. The fate of intrinsically amorphous polymers during degradation appeared more complex depending on chain structures. Two classes can be distinguished. The first class consists in polymers which can never crystallize. The second one groups polymers which can crystallize during degradation. $PLA_{37.5}GA_{25}$ is a good example of the first class, whereas PLA_{50}, $PLA_{62.5}$, PLA_{75}, $PLA_{87.5}$, $PLA_{75}GA_{25}$ and $PLA_{85}GA_{15}$ belong to

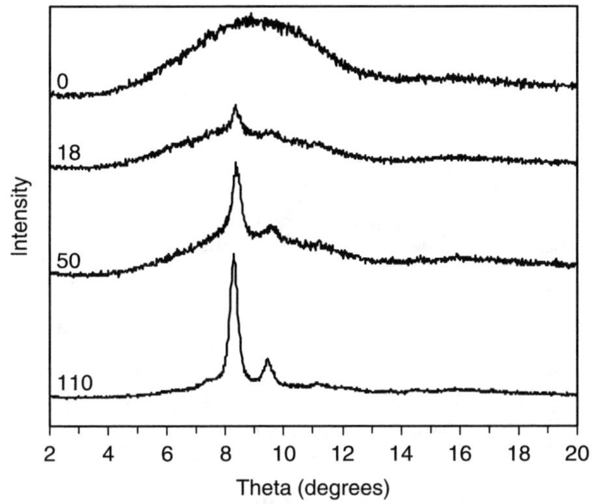

FIGURE 23.3 X-ray diffraction spectra of initially amorphous PLA_{100} after 0, 18, 50, and 110 weeks in pH = 7.4 phosphate buffer at 37°C. (From Li, S., Garreau, H., and Vert, M., Structure–property relationships in the case of the degradation of massive poly(α-hydroxy acids) in aqueous media. Part 3: influence of the morphology of poly(L-lactic acid), *J. Mater. Sci.: Mater. Med.*, 1, 198–206, 1990. With permission.)

the second class. $PLA_{37.5}GA_{25}$ cannot crystallize, even after degradation, due to the high irregularity of its chain structure. The others are crystallizable because of the presence of isotactic blocks which, once released by degradation, are susceptible to crystallization. For example, PLA_{50} is an intrinsically amorphous polymer. However, a crystalline oligomeric stereocomplex composed of L-LA and D-LA segments was detected at the later stages of PLA_{50} degradation,[63,64] as shown in Figure 23.5. Stereoregular segments were initially present within the PLA_{50} chains, which are known to present a predominantly isotactic structure resulting from the pair-addition mechanism of DL-lactide polymerization,[21] as shown in Table 23.2. $PLA_{62.5}$ led also to slightly crystalline residues of the stereocomplex type.[37] In contrast, PLA_{75} and $PLA_{87.5}$ degradation residues exhibited the same crystalline structure as PLA_{100}.[37] Insofar as $PLA_{75}GA_{25}$ and $PLA_{85}GA_{15}$ are concerned, GA units are more hydrophilic than LA ones and, as a consequence, are more inclined to

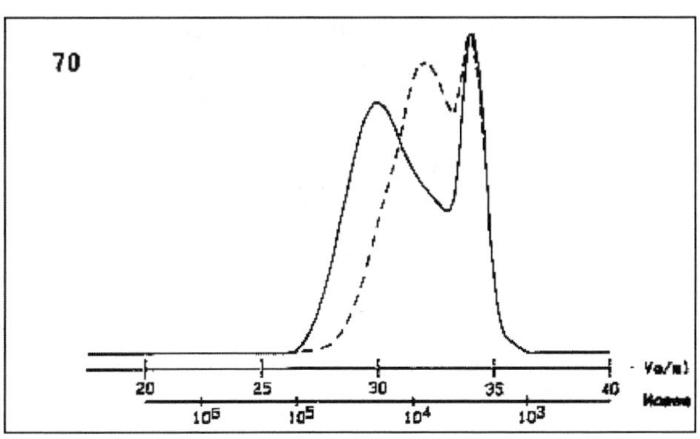

FIGURE 23.4 SEC chromatograms of initially amorphous PLA_{100} after 70 weeks in pH = 7.4 phosphate buffer at 37°C; —— is the surface, - - - - is the interior. (From Li, S., Garreau, H., and Vert, M., Structure–property relationships in the case of the degradation of massive poly(α-hydroxy acids) in aqueous media. Part 3: influence of the morphology of poly(L-lactic acid), *J. Mater. Sci.: Mater. Med.*, 1, 198–206, 1990. With permission.)

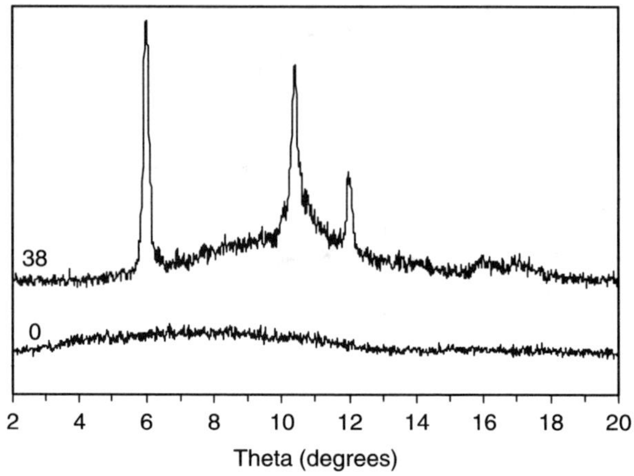

FIGURE 23.5 X-ray diffraction spectra of PLA_{50} after 0 and 38 weeks' degradation in pH = 7.4 phosphate buffer at 37°C. (From Li, S., and Vert, M., Crystalline oligomeric stereocomplex as intermediate compound in racemic poly(DL-lactic acid) degradation, *Polym. Int.*, 33, 37–41, 1994. With permission.)

hydrolysis. The preferential degradation of GA units results in L-LA-enriched segments, which can crystallize under the degradation conditions as shown by the appearance of diffraction peaks on the X-ray diffraction spectra (Figure 23.6). However, the structure of the crystalline residues is slightly different from that of PLA_{100} due to the inclusion of GA units in the crystalline domains.[38–40]

A schematic representation is given in Figure 23.7. The triangle can be divided in two types of zones: C zones are composed of intrinsically crystalline polymers and A zones of intrinsically amorphous polymers. For polymers in C zones, if they are prepared in an amorphous state by quenching, degradation will be characterized by crystallization of degradation by-products. If they are prepared in semicrystalline state by annealing, degradation will be characterized by selective degradation of amorphous regions. In both cases, no hollow structures will be obtained for large-size devices in spite of the heterogeneous degradation. In the case of amorphous polymers, three

TABLE 23.2
Probability Data of Isotactic L-LA and D-LA Sequences in Four PLA_x Stereocopolymer Chains Derived from L-Lactide and DL-Lactide in the Absence of Transesterification Reactions

Designation	Sequence	Formula	Probability			
			$PLA_{87.5}$ (p = .875)	PLA_{75} (p = .75)	$PLA_{62.5}$ (p = .625)	PLA_{50} (p = .5)
Triads	LLL(L)	p^2[a]	0.766	0.563	0.391	0.25
	DDD(D)	q^2[b]	0.016	0.063	0.141	0.25
Pentads	LLLLL(L)	p^3	0.670	0.422	0.242	0.125
	DDDDD(D)	q^3	0.002	0.016	0.053	0.125
Heptads	LLLLLLL(L)	p^4	0.586	0.316	0.153	0.0625
	DDDDDDD(D)	q^4	2×10^{-4}	0.004	0.020	0.0625
Nonads	LLLLLLLLL(L)	p^5	0.513	0.237	0.095	0.03125
	DDDDDDDDD(D)	q^5	3×10^{-5}	0.001	0.007	0.03125

[a] p = [L-LA]/([L-LA] + [D-LA]).
[b] $q = (1 - p)$.

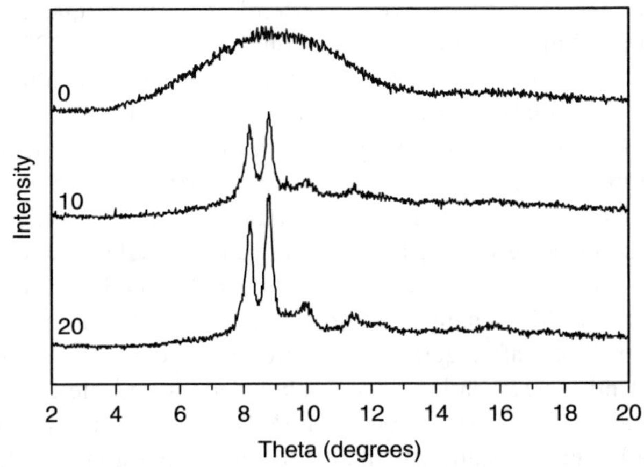

FIGURE 23.6 X-ray diffraction spectra of $PLA_{75}GA_{25}$ after 0, 10, and 20 weeks in $pH = 7.4$ phosphate buffer at 37°C. (From Li, S., Garreau, H., and Vert, M., Structure–property relationships in the case of the degradation of massive poly(α-hydroxy acids) in aqueous media. Part 2: degradation of lactide/glycolide copolymers: $PLA_{37.5}GA_{25}$ and $PLA_{75}GA_{25}$, *J. Mater. Sci.: Mater. Med.*, 1, 131–139, 1990. With permission.)

subzones can be distinguished: A1, A2 and A3. For A1 polymers, degradation will lead to hollow structures which remain amorphous due to the high irregularity of polymer chain structures. For A2 polymers, degradation will also lead to hollow structures which will partially crystallize, the crystalline structure depending on the initial composition. For A3 polymers, no hollow structures can be obtained; both the surface and interior will crystallize. It is noteworthy that degradation of small size amorphous polymers will not lead to hollow structures, but morphological changes are comparable to large size ones.[64]

All these examples and comments demonstrate that polymer morphology and morphological changes have to be taken into account to control the hydrolytic degradation of PLAGA polymers.

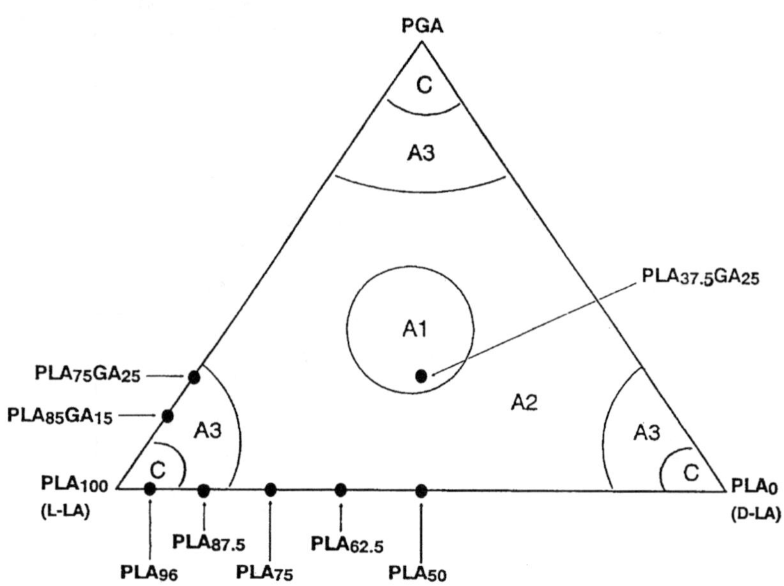

FIGURE 23.7 Schematic presentation of the degradation behaviors of PLAGA polymers with hydrolytic degradation.

PCL and its copolymers with PLA have also been largely investigated. Pitt et al. observed a steady increase in crystallinity of PCL films, from 45 to nearly 80% after 120 weeks of implantation, which was attributed to the crystallization of tie segments after chain cleavage in the amorphous phase. The low glass transition temperature of PCL ($Tg = -60°C$) facilitated the recrystallization.[51] Li et al. also observed a crystallinity increase during the *in vitro* degradation of PCL.[51] Water absorption was very limited (less than 2%) during the first 63 weeks due to the high crystallinity and hydrophobicity of PCL. Thereafter, water absorption reached 8% at 133 weeks and 12% at 200 weeks, which can be related to a decrease in molecular weight and the formation of carboxyl endgroups. On the other hand, weight loss reached 3% at 63 weeks, 11% at 133 weeks and 14% at 200 weeks. Insofar as MW changes are concerned, the molecular weight decreased steadily from an initial 58,700 to 7000 after 200 weeks. After 133 weeks, the MW distribution became trimodal due to the selective degradation of amorphous zones and the crystallite edges, the MW values of the three peaks being 2600, 5200, and 8800, respectively (Figure 23.8). In the case of a $PLA_{75}CL_{25}$ copolymer, crystallinity increased from an initial 14 to 52% at 63 weeks. The crystalline structure was of the PLA_{100}-type, showing a phase separation between the two components.[51]

B. Polymer Composition

The composition of polymer chains greatly determines the degradation rate of aliphatic polyesters. As shown in Table 23.3, the half-life in terms of weight loss of initially amorphous PLA_{100} was found to be about 110 weeks, that of initially semicrystalline PLA_{100} being much longer.[39] The introduction of 4% of D-LA units into L-LA chains considerably enhanced the degradation since initially amorphous PLA_{96} had a half-life of about 90 weeks.[40] PLA_{50} degraded much more

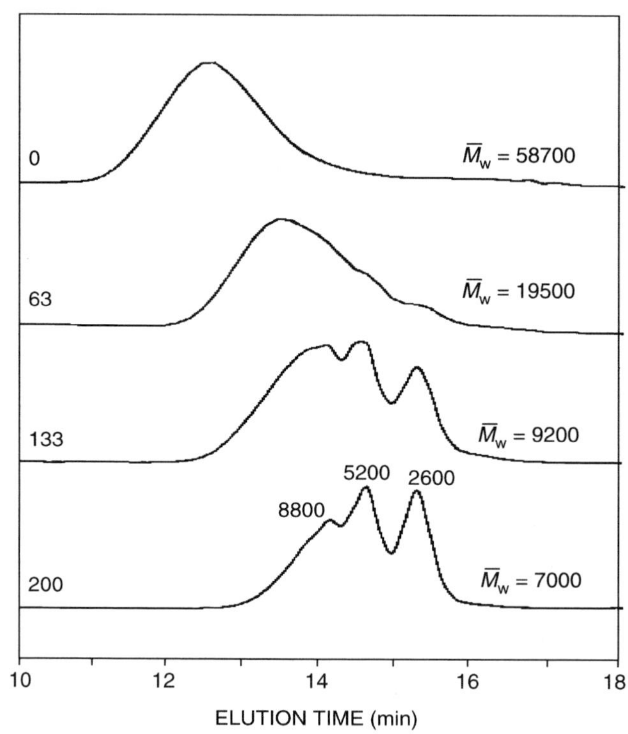

FIGURE 23.8 SEC chromatograms of PCL plates after 0, 63, 133, and 200 weeks in a pH 7.4 phosphate buffer at 37°C.

TABLE 23.3
Degradation Rates (Half-Life) of Various PLAGA Polymers Allowed to Degrade in pH = 7.4 Phosphate Buffer at 37°C

Polymer	$\bar{M}_w(\times 10^{-3})$	$\bar{M}_w/\bar{M}_n$	Half-Life[a] (weeks)
PLA_{100}	130.0	1.8	110
PLA_{96}	100.0	1.7	90
$PLA_{87.5}$	190.4	1.9	80
PLA_{75}	48.6	1.6	22
$PLA_{62.5}$	67.6	1.9	23
PLA_{50}	65.0	1.6	10
$PLA_{85}GA_{15}$	112.0	1.8	20
$PLA_{75}GA_{25}$	111.0	1.8	10
$PLA_{37.5}GA_{25}$	51.0	1.6	3

[a] Half-life corresponding to the time where half of initial material is lost.

rapidly than PLA_{100} and PLA_{96} with a half-life of 10 weeks.[35] It is also worthwhile to note that the half-life of $PLA_{87.5}$ was much longer than that of PLA_{75} because the latter led to hollow structures during degradation.[37] The copolymerization of LA with GA greatly increased the degradation rate as compared with parent homopolymers.[38,40] $PLA_{37.5}GA_{25}$ had a half-life of 3 weeks only. The half-lives of $PLA_{75}GA_{25}$ and $PLA_{85}GA_{15}$, whose degradation byproducts were capable of crystallization, were respectively, 10 and 20 weeks.

Attention has also been paid to chemical composition changes of the copolymers. The LA content increased during degradation of $PLA_{75}GA_{25}$ and $PLA_{85}GA_{15}$.[38,40] $PLA_{75}GA_{25}$ exhibited a faster increase at the early stages than $PLA_{85}GA_{15}$ due to faster degradation. At the later stages of degradation, however, the LA content tended to a limit in both cases. This preferential degradation of GA units was assigned to the higher hydrophilicity of GA units in comparison to LA ones. In consequence, the L-LA-enriched segments crystallized under the *in vitro* degradation conditions to form crystalline zones of relatively low MW,[38,40] as observed for initially amorphous PLA_{96} and PLA_{100}.[39,40] Reciprocally, the crystallization limited the preferential degradation of GA units because of their inclusion in crystalline zones. That is why, the crystalline structure of the copolymer residues was slightly different from that of PLA_{100} (Figures 23.3 and 23.6). In the case of $PLA_{37.5}GA_{25}$, no significant preferential degradation of GA units was observed, probably due to the very high degradation rate.[38]

In the case of PLACL copolymers, the degradation rate was enhanced, in comparison with the parent homopolymers PCL. For example, a $PLA_{75}CL_{25}$ copolymer exhibited a much higher weight loss rate than PCL, as shown in Figure 23.9. The copolymer lost 83% of its material after 133 weeks, while PCL lost only 11%.

C. MW and MW Distribution

MW and MW distribution are important factors in the polymer degradation because of the autocatalytic character of aliphatic polyester hydrolysis. Pitt et al. found that PLA_{50} films with an initial $\bar{M}_n$ of 14,000 were absorbed by week 28, against 60 weeks for those with an initial $\bar{M}_n$ of 49,000.[22] Chawla and Chang investigated the *in vivo* degradation of four PLA_{50} samples with different high MWs. At the end of a 48 week implantation period, samples with lower MW had degraded faster.[65] Kaetsu et al. studied the *in vivo* degradation of four PLA_{50} needle samples with relatively low MW (1100 to 2200 determined by terminal group titration).[66] In these cases, the higher the initial MW, the slower the degradation rate. Ogawa et al. observed that both the lag time and

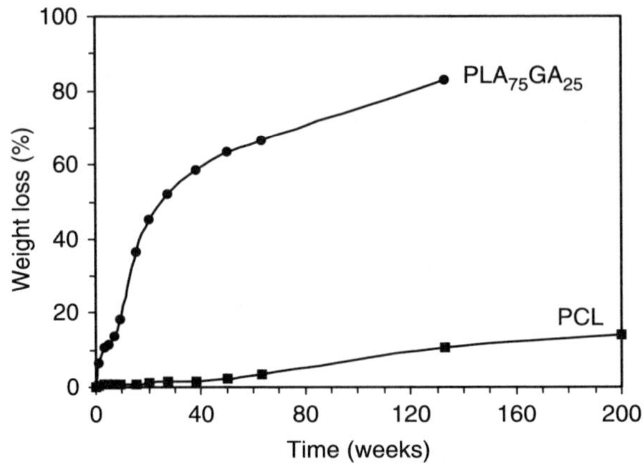

FIGURE 23.9 Weight loss profiles of PCL and $PLA_{72}CL_{25}$ with degradation in pH = 7.4 phosphate buffer at 37°C. (From Li, S., Espartero, J. L., Foch, P., and Vert, M., Structural characterization and hydrolytic degradation of a Zn metal initiated copolymer of L-lactide and ε-caprolactone, *J. Biomater. Sci. Polym. Ed.*, 8, 165–187, 1996. With permission.)

the half-life of PLA_{50} plates increased with MW during *in vivo* degradation.[67] Fukuzaki et al. studied the degradation of different PLA_{50} with relatively low MWs.[15] The authors proposed three types of degradation profile according to MW, namely, parabolic, linear, and S-types. The parabolic type observed for $\bar{M}_n = 1500$ daltons can be explained by the release of the soluble fraction present within the initial oligomers, whereas the S-type ($\bar{M}_n = 3500$ daltons) reflects a lag time due to higher MW.

The effect of MW distribution, or polydispersity ($\bar{M}_w/\bar{M}_n$), on polymer degradation has only been investigated qualitatively. The presence of low MW species and monomers was shown to accelerate the degradation. Their elimination is thus very important for the control of polymer degradation. Leeslag et al. showed that the purification of PLA_{100} led to a material more resistant to hydrolytic degradation.[61] Nakamura et al. observed a strong increase in the degradation rate of PLA stereocopolymers in the presence of DL-lactide and glycolide.[19] Recently, Mauduit et al. showed that oligomers accelerated the rate of degradation of PLA_{50} films.[68]

From these published data, one can conclude that the lower the MW, the faster the degradation rate, in agreement with the presence of more carboxylic acid catalyzing groups.

D. Size and Porosity

The size of polymer samples has been recently recognized as an important factor for the degradation of aliphatic polyesters. Mauduit et al. investigated the degradation of low and high MW PLA_{50} particles and films *in vitro*.[69] In contrast to large-size devices, corresponding SEC chromatograms remained monomodal during the whole degradation period. Zhu et al. observed the same phenomenon for microspheres of PLA and two PLAGA copolymers.[70] Grizzi et al. investigated the influence of size on the degradation of PLA_{50}. Films, powder and microspheres degraded much less rapidly as compared to large-size specimens,[71] showing conclusively that the smaller the polymer size, the slower the degradation rate. This unexpected behavior was well explained by the mechanism of autocatalytic degradation described above. In fact, if the size of polymer matrix is very small, there is no internal autocatalysis or surface/center differentiation. It is useful to keep in mind that the literature is often misleading in comparing degradation rates of microspheres, fibers, films, pellets, and other massive implants.[32,72]

The porosity of the polymer matrix is also an important factor. Lam et al. evaluated the *in vitro* and *in vivo* degradation of PLA_{100} films. The authors observed a faster degradation of nonporous films as compared with porous ones.[73] This can be assigned to the fact that in the case of porous films, no internal autocatalysis occurred due to the ionic exchanges facilitated by the porous structure.

E. Additives

Another important factor is the incorporation of acidic or basic compounds in a polymeric matrix. The understanding of the effect of incorporated compounds on matrix degradation is of particular importance in drug delivery systems. If the compound is acidic, it can accelerate the degradation of the polymer. In contrast, for basic compounds, two opposite effects can intervene simultaneously: base catalysis and neutralization of carboxyl endgroups. Whether the degradation is accelerated or slowed down depends on the relative importance of the two effects.[64,74–80]

Maulding et al. reported an acceleration of the degradation of PLA_{50} in the presence of a basic drug, thioridazine, which is a tertiary amine compound. Catalysis was attributed to the nucleophilic nature of the amino group.[74,75] Kishida et al. obtained similar results.[76] The authors found that the degradation rates of PLA_{100} matrix which normally hydrolyzes slowly in aqueous media, could be increased by incorporating basic compounds, the catalytic effect increasing with the load. The degradation rate depended also on the apparent pK_a of the conjugate acid of each basic compound. The higher the pK_a, the faster the degradation.[76] Li et al. investigated the mechanism of hydrolytic degradation of PLA_{50} matrix in the presence of a tertiary amine, namely caffeine, in order to elucidate the influence of this basic compound on the hydrolytic cleavage of polyester chains.[64] Caffeine was incorporated into PLA_{50} matrix in various contents (0 to 20%) by blending in acetone followed by solvent evaporation. The effects of caffeine on the degradation of PLA_{50} were rather complex and largely depended on the load. For low contents ($\leq$2%), caffeine was molecularly dispersed in the matrix and considerably accelerated the degradation with respect to caffeine-free devices. However, the increase of the degradation rate was not proportional to the caffeine load, due to the combined effects of base/carboxyl endgroup interaction, crystallization and matrix-controlled or channeling-controlled diffusion of caffeine. In the early stages of degradation, the overall catalytic effect was larger for devices with low caffeine contents than for highly loaded ones where caffeine was in a crystallized state and thus less available for basic catalysis.[64]

Other authors observed a decrease of the degradation rate due to interactions between polymer chain ends and additives. Mauduit et al. investigated various gentamycin/PLA_{50} blends.[68,69,77] It was found that the base form of the drug was able to neutralize polymer chain ends on blending in acetone, whereas the sulfate form required an aqueous environment. In the case of gentamycin sulfate/low MW PLA_{50} systems, interactions between drug and chain ends stabilized the matrix, the excess free drug being rapidly released.[77] Both phenomena acted against base catalysis.

The effects of sparingly soluble additives were also investigated. Verheyen et al. examined the physicochemical properties of hydroxylapatite/PLA_{100} composites in solution tests.[78] Data showed that the higher the hydroxylapatite content, the slower the decrease of MW. Li et al. investigated the hydrolytic behavior of PLA_{50}/coral blends.[79] It was found that coral, which is composed of calcium carbonate, significantly slowed down the degradation of PLA_{50} matrix and suppressed the faster internal degradation. Zhang et al. observed a decrease of degradation rate of 1.7 to 3.0 times with incorporation of salts such as $Mg(OH)_2$, $MgCO_3$, $CaCO_3$ and $ZnCO_3$ into $PLA_{25}GA_{50}$ films.[80] The stabilizing effect of hydroxylapatite, coral and other metal salts on the hydrolysis of polymer chains can be assigned to the neutralization of carboxyl endgroups by the additives, and by the degradation medium which penetrated the matrix due to the presence of the polymer/additive interfaces.

Therefore, the degradation of PLAGA polymers in the presence of basic compounds depends on a number of parameters, including, base catalysis, neutralization of carboxyl endgroups, porosity, dimensions of devices, load, and morphology of incorporated compounds. None of these factors can be considered separately if one wants to understand the effect of basic drugs on matrix degradation.

F. Others

The influence of γ irradiation, which may be used in the sterilization process of medical devices, has also been examined. Chu observed that γ irradiation of Dexon® and Vicryl® fibers resulted in an earlier appearance of the pH fall of the degradation medium and a faster loss of tensile strength,[81] in agreement with a predominant chain scission process during irradiation. Spenlehauer et al. found that γ sterilization dramatically decreased the MW of PLA_{50}, $PLA_{45}GA_{10}$ and $PLA_{37.5}GA_{25}$ microspheres and that this degradation continued on storage for GA-containing compounds.[82] Birkinshaw et al. examined the effect of γ irradiation on compression molded PLA_{50} samples.[83] The authors concluded that the primary effect of irradiation on hydrolytic degradation was associated with the initial reduction in MW, the degradation mechanism remaining the same. Volland et al. investigated the influence of γ sterilization on $PLA_{25}GA_{50}$ microspheres.[84] The MW of the polymer decreased with increasing irradiation dose, but the polydispersity remained unchanged, thus suggesting a random chain cleavage rather than an unzipping process.

It is also worth noting that no major differences were observed between the degradation rates of PLA_{50} in $pH = 7.4$ phosphate buffer and in nonbuffered saline.[35] In an initially acidic buffer medium ($pH = 3.7$), however, water absorption was limited in comparison to a $pH = 7.4$ medium.[85] In contrast, the absence of ionic strength in distilled water promoted water absorption and thus enhanced the surface/interior differentiation in the early stages.[38] On the other hand, the $pH = 7.4$ phosphate buffer enhanced solubilization of degradation byproducts in contrast to distilled water or acidic media, the carboxylate form $RCOO^-$ of organic acids being more hydrophilic than the carboxylic acid form RCOOH. Insofar as the *in vivo* degradation is concerned, it generally presented the same behaviors as the *in vitro* degradation.[42,61]

IV. CONCLUSION

In this chapter, we described the main characteristics of aliphatic polyesters during hydrolytic degradation. The surface/interior differentiation, or heterogeneous degradation of large-size polymers, is now regarded as a general phenomenon. Moreover, degradation can be the source of dramatic morphological changes depending on the initial gross composition. From the behaviors of several representative members of the PLAGA polymer family, a scheme has been proposed which can be used to qualitatively predict the fate of hydrolytically degraded PLAGA polymers, particularly morphological changes. Nevertheless, one must keep in mind that the degradation behaviors of a polymer can be very different depending on various intrinsical and extrinsical properties.

REFERENCES

1. Hoffman, A. S., Synthetic polymeric biomaterials, In *Polymeric Materials and Artificial Organs*, *ACS Symposium Series*, Gebelein, C. G., ed., Vol. 256, pp. 13–29, 1984.
2. Rosato, D. V., Polymers, processes and properties of medical plastics including markets and applications, In *Biocompatible Polymers, Metals and Composites*, Szycher, M., ed., Technomic Publishing, Lancaster, PA, pp. 1019–1067, 1983.
3. Charles, E. L., and Buffalo, N. Y., Preparation of high molecular weight polyhydroxyacetic ester, U.S. Patent 2,668,162, 1954. (February 2), (to E.I. DuPont de Nemours).

4. Schmitt, E. E., and Polistina, R. A., Surgical sutures, U.S. Patent 3,297,033, 1967 (January 10), (to Am. Cyanamid Co.).
5. Frazza, E. J., and Schmitt, E. E., A new absorbable suture, *J. Biomed. Mater. Res. Symp.*, 1, 43–58, 1971.
6. Li, S., and Vert, M., Biodegradation of aliphatic polyesters, In *Degradable Polymers: Principles and Applications*, Scott, G., ed., Kluwer Academic Publishers, Dordrecht, The Netherlands, pp. 77–131, 1995.
7. Albertsson, A. C., and Eklund, M., Influence of molecular structure on the degradation mechanism of degradable polymers: in vitro degradation of poly(trimethylene carbonate), poly(trimethylene carbonate-co-caprolactone), and poly(adipic anhydride), *J. Appl. Polym. Sci.*, 57, 87–103, 1995.
8. Mochizuki, M., and Hirami, M., Structural effects on the biodegradation of aliphatic polyesters, *Polym. Adv. Technol.*, 8, 203–209, 1996.
9. Holland, S. J., Tighe, B. J., and Gould, P. L., Polymers for biodegradable medical devices. 1. The potential of polyesters as controlled macromolecular release systems, *J. Control. Release*, 4, 155–180, 1986.
10. Vert, M., Li, S., Spenlehauer, G., and Guerin, P., Bioresorbability and biocompatibility of aliphatic polyesters, *J. Mater. Sci.: Mater. Med.*, 3, 432–446, 1992.
11. Li, S., Hydrolytic degradation characteristics of aliphatic polyesters derived from lactic and glycolic acids, *J. Biomed. Mater. Res.: Appl. Biomater.*, 48, 142–153, 1999.
12. Dunn, R. L., Clinical applications and update on the poly(α-hydroxy acids), In *Biomedical Applications of Synthetic Biodegradable Polymers*, Hollinger, J. O., ed., CRC Press, Boca Raton, FL, pp. 17–31, 1995.
13. Reed, A. M., and Gilding, D. K., Biodegradable polymers for use in surgery — poly(glycolic)/poly(lactic acid) homo- and copolymers: 2. In vitro degradation, *Polymer*, 20, 1459–1464, 1979.
14. Spinu, M., Jackson, C., Keating, M. Y., and Gardner, K. H., Material design in poly(lactic acid) systems: block copolymers, star homo- and copolymers, and stereocomplexes, *J. Macromol. Sci.: Pure Appl. Chem.*, A33, 1497–1530, 1996.
15. Fukuzaki, H., Yoshida, M., Asano, M., and Kumakura, M., Synthesis of copoly(D,L-lactic acid) with relatively low molecular weight and in vitro degradation, *Eur. Polym. J.*, 25, 1019–1026, 1989.
16. Fukuzaki, H., Yoshida, M., Asano, M., Kumakura, M., Mashimo, T., Yuasa, H., Imai, K., and Yamanaka, H., Synthesis of low molecular weight copoly(L-lactic acid/ε-caprolactone) by direct copolycondensation in the absence of catalysts, and enzymatic degradation of the polymers, *Polymer*, 31, 2006–2014, 1990.
17. Wang, N., Wu, X. S., Lujan-Upton, H., Donahue, E., and Siddiqui, A., Synthesis and characterization of lactic/glycolic acid oligomers, *Proc. Am. Chem. Soc.*, 76, 373–374, 1997.
18. Vert, M., Christel, P., Chabot, F., and Leray, J., Bioresorbable polymers for bone surgery, In *Macromolecular Biomaterials*, Hastings, G. W., and Ducheyne, P., eds., CRC Press, Boca Raton, FL, pp. 119–142, 1984.
19. Nakamura, T., Hitomi, S., Watanabe, S., Shimizu, Y., Jamshidi, K., Hyon, S. H., and Ikada, Y., Bioabsorption of polylactides with different molecular properties, *J. Biomed. Mater. Res.*, 23, 1115–1130, 1989.
20. Miller, R. A., Brady, J. M., and Cutright, D. E., Degradation rates of oral resorbable implants (polylactates and polyglycolates): rate modification with changes in PLA/PGA copolymer ratios, *J. Biomed. Mater. Res.*, 11, 711–719, 1977.
21. Chabot, F., Vert, M., Chapelle, S., and Granger, P., Configurational structures of lactic acid stereocopolymers as determined by C–H n.m.r, *Polymer*, 24, 53–60, 1983.
22. Pitt, C. G., Gratzl, M. M., Himmel, G. L., Surles, J., and Schindler, A., Aliphatic polyesters. 2. The degradation of poly(DL-lactide), poly(ε-caprolactone) and their copolymers in vivo, *Biomaterials*, 2, 215–220, 1981.
23. Pitt, C. G., Non-microbial degradation of polyesters: mechanisms and modifications, In *Biodegradable Polymers and Plastics*, Vert, M., Feijen, J., Albertsson, A. C., Scott, G., and Chiellini, E., eds., Royal Society of Chemistry, London, pp. 7–19, 1992.
24. Hutchinson, F. G., and Furr, B. J. A., Biodegradable polymers for the sustained release of polypeptides, *Biochem. Soc. Trans.*, 13, 520–523, 1985.

25. Sanders, L. M., McRae, G. I., Vitale, K. M., and Kell, B. A., Controlled delivery of an LHRH analogue from biodegradable injectable microspheres, *J. Control. Release*, 2, 187–195, 1985.
26. Kenley, R. A., Lee, M. O., Mahoney, T. R. II., and Sanders, L. M., Poly(lactide-co-glycolide) decomposition kinetics in vivo and in vitro, *Macromolecules*, 20, 2398–2403, 1987.
27. Schakenraad, J. M., Neuwenhuis, P., Molenaar, I., Helder, J., Dijkstra, P. J., and Feijen, J., In vivo and in vitro degradation of glycine/DL-lactic acid copolymers, *J. Biomed. Mater. Res.*, 23, 1271–1288, 1989.
28. Helder, J., Dijkstra, P. J., and Feijen, J., In vitro degradation of glycine/DL-lactic acid copolymers, *J. Biomed. Mater. Res.*, 24, 1005–1020, 1990.
29. Cohen, S., Yoshioka, T., Lucarelli, M., Hwang, L. H., and Langer, R., Controlled delivery systems for proteins based on poly(lactic/glycolic acid) microspheres, *Pharm. Res.*, 8, 713–720, 1991.
30. St Pierre, T., and Chiellini, E., Biodegradability of synthetic polymers for medical and pharmaceutical applications: Part 2 — Backbone hydrolysis, *J. Bioact. Comp. Polym.*, 2, 4–30, 1987.
31. Lewis, D. H., Controlled release of bioactive agents from lactide/glycolide polymers, *Drugs Pharm. Sci. (Biodegrad. Polym. Drug Deliv. Syst.)*, 45, 1–41, 1990.
32. Ginde, R. M., and Gupta, R. K., In vitro chemical degradation of poly(glycolic acid) pellets and fibres, *J. Appl. Polym. Sci.*, 33, 2411–2429, 1987.
33. Singh, M., Singh, A., and Talwar, G. P., Controlled delivery of diphtheria toxoid using biodegradable poly(D,L-lactide) microcapsules, *Pharm. Res.*, 8, 958–961, 1991.
34. Kimura, Y., Matsuzaki, Y., Yamane, H., and Kitao, T., Preparation of block copoly(ester–ether) comprising poly(lactide) and poly(oxypropylene) and degradation of its fibre in vitro and in vivo, *Polymer*, 30, 1342–1349, 1989.
35. Li, S., Garreau, H., and Vert, M., Structure–property relationships in the case of the degradation of massive poly(α-hydroxy acids) in aqueous media. part 1: poly(DL-lactic acid), *J. Mater. Sci.: Mater. Med.*, 1, 123–130, 1990.
36. Huffman, K. R., and Casey, D. J., Effects of carboxylic end groups on hydrolysis of polyglycolic acid, *J. Polym. Sci.: Polym. Chem. Ed.*, 23, 1939–1954, 1985.
37. Li, S., and Vert, M., Morphological changes resulting from the hydrolytic degradation of stereocopolymers derived from L- and DL-lactides, *Macromolecules*, 27, 3107–3110, 1994.
38. Li, S., Garreau, H., and Vert, M., Structure–property relationships in the case of the degradation of massive poly(α-hydroxy acids) in aqueous media. Part 2: Degradation of lactide/glycolide copolymers: $PLA_{37.5}GA_{25}$ and $PLA_{75}GA_{25}$, *J. Mater. Sci.: Mater. Med.*, 1, 131–139, 1990.
39. Li, S., Garreau, H., and Vert, M., Structure–property relationships in the case of the degradation of massive poly(α-hydroxy acids) in aqueous media. Part 3: influence of the morphology of poly(L-lactic acid), *J. Mater. Sci.: Mater. Med.*, 1, 198–206, 1990.
40. Vert, M., Li, S., and Garreau, H., More about the degradation of LA/GA-derived matrices in aqueous media, *J. Control. Release*, 16, 15–26, 1991.
41. Vert, M., Li, S., and Garreau, H., New insights on the degradation of bioresorbable polymeric devices based on lactic and glycolic acids, *Clin. Mater.*, 10, 3–8, 1992.
42. Thérin, M., Christel, P., Li, S., Garreau, H., and Vert, M., In vivo degradation of massive poly(α-hydroxy acids): validation of in vitro findings, *Biomaterials*, 13, 594–600, 1992.
43. Makino, K., Arakawa, M., and Kondo, T., Preparation and in vitro degradation properties of polylactide microcapsules, *Chem. Pharm. Bull.*, 33, 1195–1201, 1985.
44. Schmitt, E. A., Flanagan, D. R., and Linhardt, R. J., Importance of distinct water environments in the hydrolysis of poly(DL-lactide-*co*-glycolide), *Macromolecules*, 27, 743–748, 1994.
45. Ali, S. A. M., Doherty, P. J., and Williams, D. F., Mechanisms of polymer degradation in implantable devices. 2. Poly(DL-lactic acid), *J. Biomed. Mater. Res.*, 27, 1409–1418, 1993.
46. Zhang, X., Wyss, U. P., Pichora, D., and Goosen, M. F. A., An investigation of poly(lactic acid) degradation, *J. Bioact. Compat. Polym.*, 9, 80–100, 1994.
47. Pistner, H., Stallforth, H., Gutwald, R., Müling, J., Reuther, J., and Michel, C., Poly(L-lactide): a long-term degradation study in vivo, II. Physico-mechanical behavior of implants, *Biomaterials*, 15, 439–450, 1994.
48. Gogolewski, S., Jovanovic, M., Perren, S. M., Dillon, J. G., and Hughes, M. K., Tissue response and in vivo degradation of selected polyhydroxyacids: polylactides (PLA), poly(3-hydroxybutyrate)

(PHB), and poly(3-hydroxybutyrate-co-3-hydroxyvalerate) (PHB/HV), *J. Biomed. Mater. Res.*, 27, 1135–1148, 1993.

49. Mainil-Varlet, P., Curtis, R., and Gogolewski, S., Effect of in vivo and in vitro degradation on molecular and mechanical properties of various low molecular weight polylactides, *J. Biomed. Mater. Res.*, 36, 360–380, 1997.
50. Grijmpa, D. W., and Pennings, A. J., (Co)polymers of L-lactide. 1. Synthesis, thermal properties and hydrolytic degradation, *Macromol. Chem. Phys.*, 195, 1633–1647, 1994.
51. Li, S., Espartero, J. L., Foch, P., and Vert, M., Structural characterization and hydrolytic degradation of a Zn metal initiated copolymer of L-lactide and ε-caprolactone, *J. Biomater. Sci. Polym. Ed.*, 8, 165–187, 1996.
52. Löfgren, A., and Albertsson, A. C., Copolymers of 1,5-dioxepan-2-one with L- and DL-lactide: hydrolytic degradation behavior, *J. Appl. Polym. Sci.*, 52, 1327–1338, 1994.
53. Albertsson, A. C., and Eklund, M., Influence of molecular structure on the degradation mechanism of degradable polymers: in vitro degradation of poly(trimethylene carbonate), poly(trimethylene carbonate-*co*-caprolactone), and poly(adipic anhydride), *J. Appl. Polym. Sci.*, 57, 87–103, 1995.
54. Schliephake, H., Klosa, D., and Rahlff, M., Determination of the 3-D morphology of degradable biopolymer implants undergoing in vivo resorption, *J. Biomed. Mater. Res.*, 27, 991–998, 1993.
55. Pitt, C. G., Chaslow, F. I., Hibionada, Y. M., Klimas, D. M., and Scindler, A., Aliphatic polyesters. I. The degradation of poly(ε-caprolactone) in vivo, *J. Appl. Polym. Sci.*, 26, 3779–3787, 1981.
56. Fischer, E. W., Sterzel, H. J., and Wegner, G., Investigation of the structure of solution grown crystals of lactide copolymers by means of chemical reactions, *Kolloid-Z. u.Z. Polymere.*, 251, 980–990, 1973.
57. Carter, B. K., and Wilkes, G. L., Some morphological investigations on an absorbable copolyester biomaterial based on glycolic and lactic acid, In *Polymers as Biomaterials*, Shalaby, S. W., Hoffman, A. S., Ratner, B. D., and Horbert, T. A., eds., Plenum Press, New York, pp. 67–92, 1984.
58. Fredericks, R. J., Melveger, A. J., and Dolegiewtz, L. J., Morphological and structural changes in a copolymer of glycolide and lactide occurring as a result of hydrolysis, *J. Polym. Sci.: Polym. Phys. Ed.*, 22, 57–66, 1984.
59. Chu, C. C., Hydrolytic degradation of poly(glycolic acid): tensile strength and crystallinity study, *J. Appl. Polym. Sci.*, 26, 1727–1734, 1981.
60. Browning, A., and Chu, C. C., The effect of annealing treatments on the tensile properties and hydrolytic degradative properties of polyglycolic acid sutures, *J. Biomed. Mater. Res.*, 20, 613–632, 1986.
61. Leeslag, J. W., Pennings, A. J., Bos, R. R. M., Rozema, F. R., and Boering, G., Bioresorbable materials of poly(L-lactide). VII. In vivo and in vitro degradation, *Biomaterials*, 8, 311–314, 1987.
62. Pohjonen, T., and Törmälä, P., Hydrolytic degradation of ultra-high-strength self-reinforced poly-L-lactide. A temperature dependence study, In *Biodegradable Implants in Fracture Fixation*, Leung, K. S., Hung, L. K., and Leung, P. C., eds., The Chinese University of Hong Kong, Hong Kong, pp. 75–88, 1993.
63. Li, S., and Vert, M., Crystalline oligomeric stereocomplex as intermediate compound in racemic poly(DL-lactic acid) degradation, *Polym. Int.*, 33, 37–41, 1994.
64. Li, S., Girod-Holland, S., and Vert, M., Hydrolytic degradation of poly(DL-lactic acid) in the presence of caffeine base, *J. Control. Release*, 40, 41–53, 1996.
65. Chawla, A. S., and Chang, T. M. S., In vivo degradation of poly(lactic acid) of different molecular weights, *Biomed. Med. Dev. Art. Org.*, 13, 153–162, 1985–1986.
66. Kaetsu, I., Yoshida, M., Asano, M., Yamanaka, H., Imai, K., Yuasa, H., Mashimo, T., Susuki, K., Katakai, R., and Oya, M., Biodegradable implant composites for local therapy, *J. Control. Release*, 6, 249–263, 1987.
67. Ogawa, Y., Okada, H., Yamamoto, M., and Shimamoto, T., In vivo release profiles of leuprolide acetate from microcapsules prepared with polylactic acids or copoly(lactic/glycolic) acids and in vivo degradation of these polymers, *Chem. Pharm. Bull.*, 36, 2576–2581, 1988.
68. Mauduit, J., Bukh, N., and Vert, M., Gentamycin/poly(lactic cid) blends aimed at sustained release local antibiotic therapy administered per-operatively: III. The case of gentamycin sulfate in films of high and low molecular weight poly(DL-lactic acid), *J. Control. Release*, 25, 43–49, 1993.

69. Mauduit, J., Bukh, N., and Vert, M., Gentamycin/poly(lactic acid) blends aimed at sustained release local antibiotic therapy administered per-operatively: II. The case of gentamycin sulfate in high molecular weight poly(DL-lactic acid) and poly(L-lactic acid), *J. Control. Release*, 23, 221–230, 1993.
70. Zhu, J. H., Shen, Z. R., Wu, L. T., and Yang, S. L., In vitro degradation of polylactide and poly(lactide-co-glycolide) microspheres, *J. Appl. Polym. Sci.*, 43, 2099–2106, 1991.
71. Grizzi, I., Garreau, H., Li, S., and Vert, M., Biodegradation of devices based on poly(DL-lactic acid): size-dependence, *Biomaterials*, 16, 305–311, 1995.
72. Kwong, A. K., Chou, S., Sun, A. M., Sefton, M. V., and Goosen, M. F. A., In vitro and in vivo release of insulin from poly(lactic acid) microbeads and pellets, *J. Control. Release*, 4, 47–62, 1986.
73. Lam, K. H., Nieuwenhuis, P., Molenaar, I., Esselbrugge, H., Feijen, J., Dijkstra, P. J., and Schakenraad, J. M., Biodegradation of porous versus non-porous poly(L-lactic acid) films, *J. Mater. Sci.: Mater. Med.*, 5, 181–189, 1994.
74. Maulding, H. V., Tice, T. R., Cowsar, D. R., Fong, J. W., Pearson, J. E., and Nazareno, J. P., Biodegradable microcapsules: acceleration of polymeric excipient hydrolytic rate by incorporation of a basic medicament, *J. Control. Release*, 3, 103–117, 1986.
75. Maulding, H. V., Prolonged delivery of peptides by microcapsules, *J. Control. Release*, 6, 167–176, 1987.
76. Kishida, A., Yoshioka, S., Takeda, Y., and Uchiyama, M., Formulation-assisted biodegradable polymer matrices, *Chem. Pharm. Bull.*, 37, 1954–1956, 1989.
77. Mauduit, J., Bukh, N., and Vert, M., Gentamycim/poly(lactic acid) blends aimed at sustained release local antibiotic therapy administered per-operatively. I. The case of gentamycin base and gentamycin sulfate in poly(DL-lactic acid) oligomers, *J. Control. Release*, 23, 209–220, 1993.
78. Verheyen, C. C. P. M., Klein, C. P. A. T., De Blieckhogervorst, J. M. A., Wolke, J. G. C., Van Blitterswijn, C. A., and De Groot, K., Evaluation of hydroxylapatite/poly(L-lactide) composites: physico-chemical properties, *J. Mater. Sci.: Mater. Med.*, 4, 58–65, 1993.
79. Li, S., and Vert, M., Hydrolytic degradation of coral/poly(DL-lactic acid) bioresorbable material, *J. Biomat. Sci.: Polym. Ed.*, 7, 817–827, 1996.
80. Zhang, Y., Zale, S., Sawyer, L., and Bernstein, H., Effects of metal salts on poly(DL-lactide-co-glycolide) polymer hydrolysis, *J. Biomed. Mater. Res.*, 34, 531–538, 1997.
81. Chu, C. C., Degradation phenomena of two linear aliphatic polyester fibres used in medicine and surgery, *Polymer*, 26, 591–594, 1985.
82. Spenlehauer, G., Vert, M., Benoit, J. P., and Boddaert, A., In vitro and in vivo degradation of poly(DL-lactide/glycolide) type microspheres made by solvent evaporation method, *Biomaterials*, 10, 557–563, 1989.
83. Birkinshaw, C., Buggy, M., Henn, G. G., and Jones, E., Irradiation of poly(DL-lactide), *Polym. Degrad. Stab.*, 38, 249–253, 1992.
84. Volland, C., Wolff, M., and Kissel, T., The influence of terminal gamma-sterilization on captopril containing poly(DL-lactide-*co*-glycolide) microspheres, *J. Control. Release*, 31, 293–305, 1992.
85. Li, S., and McCarthy, S. P., Further investigations on the hydrolytic degradation of poly(DL-lactide), *Biomaterials*, 20, 35–44, 1999.

Part IV

Tissue Engineering Applications

24 Biomaterials for Genitourinary Tissue Engineering

Priyabrata Mukherjee and Anthony Atala

CONTENTS

I. INTRODUCTION

Lost or malfunctioning genitourinary tissues have traditionally been reconstructed with native nonurologic tissues (e.g., gastrointestinal segments,[1,2] skin,[3] peritoneum,[4] fascia,[5] omentum,[6] and dura[7,8]) or synthetic prostheses (silicone,[9–11] polyvinyl,[12,13] and Teflon[14–18]). Although reconstructive therapies using these materials have saved and improved countless lives, they remain imperfect solutions. Reconstruction with nonurologic native tissues rarely replaces the entire function of the original tissue and bears the risk of complications, including metabolic abnormalities, infection, perforation, and malignancy.[19–22] Furthermore, the limited amount of autologous donor tissue confines these types of reconstruction. The use of synthetic prostheses have usually failed due to the wide array of complications (e.g., device malfunction, infection, and stone formation) associated with mechanical or biocompatibility problems.[23]

Tissue engineering has emerged as a potential alternative to the current therapies for genitourinary tissue reconstruction.[24] In this approach, new functional genitourinary tissue is reconstructed by transplanting cells using biocompatible biomaterials, or by inducing tissue ingrowth from the surrounding tissue onto the biomaterials. Biomaterials play a central role in engineering functional genitourinary tissues. In this chapter, we will discuss design criteria and types of biomaterials that serve as scaffolds for engineering genitourinary tissue, and review the current concepts of genitourinary tissue engineering using various types of biomaterials.

II. IMPORTANCE OF BIOMATERIALS

Biomaterials in genitourinary tissue engineering function as an artificial extracellular matrix (ECM), and elicit biological and mechanical functions of native ECM found in tissue in the body.

Native ECM brings cells together in tissue, controls the tissue structure, and regulates the cell phenotype.[25] Biomaterials facilitate the localization and delivery of cells and bioactive factors (e.g., cell-adhesion peptides and growth factors) to desired sites in the body, define a three-dimensional space for the formation of new tissues with appropriate structure, and guide the development of new tissues with appropriate function.[26] Direct injection of cell suspensions without biomaterial matrices have been utilized in some cases,[27,28] but it is difficult to control the localization of transplanted cells. In addition, most mammalian cell types are anchorage dependent and will die if not provided with a cell-adhesion substrate. Biomaterials provide a cell-adhesion substrate and can be used to achieve cell delivery with high loading and efficiency to specific sites in the body. The configuration of the biomaterials can guide the structure of an engineered tissue. The biomaterials provide mechanical support against *in vivo* forces, thus maintaining a predefined structure during the process of tissue development. The biomaterials can be loaded with bioactive signals, such as cell-adhesion peptides and growth factors which can regulate cellular function.

III. DESIGN AND SELECTION OF BIOMATERIALS

The design and selection of the biomaterial is critical in the development of engineered genitourinary tissues. The biomaterial must be capable of controlling the structure and function of the engineered tissue in a predesigned manner by interacting with the transplanted cells and the host cells. Generally, the ideal biomaterial should be biocompatible, promote cellular interaction and tissue development, and possess proper mechanical and physical properties.

The selected biomaterial should be biodegradable and bioresorbable to support the reconstruction of a completely normal tissue without inflammation. Such behavior of the biomaterials would avoid the risk of inflammatory or foreign-body responses which may be associated with the permanent presence of a foreign material *in vivo*. The degradation products should not provoke inflammation or toxicity and must be removed from the body via metabolic pathways. The degradation rate and the concentration of degradation products in the tissues surrounding the implant must be at a tolerable level.[29]

The biomaterials should provide an appropriate regulation of cell behavior, such as adhesion, proliferation, migration, and differentiation, in order to promote the development of functional new tissue. Cell behavior in engineered tissues is regulated by multiple interactions with the microenvironment, including interactions with cell-adhesion ligands[30] and with soluble growth factors.[31] Cell adhesion-promoting factors (e.g., Arg–Gly–Asp, RGD) can be presented by the biomaterial itself or be incorporated into the biomaterial in order to control cell behavior through ligand-induced cell receptor signaling processes.[32,33] The biomaterial can also serve as a depot for the local release of growth factors and other bioactive agents that induce tissue-specific gene expression of the cells.[34]

The biomaterials should possess appropriate mechanical properties to regenerate tissue with predefined sizes and shapes. The biomaterials provide temporary mechanical support sufficient to withstand *in vivo* forces exerted by the surrounding tissue and maintain a potential space for tissue development. The mechanical support of the biomaterials should be maintained until the engineered tissue has sufficient mechanical integrity to support itself. This can be potentially achieved by an appropriate choice of mechanical and degradative properties of the biomaterials.[35]

The biomaterials need to be processed into specific configurations. A large surface-area-to-volume ratio is often desirable in order to allow the delivery of a high density of cells. A high porosity, interconnected pore structures, and specific pore sizes promote tissue ingrowth from the surrounding host tissue. Several techniques have been developed which readily control porosity, pore size, and pore structure.[26]

IV. CLASSIFICATION OF BIOMATERIALS

Generally, three classes of biomaterials have been utilized for engineering genitourinary tissues: naturally derived materials (e.g., collagen and alginate), acellular tissue matrices (e.g., bladder submucosa and small intestinal submucosa (SIS)), and synthetic polymers (e.g., polyglycolic acid (PGA), polylactic acid (PLA), and poly (lactic-*co*-glycolic acid) (PLGA)). Naturally derived materials and acellular tissue matrices have the potential advantage of biological recognition. Synthetic polymers can be manufactured reproducibly on a large scale with controlled properties of their strength, degradation rate, and microstructure.

Collagen is the most abundant and ubiquitous structural protein in the body, and may be readily purified from both animal and human tissues with enzyme treatment and salt/acid extraction.[36] Collagen has long been known to exhibit minimal inflammatory and antigenic responses,[37] and has been approved by the U.S. Food and Drug Administration (FDA) for many types of medical applications, including wound dressings and artificial skin.[38] Collagen implants degrade through a sequential attack by lysosomal enzymes. The *in vivo* resorption rate can be regulated by controlling the density of the implant and the extent of intermolecular cross-linking. The lower the density, the greater the interstitial space and generally, the larger the pores for cell infiltration, leading to a higher rate of implant degradation. Intermolecular cross-linking reduces the degradation rate by making the collagen molecules less susceptible to an enzymatic attack. Intermolecular cross-linking can be accomplished by various physical (e.g., UV radiation and dehydrothermal treatment) or chemical (e.g., glutaraldehyde, formaldehyde, and carbodiimides) techniques.[39] Collagen contains cell-adhesion domain sequences (e.g., RGD) which exhibit specific cellular interactions. This may assist to retain the phenotype and activity of many types of cells, including fibroblasts[40] and chondrocytes.[41] Collagen exhibits high tensile strength and flexibility, and these mechanical properties can be further enhanced by intermolecular cross-linking. This material can be processed into a wide variety of structures (e.g., sponges (Figure 24.1(a)), fibers, and films).[26,42,43]

Alginate, a polysaccharide isolated from sea weed, has been used as an injectable cell delivery vehicle[44] and a cell immobilization matrix,[45] owing to its gentle gelling properties in the presence of divalent ions such as calcium. Alginate is biocompatible and approved by the FDA for human use as wound dressing material. Alginate is a family of copolymers of D-mannuronate and L-guluronate. The physical and mechanical properties of alginate gel are strongly correlated with the proportion and length of polyguluronate block in the alginate chains.[44] However, alginate does not possess a biologic recognition domain. In addition, the range of mechanical properties available from alginate hydrogels is quite limited and changes in a noncontrollable manner, presumably due to the loss of ionic cross-linking. Recently, efforts were made to synthesize biodegradable alginate hydrogels with mechanical properties controllable in a wide range by intermolecular covalent cross-linking and with cell-adhesion peptides coupled to their backbones.[46]

Acellular tissue matrices are collagen-rich matrices prepared by removing cellular components from tissue (Figure 24.1b). The matrices are often prepared by mechanical and chemical manipulation of a segment of bladder tissue.[47–50] The matrices slowly degrade upon implantation, and are replaced and remodeled by ECM proteins synthesized and secreted by transplanted or ingrowing cells. Acellular tissue matrices have proven to support cell ingrowth and regeneration of genitourinary tissues, including urethra and bladder, with no evidence of immunogenic rejection.[47,50] Since the structure of the proteins (e.g., collagen and elastin) in acellular matrices are well conserved and normally arranged, the mechanical properties of the acellular matrices are not significantly different from those of native bladder submucosa.[48]

Polyesters of naturally occurring α-hydroxy acids, including PGA, PLA, and PLGA, are widely used in tissue engineering. These polymers have gained FDA approval for human use in a variety of applications, including sutures.[51] The ester bonds in these polymers are hydrolytically labile, and these polymers degrade by nonenzymatic hydrolysis. The degradation products of PGA, PLA, and

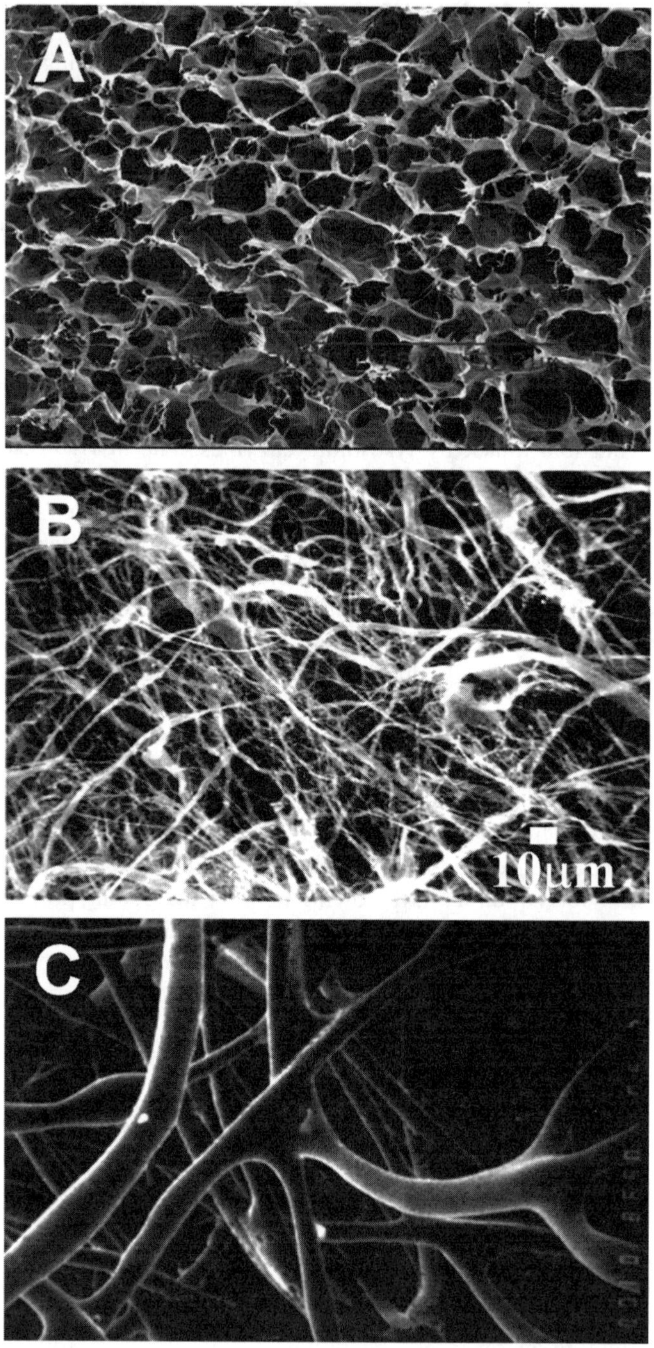

FIGURE 24.1 Representative scanning electron micrographs of three classes of biomaterials: (a) naturally derived biomaterial (collagen sponge), (b) acellular matrix (bladder submucosa), and (c) synthetic polymer (polyglycolic acid matrix).

PLGA are nontoxic, natural metabolites and are eventually eliminated from the body in the form of carbon dioxide and water.[51] The degradation rate of these polymers can be tailored from several weeks to several years by altering crystallinity, initial molecular weight, and the copolymer ratio of lactic to glycolic acid. Since these polymers are thermoplastics, they can be easily formed into a three-dimensional scaffold with a desired microstructure, gross shape and dimension by various techniques, including molding, extrusion,[52] solvent casting,[53] phase separation techniques,[54] and a gas foaming technique.[55] Many applications in genitourinary tissue engineering often require a scaffold with high porosity and ratio of surface area to volume. This has been addressed by processing biomaterials into configurations of fiber meshes (Figure 24.1(c)) and porous sponges using the techniques described above. The mechanical properties of the scaffold can be controlled by the fabrication process. A drawback of the synthetic polymers is the lack of biological recognition. As an approach toward incorporating cell recognition domain into these materials, copolymers with amino acids have been synthesized.[32,33,56,57] Other biodegradable synthetic polymers, including poly (anhydrides) and poly (ortho-esters), can also be used to fabricate scaffolds for genitourinary tissue engineering with controlled properties.[58]

V. BIOMATERIALS UTILIZED TO ENGINEER GENITOURINARY TISSUES

A. KIDNEYS

The kidney's function is dependent on the growth and differentiation of its precursor cells within the intermediate mesoderm (metanephric blastema and ureteric bud) into a mature organ consisting of many different cell types. There are at least 26 different terminally differentiated cell types in the kidney of a new born mouse that arise from at least four cell types present in the undifferentiated metanephric blastema when renal development begins.[59]

Synthetic biodegradable polymers, PGA, and polycarbonate have been utilized as cell transplantation vehicles in an attempt to reconstruct functional renal units *in vivo*. Augmentation of either isolated or total renal function with renal cell transplantation may be a feasible solution for complete replacement therapy. In two studies, renal cells were harvested, expanded *in vitro*, seeded onto polymer scaffolds, and subcutaneously implanted into athymic mice. The transplanted cells proliferated and organized into glomeruli and tubule-like structures. Neovascularity was also formed in the tissue constructs. These structures allowed for solute transport by the tubular cells across the membrane, resulting in the excretion of high levels of uric acid and creatinine through a urine-like fluid.[60] Recent success in harvesting and expanding renal cells *in vitro* and the development of biologically active scaffolds may allow the creation of functional renal units that can be applied for partial or, eventually, full replacement of kidney function.

Generation of histocompatible renal tissue using nuclear transplantation could provide unlimited source for renal cells in regenerative therapies. In our recent efforts we have demonstrated *in vivo* reconstitution and structural modeling of renal tissues from kidney cells using nuclear transplantation.[61] Renal cells were isolated from a 56-day-old cloned metanephros. After expansion, renal cells produced both erythropoietin and, 1,25-dihydroxyvitamin D_3, a key endocrine metabolite. The cloned cells were seeded onto collagen-coated cylindrical polycarbonate membranes. Units with cloned cells, without cells, and with cells from an allogenic control fetus were transplanted subcutaneously and retrieved 12 weeks after implantation into the nuclear donor animal. A straw colored fluid was seen in the reservoirs of the cloned groups. Chemical analysis of the fluid suggested the unidirectional secretion and concentration of the urea nitrogen and the creatinine was close to normal. The retrieved implants showed extensive vascularization and had selfassembled into glomeruli and tubule like structures. No immunogenic rejection response was detected in the hosts with the cloned renal cells.

B. URETERS

The major clinical stumbling blocks in the use of biomaterials for ureteral tissue reconstruction are the formation of encrustation due to calcification and the formation of bacterial bio films.[62]

Collagen tubular sponges were utilized to transplant bladder cells for replacement of ureteral segments in dogs. Five to 12 weeks following implantation, extensive regeneration of uroepithelial cell layers occurred on the luminal side with no evidence of severe hydronephrosis. However, the study showed severe stricture formation and papillary mucosal thickening at the anastomotic sites. In addition, muscle regeneration into the collagen grafts was not evident. In a urine exposure test, severe salt deposits were noted.[63]

Ureteral acellular matrices were utilized as a scaffold for the ingrowth of ureteral tissue in rats. The matrices were prepared by removing cell and lipid components from ureters. Upon implantation, the acellular matrices promoted the regeneration of all ureteral wall components with no evidence of rejection. At 3 weeks, complete epithelialization and progressive vessel infiltration occurred. At 10 weeks, smooth muscle fibers were observed. At 12 weeks, nerve fibers were first detected.[48]

Biodegradable polymer scaffolds have also been utilized as cell transplantation vehicles to reconstruct ureteral tissues. In one study, urothelial and smooth muscle cells isolated from bladders and expanded *in vitro* were seeded onto PGA scaffolds with tubular configurations and implanted subcutaneously into athymic mice. Following implantation, the urothelial cells proliferated to form a multilayered luminal lining of tubular structures, while smooth muscle cells organized into multilayered structures surrounding the urothelial cells. Abundant angiogenesis was evident. The degradation of the polymer scaffolds resulted in the eventual formation of natural urothelial tissues.[64] This study suggested that it was possible to engineer urologic tissues containing multiple cell types. This approach was expanded to replacing ureters in dogs by transplanting smooth muscle cells and urothelial cells on tubular polymer scaffolds.[65] In a pig model abundant myofibroblast regeneration was observed but there was lack of smooth muscle regeneration.[66] The problems of encrustation and biofilm formation have been addressed in many of the current reports.[66–68]

C. Bladder

Gastrointestinal tissues have been used for bladder replacement but entail inherent complications including infection, metabolic disturbances, and urolithiasis. Alternative attempts include use of matrices for tissue regeneration and cell-based transplantation.[69–71]

Collagen/Vicryl® composite membranes were utilized as a scaffold for tissue ingrowth to repair a full thickness defect in the bladder of rabbits. The collagen membranes were reinforced with meshes of Vicryl, a biodegradable polymer composed of PLGA, to strengthen the collagen membranes that are too soft to suture reliably. The results of the initial study were not encouraging due to the occurrence of severe infection.[73] However, a later study obtained a high success rate when the experiments were repeated using purification and gamma irradiation of collagen, and postoperative administration of antibiotics. At 3 weeks, normal urothelium was noted. At 6 weeks, no implanted biomaterial was identified. At 35 weeks, smooth muscle regeneration was evident. During this period, there was no evidence of urinary leakage, infection, or bladder calculi.[74]

The allogenic acellular bladder matrix has served as a scaffold for the ingrowth of host bladder wall components in rats. The matrix was prepared by mechanically and chemically removing all cellular components from bladder tissue.[50] Partial cystectomy (25 to 50%) was performed, followed by augmentation cystoplasty using acellular bladder matrices. The mucosal lining was complete within 10 days. After 4 weeks, muscular and vascular regeneration was completed. Nerve regeneration continued to improve until week 20.[50,75] The grafted bladders had significantly better capacity and compliance than the autoregenerated bladders after partial cystectomy alone. The bladders regenerated with acellular matrix grafts exhibited contractile activity to electric and carbachol stimulation.[49] Clinically relevant antigenicity was not evident.[50] However, there was a 26 to 63% incidence of bladder stone formation.[50,72,75]

In a series of studies, xenogenic SIS was utilized as a graft material to regenerate bladder tissues. SIS is a collagen-rich biomaterial prepared from the small intestine by removing the serosa, the mucosa from the inner surface, and the tunica muscularis from the outer surface of the intestine. In rat or dog models, partial cystectomy (35 to 70%) was performed, followed by bladder augmentation with SIS. All three layers of the bladder (mucosa, smooth muscle, and serosa) were regenerated with normal architecture, but with decreased muscle content.[76] Cholinergic and purinergic innervations also occurred.[77] The regenerated bladders exhibited compliance similar to that of normal bladders.[78]

Allogenic bladder submucosa was utilized as a biomaterial for bladder augmentation in dogs. The regenerated bladder tissues contained a normal cellular organization consisting of urothelium and smooth muscle, and exhibited a normal compliance. Biomaterials preloaded with cells prior to their implantation showed better tissue regeneration, compared to biomaterials implanted with no cells in which the tissue regeneration depended on the ingrowth of the surrounding tissue. The bladders showed a significant increase (100%) in capacity when augmented with scaffolds seeded with cells, compared to scaffolds without cells (30%).[79]

A viable bladder substitute was created by transplanting smooth muscle cells and urothelial cells on biodegradable polymer scaffolds into dogs. Most previous approaches have reconstructed the bladder through the replacement of partial defective walls using various biomaterials.[49,50,75–79] In this study, transplantable whole bladders were created using cultured bladder cells and polymer scaffolds. The bladder-shaped polymer scaffolds were fabricated by configuring PGA fiber-based scaffolds into a bladder-shaped mold and coating the scaffolds with a 50:50 copolymer of glycolide and lactide. Urothelial and smooth muscle cells were isolated from 1 cm^2 bladder biopsies, expanded *in vitro*, and seeded on the scaffolds. These neo-bladders were implanted in dogs that had the majority of their native bladders excised. In functional evaluations for up to 11 months, the bladder neo-organs demonstrated a normal capacity to retain urine, normal compliance, ingrowth of neural structures, and a normal histological architecture, including a normal concentration of urothelium, submucosa, and muscle. The bladder-shaped molds implanted alone, without cells, were fibrotic, with a paucity of muscle, and collapsed over time.[80]

Recently, we evaluated the *in vitro* biocompatibility of a number of naturally derived and synthetic biomaterials, namely, bladder submucosa, small intestine submucosa, collagen, alginate, polyglycolic acid, poly (L-lactic acid) and poly(L-lactic-*co*-glycolic acid) using normal human bladder smooth muscle cells.[81] All the biomaterials tested in this study, except alginate, exhibited nontoxic and bioactive effects. This study provides information on cell–biomaterial interactions and on the ability of biomaterials to support bioactive cell functions.

In another study, we investigated the phenotypic and functional characteristics of tissue engineered bladder smooth muscle derived from patients with functionally normal bladders and functionally abnormal extrophic and neuropathic bladders.[82] We demonstrated that the tissue engineered muscle from normal and diseased bladders retain their phenotype *in vitro* as well as after implantation *in vivo* in nude mice model. The cells exhibited the same degree of contractibility to electrical and chemical stimulations, regardless of their origin. These results suggest that there are no phenotypic or functional differences between muscle cells obtained from urodynamically normal or pathological bladders, and bladder muscle cells, regardless of their origin, may have the potential to be engineered into normal bladder tissues.

D. Urethra

Woven meshes of PGA (Dexon®) were used to reconstruct urethras in dogs. Three to four centimeters of the ventral half of the urethral circumference and its adjacent corpus spongiosum were excised, and the polymer mesh was sutured to the defective area. After 2 weeks, the animals were able to void through the neourethra. At 2 months, the urothelium was completely regenerated.

The polymer meshes were completely absorbed after 3 months. No complications occurred. However, the excised corpus spongiosum did not regenerate.[83]

PGA has been also used as a cell transplantation vehicle to engineer tubular urothelium *in vivo*. Cultured urothelial cells were seeded onto tubular PGA scaffolds and implanted into athymic mice. At 20 and 30 days polymer degradation was evident, and tubular urothelium formed in cells which were stained for an urothelium-associated cytokeratin.[84]

PGA mesh tubes coated with polyhydroxybutyric acid (PHB) were used to reconstruct urethras in dogs. PHB is a biodegradable thermoplastic polymer produced microbially. PHB degrades by both hydrolysis and enzyme reaction. The hydrolyzed product, 3-hydroxybutyric acid, is a natural metabolite that is contained in human blood.[85] Eight to 12 months later, complete regeneration of urothelium and adjacent connective tissue occurred. All of the polymers disappeared after 1 year. There were no anastomotic strictures or inflammatory reactions.[86]

Recently, an acellular collagen matrix has proven to be a suitable graft for repairing urethral defects, both experimentally and clinically. The acellular collagen matrix was obtained from porcine bladders for the animal studies. The neourethras demonstrated a normal urothelial luminal lining and organized muscle bundles, without any signs of strictures or complications. The animals were able to void through the neourethra.[47] These results were confirmed clinically in a series of patients whose urethral defects were repaired with human bladder acellular collagen matrices.[87] One of its advantages over nongenital tissue grafts used for urethroplasty (e.g., bladder mucosa[39] and buccal mucosa[88]) is that the material is "off-the-shelf." This eliminates the necessity of additional surgical procedures for graft harvesting, which may decrease operative time as well as the potential morbidity due to the harvest procedure.

We investigated the feasibility of using both cell seeded and nonseeded scaffolds for tubularized urethral replacement in a rabbit model.[89] The acellular collagen matrix was obtained and tubularized around a 16Fr catheter. Autologous cells were obtained by biopsy and expanded *in vitro*. The scaffolds were sequentially seeded with epithelial cells within the lumen and smooth muscle cells on the outer surface. The seeded grafts were incubated for 7 days. The unseeded grafts and controls were incubated for 24 h. Serial urethrography of the unseeded collagen matrices demonstrated the presence of strictures. In contrast, serial urethrography of the cell-seeded constructs confirmed the maintenance of a wide urethral caliber without any sign of strictures. Using antiCD-31 antibodies, angiogenesis was noted on the cell-seeded scaffolds within the organizing seromuscular layers. Unseeded grafts demonstrated a lack of vascularization in comparison to their seeded counterpart.

In another recent study we explored the feasibility of using bladder submucosa-based inert matrices as free graft substitutes for urethral stricture repair.[90] Autologous tissues from other sources have been used for urethral repair including genital or buccal tissues, tunica vaginalis and peritoneal grafts. However, complications due to hair growth, graft shrinkage, recurrent strictures, stone formation and diverticuli have been observed.[91] Nondegradable synthetic grafts have been proposed previously for urethral reconstruction but they have been associated with erosion, dislodgement, fistula, stenosis, extravasation, and calcification.[92,93] Cadaveric human bladder tissue was aseptically obtained and processed in strict compliance with state and federal guidelines. Patient ages ranging from 22 to 61 years were selected. The length of the stricture segments as determined during surgery ranged from 1.5 to 16 cm. After resection of the diseased fibrotic stricture segment, the matrix was anastomosed to the urethral plate in an onlay fashion. Out of 28 patients, 24 had a successful outcome after a single repair procedure and four had a postoperative anastomotic caliber decrease, which was treated with visual internal urethrotomies. All patients had anatomically and functionally potent urethras, as evidenced by retrograde urethrography and uroflowmetry. The repaired segments showed histological regeneration of the urethral tissue. This study confirmed the feasibility of using acellular bladder-based inert collagen matrices as free graft substitutes for patients with urethral stricture disease. The collagen matrix patch can be used as an "off-the-shelf" material in an onlay fashion.

E. Genital Tissue

Cartilage rods were created *in vivo* for penile prostheses by transplanting chondrocytes on PGA scaffolds. PGA fibers were configured into cylindrical rods. Chondrocytes isolated from the articular surface of a calf shoulder and expanded *in vitro* were seeded onto the PGA scaffolds and implanted subcutaneously into athymic mice. Cartilaginous tissues formed with the same shape as the initial implants. At 6 months, the cartilage rods were readily elastic and could withstand high degrees of pressure.[94] Additional studies demonstrated that autologous chondrocytes seeded on polymer scaffolds are able to form penile prostheses in an animal model.[95] Natural penile prostheses created from the patient's own cells may decrease the biocompatibility risks associated with the artificial prostheses.[96,97]

Corporal tissues were engineered *in vivo* by transplanting smooth muscle cells and endothelial cells on PGA scaffolds.[98] The cell-polymer constructs formed neocorporal tissue at 1 month, consisting of corporal smooth muscle and neovasculature formed by both the host and implanted endothelial cells. The availability of transplantable autologous corpus carvernosum tissue for use in reconstructive procedures would be beneficial to many patients with congenital and acquired abnormalities of the genital system.

Recently, we demonstrated that the engineering of human cartilage rods for various applications is possible. Chondrocytes were isolated from human ears, expanded *in vitro* and seeded on rod shaped biodegradable PGA scaffolds (1.2 cm diameter and 6.0 cm in length). The seeded scaffolds were maintained in stirring bioreactors for a month and then implanted subcutaneously into athymic mice. The specimens were retrieved after 2 months and analyzed. Histological analysis with hematoxylin and eosin, and alcian blue and toluidine showed the presence of mature and healthy cartilage as well as the presence of highly sulfated mucopolysaccharides in the ECMs. Most of the polymer scaffolds had degraded 2 months after implantation. The study of the mechanical behavior and the compressibility of the cartilage rods showed that the rods possessed appropriate mechanical properties to be used as penile prostheses.[99]

In another study we investigated the structural and functional integrity of the neo-corpora in a rabbit model by replacing an entire cross-sectional segment of both corporal bodies with autologous engineered tissue. Autologous acellular corporal collagen matrices were obtained from a donor rabbit penis. Autologous corpus cavernosal smooth muscle and endothelial cells were harvested, expanded *in vitro* and seeded on to the matrices. Cavernosography, cavernosometry, mating behavior, and sperm ejaculation after 6 months of implantation showed the structural and functional integrity of the neocorpora. This study demonstrated that corpus cavernosa could be created in a rabbit model by seeding autologous corpus cavernosal smooth muscle cells and endothelial cells on acellular collagen matrices.[100] We also investigated the possibility of using vaginal epithelial cells and smooth muscle cells to create vaginal tissue *in vivo*. Vaginal epithelial cells and smooth muscle cells from female rabbits were harvested, expanded *in vitro* and seeded onto PGA scaffolds. The cell seeded scaffolds were subcutaneously implanted into nude mice. After 1 week of implantation the retrieved polymer scaffolds demonstrated multilayer tissue strips of both cell types, and penetrating native vasculature was also noted. Increased organization of the smooth muscle and epithelial tissue was evident after 4 weeks. There was no evidence of tissue formation in the controls. Immunocytochemical analyses using antipancytokeratins confirmed the presence of vaginal epithelial cells in each of the constructs. Antialpha-actin smooth muscle antibodies also confirmed the presence of multilayer smooth muscle fibers and tissue at each time point. This study, for the first time, demonstrated that cell-seeded polymer scaffolds are able to form vascularized vaginal tissue *in vivo* that have phenotypic and functional properties similar to those of normal vaginal tissues.[101]

F. Injection Therapies

Endoscopic therapies using injectable bulking agents are attractive procedures to treat urinary incontinence and vesicoureteral reflux. The ideal material for an endoscopic treatment should be easily injectable, nonantigenic, nonmigratory, volume stable, and safe for human use.[102] The materials used experimentally or clinically for injection therapies include Teflon (polytetrafluoroethylene) microparticles,[16–18] polyvinyl alcohol,[53] autologous fat,[103] silicone microparticles,[10] and collagen.[104,105] However, none of these materials is ideal, due to migration, granuloma formation, and volume loss.[10,93,106,107]

Chondrocyte–alginate gel suspensions have been investigated as a possible injectable material. The gelling properties of alginate, combined with the low nutrient requirements of chondrocytes, may facilitate the formation of a stable bulking material. Chondrocytes mixed with alginate gel were injected subcutaneously into athymic mice. The injected suspensions formed cartilaginous tissues *in vivo* and showed no evidence of cartilage or alginate migration, granuloma formation, or volume loss.[108] Additional studies demonstrated that vesicoureteral reflux could be treated with an autologous chondrocyte–alginate suspension without any evidence of obstruction in a pig minimodel.[109] Clinical trials using this substance in humans are currently underway. Twenty-nine patients with vesicoureteral reflux underwent subureteric endoscopic injection of autologous chondrocytes harvested from the patient's ear. At 3 months postprocedure, 79% of the ureters were free of reflux after one or two injections.[110] Similar clinical trials with this technology have been initiated for the treatment of urinary incontinence. In addition to its use for the endoscopic treatment of reflux and incontinence, the system of injectable autologous cells may also be applicable for the treatment of other medical conditions, such as rectal incontinence, dysphonia, plastic reconstruction, and wherever an injectable biocompatible material is needed.

VI. FUTURE DIRECTIONS

Creating new materials by combining the advantages of both synthetic materials and naturally derived materials is an emerging area of tissue engineering. The hybrid materials would possess the specific biological activities of naturally derived materials as well as the favorable properties of synthetic materials, including widely controllable mechanical properties and good processability. Traditional synthetic biomaterials lack biological signals that regulate cellular functions. One interaction that directs cell behavior is cellular interaction with adhesion domains in biomaterials. The synthetic materials promote cell adhesion via indirect recognition, that is, by proteins (e.g., fibronectin and vitronectin) from the body fluids adsorbing nonspecifically to the material surface. A direct approach to promote specific cell adhesion on biomaterials would be the incorporation of cell-adhesion peptides found in natural ECM into synthetic materials. Other bioactive factors that can be incorporated into biomaterials to control engineered tissue development include growth factors[111] and DNA.[112,113]

To design a biomaterial capable of inducing tissue-specific gene expression of cells, it would be necessary to identify the key molecular components involved in the cell behaviors required for regulation of tissue function. One must also understand the related physicochemical principles that influence receptor-mediated cell regulation, since cell behavior can be influenced by quantitative aspects of the receptor–ligand interactions.[30] The expression of genes by cells in engineered tissue may also be regulated by growth factors,[31] interactions with other cells,[114] and mechanical stimuli imposed on the cells.[115,116] Once the key interactions that regulate specific cell functions are determined, the tissue-specific function of an engineered tissue could be retained by incorporating an appropriate combination of the specific signals into the biomaterials.

Engineering genitourinary tissues with multiple cell types organized in specific patterns is another challenge. This may be achieved through the development of biomaterials capable of high adhesiveness to selective cell types. Several cell-adhesion ligands with highly specific recognition sites could potentially be displayed spatially in a desirable pattern to induce specific cell organization schemes. This approach has been investigated to date in order to promote selective endothelial cell adhesion to engineered blood vessels.[117]

REFERENCES

1. Atala, A., Bauer, S. B., Hendren, W. H., and Retik, A. B., The effect of gastric augmentation on bladder function, *J. Urol.*, 149, 1099–1102, 1993.
2. Atala, A., and Hendren, W. H., Reconstruction with bowel segments, *Dial. Pediatr. Urol.*, 17, 7 1994.
3. Draper, J. W., and Stark, R. B., End results in the replacement of mucous membrane of the urinary bladder with thick-split grafts of skin, *Surgery*, 39, 434–440, 1956.
4. Hutschenreiter, G., Rumpelt, H. J., Klippel, K. F., and Hohenfellner, R., The free peritoneal transplant as a substitute for the urinary bladder wall, *Invest. Urol.*, 15, 375–379, 1978.
5. Neuhof, H., Fascia transplantation into visceral defects, *Surg. Gynecol. Obstet.*, 14, 383–427, 1917.
6. Goldstein, M. B., Dearden, L. C., and Gualtieri, V., Regeneration of subtotally cystectomized bladder patched with omentum: an experimental study in rabbits, *J. Urol.*, 97, 664–668, 1967.
7. Kelâmi, A., Ludtke-Handjery, A., Korb, G., Roll, J., Schnell, J., and Danigel, K. H., Alloplastic replacement of the urinary bladder wall with lyophilized human dura, *Eur. Surg. Res.*, 2, 195, 1970.
8. Kelâmi, A., Lyophilized human dura as a bladder wall substitute: experimental and clinical results, *J. Urol.*, 105, 518–522, 1971.
9. Bogash, M., Kohler, F. P., Scott, R. H., and Murphy, J. J., Replacement of the urinary bladder by a plastic reservoir with mechanical valves, *Urology*, 1, 900–903, 1960.
10. Henly, D. R., Barrett, D. M., Welland, T. L., O'Connor, M. K., Malizia, A. A., and Wein, A. J., Particulate silicone for use in periurethral injections: local tissue effects and search for migration, *J. Urol.*, 153, 2039–2043, 1995.
11. Small, M. P., Carrion, H. M., and Gordon, J. A., Small-Carrion penile prosthesis: new implant for management of impotence, *Urology*, 5, 479–486, 1975.
12. Kudish, H. G., The use of polyvinyl sponge for experimental cystoplasty, *J. Urol.*, 78, 232 1957.
13. Ulm, A. H., and Lo, M.-C., Total bilateral polyvinyl ureteral substitutes in the dogs, *Surgery*, 45, 313, 1959.
14. Bona, A. V., and De Gresti, A., Partial substitution of the bladder wall with Teflon tissue, *Minerva. Urol.*, 18, 43 1966.
15. Kocvara, S., and Zak, F., Ureteral substitution with Dacron and Teflon prosthesis, *J. Urol.*, 88, 365–376, 1962.
16. O'Donnell, B., and Puri, P., Treatment of vesicoureteric reflux by endoscopic injection of Teflon, *Br. Med. J.*, 289, 7–9, 1984.
17. Politano, V. A., Small, M. P., Harper, J. M., and Lynne, C. M., Periurethral Teflon injection for urinary incontinence, *J. Urol.*, 111, 180–183, 1974.
18. Politano, V. A., Periurethral polytetrafluoroethylene injection for urinary incontinence, *J. Urol.*, 127, 439–442, 1982.
19. Atala, A., Tissue engineering in the genitourinary system, In *Synthetic Biodegradable Polymer Scaffolds*, Atala, A. and Mooney, D. J., eds., Birkhäuser, Boston, MA, pp. 149–164, 1997.
20. Khoury, J. M., Timmons, S. L., Corbel, L., and Webster, G. D., Complications of enterocystoplasty, *Urology*, 40, 9–14, 1992.
21. Leong, C. H., and Ong, G. B., Gastrocystoplasty in dogs, *Aust. NZ J. Surg.*, 41, 272–279, 1972.
22. McDougal, W. S., Metabolic complications of urinary intestinal diversion, *J. Urol.*, 147, 1199–1208, 1992.
23. Atala, A., Use of non-autologous substances in VUR and incontinence treatment, *Dial. Pediatr. Urol*, 17, 11–12, 1994.
24. Kim, B. S., Mooney, D. J., and Atala, A., Tissue engineering: genitourinary system, In *Principles of Tissue Engineering*, 2nd ed., Lanza, R., Langer, R. and Vacanti, J. P., eds., Academic Press, San Diego, CA, pp. 979–983, 1999.
25. Alberts, B., Bray, D., Lewis, J., Raff, M., Roberts, K., and Watson, J. D., *Molecular Biology of the Cell*, Garland Publishing, New York, 1994, pp. 971–995.
26. Kim, B. S., and Mooney, D. J., Development of biocompatible synthetic extracellular matrices for tissue engineering, *Trends Biotechnol.*, 16, 224–230, 1998.
27. Brittberg, M., Lindahl, A., Nilsson, A., Ohlsson, C., Isaksson, O., and Peterson, L., Treatment of deep cartilage defects in the knee with autologous chondrocyte transplantation, *N. Engl. J. Med.*, 331, 889–895, 1994.

28. Ponder, K. P., Gupta, S., Leland, F., Darlington, G., Finegold, M., DeMayo, J., Ledley, F. D., Chowdhury, J. R., and Woo, S. L., Mouse hepatocytes migrate to liver parenchyma and function indefinitely after intrasplenic transplantation, *Proc. Natl Acad. Sci. USA*, 88, 1217–1221, 1991.
29. Bergsma, J. E., Rozema, F. R., Bos, R. R. M., van Rozendaal, A. W. M., de Jong, W. H., Teppema, J. S., and Joziasse, C. A. P., Biocompatibility and degradation mechanism of predegraded and non-degraded poly(lactide) implants: an animal study, *Mater. Med.*, 6, 715–724, 1995.
30. Hynes, R. O., Integrins: versatility, modulation and signaling in cell adhesion, *Cell*, 69, 11–25, 1992.
31. Deuel, T. F., Growth factors, In *Principles of Tissue Engineering*, Lanza, R. P., Langer, R. and Chick, W. L., eds., Academic Press, New York, pp. 133–149, 1997.
32. Barrera, D. A., Zylstra, E., Lansbury, P. T., and Langer, R., Synthesis and RGD peptide modification of a new biodegradable copolymer: poly(lactic acid-*co*-lysine), *J. Am. Chem. Soc.*, 115, 11010–11011, 1993.
33. Cook, A. D., Hrkach, J. S., Gao, N. N., Johnson, I. M., Pajvani, U. B., Cannizzaro, S. M., and Langer, R., Characterization and development of RGD-peptide-modified poly(lactic acid-*co*-lysine) as an interactive, resorbable biomaterial, *J. Biomed. Mater. Res.*, 35, 513–523, 1997.
34. Peters, M. C., Isenberg, B. C., Rowley, J. A., and Mooney, D. J., Release from alginate enhances the biological activity of vascular endothelial growth factor, *Biomater. Sci. Polym. Ed.*, 9, 1267–1278, 1998.
35. Kim, B. S., and Mooney, D. J., Engineering smooth muscle tissue with a predefined structure, *J. Biomed. Mater. Res.*, 41, 322–332, 1998.
36. Li, S. T., Biologic biomaterials: tissue-derived biomaterials (collagen), In *The Biomedical Engineering Handbook*, Brozino, J. D., ed., CRC Press, Boca Raton, FL, pp. 627–647, 1995.
37. Furthmayr, H., and Timpl, R., Immunochemistry of collagens and procollagens, *Int. Rev. Connect. Tissue Res.*, 7, 61 1976.
38. Pachence, J. M., Collagen-based devices for soft tissue repair, *J. Biomed. Mater. Res. (Appl. Biomater.)*, 33, 35–40, 1996.
39. Hendren, W. H., and Reda, E. F., Bladder mucosa graft for construction of male urethra, *J. Pediatr. Surg.*, 21, 189–192, 1986.
40. Silver, F. H., and Pins, G., Cell growth on collagen: a review of tissue engineering using scaffolds containing extracellular matrix, *J. Long-term Effects Med. Implants*, 2, 67–80, 1992.
41. Sam, A. E., and Nixon, A. J., Chondrocyte-laden collagen scaffolds for resurfacing extensive articular cartilage defects, *Osteoarthritis Cartilage*, 3, 47–59, 1995.
42. Cavallaro, J. F., Kemp, P. D., and Kraus, K. H., Collagen fabrics as biomaterials, *Biotechnol. Bioeng.*, 43, 781–791, 1994.
43. Yannas, I. V., and Burke, J. F., Design of an artificial skin. I. Basic design principles, *J. Biomed. Mater. Res.*, 14, 65–81, 1980.
44. Smidsrød, O., and Skjåk-Bræk, G., Alginate as an immobilization matrix for cells, *Trends Biotechnol.*, 8, 71–78, 1990.
45. Lim, F., and Sun, A. M., Microencapsulated islets as bioartificial endocrine pancreas, *Science*, 210, 908–910, 1980.
46. Rowley, J. A., Madlambayan, G., and Mooney, D. J., Alginate hydrogels as synthetic extracellular matrix materials, *Biomaterials*, 20, 45–53, 1999.
47. Chen, F., Yoo, J. J., and Atala, A., Acellular collagen matrix as a possible "off the shelf" biomaterial for urethral repair, *Urology*, 54, 407–410, 1999.
48. Dahms, S. E., Piechota, H. J., Nunes, L., Dahiya, R., Lue, T. F., and Tanagho, E. A., Free ureteral replacement in rats: regeneration of ureteral wall components in the acellular matrix graft, *Urology*, 50, 818–825, 1997.
49. Piechota, H. J., Dahms, S. E., Nunes, L. S., Dahiya, R., Lue, T. F., and Tanagho, E. A., *In vitro* functional properties of the rat bladder regenerated by the bladder acellular matrix graft, *J. Urol.*, 159, 1717–1724, 1998.
50. Probst, M., Dahiya, R., Carrier, S., and Tanagho, E. A., Reproduction of functional smooth muscle tissue and partial bladder replacement, *Br. J. Urol.*, 79, 505–515, 1997.
51. Gilding, D. K., Biodegradable polymers, In *Biocompatibility of Clinical Implant Materials*, Williams, D. F., ed., CRC Press, Boca Raton, FL, pp. 209–232, 1981.

52. Freed, L. E., Vunjak-Novakovic, G., Biron, R. J., Eagles, D. B., Lesnoy, E. C., Barlow, S. K., and Langer, R., Biodegradable polymer scaffolds for tissue engineering, *Biotechnology*, 12, 689–693, 1994.
53. Merguerian, P. A., McLorie, G. A., Khoury, A. E., Thorner, P., and Churchill, B. M., Submucosal injection of polyvinyl alcohol foam in rabbit bladder, *J. Urol.*, 144, 531–533, 1990.
54. Nam, Y. S., and Park, T. G., Porous biodegradable polymeric scaffolds prepared by thermally induced phase separation, *J. Biomed. Mater. Res.*, 47, 8–17, 1999.
55. Harris, L. D., Kim, B. S., and Mooney, D. J., Open pore biodegradable matrices formed with gas foaming, *J. Biomed. Mater. Res.*, 42, 396–402, 1998.
56. Barrera, D. A., Zylstra, E., Lansbury, P. T., and Langer, R., Copolymerization and degradation of poly(lactic acid-*co*-lysine), *Macromolecules*, 28, 425–432, 1995.
57. Intveld, P. J. A., Shen, Z. R., Takens, G. A. J., Dijkstra, P. J., and Feijen, J., Glycine glycolic acid based copolymers, *J. Polym. Sci. Polym. Chem.*, 32, 1063–1069, 1994.
58. Peppas, N. A., and Langer, R., New challenges in biomaterials, *Science*, 263, 1715–1720, 1994.
59. Al-Awqati, Q., and Oliver, J. A., Stem cells in the kidney, *Kidney Int.*, 61(2), 387–395, 2002.
60. Amiel, G. E., and Atala, A., Current and future modalities for functional renal replacement, *Urol. Clin. North Am.*, 26, 235–246, 1999.
61. Lanza, R. P., Chung, H. Y., Yoo, J. J., Wettstein, P. J., Blackwell, C., Borson, N., Hofmeister, E., Schuch, G., Soker, S., Moraes, C. T., West, M. D., and Atala, A., Generation of histocompatible tissues using nuclear transplantation, *Nat. Biotechnol.*, 20(7), 689–696, 2002.
62. Beiko, D. T., Knudsen, B. E., Watterson, J. D., and Denstedt, J. D., Biomaterials in urology, *Curr. Urol. Rep.*, 4(1), 51–55, 2003.
63. Tachibana, M., Nagamatsu, G. R., and Addonizio, J. C., Ureteral replacement using collagen sponge tube grafts, *J. Urol.*, 133, 866–869, 1985.
64. Atala, A., Freeman, M. R., Vacanti, J. P., Shepard, J., and Retik, A. B., Implantation *in vivo* and retrieval of artificial structures consisting of rabbit and human urothelium and human bladder muscle, *J. Urol.*, 150, 608–612, 1993.
65. Yoo, J. J., Satar, N., Retik, A. B., and Atala, A., Ureteral replacement using biodegradable polymer scaffolds seeded with urothelial and smooth muscle cells, *J. Urol.*, 153(Suppl.), 375A, 1995.
66. Jones, D. S., Djokic, J., McCoy, C. P., and Gorman, S. P., Poly(epsilon-caprolactone) and poly(epsilon-caprolactone)–polyvinylpyrrolidone-iodine blends as ureteral biomaterials: characterisation of mechanical and surface properties, degradation and resistance to encrustation *in vitro*, *Biomaterials*, 23(23), 4449–4458, 2002.
67. Laaksovirta, S., Valimaa, T., Isotalo, T., Tormala, P., Talja, M., and Tammela, T. L., Encrustation and strength retention properties of the self-expandable, biodegradable, self-reinforced L-lactide-glycolic acid co-polymer 80:20 spiral urethral stent *in vitro*, *J. Urol.*, 170(2 Pt 1), 468–471, 2003.
68. Gorman, S. P., Garvin, C. P., Quigley, F., and Jones, D. S., Design and validation of a dynamic flow model simulating encrustation of biomaterials in the urinary tract, *J. Pharm. Pharmacol.*, 55(4), 461–468, 2003.
69. Huard, J., Yokoyama, T., Pruchnic, R., Qu, Z., Li, Y., Lee, J. Y., Somogyi, G. T., de Groat, W. C., and Chancellor, M. B., Muscle-derived cell-mediated ex vivo gene therapy for urological dysfunction, *Gene Ther.*, 9(23), 1617–1626, 2002.
70. Atala, A., Future trends in bladder reconstructive surgery, *Semin. Pediatr. Surg.*, 11(2), 134–142, 2002.
71. Falke, G., Caffaratti, J., and Atala, A., Tissue engineering of the bladder, *World J. Urol.*, 18(1), 36–43, 2000.
72. Atala, A., Bladder regeneration by tissue engineering, *BJU Int.*, 88(7), 765–770, 2001.
73. Monsour, M. J., Mohammed, R., Gorham, S. D., French, D. A., and Scott, R., An assessment of a collagen/vicryl composite membrane to repair defects of the urinary bladder in rabbits, *Urol. Res.*, 15, 239, 1987.
74. Scott, R., Mohammed, R., Gorham, S. D., French, D. A., Monsour, M. J., Shivas, A., and Hyland, T., The evolution of a biodegradable membrane for use in urological surgery, *Br. J. Urol.*, 62, 26–31, 1988.

75. Sutherland, R. S., Baskin, L. S., Hayward, S. W., and Cunha, G. R., Regeneration of bladder urothelium, smooth muscle, blood vessels and nerves into an acellular tissue matrix, *J. Urol.*, 156, 571–577, 1996.
76. Kropp, B. P., Rippy, M. K., Badylak, S. F., Adams, M. C., Keating, M. A., Rink, R. C., and Thor, K. B., Regenerative urinary bladder augmentation using small intestine submucosa: urodynamic and histologic assessment in long-term canine bladder augmentations, *J. Urol.*, 155, 2098–2104, 1996.
77. Vaught, J. D., Kroop, B. P., Sawyer, B. D., Rippy, M. K., Badylak, S. F., Shannon, H. E., and Thor, K. B., Detrusor regeneration in the rat using porcine small intestine submucosal grafts: functional innervation and receptor expression, *J. Urol.*, 155, 374–378, 1996.
78. Kropp, B. P., Sawyer, B. D., Shannon, H. E., Rippy, M. K., Badylak, S. F., Adams, M. C., Keating, M. A., Rink, R. C., and Thor, K. B., Characterization of using small intestine submucosa regenerated canine detrusor: assessment of reinnervation, *in vitro* compliance and contractility, *J. Urol.*, 156, 599–607, 1996.
79. Yoo, J. J., Meng, J., Oberpenning, F., and Atala, A., Bladder augmentation using allogenic bladder submucosa seeded with cells, *Urology*, 51, 221–225, 1998.
80. Oberpenning, F., Meng, J., Yoo, J. J., and Atala, A., De novo reconstruction of a functional mammalian urinary bladder by tissue engineering, *Nat. Biotechnol.*, 17, 149–155, 1999.
81. Pariente, J. L., Kim, B. S., and Atala, A., *In vitro* biocompatibility assessment of naturally derived and synthetic biomaterials using normal human urothelial cells, *J. Biomed. Mater. Res.*, 55, 33–39, 2001.
82. Lai, J. Y., Yoon, C. Y., Yoo, J. J., Wulf, T., and Atala, A., Phenotypic and functional characterization of *in vivo* tissue engineered smooth muscle from normal and pathological bladders, *J. Urol.*, 168, 1853–1857, 2002.
83. Bazeed, M. A., Thüroff, J. W., Schmidt, R. A., and Tanagho, E. A., New treatment for urethral strictures, *Urology*, 21, 53–57, 1983.
84. Atala, A., Vacanti, J. P., Peters, C. A., Mandell, J., Retik, A. B., and Freeman, M. R., Formation of urothelial structures *in vivo* from dissociated cells attached to biodegradable polymer scaffolds *in vitro*, *J. Urol.*, 148, 658–662, 1992.
85. Holmes, P., Applications of PHB — a microbially produced thermoplastic, *Phys. Technol.*, 16, 32–36, 1985.
86. Olsen, L., Bowald, S., Busch, C., Carlsten, J., and Eriksson, I., Urethral reconstruction with a new synthetic absorbable device, *Scand. J. Urol. Nephrol.*, 26, 323–326, 1992.
87. Atala, A., Guzman, L., and Retik, A. B., A novel inert collagen matrix for hypospadias repair, *J. Urol.*, 162, 1148–1151, 1999.
88. Kropfl, D., Tucak, A., Prlic, D., and Verweyen, A., Using buccal mucosa for urethral reconstruction in primary and re-operative surgery, *Eur. Urol.*, 34, 216–220, 1998.
89. De Filippo, R. E., Yoo, J. J., and Atala, A., Urethral replacement using cell seeded tubularized collagen matrices, *J. Urol.*, 168, 1789–1792, 2002.
90. El-Kassaby, A. W., Retik, A. B., Yoo, J. J., and Atala, A., Urethral stricture repair with an off-the-shelf collagen matrix, *J. Urol.*, 169(1), 170–173, 2003.
91. Chen, F., Yoo, J. J., and Atala, A., Experimental and clinical experience using tissue regeneration for urethral reconstruction, *World J. Urol.*, 18(1), 67–70, 2000.
92. Wessells, H., and McAninch, J. W., Current controversies in anterior urethral stricture repair: free-graft versus pedicled skin-flap reconstruction, *World J. Urol.*, 16(3), 175–180, 1998.
93. Anwar, H., Dave, B., and Seebode, J. J., Replacement of partially resected canine urethra by polytetrafluoroethylene, *Urology*, 24(6), 583–586, 1984.
94. Yoo, J. J., Lee, I., and Atala, A., Cartilage rods as a potential material for penile reconstruction, *J. Urol.*, 160, 1164–1168, 1998.
95. Yoo, J. J., Park, H. J., Lee, I., and Atala, A., Autologous engineered cartilage rods for penile reconstruction, *J. Urol.*, 162, 1119–1121, 1999.
96. Kardar, A., and Pettersson, B. A., Penile gangrene: a complication of penile prosthesis, *Scand. J. Urol. Nephrol.*, 29, 355–356, 1995.
97. Nukui, F., Okamoto, S., Nagata, M., Kurokawa, J., and Fukui, J., Complications and reimplantation of penile implants, *Int. J. Urol.*, 4, 52–54, 1997.

98. Park, H. J., Yoo, J. J., Kershen, R. T., Moreland, R. B., Krane, R. J., and Atala, A., Reconstruction of human corpus cavernosum smooth muscle and endothelial cells *in vivo*, *J. Urol.*, 162, 1106–1109, 1999.
99. Kim, B. S., Yoo, J. J., and Atala, A., Engineering of human cartilage rods: potential applications for penile prosthesis, *J. Urol.*, 168, 1794–1797, 2002.
100. Kwon, G. T., Yoo, J. J., and Atala, A., Autologous penile corpora cavernosa replacement using tissue engineering techniques, *J. Urol.*, 168, 1754–1758, 2002.
101. De Filippo, R. E., Yoo, J. J., and Atala, A., Engineering of vaginal tissue *in vivo*, *Tissue Eng.*, 9(2), 301–306, 2003.
102. Kershen, R. T., and Atala, A., New advances in injectable therapies for the treatment of incontinence and vesicoureteral reflux, *Urol. Clin. North Am.*, 26, 81–94, 1999.
103. Palma, P. C., Riccetto, C. L., Herrmann, V., and Netto, N. R. Jr., Repeated lipoinjections for stress urinary incontinence, *J. Endourol.*, 11, 67–70, 1997.
104. Appell, R. A., Collagen injection therapy for urinary incontinence, *Urol. Clin. North Am.*, 21, 177–182, 1994.
105. Leonard, M. P., Canning, D. A., Epstein, J. I., Gearhart, J. P., and Jeffs, R. D., Local tissue reaction to the subureteral injection of glutaraldehyde cross-linked bovine collagen in humans, *J. Urol.*, 143, 1209–1212, 1990.
106. Claes, H., Stroobants, D., van Meerbeek, J., Verbeken, E., Knockaert, D., and Baert, L., Pulmonary migration following periurethral polytetrafluoroethylene injection for urinary incontinence, *J. Urol.*, 142, 821–822, 1989.
107. Malizia, A. A., Reiman, H. M., Myers, R. P., Sande, J. R., Barham, S. S., Benson, R. C., Dewanjee, M. K., and Ulz, W. J., Migration and granulomatous reaction after periurethral injection of polytef (Teflon), *JAMA*, 251, 3277–3281, 1984.
108. Atala, A., Cima, L. G., Kim, W., Paige, K. T., Vacanti, J. P., Retik, A. B., and Vacanti, C. A., Injectable alginate seeded with chondrocytes as a potential treatment for vesicoureteral reflux, *J. Urol.*, 150, 745–747, 1993.
109. Atala, A., Kim, W., Paige, K. T., Vacanti, C. A., and Retik, A. B., Endoscopic treatment of vesicoureteral reflux with a chondrocyte-alginate suspension, *J. Urol.*, 152, 641–643, 1994.
110. Diamond, D. A., and Caldamone, A. A., Endoscopic treatment of vesicoureteral reflux in children using autologous chondrocytes: preliminary results, *Pediatrics*, 102(Suppl. Part 2), 107A, 1998.
111. Bentz, H., Schroeder, J. A., and Estridge, T. D., Improved local delivery of TGF-2 by binding to injectable fibrillar collagen via difunctional polyethylene glycol, *J. Biomed. Mater. Res.*, 39, 539–548, 1998.
112. Shea, L. D., Smiley, E., Bonadio, J., and Mooney, D. J., DNA delivery from polymer matrices for tissue engineering, *Nat. Biotechnol.*, 17, 551–554, 1999.
113. Yoo, J. J., and Atala, A., A novel gene delivery system using urothelial tissue engineered neo-organs, *J. Urol.*, 158, 1066–1070, 1997.
114. Parsons-Wingerter, P. A., and Saltzman, W. M., Growth versus function in three-dimensional culture of single and aggregated hepatocytes within collagen gels, *Biotechnol. Prog.*, 9, 600–607, 1993.
115. Banes, A. J., Mechanical strain and the mammalian cell, In *Physical Forces and the Mammalian Cell*, Frangos, J. A., ed., Academic Press, New York, pp. 81–123, 1993.
116. Kim, B. S., Nikolovski, J., Bonadio, J., and Mooney, D. J., Cyclic mechanical strain regulates the development of engineered smooth muscle tissue, *Nat. Biotechnol*, 17(10), 979–983, 1999.
117. Hubbell, J. A., Massia, S. P., Desai, N. P., and Drumheller, P. D., Endothelial cell-selective materials for tissue engineering in the vascular graft via a new receptor, *Biotechnology*, 9, 568–572, 1991.

25 Engineered Blood Vessel Substitutes

Jan P. Stegemann, Shaneen L. Rowe, and Robert M. Nerem

CONTENTS

I. INTRODUCTION

The development of blood vessel substitutes using tissue engineering principles has rapidly progressed in the past decade in response to the clinical need for improved vascular grafts in surgical procedures, especially for small diameter applications. In the United States, around 40% of all deaths are caused by cardiovascular disease, and more than half of these are a result of coronary heart disease. Approximately 500,000 coronary artery bypass grafting procedures are performed annually, each requiring a small diameter vascular conduit to bypass a blocked coronary artery.[1] There has been progress in understanding the biological mechanisms behind vascular disease, and in developing pharmacological and interventional treatments, however, the *in vitro* and preclinical test-beds for evaluating these therapies and for studying vascular biology are still underdeveloped. For these reasons, there has been a strong interest in creating blood vessel analogs for use both *in vivo* as surgical grafts, and *in vitro* as models of the vasculature.

Natural blood vessels are not simply passive conduits for the flow of blood. Their ability to change their dimensions over the short and long term is important in maintaining appropriate hemodynamic conditions, and therefore in preserving patency. In addition, the smaller resistance and capacitance vessels perform an important function in the regulation of blood flow, by constricting and dilating based on neuronal, hormonal and autoregulatory mechanisms.[2] For large diameter (>6 mm inner diameter) vessel replacements, a number of synthetic materials have proven useful for producing vascular conduits. Although these approaches have worked adequately for peripheral grafting and as vascular access shunts, there are currently no acceptable synthetic blood vessel substitutes for small diameter applications. The primary reason is the difficulty in maintaining patency in purely synthetic grafts, due to the lack of an appropriate lumenal surface and the development of intimal hyperplasia at anastomotic sites. For this reason, cell-seeded grafts have been proposed as a more biologically appropriate solution. Initial efforts focused on reconstitution of the intimal layer, through endothelial cell seeding or recruitment, in order to promote hemocompatibility. In addition, more biologically compatible scaffolds were investigated as an alternative to synthetic polymers. These areas of research subsequently widened to include the incorporation of living cells in the vessel wall, in an effort to produce more fully functional grafts that mimic both the passive and active functions of the blood vessel.

The ideal blood vessel substitute would exactly mimic the structure and function of native vessels. Current tissue engineering approaches generally fall short of this ambitious goal, however, important strides have been made. A key requirement of implanted vascular grafts is their ability to remain patent for many years. Hemocompatibility is primarily determined by the lumenal lining of the vessel, and in most tissue engineering approaches the lining is expected to be a monolayer of functional endothelial cells. Another absolute requirement is that the mechanical properties of the engineered tissue must be suitable for long term implantation. This demands not only sufficient mechanical strength (suture retention, burst strength, tensile resistance), but also appropriate compliance and elasticity. In the case of biologically derived scaffolds, a key challenge is achieving appropriate polymer assembly through *in vitro* and *in vivo* processing. Additionally, in order for an implanted blood vessel substitute to fully replace the function of a native vessel, it must exhibit appropriate short and long term remodeling responses. This is largely a function of the smooth muscle cells that inhabit the vessel wall, and in most tissue engineering approaches these cells are supported in a three-dimensional scaffold material formed in the shape of a tube. Finally, any implanted material or tissue must avoid eliciting a damaging immune response. This is of particular concern in tissue engineering due to the incorporation of living cells in the engineered construct, making cell sourcing a key issue. Determining the right combination of scaffold materials and cell types, and structuring them appropriately, has remained a challenge in vascular tissue engineering.

This review of scaffolding strategies in vascular tissue engineering describes the currently most advanced techniques for producing a blood vessel substitute. The approaches are divided into those using synthetic polymer scaffolds (both permanent and degradable), those using naturally derived polymer scaffolds, and those using cell-secreted scaffolds. In all cases, the discussion is limited to approaches that involve the use of living mammalian cells, whether they are seeded on the construct initially, or recruited into the scaffold after implantation. The focus is on the current state of technology, rather than a historical review of the field, and the goal is to describe how different cell-scaffold combinations seek to address the challenges of creating mechanically sound, thrombosis resistant, immunologically safe blood vessel substitutes.

II. BLOOD VESSEL STRUCTURE AND FUNCTION

The ultimate goal of vascular tissue engineering is to recreate the function of natural blood vessels, the physical structure of which is shown schematically in Figure 25.1. The three main structural layers consist of distinct cell and matrix types, whose composition reflects their function. The intima is composed of a monolayer of endothelial cells with an underlying basement membrane consisting of loosely organized Type IV collagen and laminin. The intima lines the lumen of the blood vessel and provides the hemocompatible blood-contacting surface that is critical for the maintenance of vessel patency. The endothelium is also important in the regulation of vascular tone, by sensing flow conditions and producing signaling molecules in response. The middle layer of the natural blood vessel is the media, and is composed of smooth muscle cells in a matrix of Type I and Type III collagen, elastin and proteoglycans. The collagen fibers of the media are aligned circumferentially or helically along the axis of the vessel, reflecting the need for them to resist tension created by pulsatile blood flow. This layer has been identified as being the most significant mechanically, and is also responsible for the phasic and tonic contraction associated with regulation of blood flow by the vasculature.[3] The interaction between the endothelial cells of the intima and

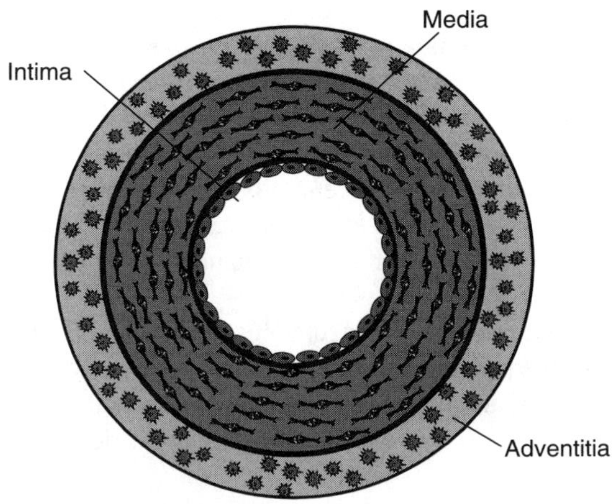

FIGURE 25.1 Schematic diagram of the structure of a normal blood vessel showing the three main tissue layers: the intima, the media, and the adventitia.

the smooth muscle cells of the media is a critical component of normal arterial function. The outer layer of the blood vessel, the adventitia, consists of loosely arranged Type I and Type III collagen, along with elastin and fibroblasts. Its structure and thickness varies with vessel size and location, and its function is as an anchoring connective tissue that maintains longitudinal tension on the blood vessel, and in larger vessels provides a bed for the ingrowth of a blood supply and nerves.

The primary function of blood vessels is to act as a conduit for the flow of blood, and they are therefore one of the main transport systems in the body. Compared to complex organs such as the liver and brain, blood vessels have a relatively simple structure and straightforward function. They are therefore good candidates for the application of a tissue engineering approach; using living cells and appropriate matrix materials, combined with defined biochemical and mechanical stimulation, to recreate tissue *in vitro*. It should be noted, however, that even a relatively simple tissue such as a muscular artery has a very specific matrix and cell organization, combined with an important and often complex set of signaling pathways, which lead to its normal biological function. Recapitulating this structure and obtaining the appropriate function is a problem that has been approached in a variety of ways.

The ideal blood vessel substitute will be able to withstand the same mechanical forces that native vessels are constantly exposed to, including cyclic radial stress, longitudinal stress, fluid shear and fluid pressure. However, it is not clear whether engineered blood vessels must have the same properties as the native vessel in order to restore function. It is likely that vessels with properties inferior to native vessels will be able to perform acceptably, at least initially, and that over time they will be further remodeled *in vivo* to better suit the environment. However, inappropriate mechanical matching over the longer term may lead to complications. Therefore, determination of the mechanical properties required for the replacement of vascular function is an important challenge in the field.

Biomechanical analysis of engineered vascular tissues can provide insight into the properties and expected behavior of these materials.[4,5] The burst or rupture strength is a critical property that indicates the pressure required to cause failure of the vessel wall. This can be determined experimentally or can be estimated from the ultimate tensile stress, which is the maximum value of stress on a conventional stress vs. strain diagram. The material modulus is a measure of stiffness that is determined by finding the slope of the linear portion of the stress vs. strain diagram. In native vessels, the major structural protein is Type I collagen, which provides tissue stiffness and high tensile strength.

Vessel compliance and elasticity are also important. Compliance is the ability of a material to dilate under pressure, defined as the change in volume in response to a change in pressure. Elasticity is the capability of a material to regain its original dimensions after deformation. In native vessels, fibrillar collagen and elastin combine to provide a matrix that is both compliant and elastic, allowing dilation without permanent deformation under the pulsatile pressures in the cardiovascular system. Native vessels and blood substitutes generally also exhibit viscoelastic behavior, meaning they exhibit properties of both a viscous fluid and an elastic solid. These materials exhibit a time-dependent response to an applied load, and their mechanical properties change as the strain rate is varied.

The complexity of natural blood vessel mechanics makes it difficult to develop an engineered substitute that fully matches the properties of native tissue. Important strides have been made in this area, mainly in the area of achieving adequate burst strengths, which, at a minimum, allows implantation of the vessel. Achievement of the appropriate elastic and viscoelastic properties has remained a more elusive goal.

III. ENDOTHELIALIZATION STRATEGIES

Since occlusion due to thrombus formation is one of the main reasons for the failure of synthetic vascular grafts,[6] there has been great emphasis on creating more hemocompatible lumenal linings, designed to shield the blood from the clot-inducing graft material. Early cell-based approaches in this area focused on coating the inner surface of permanent synthetic polymer grafts with a monolayer of living endothelial cells, as shown schematically in Figure 25.2. These efforts represented the beginning of vascular tissue engineering, in that they used living cells to provide a more physiological solution to improving graft function.[7,8] It is likely that any eventual engineered vascular tissue will require an intact, functioning endothelium in order to function well over the long term, and therefore endothelialization strategies will remain critical to the success of this field.

Recent evaluations of endothelialized synthetic grafts implanted in the lower extremities of human patients have shown promising results.[9] These studies used autologous endothelial cells harvested from a subcutaneous vein, which were expanded in culture for approximately 21 days before being used to coat the graft lumen. The cells were seeded onto expanded polytetrafluoroethylene (ePTFE) vascular prostheses that had been precoated with fibrin glue to promote cell

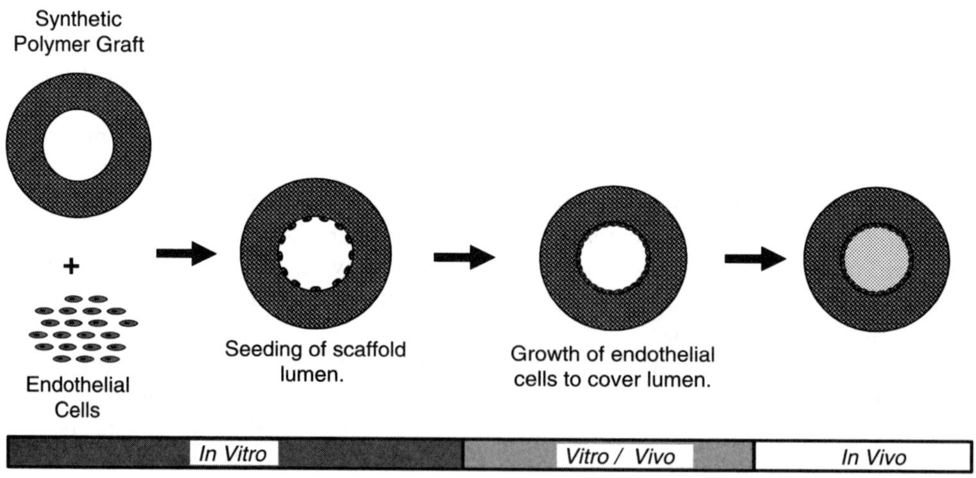

FIGURE 25.2 Schematic diagram of endothelialization of permanent synthetic scaffold materials for use as vascular grafts. Endothelial cells are seeded onto the lumen of the graft and grown to cover the inner surface, in order to provide a hemocompatible lining. See, for example Refs. 8,9,12.

attachment, and were cultured for a further 9 days to achieve confluence. Using 6 mm inner diameter grafts, the patency rate at 7 years exceeded 60%, and a shorter term evaluation of 7 mm inner diameter grafts showed greater than 80% patency at 4 years. In spite of the relatively recent success of endothelialization in peripheral (large diameter) vascular grafting, this technology has not yet produced similarly promising results in small diameter applications. Encouragingly, recent studies using 4 mm inner diameter ePTFE grafts[10] and decellularized vascular tissue[11] have suggested that endothelialization can also be used to improve the function of small caliber vessels.

A critical issue in developing practically feasible endothelialization strategies is the sourcing of cells, since endothelial cells are strongly involved in the vascular immune reaction. Initial successes using preseeded autologous endothelial cells have shown that this approach has promise, but in order to provide "off-the-shelf" availability of tissue for implantation it is likely that alternate strategies will be needed. Efforts at decreasing the time required to generate an endothelial cell lining, such that a long *in vitro* culture period is not required, have led to cell "sodding" techniques using microvascular endothelial cells extracted from liposuction fat at the time of graft implant surgery.[12] Recruitment of native endothelial cells to the graft surface after implantation is another possibility, though this is challenging in human patients due to the low proliferation and migration rate of endothelial cells *in vivo*. Immobilization of cell adhesion peptides[13] and release of growth factors[14] have been used to increase the degree graft coverage, as well as the level of endothelial cell function, and the effects of shear forces on endothelial cells seeded onto grafts also are being investigated.[15] In addition, endothelial progenitor cells are known to circulate in the blood, though in very small numbers.[16] This is another potential source of autologous cells for lumenal seeding[17] or recruitment after graft implantation. The ability to use allogeneic endothelial cells to line the lumen of implanted grafts would be a great advantage in terms of cell sourcing. In this case, however, a major challenge would be controlling the immune response in order to provide a functional, hemocompatible graft lining.

IV. SCAFFOLDING STRATEGIES

As the understanding of vascular biology has progressed, it has become increasingly clear that the smooth muscle cells that populate the medial layer of the vessel wall contribute in important ways to both the short and long term function of the blood vessel. This includes generation of contractile forces important in the vasoactive response, longer term remodeling of vessel dimensions in response to changing hemodynamic conditions, as well as pathological responses such as intimal hyperplasia and the generation of atherosclerotic plaques. In addition, the interaction between the endothelial cells of the intima and the smooth muscle cells of the media is receiving more attention as the regulatory effect of each of these cell types on the other is becoming more completely understood.

The generation of appropriate scaffolds for smooth muscle cell growth and function is a challenge that has been approached in a variety of ways. Of key importance is providing an extracellular matrix that is sufficiently strong and compliant to withstand the pulsatile pressures of the vascular system, while at the same time maintaining the cell-specific functions of the vascular smooth muscle cell. The following sections outline three general strategies for achieving these goals.

A. Degradable Synthetic Scaffolds

The archetypal tissue engineering technology of using a biodegradable synthetic polymer scaffold seeded with isolated cells[18] has also been applied to the vascular system, as shown schematically in Figure 25.3. This approach involves seeding cultured cells onto a preformed porous scaffold, which is chemically designed to degrade over time in the physiological environment. In vascular applications, the most commonly used degradable polymer scaffolds are polycaprolactone (PCL),

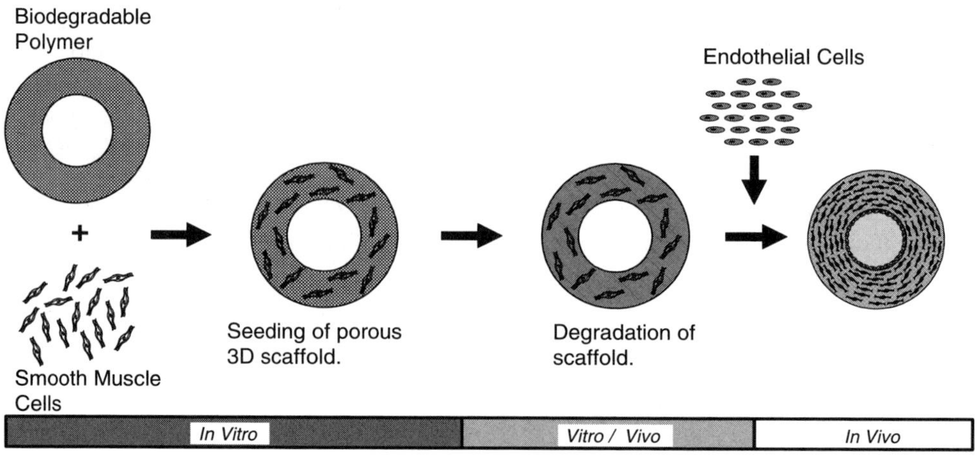

FIGURE 25.3 Schematic representation of the use of biodegradable polymer scaffolds as vascular grafts. Smooth muscle cells are seeded onto a porous degradable scaffold, which is eventually replaced by living tissue and can be lined with endothelial cells. See, for example Refs. 19,21,23.

polyglycolic acid (PGA), polylactic acid (PLA), as well as derivatives and copolymers of these materials.

The feasibility of this approach was first reported in an ovine model using PGA–polyglactin copolymers seeded in stages with autologous fibroblasts, smooth muscle cells and endothelial cells.[19] The 15 mm diameter constructs were cultured *in vitro* for 7 days, before being used to replace short segments in the low pressure pulmonary circulation. Examination of explanted grafts showed that cell-seeded grafts remained patent for up to 24 weeks, whereas control acellular polymer grafts occluded via thrombus formation. The polymer scaffold had been completely degraded by 11 weeks *in vivo*, but the resulting tissues were not deemed suitable for implantation in the higher pressure systemic circulation. A subsequent study used a modified polymer system (a PGA–polyhydroxyalkanoate composite) in order to augment the longer term mechanical properties of the construct.[20] A mixed culture of autologous ovine endothelial cells, smooth muscle cells and fibroblasts were seeded simultaneously on the lumen of the 7 mm diameter construct, and were allowed to attach and proliferate *in vitro* for 7 days prior to the construct being implanted as an infrarenal aortic graft. These engineered vessels exhibited patency up to 22 weeks, whereas acellular polymer tubes used as controls became occluded at times ranging from 1 day to 14 weeks. Evaluation of the cell seeded constructs showed collagen and elastin fiber formation in the medial layer, as well as staining for endothelial cell markers on the lumenal surface.

A similar approach using bovine and porcine cell-seeded PGA scaffolds implanted into a porcine model[21] also demonstrated very promising results. Notably, this study used a bioreactor system to impart cyclic mechanical strain to the smooth muscle cell-seeded constructs as they developed *in vitro* over a period of 8 weeks. It was found that pulsatile strain promoted homogeneous cell distribution in the construct wall, as well as more complete resorption of the polymer scaffold, leading to enhanced mechanical properties. Endothelial cells were added to the lumen of pulsed constructs, and were cultured for a period of 3 days in the presence of continuous perfusion of the vessel lumen with culture medium. Implantation of one xenogeneic construct, as well as three autologous constructs, showed that these engineered vessels could remain patent for up to 4 weeks, and that application of pulsatile strain *in vitro* improved patency rates as well as graft morphology and function.

Other investigations into the use of degradable synthetic polymer scaffolds have focused on modifying the scaffold to guide cell function and better control the biological response.

Modification of PGA–PLA scaffolds has been shown to modulate cell adhesion,[22] as well as cell phenotype,[23] and mechanical stimulation has been shown to enhance tissue properties.[24] Chemical design of other polymers has also been used to produce degradable scaffolds with potential application in the vascular system,[25] including the incorporation of tethered growth factors in order to affect cell function.[26]

Although encouraging results have been obtained using cell-seeded degradable polymer systems, significant challenges remain. Since the scaffold is designed to be replaced by cells and extracellular matrix, a high degree of cell proliferation and matrix synthesis is required to transform these synthetic constructs into fully biological tissue. Complete degradation of the polymer scaffold is difficult to achieve, especially since mass transport limitations can hamper the degradation process as the tissue becomes more densely filled with proliferating cells and new extracellular matrix. In addition, although the degradation products of most of these systems are relatively benign, the hydrolysis reaction involved can change the tissue microenvironment in such a way as to be damaging to nearby cells, for example by decreasing the local pH.[27] The inflammatory and immune responses may also be activated by polymer degradation.[28] These challenges are being approached in a variety of creative ways, by controlling scaffold properties and directing cell function.

B. Naturally-Derived Scaffolds

Another promising strategy to engineering living tissue is to provide cells with a scaffold similar to the one they are exposed to in native tissue, with the expectation that they will recognize and react appropriately to the supplied matrix and produce a competent vascular tissue. The predominant structural components of natural blood vessels are collagen (Type I and Type III) and elastin. These materials have been used to varying degrees in reconstructing vascular tissue *in vitro*. The main approaches using naturally derived matrices can be divided into two subcategories, as shown schematically in Figure 25.4 and Figure 25.5. In the first case, cells are combined with a solubilized form of the matrix proteins and a tissue is formed by subsequent gelation of the scaffold, generally using a molding process that results in the cells being embedded directly in the tissue matrix. In the second case, natural allogeneic or xenogeneic tissue is processed to remove cellular and immunogenic components, leaving an intact decellularized matrix that is subsequently used as a scaffold for repopulation with autologous cells.

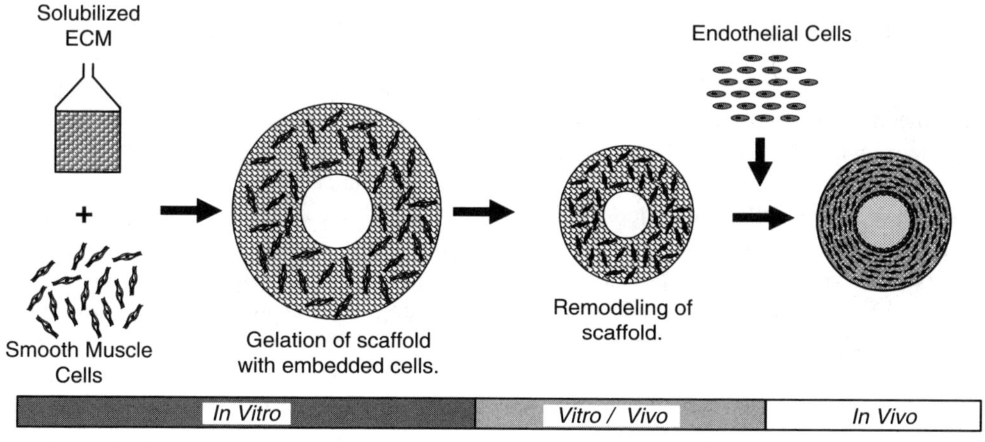

FIGURE 25.4 Schematic diagram of the use of solubilized extracellular matrix proteins, for example, Type I collagen, as scaffolds in vascular tissue engineering. Smooth muscle cells are embedded in a molded hydrogel, which is remodeled by the cellular component to produce a vascular tissue. Endothelial cells can be seeded onto the lumen to provide a hemocompatible surface. See, for example Refs. 29,33,35.

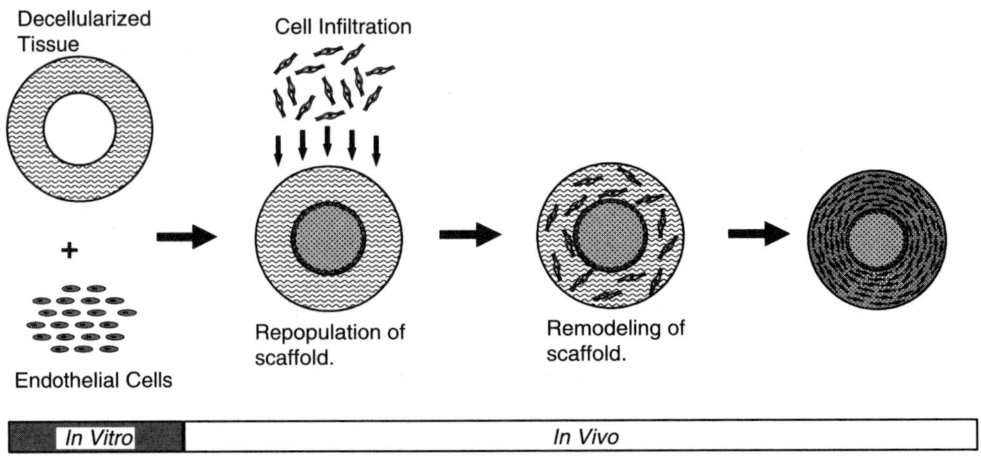

FIGURE 25.5 Schematic diagram of the use of decellularized natural tissues to produce vascular scaffolds. Endothelialization and implantation result in the recruitment of cells from the host, which repopulate and remodel the matrix. See, for example Refs. 48,50,51.

Most tissue molding approaches (Figure 25.4) have used Type I collagen as the main matrix protein, since it is difficult to obtain sufficient quantities of Type III collagen, and solubilized elastin will generally not reconstitute into a functional fibrillar form. Initial studies using tubular Type I collagen gels[29] established the feasibility of this technique, though the low mechanical strength of the matrix necessitated the use of a reinforcing synthetic polymer sleeve. More recent studies have provided important insight into how collagen gel matrices are formed, how they can be modified and controlled, and how cells interact with and remodel them. Cell-mediated gel compaction has been shown to be modulated by the presence of mechanical constraints during gelation,[30] resulting in increased fiber alignment.[31] The properties of such engineered tissues can also be improved by chemical means, for example, by promoting appropriate collagen cross-linking,[32–34] and through application of cyclic mechanical strain during the *in vitro* tissue development process.[35] Remodeling of the collagen matrix by embedded cells through the action of matrix metalloproteinases can affect the mechanical properties of collagen-based engineered tissues, and this process is also affected by the application of mechanical forces.[36,37]

A key challenge in the use of molded Type I collagen matrices as a matrix in vascular tissue engineering is the difficulty in obtaining graft mechanical properties that are appropriate for implantation. One potential solution to this problem is the use of reinforcing sleeves that provide the required mechanical strength while the cells are embedded in a reconstituted collagen matrix. The reinforcing sleeve can be made of a naturally derived polymer that has been strengthened by cross-linking,[38] or it can be a synthetic polymer with superior mechanical properties.[39] In the former case, the sleeve may be gradually remodeled and resorbed by the cellular component, whereas in the latter case a degradable or permanent polymer can be used. Reinforcement of collagen-based vascular constructs with synthetic polymers has been tested in autologous implantation studies in a canine model, and showed patency up to 27 weeks,[40] depending on the properties of the reinforcing material. Interestingly, this technique has also been used to fabricate branched vessels.[41]

In addition to Type I collagen, other naturally derived polymers that can be reconstituted in fibrillar form have been investigated as potential matrices in tissue engineering. The blood clotting protein fibrin has been suggested as an alternative to collagen matrices,[42,43] and has also been used as an additive to provide enhanced properties to collagen-based engineered tissues.[44] The production of biomimetic peptides using genetic engineering techniques is also being used to produce extracellular matrix proteins with desired mechanical and biochemical properties,[45] and methods for creating scaffolds from these are being developed. For example, electrospinning has been used to produce mesh scaffolds of collagen fibers,[46,47] although the mechanical properties and interaction with cells of these matrices has not yet been extensively characterized.

The use of decellularized natural tissue matrices as scaffolds for vascular tissue engineering also has been investigated. This method is shown schematically in Figure 25.5. Isolation and processing of intact extracellular matrix lamina from porcine tissue has been used to produce scaffold materials that are composed entirely of native structural proteins, with the associated attachment proteins also included.[48] The small intestinal submucosa (SIS), in particular, has been used to create vascular conduits, which when tested in xenogeneic implant studies[49] showed patency for up to 9 weeks as well as evidence of remodeling and improvement of graft mechanical properties. SIS has also been used in combination with gelled and cross-linked bovine collagen that has been heparinized to reduce thrombogenicity.[50] This model showed patency after 13 weeks of implantation into a xenogeneic host, as well as cellular infiltration and a recovery of vasomotor activity in response to certain contractile agonists. Decellularized vascular tissue has also been used in an effort to provide new small diameter blood vessels. Porcine carotid arteries have been processed to remove the cellular component, heparinized to reduce thrombosis and implanted in a dog model for up to 9 weeks.[51] Porcine iliac vessels have also been used in decellularized form, and have been lined with autologous endothelial progenitor cells before being implanted in a sheep model.[17] These vessels exhibited patency for up to 29 weeks, as well as a vasoactive response upon removal.

C. Cell-Secreted Scaffolds

Creation of engineered tissues does not necessarily require that a scaffold be supplied initially, since it can be argued that the most appropriate tissue matrix is that produced by the specialized tissue cells themselves. The resulting construct is fully biological, and the resident cells are well integrated into the tissue, with the ability to produce, degrade, and remodel the matrix in response to environmental cues. Although this approach is simple in concept, directing cell function to recreate appropriate tissues has remained a major challenge in tissue engineering and regenerative medicine.

In the case of vascular tissue engineering, an innovative technology has been developed that uses sequential layering of cultured tissue sheets to produce a tubular blood vessel analog,[52] as shown schematically in Figure 25.6. The robust production of extracellular matrix that is required for generation of sheets of tissue is promoted by the addition of ascorbic acid to the culture medium, a method adapted from the development of engineered dermal tissue. Fibroblasts are cultured for 35 weeks to generate an initial tissue sheet, which is then dehydrated and wrapped around a mandrel to

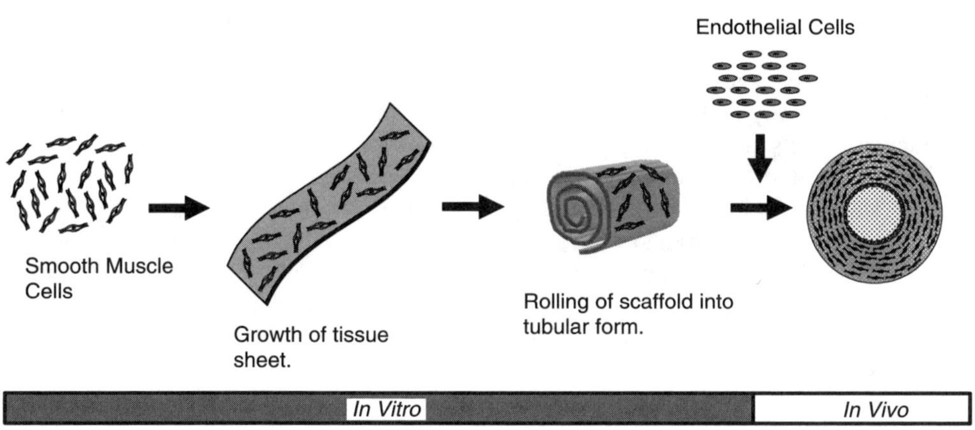

FIGURE 25.6 Schematic representation of using rolled sheets of smooth muscle cell-generated tissue to produce a vascular graft. Rolled tissues are cultured to form a conduit, which can be lined with endothelial cells. See, for example Refs. 52,63.

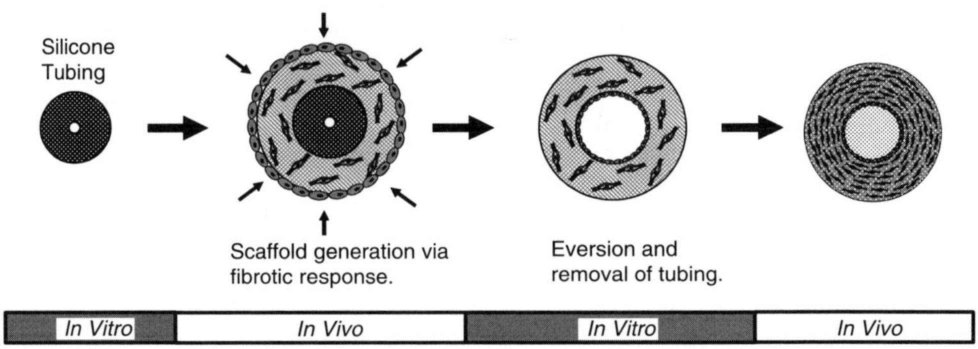

FIGURE 25.7 Schematic diagram of using the production of granulation tissue to generate a tubular conduit. A piece of silicone tubing is implanted into the peritoneal cavity and becomes encapsulated by the fibrotic host response, with an outer layer of mesothelial cells. Upon explantation, the tissue can be everted to yield a vascular graft. See Ref. 54.

produce an acellular inner sleeve. Smooth muscle cells are cultured for 3 weeks to produce sheets of tissue several cell layers thick. These sheets are then removed from the culture plate and wrapped around a mandrel to form a tubular medial structure, which is cultured in a flow-through bioreactor for 1 week. A cell-containing fibroblast sheet can then be wrapped around the medial tissue to produce an analog of the adventitia. This construct is cultured for at least an additional 8 weeks before an endothelial cell lining is seeded onto the lumen of the construct. Vascular tissues produced in this manner have exhibited remarkable burst pressures ($>$2500 mmHg), and have shown some success in xenogeneic implant studies. In addition, these vessels have exhibited evidence of contractility in response to pharmacological stimulation.[53]

The natural foreign body response, and the associated generation of granulation tissue, has also been harnessed to produce fully biological vascular tissue,[54] as shown schematically in Figure 25.7. In this approach, silicone tubing is implanted into the peritoneal cavity of the host in order to stimulate fibrous encapsulation. After 2 weeks, the resulting fibrotic response around the tubing produces a vascular structure consisting of dense connective tissue surrounded by cell-rich granulation tissue, with an outer monolayer of mesothelial cells. By removing the silicone and everting the formed tissue, the outer mesothelial cell layer becomes the inner lumenal lining and a structure similar to a blood vessel is formed. The mechanical properties of these engineered vessels have not been extensively characterized, however, autologous implantation studies showed the ability to remain patent for up to 4 months, and explanted vessels exhibited modest contractile responses to vasoactive agents.

Generation of a fully biological blood vessel substitute using only cell-secreted scaffolds has many attractive aspects, although important practical issues still need to be resolved. In the tissue-wrapping approach, the long culture times required for generation of appropriate tissue *in vitro* (at least 13 weeks, not including formation of the initial fibroblast-generated sleeve) and the extensive tissue handling required to wrap successive layers pose a challenge to producing a widely available graft. Use of the fibrotic response to generate new tissue *in vivo* has the potential to produce autologous grafts; however, it is not clear that the human fibrotic response will be appropriate for this technique. In addition, the origin and eventual function of the cells that are recruited in this approach have not been fully characterized.

V. BIOLOGICAL RESPONSE TO SCAFFOLD

As more fully biological tissue substitutes are developed, an increasing amount of attention is being directed at understanding how living cells interact with their surroundings. In tissue engineering applications, many of the cell–matrix interactions take place in a complex three-dimensional environment and cell function is further influenced by soluble biochemical as well as mechanical signals. In the case of vascular tissue, the endothelial cells that line the lumen are in the form of a monolayer, which is one of the few physiological situations in which two-dimensional culture systems are representative of the *in vivo* geometry. However, these cells are in contact with the complex biochemical environment of whole blood, and are strongly influenced by fluid shear forces.[55] In addition, it must be recognized that native endothelial cells are found on a three-dimensional substrate populated by smooth muscle cells, and that these cell types can interact directly and through biochemical signaling.[56–58] The ability of the endothelial lining to provide a continuous hemocompatible surface therefore depends on the fluid dynamic environment, as well as the composition and function of the substrate.

The smooth muscle cells of the medial layer are responsible for the agonist-stimulated contraction and dilation of blood vessels, and are also involved in longer term remodeling of the tissue. These cells are surrounded by a three-dimensional matrix that provides a variety of signals that affect cell function. The matrix itself has biochemical effects on the embedded cells, and is also involved in signaling by growth factors and mechanical forces.[59,60] It is increasingly being recognized that smooth muscle cell function in three-dimensional scaffolds is under complex control, and that studies in two-dimensional culture do not always reflect this complexity.[61] The properties of the scaffold itself can have an effect on cell phenotype,[23] as can the combined effect of the matrix, biochemical factors and mechanical forces.[62] In order to get appropriate tissue structure and remodeling, the effects of these parameters must be understood and ideally harnessed to produce vascular substitutes with the desired properties.

VI. SUMMARY AND FUTURE DIRECTIONS

The need for improved vascular grafts, particularly in small diameter applications, is clear. The trend over the past several decades has been toward more biologically appropriate and active conduits, often incorporating living cells. These engineered tissues are functionally more complex than purely synthetic or acellular materials; however, they offer the potential to more closely replicate the active functions of living native tissues. This includes the anticoagulative properties of an endothelial cell lining, as well as the remodeling and vasoactive functions of a smooth muscle cell-populated vascular wall.

There has been a great deal of progress in producing such living vascular tissues, as evidenced by the increasing number of implantation studies in both animal and human systems. However, a number of key challenges remain to be overcome before a widely usable living vascular substitute is obtained. These include achievement of appropriate mechanical properties, hemocompatibility, tissue remodeling, and vasoactivity. Careful selection and study of scaffolding strategies will certainly play an important role in achieving these goals, as cell–cell and cell–matrix interactions become more completely understood.

ACKNOWLEDGMENTS

The authors gratefully acknowledge the financial support of the National Science Foundation through ERC Award Number EEC-9731643.

REFERENCES

1. AHA. Heart Disease and Stroke Statistics — 2003 Update, 2002.
2. Morris, J. L., Gibbins, I., Kadowitz, P. J., Herzog, H., Kreulen, D. L., Toda, N., and Claing, A., Roles of peptides and other substances in cotransmission from vascular autonomic and sensory neurons, *Can. J. Physiol. Pharmacol.*, 73, 521–532, 1995.

3. Griendling, K. K., and Alexander, R. W., Cellular biology of blood vessels, In *Hurst's the Heart*, Schlant, R. C. and Alexander, R. W., eds., McGraw-Hill Inc., New York, 1994.
4. Caro, C. G., Pedley, T. J., Schroter, R. C., and Seed, W. A., *The Mechanics of the Circulation*, Oxford University Press, Oxford, 1978.
5. Fung, Y. C., *Biomechanics: Mechanical Properties of Living Tissues*, Springer, New York, 1993.
6. Cynamon, J., and Pierpont, C. E., Thrombolysis for the treatment of thrombosed hemodialysis access grafts, *Rev. Cardiovasc. Med.*, 3, S75–S82, 2002.
7. Graham, L. M., Vinter, D. W., Ford, J. W., Kahn, R. H., Burkel, W. E., and Stanley, J. C., Cultured autogenous endothelial cell seeding of prosthetic vascular grafts, *Surg. Forum*, 30, 204–206, 1979.
8. Herring, M., Gardner, A., and Glover, J., A single-staged technique for seeding vascular grafts with autogenous endothelium, *Surgery*, 84, 498–504, 1978.
9. Meinhart, J. G., Deutsch, M., Fischlein, T., Howanietz, N., Froschl, A., and Zilla, P., Clinical autologous in vitro endothelialization of 153 infrainguinal ePTFE grafts, *Ann. Thorac. Surg.*, 71, S327–S331, 2001.
10. Fields, C., Cassano, A., Makhoul, R. G., Allen, C., Sims, R., Bulgrin, J., Meyer, A., Bowlin, G. L., and Rittgers, S. E., Evaluation of electrostatically endothelial cell seeded expanded polytetrafluoroethylene grafts in a canine femoral artery model, *J. Biomater. Appl.*, 17, 135–152, 2002.
11. Carnagey, J., Hern-Anderson, D., Ranieri, J., and Schmidt, C. E., Rapid endothelialization of PhotoFix natural biomaterial vascular grafts, *J. Biomed. Mater. Res.*, 65B, 171–179, 2003.
12. Williams, S. K., Wang, T. F., Castrillo, R., and Jarrell, B. E., Liposuction-derived human fat used for vascular graft sodding contains endothelial cells and not mesothelial cells as the major cell type, *J. Vasc. Surg.*, 19, 916–923, 1994.
13. Krijgsman, B., Seifalian, A. M., Salacinski, H. J., Tai, N. R., Punshon, G., Fuller, B. J., and Hamilton, G., An assessment of covalent grafting of RGD peptides to the surface of a compliant poly(carbonate-urea)urethane vascular conduit versus conventional biological coatings: its role in enhancing cellular retention, *Tissue Eng.*, 8, 673–680, 2002.
14. Wissink, M. J. B., Beernink, R., Poot, A. A., Engbers, G. H. M., Beugeling, T., van Aken, W. G., and Feijen, J., Improved endothelialization of vascular grafts by local release of growth factor from heparinized collagen matrices, *J. Controlled Release*, 64, 103–114, 2000.
15. Braddon, L. G., Karoyli, D., Harrison, D. G., and Nerem, R. M., Maintenance of a functional endothelial cell monolayer on a fibroblast/polymer substrate under physiologically relevant shear stress conditions, *Tissue Eng.*, 8, 695–708, 2002.
16. Lin, Y., Weisdorf, D. J., Solovey, A., and Hebbel, R. P., Origins of circulating endothelial cells and endothelial outgrowth from blood, *J. Clin. Invest.*, 105, 71–77, 2000.
17. Kaushal, S., Amiel, G. E., Guleserian, K. J., Shapira, O. M., Perry, T., Sutherland, F. W., Rabkin, E., Moran, A. M., Schoen, F. J., Atala, A., Soker, S., Bischoff, J., and Mayer, J. E., Functional small-diameter neovessels created using endothelial pregenitor cells expanded ex vivo, *Nat. Med.*, 7, 1035–1040, 2001.
18. Langer, R., and Vacanti, J. P., Tissue engineering, *Science*, 260, 920–926, 1993.
19. Shinoka, T., Shum-Tim, D., Ma, P., Tanel, R., Isogai, N., Langer, R., Vacanti, J., and Mayer, J., Creation of viable pulmonary artery autografts through tissue engineering, *J. Thorac. Cardiovasc. Surg.*, 115, 536–546, 1998.
20. Shum-Tim, D., Stock, U., Hrkach, J., Shinoka, T., Lien, J., Moses, M. A., Stamp, A., Taylor, G., Moran, A. M., and Landis, W., Tissue engineering of autologous aorta using a new biodegradable polymer, *Ann. Thorac. Surg.*, 68, 2298–2304, 1999.
21. Niklason, L. E., Gao, J., Abbott, W. M., Hirschi, K. K., Houser, S., Marini, R., and Langer, R., Functional arteries grown in vitro, *Science*, 284, 489–493, 1999.
22. Nikolovski, J., and Mooney, D. J., Smooth muscle cell adhesion to tissue engineering scaffolds, *Biomaterials*, 21, 2025–2032, 2000.
23. Kim, B. S., Nikolovski, J., Bonadio, J., Smiley, E., and Mooney, D. J., Engineered smooth muscle tissues: regulating cell phenotype with the scaffold, *Exp. Cell Res.*, 251, 318–328, 1999.
24. Kim, B.-S., Nikolovski, J., Bonadio, J., and Mooney, D. J., Cyclic mechanical strain regulates the development of engineered smooth muscle tissue, *Nat. Biotechnol.*, 17, 979–983, 1999.
25. Mann, B. K., Gobin, A. S., Tsai, A. T., Schmedlen, R. H., and West, J. L., Smooth muscle cell growth in photopolymerized hydrogels with cell adhesive and proteolytically degradable domains: synthetic ECM analogs for tissue engineering, *Biomaterials*, 22, 3045–3051, 2001.

26. Mann, B. K., Schmedlen, R. H., and West, J. L., Tethered-TGF-beta increases extracellular matrix production of vascular smooth muscle cells, *Biomaterials*, 22, 439–444, 2001.
27. Agrawal, C. M., and Athanasiou, K. A., Technique to control pH in vicinity of biodegrading PLA–PGA implants, *J. Biomed. Mater. Res.*, 38, 105–114, 1997.
28. Athanasiou, K. A., Niederauer, G. G., and Agrawal, C. M., Sterilization, toxicity, biocompatibility and clinical applications of polylactic acid/polyglycolic acid copolymers, *Biomaterials*, 17, 93–102, 1996.
29. Weinberg, C. B., and Bell, E., A blood vessel model constructed from collagen and cultured vascular cells, *Science*, 231, 397–400, 1986.
30. Barocas, V. H., and Tranquillo, R. T., An anisotropic biphasic theory of tissue-equivalent mechanics: the interplay among cell traction, fibrillar network deformation, fibril alignment, and cell contact guidance, *J. Biomech. Eng.*, 119, 137–145, 1997.
31. Barocas, V. H., Girton, T. S., and Tranquillo, R. T., Engineered alignment in media equivalents: magnetic prealignment and mandrel compaction, *J. Biomech. Eng.*, 120, 660–666, 1998.
32. Brinkman, W. T., Nagapudi, K., Thomas, B. S., and Chaikof, E. L., Photo-cross-linking of type I collagen gels in the presence of smooth muscle cells: mechanical properties, cell viability, and function, *Biomacromolecules*, 4, 890–895, 2003.
33. Girton, T. S., Oegema, T. R., and Tranquillo, R. T., Exploiting glycation to stiffen and strengthen tissue equivalents for tissue engineering, *J. Biomed. Mater. Res.*, 46, 87–92, 1999.
34. van Wachem, P. B., Plantinga, J. A., Wissink, M. J., Beernink, R., Poot, A. A., Engbers, G. H., Beugeling, T., van Aken, W. G., Feijen, J., and van Luyn, M. J., In vivo biocompatibility of carbodiimide-crosslinked collagen matrices: effects of crosslink density, heparin immobilization, and bFGF loading, *J. Biomed. Mater. Res.*, 55, 368–378, 2001.
35. Seliktar, D., Black, R. A., Vito, R. P., and Nerem, R. M., Dynamic mechanical conditioning of collagen-gel blood vessel constructs induces remodeling in vitro, *Ann. Biomed. Eng.*, 28, 351–362, 2000.
36. Asanuma, K., Magid, R., Johnson, C., Nerem, R. M., and Galis, Z. S., Uniaxial strain upregulates matrix-degrading enzymes produced by human vascular smooth muscle cells, *Am. J. Physiol. Heart Circ. Physiol.*, 284, H1778–H1784, 2003.
37. Seliktar, D., Nerem, R. M., and Galis, Z. S., The role of matrix metalloproteinase-2 in the remodeling of cell-seeded vascular constructs subjected to cyclic strain, *Ann. Biomed. Eng.*, 29, 923–934, 2001.
38. Berglund, J. D., Mohseni, M. M., Nerem, R. M., and Sambanis, A., A biological hybrid model for collagen-based tissue engineered vascular constructs, *Biomaterials*, 24, 1241–1254, 2003.
39. Matsuda, T., and He, H., Newly designed compliant hierarchic hybrid vascular grafts wrapped with a microprocessed elastomeric film — I: fabrication procedure and compliance matching, *Cell Transplant.*, 11, 67–74, 2002.
40. He, H., and Matsuda, T., Arterial replacement with compliant hierarchic hybrid vascular graft: biomechanical adaptation and failure, *Tissue Eng.*, 8, 213–224, 2002.
41. Kobashi, T., and Matsuda, T., Fabrication of branched hybrid vascular prostheses, *Tissue Eng.*, 5, 515–524, 1999.
42. Grassl, E. D., Oegema, T. R., and Tranquillo, R. T., Fibrin as an alternative biopolymer to type-I collagen for the fabrication of a media equivalent, *J. Biomed. Mater. Res.*, 60, 607–612, 2002.
43. Ye, Q., Zund, G., Benedikt, P., Jockenhoevel, S., Hoerstrup, S. P., Sakyama, S., Hubbell, J. A., and Turina, M., Fibrin gel as a three dimensional matrix in cardiovascular tissue engineering, *Eur. J. Cardiothorac. Surg.*, 17, 587–591, 2000.
44. Cummings, C. L., Gawlitta, D., Nerem, R. M., and Stegemann, J. P., Collagen fibrin and collagen-fibrin mixtures as matrix materials for vascular tissue engineering (abstract), pp. 203. *Proceedings of the 2nd Joint EMBS–BMES Conference*, 2002.
45. Huang, L., McMillan, R. A., Apkarian, R. P., Pourdeyhimi, B., Conticello, V. P., and Chaikof, E. L., Generation of synthetic elastin-mimetic small diameter fibers and fiber networks, *Macromolecules*, 33, 2989–2997, 2000.
46. Huang, L., Apkarian, R. P., and Chaikof, E. L., High-resolution analysis of engineered type I collagen nanofibers by electron microscopy, *Scanning*, 23, 372–375, 2001.
47. Matthews, J. A., Wnek, G. E., Simpson, D. G., and Bowlin, G. L., Electrospinning of collagen nanofibers, *Biomacromolecules*, 3, 232–238, 2002.

48. Hodde, J. P., Record, R. D., Tullius, R. S., and Badylak, S. F., Retention of endothelial cell adherence to porcine-derived extracellular matrix after disinfection and sterilization, *Tissue Eng.*, 8, 225–234, 2002.
49. Hiles, M. C., Badylak, S. F., Lantz, G. C., Kokini, K., Geddes, L. A., and Morff, R. J., Mechanical properties of xenogeneic small-intestinal submucosa when used as an aortic graft in the dog, *J. Biomed. Mater. Res.*, 29, 883–891, 1995.
50. Huynh, T., Abraham, G., Murray, J., Brockbank, K., Hagen, P.-O., and Sullivan, S., Remodeling of an acellular collagen graft into a physiologically responsive neovessel, *Nat. Biotechnol.*, 17, 1083–1086, 1999.
51. Conklin, B. S., Richter, E. R., Kreutziger, K. L., Zhong, D.-S., and Chen, C., Development and evaluation of a novel decellularized vascular xenograft, *Med. Eng. Phys.*, 24, 173–183, 2002.
52. L'Heureux, N., Paquet, S., Labbe, R., Germain, L., and Auger, F. A., A completely biological tissue-engineered human blood vessel, *FASEB J.*, 12, 47–56, 1998.
53. L'Heureux, N., Stoclet, J. C., Auger, F., Lagaud, G. J., Germain, L., and Andriantsitohaina, R., A human tissue-engineered vascular media: a new model for pharmacological studies of contractile responses, *FASEB J.*, 15, 515–524, 2001.
54. Campbell, J. H., Efendy, J. L., and Campbell, G. R., Novel vascular graft grown within recipient's own peritoneal cavity, *Circ. Res.*, 85, 1173–1181, 1999.
55. Fisher, A. B., Chien, S., Barakat, A. I., and Nerem, R. M., Endothelial cellular response to altered shear stress, *Am. J. Physiol. Lung Cell Mol. Physiol.*, 281, L529–L533, 2001.
56. Chiu, J.-J., Chen, L.-J., Lee, P.-L., Lee, C.-I., Lo, L.-W., Usami, S., and Chien, S., Shear stress inhibits adhesion molecule expression in vascular endothelial cells induced by coculture with smooth muscle cells, *Blood*, 101, 2667–2674, 2003.
57. Imberti, B., Seliktar, D., Nerem, R. M., and Remuzzi, A., The response of endothelial cells to fluid shear stress using a co-culture model of the arterial wall, *Endothelium*, 9, 11–23, 2002.
58. Powell, R. J., Carruth, J. A., Basson, M. D., Bloodgood, R., and Sumpio, B. E., Matrix-specific effect of endothelial control of smooth muscle cell migration, *J. Vasc. Surg.*, 24, 51–57, 1996.
59. Boudreau, N. J., and Jones, P. L., Extracellular matrix and integrin signalling: the shape of things to come, *Biochem. J.*, 339, 481–488, 1999.
60. Bottaro, D. P., Liebmann-Vinson, A., and Heidaran, M. A., Molecular signaling in bioengineered tissue microenvironments, *Ann. N Y Acad. Sci.*, 961, 143–153, 2002.
61. Stegemann, J. P., and Nerem, R. M., Altered response of vascular smooth muscle cells to exogenous biochemical stimulation in two- and three-dimensional culture, *Exp. Cell Res.*, 283, 146–155, 2003.
62. Stegemann, J. P., and Nerem, R. M., Phenotype modulation in vascular tissue engineering using biochemical and mechanical stimulation, *Ann. Biomed. Eng.*, 31, 391–402, 2003.
63. Hoerstrup, S. P., Zund, G., Cheng, S., Melnitchouk, S., Kadner, A., Sodian, R., Kolb, S. A., and Turina, M., A new approach to completely autologous cardiovascular tissue in humans, *ASAIO J.*, 48, 234–238, 2002.

26 Tissue Engineering of Tendons and Ligaments

Changming Guo and Myron Spector

CONTENTS

I. INTRODUCTION

Tendon and ligaments are specialized fibrous tissues, which principally serve a mechanical function. The tendon links muscle to bone, whereas the ligament joins bone to bone. Tendons allow forces from the muscle to be transmitted to the bone for movement. Ligaments impart stability to joints. Tendons and ligaments are thus important structures that govern motion as well as share load in diarthrodial joints. Injury or loss of these structures can lead to marked changes in joint motion and significant morbidity for patients. The need for tissue engineering of tendons and ligaments is as clear and compelling as it is for other tissues. Tendons and ligaments do not spontaneously regenerate. When healing produces reparative tissue, scar-like tissue does not have the same mechanical properties as the undamaged tendon or ligament, and thus have limited functionality. While there have been promising advances in the implementation of tissue engineering strategies for tendon and ligament regeneration in tissue culture and animal experiments, no clinical treatments have yet been established. There is, however, hope that such procedures will be ready for human use in the near future.

The similarities in the structure, function, and healing of tendons and ligaments warrant their being considered together in a discussion of possible tissue engineering strategies for the regeneration of these types of tissues. The objectives of this chapter are to provide a context for treatment modalities of tendon and ligament injuries, to provide a platform from which investigators can further their own research in this field, and to stimulate new contributions to the field of tissue engineering of these important biological structures.

Tendons and ligaments sustain a wide array of injuries from simple sprains to complete rupture, which is the principal issue being addressed in this chapter. The nature of the healing response is critically dependent on whether the ligament is inside or outside of the joint (i.e., intra or extra-articular) and whether the tendon is contained within a synovial sheath. Intra-articular ligaments such as the anterior cruciate ligament (ACL) generally do not form a reparative tissue to bridge the rupture. Extra-articular ligaments and tendons may heal through the formation of scar-like tissue in the rupture gap. However, this reparative tissue does not display the same composition, structure, and mechanical properties as undamaged tissue.

Each year in the United States, at least 120,000 patients undergo tendon or ligament repairs.[1] The therapeutic options currently available are:

1. Direct suture repair
2. Reconstruction with an autograft or allograft tendon or ligament
3. Reconstruction with a synthetic prosthesis implanted alone or as an augmentation device for use with an autograft or allograft

Direct suture does not appear to be of value for certain ruptures (viz., the ruptured ACL) and is of limited use for ruptures resulting in a gap. Autograft procedures employ tissue that closely resembles the original injured tissue. The current standard of care for the ruptured ACL is implantation of a tendon autograft. However, such grafts eventually become lax in many patients and there is evidence that such treatment does not alter the progression of osteoarthritis. Moreover, there are problems of donor site morbidity, particularly when the patella tendon (PT) is employed. Allogeneic grafting of ligaments may lead to immunologic reactions and prevent adequate healing of the wound.[2] Synthetic materials may undergo wear and rupture.[3,4]

Interest in biological technologies such as tissue engineering has dramatically increased because it is now feasible to isolate living, healthy cells from tissues and organs, expand their number under cell culture conditions, combine the cells with biocompatible carrier materials in the form of scaffolds, and transplant them into patients. Tissue engineering provides the opportunity to generate living substitutes for tissues and organs, which may overcome the drawbacks of current tissue reconstruction procedures.

An understanding of the composition and properties of tendons and ligaments is important in order to: (1) demonstrate that the respective tissue has actually been regenerated, and (2) anticipate the function that might be expected of reparative tissue that falls short of complete regeneration.

II. TENDON AND LIGAMENT ANATOMY, COMPOSITION, AND FUNCTION

As organs, the tendons in the body vary in their anatomy and structural mechanical properties. The tendon consists of essentially three regions: the muscle attachment region, the midsubstance region, and the bone attachment region. An approach that may regenerate one tendon may need modification to facilitate the regeneration of another. Morphologically, tendons are comprised of a complex composite material consisting of collagen fibrils embedded in a matrix of proteoglycans, associated with a relative paucity of cells.

Ligaments are short bands of tough, flexible fibrous connective tissue that connect bones and support viscera. Grossly, ligaments appear as band- or cable-like structures with few distinguishing landmarks. Ligaments, like tendons, comprise distinguishable zones in the midsubstance region and near the bony insertion sites.

A. The Cell Types

The fibroblast is the predominant cell type in both tendons and ligaments. The cells are arranged in the spaces among the parallel collagen bundles. The cell bodies are rod or spindle-shaped and oriented in rows when seen microscopically in a longitudinal section (Figure 26.1(a)). The fibroblast, as the parenchymal cell of a tendon and ligament, has a role in maintaining the matrix structure through degenerative and formative processes comprising remodeling, and to some extent it can contribute to healing. However, tendons have a low density of cells with low mitotic activity, explaining the low turnover rate of this tissue and raising the question of the degree to which these cells can repair damaged tendons. Ligaments also have certain cells that appear chondrocytic in certain locations.[5]

Endothelial cells and nerve processes form only a small part of the cell population. There is a subpopulation of α-smooth muscle actin-containing myofibroblast-like contractile cells in normal and healing tendon[6] and ligament.[5,7] It has been hypothesized that this subpopulation of cells is involved in the “tensioning” of tendons and ligaments[7] and in the modulation of contraction–relaxation of the muscle tendon complex.

B. The Extra-Cellular Matrix (ECM)

The major constituent of tendons and ligaments is type I collagen (86% fat-free dry weight for tendons vs. 70% fat-free dry weight for ligaments). The fibrils of collagen are bound together to form fascicles that are approximately 50 to 300 μm in diameter. In a study comparing the collagen fibril diameter in tendons at different ages, Strocchi[8] found that the fibrils are small and uniform in the neonatal period, becoming large and variable from adolescence onwards. The bundles of collagen fibrils oriented along the long axis of the tendon and ligament display a crimp-like pattern with a length of about 110 μm in tendons (Figure 26.1(b)) and approximately 55 μm in ligaments.

The mechanical properties of tendons and ligaments are largely determined by the structure of collagen and its uniaxial orientation in the tissue, which impart stiffness. That type I collagen is also the principal extra-cellular matrix component of dermis, fibrocartilage, corneas, and scars, and a host of other soft tissues with diverse properties, underscores the importance of the organization and orientation of collagen, among other associated matrix molecules, as a determinant of tissue properties.

Proteoglycans make up 1 to 5% of the dry weight of the tissues. They are hydrophilic and thus, can trap and bind water, which can facilitate the resistance of the tissue to compressive

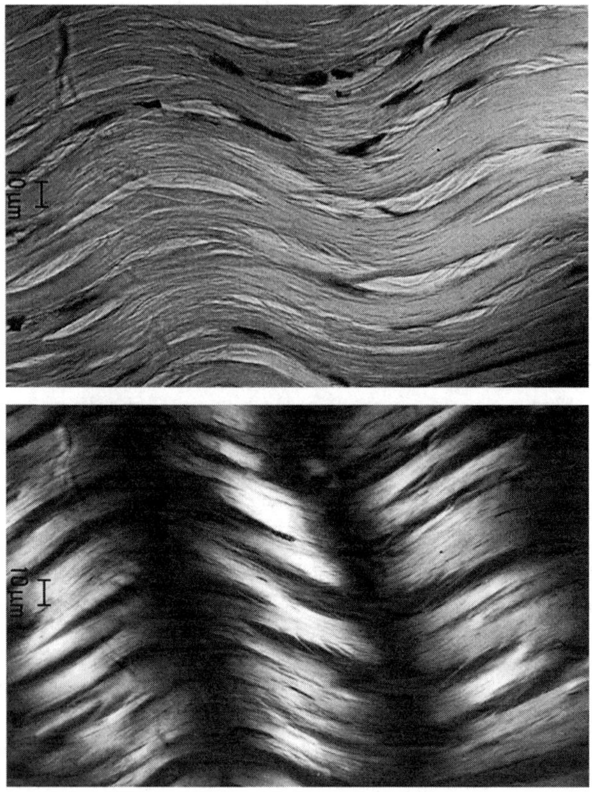

(a)

(b)

FIGURE 26.1 Photomicrographs of a longitudinal section of normal rabbit Achilles tendon under (a) normal and (b) polarized light, showing the spindle-shaped fibroblasts aligned along the collagen fibers that display a prominent crimp, highlighted by the alternating light and dark bands in the polarized light image in (b). Hematoxylin and eosin stain.

loading at certain anatomical locations. Proteoglycans are also important for their role in the interaction with collagen fibers. One of the small proteoglycans, decorin, is widely distributed and is thought to play a fundamental role in regulating collagen fiber formation *in vivo*. It has been suggested that uncontrolled lateral fusion of collagen fibrils in a decorin-deficient tendon would reduce tensile strength of the tendon. On the other hand, it also has been shown that decorin incorporation can increase the ultimate tensile strength of fibers that are not cross-linked. It has been speculated that decorin can prevent fibrillar slippage during deformation, thereby improving the tensile properties of collagen fibers. The distribution of various proteoglycans changes in response to the *in vivo* mechanical environment. Analysis of the human posterior tibial tendon from age 1.5 to 24 months showed that decorin was the predominant proteoglycan in the proximal (tension-bearing) region of the tendon. In contrast, two types of small proteoglycans (decorin and biglycan) and large proteoglycans are present in the region that experiences localized compression where the tendon passes behind the medial malleolus. The same distribution of proteoglycans for tissues under tension alone has also been documented in the human PT. This distribution showed no distinctive trends related to age after puberty.

Elastin, a fibrillar protein that affects the mechanical properties, is also present in tendons and ligaments (less then 1% dry weight). In some ligaments, such as the flaval or nuchal ligaments of the spine, elastin forms the primary structural component.

C. The Surrounding and Supporting Tissue

The fascicles within tendon are bound together by loose connective tissue, endotenon, which permits longitudinal movement of collagen fascicles and supports blood vessels, lymphatics, and nerves. A synovial sheath that allows pulley action and directs the path of the tendon encloses tendons that bend sharply, such as the flexor tendons of the hand. A bifoliate mesotenon originates on the side of the bend opposite from the pulley friction surface and joins the epitenon that covers the surface of the tendon. The sliding of this type of tendon is assisted by synovial fluid, which is extruded from the parietal synovial membrane lining the inner surface of the sheath and from the visceral synovial membrane or epitenon. Tendons not enclosed within a sheath move in a straight line and are surrounded by a loose areolar connective tissue called the paratenon, which is continuous with the tendon.

Tendons receive their blood supply from vessels in the perimysium, the periosteal insertion, and the surrounding tissue via vessels in the paratenon or mesotenon. In tendons surrounded by a paratenon, vessels enter from many points on the periphery and anastomose with a longitudinal system of capillaries. The vascular pattern of a flexor tendon within a tendon sheath is quite different. Here, blood is supplied by the proximal mesotenons and is reduced to the long and short vinculae. These tendons also receive their blood supply from their osseous insertions. Despite a large number of vessels supplying these tendons, injection studies have identified consistent areas of avascularity. These avascular regions have led several investigators to propose a dual pathway for tendon nutrition: a vascular pathway and, for the avascular regions, a synovial (diffusion) pathway. The concept of diffusional nutrition is of primary clinical significance in that it implies that tendon healing and repair can occur in the absence of adhesions and a blood supply.

Similarly, ligaments display uniform micro-vascularity, which originates from the insertion sites. Ligaments too have a variety of specialized nerve endings. Proprioception and nocioception receptors have been identified in the medial collateral ligament (MCL) of the knee.

D. Insertion into Bone

Tendon and ligament insertion into bone represents a transition from one tissue to another and can be quite complex. If tissue engineered tendons and ligaments are to function properly they have to replicate the mechanism by which the tissue is inserted into the bone. Insertions of ligaments are usually classified as either direct or indirect, the latter being more common. Direct insertions contain four morphologically distinct zones, namely, ligament, fibrocartilage, mineralized fibrocartilage, and bone. For indirect insertions, the superficial layer connects directly with periosteum while the deeper layers anchor to bone via Sharpey fibers. An example of a ligament that exhibits both types of insertion is the MCL of the knee. Its femoral insertion is direct while its tibial insertion is indirect.

Similarly, the tendon to bone insertion site is a complex transitional region that links two very different materials. The insertion site must transfer a complex loading environment effectively to prevent injury and provide proper joint function. Thomopoulos et al, performed assays at two insertion site locations: the tendon end of the insertion and the bony end of the insertion. They found that collagen was significantly more oriented at the tendon insertion compared to the bony insertion. Collagen types II, IX, and X, and aggrecan were localized only to the bony insertion, while decorin and biglycan were localized only to the tendon insertion. Thus, the tendon to bony insertion site varies dramatically along its length, in terms of its viscoelastic properties, collagen structure, and ECM composition.[9]

E. Function — Biomechanics

Tendon and ligaments both function by transmitting tensile forces, and their structure is adapted to that role. Tendon possesses one of the highest tensile strengths of any soft tissue in the body, both because its main constituent is collagen, one of the strongest fibrous proteins, and because these collagen fibers are arranged parallel to the direction of tensile force. The tensile properties of tendon can be characterized by the mechanical (material) properties of the tendon itself, as well as the structural properties of the bone–tendon–muscle structure. The mechanical properties (stress–strain relationship) of the collagen depend primarily on the architecture of the collagen fibers, and the interaction of the collagen with the ECM. The structural properties (load–elongation relationship) of the bone–tendon–muscle structure depend on the mechanical properties of the tendon substance as well as its bony insertion site and the myotendinous junction. It has been shown in animals that the elastic modulus of tendon ranges from 500 to 1200 MPa, whereas the ultimate tensile strength ranges from 45 to 125 MPa. In humans, the elastic modulus ranges from 1200 to 1800 MPa. The ultimate tensile strength of tendon ranges from 50 to 105 MPa, and the ultimate strain ranges from 9 to 35%. Structural properties of bone–tendon–muscle complexes have also been studied extensively, but vary greatly with different complexes from various anatomic locations.

F. Differences between Tendons and Ligaments

Ligaments and tendons are similar in structural composition as well as mechanical behavior; however, they differ in a number of ways. Physically, ligaments are short, wide, load-bearing tissue structures that connect bones, whereas tendons are longer and narrower structures that connect muscle to bone. Biochemically, ligaments contain a lower percentage of collagen and a higher percentage of glycosaminoglycans than tendons.[10]

Biomechanically, the collagen fibers in tendons are more longitudinally organized than those in ligaments and display a longer crimp length. Collectively, these morphological features explain why the load–deformation behavior of tendon displays less of a "toe region" than ligament during the initial stages of loading. Ligaments experience a more varied loading because of their role in joint stability and, therefore, display a broader distribution of fiber directions.

Studies suggest that ligaments are more metabolically active than tendons, having more plump cellular nuclei, higher DNA content, larger amounts of reducible cross-links, and the presence of more type III collagen.

Despite the differences in the make-up of tendons and ligaments, the similarities are such that one might expect that tissue-engineering strategies that work for one would show promise for the other.

III. HEALING PROCESSES

The ability of tendon and ligament to heal depends on factors such as anatomical location, vascularity, skeletal maturity, and tissue loss. Factors that favor the healing of ligaments and tendons relate to their relatively simple structure (viz., uniaxially aligned fibrils of type I collagen) and the fact that the parenchymal cell is the ubiquitous fibroblast. Factors that disfavor healing include the poor vascularity, the relatively low number density of cells, the high loading of the tissues that is difficult to control during healing, and the fibrinolytic effect of synovial fluid acting on intra-articular ligaments and tendons contained in a synovial sheath. It may be the absence of a provisional fibrin clot bridging the ends of a ruptured tendon or ligament that is the single most important factor disfavoring healing. Recent studies have shown that ruptured human ACL displays a proliferative response of cells in the tissue that forms on the surface of the ruptured ends. However, these cells are not able to bridge the ruptured ends because there is no provisional scaffold (i.e., the fibrin scaffold) to facilitate their migration. This underscores the importance of a biomaterial scaffold in the tissue engineering strategy employed when the goal is the regeneration of the tissue *in vivo*.

The fibrocollagenous tissue that we call "scar" (that is, the end stage of healing in cases of repair versus regeneration) comprises uniaxially aligned type I collagen, similar to that comprising tendons and ligaments. This scar tissue may be of some functional value for certain ligaments and tendons, and therefore raises the issue that complete regeneration of these tissues may not be necessary to recover meaningful function. Differences among scars, tendons, and ligaments include differences in the collagen fiber diameter distribution and crimp, and the percentage of other components such as proteoglycans and elastin. These differences may prove to be important for the long-term function of the tissue and, therefore, compel the development of tissue engineering strategies for the complete regeneration of these tissues.

Spontaneous healing of the tendon has been extensively studied in the flexor tendons of the hand and the Achilles tendon. Much of the following that focuses on the healing of tendons also applies to ligament healing. Of particular interest in ligament healing is that while extra-articular ligaments, such as the collateral ligament of the knee joint, heal producing reparative tissue that can provide some of the function of the natural ligament (i.e., up to 80% of the tensile strength after 1 year), the intra-articular ligaments generally display no sign of healing (i.e., no reparative tissue).

In the following discussion about issues related to the healing of tendon and ligament, what can be said about one of the tissues often relates to findings related to the other. The following is not an exhaustive review of the literature, but rather a presentation of selected points that provide a context in which to consider tendon and ligament tissue engineering.

A. Is there Any Part of the Tendon that Can Self-Regenerate?

Regeneration in this chapter shall refer to the recreation of a morphological, biochemical and biomechanical tendon or ligament that is identical to the preinjured tissue. Several studies have claimed the regeneration of tendon. However, on close inspection, the "regenerated" tendon differs from normal, undamaged tissue with respect to one or more of three main parameters: histological, biochemical, or biomechanical.

Eriksson et al.[11] and Ferretti et al.[12] have described the regeneration of the semitendinosus tendon after resection. Macroscopically, histologically, and immunohistochemically the regenerated tendon closely resembled normal ones. There were, however, focal scar-like areas. Martini et al.[13] have shown that the synovium can self regenerate if the visceral sheet around the flexor tendon remains undamaged. However, other authors have demonstrated that the synovial sheath does not maintain its characteristics, as demonstrated histologically.

B. Embryological vs. Adult Healing

Studies on healing characteristics of the sheep fetus demonstrate that the fetal tendon exhibits scar less healing. Of interest was that the fetal tissue was much more cellular than adult tissue with fibroblasts that were considerably more round than those typical for normal adult tissue. Also, the fiber bundles were less eosinophilic with wider spaces among the bundles, filled with endotenon tissue and cells.

Similarly, the transected fetal lamb tendon displayed no scarring. The tendon healed at 6 weeks and there were no adhesions or ruptures.[14] Farata et al. also compared tendon healing between 80 and 110 days of development in the sheep embryo. They found that the regenerated 80-day sheep tendon appeared histologically and geometrically the same as normal tendons. However, the mechanical strength was lower than a normal tendon.

The response of the adult Achilles tendon to injury by full transection, reviewed by several articles, follows a sequence similar to that in other connective tissues such as ligament and skin. This sequence is generally considered to consist of three overlapping phases: inflammation, repair, and remodeling. Following is a summary of the response of a tendon to a lesion produced by a full transection of the tendon, as reported by several authors. After complete transection, the ends of the tendon retract spontaneously. The gap created ranges from 2–4 mm, in the MCL of a rabbit, to 9–12 mm, in a rabbit tendoarchilles tendon. Additional retraction of the tendon ends can occur with movement of the knee or calcaneal joints.

Within the first few hours of injury, the collagenous matrix is disrupted and tendon and blood cells die. A hemorrhagic exudate fills the entire lesion site and a fibrin clot forms and seals the wound within minutes. The clot contains inflammatory products (fibrin, platelets, red cells, nuclear, and matrix debris). This clot has little tensile strength, but is important as the provisional scaffold which allows cells to migrate into the rupture gap.

The inflammatory stage of healing generally starts within hours of injury and can take from 3 to 10 days to complete. This stage is associated with "clean-up" of the lesion site. Polymorphonuclear neutrophils (PMNs) and lymphocytes, and other inflammatory cells, invade and populate the wound site within hours of injury. Monocytes and macrophages appear soon after to continue the phagocytosis of the cell and matrix debris.

The repair phase, starting as early as 10 days postinjury, can take up to 2 to 5 weeks to complete. Undifferentiated and disorganized fibroblasts containing well-developed endoplasmic reticulum, infiltrating from the wound edge and paratenon, begin to proliferate in the wound site within the fibrin mesh of the clot. Simultaneously, endothelial cells of surrounding vessels enlarge and proliferate forming capillary buds that follow the migrating fibroblasts. Together the fibroblasts, macrophages, and capillaries form the granulation tissue in the wound site. This early stage of the repair phase is characterized by increased cellularity. A collagen bridge joining the ruptured ends, which initially comprises predominantly type III collagen, gradually replaces the fibrin clot. The type III collagen fibers, which are smaller in diameter than the type I collagen to be deposited in the next stage of healing, are referred to as reticular fibers because of the network-like pattern of their deposition; the collagen type III fibers do not aggregate in a preferential direction like the type I fibers. Both collagen production and fibroblast proliferation peak during this phase (characterized by loosely organized fibrous tissue), and subsequently decrease over the next several months.

The remodeling phase begins as early as 3 weeks and lasts for over 1 to 2 years. It is marked by a reduction in the production of type III collagen and reorganization of the type I collagen fibers. During the remodeling stage, the matrix fibers reorient themselves along the long direction of the tendon. This direction coincides with the direction of the tensile stress in the tendon. The remodeling stage is also marked by a decrease in the number of fibroblasts present in the tissue and a decrease in the overall volume of the scar tissue. The tensile strength of the tendon increases through this period of remodeling, even though the total volume is decreasing. This increase has been explained by the reorganization of the collagen fibers, which has been observed to occur during the same period. In the analogous case of ligament healing, however, it has been shown that the tensile strength of the reparative tissue did not return to normal ligament levels even after a year. This data can probably be extended to the case of the tendon.

In summary, the response of the mature Achilles tendon to an injury involving a full transection of the tendon, results in reparative fibrous tissue that lacks the structure of normal tendon. During the process of remodeling, collagen fibers initially arranged in random directions become reoriented in the longitudinal direction of the tendon. However, there is never a return to normal composition and architecture, thus demonstrating the need for tissue engineering approaches.

C. Source of Reparative Cells

There is controversy surrounding the location of the cells responsible for collagen synthesis during tendon repair. On one side of the controversy is the concept that the tendon has the necessary cells to produce collagenous tissue (the intrinsic mechanism), while on the other side, there is the belief that the source of the collagen-producing cells is outside the tendon (i.e., an extrinsic source such as the surrounding tissues or from the tendon sheath). Some believe that both intrinsic and extrinsic sources of collagen-producing cells contribute to the healing process.

In vitro studies have shown that in response to tendon injury, cells within the tendon had the ability to migrate to and proliferate in the wound site. In these studies, by 6 weeks, the injury site appeared to be filled with collagen. *In vivo* data appear to parallel these *in vitro* studies. Matthews and Richards[15] showed that cells within the rabbit flexor tendon participated in the wound healing process when the synovial sheath was not violated, and the tendon was immobilized, with cells from external sources (e.g., tendon sheath, blood vessels, and other neighboring tissue) migrating and proliferating into the wound site. Tendon repair, in that case, involved the participation of all surrounding tissue in the healing of the entire wound.[15]

Murray and Spector et al.[16] demonstrated that the proximal and distal human ACL stumps not only had different morphology but demonstrated different outgrowth behaviors. They also showed that TGF-β1 had an inhibitory effect on the outgrowth of ACL explants.[16] The presence of a collagen–GAG scaffold encouraged earlier migration and proliferation of the cells from the explant into the matrices.[17]

Doubts still remain about whether there are sufficient pools of fibroblasts in the tendon and ligament to adequately populate the defect. This uncertainty is rationale for the use of exogenous cultured cells in tissue engineering modalities. However, it has been demonstrated that allograft cells perish in the transplanted tendon and ligament allografts.[18] The host cells replaced the donor cells in a rapid manner. Thus autologous cells may be more suitable for tissue engineering.

D. Normal Tendons and Ligaments vs. Scar Tissue

Regeneration should result in tissue that is indistinguishable from the original tissue, that is, the newly formed tissue has morphological, ultrastructural, biochemical, and biomechanical functions identical to those of the original tissue. Repair, in the classical use of the term, results in fibrocollagenous tissue which is distinguishable from the original tissue and is known as "scar." Scar tissue tends to be more vascular and have increased cell density, in comparison to tendon and ligament. Reparative scar has a higher cell metabolism and the presence of inflammatory cells can be seen. The collagen fibrils are smaller in scar and there are defects in collagen packing, when compared to tendon and ligament. The cross-links between the collagen fibers are also immature. Biochemically, repair tissue contains more type III[19] and type V collagen than normal tendons (i.e., higher collagen type III: type I ratio). Scar tissue also has larger proteoglycans, an excess of other glycoproteins and immature elastin in comparison to normal ligaments.

Some studies have claimed regeneration of tendon; however, there were still differences from the original tendon. In this ongoing attempt to regenerate tendon, we would inevitably move from a spectrum from complete scar to 100% regenerated tendon. Controversy, however, exists as to which parameters serve as the best measure for tendon regeneration. Some authors have measured differences such as crimp pattern, average fibril diameter, and distribution of the collagen fibrils within the tissue. Other measures include using polarized light to determine the orientation of the collagen fibers. For example, Kato et al.[20] found the crimp length of the healing tendon in their animal model was smaller than that of the original tendon. Collagen fibril diameters were also significantly smaller than that of normal tendon. Biomechanically, the repair tendon has a mechanical strength of 40 to 60% of the normal tendon.

IV. TENDON AND LIGAMENT TISSUE ENGINEERING: BASIC CONSIDERATIONS

Tissue engineering approaches applied to tendon and ligament, as with other tissues, are founded on the use of biomaterial scaffolds (matrices), cells, and soluble regulators, alone or in combination. Also, as with other applications, one of the two goals might be set: to engineer tendon *in vitro* for subsequent implantation or to develop an implant to facilitate regeneration *in vivo*. A problematic issue related to the former approach is the need to have the tendon or ligament tissue, engineered *in vitro*, becomes incorporated in the host tissue when it is implanted. Work to date has focused on the implementation of absorbable synthetic and natural matrices, alone, as implants for the engineering of the tissue by the host. While the results of this approach have been promising, there are indications that exogenous cells will be necessary for more complete regeneration. This has led to preliminary studies of methods for seeding certain matrices with tendon and mesenchymal stem cells *in vitro*.

A. Candidate Cells

As with several types of tissue and organs, tendon and ligament tissue engineering strategies may require the use of exogenous cells for implantation. The practical challenges of obtaining allogeneic cells for human implantation and the potential problems associated with the use of allogeneic cells, has led to the exploration of other cell sources. Autologous tenocytes and ligaments cells are being employed for experimental purposes. Another approach has focused on adult stem cells derived from bone marrow. The adult bone marrow stroma contains a subset of non-hematopoietic cells referred to as mesenchymal stem cells (MSCs), or mesenchymal progenitor cells. These cells have the capacity to undergo extensive replication in an undifferentiated state *ex vivo*. In addition, MSCs have the potential to develop either *in vitro* or *in vivo* into distinct mesenchymal tissues, including bone, cartilage, fat, tendon, muscle, and marrow stroma, which suggest these cells are an attractive cell source for tendon and ligament tissue engineering approaches. In the future other cell sources may also be available, including embryonic stem cells and embryonic tendon cells.[21]

B. Roles of the Scaffold

The biomaterials used for the fabrication of fibrous and porous sponge-like scaffolds for tissue engineering are resorbable. Numerous previous investigations have employed non-resorbable biomaterials for the reconstruction of tendons and ligaments, and several of these devices found their way into clinical use. In most cases the generally poor clinical performance led to the abandonment of these approaches.

The principal role of the biomaterial scaffold in tissue engineering is to facilitate cellular processes that will result in regeneration of the tissue. Therefore, the critical attributes of the biomaterial relate to its biological interaction with the infiltrating or seeded cells, and not its mechanical properties that must permanently substitute for the tendon or ligament over the long term.

Differentiation and cell behavior are continuously regulated processes. Interactions between the cell and its environment play a major role in maintaining the stable expression of differentiation-specific genes. An important component of the cellular environment is the ECM. Synthetic and natural scaffolds, acting in some ways as analogs of ECM, can thus influence cell behavior. The ECM, and tissue engineering scaffolds, offer structural support for the cells, and can also act as a physical barrier or selective filter to soluble molecules. The ECM and scaffold also plays a role in regulating the differentiated phenotype of cells. There are at least three mechanisms by which the ECM can regulate cell behavior. One is through the composition of the ECM. The second is through synergistic interactions between growth factors and matrix molecules. The third is through the cell surface receptors that mediate adhesion to ECM components.

There is, therefore, no doubt that the choice of scaffold plays an important role in the tissue engineering of the tendon and ligament. Many varied scaffold materials have been evaluated. However, some of the key factors for a suitable scaffold include biomechanical strength, especially during the initial stages. This would allow early mobilization and faster recovery of function. The biomaterial should preferably be biodegradable. This would allow the host natural tissues to totally replace the temporary scaffold. There would be less inflammatory reaction and consequent scarring. It is important that the degradation products be non-toxic and non-immunogenic. Other bioactive properties in terms of architecture, spatial organization of adhesion, growth, and differentiation signals are also important. A summary of different scaffolds used is given in Table 26.2.

(a)
(b)

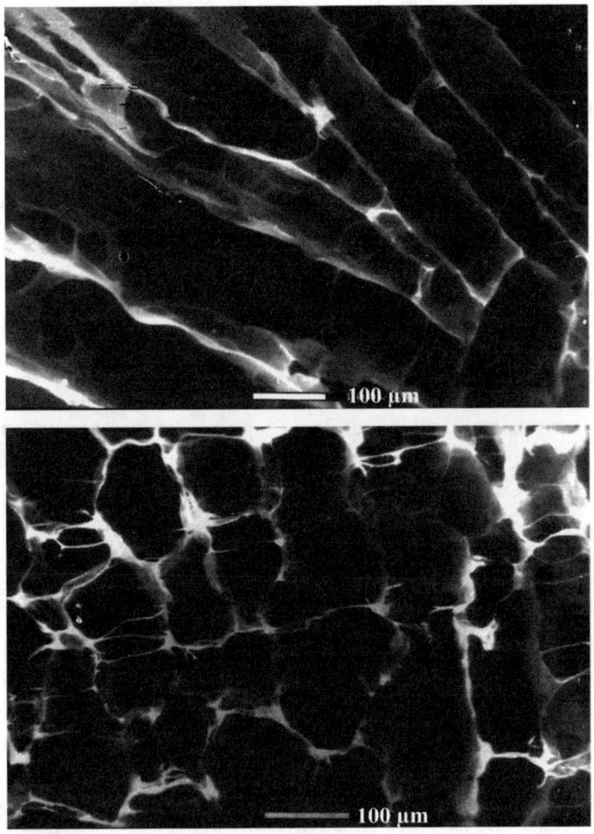

FIGURE 26.2 Scanning electron micrographs of a collagen–glycosaminoglycan scaffold fabricated as a cylinder with pore channel aligned along the long axis of the rod. Figure 26.2(a) shows a longitudinal section through the scaffold, and Figure 26.2(b) depicts a transverse section. The average pore diameter is approximately 100 μm in cross-section.

A variety of methods have been employed for the fabrication of scaffolds for tissue engineering. In the case of oriented tissues such as ligaments and tendons, a desired attribute of a sponge-like matrix could be the uniaxial orientation of pore channels. Such an approach has been employed for the fabrication of scaffolds for nerve regeneration.[22] One method of scaffold fabrication that allows the control of pore orientation is freeze-drying. In this technique the pore diameter distribution and orientation are determined by ice crystal formation, which can be controlled by the design of the mold and freezing parameters. This technique has been applied to the production of collagen–glycosaminoglycan scaffolds for tendon and ligament tissue engineering.[23] These scaffolds demonstrate a preferred orientation of the channels along the long axis of the porous rod (Figure 26.2). Such geometries may help to guide the architecture of newly synthesized matrix by the infiltrating or seeded cells.

V. *IN VITRO* TISSUE ENGINEERING INVESTIGATIONS EMPLOYING BIOMATERIAL SCAFFOLDS

That the midsubstance of tendon and ligament is poorly vascular, with a relatively low density of parenchymal cells, suggests that matrices seeded with exogenous cells may be necessary to facilitate regeneration. This has led to several *in vitro* investigations in which candidate cells were seeded into scaffolds for the purpose of assessing the influence of the scaffold material on the proliferation and biosynthetic activity of the cells. Such studies are of use for the screening of candidate biomaterials and cell types as well as for the investigation of *in vitro* conditions most suitable for the preparation of cell-seeded constructs as implants.

A. Evaluation of Biomaterials as Cell-Seeded Constructs

Recent preliminary investigations have seeded fibers of a synthetic polymer and a collagen–GAG sponge with tenocytes *in vitro* in order to ultimately develop cell-seeded implants for tendon engineering.

Bioresorbable prostheses currently being investigated to facilitate tendon healing include: collagen fibers, reconstituted collagen type I, di-catechol nordihydroguaiaretic acid (NDGA) collagen composite, resorbable fibers of dimethltrimenthylene carbonate–trimethylene carbonate copolymer, and a composite artificial tendon of poly (2-hydroxyethylmethylcrylate)/poly(caprolactone) blend hydrogel matrix, poly(lactic acid) fibers and silk–fiber matrix.

B. Collagen Fibers

Collagen fibers composed of type I collagen molecules were studied for biocompatibility and mechanical properties. These fibers were cross-linked using two processes; glutaraldehyde or dehydration followed by exposure to cyanamide (DHT/C). The latter method produces urea only as a by-product of the cross-linking process and is postulated to be more biocompatible. An *in vitro* model using rat tendon fibroblasts growing on individual fibers was used to evaluate outgrowth rates, cell and fiber interactions, and cell morphology. These studies showed an advantage with DHT/C cross-linking, relative to glutaraldehyde cross-linking, in promoting fibroblast growth. *In vivo* intra-muscular implantation in rats showed excellent biocompatibility for both kinds of collagen implants and aligned in-growth of cells from the MCL into the implant. Mechanical testing demonstrated the higher strength of dry fibers; however, upon hydration, there was a marked decrease in stress to failure. This reduction in strength was principally due to swelling which results in an increase of the cross-sectional area. These collagen fibers appear to be very biocompatible, even in the presence of low concentrations of glutaraldehyde. They promote fibrous aligned in-growth in a setting of ligament healing. Thus, they represent a promising candidate as a scaffold ligament or tendon prosthesis if cross-link density can be increased.[24]

Dunn evaluated high-strength resorbable collagen fiber scaffolds with intra-articular ACL or extra-articular PT rabbit fibroblasts. Fibroblasts attached, proliferated, and secreted new collagen onto the ligament analogs *in vitro*. Fibroblast function depended on the tissue culture substrate (ligament analog versus tissue culture plate) and the origin of the fibroblasts (ACL vs. PT). PT fibroblasts proliferated more rapidly than ACL fibroblasts when cultured on ligament analogs. Collagen synthesis by ACL and PT fibroblasts was approximately tenfold greater on ligament analogs than on tissue culture plates. The composition, structure, and geometry of the collagen fiber scaffolds may promote collagen synthesis within ligament analogs *in vitro*.[25]

Methods for stabilizing collagen-based materials with catechol-containing monomers were developed in order to produce fibers with mechanical properties in tension comparable to those of normal tendons. Fibers produced from pepsin were solubilized and bovine tendon type I collagen were polymerized with the NDGA. Polymerization was based on the chemical oxidation of the constituent *o*-catechols to reactive *o*-quinone functionalities. NDGA caused a dose dependent increase in the tensile strength and stiffness of the type I collagen fibers. A second treatment with NDGA improved the tensile properties significantly. Comparison of the effects of NDGA with those of biologically relevant mono-catechols indicated that the bicatechol functionality of NDGA was responsible for generation of the superior tensile properties. Elimination of unreacted intermediates from the treated fibers with ethanol increased the effectiveness of the cross-linking process while simultaneously sterilizing the material. Catalyzing oxidation by saturating the reaction buffer with oxygen increased the effectiveness of polymerization and the resulting tensile properties of the treated fibers. The ultimate tensile strength of the optimized NDGA-treated fibers averaged 90 MPa; the elastic modulus of these fibers averaged 580 MPa. Both values are comparable to undamaged tendons. The material properties of the NDGA cross-linked fibers exceed the properties of collagen fibers treated with other cross-linking strategies such as glutaraldehyde and carbodiimide. These results indicate that NDGA cross-linking may provide a viable approach to stabilizing collagenous materials for use in repair of ruptured, lacerated or surgically transected tendons, as well as other biomaterial constructs for surgical repair of musculoskeletal injuries and disease.[26]

Collagen fibers from different sources also show varying mechanical properties. Extrusion of bovine Achilles tendon collagen fibers significantly affected fiber diameter. This increases the percentage of smaller fibers which display higher tangent moduli and peak stresses. Mechanical properties of 125 μm diameter extruded fibers were similar to the properties reported for human ligaments. Collagen fibers from a rat tail tendon displayed characteristic strain-softening behavior, and scaffolds of rat tail fibers demonstrated a nonintuitive relationship between tangent modulus and specimen length. Gentleman et al. used a composite scaffold (extruded collagen fibers cast within a gel of Type I rat tail tendon collagen) and showed that the fibroblast seeded scaffolds displayed significantly higher tangent moduli and peak stresses than those without cells.[27]

C. Silk–Fiber Matrix

Altman studied a silk–fiber matrix as a candidate material for tissue engineering ACL.[28] The matrix was successfully designed to match the complex and demanding mechanical requirements of a normal human ACL, including adequate fatigue performance. This protein matrix supported the attachment, expansion, and differentiation of adult human progenitor bone marrow stromal cells based on scanning electron microscopy, DNA quantification, and the expression of collagen types I and III and tenascin-C markers. The results warrant additional investigation of selectively prepared silkworm fiber matrices that, aside from providing unique benefits in terms of mechanical properties as well as biocompatibility and slow degradability, can provide suitable biomaterial matrices for the support of adult stem cell differentiation toward ligament lineages.

D. Seeding of Cultured Tenocytes into Porous Collagen–GAG Matrices

Tenocytes, recovered by collagenase digestions of the Achilles and plantaris tendon of adult New Zealand white rabbits, were grown to confluence, and subsequently passaged to increase cell numbers. In a preliminary experiment, collagen–GAG matrices (described in a previous section) were seeded with 0.75 to 1.5 million cells.

The number of cells and their distribution in collagen–GAG matrices varied from a small number of cells uniformly distributed throughout the matrix to cells concentrated on the surface of the matrix. Cells were either spread out along the surface of the collagen–GAG material or aggregated. A cell count of the medium and trypsinized contents of the wells, in which the samples were seeded, revealed that from 2 to 50% of the seeded tendon cells were not incorporated into the collagen–GAG matrices after one day. In general, the tenocytes appeared to infiltrate to a depth of 0.35 mm into the 1 mm thick samples. In many of the matrices, the interior of the matrix was devoid of cells. The majority of cells were near or on the surface of the cells. The degree of infiltration appeared to be dependent on the pore diameter of the collagen–GAG matrices. Matrices with a larger pore diameter (120 μm) allowed for the infiltration of cells into the interior of the matrices and produced, in general, matrices with evenly distributed cells throughout the devise. While matrices with larger pore diameters would be desirable as cell transplantation devices because of the deeper cell infiltration and the uniformity of cell distribution, there might be drawbacks to using too large of a pore diameter. For matrices with pore sizes averaging 120 μm, the cells attached and spread out along the surface of collagen–GAG fibers. The surface area of collagen–GAG sponges is inversely proportional to the matrix pore diameter; therefore, with the larger pore diameters, there is less surface area to which the cells can attach. Future studies need to systematically investigate the matrix characteristics that yield optimal cell seeding.[29]

VI. ANIMAL STUDIES IMPLEMENTING TISSUE ENGINEERING SCAFFOLDS

A. ANIMAL MODELS: ENTUBULATION

Numerous animal models have been employed for the investigation of tendon and ligament tissue engineering. (See Table 26.1) The model employed should be appropriate for the tissue studied. Ligaments could be differentiated into intra-articular and extra-articular models whereas tendons should be divided into intra-synovial and extra-synovial models. Important factors of consideration include size of the animal, anatomic features and techniques available for tissue analysis.[30] Ligament healing models have generally included the MCLs because of their accessibility, in contrast to the ACL which even in larger animals is relatively short and difficult to access. Tendon healing models frequently include the flexor tendons of the hand and the Achilles tendon.

One aspect of the healing of tendons and ligaments, as with many other tissues, is the negative influence of surrounding tissues on healing as they collapse and infiltrate into the defect (i.e., rupture gap). One approach to overcome this phenomenon is to entabulate the ruptures ends. This approach has been successfully used to isolate the defect in other tubular tissues, such as peripheral nerve and spinal cord. While impractical and perhaps problematic as long term implants for this application, silicone tubes have been of some value for experimental purposes.

As an example of one approach that was employed in an animal investigation, a 1 cm gap in the rabbit Achilles tendon was treated by "entubulating" the tendon stumps in a silicon tube, as was done in an earlier study of flexor tendon healing and in prior studies of bone and peripheral nerve regeneration. In this model, the Achilles tendon was transected at its midpoint, after which the tendon

TABLE 26.1
Different Animal Models Employed

Animal	Lesion (Size)	Immobilization Method (Time)	Reference
Rabbit	Transection	None	63
Rabbit	Excision of proximal third	None	64
Rabbit	Transection	None	65
Rabbit	Excision of 1.5 cm	None	66
Rabbit	Excision of 2.0 cm midsection	Long plaster cast (6 weeks)	67
Rabbit	Excision of 3.0 cm	None	20
Sheep	Excision of 2.5–4 cm	Plaster (6 weeks)	68
Sheep	Transection	None	69
Rat	Transection	External fixation (12 days)	70
Rat	Transection	Transection of skin, nerve, muscle and bone	71
Dog	Transection (sheath and tendon)	Shoulder — spica cast or controlled passive motion	72
Chicken	Transection (sheath and tendon)	None	73
Chicken	Transection	None	74

stumps retracted about 8 to 10 mm. The stumps were subsequently inserted into a silicone tube. A 10 mm gap in the tendon was maintained by stitches through each stump and the tube, which allowed any mechanical stresses imparted to the system to be taken up by the sutures. In this model, the tendons of the peroneus longus, brevis, and tertius were cut and a portion of each tendon dissected away to immobilize the tendon gap site. Moreover, the plantaris tendon was cut with a Z-plasty, the ends separated, and then sutured under minimum tension. The knee joint was immobilized by external fixation to reduce loading to the tendon. The spontaneous healing in the entubulated 1 cm gap after 6 weeks comprised a thin continuous cable of the fibrous tissue, less than 1 cm in diameter. There was no significant increase in the diameter of the reparative tissue after 12 weeks. The tissue comprised dense aggregates of crimped collagen fibers, with a wavelength (12 μm) and fiber bundle thickness that were both significantly shorter than those in normal tendon. Future studies need to be directed toward answering the question: is this crimped tissue end-stage scar?

Findings with the entubulated tendon gap model in the rabbit demonstrate that a tendon wound site can be isolated from the effect of extrinsic factors during healing, and that there is only limited capability for spontaneous intrinsic healing, with no indication of regeneration of tendon. An *in vivo* model of this type could thus be valuable for the evaluation of constructs developed for engineering tendons. Future studies need to consider how certain characteristics of the tube affect the healing process. One potentially important feature is permeability. As there is limited blood supply to the midsubstance of many tendons, regenerating tissue in the tube needs to derive nutrients from the surrounding environment. While the permeability of the tube may facilitate this nutrition, from the tissue forming in the entubulated gap, it might allow the loss of intrinsic regulators of growth and remodeling from the lesion. These considerations have been addressed in entubulated peripheral nerve regeneration, and may be relevant for tendon and ligament healing.

B. Fibrous Scaffolds

1. Collagen Fiber Scaffold

Kato et al., reported the use of carbodiimide-cross-linked and glutaraldehyde-cross-linked collagen–fiber implants for the Achilles tendon of rabbits. They found that healing in gaps in the tendon, bridged by the devices, was affected by the rate of the implant degradation. The carbodiimide-cross-linked implant was resorbed within 10 weeks and was replaced with a "neotendon." This reparative tissue was characterized by "aligned, crimped collagen fiber bundles" as early as within 20 weeks. The slower degrading glutaraldehyde-cross-linked implant induced tissue was "not as developed" (was not as aligned and was not crimped) as the carbodiimide-cross-linked implant. While these results are promising, it should be noted that Kato et al. stated that the neotendon was "similar, but not identical, to normal tendon" 1 year after implantation of the prosthesis. The tissue in Kato's tendon lesion site was described to have a crimp wavelength of 10 μm. In Kato's study, it was also observed that this crimp pattern was present from 3 to 52 weeks with minimal change in the crimped characteristics. The fact that fibrous tissue with this crimp pattern did not appear to remodel significantly after 52 weeks may suggest that it is a terminal "scar." This raises the question of whether complete regeneration is necessary for the reparative tissue to be of functional value (see Table 26.2; Figure 26.3).

2. Mesenchymal Stem Cell-Seeded Knitted PLGA Fiber Scaffold

It is well known that D,L-lactide-*co*-glycolide (PLGA, 10:90) is commonly used as a surgical suture material (Figure 26.4(a)). In one study, MSCs were found to adhere and proliferate on PLGA films.[31] As for scaffold structure, a knitted technique has been used for ligament reconstruction and showed promising results.[32] MSCs were seeded into PLGA scaffold and implanted into 1 cm long gap defects in the Achilles tendon of New Zealand White rabbits. They found an increased volume of regenerated tissue with type I and III collagen fibers present in the neotendon 4 weeks

TABLE 26.2
Types of Scaffolds Utilized

Tendon or Ligament	Nonsoluble Regulator	Soluble Regulator	Cell Type	Reference
Ligament (MCL)	Medial collateral ligament	None	Bone marrow cells	75
Ligament (ACL)	Collagen matrix	None	Fibroblast	25
Ligament (ACL)	Extruded collagen fibers	None	None	76
Ligament (ACL and MCL)	Biodegradable polymer fiber scaffold	TGF	ACL and MCL fibroblast	77
Ligament	Woven carbon fibers and PGA	None	Embryonic tendon cells	78
Ligament (ACL)	Silk–fiber matrix	None	Bone marrow cells	28
Tendon	Collagen matrix	None	Mesenchymal stem cell	79
Tendon	Collagen matrix	None	Mesenchymal stem cell	80
Tendon	Collagen scaffold	None	Mesenchymal stem cell	81
Tendon	Reconstituted collagen tendon scaffold	None	None	82
Tendon	Reconstituted type I collagen	None	None	83
Tendon	Collagen tendon prosthesis	None	None	20
Tendon	Reconstituted type I collagen	None	Rat tendon fibroblast	24
Tendon	Marlex mesh	None	None	84
Tendon	Small intestine sub mucosa	None	None	85,86
Tendon	Dacron vascular graft	None	None	87
Tendon	Chitin, p-CL, PLA	None	None	88
Tendon	Polyglycolic acid fibers	None	Autologous tenocyte	89
Tendon	Knitted PLGA fiber	None	Bone marrow stromal cells	31
Tendon	NDGA–collagen composite fibers	None	None	26
Tendon	Carbon Fiber	None	Rat tendon fibroblast	90
Tendon (Achilles)	Carbon fiber	None	Human achilles tendon	91

postoperatively. The histology of the neotendon at 12 weeks showed organized bundles of highly crimped fibers (Figure 26.4(b)). Compared to normal tendons, the length of the crimp patterns were shorter and the cells numbers higher.[31]

3. Tenocyte-Seeded Polyglycolic Acid Fiber Mesh

In a recent investigation, samples of non-woven mesh of polyglycolic acid fibers, with interstitial spaces from 75 to 100 μm in diameter, were seeded with tendon cells isolated from newborn calves by collagenase treatment. After 1 week in culture, the cell-seeded specimens were implanted subcutaneously in nude mice for up to 10 weeks. Histological evaluation of the 10 week samples showed "parallel linear organization of collagen bundles throughout the specimens." Mechanical testing revealed that the tissue engineered neotendon structures had approximately one third the tensile strength of normal tendon (11 vs. 32 MPa), 8 weeks postimplantation. These promising findings are serving as the basis for efforts to further improve engineering tendon. Additional studies will be required to determine if comparable results can be achieved with adult cells.

C. Sponge-Like Scaffolds

1. Collagen–GAG Matrices Contained within a Tube

A prior study investigated the regeneration of rabbit Achilles tendons, induced by a collagen–GAG sponge-like scaffold contained within a silicone tube in the animal model described earlier. The presence of the collagen–GAG matrix altered the process of tendon healing when compared to the

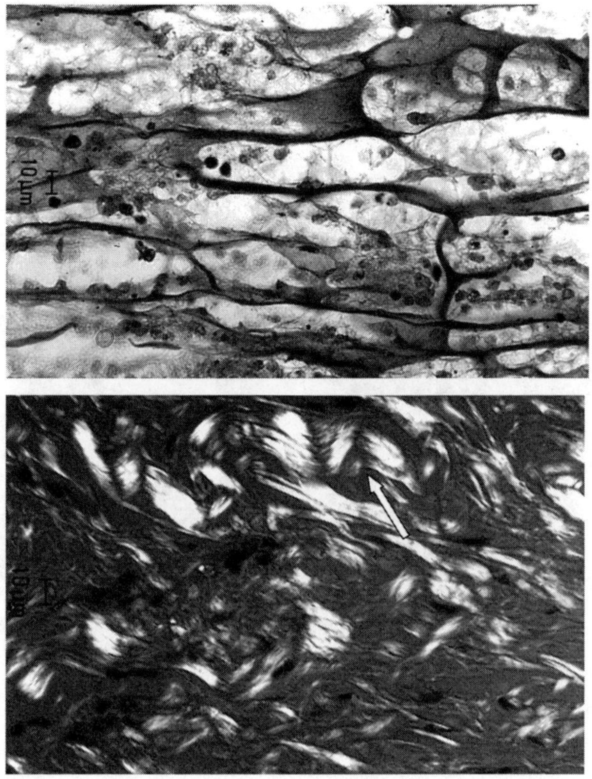

FIGURE 26.3 Histology showing the cells and extra-cellular matrix filling a porous collagen–GAG scaffold: (a) 3 weeks after being implanted inside a silicone tube bridging a rupture of the Achilles tendon in a rabbit, and (b) 12 weeks postoperatively viewed under polarized light. The black arrows in (a) show the cells that have infiltrated the scaffold and the white arrows show the fine fibrillar matrix representing residual fibrin clot and newly synthesized reticular matrix. The white arrow in (b) shows the crimped collagen. Hematoxylin and eosin stain.

control sites implanted with the tube alone. Tubes filled with collagen–GAG matrix contained a significantly greater volume of tissue at the time periods of evaluation: 3, 6, and 12 weeks. Collagen–GAG scaffolds obtained 3 weeks after implantation were filled with undifferentiated mesenchymal cells and a fine fibrillar matrix consisting of the residual fibrin clot and a newly synthesized reticular matrix (Figure 26.3(a)). These findings demonstrated the capability of host cells to infiltrate the scaffold with its interconnecting pores. Moreover, it demonstrated the potential value of the scaffold in stabilizing the fibrin clot, which serves as a provisional scaffold for cell migration into the defect site. The longitudinal orientation of the collagen–GAG struts of the implanted scaffold could still be seen 3 weeks after implantation (Figure 26.3(a)). Granulation tissue persisted for a longer period of time in the lesion site of collagen–GAG-filled defects, and the amount of dense fibrous tissue increased continuously during the period of study in defects filled with collagen–GAG matrix. In contrast, the amount of dense fibrous tissue decreased after 6 weeks in originally empty tubes. In tubes that did not contain the CG matrix, the new tissue consisted of dense aggregates of crimped fibers with a wavelength and fiber bundle thickness that were significantly shorter (10 μm) than those in normal tendon, and consistent with the type of scar that is the end result of repair of many connective tissues. The newly synthesized collagen bundles in the collagen–GAG scaffolds after 12 weeks also occasionally displayed crimp (Figure 26.3(b) white arrow), but the wavelength appeared to be longer than in the controls and there were generally fewer regions displaying crimp. These

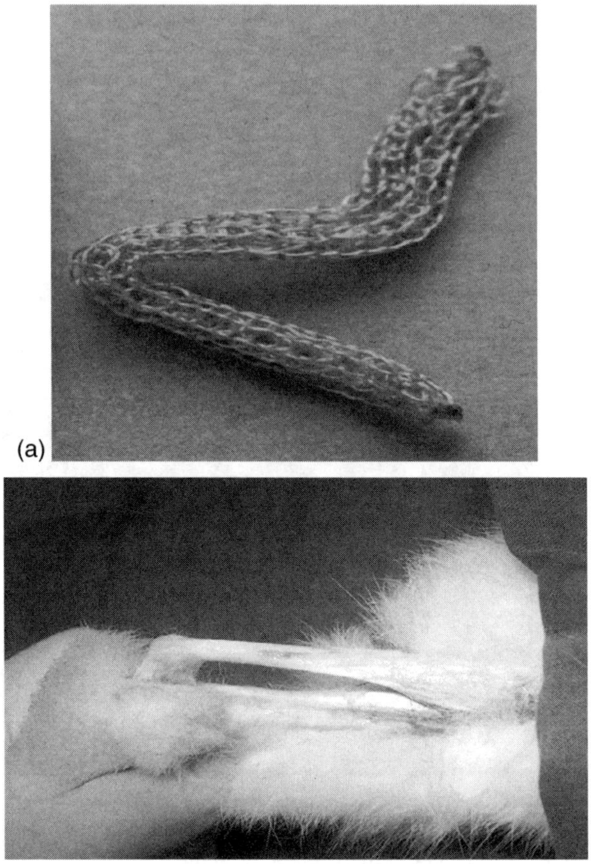

FIGURE 26.4 (a) Unseeded knitted PLGA scaffold and (b) Macro morphology of mesenchymal stem cell-seeded scaffold in the rabbit Achilles tendon at 12 weeks postimplantation. (Courtesy of Dr. James Goh.)

results demonstrate that the presence of a collagen–GAG scaffold can modify the reparative tissue architecture. Additional studies will be required to demonstrate the potential value of this approach.

D. Gel-Like Scaffolds

In a recently reported study, autologous MSCs seeded in a collagen gel were implanted into 1 cm long defects in the lateral gastrocnemius tendons of rabbit for 3 months. The cells were mixed with the collagen solution and incubated for 36 to 40 h in the presence of biodegradable sutures, such that the gel contracted around the suture to form an "integrated implant." During the culture period, tensile loading was applied to the sutures in order to align the cells seeded in the incorporated gel. The implant was then sutured into the gap in the rabbit tendon.

After 3 months, the reparative tissue comprised "dense bands of matrix organized in the axial direction of the tendon." Controls with sutures only (but without the unseeded gel) "demonstrated similar attributes but with less volume and less organization than those seen in MSC-treated repair tissue." While the mechanical properties of the cell-treated defects were improved relative to the suture-only group, they were still significantly below normal levels. Further investigation of this novel approach will be required to determine how the system can be modified to achieve a more physiological neotendon. Hildebrand et al.[33] describes some of the responses of the donor and recipient cells when transplanted to a ligament or tendon.

VII. EFFECTS OF GROWTH FACTORS

The biology of the healing tendon and the growth factors and receptors produced are slowly being understood (see Table 26.3). The healing cells and the transplanted cells show responsiveness to growth factor stimulation. Growth factors have been introduced either in protein or gene format. Basic FGF (FGF-2) has been shown to increase the proliferation rate and type III collagen in a dose dependent manner for up to 7 days. However, it did not influence the ultimate stress and the pyridinoline content of the healing tendons.[34] Abrahamsson et al.[35] showed that rhIGF-1 increased cellular proliferation and increased matrix synthesis of tendon explant cultures. Dahlgren et al.[36] injected rhIGF-1 into a equine model of flexor tendinitis and showed that it resulted in a tendon with increased cellularity and collagen content. The IGF treated tendon was also stiffer than the saline treated controls. Spindler et al. attempted injection of TGF-β2 into rabbit MCL at 0.1, 1, or 5 μg doses and found that all treatment groups demonstrated an increase in scar mass, but no group had a significant increase in load to failure at 6 weeks. The addition of 0.1 μg TGF-β2 led to a significant increase in scar stiffness. The addition of PDGF had no significant effect on any of the parameters studied. This study suggests that the mechanical stiffness, but not the load at failure, of ligament scar can be significantly altered by the administration of TGF-β2.[37]

Collagen diameter and its organization play an important role in the ultimate strength of the tendon. Decorin, a proteoglycan that inhibits fibrillogenesis, may be inhibited with antisense decorin oligodeoxynucleotides. This was shown to increase the final diameter of the collagen fibrils in young scars and improve scar failure strength and scar creep elongation.[38] Growth differentiation factors 5 and 6 have been shown to increase the tensile strength of the regenerate tendon.[39,40]

The response of ligaments appears different compared to the tendon.[41] Epidermal growth factor (EGF), basic fibroblast growth factor (bFGF) and platelet derived growth factor-BB (PDGF-BB) appear to significantly increase fibroblast proliferation,[42–45] while TGF-β1 superiorly promotes extracellular matrix synthesis.[42] *In vivo* studies, ruptured rat MCL exposed to PDGF-BB have ultimate load, ultimate elongation, and energy absorbed to failure that is significantly greater than those from other groups.

TABLE 26.3
Effect of Growth Factors on Healing and Repair of Tendon

Tendon or Ligament	Soluble Regulator	Reference
Ligament (MCL)	EGF, PDGF-BB, TGF-β1	42
Ligament (ACL)	TGF-β1, PDGF-AB, EGF, FGF2	92
Ligament (ACL)	EGF, PDGF-AB, TGF-β1, IGF-1, insulin	43
Ligament (MCL and ACL)	EGF, PDGF-BB, bFGF, TGF-β1, IGF-1	44
Ligament (MCL and ACL)	EGF, bFGF, PDGF-AA and -BB, IGF-1 and interleukin-1-alpha	45
Tendon	IGF-1	36
Patella tendon and ACL	PDGF-AB and TGF-β1	41
Tendon	BMP-12 gene	52
Tendon	GDF-5	40
Tendon (rat patella)	bFGF	34
Tendon	rhIGF-1	35
Tendon	bFGF	93
Tendon	GDF 5,6	39
Tendon	rEGH	94
Tendon	Decorin antisense gene therapy	38
Tendon	CDMP-2	95
Tendon	EGF, IGF	96
Tendon	FGF, PDGF	97

It is currently not clear if using a single growth factor or a combination of growth factors either together or in temporal fashion would yield the best results. It is also not yet known whether the best mode for delivering these growth factors is in protein molecules or via genetic alteration.

VIII. USE OF GENE THERAPY FOR TENDON/LIGAMENT TISSUE ENGINEERING

Recent advances in gene transfer technology permit the design of strategies to transfer various genes encoding modulatory species of ribonucleic acid or proteins such as growth factors, receptors, and transcription factors to genetically augment tissue engineering of tendons and ligaments. Several investigators have attempted transfecting cell cultures or local tissues directly. Various vehicles and techniques have been developed to deliver therapeutic genes into cells, such as viral vectors, and physical/chemical gene delivery methods including naked DNA and particle-mediated gene transfer, the micro-seeding technique, and the application of lipids. Goomer et al.[46] used a non-viral *in vivo* gene therapy to transfect a marker gene into flexor tendons with efficacy approaching 100%. Hildebrand et al.[47] showed that when normal and injured rabbit knee ligaments (viz., the MCL and ACL) were transfected with transduced ligament fibroblast or adenovirus particles, there were similar lengths of gene expression in the ruptured and uninjured MCLs. Pelinkovic et al.[48] injected genetically engineered, highly purified muscle derived muscle cell (MDC) to the supraspinatus tendons of nude rats. The results indicate that the rotator cuff tendon matrix and its original cellular components modulated the injected MDCs towards a fibroblastic phenotype. Gerich et al.[49] studied various retroviral vectors; MFG lacZ and BAG lacZ neo(r) and adenovirus LacZ were evaluated for their ability to deliver genes to cells of ligamentous origin. They found that the adenovirus was the most effective vector in short term experiments. However, expression was transient. Although retrovirus gave lower initial transduction efficiencies, the percentage of transduced cells could be increased by the use of the selectable marker gene neo(r). Menetrey et al.[50] compared direct, fibroblast, and myoblast mediated gene transfer into the ACL. They found that all three methods showed the presence of marker genes within the ligament. Nakamura et al. compared the efficacy of two different cationic liposomes, Lipofectin, and hemagglutinating virus of Japan (HVJ)-cationic liposomes, on nuclear uptake of fluorescence-labeled phosphorothioate oligodeoxyribonucleotide (S-ODN) by ligament scar fibroblasts and suppression of decorin mRNA expression when antisense decorin S-ODN was transferred. They found that there was no significant difference in nuclear uptake of fluorescent ODN between the two methods. However, only HVJ-cationic liposomes had a significant effect on suppression of decorin mRNA expression levels. To address the discrepancy, the molecular integrity of the transferred ODN in the cells was assessed by analysis of fluorescence resonance energy transfer (FRET) within double-fluorescence-labeled S-ODN. More than 70% of the ODN transfected by HVJ-cationic liposomes remained intact within the nucleus at 20 hours after transfection, while the majority of the ODN transferred by Lipofectin was degraded at this point. These results suggest a strong relationship between the nuclear integrity of transfected antisense ODN and its suppression of target mRNA expression.[51]

Lou et al. used BMP-12, a human homologue of mouse growth/differentiation factor (GDF)-7. They had previously reported that injection of mesenchymal progenitor cells transferred with the BMP-12 gene into the muscles of nude mice induced tendon-like tissue formation. They now investigated the effect of BMP-12 gene transfer on tendon cells and observed that adenovirus-mediated BMP-12 *in vitro* gene transfer into chicken tendon cells increased type I collagen synthesis but did not induce change in alkaline phosphatase activity. BMP-12 gene transfer into a complete tendon laceration chicken model resulted in a twofold increase of tensile strength and stiffness of repaired tendons, indicating improved tendon healing *in vivo*.[52] Collectively, these laboratory studies show that gene therapy shows some promise for tendon and ligament tissue engineering.

IX. EFFECT OF EXTERNAL FACTORS ON TENDON REGENERATION

There is increasing evidence that the mechanical stress of mobilizing healing tendons not only decreases adhesions, but improves the biomechanical attributes of the healing tendon.

Palmes et al.[53] showed that postoperative mobilization resulted in a continuous and significantly more rapid restoration of load to failure in comparison to the immobilized group. Tendon deflection decreased with postoperative mobilization, whereas it increased with immobilization. After 112 days the tendons of the mobilization group had regained their original tendon stiffness, whereas the immobilized tendons only reached only about half the tendon stiffness observed in the control tendons. Histologically, postoperative mobilization led to increased immigration of inflammatory cells in the early phase. In the late phase, as compared to immobilization, tendon structure was more mature with fiber bundles arranged in parallel and interposed tenocytes.

This histological and biomechanical improvement is probably partially due to the orientation of the fibroblast during stretching. Fibroblast orientation to stretch begins at 3 hours and appears nearly complete by 24 hours. Cultures stretched for 2 to 3 hours continue to exhibit greater degrees of orientation compared to controls. The fibroblast begins to orient within 3 hours of initiation of stretch and they continue to orient for several hours after cessation of stretch.[54] Altman et al.[55] demonstrated that the application of mechanical stress over a period of 21 days up-regulated ligament fibroblast markers, including collagen types I and III and tenascin-C, fostered statistically significant cell alignment and density, and resulted in the formation of oriented collagen fibers, all features characteristic of ligament cells.

In addition to improving orientation, external forces can also influence collagen production. Qin et al.[56] showed that a cyclic mechanical strain applied in three-dimensional culture stimulated tendon cell growth and increased collagen type I production, compared to unstretched controls. Murray et al.[57] also showed that a pulsed magnetic field can specifically increase the production of collagen, the major differentiated function of fibroblast, possibly by altering cyclic-AMP metabolism.

However, at the end of the spectrum, too much loading may increase interleukin 6 production, causing increased inflammatory response. Skutek experimentally stretched human tendon fibroblast of healthy human tendons for 15 and 60 min with 1 Hz and an elongation of 5%. He observed that interleukin 6 production was significantly increased after 15 min of cyclic biaxial mechanical stretching for 4 to 8 h, and for 2 h after 60 min stretching.[58]

Despite these informative studies, the amount of stress, the length of application, and the type of loading (constant or cyclic) are still unknown parameters necessary for tissue regeneration (see Table 26.4).

Other modalities of treatment that have shown promise include using ultrasound, electrical stimulation and laser phototherapy or a combination of them. Enwemeka et al. used continuous waves at space-average intensity of 0.5 W cm^2 for 5 min everyday for 9 days. He found that sonication induced a significant increase in the tensile strength, tensile stress and energy absorption capacity of the tendons.[59] Gum used a multimodality protocol with low intensity ultrasound, electrical stimulation, and low intensity Ga:As laser photo stimulation. The multimodality treatment produced a 14% increase in maximal strength, a 42% increase in load-at-break, a 20% increase in maximal stress, a 45% increase in stress-at-break, a 21% increase in maximal strain, and a 14% increase in strain-at-break.[60] The improvements measured in multitherapy were consistent but less remarkable, compared to their earlier work with single modality protocols.

TABLE 26.4
Effect of Physical Environment on Healing and Repair of Tendon

Tendon or Ligament	Physical Factor	Cell Type	Reference
Ligament	Cyclic hydrostatic compression and cyclic tension	MCL scar	98
Tendon	Cyclic mechanical stretching	Tendon fibroblast	58
Tendon	Intermittent hyperbaric oxygen	Tendon fibroblast	61
Tendon	Tensile loading	Achilles tendon	53
Tendon	*Ex vivo* cyclic loading	Canine flexor tendons	99
Tendon	Constant mechanical tension	Rabbit flexor tendons	100
Tendon	Variation of rehabilitation force	Canine flexor profundus tendon	101
Tendon	Stress shielding	Rabbit patella tendon	102
Tendon	Ultrasound, electrical stimulation, laser	Rabbit Achilles tendon	60
Tendon	Low intensity ultrasound	Rabbit Achilles tendon	59
Tendon	Hyaluronic acid	Rabbit profundus flexor tendon	103
Tendon	Temperature and time	Bovine Achilles tendon	104
Tendon	Cyclic mechanical stretch	Hen tenocytes	56

They hypothesize that the beneficial effects of ultrasound and laser photo stimulation on tendon healing may counteract one another when applied simultaneously.

Another study evaluated the effect of hyperbaric oxygen on procollagen messenger RNA levels and collagen synthesis on healing rat tendon laceration. Ishii et al. found that healing tendons treated with two atmospheres, 100% oxygen for 60 min, once a day increases pro-a1(I) mRNA expression.[61]

X. USE OF BIOREACTORS FOR TENDON/LIGAMENT TISSUE ENGINEERING

As research unveils increasing number of factors that affect tendon regeneration, we are faced with compelling reasons to control these factors. Advance bioreactors are essential for meeting the complex requirements of *in vitro* engineering of a tendon or ligament. Altman[62] devised an advanced bioreactor that could dissolve oxygen tension, and control for mechanical strain, temperature and pH. The bioreactor would ideally offer a favorable mechanical and biochemical environment to promote tissue regeneration.

ACKNOWLEDGMENTS

This work was supported, in part, by the Department of Veterans Affairs and the National Medical Research Council of Singapore.

REFERENCES

1. Langer, R., and Vacanti, J. P., Tissue engineering, *Science*, 260(5110), 920–926, 1993.
2. Sabiston, P., Frank, C., Lam, T., Shrive, N., Allograft ligament transplantation. A morphological and biochemical evaluation of a medial collateral ligament complex in a rabbit model, *Am. J. Sports Med.*, 18(2), 160–168, 1990.
3. Olson, E. J., Kang, J. D., Fu, F. H., Georgescu, H. I., Mason, G. C., and Evans, C. H., The biochemical and histological effects of artificial ligament wear particles: in vitro and in vivo studies, *Am. J. Sports Med.*, 16(6), 558–570, 1988.

4. Paulos, L. E., Rosenberg, T. D., Grewe, S. R., Tearse, D. S., and Beck, C. L., The GORE–TEX anterior cruciate ligament prosthesis. A long-term followup, *Am. J. Sports Med.*, 20(3), 246–252, 1992.
5. Murray, M. M., and Spector, M., Fibroblast distribution in the anteromedial bundle of the human anterior cruciate ligament: the presence of α-smooth muscle actin positive cells, *J. Orthop. Res.*, 17, 18–27, 1999.
6. Ippolito, E., Natali, P. G., Postacchini, F., Accinni, L., and De Martino, C., Ultrastructural and immunochemical evidence of actin in the tendon cells, *Clin. Orthop.*, 126, 282–284, 1977.
7. Faryniarz, D. A., Chaponnier, C., Gabbiani, G., Yannas, I. V., and Spector, M., Myofibroblasts in the healing lapine medial collateral ligament: possible mechanisms of contraction, *J. Orthop. Res.*, 14, 228–231, 1996.
8. Strocchi, R., De Pasquale, V., Guizzardi, S., Govoni, P., Facchini, A., Raspanti, M., Girolami, M., and Giannini, S., Human Achilles tendon: morphological and morphometric variations as a function of age, *Foot Ankle*, 12(2), 100–104, 1991.
9. Thomopoulos, S., Williams, G. R., Gimbel, J. A., Favata, M., and Soslowsky, L. J., Variation of biomechanical, structural, and compositional properties along the tendon to bone insertion site, *J. Orthop. Res.*, 21(3), 413–419, 2003.
10. Amiel, D., Frank, C., Harwood, F., Fronek, J., and Akeson, W., Tendons and ligaments: a morphological and biochemical comparison, *J. Orthop. Res.*, 1(3), 257–265, 1984.
11. Eriksson, K., Kindblom, L. G., Hamberg, P., Larsson, H., and Wredmark, T., The semitendinosus tendon regenerates after resection: a morphologic and MRI analysis in 6 patients after resection for anterior cruciate ligament reconstruction, *Acta Orthop. Scand.*, 72(4), 379–384, 2001.
12. Ferretti, A., Conteduca, F., Morelli, F. and Masi, V., Regeneration of the semitendinosus tendon after its use in anterior cruciate ligament reconstruction: a histologic study of three cases, *Am. J. Sports Med.*, 30(2), 204–207, 2002.
13. Martini, A. K., Animal experiment studies on the topic of regeneration of the synovial membrane following synovectomy of flexor tendons in the so-called no man's land, *Handchir. Mikrochir. Plast. Chir.*, 18(4), 199–203, 1986.
14. Al-Qattan, M. M., Posnick, J. C., Lin, K. Y., Thorner, P., Fetal tendon healing: development of an experimental model, *Plast. Reconstr. Surg.*, 92(6), 1155–1160, 1993, discussion 1161.
15. Matthews, P., and Richards, H., Factors in the adherence of flexor tendon after repair: an experimental study in the rabbit, *J. Bone Joint Surg. Br.*, 58(2), 230–236, 1976.
16. Murray, M. M., Bennett, R., Zhang, X., and Spector, M., Cell outgrowth from the human ACL in vitro: regional variation and response to TGF-beta1, *J. Orthop. Res.*, 20(4), 875–880, 2002.
17. Murray, M. M., and Spector, M., The migration of cells from the ruptured human anterior cruciate ligament into collagen–glycosaminoglycan regeneration templates in vitro, *Biomaterials*, 22(17), 2393–2402, 2001.
18. Jackson, D. W., and Simon, T. M., Donor cell survival and repopulation after intraarticular transplantation of tendon and ligament allografts, *Microsc. Res. Tech.*, 58(1), 25–33, 2002.
19. Maffulli, N., Ewen, S. W., Waterston, S. W., Reaper, J., and Barrass, V., Tenocytes from ruptured and tendinopathic achilles tendons produce greater quantities of type III collagen than tenocytes from normal achilles tendons. An in vitro model of human tendon healing, *Am. J. Sports Med.*, 28(4), 499–505, 2000.
20. Kato, Y. P., Dunn, M. G., Zawadsky, J. P., Tria, A. J., and Silver, F. H., Regeneration of Achilles tendon with a collagen tendon prosthesis. Results of a one-year implantation study, *J. Bone Joint Surg. Am.*, 73(4), 561–574, 1991.
21. Yang, Z. M., Xie, H. Q., and Xiang, Z., Culture of the transformed human embryonic tendon cells and its biological characteristics in vitro, *Zhongguo Xiu Fu Chong Jian Wai Ke Za Zhi*, 13(2), 99–104, 1999.
22. Yannas, I. V., Biologically active analogues of the extracellular matrix: artificial skin and nerves, *Angew. Chem. Int. Ed. Eng.*, 29(1), 20–35, 1990.
23. Louie, L. K., Yannas, I. V., and Spector, M., Development of a collagen–GAG copolymer implant for the study of tendon regeneration, In *Biomaterials for Drug and Cell Delivery*, Mikos, A. G., ed., MRS, Pittsburgh, PA, pp. 19–24, 1994.
24. Law, J. K., Parsons, J. R., Silver, F. H., Weiss, A. B., An evaluation of purified reconstituted type 1 collagen fibers, *J. Biomed. Mater. Res.*, 23(9), 961–977, 1989.

25. Dunn, M. G., Liesch, J. B., Tiku, M. L., and Zawadsky, J. P., Development of fibroblast-seeded ligament analogs for ACL reconstruction, *J. Biomed. Mater. Res.*, 29(11), 1363–1371, 1995.
26. Koob, T. J., and Hernandez, D. J., Material properties of polymerized NDGA–collagen composite fibers: development of biologically based tendon constructs, *Biomaterials*, 23(1), 203–212, 2002.
27. Gentleman, E., Lay, A. N., Dickerson, D. A., Nauman, E. A., Livesay, G. A., Dee, K. C., Mechanical characterization of collagen fibers and scaffolds for tissue engineering, *Biomaterials*, 24(21), 3805–3813, 2003.
28. Altman, G. H., Horan, R. L., Lu, H. H., Moreau, J., Martin, I., Richmond, J. C., and Kaplan, D. L., Silk matrix for tissue engineered anterior cruciate ligaments, *Biomaterials*, 23(20), 4131–4141, 2002.
29. Louie, I. V. Y., Hsu, H. P., and Spector, M., Healing of tendon defects implanted with a porous collagen–GAG matrix: histological evaluation, *Tissue Eng.*, 3(2), 187–195, 1997.
30. Carpenter, J. E., Thomopoulos, S., and Soslowsky, L. J., Animal models of tendon and ligament injuries for tissue engineering applications, *Clin. Orthop.*, 367(Suppl), S296–S311, 1999.
31. Ouyang, H. W., Goh, J. C., Mo, X. M., Teoh, S. H., Lee, E. H., The efficacy of bone marrow stromal cell-seeded knitted PLGA fiber scaffold for Achilles tendon repair, *Ann. N Y Acad. Sci.*, 961, 126–129, 2002.
32. Urban, J., Rutowski, R., and Staniszewska-Kus, J., Experimental research on the usefulness of the Dallos polyester prostheses in reconstructive operations of knee joint ligaments, *Polim. Med.*, 30(1–2), 3–9, 2000.
33. Hildebrand, K. A., Jia, F., and Woo, S. L., Response of donor and recipient cells after transplantation of cells to the ligament and tendon, *Microsc. Res. Tech.*, 58(1), 34–38, 2002.
34. Chan, B. P., Fu, S., Qin, L., Lee, K., Rolf, C. G., and Chan, K., Effects of basic fibroblast growth factor (bFGF) on early stages of tendon healing: a rat patellar tendon model, *Acta Orthop. Scand.*, 71(5), 513–518, 2000.
35. Abrahamsson, S. O., Matrix metabolism and healing in the flexor tendon. Experimental studies on rabbit tendon, *Scand. J. Plast. Reconstr. Surg. Hand Surg. Suppl.*, 23, 1–51, 1991.
36. Dahlgren, L. A., van der Meulen, M. C., Bertram, J. E., Starrak, G. S., and Nixon, A. J., Insulin-like growth factor-I improves cellular and molecular aspects of healing in a collagenase-induced model of flexor tendinitis, *J. Orthop. Res.*, 20(5), 910–919, 2002.
37. Spindler, K. P., Murray, M. M., Detwiler, K. B., Tarter, J. T., Dawson, J. M., Nanney, L. B., and Davidson, J. M., The biomechanical response to doses of TGF-beta2 in the healing rabbit medial collateral ligament, *J. Orthop. Res.*, 21(2), 245–249, 2003.
38. Nakamura, N., Hart, D. A., Boorman, R. S., Kaneda, Y., Shrive, N. G., Marchuk, L. L., Shino, K., Ochi, T., and Frank, C. B., Decorin antisense gene therapy improves functional healing of early rabbit ligament scar with enhanced collagen fibrillogenesis in vivo, *J. Orthop. Res.*, 18(4), 517–523, 2000.
39. Aspenberg, P., and Forslund, C., Enhanced tendon healing with GDF 5 and 6, *Acta Orthop. Scand.*, 70(1), 51–54, 1999.
40. Rickert, M., Jung, M., Adiyaman, M., Richter, W., and Simank, H. G., A growth and differentiation factor-5 (GDF-5)-coated suture stimulates tendon healing in an Achilles tendon model in rats, *Growth Factors*, 19(2), 115–126, 2001.
41. Spindler, K. P., Imro, A. K., Mayes, C. E., and Davidson, J. M., Patellar tendon and anterior cruciate ligament have different mitogenic responses to platelet-derived growth factor and transforming growth factor beta, *J. Orthop. Res.*, 14(4), 542–546, 1996.
42. Woo, S. L., Smith, D. W., Hildebrand, K. A., Zeminski, J. A., and Johnson, L. A., Engineering the healing of the rabbit medial collateral ligament, *Med. Biol. Eng. Comput.*, 36(3), 359–364, 1998.
43. DesRosiers, E. A., Yahia, L., and Rivard, C. H., Proliferative and matrix synthesis response of canine anterior cruciate ligament fibroblasts submitted to combined growth factors, *J. Orthop. Res.*, 14(2), 200–208, 1996.
44. Scherping, S. C. Jr., Schmidt, C. C., Georgescu, H. I., Kwoh, C. K., Evans, C. H., and Woo, S. L., Effect of growth factors on the proliferation of ligament fibroblasts from skeletally mature rabbits, *Connect. Tissue Res.*, 36(1), 1–8, 1997.
45. Schmidt, C. C., Georgescu, H. I., Kwoh, C. K., Blomstrom, G. L., Engle, C. P., Larkin, L. A., Evans, C. H., and Woo, S. L., Effect of growth factors on the proliferation of fibroblasts from the medial collateral and anterior cruciate ligaments, *J. Orthop. Res.*, 13(2), 184–190, 1995.

46. Goomer, R. S., Maris, T. M., Gelberman, R., Boyer, M., Silva, M., and Amiel, D., Nonviral in vivo gene therapy for tissue engineering of articular cartilage and tendon repair, *Clin. Orthop.*, 379(Suppl), S189–S200, 2000.
47. Hildebrand, K. A., Deie, M., Allen, C. R., Smith, D. W., Georgescu, H. I., Evans, C. H., Robbins, P. D., and Woo, S. L., Early expression of marker genes in the rabbit medial collateral and anterior cruciate ligaments: the use of different viral vectors and the effects of injury, *J. Orthop. Res.*, 17(1), 37–42, 1999.
48. Pelinkovic, D., Lee, J. Y., Engelhardt, M., Rodosky, M., Cummins, J., Fu, F. H., Huard, J., Muscle cell-mediated gene delivery to the rotator cuff, *Tissue Eng.*, 9(1), 143–151, 2003.
49. Gerich, T. G., Kang, R., Fu, F. H., Robbins, P. D., and Evans, C. H., Gene transfer to the patellar tendon, *Knee Surg. Sports Traumatol. Arthrosc.*, 5(2), 118–123, 1997.
50. Menetrey, J., Kasemkijwattana, C., Day, C. S., Bosch, P., Fu, F. H., Moreland, M. S., and Huard, J., Direct-, fibroblast- and myoblast-mediated gene transfer to the anterior cruciate ligament, *Tissue Eng.*, 5(5), 435–442, 1999.
51. Nakamura, N., Hart, D. A., Frank, C. B., Marchuk, L. L., Shrive, N. G., Ota, N., Taira, K., Yoshikawa, H., and Kaneda, Y., Efficient transfer of intact oligonucleotides into the nucleus of ligament scar fibroblasts by HVJ-cationic liposomes is correlated with effective antisense gene inhibition, *J. Biochem. (Tokyo)*, 129(5), 755–759, 2001.
52. Lou, J., Tu, Y., Burns, M., Silva, M. J., and Manske, P., BMP-12 gene transfer augmentation of lacerated tendon repair, *J. Orthop. Res.*, 19(6), 1199–1202, 2001.
53. Palmes, D., Spiegel, H. U., Schneider, T. O., Langer, M., Stratmann, U., Budny, T., and Probst, A., Achilles tendon healing: long-term biomechanical effects of postoperative mobilization and immobilization in a new mouse model, *J. Orthop. Res.*, 20(5), 939–946, 2002.
54. Neidlinger-Wilke, C., Grood, E., Claes, L., and Brand, R., Fibroblast orientation to stretch begins within three hours, *J. Orthop. Res.*, 20(5), 953–956, 2002.
55. Altman, G. H., Horan, R. L., Martin, I., Farhadi, J., Stark, P. R., Volloch, V., Richmond, J. C., Vunjak-Novakovic, G., and Kaplan, D. L., Cell differentiation by mechanical stress, *FASEB J.*, 16(2), 270–272, 2002.
56. Qin, T., Yang, Z., Xie, H., Li, X., Li, S., Ye, G., and Li, S., Initial study on three-dimensional culture of tenocytes under cyclic mechanical stretch, *Hua Xi Yi Ke Da Xue Xue Bao*, 33(1), 1–4, 2002.
57. Murray, J. C., and Farndale, R. W., Modulation of collagen production in cultured fibroblasts by a low-frequency, pulsed magnetic field, *Biochim. Biophys. Acta*, 838(1), 98–105, 1985.
58. Skutek, M., van Griensven, M., Zeichen, J., Brauer, N., and Bosch, U., Cyclic mechanical stretching enhances secretion of Interleukin 6 in human tendon fibroblasts, *Knee Surg. Sports Traumatol. Arthrosc.*, 9(5), 322–326, 2001.
59. Enwemeka, C. S., Rodriguez, O., and Mendosa, S., The biomechanical effects of low-intensity ultrasound on healing tendons, *Ultrasound Med. Biol.*, 16(8), 801–807, 1990.
60. Gum, S. L., Reddy, G. K., Stehno-Bittel, L., and Enwemeka, C. S., Combined ultrasound, electrical stimulation, and laser promote collagen synthesis with moderate changes in tendon biomechanics, *Am. J. Phys. Med. Rehabil.*, 76(4), 288–296, 1997.
61. Ishii, Y., Miyanaga, Y., Shimojo, H., Ushida, T., and Tateishi, T., Effects of hyperbaric oxygen on procollagen messenger RNA levels and collagen synthesis in the healing of rat tendon laceration, *Tissue Eng.*, 5(3), 279–286, 1999.
62. Altman, G. H., Lu, H. H., Horan, R. L., Calabro, T., Ryder, D., Kaplan, D. L., Stark, P., Martin, I., Richmond, J. C., and Vunjak-Novakovic, G., Advanced bioreactor with controlled application of multi-dimensional strain for tissue engineering, *J. Biomech. Eng.*, 124(6), 742–749, 2002.
63. Postacchini, F., Accinni, L., Natali, P. G., Ippolito, E., and DeMartino, C., Regeneration of rabbit calcaneal tendon: a morphological and immunochemical study, *Cell Tissue Res.*, 195(1), 81–97, 1978.
64. Aragona, J., Parsons, J. R., Alexander, H., and Weiss, A. B., Soft tissue attachment of a filamentous carbon-absorbable polymer tendon and ligament replacement, *Clin. Orthop.*, 160, 268–278, 1981.
65. Shieh, S. J., Zimmerman, M. C., and Parsons, J. R., Preliminary characterization of bioresorbable and nonresorbable synthetic fibers for the repair of soft tissue injuries, *J. Biomed. Mater. Res.*, 24(7), 789–808, 1990.
66. Tauro, J. C., Parsons, J. R., Ricci, J., and Alexander, H., Comparison of bovine collagen xenografts to autografts in the rabbit, *Clin. Orthop.*, (266), 271–284, 1991.

67. McMaster, W. C., Kouzelos, J., Liddle, S., and Waugh, T. R., Tendon grafting with glutaraldehyde fixed material, *J. Biomed. Mater. Res.*, 10(2), 259–271, 1976.
68. Howard, C. B., McKibbin, B., and Ralis, Z. A., The use of Dexon as a replacement for the calcaneal tendon in sheep, *J. Bone Joint Surg. Br.*, 67(2), 313–316, 1985.
69. Siebert, H. R., Rueger, J. M., and Pannike, A., Experimental studies of ligament replacement, *Unfallchirurgie*, 11(5), 247–250, 1985.
70. Murrell, G. A., Lilly, E. G., 3rd, Goldner, R. D., Seaber, A. V., and Best, T. M., Effects of immobilization on Achilles tendon healing in a rat model, *J. Orthop. Res.*, 12(4), 582–591, 1994.
71. Drzewiecki, A. E., Sarkar, K., Wu, Y., and Uhthoff, H. K., Usefulness of a new technique for hind limb immobilization in rats for the study of tendon healing, *Arch. Orthop. Trauma Surg.*, 111(1), 39–42, 1991.
72. Gelberman, R. H., Vande Berg, J. S., Lundborg, G. N., and Akeson, W. H., Flexor tendon healing and restoration of the gliding surface. An ultrastructural study in dogs, *J. Bone Joint Surg. Am.*, 65(1), 70–80, 1983.
73. Garner, W. L., McDonald, J. A., Koo, M., Kuhn, C., 3rd, and Weeks, P. M., Identification of the collagen-producing cells in healing flexor tendons, *Plast. Reconstr. Surg.*, 83(5), 875–879, 1989.
74. Siddiqi, N. A., Hamada, Y., and Noryia, A., The healing of flexor tendons in chickens, *Int. Orthop.*, 16(4), 363–368, 1992.
75. Watanabe, N., Woo, S. L., Papageorgiou, C., Celechovsky, C., and Takai, S., Fate of donor bone marrow cells in medial collateral ligament after simulated autologous transplantation, *Microsc. Res. Tech.*, 58(1), 39–44, 2002.
76. Dunn, M. G., Avasarala, P. N., and Zawadsky, J. P., Optimization of extruded collagen fibers for ACL reconstruction, *J. Biomed. Mater. Res.*, 27(12), 1545–1552, 1993.
77. Lin, V. S., Lee, M. C., O'Neal, S., McKean, J., and Sung, K. L., Ligament tissue engineering using synthetic biodegradable fiber scaffolds, *Tissue Eng.*, 5(5), 443–452, 1999.
78. Xie, H. Q., Yang, Z. M., and Xin, J. P., Short tandem repeat loci examination after repair of coracoclavicular ligament injury by tissue engineered tendon, *Zhongguo Xiu Fu Chong Jian Wai Ke Za Zhi*, 14(4), 237–240, 2000.
79. Young, R. G., Butler, D. L., Weber, W., Caplan, A. I., Gordon, S. L., and Fink, D. J., Use of mesenchymal stem cells in a collagen matrix for Achilles tendon repair, *J. Orthop. Res.*, 16(4), 406–413, 1998.
80. Awad, H. A., Butler, D. L., Harris, M. T., Ibrahim, R. E., Wu, Y., Young, R. G., Kadiyala, S., and Boivin, G. P., In vitro characterization of mesenchymal stem cell-seeded collagen scaffolds for tendon repair: effects of initial seeding density on contraction kinetics, *J. Biomed. Mater. Res.*, 51(2), 233–240, 2000.
81. Awad, H. A., Butler, D. L., Boivin, G. P., Smith, F. N., Malaviya, P., Huibregtse, B., and Caplan A. I., Autologous mesenchymal stem cell-mediated repair of tendon, *Tissue Eng.*, 5(3), 267–277, 1999.
82. Goldstein, J. D., Tria, A. J., Zawadsky, J. P., Kato, Y. P., Christiansen, D., and Silver, F. H., Development of a reconstituted collagen tendon prosthesis. A preliminary implantation study, *J. Bone Joint Surg. Am.*, 71(8), 1183–1191, 1989.
83. Wasserman, A. J., Kato, Y. P., Christiansen, D., Dunn, M. G., and Silver, F. H., Achilles tendon replacement by a collagen fiber prosthesis: morphological evaluation of neotendon formation, *Scanning Microsc.*, 3(4), 1183–1197, 1989, discussion 1197–1200.
84. Ozaki, J., Fujiki, J., Sugimoto, K., Tamai, S., and Masuhara, K., Reconstruction of neglected Achilles tendon rupture with Marlex mesh, *Clin. Orthop.*, 238, 204–208, 1989.
85. Badylak, S. F., Tullius, R., Kokini, K., Shelbourne, K. D., Klootwyk, T., Voytik, S. L., Kraine, M. R., and Simmons, C., The use of xenogeneic small intestinal submucosa as a biomaterial for Achilles tendon repair in a dog model, *J. Biomed. Mater. Res.*, 29(8), 977–985, 1995.
86. Badylak, S., Arnoczky, S., Plouhar, P., Haut, R., Mendenhall, V., Clarke, R., and Horvath, C., Naturally occurring extracellular matrix as a scaffold for musculoskeletal repair, *Clin. Orthop.*, 367(Suppl), S333–S343, 1999.
87. Lieberman, J. R., Lozman, J., Czajka, J., and Dougherty, J., Repair of Achilles tendon ruptures with Dacron vascular graft, *Clin. Orthop.*, 234, 204–208, 1988.

88. Sato, M., Maeda, M., Kurosawa, H., Inoue, Y., Yamauchi, Y., and Iwase, H., Reconstruction of rabbit Achilles tendon with three bioabsorbable materials: histological and biomechanical studies, *J. Orthop. Sci.*, 5(3), 256–267, 2000.
89. Cao, Y., Liu, Y., Liu, W., Shan, Q., Buonocore, S. D., and Cui, L., Bridging tendon defects using autologous tenocyte engineered tendon in a hen model, *Plast. Reconstr. Surg.*, 110(5), 1280–1289, 2002.
90. Ricci, J. L., Gona, A. G., and Alexander, H., In vitro tendon cell growth rates on a synthetic fiber scaffold material and on standard culture plates, *J. Biomed. Mater. Res.*, 25(5), 651–666, 1991.
91. Parsons, J. R., Weiss, A. B., Schenk, R. S., Alexander, H., and Pavlisko, F., Long-term follow-up of Achilles tendon repair with an absorbable polymer carbon fiber composite, *Foot Ankle*, 9(4), 179–184, 1989.
92. Meaney Murray, M., Rice, K., Wright, R. J., and Spector, M., The effect of selected growth factors on human anterior cruciate ligament cell interactions with a three-dimensional collagen–GAG scaffold, *J. Orthop. Res.*, 21(2), 238–244, 2003.
93. Chan, B. P., Chan, K. M., Maffulli, N., Webb, S., and Lee, K. K., Effect of basic fibroblast growth factor. An in vitro study of tendon healing, *Clin. Orthop.*, 342, 239–247, 1997.
94. Dowling, B. A., Dart, A. J., Hodgson, D. R., Rose, R. J., and Walsh, W. R., Recombinant equine growth hormone does not affect the in vitro biomechanical properties of equine superficial digital flexor tendon, *Vet. Surg.*, 31(4), 325–330, 2002.
95. Forslund, C., and Aspenberg, P., Tendon healing stimulated by injected CDMP-2, *Med. Sci. Sports Exerc.*, 33(5), 685–687, 2001.
96. Jann, H. W., Stein, L. E., and Slater, D. A., In vitro effects of epidermal growth factor or insulin-like growth factor on tenoblast migration on absorbable suture material, *Vet. Surg.*, 28(4), 268–278, 1999.
97. Stein, L. E., Effects of serum, fibroblast growth factor, and platelet-derived growth factor on explants of rat tail tendon: a morphological study, *Acta Anat. (Basel)*, 123(4), 247–252, 1985.
98. Majima, T., Marchuk, L. L., Sciore, P., Shrive, N. G., Frank, C. B., and Hart, D. A., Compressive compared with tensile loading of medial collateral ligament scar in vitro uniquely influences mRNA levels for aggrecan, collagen type II, and collagenase, *J. Orthop. Res.*, 18(4), 524–531, 2000.
99. Ditsios, K. T., Burns, M. E., Boyer, M. I., Gelberman, R. H., and Silva, M. J., The rigidity of repaired flexor tendons increases following ex vivo cyclic loading, *J. Biomech.*, 35(6), 853–856, 2002.
100. Mass, D. P., Tuel, R. J., Labarbera, M., and Greenwald, D. P., Effects of constant mechanical tension on the healing of rabbit flexor tendons, *Clin. Orthop.*, 296, 301–306, 1993.
101. Goldfarb, C. A., Harwood, F., Silva, M. J., Gelberman, R. H., Amiel, D., and Boyer, M. I., The effect of variations in applied rehabilitation force on collagen concentration and maturation at the intrasynovial flexor tendon repair site, *J. Hand Surg. [Am.]*, 26(5), 841–846, 2001.
102. Yamamoto, E., Hayashi, K., and Yamamoto, N., Effects of stress shielding on the transverse mechanical properties of rabbit patellar tendons, *J. Biomech. Eng.*, 122(6), 608–614, 2000.
103. Salti, N. I., Tuel, R. J., and Mass, D. P., Effect of hyaluronic acid on rabbit profundus flexor tendon healing in vitro, *J. Surg. Res.*, 55(4), 411–415, 1993.
104. Drew, P. J. et al., The effects of temperature and time on thermal bond strength in tendons, *Lasers Med. Sci.*, 16(4), 291–298, 2001.

27 Tissue Engineering of the Cornea

May Griffith, Fengfu Li, Chris Lohmann, Heather D. Sheardown, Shigeto Shimmura, and David J. Carlsson

CONTENTS

I. INTRODUCTION

Over the years, there have been reports of tissue engineered (TE) corneas that range in complexity and functionality from materials and devices that are designed to restore solely the ocular function, to those that are designed to do this by integration with part or the full thickness of damaged or diseased corneas and promoting regeneration of the host tissue. TE lenticules are also being designed to be implanted into the cornea in order to alter the refractive properties of the eye, thereby improving vision. Beyond the need for vision restoration, there are other critical questions of bonding or integration with the host tissue and epithelial overgrowth to restore the cornea's protective surface layer. Even more demanding, but very desirable, is the regeneration of innervation to restore touch and hydration sensitivity.

II. THE NEED FOR TE CORNEAS

According to the World Health Organization (WHO), corneal diseases are a major cause of vision loss and blindness, second only to cataracts in overall importance.[1] Ocular trauma and cornea ulceration is estimated to result in 1.5 to 2 million new cases of corneal blindness annually,[1] while corneal scarring from measles is a major cause of vision loss in children.[2] Corneal blindness affects over 10 million individuals worldwide (estimates from the Vision Share Consortium of Eye Banks, U.S.A.), and the only widely accepted treatment is transplantation with human donor tissue. At present, over 45,000 corneal transplants per year are performed in the U.S.A. Under ideal circumstances, traditional allograft cornea transplantation has a quite high success rate, with *ca.* 80% of grafts still clear after 2 years, dropping to *ca.* 65% after 5 years.[3,4] Success is dependent upon several parameters. Most importantly, the patient's condition needs to be amenable to transplantation. For example, problems such as inactive central scars or keratoconus have a good prognosis while others, such as alkali burns or neurotrophic scars secondary to Herpes zoster ophthalmicus, have a poor prognosis.

Because of the aging population within the U.S. and Canada and people's longer life expectancies, it is expected that the North American cornea donor pool will shrink as the demand increases (older corneas are less acceptable and the elderly are more likely to need transplants). Furthermore, the availability of acceptable donor tissue is expected to decrease with the increasing incidence of infectious diseases, including HIV and hepatitis, as well as the growing popularity of laser *in situ* keratomileusis (LASIK) for correcting refractive errors. These surgically treated corneas are unacceptable donor tissue. Another serious disadvantage of cornea allograft transplantation is the possible transmission of infectious agents. Person-to-person transmisssion of the rabies virus[5] and at least one case of Creutzfelst–Jakob disease[6] has been reported. Hepatitis B and C and HIV can be isolated from tears and there is concern about their possible transmission. Given the knowledge of infectious transmission being possible in prion form, it also conceivable that transmission of as yet unknown pathogens can occur also.

An alternative to human tissue donations are artificial substitutes. At present, however, widely accepted corneal substitutes are not available.[7] Prostheses have been developed, but none integrates seamlessly into the host tissue.[7] Furthermore, no artificial substitutes proposed to date have addressed the reinnervation issue, although loss of corneal innervation has been shown to lead to visual loss.[8]

Recently, refractive surgery using an excimer laser has become a very popular means of correcting refractive errors of the cornea and thereby improving vision without the use of external devices such as spectacles or contact lenses. While refractive procedures are expected to be reproducible and show long-term stability, regression of the corrective effects of the surgery, (e.g., by hypertrophy of the epithelium) has been documented and is one of the more common problems, and secondary procedures or touch-ups, are now routinely accepted.[9] Other competitive methods of refractive correction are being developed, including phakic intraocular lenses (IOLs), intrastromal inlays, intrastromal corneal rings (INTACS; which do not compromise the central optical area and are reversible), and lenticules that are embedded between the corneal stroma and epithelium (corneal onlays). In this chapter, however, we will focus primarily on the development of scaffolding for transplantable, engineered corneal replacements.

III. REPLACEMENTS FOR HUMAN CORNEAS

Human corneas are extremely tough and tear resistant, but perfect reproduction of these mechanical properties may not be necessary for a substitute or replacement to be functional, provided that it is tough enough to survive handling, implantation stresses, and wear and tear, and is also functionally comparable in other key areas.[10] Critical functions of the cornea include high optical transparency with minimal scatter for vision, and toughness to protect the more delicate inner parts of the eye. A functional, sensitive nerve network is also important, both as a highly effective warning of potential injury, and as a key link in eye humidification through signaling when blinking must occur to prevent the potentially dangerous situation of "dry eye" that can occasionally lead to ulceration and vision loss.[11]

The human cornea comprises some 300 layers of collagen fibrils arranged in offset sheets, parallel to the plane of the cornea. This precise, ordered structure was originally proposed to be essential for the high optical clarity of the cornea[12] and this proposal is still widely accepted. However, clarity has been shown to result from a combination of refractive index matching and, more importantly, the presence of structural components with diameters less than the shortest wavelength of visible light (*ca.* 380 nm).[13,14] This is graphically illustrated by atomic force microscopy (AFM) images of fully hydrated scleral (white of eye) and stromal (corneal) layers (Figure 27.1).[15] Optical clarity (freedom from absorption and scatter in the visible region) requires a synthetic polymer or biopolymer that is free of chromophores absorbing in the visible region and also free of subunits that will scatter light. These scattering centers may include crystalline domains or large aggregates of microfibrils. Amorphous synthetic polymers, free of crystallinity, such as PMMA, are widely used when transparency is required. Biopolymers, such as collagen I and fibrin,

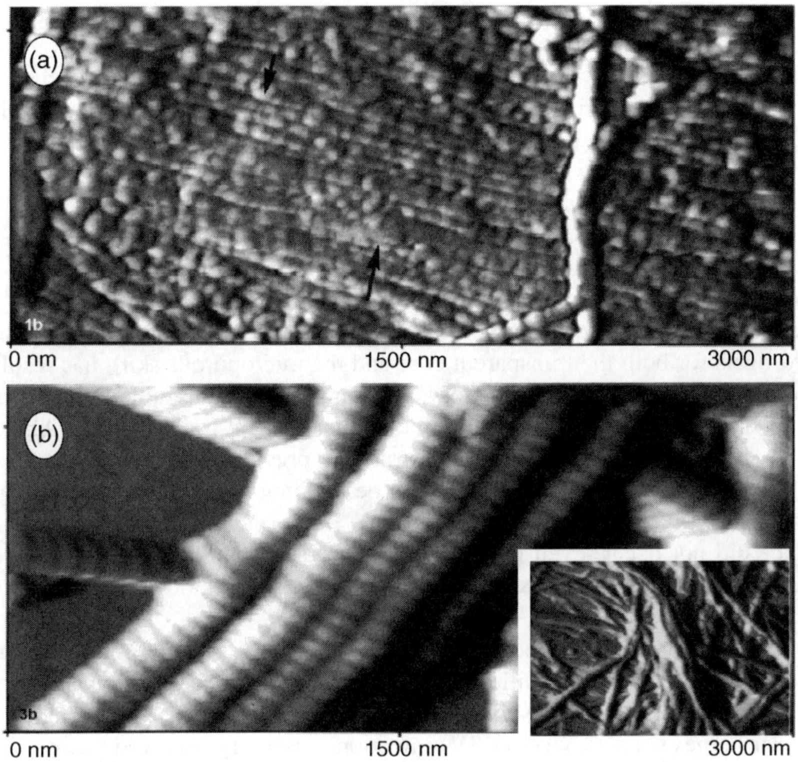

FIGURE 27.1 AFM images of hydrated human cornea (a). Arrows indicate regular, fine plies of collagen fibrils in the cornea, compared to the much thicker fibrils of the sclera (b). Inset in (b) shows a lower magnification of scleral collagen fibrils, which are disorganized compared to that of the cornea. (From Meller, D., Peters, K., and Meller, K., *Cell Tissue Res.*, 288, 111, 1997. With permission.)

are inherently fibrous in nature because of the ready association of their nanofibrilar subunits. However, by careful control of cross-linking conditions, both biopolymers can be locked in a form where their microfibrils are below *ca.* 300 nm diameter, which is well below the lowest visible wavelengths (Li et al.,[16] unpublished observations). However, because these clear materials have a disordered array of microfibrilar structures, they are less tough than the composite structure of the natural cornea that is based on thin plies of roughly parallel collagen filaments, with glycosaminoglycans and water contributing to inter filament bonding. TE corneal materials that are in use, or being developed, are based on both amorphous synthetic polymers and microfibrilar biopolymers.

IV. SYNTHETIC SCAFFOLDS AND KERATOPROSTHESES

Keratoprostheses (KPros, commonly referred to as artificial corneas) are usually completely synthetic constructs designed to replace the central portion of an opaque cornea. Early devices with rigid components required complex surgery and led to high incidences of complications, such as extrusion, melting, aqueous leakage, infection, retroprosthetic membrane formation, and glaucoma. Rigidity, lack of free flow of nutrients (oxygen, glucose), and lack of biointegration contributed to device failure. The "core and skirt" concept has now been widely adopted.[7] This design is based upon a biointegratable "skirt" (containing interconnected pores in the 10 to 30 μm diameter range) that surrounds a clear, central optic. Tissue breakdown around the anchoring skirt and extrusion of the KPro is still a major cause of its failure. The clear core allows light transmission into the eye, while the porous skirt is strong enough for suture placement and eventually anchors the device to the corneal stroma via fibroblast ingrowth and collagen deposition sufficient to provide firm anchorage. The anterior surface of the device should allow epithelialization to occur, although the inhibition of epithelial downgrowth is necessary. The posterior surface should inhibit cellular attachment and proliferation to prevent the formation of opaque retroprosthetic membranes. The optic portion of the device, aside from being transparent, should provide a refractive power similar to that of the normal cornea and the optic should have a diameter sufficient to allow posterior segment visualization and a reasonable field of view. In order to maintain cellular activity in the skirt, there must be adequate oxygen and nutrient permeability. None of the materials used in device fabrication should elicit any immune, inflammatory, or mutagenic responses.

The introduction of highly hydrophilic constructs, based on poly(2-hydroxyethyl methacrylate; pHEMA) hydrogels for both the transparent core and the microporous skirt, has resolved many of the initial problems of extrusion, inflammatory, and immune reactions, although calcification was a problem in the earlier iterations.[17] Recently, multicenter clinical trials with the AlphaCor™ KPro (based on the pHEMA hydrogel skirt core construction, previously known as the Chirila KPro)[18] showed that a full thickness, synthetic device can be maintained in the cornea through anchorage via fibroblast ingrowth into the peripheral portion. However, epithelialization of the pHEMA anterior surface did not occur, even though it is realized that epithelialization would be ideal.[18] Nerve regeneration in these prostheses has not been reported, but would be essential to more normal corneal function.

The AlphaCor KPro device has been used as an alternative to donor corneal tissue in patients who would be at high risk of conventional corneal graft failure, a very demanding situation for KPro retention. Hicks et al. have described a study of AlphaCor KPros in patients, some with a history of ocular herpes simplex virus (HSV) infection.[19] Surgery involved a two-stage procedure in which the device was placed within an intrastromal pocket closed by suturing a conjunctival flap over the anterior surface of the cornea. After a period of 12 weeks, the device optic was exposed by removing the conjunctival flap. The aesthetics of this optically functional Kpro have raised some problems.[18] Corneal melt related complications occurred in 30% of the cases, with loss of stromal corneal tissue adjacent or anterior to the sponge skirt of the device. Of these failures, 75% had a history of ocular HSV, with a mean time to melt onset of 7.5 months. Although use of a putative anticollagenase agent (guttae medroxyprogesterone) retarded melt onset, this nonapproved treatment was stopped at the request of regulatory authorities because the drug was not approved for use with keratoprostheses. AlphaCor implantation may precipitate reactivation of latent HSV such that reactivation and resultant inflammation reduce device biointegration and facilitate melting of corneal stromal tissue anterior to the device. It was concluded that a history of HSV should be an exclusion factor for AlphaCor surgery.

Other core skirt KPros have been proposed, such as those based on a porous semitransparent poly tetrafluoroethylene (PTFE) skirt and a central optic of poly vinyl pyrrolidone (PVP) coated silicone rubber (PDMS)[20,21] or on poly(butyl methacrylate), hexaethyleneglycolmethacrylate with a dimethacrylate cross-linker.[22] These latter copolymers allowed cultured human epithelial cells to proliferate readily and to migrate on their surfaces. There is wide agreement that the long-term stability of these devices would be greatly improved by the presence of an epithelial layer. Surface modification of the various materials, using many different techniques and many different molecules, has been extensively tried. Surface modification with macromolecules of the extracellular matrix (ECM) or bioactive fragments such as cell adhesion oligopeptides including the fibronectin-based Arg–Gly–Asp–(Ser) (RGD(S)),[23] laminin-based Tyr–Ile–Gly–Ser–Arg (YIGSR)[24] and a novel collagen-based peptide Gly–Pro–Nleu[25] has also been examined. Surface modification with combinations of peptides, including the cell adhesion peptides RGDS and YIGSR as well as synergistic counterparts Pro–His–Ser–Arg–Asn (PHSRN) and Pro–Asp–Ser–Gly–Arg (PDSGR), demonstrated that corneal epithelial cell adhesion is greatly improved on surfaces with the cell adhesion peptides and at least one of the counterparts.[26]

A dramatic example of the importance of ECM components on corneal epithelial cell adhesion and differentiation comes from the development of corneal onlays (lenticules that would be implanted between the corneal stroma and epithelium to correct refractive errors). Evans et al.[27] have recently reported a porous, transparent, perfluoropolyether (PFPE) based polymer that was made more hydrophilic by production of a zwitterion copolymer through UV-initiated copolymerization of the PFPE-dimethacrylate and a zwitterionic monomer [(2-(methacryloyloxy)ethyl)-dimethyl(3-sulfopropyl)ammonium hydroxide inner salt]. To enhance epithelial overgrowth, a thin layer of collagen I was covalently immobilized on the anterior surface of each lenticule by a two-step process. The 5 to 10 nm layer of collagen I was covalently immobilized by reductive amination of the surfaces, following exposure to a radio frequency glow discharge. In a clinical comparison of implanted lenticules and sham wounds, epithelium completely covered the feline, corneal wound bed (sham) by Days 3 to 9 and the exposed lenticule surface (implanted) by Days 5 to 11 in six of the seven implanted corneas. Overall, the corneas in both series were quiet with no signs of thinning, remained transparent by slit lamp examination, maintained multilayered epithelial cover, and supported a stable tear film during the observation period. Light microscopy of the sham-wounded corneas revealed six to seven layers of cells constituting the epithelium on the central portion of the original wound bed at 4 weeks that had increased to eight to ten layers by 8 weeks. This was slightly less than the 12 layers of cells in normal feline corneal epithelium. An overall examination of the stromal tissue in the implanted corneas at both time points revealed that the stroma was relatively normal in appearance with no evidence of thinning (stromal melting). The steady initial growth of epithelium over the lenticules was undoubtedly associated with the presence of the covalently immobilized collagen I on the anterior surface of the lenticules.

V. CELL-BASED CORNEAL MODEL STRUCTURES

In the late 1990s, several prototype corneas were described[28–30] that depended upon the propensity of ECM molecules to self-organize and for cells to integrate into such scaffolds. Griffith et al. used either bovine dermis or rat tail collagen, blended with chondroitin 5-sulfate and with immortalised stromal cells.[29] Human epithelial cells were grown over a synthetic stroma containing premixed cells to form multilayers. These matrices were further optimized by growing a single layer of human endothelial cells on the posterior surface of their stromal layer, with fabricated Descemet's and Bowman's layers between the posterior and anterior cell layers, respectively. These lightly cross-linked hydrogel structures had some integrity, and were proposed as systems for the study of wound healing *in vitro* and as replacements for the animal eye tests (Draize test)[28–30] but were too fragile for transplantation.

Several groups have been developing corneal equivalents using completely natural materials as potentially implantable replacements. The model developed by the Laboratoire d'Organogenese Experimentale (LOEX)[30] used a self-assembly approach whereby stromal cells were provided with the nutrients and appropriate factors such as ascorbic acid to induce production of collagen and other ECM macromolecules in sheets.[31] The sheets were then stacked together and an epithelium was seeded on top of the stack. In their previous study of tissue engineered blood vessels, tensile strength achieved by this method was demonstrated to be high,[32] suggesting that this might eventually be achieved in the corneal models as well. However, no optical data from this corneal model was reported.

In a different approach, Han et al.[33] have prepared a bioengineered ocular surface tissue replacement consisting of human limbal cells, believed to be corneal epithelial stem cells in a cross-linked, human, fibrin gel. The cells were suspended in a human fibronectin/fibrin gel cross-linked by human thrombin and factor XIII, derived from a fibrinogen rich cryoprecipitate of human plasma and proliferated in the fibrin gel to near confluence over the 15 days. This bioengineered corneal surface tissue created a transportable, pliable, and stable tissue replacement. Because both the cells and the plasma components of the fibrin gel are of human origin (potentially derived from the patient), this tissue replacement represents a totally autologous bioengineered replacement tissue. However, production of the normal, stratified, epithelial architecture and adequate optical clarity was not discussed.

VI. BIOSYNTHETIC MATRIX REPLACEMENTS AS SCAFFOLDS

The ideal biomaterial scaffold for achieving regeneration should be "smart," duplicating the environmental conditions that direct the development of the original tissue. The "tissue template" properties of different ECM macromolecules allow for specification of different cell attractive environments, and these can be exploited to create engineered tissues for regeneration of specific tissues and organs. However, in tissues such as the cornea, high optical clarity and adequate tensile strength and toughness are required. Hence, enhancement of the properties of the natural polymers with synthetic components is necessary.

Cells grow well on the surfaces (2D growth) of many synthetic polymers. However, the successful ingrowth or encapsulation (3D growth) of living cells has only been demonstrated in a few, fully synthetic polymers, such as those based on polyethylene oxide, polypropylene oxide and poly(*N*-isopropylacrylamide) (PNiPAAm).[34,35] In contrast, many natural biopolymer hydrogels, such as those based on alginate, fibrinogen–fibrin, chitosan, agarose, albumin, collagens, and their derivatives, are widely used to encapsulate living cells. Hydrogels of collagen I, the dominant biopolymer in the human cornea, are particularly attractive as matrix replacement type scaffolds, partly because of their strength at relatively low concentrations, resulting from the virtually rigid rod properties of the collagen Type I triple helix.[36] In addition, collagen brings the cell attachment motif RGD.[37] However, both the biodegradation resistance of collagen I and the strength of hydrogels in general at low concentrations (<10 wt/vol.%) need to be enhanced by chemical cross-linking.[35]

Based on our experience with the development of soft prototype, hydrogel scaffolds (0.3 wt% collagen),[29] we have now prepared much more robust, functional, cornea matrices based on higher collagen concentrations (2 to 5 wt/vol.%) and higher cross-link densities.[38] Fibrillogenesis (self-association of collagen I triple helices at neutral pH) to give opaque gels during storage or incubation at 37°C was overcome by the use of pepsin digested (atelo)collagen and by careful pH control during neutralization at 0°C followed by collagen cross-linking through its ε-NH_2 groups with a NiPAAm-based copolymer [poly(*N*-isopropylacrylamide-*co*-acrylic acid-*co*-acryloxysuccinimide] or its YIGSR-modified analog (copolymers abbreviated to TERP and TERP5, respectively). These composite collagen–copolymer matrices could be molded to the curvature and dimensions of a cornea. They also showed high optical clarity (Figure 27.2(a)) with direct transmission and back scatter of visible light comparable to that of human corneas as measured by the same optical method. Visually, the composite matrices are superior to the collagen-only gels that are translucent at best (Figure 27.2(b)). These collagen–copolymer matrices had a glucose diffusion permeability coefficient (2.7×10^{-6} cm^2/s) higher than the natural stroma,[39] although adequate inulin and albumin transport[28] is still to be measured. Furthermore, they were adequately robust for suturing during surgery (Figure 27.2(c)).

Currently, transplantation of human corneas usually involves a full thickness replacement by a surgical technique called penetrating keratoplasty (PK). Lamellar keratoplasty (LKP) is an alternative surgical procedure that requires removal of only the damaged or diseased epithelium and stroma, leaving the endothelium intact. Nonpenetration of the aqueous humor reduces the rate of rejection and postoperative complications such as leakage, improving long-term graft stability.[40,41] Consequently, we used our collagen–TERP5 hydrogels as corneal LKP replacements, sutured into one cornea of each of a series of Yucatan microswine (Figure 27.2(c)). No adverse inflammatory or immune reaction was found after implantation of either biosynthetic matrices or pig cornea

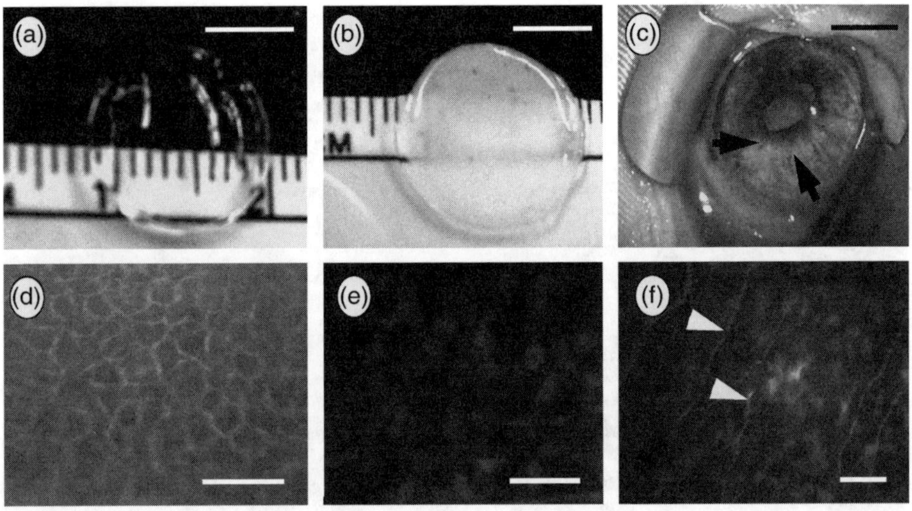

FIGURE 27.2 Optical clarity of corneal implant composed of composite collagen–TERP5 (a) compared to the translucent collagen only hydrogel (b). (c) Shows an implant grafted within the host tissue. Sutures are shown (arrows). Scale bar, 5 mm. (d)–(f) show clinical *in vivo* confocal microscopic images of 3-week implants. (d) Regenerated corneal epithelium on the implant surface. Scale bar, 25 μm. (e) Stromal cells and within the implant. Scale bar, 25 μm. (f) Ingrowing nerve fibers (arrowheads) within the implant at the epithelial–stromal interface. Stroma cells are present in this image. Scale bar, 25 μm.

allografts that served as controls. Epithelial cell growth over the implant (Figure 27.2(d)) was observed in all animals at 3 weeks postsurgery. Clinical *in vivo* confocal microscopy of the implanted matrices at 3 weeks postsurgery also showed ingrowth of stromal cells (Figure 27.2(e)) and regeneration of corneal nerves (Figure 27.2(f)).

Histological sections through corneas with implants showed a distinct but smooth, implant–host tissue interface (Figure 27.3(a)). In both corneas with implants or allografts, the regenerated epithelium was stratified (Figure 27.3(b)), overlying a regenerated basement membrane that showed the presence of hemidesmosomes by Type VII collagen immunohistochemistry[42] (Figure 27.3(c)) transmission electron microscopy (Figure 27.3(d)). Neurofilament positive subepithelial nerves were also observed.

In laser refractive surgical procedures, postsurgical nerve ingrowth can only occur radially from the patient's residual corneal rim, with some touch sensitivity of the central cornea only registered after approximately 18 months.[43] Previous studies of restoration of touch sensitivity have indicated that only minimal function is detected even 10 years after partial thickness lenticule transplantation from a human donor's cornea.[44] From touch sensitivity measurements at five points on the corneal implant in three pigs repre and postoperatively by esthesiometery,[38] a dramatic drop in sensitivity was found immediately after surgery (Figure 27.4). However by 21 days postoperative, sensitivity had returned to preoperative levels. The rapid innervation after implantation could occur either from the stromal rim around the incision or from the remaining stromal bed posterior to the implant. Both mechanisms are possible as touch sensitivity returned quite uniformly across the whole implant area (Figure 27.4).

Innervation is crucial to healthy corneal function. LKP and inlays can be expected to cause dry eye (loss of the normal reflex action of blinking to restore corneal surface hydration) because of the severing of the nerve plexus during implantation, as occurs also during LASIK. Battat et al.[45] have confirmed that denervation resulting from a microkeratome pass causes a disruption of the ocular surface tear dynamics persisting for as long as 16 months after surgery. In addition, Goren et al.

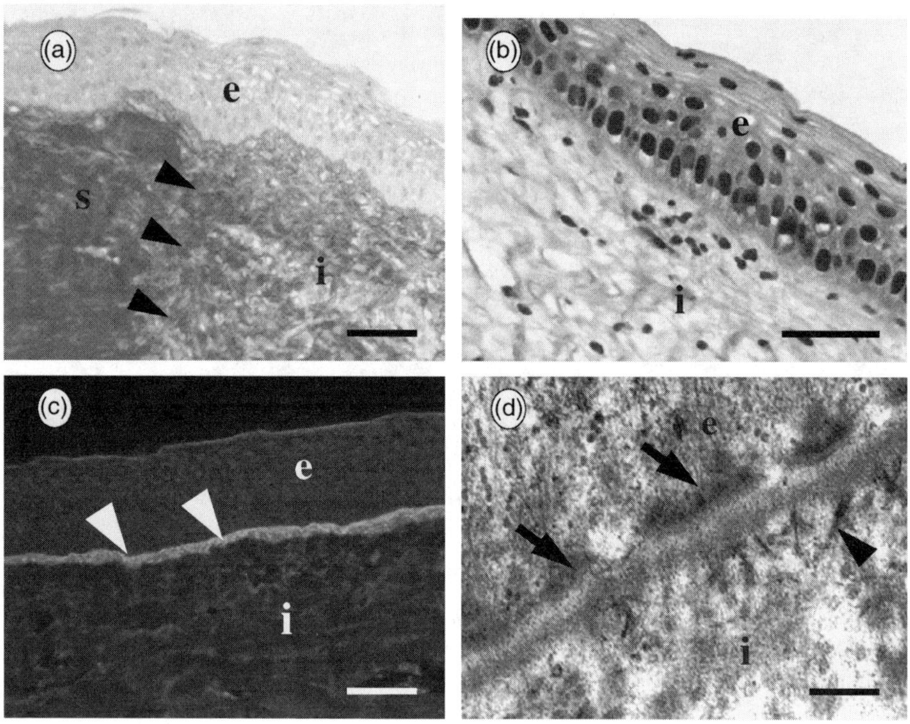

FIGURE 27.3 Postsurgical implant–host integration at 6 weeks postsurgery. (a) Picrosirius red stained section showing the interface (arrowheads) between host corneal stroma (s) and implant (i). The implant surface is covered by regenerated, stratified epithelium (e). Scale bar, 40 μm. (b) H&E stained section, showing stratified epithelium (e) that has regrown over the implant (i). Stromal cells have moved into the implant. Scale bar, 40 μm. (c) Immunolocalization of Type VII collagen, a marker for hemidesmosomes, at the epithelium–implant interface (arrowheads). Scale bar, 40 μm. (d) TEM of epithelium–implant interface showing morphological features consistent with hemidesmosomes (arrows) attached to the underlying basement membrane. Scale bar, 200 nm.

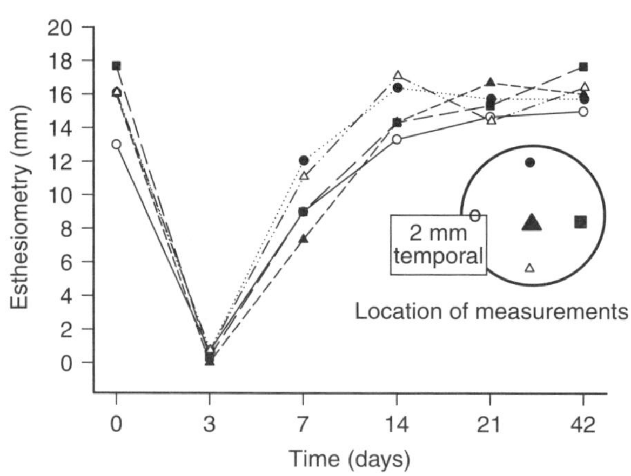

FIGURE 27.4 Esthesiometry testing of touch sensitivity of pig implants. Touch sensitivity was measured at five points on each cornea, as indicated in the diagram; $n = 3$.

have reported that the quality of tear fluid in eyes with even slightly diminished Cochet–Bonnet corneal sensitivity scores produce tears with a significantly decreased concentration of the antimicrobial protein lactoferrin.[46]

VII. TISSUE ENGINEERING OF THE CORNEA: ARE WE THERE YET?

The human cornea is a complex structure dependent on the functional co-operation of epithelial, stromal, and endothelial cells, as well as ingrowing corneal nerves. Sustained, systematic progress towards a TE analog has been achieved over the last 15 years. The latest collagen–TERP5 implants were designed with the objective of promoting host tissue repair and regeneration by acting as a tissue templates to encourage the ingrowth of host corneal cells and nerves.[38] Overall, it is now possible to control the strength and optical clarity of noncytotoxic, biosynthetic composites to the point that TE corneal replacements could, with further development, address future shortages of donor corneas. Furthermore, such TE corneal implants would circumvent potential problems resulting from the lack of nerve regeneration after surgery, found both in human donor tissue.[43,44] Preliminary, short-term animal testing has been successful, but extending this work to longer term studies of implant performance and remodeling by the cell occupants will be pivotal to demonstrating if this approach is viable.

REFERENCES

1. Whitcher, J. P., Srinivasan, M., and Upadhyay, M. P., Corneal blindness: a global perspective, *Bull. World Health Org.*, 79, 214–221, 2001.
2. Gilbert, C., and Foster, A., Childhood blindness in the context of VISION 2020 — the right to sight, *Bull. World Health Org.*, 79, 227–232, 2001.
3. Beekhuis, W. H., Current clinicians' opinions on risk factors in corneal grafting. Results of a survey among surgeons in the eurotransplant area, *Cornea*, 14(1), 39–42, 1995.
4. Sit, M., Weisbrod, D. J., Naor, J., and Slomovic, A. R., Corneal graft outcome study, *Cornea*, 20(2), 129–133, 2001.
5. Houff, S. A., Burton, R. C., Wilson, R. W., Henson, T. E., London, W. T., Baer, G. M., Anderson, L. J., Winkler, W. G., Madden, D. L., and Sever, J. L., Human-to-human transmission of rabies virus by corneal transplant, *N. Engl. J. Med.*, 300, 603–604, 1979.
6. Duffy, P., Wolf, J., Collins, G., DeVoe, A. G., Streeten, B., and Cowen, D., Possible person-to-person transmission of Creutzfeldt–Jakob disease, *N. Engl. J. Med.*, 290(12), 692–693, 1974.
7. Chirila, T. V., An overview of the development of artificial corneas with porous skirts and the use of pHEMA for such application, *Biomaterials*, 22, 3311–3317, 2001.
8. Lambiase, A., Rama, P., Bonini, S., Caprioglio, G., and Aloe, L., Topical treatment with nerve growth factor for corneal neurotrophic ulcers, *N. Engl. J. Med.*, 338, 1174–1180, 1998.
9. Belin, M. W., Evaluating emerging refractive technologies, *Int. Ophthalmol. Clin.*, 42, 1–18, 2002.
10. Chapekar, M. S., Tissue engineering: challenges and opportunities, *J. Biomed. Mater. Res. (Appl. Biomater.)*, 53, 617–620, 2000.
11. Stern, M. E., Beuerman, R. W., Fox, R. I., Gao, J., Mircheff, A. K., and Pflugfelder, S. C., A unified theory of the role of the ocular surface in dry eye, *Adv. Exp. Med. Biol.*, 438, 643–651, 1988.
12. Maurice, D. M., The structure and transparency of the cornea, *J. Physiol. (London)*, 136, 263–286, 1957.
13. Freegard, T. J., The physical basis of transparency of the normal cornea, *Eye*, 11, 465–471, 1997.
14. Benedek, G. B., Theory of transparency of the eye, *Appl. Opt.*, 10, 459, 1971.
15. Meller, D., Peters, K., and Meller, K., Human cornea and sclera studied by atomic force microscopy, *Cell Tissue Res.*, 288, 111–118, 1997.
16. Nells, V., and Herrmann, R., The configuration of fibrin clots determines capillary morphogenesis and endothelial cell migration, *Microvasc. Res.*, 51, 347–364, 1996.

17. Vijayasekaran, S., Chirila, T. V., Robertson, T. A., Lou, X., Fitton, J. H., Hicks, C. R., and Constable, I. J., Calcification of poly(2-hydroxyethyl methacrylate) hydrogel sponges implanted in the rabbit cornea: a 3-month study, *J. Biomater. Sci. Polym. Ed.*, 11(6), 599–615, 2000.
18. Crawford, G. J., Hicks, C. R., Lou, X., Vijayasekaran, S., Tan, D., Mulholland, B., Chirila, T. V., and Constable, I. J., The Chirila Keratoprosthesis: phase I human clinical trial, *Ophthalmology*, 109(5), 883–889, 2002.
19. Hicks, C. R., Crawford, G. J., Tan, D. T., Snibson, G. R., Sutton, G. L., Gondhowiardjo, T. D., Lam, D. S., and Downie, N., Outcomes of implantation of an artificial cornea. AlphaCor: effects of prior ocular herpes simplex infection, *Cornea*, 21(7), 685–690, 2002.
20. Legeais, J. M., Drubaix, I., Briat, B., Renard, G., and Pouliquen, Y., Second generation bio-integrated keratoprosthesis. Implantation in animals, *J. Fr. Ophthalmol.*, 20, 42–48, 1997.
21. Legeais, J. M., and Renard, G., A second generation of artificial cornea (BiokroII), *Biomaterials*, 19, 1517–1522, 1998.
22. Bruining, M. J., Pijpers, A. P., Kingshott, P., and Koole, L. H., Studies on new polymeric biomaterials with tunable hydrophilicity, and their possible utility in corneal repair surgery, *Biomaterials*, 23, 1213–1219, 2002.
23. Kobayashi, H., and Ikada, Y., Corneal cell adhesion and proliferation on hydrogel sheets bound with cell-adhesive proteins, *Curr. Eye Res.*, 10, 899–908, 1991.
24. Merrett, K., Griffith, C. M., Deslandes, Y., Pleizier, G., and Sheardown, H., Adhesion of corneal epithelial cells to cell adhesion peptide modified pHEMA surfaces, *J. Biomater. Sci. Polymer. Ed.*, 12, 647–671, 2001.
25. Johnson, G., Jenkins, M., McLean, K. M., Griesser, H. J., Kwak, J., Goodman, M., and Steele, J. G., Peptoid-containing collagen mimetics with cell binding activity, *J. Biomed. Mater. Res.*, 51, 612–624, 2000.
26. Aucoin, L., Griffith, C. M., Pleizier, G., Deslandes, Y., and Sheardown, H., Interactions of corneal epithelial cells and surfaces modified with cell adhesion peptide combinations, *J. Biomater. Sci. Polym. Ed.*, 13, 447–462, 2002.
27. Evans, M. D., Xie, R. Z., Fabbri, M., Bojarski, B., Chaouk, H., Wilkie, J. S., McLean, K. M., Cheng, H. Y., Vannas, A., and Sweeney, D. F., Progress in the development of a synthetic corneal onlay, *Invest. Ophthalmol. Vis. Sci.*, 43(10), 3196–3201, 2002.
28. Schneider, A. I., Maier-Reif, K., and Graeve, T., Constructing an in vitro cornea from cultures of the three specific corneal cell types, *In Vitro Cell. Dev. Biol. Anim.*, 35(9), 515–526, 1999.
29. Griffith, M., Osborne, R., Munger, R., Xiong, X., Doillon, C. J., Laycock, N. I. C., Hakim, M., Song, Y., and Watsky, M. A., Functional human corneal equivalents from cell lines, *Science*, 286, 2169–2172, 1999.
30. Germain, L., Auger, F. A., Grandbois, E., Guignard, R., Giasson, M., Boisjoly, H., and Guérin, S. L., Reconstructed human cornea produced in vitro by tissue engineering, *Pathobiology*, 67(3), 140–147, 1999.
31. Gaudreault, M., Carrier, P., Larouche, K., Leclerc, S., Giasson, M., Germain, L., and Guerin, S. L., Influence of Sp1/Sp3 expression on corneal epithelial cells proliferation and differentiation properties in reconstructed tissues, *Invest. Ophthalmol. Vis. Sci.*, 44, 1447–1457, 2003.
32. Auger, F. A., Remy-Zolghadri, M., Grenier, G., and Germain, L., A truly new approach for tissue engineering: the LOEX self-assembly technique, *Ernst Schering Res. Found. Workshop*, 35, 73–88, 2002.
33. Han, B., Schwab, I. R., Madsen, T. K., and Isseroff, R. R., A fibrin-based bioengineered ocular surface with human corneal epithelial stem cells, *Cornea*, 21, 505–510, 2002.
34. Lee, K. Y., and Mooney, D. J., Hydrogels for tissue engineering, *Chem. Rev.*, 101, 1869–1879, 2001.
35. Hoffman, A. S., Hydrogels for biomedical applications, *Adv. Drug Deliv. Rev.*, 43, 3–12, 2002.
36. Amis, E. J., Carriere, C. J., Ferry, J. D., and Veis, A., Effect of pH on collagen flexibility determined from dilute solution viscoelastic measurements, *Int. J. Biol. Macromol.*, 7, 130–134, 1985.
37. Pierschbacher, M. D., and Ruoslahti, E., Influence of stereochemistry of the sequence Arg–Gly–Asp–Xaa on binding specificity in cell adhesion, *J. Biol. Chem.*, 262, 17294–17298, 1987.
38. Li, F., Carlsson, D. J., Lohmann, C. P., Suuronen, E. J., Vascotto, S., Kobuch, K., Sheardown, H., Munger, R., and Griffith, M., Cellular and nerve regeneration within a biosynthetic extracellular matrix: corneal implantation, *Proc. Natl Acad. Sci. USA*, 100, 15346–15351, 2003.

39. McCarey, B. E., and Schmidt, F. H., Modelling glucose distribution in the cornea, *Curr. Eye Res.*, 9, 1025–1039, 1990.
40. Archila, E. A., Deep lamellar keratoplasty dissection of host tissue with intrastromal air injection, *Cornea*, 3, 217–218, 1984.
41. Terry, M. A., The evolution of lamellar grafting techniques over twenty-five years, *Cornea*, 19, 611–616, 2000.
42. Gipson, I. K., Spurr-Michaud, S. J., and Tisdale, A. S., Hemidesmosomes and anchoring fibril collagen appear synchronously during development and wound healing, *Dev. Biol.*, 126, 253–262, 1988.
43. Mathers, W. D., Jester, J. V., and Lemp, M. A., Return of human corneal sensitivity after penetrating keratoplasty, *Arch. Ophthalmol.*, 106, 210–211, 1988.
44. Kaminski, S. L., Biowski, R., Lucas, J. R., Koyuncu, D., and Grabner, G., Corneal sensitivity 10 years after epikeratoplasty, *J. Refract. Surg.*, 18, 731–736, 2002.
45. Battat, L., Macri, A., Dursun, D., and Pflugfelder, S., Effects of laser in situ keratomileusis on tear production, clearance, and the ocular surface, *Ophthalmology*, 108, 1230–1235, 2001.
46. Goren, M. B., LASIK and dry eye, *Ophthalmology*, 109, 1947–1948, 2002.

28 Materials Employed for Breast Augmentation and Reconstruction

Parul Natvar Patel and Charles W. Patrick, Jr.

CONTENTS

I. INTRODUCTION

The augmentation of the breast, or restoration of the breast following tumor resection in breast cancer patients, presents a challenge in plastic surgery.[1–4] The numerous disadvantages of current techniques have served as a driving force for the development of new techniques in tissue engineering involving the formation of autologous adipose tissue within sophisticated scaffolds. This chapter discusses the inadequacies of current methods of breast augmentation and restoration, and the innovative adipose tissue engineering strategies being developed to abrogate these limitations to improve patient outcomes and quality of life. A brief description of breast anatomy is first described.

II. ANATOMY OF THE BREAST

In order to translate practically tissue engineering into the clinical setting, a number of parameters must be defined. The mass, organization, and function of the breast must be addressed for individual patients, since the size and shape of the breast varies from person to person, between races, and within the same individual at different stages in their life.[5] This necessitates a clear understanding of breast architecture and composition. The breast is a specialized organ composed of glandular, ductal, connective and adipose tissue.[6] Each breast contains 15 to 25 lobes of compound glands that are embedded in fibrous and adipose tissue (Figure 28.1).[7] These lobes, each containing an

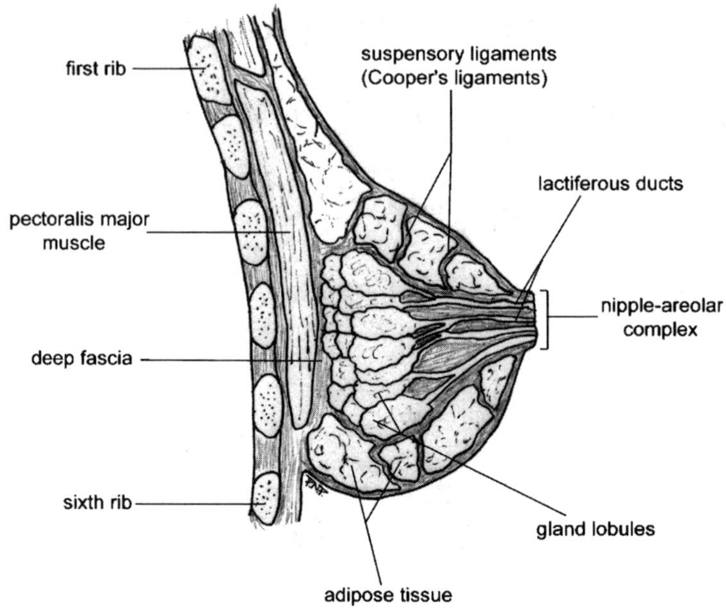

FIGURE 28.1 Anatomy of the breast: a cross-section through the sagittal plane.

excretory duct that drains into the lactiferous sinus, radiate from a central nipple–areolar complex. The breasts are typically the most prominent superficial structures of the anterior thoracic wall, overlying the pectoral muscles and extending from the clavicle superiorly to the sixth rib inferiorly, and from the midsternal line medially to the axilla laterally, forming the axillary tail of Spence. The size and shape of a breast is determined by the overlying skin envelope and the adipose tissue surrounding the glandular structures. The presence of adipose tissue lobules interspersed between fibrous and glandular components of the breast help determine the bulk, softness, and contour of the breast mound. The deep aspect of the breast is separated from the underlying pectoralis major muscle by the deep fascia. Between the breast and this deep fascia is a thin layer of loose connective tissue (within the retromammary space) devoid of adipose tissue that allows the breast to move freely over the deep fascia. The breast is firmly attached to the skin and underlying structures by fibrous bands referred to as suspensory ligaments (Cooper's ligaments), which provide additional support and also contribute to determining the shape and contour of the breast.[5]

III. CLINICAL IMPETUS

Breast cancer is the second leading cause of cancer in women (after skin cancer), affecting one in every eight women in America and one in every three women with cancer.[8] Owing to the large number of clinical occurrences, breast reconstruction following lumpectomy (partial removal of breast tissue) or radical mastectomy (total removal of the breast) has become the sixth most common reconstructive procedure performed in the United States.[9] Lumpectomy defects that are less than 25% of the total breast volume can be corrected by rearranging the local breast tissue.[10] Larger lumpectomies and mastectomies require more drastic reconstruction modalities. Studies show that many women who have had a mastectomy tend to suffer from a syndrome "marked by anxiety, insomnia, depressive attitudes, occasional ideas of suicide, and feelings of shame and worthlessness."[11] Reconstruction of the breast mound following a mastectomy has proven to alleviate the sense of mutilation and suffering that women experience postsurgery.[12] As a result, breast reconstruction is offered as an option to any woman undergoing surgery for breast cancer. Breast reconstruction was performed on 74,090 patients in 2002 with 50% of them having the procedure at the same time as a mastectomy. Furthermore, 236,888 breast augmentation procedures were performed in 2002, being the third most common surgical cosmetic procedure.[9]

IV. STANDARD OF CARE

When planning breast augmentation or reconstruction, one of three options is currently considered. First, either a fixed-volume or expandable prosthesis is placed alone for breast augmentation. This option is most popular with women desiring an increase in breast size.[13] Second, autologous tissue is used to reconstruct the breast. This option is most commonly performed for patients following a mastectomy. Finally, autologous tissue in combination with a prosthetic device is used. If the patient's own tissue does not contribute an adequate volume for reconstruction, then a prosthetic device can be additionally implanted to fill the defect.[10,14] Each of these clinical strategies is briefly discussed below.

A. Implants and Expanders

There are two basic types of prosthetic devices available: fixed-volume breast implants and tissue expanders. The advantage of using an implant or expander is that these devices can be manufactured in a broad range of sizes, contours, profiles, and textures. Fixed-volume implants have an outer layer or envelope of vulcanized silicone and can be filled with saline or silicone.[10,14] Implants filled with silicone gel are available on a limited basis due to the controls placed by the FDA in 1992.[13,15] Tissue expanders are implants in which the volume can be altered after implantation. Breast reconstruction using tissue expanders involves a temporary expander being inserted and inflated to stretch the skin to a slightly larger than necessary state, at which point the device is removed and replaced with a permanent implant of desired size. Expanders possess an outer envelope of silicone elastomer and serial injections of saline are made over time, usually once per week, to inflate the implant.[10,14] Both implants and expanders are available with a smooth or textured silicone surface. Textured surfaces result in a lower incidence of capsular contracture after implantation than smooth surface implants. Textured surfaces also serve to stabilize implant movement and facilitate tissue expansion. One drawback to having a textured envelope is that these implants have a slightly thicker vulcanized silicone shell that may be more visible in some patients by exhibiting a rippled appearance through the skin.[10,14] The thicker shell may also contribute to the implant folding and potentially deflating following insertion.[10,14]

The most common complication resulting from implants and expanders is capsular contracture (Figure 28.2).[10,14] Inserting an implant into the body inevitably leads to the formation of a capsule of firm, fibrous scar tissue around the implant due to a foreign body reaction.[16–19] The implant capsule constricts over time, making the augmented breast feel harder and firmer than desired.[10,14,15,20] Capsular contracture often results in a spherical breast appearance,[10,14,15] chronic chest wall discomfort, and restricted shoulder or arm movement.[10,14] Although capsule formation is normal after breast augmentation, not all capsules result in severe contraction.[21] Capsular contracture is graded on the Baker scale of I to IV, where I denotes a breast that looks and feels soft and IV denotes a breast that feels firm, is distorted in shape, and causes pain to the patient. The potential for calcium deposits accumulating in the capsule can also interfere with tumor detection.[10,14]

Implants are also subject to other local complications such as rupture, possible silicone or saline leakage,[10,14] displacement, deformation,[10,14,22] chronic seroma, implant exposure,[10,14] hematoma, and loss of nipple sensation.[19] Breast augmentation using tissue expanders also results in more complications than using fixed-volume implants alone. Infection is most common with expanders due to the bacteria that can be introduced from the serial saline injections. Fill-valve problems and skin loss are also problems associated with using tissue expanders.[14]

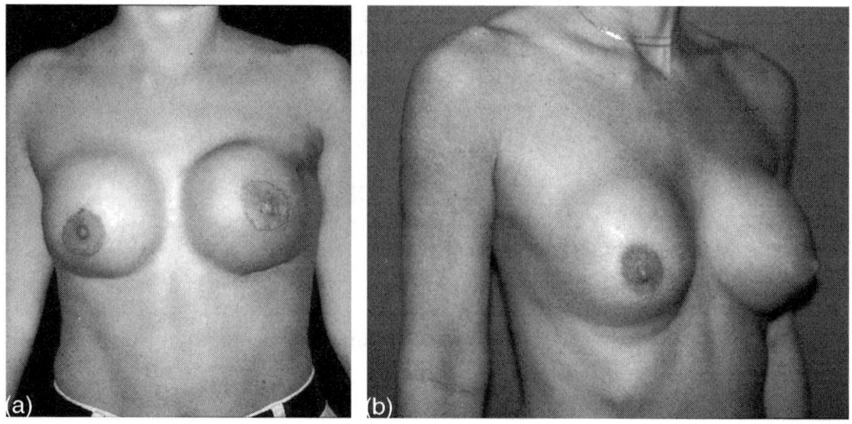

FIGURE 28.2 Examples of capsular contracture around breast implants. (a) Patient possesses capsular contracture of both breasts, resulting in breast hardness, an unnatural spherical profile, and breast asymmetry. (b) Similarly, this patient possesses capsular contracture of her right breast (Photographs are courtesy of Elisabeth Beahm, M.D., Department of Plastic Surgery, M.D. Anderson Cancer Center, The University of Texas).

B. Autologous Tissue

Breast reconstruction considers the aesthetic and anatomic properties of the breast mound in order to achieve optimal clinical results. Early on, autologous adipose tissue was investigated as a virtually limitless source of material for soft tissue repair.[23] It is easily harvested, readily available, and most patients possess excessive amounts that can be harvested without producing significant contour defects. However, despite the theoretical advantages, autologous adipose tissue injections have demonstrated poor results,[24] with a 50 to 70% reduction in graft volume due to resorption.[25–28] Moreover, mature adipocytes cannot be expanded *ex vivo* because they are terminally differentiated.[29] Additional problems with autologous adipose tissue injections include microdeposits of serum calcium from fat resorption that result in obscure mammographies and interfere with tumor detection.[23,30]

Breast reconstruction using the patient's own tissues, rather than implantable devices, tends to produce better results[10] with fewer complications.[14] The use of autologous tissues tends to produce a breast mound that better recreates the shape, contour, softness, and fullness of the natural breast than the use of implants.[10] There are a number of sites for donor tissue for breast reconstruction. The primary areas include the abdomen (transverse rectus abdominis musculocutaneous (TRAM) flap)[10,14] and the back (latissimus dorsi musculocutaneous flap). The buttocks (inferior gluteus maximum musculocutaneous flap) and the hips (Rubens flap) are seldom used to reconstruct the breast mound. Tissue availability must be considered for each donor site based on the patient.[14]

Abdominal flap reconstruction is the most common, yet technically demanding, procedure performed for breast reconstruction using autologous tissues.[10] The TRAM flap conveniently provides abdominal wall skin and adipose tissue, along with the blood vessels that supply these tissues to reconstruct the breast. Breasts reconstructed with the TRAM flap have shown to maintain symmetry over time. Patients also view the accompanying abdominoplasty, or "tummy tuck," as an added benefit of having reconstruction using the TRAM flap. A disadvantage to having this surgery performed is that additional procedures to reconstruct the nipple–areola and to refine the breast mound are required. Additionally, due to the complexity of the TRAM flap procedure, complications such as prolonged pain, abdominal weakness, an extended period of recovery, and occasional areas of necrosis accompany the surgery.[14] Finally, contour abnormalities of the abdomen may result after TRAM flap reconstruction, including upper and lower abdominal bulge, upper fullness, and hernia.[31–33] The latissimus dorsi muscle is a large triangular muscle that extends over the lower half of the thorax and lumbar region. Although the procedure involving the latissimus dorsi flap is less involved than the TRAM flap procedure, the long-term results of the latissimus dorsi flap procedure are inferior. The muscle tends to atrophy over time and there is typically a volume discrepancy in the back from the donor site.[14]

Breast augmentations and reconstructions are adversely affected by radiation therapy. A high percentage of breast cancer patients need radiation therapy, which causes an increase in the risk of capsular contracture around the implant and eventual loss of the implant. Radiation therapy also increases the incidence of fibrosis and, hence, the incidence of capsular contraction around breasts reconstructed with implants or autologous tissues.[10]

V. OVERALL TISSUE ENGINEERING STRATEGY

Due to the complexity of breast augmentation and reconstruction measures, and the complications associated with these procedures, various tissue engineering strategies are being investigated as a means to repair and restore the breast. Tissue engineering strategies in general involve seeding cells and introducing appropriate tissue induction and differentiation growth factors in a three-dimensional natural, synthetic, or hybrid scaffold in order to develop "biological substitutes that restore, maintain, or improve tissue function."[1,2,34] Numerous scaffold materials, ranging from non-woven fiber and hydrogel extracellular matrix structures to biodegradable polymers in the form of foams, non-woven fibers, and hydrogels are being investigated (Table 28.1). Scaffold materials must be able to mechanically support and guide tissue formation, such as adipogenesis. Materials must also be biocompatible, biodegradable, easily processed,[35,36] resistant to mechanical strain, and easily shaped to the surgeon's specifications.[12] Also, materials must permit variability in shape and volume in order to personalize the scaffold to meet the patient's specific contour and size needs. Ideally, the scaffold should recapitulate the endogenous extra-cellular matrix as much as possible. Acellular tissue engineering constructs are implanted in patients and rely on recruitment of surrounding cells or are tailored to remain acellular, whereas cellular tissue engineering constructs are either grown *ex vivo* in sophisticated bioreactors and then implanted in the patient or are placed directly *in vivo* with the patient serving as a bioreactor. Preadipocytes, precursor cells that differentiate into mature adipocytes, can be seeded onto a scaffold and allowed to proliferate and differentiate to promote the formation of adipose tissue. The materials that are under investigation as potential scaffolds for adipose tissue growth are discussed in Section VI (Figure 28.3).

TABLE 28.1
References for Natural and Synthetic Materials Employed in Adipose Tissue Engineering Strategies

Natural Materials	Synthetic Materials
Alginate[43,54,55]	Polyethylene glycol[35]
Collagen[40–42,44,45]	Poly(glycolic) acid[4,47,48]
Fibrin[49]	Poly(L-lactic-co-glycolic) acid[37,38,61,62]
Hyaluronic acid[42,43]	Polytetrafluoroethylene[39]
Matrigel[50–52]	–

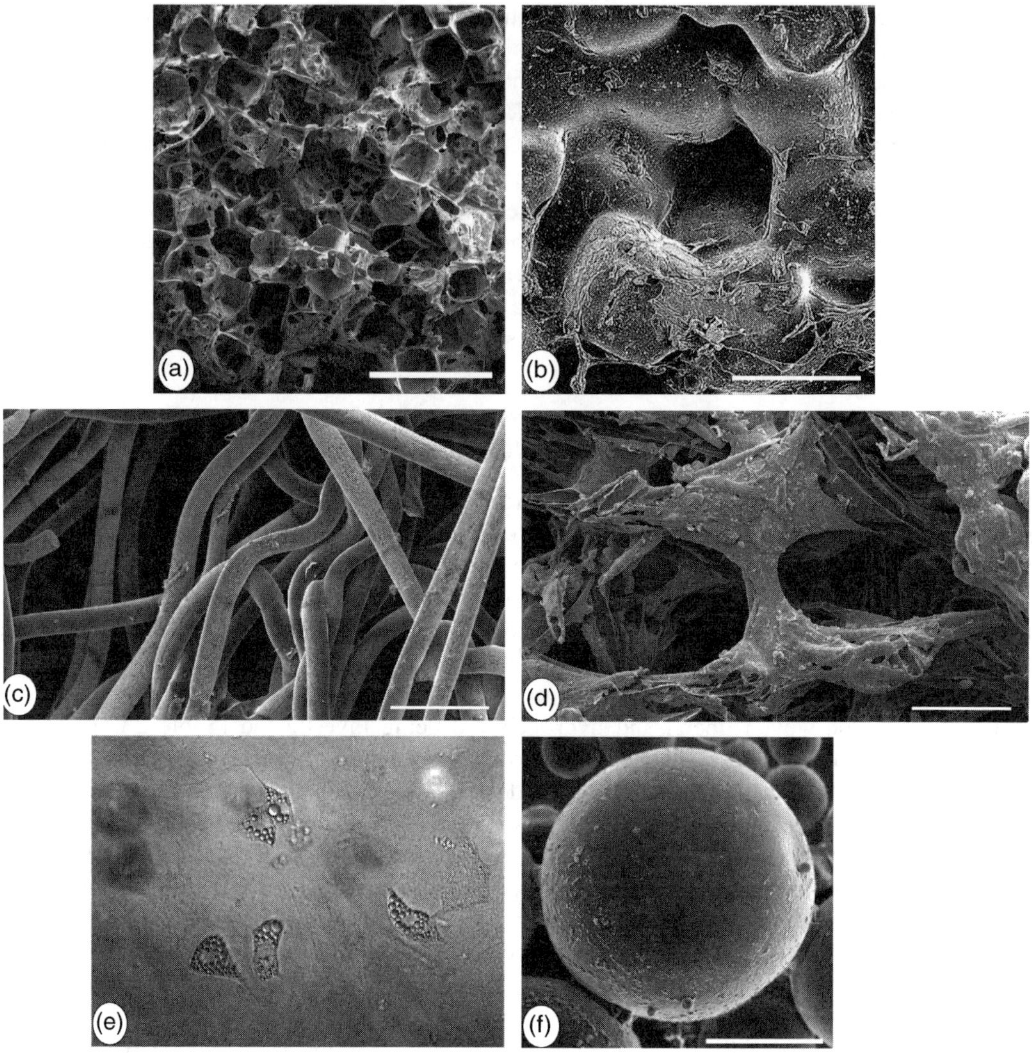

FIGURE 28.3 Examples of materials employed in adipose tissue engineering: (a) PLGA foam (bar denotes 600 μm), (b) PLGA foam with preadipocytes within the pores (bar denotes 300 μm), (c) non-woven polymer fiber (bar denotes 100 μm), (d) non-woven polymer fiber with preadipocytes spanning the fibers (bar denotes 100 μm), (e) preadipocytes differentiating within a fibrin hydrogel (300 × magnification), and (f) PLGA/PEG microsphere used to deliver bioactive factors (bar denotes 30 μm).

VI. MATERIALS AS SCAFFOLDS

A. Natural and Synthetic Polymer Foams and Non-Woven Meshes

Rigid poly(L-lactic-*co*-glycolic-acid) (PLGA) polymer foams seeded with rat preadipocytes have been implanted subcutaneously into male Lewis rats in and effort to generate *de novo* adipose tissue (Figure 28.3(a and b)). Results from short-term (2 to 5 weeks) and long-term (1 to 12 months) studies demonstrated the successful formation of adipose tissue for up to 2 months.[37,38] The volume of generated adipose tissue then began to decrease and was completely resorbed by 5 months. The resorption of adipose tissue after 2 months may be due to factors such as a lack of adequate vascularization, lack of support structure after the PLGA degraded, the anatomical site of implantation not being conducive to long-term maintenance of adipose tissue, or a limitation related to the small animal model used. The long-term maintenance or persistence of generated tissue is a challenge for all tissue engineering applications.

In addition to foams or sponges, non-woven fiber meshes have been studied as potential scaffolds for adipose tissue engineering (Figure 28.3(c) and (d)). Fluortex monofilament-expanded polytetrafluoroethylene (52 μm pore size) is one such material. Since preadipocytes do not attach to polytetrafluoroethylene, the material was coated with fibronectin to optimize cell seeding efficiency. Fibronectin coating resulted in a significantly higher number of attached human preadipocytes than collagen or albumin coatings. Human preadipocytes were able to proliferate and differentiate into adipocytes on the fibronectin-coated expanded polytetrafluoroethylene *in vitro* over a period of 120 h.[39] *In vivo* studies have not been conducted.

von Heimburg and colleagues[40–42] have studied porous collagen scaffolds and two types of hyaluronan-based materials[42,43] for adipose tissue engineering. The collagen scaffolds are constructed based on a directional solidification method followed by freeze-drying to obtain a uniformly porous structure.[44,45] The hyaluronic acid-based devices were manufactured into sponges or scaffolds composed of non-woven fibers. The sponges demonstrated open, interconnecting pores ranging from 50 to 340 μm while the non-woven mesh had an interfiber distance of 100 to 300 μm. *In vivo* studies conducted in mice showed that, due to the larger, interconnected pores, the hyaluronan sponge proved to be a better scaffold than the freeze-dried collagen construct or the hyaluronan non-woven carrier. After 8 months, a greater number of adipocytes were found in the sponge compared to the non-woven mesh. The large pore size is important for the preadipocytes to incorporate lipids and enlarge during differentiation. Pore size was also found to be a factor in adipogenesis in several proprietary Johnson & Johnson biodegradable polymer sponges and non-woven fibers.[46]

Beahm and colleagues[4,47,48] have demonstrated recruitment of resident preadipocytes and formation of *de novo* adipose tissue in silicone molds of various shapes (sheet, hemisphere, sphere). Packed with poly(glycolic acid) (PGA) fibers, Matrigel, and bFGF, the molds were sutured to the superficial inferior epigastric vessel of nude rats to create a vascular pedicle model. Qualitative assessment over 4 to 20 weeks demonstrated *de novo* adipogenesis. This approach is currently being translated to a large animal model (Yucatan micropig) to assess adipose tissue engineering strategies.

B. Hydrogels

Numerous natural and synthetic hydrogels are being investigated for primarily small breast defects, soft tissue repair, and cosmetic applications. As a natural wound healing matrix, fibrin is often studied as a potential injectable, cell-seeded scaffold. Investigators in Germany have seeded human preadipocytes within defined fibrin gels and demonstrated cell proliferation and differentiation (Figure 28.3(e)).[49] Similarly, reconstituted basement membrane of a mouse tumor, or Matrigel, has been shown by several investigators to induce migration, proliferation, and differentiation of preadipocytes when supplemented with basic fibroblast growth factor (bFGF).[50–52] It is speculated that when Matrigel plus bFGF is injected into a subcutaneous space, preadipocytes residing in the adjacent connective tissue migrate into the extra-cellular matrix hydrogel. Results demonstrate that endothelial cells are also recruited.[52] It remains unclear whether the observed angiogenesis drives the adipogenesis or *vice versa*. When gelatin microspheres loaded with bFGF were coimplanted with Matrigel, similar results were obtained.[53]

Porous alginate material (a naturally derived hydrogel) has also been investigated as a construct for soft tissue engineering. Halberstadt et al.[43] modified the alginate material with the peptide sequence; arginine, glycine, and aspartic acid (RGD), to allow cells to adhere to the construct. *In vitro* studies demonstrated that the porous alginate–RGD material supported cell attachment, adhesion, and proliferation.[54] Small animal studies performed over 6 months showed that the

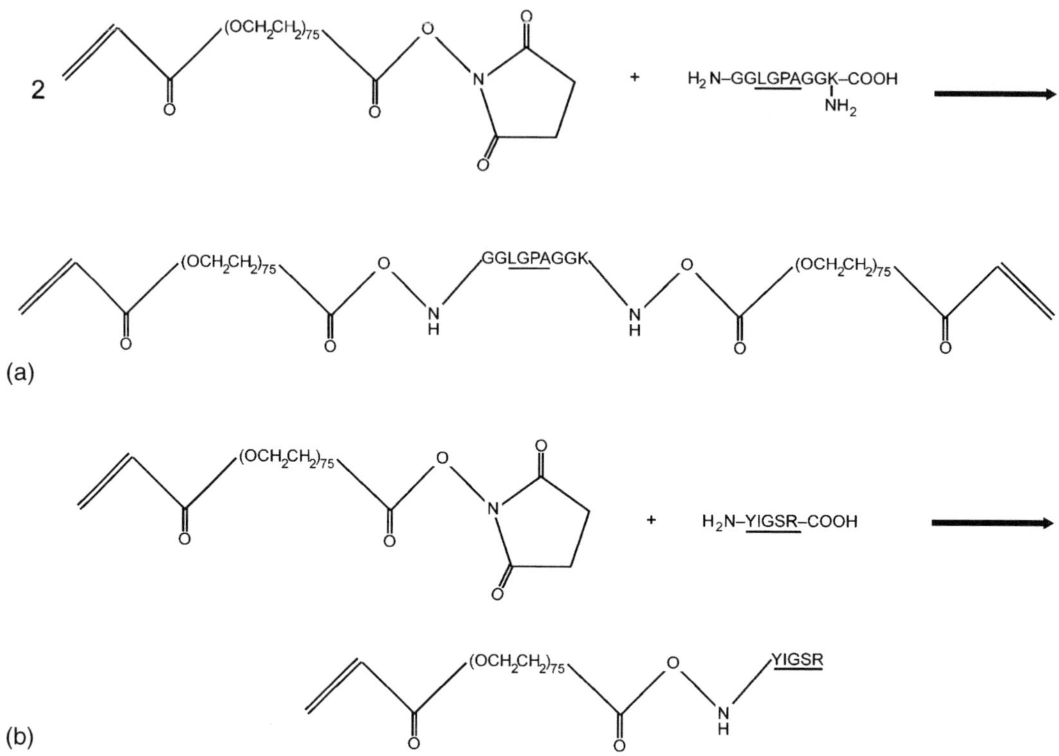

FIGURE 28.4 (a) Coupling of collagenase sensitive peptide, LGPA, with acrl–PEG–NHS to form a degradable hydrogel; (b) coupling of cell adhesive peptide, YIGSR, with acrl–PEG–NHS to facilitate preadipocyte binding to hydrogel.

implanted material was also conducive to tissue ingrowth and did not elicit major inflammatory responses.[55] In a large animal model (sheep), the material was seeded with preadipocytes and injected into the nape of the neck. Well-defined adipose tissue was identified within the hydrogel at 1 and 3 months. Unfortunately, it cannot be determined whether the adipose tissue growth resulted from the previously seeded preadipocytes in the material or from resident preadipocytes.[43]

Collagen hydrogels have also been investigated as a three-dimensional biological matrix upon which preadipocytes are cocultured with human mammary epithelial cells in order to form tissue that closely resembles the normal human breast. Histologic analysis of the collagen gels from *in vitro* studies indicates a pattern of ductal structures of human mammary epithelial cells within clusters of adipocytes, similar to the architecture of breast tissue. These findings indicate that there is a potential for breast tissue to be regenerated *in vitro* on a three-dimensional scaffold and later implanted for breast reconstruction purposes.[56]

Synthetic polymer hydrogels are also actively being studied as potential tissue engineering scaffolds because of the ability to derivatize the polymers with bioactive functional groups and, thereby, control molecular and cell function. Hydrogels are viscoelastic polymeric structures that contain a significant volume fraction of water (usually $>90\%$). The three-dimensional polymeric structures of the hydrogel are held together primarily by cross-linking. Currently, hydrogels are being employed in biomedicine for controlled drug release, soft tissue augmentation, cell separations, and biosensors.[57,58] In the tissue engineering arena, hydrogels can be used as nascent materials, or they can be modified with bioactive peptides that aid them in mimicking cell adhesion properties of the extra-cellular matrix. Proteolytically degradable peptides can be incorporated into the backbone of the polymer to form hydrogels that are degraded by cell-secreted enzymes.[35] For instance, polyethylene glycol (PEG) can be modified with the peptide sequence; leucine, glycine, proline, and alanine (LGPA), to form a polymer degradable by collagenase (Figure 28.4(a)). Preadipocyte adhesion sites can also be coupled to polyethylene glycol using the peptide; tyrosine, isoleucine, glycine, serine, and arginine (YIGSR) (Figure 28.4(b)). YIGSR is a cell binding peptide found on laminin. Wu and Patrick[59] have shown that preadipocytes bind preferentially to laminin-1 and that cell adhesion to and migration on laminin-1 is mediated, in part, by the $\alpha_1\beta_1$ integrin. Combining the degradable polyethylene glycol with the polymer coupled with cell adhesion sites produces a synthetic hydrogel that has been shown to be adequate scaffold material *in vitro*. The preadipocytes can be mixed into a polyethylene glycol solution and then photopolymerized into a hydrogel. The clinical benefit of having a polymerizable hydrogel such as polyethylene glycol is that the cell–polymer solution can easily be injected into the defect to be corrected and then photopolymerized *in situ* into a hydrogel. The need for complex surgical intervention would thus be eliminated.[60]

VII. MATERIALS AS DELIVERY VEHICLES FOR ADIPOGENIC FACTORS

Sophisticated biomaterials can be modified to present factors that promote adipogenesis. Modifications include encapsulating a factor or tethering a factor to polymer chains. PLGA/polyethylene glycol (PLGA/PEG) microspheres have been investigated as growth factor delivery vehicles in the absence of Matrigel (Figure 28.3(f)).[61] Insulin, insulin-like growth factor-1 (IGF-1), and bFGF have been administered to the muscular fascia of the rat abdominal wall via these PLGA/PEG microspheres.[62] At the 4 weeks harvest, adipose tissue was grossly observed at the site of implantation. Histologic and image analysis showed that the microspheres treated with both insulin and IGF-1 demonstrated statistically greater increases in adipose tissue (composed of fibroblasts and adipocytes) than did empty microspheres or microspheres treated with insulin, IGF-1, or bFGF alone or with all three growth factors. This *de novo* generation of adipose tissue may result from one of three mechanisms. Stem cells may be present in the fascia and stimulated to differentiate into preadipocytes and, subsequently, adipocytes. Another possible explanation is that preadipocytes residing in the fascia are stimulated to differentiate into adipose tissue. The third explanation, and the most speculative, is that the fibroblasts in the fascia dedifferentiate and then differentiate into preadipocytes and subsequently adipocytes. Regardless of the means, results indicate that adipose tissue can be generated in fascia within a short period (4 weeks) via delivery of microspheres treated with insulin and IGF-1.[62]

VIII. FUTURE CONSIDERATIONS

The generation of adipose tissue using novel biomaterials has the potential to change the way reconstructive surgery is practiced, as well as increase patients' quality of life. The fabrication of materials specific to adipose tissue is in its infancy. Current scaffolding strategies show promise for adipose tissue growth, but are limited by the level of vascular support for the scaffold. An adequate blood supply must be established to accompany *de novo* adipose tissue formation. The scaffold materials must also be able to withstand possible radiation exposure. Future studies must also address the long-term maintenance of newly generated adipose tissue, as well as the development of appropriate large animal models to assess adipose tissue engineering strategies.

REFERENCES

1. Patrick, C. W. Jr., Tissue engineering of fat, *Surg. Oncol.*, 19, 302–311, 2000.
2. Patrick, C. W. Jr., Tissue engineering strategies for soft tissue repair, *Anat. Rec.*, 263, 361–366, 2001.

3. Patel, P., Robb, G. L., and Patrick, C. W. Jr., Soft tissue restoration using tissue engineering, *Semin. Plast. Surg.*, 17, 99–105, 2003.
4. Beahm, E., Walton, R., and Patrick, C. W. Jr., Progress in adipose tissue construct development, *Clin. Plast. Surg.*, 17, 547–558, 2003.
5. Moore, K. L., *Clinically Oriented Anatomy*, 3rd ed., Williams & Wilkins, Baltimore, MD, pp. 45–48, 1992.
6. Dunnington, G. L., Andriole, D. A., and Kaiser, S., Breast, In *Essentials of General Surgery*, Lawrence, S., ed., Lippincott Williams & Wilkins, Philadelphia, PA, pp. 368–385, 2000.
7. Junqueira, L. C., Carneiro, J., and Kelley, R. O., *Basic Histology*, Appleton & Lange, Norwalk, CT, pp. 461–465, 1989.
8. American Cancer Society. *Breast Cancer Statistics for 2002*, American Cancer Society, 2002.
9. American Society of Plastic Surgeons. *2001 National Clearinghouse of Plastic Surgery Statistics*, American Society of Plastic Surgeons, 2002.
10. Robb, G. L., Reconstructive surgery, In *Breast Cancer*, Hunt, K. K., Robb, G. L., Strom, E. A. and Ueno, N. T., eds., Springer, New York, pp. 223–253, 2001.
11. Renneker, R., and Cutler, M., Psychological problems of adjustment to cancer of the breast, *JAMA*, 148, 834, 1952.
12. Jacobson, N., The socially constructed breast: breast implants and the medical construction of need, *Am. J. Public Health*, 88, 1254–1261, 1998.
13. Zuckerman, D., Commentary: are breast implants safe?, *Plast. Surg. Nurs.*, 22, 66–71, 2002.
14. Bostwick, J. III, *Plastic and Reconstructive Surgery*, 2nd ed., Quality Medical Publishing, Inc., St. Louis, MO, p. 1606, 2000.
15. Gerszten, P. C., A formal risk assessment of silicone breast implants, *Biomaterials*, 20, 1063–1069, 1999.
16. Pollock, H., Breast capsular contracture: a retrospective study of textured versus smooth silicone implants, *Plast. Reconstr. Surg.*, 91, 404–407, 1993.
17. Peters, W., Pritzker, K., Smith, D., Fornasier, V., Holmyard, D., Lugowski, S., Kamel, M., and Visram, F., Capsular calcification associated with silicone breast implants: incidence, determinants, and characterization, *Ann. Plast. Surg.*, 41, 348–360, 1998.
18. Coleman, D. J., Foo, I. T. H., and Sharpe, D. T., Textured of smooth implants for breast augmentation? A prospective controlled trial, *Br. J. Plast. Surg.*, 44, 444–448, 1991.
19. Baran, C. N., Peker, F., Ortak, T., Sensoz, O., and Baran, N. K., A different strategy in the surgical treatment of capsular contracture: leave capsule intact, *Aesthetic Plast. Surg.*, 25, 427–431, 2001.
20. Bosetti, M., Navone, R., Rizzo, E., and Cannas, M., Histochemical and morphometric observations on the new tissue formed around mammary expanders coated with pyrolytic carbon, *J. Biomed. Mater. Res.*, 40, 307–313, 1998.
21. Pandya, A. N., and Dickson, M. G., Capsule within a capsule: an unusual entity, *Br. J. Plast. Surg.*, 55, 455–456, 2002.
22. Clough, K. B., O'Donoghue, J. M., Fitoussi, A. D., Nos, C., and Falcou, M. C., Prospective evaluation of late cosmetic results following breast reconstruction: I. Implant reconstruction, *Plast. Reconstr. Surg.*, 107, 1702–1709, 2001.
23. Coleman, W. P. III, Fat transplantation, *Dermatol. Clin.*, 17, 891–898, 1999.
24. Ellenbogen, R., Invited comment, *Aesthetic Plast. Surg.*, 24, 197, 1990.
25. de la Fuente, A., and Tavora, T., Fat injection for the correction of facial lipodistrophies: a preliminary report, *Aesthetic Plast. Surg.*, 12, 39–43, 1988.
26. Chajchir, A., and Benzaquen, I., Liposuction fat grafts in face wrinkles and hemifacial atrophy, *Aesthtic Plast. Surg.*, 10, 115–117, 1986.
27. Matsudo, P., and Toledo, L., Experience of injected fat grafting, *Aesthetic Plast. Surg.*, 12, 35–38, 1988.
28. Niechajev, I., and Sevcuk, O., Long term results of fat transplantation: clinical and histologic studies, *Plast. Reconstr. Surg.*, 94, 496–506, 1994.
29. Robb, G., Miller, M., and Patrick, C. W. Jr., Breast reconstruction, In *Methods in Tissue Engineering*, Atala, A. and Lanza, R., eds., Academic Press, San Diego, pp. 881–889, 2001.
30. Ersek, R. A., Chang, P., and Salisbury, M. A., Lipo layering of autologous fat: an improved technique with promising results, *Plast. Reconstr. Surg.*, 101, 820–826, 1998.

31. Hartrampf, C. R. Jr., Abdominal wall competence in transverse abdominal island flap operations, *Ann. Plast. Surg.*, 12, 139–146, 1984.
32. Lejour, M., and Dome, M., Abdominal wall function after rectus abdominis transfer, *Plast. Reconstr. Surg.*, 87, 1054–1068, 1991.
33. Feller, A. M., Free TRAM: results and abdominal wall function, *Clin. Plast. Surg.*, 21, 223–232, 1994.
34. Langer, R., and Vacanti, J. P., Tissue engineering, *Science*, 260, 920–926, 1993.
35. Mann, B., Gobin, A., Tsai, A., and West, J., Smooth muscle cell growth in photopolymerized hydrogels with cell adhesive and proteolytically degradable domains: synthetic ECM analogs for use in tissue engineering, *Biomaterials*, 22, 3045–3051, 2001.
36. West, J. L., and Hubbell, J. A., Bioactive polymers, In *Synthetic Biodegradable Polymer Scaffolds*, Atala, A. and Mooney, D. J., eds., Springer, Berlin, p. 348, 1997.
37. Patrick, C. W. Jr., Chauvin, P. B., and Reece, G. P., Preadipocyte seeded PLGA scaffolds for adipose tissue engineering, *Tissue Eng.*, 5, 139–151, 1999.
38. Patrick, C. W. Jr., Zheng, B., Johnston, C., and Reece, G. P., Long-term implantation of preadipocyte seeded PLGA scaffolds, *Tissue Eng.*, 8, 283–293, 2002.
39. Kral, J. G., and Crandall, D. L., Development of a human adipocyte synthetic polymer scaffold, *Plast. Reconstr. Surg.*, 104, 1732–1738, 1999.
40. von Heimburg, D., Lemperle, G., Dippe, B., and Kruger, S., Free transplantation of fat autografts expanded by tissue expanders in rats, *Br. J. Plast. Surg.*, 47, 470–476, 1994.
41. von Heimburg, D., Zachariah, S., Heschel, I., Kuhling, H., Schoof, H., Hafemann, B., and Pallua, N., Human preadipocytes seeded on freeze–dried collagen scaffolds investigated in vitro and in vivo, *Biomaterials*, 22, 429–438, 2001.
42. von Heimburg, D., Zachariah, S., Low, A., and Pallua, N., Influence of different biodegradable carriers on the in vivo behavior of human adipose precursor cells, *Plast. Reconstr. Surg.*, 108, 411–420, 2001.
43. Halberstadt, C., Austin, C., Rowley, J., Culberson, C., Loebsack, A., Wyatt, S., Coleman, S., Blacksten, L., Burg, K., Mooney, D., and Holder, W. Jr., A hydrogel material for plastic and reconstructive applications injected into the subcutaneous space of a sheep, *Tissue Eng.*, 8, 309–319, 2002.
44. Heschel, I., and Rau, G., *Method for Producing Porous Structures*, 98/03403, 1997.
45. Heschel, I., Luckge, C., Rodder, M., Garbering, C., and Rau, C., Possible applications of directional solidification techniques in cryobiology, In *Advances in Cyrogenic Engineering*, Kittel, P., ed., Plenum Press, New York, pp. 13–19, 1996.
46. Roweton, S., Freeman, L., Patrick, C. W. Jr., Dempsy, K., and Zimmerman, M., *Preadipocyte-Seeded Absorbable Matrices*, Johnson & Johnson Excellence in Science Symposium, New Jersey, 2000.
47. Beahm, E., Wu, L., and Walton, R. L., *Lipogenesis in a Vascularized Engineered Construct*, American Society of Reconstructive Microsurgeons, San Diego, CA, 2000.
48. Walton, R. L., Beahm, E. K., and Wu, L., De novo adipose formation in a vascularized construct. *Microsurgery*, 24, 378–384, 2004.
49. Stark, B., personal communication, 2003.
50. Kawaguchi, N., Toriyama, K., Nicodemou-Lena, E., Inou, K., Torii, S., and Kitagawa, Y., De novo adipogenesis in mice at the site of injection of basement membrane and basic fibroblast growth factor, *PNAS*, 95, 1062–1066, 1998.
51. Passaniti, A., Taylor, R. M., Pili, R., Guo, Y., Long, P. V., Haney, J. A., Pauley, R. R., Grant, D. S., and Martin, G. R., A simple, quantitative method for assessing angiogenesis and antiangiogenic agents using reconstituted basement membrane, heparin, and fibroblast growth factor, *Lab Invest.*, 67, 519–528, 1992.
52. Toriyama, K., Kawaguchi, N., Kitoh, J., Tajima R., Inou K., Kitagawa Y., and Torii S., Endogenous adipocytes precursor cells for regenerative soft-tissue engineering, *Tissue Eng.*, 8, 157–165, 2002.
53. Kimura, Y., Ozeki, M., Inamoto, T., and Tabata, Y., Time course of de novo adipogenesis in Matrigel by gelating microspheres incorporating basic fibroblast growth factor, *Tissue Eng.*, 8, 603–613, 2002.
54. Eiselt, P., Yeh, J., Latvala, R. K., Shea, L. D., and Mooney, D. J., Porous carriers for biomedical applications based on alginate hydrogels, *Biomaterials*, 21, 1921–1927, 2000.
55. Halberstadt, C. R., Mooney, D. J., and Burg, K. J. L., Eislet, P., Rowley, J. A., Beiler, R. J., Roland, W. D., Culberson, C. R., Greene, K. G., Wyatt, S., Loebsack, C. E., and Holder, W. D., The design and implementation of an alginate material for soft tissue engineering. *Sixth World Biomaterial Congress*, 2000.

56. Huss, F. R. M., and Kratz, G., Mammary epithelial cell and adipocytes co-culture in a 3-D matrix: the first step towards tissue-engineered human breast tissue, *Cells Tissues Organs*, 169, 361–367, 2001.
57. Andrade, J. D., 1976. Hydrogels for medical and related applications. *ACS Symposium Series*, Washington, DC.
58. Peppas, N. A., ed., *Hydrogels in Medicine and Pharmacy*, CRC Press, Boca Raton, FL, p. 180, 1987.
59. Patrick, C. W., Jr., and Wu, X., Integrin-mediated preadipocyte adhesion and migration on laminin. *Ann. Biomed. Eng.*, 31, 505–515, 2003.
60. Elisseeff, J., McIntosh, W., Anseth, K., Riley, S., Ragan, P., and Langer, R., Photoencapsulation of chondrocytes in poly(ethylene oxide)-based semi-interpenetrating networks, *J. Biomed. Mater. Res.*, 51, 164–171, 2000.
61. King, T. W., and Patrick, C. W. Jr., Development and in vitro characterization of vascular endothelial growth factor (VEGF)-loaded poly(DL-lactic-co-glycolic acid)/poly(ethylene glycol) microspheres using a solid encapsulation/single emulsion/solvent extraction technique, *J. Biomed. Mater. Res.*, 51, 383–390, 2000.
62. Yuksel, E., Weinfeld, A. B., Cleek, R., Waugh, J. M., Jensen, J., Boutros, S., Shenaq, S. M., and Spira, M., *De novo* adipose tissue generation through long-term, local delivery of insulin and insulin-like growth factor-1 by PLGA/PEG microspheres in an *in vivo* rat model: a novel concept and capability, *Plast. Reconstr. Surg.*, 105, 1721–1729, 2000.

29 Scaffolding in Periodontal Engineering

Orasa Anusaksathien, Qi-ming Jin, Peter X. Ma, and William V. Giannobile

CONTENTS

I. CHALLENGES IN THE TREATMENT OF PERIODONTAL ALVEOLAR BONE DEFECTS

Periodontal diseases are initiated by inflammatory responses to bacterial plaque biofilms, and continue slowly, but progressively to destroy the structures surrounding the tooth root surface, including cementum, alveolar bone, and periodontal ligament. If the disease is left untreated, tooth loss will subsequently occur. It has been estimated that periodontal disease afflicts approximately 15% of the U.S. population.[1] The conventional periodontal treatment results in repair, whereby the restoration of the lost tooth support is rarely achieved in a predictable fashion. The challenges in regenerating periodontal tissues include several limiting factors such as: the lack of blood supply on the diseased root surface, a complex microbiota that can contaminate wounds at the soft–hard tissue interface, and occlusal forces on the tooth complex in traverse and axial plans.[2] Thus, the development of predictable therapies to regenerate large "critical size" periodontal bone defects is needed.

In order to attain maximal regeneration of periodontal structures, several factors are important to optimize healing: control of microbial infection, stabilization and maintenance of the wound compartment to allow migration of mesenchymal cells to reoccupy the defect, and release of chemotactic, inductive, and growth factors (GFs) to stimulate proliferation and attachment of putative regenerative cells. Several procedures have been performed surgically including

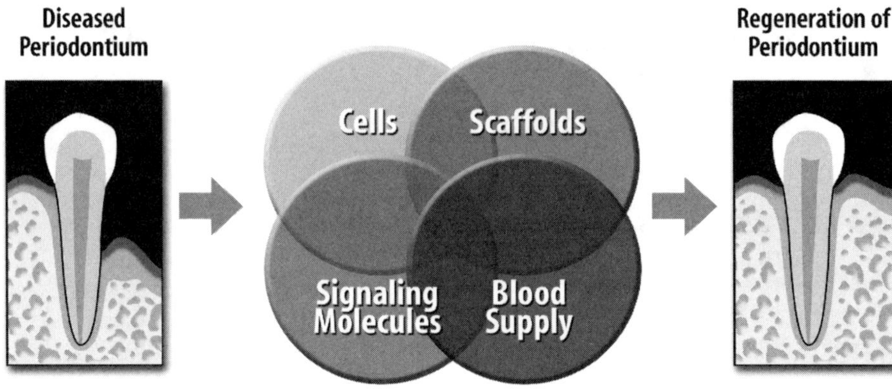

FIGURE 29.1 Schematic representation of critical factors required in periodontal tissue engineering. Reconstruction of lost periodontal tissue after root decontamination of bacterial plaque requires the combination of cells, scaffolds, signaling molecules, and vasculature.

root-conditioning agents, osteoconductive and osteoinductive grafting materials such as autogenous, allograft, and xenograft bone materials, guided tissue regeneration (GTR) using cell occlusive barriers, and combinations of these treatments. In general, existing therapies utilized in an attempt to stimulate periodontal regeneration are limited in both predictability and extent of outcome. Therefore, further development of novel treatments requires the consideration of many factors which dictate the degree of periodontal regeneration, including cells, scaffolds, bioactive molecules, and vasculature (Figure 29.1).

Over the past 20 years researchers have used advances made in material sciences and molecular biology to apply to periodontal tissue engineering, including the development of scaffolding devices, the use of regenerative molecules, for example, platelet-derived growth factor (PDGF) or bone morphogenetic proteins (BMPs), to treat advanced periodontal disease. These devices and regenerative molecules are in different stages of development from preclinical studies to FDA-approved therapies. This chapter will focus on the emerging therapies in the areas of material science, growth factor delivery, and cell/gene therapy for the restoration of periodontal structures.

II. IMPORTANCE OF DELIVERY DEVICES FOR PERIODONTAL TISSUE ENGINEERING

The use of scaffolds serves not only as space maintainers, but also as delivery vehicles for targeting bioactive molecules to the wound site. A list of scaffolds utilized in periodontal tissue regenerative procedures is shown in Table 29.1. These scaffolds may function as osteoconductive and/or osteoinductive devices depending on the type of material used. There are two basic types of scaffolds used in periodontology: naturally derived scaffolds and synthetic biomimetic scaffolds. Autogenous, allograft, and xenograft bone are naturally derived scaffold materials that function as space maintainers and as lattices for tissue ingrowth within periodontal lesions. Alloplasts and other polymers as scaffolds are also used for a variety of purposes similar to natural scaffolds, but often times lead to fibrous tissue encapsulation when devoid of bioactive molecules.

Three-dimensional scaffolds act as templates that guide and accelerate new tissue formation. The physical microstructure of the biomaterials is one of the important factors governing the osteoconductive nature of the scaffold. The internal architecture of the scaffold controls the quantity and shape of the substratum surfaces available for cellular ingrowth. The density, numbers, sizes of pores, and total void volume of the construct determine the type of cell that migrates into

TABLE 29.1
Scaffold Materials for Periodontal/Craniofacial Repair

Biomaterial	Tradename	References
Allografts		
Calcified freeze dried bone, decalcified freeze dried bone	Grafton®, Lifenet®, Musculoskeletal Transplant Foundation®	14,15,117–120
Xenografts		
Bovine mineral matrix, bovine-derived HA	Bio-Oss, OsteoGraf®, Pep-Gen P-15®	20,21,24,28,29,121,122
Alloplasts		
Hydroxyapaptite dense HA, porous HA, resorbable HA	Osteogen®, Periograf®, ProOsteone®	59–62
Tricalcium phosphate (TCP), calcium phosphate cement	Synthograft®, α-BSM®	65,66,117,123–125
Hard tissue replacement polymers	Bioplant®	126,127
Bioactive glass (SiO_2, CaO, Na_2O, P_2O_5)	PerioGlas®, BioGran®	25,71
Coral-derived calcium carbonate	Biocoral®	128–130
Polymers & collagens		
Collagen	Helistat®, Collacote®, Colla-Tec®, Gelfoam®	110,131–135
Poly(lactide-copolyglycolide, PLGA)		92,115,116
Methylcellulose		106,136–138
Hyaluronic acid ester	Hy®	52
Chitosan		139
Enamel matrix derivative (EMD)	Emdogain®	22,140–145

the scaffold.[3] In bone tissue engineering, it has been shown that pore sizes less than 15 to 50 μm encourages fibrous tissue invasion, pore sizes 50 to 150 μm leads to osteoid formation, and pore sizes greater than 150 μm promote the ingrowth of bone mineral.[4] Currently, it has been accepted that pore sizes that range from 150 to 500 μm are optimal for bone and mineral tissue formation.[5] Ideally, the scaffold should be biodegradable, biocompatible, and highly porous with a high surface area/volume ratio, allowing the transportation of nutrients and metabolites, and permitting cell and tissue ingrowth, as well as serving as a reservoir for bioactive molecules.[6–8] Alternatively, bioactive molecules could be incorporated in the forms of microparticles, nanoparticles, fibers, or injectable devices.[9,10]

The controlled-release device carrying bioactive molecules could be fabricated to release such bioactive molecules at optimal doses in a timely manner depending on the biological demand of the target tissue. Growth factors can be released at constant, pulsatile, and/or time programmed patterns. The constant release of bioactive molecules is the basic approach to deliver such molecules to the target tissue. The rate of release depends on the type of biomaterial used and the topography of the scaffold. The pulsatile release polymer devices can be fabricated to release bioactive molecules in response to either external stimuli such as magnetic, ultrasound, thermal, and electrical or self-regulated stimuli.[11] A time-programmed release device has been developed as a so-called "controlled-release microchip" that incorporates bioactive molecules in different reservoirs with the attached electrochemical sensor using microfabrication technology.[12] The device consists of micrometer-scale pumps, valves, and flow channels to deliver liquid solutions of single or multiple substances on demand in a continuous or pulsatile manner. The future development of controlled-release devices to deliver multiple bioactive molecules in periodontal wounds warrants further investigation.

III. SCAFFOLDING DEVICES USED FOR PERIODONTAL TISSUE ENGINEERING

A. Naturally Derived Scaffolds

1. Allograft–Xenograft Biomaterials

Allograft bone materials such as decalcified freeze-dried bone allograft (DFDBA) have been shown to contain multiple bioactive molecules (e.g., osteoinductive molecules such as BMPs).[13] The materials have been manufactured in many forms such as powders, blocks, moldable forms that maintain cohesiveness (putty), flowable forms, or gels.[14] Allograft bone materials have been used widely in treating periodontal defects and have resulted in the promotion of periodontal tissue regeneration.[15] However, the use of these allograft materials possess potential, although only highly remote risks for transmissible diseases such as HIV, hepatitis B and C.[16,17] In addition, the osteoinductive ability of these allografts varies among bone banks raising the question of predictability of the results.[18] Several bone banks have established the test for osteoinductive capacity of each lot and employ a battery of tests to confirm the elimination of transmissible diseases.

Xenografts derived from bovine bone have been utilized as an alternative grafting material in periodontology and oral surgery to avoid the volume restriction and the possibility of disease transmission to humans. Natural bone mineral matrix (Bio-Oss® Geistlich/Osteohealth Co., Wolhusen, Switzerland) and bovine derived hydroxyapatite (anorganic bone matrix, ABM) are commercially available forms used in oral and craniofacial reconstruction procedures. Bio-Oss, a commercially available product in clinic, is a highly purified natural bovine bone mineral matrix, which not only has the anorganic composition similar to human bone, but also maintains the bony architecture. Bio-Oss used alone or with collagen barrier membranes promotes new cementum and new bone formation in humans.[19–21] Bio-Oss has also been used as a delivery vehicle to reconstruct mandibular defects in combination with BMP-7 as well as in the repair of human intrabony defects with enamel matrix derivative (EMD).[22,23] These studies showed that this material is highly biocompatible, osteoconductive, and resorbable. A inorganic bone matrix has been used as a grafting material in periodontal intrabony defects and showed evidence of promoting periodontal tissue regeneration.[24,25]

Bovine anorganic bone mineral (ABM) coated with synthetic collagen type 1 containing 15 amino acids residues (P-15) has been well-studied over the last 5 years due to its *in vitro* and *in vivo* data, suggesting that the P-15 peptide promotes cell attachment and proliferation on bone mineral. Qian and Bhatnagar[26] first reported the use of synthetic P-15 collagen peptide molecule located in collagen type I molecule between amino acid residues 766 and 780. They have shown enhanced attachment and three-dimensional colony formation on ABM-P-15, by human PDL fibroblasts.[27] Short- and long-term clinical studies showed that a single application of ABM-P-15 to treat advanced periodontal defects resulted stimulation of bone neogenesis.[28,29] In addition, histological evaluation in an animal model demonstrated encouraging results that ABM-P-15 enhances bone formation.[30]

2. Cell Occlusive Devices and Collagen-Based Materials

The purpose of using barriers as cell occlusive devices for GTR procedures is to exclude or restrict the repopulation of epithelial and gingival connective cells, and to maximize PDL, cementoblast, and osteoblast cell growth into periodontal defects.[31] A list of cell occlusive devices including nonabsorbable and absorbable barriers is shown in Table 29.2. The nonabsorbable barriers are biocompatible materials manufactured from cellulose or expanded polytetrafluoroethylene (ePTFE). The absorbable barriers are polymers made of polylactic acid, polyglycolic acid, polyglactin, and both soluble and nonsoluble collagen.[32] These membrane barriers were designed to function as cell-occlusive, space maintaining, and clot stabilizing for a period of time, usually 5 to 6 weeks after placement. This timeframe is to allow for endothelial,

TABLE 29.2
Cell Occlusive Barriers in Periodontal/Craniofacial Repair

Cell Occlusive Barrier	Tradename	References
Nonresorbable		
Cellulose, ePTFE	Millipore filter®, Gore-Tex®	118,145–149
Resorbable		
Polylactic acid and polyglycolic acid (PLGA), Polyglactin-910, Poly(L-lactide) (PLLA)	Resolut®, Atrisorb®, Vicryl-Netz®	91,119,145,146,150,151
Collagen bovine tendon Type I, porcine dermis Type I + III	Biomend®, BioGide®, Ossix®	152–155
Plaster of Paris calcium sulfate	Cap-Set®, Hap-Set®	155–158

PDL, cementoblast and osteoblast migration, differentiation, and proliferation in the wound area for the formation of new periodontal attachment. The design of such barriers needs to be porous, and the pore size needs to be large enough for nutrient flow, but small enough to restrict epithelial and gingival connective tissue cell migration (Figure 29.2). The use of nonabsorbable barriers requires a second surgical procedure for barrier removal. In addition, prolonged retention of barriers may cause infection and tissue tear.[33] Absorbable barriers made from synthetic polymers are biocompatible, and biodegradable through physicochemical hydrolysis. Collagen barriers are degraded through an enzymatic process that may take from 6 to 8 weeks to as long as 4 to 8 months. Collagen membranes are fabricated to ease clinical manageability and shorten chairtime without the need for removal.

Clinical and histological data have shown modest improvements of GTR plus graft compared to either treatment alone.[34] There is insufficient data to support the use of grafting materials for enhancement of periodontal healing.[34,35] GTR is costly and technically sensitive and requires frequent follow-up maintenance visits. The predictability of the treatment outcomes varies widely depending on patient oral hygiene, exposure of barrier leading to the contamination of micro-organisms, smoking, and patients' compliance, as well as the topography of the defects.[36,37]

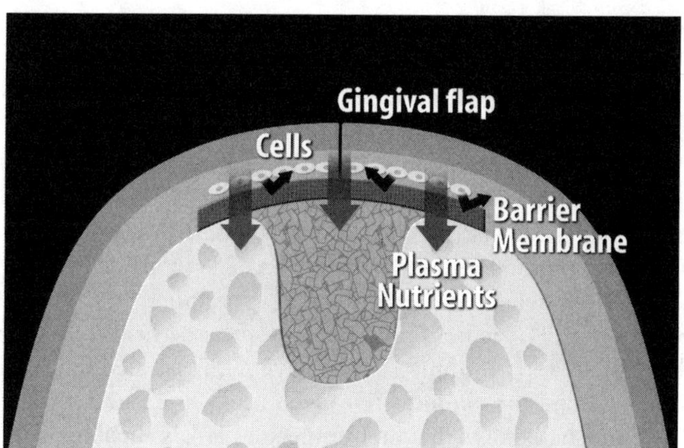

FIGURE 29.2 Barrier membrane used to compartmentalize a periodontal bone defect by guided tissue regeneration. The microstructure of the barrier membrane is designed to block gingival epithelial and connective tissue cell migration through the barrier and allow plasma nutrients from the flap to flow through. The membrane is placed between gingival flap and overlying alveolar bone defect to permit osteogenic cells to migrate into the defect for tissue regeneration.

Therefore, it is important for the development of alternative treatment modalities in treating critical size periodontal defects.

Collagen barrier membranes, made mostly from type I collagen extracted from tissues of bovine or porcine origin, have also been used in GTR. The collagen membrane has many properties:[32,38,39]

1. *Bioabsorbable*. Collagen membrane can be degraded *in vivo* enzymatically and its degradation rate can be controlled by cross-linking. Several collagen membranes with various degradation rates are available clinically. In addition, there is no need for second stage surgery.
2. *Hemostasis*. Collagen initiates clot formation and wound stability stimulating platelet attachment and enhancing fibrin linkage.
3. *Chemotaxis*. Collagen promotes primary wound coverage that is critical for the success of GTR via its ability to enhance fibroblast cell migration.
4. *Clinical manageability*. Collagen can be easily manipulated and adapted to various anatomical bony defects.
5. *Safety and biocompatibility*. Collagen elicits only a weak immune response, and cross-linking agents and sterilization can lower its host immune reaction.[40]
6. *Barrier function*. Collagen barriers have the ability to exclude gingival fibroblasts and epithelial cells from invading into periodontal wound sites.

Due to these properties, collagen membranes have been widely utilized to regenerate deep intrabony defects and augment horizontal bone formation. As a result, clinical attachment level is improved and periodontal pocket depth is reduced.[41] Moreover, collagen barriers are also used to repair dehiscence and fenestration defects associated with endosseous oral titanium implants, and consequently improves implant osseointegration.[41,42] More recently, Lee et al.[43] have shown that polymer/βTCP barriers impregnated with BMP-2 greatly improved craniofacial bone regeneration *in vivo*. This approach of combining barriers with bioactive molecules offers tremendous possibilities for bone tissue engineering.

3. Hyaluronic Acid-Based Scaffolds

Hyaluronic acid (HA) is a natural abundant nonsulfated glycosaminoglycan (GAG) found in the extracellular matrix, which is composed of (β-1,4)-linked D-glucuronic acid and (β-1,3)-*N*-acetyl-D-glucosamine. HA is nontoxic and biocompatible. HA can be degraded enzymatically *in vivo* and the rate of biodegradation can be controlled by cross-linking.[44,45] Because HA has three reaction groups, hydroxyl, carboxyl, and acetamido, HA is easily modified chemically according to various clinical requirements.[46] By binding its receptors such as CD44 and RHAMM, HA regulates cell adhesion, motility, proliferation, and differentiation.[47,48] Therefore, HA acts as a bioactive scaffold.

As a scaffold biomaterial, HA is used not only in soft tissue healing such as skin, cornea, and nerve, but also in hard tissue repair. Some studies have demonstrated that high molecular weight HA (200 to 400 kDa) is chondro-osteoinductive, while low molecular weight HA oligomers are angiogenic.[49] Sasaki et al.[50] has shown that high-molecular HA is capable of accelerating new bone formation. In the treatment of osteochondral defects, Caplan[51] revealed that cross-link modified high molecular HA accelerated cartilage formation better than the more stable benzylated high molecular weight HA. Hyaluronic acid scaffolds have also been used in dentistry to restore periodontal defects and to repair tooth molar pulpal defects while demonstrating good biocompatibility.[52,53] In addition, HA scaffolds have been used as carriers to delivery GFs such as BMPs and FGF-2.[54,55]

4. Calcium Phosphate-Based Ceramics

Calcium and phosphorus are the principal inorganic constituents of hard tissues (bone and teeth), which exist in the form of crystallized sodium-, magnesium-, and carbonate-hydroxyapatite (HAP). Hydroxyapatite has a high ability to adsorb various proteins in serum, especially to bind fibronectin and vitronectin.[56,57] This property improves cell attachment to its surface and thus has excellent biocompatibility. HAPs biodegradation depends on its origin, composition, and synthetic formulation.[58] Hydroxyapatite has been used as an osteoconductive device for a variety of craniofacial and periodontal applications.[59–62]

In addition to natural HAPs, there are several synthetic calcium phosphate scaffolds including resorbable hydroxyapatite, β-TCP (β-tricalcium phosphate), and the mixture of HAP and β-TCP (biphasic calcium phosphate: BCP). HAP is stable under physiological conditions and slow to resorb, while β-TCP and BCP resorb more rapidly.[58] Dependent on the application, calcium phosphate is prepared in powder, granule, or block forms. The powders of HAP and β-TCP can be used to coat the surfaces of titanium implants to accelerate dental implant osteointegration.[63,64] Calcium phosphate granules are usually used to prepare injectable bone cement. Most block forms of calcium phosphate are porous, which allows for bone ingrowth. Porous HAP or β-TCP is widely utilized as a bone substitute to repair osseous defects around teeth or dental implants[65,66] or as a carrier to deliver GFs.[67–69] In addition, HAP can be used as a cell substratum for osteoblasts in order to restore bone defects by tissue engineering methods.[70]

Bioactive glass is an alloplast material consisting of SiO_2, CaO, Na_2O, and P_2O_5 with grain sizes of 90 to 170 μm. Bioactive glass serves as a biocompatible and osteoconductive scaffold to treat intrabony and tooth furcation periodontal defects with modest success.[71–74] It has been shown that following the migration of osteogenic cells into the defect area and collagen incorporation, a firm bonding of bioactive glass with bone and connective tissue is achieved within 7 to 10 days.[75] Histological studies in animals showed evidence of true regeneration with newly formed cementum, PDL, and alveolar bone.[76,77] However, a human histological study showed that the graft materials were embedded in the dense connective tissue stroma with minimal new bone formation.[25]

B. Synthetic Scaffolds

1. Poly(lactide-coglycolide)

Poly(lactide-coglycolide) (PLGA) is a family of synthetic copolymers of lactide and glycolide. This type of scaffold can be manufactured precisely and is easy to fabricate into any shape according to clinical and experimental requirements. PLGA is biodegradable; the mechanism of which includes physicochemical hydrolysis and enzymatic degradation.[78,79] The degradation rate is associated with various characteristics such as molecular structure, crystallinity, and copolymer ratio.[80,81]

PLGA is a common polymer scaffold widely used in pharmaceutical science and tissue engineering. PLGA not only as a carrier delivers DNA[82] and growth factors such as BMP, VEGF, and NGF,[83–85] but also acts as an artificial extracellular matrix to engineer skin, liver, cartilage, and bone.[10,86–88] For periodontal applications, PLGA microspheres have been used as a local drug delivery system to deliver antibiotics such as minocycline.[89] Secondly, PLGA acts as a carrier to deliver growth factors such as BMP to periodontal defects and promote periodontal regeneration.[90] Moreover, PLGA as an occlusive membrane has been used in periodontal tissue guided regeneration and improves alveolar bone formation and cementogenesis.[91] PLGA has also been used to engineer cementum and complex tooth structures.[92,93] Combining the advantages of biodegradable polymers with the excellent osteoconductivity, highly porous polymer/HAP scaffolds have been developed.[94,95] The osteoconductive composite scaffolds were shown to be superior to the plain polymer scaffolds in osteoblastic cell seeding uniformity, proliferation, and

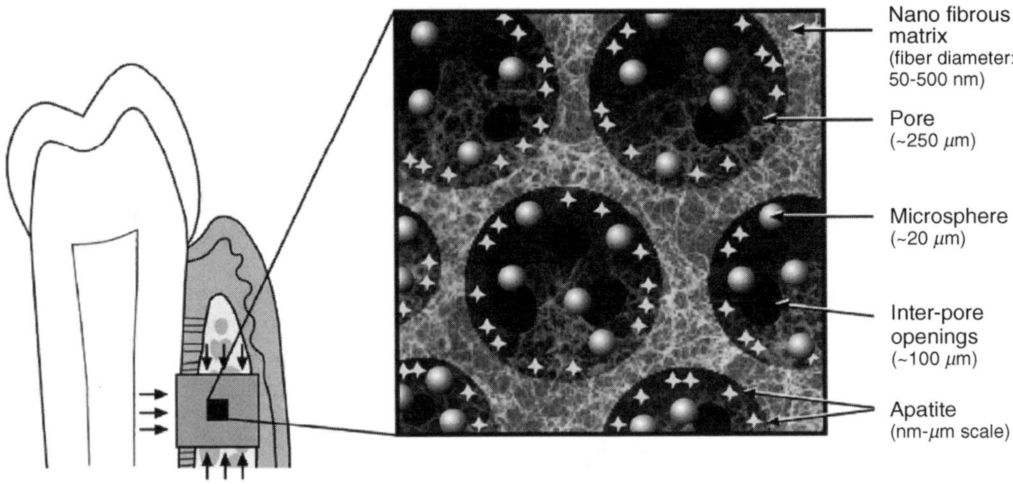

FIGURE 29.3 Schematic representation of "biomimetic nanoscaffolds." The scaffold design proposed for periodontal tissue engineering combines novel nanofibrous scaffolds with microspheres for controlled release of putative regenerative factors. The nanofibrous scaffolding design will utilize the architectural features of collagen, providing a high surface area for cell attachment and new matrix deposition, and an open structure allowing an interactive environment for cell–cell, cell–nutrient, and cell–signal molecule interactions. The bone mineral mimic apatite modification will enhance osteoconductivity of the scaffold. The biodegradable microspheres will release the regenerative factors in a controlled fashion thereby enhancing the duration that the factors are present in the local environment. The scaffold design coupled with inductive factors will provide the environment needed for cell recruitment (arrows indicate cell migration) and subsequent cell differentiation as required for regenerating periodontal tissues.

differentiated marker expression.[96] Recent work by Ma and coworkers suggests potential of using nanofibrous scaffolds combined with PLGA microspheres for periodontal repair (Figure 29.3)

IV. PROTEIN DELIVERY FOR PERIODONTAL TISSUE ENGINEERING

In order to enhance the *in vivo* efficacy, the incorporation of bioactive molecules into scaffolding materials may facilitate sustained factor release for periods of time. There are two basic modes of incorporating bioactive molecules into the scaffolds at the time of[97,98] or after the fabrication.[99] Bioactive molecules incorporated directly into a bioresorbable scaffold are generally released by a diffusion-controlled mechanism that is regulated by the pore sizes such that different pore sizes affect the tortuosity of the scaffold and thereby control the release of protein.[100] The rate of bioactive molecules released depends on the type and rate of degradation of the delivery device, and the rate of bioactive molecules diffusion through pores of the scaffolds. Wei et al. has incorporated human parathyroid hormone (PTH 1–34) into biodegradable PLGA microspheres and demonstrated that PTH could be released in a controlled fashion.[9] These polymer microspheres were able to preserve the bioactivity of the PTH as evidenced by the stimulated release of cAMP by rat osteosarcoma cells *in vitro*, as well as increased serum calcium levels when injected subcutaneously into mice. Murphy et al.[8] has shown the release of vascular endothelial growth factor (VEGF, a potent mitogen specific for endothelial cells) from PLGA scaffolds for up to 15 days. Furthermore, enhanced vascularization in the scaffold has been observed when the VEGF incorporated polymers seeded with endothelial cells were implanted *in vivo*.[101] In addition, it has been shown that IGF-1/TGFβ1 incorporated in PLGA microspheres is gradually released over 15 days.[102] Furthermore, *in vitro* culture of IGF-1/TGFβ1 containing microspheres photoencapsulated with chondrocytes in a hydrogel system resulted in increased cell numbers and glycosaminoglycan production compared to the control gels either without microspheres at 2 weeks.[102] Recently Lutolf et al.[103] have shown that hydrogels that contain cell adhesive molecule motif (RGD) and substrates for matrix metalloproteinases (MMPs) as linkers to polymer chains work as biomimetic scaffolds. This approach works well since cell invasion into the scaffolds regulates the release of BMPs to heal craniofacial defects. However, in periodontal application, it has been shown that delivery of PDGF/insulin-like growth factor-I (IGF-I) using methylcellulose carriers in treating periodontal defects had a shorter half-life (3 to 4 h).[104] Development of a suitable controlled release of the bioactive molecules devices and/or an alternative mode of delivery is needed.

Several bioactive molecules have been utilized in periodontal wounds in preclinical and clinical studies. These bioactive molecules include PDGF, IGF-I, basic fibroblast growth factor (FGF-2), TGF-β1, BMPs-2 and -7, and EMD. The details of preclinical and clinical studies have been reviewed.[105] In general, the results from the *in vivo* studies have shown that all of the above mentioned bioactive molecules, except TGF-β exhibited an ability to promote periodontal tissue regeneration. BMPs showed the greatest potential, since they are potent osteoinductive mediators. In addition, combination of PDGF/IGF-I accelerated more pronounced regenerative outcomes compared to the use of each GF alone.[106] Future therapy using various combinations of GFs promoting optimal periodontal tissue regeneration warrant further studies.

V. CELL DELIVERY FOR PERIODONTAL ENGINEERING

To date, cell therapy has just recently been employed to periodontology. Recent applications for the use of development and characterization of tissue-engineered human oral mucosa equivalent have been described by Izumi et al.[107] Thus the premise that expansion of cells in an *ex vivo* environment can be propagated *in vitro* for subsequent transplantation in areas of insufficient donor tissue.

Somerman and colleagues[108] have demonstrated the potential usage of cell therapy with cloned cementoblasts. In a series of reports, the group has cloned and characterized a cementoblast cell line which possesses many of the phenotypic characteristics of tooth lining cells *in vivo*. More recently it has been demonstrated that transplantation of cloned cementoblasts in PLGA carriers leads to the repair of large periodontal alveolar bone defects in rodents.[109] The regeneration achieved was only limited by the size of the transplanted PLGA carrier. Jin et al. has also shown that skin fibroblasts transduced by the BMP-7 gene promoted the tissue engineering of periodontal bone defects including new bone, functional PDL, and tooth root cementum[110] (Figure 29.4). By utilizing

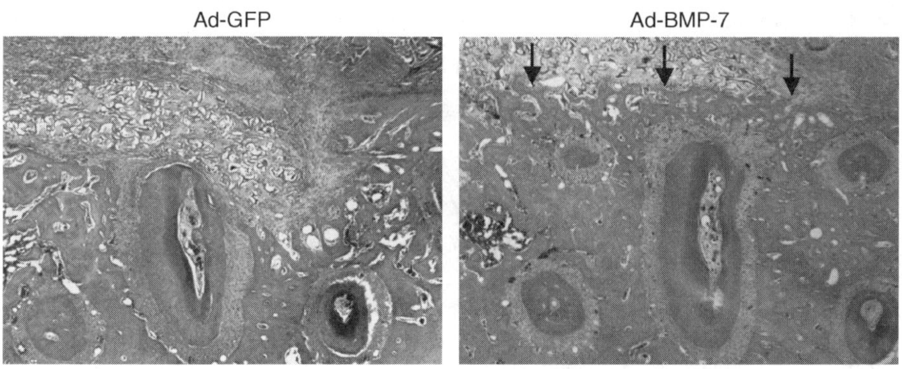

FIGURE 29.4 The microphotographs of rat large periodontal defects treated with skin fibroblasts transduced by adenovirus encoding a control (green fluorescent protein [Ad-GFP]) and bone morphogenetic protein-7 (Ad-BMP-7) for 35 days. In Ad-GFP treated specimens, little bone formation was seen (left panel), while bone formation bridging the defect was found (arrow) in Ad-BMP-7 treated group (right panel). In addition, Ad-BMP-7 treatment also induced cementum formation and vertical periodontal ligament formation.

strategies to better understand the basic biologic mechanisms involved in cementogenesis, optimal therapies for periodontal wound healing may be developed.

VI. GENE DELIVERY IN PERIODONTAL TISSUE ENGINEERING

Gene therapy is an alternative mode of targeting bioactive molecules via incorporating DNA in the delivery system in order to prolong the bioavailability of such molecules for longer time periods than those delivered by protein.[111,112] The DNA sequences of genes of interest can be inserted in many vector systems, such as plasmids, adenovirus, adeno-associated virus, and retrovirus. Application of each type of vector is determined by the duration of delivery of such biomolecules and the adverse effects of each vector. Development of DNA incorporated in the scaffolds has to allow for transfection of sufficient cells to produce inductive doses of the desired protein.[82] For example, the incorporation of plasmid DNA encoding PDGF in PLGA scaffolds showed encouraging results in the sustained release of plasmid DNA from matrices and led to cell transfection and sustained production of PDGF for up to 28 days or longer. In addition, *in vivo* delivery of the PDGF plasmid DNA promoted matrix deposition and blood vessel formation in the developing tissue for 4 weeks.[82] Moreover, combination DNA and cell therapy could be performed by transducing cells with DNA to obtain optimal numbers of transduced cells and seeding the cells in the scaffolds. The cell/gene scaffolds could then be implanted in the wound site to provide a sustained released for regenerating of tissues and organs. We have successfully used direct gene delivery to repair *ex vivo* gingival wounds,[113] and periodontal defects.[114] In addition, the use of combined cell and gene therapy (*ex vivo* gene transfer) has also shown significant benefits in cementum engineering[115,116] and for periodontal tissue engineering.[110]

VII. FUTURE APPLICATIONS

The future development of biomimetic scaffolds for periodontal tissue engineering appears to be an important area in periodontology for new therapies to treat patients. Strategies to develop "smart" devices that are osteoconductive, space maintaining, and bioactive may be the combination needed to optimize therapy. Exciting recent studies using biomimetic scaffolds for protein, gene, and cell therapy offer a bright future for periodontology.

ACKNOWLEDGMENTS

This study was supported by NIH/NIDCR grants DE 11960, DE 13397, DE 14755, and DE 15384.

REFERENCES

1. Williams, R. C., Periodontal disease, *N. Engl. J. Med.*, 322, 373–382, 1990.
2. Saygin, N. E., Giannobile, W. V., and Somerman, M. J., Molecular and cell biology of cementum, *Periodontology 2000*, 24, 73–98, 2000.
3. Warren, S. M., Fong, K. D., Chen, C. M., Loboa, E. G., Cowan, C. M., Lorenz, H. P., and Longaker, M. T., Tools and techniques for craniofacial tissue engineering, *Tissue Eng.*, 9, 187–200, 2003.
4. Hulbert, S. F., Young, F. A., Mathews, R. S., Klawitter, J. J., Talbert, C. D., and Stelling, F. H., Potential of ceramic materials as permanently implantable skeletal prostheses, *J. Biomed. Mater. Res.*, 4, 433–456, 1970.
5. Daculsi, G., and Passuti, N., Effect of the macroporosity for osseous substitution of calcium phosphate ceramics, *Biomaterials*, 11, 86–87, 1990.
6. Ma, P. X., and Choi, J. W., Biodegradable polymer scaffolds with well-defined interconnected spherical pore network, *Tissue Eng.*, 7, 23–33, 2001.

7. Ma, P. X., and Zhang, R., Synthetic nano-scale fibrous extracellular matrix, *J. Biomed. Mater. Res.*, 46, 60–72, 1999.
8. Murphy, W. L., Peters, M. C., Kohn, D. H., and Mooney, D. J., Sustained release of vascular endothelial growth factor from mineralized poly(lactide-co-glycolide) scaffolds for tissue engineering, *Biomaterials*, 21, 2521–2527, 2000.
9. Wei, G., Pettway, G. J., McCauley, L. K., and Ma, P. X., The release profiles and bioactivity of parathyroid hormone from poly(lactic-co-glycolic acid) microspheres, *Biomaterials*, 25, 345–352, 2004.
10. Mooney, D. J., Sano, K., Kaufmann, P. M., Majahod, K., Schloo, B., Vacanti, J. P., and Langer, R., Long-term engraftment of hepatocytes transplanted on biodegradable polymer sponges, *J. Biomed. Mater. Res.*, 37, 413–420, 1997.
11. Kost, J., and Langer, R., Responsive polymeric delivery systems, *Adv. Drug Deliv. Rev.*, 46, 125–148, 2001.
12. Santini, J. T. Jr., Cima, M. J., and Langer, R., A controlled-release microchip, *Nature*, 397, 335–338, 1999.
13. Urist, M. R., Bone: formation by autoinduction, *Science*, 150, 893–899, 1965.
14. Russell, J. L., Grafton demineralized bone matrix: performance consistency, utility, and value, *Tissue Eng.*, 6, 435–440, 2000.
15. Bowers, G. M., Chadroff, B., Carnevale, R., Mellonig, J., Corio, R., Emerson, J., Stevens, M., and Romberg, E., Histologic evaluation of new attachment apparatus formation in humans. Part III, *J. Periodontol.*, 60, 683–693, 1989.
16. Mellonig, J. T., Prewett, A. B., and Moyer, M. P., HIV inactivation in a bone allograft, *J. Periodontol.*, 63, 979–983, 1992.
17. Scarborough, N. L., White, E. M., Hughes, J. V., Manrique, A. J., and Poser, J. W., Allograft safety: viral inactivation with bone demineralization, *Contemp. Orthop.*, 31, 257–261, 1995.
18. Schwartz, Z., Mellonig, J. T., Carnes, D. L. Jr., de la Fontaine, J., Cochran, D. L., Dean, D. D., and Boyan, B. D., Ability of commercial demineralized freeze-dried bone allograft to induce new bone formation, *J. Periodontol.*, 67, 918–926, 1996.
19. Houser, B. E., Mellonig, J. T., Brunsvold, M. A., Cochran, D. L., Meffert, R. M., and Alder, M. E., Clinical evaluation of anorganic bovine bone xenograft with a bioabsorbable collagen barrier in the treatment of molar furcation defects, *Int. J. Periodont. Restorative Dent.*, 21, 161–169, 2001.
20. Mellonig, J. T., Human histologic evaluation of a bovine-derived bone xenograft in the treatment of periodontal osseous defects, *Int. J. Periodont. Restorative Dent.*, 20, 19–29, 2000.
21. Richardson, C. R., Mellonig, J. T., Brunsvold, M. A., McDonnell, H. T., and Cochran, D. L., Clinical evaluation of Bio-Oss: a bovine-derived xenograft for the treatment of periodontal osseous defects in humans, *J. Clin. Periodontol.*, 26, 421–428, 1999.
22. Scheyer, E. T., Velasquez-Plata, D., Brunsvold, M. A., Lasho, D. J., and Mellonig, J. T., A clinical comparison of a bovine-derived xenograft used alone and in combination with enamel matrix derivative for the treatment of periodontal osseous defects in humans, *J. Periodontol.*, 73, 423–432, 2002.
23. Terheyden, H., Knak, C., Jepsen, S., Palmie, S., and Rueger, D. R., Mandibular reconstruction with a prefabricated vascularized bone graft using recombinant human osteogenic protein-1: an experimental study in miniature pigs. Part I: prefabrication, *Int. J. Oral Maxillofac. Surg.*, 30, 373–379, 2001.
24. Camelo, M., Nevins, M. L., Schenk, R. K., Simion, M., Rasperini, G., Lynch, S. E., and Nevins, M., Clinical, radiographic, and histologic evaluation of human periodontal defects treated with Bio-Oss and Bio-Gide, *Int. J. Periodont. Restorative Dent.*, 18, 321–331, 1998.
25. Nevins, M. L., Camelo, M., Nevins, M., King, C. J., Oringer, R. J., Schenk, R. K., and Fiorellini, J. P., Human histologic evaluation of bioactive ceramic in the treatment of periodontal osseous defects, *Int. J. Periodont. Restorative Dent.*, 20, 458–467, 2000.
26. Qian, J. J., and Bhatnagar, R. S., Enhanced cell attachment to anorganic bone mineral in the presence of a synthetic peptide related to collagen, *J. Biomed. Mater. Res.*, 31, 545–554, 1996.
27. Bhatnagar, R. S., Qian, J. J., Wedrychowska, A., Sadeghi, M., Wu, Y. M., and Smith, N., Design of biomimetic habitats for tissue engineering with P-15, a synthetic peptide analogue of collagen, *Tissue Eng.*, 5, 53–65, 1999.

28. Yukna, R., Salinas, T. J., and Carr, R. F., Periodontal regeneration following use of ABM/P-1 5: a case report, *Int. J. Periodont. Restorative Dent.*, 22, 146–155, 2002.
29. Yukna, R. A., Callan, D. P., Krauser, J. T., Evans, G. H., Aichelmann-Reidy, M. E., Moore, K., Cruz, R., and Scott, J. B., Multi-center clinical evaluation of combination anorganic bovine-derived hydroxyapatite matrix (ABM)/cell binding peptide (P-15) as a bone replacement graft material in human periodontal osseous defects. 6-month results, *J. Periodontol.*, 69, 655–663, 1998.
30. Barboza, E. P., de Souza, R. O., Caula, A. L., Neto, L. G., Caula Fde, O., and Duarte, M. E., Bone regeneration of localized chronic alveolar defects utilizing cell binding peptide associated with anorganic bovine-derived bone mineral: a clinical and histological study, *J. Periodontol.*, 73, 1153–1159, 2002.
31. Melcher, A. H., On the repair potential of periodontal tissues, *J. Periodontol.*, 47, 256–260, 1976.
32. Wang, H.-L., and Giannobile, W. V., Guided bone regeneration utilizing absorbable collagen membranes. *Proceedings of Fifth World Congress for Oral Implantology: Biomechanics and Tissue Engineering*, Tokyo, pp. Ke-4, 2001.
33. Murphy, W. L., and Mooney, D. J., Controlled delivery of inductive proteins, plasmid DNA and cells from tissue engineering matrices, *J. Periodontal Res.*, 34, 413–419, 1999.
34. Needleman, I., Tucker, R., Giedrys-Leeper, E., and Worthington, H., A systematic review of guided tissue regeneration for periodontal infrabony defects, *J. Periodontal Res.*, 37, 380–388, 2002.
35. Jepsen, S., Eberhard, J., Herrera, D., and Needleman, I., A systematic review of guided tissue regeneration for periodontal furcation defects. What is the effect of guided tissue regeneration compared with surgical debridement in the treatment of furcation defects?, *J. Clin. Periodontol.*, 29(Suppl. 3), 103–116, 2002.
36. Machtei, E. E., and Schallhorn, R. G., Successful regeneration of mandibular Class II furcation defects: an evidence-based treatment approach, *Int. J. Periodont. Restorative Dent.*, 15, 146–167, 1995.
37. Sanz, M., and Giovannoli, J. L., Focus on furcation defects: guided tissue regeneration, *Periodontology 2000*, 22, 169–189, 2000.
38. Bunyaratavej, P., and Wang, H. L., Collagen membranes: a review, *J. Periodontol.*, 72, 215–229, 2001.
39. Wang, H. L., and Carroll, W. J., Using absorbable collagen membranes for guided tissue regeneration, guided bone regeneration, and to treat gingival recession, *Compend Contin Educ. Dent.*, 21, 399–402, 2000, see also p. 404 and 406 passim; quiz 414.
40. Sato, K., Radiation sterilization of medical equipment, *Radioisotopes*, 32, 431–440, 1983.
41. Cortellini, P., and Tonetti, M. S., Focus on intrabony defects: guided tissue regeneration, *Periodontology*, 22, 104–132, 2000.
42. Oh, T. J., Meraw, S. J., Lee, E. J., Giannobile, W. V., and Wang, H. L., Comparative analysis of collagen membranes for the treatment of implant dehiscence defects, *Clin. Oral Implants Res.*, 14, 80–90, 2003.
43. Lee, Y. M., Nam, S. H., Seol, Y. J., Kim, T., Lee, S. J., Ku, Y., Rhyu, I. C., Chung, C. P., Han, S. B., and Choi, S. M., Enhanced bone augmentation by controlled release of recombinant human bone morphogenetic protein-2 from bioabsorbable membranes, *J. Periodontol.*, 74, 865–872, 2003.
44. Baier Leach, J., Bivens, K. A., Patrick, C. W. Jr., and Schmidt, C. E., Photocrosslinked hyaluronic acid hydrogels: natural, biodegradable tissue engineering scaffolds, *Biotechnol. Bioeng.*, 82, 578–589, 2003.
45. Collier, J. H., Camp, J. P., Hudson, T. W., and Schmidt, C. E., Synthesis and characterization of polypyrrole-hyaluronic acid composite biomaterials for tissue engineering applications, *J. Biomed. Mater. Res.*, 50, 574–584, 2000.
46. Bulpitt, P., and Aeschlimann, D., New strategy for chemical modification of hyaluronic acid: preparation of functionalized derivatives and their use in the formation of novel biocompatible hydrogels, *J. Biomed. Mater. Res.*, 47, 152–169, 1999.
47. Laurent, T. C., and Fraser, J. R., Hyaluronan, *FASEB J.*, 6, 2397–2404, 1992.
48. Knudson, W., Casey, B., Nishida, Y., Eger, W., Kuettner, K. E., and Knudson, C. B., Hyaluronan oligosaccharides perturb cartilage matrix homeostasis and induce chondrocytic chondrolysis, *Arthritis Rheum.*, 43, 1165–1174, 2000.

49. Chen, W. Y., and Abatangelo, G., Functions of hyaluronan in wound repair, *Wound Repair Regen.*, 7, 79–89, 1999.
50. Sasaki, T., and Watanabe, C., Stimulation of osteoinduction in bone wound healing by high-molecular hyaluronic acid, *Bone*, 16, 9–15, 1995.
51. Caplan, A. I., Tissue engineering designs for the future: new logics, old molecules, *Tissue Eng.*, 6, 1–8, 2000.
52. Wikesjo, U. M., Lim, W. H., Thomson, R. C., Cook, A. D., Wozney, J. M., and Hardwick, W. R., Periodontal repair in dogs: evaluation of a bioabsorbable space-providing macroporous membrane with recombinant human bone morphogenetic protein-2, *J. Periodontol.*, 74, 635–647, 2003.
53. Sasaki, T., and Kawamata-Kido, H., Providing an environment for reparative dentine induction in amputated rat molar pulp by high molecular-weight hyaluronic acid, *Arch. Oral Biol.*, 40, 209–219, 1995.
54. Radomsky, M. L., Aufdemorte, T. B., Swain, L. D., Fox, W. C., Spiro, R. C., and Poser, J. W., Novel formulation of fibroblast growth factor-2 in a hyaluronan gel accelerates fracture healing in nonhuman primates, *J. Orthop. Res.*, 17, 607–614, 1999.
55. Kim, H. D., and Valentini, R. F., Retention and activity of BMP-2 in hyaluronic acid-based scaffolds in vitro, *J. Biomed. Mater. Res.*, 59, 573–584, 2002.
56. Kilpadi, K. L., Chang, P. L., and Bellis, S. L., Hydroxylapatite binds more serum proteins, purified integrins, and osteoblast precursor cells than titanium or steel, *J. Biomed. Mater. Res.*, 57, 258–267, 2001.
57. Rosengren, A., Pavlovic, E., Oscarsson, S., Krajewski, A., Ravaglioli, A., and Piancastelli, A., Plasma protein adsorption pattern on characterized ceramic biomaterials, *Biomaterials*, 23, 1237–1247, 2002.
58. Dorozhkin, S. V., and Epple, M., Biological and medical significance of calcium phosphates, *Angew. Chem. Int. Ed. Engl.*, 41, 3130–3146, 2002.
59. Yukna, R. A., Mayer, E. T., and Amos, S. M., 5-year evaluation of durapatite ceramic alloplastic implants in periodontal osseous defects, *J. Periodontol.*, 60, 544–551, 1989.
60. Stahl, S. S., and Froum, S. J., Histologic and clinical responses to porous hydroxylapatite implants in human periodontal defects. Three to twelve months postimplantation, *J. Periodontol.*, 58, 689–695, 1987.
61. Yukna, R. A., Harrison, B. G., Caudill, R. F., Evans, G. H., Mayer, E. T., and Miller, S., Evaluation of durapatite ceramic as an alloplastic implant in periodontal osseous defects. II. Twelve month reentry results, *J. Periodontol.*, 56, 540–547, 1985.
62. Rabalais, M. L. Jr., Yukna, R. A., and Mayer, E. T., Evaluation of durapatite ceramic as an alloplastic implant in periodontal osseous defects. I. Initial six-month results, *J. Periodontol.*, 52, 680–689, 1981.
63. Jinno, T., Davy, D. T., and Goldberg, V. M., Comparison of hydroxyapatite and hydroxyapatite tricalcium-phosphate coatings, *J. Arthroplasty.*, 17, 902–909, 2002.
64. Zechner, W., Tangl, S., Furst, G., Tepper, G., Thams, U., Mailath, G., and Watzek, G., Osseous healing characteristics of three different implant types, *Clin. Oral Implants Res.*, 14, 150–157, 2003.
65. Trombelli, L., Heitz-Mayfield, L. J., Needleman, I., Moles, D., and Scabbia, A., A systematic review of graft materials and biological agents for periodontal intraosseous defects, *J. Clin. Periodontol.*, 29(Suppl. 3), 117–135, 2002, discussion 160–112.
66. Rafter, M., Baker, M., Alves, M., Daniel, J., and Remeikis, N., Evaluation of healing with use of an internal matrix to repair furcation perforations, *Int. Endod. J.*, 35, 775–783, 2002.
67. Gille, J., Dorn, B., Kekow, J., Bruns, J., and Behrens, P., Bone substitutes as carriers for transforming growth factor-beta(1) (TGF-beta(1)), *Int. Orthop.*, 26, 203–206, 2002.
68. Jin, Q. M., Takita, H., Kohgo, T., Atsumi, K., Itoh, H., and Kuboki, Y., Effects of geometry of hydroxyapatite as a cell substratum in BMP-induced ectopic bone formation, *J. Biomed. Mater. Res.*, 52, 491–499, 2000.
69. Arm, D. M., Tencer, A. F., Bain, S. D., and Celino, D., Effect of controlled release of platelet-derived growth factor from a porous hydroxyapatite implant on bone ingrowth, *Biomaterials*, 17, 703–709, 1996.

70. Takebe, J., Itoh, S., Okada, J., and Ishibashi, K., Anodic oxidation and hydrothermal treatment of titanium results in a surface that causes increased attachment and altered cytoskeletal morphology of rat bone marrow stromal cells in vitro, *J. Biomed. Mater. Res.*, 51, 398–407, 2000.
71. Yukna, R. A., Evans, G. H., Aichelmann-Reidy, M. B., and Mayer, E. T., Clinical comparison of bioactive glass bone replacement graft material and expanded polytetrafluoroethylene barrier membrane in treating human mandibular molar class II furcations, *J. Periodontol.*, 72, 125–133, 2001.
72. Froum, S. J., Weinberg, M. A., and Tarnow, D., Comparison of bioactive glass synthetic bone graft particles and open debridement in the treatment of human periodontal defects. A clinical study, *J. Periodontol.*, 69, 698–709, 1998.
73. Lovelace, T. B., Mellonig, J. T., Meffert, R. M., Jones, A. A., Nummikoski, P. V., and Cochran, D. L., Clinical evaluation of bioactive glass in the treatment of periodontal osseous defects in humans, *J. Periodontol.*, 69, 1027–1035, 1998.
74. Zamet, J. S., Darbar, U. R., Griffiths, G. S., Bulman, J. S., Bragger, U., Burgin, W., and Newman, H. N., Particulate bioglass as a grafting material in the treatment of periodontal intrabony defects, *J. Clin. Periodontol.*, 24, 410–418, 1997.
75. Greenspan, D. C., Bioactive glass: mechanisms of bone bonding, *Swed. Dent. J.*, 91, 31–35, 1999.
76. Fetner, A. E., Hartigan, M. S., and Low, S. B., Periodontal repair using PerioGlas in nonhuman primates: clinical and histologic observations, *Compendium*, 15, 932 1994, see also pp. 935–938; quiz 939.
77. Karatzas, S., Zavras, A., Greenspan, D., and Amar, S., Histologic observations of periodontal wound healing after treatment with PerioGlas in nonhuman primates, *Int. J. Periodont. Restorative Dent.*, 19, 489–499, 1999.
78. Agrawal, C. M., McKinney, J. S., Lanctot, D., and Athanasiou, K. A., Effects of fluid flow on the in vitro degradation kinetics of biodegradable scaffolds for tissue engineering, *Biomaterials*, 21, 2443–2452, 2000.
79. Cai, Q., Shi, G., Bei, J., and Wang, S., Enzymatic degradation behavior and mechanism of poly(lactide-co-glycolide) foams by trypsin, *Biomaterials*, 24, 629–638, 2003.
80. Athanasiou, K. A., Agrawal, C. M., Barber, F. A., and Burkhart, S. S., Orthopaedic applications for PLA-PGA biodegradable polymers, *Arthroscopy*, 14, 726–737, 1998.
81. Hutmacher, D. W., Scaffolds in tissue engineering bone and cartilage, *Biomaterials*, 21, 2529–2543, 2000.
82. Shea, L. D., Smiley, E., Bonadio, J., and Mooney, D. J., DNA delivery from polymer matrices for tissue engineering, *Nat. Biotechnol.*, 17, 551–554, 1999.
83. Hadlock, T. A., Sheahan, T., Cheney, M. L., Vacanti, J. P., and Sundback, C. A., Biologic activity of nerve growth factor slowly released from microspheres, *J. Reconstr. Microsurg.*, 19, 179–184, 2003.
84. Matin, K., Senpuku, H., Hanada, N., Ozawa, H., and Ejiri, S., Bone regeneration by recombinant human bone morphogenetic protein-2 around immediate implants: a pilot study in rats, *Int. J. Oral Maxillofac. Implants*, 18, 211–217, 2003.
85. Sheridan, M. H., Shea, L. D., Peters, M. C., and Mooney, D. J., Bioabsorbable polymer scaffolds for tissue engineering capable of sustained growth factor delivery, *J. Control. Release*, 64, 91–102, 2000.
86. Bryan, D. J., Tang, J. B., Holway, A. H., Rieger-Christ, K. M., Trantolo, D. J., Wise, D. L., and Summerhayes, I. C., Enhanced peripheral nerve regeneration elicited by cell-mediated events delivered via a bioresorbable PLGA guide, *J. Reconstr. Microsurg.*, 19, 125–134, 2003.
87. Shea, L. D., Wang, D., Franceschi, R. T., and Mooney, D. J., Engineered bone development from a pre-osteoblast cell line on three-dimensional scaffolds, *Tissue Eng.*, 6, 605–617, 2000.
88. Sherwood, J. K., Riley, S. L., Palazzolo, R., Brown, S. C., Monkhouse, D. C., Coates, M., Griffith, L. G., Landeen, L. K., and Ratcliffe, A., A three-dimensional osteochondral composite scaffold for articular cartilage repair, *Biomaterials*, 23, 4739–4751, 2002.
89. Williams, R. C., Paquette, D. W., Offenbacher, S., Adams, D. F., Armitage, G. C., Bray, K., Caton, J., Cochran, D. L., Drisko, C. H., Fiorellini, J. P., Giannobile, W. V., Grossi, S., Guerrero, D. M., Johnson, G. K., Lamster, I. B., Magnusson, I., Oringer, R. J., Persson, G. R., Van Dyke, T. E., Wolff, L. F., and Santucci, E. A., Treatment of periodontitis by local administration of minocycline microspheres: a controlled trial, *J. Periodontol.*, 72, 1535–1544, 2001.
90. Kinoshita, A., Oda, S., Takahashi, K., Yokota, S., and Ishikawa, I., Periodontal regeneration by application of recombinant human bone morphogenetic protein-2 to horizontal circumferential defects created by experimental periodontitis in beagle dogs, *J. Periodontol.*, 68, 103–109, 1997.

91. Kurtis, B., Unsal, B., Cetiner, D., Gultekin, E., Ozcan, G., Celebi, N., and Ocak, O., Effect of polylactide/glycolide (PLGA) membranes loaded with metronidazole on periodontal regeneration following guided tissue regeneration in dogs, *J. Periodontol.*, 73, 694–700, 2002.
92. Jin, Q., Zhao, M., Webb, S. A., Berry, J. E., Somerman, M. J., and Giannobile, W. V., Cementum engineering using three-dimensional polymer scaffolds, *J. Biomed. Mater. Res.*, 67A, 54–60, 2003.
93. Young, C. S., Terada, S., Vacanti, J. P., Honda, M., Bartlett, J. D., and Yelick, P. C., Tissue engineering of complex tooth structures on biodegradable polymer scaffolds, *J. Dent. Res.*, 81, 695–700, 2002.
94. Zhang, R., and Ma, P. X., Poly(alpha-hydroxyl acids)/hydroxyapatite porous composites for bone-tissue engineering. I. Preparation and morphology, *J. Biomed. Mater. Res.*, 44, 446–455, 1999.
95. Zhang, R., and Ma, P. X., Porous poly(L-lactic acid)/apatite composites created by biomimetic process, *J. Biomed. Mater. Res.*, 45, 285–293, 1999.
96. Ma, P. X., Zhang, R., Xiao, G., and Franceschi, R., Engineering new bone tissue in vitro on highly porous poly(alpha-hydroxyl acids)/hydroxyapatite composite scaffolds, *J. Biomed. Mater. Res.*, 54, 284–293, 2001.
97. Lo, H., Kadiyala, S., Guggino, S. E., and Leong, K. W., Poly(L-lactic acid) foams with cell seeding and controlled-release capacity, *J. Biomed. Mater. Res.*, 30, 475–484, 1996.
98. Whang, K., Tsai, D. C., Nam, E. K., Aitken, M., Sprague, S. M., Patel, P. K., and Healy, K. E., Ectopic bone formation via rhBMP-2 delivery from porous bioabsorbable polymer scaffolds, *J. Biomed. Mater. Res.*, 42, 491–499, 1998.
99. Fournier, N., and Doillon, C. J., Biological molecule-impregnated polyester: an in vivo angiogenesis study, *Biomaterials*, 17, 1659–1665, 1996.
100. Babensee, J. E., McIntire, L. V., and Mikos, A. G., Growth factor delivery for tissue engineering, *Pharm. Res.*, 17, 497–504, 2000.
101. Peters, M. C., Polverini, P. J., and Mooney, D. J., Engineering vascular networks in porous polymer matrices, *J. Biomed. Mater. Res.*, 60, 668–678, 2002.
102. Elisseeff, J., McIntosh, W., Fu, K., Blunk, B. T., and Langer, R., Controlled-release of IGF-I and TGF-beta1 in a photopolymerizing hydrogel for cartilage tissue engineering, *J. Orthop. Res.*, 19, 1098–1104, 2001.
103. Lutolf, M. P., Weber, F. E., Schmoekel, H. G., Schense, J. C., Kohler, T., Muller, R., and Hubbell, J. A., Repair of bone defects using synthetic mimetics of collagenous extracellular matrices, *Nat. Biotechnol.*, 21, 513–518, 2003.
104. Lynch, S. E., de Castilla, G. R., Williams, R. C., Kiritsy, C. P., Howell, T. H., Reddy, M. S., and Antoniades, H. N., The effects of short-term application of a combination of platelet-derived and insulin-like growth factors on periodontal wound healing, *J. Periodontol.*, 62, 458–467, 1991.
105. Anusaksathien, O., and Giannobile, W. V., Growth factor delivery to re-engineer periodontal tissues, *Curr. Pharm. Biotechnol.*, 3, 129–139, 2002.
106. Giannobile, W. V., Hernandez, R. A., Finkelman, R. D., Ryan, S., Kiritsy, C. P., D'Andrea, M., and Lynch, S. E., Comparative effects of platelet-derived growth factor-BB and insulin-like growth factor-I, individually and in combination, on periodontal regeneration in Macaca fascicularis, *J. Periodontal Res.*, 31, 301–312, 1996.
107. Izumi, K., Feinberg, S. E., Iida, A., and Yoshizawa, M., Intraoral grafting of an ex vivo produced oral mucosa equivalent: a preliminary report, *Int. J. Oral Maxillofac. Surg.*, 32, 188–197, 2003.
108. Somerman, M. J., Ouyang, H. J., Berry, J. E., Saygin, N. E., Strayhorn, C. L., D'Errico, J. A., Hullinger, T., and Giannobile, W. V., Evolution of periodontal regeneration: from the roots' point of view, *J. Periodontal Res.*, 34, 420–424, 1999.
109. Zhao, M., Jin, Q.-M., Berry, J. E., Nociti, F. H. J., Giannobile, W. V., and Somerman, M., Periodontal tissue engineering by cementoblastic cell therapy, *J. Periodontol.*, 75, 154–161, 2004.
110. Jin, Q.-M., Anusaksathien, O., Webb, S. A., Rutherford, R. B., and Giannobile, W. V., Gene therapy of bone morphogenetic protein for periodontal tissue engineering, *J. Periodontol.*, 74, 202–213, 2003.
111. Doukas, J., Chandler, L. A., Gonzalez, A. M., Gu, D., Hoganson, D. K., Ma, C., Nguyen, T., Printz, M. A., Nesbit, M., Herlyn, M., Crombleholme, T. M., Aukerman, S. L., Sosnowski, B. A., and Pierce, G. F., Matrix immobilization enhances the tissue repair activity of growth factor gene therapy vectors, *Hum. Gene Ther.*, 12, 783–798, 2001.

112. Lieberman, J. R., Le, L. Q., Wu, L., Finerman, G. A., Berk, A., Witte, O. N., and Stevenson, S., Regional gene therapy with a BMP-2-producing murine stromal cell line induces heterotopic and orthotopic bone formation in rodents, *J. Orthop. Res.*, 16, 330–339, 1998.
113. Anusaksathien, O., Webb, S. A., Jin, Q., and Giannobile, W. V., PDGF gene delivery stimulates ex vivo gingival repair, *Tissue Eng.*, 9, 745–756, 2003.
114. Jin, Q.-M., Anusaksathien, O., Webb, S. A., and Giannobile, W. V., Platelet-derived growth factor gene delivery promotes periodontal tissue repair (abstr), *Mol. Ther.*, 7(Part 2), S399 2003.
115. Anusaksathien, O, Jin, Q-M, Zhao, M, Somerman, MJ, and Giannobile, WV. Effect of sustained gene delivery of platelet-derived growth factor (PDGF) or its antagonist (PDGF-1308) on tissue-engineered cementum, *J. Periodontol.,* 75, 429–440, 2004.
116. Jin, Q-M, Zhao, M, Economides, AN, Somerman, MJ, and Giannobile, WV. Noggin gene delivery inhibits cementoblast-induced mineralization, *Connect. Tissue Res.*, 45, 50–59, 2004.
117. Blumenthal, N. M., Koh-Kunst, G., Alves, M. E., Miranda, D., Sorensen, R. G., Wozney, J. M., and Wikesjo, U. M., Effect of surgical implantation of recombinant human bone morphogenetic protein-2 in a bioabsorbable collagen sponge or calcium phosphate putty carrier in intrabony periodontal defects in the baboon, *J. Periodontol.*, 73, 1494–1506, 2002.
118. Anderegg, C. R., Martin, S. J., Gray, J. L., Mellonig, J. T., and Gher, M. E., Clinical evaluation of the use of decalcified freeze-dried bone allograft with guided tissue regeneration in the treatment of molar furcation invasions, *J. Periodontol.*, 62, 264–268, 1991.
119. Gager, A. H., and Schultz, A. J., Treatment of periodontal defects with an absorbable membrane (polyglactin 910) with and without osseous grafting: case reports, *J. Periodontol.*, 62, 276–283, 1991.
120. Rummelhart, J. M., Mellonig, J. T., Gray, J. L., and Towle, H. J., A comparison of freeze-dried bone allograft and demineralized freeze-dried bone allograft in human periodontal osseous defects, *J. Periodontol.*, 60, 655–663, 1989.
121. Camelo, M., Nevins, M. L., Lynch, S. E., Schenk, R. K., Simion, M., and Nevins, M., Periodontal regeneration with an autogenous bone-Bio-Oss composite graft and a Bio-Gide membrane, *Int. J. Periodont. Restorative Dent.*, 21, 109–119, 2001.
122. Nevins, M. L., Camelo, M., Lynch, S. E., Schenk, R. K., and Nevins, M., Evaluation of periodontal regeneration following grafting intrabony defects with Bio-Oss collagen: a human histologic report, *Int. J. Periodont. Restorative Dent.*, 23, 9–17, 2003.
123. Saffar, J. L., Colombier, M. L., and Detienville, R., Bone formation in tricalcium phosphate-filled periodontal intrabony lesions. Histological observations in humans, *J. Periodontol.*, 61, 209–216, 1990.
124. Stahl, S. S., and Froum, S., Histological evaluation of human intraosseous healing responses to the placement of tricalcium phosphate ceramic implants. I. Three to eight months, *J. Periodontol.*, 57, 211–217, 1986.
125. Wikesjo, U. M., Sorensen, R. G., Kinoshita, A., and Wozney, J. M., RhBMP-2/alphaBSM induces significant vertical alveolar ridge augmentation and dental implant osseointegration, *Clin. Implant Dent. Relat. Res.*, 4, 174–182, 2002.
126. Stahl, S. S., Froum, S. J., and Tarnow, D., Human clinical and histologic responses to the placement of HTR polymer particles in 11 intrabony lesions, *J. Periodontol.*, 61, 269–274, 1990.
127. Yukna, R. A., and Yukna, C. N., Six-year clinical evaluation of HTR synthetic bone grafts in human grade II molar furcations, *J. Periodontal Res.*, 32, 627–633, 1997.
128. Yukna, R. A., and Yukna, C. N., A 5-year follow-up of 16 patients treated with coralline calcium carbonate (BIOCORAL) bone replacement grafts in infrabony defects, *J. Clin. Periodontol.*, 25, 1036–1040, 1998.
129. Kim, C. K., Choi, E. J., Cho, K. S., Chai, J. K., and Wikesjo, U. M., Periodontal repair in intrabony defects treated with a calcium carbonate implant and guided tissue regeneration, *J. Periodontol.*, 67, 1301–1306, 1996.
130. Mora, F., and Ouhayoun, J. P., Clinical evaluation of natural coral and porous hydroxyapatite implants in periodontal bone lesions: results of a 1-year follow-up, *J. Clin. Periodontol.*, 22, 877–884, 1995.
131. Hanisch, O., Sorensen, R. G., Kinoshita, A., Spiekermann, H., Wozney, J. M., and Wikesjo, U. M., Effect of recombinant human bone morphogenetic protein-2 in dehiscence defects with non-submerged immediate implants: an experimental study in Cynomolgus monkeys, *J. Periodontol.*, 74, 648–657, 2003.

132. Selvig, K. A., Sorensen, R. G., Wozney, J. M., and Wikesjo, U. M., Bone repair following recombinant human bone morphogenetic protein-2 stimulated periodontal regeneration, *J. Periodontol.*, 73, 1020–1029, 2002.
133. Choi, S. H., Kim, C. K., Cho, K. S., Huh, J. S., Sorensen, R. G., Wozney, J. M., and Wikesjo, U. M., Effect of recombinant human bone morphogenetic protein-2/absorbable collagen sponge (rhBMP-2/ACS) on healing in 3-wall intrabony defects in dogs, *J. Periodontol.*, 73, 63–72, 2002.
134. Giannobile, W. V., Ryan, S., Shih, M. S., Su, D. L., Kaplan, P. L., and Chan, T. C., Recombinant human osteogenic protein-1 (OP-1) stimulates periodontal wound healing in class III furcation defects, *J. Periodontol.*, 69, 129–137, 1998.
135. Ripamonti, U., Heliotis, M., Rueger, D. C., and Sampath, T. K., Induction of cementogenesis by recombinant human osteogenic protein-1 (hop-1/bmp-7) in the baboon (Papio ursinus), *Arch. Oral Biol.*, 41, 121–126, 1996.
136. Howell, T. H., Fiorellini, J. P., Paquette, D. W., Offenbacher, S., Giannobile, W. V., and Lynch, S. E., A phase I/II clinical trial to evaluate a combination of recombinant human platelet-derived growth factor-BB and recombinant human insulin-like growth factor-I in patients with periodontal disease, *J. Periodontol.*, 68, 1186–1193, 1997.
137. Rutherford, R. B., Niekrash, C. E., Kennedy, J. E., and Charette, M. F., Platelet-derived and insulin-like growth factors stimulate regeneration of periodontal attachment in monkeys, *J. Periodontal Res.*, 27, 285–290, 1992.
138. Lynch, S. E., Williams, R. C., Polson, A. M., Howell, T. H., Reddy, M. S., Zappa, U. E., and Antoniades, H. N., A combination of platelet-derived and insulin-like growth factors enhances periodontal regeneration, *J. Clin. Periodontol.*, 16, 545–548, 1989.
139. Park, Y. J., Lee, Y. M., Park, S. N., Sheen, S. Y., Chung, C. P., and Lee, S. J., Platelet derived growth factor releasing chitosan sponge for periodontal bone regeneration, *Biomaterials*, 21, 153–159, 2000.
140. Okuda, K., Momose, M., Miyazaki, A., Murata, M., Yokoyama, S., Yonezawa, Y., Wolff, L. F., and Yoshie, H., Enamel matrix derivative in the treatment of human intrabony osseous defects, *J. Periodontol.*, 71, 1821–1828, 2000.
141. Parashis, A., and Tsiklakis, K., Clinical and radiographic findings following application of enamel matrix derivative in the treatment of intrabony defects. A series of case reports, *J. Clin. Periodontol.*, 27, 705–713, 2000.
142. Parodi, R., Liuzzo, G., Patrucco, P., Brunel, G., Santarelli, G. A., Birardi, V., and Gasparetto, B., Use of Emdogain in the treatment of deep intrabony defects: 12-month clinical results. Histologic and radiographic evaluation, *Int. J. Periodont. Restorative Dent.*, 20, 584–595, 2000.
143. Rasperini, G., Silvestri, M., Schenk, R. K., and Nevins, M. L., Clinical and histologic evaluation of human gingival recession treated with a subepithelial connective tissue graft and enamel matrix derivative (Emdogain): a case report, *Int. J. Periodont. Restorative Dent.*, 20, 269–275, 2000.
144. Sculean, A., Chiantella, G. C., Windisch, P., and Donos, N., Clinical and histologic evaluation of human intrabony defects treated with an enamel matrix protein derivative (Emdogain), *Int. J. Periodont. Restorative Dent.*, 20, 374–381, 2000.
145. Pontoriero, R., Wennstrom, J., and Lindhe, J., The use of barrier membranes and enamel matrix proteins in the treatment of angular bone defects. A prospective controlled clinical study, *J. Clin. Periodontol.*, 26, 833–840, 1999.
146. Caffesse, R. G., Mota, L. F., Quinones, C. R., and Morrison, E. C., Clinical comparison of resorbable and non-resorbable barriers for guided periodontal tissue regeneration, *J. Clin. Periodontol.*, 24, 747–752, 1997.
147. Metzler, D. G., Seamons, B. C., Mellonig, J. T., Gher, M. E., and Gray, J. L., Clinical evaluation of guided tissue regeneration in the treatment of maxillary class II molar furcation invasions, *J. Periodontol.*, 62, 353–360, 1991.
148. Gottlow, J., Nyman, S., Karring, T., and Lindhe, J., New attachment formation as the result of controlled tissue regeneration, *J. Clin. Periodontol.*, 11, 494–503, 1984.
149. Nyman, S., Lindhe, J., Karring, T., and Rylander, H., New attachment following surgical treatment of human periodontal disease, *J. Clin. Periodontol.*, 9, 290–296, 1982.

150. Christgau, M., Schmalz, G., Wenzel, A., and Hiller, K. A., Periodontal regeneration of intrabony defects with resorbable and non-resorbable membranes: 30-month results, *J. Clin. Periodontol.*, 24, 17–27, 1997.
151. Park, Y. J., Ku, Y., Chung, C. P., and Lee, S. J., Controlled release of platelet-derived growth factor from porous poly(L-lactide) membranes for guided tissue regeneration, *J. Control. Release*, 51, 201–211, 1998.
152. Chung, K. M., Salkin, L. M., Stein, M. D., and Freedman, A. L., Clinical evaluation of a biodegradable collagen membrane in guided tissue regeneration, *J. Periodontol.*, 61, 732–736, 1990.
153. Blumenthal, N., and Steinberg, J., The use of collagen membrane barriers in conjunction with combined demineralized bone-collagen gel implants in human infrabony defects, *J. Periodontol.*, 61, 319–327, 1990.
154. Pitaru, S., Tal, H., Soldinger, M., Azar-Avidan, O., and Noff, M., Collagen membranes prevent the apical migration of epithelium during periodontal wound healing, *J. Periodontal Res.*, 22, 331–333, 1987.
155. Orsini, M., Orsini, G., Benlloch, D., Aranda, J. J., Lazaro, P., Sanz, M., De Luca, M., and Piattelli, A., Comparison of calcium sulfate and autogenous bone graft to bioabsorbable membranes plus autogenous bone graft in the treatment of intrabony periodontal defects: a split-mouth study, *J. Periodontol.*, 72, 296–302, 2001.
156. Setya, A. B., and Bissada, N. F., Clinical evaluation of the use of calcium sulfate in regenerative periodontal surgery for the treatment of Class III furcation involvement, *Periodontal Clin. Invest.*, 21, 5–14, 1999.
157. Cho, K. S., Choi, S. H., Han, K. H., Chai, J. K., Wikesjo, U. M., and Kim, C. K., Alveolar bone formation at dental implant dehiscence defects following guided bone regeneration and xenogeneic freeze-dried demineralized bone matrix, *Clin. Oral Implants Res.*, 9, 419–428, 1998.
158. Anson, D., Calcium sulfate: a 4-year observation of its use as a resorbable, *Educ. Dent.*, 17, 895–899, 1996.

30 Tissue Engineering of Craniofacial Structure

Kacey G. Marra, Mark P. Mooney, and Jeffrey O. Hollinger

CONTENTS

I. FDA-APPROVED BIODEGRADABLE POLYMERS

Biodegradable polyesters have made a significant impact in medicine during the last century. Particularly, the usefulness of poly(lactic acid) (PLA) and poly(glycolic acid) (PGA), and copolymers thereof, has been demonstrated in myriad clinical applications. Their structures are depicted in Figure 30.1. These polymers are degraded hydrolytically and enzymatically, resulting in a breakdown of the polymer chain.

A. Poly(glycolic Acid)

Poly(glycolic acid) (PGA) is synthesized by the ring opening of glycolide; hence, PGA is also referred to as poly(glycolide). PGA is a highly insoluble polymer; hexafluoroisopropanol is one of PGA's only solvents, and this solvent is highly toxic. Therefore, PGA is more commonly utilized as a copolymer with PLA.

B. Poly(lactic Acid)

Poly(lactic acid) (PLA) is synthesized by the ring opening of lactide. As two enantiomeric isomers of lactide exist (e.g., D and L), both poly(L-lactic acid) (PLLA) and poly(D-lactic acid) (PDLA)

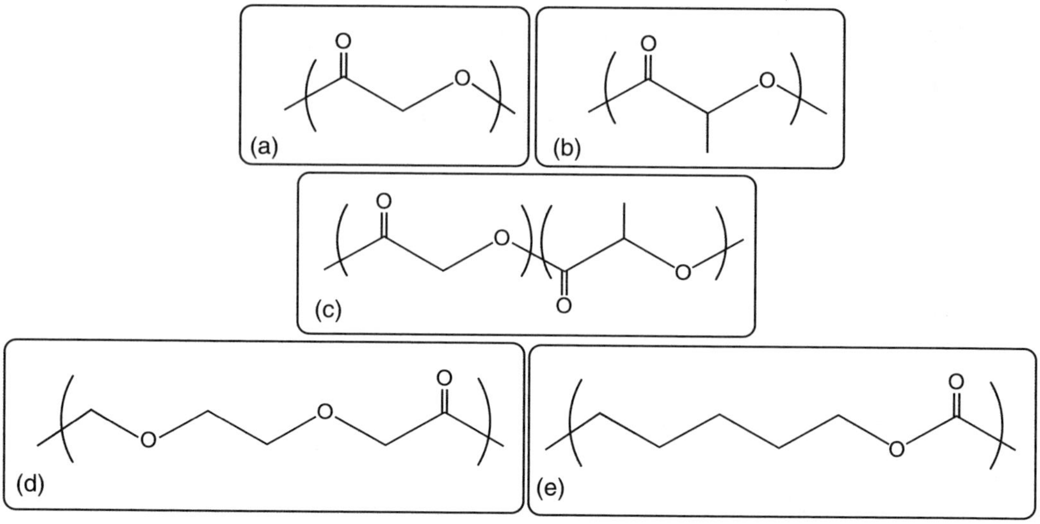

FIGURE 30.1 Structures of (a) PGA, (b) PLA, (c) PLGA, (d) polydioxanone, and (e) polycaprolactone.

have been examined as biomaterials. PLLA is more crystalline, and degrades more slowly than PDLA. PLLA was first used as plates and screws by Getter et al.[1] in the early 1970s in a canine mandibular fracture model, with promising results. Clinically, PLLA screws and polydioxanone (Figure 30.2) bands were examined in pediatric cranioplasties and cranial reconstructions by Illi et al.[2–4] in the early 1990s. Although initial clinical results of PLLA in zygomatic fractures was promising, patients exhibited swelling at the site of implantation as long as 6 years after implantation.[5]

A copolymerization of PDLA with PLLA results in PDLLA, which is being used clinically in craniofacial fixation applications. PDLLA is comprised of 70% PLLA and 30% PDLA, and is manufactured by Bionx Implants (as Biosorb), by Synthes Maxillofacial (as the Synthes Resorbable Fixation System), and by Macropore (as Macrosorb FX). These implants degrade *in vivo* within 1 to 6 years.[6] PDLLA is manufactured in sheets, plates, screws, and pins.

C. Poly(lactic-*co*-glycolic Acid)

The copolymer of PLA and PGA, poly(lactic-*co*-glycolic acid) (PLGA), has been extensively studied in a variety of tissue engineering applications. Of interest to craniofacial applications, particularly pediatric zygomatic, midfacial, and periorbital reconstruction, is the PLGA copolymer that consists of 82% PLLA and 18% PGA, referred to as LactoSorb. LactoSorb was introduced in 1996 and is being sold by Lorenz Surgical Inc. LactoSorb is completely resorbed in 12 months,[7] and has been used clinically in pediatric craniofacial reconstruction for 6 years.

Another clinically useful PLGA polymer is DeltaSystem, manufactured by Stryker-Leibinger. This tripolymer consists of 85% PLLA, 5% PDLLA, and 10% PGA, and maintains 81% of its mechanical strength 8 weeks after implantation. Complete resorption is between 1.5 and 3 years.[6]

D. Other Biodegradable and Native Polymers

Other biodegradable polymers that are being examined for craniofacial reconstruction include polydioxanone (Figure 30.1(d)) and polycaprolactone (Figure 30.1(e)). Polydioxanone (manufactured by Ethicon) has been examined clinically in cranial vault procedures with promising results.[8] Polycaprolactone is a polyester that has been examined in a porcine orbital defect model.[9]

(a)

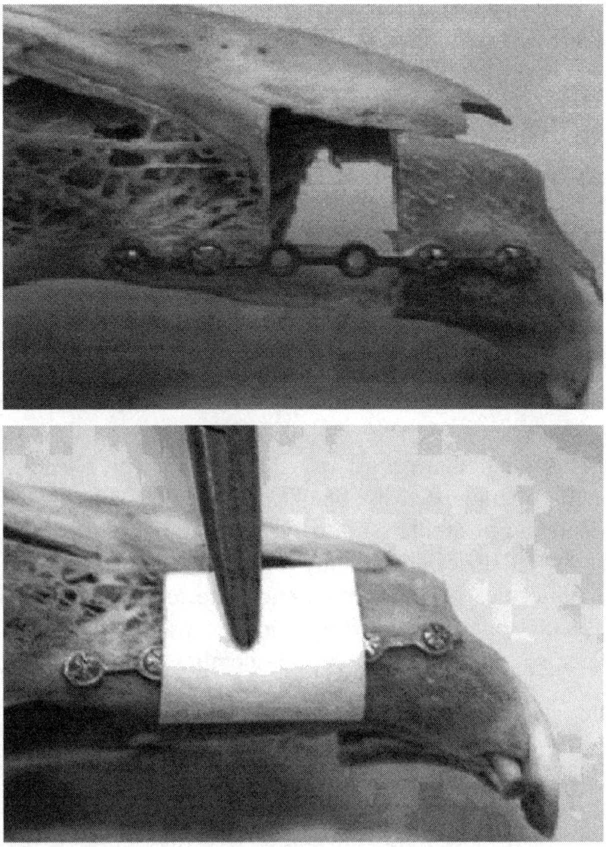

(b)

FIGURE 30.2 Cleaned and dried adult rabbit skull showing (a) the modified Le Fort I osteotomy procedure and rigid fixation, and (b) positioning of the collagen membrane used for guided tissue regeneration.

However, the most bone growth was observed when the polycaprolactone scaffold was seeded with bone marrow cells. The authors have recently described the seeding of autologous bone marrow in biodegradable scaffolds.[10] Caprotite, a physical blend of polycaprolactone, PLGA and hydroxyapatite, was fabricated into 80% porous constructs. The pore size range was 150 to 250 μm, and the 8 mm diameter disks were 1 mm thick. Caprotite seeded with autologous bone marrow was compared to unseeded Caprotite, autograft, and empty defect. Results revealed that defects repaired with seeded disks did not show significantly different bone area regeneration compared with negative control defects and unseeded disks, via three-dimensional computed tomography and histological analysis. However, defects repaired with seeded disks demonstrated small bony islands throughout the scaffold and no fibrous nonunions, while negative defects and defects repaired with unseeded disks demonstrated bony ingrowth only from the defect margins and fibrous nonunions in the center. These findings suggest that scaffold seeding promoted osteogenesis in the defects, although at a very slow rate, and future studies include modifying the concentration of seeded cells.

Bovine collagen membranes have been used for guided tissue regeneration (GTR) to retard fibrous tissue in growth and facilitate osseous regeneration following craniofacial reconstruction in clinical[11] and experimental models.[12–15] The development of fibrous nonunions after orthognathic surgery is thought to result from an interaction of biomechanical stress and the differential and more rapid migration of fibroblasts (compared with osteoblasts) into the wound site during healing. GTR is a technique used to facilitate wound healing by excluding

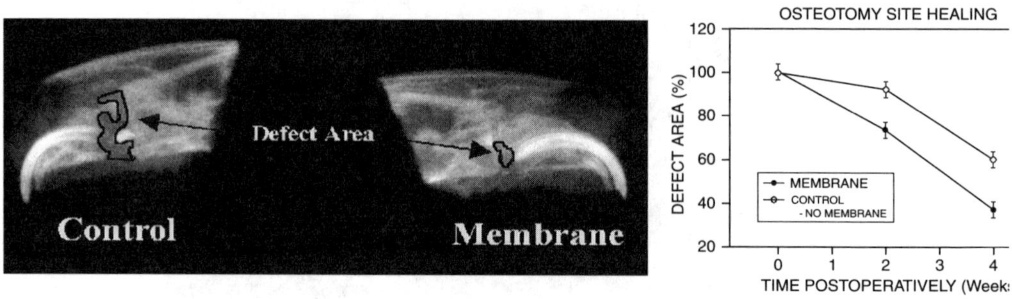

FIGURE 30.3 Lateral radiographs of the remaining defects in the Le Fort I osteotomy sites in rabbits at 4 weeks postop. Note the significantly enhanced new bone growth in the side protected with a collagen membrane. Graph shows defect area percentage in control defects and defects covered with membrane.

undesirable tissues. This method involves using a barrier, which excludes the unwanted cells but still allows the cells of interest to repopulate the healing site. GTR is used at wound sites and is inserted between the desired tissue to be regenerated and the surrounding tissues. The membrane functions as a barrier, depending on the pore size of the membrane used and the size of the local molecules or cells. In essence, the membrane works as a filter to prevent the flow of unwanted cells but still allows the flow of wanted cells and molecules that are of small enough diameter to migrate through the membranes pore size. This filtering process

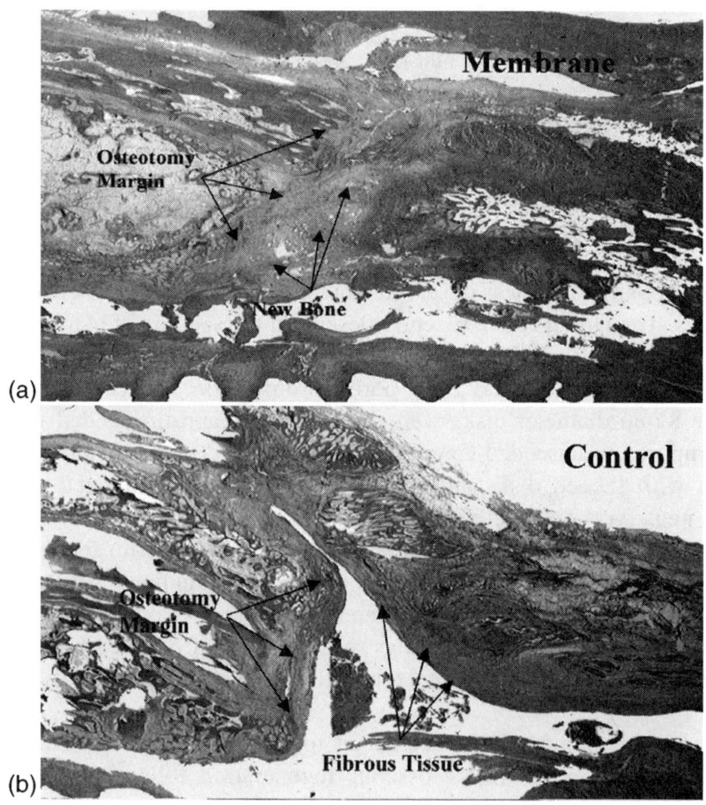

FIGURE 30.4 Histophotomicrographs (original magnification: × 32) of Le Fort I ostetotomies (a) covered, or (b) not covered with a resorbable collagen membrane in adult rabbits. Note the significant new bone formation in the membrane defect compared to the uncovered control side.

allows the preferred tissue to thrive and flourish at the wound site with out the impingement of unwanted and possibly inhibitory cells and tissues. This results in the repopulation of the wound site with the preferred tissue to complete the wound healing process at an accelerated rate.[11–15] A series of studies by Mooney and colleagues has shown that rigid fixation of mobilized zygomatic or maxillary Le Fort I osteotomies in rabbits (Figure 30.2), in combination with GTR using a collagen membrane, significantly improved new bone regeneration and bony defect healing by approximately 20% compared to nonGTR controls (Figures 30.3 and 30.4).[13–15] Synthetic membranes, such as Gore-Tex or silastic, have been previously shown to be efficient as GTR membranes. However, immunological and foreign body reactions have been observed postoperatively with the use of these membranes and require additional surgical procedures to retrieve and remove the membranes.[11–15] With the recent technological improvements in collagen harvesting and processing, past problems with immunological reactions and inflammatory responses to collagen have been reduced, thus making this substance a possible alternative to synthetic polymers. Collagen can also support and bind stem cells and growth factors, which will improve bone healing rates in the craniofacial complex.

II. GROWTH FACTORS AND BONE REGENERATION

The use of growth factors in tissue-engineered bone substitutes has been well-documented. Recently, a comprehensive review on the role of growth factors in maxillofacial reconstruction was written by Schilephake,[16] and a review on the role of growth factors in bone repair was written by Lieberman et al.[17] We will describe the growth factors that are most significant in craniofacial reconstruction.

A. Bone Morphogenetic Proteins

The importance of the bone morphogenetic protein family (BMPs) in bone regeneration is well known, first identified by Urist.[18] Comprehensive overviews of BMPs were recently written by Wozney,[19] and Ducy and Karsenty.[20] BMPs 2 to 9 are members of the transforming growth factor (TGF-β) family, and BMPs 2, 4, and 7 have been most studied in bone repair, utilizing carriers such as collagen, PLGA, PLA, and calcium phosphates.[21–25] Particularly, Hollinger's laboratory has developed a hybrid device for BMP2 and osteoblast precursor cells that is a biomimetic matrix consisting of Types I and IV collagens and PLGA.[26] This composition was successful in restoring critical-sized defects in rats' calvariae. Hollinger's laboratory also incorporated BMP2 and rhTGF-β_1 into PLA implants as well as PLGA sponges and implanted into canine spinal fusion models and rabbit calvarial defects.[27–30] Results indicated that the incorporation of BMP2 improved new bone formation, and enhancement with rhTGF-β_1 demonstrated an improved bone ingrowth and osseous regeneration. Meikle et al.[31] has examined PLGA and bone matrix proteins in a noncritical size defect in the rabbit model. There was a nonunion between the polymer (without protein) and bone, and the PLGA did not out-perform the control (autograft). When bone matrix proteins extracted from bovine cortical bone were incorporated into the implants, there was a significant inhibitory effect on bone repair due to a cellular and humoral immune response.

Although results with BMPs are promising, in published clinical studies milligram doses of BMP were required in healthy, adult populations (1.7 to 3.4 mg).[32,33] Dosing patients with high levels of BMP can be a clinical concern. Furthermore, rhBMP-2 has had limited success for maxillary sinus lifts[33] and alveolar bone reservation.[34] However, Friedlaender et al.[35] demonstrated that BMP7, delivered via Type I collagen, was a safe and effective treatment for human tibial nonunions. The BMPs continue to be extensively studied and their potential in bone tissue engineering remains promising.

B. Insulin-Like Growth Factors

The insulin-like growth factor family is also well studied. IGF-I is the somatomedin of growth hormone and is essential for normal skeletal growth and development. Liu et al.[36] demonstrated the severe impairment of bone growth in IGF-I knockout mice. Both IGF-I and IGF-II have a significant role in the growth of bone, brain, kidney, liver, and muscle.[37,38] There is a rich literature base correlating decreased IGF-I levels with increased bone mass loss in the elderly.[39–48] Variations in hormonal status can be associated with gender, age, anatomic location, and pathophysiological state, and is likely involved with the clinical differences in repair potential between these conditions.

There are several studies indicating that local application IGF-I can positively influence calvarial and fracture healing.[49–54] Prisell et al.[51] has examined hyaluronan sponges containing IGF-I in both young and aged rats, and found that IGF-I did not increase bone formation in the young (6 to 7 weeks) rats, but did increase in the aged (19 to 27 month) rats. Interestingly, Busch et al.[53] demonstrated that bone formation in young rat calvarial defects was enhanced with IGF-I delivery from polyorthoester membranes. Wakisaka et al.[52] locally delivered IGF-I for 14 days to the femurs of 24-month-old male rats and found that bone formation was stimulated without any appreciable effect on bone resorption. Recently, Schmidmaier et al.[50] reported that IGF-I and TGF-β_1 incorporated into polylactide implants accelerated fracture healing significantly in rats without systemic side effects.

The concentration of IGF-II in bone and the stimulatory effects of IGF-II on bone formation have been examined.[48,55–58] Although IGF-II is more abundant in bone than IGF-I, IGF-I has been found to be 4 to 7 times more potent than IGF-II.[59] Furthermore, it has been demonstrated that IGF-II mRNA levels remain constant during *in vitro* aging, in comparison to the decline in IGF-I levels.[55,60]

The stimulatory effects of IGF-I and IGFBP-5 on osteoclasts has been examined.[58,61–66] Although numerous studies indicating the stimulatory effects of IGF-I and IGFBP-5 on bone formation have been cited, it must be stated that many growth factors may be involved in bone resorption.[63] However, Wakisaka et al.[52] locally delivered IGF-I for 14 days to the femurs of 24-month-old male rats and found that bone formation was stimulated without any appreciable effect on bone resorption. The role of IGF-I in bone healing remains controversial.

C. Transforming Growth Factors

The TGF-β and fibroblast growth factor (FGF) families have been examined as key growth factors in bone regeneration.[67–70] TGF-βs are a superfamily of growth factors comprising over two dozen related polypeptides that include the TGF-βs and BMPs.[71–73] The various TGF-β isoforms are differentially expressed in the post cranial and cranial bones, dura mater, and cranial vault sutures. They play an essential role in many biological processes, including collagen synthesis, wound healing, bone regeneration, suture patency, and eventual suture fusion.[71–74] Experimental studies have shown that recombinant human TGF beta one (rhTGF-β_1) induces bone formation in rat parietal bone,[75] as well as in rabbit skull defects.[70,76] The combination of TGF-β_1 and IGF-I in aged female rat femoral defects resulted in enhanced bone healing.[77] Mooney and colleagues have shown that TGF-β_3 or anti-TGF-β_2 antibody, delivered in a slow release collagen vehicle over a 6 week period, were found to rescue pathologically fusing coronal sutures (Figure 30.5) and prevent resynostosis and reossification following coronal suturectomy (Figure 30.6 and Figure 30.7) in young rabbits with familial craniosynostosis.[78–81] These findings suggest that exogenous TGF-βs may be used to regulate osseous growth and wound healing by affecting bone cell proliferation and apoptosis through interactions with other TGF-β isoforms and their receptors[74,81] and may also be used to engineer craniofacial structures.

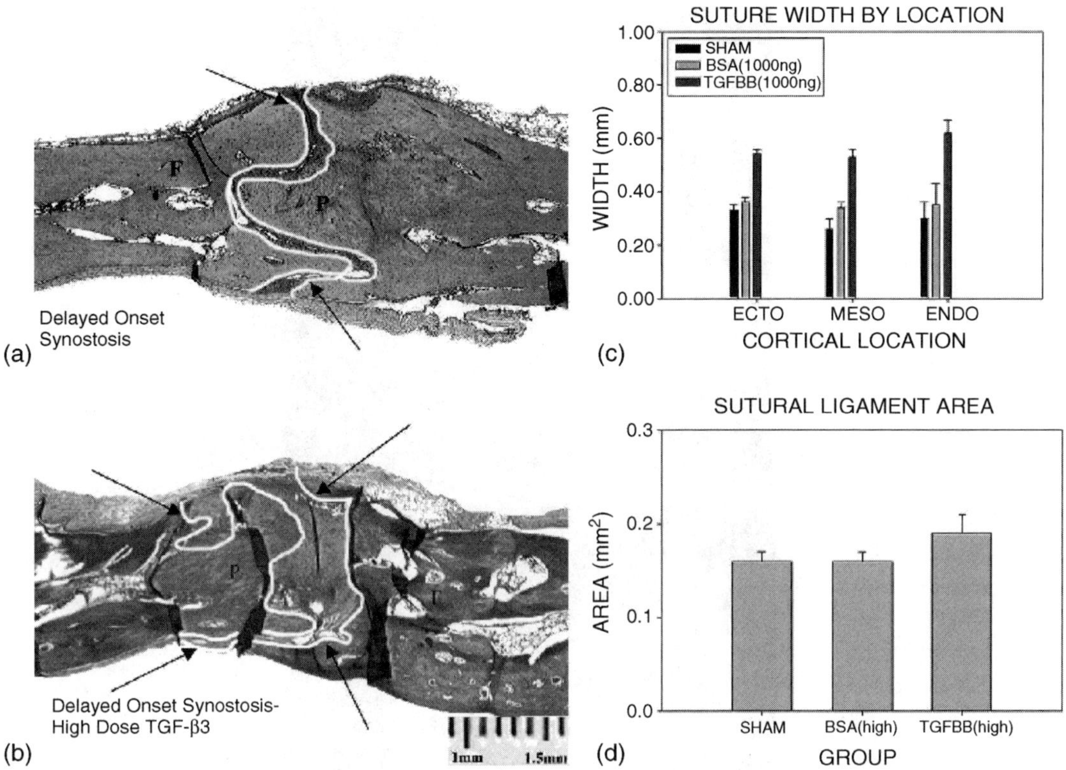

FIGURE 30.5 Histophotomicrograph (original magnification × 25) of (a) an untreated coronal suture and (b) a 25-day-old coronal suture treated with 1000 ng of TGF-β3 from 84-day-old craniosynostotic rabbits. Note the thickened osteogenic fronts, bony bridge, and narrowed coronal suture (arrows and traced in white) in the untreated rabbit, and the widened, patent suture and thinner osteogenic fronts in the treated rabbit (F = frontal bone; P = parietal bone). (c) Suture width, and (d) suture area by group.

D. Fibroblast Growth Factors

Basic FGF (or FGF-2), is part of a family of at least 19 structurally related growth factors. This family is a potent mitogen and has been implicated in normal and pathological processes ranging from embryogenesis, tumor angiogenesis, and skeletal and soft tissue wound healing.[82,83] In bone, FGF-2 is produced intracellularly and secreted into the extracellular matrix; it also localizes in the nucleus of osteoblasts and regulates cells function at the transcriptional level.[74,83,84] The bone anabolic effects of FGF-2 in 5-month-old ovariectomized rats was examined by Liang et al.[85] Results indicated that systemic delivery of FGF-2 for 2 weeks resulted in increased bone mass in the tibia. FGF-2 is stored in bone in the active form,[86] and has also been found to enhance demineralized bone-induced osteogenesis.[87] Tabata et al.[88] found that the delivery of FGF-2 via a biodegradable gelatin hydrogel to monkey skull defects resulted in increased bone mineral density. In a different primate model, but also using gelatin as the delivery vehicle, Takayama et al.[89] demonstrated improved periodontal regeneration with increasing FGF-2 dosage. Using a different carrier, hyaluronan gel, Radomsky et al.[90] demonstrated accelerated fracture healing in a primate model. Perturbation of FGF-2 expression also has significant effects on craniofacial sutural and skeletal formation. Homozygote FGF-2 knock out mice (−/−) showed decreased trabecular bone volume, mineral apposition rate, and bone formation rates[83,91] while FGF-2 overexpression transgenic mice showed macrocephaly, increased maturation and proliferation of osteoprogenitor cells, an increase in mineralized nodules, and premature fusion of the cranial vault sutures.[83,92,93]

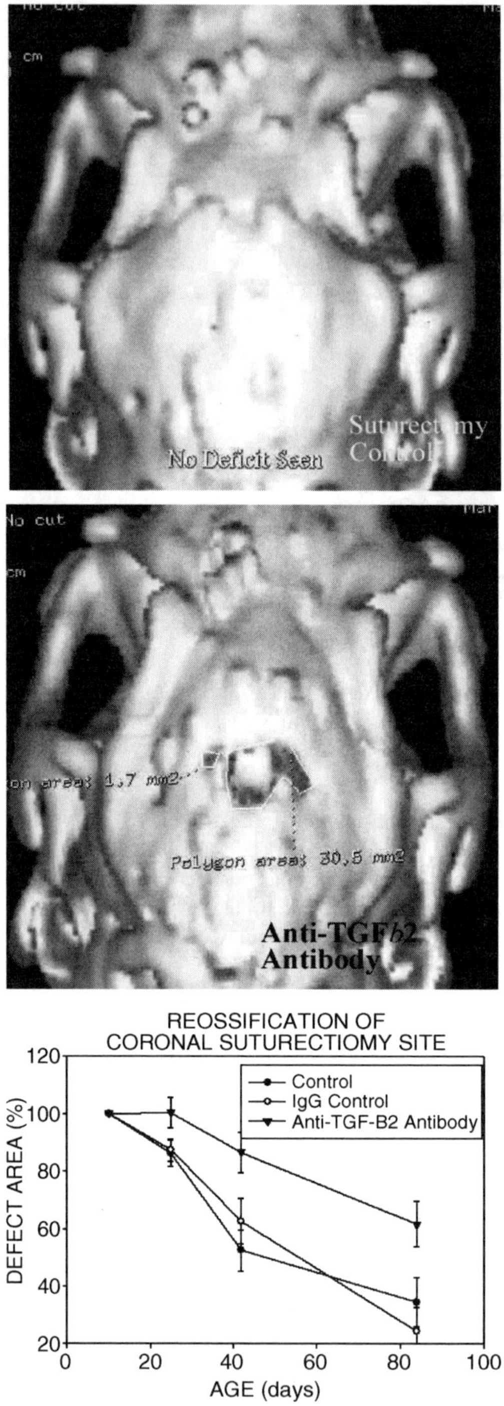

FIGURE 30.6 Dorsal 3D-CT scans of suturectomy sites in 84-day-old craniosynostotic rabbits. Note the complete resossification of the defect in the control rabbit (top) compared to patent defect in the rabbit treated with anti-TGF-β_2 antibody at 10 days of age (middle). Graph of mean suturectomy defect area by group and age (bottom). Note the significantly increased defect area in the synostotic rabbits receiving anti-TGF-β_2 antibody compared to controls.

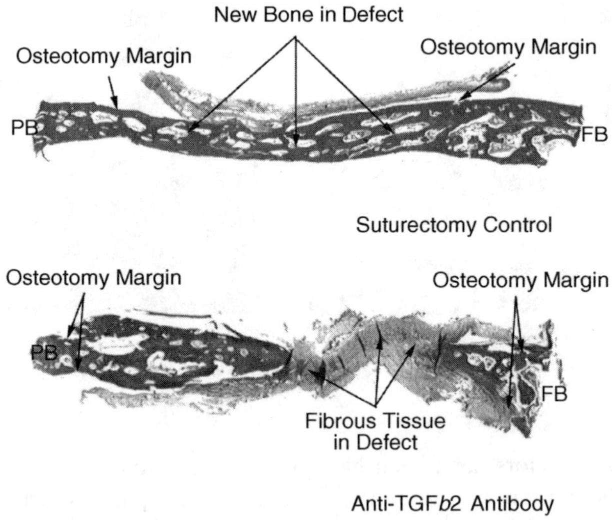

FIGURE 30.7 Histophotomicrographs (original magnification ×15) of an untreated control coronal suturectomy site (top) and a suturectomy sited treated at 10 days of age with anti-TGF-β_2 antibody (bottom) from 84-day-old craniosynostotic rabbits. Note the extensive reossification of the suturectomy site in the control compared to the fibrous nonunion in the suturectomy site treated with anti-TGF β_2 antibody. (PB = parietal bone; FB = frontal bone.)

FGF-2 is thought to interact with various TGF-β isoforms and developmental genes (i.e., Noggin, Twist, Runx, Msx2) and work through similar bone producing pathways and intracellular signaling systems.[74,83,84]

E. Platelet-Derived Growth Factors

Platelet-derived growth factor (PDGF) has been examined as a stimulator of bone formation as well as a recruiter of periodontal ligament cells.[67,94,95] PDGF was discovered by Lynch et al.[96] in the 1980s to promote bone regeneration, cementum formation, and periodontal ligament growth. PDGF has been shown to enhance bone formation in older rats, as demonstrated in ectopic bone formation in a subcutaneous model, with demineralized bone matrix as the delivery vehicle.[97] A stimulatory effect on bone healing in rabbit tibial osteotomies was demonstrated, using collagen as the carrier.[95] Combined with IGF-I, PDGF has demonstrated enhanced bone formation in periodontal applications in canines.[98–100] Recently, rhPDGF-BB has been examined in a human class II furcation model. Using bone allograft as the PDGF carrier, new bone, cementum, and periodontal regeneration was demonstrated.[101] PDGF is one of the more promising growth factors in craniofacial tissue engineering.

III. GROWTH FACTOR ENCAPSULATION

Bone healing is a co-ordinated series of cellular and biochemical events and therapeutic interventions to augment healing need to be in register with these elements. Predictable spatial and temporal delivery of growth factors for craniofacial bone repair can be accomplished by controlled, local delivery using several strategies.

The delivery of growth factors can be controlled by drug loading, polymer composition, and processing techniques. Babensee et al.[102] has written a comprehensive review on drug and growth factor delivery, as have Saltzman and Olbricht.[103] Growth factors can be encapsulated in microspheres (mean diameter 1 to 1000 μm) or nanospheres (mean diameter 1 to 1000 nm).

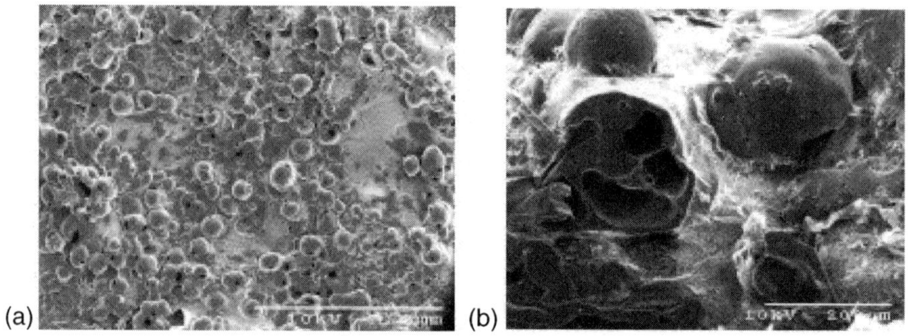

FIGURE 30.8 (a) SEM of PLGA microsphere within a PLGA scaffold, and (b) PLGA microspheres within a fibrin gel.

Drugs and other growth factors have been incorporated into biodegradable PLGA microspheres, such as BMP-2, [104] human growth hormone,[105] Japanese encephalitis virus vaccine,[106] vascular endothelial growth factor (VEGF),[107] IGF-I,[108] PDGF,[109] TGF-β_1,[110,111] FGF-2,[112] cyclosporine,[113] and cisplatin.[114] DeLuca's group has published extensively on controlled drug release from PLGA microspheres,[115–121] including a recent study that demonstrated that microsphere modification of polymer ratios results in a controlled drug release.[116] Growth factor incorporation into polymer nanospheres has been less widely studied.[122–127] Although the encapsulation efficiency of nanospheres is usually lower than that of microspheres, the release rate is usually faster.[126–128] It has been found that the rate of release decreases and the duration increases with increasing sphere size.[129] Other methods of delivery include adsorption of growth factors to the surface prior to implantation[26] and incorporation of growth factors during the scaffold fabrication process.[130] Additionally, microspheres containing growth factors can be injected into a porous scaffold, postfabrication; one relevant study describes PLGA microspheres encapsulating epidermal growth factor that were injected into polymer sponges and implanted into the mesentery of Lewis rats for 1 week.[131] More recently, Richardson et al. reported the dual delivery of two angiogenic growth factors from a tissue-engineered scaffold.[109] VEGF was adsorbed to the surface of the scaffold while PDGF was encapsulated in PLGA microspheres and mixed within the scaffold.

Marra's laboratory has recently examined the incorporation of polymer microspheres into polymer scaffolds[132,133] as well as into fibrin scaffolds.[134] By protecting the PLGA microspheres with a membrane that resists organic solvents, microspheres were able to be incorporated into tissue-engineered scaffold during fabrication. This technique should result in a more homogeneous and quantifiable distribution of the microspheres, as well as the possibility of incorporating more than one growth factor (Figure 30.8). Fibrin has been examined as a drug or growth factor delivery vehicle,[135–145] but not in relation to the incorporation of polymeric microspheres.

IV. CONCLUSIONS

The combination of scaffolds, cells, and growth factors for tissue-engineered craniofacial reconstruction is a promising area of research. Although this chapter focuses on polymeric scaffold material, it must be noted that calcium phosphates are also being examined as bone substitutes in craniofacial repair.[146] Additionally, osteogenesis in craniofacial applications using demineralized bone implants has been examined by Glowacki et al.[147] Chapter 20 describes the role of ceramics in bone tissue engineering. Finally, the use of cells, particularly mesenchymal stem cells, has been studied with promising results.[148] See Chapter 20 for a more comprehensive review of the incorporation of cells into tissue-engineered scaffolds. The most prevalent growth factors for craniofacial reconstruction include FGF-2, PDGF-BB, TGF-βs and BMPs.

The future of craniofacial tissue engineering will likely include the synthesis of new polymeric materials that are conducive to stem cell seeding, as well as biomaterials that can deliver growth factors in a controlled fashion.

ACKNOWLEDGMENTS

The authors thank Phil Campbell for helpful discussion concerning the role of IGF-I in bone repair. Partial support for this work is from NIH RO1 DE13108 (JOH), NIH RO1 DE11416 (JOH), NIH/NIDCR DE13078 (MPM), the Pennsylvania Infrastructure Technology Alliance (KGM), and the Pittsburgh Tissue Engineering Initiative (KGM).

REFERENCES

1. Getter, L., Curtright, D. E., Bhaskar, S. N., and Augsburg, J. K., A biodegradable intraosseous appliance in the treatment of mandibular fractures, *J. Oral Surg.*, 30, 344, 1972.
2. Illi, O. E., Stauffer, U. G., Sailer, H. F., and Weigum, H., [Resorbable implants in craniofacial surgery in childhood. A contribution to the development of poly(lactide) implants], *Helv. Chir. Acta*, 58, 123, 1991.
3. Illi, O. E., Stauffer, U. G., Sailer, H., and Beck, P., [Biodegradable osteosynthesis applied to craniofacial surgery], *Chir. Pediatr.*, 31, 240, 1990.
4. Illi, O. E., Weigum, H., and Misteli, F., Biodegradable implant materials in fracture fixation, *Clin. Mater.*, 10, 69, 1992.
5. Goldstein, J. A., The use of bioresorbable material in craniofacial surgery, *Clin. Plast. Surg.*, 28, 653, 2001.
6. Imola, M. J. and Schramm, V. L., Resorbable internal fixation in pediatric cranial base surgery, *Laryngoscope*, 112, 1897, 2002.
7. Eppley, B. L. and Reilly, M., Degradation characteristics of PLLA–PGA bone fixation devices, *J. Craniofac. Surg.*, 8, 116, 1997.
8. Fearon, J. A., Rigid fixation of the calvaria in craniosynostosis without using "rigid" fixation, *Plast. Reconstr. Surg.*, 111, 27, 2003.
9. Rohner, D., Hutmacher, D. W., See, P., Tan, K. C., Yeow, V., Tan, S. Y., Lee, S. T., and Hammer, B., [Individually CAD-CAM technique designed, bioresorbable 3-dimensional polycaprolactone framework for experimental reconstruction of craniofacial defects in the pig], *Mund. Kiefer Gesichtschir.*, 6, 162, 2002.
10. Bidic, S. M., Calvert, J. W., Marra, K., Kumta, P., Campbell, P., Mitchell, R., Wigginton, W., Hollinger, J. O., Weiss, L., and Mooney, M. P., Rabbit calvarial wound healing by means of seeded Caprotite scaffolds, *J. Dent. Res.*, 82, 131, 2003.
11. Lorenzoni, M., Pertl, C., Polansky, R., and Wegscheider, W. A., Guided bone regeneration with barrier membranes — a clinical and radiographic follow-up study after 24 months, *Clin. Oral Implants Res.*, 10, 16, 1999.
12. Mundell, R. D., Mooney, M. P., Siegel, M. I., and Losken, A., Osseous guided tissue regeneration using a collagen barrier membrane, *J. Oral Maxillofac. Surg.*, 51, 1004, 1993.
13. Mooney, M. P., Mundell, R. D., Stetzer, K., Ochs, M. W., Milch, E. A., Buckley, M. J., and Siegel, M. I., The effects of guided tissue regeneration and fixation technique on osseous wound healing in rabbit zygomatic arch osteotomies, *J. Craniofac. Surg.*, 7, 46, 1996.
14. Stetzer, K., Cooper, G., Gassner, R., Kapucu, R., Mundell, R., and Mooney, M. P., Effects of fixation type and guided tissue regeneration on maxillary osteotomy healing in rabbits, *J. Oral Maxillofac. Surg.*, 60, 427, 2002.
15. Verschueren, D., Gassner, R., Mitchell, R., and Mooney, M. P., Le Fort I osteotomy healing in rabbits with the use of guided tissue regeneration (GTR), *J. Dent. Res.*, 82, 1574, 2003.
16. Schilephake, H., Bone growth factors in maxillofacial skeletal reconstruction, *Int. J. Oral Maxillofac. Surg.*, 31, 469, 2002.

17. Lieberman, J. R., Daluiski, A., and Einhorn, T. A., The role of growth factors in the repair of bone. Biology and clinical applications, *J. Bone Joint Surg. Am.*, 84-A, 1032, 2002.
18. Urist, M. R., Bone: formation by autoinduction, *Science*, 150, 893, 1965.
19. Wozney, J. M., Overview of bone morphogenetic proteins, *Spine*, 27, S2, 2002.
20. Ducy, P. and Karsenty, G., The family of bone morphogenetic proteins, *Kidney Int.*, 57, 2207, 2000.
21. Boyne, P. J., Animal studies of application of rhBMP-2 in maxillofacial reconstruction, *Bone*, 19, 83, 1996.
22. Gao, T. J., Lindholm, T. S., Marttinen, A., and Puolakka, T., Bone inductive potential and dose-dependent response of bovine bone morphogenetic protein combined with type IV collagen carrier, *Ann. Chir. Gynaecol. Suppl.*, 207, 77, 1993.
23. Kawai, T., Mieki, A., Ohon, Y., Umemura, M., Kataoka, H., Kurita, S., Koie, M., Jinde, T., Hasegawa, J., and Urist, M. R., Osteoinductive activity of composites of bone morphogenetic protein and pure titanium, *Clin. Orthop. Rel. Res.*, 290, 296, 1993.
24. Kawamura, M. and Urist, M. R., Induction of callus formation by implants of bone morphogenetic protein and associated bone matrix noncollagenous proteins, *Clin. Orthop. Rel. Res.*, 236, 240, 1988.
25. Murata, M., Maki, F., Sato, D., Shibata, T., and Arisue, M., Bone augmentation by onlay implant using recombinant human BMP-2 and collagen on adult rat skull without periosteum, *Clin. Oral Implants Res.*, 11, 289, 2000.
26. Winn, S. R., Schmitt, J. M., Buck, D., Hu, Y., Grainger, D., and Hollinger, J. O., Tissue-engineered bone biomimetic to regenerate calvarial critical-sized defects in athymic rats, *J. Biomed. Mater. Res.*, 45(4), 414, 1999.
27. McKinney, L. and Hollinger, J. O., A bone regeneration study: transforming growth factor-beta 1 and its delivery, *J. Craniofac. Surg.*, 7, 36, 1996.
28. Hollinger, J. O. and Leong, K., Poly(alpha-hydroxy acids): carriers for bone morphogenetic proteins, *Biomaterials*, 17, 187, 1996.
29. Hollinger, J. O., Mayer, M., Buck, D., Zegzula, H. D., Ron, E., Smith, J., Lin, L., and Wozney, J., Poly(alpha-hydroxy acid) carrier for delivering recombinant human bone morphogenetic protein-2 for bone regeneration, *J. Control. Release*, 39, 287, 1996.
30. Hollinger, J. O., Schmitt, J. M., Buck, D. C., Shannon, R., Joh, S. P., Zegzula, H. D., and Wozney, J., Recombinant human bone morphogenetic protein-2 and collagen for bone regeneration, *J. Biomed. Mater. Res.*, 43(4), 356, 1998.
31. Meikle, M. C., Papioannou, S., Ratledge, T. J., Speight, P. M., Watt-Smith, S. R., Hill, P. A., and Reynolds, J. J., Effect of poly-DL-lactide-*co*-glycolide implants and xenogeneic bone matrix-derived growth factors on calvarial bone repair in the rabbit, *Biomaterials*, 15, 513, 1994.
32. Geesink, R. G., Hoefnagels, N. H., and Bulstra, S. K., Osteogenic activity of PO-1 bone morphogenetic protein (BMP-7) in a human fibular defect, *J. Bone Joint Surg.*, 81-B, 710, 1999.
33. Boyne, P. J., Marx, R. E., Nevins, M., Triplett, G., Lazaro, E., Lilly, L. C., Alder, M., and Nummikoski, P., A feasibility study evaluation rhBMP-2/absorbable collagen sponge for maxillary sinus augmentation, *Int. J. Periodontal Restor. Dent.*, 17, 11, 1997.
34. Howell, T. H., Fiorellini, J., Jones, A., Alder, M., Nummikoski, P., Lazaro, M., Lilly, L., and Cochran, D., A feasibility study evaluating rhBMP-2/absorbable collagen sponge device for local alveolar ridge preservation of augmentation, *Int. J. Periodontal Restor. Dent.*, 17, 125, 1997.
35. Friedlaender, G. E., Perry, C. R., Cole, J. D., Cook, S. D., Cierny, G., Muschler, G. F., Zych, G. A., Calhoun, J. H., LaForte, A. J., and Yin, S., Osteogenic protein-1 (bone morphogenetic protein-7) in the treatment of tibial nonunions, *J. Bone Joint Surg. Am.*, 83-A(Suppl. 1), S151, 2001.
36. Liu, J. P., Baker, J., Perkins, A. S., Robertson, E. J., and Efstratiadis, A., Mice carrying null mutations of the genes encoding insulin-like growth factor I (Igf-1) and type 1 IGF receptor (Igf1r), *Cell*, 75, 59, 1993.
37. Mohan, S., and Baylink, D. J., Development of a simple valid method for the complete removal of insulin-like growth factor (IGF)-binding proteins from IGFs in human serum and other biological fluids: comparison with acid–ethanol treatment and C18 Sep–Pak separation, *J. Clin. Endocrinol. Metab.*, 80, 637, 1995.
38. Bautista, C., Mohan, S., and Baylink, D., Insulin-like growth factors I and II are present in the skeletal tissue of ten vertebrates, *Metabolism*, 39, 96, 1993.

39. Seck, T., Bretz, A., Krempien, R., Krempien, B., Ziegler, R., and Pfeilschifter, J., Age-related changes in insulin-like growth factor I and II in human femoral cortical bone: lack of correlation with bone mass, *Bone*, 24, 387, 1999.
40. Calo, L., Castrignano, R., Davis, P. A., Carraro, G., Pagnin, E., Giannini, S., Semplicini, A., and D'Angelo, A., Role of insulin-like growth factor-I in primary osteoporosis: a correlative study, *J. Endocrinol. Invest.*, 23, 223, 2000.
41. Langlois, J. A., Rosen, C. J., Visser, M., Hannan, M. T., Harris, T., Wilson, P. W., and Kiel, D. P., Association between insulin-like growth factor I and bone mineral density in older women and men: the Framingham heart study, *J. Clin. Endocrinol. Metab.*, 83, 4257, 1998.
42. Kim, J. G., Shin, C. S., Choi, Y. M., Moon, S. Y., Kim, S. Y., and Lee, J. Y., The relationship among circulating insulin-like growth factor components, biochemical markers of bone turnover and bone mineral density in postmenopausal women under the age of 60, *J. Nutr. Health Aging*, 2, 39, 1998.
43. Rosen, C. J., Glowacki, J., and Craig, W., Sex steroids, the insulin-like growth factor regulatory system, and aging: implications for the management of older postmenopausal women, *J. Nutr. Health Aging*, 2, 39, 1998.
44. Wuster, C., Harle, U., Rehn, U., Muller, C., Knauf, K., Koppler, D., Schwabe, C., and Ziegler, R., Benefits of growth hormone treatment on bone metabolism, bone density and bone strength in growth hormone deficiency and osteoporosis, *Growth Horm. IGF Res.*, 8(Suppl. A), 87, 1998.
45. Florini, J. R., Prinz, P. N., Vitiello, M. V., and Hintz, R. L., Somatomedin-C levels in healthy young and old men: relation to peak and 24-hour integrated levels of growth hormone, *J. Gerontol.*, 40, 2, 1985.
46. Muñoz-Torres, M., Mezquita-Raya, P., Lopez-Rodriguez, F., Torres-Vela, E., de Dios Luna, J., and Escobar-Jimenez, F., The contribution of IGF-I to skeletal integrity in postmenopausal women, *Clin. Endocrinol.*, 55, 759, 2001.
47. Ammann, P., Bourrin, S., Bonjour, J. P., Meyer, J. M., and Rizzoli, R., Protein undernutrition-induced bone loss is associated with decreased IGF-I levels and estrogen deficiency, *J. Bone Miner. Res.*, 15, 683, 2000.
48. Boonen, S., Mohan, S., Dequeker, J., Aerssens, J., Vanderschueren, D., Verbeke, G., Broos, P., Bouillon, R., and Baylink, D. J., Down-regulation of the serum stimulatory components of the insulin-like growth factor (IGF) system (IGF-I, IGF-II, IGF binding protein [BP]-3, and IGFBP-5) in age-related (type II) femoral neck osteoporosis, *J. Bone Miner. Res.*, 14, 2150, 1999.
49. Isgaard, J., Nilsson, A., Lindahl, A., Jansson, J.-O., and Isaksson, O. G. P., Effects of local administration of GH and IGF-1 on longitudinal bone growth in rats, *Am. J. Physiol.*, 250, E367, 1986.
50. Schmidmaier, G., Wildemann, B., Bail, H., Lucke, M., Fuchs, T., Stemberger, A., Flyvbjerg, A., Haas, N. P., and Raschke, M., Local application of growth factors (insulin-like growth factor-1 and transforming growth factor-beta1) from a biodegradable poly(D,L-lactide) coating of osteosynthetic implants accelerates fracture healing in rats, *Bone*, 28, 341, 2001.
51. Prisell, P. T., Aspenberg, P., Wikstrom, B., Wredmark, T., and Norstedt, G., Insulin-like growth factor I increases bone formation in old or corticosteroid treated rats, *Acta Orthop. Scand.*, 68, 586, 1997.
52. Wakisaka, A., Tanaka, H., Barnes, J., and Liang, C. T., Effect of locally infused IGF-I on femoral gene expression and bone turnover activity in old rats, *J. Bone Miner. Res.*, 13, 13, 1998.
53. Busch, O., Solheim, E., Bang, G., and Tornes, K., Guided tissue regeneration and local delivery of insulin-like growth factor I by bioerodible polyorthoester membranes in rat calvarial defects, *Int. J. Oral Maxillofac. Implants*, 11, 498, 1996.
54. Thaller, S. R., Salzhauer, M. A., Rubinstein, A. J., Thion, A., and Tesluk, H., Effect of insulin-like growth factor type I on critical size calvarial bone defects in irradiated rats, *J. Craniofac. Surg.*, 9, 138, 1998.
55. Kveiborg, M., Flyvbjerg, A., Rattan, S. I., and Kassem, M., Changes in the insulin-like growth factor-system may contribute to in vitro age-related impaired osteoblast functions, *Exp. Gerontol.*, 35, 1061, 2000.
56. Jia, D. and Heersche, J. N. M., Insulin-like growth factor-1 and -2 stimulate osteoprogenitor proliferation and differentiation and adipocyte formation in cell populations derived from adult rat bone, *Bone*, 27, 785, 2000.

57. Mohan, S. and Baylink, D. J., The role of insulin-like growth factor-II in the coupling of bone formation to resorption, In *Modern Concepts of Insulin-like Growth Factors*, Spencer, E. M., ed., Elsevier, Amsterdam, p. 169, 1991.
58. Nicolas, V., Mohan, S., Honda, Y., Prewett, A., Finkelman, R. D., Baylink, D. J., and Farley, J. R., An age-related decrease in the concentration of insulin-like growth factor binding protein-5 in human cortical bone, *Calcif. Tissue Int.*, 57, 206, 1995.
59. Lind, M., Growth factors, possible new clinical tools: a review, *Acta Orthop. Scand.*, 67, 407 1996.
60. Seck, T., Scheppach, B., Scharla, S., Diel, I., Blum, W. F., Bismar, H., Schmid, G., Krempien, B., Ziegler, R., and Pfeilschifter, J., Concentration of insulin-like growth factor (IGF)-I and -II in iliac crest bone matrix from pre- and postmenopausal women: relationship to age, menopause, bone turnover, bone volume, and circulating IGFs, *J. Clin. Endocrinol. Metab.*, 83, 2331, 1998.
61. Kanatani, M., Sugimoto, T., Nishiyama, K., and Chihara, K., Stimulatory effect of insulin-like growth factor binding protein-5 on mouse osteoclast cell formation and osteoclastic bone-resorbing activity, *J. Bone Miner. Res.*, 15, 902 2000.
62. Santhanagopal, A. and Dixon, S. J., Insulin-like growth factor I rapidly enhances acid efflux from osteoblastic cells, *Am. J. Physiol.*, 277, E423, 1999.
63. Canalis, E., Pash, J., and Varghese, S., Skeletal growth factors, *Crit. Rev. Eukaryot. Gene. Expr.*, 3, 155, 1993.
64. Ghiron, L. J., Thompson, J. L., Holloway, L., Hintz, R. L., Butterfield, G. E., Hoffam, A. R., and Marcus, R., Effects of recombinant insulin-like growth factor-I and growth hormone on bone turnover in elderly women, *J. Bone Miner. Res.*, 12, 1844, 1995.
65. Guicheux, J., Heymann, D., Rousselle, A. V., Gouin, F., Pilet, P., Yamada, S., and Daculsi, G., Growth hormone stimulatory effects on osteoclastic resorption are partly mediated by insulin-like growth factor I: an in vitro study, *Bone*, 22, 25, 1998.
66. Hill, P. A., Reynolds, J. J., and Meikle, M. C., Osteoblasts mediate insulin-like growth factor-I and -II stimulation of osteoclast formation and function, *Endocrinology*, 136, 124, 1995.
67. Solheim, E., Growth factors in bone, *Int. Orthop.*, 22, 410, 1998.
68. Bostrom, M. P., Saleh, K. J., and Einhorn, T. A., Osteoinductive growth factors in preclinical fracture and long bone defect models, *Orthop. Clin. North Am.*, 30, 647, 1999.
69. Nielsen, H. M., Andreassen, T. T., Ledet, T., and Oxlund, H., Local injection of TGF-beta increases the strength of tibial fractures in the rat, *Acta Orthop. Scand.*, 65, 37, 1994.
70. Hong, L., Tabata, Y., Miyamoto, S., Yamada, K., Aoyama, I., Tamura, M., Hashimoto, N., and Ikada, Y., Promoted bone healing at a rabbit skull gap between autologous bone fragment and the surrounding intact bone with biodegradable microspheres containing transforming growth factor-beta1, *Tissue Eng.*, 6, 331, 2000.
71. Massague, J., Attisano, L., and Wrana, J. L., The TGF-β family and its composite receptors, *Trends Cell Biol.*, 4, 172, 1994.
72. Centrella, M., Horowitz, M., Wozney, J. M., and McCarthy, T. L., Transforming growth factor beta (TGF-beta) family members of bone, *Endocr. Rev.*, 15, 27, 1994.
73. Cohen, M. M. J., TGFβ and suture biology, In *Craniosynostosis: Diagnosis, Evaluation, and Management*, MacLean, R. E., Ed., Oxford University Press, New York, p. 69, 2000.
74. Opperman, L. A. and Ogle, R. C., Molecular studies of craniosynostosis: factors affecting cranial suture morphogenesis and patency, In *Understanding Craniofacial Anomalies: The Etiopathogenesis of Craniosynostosis and Facial Clefting*, Siegel, M. I., Ed., Wiley, New York, p. 497, 2002.
75. Tanaka, T., Taniguchi, Y., Gotoh, K., Satoh, R., Inazu, M., and Ozawa, H., Morphological study of recombinant human transforming growth factor β1-induced intramembranous ossification in neonatal rat parietal bone, *Bone*, 14, 117, 1993.
76. Beck, L. S., Deguzman, L., Lee, W. P., Xu, Y., McFatridge, L. A., Gillett, N. A., and Amento, E. P., TGF-β1 induces bone closure of skull defects, *J. Bone Miner. Res.*, 6, 1257 1991.
77. Blumenfeld, I., Srouji, S., Lanir, Y., Laufer, D., and Livne, E., Enhancement of bone defect healing in old rats by TGF-beta and IGF-1, *Exp. Gerontol.*, 37, 553, 2002.
78. Mooney, M. P., Siegel, M. I., and Opperman, L. A., Animal models of craniosynostosis: experimental, congenital, and transgenic, In *Understanding Craniofacial Anomalies: The Etiopathogenesis of Craniosynostosis and Facial Clefting*, Siegel, M. I., Ed., Wiley, New York, p. 209, 2002.

79. Mooney, M. P., Losken, H. W., Moursi, A., Mitchell, R., Winnard, P., Ozerdem, O., Bradley, J., Azari, K., Acarturk, O., Chang, T., Opperman, L. A., Stelnicki, E. J., and Siegel, M. I., Inhibition of postoperative resynostosis with anti-TGF-β2 antibody, *J. Dent. Res.*, 81, 221, 2002.
80. Mooney, M. P., Losken, H. W., Moursi, A., Mitchell, R., Wigginton, W., Winnard, P., Bradley, J., Azari, K., Acarturk, O., Opperman, L. A., Stelnicki, E. J., and Siegel, M. I., Postoperative cytokine therapy allows for normal intracranial volume in craniosynostotic rabbits, *J. Dent. Res.*, 83, 1517, 2003.
81. Chong, S. L. , Mitchell, R., Moursi, A., Winnard, P., Losken, H. W., Bradley, J., Ozerdem, O., Azari, K., Acarturk, O., Opperman, L. A., Siegel, M. I., and Mooney, M. P., Rescue of coronal suture fusion using transforming growth factor-beta 3 (Tgf-β3) in rabbits with delayed-onset craniosynostosis, *Anat. Rec.*, 274, 962, 2003.
82. Xu, J., Liu, Z., and Ornitz, D. M., Temporal and spatial gradients of FGF-8 and FGF-17 regulate proliferation and differentiation of midline cerebellar structures, *Development*, 127, 1833, 2000.
83. Moursi, A. M., Winnard, P. L., Winnard, A. V., Rubenstrunk, J. M., and Mooney, M. P., Fibroblast growth factor 2 induces increased calvarial osteoblast proliferation and cranial suture fusion, *Cleft Palate Craniofac. J.*, 39, 487, 2002.
84. Marie, P. J., Debiais, F., and Hay, E., Regulation of human cranial osteoblast phenotype by FGF-2, FGFR-2 and BMP-2 signaling, *Histol. Histopathol.*, 17, 877, 2002.
85. Liang, H., Pun, S., and Wronski, T. J., Bone anabolic effects of basic fibroblast growth factor in ovariectomized rats, *Endocrinology*, 140, 5780, 1999.
86. Gospodarowicz, D., Fibroblast growth factor: chemical structure and biological function, *Clin. Orthop.*, 257, 231, 1990.
87. Aspenberg, P. and Lohmander, L. S., Fibroblast growth factor stimulates bone formation: bone induction studied in rats, *Acta Orthop. Scand.*, 60, 473, 1989.
88. Tabata, Y., Yamada, K., Hong, L., Miyamoto, S., Hashimoto, N., and Ikada, Y., Skull bone regeneration in primates in response to basic fibroblast growth factor, *J. Neurosurg.*, 91, 851, 1999.
89. Takayama, S., Murakami, S., Shimabukuro, Y., Kitamura, M., and Okada, H., Periodontal regeneration by FGF-2 (bFGF) in primate models, *J. Dent. Res.*, 80, 2075, 2001.
90. Radomsky, M., Aufdemorte, T., Swain, L., Fox, C., Spiro, R., and Poser, J., Novel formulation of fibroblast growth factor-2 in a hyaluronan gel accelerates fracture healing in nonhuman primates, *J. Orthop. Res.*, 17, 607 1999.
91. Montero, A., Okada, Y., Tomita, M., Ito, M., Tsurukami, H., Nakamura, T., Doetschman, T., Coffin, J. D., and Hurley, M. M., Disruption of the fibroblast growth factor-2 gene results in decreased bone mass and bone formation, *J. Clin. Invest.*, 105, 1085, 2000.
92. Scutt, A. and Bertram, P., Basic fibroblast growth factor in the presence of dexamethasone stimulates colony formation expansion, and osteoblastic differentiation by rat bone marrow stromal cells, *Calcif. Tissue Int.*, 64, 69, 1999.
93. Greenwald, J. A., Mehrara, B. J., Spector, J. A., Warren, S. M., Fagenholz, P. J., Smith, L. E., Bouletreau, P. J., Crisera, F. E., Ueno, H., and Longaker, M. T., In vivo modulation of FGF biological activity alters cranial suture fate, *Am. J. Pathol.*, 158, 441, 2001.
94. Bolander, M. E., Regulation of fracture repair by growth factors, *Proc. Soc. Exp. Biol. Med.*, 200, 165, 1992.
95. Nash, T. J., Howlett, C. R., Martin, C., Steele, J., Johnson, K. A., and Hicklin, D. J., Effect of platelet-derived growth factor on tibial osteotomies in rabbits, *Bone*, 15, 203, 1994.
96. Lynch, S. E., Polson, A. M., Howell, T. H., Reddy, M. S., Zappa, U. E., and Antoniades, H. N., A combination of platelet derived and insulin-like growth factors enhances periodontal regeneration, *J. Clin. Periodontol.*, 16, 545, 1989.
97. Howes, R., Bowness, J. M., Grotendorst, G. R., Martin, G. R., and Reddi, A. H., Platelet-derived growth factor enhances demineralized bone matrix-induced cartilage and bone formation, *Calcif. Tissue Int.*, 42, 34, 1988.
98. Lynch, S. E., Buser, D., Hernandez, R. A., Weber, H. P., Stich, H., Fox, C. H., and Williams, R. C., Effects of the platelet-derived growth factor/insulin-like growth factor-I combination on bone regeneration around titanium dental implants. Results of a pilot study in beagle dogs, *J. Periodontol.*, 62, 710, 1991.

99. Lynch, S. E., Ruiz, G., Williams, R. C., Kiritsy, C. P., Howell, T. H., Reddy, M. S., and Antoniades, H. N., The effects of short term application of a combination of platelet-derived growth factors and insulin-like growth factors on periodontal wound healing, *J. Periodontol.*, 62, 458 1991.
100. Giannobile, W. V., Hernandez, R. A., Finkelman, R. D., Ryan, S., Kiritsy, C. P., D'Andrea, M., and Lynch, S. E., Comparative effects of platelet-derived growth factor-BB and insulin-like growth factor-I, individually and in combination, on periodontal regeneration in *Macaca fascicularis*, *J. Periodontal Res.*, 31, 301, 1996.
101. Nevins, M., Camelo, M., Nevins, M. L., Schenk, R. K., and Lynch, S. E., Periodontal regeneration in humans using recombinant human platelet-derived growth factor BB (rhPDGF-BB) and allogeneic bone, *J. Periodontal Res.*, 31, 301, 2003.
102. Babensee, J. E., McIntire, L. V., and Mikos, A. G., Growth factor delivery for tissue engineering, *Pharm. Res.*, 17(5), 497, 2000.
103. Saltzman, W. M. and Olbricht, W. L., Building drug delivery into tissue engineering, *Nat. Rev. Drug Discov.*, 1, 177, 2002.
104. Weber, F. E., Eyrich, G., Gratz, K. W., Maly, F. E., and Sailer, H. F., Slow and continuous application of human recombinant bone morphogenetic protein via biodegradable poly(lactide-*co*-glycolide) foamspheres, *Int. J. Oral Maxillofac. Surg.*, 31, 60, 2002.
105. Cleland, J. L., Mac, A., Boyd, B., Yang, J., Duenas, E. T., Yeung, D., Brooks, D., Hsu, C., Chu, H., Mukku, V., and Jones, A. J., The stability of recombinant human growth hormone in poly(lactic-*co*-glycolic) (PLGA) microspheres, *Pharm. Res.*, 14(4), 420, 1997.
106. Khang, G., Cho, J. C., Lee, J. W., Rhee, J. M., and Lee, H. B., Preparation and characterization of Japanese encephalitis virus vaccine loaded poly(L-lactide-*co*-glycolide) microspheres for oral immunization, *Biomed. Mater. Eng.*, 9(1), 49, 1999.
107. King, T. W. and Patrick, C. W. Jr., Development and in vitro characterization of vascular endothelial growth factor (VEGF)-loaded poly(D,L-lactic-*co*-glycolic acid)/poly(ethylene glycol) microspheres using a solid encapsulation/single emulsion/solvent extraction technique, *J. Biomed. Mater. Res.*, 51, 383, 2000.
108. Lam, X. M., Duenas, E. T., Daugherty, A. L., Levin, N., and Cleland, J. L., Sustained release of recombinant human insulin-like growth factor-I for treatment of diabetes, *J. Control. Release*, 67(2–3), 281, 2000.
109. Richardson, T. P., Peters, M. C., Ennet, A. B., and Mooney, D. J., Polymeric system for dual growth factor delivery, *Nat. Biotechnol.*, 19, 1029, 2001.
110. Lu, L., Stamatas, G. N., and Mikos, A. G., Controlled release of transforming growth factor beta-1 from biodegradable polymer microparticles, *J. Biomed. Mater. Res.*, 50, 440, 2000.
111. Peter, S. J., Lu, L., Kim, D. J., Stamatas, G. N., Miller, M. J., Yaszemski, M. J., and Mikos, A. G., Effects of transforming growth factor beta-1 released from biodegradable polymer microparticles on marrow stromal osteoblasts cultured on poly(propylene fumarate) substrates, *J. Biomed. Mater. Res.*, 50, 452, 2000.
112. Nugent, M. A., Chen, O. S., and Edelman, E. R., Controlled release of fibroblast growth factor: activity in cell culture, *Mater. Res. Soc. Symp. Proc.*, 252, 273, 1992.
113. Chacon, M., Molpeceres, J., Berges, L., Guzman, M., and Aberturas, M. R., Stability and freeze-drying of cyclosporine loaded poly(D,L-lactide-glycolide) carriers, *Eur. J. Pharm. Sci.*, 8(2), 99, 1999.
114. Verrijk, R., Smolders, I. J., McVie, J. G., and Begg, A. C., Polymer-coated albumin microspheres as carriers for intravascular tumour targeting of cisplatin, *Cancer Chemother. Pharmacol.*, 29, 117, 1991.
115. Hausberger, A. G. and DeLuca, P. P., Characterization of biodegradable poly(D,L-lactide-*co*-glycolide) polymers and microspheres, *J. Pharm. Biomed. Anal.*, 13, 747, 1995.
116. Burton, K. W., Shameem, M., Thanoo, B. C., and DeLuca, P. P., Extended release peptide delivery systems through the use of PLGA microsphere combinations, *J. Biomater. Sci. Polym. Ed.*, 11, 715, 2000.
117. Capan, Y., Woo, B. H., Gebrekidan, S., Ahmed, S., and DeLuca, P. P., Influence of formulation parameters on the characteristics of poly(D,L-lactide-*co*-glycolide) microspheres containing poly(L-lysine) complexed plasmid DNA, *J. Control. Release*, 60, 279, 1999.
118. Ravivarapu, H. B., Burton, K., and DeLuca, P. P., Polymer and microsphere blending to alter the release of a peptide from PLGA microspheres, *Eur. J. Pharm. Biopharm.*, 50, 263, 2000.

119. Ravivarapu, H. B., Lee, H., and DeLuca, P. P., Enhancing initial release of peptide from poly(D,L-lactide-*co*-glycolide) (PLGA) microspheres by addition of a porosigen and increasing drug load, *Pharm. Dev. Technol.*, 5, 287, 2000.
120. Schrier, J. A. and DeLuca, P. P., Recombinant human bone morphogenetic protein-2 binding and incorporation in PLGA microsphere delivery systems, *Pharm. Dev. Technol.*, 4, 611, 1999.
121. Rodgers, J. B., Vasconez, H. C., Wells, M. D., DeLuca, P. P., Faugere, M. C., Fink, B. F., and Hamilton, D., Two lyophilized polymer matrix recombinant human bone morphogenetic protein-2 carriers in rabbit calvarial defects, *J. Craniofac. Surg.*, 9, 147, 1998.
122. Polakovic, M., Gorner, T., Gref, R., and Dellacherie, E., Lidocaine loaded biodegradable nanospheres. II. Modeling of drug release, *J. Control. Release*, 60(2–3), 169, 1999.
123. Suh, H., Jeong, B., Rathi, R., and Kim, S. W., Regulation of smooth muscle cell proliferation using paclitaxel-loaded poly(ethylene oxide)–poly(lactide/glycolide) nanospheres, *J. Biomed. Mater. Res.*, 42(2), 331, 1998.
124. Barichello, J. M., Morishita, M., Takayama, K., and Nagai, T., Encapsulation of hydrophilic and lipophilic drugs in PLGA nanoparticles by the nanoprecipitation method, *Drug. Dev. Ind. Pharm.*, 25(4), 471, 1999.
125. Carino, G. P., Jacobs, J. S., and Mathiowitz, E., Nanosphere based oral insulin delivery, *J. Control. Release*, 65(1–2), 261, 2000.
126. Feng, S. S., Huang, G. F., and Mu, L., Nanospheres of biodegradable polymers: a system for clinical administration of an anticancer drug paclitaxel (Taxol), *Ann. Acad. Med. Singapore*, 29, 633, 2000.
127. Görner, T., Gref, R., Michenot, D., Sommer, F., Tran, M. N., and Dellacherie, E., Lidocaine-loaded biodegradable nanospheres. I. Optimization of the drug incorporation into the polymer matrix, *J. Control. Release*, 57, 259, 1999.
128. Wang, Y. M., Sato, H., Adachi, I., and Horikoshi, I., Preparation and characterization of poly(lactic-*co*-glycolic acid) microspheres for targeted delivery of a novel anticancer agent, taxol, *Chem. Pharm. Bull.*, 44, 1935, 1996.
129. Berkland, C., King, M., Cox, A., Kim, K., and Pack, D. W., Precise control of PLG microsphere size provides enhanced control of drug release rate, *J. Control. Release*, 82, 137, 2002.
130. Sheridan, M. H., Shea, L. D., Peters, M. C., and Mooney, D. J., Bioabsorbable polymer scaffolds for tissue engineering capable of sustained growth factor delivery, *J. Control. Release*, 64(1–3), 91, 2000.
131. Mooney, D. J., Kaufmann, P. M., Sano, K., Schwendeman, S. P., Majahod, K., Schloo, B., Vacanti, J. P., and Langer, R., Localized delivery of epidermal growth factor improves the survival of transplanted hepatocytes, *Biotech. Bioeng.*, 50, 422, 1996.
132. Meese, T. M., Hu, Y., Nowak, R. W., and Marra, K. G., Surface studies of coated polymer microspheres and protein release from tissue-engineered scaffolds, *J. Biomater. Sci. Polym. Ed.*, 13, 141, 2002.
133. Hu, Y., Hollinger, J. O., and Marra, K. G., Controlled release from coated polymer microparticles embedded in tissue-engineered scaffolds, *J. Drug Target.*, 9, 431, 2001.
134. Royce, S. M. and Marra, K. G., Release of FGF-1 from PLGA microspheres incorporated in fibrin scaffolds, *J. Control. Release*, 15, 1327, 2004.
135. Redl, H., Stanek, G., Hirchl, A., and Seelich, T., In vitro properties of mixtures of fibrin seal and antibiotics, *Biomaterials*, 4, 29, 1983.
136. Greco, F., De Palma, L., Spagnolo, N., Rossi, A., Specchia, N., and Gigante, A., Fibrin antibiotic mixtures: an in vitro study assessing the possibility of using a biologic carrier for local drug delivery, *J. Biomed. Mater. Res.*, 5, 39, 1991.
137. Lasa, C. Jr., Hollinger, J., Drohan, W., and MacPhee, M., Delivery of demineralized bone powder by fibrin sealant, *J. Plast. Reconstr. Surg.*, 96, 1409, 1995.
138. Currie, L. J., Sharpe, J. R., and Martin, R., The use of fibrin glue in skin grafts and tissue-engineered skin replacements: a review, *Plast. Reconstr. Surg.*, 108, 1713, 2001.
139. Yin, Q. W., Kemp, G. J., Yu, L. G., Wagstaff, S. C., and Frostick, S. P., Neurotrophin-4 delivered by fibrin glue promotes peripheral nerve regeneration, *Muscle Nerve*, 24, 345, 2001.
140. Sakiyama-Elbert, S. E. and Hubbell, J. A., Controlled release of nerve growth factor from a heparin-containing fibrin-based cell ingrowth matrix, *J. Control. Release*, 69, 149, 2000.

141. Hildebrand, K. A., Woo, S. L., Smith, D. W., Allen, C. R., Deie, M., Taylor, B. J., and Schmidt, C. C., The effects of platelet-derived growth factor-BB on healing of the rabbit medial collateral ligament. An in vivo study, *Am. J. Sports Med.*, 26, 549, 1998.
142. Iwaya, K., Mizoi, K., Tessler, A., and Itoh, Y., Neurotrophic agents in fibrin glue mediate adult dorsal root regeneration into spinal cord, *Neurosurgery*, 44, 589, 1999.
143. Pandit, A. S., Feldman, D. S., Caulfield, J., and Thompson, A., Stimulation of angiogenesis by FGF-1 delivered through a modified fibrin scaffold, *Growth Factors*, 15, 113, 1998.
144. Ye, Q., Zund, G., Benedikt, P., Jockenhoevel, S., Hoerstrup, S. P., Sakyama, S., Hubbell, J. A., and Turina, M., Fibrin gel as a three dimensional matrix in cardiovascular tissue engineering, *Eur. J. Cardiothorac. Surg.*, 17, 587, 2000.
145. Schense, J. C. and Hubbell, J. A., Cross-linking exogenous bifunctional peptides into fibrin gels with factor XIIIa, *Bioconjug. Chem.*, 10, 75, 1999.
146. Baker, S. B., Weinzweig, J., Kirscier, R. E., and Bartlett, S. P., Applications of a new carbonated calcium phosphate bone cement: early experience in pediatric and adult craniofacial reconstruction, *Plast. Reconstr. Surg.*, 109, 1789, 2002.
147. Glowacki, J., Kaban, L. B., Murray, J. E., Folkman, J., and Mulliken, J. B., Application of the biological principle of induced osteogenesis for craniofacial defects, *Lancet*, 1, 959, 1981.
148. Boo, J. S., Yamada, Y., Okazaki, Y., Hibino, Y., Okada, K., Hata, K., Yoshikawa, T., Sugiura, Y., and Ueda, M., Tissue-engineered bone using mesenchymal stem cells and a biodegradable scaffold, *J. Craniofac. Surg.*, 13, 231, 2002.

31 Hemoglobin-Based Red Blood Cell Substitutes

Thomas Ming Swi Chang

CONTENTS

I. INTRODUCTION

The first report on artificial cells, including artificial red blood cells, was published as early as 1964.[1] Our early research shows the feasibility of artificial cells in biotechnology including blood substitutes, enzyme therapy, cell encapsulation, drug delivery, and other applications (Figure 31.1).[1–3] Increasing interest about this approach only developed in the early 1980s, which follows the explosive international interests in all areas of biotechnology around this same time. This has resulted in increasing progress globally on the use of artificial cells in biotechnology and medicine.[4–8] It was the concerns regarding HIV in donor blood that stimulated the development of hemoglobin-based blood substitutes as red blood cell replacements starting in the early 1990s.[9–15] These can be stored at 4°C or room temperature for more than 1 year. They have no blood group antigen and can be used as universal donors. They can also be sterilized to remove and inactivate infectious agents like HIV and hepatitis viruses. This review summarizes the present status on hemoglobin-based blood substitutes.

II. FIRST-GENERATION BLOOD SUBSTITUTES

A. POLYHEMOGLOBIN

Inside red blood cells, hemoglobin (Hb) is a tetramer of four subunits with a 2,3-GPD pocket. When infused, the tetramer breaks down into toxic dimers. Our basic research shows that bifunctional

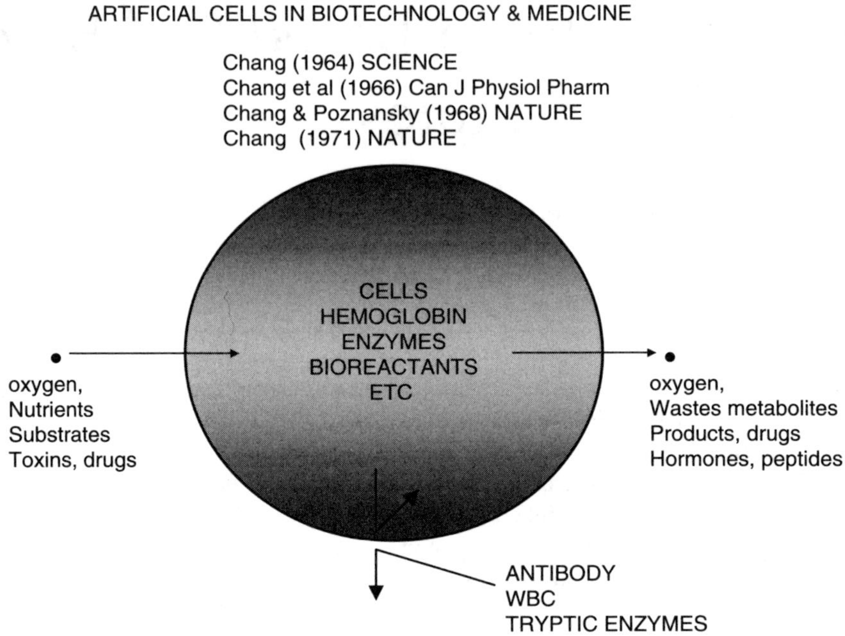

FIGURE 31.1 Basic principle of artificial cells. (From *Artificial Cells, Blood Substitutes & Biotechnology. An International Journal,* Vol. 32, pp. 1–16, 2004, with permission from Marcel Dekker, publisher.[16])

agents can be used to cross-link hemoglobin to prevent its break down.[1,3] We first used sebacyl chloride to cross-link hemoglobin into polyhemoglobin.[1,3] The reaction is as follows:

$$\underset{\text{Sebacyl Chloride}}{Cl(CH_2)_8Cl} + \underset{\text{Hemoglobin}}{HBNH_2} = \underset{\text{Cross-linked polyhemoglobin}}{HB(CH_2)_8HB}$$

We then used another bifunctional agent, glutaraldehyde, to cross-link hemoglobin and an rbc enzyme, catalase, into soluble polyhemoglobin.[17] The reaction is as follows:

$$\underset{\text{Glutaraldehyde}}{H{-}CO{-}(CH_2)_3{-}CO{-}H} + \underset{\text{Hemoglobin}}{HBNH_2} = \underset{\text{Cross-linked polyhemoglobin}}{HB{-}NH{-}CO{-}(CH_2)_3{-}CO{-}NH{-}HB}$$

This glutaraldehyde cross-linked polyhemoglobin approach has been developed by a number of groups to its present state suitable for clinical use as described below.

B. Glutaraldehyde Cross-Linked Human Pyridoxalated Polyhemoglobin

Gould et al.[18,19] have reported their clinical trial using glutaraldehyde polymerized pyridoxalated human polyhemoglobin with the following characteristics. They have removed most of the single cross-linked tetrameric hemoglobin ($<1\%$ remaining) leaving $>99\%$ of hemoglobin in the form of polyhemoglobin. Each unit (500 ml) contains 50 g of the polymerized hemoglobin in lactated Ringer solution. It has a P_{50} of 28 to 30 mmHg and its half-life in the circulation after infusion is 24 h and shelf life is more than 1 year. Their phase I clinical trial shows that the infusion of 500 ml did not produce any gastrointestinal smooth muscle spasm, or vasoconstriction, nor had any effects on the kidney or other organs.

In their phase II clinical trial in trauma surgery,[20] 44 trauma patients aged 19 to 75 with an injury severity score (ISS) of 21 ± 10 were randomized to receive red cells ($N = 23$) or up to six units of PolyHeme ($N = 21$) after trauma and during emergent operations. After six units of PolyHeme, any further transfusion needed would be given as allogeneic blood. This way, 12 patients received six units, three patients received four units, one patient received three units, two patients received two units, and one patient received one unit of PolyHeme. There was no significant difference between the control group using rbc and the experimental group using polyhemoglobin in regard to any adverse events. There were no vasoactive properties, renal dysfunction, fever, or other measurements of organ functions. Gould et al. followed the total hemoglobin as: total (Hb) = rbc(Hb) + poly(Hb). They showed that each 500 ml of PolyHeme (50 g Hb) raises the poly (Hb) of the plasma by about 1 g/dl. In the experimental group, at the end of infusion: total (Hb) was 9.0 ± 2.0 g/dl with 5.8 ± 2.8 g/dl from red blood cells and 3.9 ± 1.3 g/dl from the infused poly (Hb). The authors calculated this to represent the replacement of greater than 40% of the total circulating (Hb) by PolyHeme. They also found that PolyHeme with a circulation half-life of 24 h was able to significantly decrease the need for blood transfusion during the first 24 h. They recently reported the results of their phase III clinical trial with a much larger number of patients.[19] Infusion of PolyHeme maintains adequate total (Hb) even in those with rbc (Hb) of less than 3 g/dl. It is safe during rapid, massive infusion, to the amount of 20 units studied. Their data shows that this approach significantly decreases the mortality of trauma patients including those with less than 3 g/dl total red blood cell hemoglobin when compared to historical controls.[19]

C. Glutaraldehyde Cross-Linked Bovine Polyhemoglobin

Biopure Corporation has developed glutaraldehyde cross-linked polyhemoglobin using bovine hemoglobin instead of human hemoglobin, and they have described this in detail elsewhere.[21–23] Bovine hemoglobin is available in large amounts, furthermore, unlike human hemoglobin, even without 2-3-diphosphoglycerate or its analogue, pyridoxal-phosphate, bovine hemoglobin has much higher P_{50} than human hemoglobin. They also remove as much single cross-linked tetrameric hemoglobin as possible resulting in a preparation with less than 5% of cross-linked tetrameric hemoglobin. Single cross-linked tetrameric hemoglobin, unlike polyhemoglobin, can cross the intercellular junction and may result in vasoconstriction. Each unit (500 ml) contains 6 to 7 g of hemoglobin in lactated Ringer with a P_{50} of 38 mmHg and a half-life in the circulation of 24 h. One of the most surprising findings is that its shelf life at room temperature is more than 2 years. LaMuraglia et al.[22] reported their phase II single blind, multicenter trial in infrarenal aortic aneurysm surgery. A group of 72 patients were randomized to receive a blood substitute ($N = 48$) or allogenic red blood cells ($N = 24$) at the time of the first transfusion decision. Those randomized to the blood substitute group received 60 g of polyhemoglobin for the initial transfusion, and had the option to receive three further doses of 30 g each within 96 h. After this, allogenic red blood cells were given for any further requirements in the 28 days following randomization. The two groups were comparable for all baseline characteristics. All patients in the control group required allogenic red blood cell transfusions. Of those in the blood substitute group, 27% did not require allogenic red blood cell transfusion in the 28-day period. Antibody levels of IgE to bovine hemoglobin were not induced in any of the patients in this study. There was a transient 15% increase in blood pressure, but the cardiac index and pulmonary pressures were not affected. The transfusions were well tolerated and there were no significant differences in complications, morbidity, or mortality rates between the two groups. Mullon et al. reported in the *New England Journal of Medicine* the use of this in a patient with acute hemolytic anemia.[23] Hemolysis was so severe that they were unable to maintain the hemoglobin level that was less than 4 g/dl despite standard treatment plus a daily infusion of eight units of donor blood. Incremental infusion of their polyhemoglobin in the amounts of 5, 3 and 3 units, a total of 11 units over 7 days, were able to maintain the total hemoglobin level at 5.5 g/dl. The patient eventually recovered and was discharged. This is now approved for routine clinical uses in South Africa. Sprung et al. recently reported their results of a multicenter, randomized, single-blinded trial in surgical patients,[24] and have subsequently applied to the U.S. Food and Drug Administration (FDA) for approval.

D. Other Types of Modified Hemoglobin

The above two polyhemoglobin preparations are described in detail since they are in the most advanced stage of development and clinical trial pending FDA approval. There are a number of other types of first-generation modified hemoglobin (Figure 31.2). One is based on the use of *o*-raffinose to cross-link human hemoglobin into a mixture of polyhemoglobin and cross-linked tetrameric hemoglobin.[25] Another type is to intramolecularly cross-link hemoglobin to form a single cross-linked tetrameric hemoglobin.[26] The results of their phase III clinical trials have led them to develop a second-generation recombinant human hemoglobin. Recombinant tetrameric human hemoglobin can be prepared using genetically engineered *Escherichia coli*.[27] Results from clinical trials[28] led to the development of a second-generation recombinant human hemoglobin that no longer has vasoactivity in animal studies.[29] Another group of hemoglobin-based blood substitutes is conjugated hemoglobin formed by cross-linking hemoglobin to polymer.[1,3] This has been extended to form soluble conjugated hemoglobin by cross-linking individual hemoglobin molecule to a soluble polymer and has been used in clinical trials.[30] More recently, a new Maleimide PEG-hemoglobin has been developed and is now in a phase II clinical trial.[14]

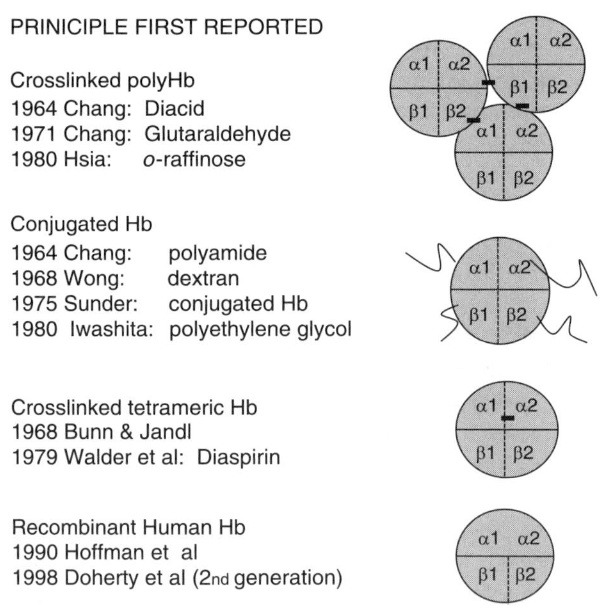

FIGURE 31.2 Different types of first-generation hemoglobin-based blood substitutes (from *Monograph on Blood Substitutes: Principles, Methods, Products and Clinical Trials*. Karger Landes Systems, 1997, with permission from copyright holder).

III. SECOND-GENERATION BLOOD SUBSTITUTES

A. Polyhemoglobin–Superoxide Dismutase–Catalase in Ischemia Reperfusion

The two polyhemoglobin preparations described above have a number of advantages when compared to donor red blood cells, and they are particularly useful for surgery. However, these are only oxygen carriers and do not have the full functions of red blood cells that may be needed for certain clinical conditions.[15] For example, reperfusion with oxygen carrying fluids like polyhemoglobin to sustain severely ischemic tissues as in strokes, myocardial infarction, organ transplantation, severe sustained hemorrhagic shock, and other conditions, can result in the release of superoxide, oxygen radicals, and other reactions leading to tissue injuries. First-generation modified hemoglobin usually does not contain antioxidant enzymes normally present in red blood cells. To solve this problem we cross-linked catalase and superoxide dismutase (SOD) to hemoglobin to form PolyHb–SOD-catalase (CAT).[31] Compared to PolyHb, PolyHb–SOD-catalase removes significantly more oxygen radicals and peroxides and stabilizes the cross-linked hemoglobin resulting in decrease oxidative iron and heme release.[32] In the reperfusion of ischemic rat intestines, PolyHb–SOD-CAT also significantly reduces the increase in oxygen radicals as measured by an increase in 3,4 dihydroxybenzoate when compared to using PolyHb.[33] Unlike PolyHb, PolyHb–SOD-CAT did not cause a significant increase in oxygen radicals when used to reperfuse ischemic rat intestines.[33] More recently, in a transient global cerebral ischemia rat model, we found that after 60 min of ischemia, reperfusion with polyHb resulted in significant increases in brain edema due to the breakdown of blood–brain barrier caused by the damaging oxygen radicals formed from ischemia reperfusion.[34] On the other hand, polyHb–SOD-CAT supplies oxygen and at the same time removes oxygen radicals, and as a result does not cause this adverse effect.[34]

IV. THIRD-GENERATION BLOOD SUBSTITUTES

The above first- and second-generation hemoglobin-based blood substitutes are not covered by a membrane as in red blood cells. Thus, hemoglobin are exposed to the external environment with more potential for oxidant activity. Furthermore, the circulation half-life is rather short: 24 h. As a result, Chang's original idea of a complete artificial rbc[1,3] is now being developed as a third-generation blood substitute in two forms.

A. Hemoglobin Lipid Vesicles

The first type is in the form of lipid vesicles containing hemoglobin.[35] This has been explored by Rudolph in the U.S.,[11] and more recently by Tsuchida's group in Japan.[10] The U.S. group has modified the surface properties to result in a circulation half-life of about 50 h.[36] The Japanese group is completing their animal studies towards a clinical trial.

B. Nanodimension Artificial Red Blood Cells with Biodegradable Polymeric Membrane

We are developing another type based on biodegradable polymer (polyethylene glycol polylactide copolymer) and nanotechnology resulting in nanocapsules of 100 to 150 nm diameter containing hemoglobin and rbc enzymes (Figure 31.2).[37–40] Unlike lipids, biodegradable membranes are converted to water and carbon dioxide in the body, and therefore do not accumulate in the reticuloendothelial system (Figure 31.3). We have included superoxide dismutase, catalase, carbonic anhydrase, and in addition, multienzyme systems to prevent the accumulation of methemoglobin. Our recent studies show that by using a polyethylene-glycol-polylactide copolymer membrane we are able to increase the circulation time of these nanoartificial rbc to double that of polyHb.[40]

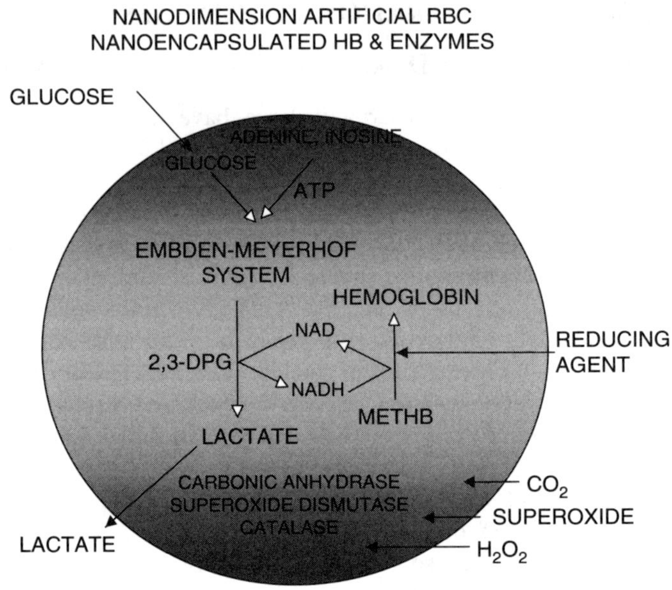

FIGURE 31.3 Nanodimension artificial red blood cells with polyethylene-glyco-polylactide membrane. In addition to hemoglobin, this contains the same enzymes that are normally present in red blood cells. Thus, it has the complete function of the red blood cells. (From *Artificial Cells, Blood Substitutes & Biotechnology. An International Journal,* Vol. 31(3), pp. 231–248, 2003, with permission from Marcel Dekker, publisher.)

ACKNOWLEDGMENTS

The support of the Canadian Institute for Health Research (previously the Medical Research Council of Canada), the "Virage" Centre of Excellence in Biotechnology from the Quebec Ministry of Education, Science and Technology, the Quebec Ministry of Health FRSQ Research Group on Blood Substitutes in Transfusion Medicine, and the Bayer-Canadian Blood Agency-Hema Quebec Research and Development Fund (previously Bayer-Canadian Red Cross) are all gratefully acknowledged.

REFERENCES

1. Chang, T. M. S., Semipermeable microcapsules, *Science*, 146(3643), 524–525, 1964.
2. Chang, T. M. S., MacIntosh, F. C., and Mason, S. G., Semipermeable aqueous microcapsules: I. Preparation and properties, *Can. J. Physiol. Pharmacol.*, 44, 115–128, 1966.
3. Chang, T. M. S., *Artificial Cells*, Monograph. Charles C Thomas, Springfield, IL, 1972 (available for free online viewing at www.artcell.mcgill.ca).
4. Chang, T. M. S., *Artificial Cells "Encyclopedia of Human Biology"* 2nd ed., Dulbecco, Renalo [editor-in-chief], Academic Press, Inc., San Diego, CA, pp. 457–463, 1997.
5. Kuhreibez, W. M., Lauza, P. P., Cuicks, W. L., eds., *Cell Encapsulation Technology and Therapy*, Burkhauser, Boston, 1999.
6. Hunkeler, D., Prokop, A., Cherrington, A. D., Rajotte, R., Sefton, M., eds., *Bioartificial Organs A: Technology, Medicine and Material, Ann. NY Acad. Sci.*, 875, 1999; *Acad. Sci.* 831, 271–279.
7. Chang, T. M. S., Artificial cells bioencapsulation in macro, micro, nano and molecular dimensions, *Artif. Cells Blood Substit. Biotechnol. Int. J.*, 32, 1–14, 2004.
8. Chang, T. M. S., Therapeutic applications of polymeric artificial cells. *Nat. Rev. Drug Discov.*, 4, 221–235, 2005.

9. Chang, T. M. S., *Blood Substitutes: Principles, Methods, Products and Clinical Trials*, Vol. 1, Karger/Landes, Basel/Texas, 1997.
10. Tsuchida, E., ed., *Blood Substitutes: Present and Future Perspectives*, Elservier, Amsterdam, 1998.
11. Rudolph, A. S., Rabinovici, R., Feuerstein, G. Z., eds., *Rbc Substitutes*, Marcel Dekker, Inc., New York, 1997.
12. Chang, T. M. S., "Red Blood Cell Substitutes" Best Practice & Research, *Clin. Haematol.*, 13(4), 651–668, 2000.
13. Chang, T. M. S., Oxygen Carriers, *Curr. Opin. Investig. Drugs*, 3(8), 1187–1190, 2002.
14. Winslow, R., Current status of blood substitute research: towards a new paradigm, *J. Int. Med.*, 253, 508–517, 2003.
15. Chang, T. M. S., New generations of red blood cell substitutes, *J. Int. Med.*, 253, 527–535, 2003.
16. Chang, T. M. S., Artificial cells bioencapsulation in macro, micron, nano and molecular dimensions, *Artif. Cells Blood Substit. Biotechnol. Int. J.*, 32, 1–16, 2004.
17. Chang, T. M. S., Stabilization of enzyme by microencapsulation with a concentrated protein solution or by crosslinking with glutaraldehyde, *Biochem. Biophys. Res. Commun.*, 44, 1531–1533, 1971.
18. Gould, S. A., Sehgal, L. R., Sehgal, H. L., DeWoskin, R., and Moss, G. S., The clinical development of human polymerized hemoglobin, *Blood Substitutes: Principles, Methods, Products and Clinical Trials*, Vol. 2, Chang, T. M. S., ed., Karger/Landes, Basel/Texas, pp. 12–28, 1998.
19. Gould, S. A., Moore, E. E., Hoyt, D. B., Burch, J. M., Haenel, J. B., Garcia, J., DeWoskin, R., and Moss, G. S., The first randomized trial of human polymerized hemoglobin as a blood substitute in acute trauma and emergent surgery, *J. Am. Coll. Surg.*, 187, 113–120, 1998.
20. Gould, S. A., Moore, E. E., Hoyt, D. B., Ness, P. M., Norris, E. J., Carson, J. L., Hides, G. A., Freeman, I. H. G., DeWoskin, R., and Moss, G. S., The life-sustaining capacity of human polymerized hemoglobin when red cells might be unavailable, *J. Am. Coll. Surg.*, 195, 445–452, 2002.
21. Pearce, L. B., and Gawryl, M. S., Overview of preclinical and clinical efficacy of Biopure's HBOCs, *Blood Substitutes: Principles, Methods, Products and Clinical Trials*, Vol. 2, Chang, T. M. S., ed., Karger/Landes, Basel/Texas, pp. 82–98, 1998.
22. LaMuraglia, G. M., O'Hara, P. J., Baker, W. H., Naslund, T. C., Norris, E. J., Li, J., and Vandermeersch, E., The reduction of the allogenic transfusion requirement in aortic surgery with a hemoglobin-based solution, *J. Vasc. Surg.*, 31(2), 2000.
23. Mullon, J., Giacoppe, G., Clagett, C., McCune, D., and Dillard, T., Transfusions of polymerized bovine hemoglobin in a patient with severe autoimmune hemolytic anemia, *N. Engl. J. Med.*, 342, 1638–1643, 2000.
24. Sprung, J., Kindscher, J. D., Wahr, J. A., Levy, J. H., Monk, T. G., Moritz, M. W., and O'Hara, P. J., The use of bovine hemoglobin glutamer-250 (Hemopure) in surgical patients: results of a multicenter, randomized, single-blinded trial, *Anesth. Analg.*, 94, 799–808, 2002.
25. Adamson, J. G., Moore, C., and Hemolink, T. M., An *o*-Raffinose crosslinked hemoglobin-based oxygen carrier, In *Blood Substitutes: Principles, Methods, Products and Clinical Trials*, Chang, T. M. S., ed., Karger/Landes, Basel/Texas, pp. 62–79, 1998.
26. Nelson, D. J., Blood and HemAssistTM (DCLHb): potentially a complementary therapeutic team, *Blood Substitutes: Principles, Methods, Products and Clinical Trials*, Vol. 2, Chang, T. M. S., ed., Karger/Landes, Basel/Texas, pp. 39–57, 1998.
27. Hoffman, S. J., Looker, D. L., Roehrich, J. M., Cozart, P. E., Durfee, S. L., Tedesco, J. L., and Stetler, G. L., Expression of fully functional tetrameric human hemoglobin in escherichia coli, *Proc. Natl Acad. Sci. USA*, 87, 8521–8525, 1990.
28. Freytag, J. W., and Templeton, D., OptroTM (recombinant human hemoglobin): a therapeutic for the delivery of oxygen and the restoration of blood volume in the treatment of acute blood loss in trauma and surgery, In *Red Cell Substitutes: Basic Principles and Clinical Application*, Rudolph, A. S., Rabinovici, R. and Feuerstein, G. Z., eds., Marcel Dekker, Inc., New York, pp. 325–334, 1997.
29. Doherty, D. H., Doyle, M. P., Curry, S. R., Vali, R. J., Fattor, T. J., Olson, J. S., and Lemon, D. D., Rate of reaction with nitric oxide determines the hypertensive effect of cell-free hemoglobin, *Nat. Biotechnol.*, 16, 672–676, 1998.
30. DeAngelo, J., Nitric oxide scavengers in the treatment of shock associated with systemic inflammatory response syndrome, *Exp. Opin. Pharmacother.*, 1, 19–29, 1999.

31. D'Agnillo, F., and Chang, T. M. S., Polyhemoglobin-superoxide dismutase catalase as a blood substitute with antioxidant properties, *Nat. Biotechnol.*, 16(7), 667–671, 1998.
32. D'Agnillo, F., and Chang, T. M. S., Absence of hemoprotein-associated free radical events following oxidant challenge of crosslinked hemoglobin-superoxide dismutase-catalase, *Free Radic. Biol. Med.*, 24(6), 906–912, 1998.
33. Razack, S., D'Agnillo, F., and Chang, T. M. S., Effects of polyhemoglobin-catalase-superoxide dismutase on oxygen radicals in an ischemia-reperfusion rat intestinal model, *Artif. Cells Blood Substit. Immobil. Biotechnol.*, 25, 181–192, 1997.
34. Powanda, D., and Chang, T. M. S., Cross-linked polyhemoglobin-superoxide dismutase-catalase supplies oxygen without causing blood brain barrier disruption or brain edema in a rat model of transient global brain ischemia-reperfusion, *Artif. Cells Blood Substit. Immobil. Biotechnol. Int. J.*, 30, 25–42, 2002.
35. Djordjevich, L., and Miller, I. F., Synthetic erythrocytes from lipid encapsulated hemoglobin, *Exp. Hematol.*, 8, 584, 1980.
36. Philips, W. T., Klpper, R. W., Awasthi, V. D., Rudolph, A. S., Cliff, R., Kwasiborski, V. V., and Goins, B. A., Polyethylene glyco-modified liposome-encapsulated hemoglobin: a long circulating red cell substitute, *J. Pharm. Exp. Ther.*, 288, 665–670, 1999.
37. Yu, W. P., and Chang, T. M. S., Submicron biodegradable polymer membrane hemoglobin nanocapsules as potential blood substitutes: a preliminary report, *J. Artif. Cells Blood Substit. Immobil. Biotechnol.*, 22, 889–894, 1994.
38. Yu, W. P., and Chang, T. M. S., Submicron polymer membrane hemoglobin nanocapsules as potential blood substitutes: preparation and characterization, *Artif. Cells Blood Substit. Immobil. Biotechnol. Int. J.*, 24, 169–184, 1996.
39. Chang, T. M. S., and Yu, W. P., Nanoencapsulation of hemoglobin and rbc enzymes based on nanotechnology and biodegradable polymer, *Blood Substitutes: Principles, Methods, Products and Clinical Trials*, Vol. 2, Chang, T. M. S., ed., Karger, Basel, pp. 216–231, 1998.
40. Chang, T. M. S., Powanda, D., and Yu, W. P., Analysis of polyethylene-glycol-polylactide nano-dimension artificial red blood cells in maintaining systemic hemoglobin levels and prevention of methemoglobin formation, *Artif. Cells Blood Substit. Biotechnol. Int. J.*, 31(3), 231–248, 2003.

32 Nerve Regeneration

Erin Lavik and Robert Langer

CONTENTS

This chapter will focus primarily on scaffolds for regeneration in the central nervous system (CNS) since there is little endogenous regeneration in the CNS following injury in higher order mammals, whereas, there is significant regeneration in the peripheral nervous system (PNS) following injury. The CNS consists of the brain, spinal cord, and retina. The PNS consists of the sensory neurons which carry impulses from the receptors to the CNS and the motor neurons which carry impulses to the muscles and glands. There is a plethora of texts on the anatomy of the nervous system.[1–4]

The major cells types of the nervous system are the neurons and glia. Neurons are the cells which carry the impulses throughout the nervous system. Each has a nucleus called a cell body or soma, and extends axons, or neurites, which form synapses with the dendrites of other neurons.

The majority of the cell bodies of the neurons lie within the CNS. Some of the sensory neurons, such as the olfactory neurons, have their cell bodies outside the CNS. Glia are all the nonneural cells of the nervous system and include oligodendrocytes, astrocytes, and microglia in the CNS and Schwann cells in the PNS.

Oligodendrocytes produce the myelin sheaths which insulate the axons of the neurons. Their equivalent within the PNS are the Schwann cells. Schwann cells differ from oligodendrocytes in that they surround and myelinate individual axons, whereas oligodendrocytes extend processes to myelinate several axons. Since there are fewer cells responsible for myelination in the CNS, more efficient packing of axons is possible.[5] Astrocytes are less well understood but are believed to be involved in potassium ion and neurotransmitter concentration regulation, as well as in nutritional support for the neurons. The microglia are phagocytic cells that respond to infection and injury in the CNS.

I. SCAFFOLDS FOR THE PNS

A. The Structure of the PNS

The PNS is defined as the portion of the nervous system which connects the CNS with the tissue and organs of the body. The PNS consists of neurons and glia; however, the cell bodies of most of the neurons lie within the CNS. The PNS, then, is primarily axons and glia. The myelinating cells of the PNS are the Schwann cells which are thought to be critical for the regenerative capacity of the PNS. Ide provides an excellent review of the PNS in particular elucidating the structure and role of the basal lamina, which is the basement membrane extracellular matrix of the PNS.[6]

The sciatic nerve of the PNS which connects the lumbar portion of the spinal cord with the hind limbs has been one of the most studied nerves with respect to regeneration, particularly in rodent models, in great part because of its size, accessibility, and functional characterization. It connects the legs or hind limbs with the CNS in the lumbar region and is responsible for motor control of the hind limbs.

B. Previous Treatments for Repair of the PNS

Axons in the PNS are capable of regeneration after injury. However, when an axon is cut or damaged over a significant distance, intervention is often needed to restore function. When peripheral nerves are severed and the ends are adjacent, the ends may be sutured with some success. More often, however, the ends are damaged and tissue must be removed so that direct suturing results in the axon being under tension. Therefore, some sort of intermediate section which permits regeneration is needed.[7]

The most common clinical approach is to use autografts. These are autologous peripheral nerve segments taken from other, less critical, locations and sutured between the damaged ends of the nerve.[8] There are several limitations to the use of autografts including diameter matching between the autograft and host nerve, availability of nerves for harvest, invasive obtainment, and only partial recovery. These limitations have motivated other approaches for repair.

Allografts of nerves harvested from other sources, including cadavers, has been explored with limited success.[9] Venous conduits have also been employed with limited success. The conduits appear to act as substrates which promote the ingrowth of Schwann cells which, in turn, promote axonal growth and guidance.[10] It is no surprise that the logical extension of this was to the use of hollow tubes to promote repair.

Hollow tubes have been widely studied in the field. Their initial use was motivated in part by the need for efficient means of dealing with traumatic injuries during wars when autografts and allografts were not feasible.[11] Silicone tubes have been widely studied and have shown some success. They do promote the formation of a nerve cable across gaps of 10 mm or less.[12,13]

One possibility for the success of such hollow tubes may be confinement of growth factors released in response to the injury from the tissue near the nerve stumps. Essentially, the tube may limit diffusion of growth factors from the nerve stumps, keeping the local concentration at a high enough level to stimulate repair, higher than in the same injury case without the tube. Whatever the mechanism, silicone tubes have not been as successful as autografts. Tubes made from degradable materials[7,14] have led to thicker cables containing more myelinated axons than their nondegradable counterparts. In some cases they have produced similar functional outcomes to autografts for limited gap lengths.[7] However, regeneration has been only partially successful with tubes alone.

C. Scaffolds for Repair in the PNS

Limitations on the availability of tissue as well as the limited success of conduits has prompted the development of other approaches to repair including fibers, sutures, gels, and porous scaffolds to aid in physical guidance and growth factor presentation for regenerating axons (Figure 32.1).[15–18]

During development, axons sprout and seek targets. The axons are guided by chemical gradients and surface molecules.[19,20] Scaffolds have the potential to simulate these targeting cues through their surface chemistries and structures. Scaffolds also have the potential to offer directional guidance through their pore structures, as well as anchorage points for growth cones during regeneration. One of the simplest scaffolds with an oriented structure to guide axons is that of a series of filaments. Poly(L-lactic acid) filaments incorporated into hollow tubes succeeded in promoting regeneration of a limited number of axons across 14 and 18 mm gaps.[21] Degradable sutures, which are essentially degradable polymer filaments have shown similar results in 7 and 15 mm gaps,[22] but regeneration along the longer gap remained poor.

The next level of complexity for scaffolds is the natural and synthetic gels. While gels may seem to be very simple structures on the macroscopic level, they have the potential to be extremely complex and highly tailored at the molecular level. The incorporation of gels consisting of laminin

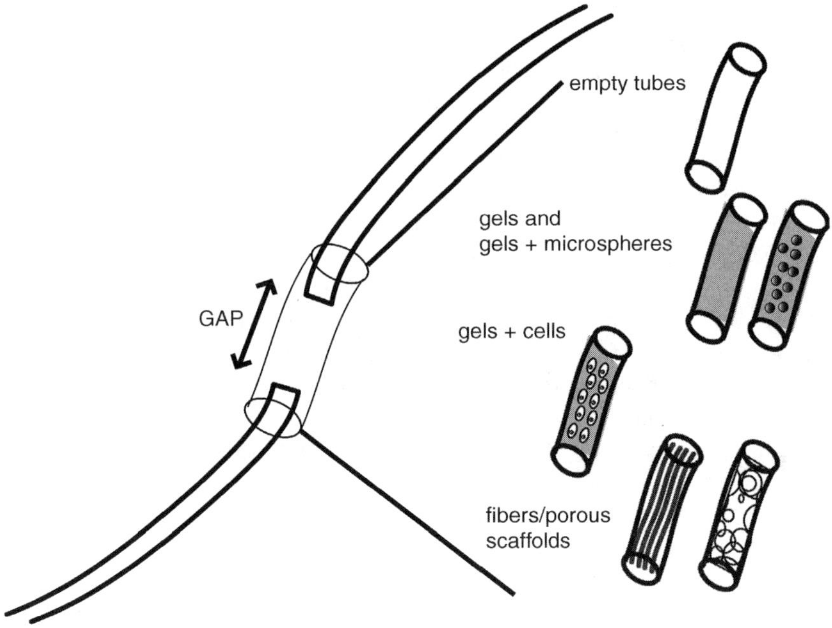

FIGURE 32.1 A schematic summarizing the approaches using tubes and scaffolds to promote repair in the PNS.

or collagen have led to the formation of nerve cables in the gap which are larger in diameter than their empty tube counterparts.[23] Laminin functionalized gels coupled with sustained delivery of nerve growth factor (NGF) performed comparably to autografts for injuries of 10 mm.[24] Again, though, the challenge has been to bridge larger gaps.

Porous collagen scaffolds have also been successful in bridging gaps and have been shown to favor the formation of high diameter axons as compared with empty tubes.[18] The high diameter axons are important for normal physiological performance. These scaffolds can provide physical cues to direct regenerating axons through their oriented pore structures.

One of the major limitations of the above approaches appears to have been the limited ingrowth of Schwann cells to provide a permissive environment for repair and to remyelinate the regenerating axons. In cases where Schwann cell migration was promoted, regeneration was improved.[22]

Microspheres which provide controlled release of growth factors,[25] Schwann cells,[26] and neural progenitors[27] have been incorporated in gels within the tubes to provide a similar environment to the autografts as well as the capacity to myelinate regenerating fibers. These have led to significant improvements in terms of recovery, but the autograft remains the choice in the clinic.

Scaffolds have shown promise for peripheral nerve regeneration. For the next generation of scaffolds, the question arises as to what factors and characteristics would make the scaffold most permissive for repair. A great deal is known about the structure, surface chemistry, and neurotrophic factors involved in endogenous peripheral regeneration[6] as well as the cues that permit axons to find their targets.[19] The key now is to incorporate these cues and characteristics into scaffolds to promote regeneration.[25,28] By displaying the appropriate surface chemistry and growth factors a trail may be placed to guide regeneration.

II. SCAFFOLDS FOR THE CNS: THE SPINAL CORD

A. Basic Anatomy of the Spinal Cord (an Engineering View)

It is necessary to first understand the anatomy of a healthy spinal cord before one can consider methods for the repair of an injured one. There is a plethora of texts on the anatomy of the nervous system, and the spinal cord in particular.[1–4] The spinal cord is the section of the CNS which runs down the vertebral column containing neurons and glia. As noted previously, glia are all the nonneural cells of the CNS, and include oligodendrocytes, astrocytes, and microglia.

The spinal cord is composed of two primary regions, the white matter and the gray matter. The cell bodies of the CNS lie within the gray matter. The white matter contains the long tracts of myelinated axons. The myelination gives the white matter its color, and hence its name.

There are several kinds of neurons with particular tracts in the spinal cord. Sensory, or afferent neurons, have cell bodies in the dorsal roots of the spinal cord. The axons of the motor neurons make up the descending pathways. Of particular note regarding the motor neurons is the corticospinal tract which has cell bodies in the cerebral cortex of the brain. It was once believed to be responsible for all voluntary motor control but is now believed to have a more specific role involving "skilled movements of the distal muscles of the limbs."[1] This pathway is of particular interest with regard to regeneration because it plays such a tremendous role in motor control and is easily identified via tracers for anatomical characterization.

Regeneration of a large number of particular tracts may not be necessary, though. There have been cases of return of sensory function following severance of one of the sensory tracts, the lateral spinothalmic tract. The procedure, a tractotomy, is performed to relieve severe intractable pain, and results in the loss of the ability to feel pain on the contralateral side below the cut. However, in some cases the effect is only temporary, which suggests rerouting of the signal through other pathways.[1] Thus, it may be possible to create new pathways which are able to function in the place of damaged ones.

B. Results of Spinal Cord Injury

Traumatic spinal cord injury results in axonal damage or severance, and in the loss of neurons, supporting cells, and myelin. While primary injury accounts for a portion of the damage, secondary injury processes, which occur for hours and days following injury, lead to further damage.[29] The majority of spinal cord injuries are contusion injuries in which the spinal cord sustains an impact without actual severance. In a severe injury, hemorrhages are seen at the injury epicenter in the gray matter within 30 min of the initial impact.[30]

Depending on the severity of impact, there is hemorrhaging within the gray matter at the injury epicenter followed by hemorrhaging in the white matter, axonal swelling, and edema formation leading to ischemia and hypoxia.[31] The release of excitatory amino acids and cytokines, lipid peroxidation, the invasion of immune cells, necrosis and apoptosis of neurons and glia, and degeneration of axons contribute to the degeneration, demyelination, and lesion expansion which continues for days following injury.[29,32]

The expansion of the lesion has been directly correlated with functional deficits. As the lesion encompasses a greater degree of the white matter, motor performance deteriorates.[33–35] However, it is important to note that with even a small degree of white matter remaining, there is still a significant degree of functional recovery. For example, Basso et al.[36] report that in a group of five rats where 9.9 ± 4.8% of the tissue at the injury epicenter was spared, the average BBB score (the Basso, Beattie, Bresnahan score which is a detailed 21 point scoring system used to quantify locomotion following SCI) following injury was 10.6 ± 0.6%, which correlates to occasional weight supported plantar steps. Similar results have been seen by other groups including Noble and Wrathall[35] who report that, following a moderate contusion injury using a weight drop method, at 8 weeks the animals retained no gray matter at the injury epicenter and only 3.1 ± 2.2% of their white matter, yet they exhibited a 75% deficit in a combined behavioral score calculated from the results of a battery of tests including analysis of locomotion, sensory reflexes, and swimming.[37] With minimal tissue remaining, the Sprague–Dawley rats used in the experiment still retained a significant amount of function. This suggests that if one can stem the secondary injury processes and promote even a small amount of regeneration following a severe spinal cord injury, significant functional recovery may occur.

Unlike the PNS, regeneration is essentially absent in the CNS following injury. In 1988, Schwab and Caroni[38] showed that myelin in the CNS appeared to inhibit regeneration, and in 1990, Schnell and Schwab[39] showed that regeneration was promoted in the spinal cord by the introduction of an antibody against myelin-inhibitory protein. Subsequent work suggested the administration of the antibody also promoted functional recovery.[40] Recently, the gene and the protein it encodes have been identified, and the protein has been shown to inhibit neurite growth.[41,42] The myelin inhibitory protein is not the only factor that inhibits regeneration following spinal cord injury. The upregulation of activated astrocytes, which appear to contribute to lesion expansion, may also participate in the formation of glial scarring which impedes regeneration.[43–45]

While the impediments are great, there is evidence that the CNS, even the white matter, can support regeneration. Davies et al. showed that axons could regenerate in uninjured white matter.[46] They further showed axons regenerated not only in uninjured white matter but in degenerating white matter of the spinal cord.[47] It appears that the injury site itself is inhibitory, but if axons traverse the injury site they may regenerate beyond the injury. Glial scar formation occurs, even in severe spinal cord injury, after the initial impact during the secondary injury processes.[30] If one can overcome the inhibitory environment of the injury site and suppress glial scar formation, at least temporarily, one may be able promote regeneration and functional recovery. Regeneration of a large number of particular tracts may not be necessary. There have been cases of return of sensory function following severance of one of the sensory tracts, the lateral spinothalmic tract. The procedure, a tractotomy, is performed to relieve severe intractable pain, and results in the loss of the ability to feel pain on the contralateral side below the cut. However, in some cases the effect is only temporary, which suggests rerouting of the signal through other pathways.[1] Thus, it may be possible to create new pathways which are able to function in the place of damaged ones.

III. APPROACHES TO TREATMENT, REPAIR, AND REGENERATION

A. Drug Interventions

1. Methylprednisolone

The current treatment for severe spinal cord injury is the administration of high doses of methylprednisolone within 8 h of injury.[48] It has drawn serious criticism and is not practiced at many hospitals. There are serious side effects to the treatment, and the efficacy is still debated. Clinical trials of the steroid have shown functional improvement when administrated immediately after injury.[49–51] While the exact mechanism of methylprednisolone is not known, it does inhibit lipid peroxidation and inflammation[29] and appears to reduce secondary injury. Although administration of methylprednisolone has led to improved functional outcomes as compared to a placebo treatment in preliminary trials, significant functional deficits remained.

2. Growth Factors

Growth factors are molecules that regulate protein synthesis which in turn regulates cell division and growth. Neurotrophic factors or neurotrophins are a specific class of growth factors which affect the development of the nervous system.[52] Among other regulatory behaviors, neurotrophins provide critical guidance for extending axons. Retrograde transport studies have been used to identify some of the neuronal cell populations in the CNS which respond to the various neurotrophins. DiStefano et al.[53] identified brain derived neurotrophic factor (BDNF) and neurotrophic factor 3 (NT-3) as being retrogradely transported primarily by motor neurons, whereas NGF was the primary factor retrogradely transported by populations of sensory neurons. This suggests that the axons of sensory neurons may be stimulated and guided by NGF, and likewise, axons of motor neurons may be stimulated and guided by NT-3 and BDNF.

Experimental studies of axonal extension in the CNS utilizing these neurotrophic factors correlate well with the above work. When NT-3 and BDNF were combined with Schwann cell-seeded guidance tubes,[54] the neurotrophic factors stimulated the ingrowth of axons from certain motor neurons including some distant neurons from the brain stem. Studies on the extension of axons in the corticospinal tract, a tract believed to be critical for voluntary motor control, suggest that NT-3 leads to axon sprouting and extension following severance,[55–57] but that BDNF and NGF have little effect.[57] NGF, however, appears to have an effect on the extension and ingrowth of sensory neurons. NGF has been shown to induce not only extension of sensory axons *in vivo*[58,59] but also appears to stimulate the ingrowth of sensory axons into the caudal end of a transected spinal cord, overcoming the effects of myelin-associated inhibitory proteins.[59]

While the use of neurotrophins only has been shown to improve functional recovery of sensory axons following compression injuries,[60] they appear to be most effective when used in concert with other therapeutic approaches such as implants or grafts.[54,61,62]

3. Role of Other Drugs and Antibodies

As discussed above, the introduction of the antibody to the myelin-associated inhibitory molecule leads to regrowth of axons following transection[39] and associated functional improvement.[40] More recent work has focused on identifying the specific sections of the antigens on which the antibody acts.[41,42,63] This work may lead to the development of new drugs to promote the regrowth of axons. Similar effects regarding axonal sprouting and regrowth have also been seen using drugs such as inosine. Specifically, inosine has been shown to induce significant outgrowth of the corticospinal tract following transection in an adult rat model.[64] Inosine may be attractive because it is more easily obtained than the inhibitory antibody. Another drug, Cyclosporin A has also been shown to be potentially therapeutic. Rather than inducing axonal outgrowth, it appears to be neuroprotective, mitigating secondary injury.[65,66] It is also one of the most common immunosuppressive drugs used during spinal cord injury studies, particularly with the implantation of cells or xenografts. Its use may, then, add a degree of complexity to results due to its neuroprotective effects.

B. Cell Therapies for the Spinal Cord

1. Schwann Cells

Schwann cells appear to be crucial for regeneration in the PNS. Thus, they have drawn great attention as a means for regeneration in the CNS. The Bunge group at the Miami Project to Cure Paralysis has pursued the use of guidance tubes seeded with Schwann cells in Matrigel to promote the regeneration of axons in the spinal cord. First, Xu et al. (1995)[67] showed that axon regeneration was greatly improved when Schwann cells were seeded into a guidance tube filled with Matrigel as compared with just the guidance tube and Matrigel alone. Extending this work, Xu et al. (1995)[54] showed that the ingrowth of CNS axons was virtually doubled by the introduction of the neurotrophins, NT-3 and BDNF, and the treatment stimulated regeneration of the axons associated with the vestibulospinal and raphespinal tracts, which are important in motor control. Chen et al.[68] in a related study, found that similar results could be achieved with the same grafts and the introduction of methylprednisolone.

In the above studies, the distal end of the guidance tube was blocked, and only ingrowth was studied. However, in more recent work, the Bunge group used a guidance tube which was open on both ends and repeated the initial study using Schwann cells. They found that the axons formed a bridge across the transection, but exhibited no ingrowth into the distal end of the spinal cord.[69]

The lack of ingrowth into the distal end of the cord is not surprising based on an understanding of the inhibitory nature of the damaged white matter.[38,39] However, these inhibitory effects appear to be quite complicated. While regeneration in very limited lesions, such as the transection of the corticospinal tract, has been achieved using IN-1,[39,40,70] even in a chronic injury model,[71] the antibody appeared to be incapable of producing a similar response in the larger injury model. Introduction of the antibody to the myelin inhibitory protein, IN-1, in this model induced sprouting of the corticospinal tract, but no ingrowth into the distal end of the cord.[72] It is possible that while the antibody shows promise for limited injuries, it is less effective for larger injuries which involve the damage to or loss of several millimeters of the cord. The reasons for this have yet to be determined. Interestingly, the implantation of the same system of Schwann cells seeded in a PAN/PVC tube filled with Matrigel in a hemisection model system was more successful. Without growth factors, or antibodies to inhibitory molecules, nerve fibers were seen extending into the gray matter distal to the lesion site.[73]

2. Olfactory Ensheathing Cells

While peripheral nerve bridges and Schwann cell seeded guidance channels have been shown to facilitate extensive ingrowth, there is still limited outgrowth of axons beyond the lesion site. There is also the issue that these cells and grafts are not native to the CNS. Therefore, in the hopes of facilitating greater regeneration with cells derived from the CNS, olfactory ensheathing cells (OECs) were isolated and injected along with the implantation of a Schwann cell seeded PAN/PVC guidance channel filled with Matrigel.[74] The addition of the OECs led to axons extending over long distances into the caudal cord. Previously, OECs had been shown to lead to improvements in function when injected to a very limited lesion of the corticospinal tract in rats.

Axons sprouted and grew through the lesion into the caudal end of the tract, and several of the treated animals exhibited improvements in behavior involving forepaw reaching.[75] OECs alone have also been shown to promote long distance regeneration of the corticospinal tract and accompanying functional improvements, including weight support while stationary, following a full transection.[76]

While the work is promising, there are inherent disadvantages to this approach. Solely reconnecting axons does not address the issue of the loss of cells in the gray matter due to the injury. The neurons in the gray matter play a tremendous role in the function of the system. Second, it is not clear that one will need to bridge long gaps in the spinal cord to achieve functional regeneration; one may very well succeed in restoring function through the development and activation of interneurons.[5]

3. Stem Cells

Ideally, one would like to create a permissive environment for regeneration and introduce cells which are capable of replacing lost ones, as well as form new connections. In general, adult neurons do not proliferate.[77] However, stem cells, neural precursors, and neural progenitors do.

While the definition of a stem cell remains a topic of distinct debate, a simple definition is that they are cells which are pluripotent, meaning that they are capable of forming all of the other cells types, and indefinitely self-renewing.[78] The most primordial, pluripotent stem cell is the embryonic stem cell, capable of forming all of the other cell types. There are also multipotent stem cells which have, until recently, been believed to be more committed to form a particular set of cells. These include hematopoetic stem cells which can form blood cells, mesenchymal stem cells which can form muscle, bone, cartilage, and tendon cell types, and neural stem cells (NSCs), which can form the cells of the nervous system.[79] These stem cells may be less committed than originally believed. There is evidence which suggests that some of these cells might be capable of transdifferentiation.[80–86] This work is extremely preliminary but very intriguing. Stem cells hold great promise for regeneration of the spinal cord after injury, but there are great challenges to capitalize on this promise.

a. *Embryonic Stem Cells*

Mouse embryonic stem cells have been studied for more than 20 years, but it was the isolation of human embryonic stem cells (ES cells) and embryonic germ cells (EG cells) from blastocysts and primordial germ cells of the developing gonadal ridge in 1998[87,88] which stimulated public and political consciousness. Although murine ES cells have been studied for more than 20 years, very little is still known about how to control their proliferation and differentiation.[89,90] What is known is that when murine embryonic stem cells, which have been cultured *in vitro,* are reintroduced to embryos which are then allowed to develop *in vivo*, the ES cells integrate and differentiate such that the resulting mice are chimeric.[91,92] The introduction of ES cells, which have been grown in a medium that favors differentiation, into neural precursors in the ventricles of embryonic rats leads to their incorporation and differentiation into neurons and glia.[93,94]

These cells have also been shown to have impressive therapeutic potential in spinal cord injuries. When ES cells, treated with retinoic acid, were injected 9 days after crush injuries in a rat spinal cord model, the cells integrated and differentiated into neurons and glia. More importantly though, the treated animals showed statistically significant improvement in open field locomotion (walking) in comparison to the control group which did not receive cells.[95] ES cells are also capable of becoming oligodendrocytes and remyelinating a demyelinated spinal cord.[96]

More must be learned about ES cells, including their tumorigenic potential. The cells must be differentiated completely and terminally. There is evidence that murine ES cells are tumorigenic[90] and certainly further work must be done to assess this. In the above spinal cord studies, no tumors were seen.[95,96]

b. Neural Stem Cells

Adult NSCs were first isolated and propagated in 1992 by Reynolds and Weiss.[97] They found that cells obtained from adult mouse striata and cultured in epidermal growth factor (EGF) could survive and proliferate *in vitro* and, with the proper chemical cues, differentiate into neurons and astrocytes. Previously, it was known that such neural progenitors existed during the development of the embryo, but such cells were thought to be absent from the adult CNS.[77] Now progenitor cells have been found throughout the adult brain and spinal cord.[98–101] There are several excellent reviews of neural stem cell biology which address the current state of the field and the therapeutic potential of the cells.[78,102–106]

The cells of the Reynolds and Weiss paper have been termed "stem-like" by some investigators because they were not shown to differentiate into oligodendrocytes.[103] In subsequent work, Weiss et al. showed that when propagated in EGF and basic fibroblast growth factor (bFGF), the cells isolated from the adult brain but and spinal cord were stem cells capable of both self-renewal and differentiation into neurons, astrocytes, and oligodendrocytes.[107]

The conditions under which the cells are isolated and propagated appear to play a critical role in their subsequent potential *in vitro* and *in vivo*. Briefly, a number of growth factors and substrates have been identified to facilitate the isolation, propagation, and differentiation of NSCs and neural progenitors *in vitro*. bFGF has been shown to be sufficient to maintain the proliferation and self-renewal of stem cells *in vitro*[101,108,109] while NT-3 and BDNF induce differentiation to neurons *in vitro*.[110] Platelet-derived growth factor (PDGF) has been shown to induce neuronal differentiation of both embryonic and adult NSCs, while ciliary neurotrophic factor (CNTF) induces astrocytes, and the thyroid hormone triiodothyronine (T3) induces oligodendrocytes.[111] It has further been shown that substrates can play a very important role in the fate of NSCs.[112] Specifically, fibronectin induced Schwann cells from NSCs while poly-D-lysine promoted neurons. Cells which have differentiated *in vitro* are capable of being functionally active. NSCs which differentiate to neurons have been shown to form active synapses in culture.[113]

NSCs and the more differentiated progeny, neural progenitors, have been shown to integrate, differentiate, and lead to functional improvement following a variety of injuries to the CNS. When implanted into the brains of healthy adult rats, both neural progenitors[114] and neural precursors[115] integrated and differentiated to form the appropriate cell types for each specific region with appropriate morphologies. In the damaged brain, NSCs and progenitors have been shown to integrate and differentiate to replace the cell types lost due to disease or injury.[116–118]

Following spinal cord injury, the behavior of NSCs is less clear. There are reports suggesting that NSCs are capable of forming glia but not neurons when introduced following a crush injury to the spinal cord.[119] Neural progenitors have likewise been found to form multipolar neurons in a healthy spinal cord but only bipolar neurons when injected after transection of the cord.[120] It was hypothesized the cell–cell contacts are essential for differentiation into complex morphologies. Other work has suggested that NSCs are capable of forming neurons, even motor neurons, following injury.[121]

Beyond the possibility that NSCs are capable of becoming functional neurons following spinal cord injury, they are attractive because they can form glia. The formation of glia may be an important aspect of regeneration. Not only can they insulate the axons and provide the appropriate supporting structures and trophic factors, but it has been shown that glia appear to be essential to the formation of functional synaptic connections for developing neurons *in vitro*.[122] Remyelination following spinal cord injury has also been shown to lead to functional improvement.[96,123,124]

C. Role of Surface Chemistry: The Extra Cellular Matrix

Cells are sensitive to their environment. They may respond differently to chemical signals and events as a result of their interaction with the extracelluar matrix (ECM),[125] or as a result of cell–cell contacts. The effects of cell–cell contact lead to cells behaving differently under two-dimensional and three-dimensional culture conditions.[126] Thus, cell–surface interactions and cell–scaffold architecture interactions are not easily deconvoluted. As noted above, with regard to NSCs, it has been shown that cell–surface interactions not only affect CNS stem cell proliferation and survival but neuronal differentiation as well.[112]

One particular ECM adhesion protein that has drawn a great deal of interest with respect to neural cell adhesion and neurite extension is laminin. Laminin has been shown to counter the inhibitory effects of the myelin-associated proteins with respect to neurite growth.[127] Laminin, and its homologue, S-laminin, are expressed in the CNS during development,[128] but they are essentially absent from the adult CNS. However, laminin and S-laminin are still found in the adult PNS and there is *in vitro* evidence to suggest that they play a role in creating the permissive environment for regeneration in the PNS.[127]

Laminin consists of an A chain, a B1 chain, and a B2 chain. It stimulates neural adhesion and neurite extension, but its homologue, S-laminin, which is very similar to the B1 chain of laminin, appears to only elicit neural adhesion and does not promote neurite extension *in vivo*.[129] S-Laminin is a homolog of the B1 chain of laminin and differs with respect to specific amino acid sequences,[130] which appear to be the active sequences for neurite adhesion.[131] The sequences of the B1 chain of laminin have been studied with respect to cell attachment, migration, and receptor binding, and the sequence, CDPGYIGSR has been shown to exhibit the highest activity. Likewise, Tashiro et al.[132] found that the IKVAV sequence of the A chain of laminin promoted cellular attachment and neurite outgrowth in a two-dimensional culture dish with PC 12 cells, an immortalized cell line from rat adrenal sarcoma which extends neurite-like processes when exposed to NGF. However, when the same sequence was studied in a three-dimensional culture system by Bellamokonda et al.,[133] they found that while it promoted extension in PC12 cells, it did not promote substantial neurite extension in dorsal root ganglion (DRG) neurons. Instead, they found that the YIGSR sequence from the B1 chain was far more effective for neurite extension for the DRG cells. A comparison of the papers suggest that PC12 cells are not a very good model for neurite extension studies involving surface interactions, and the B1 chain may play a critical role in neurite adhesion and extension. This is not entirely surprising considering the apparent relationship between S-laminin and laminin. Based on the developmental work and on the three-dimensional culture experiments, it would appear that one should focus on the study of laminin and of the specific peptide sequences, including YIGSR of the B1 chain in particular, for neural regeneration.

D. Scaffolds in the Spinal Cord

1. Axonal Guidance: Utilizing the Regenerative Potential of the PNS to Induce Regeneration in the CNS

Autographs and guidance channels have had a great deal of success with respect to regeneration in the PNS, and several groups have sought to apply these techniques to the CNS. Techniques involving carbon fibers,[134] collagen,[135] and Matrigel, an ECM derived from mouse sarcoma, in a poly(acrylonitrile):poly(vinylchloride) (PAN/PVC) tube[54,67,69,72–74,136,137] have been studied with respect to axon extension and guidance in the spinal cord.

The permissive nature of the gray matter may provide another route for overcoming the inhibitory myelinating proteins of the white matter. Cheng et al.[138] capitalized on this using peripheral nerve grafts to bridge the gap in their adult rat model by connecting grafts from the white matter to the gray matter in an effort to avoid the myelin-associated inhibitory molecules. They found that choice of graft connection sites was critical to their success. When they connected white matter to white matter and gray matter to gray matter with their bridging PNS axons, they did not achieve regeneration, even when they used the aFGF containing fibrin glue. They were only successful when they connected white matter to gray matter. In other promising studies, similar nerve bridges were shown to induce sprouting several millimeters below the injury site when coupled with neurotrophic factors,[61,139] though the degree of functional recovery was limited.

Axonal guidance using nerve grafts, carbon fibers, collagen or Schwann cell seeded channels have all shown promise, yet, with all the possible permutations of the designs, significant, reproducible functional recovery has not yet been achieved.

E. A New Approach to Spinal Cord Repair: Tissue Engineering

The principles of tissue engineering have been applied to model[140] and treat damage to the CNS[141,142] utilizing scaffolds alone[134,135,143] or with Schwann cells,[54,67,69,73] olfactory ensheathing cells,[74] and neyral stem cells.[144,145] However, this work has relied on using very simple scaffolds, made primarily of synthetic or natural hydrogels, which provide support but no guidance or direction to the donor cells or host tissue (Figure 32.2).

The intelligent design of the scaffold affords one the opportunity to not only support but direct cells and guide repair. Rather than simply taking a gel or a hollow tube, one can create a scaffold whose architecture is designed to mimic that of a healthy spinal cord so that as the scaffold is resorbed and new tissue forms, it is more likely that the tissue will be as physiologically normal as possible. A highly simplified view of the spinal cord is that of a coaxial cylindrical scaffold with the inner cylinder being the gray matter and the outer being the white matter. By designing a scaffold that mimics this structure, axon extension and guidance may be coupled with cellular replacement to facilitate regeneration.

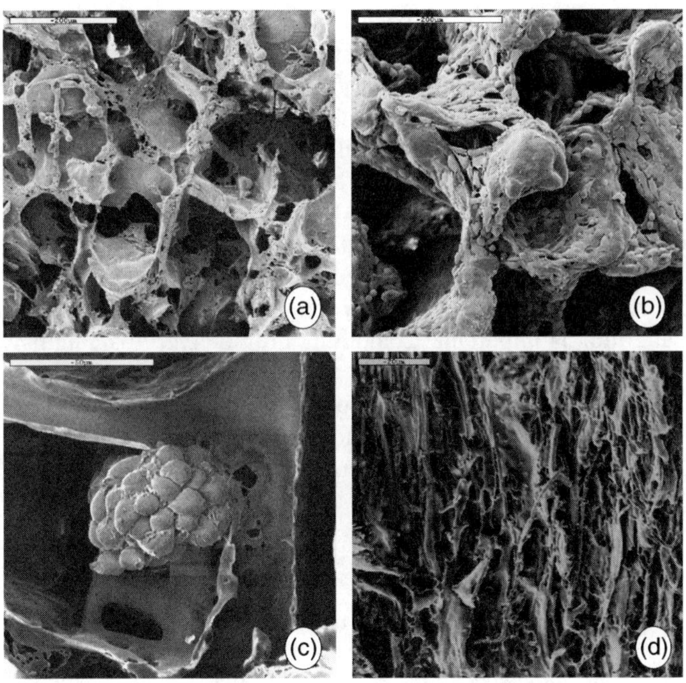

FIGURE 32.2 SEM micrographs showing a number of PLGA scaffolds developed for spinal cord repair: (a) a PLGA sponge fabricated using a salt-leaching technique used for seeding neural stem cells (NSCs) for implantation into the spinal cord, (b) a similar scaffold seeded with NSCs, (c) a number of NSCs on a section of a scaffold immediately following seeding, and (d) an oriented pore structure in a PLGA scaffold used to guide regenerating axons.

Through the control of scaffold architecture, surface chemistry, and degradation rate, a scaffold which supports the integration of NSCs and facilitates regeneration in an adult rat hemisection model has been created. The scaffold is a first step in the development of a novel therapy which may, one day, lead to a cure for traumatic spinal cord injury and other injuries and diseases of the spinal cord.

IV. SCAFFOLDS FOR THE CNS: THE BRAIN

There has been less work surrounding the development of scaffolds for the brain than for the PNS or the spinal cord. Poly(glycolic acid) meshes[146] and gels[143,147] have been used as a means for transplanting NSCs in cavities created in the brain. Microspheres releasing NGF have also been used as scaffolds for NSCs in lesions created in the brain.[148] The latter approach combines the injectable nature of certain hydrogels with growth factor delivery and three-dimensional surfaces for neural stem cell attachment and differentiation. It marks a major step in the development of scaffolding in the brain.

V. THE FUTURE FOR SCAFFOLDS IN THE NERVOUS SYSTEM

Polymer scaffolds have the potential to provide a new means to control the physical and chemical environment of the nervous system. By controlling this environment, one may be able to promote regeneration, the replacement of cells, and, perhaps, promote functional recovery. There are immense challenges to all of this. The successes thus far have been small but important.

Polymer scaffolds may control the physical environment by their pore architecture in the case of hydrophobic polymers such as PLGA. In the case of hydrogels, the architecture may be controlled via the location and density of cross-links. One may also chosen to create a composite system in which different polymers provide complimentary physical cues such as a very permissive polymer for growth coupled with one which guides the direction. The physical environment may also be controlled by the mechanical properties of the polymer structure. Through the choice of polymer or polymer blend, one may create scaffolds which have the same compliance as nervous tissue.

The chemical environment may be controlled through the surface chemistry of the scaffold, as well as the incorporation of drug delivery paradigms. As noted above, growth factors have been used extensively to promote repair with limited success. The body, however, does not simply deliver growth factors. It does so in a particular temporal and spatial manner. It is reasonable to believe that this patterning is important. Combining drug delivery paradigms with scaffolds may provide a means to create such patterning.

Ultimately, while scaffolds may or may not provide the cure for injuries to and diseases of the nervous system, they will provide a critical tool for understanding the nervous system and, hopefully, will lead to the development of new treatment paradigms.

REFERENCES

1. Gilman, S., and Newman, S. W., *Manter and Gatz's Essentials of Clinical Neuroanatomy and Neurophysiology*, 8th ed., F.A. Davis Company, Philadelphia, 1992.
2. Davidoff, R. A., ed., *Handbook of the Spinal Cord, Anatomy and Physiology*, Vol. 2&3, Marcell Dekker, Inc., New York, 1984.
3. Butler, A. B., and Hodos, W., *Comparative Vertebrate Neuroanatomy: Evolution and Adaptation*, Wiley-Liss, New York, 1996.
4. Paxinos, G., ed., *Hindbrain and Spinal Cord, The Rat Nervous System*, Vol. 2, Academic Press, Sydney, 1985.
5. Whittemore, S., Lecture at the Kentucky Trust for Spinal Cord Regeneration Conference, 1997.

6. Ide, C., Peripheral nerve regeneration, *Neurosci. Res.*, 25(2), 101–121, 1996.
7. Ljungberg, C., Johansson-Ruden, G., Bostrom, K. J., Novikov, L., and Wiberg, M., Neuronal survival using a resorbable synthetic conduit as an alternative to primary nerve repair, *Microsurgery*, 19(6), 259–264, 1999.
8. Buntic, R. F., Buncke, H. J., Kind, G. M., Chin, B. T., Ruebeck, D., and Buncke, G. M., The harvest and clinical application of the superficial peroneal sensory nerve for grafting motor and sensory nerve defects, *Plast. Reconstr. Surg.*, 109(1), 145–151, 2002.
9. Matsuyama, T., Midha, R., Mackinnon, S. E., Munro, C. A., Wong, P. Y., and Ang, L. C., Long nerve allografts in sheep with cyclosporin A immunosuppression, *J. Reconstr. Microsurg.*, 16(3), 219–225, 2000.
10. Strauch, B., Rodriguez, D. M., Diaz, J., Yu, H. L., Kaplan, G., and Weinstein, D. E., Autologous schwann cells drive regeneration through a 6-cm autogenous venous nerve conduit, *J. Reconstr. Microsurg.*, 17(8), 589–595, 2001.
11. Weiss, P., The technology of nerve regeneration: a review. Sutureless tubulation and related methods of nerve repair, *J. Neurosurg.*, 1, 400–450, 1944.
12. Lundborg, G., Dahlin, L. B., Danielsen, N., Gelberman, H., Longo, F., and Powell, H., Nerve regeneration in silicone chambers, influence of gap length and of distal stump component, *Exp. Neurol.*, 76, 361–375, 1982.
13. Ikeguchi, R., Kakinoki, R., Matsumoto, T., Tsuji, H., Ishikawa, J., and Nakamura, T., Rat nerve regeneration through a silicone chamber implanted with negative carbon ions, *Dev. Brain Res.*, 140(1), 127–131, 2003.
14. Valero-Cabre, A., Tsironis, K., Skouras, E., Perego, G., Navarro, X., and Neiss, W. F., Superior muscle reinnervation after autologous nerve graft or poly-L-lactide-epsilon-caprolactone (PLC) tube implantation in comparison to silicone tube repair, *J. Neurosci. Res.*, 63(2), 214–223, 2001.
15. Maquet, V., Martin, D., Malgrange, B., Franzen, R., Schoenen, J., Moonen, G., and Jerome, R., Peripheral nerve regeneration using bioresorbable macroporous polylactide scaffolds, *J. Biomed. Mater. Res.*, 52, 639–651, 2000.
16. Evans, G. R. D., Peripheral nerve injury: a review and approach to tissue engineered constructs, *Anat. Rec.*, 263(4), 396–404, 2001.
17. Rodríguez, F. J., Gómez, N., Perego, G., and Navarro, X., Highly permeable polylactide-caprolactone nerve guides enhance peripheral nerve regeneration through long gaps, *Biomaterials*, 20, 1489–1500, 1999.
18. Chamberlain, L. J., Yannas, I. V., Hsu, H. P., Strichartz, G., and Spector, M., Collagen-GAG substrate enhances the quality of nerve regeneration through collagen tubes up to level of autograft, *Exp. Neurol.*, 154(2), 315–329, 1998.
19. Holt, C. E., and Harris, W. A., Target selection: invasion, mapping and cell choice, *Curr. Opin. Neurobiol.*, 8(1), 98–105, 1998.
20. Key, B., and St John, J., Axon navigation in the mammalian primary olfactory pathway: where to next?, *Chem. Senses*, 27(3), 245–260, 2002.
21. Ngo, T.-T. B., Waggoner, P., Romero, A., and Nelson, K., Poly(L-lactide) microfilaments enhance peripheral nerve regeneration across extended nerve lesions, *J. Neurosci. Res.*, 72, 227–238, 2003.
22. Scherman, P., Lundborg, G., Kanje, M., and Dahlin, L. B., Neural regeneration along longitudinal polyglactin sutures across short and extended defects in the rat sciatic nerve, *J. Neurosurg.*, 95(2), 316–323, 2001.
23. Madison, R., Da Silva, C., and Dikkes, P., Entubulation repair with protein additives increases the maximum nerve gap distance successfully bridged with tubular prostheses, *Brain Res.*, 447, 325–334, 1988.
24. Yu, X. J., and Bellamkonda, R. V., Tissue-engineered scaffolds are effective alternatives to autografts for bridging peripheral nerve gaps, *Tissue Eng.*, 9(3), 421–430, 2003.
25. Xu, X. Y., Yee, W. C., Hwang, P. Y. K., Yu, H., Wan, A. C. A., Gao, S. J., Boon, K. L., Mao, H. Q., Leong, K. W., and Wang, S., Peripheral nerve regeneration with sustained release of poly (phosphoester) microencapsulated nerve growth factor within nerve guide conduits, *Biomaterials*, 24(13), 2405–2412, 2003.

26. Evans, G. R. D., Brandt, K., Katz, S., Chauvin, P., Otto, L., Bogle, M., Wang, B., Meszlenyi, R. K., Lu, L. C., Mikos, A. G., and Patrick, C. W., Bioactive poly(L-lactic acid) conduits seeded with Schwann cells for peripheral nerve regeneration, *Biomaterials*, 23(3), 841–848, 2002.
27. Murakami, T., Fujimoto, Y., Yasunaga, Y., Ishida, O., Tanaka, N., Ikuta, Y., and Ochi, M., Transplanted neuronal progenitor cells in a peripheral nerve gap promote nerve repair, *Brain Res.*, 974(1–2), 17–24, 2003.
28. Bryan, D. J., Tang, J. B., Holway, A. H., Rieger-Christ, K. M., Trantolo, D. J., Wise, D. L., and Summerhayes, I. C., Enhanced peripheral nerve regeneration elicited by cell-mediated events delivered via a bioresorbable PLGA guide, *J. Reconstr. Microsurg.*, 19(2), 125–134, 2003.
29. Beattie, M. S., Farooqui, A. A., and Bresnahan, J. C., Review of current evidence for apoptosis after spinal cord injury, *J. Neurotrauma*, 17, 915–925, 2000.
30. Osterholm, J. L., The pathological response to spinal cord injury, *J. Neurosurg.*, 40, 5–33, 1974.
31. Ducker, T. B., Kindt, G. W., and Kempe, L. G., Pathological findings in acute experimental spinal cord trauma, *J. Neurosurg.*, 35, 700–708, 1971.
32. Liu, X. Z., Xu, X. M., Hu, R., Du, C., Zhang, S. X., McDonald, J. W., Dong, H. X., Wu, Y. J., Fan, G. S., Jacquin, M. F., Hsu, C. Y., and Choi, D. W., Neuronal and glial apoptosis after traumatic spinal cord injury, *J. Neurosci.*, 17, 5395–5406, 1997.
33. Fehlings, M. G., and Tator, C. H., The relationship among severity of spinal cord injury, residual neurological function, axon counts, and counts of retrogradely labeled neurons after experimental spinal cord injury, *Exp. Neurol.*, 132, 220–228, 1995.
34. Noble, L. J., and Wrathall, J. R., Spinal cord contusion in the rat: morphometric analyses of alterations in the spinal cord, *Exp. Neurol.*, 88, 1985.
35. Noble, L. J., and Wrathall, J. R., Correlative analyses of lesion development and functional status after graded spinal cord contusive injuries in the rat, *Exp. Neurol.*, 103, 34–40, 1989.
36. Basso, D. M., Beattie, M. S., and Bresnahan, J. C., Graded histological and locomotor outcomes after spinal cord contusion using the NYU weight-drop device versus transection, *Exp. Neurol.*, 139, 244–256, 1996.
37. Kerasidis, H., Wrathall, J. R., and Gale, K., Behavioral assessment of functional deficit in rats with contusive spinal cord injury, *J. Neurosci. Methods*, 20, 167–189, 1987.
38. Schwab, M. E., and Caroni, P., Oligodendrocytes and CNS myelin are nonpermissive substrates for neurite growth and fibroblast spreading in vitro, *J. Neurosci.*, 8, 2381–2393, 1988.
39. Schnell, L., and Schwab, M. E., Axonal regeneration in the rat spinal cord produced by an antibody against myelin-associated neurite growth inhibitors, *Nature*, 343, 269–272, 1990.
40. Bregman, B. S., Kunkel-Bagden, E., Schnell, L., Dai, H. N., Gao, D., and Schwab, M. E., Recovery from spinal cord injury mediated by antibodies to neurite growth inhibitors, *Nature*, 378, 498–501, 1995.
41. Grandpré, T., Nakamura, F., Vartanian, T., and Strittmatter, S. M., Identification of the nogo inhibitor of axon regeneration as a reticulon protein, *Nature*, 403, 439–444, 2000.
42. Chen, M. S., Huber, A. B., Haar, M. E. v. d., Frank, M., Schnell, L., Spillmann, A. A., Christ, F., and Schwab, M. E., Nogo-A is a myelin-associated neurite outgrowth inhibitor and an antigen for monoclonal antibody IN-1, *Nature*, 403, 434–439, 2000.
43. Fitch, M. T., and Silver, J., Activated macrophages and the blood-brain barrier: inflammation after CNS injury leads to increases in putative inhibitory molecules, *Exp. Neurol.*, 148, 587–603, 1997.
44. Fitch, M. T., and Silver, J., Glial cell extracellular matrix: boundaries for axon growth in development and regeneration, *Cell Tissue Res.*, 290, 379–384, 1997.
45. Fitch, M. T., Doller, C., Combs, C. K., Landreth, G. E., and Silver, J., Cellular and molecular mechanisms of glial scarring and progressive cavitation: in vivo and in vitro analysis of inflammation-induced secondary injury after CNS trauma, *J. Neurosci.*, 19, 8182–8198, 1999.
46. Davies, S. J., Fitch, M. T., Memberg, S. P., Hall, A. K., Raisman, G., and Silver, J., Regeneration of adult axons in white matter tracts of the central nervous system, *Nature*, 390, 680–683, 1997.
47. Davies, S. J. A., Goucher, D. R., Doller, C., and Silver, J., Robust regeneration of adult sensory axons in degenerating white matter of the adult rat spinal cord, *J. Neurosci.*, 19, 5810–5822, 1999.
48. Hall, E. D., Yonkers, P. A., Taylor, B. R., and Sun, F. F., Lack of effect of postinjury treatment with methylprednisolone or tirilazad mesylate on the increase in eicosanoid levels in the acutely injured cat spinal cord, *J. Neurotrauma*, 12, 245–256, 1995.

49. Bracken, M. B., Shepard, M., Collins, W., Holford, T., Young, W., Baskin, D., Eisenberg, H., Flamm, E., Leosummers, L., Maroon, J., Marshall, L., Perot, P., Piepmeier, J., Sonntag, V., Wagner, F., Wilberger, J., and Winn, H., A randomized controlled trial of methylprednisolone of naloxone in the treatment of acute spinal-cord injury study, *N Engl. J. Med.*, 20, 1405–1411, 1990.
50. Bracken, M. B., Sherpard, M. J., Collins, W. F., Holford, T. R., Baskin, D. S., Eisenberg, H. M., Flamm, E., Leo-Summers, L., Maroon, J., Marshall, L. F., Perot, P. L., Piepmeier, J., Sonntag, V. R. H., Wagner, F. C., Wilberger, J. E., Winn, H. R., and Young, W., Methylprednisolone or naxolone treatment after acute spinal cord injury: 1-year follow-up data: results of the second national acute spinal cord injury study, *J. Neurosurg.*, 76, 23–31, 1992.
51. Bracken, M. B., Shepard, M., Holford, T., Leosummers, L., Aldrich, E., Fazl, M., Fehlings, M., Herr, D., Hitchon, P., Marshall, L., Nockels, R., Pascal, V., Perot, P., Piepmeier, J., Sonntag, V., Wagner, F., Wilberger, J., Winn, H., and Young, W., Administration of methylprednisolone for 24 or 48 h or tirilazad mesylate for 48 h in the treatment of acute spinal cord injury — results of the third national acute spinal cord injury randomized controlled trial, *J. Am. Med. Assoc.*, 277, 1597–1604, 1997.
52. Alberts, B., Bray, D., Lewis, J., Raff, M., Roberts, K., and Watson, J. D., *Molecular Biology of the Cell*, Garland Publishing, Inc., New York, 1994.
53. DiStefano, P. S., Friedman, B., Radziejewski, C., Alexander, C., Boland, P., Schick, C. M., Lindsay, R. M., and Wiegand, S. J., The neurotrophins BNDF, NT-3, and NGF display distinct patterns of retrograde axonal transport in peripheral and central neurons, *Neuron*, 8, 983–993, 1992.
54. Xu, X. M., Guénard, V., Kleitman, N., Aebischer, P., and Bunge, M. B., A combination of BDNF and NT-3 promotes supraspinal axonal regeneration into Schwann cell grafts in adult rat thoracic spinal cord, *Exp. Neurol.*, 134, 261–272, 1995.
55. Grill, R. J., Blesch, A., and Tuszynski, M. H., Robust growth of chronically injured spinal cord axons induced by grafts of genetically modified NGF-secreting cells, *Exp. Neurol.*, 148, 444–452, 1997.
56. Grill, R., Murai, K., Blesch, A., Gage, F. H., and Tuszynski, M. H., Cellular delivery of neurotrophin-3 promotes corticospinal axonal growth and partial functional recovery after spinal cord injury, *J. Neurosci.*, 17, 5560–5572, 1997.
57. Schnell, L., Schneider, R., Kolbeck, R., Barde, Y.-A., and Schwab, M. E., Neurotrophin-3 enhances sprouting of corticospinal tract during development and after adult spinal cord lesion, *Nature*, 367, 170–173, 1994.
58. Tuszynski, M. H., Peterson, D. A., Ray, J., Baird, A., Nakahara, Y., and Gage, F. H., Fibroblasts genetically modified to produce nerve growth factor induce robust neuritic ingrowth after grafting to the spinal cord, *Exp. Neurol.*, 126, 1–14, 1994.
59. Oudega, M., and Hagg, T., Nerve growth factor promotes regeneration of sensory axons into adult rat spinal cord, *Exp. Neurol.*, 140, 218–229, 1996.
60. Ramer, M. S., Priestley, J. V., and McMahon, S. B., Functional regeneration of sensory axons into the adult spinal cord, *Nature*, 403, 312–316, 2000.
61. Blits, B., Dijkhuizen, P. A., Boer, G. J., and Verhaagen, J., Intercostal nerve implants transduced with an adenoviral vector encoding neurotrophin-3 promote regrowth of injured rat corticospinal tract fibers and improve hindlimb function, *Exp. Neurol.*, 164, 25–37, 2000.
62. Bregman, B. S., McAtee, M., Dai, H. N., and Kuhn, P. L., Neurotrophic factors increase axonal growth after spinal cord injury and transplantation in the adult rat, *Exp. Neurol.*, 148, 475–494, 1997.
63. Fournier, A. E., GrandPre, T., and Strittmatter, S. M., Identification of a receptor mediating nogo-66 inhibition of axonal regeneration, *Nature*, 409, 341–346, 2001.
64. Benowitz, L. I., Goldberg, D. E., Madsen, J. R., Soni, D., and Irwin, N., Inosine stimulates extensive axon collatoral growth in the rat corticospinal tract after injury, *Proc. Natl Acad. Sci.*, 96, 13486–13490, 1999.
65. Sullivan, P. G., Rabchevsky, A. G., Hicks, R. R., Gibson, T. R., Fletcher-Turner, A., and Scheff, S. W., Dose-response curve and optimal dosing regimen of cyclosporin A after traumatic brain injury in rats, *Neuroscience*, 101, 289–295, 2000.
66. Sullivan, P. G., Thimpson, M., and Scheff, S. W., Continuous infusion of cyclosporin A postinjury significantly ameliorates cortical damage following traumatic brain injury, *Exp. Neurol.*, 161, 631–637, 2000.

67. Xu, X. M., Guénard, V., Kleitman, N., and Bunge, M. B., Axonal regeneration into Schwann cell-seeded guidance channels grafted into transected adult rat spinal cord, *J. Comp. Neurol.*, 351, 145–160, 1995.
68. Chen, A., Xu, X. M., Kleitman, N., and Bunge, M. B., Methylprednisolone administration improves axonal regeneration into Schwann cell grafts in transected adult rat thoracic spinal cord, *Exp. Neurol.*, 138, 261–276, 1996.
69. Xu, X. M., Chen, A., Guénard, V., Kleitman, N., and Bunge, M. B., Bridging Schwann cell transplants promotes axonal regeneration from both the rostral and caudal stumps of the transected adult rat spinal cord, *J. Neurocytol.*, 26, 1–16, 1997.
70. Brosamle, C., Huber, A. B., Fiedler, M., Skerra, A., and Schwab, M. E., Regeneration of lesioned corticospinal tract fibers in the adult rat induced by a recombinant, humanized IN-1 antibody fragment, *J. Neurosci.*, 20, 8061–8068, 2000.
71. Meyenburg, J. v., Brøsamle, C., Metz, G. A. S., and Schwab, M. E., Regeneration and sprouting of chronically injured corticospinal tract fibers in adult rats promoted by NT-3 and the mAb IN-1, which neutralizes myelin-associated neurite growth inhibitors, *Exp. Neurol.*, 154, 583–594, 1998.
72. Guest, J. D., Hesse, D., Schnell, L., Schwab, M. E., Bunge, M. B., and Bunge, R. P., Influence of IN-1 antibody and acidic FGF-fibrin glue on the response of injured corticospinal tract axons to human Schwann cell grafts, *J. Neurosci. Res.*, 50, 888–905, 1997.
73. Xu, X., Zhang, S.-X., Li, H., Aebischer, P., and Bunge, M., Regrowth of axons into the distal spinal cord through a Schwann-cell-seeded mini-channel implanted into hemisected adult rat spinal cord, *Eur. J. Neurosci.*, 11, 1723–1740, 1999.
74. Ramón-Cueto, A., Plant, G. W., Avila, J., and Bunge, M. B., Long-distance axonal regeneration in the transected adult rat spinal cord is promoted by olfactory ensheathing glia transplants, *J. Neurosci.*, 18, 3803–3815, 1998.
75. Li, Y., Field, P. M., and Raisman, G., Repair of adult rat corticospinal tract by transplanting of olfactory ensheathing cells, *Science*, 277, 2000–2002, 1997.
76. Ramón-Cueto, A., Cordero, M. I., Santos-Benito, F. F., and Avila, J., Functional recovery of paraplegic rats and motor axon regeneration in their spinal cords by olfactory ensheathing glia, *Neuron*, 25, 425–435, 2000.
77. Barinaga, M., Challenging the 'No New Neurons' dogma, *Science*, 255, 1646, 1992.
78. Morrison, S. J., Shah, N. M., and Anderson, D. J., Regulatory mechanisms in stem cell biology, *Cell*, 88, 287–298, 1997.
79. Vogel, G., Harnessing the power of stem cells, *Science*, 283, 1432–1434, 1999.
80. Bjornson, C. R. R., Rietze, R. L., Reynolds, B. A., Magli, M. C., and Vescovi, A. L., Turning brain into blood: a hematopoietic fate adopted by adult neural stem cells in vivo, *Science*, 283, 534–537, 1999.
81. Woodbury, D., Schwartz, E., Prockop, D., and Black, I., Adult rat and human bone marrow stromal cells differentiate into neurons, *J. Neurosci. Res.*, 61, 364–370, 2000.
82. Mezy, E., Chandross, K., Harta, G., Maki, R., and McKercher, S., Turning blood into brain: cells bearing neuronal antigens generated in vivo from bone marrow, *Science*, 290, 1779–1782, 2000.
83. Sanchez-Ramos, J., Song, S., Cardozo-Pelaez, F., Hazzi, C., Stedeford, T., Willing, A., Freeman, T., Saporta, S., Janssen, W., Patel, N., Cooper, D., and Sanberg, P., Adult bone marrow stromal cells differentiate into neural cells in vitro, *Exp. Neurol.*, 164, 247–256, 2000.
84. Pittenger, M., Mackay, A., Beck, S., Jaiswal, R., Douglas, R., Mosca, J., Moorman, M., Simonetti, D., Craig, S., and Marshak, D., Multilineage potential of adult human mesenchymal stem cells, *Science*, 284, 143–147, 1999.
85. Clarke, D. L., Johansson, C. B., Wilbertz, J., Veress, B., Nilsson, E., Karlstrøm, H., Lendahl, U., and Frién, J., Generalized potential of adult neural stem cells, *Science*, 288, 1660–1663, 2000.
86. Kondo, T., and Raff, M., Oligodendrocyte precursor cells reprogrammed to become multipotential CNS stem cells, *Science*, 289, 1754–1757, 2000.
87. Thomson, J. A., Itskovitz-Eldor, J., Shapiro, S. S., Waknitz, M. A., Swiergiel, J. J., Marshall, V. S., and Jones, J. M., Embryonic stem cell lines derived from human blastocysts, *Science*, 282, 1145–1147, 1998.
88. Shamblott, M., Axelman, J., Wang, S., Bugg, E., Littlefield, J., Donovan, P., Blumenthal, P., Huggins, G., and Gearhart, J., Derivation of pluripotent stem cells from cultured human primordial germ cells, *Proc. Natl Acad. Sci.*, 95, 13726–13731, 1998.

89. Watt, F., and Hogan, B., Out of Eden: stem cells and their niches, *Science*, 287, 1427–1430, 2000.
90. Solter, D., and Gearhart, J., Putting stem cells to work, *Science*, 283, 1468–1470, 1999.
91. Nagy, A., Rossant, J., Nagy, R., Abramow-Newerly, W., and Roder, J., Derivation of completely cell culture-derived mice from early-passage embryonic stem cells, *Proc. Natl Acad. Sci.*, 90, 8424–8428, 1993.
92. Brook, F., and Gardner, R., The origin and efficient derivation of embryonic stem cells in the mouse, *Proc. Natl Acad. Sci.*, 94, 5709–5712, 1997.
93. Brüstle, O., Spiro, A. C., Karram, K., Choudhary, K., and Okabe, S., In vitro-generated neural precursors participate in mammalian brain development, *Proc. Natl Acad. Sci.*, 94, 14809–14814, 1997.
94. Brüstle, O., Jones, K., Learish, R., Karram, K., Choudhary, K., Wiesler, O., Duncan, I., and McKay, R., Embryonic stem cell-derived glial precursors: a source of myelinating transplants, *Science*, 285, 754–756, 1999.
95. McDonald, J. W., Liu, X.-Z., Qu, Y., Liu, S., Mickey, S. K., Turetsky, D., Gottleib, D. I., and Choi, D. W., Transplanted embryonic stem cells survive, differentiate and promote recovery in injured rat spinal cord, *Nat. Med.*, 5(12), 1410–1412, 1999.
96. Liu, S., Qu, Y., Stewart, T. J., Howard, M. J., Chakrabortty, S., Holekamp, T. F., and McDonald, J. W., Embryonic stem cells differentiate into oligodendrocytes and myelinate in culture and after spinal cord transplantation, *Proc. Natl Acad. Sci.*, 97, 6126–6131, 2000.
97. Reynolds, B. A., and Weiss, S., Generation of neurons and astrocytes from isolated cells of the adult mammalian central nervous system, *Science*, 255, 1707–1710, 1992.
98. Horner, P. J., Power, A. E., Kempermann, G., Kuhn, H. G., Palmer, T. D., Winkler, J., Thal, L. J., and Gage, F. H., Proliferation and differentiation of progenitor cells throughout the intact adult rat spinal cord, *J. Neurosci.*, 20, 2218–2228, 2000.
99. Mathis, L., and Nicolas, J. F., Different clonal dispersion in the rostral and caudal mouse central nervous system, *Development*, 127, 1277–1290, 2000.
100. Kehl, L. J., Fairbanks, C. A., Laughlin, T. M., and Wilcox, G. L., Neurogenesis in postnatal rat spinal cord: a study in primary culture, *Science*, 276, 586–589, 1997.
101. Shihabuddin, L. S., Ray, J., and Gage, F. H., FGF-2 is sufficient to isolate progenitors found in the adult mammalian spinal cord, *Exp. Neurol.*, 148, 577–586, 1997.
102. McKay, R., Stem cells in the central nervous system, *Science*, 276, 66–71, 1997.
103. Temple, S., and Alvarez-Buylla, A., Stem cells in the adult mammalian central nervous system, *Curr. Opin. Neurobiol.*, 9, 135–141, 1999.
104. Morshead, C. M., and Kooy, D. v. d., A new 'spin' on neural stem cells?, *Curr. Opin. Neurobiol.*, 11, 59–65, 2001.
105. Gage, F. H., Mammalian neural stem cells, *Science*, 287, 1433–1438, 2000.
106. Gage, F. H., Cell therapy, *Nature*, 392, 18–23, 1998.
107. Weiss, S., Dunne, C., Hewson, J., Whol, C., Wheatley, M., Peterson, A., and Reynolds, B., Multipotent CNS stem cells are present in the adult mammalian spinal cord and ventricular neuroaxis, *J. Neurosci.*, 16, 7599–7609, 1996.
108. Gritti, A., Parati, E. A., Cova, L., Frolichsthal, P., Galli, R., Wanke, E., Faravelli, L., Morassutti, D. J., Roisen, F., Nickel, D. D., and Vescovi, A. L., Multipotent stem cells from the adult mouse brain proliferate and self-renew in response to basic fibroblast growth factor, *J. Neurosci.*, 16, 1091–1100, 1996.
109. Kitchens, D., Snyder, E., and Gottlieb, D., FGF and EGF are mitogens for immortalized neural progenitors, *J. Neurobiol.*, 25, 797–807, 1994.
110. Vicario-Abejón, C., Johe, K. K., Hazel, T. G., Collazo, D., and McKay, R. D. G., Functions of basic fibroblast growth factor and neurotrophins in the differentiation of hippocampal neurons, *Neuron*, 15, 105–114, 1995.
111. Johe, K. K., Hazel, T. G., Muller, T., Dugich-Djordjevic, M. M., and McKay, R. D. G., Single factors direct the differentiation of stem cells from the fetal and adult central nervous system, *Genes Dev.*, 10, 3129–3140, 1996.
112. Stemple, D. L., and Anderson, D. J., Isolation of a stem cell for neurons and glia from the mammalian neural crest, *Cell*, 71, 973–985, 1992.

113. Toda, H., Takahashi, J., Mizoguchi, A., Koyano, K., and Hashimoto, N., Neurons generated from adult rat hippocampal stem cells form glutamatergic and GABAergic synapses in vitro, *Exp. Neurol.*, 165, 66–76, 2000.
114. Suhonen, J. O., Peterson, D. A., Ray, J., and Gage, F. H., Differentiation of adult hippocampus-derived progenitors into olfactory neurons in vivo, *Nature*, 383, 624–627, 1996.
115. Shihabuddin, L. S., Hertz, J. A., Holets, V. R., and Whittemore, S. R., The adult CNS retains the potential to direct region-specific differentiation of a transplanted neuronal precursor cell line, *J. Neurosci.*, 15, 6666–6678, 1995.
116. Isacson, O., Deacon, T. W., Pakzaban, P., Galpern, W. R., Dinsmore, J., and Burns, L. H., Transplanted xenogenic neural cells in neurodegenerative disease models exhibit remarkable axonal target specificity and distinct growth patterns of glial and axonal fibres, *Nat. Med.*, 1, 1189–1194, 1995.
117. Snyder, E., Yoon, C., Flax, J., and Macklis, J., Multipotent neural precursors can differentiate toward replacement of neurons undergoing targeted apoptotic degeneration in adult mouse neocortex, *Proc. Natl Acad. Sci.*, 94, 11663–11668, 1997.
118. Yandava, B., Billinghurst, L., and Snyder, E., "Global" cell replacement is feasible via neural stem cell transplantation: evidence from the dysmyelinated shiverer mouse brain, *Proc. Natl Acad. Sci.*, 96, 7029–7034, 1999.
119. Cao, Q., Zhang, Y. P., Howard, R. M., Walters, W. M., Tsoulfas, P., and Wittemore, S. R., Pluripotent stem cells engrafted into the normal or lesioned adult rat spinal cord are restricted to a glial lineage, *Exp. Neurol.*, 167, 48–58, 2001.
120. Onifer, S. M., Cannon, A. B., and Whittemore, S. R., Altered differentiation of CNS neural progenitor cells after transplantation into the injured adult rat spinal cord, *Cell Transplant.*, 6, 327–338, 1997.
121. Park, K., Liu, S., Flax, J., Nissim, S., Stieg, P., and Snyder, E., Transplantation of neural progenitor and stem cells: developmental insights may suggest new therapies for spinal cord and other CNS dysfunction, *J. Neurotrauma*, 16, 675–687, 1999.
122. Pfrieger, F. W., and Barres, B. A., Synaptic efficacy enhanced by glial cells in vitro, *Science*, 277, 1684–1687, 1997.
123. Jeffery, N. D., Crang, A. J., O'Leary, M. T., Hodge, S. J., and Blakemore, W. F., Behavioral consequences of oligodendrocyte progenitor cell transplantation into experimental demyelinating lesions in the rat spinal cord, *Eur. J. Neurosci.*, 11, 1508–1514, 1999.
124. Akiyama, Y., Honmou, O., Kato, T., Uede, T., Hashi, K., and Kocsis, J. D., Transplantation of clonal neural precursor cells derived from adult human brain establishes functional peripheral myelin in the rat spinal cord, *Exp. Neurol.*, 167, 27–39, 2001.
125. Gospodarowicz, D., Greenburg, G., and Birdwell, C. B., Determination of cellular shape by the extracellular matrix and its correlation with the control of cellular growth, *Cancer Res.*, 38, 4155–4171, 1978.
126. Freshney, R. I., *Culture of Animal Cells: A Manual of Basic Technique*, 3rd ed., Wiley-Liss, New York, 1994.
127. David, S., Braun, P. E., Jackson, D. L., Kottis, V., and McKerracher, L., Laminin overrides the inhibitory effects of peripheral nervous system and central nervous system myelin-derived inhibitors of neurite growth, *J. Neurosci. Res.*, 42, 594–602, 1995.
128. Hunter, D. D., Llinas, R., Ard, M., Merlie, J. P., and Sanes, J. R., Expression of S-laminin and laminin in the developing rat central nervous system, *J. Comp. Neurol.*, 323, 238–251, 1992.
129. Letourneau, P. C., Condic, L., and Snow, D. M., Interactions of developing neurons with the extracelluar matrix, *J. Neurosci.*, 14, 915–938, 1994.
130. Hunter, D. D., Shah, V., Merlie, J. P., and Sanes, J. R., A laminin-like adhesive protein concentrated in the synaptic cleft of the neuromuscular junction, *Nature*, 338, 229–234, 1989.
131. Hunter, D. D., Porter, B. E., Bulock, J. W., Adams, S. P., Merlie, J. P., and Sanes, J. R., Primary sequence of a motor neuron-selective adhesive site in the synaptic basal lamina protein S-laminin, *Cell*, 59, 905–913, 1989.
132. Tashiro, K.-I., Sephel, G. C., Weeks, B., Sasaki, M., Martin, G. R., Kleinman, H. K., and Yamada, Y., A synthetic peptide containing the IKVAV sequence from the a chain of laminin mediates cell attachment, migration, and neurite outgrowth, *J. Biol. Chem.*, 264, 16174–16182, 1989.

133. Bellamkonda, R., Ranieri, J. P., and Aebischer, P., Laminin oligopeptide derivatized agarose gels allow three-dimensional neurite extension in vitro, *J. Neurosci. Res.*, 41, 501–509, 1995.
134. Khan, T., Dauzvardis, M., and Sayers, S., Carbon filament implants promote axonal growth across the transected rat spinal cord, *Brain Res.*, 541, 139–145, 1991.
135. Marchand, R., and Woerly, S., Transected spinal cords grafted with in situ self-assembled collagen matrices, *Neuroscientist*, 36, 45–60, 1990.
136. Guest, J. D., Rao, A., Olson, L., Bunge, M. B., and Bunge, R. P., The ability of human Schwann cell grafts to promote regeneration in the transected nude rat spinal Cord, *Exp. Neurol.*, 148, 502–522, 1997.
137. Gautier, S. E., Oudega, M., Fragoso, M., Chapon, P., Plant, G. W., Bunge, M. B., and Parel, J.-M., Poly(α-hydroxyacids) for application in the spinal cord: resorbability and biocompatability with adult rat Schwann cells and spinal cord, *J. Biomed. Mater. Res.*, 42, 642–654, 1998.
138. Cheng, H., Cao, Y., and Olson, L., Spinal cord repair in adult paraplegic rats: partial restoration of hind limb function, *Science*, 273, 510–513, 1996.
139. Oudega, M., and Hagg, T., Neurotrophins promote regeneration of sensory axons in the adult rat spinal cord, *Brain Res.*, 818, 431–438, 1999.
140. Bellamkonda, R., Ranieri, J. P., Bouche, N., and Aebischer, P., Hydrogel-based three dimensional matrix for neural cells, *J. Biomed. Mater. Res.*, 29, 663–671, 1995.
141. Bellamkonda, R., and Aebischer, P., Review: tissue engineering in the nervous system, *Biotechnol. Bioeng.*, 43, 543–554, 1994.
142. Borkenhagen, M., and Aebischer, P., Tissue-engineering approaches for central and peripheral nervous-system regeneration, *MRS Bull.*, 21, 59–61, 1996.
143. Woerly, S., Petrov, P., Sykova, E., Roitbak, T., Simonova, Z., and Harvey, A., Neural tissue formation within porous hydrogels implanted in brain and spinal cord lesions: ultrastructural, immunohistochemical, and diffusion studies, *Tissue Eng.*, 5, 467–488, 1999.
144. Vacanti, M., Leonard, J., Dore, B., Bonassar, L., Cao, Y., Stachelek, S., Yu, C., O'Connell, F., Vacanti, J., and Vacanti, C., Tissue engineered spinal cord, *FASEB J.*, 14(4), A446–A446, 2000.
145. Vacanti, M., Leonard, J., Dore, B., Bonassar, L., Cao, Y., Stacheleck, S., Vacanti, J., O'Connell, F., Yu, C., Farwell, A., and Vacanti, C., Tissue-engineered spinal cord, *Transplant. Proc.*, 33, 592–598, 2001.
146. Park, K., Teng, Y., and Snyder, E., The injured brain interacted reciprocally with neural stem cells supported by scaffolds to reconstitute lost tissue, *Nat. Biotechnol.*, 20, 1111–1117, 2002.
147. Tate, M. C., Shear, D. A., Hoffman, S. W., Stein, D. G., and LaPlaca, M. C., Biocompatibility of methylcellulose-based constructs designed for intracerebral gelation following experimental traumatic brain injury, *Biomaterials*, 22(10), 1113–1123, 2001.
148. Mahoney, M., and Saltzman, W., Transplantation of brain cells assembled around a programmable synthetic microenvironment, *Nat. Biotechnol.*, 19, 934–939, 2001.

33 Functional Tissue Engineering of Cartilage and Myocardium: Bioreactor Aspects

Milica Radisic, Bojana Obradovic, and Gordana Vunjak-Novakovic

CONTENTS

I. INTRODUCTION

Tissue engineering combines the principles of biology, engineering, and medicine to create functional grafts capable of repairing native tissue following a congenital deformity, disease, or trauma. Engineered tissues can provide high-fidelity models for basic studies of cell function and tissue development, and responses to genetic alterations, drugs, hypoxia, and physical stimuli. The overall objective of tissue engineering is the restoration of normal tissue function. Ideally, lost or damaged tissue should be replaced by an engineered graft that can reestablish appropriate structure, composition, cell signaling, and key function(s) of the native tissue. In light of this paradigm, the clinical utility of tissue engineering will likely depend on our ability to replicate the site-specific properties of the tissue being replaced across different size scales and establish the specific differentiated cell phenotype, the composition, architectural organization, and biomechanical properties of the extracellular matrix (ECM), and provide the continuity and strength of the interface with the neighboring host tissues.[1]

It is thought that the cell function *in vitro* can be modulated by the same factors known to play a role during embryogenesis. This involves the utilization of: (1) scaffolds for cell attachment and tissue formation, (2) a physiological milieu, (3) supply of nutrients, oxygen and growth factors, and (4) application of physical regulatory signals (Figure 33.1). *In vivo*, the processes of cell differentiation and tissue assembly are directed by multiple factors acting in concert and according to specific spatial and temporal sequences. *In vitro,* biophysical regulation of cultured cells can be achieved by an integrated use of biomaterial scaffolds and bioreactors (Figure 33.2).

Based on the above principles, a "biomimetic" approach has been established that involves the *in vitro* creation of immature but functional tissues by an integrated use of: (1) cells that can be selected, expanded, and transfected to express the genes of interest, (2) biomaterial scaffolds that serve as a structural and logistic template for tissue development and biodegrade at a controlled rate, and (3) bioreactors that provide environmental conditions necessary for the cells to regenerate functional tissue.

In this chapter, we summarize some of the design requirements for tissue engineering bioreactors, and focus on bioreactor cultivations of articular cartilage and cardiac muscle. These two tissues were selected because of their high clinical relevance and distinctly different structural and functional properties. Articular cartilage is avascular, and contains only one cell type (chondrocyte) in abundant ECM. Chondrocytes are present at low concentration and are supplied with oxygen and nutrients by a combination of diffusion and fluid flow during joint loading. Although the biosynthesis rates depend on gas exchange,[2] cells remain viable under hypoxic conditions, which are inherent for adult cartilage in an articular joint.[3] In contrast, cardiac muscle is a highly differentiated, vascularized tissue which contains several cell types (cardiac myocytes, fibroblasts, endothelial cells).[4,5] Cardiac myocytes are highly metabolically active, and cannot tolerate hypoxia for prolonged periods of time.[6,7] Also, cardiac cells do not proliferate in culture, such that the establishment and maintenance of a high density of viable cells is essential for engineering cardiac constructs with physiological cell density. These differences in cell metabolism, and in particular in cell respiration rates and tolerance to hypoxia, impose different

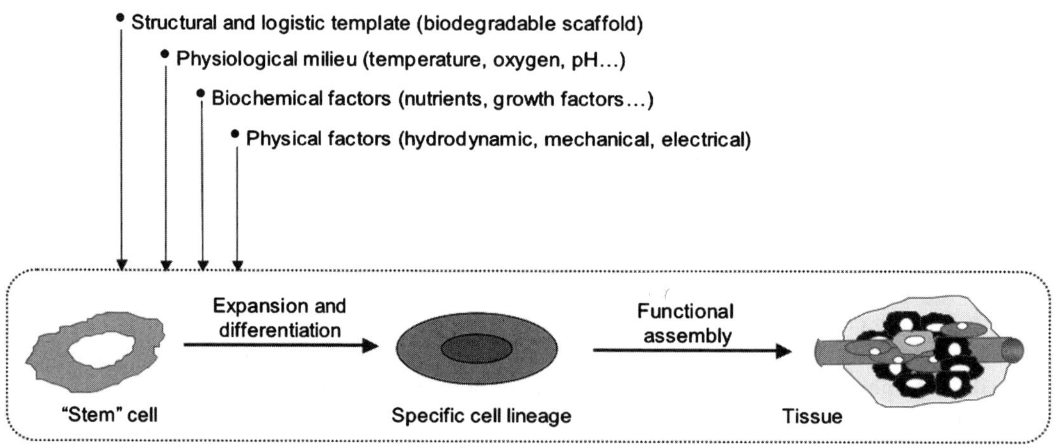

FIGURE 33.1 Developmental paradigm. Tissue development and remodeling, *in vivo* and *in vitro*, involves the proliferation and differentiation of stem/progenitor cells and their subsequent assembly into a tissue structure. Cell function and the progression of tissue assembly depend on: (a) the availability of a scaffold for cell attachment and tissue formation, (b) the maintenance of physiological conditions in cell/tissue environment, (c) the supply of nutrients, oxygen, metabolites, and growth factors, and (d) the presence of physical regulatory factors.

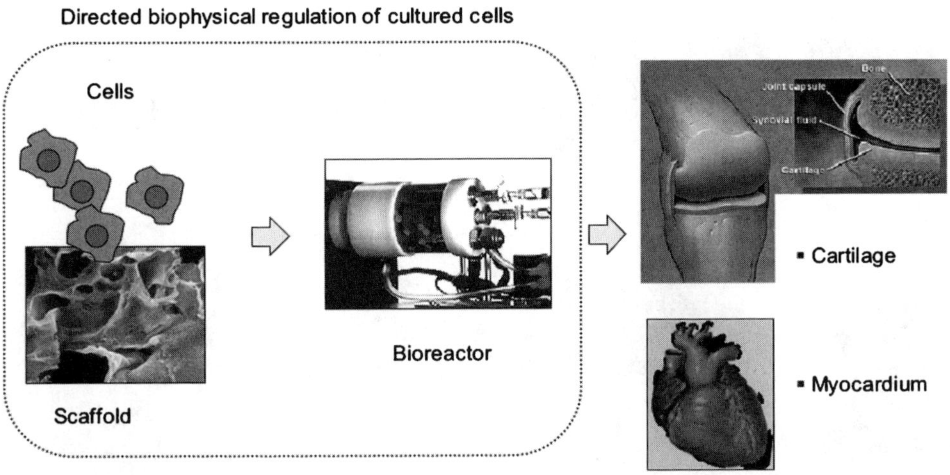

FIGURE 33.2 Tissue engineering paradigm. The regulatory factors of cell differentiation and tissue assembly depicted in Figure 33.1 can be utilized *in vitro* to engineer functional tissues by an integrated use of isolated cells, biomaterial scaffolds, and bioreactors. The cells themselves (either differentiated or progenitor/stem cells seeded onto a scaffold and cultured in a bioreactor) carry out the process of tissue formation, in response to regulatory signals. The scaffold provides a structural, mechanical, and logistic template for cell attachment and tissue formation. The bioreactor provides the environmental conditions and regulatory signals (biochemical and physical) that induce, enhance, or at least support the development of functional tissue constructs.

requirements to the design and operation of bioreactors that are optimal for tissue engineering of cartilage and cardiac muscle.

II. TISSUE ENGINEERING BIOREACTORS

Ideally, a bioreactor should provide an *in vitro* environment for rapid and orderly tissue development starting from isolated cells and three-dimensional scaffolds. Bioreactors are designed to perform one or more of the following functions:

- establish spatially uniform concentrations of cells seeded onto clinically sized biomaterial scaffolds,
- control conditions in culture medium (e.g., temperature, pH, osmolality, levels of oxygen, nutrients, metabolites, regulatory molecules),
- facilitate mass transfer between the cells and the culture environment, and
- provide physiologically relevant physical signals (e.g., interstitial fluid flow, shear, pressure, compression, stretch).[8]

An overview of representative tissue engineering bioreactors is shown in Figure 33.3. Bioreactors were shown to enhance both the effectiveness of tissue engineering scaffolds,[9] and the maintenance of differentiated cell function.[8,10,11] The regime of flow (laminar and dynamic rather than turbulent and steady) markedly affected the cultivation of clinically sized, mechanically functional constructs. Importantly, hydrodynamically active environments improved construct compositions and mechanical properties as compared to the static culture.

In the rotating bioreactor (Figure 33.3a), up to 12 tissue constructs are suspended in the 110 ml annular space between two cylinders, gas is exchanged via a silicone membrane, and mixing is provided by construct settling in the rotating flow (Figure 33.3b).[12]

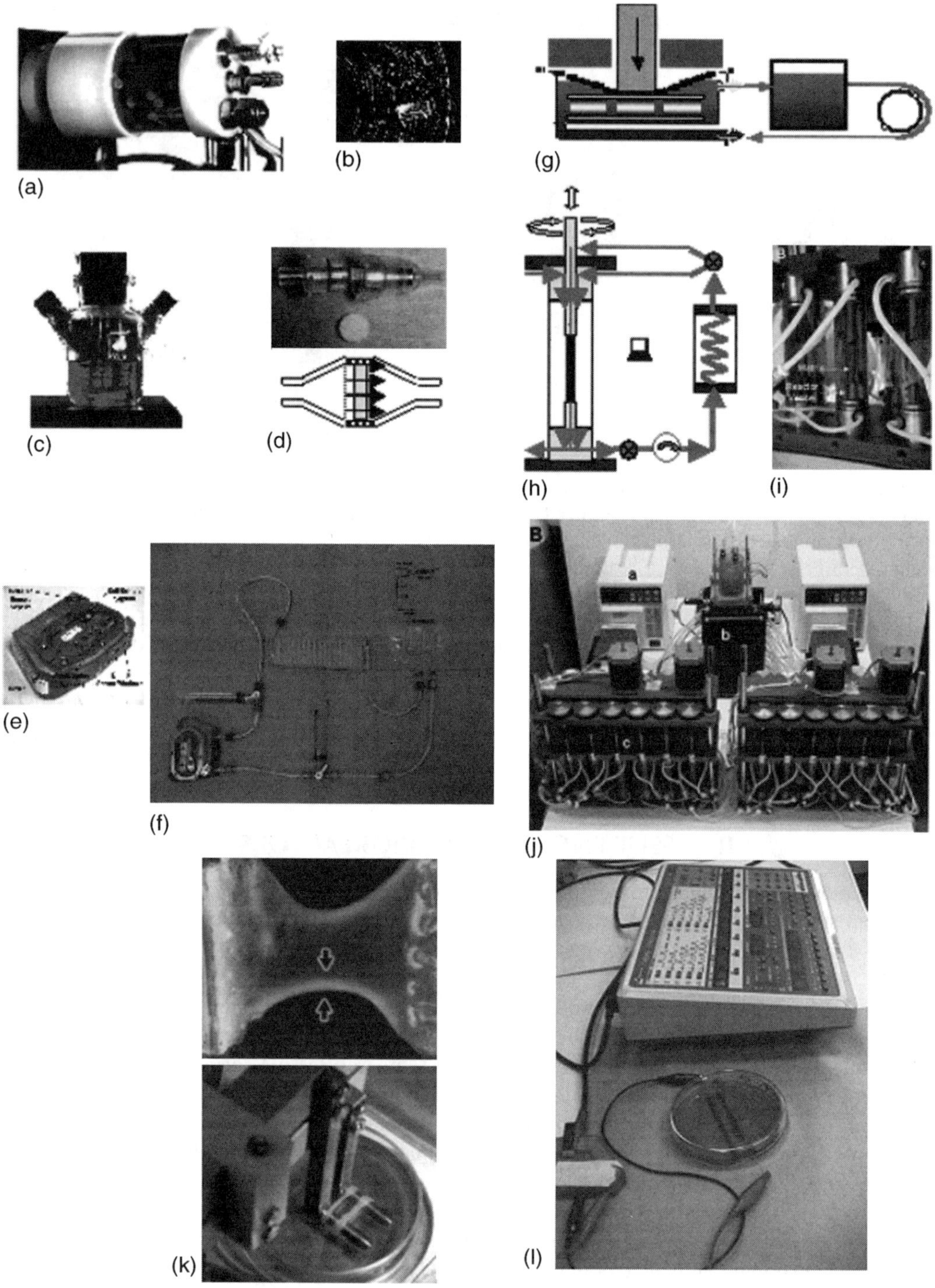
(a)
(b)
(g)
(c)
(d)
(h)
(i)
(e)
(f)
B
a
b
c
(j)
(k)
(l)

Spinner flasks (Figure 33.3c) are 120 ml vessels in which constructs are fixed in place, gas is exchanged via surface aeration, and mixing is provided by magnetic stirring.[13]

Perfused cartridges are 1.5 ml vessels designed to provide interstitial fluid flow through the construct within; gas is exchanged by medium recirculation through an external gas exchanger (Figure 33.3d).[14]

Perfused chambers (Figure 33.3f) are 3, 10, or 30 ml volume vessels designed to provide medium flow around tissue constructs or cultured cells. They are connected to an external loop containing a gas exchanger, a miniature pump, and a small reservoir with a fresh medium, and designed for use either in an incubator or as a "portable" system for transferring engineered tissue constructs from one laboratory to another.[15]

The bioreactor system with medium perfusion and mechanical loading contains five chambers that can accommodate up to five engineered constructs apiece (Figure 33.3g); gas is exchanged via an external device, mixing is provided by recirculation, and loading is provided by a mechanical spectrometer capable of applying a variety of mechanical stimulation regimes.[16]

One of the most advanced bioreactor systems (Figure 33.3h,i and j) is a bench-top device that contains 24 tissue culture vessels with medium perfusion and mechanical loading. Each cartridge (Figure 33.3j) is connected to an independent perfusion loop and can be subjected to multi-dimensional mechanical loading (dynamic compression/tension and torsion).

A bioreactor system for mechanical stimulation of engineered heart tissue involves casting a cell suspension in collagen between two Velcro-coated silicone tubes. Cyclic stretch is applied at the frequency of 1.5 Hz and the strain rate of up to 20% using a custom made stretching device[17] (Figure 33.3k, top). In an improved setup, neonatal rat cardiac cells are resuspended in the collagen/ Matrigel® mix and cast into circular molds[18] (Figure 33.3k, bottom). After 7 days of static culture, the strips of cardiac tissue are placed around two rods of a custom made mechanical stretcher and subjected to either unidirectional or cyclic stretch.

Electrical stimulation during culture is provided by placing tissue constructs between two parallel electrodes (carbon rods) connected to a cardiac stimulator via platinum wires.[19] The stimulator is programmed to apply supra threshold, square biphasic pulses 2 msec duration and 60 bpm (Figure 33.3l).

Seeding of three-dimensional scaffolds with isolated cells (e.g., chondrocytes, bone-marrow derived progenitors, cardiac myocytes) is the first step of bioreactor cultivation of engineered tissues. Seeding requirements include (1) high cell yield (to maximize cell utilization), (2) high kinetic rate of cell attachment (to minimize the time in suspension for anchorage-dependent cells), (3) high and spatially uniform distribution of attached cells (to promote rapid and spatially uniform tissue development),[8,20] and (4) immediate establishment of interstitial flow through scaffolds seeded with cells sensitive to hypoxia (e.g., cardiac myocytes) to prevent cell damage and death during seeding.

FIGURE 33.3 Overview of tissue engineering bioreactors. (a) Rotating bioreactor in which tissue constructs are cultured freely suspended in dynamic laminar flow;[12] (b) flow visualization of fluid flow around a suspended construct (cross-sectional view) (8, photo courtesy of P. Neitzel); (c) spinner flask in which constructs are fixed in place and cultured either statically or in a well mixed medium;[13] (d) photo and schematic of a perfused cartridge with interstitial flow of culture medium that provides mass transport throughout the construct interior (14, photo courtesy of N. Dunkelman); (e, f) perfused chamber within a "portable" single loop[15]; (g) perfused chamber system that provides mechanical loading to a construct[16]; (h, i) schematic and photo of a perfused column that provides interstitial flow of culture medium and mechanical loading of the construct[85]; (j) photo of 24 columns within a modular bioreactor system that provides medium perfusion and multi-dimensional mechanical loading (two sets of $N = 12$ columns each);[85] (k, top) photo of an engineered heart tissue attached to Velcro-coated silicone tubes for mechanical stimulation and a setup for cyclic mechanical stimulation;[18] (l) photo of a setup for electrical stimulation consisting of a Petri dish fitted with two carbon rods, connected to a commercial cardiac stimulator.[19]

For engineered cartilage, cell densities are commonly in the range of 10 to 100 × 10^6 cells/cm^3 of scaffold volume. Highly porous scaffolds are best seeded using mixed flasks in which scaffolds are fixed in place and exposed to well-mixed cell suspension (typically, up to 12 scaffolds per flask in 120 ml volume of culture medium containing 5 × 10^5 cells/cm^3 and magnetically stirred at 50 to 80 rpm). Mixing maintains the uniformity of suspension and provides relative velocity between the cells and the scaffolds, and essentially all cells attach throughout the scaffold volume in less than 24 h.[20] The kinetics and possible mechanisms of cell seeding were rationalized using a simple mathematical model, which can be used to select seeding conditions for specific scaffold sizes and cell seeding densities.[20] The diffusional transport of oxygen, nutrients, and metabolites within the scaffold are acceptable for seeding of chondrocytes due to their relatively moderate oxygen requirements and ability to survive hypoxic and acidic conditions. The main advantages of dynamic seeding in spinner flasks are the high yield and high spatial uniformity of cell attachment, the conditions known to promote rapid chondrogenesis.[20]

For engineered cardiac muscle, cell densities that need to be established on seeded scaffolds are ≥100 × 10^6 cells/cm^3 of scaffold volume. In addition to the general requirements listed above, seeding of cardiac myocytes requires the efficient supply of nutrients and oxygen to the cells at all times during seeding and cultivation, in order to maintain cell viability. The technique of seeding that was specifically developed for cardiac myocytes involved (1) the rapid inoculation of cardiac myocytes into collagen sponges using Matrigel as a cell delivery vehicle, and (2) the transfer of inoculated scaffolds into perfused cartridges with the immediate establishment of the interstitial flow of culture medium. Forward-reverse flow was used for the initial period of 1.5–4.5 h in order to further increase the rate and spatial uniformity of cell seeding.[21] In this system, cells were "locked" into the scaffold during a short (10 min) gelation period, and evenly distributed by medium perfusion.

Cultivation of cell–polymer constructs for cartilage and cardiac tissue engineering is routinely done in static flasks, mixed flasks, rotating vessels, and perfused cartridges, under conditions summarized in Figure 33.4. The hydrodynamic environment of a bioreactor is important as it can affect cell function in engineered tissues in at least two ways: via associated effects on mass transport between the tissue and culture medium, and by direct physical stimulation of the cells.

In static flasks, the cultured tissues are fixed in place and exposed to static medium, such that the mass transport between the tissue and culture medium, as well as within the tissue, is governed by molecular diffusion, and there is no hydrodynamic shear acting at the cells.

In mixed flasks, tissues are fixed in place like in static flasks, and exposed to a well mixed medium. The flow conditions were characterized as turbulent, with the associated hydrodynamic shear that was below the level causing cell death or damage, but sufficient to affect the function of cells at construct surfaces.[13] Mass transport between the tissue and culture medium is enhanced by convection, whereas the transport within the tissue remains governed by molecular diffusion like in static flasks.

In rotating vessels, cultured tissues are dynamically suspended in the rotating flow without external fixation. The flow conditions were characterized as dynamic and laminar, with tissue constructs settling in a tumble-slide regime associated with fluctuations in fluid pressure, velocity, and shear.[12] Mass transport between the tissue and culture medium is enhanced by dynamic laminar convection, a flow regime that is stimulatory to the cells, whereas the transport within the tissue remains governed by molecular diffusion.

In perfused cartridges, cultured tissues are fixed in place and perfused with a culture medium, at interstitial velocities comparable to those of blood flow in native tissues. Medium flow is within the laminar regime, and can be either steady or pulsatile. The enhancement of mass transport is achieved by interstitial flow of medium in conjunction with external gas and medium exchange, resulting in short diffusional distances and better control of the microenvironmental conditions than in any of the above three systems.

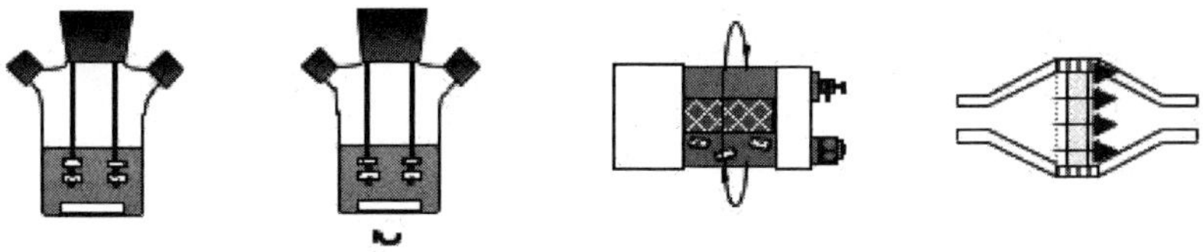

Cultivation vessel	Static flask	Mixed flask	Rotating vessel	Perfused cartridge
Medium volume (cm^3)	120	120	110	1.5
Engineered tissues (5 mm diameter x 2 mm thick discs)	Fixed in place; up to 12 per vessel	Fixed in place; up to 12 per vessel	Freely settling; up to 12 per vessel	Fixed in place; 1 per cartridge
Medium exchange	Batch-wise (3 cm^3 per construct per day)	Batch-wise (3 cm^3 per construct per day)	Batch-wise (3 cm^3 per construct per day)	Continuous recirculation (0.01–3 mL/min); periodic exchange (3 cm^3 per construct per day)
Gas exchange	Continuous, via surface aeration	Continuous, via surface aeration	Continuous, via an internal membrane	Continuous, via an external gas exchange
Stirring/rotation rate (s^{-1})	0	0.83–1.25	0.25–0.67	0
Flow conditions	Static fluid	Turbulent(1)	Laminar(2)	Laminar(2)
Mixing mechanism	None	Magnetic stirring	Settling in rotational flow	Recirculating flow
Mass transfer in bulk medium	Molecular diffusion	Convection (due to medium stirring)	Convection (due to tissue settling)	Convection (due to medium flow)
Fluid shear at construct surfaces	None	Steady, turbulent	Dynamic, laminar	Steady, laminar
References	[8, 13, 25, 62, 70]	[8, 11, 13, 20, 25, 48, 62, 70]	[8–12, 25, 26, 28, 51, 52, 62, 65, 69, 70, 87]	[14, 30, 61, 62]

(1) The smallest turbulent eddies had a diameter of 250 μm and velocity of 0.4 cm/s; estimated according to Cherry and Papoutsakis [86] by Vunjak-Novakovic et al. [13]
(2) Tissues were settling in a laminar tumble-slide regimen in a rotational field; estimated according to Clift et al. [88] by Freed and Vunjak-Novakovic [12] and Neitzel et al. [87]

FIGURE 33.4 Representative bioreactors. (top) Schematic presentation of construct cultivation in static flasks, mixed flasks, rotating vessels and perfused cartridges. (bottom) Overview of the operating conditions for each vessel type.[8,12,13,86–88]

III. BIOREACTOR CULTIVATION OF ENGINEERED CARTILAGE

Table 33.1 gives an overview of representative studies of cartilage tissue engineering that involved the use of bioreactors (i.e., various regimes of flow, mixing, medium, and gas exchange, and/or the application of biochemical and physical regulatory signals) and demonstrated the functionality of engineered tissue (e.g., by measuring compressive, tensile or shear properties, or integration with adjacent host tissues).

A. Orbitally Mixed Dishes

The structural and functional properties of engineered constructs based on bovine calf chondrocytes and PGA scaffolds improved progressively and concomitantly over 12 weeks of cultivation in orbitally mixed dishes.[22] The unconfined compression aggregate modulus increased 40-fold to 10% of normal levels, and the hydraulic permeability decreased four orders of magnitude to the range of values measured for native cartilage.[22] In separate related studies, shear modulus of 8-week constructs positively correlated with GAG and collagen fractions but remained subnormal.[23]

Engineered constructs based on bovine chondrocytes and PGA scaffolds and cartilage explants cultured for 9 weeks in orbitally mixed dishes contained comparable amounts of proteoglycans, whereas the volumetric fractions of collagen in constructs were only one half of those in explants.[24] In confined compression, engineered constructs, cultured explants, and freshly explanted cartilage had similar stress–strain relationships, consistent with the model of prestressed collagen fibers balanced by swelling pressure of proteoglycans. The aggregate and shear moduli decreased from cartilage to explants and constructs, and correlated with the tissue fractions of proteoglycans and collagen.[24]

TABLE 33.1
Bioreactor Studies of Cartilage Tissue Engineering

Type of Bioreactor (Duration of Culture)	Cell Source	Scaffold (Material, Dimensions)	Regulatory Factors (Biochemical)	Regulatory Factors (Physical)	References
Static dish (7 weeks)	Articular chondrocyte (bovine, 1–2 weeks old, expanded)	Agarose gel, 16 mm dia × 1 mm thick	10% FBS	Mechanical compression static: up to 50% strain; dynamic: 3% strain, 0.01–1 Hz)	35, 38
Static dish (4 weeks)	Articular chondrocyte (bovine, 3–5 months old)	Agarose gel, 6.76 mm dia × 1.7 mm thick; 4.76 mm × 1.6 mm thick	10% FBS	Mechanical compression; dynamic; unconfined; 3 cycles per day; 1 h on 1 h off; 5% strain amplitude; 1 Hz	36, 39
Static dish (4 weeks)	Articular chondrocyte	Agarose gel, 4.76 mm × 1.6 mm thick	10% FBS plus 10 ng/ml TGF-b, or 300 ng/ml IGF-I	Mechanical compression; (dynamic; unconfined; 2% offset; 5% strain; 1 Hz; 1 h on, 1 h off, 3 cycles/day)	49
Static and mixed dishes (2–12 weeks)	Articular chondrocyte (bovine, 2–4 weeks old)	Porous PLLA	10% FBS during cell expansion and construct cultivation	Convective mixing; convective flow	12, 23, 84–86
Perfused bag (5 weeks) Perfused cartridges (4 weeks) Perfused chambers (2 weeks)	Articular chondrocyte (bovine, 2–4 weeks old); cartilage explants	10 mm × (5–10) mm × (2–3) mm; Fibrous PGA (10 mm dia × (1–5) mm thick)			
Mixed dishes (9 ± 1 weeks)		Fibrous PGA 5 mm dia × 2 mm thick	10% FBS	Convective mixing; osmotic pressure	24
Compression chamber (5 weeks) preculture in mixed flasks	Articular chondrocyte (equine; young and adult)	Fibrous PGA, 10 mm × 10 mm × 1 mm	10% FBS	Flow; intermittent pressure; (3.4 or 6.9 MPa; 5 sec on, 30 sec off)	42, 45, 46

Perfused bioreactor (4 weeks)	Articular chondrocyte (bovine, 2–4 weeks old); human (30–65 year old) rabbit	PLGA-coated fibrous PGA, 12.7 mm dia × 1 mm thick; porous collagen (2 mm thick), PLGA or polydioxanon mesh	10% FBS 2% FCS	Medium perfusion (interstitial velocity: 0.1–10 μm/sec or around constructs)	29–31
Static dish (3 weeks);	Articular chondrocyte (bovine, 2–4 weeks old)	fibrous PGA, 10 mm dia × 2 mm thick	10% FBS	Mechanical compression (static and dynamic; confined) (static offset: 0, 1, 300 or 50%; 5% strain; 0.001 or 0.1 Hz)	47
Rotating vessel (7 months)	Articular chondrocyte (bovine, 2–4 weeks old)	Fibrous PGA, 5 mm dia × 2 mm thick	10% FBS		25, 28, 87
Seeding in mixed flasks (3 days) Static and mixed flasks Rotating vessel (6 weeks)	Articular chondrocyte (bovine, 2–4 weeks old)	Fibrous PGA, 5 mm dia × 2 mm thick	10% FBS	Flow and mixing	13, 26, 70, 87
Rotating vessel (5 weeks)	Articular chondrocyte (bovine, 2–4 weeks old)	Fibrous PGA, 5 mm dia × 2 mm thick	10% FBS oxygen (40 or 80 mg Hg)	Flow and mixing	2, 52
Static and mixed dishes Static and mixed flasks Rotating vessel (4 weeks)	Articular chondrocyte (bovine, 2–4 weeks old)	Fibrous PGA, 5 mm dia × 2 mm thick	10% FBS IGF-I	Flow and mixing	48
Rotating vessel (4 weeks)	Articular chondrocyte (bovine, 2–4 weeks old) transfected with human IGF-I	Fibrous PGA, 5 mm dia × 2 mm thick	None	Flow and mixing	88
Rotating vessel (4 weeks)	Articular chondrocyte (bovine, 2–4 weeks old)	Fibrous PGA, 5 mm dia × 2 mm thick	10% FBS and either TGF-b + FGF-2, or FGF-2 or IGF-I	Flow and mixing	50

Continued

TABLE 33.1
Continued

Type of Bioreactor (Duration of Culture)	Cell Source	Scaffold (Material, Dimensions)	Regulatory Factors (Biochemical)	Regulatory Factors (Physical)	References
Rotating vessel (4 weeks)	Articular chondrocyte (bovine, 2–4 weeks old)	PGA (fibrous mesh; composite) Hyaff-11 (benzylated HA, porous sponge; fibrous mesh) 5 mm dia × 2 mm thick	10% FBS	Flow and mixing	9
Constructs: rotating vessels (5 d; 5 weeks) Construct/explant composites: rotating vessels (1–8 weeks)	Articular chondrocyte (bovine, 2–4 weeks old); Cartilage explants (bovine, 2–4 weeks old)	Constructs: fibrous PGA, 5 mm dia × 2 mm thick Explants: 10/5 mm dia × 2 mm thick rings (intact or trypsin treated)	10% FBS	Flow and mixing	89

B. Static Flasks, Mixed Flasks, Rotating Vessels, and Perfused Cartridges

Three representative types of bioreactors used for cartilage tissue engineering include static flasks, mixed flasks, and rotating bioreactors. The effects of bioreactor flow and mixing on construct structure and function are described in more detail in the last section of this chapter. Construct compositions and mechanical properties improved from static to mixed flasks and rotating vessels. Constructs cultured in rotating vessels were uniformly cartilaginous, contained 75% as much GAG, and 40% as much total collagen (hydroxyproline) per unit wet weight as normal cartilage, and had normal tissue morphology except for the lack of zonal organization.[25] The collagen network of 6-week constructs had normal fibril density and diameter, and the fractions of collagen types II, IX, and X, but the number of pyridinium cross-links per collagen molecule was only one third of normal.[26]

The compressive modulus of 6-week constructs was 175 kPa, as compared to 270 kPa measured for the articular surface of bovine articular cartilage,[27] and 710 or 950 kPa measured for the deep zone.[27,28] The dynamic stiffness and hydraulic permeability was both subnormal, although within the order of magnitude of values measured for bovine articular cartilage. Prolonged (7-month) cultivation yielded constructs with normal wet weight fractions of GAG and normal aggregate modulus and hydraulic permeability, but the fraction of collagen and dynamic stiffness remained subnormal.[28] These studies demonstrated that medium flow and mixing can be utilized to achieve spatially uniform cell distributions in thick constructs and to increase the rates of mass transport at construct surfaces, but also that the regime of flow (laminar or turbulent) can markedly affect cell function and thereby the structure and mechanical function of engineered constructs.

Direct perfusion through cultured tissue constructs also stimulated chondrogenesis, presumably due to combined effects of enhanced mass transport, pH regulation, and fluid shear in the cell microenvironment,[29–31] in particular at physiological interstitial flow velocities, approximately 1 μm/sec.[32–34] It is possible that the observed effects of dynamic mechanical compression *in vitro* (e.g., [35,36]) and *in vivo*[37] were due in part to the increased fluid flow within cultured tissues.

C. Mechanical Stimulation

During habitual loading, articular cartilage is exposed to fluctuating mechanical forces, which provide interstitial fluid flow and tissue nutrition. Effects of static and dynamic compression were studied for articular bovine calf chondrocytes cultured statically in agarose gel for up to 7 weeks.[35,38] Constructs responded to mechanical forces in a manner similar to that of native cartilage: static compression suppressed ECM synthesis by an amount that increased with increasing compression amplitude and culture time, whereas dynamic compression stimulated ECM synthesis by an amount that increased with ECM accumulation and culture time. An increase in cell seeding density was associated with improved construct compositions and mechanical properties, an improvement that was counterbalanced with decreased beneficial effects of mechanical loading.[39]

Dynamic compression enhanced chondrogenesis of chick-limb bud cells in a manner dependent on the frequency and duration of loading.[40,41] Intermittent hydrostatic pressure improved compositions and mechanical properties of young but not old chondrocytes cultured on a fibrous PGA scaffold in a perfused bioreactor for 5 weeks.[42–46] Short term static compression suppressed while dynamic compression enhanced the synthesis and retention of GAGs and protein in bovine chondrocytes cultured statically on PGA scaffolds for 3 weeks.[47]

D. Interactions of Growth Factors and Hydrodynamic or Mechanical Stimulation

The interactive effects of the insulin-like growth factor-I IGF-I and mechanical environment (static and orbitally mixed dishes, static and mixed flasks, rotating vessels) on structure, biochemical composition, and mechanical properties of engineered cartilage were studied using the chondrocyte–PGA model system.[48] The IGF-I and flow modulated tissue properties independently in a manner consistent with previous studies, and interacted to produce tissues superior to those obtained by utilizing the two factors independently. After 4 weeks of culture, the best constructs had wet weight fractions of GAG and collagen and equilibrium moduli that were approximately 50% of values measured for the middle zone of native articular cartilage. TGF-β and IGF-I interacted with mechanical loading in a synergetic manner and improved the compositions and mechanical properties of constructs cultured for 5 weeks in agarose gel, to the extent greater than the sum of effects of either stimulus applied alone.[49] These studies demonstrated that the beneficial effects of growth factors can be amplified by the application of dynamic mechanical loading.

In separate related studies, growth factors supplemented sequentially to culture medium (TGF-β/FGF-2 early, IGF-I later during culture) in a hydrodynamically active bioreactor environment, markedly and significantly improved the compositions and mechanical properties of engineered cartilage.[50] After 4 weeks of culture, constructs contained up to 4.5% wet weight GAG, up to 4.5% wet weight collagen and had equilibrium moduli of up to 400 kPa. Likewise, a hydrodynamically active environment present in rotating bioreactors amplified the beneficial effects of polymer scaffolds on construct compositions and mechanical properties.[9] Bioreactor hydrodynamics and scaffold structure acted in concert to yield engineered cartilage that contained high fractions of GAG and type II collagen, and had equilibrium moduli of 400 to 540 kPa after only 4 weeks of bioreactor cultivation. Taken together, these recent studies demonstrated that engineered cartilage with the composition and mechanical behavior within the range of values measured for immature (fetal-like) native cartilage can be grown in bioreactors within approximately 4 weeks.

Gene transfer of the human IGF-I was also utilized for cartilage tissue engineering in an attempt to further enhance construct compositions and functional properties.[51] Calf articular chondrocytes (unmodified, genetically modified to overexpress the *Escherichia coli* β-galactosidase gene or the human IGF-I gene) were cultured on PGA scaffolds in bioreactors and evaluated structurally and functionally, *in vitro* and *in vivo*. Transgene expression was maintained both *in vitro* and *in vivo* and resulted in rapid and progressive chondrogenesis. After 4 weeks of culture, IGF-I constructs contained markedly larger amounts of GAGs and collagen, and had a fourfold higher equilibrium moduli as compared to nontransfected or lacZ constructs. The observed enhancement of *in vitro* chondrogenesis by spatially defined overexpression of human IGF-I suggested that cartilage tissue engineering based on genetically modified chondrocytes may be advantageous compared to either gene transfer or tissue engineering alone.

E. Controlled Bioreactor Studies

One important advantage of bioreactors is that they can be used in controlled *in vitro* studies of engineered tissues. Two examples listed here include (1) mathematical modeling of the progression of chondrogenesis, and (2) quantitative studies of the effects of culture duration on integrative properties of engineered cartilage.

The progression of chondrogenesis in constructs based on articular chondrocytes and fibrous scaffolds has been associated with temporal and spatial changes in local concentrations of the cells and cartilaginous matrix. Quantitatively, the development of tissue engineered cartilage has been modeled using a spatially varying, deterministic, continuum model.[52] The model accounted for the deposition and diffusion of oxygen and GAG as a function of the spatial position within the construct and the duration of tissue cultivation. Here, GAG was taken as a marker of chondrogenesis in light of prior association of its deposition with that of collagen type II.[25] Model predictions for concentration profiles of GAGs were qualitatively and quantitatively consistent with those measured via high-resolution (40 μm) image processing of histological sections of tissue samples harvested at timed intervals.[53]

To study integration, disc-shaped constructs cultured for 5 days or 4 weeks, or cartilage explants (intact or trypsin-treated) were sutured into cartilage rings (intact or trypsin-treated), cultured for 1 to 8 weeks and evaluated structurally and functionally (compressive stiffness of the central disk, adhesive strength of the integration interface). Immature constructs integrated better than either mature constructs or cartilage explants. Integration of immature constructs involved cell proliferation and the progressive formation of cartilaginous tissue, in contrast to the integration of more mature constructs or native cartilage that involved only the secretion of matrix components.

IV. BIOREACTOR CULTIVATION OF ENGINEERED CARDIAC TISSUE

Table 33.2 gives an overview of representative studies of cardiac tissue engineering that involved the use of bioreactors (i.e., various regimes of flow, mixing, medium, and gas exchange, and/or the application of biochemical and physical regulatory signals) and demonstrated the functionality of engineered tissue (e.g., by measuring signal propagation, contractile properties, or ability to survive *in vivo*).

A. Static and Orbitally Mixed Dishes

Cardiac tissue constructs were successfully cultivated in dishes using a variety of scaffolds and cell sources. Fetal rat ventricular cardiac myocytes were expanded, inoculated into collagen sponges,

TABLE 33.2
Bioreactor Studies of Engineered Cardiac Tissue

Type of Bioreactor	Heart Cell Source	Scaffold	Construct Size	Initial Cell Number	Reference
Static dish	Fetal rat	Alginate	6 mm dia × 1 mm	3×10^5	58
Static dish	Fetal rat	Bovine collagen (Gelfoam sponge)	$5 \times 5 \times 1$ mm^3	*ca.* 10^4	54
Static dish	Neonatal rat	Bovine collagen (Tissue Fleece mesh)	$20 \times 15 \times 2.5$ mm^3	2×10^6	57
Static dish	Neonatal rat	Bovine collagen	$5 \times 5 \times 3$ mm^3	0.5×10^6	84
Mixed dish (20 rpm)	Neonatal rat	Bovine collagen (Ultrafoam sponge)	10 mm dia × 1.5 mm thick	$6–12 \times 10^6$	14, 60
Spinner flask (static, 50, 90 rpm)	Embryonic chick Neonatal rat	PGA mesh	5 mm dia × 2 mm thick	$1.3–8 \times 10^6$	62
Spinner flask (50 rpm)	Neonatal rat	PGA, sPGA mesh	5 mm dia × 2 mm thick	8×10^6	63, 64
HARV, STLV (11 rpm)	Neonatal rat	PGA, sPGA, lPGA mesh	5 mm dia × 2 mm thick	8×10^6	62–64
HARV	Neonatal rat	Bovine collagen	$5 \times 5 \times 3$ mm^3	0.5×10^6	84
Perfused cartridge (0.2–3 ml/min)	Neonatal rat	PGA mesh	11 mm dia × 2 mm thick	24×10^6	61, 66
Perfused cartridge (0.2–3 ml/min)	Neonatal rat	Bovine collagen (Ultrafoam sponge)	10 mm dia × 1.5 mm thick	$6–12 \times 10^6$	14, 60
Cyclic stretch (1.5–2 Hz, 1–20% strain)	Embryonic chick Neonatal rat	Collagen gel		$1–2.5 \times 10^6$	17, 18
Cyclic stretch (1.3 Hz)	Human heart	Bovine collagen (Gelfoam sponge)	$30 \times 20 \times 3$ mm^3 $20 \times 20 \times 3$ mm^3	$3–30 \times 10^6$	67
Electrical stimulation (2 msec, 1 Hz, 5 V)	Neonatal rat	Bovine collagen (Ultrafoam sponge)	$8 \times 6 \times 1.5$ mm^3	5×10^6	19

and cultivated in static dishes for up to 4 weeks.[54] The cells proliferated in culture and expressed multiple sarcomeres. Adult human ventricular cells were used in a similar system, but they exhibited no proliferation.[55] Neonatal rat cardiac myocytes cultured in collagen sponges formed spontaneously contracting constructs within 36 h[56,57] and maintained their activity for up to 12 weeks. The contractile force increased upon addition of calcium and epinephrine.

Fetal cardiac myocytes cultivated statically in alginate formed spontaneously beating aggregates in the scaffold pores after 4 days of culture.[58] High density seeding of cells in alginate (on the order of 10^8 cells/cm^3) was achieved using centrifugal forces.[59] Neonatal rat cardiac myocytes were inoculated into collagen sponges (Ultrafoam) at high density (*ca.* 10^8 cells/cm^3) using Matrigel and cultivated in orbitally mixed dishes. The constructs contracted synchronously in response to electrical stimulation generated by a pace-maker, and expressed the full set of cardiac differentiation markers: cardiac troponin I, sarcomeric α-actin, and tropomyosin.[14,60]

Despite enhanced mass transport at construct surfaces achieved in orbitally mixed dishes, the main mode of oxygen and nutrient transport within the construct was molecular diffusion, like in static cultures. Diffusion alone is capable of satisfying oxygen demand of only about 100 μm thick surface layer of compact tissue.[61] Therefore, the interior of the constructs remained mostly acellular. Consistent with the diffusional limitations of oxygen transport to the cells is the anaerobic glucose metabolism exhibited by the constructs cultivated in orbitally mixed dishes.[14]

B. Mixed Flasks

To improve cell survival and assembly on all surfaces of the engineered tissue, cardiac constructs were seeded and cultured in mixed flasks (Figure 33.4). Neonatal rat or embryonic chick ventricular myocytes were seeded onto PGA scaffolds at the density of 2×10^6 cells/cm^3 by placing a dilute cell suspension in the spinner flasks and mixing for 3 days (50 rpm).[62] After 2 weeks in culture, constructs based on neonatal rat cardiac myocytes increased in size and exhibited spontaneously beating areas, whereas constructs based on embryonic chick myocytes exhibited no contractions and reduced in size by 60%. Immunohistochemistry, revealed presence of large number of nonmyocytes in the constructs based on embryonic chick heart cells, while constructs based on neonatal rat cells consisted mostly of elongated cardiomyocytes.

Mixing had significant effects on the construct metabolism and cellularity.[62] Constructs cultivated in mixed flasks (90 or 50 rpm) had significantly higher cellularity index (*ca.* 20 μg DNA/construct) and metabolic activity (*ca.* 150 MTT units/mg DNA), compared to the constructs cultivated in static flasks (*ca.* 5 μg DNA/construct, and *ca.* 50 MTT units/mg DNA, respectively). Mixing maintained medium gas and pH levels within the physiological range yielding a more aerobic glucose metabolism (L/G $\approx$ 1.5) in mixed flasks compared to the anaerobic glucose metabolism in static flaks (L/G $>$ 2).[62]

After 1 week of culture, constructs seeded in mixed flasks with 8×10^6 neonatal rat cardiac myocytes contained a peripheral tissue-like region (50 to 70 μm thick) in which cells stained positive for tropomyosin and organized in multiple layers in a three-dimensional configuration.[63] Electrophysiological studies conducted using a linear array of extracellular electrodes showed that the peripheral layer of the constructs exhibited relatively homogeneous electrical properties and sustained macroscopically continuous impulse propagation on a centimeter-size scale.[63] Constructs based on cell preparations that were further enriched with cardiomyocytes by cell preplating exhibited lower excitation threshold (ET), higher conduction velocity, higher maximum capture rate (MCR), and higher maximum and average amplitude.[63]

As compared to orbitally mixed dishes, where the bottom construct surface was in contact with the dish surface, in spinner flaks mixing improved the delivery of oxygen and nutrients to the entire surface of the tissue construct and yielded more uniform cell distribution in the peripheral tissue layer. However, the interior of these constructs remained largely acellular due to the diffusional limitations of oxygen transport to the cells in the construct interior, and the construct cellularity remained significantly lower than in the neonatal rat ventricle.[63] An additional drawback of the cultivation in spinner flasks is that turbulent flow conditions may induce cell dedifferentiation and the formation of a fibrous capsule at construct surfaces.

C. Rotating Vessels

Two types of rotating bioreactors: high aspect ratio vessel (HARV) and slow turning lateral vessel (STLV, shown in Figures 33.3 and 33.4) were utilized for cardiac tissue engineering.[62,64] Both vessels enable the cultivation of constructs that are suspended in laminar flow of medium without external fixation. The main difference is that HARV has larger gas exchange membrane and therefore better oxygen supply to the constructs.[8,62,64]

Seeding of neonatal rat cardiac myocytes from a dilute cell suspension onto PGA scaffolds in HARVs resulted in a higher yield of cell attachment and less cell damage (as assessed by medium lactate dehydrogenase [LDH] levels) as compared to the seeding in spinner flasks. This difference was likely due to the laminar flow conditions that exert less cell damage compared to the turbulent flow conditions of the spinner flasks.[62] Medium gas composition and pH were maintained within the physiological ranges and resulted in aerobic cell metabolism (as assessed by lactate yield on glucose that was approximately 1 mol/mol).

Metabolic activity of the constructs cultivated in HARVs was comparable to those of neonatal rat ventricles (*ca.* 250 units MTT/mg DNA) and significantly higher than for those cultivated in mixed flasks (*ca.* 150 units MTT/mg DNA, respectively). Hyperthrophy index was comparable to the neonatal tissue (18 mg protein/mg DNA) but the cellularity remained subnormal.[62]

Cultivation of cells enriched by preplating for cardiac myocytes in HARVs on laminin coated PGA in low serum medium yielded formation of up to 160 μm thick layer of peripheral cardiac tissue with electrophysiological properties that were better than for constructs cultivated in spinner flasks and comparable to native ventricles.[64] The expression levels of cardiac proteins connexin-43, creatin kinase-MM, and sarcomeric myosin heavy chain were lower in HARV cultivated constructs than in neonatal rat ventricle, but higher than in the spinner flask cultivated constructs.[64]

Action potential duration in the tissue constructs cultivated in HARVs was 1.8 to 2.4 times longer than those of native ventricle and the MCR was 68% as high. Pharmacological studies done with 4-aminopyridine indicated that a decrease in the transient outward potassium current could be responsible for the observed differences in APD and MCR.[65]

D. Perfused Cartridges with Interstitial Flow of Medium

Cardiac tissue constructs (11 mm in diameter × 2 mm thick) based on neonatal rat ventricular cardiac myocytes and PGA scaffold were placed in a perfusion chamber and pulsatile flow of medium through the construct was provided by a peristaltic pump (0.2 to 3 ml/min, corresponding to interstitial velocities of 35 to 500 μm/sec). A coil of silicone tubing in the external loop served as a gas exchanger. The resulting constructs had uniform distribution of cells expressing sarcomeric α-actin and cardiac specific ultrastructural features (sarcomeres, gap junctions), but the cell density remained low due to the diffusional limitations of oxygen transport during seeding that was carried out in tissue culture dishes for 48 h and resulted in substantial cell loss.[61,66]

In recent studies, the rate, yield, and uniformity of cell seeding were significantly improved by utilizing rapid gel inoculation of cells into collagen scaffold in conjunction with the immediate establishment of interstitial medium flow.[14] Neonatal rat cardiomyocytes suspended in Matrigel were inoculated on collagen sponges (10 mm diameter × 1.5 mm thick discs) at a physiologic cell density (1.35×10^8 cells/cm^3) and cultured with interstitial medium flow (0.5 ml/min, corresponding to an interstitial velocity within the construct of 425 μm/sec). During the initial 1.5 h period of cell seeding, the direction of medium flow was alternated every 5 min to prevent the washout of loosely bound cells; unidirectional medium flow was applied during construct cultivation.

Medium perfusion markedly improved construct properties according to all tested parameters.[60] After 7 days of cultivation, the viability of cells in perfused constructs was indistinguishable from that of freshly isolated cells (*ca.* 85%), as compared to rather low viability of dish-grown constructs (*ca.* 45%). The molar ratio of lactate produced to glucose consumed in perfused constructs (L/G $\approx$ 1) indicated aerobic cell metabolism, in contrast to anaerobically metabolizing (L/G $\approx$ 2) dish-grown constructs. Perfused constructs and native ventricles had more cells in the S phase than in the G2/M phases, whereas the cells from dish-grown constructs appeared unable to complete the cell cycle and accumulated in the G2/M phase. Cells expressing cardiac-specific differentiation markers (sarcomeric α-actin, cardiac troponin I, sarcomeric tropomyosin) were present throughout the perfused constructs, while the dish-grown constructs had most cells located within a 100 to 300 μm thick surface layer, around an empty interior. In response to electrical stimulation, perfused constructs contracted synchronously, had lower excitation thresholds, and recovered their baseline function levels following treatment with a gap junctional blocker; dish-grown constructs exhibited arrhythmic contractile patterns and failed to recover their baseline levels.

Although interstitial medium flow enabled engineering of compact tissue that had the physiologic density of viable aerobically metabolizing cells, most cells were round and mononucleated, indicating the lack of certain physical stimuli.

E. Mechanical Stimulation

To provide appropriate mechanical stimulation during construct cultivation, neonatal rat cardiac myocytes were reconstituted in collagen gel and cultivated in the presence of cyclic stretch. In one setup, neonatal rat ventricular myocytes were resuspended in a gel consisting of collagen I and Matrigel in culture medium[17] and cast into wells containing Velcro coated silicone tubes. After 4 days in culture a strip of biconcave tissue was formed and fixed at each end to a piece of silicone tubing. Unidirectional and cyclic stretch at the frequency of 1.5 Hz and the strain rate of up to 20% was employed for 6 days. Stretched constructs exhibited improved morphology and organization of cardiac myocytes, and higher RNA/DNA and protein/cell ratios. The force of contraction was higher in stretched constructs both under basal conditions and after stimulation with isoprenaline.[17]

In an improved setup, neonatal rat cardiac cells were resuspended in the collagen/Matrigel mix and cast into circular molds.[18] After 7 days of static culture, the strips of cardiac tissue were placed around two rods of a custom-made mechanical stretcher and subjected to unidirectional or cyclic stretch. Histology and immunohistochemistry revealed a formation of intensively interconnected, longitudinally oriented cardiac muscle bundles with morphological features resembling adult rather than immature native tissue. Fibroblasts, macrophages, and capillary structures positive for CD31 were detected. Cardiomyocytes exhibited well developed ultrastructural features: sarcomeres arranged in myofibrils, with well developed Z, I, A, H, and M bands, specialized cell–cell junctions, T tubules as well as basement membrane. Contractile properties were similar to those measured for the native tissue, with high ratio of twitch to resting tension and strong β-adrenergic response. Action potentials characteristic of rat ventricular myocytes were recorded.

Cyclic mechanical stretch (1.33 Hz) was applied to the constructs based on collagen scaffold (Gelfoam) and human heart cells (isolated from children undergoing repair of Tetralogy of Fallot[67]). A rectangular piece of tissue was fixed at one end to the bottom of a square dish; the other end attached to a freely moving steel rod. The cyclic movement of a rod is induced by a dynamically changing magnetic field. Constructs subjected to chronic stretch had improved cell distributions, collagen matrix formation, and increased cell proliferation.

F. Electrical Stimulation

Constructs prepared by seeding collagen sponges ($6 \times 8 \times 1.5$ mm^3) with neonatal rat ventricular cells (5×10^6) were stimulated using suprathreshold square biphasic pulses (2-msec duration, 1 Hz, 5 V/cm). The stimulation was initiated after 1 to 5 days of scaffold seeding (a 3-day period was optimal) and applied for up to 8 days. Stimulation resulted in significantly better contractile responses to pacing as compared to unstimulated controls, as evidenced by the sevenfold higher amplitude of contractions, lower excitation threshold, and higher MCR. Excitation–contraction coupling of cardiac myocytes in stimulated constructs was also evidenced by transmembrane potentials that were similar to the action potentials reported previously for cells from mechanically stimulated constructs.[18] Importantly, the improved contractile properties were not the result of increased cellularity.

Stimulated constructs exhibited higher levels of α-myosin heavy chain (α-MHC), Connexin 43, creatin kinase-MM, and cardiac troponin I expression, and contained thick aligned myofibers that resembled myofibers in the native heart. At the ultrastructural level, cells in stimulated constructs exhibited specialized features characteristic of native myocardium. Gap junctions, intercalated discs, and microtubules were all markedly more frequent in stimulated group compared to the nonstimulated group. Cells in stimulated constructs contained aligned myofibrils, and well developed sarcomeres with clearly visible M, Z lines, H, I, and A bands. In most cells, Z lines were aligned, and the intercalated discs were positioned between two Z lines. In contrast, nonstimulated constructs had poorly developed cardiac-specific organelles and poor organization of ultrastructural features.

These studies suggest that electrical stimulation of construct contractions during cultivation progressively enhanced the excitation–contraction coupling and improved the properties of engineered myocardium at the cellular, ultrastructural, and tissue levels.

V. CASE STUDIES

(1) **Cartilage**. Hydrodynamic factors in bioreactors can modulate chondrogenesis in at least two ways: via associated effects on mass transport of biochemical factors between the developing tissue and culture medium and by direct physical stimulation of the cells. The effects of three different culture environments: static flasks (tissues fixed in place, static medium), mixed flasks (tissues fixed in place, turbulent flow) and rotating bioreactors (tissues suspended in laminar flow) were studied for engineered cartilage constructs and native cartilage explants.

Average steady-state concentrations of metabolites measured in culture medium over 6 weeks of cultivation are shown in Table 33.3. Static cultures of cartilage constructs and explants had comparable metabolic parameters, most of which were significantly different from those measured in mixed flasks and rotating bioreactors. Average steady-state levels of pO_2, ammonia and pH were lower, and p_{CO_2}, and glucose level were higher in static than in mixed cultures. Higher yields of lactate on glucose ($Y_{L/G}$) in static than in mixed cultures imply more anaerobic metabolism in cells from statically cultured constructs and explants.

In static flasks, cartilaginous matrix accumulated mostly at the periphery of constructs and explants (Figure 33.5d and a, respectively), in contrast to constructs and explants cultured in mixed flasks (Figure 33.5e and b, respectively) where cartilaginous matrix accumulated in the inner tissue phase but was surrounded by a thick fibrous capsule (*ca.* 450 μm after 6 weeks of culture[53]). Only in rotating vessels were constructs and explants cartilaginous over their entire cross-sections and had normal tissue morphology except for the lack of zonal organization (Figure 33.5f and c, respectively). The gradients of GAG concentration at the tissue surfaces were consistent with the measured fractional release of newly synthesized GAG into the culture medium: 10 to 30% for constructs and explants cultured in static flasks and rotating vessels, and 40 to 60% for constructs and explants cultured in mixed flasks.[68]

TABLE 33.3
Average Steady-State Levels of Metabolic Parameters in Cultures of Engineered and Native Cartilage in Static Flasks, Mixed Flasks and Rotating Vessels (Reproduced with Permission from Obradovic et al.[68])

	Static Flask		Mixed Flask		Rotating Vessel	
	Constructs	Explants	Constructs	Explants	Constructs	Explants
pH	6.85 ± 0.04^{1}	6.91 ± 0.03^{1}	6.97 ± 0.05^{2}	7.03 ± 0.03^{2}	6.87 ± 0.04^{1}	7.07 ± 0.01^{2}
P_{CO_2} (mmHg)	106.2 ± 4.0^{1}	103.7 ± 5.0^{1}	80.5 ± 3.0^{2}	79.5 ± 3.0^{2}	85.9 ± 2^{3}	79.6 ± 1.0^{2}
P_{O_2} (mmHg)	81.2 ± 3.3^{1}	84.5 ± 4.0^{1}	114.2 ± 4.0^{2}	124.0 ± 3.0^{3}	77.6 ± 3^{1}	114.1 ± 3.0^{2}
Lactate (g/l)	1.28 ± 0.15^{1}	1.08 ± 0.16^{1}	1.07 ± 0.24^{1}	1.04 ± 0.13^{1}	1.56 ± 0.18^{2}	0.74 ± 0.13^{2}
Glucose (g/l)	2.46 ± 0.15^{2}	2.74 ± 0.15^{1}	2.23 ± 0.31^{2}	2.62 ± 0.18^{1}	1.94 ± 0.22^{2}	3.08 ± 0.16^{2}
$Y_{L/G}$ (mol L/mol G)	1.55	1.67	1.30	1.22	1.28	1.38
Ammonia (g/l)	0.013 ± 0.002^{1}	0.012 ± 0.002^{1}	0.036 ± 0.004^{2}	0.034 ± 0.004^{2}	0.034 ± 0.003^{2}	0.022 ± 0.005^{3}

Superscripts indicate significantly different levels of a given variable among groups, as assessed by ANOVA in conjunction with Tukey's-HSD test ($p < .05$).

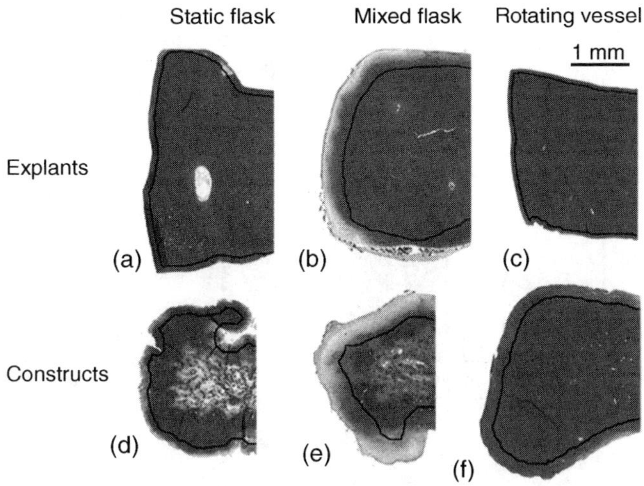

FIGURE 33.5 Histological structures of cultured cartilage constructs and explants. Explants of bovine articular cartilage and engineered constructs based on fibrous PGA scaffolds and bovine chondrocytes were prepared in form of 5 mm diameter × 2 mm thick discs and cultured for 6 weeks in three different environments. Representative histological sections are shown for cartilage explants (a, b, c) and constructs (d, e, f) cultured in static flasks (a, d), mixed flasks (b, e), and rotating vessels (c, f). Stain: safranin O. (Images from Vunjak-Novakovic, G., Obradovic, B., Martin, I., and Freed, L. E., Bioreactor studies of native and tissue engineered cartilage, *Biorheology*, 39, 259–268, 2002. With permission.)

The observed differences in tissue morphology could be related to the respective differences in flow and mass transport conditions in the three culture vessels (Figure 33.4). In addition, the modeling of GAG distribution in native cartilage explants quantified the effects of hydrodynamic conditions on GAG synthesis and diffusion rates.[69] In static flasks, mass transport in a culture medium occurs by molecular diffusion only (therefore tissue formation on the construct periphery with low GAG release). In mixed flasks, mechanical stirring generates convective flow that enhances mass transport in bulk medium, but also generates turbulent shear (therefore formation of a GAG depleted capsule with elongated fibroblast-like cells at construct surfaces). In rotating vessels, construct settling generates laminar fluid flow with dynamic fluctuations in fluid velocity, shear, and pressure that enhance mass transport at construct surfaces without adverse hydrodynamic effects (therefore spatially uniform chondrogenesis throughout the construct volume and GAG release comparable to that measured in static cultures). Notably, the effects of bioreactor environment on tissue morphology were similar for cartilage constructs and explants (Figure 33.5, compare a and d, b and e, c and f).

Over 6 weeks of *in vitro* culture, the wet weights of constructs and explants increased four- to fivefold (Figure 33.6a and b, respectively), due to the progressive accumulation of newly synthesized GAG and collagen. This finding is important as it documents that bioreactors can support the growth of both native and engineered tissues. In both cases, dynamic laminar flow in rotating vessels was superior to either static or turbulent flow. In constructs from rotating vessels, the fractions of GAG (Figure 33.6c) and total collagen (Figure 33.6e) were fivefold higher than in the initial constructs, and markedly higher than in either static or mixed flasks. In explants cultured in rotating vessels, the fractions of GAG (Figure 33.6d) and total collagen (Figure 33.6f) were maintained at their initial levels, in contrast to those cultured in static and mixed flasks where the total amounts of GAG and collagen increased but their fractions decreased to 50 to 70% of initial levels. Therefore, the dynamic laminar flow in rotating vessels supported the synthesis and assembly of cartilage matrix resulting in a balanced growth of engineered and native cartilage and the establishment or maintenance of normal tissue compositions. This was in sharp contrast with the cultivations in static and mixed flasks which resulted in inferior growth rates and tissue compositions.

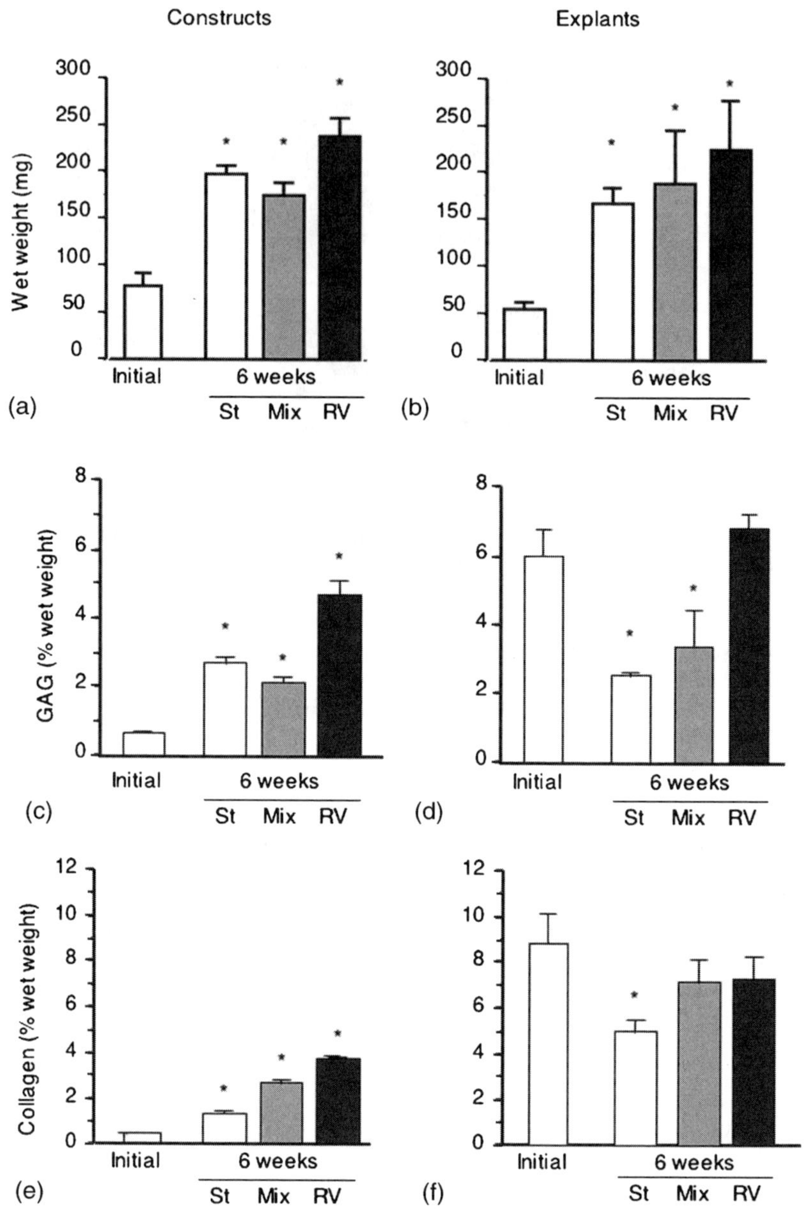

FIGURE 33.6 Biochemical compositions of (a, c, e) cultured cartilage constructs and (b, d, f) explants. (a, b) tissue wet weight (mg), (c, d) glycosaminoglycans (% wet weight), (e, f) total collagen (% wet weight). Data represents the ave ± *SD* for $N = 3–6$ samples per group at time zero (initial) and after 6 weeks of culture. ST, MIX, and RV, respectively, refer to static flasks, mixed flasks and rotating bioreactor vessels. (*) significantly different levels of a given variable among groups, as assessed by ANOVA in conjunction with Tukey's-HSD test ($p < .05$). (Data from Vunjak-Novakovic, G., Obradovic, B., Martin, I., and Freed, L. E., Bioreactor studies of native and tissue engineered cartilage, *Biorheology*, 39, 259–268, 2002. With permission.)

Biomechanical construct properties correlated with their biochemical compositions, both for constructs cultured in different hydrodynamic environments and constructs cultured for various periods of time (Figure 33.7). After 6 weeks of cultivation, the equilibrium modulus of constructs increased from static and mixed flasks and rotating vessels (53, 51, and 172 kPa, respectively,

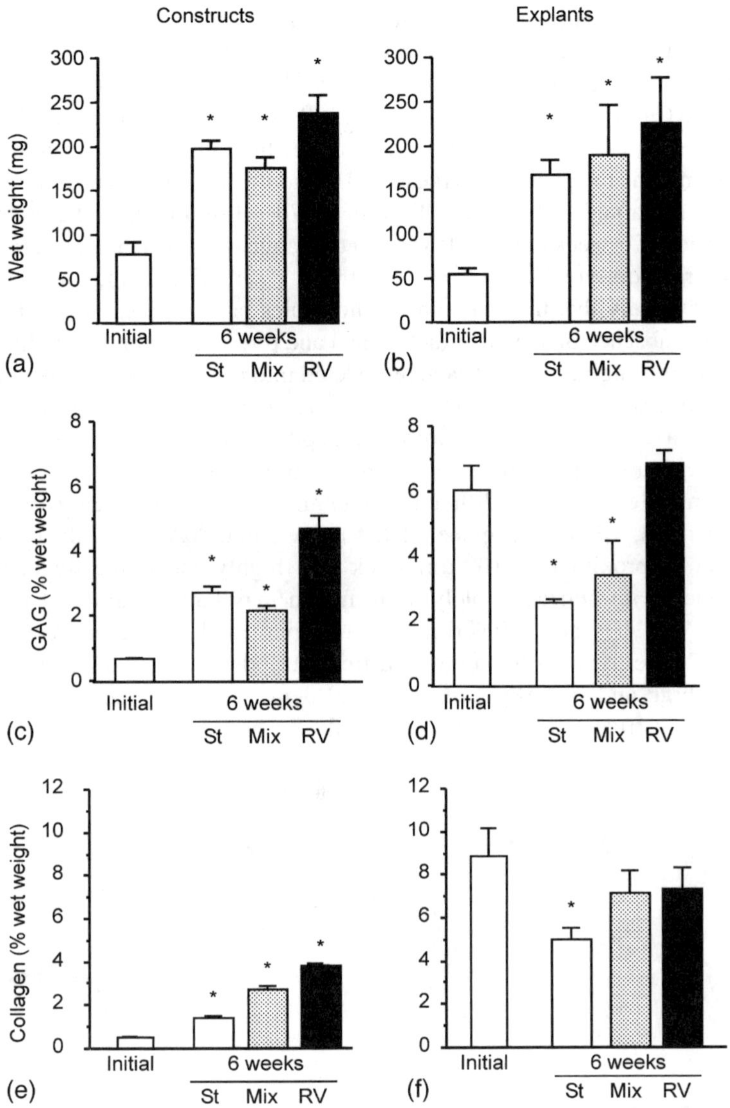

FIGURE 33.7 Effects of bioreactor hydrodynamics and culture duration on mechanical properties of engineered cartilage. (a, c, e) Constructs cultured in static flasks, mixed flasks, and rotating vessels. (b, d, f) Constructs cultured rotating vessels for 3 days, 6 weeks, and 7 months, respectively. Data for cartilage explants are shown for comparison. (a, b) Equilibrium modulus (kPa) measured in radially confined static compression. (c, d) Dynamic stiffness measured in radially confined dynamic compression. (e, f) Hydraulic permeability ($10^{15}m^4/Ns$) calculated from compression data. Data represent the ave $\pm$ *SD* for $N = 3$–6 samples per group. ST, MIX, and RV, respectively, refer to static flasks, mixed flasks and rotating bioreactor vessels. (Data from Vunjak-Novakovic, G., Obradovic, B., Martin, I., and Freed, L. E., Bioreactor studies of native and tissue engineered cartilage, *Biorheology*, 39, 259–268, 2002. With permission.)

Figure 33.7a),[27] in parallel with the increases in GAG and collagen concentrations (Figure 33.6c and e). The dynamic stiffness of constructs from rotating vessels was also approximately three times as high as in either static or mixed flasks (Figure 33.7c). Consistently, the hydraulic permeability decreased from static and mixed flasks to rotating vessels (Figure 33.7e). All biomechanical parameters were inferior to those measured for freshly explanted cartilage.

For engineered cartilage, the optimal duration of *in vitro* cultivation has not yet been determined. With time in culture, the constructs more closely approximated articular cartilage, both structurally and functionally.[28,70] Compared to adult native cartilage, 6-week constructs had subnormal equilibrium modulus (Figure 33.7b), dynamic stiffness (Figure 33.7d), and fourfold higher hydraulic permeability (Figure 33.7f). After 7 months of culture, the equilibrium modulus and hydraulic permeability came into the range of values measured for adult cartilage (Figure 33.7b and f), whereas dynamic stiffness remained subnormal (Figure 33.7d). Importantly, the compositions and mechanical properties of 6-week constructs were in the range of values measured for fetal cartilage, suggesting that bioreactors yield engineered constructs resembling immature rather than mature native cartilage. It is likely that the functional deficiencies present in engineered cartilage grown *in vitro* are due to the absence of specific biochemical and physical factors normally present *in vivo*. Following implantation, engineered cartilage remodeled into physiologically stiff cartilage and new subchondral bone, in response to local and systemic regulatory factors (e.g., [71–74]). Notably, columnar cells and a tidemark at an appropriate depth were observed in engineered cartilage following *in vivo* exposure to physiological loading,[72] but not *in vitro*.

(2) **Cardiac muscle**. A major difficulty in engineering most tissue types is that native tissues are vascularized, whereas engineered tissues are not.[75] As a result, engineered tissues that are more than approximately 100 μm thick and highly metabolically active (e.g., heart muscle) and supplied with nutrients solely by diffusion may have insufficient mass transport to and from the cells.[76–79] Tissue engineering of 1 to 5 mm thick, functional constructs based on cells that cannot tolerate hypoxia for prolonged time periods critically depends on our ability to seed the cells at a high and spatially uniform initial density, and to maintain their viability and function. Cardiac constructs cultured in mixed flasks had an approximately 100 μm thick

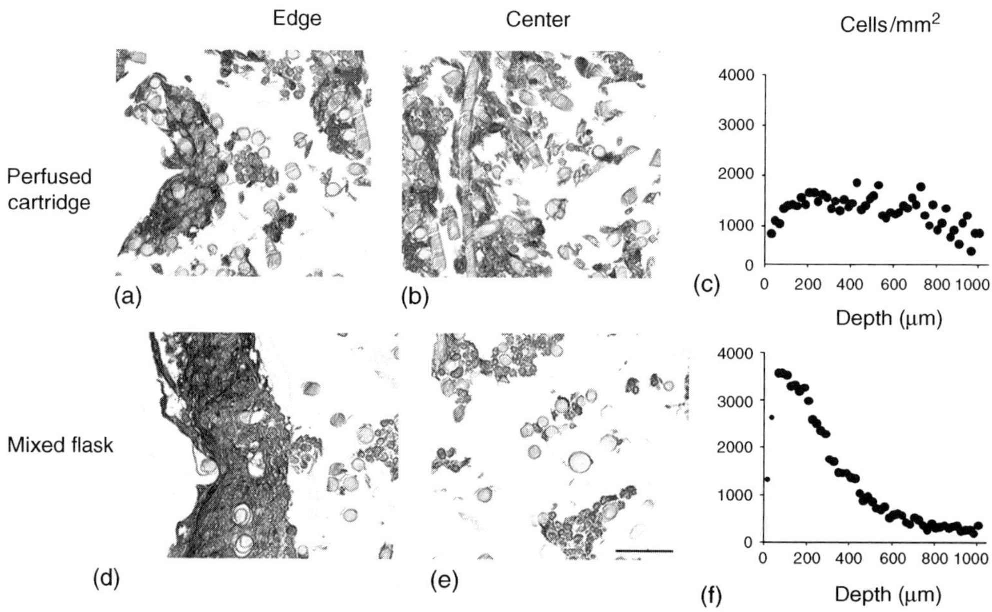

FIGURE 33.8 Effects of perfusion on cell distributions in cardiac constructs. Data are shown for PGA based constructs cultured for 10 days in (a, b, c) perfused cartridges at 0.6 ml/min or (d, e, f) mixed flasks at 50 rpm. Histomorphology (α-sarcomeric actin, scale bar 100 μm) for constructs cultivated in (a, b) perfused cartridges and (f) mixed flasks. Cross sectional cell distributions determined by image analysis for (c) constructs cultivated in perfused cartridges and (f) mixed flasks. (From Carrier, R. L., Rupnick, M., Langer, R., Schoen, F. J., Freed, L. E., and Vunjak-Novakovic, G., Perfusion improves tissue architecture of engineered cardiac muscle, *Tissue Eng.*, 8, 175–188, 2002. With permission.)

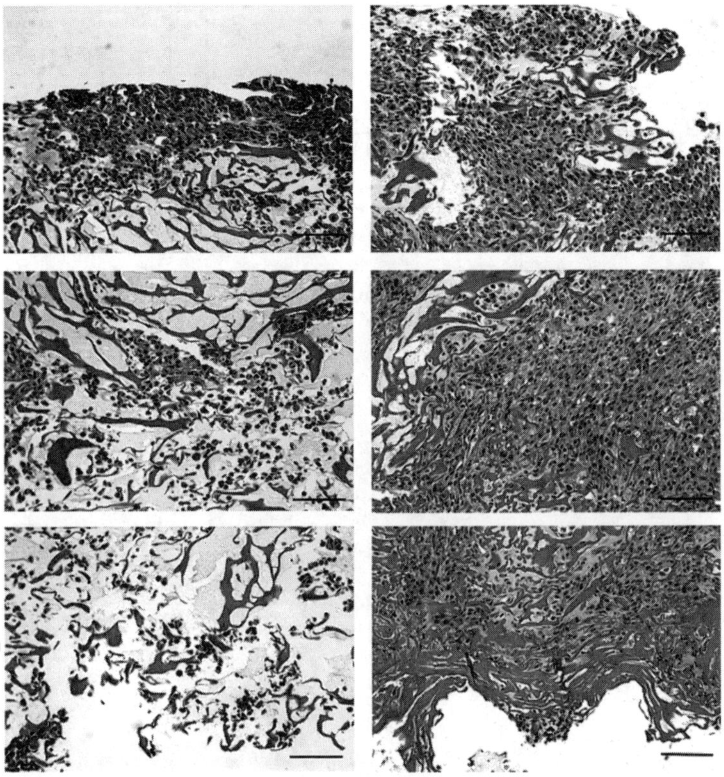

FIGURE 33.9 Effects of perfusion during seeding and cultivation on cardiac cell distribution. Cross-sections of constructs inoculated with 12 million cells and then transferred for a period of 4.5 h either into dishes (25 rpm, left) or into perfused cartridges (1.5 ml/min, right). The top, center and bottom areas of a 650-μm wide strip extending from one construct surface to the other are shown. Scale bar 100 μm. (From Radisic, M., Euloth, M., Yang, L., Langer, R., Freed, L. E., and Vunjak-Novakovic, G., High density seeding of myocyte cells for tissue engineering, *Biotechnol. Bioeng.*, 82, 403–414, 2003. With permission.)

peripheral tissue-like region around a relatively cell-free interior (Figure 33.8d and e). Cardiac myocytes are densely packed and highly metabolically active, therefore nutrients, particularly oxygen, are depleted within a relatively thin layer of viable tissue, and viable cells are concentrated mostly in the peripheral region of the construct (Figure 33.8f). Natural myocardium obviates this difficulty through a rich vasculature, with average capillary-to-capillary distances in rat heart of only 17 to 19 μm, approximately the width of an individual cardiac myocyte.[80–83]

One way to reduce diffusional distances for mass transport, improve control of oxygen, pH, nutrients, and metabolites in the cell microenvironment, and thereby increase the thickness and spatial uniformity of engineered cardiac tissue is to perfuse constructs with culture medium. Perfusion improved the spatial uniformity of cell distribution in 9.5 mm in diameter, 2 mm thick constructs based on neonatal rat cardiac myocytes and fibrous PGA scaffolds (Figure 33.8a, b, and c), However, the total cell numbers were comparable for constructs cultured in perfused cartridges and mixed flasks (Figure 33.8c and f). During cultivation, the flow of medium redistributed the cells across the entire volume of the construct, but the cell density remained low due to the limitations in oxygen transport during cell seeding.

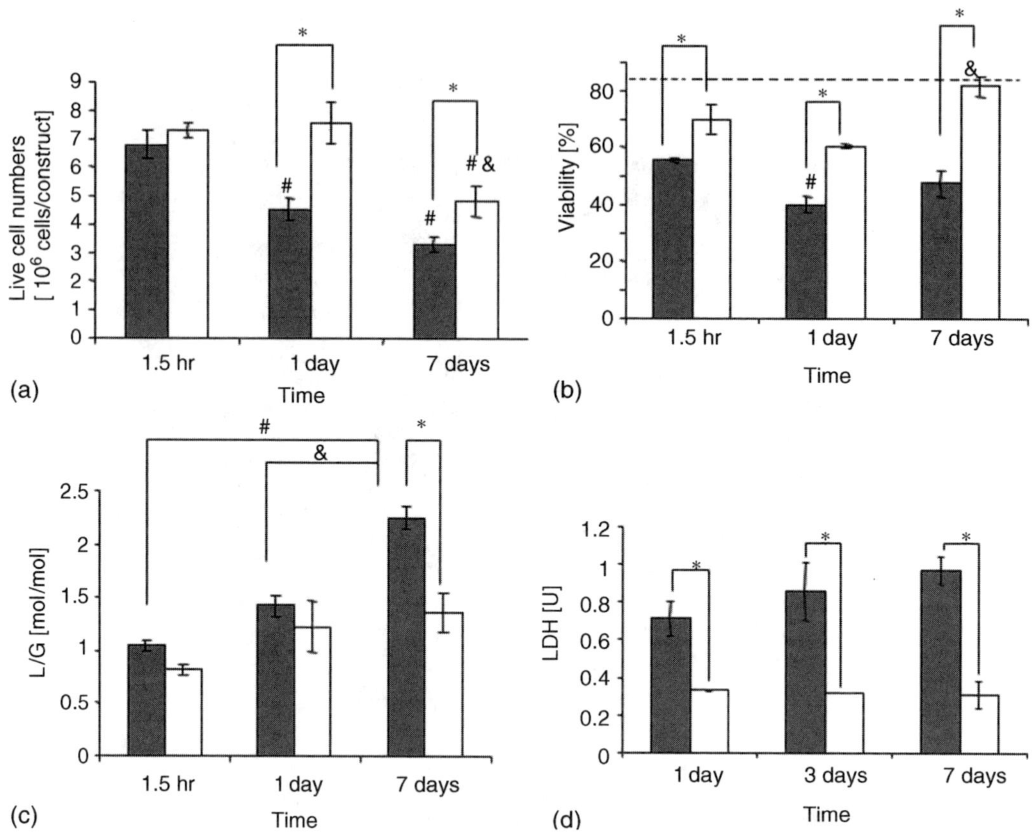

FIGURE 33.10 Effects of perfusion on cardiac cell viability and metabolism. (a) Live cell number per construct. (b) Viability of cells on the construct. Dashed line represents viability of freshly isolated cells: 83.8 ± 2.0 ($N = 6$). (c) Molar ratio of lactate produced to glucose consumed. (d) Lactate dehydrogenase (LDH) content (U) in culture medium. Open bars represent perfused constructs; filled bars represent dish-grown controls. Data are expressed as average ± standard error; p values were calculated by one-way ANOVA in conjunction with Tukey's test for pair-wise multiple comparisons ($N = 3$–9 samples per data point). Differences were considered significant if $p < .05$. * Significant differences between perfused and dish-grown constructs; # significant difference between 1 and 7 day, and 1.5 h, constructs; & significant difference between 1 and 7 day constructs. (From Radisic, M., Yang, L., Boublik, J., Cohen, R. J., Langer, R., Freed, L. E., and Vunjak-Novakovic, G., Medium perfusion enables engineering of compact and contractile cardiac tissue, *Am. J. Physiol. Heart Circ. Physiol.*, 286, H507–H516, 2004, with modification. With permission.)

To overcome this limitation, and establish physiologic densities of viable cardiac myocytes, a new method for cell seeding was developed that combined rapid cell inoculation into a porous scaffold using Matrigel with the establishment of the perfusion of culture medium through the construct immediately after gel hardening (15 min).[14] Cell distributions in the top, center, and bottom areas of a 0.65 μm wide strip extending from one construct surface to the other are shown in Figure 33.9. Constructs seeded in dishes had most cells located in the 100 to 200 μm thick layer at the top surface, and only a small number of cells penetrated the entire construct depth (Figure 33.9). Constructs seeded in perfusion exhibited high and spatially uniform cell density throughout the perfused volume of the construct (Figure 33.9). Clearly, medium perfusion during seeding was important for engineering thick constructs with high densities of viable cells.

To engineer thick, compact, functional cardiac constructs, we developed a biomimetic culture system that can mimic the convective-diffusive oxygen transport present *in vivo* and maintain

TABLE 33.4
Contractile Properties of Neonatal Ventricles, 7-Day Perfused Constructs and 7-Day Dish-Grown Constructs

	Neonatal Rat Ventricle	Perfused Construct (7 days @ 0.5 ml/min)	Dish-Grown Construct (7 days @ 25 rpm)
		ET (V)	
Before PA	1.6 ± 0.1	3.3 ± 0.2*	4.5 ± 0.4*&
After PA	1.5 ± 0.1	3.5 ± 0.1*	4.4 ± 0.1*&
		MCR at 150% ET (bpm)	
Before PA	413 ± 7	420 ± 30	502 ± 32
After PA	465 ± 15	415 ± 35	378 ± 31§
		MCR at 200% ET (bpm)	
Before PA	427 ± 40	415 ± 45	523 ± 14
After PA	427 ± 58	435 ± 45	380 ± 31§

Excitation threshold (ET) was measured at a stimulation frequency of 60 bpm. Maximum capture rate (MCR) was measured at voltages equivalent to 150 and 200% ET. Data were collected before the treatment with palmitoleic acid (before PA) and after the washout of palmitoleic acid (after PA). Constructs could not be induced to contract in the solution containing PA. Data are expressed as averages ± standard error. p values were calculated by one-way ANOVA in conjunction with Tukey's test for pair-wise multiple comparisons. ($N = 2\text{–}6$). Differences were considered statistically significant if $p < .05$. (From Radisic, M., Yang, L., Boublik, J., Cohen, R. J., Langer, R., Freed, L. E., and Vunjak-Novakovic, G., Medium perfusion enables engineering of compact and contractile cardiac tissue, *Am. J. Physiol. Heart Circ. Physiol.*, 286, H507–H516, 2004. With permission.) *Significant difference between constructs and neonatal rat ventricles, &significant difference between perfused and dish-grown constructs, §significant difference before and after PA treatment.

oxygen supply to the cells at all times during the *in vitro* cultivation. Neonatal rat cardiomyocytes suspended in Matrigel were cultured on collagen sponges at physiologically high initial density (1.35×10^8 cells/cm^3) for 7 days with interstitial flow of medium; constructs cultured in orbitally mixed dishes and neonatal rat ventricles served as controls.

To provide convective-diffusive oxygen supply to the cells, interstitial medium flow was maintained for 7 days during cultivation; constructs seeded and cultivated in dishes served as a control. Throughout the cultivation, the number of live cells in perfused constructs was significantly higher than in dish-grown constructs (Figure 33.10a). Notably, the number of live cells in dish-grown constructs decreased rapidly during the first day of culture and continued to decrease between Days 1 and 7. In contrast, live cell numbers in perfused constructs were constant during Day 1, and decreased only slowly with time in culture (Figure 33.10a). Cell viability was significantly higher in perfused than in dish-grown constructs at all time points (Figure 33.10b). Importantly, the final cell viability in perfused constructs (81.6 ± 3.7%, Figure 33.10b) was not significantly different than the viability of the freshly isolated cells (83.8 ± 2.0) and it was markedly higher than the cell viability in dish-grown constructs (47.4 ± 7.8%, Figure 33.10b). The molar ratio of lactate produced to glucose consumed (L/G) was approximately 1 for perfused constructs throughout the duration of culture, indicating aerobic cell metabolism (Figure 33.10c). In dishes, L/G increased progressively from 1 to approximately 2, indicating a transient to anaerobic cell metabolism. (Figure 33.10c). Cell damage was assessed by monitoring the activity of LDH in culture medium. At all time points, the levels of LDH were significantly lower in perfusion than in dish cultures (Figure 33.10d), indicating that medium perfusion reduced cell damage.

Spontaneous contractions were observed in some constructs early in culture (dish-grown constructs 2 to 3 days after seeding), and ceased after approximately 5 days of cultivation, indicating the maturation of engineered tissue. In response to electrical stimulation (e.g., at 5 V and 60 bpm), all constructs were reproducibly induced to contract synchronously. However, in perfused constructs the contraction frequency was constant, whereas in dish-grown constructs the contraction frequency spontaneously increased every 1 to 2 min and the contraction pattern appeared arrhythmic. Overall, medium perfusion during cell seeding and construct cultivation markedly improved the contractile behavior of engineered cardiac constructs (Table 33.4).

VI. SUMMARY

The paradigm discussed in this chapter is that the restoration of normal tissue function can be achieved by using *in vitro* constructs that are immature but functional and thus have the ability to regenerate the molecular, structural, and functional properties of a compromised native tissue. Engineered tissues can also serve as high-fidelity models for basic studies of cells and tissues. Functional tissue constructs for scientific studies and tissue repair can be engineered *in vitro* by using differentiated or progenitor cells, biomaterial scaffolds, and bioreactors. We have discussed the design and operation of representative bioreactors that have been used to engineer articular cartilage and cardiac muscle, two distinctly different tissues of high clinical interest.

ACKNOWLEDGMENTS

The work presented in this chapter has been supported by the National Aeronautics and Space Administration (GV-N, MR), the National Institutes of Health (GV-N), Poitras Fellowship (MR), and Ministry of Science, Technology, and Development of the Republic of Serbia (Grant, 1776) (BO). The authors would like to thank Sue Kangiser for her help with the manuscript preparation.

REFERENCES

1. Vunjak-Novakovic, G., and Goldstein, S. A., Biomechanical principles of cartilage and bone tissue engineering, In *Basic Orthopaedic Biomechanics and Mechanobiology*, Mow, V. C. and Huiskes, R., eds., Lippincott-Williams and Wilkens, Baltimore, MD, pp. 343–408, chap. 8, 2004.
2. Obradovic, B., Carrier, R. L., Vunjak-Novakovic, G., and Freed, L. E., Gas exchange is essential for bioreactor cultivation of tissue engineered cartilage, *Biotechnol. Bioeng.*, 63, 197–205, 1999.
3. Buckwalter, J. A., and Mankin, H. J., Articular cartilage, part I: tissue design and chondrocyte–matrix interactions, *J. Bone Joint Surg. Am.*, 79A, 600–611, 1997.
4. MacKenna, D. A., Omens, J. H., McCulloch, A. D., and Covell, J. W., Contribution of collagen matrix to passive left ventricular mechanics in isolated rat heart, *Am. J. Physiol.*, 266, H1007–H1018, 1994.
5. Brilla, C. G., Maisch, B., Rupp, H., Sunck, R., Zhou, G., and Weber, K. T., Pharmacological modulation of cardiac fibroblast function, *Herz*, 20, 127–135, 1995.
6. Vander, A. J., Sherman, J. H., and Luciano, D. S., *Human Physiology*, McGraw-Hill, New York, 1985.
7. Schoen, F. J., The heart, In *Robbins Pathologic Basis of Disease*, Cotran, R. S., Kumar, V., Collins, T., and Robbins, S. L., eds., W.B. Saunders, Philadelphia, PA, pp. 543–599, 1999.
8. Freed, L. E., and Vunjak-Novakovic, G., Tissue engineering bioreactors, In *Principles of Tissue Engineering*, Lanza, R. P., Langer, R. and Vacanti, J., eds., Academic Press, San Diego, CA, pp. 143–156, 2000.
9. Pei, M., Solchaga, L. A., Seidel, J., Zeng, L., Vunjak-Novakovic, G., Caplan, A. I., and Freed, L. E., Bioreactors mediate the effectiveness of tissue engineering scaffolds, *FASEB J.*, 16, 1691–1694, 2002.
10. Vunjak-Novakovic, G., Fundamentals of tissue engineering: scaffolds and bioreactors, In *Tissue Engineering of Cartilage and Bone*, Caplan, A. I., ed., Wiley, London, pp. 34–51, 2003.

11. Vunjak-Novakovic, G., Obradovic, B., Madry, H., Altman, G., and Kaplan, D., Bioreactors for orthopaedic tissue engineering, In *Orthopaedic Tissue Engineering: Basic Science and Practice*, Caplan, A. I. and Goldberg, V., eds., Marcel Dekker, New York, pp. 123–147, 2004.
12. Freed, L. E., and Vunjak-Novakovic, G., Cultivation of cell–polymer constructs in simulated microgravity, *Biotechnol. Bioeng.*, 46, 306–313, 1995.
13. Vunjak-Novakovic, G., Freed, L. E., Biron, R. J., and Langer, R., Effects of mixing on the composition and morphology of tissue-engineered cartilage, *AIChE J.*, 42, 850–860, 1996.
14. Radisic, M., Euloth, M., Yang, L., Langer, R., Freed, L. E., and Vunjak-Novakovic, G., High density seeding of myocyte cells for tissue engineering, *Biotechnol. Bioeng.*, 82, 403–414, 2003.
15. Vunjak-Novakovic, G., Searby, N. D., de Luis, J., and Freed, L. E., Microgravity studies of cells and tissues, *Ann. NY Acad. Sci.*, 974, 504–517, 2002.
16. Seidel, J. O., Pei, M., Gray, M. L., Langer, R., Freed, L. E., and Vunjak-Novakovic, G., Long-term culture of tissue engineered cartilage in a perfused chamber with mechanical stimulation, *Biorheology*, 41(3–4), 445–458, 2004.
17. Fink, C., Ergun, S., Kralisch, D., Remmers, U., Weil, J., and Eschenhagen, T., Chronic stretch of engineered heart tissue induces hypertrophy and functional improvement, *FASEB J.*, 14, 669–679, 2000.
18. Zimmermann, W. H., Schneiderbanger, K., Schubert, P., Didie, M., Munzel, F., Heubach, J. F., Kostin, S., Nehuber, W. L., and Eschenhagen, T., Tissue engineering of a differentiated cardiac muscle construct, *Circ. Res.*, 90, 223–230, 2002.
19. Functional assembly of engineered myocardium by electrical stimulation of cardiac myocytes cultured on scaffolds, *Proc. Natl Acad. Sci.*, 101(52), 18129–18134, 2004.
20. Vunjak-Novakovic, G., Obradovic, B., Bursac, P., Martin, I., Langer, R., and Freed, L. E., Dynamic cell seeding of polymer scaffolds for cartilage tissue engineering, *Biotechnol. Prog.*, 14, 193–202, 1998.
21. Vunjak-Novakovic, G., and Radisic, M., Cell seeding of polymer scaffolds, *Methods Mol. Biol.*, 238, 131–146, 2004.
22. Ma, P. X., Schloo, B., Mooney, D., and Langer, R., Development of biomechanical properties and morphogenesis of in vitro tissue engineered cartilage, *J. Biomed. Mater. Res.*, 29, 1587–1595, 1995.
23. Stading, M., and Langer, R., Mechanical shear properties of cell–polymer cartilage constructs, *Tissue Eng.*, 5, 241–250, 1999.
24. Bursac, P. M., *Collagen Network Contributions to Structure–Function Relationships in Cartilaginous Tissues in Compression*, Boston, MA. 2001.
25. Freed, L. E., Hollander, A. P., Martin, I., Barry, J. R., Langer, R., and Vunjak-Novakovic, G., Chondrogenesis in a cell–polymer–bioreactor system, *Exp. Cell Res.*, 240, 58–65, 1998.
26. Riesle, J., Hollander, A. P., Langer, R., Freed, L. E., and Vunjak-Novakovic, G., Collagen in tissue-engineered cartilage: types, structure and crosslinks, *J. Cell. Biochem.*, 71, 313–327, 1998.
27. Chen, A. C., Bae, W. C., Schinagl, R. M., and Sah, R. L., Depth- and strain-dependent mechanical and electromechanical properties of full-thickness bovine articular cartilage in confined compression, *J. Biomech.*, 34, 1–12, 2001.
28. Freed, L. E., Langer, R., Martin, I., Pellis, N., and Vunjak-Novakovic, G., Tissue engineering of cartilage in space, *Proc. Natl Acad. Sci. USA*, 94, 13885–13890, 1997.
29. Pazzano, D., Mercier, K. A., Moran, J. M., Fong, S. S., DiBiasio, D. D., Rulfs, J. X., Kohles, S. S., and Bonassar, L. J., Comparison of chondrogensis in static and perfused bioreactor culture, *Biotechnol. Prog.*, 16, 893–896, 2000.
30. Dunkelman, N. S., Zimber, M. P., Lebaron, R. G., Pavelec, R., Kwan, M., and Purchio, A. F., Cartilage production by rabbit articular chondrocytes on polyglycolic acid scaffolds in a closed bioreactor system, *Biotechnol. Bioeng.*, 46, 299–305, 1995.
31. Sittinger, M., Bujia, J., Minuth, W. W., Hammer, C., and Burmester, G. R., Engineering of cartilage tissue using bioresorbable polymer carriers in perfusion culture, *Biomaterials*, 15, 451–456, 1994.
32. Maroudas, A., Physiochemical properties of articular cartilage, In *Adult Articular Cartilage*, Freeman, M. A. R., ed., Pitman Medical, London, pp. 215–290, 1979.
33. Mow, V. C., Kuei, S. C., Lai, W. M., and Armstrong, C. G., Biphasic creep and stress relaxation of articular cartilage in compression: theory and experiments, *J. Biomech. Eng.*, 102, 73–84, 1980.

34. Mow, V. C., Ratcliffe, A., Rosenwasser, M. P., and Buckwalter, J. A., Experimental studies on the repair of large osteochondral defects at a high weight bearing area of the knee joint: a tissue engineering study, *Trans. Am. Soc. Mech. Eng.*, 113, 198–207, 1991.
35. Buschmann, M. D., Gluzband, Y. A., Grodzinsky, A. J., and Hunziker, E. B., Mechanical compression modulates matrix biosynthesis in chondrocyte/agarose culture, *J. Cell Sci.*, 108, 1497–1508, 1995.
36. Mauck, R. L., Soltz, M. A., Wang, C. C. B., Wong, D. D., Chao, P. G., Valhmu, W. B., Hung, C. T., and Ateshian, G. A., Functional tissue engineering of articular cartilage through dynamic loading of chondrocyte-seeded agarose gels, *J. Biomech. Eng.*, 122, 252–260, 2000.
37. Takahashi, I., Nuckolis, G. H., Takahashi, K., Tanaka, O., Semba, I., Dashner, R., Shum, L., and Slavkin, H. C., Compressive force promotes Sox9, type II collagen and aggrecan and inhibits IL-1b expression resulting in chondrogenesis in mouse embryonic limb bud mesenchymal cells, *J. Cell Sci.*, 111, 2067–2076, 1998.
38. Buschmann, M. D., Gluzband, Y. A., Grodzinsky, A. J., Kimura, J. H., and Hunziker, E. B., Chondrocytes in agarose culture synthesize a mechanically functional extracellular matrix, *J. Orthop. Res.*, 10, 745–758, 1992.
39. Mauck, R. L., Seyhan, S. L., Ateshian, G. A., and Hung, C. T., Influence of seeding density and dynamic deformational loading on the developing structure/function relationships of chondrocyte-seeded agarose hydrogels, *Ann. Biomed. Eng.*, 30, 1046–1056, 2002.
40. Elder, S. H., Kimura, J. H., Soslowsky, L. J., Lavagnino, M., and Goldstein, S. A., Effect of compressive loading on chondrocyte differentiation in agarose cultures of chick limb-bud cells, *J. Orthop. Res.*, 18, 78–86, 2000.
41. Elder, S. H., Goldstein, S. A., Kimura, J. H., Soslowsky, L. J., and Spengler, D. M., Chondrocyte differentiation is modulated by frequency and duration of cyclic compressive loading, *Ann. Biomed. Eng.*, 29, 476–482, 2001.
42. Heath, C. A., and Magari, R., Mini-review: mechanical factors affecting cartilage regeneration in vitro, *Biotechnol. Bioeng.*, 50, 430–437, 1996.
43. Carver, S. E., and Heath, C. A., Influence of intermittent pressure, fluid flow, and mixing on the regenerative properties of articular chondrocytes, *Biotechnol. Bioeng.*, 65, 274–281, 1999.
44. Carver, S. E., and Heath, C. A., Increasing extracellular matrix production in regenerating cartilage with intermittent physiological pressure, *Biotechnol. Bioeng.*, 62, 166–174, 1999.
45. Carver, S. E., and Heath, C. A., Semi-continuous perfusion system for delivering intermittent physiological pressure to regenerating cartilage, *Tissue Eng.*, 5, 1–11, 1999.
46. Heath, C. A., The effects of physical forces on cartilage tissue engineering, *Biotechnol. Genet. Eng. Rev.*, 17, 533–551, 2000.
47. Davisson, T., Kunig, S., Chen, A., Sah, R., and Ratcliffe, A., Static and dynamic compression modulate matrix metabolism in tissue engineered cartilage, *J. Orthop. Res.*, 20, 842–848, 2002.
48. Gooch, K. J., Blunk, T., Courter, D. L., Sieminski, A. L., Bursac, P. M., Vunjak-Novakovic, G., and Freed, L. E., IGF-I and mechanical environment interact to modulate engineered cartilage development, *Biochem. Biophys. Res. Commun.*, 286, 909–915, 2001.
49. Mauck, R. L., Nicoll, S. B., Seyhan, S. L., Ateshian, G. A., and Hung, C. T., Synergistic action of growth factors and dynamic loading for articular cartilage tissue engineering, *Tissue Eng.*, 9, 597–611, 2003.
50. Pei, M., Seidel, J., Vunjak-Novakovic, G., and Freed, L. E., Growth factors for sequential cellular de- and re-differentiation in tissue engineering, *Biochem. Biophys. Res. Commun.*, 294, 149–154, 2002.
51. Madry, H., Padera, R., Seidel, J., Freed, L. E., Langer, R., Trippel, S. B., and Vunjak-Novakovic, G., Tissue engineering of cartilage enhanced by the transfer of a human insulin-like growth factor-I gene, *Trans. Orthop. Res. Soc.*, 26, 289 2001.
52. Obradovic, B., Meldon, J. H., Freed, L. E., and Vunjak-Novakovic, G., Glycosaminoglycan deposition in engineered cartilage: experiments and mathematical model, *AIChE J.*, 46, 1860–1871, 2000.
53. Martin, I., Obradovic, B., Freed, L. E., and Vunjak-Novakovic, G., A method for quantitative analysis of glycosaminoglycan distribution in cultured natural and engineered cartilage, *Ann. Biomed. Eng.*, 27, 656–662, 1999.
54. Li, R.-K., Jia, Z. Q., Weisel, R. D., Mickle, D. A. G., Choi, A., and Yau, T. M., Survival and function of bioengineered cardiac grafts, *Circulation*, 100, II63–II69, 1999.

55. Li, R.-K., Yau, T. M., Weisel, R. D., Mickle, D. A. G., Sakai, T., Choi, A., and Jia, Z.-Q., Construction of a bioengineered cardiac graft, *J. Thorac. Cardiovasc. Surg.*, 119, 368–375, 2000.
56. Kofidis, T., Akhyari, P., Boublik, J., Theodorou, P., Martin, U., Ruhparwar, A., Fischer, S., Eschenhagen, T., Kubis, H. P., Kraft, T., Leyh, R., and Haverich, A., In vitro engineering of heart muscle: artificial myocardial tissue, *J. Thorac. Cardiovasc. Surg.*, 124, 63–69, 2002.
57. Kofidis, T., Akhyari, P., Wachsmann, B., Mueller-Stahl, K., Boublik, J., Ruhparwar, A., Mertsching, H., Balsam, L., Robbins, R., and Haverich, A., Clinically established hemostatic scaffold (tissue fleece) as biomatrix in tissue- and organ-engineering research, *Tissue Eng.*, 9, 517–523, 2003.
58. Leor, J., Aboulafia-Etzion, S., Dar, A., Shapiro, L., Barbash, I. M., Battler, A., Granot, Y., and Cohen, S., Bioengineerred cardiac grafts: a new approach to repair the infarcted myocardium?, *Circulation*, 102, III56–III61, 2000.
59. Dar, A., Shachar, M., Leor, J., and Cohen, S., Cardiac tissue engineering optimization of cardiac cell seeding and distribution in 3D porous alginate scaffolds, *Biotechnol. Bioeng.*, 80, 305–312, 2002.
60. Radisic, M., Yang, L., Boublik, J., Cohen, R. J., Langer, R., Freed, L. E., and Vunjak-Novakovic, G., Medium perfusion enables engineering of compact and contractile cardiac tissue, *Am. J. Physiol. Heart Circ. Physiol.*, 286, H507–H516, 2004.
61. Carrier, R. L., Rupnick, M., Langer, R., Schoen, F. J., Freed, L. E., and Vunjak-Novakovic, G., Perfusion improves tissue architecture of engineered cardiac muscle, *Tissue Eng.*, 8, 175–188, 2002.
62. Carrier, R. L., Papadaki, M., Rupnick, M., Schoen, F. J., Bursac, N., Langer, R., Freed, L. E., and Vunjak-Novakovic, G., Cardiac tissue engineering: cell seeding, cultivation parameters and tissue construct characterization, *Biotechnol. Bioeng.*, 64, 580–589, 1999.
63. Bursac, N., Papadaki, M., Cohen, R. J., Schoen, F. J., Eisenberg, S. R., Carrier, R., Vunjak-Novakovic, G., and Freed, L. E., Cardiac muscle tissue engineering: toward an in vitro model for electrophysiological studies, *Am. J. Physiol. Heart Circ. Physiol.*, 277, H433–H444, 1999.
64. Papadaki, M., Bursac, N., Langer, R., Merok, J., Vunjak-Novakovic, G., and Freed, L. E., Tissue engineering of functional cardiac muscle: molecular, structural and electrophysiological studies, *Am. J. Physiol. Heart Circ. Physiol.*, 280, H168–H178, 2001.
65. Bursac, N., Papadaki, M., White, J. A., Eisenberg, S. R., Vunjak-Novakovic, G., and Freed, L. E., Cultivation in rotating bioreactors promotes maintenance of cardiac myocyte electrophysiology and molecular properties, *Tissue Eng.*, 9, 1243–1253, 2003.
66. Carrier, R. L., Rupnick, M., Langer, R., Schoen, F. J., Freed, L. E., and Vunjak-Novakovic, G., Effects of oxygen on engineered cardiac muscle, *Biotechnol. Bioeng.*, 78, 617–625, 2002.
67. Akhyari, P., Fedak, P. W. M., Weisel, R. D., Lee, T. Y. J., Verma, S., Mickle, D. A. G., and Li, R. K., Mechanical stretch regimen enhances the formation of bioengineered autologous cardiac muscle grafts, *Circulation*, 106, I137–I142, 2002.
68. Obradovic, B., Martin, I., Freed, L. E., and Vunjak-Novakovic, G., Bioreactor studies of natural and tissue engineered cartilage, *Ortop. Traumatol. Rehabilitacja*, 3, 181–189, 2001.
69. Obradovic, B., Bugarski, D., Petakov, M., Bugarski, B., Meinel, L., and Vunjak-Novakovic, G., Effects of bioreactor hydrodynamics on native and tissue engineered cartilage. In *First International Congress on Bioreactor Technology in Cell, Tissue Culture and Biomedical Applications*, Sorvari, S., ed., Tampere, Finland, pp. 61–70, 2003.
70. Vunjak-Novakovic, G., Martin, I., Obradovic, B., Treppo, S., Grodzinsky, A. J., Langer, R., and Freed, L. E., Bioreactor cultivation conditions modulate the composition and mechanical properties of tissue engineered cartilage, *J. Orthop. Res.*, 17, 130–138, 1999.
71. Wakitani, S., Goto, T., Pineda, S. J., Young, R. G., Mansour, J. M., Caplan, A. I., and Goldberg, V. M., Mesenchymal cell-based repair of large, full-thickness defects of articular cartilage, *J. Bone Joint Surg. Am.*, 76A, 579–592, 1994.
72. Schaefer, D., Martin, I., Jundt, G., Seidel, J., Heberer, M., Grodzinsky, A. J., Bergin, I., Vunjak-Novakovic, G., and Freed, L. E., Tissue engineered composites for the repair of large osteochondral defects, *Arthritis Rheum.*, 46, 2524–2534, 2002.
73. Sellers, R. S., Zhang, R., Glasson, S. S., Kim, H. D., Peluso, D., D'Augusta, D. A., Beckwith, K., and Morris, E. A., Repair of articular cartilage defects one year after treatment with recombinant human bone morphogenetic protein-2 (rhBMP-2), *J. Bone Joint Surg. Am.*, 82, 151–160, 2000.

74. Caplan, A. I., Elyaderani, M., Mochizuki, Y., Wakitani, S., and Goldberg, V. M., Principles of cartilage repair and regeneration, *Clin. Orthop. Relat. Res.*, 342, 254–269, 1997.
75. Putnam, A. J., and Mooney, D. J., Tissue engineering using synthetic extracellular matrices, *Nat. Med.*, 2, 824–826, 1996.
76. Cima, L. G., and Langer, R., Engineering human tissue, *Chem. Eng. Prog.*, June, 46–54, 1993.
77. Colton, C. K., Implantable biohybrid artificial organs, *Cell Transplant.*, 4, 415–436, 1995.
78. Mooney, D. J., Mazzoni, C. L., Breuer, C., McNamara, K., Hern, D., Vacanti, J. P., and Langer, R., Stabilized polyglycolic acid fiber-based tubes for tissue engineering, *Biomaterials*, 17, 115–124, 1996.
79. Peters, M. C., Isenberg, B. C., Rowley, J. A., and Mooney, D. J., Release from alginate enhances the biological activity of vascular endothelial growth factor, *J. Biomater. Sci. Polym. Ed.*, 9, 1267–1278, 1998.
80. Korecky, B., Hai, C. M., and Rakusan, K., Functional capillary density in normal and transplanted rat hearts, *Can. J. Physiol. Pharmacol.*, 60, 23–32, 1982.
81. Rakusan, K., and Korecky, B., The effect of growth and aging on functional capillary supply of the rat heart, *Growth*, 46, 275–281, 1982.
82. Steinhausen, M., Tillmanns, H., and Thederan, H., Microcirculation of the epimyocardial layer of the heart. I. A method for in vivo observation of the microcirculation of superficial ventricular myocardium of the heart and capillary flow pattern under normal and hypoxic conditions, *Pflugers Arch. Eur. J. Phys.*, 374, 57–66, 1978.
83. Zadeh, B. J., Gonzalez-Sanchez, A., Fischman, D. A., and Bader, D. M., Myosin heavy chain expression in embryonic cardiac cell culture, *Dev. Biol.*, 115, 204–214, 1986.
84. van Luyn, M. J. A., Tio, R. A., van Seijen, X. J. G. Y., Plantinga, J. A., de Leij, L. F. M. H., DeJongste, M. J. L., and van Wachem, P. B., Cardiac tissue engineering: characteristics of in unison contracting two- and three-dimensional neonatal rat ventricle cell (co)-cultures, *Biomaterials*, 23, 4793–4801, 2002.
85. Altman, G. H., Lu, H. H., Horan, R. L., Calabro, T., Ryder, D., Kaplan, D. L., Stark, P., Martin, I., Richmond, J. C., and Vunjak-Novakovic, G., Advanced bioreactor with controlled application of multi-dimensional strain for tissue engineering, *J. Biomech. Eng.*, 124, 742–749, 2002.
86. Cherry, R. S., and Papoutsakis, T., Physical mechanisms of cell damage in microcarrier cell culture bioreactors, *Biotechnol. Bioeng.*, 32, 1001–1014, 1988.
87. Neitzel, G. P., Nerem, R. M., Sambanis, A., Smith, M. K., Wick, T. M., Brown, J. B., Hunter, C., Jovanovic, I. P., Malaviya, P., Saini, S., and Tan, S., Cell function and tissue growth in bioreactors: fluid mechanical and chemical environments, *J. Jpn Soc. Microgr. Appl.*, 15, 602–607, 1998.
88. Clift, R., Grace, J. R., and Weber, M. F., *Bubbles, Drops, and Particles*, Academic Press, New York, pp. 16–29, 97–105, 142–148, 1978.
89. Vunjak-Novakovic, G., Obradovic, B., Martin, I., and Freed, L. E., Bioreactor studies of native and tissue engineered cartilage, *Biorheology*, 39, 259–268, 2002.

34 Stem Cells in Tissue Engineering

Victor Prisk and Johnny Huard

CONTENTS

I. INTRODUCTION

Tissue engineering is a rapidly evolving field involved in creating living tissue to repair, replace, or augment diseased tissue. The main goal of tissue engineering is to construct biomaterials that are capable of integrating bioactive molecules (e.g., growth factors) or cells.[1–4] Tissues can be synthesized via both *in vitro* and *in vivo* techniques and may be made to resemble components of virtually every mammalian organ system. Toward this end, different polymers, ceramics, proteins, and cells have been tested.[5] Additionally, gene therapy techniques can further enhance this developmental process, making more therapeutic tissues. No matter which technique is used or which organ system is mimicked, tissue engineering requires four critical components (Figure 34.1):

(1) *Production*: stem cells, progenitor/precursor cells;
(2) *Conduction*: matrices or scaffolds;
(3) *Induction*: signaling proteins or growth factors;
(4) *Mechanical Stimulation*: biomechanical forces such as shear or strain.

The "production" component includes the cellular precursors which may include single or multiple cell types and cells at various levels of maturity, from stem cells to fully differentiated cells. These cells include stem cells and the progenitors that they become. This chapter will focus on stem cells and their use in tissue engineering. To illustrate, we will focus on the muscle-derived stem cell and its applications to various organ systems and gene therapy techniques.

II. STEM CELLS

Stem cells are responsible for the development and regeneration of tissues and organs. Biochemical and biomechanical signals trigger the proliferation and differentiation of stem cells in early

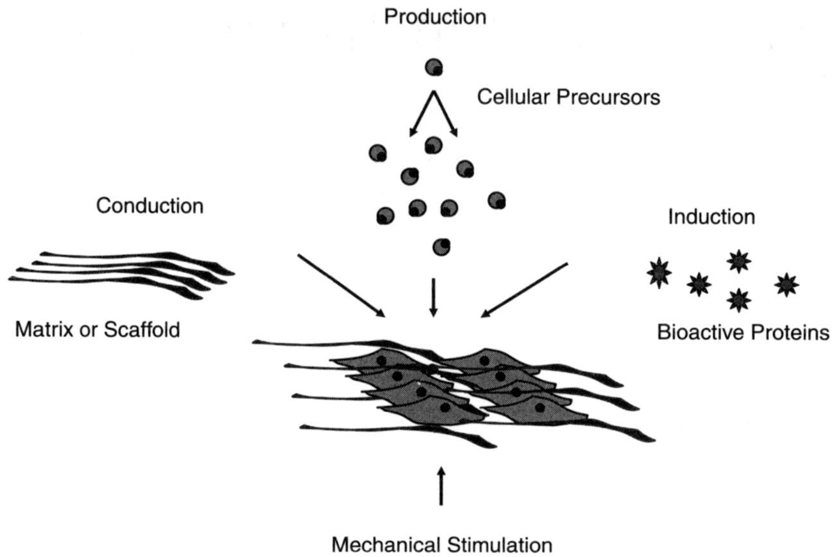

FIGURE 34.1 Components critical to tissue engineering. Production includes cellular precursors, like stem cells. Conduction refers to the matrices and scaffolds on which cells grow. Induction refers to bioactive proteins, such as growth factors. Mechanical stimulation refers to forces which aid in the induction and alignment of the conducting scaffold.

development and in regeneration after injury or disease. Stem cells may be of postnatal or embryonic origin.[6] Postnatal stem cells have been found in many tissues throughout the body, including skin, muscle, bone marrow, brain, and liver. Pluripotent and self-renewing embryonic stem cells are found in the inner cell mass of the early blastocyst. As research pertaining to embryonic stem cells is the center of much debate, this chapter will primarily focus on postnatal stem cells.

Postnatal stem cells are defined by their two major characteristics: multilineage differentiation and selfrenewal ability. The hierarchy of the multilineage differentiation of cells is defined by the following terms:

(1) *Totipotent*: cells capable of giving rise to an entire organism;
(2) *Pluripotent*: cells capable of giving rise to all three germ layers;
(3) *Multipotent*: cells specific to a particular germ layer giving rise to organ-specific progenitor and precursor cells.

Totipotency is lost very early in development at the zygote and early blastocyst stages. Thereafter, pluripotent cells may give rise to all three germ layers but are incapable of giving rise to an entire organism. Multipotent cells, which emerge later in development, are present in postnatal tissues to repopulate and regenerate those tissues in response to biochemical and biomechanical signals. The further differentiated organ-specific progenitor and precursor cells give rise to more differentiated cells with functions specific to that organ.[7]

Although postnatal stem cells reside in specialized tissues throughout the body, they retain plasticity in their ability to differentiate into multiple tissue types. For example, bone marrow-derived stem cells have been shown to differentiate into skeletal muscle and muscle-derived stem cells have been shown to differentiate into hematopoietic lineages.[8–11] Some have argued against the multipotency of postnatal stem cells by suggesting that these cells are only fusing with native specialized cells.[12,13] With conflicting data and many unresolved issues regarding differentiation, transdifferentiation, or dedifferentiation of postnatal stem cells and their progeny, a great deal of research is warranted.

The characteristic of selfrenewal is very important, especially in organs and tissues with rapid turnover due to either apoptosis or injury. Selfrenewal refers to a stem cell's ability to repopulate organ-specific progenitor cells while preserving the ability of that tissue to undergo future rounds of the same process. Two mechanisms are suggested for how stem cells may accomplish this.[14] First, perhaps asymmetric cellular divisions give rise to one cell destined to become a differentiated, mature tissue cell and one renewed stem cell. Second, the individual stem cells may respond stochastically by either differentiating or selfrenewing. In general, most of the stem cells in adult tissues remain quiescent and only a small proportion fulfills the need of homeostatic tissue turnover.

III. GENE THERAPY

The optimization and acceleration of tissue repair after injury has been accomplished through the use of minimally invasive surgical techniques, novel instruments, advanced pharmaceuticals, and modern rehabilitation. Although some tissues, such as liver parenchyma, have incredible regenerative capacity, other tissues, such as articular cartilage, have very limited healing capacity. Diseased tissues with massive fibrosis or segmental defects also present great challenges to the modern clinician. Often, this lack of tissue repair results from poor blood supply, paucity of bioactive inductive proteins, and reduced cell turnover.

In vivo and *ex vivo* gene therapy techniques have been utilized in an attempt to stimulate tissue regeneration through delivery of various cytokines and growth factors. Growth factors are small peptides that act locally to stimulate cell proliferation, migration, differentiation, and matrix synthesis. Growth factors may be synthesized by resident cells at the injury site and infiltrating inflammatory or reparatory cells. Using recombinant DNA technology, large quantities of many known growth factors can be produced for the purpose of treatment. However, the use of recombinant proteins to stimulate tissue regeneration presents therapeutic challenges. Owing to their very short biological half-lives, very high doses and repeated injections are often required to produce the desired therapeutic effect.

Combining stem cell therapies with gene therapy techniques may allow for improved delivery of the therapeutic protein. Since gene therapy provides the protein's genetic code, rather than the rapidly degradable protein itself, this technique may result in higher and more sustained protein levels. For example, Lewthwaite et al.[15] reported that systemic injection of recombinant interleukin-1 receptor antagonist protein in rabbits could not inhibit the induction of arthritis. Conversely, Makarov et al.[16] showed, through use of gene therapy to synoviocytes, that retroviral delivery of a secreted interleukin-1 receptor antagonist cDNA resulted in a reduction of experimentally induced arthritis. The need for highly concentrated protein dosing is particularly well illustrated by effective dosages for recombinant human bone morphogenetic protein-2 (rhBMP-2). As one moves up the mammalian chain from rat to sheep, the effective dosages of rhBMP-2 can increase a thousand- to a million-fold.[17–19] Gene therapy allows the targeted delivery of higher concentrations of proteins to specifically targeted tissues, thus improving therapeutic effect.

Using gene therapy requires that one choose the therapeutic protein of interest, the target cell or tissue, the vector to deliver the protein, the promoter to regulate gene expression, and the vector delivery method.[20] After the protein, target cell, and vector have been chosen, the vector needs to be inserted into the target cell. This may occur through either systemic or local delivery to the targeted tissue. Systemic delivery has the advantage of targeting all affected cells throughout the body. This technique would be of particular interest in congenital metabolic diseases or metastatic disease. Nonetheless, widespread expression of the transgene may result in unwanted and life threatening side effects.

Local delivery of protein can potentially avoid such systemic side effects. A gene construct can be delivered to a specified target tissue by either direct or indirect injection of the vector. The direct approach involves injection of the vector into the desired tissue. Limitations to this method include

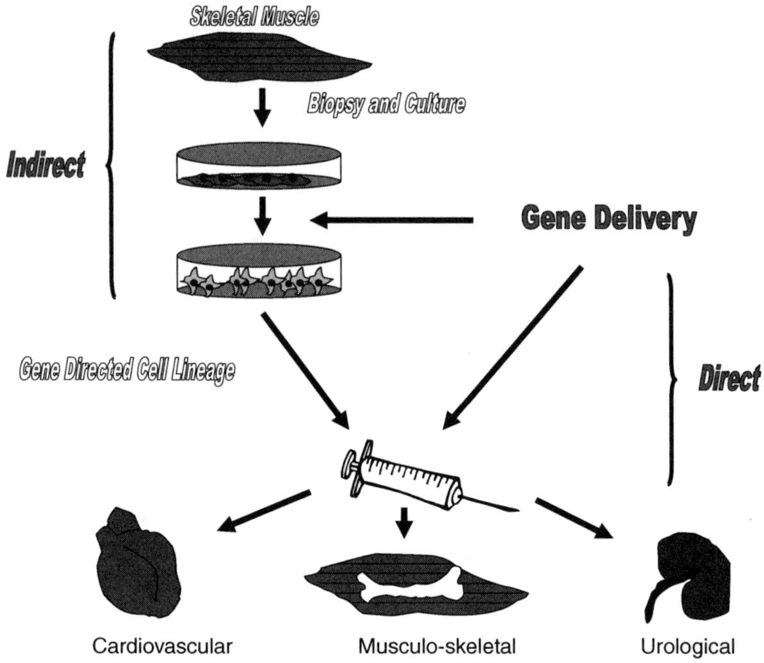

FIGURE 34.2 The *ex vivo* gene therapy approach involves harvesting and culturing stem cells and subsequently inserting a gene of interest through gene therapy techniques. This process allows for better control of gene transfer and induction of a desired cell lineage prior to reinsertion into the diseased tissue. Otherwise, a gene vector can be injected directly into the tissue of interest *in vivo*.

an inability to test the selected tissue for transgene expression and an inability to target specific cell types. The indirect, *ex vivo*, gene therapy approach involves removing the target cell of interest from the body and subsequently inserting the transgene *in vitro* (Figure 34.2). After the cells are stably transduced and tested for transgene expression, they can then be reimplanted into the host. Although indirect gene therapy is technically more demanding it has the advantage of *in vitro* testing to ensure expression and evaluate for unpredicted effects of gene expression on the cells of interest (e.g., mutagenesis). Indirect gene therapy has been performed effectively on various cell types, including bone marrow, mesenchymal, and muscle-derived stem cells.[21] Likewise, many tissues have been successfully targeted with *ex vivo* gene therapy, including the spine, articular cartilage, skeletal muscle, and bone.[22–25] Importantly, indirect gene therapy has been successful in the targeting of the human metacarpophalangeal joint.[26]

IV. STEM CELL BASED THERAPY

Stem cells display some characteristics that make them particularly useful for *ex vivo* gene therapy techniques. With their selfrenewing ability, genetically engineered stem cells may be able to eliminate the need for repeated administrations of the therapeutic gene. The first step in utilizing stem cells for cell-based gene therapy techniques is to harvest the cells and grow them in culture. Stem cells typically represent only a very small percentage of cells in postnatal tissues.[14,27] Thus their *ex vivo* expansion is crucial for cell transplantation and stem cell mediated gene therapy application. Research on cytokine and growth factor-induced expansion of both hematopoietic and muscle-derived stem cells has shown promise.[14] Hematopoietic stem cells have been expanded through culture with several individual and mixed cytokines, including interleukins, Flt-3 ligand, hepatocyte growth factor, and stem cell factor.[28,29] Likewise, muscle-derived stem cells have been successfully expanded using insulin-like growth factor-1, hepatocyte growth factor, epidermal growth factor, fibroblast growth factor-2, and stem cell factor.[14] These proteins may act on the stem cells to either reduce division time or to recruit nondividing cells into the mitotic cycle. The greatest challenge to the *ex vivo* expansion of stem cells with cytokines is ensuring that the stem cells are retaining their selfrenewal ability without exhibiting lineage differentiation.

A population of muscle-derived stem cells has been identified as being capable of prolonged production of multipotent progeny, and lacking the markers of mature muscle tissue.[27] Muscle-derived stem cells which express both myogenic markers and stem cell markers, such as desmin, c-met, stem cell antigen-1 (Sca-1), and CD34, are capable of fusing and forming myotubes under certain culture conditions while a subset remains undifferentiated and multipotent.[27,30] However, when stimulated *in vitro* by bone morphogenetic protein-2 (BMP-2) or bone morphogenic protein-4 (BMP-4), these cells show morphological changes suggesting their differentiation into osteoblasts.[31]

Muscle-derived stem cells display several attributes that make them ideal for gene therapeutics and tissue engineering:

1. Tissue harvesting via muscle biopsy is relatively easy and can be performed without compromising patient health.
2. Muscle-derived cells tolerate *ex vivo* manipulation well and are easily transducible by a variety of viral vectors.[21,32,33]
3. Muscle-derived stem cells display multilineage differentiation and selfrenewal capabilities when transplanted *in vivo*.[27]
4. Muscle-derived stem cells display an enhanced regenerative capacity.[27]

Muscle cells have been used as a vehicle for gene delivery in both muscle-related diseases and nonmuscle-related diseases. With the muscle-derived cells' ability to fuse with other muscle cells, their potential to act as gene delivery vehicles in diseases like Duchenne muscular dystrophy is clear. However, muscle-derived cells also have shown promise in treatment of many nonmuscle diseases. Muscle-derived cells, such as primary myoblasts, have been used experimentally to provide long-term expression of factor IX protein after implantation and show potential for treatment of hemophilia B.[34] Likewise, myoblasts have been used for delivery of human growth hormone, human proinsulin, and tyrosine hydroxylase.[35–37] Muscle-derived stem cells may theoretically act as even better gene delivery vehicles due to their selfrenewal, multilineage differentiation, and immune privileged transplantation behavior. Muscle-derived cells, including stem cells, have been used to improve the healing of multiple tissues within the musculoskeletal, cardiovascular, and urogenital systems. In the remainder of this chapter, we review the use of these cells in the treatment of bone, intraarticular, cardiac, and urological disorders.

V. MUSCULOSKELETAL SYSTEM

A. Bone Healing

Nonunion, or delayed healing, of fractures can be severely debilitating to a patient and presents a challenge to the orthopedic surgeon. Fractures resulting from high-energy trauma with extensive soft tissue stripping or segmental bone loss often exhibit delay or failure of bony union. Further compromise by infection, poor bone quality (i.e., osteoporosis), or inadequate fixation can complicate the management of fracture nonunions. Treatment options, especially in cases with segmental bone loss, include autogenic or allogenic bone-grafting, allograft supplemented with osteogenic proteins, bone transport, vascularized bone-grafting, or amputation.[38–41]

Similarly, difficulties in bone healing are experienced in oncology or craniofacial reconstruction where patients must endure multiple procedures and lengthy recovery periods.

For optimal bone healing, bone substitutes must contain osteoconductive, osteoinductive, and osteoproductive elements. Various scaffolds have been used as an osteoconductive element for bone ingrowth.[5] Likewise, multiple osteoinductive growth factors have been used to trigger osteogenic precursors, such as local stem cells, to initiate the healing cascade, including chemotaxis, mitosis, and differentiation.[42] The final element, osteoproductive cells, directly participates in bone formation after stimulation by the osteoinductive factors. Native precursors or even muscle-derived stem cells will manufacture osteoid in the osteoconductive scaffold when stimulated by osteoinductive growth factors.[21,31,43–45]

Numerous studies have documented the ability of muscle-derived stem cells to effectively deliver biologically active doses of bone morphogenetic proteins (BMPs) to a fracture site.[21,31,43–45] However, these cells not only act as delivery vehicles for osteoinductive proteins, they act as osteoproductive components by differentiating into an osteogenic lineage.[31] This osteogenic potential of muscle-derived cells was explored through a series of studies.[31,43,45–47] Day et al.[48] reported that muscle-derived cells carrying a LacZ reporter gene were capable of mediating gene transfer to bone defects. Bosch and co-workers[43,45] further examined the ability of muscle-derived cells transduced by adenoviral BMP-2 to produce heterotopic bone within 2 weeks of cell injection into the tricep surae of immunodeficient mice. Further examination of the fate of the transplanted and transduced muscle-derived cells showed that these cells had become osteoblastic and were capable of expressing bone proteins like osteocalcin.[43,46] Lee et al.[31] and Wright et al.[49] reported that retrovirus transduced muscle cells and muscle-derived stem cells are capable of improving the healing of critical sized calvarial defects by fulfilling both osteoinductive and osteoproductive roles. Moreover, Peng and colleagues[50] demonstrated that multiple growth factors can be beneficial to bone healing when delivered in the right proportions of independently transduced cells. Using muscle-derived stem cells retrovirally transduced to express either BMP-4 or vascular endothelial growth factor (VEGF), a ratio of 5 BMP-4 expressing cells to every 1 VEGF expressing cell was shown to synergistically improve bone healing better than treatment without VEGF or with higher concentrations of VEGF after implantation of the cells together within a critical-sized calvarial defect.[50] Concerning bone healing, it appears that cells that not only act as a gene delivery vehicle, but also have osteogenic potential, may prove best in healing critical defects. Many more studies need to be done to explore the osteogenic potential of stem cells and how delivery of different growth factors via stem cells can further improve bone healing.

B. Intraarticular Disorders

There are many articular disorders which can be candidates for gene therapy and tissue engineering. Arthritis, trauma, and aging can result in structural damage to articular cartilage, menisci, and ligaments. This section focuses on using gene therapy techniques for articular cartilage and meniscus tissue engineering.

Perhaps the most challenging problem in the treatment of musculoskeletal disorders is the repair and regeneration of articular cartilage. In contrast to bone, articular cartilage, with its lack of blood supply and chondrogenic potential, heals very poorly. Multiple surgical techniques including debridement and resurfacing, subchondral drilling and microfracture, and abrasion techniques attempt to allow osteochondral progenitors from the marrow to participate in the repair process.[51–56] However, the cartilage formed after utilization of these techniques often lacks proteoglycans and the Type II collagen that makes native cartilage structurally sound.[57] Newer strategies include the use of autologous chondrocyte transplantation to the site of the cartilage defect.[58] The difficulty with this technique is the harvesting of chondrocytes by an arthroscopic procedure and, consequently, expanding the small numbers obtained in culture. Investigators have tried to overcome these limitations by attempting to generate chondrocytes from mesenchymal stem cells in culture.[59] Adachi et al.[60] suggested that muscle-derived cells may also have potential for use in the healing of articular cartilage defects. When muscle-derived cells were transplanted into full-thickness articular cartilage defects in rabbits, these cells improved the healing of the defect with an efficacy equivalent to the chondrocyte transplantation, and even showed better incorporation and Type II collagen expression for up to 24 weeks.[60] In addition, muscle cells are easier to harvest and culture than chondrocytes and thus may represent a more attractive source chondroproductive cells.

The intraarticular meniscus of the knee also lacks healing potential and appears to be amenable to application of gene therapy techniques and tissue engineering. The meniscus, which plays multiple roles in protecting the joint surfaces, is often debrided or removed when severely damaged. Because many studies have documented the high incidence of degenerative changes that occur after total meniscectomy, orthopedic surgeons have focused on mensical repair techniques.[61] However, because only the outer one-third of the meniscus is vascularized, its healing potential is very limited in the inner two-thirds. The prospect of producing a biological meniscus replacement using tissue engineering techniques or modulating the meniscal cells by both *in vivo* and *ex vivo* gene therapy techniques is quite fascinating. Direct and indirect gene delivery techniques have proven effective in the meniscus. Both muscle-derived stem cells and meniscus-derived cells have been successfully genetically manipulated via *ex vivo* gene transfer and retransplanted into the menisci with stable gene expression.[62,63] Much research remains to be done in order to decide which cells and which growth factors will better promote meniscus regeneration.

C. Skeletal Muscle

The discovery of the muscle-derived stem cell, and perhaps the most promising application for their use, resulted from studies investigating myoblast transplantation for the treatment of muscular dystrophies. Only a small percentage of donor cells survive the initial transplantation process, due to a nonspecific inflammatory reaction and necrosis along with other yet to be defined events.[33,64–66] Interestingly, the survival of a portion of transplanted cells does not appear to be by default. Specific populations of muscle precursor cells, including stem cells, have been investigated for their improved transplantation behavior. Qu-Petersen et al.[27] wrote that muscle-derived stem cell populations survived transplantation, proliferated, and fused better than differentiated myoblast populations in immunocompetent mice. The immune privileged behavior, selfrenewal ability, and multilineage differentiation ability likely confer muscle-derived stem cells with their improved transplantation survival behavior.[27] Whether or not cell transplantation techniques can deliver and distribute dystrophin adequately throughout the muscle and body for therapeutic benefit has yet to be determined.

Muscle injuries are extremely common and have a tendency to reoccur.[67,68] Limitations in force production and reinjury are often attributed to the regeneration of restrictive fibrotic tissue which develops at the original site of injury.[24] Gene therapy may prove useful for development of techniques to improve the healing of muscle injuries. Li and Huard have documented that transforming growth factor-beta 1 (TGF-β1) plays a role in inducing formation of the fibrosis that limits muscle healing after severe injury.[24] They also documented that the muscle-derived stem cell is capable of differentiating into a myofibroblast-like lineage *in vitro* and is capable of contributing to scar formation after muscle injury *in vivo*.[24] It has also been reported that various inhibitors of TGF-β1 are capable of improving muscle regeneration and force production by limiting muscle fibrosis.[69–71] *Ex vivo* gene therapy techniques may provide a better method to deliver inhibitors of TGF-β1 to injured muscle. The proteoglycan decorin appears to be beneficial not only in reducing fibrosis, but also in improving muscle regeneration.[70] Delivery of a muscle-derived stem cell expressing decorin or a viral vector encoding for decorin to injured muscle may help improve long-term outcomes by reducing muscle fibrosis and, thus, recurrence of injury.[70]

VI. CARDIOVASCULAR SYSTEM

Cardiovascular disease is the major cause of morbidity and mortality in the U.S. Congestive heart failure due to either ischemic heart disease or other causes (i.e., hypertension, infection, autoimmune disease) results from a loss of functional heart muscle. Despite new drug therapies and surgical techniques, outcomes are still disappointing.[72] As with other tissues, the heart muscle displays very poor regenerative capacity. After ischemic injury (i.e., myocardial infarction) the necrotic area is replaced by an akinetic fibrotic scar. Likewise, many cardiomyopathies result in reduced myocyte viability, decreased contractility, and resultant heart failure. Recent studies now support cell transplantation to the heart as a potential solution to restore myocardial cells.[73] Studies in rodents, large animals, and even humans now demonstrate successful cell transplantation to diseased heart tissue.[74–79] Additionally, one might consider cell-mediated gene therapy to deliver antifibrotic agents to injured heart muscle.

A critical issue with cell transplantation to cardiovascular tissue is deciding which cell types to use. To date, the fetal cardiomyocytes showed promise with both contractile proteins and synchronous contraction behavior both *in vitro* and *in vivo*.[75,80,81] However, as with human embryonic stem cells, fetal cardiomyocytes are impractical for clinical use because of ethical concerns and immunorejection.[82] This leaves the muscle-derived cells, including myoblasts and stem cells, and bone marrow-derived cells.[77,79,83,84–87] Oshima et al.[86] recently described the stable delivery of a β-galactosidase reporter gene to cardiac muscle by transplantation of muscle-derived stem cells retrovirally transduced to express β-galactosidase. Furthermore, Payne et al.[87] recently demonstrated delivery of dystrophin expression using muscle-derived stem cell transplantation into a dystrophin-deficient heart. It is also conceivable that muscle-derived stem cells could deliver proteins to improve angiogenesis (e.g., VEGF) or inhibit fibrosis (e.g., decorin) in a cardiomyopathic or infarcted heart. Future studies of stem cell transplantation to cardiac muscle will need to focus on the differentiation of the stem cells into a functional cardiomyocyte lineage, as well as gene therapy techniques for improving regeneration and limiting fibrotic scar formation.

VII. UROGENITAL SYSTEM

Urinary incontinence is a serious medical and social condition with a tremendous number of sufferers world-wide. Stress incontinence, due to a weak or stretched urethral sphincter, results in leakage of urine with coughing, sneezing, laughing, or jumping. Treatment of this condition typically involves the conservative prescription of Kegel exercises or the more invasive surgical techniques or injection of bulking agents. Glutaraldehyde cross-linked collagen injections into the urethral sphincter is a quick, outpatient procedure that provided excellent short-term improvement in incontinence symptoms. However, some patients are allergic to the bovine collagen and the collagen is often reabsorbed, thus multiple injections are often required. In light of this, tissue engineering and use of muscle-derived cells has been explored as an alternative treatment for stress incontinence.[88–90]

Cell based *ex vivo* gene therapy techniques appear to be the most reasonable technique for building up a deficient urethral sphincter. Yokoyama et al. demonstrated that β-galactosidase expressing muscle-derived stem cells could survive longer than bovine collagen after injection into the bladder and urethral wall of immunodeficient mice.[89] In fact, they demonstrated that 88% of the β-galactosidase expression seen on Day 3 remained at Day 30. Further studies by Lee et al. reported an improvement in leak point pressures after muscle-derived stem cell injection in a denervated female rat model of stress urinary incontinence.[88] Allogenic muscle-derived stem cells significantly improved the incontinence threshold in sciatic nerve-transected animals compared to denervated animals injected with saline, after both one and four weeks.[88] Future studies will need to be performed to determine if any induction agents (i.e., growth factors) will be useful for treatment of urologic dysfunction.

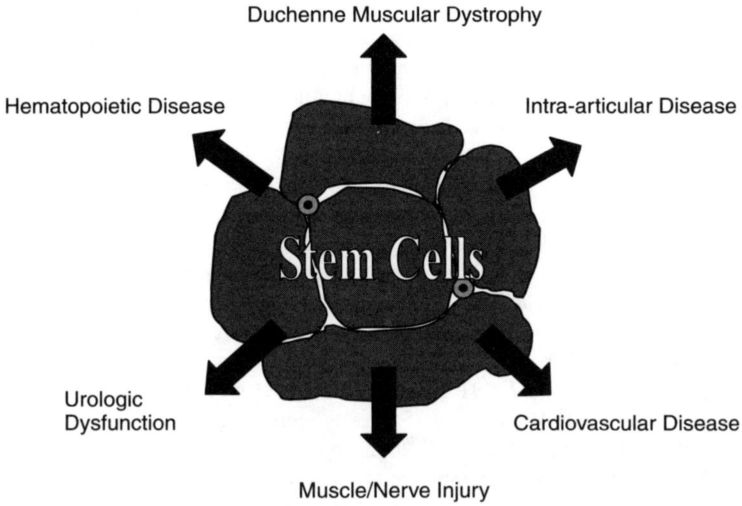

FIGURE 34.3 Schematic representation showing the use of muscle-derived stem cells for the treatment of various diseases and conditions throughout the body.

VIII. CONCLUSION

Gene therapy and tissue engineering, particularly through the *ex vivo* approach, show a great deal of promise for the future of medical therapies (Figure 34.3). It appears as though stem cells demonstrate behaviors that make them ideal for transplantation therapies. Of note there are many questions and issues that must be addressed in relation to using any type of stem cell, including: choice of cell origin, optimal isolation techniques, expansion to obtain therapeutic quantities of cells, differentiation and lineage progression signals, identification of the cells of interest, and the effects of genetic manipulation on transplantation and therapeutic behavior. Furthermore, a relative paucity of data exists examining the effects of various materials and their properties on the differentiation capacity of stem cells. As more and more of these questions and issues are explored in the laboratory, the therapeutic potential of stem cell mediated tissue engineering will be clarified.

REFERENCES

1. Solchaga, L. A., Dennis, J. E., Goldberg, V. M., and Caplan, A. I., Hyaluronic acid-based polymers as cell carriers for tissue-engineered repair of bone and cartilage, *J. Orthop. Res.*, 17(2), 205–213, 1999.
2. Wright, V. J., Peng, H., and Huard, J., Muscle-based gene therapy and tissue engineering for the musculoskeletal system, *Drug Discov. Today*, 6(14), 728–733, 2001.
3. Martinek, V., Fu, F. H., Lee, C. W., and Huard, J., Treatment of osteochondral injuries, *Genet. Eng. Clin. Sports Med.*, 20(2), 403–416, 2001.
4. Musgrave, D. S., Fu, F. H., and Huard, J., Gene therapy and tissue engineering in orthopaedic surgery, *J. Am. Acad. Orthop. Surg.*, 10(1), 6–15, 2002.
5. Christenson, L., Mikos, A. G., Gibbons, D. F., and Picciolo, G. L., Biomaterials for tissue engineering: summary, *Tissue Eng.*, 3(1), 71–73, 1997.
6. Henningson, C. T. Jr., Stanislaus, M. A., and Gewirtz, A. M., 28. Embryonic and adult stem cell therapy, *J. Allergy Clin. Immunol.*, 111(Suppl. 2), S745–S753, 2003.
7. Ham, R., and Veomett, M., *Mechanisms of Development*, CV Mosby, St Louis, MO, pp. 5–107, 1980.
8. Jackson, K. A., Mi, T., and Goodell, M. A., Hematopoietic potential of stem cells isolated from murine skeletal muscle, *Proc. Natl Acad. Sci. USA*, 96(25), 14482–14486, 1999.

9. Corti, S., Strazzer, S., Del Bo, R., Salani, S., Bossolasco, P., Fortunato, F., Locatelli, F., Soligo, D., Moggio, M., Ciscato, P., Prelle, A., Borsotti, C., Bresolin, N., Scarlato, G., and Comi, G. P., A subpopulation of murine bone marrow cells fully differentiates along the myogenic pathway and participates in muscle repair in the mdx dystrophic mouse, *Exp. Cell Res.*, 277(1), 74–85, 2002.
10. Torrente, Y., Tremblay, J. P., Pisati, F., Belicchi, M., Rossi, B., Sironi, M., Fortunato, F., El Fahime, M., D'Angelo, M. G., Caron, N. J., Constantin, G., Paulin, D., Scarlato, G., and Bresolin, N., Intraarterial injection of muscle-derived CD34(+)Sca-1(+) stem cells restores dystrophin in mdx mice, *J. Cell Biol.*, 152(2), 335–348, 2001.
11. Cao, B., Zheng, B., Jankowski, R. J., Kimura, S., Ikezawa, M., Deasy, B., Cummins, J., Epperly, M., Qu-Petersen, Z., and Huard, J., Muscle stem cells differentiate into haematopoietic lineages but retain myogenic potential, *Nat. Cell Biol.*, 5(7), 640–646, 2003.
12. Ying, Q. L., Nichols, J., Evans, E. P., and Smith, A. G., Changing potency by spontaneous fusion, *Nature*, 416(6880), 545–548, 2002.
13. Terada, N., Hamazaki, T., Oka, M., Hoki, M., Mastalerz, D. M., Nakano, Y., Meyer, E. M., Morel, L., Petersen, B. E., and Scott, E. W., Bone marrow cells adopt the phenotype of other cells by spontaneous cell fusion, *Nature*, 416(6880), 542–545, 2002.
14. Deasy, B. M., Qu-Peterson, Z., Greenberger, J. S., and Huard, J., Mechanisms of muscle stem cell expansion with cytokines, *Stem Cells*, 20(1), 50–60, 2002.
15. Lewthwaite, J., Blake, S. M., Hardingham, T. E., Warden, P. J., and Henderson, B., The effect of recombinant human interleukin 1 receptor antagonist on the induction phase of antigen induced arthritis in the rabbit, *J. Rheumatol.*, 21(3), 467–472, 1994.
16. Makarov, S. S., Olsen, J. C., Johnston, W. N., Anderle, S. K., Brown, R. R., Baldwin, A. S. Jr., Haskill, J. S., and Schwab, J. H., Suppression of experimental arthritis by gene transfer of interleukin 1 receptor antagonist cDNA, *Proc. Natl Acad. Sci. USA*, 93(1), 402–406, 1996.
17. Yasko, A. W., Lane, J. M., Fellinger, E. J., Rosen, V., Wozney, J. M., and Wang, E. A., The healing of segmental bone defects, induced by recombinant human bone morphogenetic protein (rhBMP-2). A radiographic, histological, and biomechanical study in rats, *J. Bone Joint Surg. Am.*, 74(5), 659–670, 1992.
18. Bostrom, M., Lane, J. M., Tomin, E., Browne, M., Berberian, W., Turek, T., Smith, J., Wozney, J., and Schildhauer, T., Use of bone morphogenetic protein-2 in the rabbit ulnar nonunion model, *Clin. Orthop.*, (327), 272–282, 1996.
19. Kirker-Head, C. A., Gerhart, T. N., Schelling, S. H., Hennig, G. E., Wang, E., and Holtrop, M. E., Long-term healing of bone using recombinant human bone morphogenetic protein 2, *Clin. Orthop.*, (318), 222–230, 1995.
20. Hannallah, D., Peterson, B., Lieberman, J. R., Fu, F. H., and Huard, J., Gene therapy in orthopaedic surgery, *Instr. Course Lect.*, 52, 753–768, 2003.
21. Musgrave, D. S., Bosch, P., Lee, J. Y., Pelinkovic, D., Ghivizzani, S. C., Whalen, J., Niyibizi, C., and Huard, J., Ex vivo gene therapy to produce bone using different cell types, *Clin. Orthop.*, (378), 290–305, 2000.
22. Riew, K. D., Wright, N. M., Cheng, S., Avioli, L. V., and Lou, J., Induction of bone formation using a recombinant adenoviral vector carrying the human BMP-2 gene in a rabbit spinal fusion model, *Calcif. Tissue Int.*, 63(4), 357–360, 1998.
23. Mason, J. M., Grande, D. A., Barcia, M., Grant, R., Pergolizzi, R. G., and Breitbart, A. S., Expression of human bone morphogenic protein 7 in primary rabbit periosteal cells: potential utility in gene therapy for osteochondral repair, *Gene Ther.*, 5(8), 1098–1104, 1998.
24. Huard, J., Li, Y., and Fu, F. H., Muscle injuries and repair: current trends in research, *J. Bone Joint Surg. Am.*, 84-A(5), 822–832, 2002.
25. Lee, J. Y., Musgrave, D., Pelinkovic, D., Fukushima, K., Cummins, J., Usas, A., Robbins, P., Fu, F. H., and Huard, J., Effect of bone morphogenetic protein-2-expressing muscle-derived cells on healing of critical-sized bone defects in mice, *J. Bone Joint Surg. Am.*, 83-A(7), 1032–1039, 2001.
26. Evans, C. H., Robbins, P. D., Ghivizzani, S. C., Herndon, J. H., Kang, R., Bahnson, A. B., Barranger, J. A., Elders, E. M., Gay, S., Tomaino, M. M., Wasko, M. C., Watkins, S. C., Whiteside, T. L., Glorioso, J. C., Lotze, M. T., and Wright, T. M., Clinical trial to assess the safety, feasibility, and efficacy of transferring a potentially anti-arthritic cytokine gene to human joints with rheumatoid arthritis, *Hum. Gene Ther.*, 7(10), 1261–1280, 1996.

27. Qu-Petersen, Z., Deasy, B., Jankowski, R., Ikezawa, M., Cummins, J., Pruchnic, R., Mytinger, J., Cao, B., Gates, C., Wernig, A., and Huard, J., Identification of a novel population of muscle stem cells in mice: potential for muscle regeneration, *J. Cell Biol.*, 157(5), 851–864, 2002.
28. Goff, J. P., Shields, D. S., and Greenberger, J. S., Influence of cytokines on the growth kinetics and immunophenotype of daughter cells resulting from the first division of single CD34(+)Thy-1(+)lin-cells, *Blood*, 92(11), 4098–4107, 1998.
29. Shih, C. C., Hu, M. C., Hu, J., Medeiros, J., and Forman, S. J., Long-term ex vivo maintenance and expansion of transplantable human hematopoietic stem cells, *Blood*, 94(5), 1623–1636, 1999.
30. Baroffio, A., Hamann, M., Bernheim, L., Bochaton-Piallat, M. L., Gabbiani, G., and Bader, C. R., Identification of self-renewing myoblasts in the progeny of single human muscle satellite cells, *Differentiation*, 60(1), 47–57, 1996.
31. Lee, J. Y., Qu-Petersen, Z., Cao, B., Kimura, S., Jankowski, R., Cummins, J., Usas, A., Gates, C., Robbins, P., Wernig, A., and Huard, J., Clonal isolation of muscle-derived cells capable of enhancing muscle regeneration and bone healing, *J. Cell Biol.*, 150(5), 1085–1100, 2000.
32. van Deutekom, J. C., Floyd, S. S., Booth, D. K., Oligino, T., Krisky, D., Marconi, P., Glorioso, J. C., and Huard, J., Implications of maturation for viral gene delivery to skeletal muscle, *Neuromuscular Disord.*, 8(3–4), 135–148, 1998.
33. Qu, Z., Balkir, L., van Deutekom, J. C., Robbins, P. D., Pruchnic, R., and Huard, J., Development of approaches to improve cell survival in myoblast transfer therapy, *J. Cell Biol.*, 142(5), 1257–1267, 1998.
34. Dai, Y., Roman, M., Naviaux, R. K., and Verma, I. M., Gene therapy via primary myoblasts: long-term expression of factor IX protein following transplantation in vivo, *Proc. Natl Acad. Sci. USA*, 89(22), 10892–10895, 1992.
35. Dhawan, J., Pan, L. C., Pavlath, G. K., Travis, M. A., Lanctot, A. M., and Blau, H. M., Systemic delivery of human growth hormone by injection of genetically engineered myoblasts, *Science*, 254(5037), 1509–1512, 1991.
36. Simonson, G. D., Groskreutz, D. J., Gorman, C. M., and MacDonald, M. J., Synthesis and processing of genetically modified human proinsulin by rat myoblast primary cultures, *Hum. Gene Ther.*, 7(1), 71–78, 1996.
37. Jiao, S., Williams, P., Safda, N., Schultz, E., and Wolff, J. A., Co-transplantation of plasmid-transfected myoblasts and myotubes into rat brains enables high levels of gene expression long-term, *Cell Transplant.*, 2(3), 185–192, 1993.
38. Einhorn, T. A., Enhancement of fracture-healing, *J. Bone Joint Surg. Am.*, 77(6), 940–956, 1995.
39. Lowenberg, D. W., Feibel, R. J., Louie, K. W., and Eshima, I., Combined muscle flap and Ilizarov reconstruction for bone and soft tissue defects, *Clin. Orthop.*, (332), 37–51, 1996.
40. Prokuski, L. J., and Marsh, J. L., Segmental bone deficiency after acute trauma. The role of bone transport, *Orthop. Clin. North Am.*, 25(4), 753–763, 1994.
41. Betz, R. R., Limitations of autograft and allograft: new synthetic solutions, *Orthopedics*, 25(Suppl. 5), s561–s570, 2002.
42. Reddi, A. H., Bone morphogenetic proteins, bone marrow stromal cells, and mesenchymal stem cells. Maureen Owen revisited, *Clin. Orthop.*, (313), 115–119, 1995.
43. Bosch, P., Musgrave, D. S., Lee, J. Y., Cummins, J., Shuler, T., Ghivizzani, T. C., Evans, T., Robbins, T. D., and Huard, J., Osteoprogenitor cells within skeletal muscle, *J. Orthop. Res.*, 18(6), 933–944, 2000.
44. Musgrave, D. S., Pruchnic, R., Bosch, P., Ziran, B. H., Whalen, J., and Huard, J., Human skeletal muscle cells in ex vivo gene therapy to deliver bone morphogenetic protein-2, *J. Bone Joint Surg. Br.*, 84(1), 120–127, 2002.
45. Bosch, P., Musgrave, D., Ghivizzani, S., Latterman, C., Day, C. S., and Huard, J., The efficiency of muscle-derived cell-mediated bone formation, *Cell Transplant.*, 9(4), 463–470, 2000.
46. Musgrave, D. S., Pruchnic, R., Wright, V., Bosch, P., Ghivizzani, S. C., Robbins, P. D., and Huard, J., The effect of bone morphogenetic protein-2 expression on the early fate of skeletal muscle-derived cells, *Bone*, 28(5), 499–506, 2001.
47. Musgrave, D. S., Bosch, P., Ghivizzani, S., Robbins, P. D., Evans, C. H., and Huard, J., Adenovirus-mediated direct gene therapy with bone morphogenetic protein-2 produces bone, *Bone*, 24(6), 541–547, 1999.

48. Day, C. S., Bosch, P., Kasemkijwattana, C., Menetrey, J., Moreland, M. S., Fu, F. H., Ziran, B., and Huard, J., Use of muscle cells to mediate gene transfer to the bone defect, *Tissue Eng.*, 5(2), 119–125, 1999.
49. Wright, V., Peng, H., Usas, A., Young, B., Gearhart, B., Cummins, J., and Huard, J., BMP4-expressing muscle-derived stem cells differentiate into osteogenic lineage and improve bone healing in immunocompetent mice, *Mol. Ther.*, 6(2), 169–178, 2002.
50. Peng, H., Wright, V., Usas, A., Gearhart, B., Shen, H. C., Cummins, J., and Huard, J., Synergistic enhancement of bone formation and healing by stem cell-expressed VEGF and bone morphogenetic protein-4, *J. Clin. Invest.*, 110(6), 751–759, 2002.
51. Insall, J. N., Intra-articular surgery for degenerative arthritis of the knee. A report of the work of the late K.H. Pridie, *J. Bone Joint Surg. Br.*, 49(2), 211–228, 1967.
52. Insall, J., The Pridie debridement operation for osteoarthritis of the knee, *Clin. Orthop.*, 101(01), 61–67, 1974.
53. Mitchell, N., and Shepard, N., The resurfacing of adult rabbit articular cartilage by multiple perforations through the subchondral bone, *J. Bone Joint Surg. Am.*, 58(2), 230–233, 1976.
54. Johnson, L. L., Arthroscopic abrasion arthroplasty historical and pathologic perspective: present status, *Arthroscopy*, 2(1), 54–69, 1986.
55. Bert, J. M., Role of abrasion arthroplasty and debridement in the management of osteoarthritis of the knee, *Rheum. Dis. Clin. North Am.*, 19(3), 725–739, 1993.
56. Steadman, J. R., Rodkey, W. G., and Briggs, K. K., Microfracture to treat full-thickness chondral defects: surgical technique, rehabilitation, and outcomes, *J. Knee Surg.*, 15(3), 170–176, 2002.
57. Shapiro, F., Koide, S., and Glimcher, M. J., Cell origin and differentiation in the repair of full-thickness defects of articular cartilage, *J. Bone Joint Surg. Am.*, 75(4), 532–553, 1993.
58. Brittberg, M., Lindahl, A., Nilsson, A., Ohlsson, C., Isaksson, O., and Peterson, L., Treatment of deep cartilage defects in the knee with autologous chondrocyte transplantation, *N. Engl. J. Med.*, 331(14), 889–895, 1994.
59. Wakitani, S., Kimura, T., Hirooka, A., Ochi, T., Yoneda, M., Yasui, N., Owaki, H., and Ono, K., Repair of rabbit articular surfaces with allograft chondrocytes embedded in collagen gel, *J. Bone Joint Surg. Br.*, 71(1), 74–80, 1989.
60. Adachi, N., Sato, K., Usas, A., Fu, F. H., Ochi, M., Han, C. W., Niyibizi, C., and Huard, J., Muscle derived, cell based ex vivo gene therapy for treatment of full thickness articular cartilage defects, *J. Rheumatol.*, 29(9), 1920–1930, 2002.
61. Bonneux, I., and Vandekerckhove, B., Arthroscopic partial lateral meniscectomy long-term results in athletes, *Acta Orthop. Belg.*, 68(4), 356–361, 2002.
62. Martinek, V., Usas, A., Pelinkovic, D., Robbins, P., Fu, F. H., and Huard, J., Genetic engineering of meniscal allografts, *Tissue Eng.*, 8(1), 107–117, 2002.
63. Goto, H., Shuler, F. D., Lamsam, C., Moller, H. D., Niyibizi, C., Fu, F. H., Robbins, P. D., and Evans, C. H., Transfer of lacZ marker gene to the meniscus, *J. Bone Joint Surg. Am.*, 81(7), 918–925, 1999.
64. Beauchamp, J. R., Morgan, J. E., Pagel, C. N., and Partridge, T. A., Dynamics of myoblast transplantation reveal a discrete minority of precursors with stem cell-like properties as the myogenic source, *J. Cell Biol.*, 144(6), 1113–1122, 1999.
65. Guerette, B., Asselin, I., Skuk, D., Entman, M., and Tremblay, J. P., Control of inflammatory damage by anti-LFA-1: increase success of myoblast transplantation, *Cell Transplant.*, 6(2), 101–107, 1997.
66. Cossu, G., and Mavilio, F., Myogenic stem cells for the therapy of primary myopathies: wishful thinking or therapeutic perspective?, *J. Clin. Invest.*, 105(12), 1669–1674, 2000.
67. Hurme, T., Kalimo, H., Lehto, M., and Jarvinen, M., Healing of skeletal muscle injury: an ultrastructural and immunohistochemical study, *Med. Sci. Sports Exerc.*, 23(7), 801–810, 1991.
68. Garrett, W. E. Jr., Muscle strain injuries, *Am. J. Sports Med.*, 24(Suppl. 6), S2–S8, 1996.
69. Foster, W., Li, Y., Usas, A., Somogyi, G., and Huard, J., Gamma interferon as an antifibrosis agent in skeletal muscle, *J. Orthop. Res.*, 21(5), 798–804, 2003.
70. Fukushima, K., Badlani, N., Usas, A., Riano, F., Fu, F., and Huard, J., The use of an antifibrosis agent to improve muscle recovery after laceration, *Am. J. Sports Med.*, 29(4), 394–402, 2001.
71. Chan, Y. S., Li, Y., Foster, W., Horaguchi, T., Somogyi, G., Fu, F. H., and Huard, J., The antifibrotic effects of suramin in injured skeletal muscle after laceration, *J. Appl. Physiol.*, 95(2), 771–780, 2003.

72. Flather, M. D., Yusuf, S., Kober, L., Pfeffer, M., Hall, A., Murray, G., Torp-Pedersen, C., Ball, S., Pogue, J., Moye, L., and Braunwald, E., Long-term ACE-inhibitor therapy in patients with heart failure or left-ventricular dysfunction: a systematic overview of data from individual patients. ACE-inhibitor myocardial infarction collaborative group, *Lancet*, 355(9215), 1575–1581, 2000.
73. Reinlib, L., and Field, L., Cell transplantation as future therapy for cardiovascular disease?: A workshop of the National Heart, Lung, and Blood Institute, *Circulation*, 101(18), E182–E187, 2000.
74. Jain, M., DerSimonian, H., Brenner, D. A., Ngoy, S., Teller, P., Edge, A. S., Zawadzka, A., Wetzel, K., Sawyer, D. B., Colucci, W. S., Apstein, C. S., and Liao, R., Cell therapy attenuates deleterious ventricular remodeling and improves cardiac performance after myocardial infarction, *Circulation*, 103(14), 1920–1927, 2001.
75. Li, R. K., Jia, Z. Q., Weisel, R. D., Mickle, D. A., Zhang, J., Mohabeer, M. K., Rao, V., and Ivanov, J., Cardiomyocyte transplantation improves heart function, *Ann. Thorac. Surg.*, 62(3), 654–660, 1996.
76. Scorsin, M., Hagege, A., Vilquin, J. T., Fiszman, M., Marotte, F., Samuel, J. L., Rappaport, L., Schwartz, K., and Menasche, P., Comparison of the effects of fetal cardiomyocyte and skeletal myoblast transplantation on postinfarction left ventricular function, *J. Thorac. Cardiovasc. Surg.*, 119(6), 1169–1175, 2000.
77. Rajnoch, C., Chachques, J. C., Berrebi, A., Bruneval, P., Benoit, M. O., and Carpentier, A., Cellular therapy reverses myocardial dysfunction, *J. Thorac. Cardiovasc. Surg.*, 121(5), 871–878, 2001.
78. Taylor, D. A., Atkins, B. Z., Hungspreugs, P., Jones, T. R., Reedy, M. C., Hutcheson, K. A., Glower, D. D., and Kraus, W. E., Regenerating functional myocardium: improved performance after skeletal myoblast transplantation, *Nat. Med.*, 4(8), 929–933, 1998.
79. Menasche, P., Hagege, A. A., Scorsin, M., Pouzet, B., Desnos, M., Duboc, D., Schwartz, K., Vilquin, J. T., and Marolleau, J. P., Myoblast transplantation for heart failure, *Lancet*, 357(9252), 279–280, 2001.
80. Soonpaa, M. H., Koh, G. Y., Klug, M. G., and Field, L. J., Formation of nascent intercalated disks between grafted fetal cardiomyocytes and host myocardium, *Science*, 264(5155), 98–101, 1994.
81. Yokomuro, H., Li, R. K., Mickle, D. A., Weisel, R. D., Verma, S., and Yau, T. M., Transplantation of cryopreserved cardiomyocytes, *J. Thorac. Cardiovasc. Surg.*, 121(1), 98–107, 2001.
82. Li, R. K., Mickle, D. A., Weisel, R. D., Mohabeer, M. K., Zhang, J., Rao, V., Li, G., Merante, F., and Jia, Z. Q., Natural history of fetal rat cardiomyocytes transplanted into adult rat myocardial scar tissue, *Circulation*, 96(Suppl. 9), II-179–II-186, 1997.
83. Fuchs, S., Baffour, R., Zhou, Y. F., Shou, M., Pierre, A., Tio, F. O., Weissman, N. J., Leon, M. B., Epstein, S. E., and Kornowski, R., Transendocardial delivery of autologous bone marrow enhances collateral perfusion and regional function in pigs with chronic experimental myocardial ischemia, *J. Am. Coll. Cardiol.*, 37(6), 1726–1732, 2001.
84. Tomita, S., Li, R. K., Weisel, R. D., Mickle, D. A., Kim, E. J., Sakai, T., and Jia, Z. Q., Autologous transplantation of bone marrow cells improves damaged heart function, *Circulation*, 100(Suppl. 19), II-247–II-256, 1999.
85. Sakai, T., Ling, Y., Payne, T. R., and Huard, J., The use of ex vivo gene transfer based on muscle-derived stem cells for cardiovascular medicine, *Trends Cardiovasc. Med.*, 12(3), 115–120, 2002.
86. Oshima, H., Sakai, T., Payne, T., Ling, Y., Qu-Petersen, Z., and Huard, J., Long-term survival of novel muscle-derived stem cells after transplantation into myocardium, *Heart Surg. Forum*, 6(1), 7 2002.
87. Payne, T., Sakai, T., Oshima, H., Ling, Y., Qu-Petersen, Z., Cummins, J., and Huard, J., Novel muscle-derived stem cells deliver dystrophin into a dystrophin-deficient murine heart, *Heart Surg. Forum*, 6(1), 7 2002.
88. Lee, J. Y., Cannon, T. W., Pruchnic, R., Fraser, M. O., Huard, J., and Chancellor, M. B., The effects of periurethral muscle-derived stem cell injection on leak point pressure in a rat model of stress urinary incontinence, *Int. Urogynecol. J. Pelvic Floor Dysfunct.*, 14(1), 31–37, 2003.
89. Yokoyama, T., Yoshimura, N., Dhir, R., Qu, Z., Fraser, M. O., Kumon, H., de Groat, W. C., Huard, J., and Chancellor, M. B., Persistence and survival of autologous muscle derived cells versus bovine collagen as potential treatment of stress urinary incontinence, *J. Urol.*, 165(1), 271–276, 2001.
90. Huard, J., Yokoyama, T., Pruchnic, R., Qu, Z., Li, Y., Lee, J. Y., Somogyi, G. T., de Groat, W. C., and Chancellor, M. B., Muscle-derived cell-mediated ex vivo gene therapy for urological dysfunction, *Gene Ther.*, 9(23), 1617–1626, 2002.

35 Osteochondral Tissue Engineering — Regeneration of Articular Condyle from Mesenchymal Stem Cells

Adel Alhadlaq and Jeremy J. Mao

CONTENTS

I. INTRODUCTION

Human motion is made possible by synovial joints. Not only do synovial joints allow the movement of articulating bones, but they serve to distribute functional loading during daily activities. A typical synovial joint is composed of two articular condyles enclosed in a joint capsule, stabilized by ligaments, and powered by skeletal muscles that are attached to bones via tendons. Unequivocally, synovial joint components must function in unison for a wide variety of human motions. Structural deficiency of joint components may result in accumulative failure of the entire joint. Arthritis, which encompasses more than 100 diseases and conditions, is now recognized as the leading cause of physical disability worldwide. Approximately 70 million adults in the United States alone suffer from some form of arthritis or chronic joint symptoms;[1] a substantial increase over previously estimated figures, likely due to enhanced diagnostic procedures and increased life expectancy. In addition to pathological conditions and their complications, traumatic failures of articular joints due to sports-related injuries and various other causes have contributed to the substantial increase in the number of total joint replacement procedures during the last decade.[2] Extensive demand for joint tissue replacement has motivated several meritorious approaches to tissue engineer the various phenotypic components of the articular joint.

The articular condyle consists of articular cartilage and subchondral bone. Despite their common mesenchymal origin, articular cartilage and subchondral bone are two distinct adult tissue phenotypes with few common morphological features, and yet they are structurally integrated and function in harmony to withstand a mechanical loading of up to several times the weight of the body.[3,4] Whereas bone regeneration readily occurs in the presence of angiogenesis up to a certain bony defect size, articular cartilage has poor capacity for self regeneration. Once damaged, an articular cartilage undergoes degenerative events such as the loss and destruction of key structural components, including Type II collagen and proteoglycans. Cartilage's poor capacity for self-regeneration is likely attributed to the paucity of tissue-forming cells, chondrocytes, that account for approximately 5% of the volume of articular cartilage. Since articular cartilage is avascular, its regeneration cannot rely on blood flow for the supply of cartilage tissue-forming cells. Thus, the self-regenerating capacity of articular cartilage depends upon the sparsely available chondroprogenitor cells, or perhaps mesenchymal stem cells (MSCs) that habitually reside in cartilage. Furthermore, articular cartilage is devoid of nerve supplies. Consequently, articular cartilage injuries are often not detected as joint pain until subchondral bone, which contains rich innervation, is involved. Therefore, articular cartilage defects, seen in arthritis and injuries, are often osteochondral defects.

Previous creditable attempts using tissue engineering approaches have significantly contributed, with noticeable clinical success, to the treatment of localized osteochondral defects.[5–10] Nonetheless, total condylar replacement remains a significant orthopedic challenge. As a result of severe arthritis or injuries, total joint replacement is associated with amplified structural and proportional complexity of the articular condyle relative to localized osteochondral lesions.[11–15] Bone and cartilage tissue engineering attempts utilizing isolated mature osteoblasts and chondrocytes have demonstrated considerable achievements using both *ex vivo* and *in vivo* approaches.[10,16–19] Tissue engineered articulations resembling human phalanges were created by seeding biodegradable polymer matrices with bovine articular chondrocytes and periosteal cells.[20] However, donor site morbidity and limited tissue supply associated with autologous grafting and potential immunorejection and transmission of pathogens associated with allo or xenogenic grafts exemplify the deficiencies associated with these approaches.[21–27]

Adult MSCs are advantageous for tissue engineering of osteochondral tissues due to their autogenic property and ability to readily differentiate into chondrogenic and osteogenic lineages.[28,29] Built upon previous meritorious experimental approaches in bone and cartilage tissue engineering, this chapter emphasizes cellular and scaffolding approaches for tissue engineering human-sized articular condyles from adult MSCs with a motivation to further advance the field of osteochondral tissue engineering potential for therapeutic applications in total condylar replacement.

II. COMPUTER-AIDED SCAFFOLDING AND MOLDING APPROACHES

Computer-aided tissue engineering and organ printing aim at accurate biomimicking and reconstruction of the structural complexity of normal tissues and organs.[30–32] The computer-aided designing approach for tissue engineering scaffolds offers the ability to control not only the external anatomical shape of the construct, but also the detailed architectural design of the interior. The accuracy of replication of the anatomical shape of the missing or degenerated organ positively correlates with the functional performance of tissue-engineered replacement and the host adaptability to the newly introduced implant. Simultaneously, optimization of the internal architecture of the scaffold, such as pore size and shape, channel orientation, and surface texture can potentially influence tissue formation within the construct, enhance vascularization, and affect the rate of extracellular matrix synthesis and tissue maturation stage.

The first step towards biomimetic tissue engineering of a degenerated, missing, or deformed articular condyle is the development of a "blueprint", or three-dimensional model, of the condyle for proper design and fabrication of a tissue engineered replacement. One successful approach for fabrication of multiple tissue engineering scaffolds utilizes commercially available software programs such as PV-Wave (Visual Numerics, Incorporated, San Ramon, CA, USA) or IDL (Interactive Data Language, Research Systems, Incorporated, Boulder, CO, USA) to read and digitize computerized tomography (CT) or magnetic resonance imaging (MRI) data for reconstruction of the external shape of the tissue engineering scaffold.[33,34] Since most synovial joints are bilaterally symmetric structures, the mirror CT or MRI image of the condyle in the normal side can be used as a template to construct a three-dimensional image for the scaffold replacing the contralateral abnormal condyle. In addition to defining and generating the external contour of the scaffold, computer-aided engineering and organ printing may reproduce the internal scaffold architecture such as pore size and shape, interconnectivity, and channel orientation.[12,33–35] By digitally combining the interior architecture image with the external scaffold image, a representative three-dimensional image of the scaffold can be used to fabricate the actual scaffold using solid free-form fabrication (SFF) technology.[36] SFF utilizes layer-manufacturing strategies to build three-dimensional physical engineering models from computer-aided design files [34,37] and has the potential to improve the scaffold design by controlling architectural scaffold parameters such as pore size and geometry, interconnectivity, and overall scaffold dimensional properties. An example for creating a three-dimensional scaffolding model for an articular condyle using SFF technology is shown in Figure 35.1. In this example, a polyurethane condylar–ramus unit of a human mandible was created from CT data utilizing a special SFF process known as stereolithography. Stereolithography utilizes a computer-controlled ultraviolet laser to photopolymerize consecutive layers of liquid resin to build solid objects.[34]

Upon construction of the three-dimensional scaffold resembling the articular condyle, the following step is to transform this scaffold into a biologically interactive scaffold by incorporating cells and/or growth factors that will result in the new tissue formation. A potential application of SFF technique is to create a solid three-dimensional model for the condyle from which a

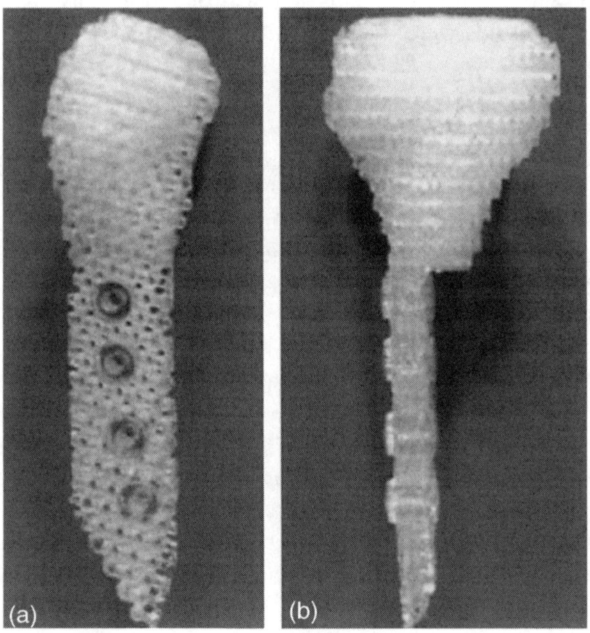

FIGURE 35.1 Fabrication of three-dimensional scaffolding model using solid SFF. (a) Sagittal view of a condylar-ramus unit made of polyurethane constructed using stereolithographic apparatus (SLA250/40, 3D Systems, Incorporated, Valencia, CA, USA) starting from CT images of human temporomandibular joint area. Sleeves in the ramus region have been added to the scaffold for the fixation screws. (b) Coronal view of the scaffold. (From Feinberg, S. E., Hollister, S. J., Halloran, J. W., Chu, T. M., and Krebsbach, P. H., Image-based biomimetic approach to reconstruction of the temporomandibular joint, *Cells Tissues Organs*, 169, 309–321, 2001. With permission.)

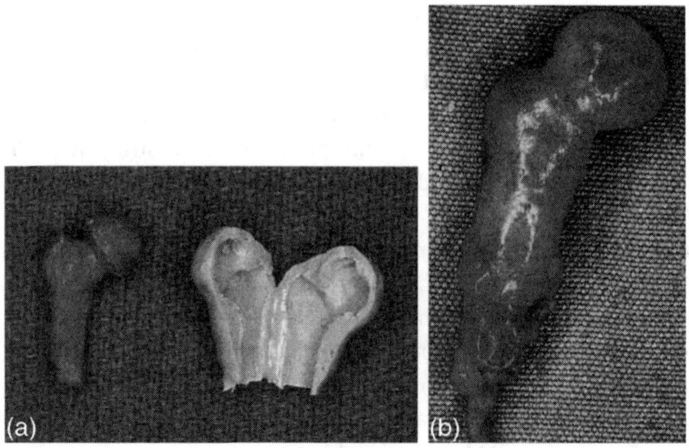

FIGURE 35.2 Molding and tissue engineering of a rat femoral head. (a) Silicone rubber mold (right) of a rat femur cast. (b) The resultant tissue engineered structure in the shape of a rat femoral head after 10 days of *in vivo* implantation in rat. (From Khouri, R. K., Koudsi, B., and Reddi, H., Tissue transformation into bone in vivo. A potential practical application, *JAMA*, 266, 1953–1955, 1991. With permission.)

negative mold can be obtained and used to fabricate the final scaffold.[11,13,14,38] As early as 1991, Khouri et al.[38] have reported the use of silicone rubber molds of rat femur casts (Figure 35.2(a)) to tissue engineer a bony structure in the shape of a femoral head (Figure 35.2(b)) from muscle flaps. Abukawa et al. [14] utilized a negative mold created by a rubber-based impression of a minipig mandibular condyle to fabricate a porous polymer scaffold from poly DL-lactic-*co*-glycolic acid (PLGA). The fabricated scaffolds were then seeded with osteogenic-differentiated minipig MSCs and the cell-scaffold constructs were incubated *ex vivo* in a rotational oxygen-permeable bioreactor system. Following a 6-week incubation period, bone formation was demonstrated in the experimental condyles, mainly at the outer surface of the tissue engineered constructs.[14] Using a different scaffolding strategy, Alhadlaq and Mao[13] created a bilayered osteochondral construct in the shape and dimensions of a human mandibular condyle using a photopolymerizable poly(ethylene glycol)-based hydrogel system encapsulating chondrogenic- and osteogenic-differentiated adult rat MSCs. The fabrication of these articular condyles was initiated by making a positive mold for a human cadaver mandibular condyle (Figure 35.3(a)) (similar to the positive mold that can be obtained for a patient's condyle with SFF technique in a clinical setting) and fabricating a negative mold using polysiloxane impression material (Memosil; Heraeus Kulzer, Incorporated, Armonk, NY, USA) (Figure 35.3(b)–(d)). The hollow negative molds were subsequently loaded in two stratified layers of hydrogel solution containing the chondrogenic- and osteogenic-differentiated MSCs, respectively (detailed experimental procedure discussed below in Section V). Each loading process was followed by a photopolymerization protocol using long-wave, 365 nm ultraviolet lamp (Glowmark, Upper Saddle River, NJ, USA) at an intensity of $\sim$4 mW/cm^2 for 5 min.[13] In comparison with solid scaffold material such as PLGA, the aqueous phase of the PEG allows cell encapsulation throughout the hydrogel. Consequently, tissue growth likely occurs both inside out and outside in.

These molding approaches may potentially function as a platform for therapeutic applications in total condylar replacement. The patient's own MSCs can be isolated via minimally invasive orthopedic procedures such as needle aspiration, culture-expanded *ex vivo*, induced to differentiate into cartilage- and bone-forming cells (chondrocytes and osteoblasts, respectively), and used in combination with the scaffolding materials to fill the negative mold that resembles the patient's original defect. However, several practical challenges, such as mechanical enablement of tissue engineered constructs to withstand the mechanical stresses experienced by normal joint function, must be taken into consideration before a patient's own articular condyle can be tissue engineered.

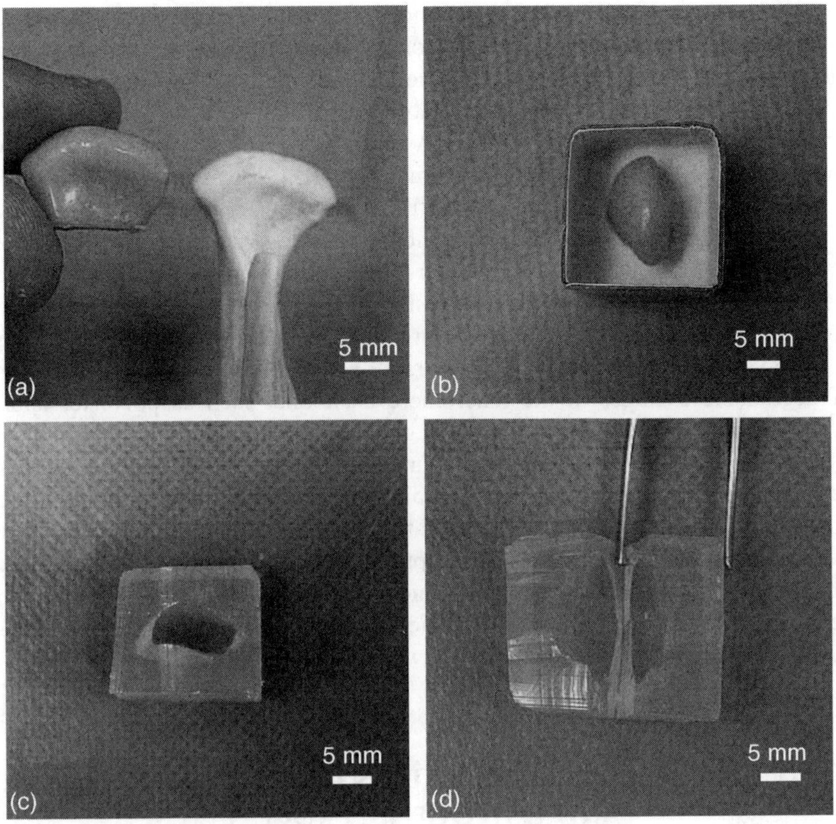

FIGURE 35.3 Fabrication of a negative mold for a human cadaver mandibular condyle. (a) An acrylic positive mold made from an alginate impression of the cadaver mandibular condyle. (b) The acrylic positive mold is attached to the bottom of a suitable container. (c) Negative mold for the acrylic model is created by filling the container shown in (b) with polysiloxane clear impression material. (d) The final negative mold is shaped as a bivalve to allow for construct removal after solidification.

III. SCAFFOLD MATERIALS AND DESIGN

The optimal scaffolding material should possess certain inherent properties such as being biocompatible, having no toxic degradation products, supporting cell viability and tissue formation, and having a degradation rate that corresponds to the rate of tissue formation and mineralization within the scaffold. In addition, due to the load-bearing nature of normal articular condyles, the tissue engineered condyle should ideally have initial physical strength to withstand joint loading. Scaffolding materials can be classified as naturally occurring or synthetic and they can be either biodegradable or nonbioresorbable. Materials that are resorbable, allow tissue formation, and facilitate predictable and specific cellular responses represent the third generation of biomaterials.[39] Poly(glycolic acid) (PGA), poly(lactic acid) (PLA), and their copolymers are the most widely used scaffold materials for bone and cartilage tissue engineering.[40] Owing to their known biocompatibility and controllable degradation and physical properties, PGA/PLA and copolymers have dominated in tissue engineering of osteochondral constructs (e.g., [11,14,23,41,42]). With continued progression toward improving the physical and chemical properties of PLGA polymers (e.g., [19,43–46]), the current relatively fast degradation rate with subsequent inferior mechanical properties, and the release of acidic degradation products with the risk of inflammatory reaction development,[47,48] necessitated persistent exploration of alternative scaffolding materials. Injectable polymers have gained increasing popularity among the next generation candidates for bone and cartilage tissue engineering,[49–51] mainly due to the potential of implanting the cell/polymer combination in a minimally invasive manner. Among the injectable polymers, poly(ethylene oxide) (PEO) hydrogels have been approved by the FDA for several medical applications due to their low toxicity and biocompatibility.[52,53] Poly(ethylene glycol) (PEG)-based hydrogels have been demonstrated to maintain cell viability and matrix release of encapsulated chondrocytes,[19,50,54] osteoblasts,[55,56] and MSC-derived chondrogenic and osteogenic cells.[13,24,57,58] The physical properties and degradation rate of these polymers are correlated with the choice of monomer(s). While poly(anhydrides) degrade by surface erosion, polyesters' degradation tends to be vast in nature with reduced control over degradation products.[40,59] Ongoing efforts are being made to investigate the enhancement of the degradation rate of PEG-based hydrogel systems through the addition of degradable linkages to the macromere backbone, such as polyester and phosphate groups.[60,62] Understanding and monitoring the degradation process of the scaffold is central to the long term success of the implant. Because mechanical properties of the implant decrease with the scaffold degradation, the degradation rate should be tailored to allow for cell proliferation, migration, and release of extracellular matrix that will eventually grant the implant with its own inherent mechanical strength. Many recent attempts have focused on enhancing the osteogenic and chondrogenic process via well-controlled mechanical modulation protocols, with the eventual goal of reducing the time period in which the mechanical strength of the implant is derived solely from the scaffold material. Some of these approaches will be summarized in Section IV.

IV. MECHANICAL MODULATION OF CHONDROGENIC AND OSTEOGENIC PROCESS

Following extensive experimental efforts of tissue engineering osteochondral tissues, the most challenging task appears to be functionally enabling the tissue engineered construct to withstand the mechanical loads that are experienced by normal osteochondral tissue, especially the articular condyle.[3,63] Mechanical loading, especially with rapid oscillating in peak amplitude, has been shown to favorably modulate chondrogenesis and osteogenesis.[63–68] Tailored mechanical strain enhances osteoblast differentiation and matrix mineralization, and affects reorganization of molecules involved in cell–matrix interaction.[69–72] Likewise, both static and dynamic compressions alter gene expression and modulate metabolic behavior of mature chondrocytes seeded in collagen scaffolds.[73,74] Cyclic strain of adult human MSCs in monolayer culture enhances matrix mineralization.[75] In our recent work, using a Flexercell Tension Plus System (FX-4000; Flexcell, Hillsborough, NC, USA), we applied a uniform 3% equibiaxial strain at 0.25 Hz on monolayer culture of human MSCs cultured for 4 weeks in basic culture medium Dulbecco's Modified Eagle's Medium-Low Glucose (DMEM-LG; Sigma, St Louis, MO, USA) supplemented with 10% fetal bovine serum (FBS) (Biocell, Rancho Dominguez, CA, USA) and 1% antibiotic–antimycotic (Gibco, Carlsbad, CA, USA). This resulted in osteogenic differentiation and maturation, as assayed by alkaline phosphatase activity (Figure 35.4(a)) and calcium deposition (Figure 35.4(b)), respectively, that correlates with monolayer cultures incubated with specially formulated osteogenic culture medium consisting of 100 nM dexamethasone, 10 mM β-glycerophosphate, and 0.05 mM ascorbic acid-2-phosphate (all from Sigma-Aldrich, St Louis, MO, USA). Increasing evidence of the potential enhancement of extracellular matrix properties of both bone and cartilage through mechanical modulation is recognized as functional tissue engineering, a field with a current focus on the responses of progenitor cells to mechanical stresses with the ultimate goal of enhancing the maturation and mechanical strength of tissue engineered structures.[76–79]

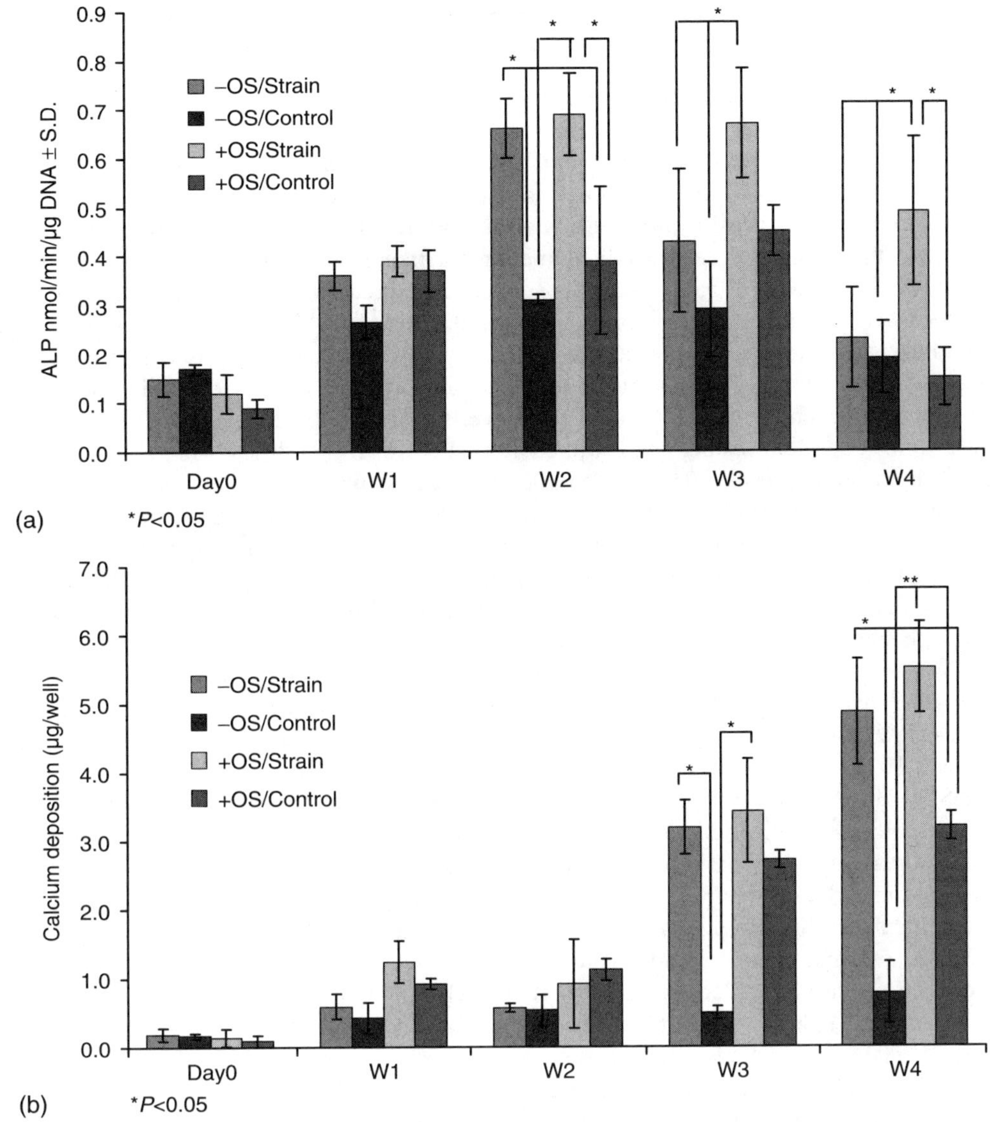

FIGURE 35.4 Alkaline phosphatase activity (ALP) and calcium deposition in MSC monolayer culture after application of cyclic strain. (a) ALP activity, indicative of osteogenic differentiation, in four groups of MSC monolayer cultures subjected to 0.25 Hz of equibiaxial cyclic strain using Flexercell system for 4 weeks. After 2 weeks, MSC monolayer cultures grown with basic medium (−OS) and subjected to strain showed ALP activity that correlates with that of cultures incubated with osteogenic supplement (+OS) and subjected to strain. The ALP in both cultures was significantly higher ($P < 0.05$, $N = 6$ per condition) than cultures incubated with similar corresponding culture media but not subjected to strain (control). (b) Calcium deposits in four groups of MSC monolayer cultures subjected to 0.25 Hz of equibiaxial cyclic strain using Flexercell system for 4 weeks. After 4 weeks, MSC monolayer cultures grown with basic medium (−OS) and subjected to strain had calcium deposits that correlate with that of cultures incubated with osteogenic supplement (+OS) and subjected to strain. The calcium deposition in both cultures was significantly higher ($P < 0.05$, $N = 6$ per condition) than cultures incubated with similar corresponding culture media but not subjected to strain (control).

V. REPRESENTATIVE TISSUE ENGINEERING OF HUMAN ARTICULAR CONDYLES FROM ADULT STEM CELLS

A wide variety of biocompatible polymers have been utilized as scaffolds for the delivery of cells and growth factors to heal various osteochondral defects (For a review see Refs. 5,40,80–83). Hydrogels are hydrophilic, three-dimensional, polymer networks that absorb large amounts of water or biological fluids while maintaining their distinct three-dimensional structure.[52,84] Due to their water content and diffusion properties, hydrogel scaffolds provide tissue-forming cells with an environment that mimics the extracellular matrix.[85] Among the hydrogel family, poly(ethylene glycol)- (PEG)-based hydrogel system has evoked a great deal of interest as a potential suitable carrier not only for cartilage tissue engineering,[13,19,24,50,54] but also for bone tissue engineering.[13,55,56] Additional advantages offered by PEG hydrogel and its derivatives are the availability of the hydrogel system in an initial liquid form that allows for mixing and uniform distribution of the encapsulated tissue-forming cells and the ability to form the cell/polymer mixture into complex shapes and dimensions before it solidifies via a biocompatible photopolymerization process. These properties make this hydrogel system and its derivatives one of the top potential candidates as a bone and cartilage tissue engineering scaffolding material. In order to further investigate the potential application of poly(ethylene glycol)-based polymer encapsulating *in vitro* expanded and differentiated MSCs for the replacement of missing or degenerated articular condyles, we have evaluated the feasibility of tissue engineering bone and cartilage phenotypes in bilayered constructs fabricated in the shape of human articular condyle following an extended *in vivo* implantation in immunodeficient mice for up to 12 weeks.[13]

Rat bone marrow-derived MSCs were harvested from 2- to 4-month-old (200 to 250 g) male Sprague–Dawley rats (Harlan, Indianapolis, IN, USA).[13] Whole bone marrow plugs were flushed out using a 10 ml syringe with Dulbecco's Modified Eagle's Medium-Low Glucose (DMEM-LG; Sigma, St Louis, MO, USA) supplemented with 10% FBS (Biocell, Rancho Dominguez, CA, USA) and 1% antibiotic–antimycotic (Gibco, Carlsbad, CA, USA). Marrow samples were collected and mechanically disrupted by passage through 16, 18, and 20 gauge needles. Cells were centrifuged, resuspended in serum-supplemented medium, counted and plated at 5×10^7 cells/100 mm culture dish, and incubated in 95% air/5% CO_2 at 37°C with a fresh medium change every 3 to 4 days. When large colonies were formed (typically 2 to 3 weeks), primary MSCs were trypsinized, counted, and passaged at a density of 5 to 7×10^5 cells/100 mm dish.

First-passage MSCs were cultured separately in a chondrogenic or osteogenic medium for 1 week.[13] The chondrogenic medium contained a supplement of 10 ng/ml TGF-β1 (RDI, Flanders, NJ, USA), whereas the osteogenic medium contained 100 nM dexamethasone, 10 mM β-glycerophosphate, and 0.05 mM ascorbic acid-2-phosphate (all from Sigma-Aldrich, St Louis, MO, USA). Cultures were incubated in 95% air/5% CO_2 at 37°C with medium changes every three to 4 days.

Poly(ethylene glycol) diacrylate (PEGDA) (MW 3400; Shearwater Polymers, Huntsville, AL, USA) was dissolved in sterile PBS supplemented with 100 units/ml penicillin and 100 μg/ml streptomycin (Gibco, Carlsbad, CA, USA) to a final solution of 10% w/v. A photoinitiator, 2-hydroxy-1-[4-(hydroxyethoxy) phenyl]-2-methyl-1-propanone (Ciba, Tarrytown, NY, USA), was added to the PEGDA solution to obtain a final photoinitiator concentration of 0.05% w/v.[13]

After 1 week of incubation in either the chondrogenic or osteogenic supplemented medium, MSC-derived chondrogenic and osteogenic cells were trypsinized, counted, and resuspended in PEGDA polymer/photoinitiator solution at a density of 5×10^6 cells/ml for the 4 week group or 20×10^6 cells/ml for the 12 week group.[13] A 150 μl aliquot of cell/polymer suspension with MSC-derived chondrogenic cells was loaded into a hollow polysiloxane mold of a cadaver human mandibular condyle (Figure 35.3). The chondrogenic layer was photopolymerized by a long-wave, 365 nm ultraviolet lamp (Glowmark, Upper Saddle River, NJ, USA) at an intensity of $\sim$4 mW/cm^2 for 5 min. Cell/polymer suspension containing MSC-derived osteogenic cells (approximately 600 μl) was then loaded to occupy the remainder of the mold, followed by the same photopolymerization protocol. The polymerized osteochondral constructs were then removed from the mold, washed twice with PBS, and implanted in subcutaneous pockets in the dorsum of severe combined immunodeficient (SCID) mice (Harlan, Indianapolis, IN, USA).

De novo formation of the articular condyle-like structures was observed after 4 and 12 week *in vivo* implantation of the osteochondral constructs in the subcutaneous pockets of the dorsum of SCID mice (Figure 35.5 and Figure 35.7).[13] Following 4 weeks *in vivo* implantation, the harvested articular condyles (Figure 35.5(e)) resembled the macroscopic shape and dimensions of the cell–hydrogel construct (Figure 35.5(c)) as well as the positive and negative condylar molds (Figure 35.5(a) and 5(b)). There was a superficial transparent portion and an inner photo-opaque portion in the superior view of the harvested articular condyle (Figure 35.5(e)), representing chondrogenic and osteogenic elements, respectively.

The interface between the upper-layer PEG hydrogel encapsulating MSC-derived chondrogenic cells and the lower-layer encapsulating MSC-derived osteogenic cells (above and below the red line in Figure 35.5(c)) demonstrated distinctive microscopic characteristics (Figure 35.6). The chondrogenic and osteogenic portions of the PEG hydrogel remained in their respective layers without substantial migration into each other's territories (observe the interface in Figure 35.6(a)). The chondrogenic layer contained sparse chondrocyte-like cells surrounded by abundant intercellular matrix (Figure 35.6(b)). The intercellular matrix of the chondrogenic layer showed an intense reaction to safranin O (Figure 35.6(b)), a cationic chondrogenic marker that binds to cartilage-specific glycosaminoglycans (GAG) such as chondroitin sulfate and keratan sulfate.[68,86]

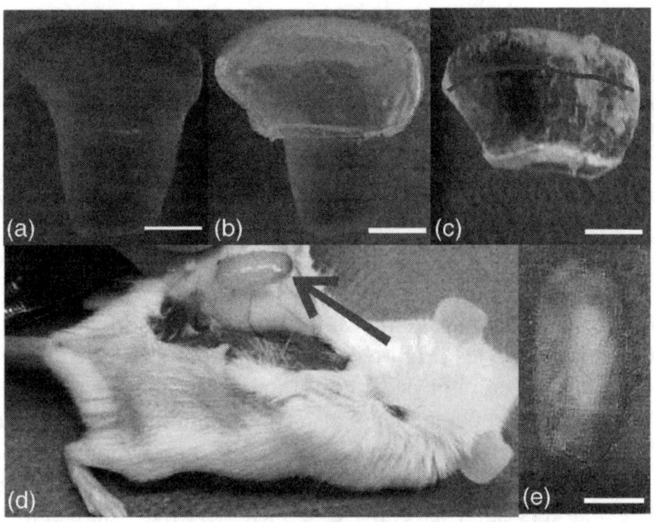

FIGURE 35.5 Fabrication and tissue engineering of a human-shaped articular condyle from rat bone-marrow derived MSC. (a) Acrylic model made from alginate impression of a human cadaver mandibular condyle. (b) Polyurethane negative mold fits the acrylic human articular condyle model. (c) A poly(ethylene glycol)-based hydrogel construct of the human mandibular condyle was fabricated in a two-phase process: (1) PEG-hydrogel solution encapsulating MSC-derived chondrogenic cells was loaded to occupy the top part of the negative mold (above the thin red line), followed by photopolymerization, and (2) additional PEG-hydrogel solution encapsulating MSC-derived osteogenic cells was loaded to occupy the remainder of the mold below the red line, followed by photopolymerization. (d) Harvest of tissue engineered articular condyle after 4 week subcutaneous implantation of the osteochondral construct in the dorsum of immunodeficient mice. (e) Harvested osteochondral construct retained the shape and dimensions of the molded articular condyle. Scale bar: 3 mm.

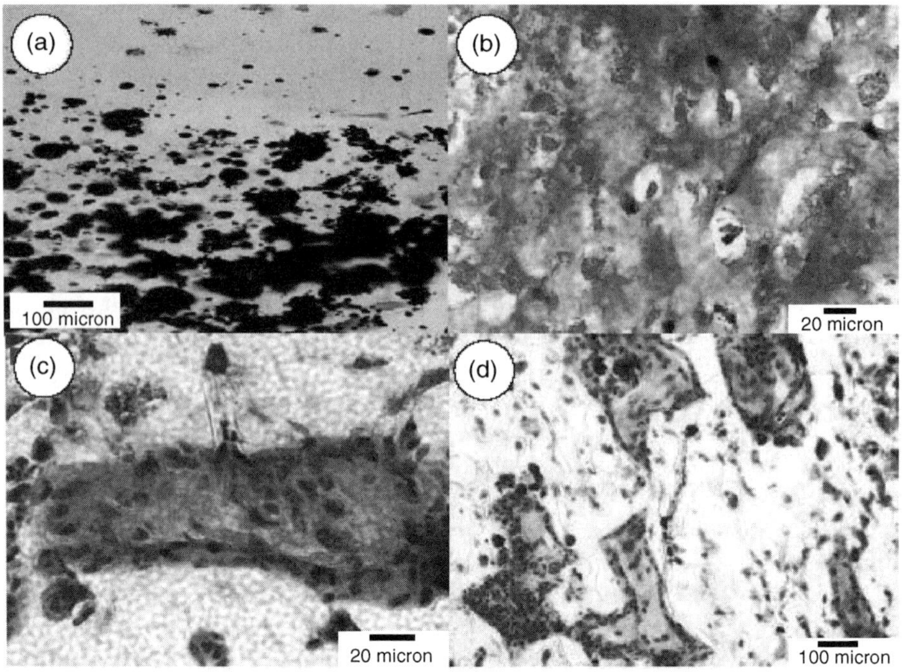

FIGURE 35.6 Representative photomicrographs of a tissue engineered articular condyle from rat bone-marrow derived MSCs harvested after 4 weeks of *in vivo* implantation in immunodeficient mice. (a) Von Kossa stained section of the osteochondral interface of tissue engineered articular condyle. The upper third of (a) represents the chondrogenic portion characterized by abundant intercellular space between MSC-derived chondrocyte-like cells. The lower two thirds of (a) represents the osteogenic portion characterized by abundant mineralization nodules. (b) Positive safranin O red stain indicates the presence of cartilage-specific glycosaminoglycans in the intercellular matrix of the chondral portion of the tissue engineered articular condyle. (c) H&E stained section from the osseous portion of the construct showing a representative island of bone trabecula-like structure in which MSC-derived osteoblast-like cells reside. (d) Positive toluidine blue staining of trabecula-like structures in the osseous portion of the tissue engineered articular condyle.

Some of the MSC-derived chondrogenic cells were surrounded by pericellular matrix, characteristic of natural chondrocytes (Figure 35.6(b)). By contrast, the osteogenic layer contained mineralization nodules that were confirmed to be mineral crystals by von Kossa staining (lower portion of Figure 35.6(a)). The osteogenic layer also showed multiple island structures occupied by osteoblast-like cells as exemplified in Figure 35.6(c). These island structures demonstrated a positive reaction to toluidine blue (Figure 35.6(d)). The chondrogenic layer showed negative staining to osteogenic markers such as von Kossa, whereas the osteogenic layer demonstrated negative reactions to chondrogenic markers such as safranin O (data not shown).

Similarly, following 12 weeks of *in vivo* implantation in the dorsum of SCID mice, articular condyles in the shape and dimensions of the molded human mandibular condyle formed *de novo* (Figure 35.7). The tissue engineered articular condyles were proportionally analogous to the acrylic model of the human cadaver mandibular condyle (Figure 35.7(a)). The constructs were firm and unyielding upon physical manipulation. The chondrogenic and osteogenic layers of the tissue engineered constructs were inseparable with no observable seam at the interface, signifying material integration between the two stratified layers.

The chondrogenic and osteogenic layers of the tissue engineered articular condyles demonstrated distinctive histologic characteristics (Figure 35.8 and Figure 35.9).[13] Microscopic examination of the interface between the top PEG hydrogel layer, encapsulating MSC-derived chondrogenic cells,

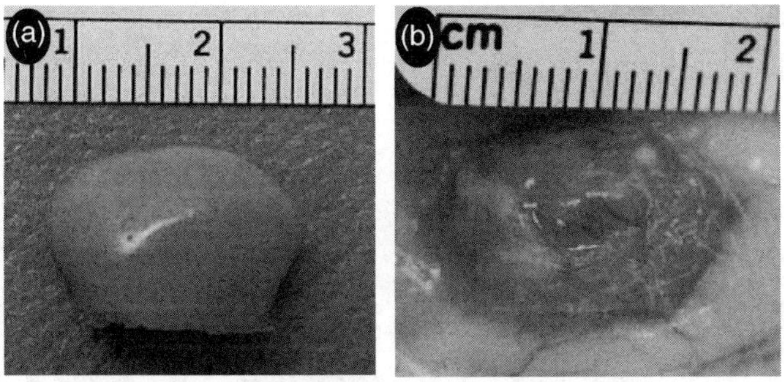

FIGURE 35.7 Tissue engineered osteochondral construct in the shape and dimensions of a cadaver human mandibular condyle. (a) Acrylic model represents a positive mold of an alginate impression of a human cadaver mandibular condyle. (b) Harvest of a tissue engineered condyle after 12 week *in vivo* subcutaneous implantation in the dorsum of immunodeficient mouse. The tissue engineered articular condyle retained the shape and dimensions of the cadaver human mandibular condyle.

and the bottom PEG hydrogel layer, encapsulating MSC-derived osteogenic cells, revealed distinctive phenotype-specific characteristics of cartilage- and bone-like tissues, respectively. The chondrogenic and osteogenic portions of the tissue engineered articular constructs largely remained in their respective layers (Figure 35.8(a)–(c)). However, in comparison with our previous work of 4 week and 8 week [13] *in vivo* incubation, there was detectable mutual infiltration of the cartilaginous and osseous components into each other's territories (Figure 35.8(a)–(c) and Figure 35.9). The chondrogenic layer consisted of chondrocyte-like cells, resident in lacunae and surrounded by abundant intercellular matrix, that showed intense reactions to cartilage-specific safranin O histochemical staining (Figure 35.8(b)). Type II collagen was immunolocalized throughout the chondrogenic layer (Figure 35.8(c)). Sparse areas of positive reaction to safranin O and immunolocalized Type II collagen were also observed within the osteogenic layer, near the osteochondral interface (Figure 35.8(b) and (c)). Type X collagen was immunolocalized to the deep chondrogenic layer adjacent to the tissue engineered osteochondral junction (Figure 35.8(d)), characteristic of the extracellular matrix of the hypertrophic and degenerating chondrocytes. By contrast, the osteogenic layer mostly showed negative reaction to safranin O (Figure 35.8(b)) and a lack of immunolocalization of Type II collagen (Figure 35.8(c)). Multiple islands of bone trabecula-like structures were observed in the osteogenic layer that were occupied by cells both on the surface and in the center, embedded within an abundant extracellular matrix (Figure 35.8 and Figure 35.9). Cells in the osteogenic layer clustered within island structures and at the interface between the chondrogenic and osteogenic layers (Figure 35.9). The osteogenic layer expressed Type I collagen, osteopontin, and osteonectin, while the chondrogenic layer lacked positive immunolocalization of these bone markers (Figure 35.9(a), (c) and (d)). Alkaline phosphatase, an early osteogenic marker was observed within the chondrogenic layer at the interface, possibly suggesting an early mineralization phase at the interface between the chondrogenic and the osteogenic portion (Figure 35.9(b)).

Our work demonstrates the feasibility of *de novo* formation of human-shaped small articular condyles with two stratified layers of chondrogenic and osteogenic histogenesis from a single population of bone marrow-derived chondrogenic and osteogenic cells encapsulated in a hydrogel scaffold.[13] The presence of abundant matrix synthesis and the immunolocalization of chondrogenic and osteogenic markers in corresponding layers of the tissue engineered articular condyles indicates survival, matrix synthesis and continuing phenotypic differentiation of encapsulated MSC-derived chondrogenic and osteogenic cells. Qualitatively, it appears that more matrix maturation has taken place by comparing the present data obtained after a 12 week *in vivo* implantation with our previous work of 4 and 8 week implantations.[13] The reciprocal expression of cartilaginous and osseous

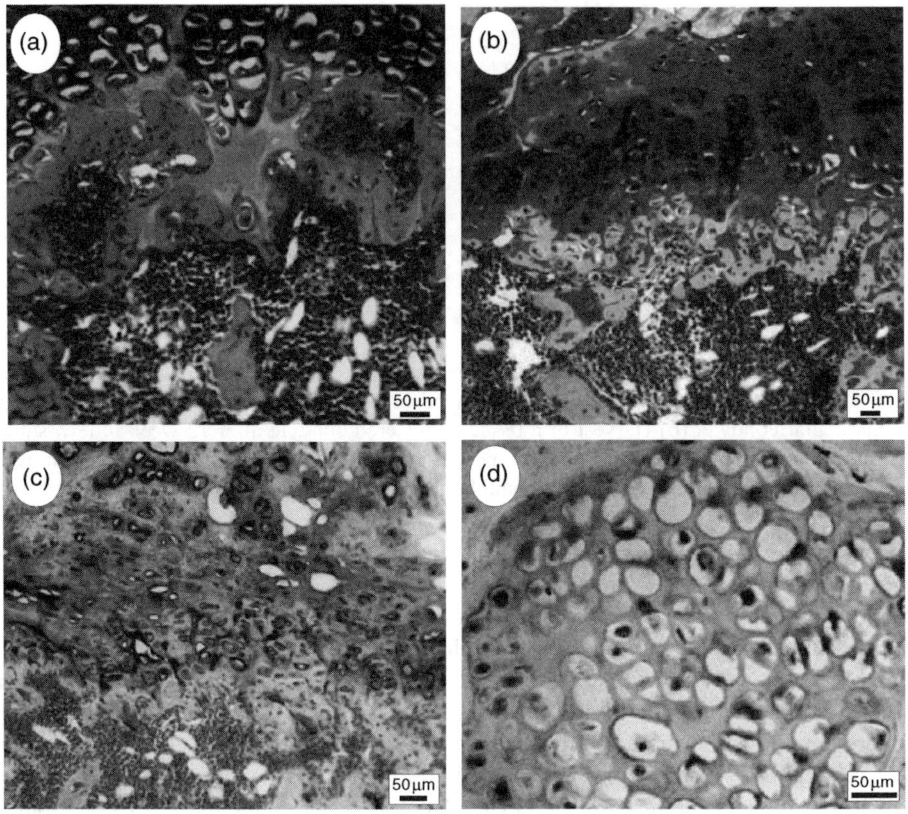

FIGURE 35.8 Histological and immunohistochemical characterization of cartilage- and bone-like tissues of tissue engineered articular condyle. (a) H&E stained section of the osteochondral interface of the tissue engineered articular condyle. The upper third of the photomicrograph represents the chondrogenic component of the tissue engineered osteochondral construct, and is characterized by relatively large chondrocyte-like cells housed in lacunae-like structures and surrounded by abundant extracellular matrix. The lower two thirds of the micrograph represents the osteogenic component of the tissue engineered osteochondral construct, and is characterized by isolated islands occupied by few cells and sparsely distributed between large clusters of cells. (b) Positive safranin O red staining of the chondral portion of the tissue engineered osteochondral construct indicating high concentration of cartilage-specific glycosaminoglycans in the extracellular matrix. In contrast, the osseous portion of the tissue engineered osteochondral construct showed negative reaction to safranin O staining. (c) Positive immunohistochemical localization of collagen Type II in the chondral portion of the tissue engineered osteochondral construct relative to the lower osseous component. (d) Positive immunohistochemical localization of Type X collagen in a cluster of chondrocytes-like cells in deeper region of the chondrogenic component (next to osseous tissue) of the tissue engineered osteochondral construct.

markers within the chondrogenic and osteogenic layers, along with the appearance of a wavy osteochondral interface, suggests mutual infiltration by cartilaginous and osseous tissues. A potential explanation for this phenomenon is the possibility of mutual diffusion of growth factors along the interface between the two layers, influencing the phenotypic commitment of MSCs toward a specific lineage. The appearance of hypertrophic chondrocyte-like cells, with immunolocalization of both Type X collagen and alkaline phosphatase, in the deep region of the chondrogenic layer likely indicates an early mineralization phase toward transition, from chondrocyte hypertrophy to osteogenic phenotype. In support of this assumption is the continuing expression, although sparsely and less intensely, of chondrogenic markers such as Type II collagen

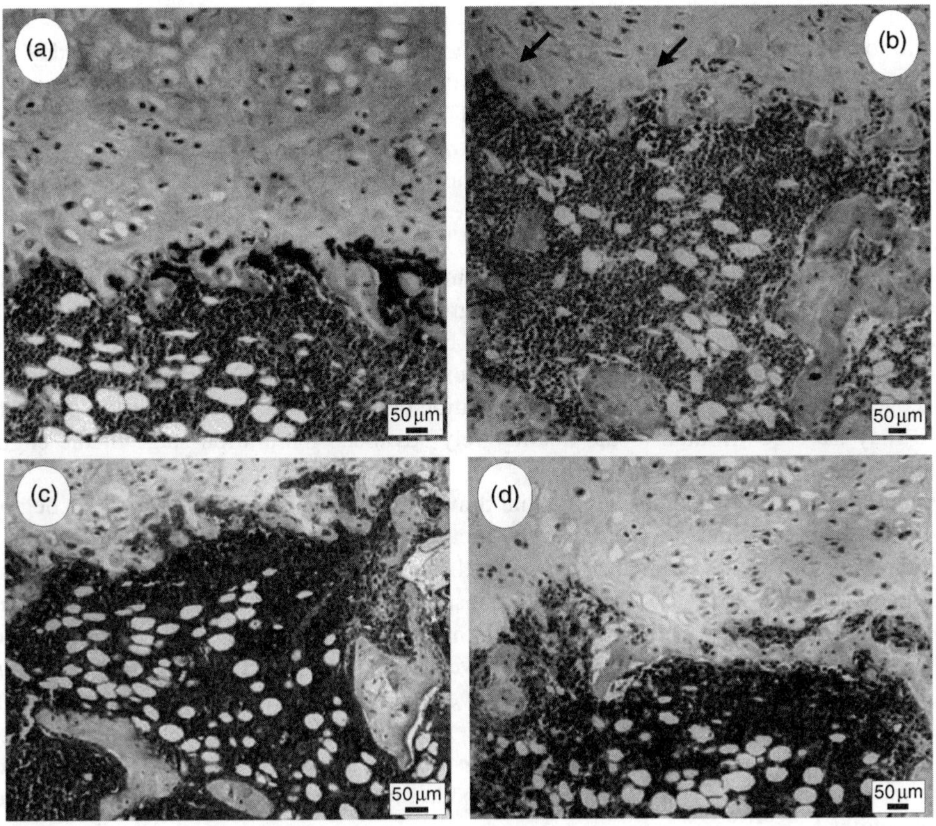

FIGURE 35.9 Immunohistochemical localization of osteogenic markers in the tissue-engineered articular condyle. (a) Positive immunohistochemical localization of Type I collagen at the osteochondral interface and within the osseous portion of the osteochondral construct. (b) General negative reaction to antibody against alkaline phosphatase except for slight and sparse reaction within the chondral portion and at the osteochondral junction (arrows). (c) Positive immunolocalization of ostepontin within the lower osseous portion of the tissue engineered osteochondral construct. By contrast, the chondral potion of the tissue engineered osteochondral construct lacked these osteogenic markers. (d) Positive immunolocalization of osteonectin within the lower osseous portion of the tissue engineered osteochondral construct. By contrast, the chondral potion of the tissue engineered osteochondral construct lacked these osteogenic markers.

and safranin O within the osteogenic portion of the tissue engineered osteochondral construct, pointing to a phenotypic transition at the osteochondral interface.

The approach of tissue engineering articular condyles with both cartilaginous and osseous components from a single population of adult MSCs represents another step toward the eventual therapeutic goal of condylar replacement in comparison with previous utilization of isolated mature chondrocytes and osteoblasts,[6,7,20,87,88] although these previous approaches have yielded excellent data on which the present work has been based. The imbricative design of osteochondral constructs in the dimensions of a human articular condyle for total condylar replacement is likely a key parameter since cell survival and viability are increasingly challenging for large three-dimensional scaffolds utilized for bone and cartilage tissue engineering.[89,90] The fabrication of uniform yet stratified osteochondral constructs in the shape and dimensions of human articular condyles from a single hydrogel system has the potential to increase the likelihood of physical integration between the cartilaginous and osseous components of the constructs, similar to other tissue engineered composite constructs.[16,20] An empirical concern coupled with seeding cells in prefabricated three-dimensional scaffolds is the tendency for the seeded cells and the synthesized extracellular matrix to localize towards the outer surface of the scaffold, where most cells are seeded.[14,91] We adopted several estimable approaches to encapsulate MSC-derived chondrogenic and osteogenic cells in the aqueous phase of PEG-based hydrogel (e.g., [50,56]), thus maximizing the possibility of uniform cell distribution and encouraging *de novo* tissue formation from within the photopolymerized constructs. The ease of fabrication, and the ability to mold and shape the photopolymerizable hydrogel systems into given proportions, are advantages for their use in the tissue engineering of osteochondral constructs.[84,92]

The initial encapsulation density has been increased from 5×10^6 cells/ml for the 4 week constructs to 20×10^6 cells/ml in the 12 week group. Both proliferation and matrix synthesis of chondrocytes in three-dimensional cultures have been shown to be influenced by the cell seeding density.[93,94] Moreover, clustering of marrow stromal cells enhances their differentiation towards the osteogenic lineage.[95] Thus, greater encapsulation density and longer incubation period than in Ref. 13 may have contributed to the qualitatively advanced maturation stage of the engineered formation of bone- and cartilage-like tissues.

Despite the above advantages, a potential drawback of PEG-based hydrogel is its degradation behavior. Research efforts are increasingly focused on investigating the enhancement of the degradation rate of PEG-based hydrogel system through the addition of degradable linkages to the macromere backbone such as polyester and phosphate groups.[60–62] Our experience with *in vivo* implantation of the PEGDA hydrogel encapsulating rat MSC-derived chondrogenic and osteogenic cells has revealed a positive relationship between its degradation and the period of *in vivo* implantation.[13] Nonetheless, the experimental divergence in our approach by *ex vivo* incubation of MSCs with chondrogenic and osteogenic supplements prior to the hydrogel encapsulation, the environmental differences between *in vivo* and *in vitro* incubation, and the extended implantation periods may contribute to the differences observed in the degradation behavior of PEGDA relative to other studies.[19,24,62,96]

The ultimate goal of synovial joint engineering is to tissue engineer autologous articular condyles *ex vivo* that are structurally and functionally sound to serve as replacements for missing or degenerated articular condyles. However, a number of challenging issues need to be addressed before reaching this ambitious goal. Zonal organization of articular cartilage is of structural and functional importance.[97–102] Thus, tissue engineered articular cartilage may need to incorporate appropriate molecular cues to recapitulate the developmental process of normal articular cartilage. The selection of a photopolymerizable hydrogel system facilitates this goal, especially since the PEG hydrogel system successfully generates multilayered articular cartilage constructs with zonal organization by encapsulating bovine chondrocytes from corresponding zones of the femur articular cartilage.[19] An additional approach of encapsulating growth factors for *in vivo* modulation of chondrogenesis and osteogenesis is likely necessary for continuing differentiation and phenotypic maintenance of MSC-differentiated chondrogenic and osteogenic cells in an *in vivo* environment.[103–107] However, as discussed previously, the most challenging task may prove to be the functional enablement of tissue engineered articular condyles to withstand the mechanical loads that are experienced by normal articular condyles. In addition to the possibility of enhancing the mechanical properties of the scaffolding material or the utilization of a composite structure, the mechanical modulation approaches to accelerate and enhance the chondrogenic and osteogenic processes within the tissue engineered constructs may prove to be more influential toward the long-term success of the implant, since the ultimate mechanical outcome will be a function of the newly formed tissues rather than the temporary degrading scaffold. Nevertheless, the presented examples of osteochondral tissue engineering may serve as initial proof of the technical and biological feasibility of tissue engineering human-sized articular condyles with chondral and osseous components from a single population of adult stem cells. Much additional work is required along several fronts before the presently pursued approaches can be utilized for therapeutic applications.

VI. CONCLUSIONS AND FUTURE PERSPECTIVES

The field of tissue engineering originated with commendable approaches to engineering single structures, such as skin, blood vessels, cartilage, and bone. Recent attempts have been made to fabricate more complex tissues, such as osteochondral constructs for potential regeneration of injured articular joint or joint components. Clinically, initial cartilage injuries are rarely associated with pain and hence, seldom detected by physicians. Clinically detectable joint pathologies, as in osteoarthritis and rheumatoid arthritis, are often osteochondral defects. The emerging field of osteochondral tissue engineering meets the challenge to study the phenotypic interaction of chondrocytes and osteoblasts, which despite both originating from MSC, behave rather differently and function in contrasting environments. The use of MSC, obtainable with minimally invasive and clinically accepted procedures from several sources in the patient's own body (such as from bone marrow or adipose tissue), is more clinically desirable than the use of mature chondrocytes or osteoblasts. Hydrogels, whether natural or synthetic, are promising candidates as the primary constituents of either one or both components of osteochondral constructs; not only as suitable carriers for the tissue-forming cells but also as reservoirs for controlled growth factor delivery. A number of challenges must be overcome to regenerate an articular condyle for the bioengineered replacement of a defective or missing condyle. The influential dynamic environment that articular joints habitually exist in will be the major hurdle. Even recently utilized polymers do not possess appropriate mechanical properties to withstand such a mechanically challenging environment. Other legitimate concerns, especially in light of complicated and sizable osteochondral defects such as the entire articular condyle, include the biological maintenance of tissue phenotype, the establishment of an interactive and well-integrated interface between the cartilagenous and osseous components, and the development of architecturally sound tissue structure throughout the construct. Even so, a multitude of approaches, some of which have been presented in this chapter, are anticipated to yield significant advances toward tissue engineering of optimal osteochondral constructs for the regeneration of localized osteochondral lesions and complete articular condyles for total joint replacement.

ACKNOWLEDGMENTS

We are grateful to Dr. Arnold Caplan's laboratory at the Case Western Reserve University and Dr. Jennifer Elisseeff's laboratory at the Johns Hopkins University for sharing their experimental protocol. We thank Ross Kopher and Dr. L. Hong for technical assistance. A. Lopez, S. Han, and T. Joshi are gratefully acknowledged for capable processing of histologic specimens. This research was supported in part by a Biomedical Engineering Research Grant from the Whitaker Foundation RG-01-0075, IRIB Grant on Biotechnology jointly from the University of Illinois at Chicago (UIC) and University of Illinois at Urbana-Champaign (UIUC), and by USPHS Research Grants DE13964 and DE15391 from the National Institute of Dental and Craniofacial Research, and EB02332 from the National Institute of Biomedical Imaging and Bioengineering, National Institutes of Health, Bethesda.

REFERENCES

1. Centers for Disease Control and Prevention (CDC), Prevalence of self-reported arthritis or chronic joint symptoms among adults — United States, *MMWR*, 51, 948–950, 2002.
2. American Academy of Orthopedic Surgeons, Arthroplasty and total joint replacement procedures 1991–2000, American Academy of Orthopedic Surgeons, United States, 2001.
3. Martin, R. B., Burr, D. B., and Sharkey, N. A., *Skeletal Tissue Mechanics*, 1st ed., Springer, New York, pp. 79–126, 1998.

4. Poole, A. R., Kojima, T., Yasuda, T., Mwale, F., Kobayashi, M., and Laverty, S., Composition and structure of articular cartilage: a template for tissue repair, *Clin. Orthop.*, 391(Suppl.), 26–33, 2001.
5. Temenoff, J. S., and Mikos, A. G., Injectable biodegradable materials for orthopedic tissue engineering, *Biomaterials*, 21, 2405–2412, 2000.
6. Schaefer, D., Martin, I., Shastri, P., Padera, R. F., Langer, R., Freed, L. E., and Vunjak-Novakovic, G., In vitro generation of osteochondral composites, *Biomaterials*, 21, 2599–2606, 2000.
7. Schaefer, D., Martin, I., Jundt, G., Seidel, J., Heberer, M., Grodzinsky, A., Bergin, I., Vunjak-Novakovic, G., and Freed, L. E., Tissue-engineered composites for the repair of large osteochondral defects, *Arthritis Rheum.*, 46, 2524–2534, 2002.
8. Solchaga, L. A., Gao, J., Dennis, J. E., Awadallah, A., Lundberg, M., Caplan, A. I., and Goldberg, V. M., Treatment of osteochondral defects with autologous bone marrow in a hyaluronan-based delivery vehicle, *Tissue Eng.*, 8, 333–347, 2002.
9. Garfein, E. S., Orgill, D. P., and Pribaz, J. J., Clinical applications of tissue engineered constructs, *Clin. Plast. Surg.*, 30, 485–498, 2003.
10. Hunziker, E. B., Tissue engineering of bone and cartilage. From the preclinical model to the patient, Novartis Found. Symp., 249, 70–78, 2003.
11. Weng, Y., Cao, Y., Silva, C. A., Vacanti, M. P., and Vacanti, C. A., Tissue-engineered composites of bone and cartilage for mandible condylar reconstruction, *J. Oral Maxillofac. Surg.*, 59, 185–190, 2001.
12. Hollister, S. J., Maddox, R. D., and Taboas, J. M., Optimal design and fabrication of scaffolds to mimic tissue properties and satisfy biological constraints, *Biomaterials*, 23, 4095–4103, 2002.
13. Alhadlaq, A., and Mao, J. J., Tissue-engineered neogenesis of human-shaped mandibular condyle from rat mesenchymal stem cells, *J. Dent. Res.*, 82, 951–956, 2003.
14. Abukawa, H., Terai, H., Hannouche, D., Vacanti, J. P., Kaban, L. B., and Troulis, M. J., Formation of a mandibular condyle in vitro by tissue engineering, *J. Oral. Maxillofac. Surg.*, 61, 94–100, 2003.
15. DeWal, H., Su, E., and DiCesare, P. E., Instability following total hip arthroplasty, *Am. J. Orthop.*, 32, 377–382, 2003.
16. Alsberg, E., Anderson, K. W., Albeiruti, A., Rowley, J. A., and Mooney, D. J., Engineering growing tissues, *Proc. Natl Acad. Sci. USA*, 17, 12025–12030, 2002.
17. Hung, C. T., Lima, E. G., Mauck, R. L., Taki, E., LeRoux, M. A., Lu, H. H., Stark, R. G., Guo, X. E., and Ateshian, G. A., Anatomically shaped osteochondral constructs for articular cartilage repair, *J. Biomech.*, 36, 1853–1864, 2003.
18. Brittberg, M., Peterson, L., Sjogren-Jansson, E., Tallheden, T., and Lindahl, A., Articular cartilage engineering with autologous chondrocyte transplantation. A review of recent developments, *J. Bone Joint Surg. Am.*, 85-A(Suppl. 3), 109–115, 2003.
19. Kim, T. K., Sharma, B., Williams, C. G., Uffner, R. M. A., Malik, A., McFarland, E. G., and Elisseeff, J. H., Experimental model for cartilage tissue engineering to regenerate the zonal organization of articular cartilage, *Osteoarthritis Cartilage*, 11, 653–664, 2003.
20. Isogai, N., Landis, W., Kim, T. H., Gerstenfeld, L. C., Upton, J., and Vacanti, J. P., Formation of phalanges and small joints by tissue-engineering, *J. Bone Joint Surg. Am.*, 81, 306–316, 1999.
21. Mooney, D. J., and Mikos, A. G., Growing new organs, *Sci. Am.*, 280, 60–65, 1999.
22. Lu, L., Zhu, X., Valenzuela, R. G., Currier, B. L., and Yaszemski, M. J., Biodegradable polymer scaffolds for cartilage tissue engineering, *Clin. Orthop.*, 391(Suppl.), 251–270, 2001.
23. Sherwood, J. K., Riley, S. L., Palazzolo, R., Brown, S. C., Monkhouse, D. C., Coates, M., Griffith, L. G., Landeen, L. K., and Ratcliffe, A., A three-dimensional osteochondral composite scaffold for articular cartilage repair, *Biomaterials*, 23, 4739–4751, 2002.
24. Williams, C. G., Kim, T. K., Taboas, A., Malik, A., Manson, P., and Elisseeff, J., In vitro chondrogenesis of bone marrow-derived mesenchymal stem cells in a photopolymerizing hydrogel, *Tissue Eng.*, 9, 679–688, 2003.
25. Ahmad, C. S., Guiney, W. B., and Drinkwater, C. J., Evaluation of donor site intrinsic healing response in autologous osteochondral grafting of the knee, *Arthroscopy*, 18, 95–98, 2002.
26. Sikavitsas, V. I., Bancroft, G. N., and Mikos, A. G., Formation of three-dimensional cell/polymer constructs for bone tissue engineering in a spinner flask and a rotating wall vessel bioreactor, *J. Biomed. Mater. Res.*, 62, 136–148, 2002.

27. Lutolf, M. P., Weber, F. E., Schmoekel, H. G., Schense, J. C., Kohler, T., Muller, R., and Hubbell, J. A., Repair of bone defects using synthetic mimetics of collagenous extracellular matrices, *Nat. Biotechnol.*, 21, 513–518, 2003.
28. Caplan, A. I., Mesenchymal stem cells, *J. Orthop. Res.*, 9, 641 1991.
29. Pittenger, M. F., Mackay, A. M., Beck, S. C., Jaiswal, R. K., Douglas, R., Mosca, J. D., Moorman, M. A., Simonetti, D. W., Craig, S., and Marshak, D. R., Multilineage potential of adult human mesenchymal stem cells, *Science*, 284, 143–147, 1999.
30. Sun, W., and Lal, P., Recent development on computer aided tissue engineering — a review, *Comput. Methods Programs Biomed.*, 67, 85–103, 2002.
31. Hutmacher, D. W., Scaffold design and fabrication technologies for engineering tissues-state of the art and future perspectives, *J. Biomater. Sci. Polym. Ed.*, 12, 107–124, 2001.
32. Mironov, V., Boland, T., Trusk, T., Forgacs, G., and Markwald, R. R., Organ printing: computer-aided jet-based 3D tissue engineering, *Trends Biotechnol.*, 21, 157–161, 2003.
33. Hollister, S. J., Levy, R. A., Chu, T. M., Halloran, J. W., and Feinberg, S. E., An image-based approach for designing and manufacturing craniofacial scaffolds, *Int. J. Oral Maxillofac. Surg.*, 29, 67–71, 2000.
34. Feinberg, S. E., Hollister, S. J., Halloran, J. W., Chu, T. M., and Krebsbach, P. H., Image-based biomimetic approach to reconstruction of the temporomandibular joint, *Cells Tissues Organs*, 169, 309–321, 2001.
35. Chu, T. M., Hollister, S. J., Halloran, J. W., Feinberg, S. E., and Orton, D. G., Manufacturing and characterization of 3-d hydroxyapatite bone tissue engineering scaffolds, *Ann. NY Acad. Sci.*, 961, 114–117, 2002.
36. Taboas, J. M., Maddox, R. D., Krebsbach, P. H., and Hollister, S. J., Indirect solid free form fabrication of local and global porous, biomimetic and composite 3D polymer–ceramic scaffolds, *Biomaterials*, 24, 181–194, 2003.
37. Sachlos, E., and Czernuszka, J. T., Making tissue engineering scaffolds work. Review: the application of solid freeform fabrication technology to the production of tissue engineering scaffolds, *Eur. Cell Mater.*, 5, 29–39, 2003.
38. Khouri, R. K., Koudsi, B., and Reddi, H., Tissue transformation into bone in vivo. A potential practical application, *JAMA*, 266, 1953–1955, 1991.
39. Hench, L. L., and Polak, J. M., Third-generation biomedical materials, *Science*, 295, 1014–1017, 2002.
40. Gunatillake, P. A., and Adhikari, R., Biodegradable synthetic polymers for tissue engineering, *Eur. Cell Mater.*, 5, 1–16, 2003.
41. Karp, J. M., Shoichet, M. S., and Davies, J. E., Bone formation on two-dimensional poly(DL-lactide-co-glycolide) (PLGA) films and three-dimensional PLGA tissue engineering scaffolds in vitro, *J. Biomed. Mater. Res.*, 64, 388–396, 2003.
42. Chen, G., Sato, T., Ushida, T., Hirochika, R., Shirasaki, Y., Ochiai, N., and Tateishi, T., The use of a novel PLGA fiber/collagen composite web as a scaffold for engineering of articular cartilage tissue with adjustable thickness, *J. Biomed. Mater. Res.*, 67, 1170–1180, 2003.
43. Nelson, K. D., Romero, A., Waggoner, P., Crow, B., Borneman, A., and Smith, G. M., Technique paper for wet-spinning poly(L-lactic acid) and poly(DL-lactide-co-glycolide) monofilament fibers, *Tissue Eng.*, 9, 1323–1330, 2003.
44. Oh, S. H., Kang, S. G., Kim, E. S., Cho, S. H., and Lee, J. H., Fabrication and characterization of hydrophilic poly(lactic-*co*-glycolic acid)/poly(vinyl alcohol) blend cell scaffolds by melt-molding particulate-leaching method, *Biomaterials*, 24, 4011–4021, 2003.
45. Lieb, E., Tessmar, J., Hacker, M., Fischbach, C., Rose, D., Blunk, T., Mikos, A. G., Gopferich, A., and Schulz, M. B., Poly(D,L-lactic acid)-poly(ethylene glycol)-monomethyl ether diblock copolymers control adhesion and osteoblastic differentiation of marrow stromal cells, *Tissue Eng.*, 9, 71–84, 2003.
46. Lu, H. H., El-Amin, S. F., Scott, K. D., and Laurencin, C. T., Three-dimensional, bioactive, biodegradable, polymer-bioactive glass composite scaffolds with improved mechanical properties support collagen synthesis and mineralization of human osteoblast-like cells in vitro, *J. Biomed. Mater. Res.*, 64, 465–474, 2003.
47. Fu, K., Pack, D. W., Klibanov, A. M., and Langer, R., Visual evidence of acidic environment within degrading poly(lactic-*co*-glycolic acid) (PLGA) microspheres, *Pharm. Res.*, 17, 100–106, 2000.

48. Lu, L., Peter, S. J., Lyman, M. D., Lai, H. L., Leite, S. M., Tamada, J. A., Uyama, S., Vacanti, J. P., Langer, R., and Mikos, A. G., In vitro and in vivo degradation of porous poly(DL-lactic-*co*-glycolic acid) foams, *Biomaterials*, 21, 1837–1845, 2000.
49. Peter, S. J., Miller, M. J., Yasko, A. W., Yaszemski, M. J., and Mikos, A. G., Polymer concepts in tissue engineering, *J. Biomed. Mater. Res.*, 43, 422–427, 1998.
50. Elisseeff, J., Anseth, K., Sims, D., McIntosh, W., Randolph, M., and Langer, R., Transdermal photopolymerization for minimally invasive implantation, *Proc. Natl Acad. Sci. USA*, 96, 3104–3107, 1999.
51. Drury, J. L., and Mooney, D. J., Hydrogels for tissue engineering: scaffold design variables and applications, *Biomaterials*, 24, 4337–4351, 2003.
52. Lee, K. Y., and Mooney, D. J., Hydrogels for tissue engineering, *Chem. Rev.*, 101, 1869–1879, 2001.
53. Poshusta, A. K., and Anseth, K. S., Photopolymerized biomaterials for application in the temporomandibular joint, *Cells Tissues Organs*, 169, 272–278, 2001.
54. Elisseeff, J., McIntosh, W., Anseth, K., Riley, S., Ragan, P., and Langer, R., Photoencapsulation of chondrocytes in poly(ethylene oxide)-based semi-interpenetrating networks, *J. Biomed. Mater. Res.*, 51, 164–171, 2000.
55. Lee, K. Y., Alsberg, E., and Mooney, D. J., Degradable and injectable poly(aldehyde guluronate) hydrogels for bone tissue engineering, *J. Biomed. Mater. Res.*, 56, 228–233, 2001.
56. Burdick, J. A., and Anseth, K. S., Photoencapsulation of osteoblasts in injectable RGD-modified PEG hydrogels for bone tissue engineering, *Biomaterials*, 23, 4315–4323, 2002.
57. Shin, H., Jo, S., and Mikos, A. G., Modulation of marrow stromal osteoblast adhesion on biomimetic oligo[poly(ethylene glycol) fumarate] hydrogels modified with Arg–Gly–Asp peptides and a poly(ethyleneglycol) spacer, *J. Biomed. Mater. Res.*, 61, 169–179, 2002.
58. Behravesh, E., and Mikos, A. G., Three-dimensional culture of differentiating marrow stromal osteoblasts in biomimetic poly(propylene fumarate-*co*-ethylene glycol)-based macroporous hydrogels, *J. Biomed. Mater. Res.*, 66, 698–706, 2003.
59. Suggs, L. J., and Mikos, A. G., Synthetic biodegradable polymers for medical applications, In *Physical Properties of Polymers Handbook*, Mark, J. E., ed., American Institute of Physics, New York, 1996.
60. Anseth, K. S., Metters, A. T., Bryant, S. J., Martens, P. J., Elisseeff, J. H., and Bowman, C. N., In situ forming degradable networks and their application in tissue engineering and drug delivery, *J. Control. Release*, 78, 199–209, 2002.
61. Halstenberg, S., Panitch, A., Rizzi, S., Hall, H., and Hubbell, J. A., Biologically engineered protein-*graft*-poly(ethylene glycol) hydrogels: a cell adhesive and plasmin-degradable biosynthetic material for tissue repair, *Biomacromolecules*, 3, 710–723, 2002.
62. Wang, D. A., Williams, C. G., Li, Q., Sharma, B., and Elisseeff, J. H., Synthesis and characterization of a novel degradable phosphate-containing hydrogel, *Biomaterials*, 24, 3969–3980, 2003.
63. Burr, D. B., Robling, A. G., and Turner, C. H., Effects of biomechanical stress on bones in animals, *Bone*, 30, 781–786, 2002.
64. Carter, D. R., Beaupre, G. S., Giori, N. J., and Helms, J. A., Mechanobiology of skeletal regeneration, *Clin. Orthop.*, 355(Suppl.), 41–55, 1998.
65. Cheline, A. J., Reddi, A. H., and Martin, R. B., Bone morphogenetic protein-7 selectively enhances mechanically induced bone formation, *Bone*, 31, 570–574, 2002.
66. Grodzinsky, A. J., Levenston, M. E., Jin, M., and Frank, E. H., Cartilage tissue remodeling in response to mechanical forces, *Ann. Rev. Biomed. Eng.*, 2, 691–713, 2000.
67. Kopher, R. A., and Mao, J. J., Sutural growth modulated by the oscillatory component of micromechanical strain, *J. Bone Miner. Res.*, 18, 521–528, 2003.
68. Wang, X., and Mao, J. J., Accelerated chondrogenesis of the rabbit cranial base growth plate upon oscillatory mechanical stimuli, *J. Bone Miner. Res.*, 17, 457–462, 2002.
69. Neidlinger-Wilke, C., Wilke, H. J., and Claes, L., Cyclic stretching of human osteoblasts affects proliferation and metabolism: a new experimental method and its application, *J. Orthop. Res.*, 12, 70–78, 1994.
70. Wozniak, M., Fausto, A., Carron, C. P., Meyer, D. M., and Hruska, K. A., Mechanically strained cells of the osteoblast lineage organize their extracellular matrix through unique sites of alphavbeta3-integrin expression, *J. Bone Miner. Res.*, 15, 1731–1745, 2000.

71. Di Palma, F., Douet, M., Boachon, C., Guignandon, A., Peyroche, S., Forest, B., Alexandre, C., Chamson, A., and Rattner, A., Physiological strains induce differentiation in human osteoblasts cultured on orthopaedic biomaterial, *Biomaterials*, 24, 3139–3151, 2003.
72. Guignandon, A., Akhouayri, O., Usson, Y., Rattner, A., Laroche, N., Lafage-Proust, M. H., Alexandre, C., and Vico, L., Focal contact clustering in osteoblastic cells under mechanical stresses: microgravity and cyclic deformation, *Cell Commun. Adhes.*, 10, 69–83, 2003.
73. Hunter, C. J., Imler, S. M., Malaviya, P., Nerem, R. M., and Levenston, M. E., Mechanical compression alters gene expression and extracellular matrix synthesis by chondrocytes cultured in collagen I gels, *Biomaterials*, 23, 1249–1259, 2002.
74. Lee, C. R., Grodzinsky, A. J., and Spector, M., Biosynthetic response of passaged chondrocytes in a type II collagen scaffold to mechanical compression, *J. Biomed. Mater. Res.*, 64, 560–569, 2003.
75. Simmons, C. A., Matlis, S., Thornton, A. J., Chen, S., Wang, C. Y., and Mooney, D. J., Cyclic strain enhances matrix mineralization by adult human mesenchymal stem cells via the extracellular signal-regulated kinase (ERK1/2) signaling pathway, *J. Biomech.*, 36, 1087–1096, 2003.
76. Goldstein, S. A., Tissue engineering: functional assessment and clinical outcome, *Ann. NY Acad. Sci.*, 961, 183–192, 2002.
77. Mow, V. C., and Wang, C. C., Some bioengineering considerations for tissue engineering of articular cartilage, *Clin. Orthop.*, 367(Suppl.), 204–223, 1999.
78. Shieh, A. C., and Athanasiou, K. A., Principles of cell mechanics for cartilage tissue engineering, *Ann. Biomed. Eng.*, 31, 1–11, 2003.
79. Caplan, A. I., Embryonic development and the principles of tissue engineering, *Novartis Found. Symp.*, 249, 17–25, 2003.
80. Athanasiou, K. A., Agrawal, C. M., Barber, F. A., and Burkhart, S. S., Orthopaedic applications for PLA–PGA biodegradable polymers, *Arthroscopy*, 14, 726–737, 1998.
81. Behravesh, E., Yasko, A. W., Engel, P. S., and Mikos, A. G., Synthetic biodegradable polymers for orthopaedic applications, *Clin. Orthop.*, 367(Suppl.), 118–129, 1999.
82. Temenoff, J. S., and Mikos, A. G., Review: tissue engineering for regeneration of articular cartilage, *Biomaterials*, 21, 431–440, 2000.
83. Middleton, J. C., and Tipton, A. J., Synthetic biodegradable polymers as orthopedic devices, *Biomaterials*, 21, 2335–2346, 2000.
84. Hoffman, A. S., Hydrogels for biomedical applications, *Adv. Drug Deliv. Rev.*, 54, 3–12, 2002.
85. Peppas, N. A., Huang, Y., Torres-Lugo, M., Ward, J. H., and Zhang, J., Physicochemical foundations and structural design of hydrogels in medicine and biology, *Annu. Rev. Biomed. Eng.*, 2, 9–29, 2000.
86. Lammi, M., and Tammi, M., Densitometric assay of nanogram quantities of proteoglycans precipitated on nitrocellulose membrane with safranin O, *Anal. Biochem.*, 168, 352–357, 1988.
87. Isogai, N., Landis, W. J., Mori, R., Gotoh, Y., Gerstenfeld, L. C., Upton, J., and Vacanti, J. P., Experimental use of fibrin glue to induce site-directed osteogenesis from cultured periosteal cells, *Plast. Reconstr. Surg.*, 105, 953–963, 2000.
88. Liu, Y., Chen, F., Liu, W., Cui, L., Shang, Q., Xia, W., Wang, J., Cui, Y., Yang, G., Liu, D., Wu, J., Xu, R., Buonocore, S. D., and Cao, Y., Repairing large porcine full-thickness defects of articular cartilage using autologous chondrocyte-engineered cartilage, *Tissue Eng.*, 8, 709–721, 2002.
89. Yang, S., Leong, K. F., Du, Z., and Chua, C. K., The design of scaffolds for use in tissue engineering. Part I. Traditional factors, *Tissue Eng.*, 7, 679–689, 2001.
90. Yang, S., Leong, K. F., Du, Z., and Chua, C. K., The design of scaffolds for use in tissue engineering. Part II. Rapid prototyping techniques, *Tissue Eng.*, 8, 1–11, 2002.
91. Botchwey, E. A., Dupree, M. A., Pollack, S. R., Levine, E. M., and Laurencin, C. T., Tissue engineered bone: measurement of nutrient transport in three-dimensional matrices, *J. Biomed. Mater. Res.*, 67, 357–367, 2003.
92. Langer, R. S., and Vacanti, J. P., Tissue engineering: the challenges ahead, *Sci. Am.*, 280, 86–89, 1999.
93. Puelacher, W. C., Wisser, J., Vacanti, C. A., Ferraro, N. F., Jaramillo, D., and Vacanti, J. P., Temporomandibular joint disc replacement made by tissue-engineered growth of cartilage, *J. Oral Maxillofac. Surg.*, 52, 1172–1177, 1994.

94. Iwasa, J., Ochi, M., Uchio, Y., Katsube, K., Adachi, N., and Kawasaki, K., Effects of cell density on proliferation and matrix synthesis of chondrocytes embedded in atelocollagen gel, *Artif. Organs*, 27, 249–255, 2003.
95. Goldstein, A. S., Effect of seeding osteoprogenitor cells as dense clusters on cell growth and differentiation, *Tissue Eng.*, 7, 817–827, 2001.
96. Bryant, S. J., and Anseth, K. S., Controlling the spatial distribution of ECM components in degradable PEG hydrogels for tissue engineering cartilage, *J. Biomed. Mater. Res.*, 64, 70–79, 2003.
97. Osborn, K. D., Trippel, S. B., and Mankin, H. J., Growth factor stimulation of adult articular cartilage, *J. Orthop. Res.*, 7, 35–42, 1989.
98. Buckwalter, J. A., and Mankin, H. J., Articular cartilage: tissue design and chondrocyte-matrix interactions, *Instr. Course Lect.*, 47, 477–486, 1998.
99. Hu, K., Radhakrishnan, P., Patel, R. V., and Mao, J. J., Regional structural and viscoelastic properties of fibrocartilage upon dynamic nanoindentation of the articular condyle, *J. Struct. Biol.*, 136, 470–475, 2001.
100. Hunziker, E. B., Articular cartilage repair: basic science and clinical progress. A review of the current status and prospects, *Osteoarthritis Cartilage*, 10, 432–463, 2002.
101. Hunziker, E. B., Quinn, T. M., and Hauselmann, H. J., Quantitative structural organization of normal adult human articular cartilage, *Osteoarthritis Cartilage*, 10, 564–572, 2002.
102. Patel, R. V., and Mao, J. J., Microstructural and elastic properties of the extracellular matrices of the superficial zone of neonatal articular cartilage by atomic force microscopy, *Front. Biosci.*, 8, 18–25, 2003.
103. Sah, R. L., Trippel, S. B., and Grodzinsky, A. J., Differential effects of serum, insulin-like growth factor-I, and fibroblast growth factor-2 on the maintenance of cartilage physical properties during long-term culture, *J. Orthop. Res.*, 14, 44–52, 1996.
104. Tabata, Y., Hong, L., Miyamoto, S., Miyao, M., Hashimoto, N., and Ikada, Y., Bone formation at a rabbit skull defect by autologous bone marrow cells combined with gelatin microspheres containing TGF-beta1, *J. Biomater. Sci. Polym. Ed.*, 11, 891–901, 2000.
105. Elisseeff, J., McIntosh, W., Fu, K., Blunk, B. T., and Langer, R., Controlled-release of IGF-I and TGF-beta1 in a photopolymerizing hydrogel for cartilage tissue engineering, *J. Orthop. Res.*, 19, 1098–1104, 2001.
106. Hanada, K., Solchaga, L. A., Caplan, A. I., Hering, T. M., Goldberg, V. M., Yoo, J. U., and Johnstone, B., BMP-2 induction and TGF-beta1 modulation of rat periosteal cell chondrogenesis, *J. Cell. Biochem.*, 81, 284–294, 2001.
107. Hong, L., Miyamoto, S., Hashimoto, N., and Tabata, Y., Synergistic effect of gelatin microspheres incorporating TGF-beta1 and a physical barrier for fibrous tissue infiltration on skull bone formation, *J. Biomater. Sci. Polym. Ed.*, 11, 1357–1369, 2000.

36 Tissue Engineered Meniscal Tissue

Mark A. Randolph and Thomas J. Gill

CONTENTS

I. INTRODUCTION

Once described as "functionless remnants of leg muscle origin,"[1] the menisci in the knee are now regarded as important structures for knee stability, shock absorption, and load distribution.[2–6] Menisci are semilunar or C-shaped fibrocartilage tissues with specific geometry, composition, and material properties related to their function in the knee.[2] Their location in the knee, and the extreme forces that the menisci can be subjected to, make them frequently susceptible to injury — common in rigorous sports related activities. Trauma often manifests as lesions, such as radial tears or bucket-handle tears, in the fibrocartilage of the meniscus. Due in large part to the limited vascularity of the meniscus, which is restricted to the outer third, this tissue has little innate ability to heal spontaneously.[7,8] Although common in the past, the practice of total meniscectomy has been largely abandoned, except in extreme cases, because there is degeneration of the joint over the long term resulting in pain and discomfort.[9–18] Partial meniscectomy is still practiced if the lesion cannot be satisfactorily stabilized with sutures or anchors.[19,20] Particularly troublesome are lesions to the inner portions of the meniscus where the tissue is not vascularized.

Engineering tissue to repair part or all of an injured meniscus is an emerging strategy for restoring the form and function of meniscal fibrocartilage. Specific consideration must be given to the type of cells necessary, the scaffold, and the physical forces within the microenvironment in which the meniscus is located. Although many paradigms have emerged for engineering cartilage for the articular surface, few attempts have been reported for engineering meniscal tissue or the entire menisci. Important considerations for engineering meniscal tissue include the material properties of the meniscal substance, the geometry of the construct, and the functional integrity of meniscal attachments. Although many attempts to repair meniscus involve transplanting allogeneic menisci or using solid implants made from materials like polyethylene, this chapter will focus on options in which cells are part of the composite structure. This chapter will summarize the anatomy and function of the knee meniscus, explore several possibilities for obtaining cells to engineer meniscal tissue, discuss both biological and synthetic scaffolds, and provide a perspective on the potential for engineering tissue to repair torn menisci or replace the entire meniscus.

II. ANATOMY AND COMPOSITION

The menisci are two semilunar or C-shaped fibraocartilage structures interposed medially and laterally between the femoral chondyle and tibial plateau.[3] In cross-section, they appear wedge-shaped and slightly concave on the anterior or femoral surface and have a thick peripheral region. The menisci first appear as condensations of mesenchymal cells in the human fetus at about 8 weeks gestation while joint cavitation occurs in the developing lower limb.[21] Together, the two menisci cover about two thirds of the tibial plateau. The menisci are nearly devoid of innervation, with the exception of the anterior and posterior meniscal horns which may be related to joint proprioception.[22,23]

The menisci are held in position through the attachments to the joint capsule and ligaments surrounding the knee.[1,3,24,25] These attachments maintain correct positioning of the meniscus during radial and anteroposterior displacement of the knee and provide crucial blood supply to the peripheral regions of the meniscus. Injury to the ligamentous attachments can often lead to malpositioning of the meniscus, subjecting it to injury when the knee is torqued into abnormal positions. Disruption of the ligamentous attachments are a challenging surgical problem in knee reconstruction or with the placement of allogeneic implants. Although the mechanical function and integrity of the ligaments are important to knee function, little attention has been focused on recreating them through engineering approaches.

Blood supply. The human meniscus is only vascularized in the outer 10 to 25% (Figure 36.1).[8] These vascularized outer regions have greater propensity to heal when stabilized with sutures or anchors than the inner avascular regions. The primary blood supply to the medial meniscus is largely derived from the superior and inferior medial geniculate arteries, and the lateral meniscus is supplied by the lateral geniculate arteries. At the microvascular level, a capillary plexis that

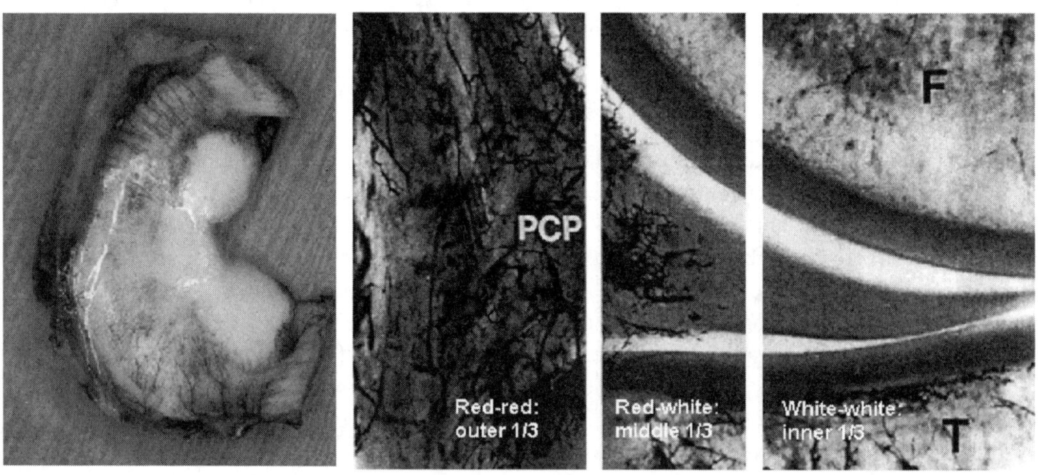

FIGURE 36.1 An India ink injection (left) shows that vessels penetrate the peripheral regions of the medial meniscus in a 3-month-old swine, but that the inner region is nonvascularized. Specimen on the right is a human specimen showing the relative zones of the meniscus and the extent of vessel penetration. (From Arnoczky, S. P., and Warren, R. F., Microvasculature of the human meniscus, *Am. J. Sports Med.*, 10, 90–95, 1982. With permission.)

originates in the joint capsule and synovial tissues surrounding the joint supply the menisci.[8,25–27] India ink injection studies of swine meniscus in our laboratory have shown that vascularization is restricted to the outer half of the young animals (Figure 36.1). In addition to its location in the avascular environment of the knee joint, the limited vascularity of the meniscus probably has a negative effect on its ability to recruit cells to a lesion for normal wound repair. This may be especially true for the nonvascular inner regions.

Composition. About 70 to 75% of the wet weight of the meniscus consists of water.[21,28–30] The dry weight is comprised of 60 to 70% collagen, 1% proteoglycans, and 8 to 13% noncollagenous proteins such as elastin. Collagens are primarily Type I (90%) with smaller amounts of II, III, V, and VI.[28,30] The collagen fibers are predominantly oriented circumferentially, which may be related to the mechanical forces on the menisci, with some radially oriented fibers (Figure 36.2). The meniscus, like auricular cartilage, has elastin fibers throughout that bridge the collagen fibers. There are also inhomogeneities in tissue composition from the peripheral vascular regions to the inner avascular regions. This is appreciated by the histological appearance of the inner portion that resembles articular cartilage, whereas the outer portions are more fibrocartilage-like (Figure 36.3).

III. FUNCTION

The interposition of the menisci between major weight-bearing bones of the body, the tibia and femur, are important constraints for engineering meniscal tissues. The complex geometry of the

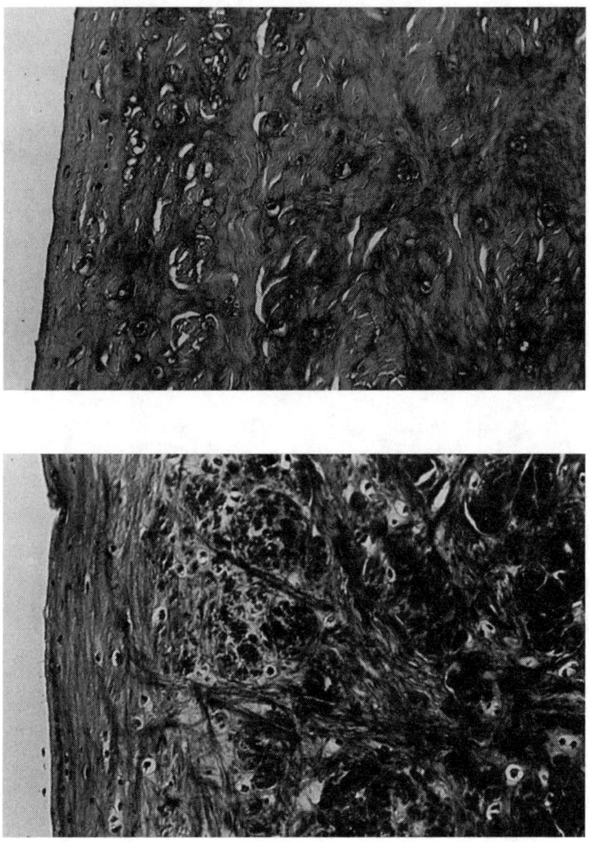

FIGURE 36.2 Photomicrographs of a swine meniscus showing the collagen fiber distribution in the cartilage matrix. The cells on the meniscal surface to the left are fusiform whereas cells deeper in the matrix appear more rounded and some are in distinct lacunae (top: H&E, ×100; bottom: Massons Trichrome, ×100).

FIGURE 36.3 Photomicrograph of the entire swine meniscus showing the spatial differences in the cells and matrix from the peripheral regions on the left to the inner, avascular regions on the right (H&E, ×5).

menisci provides congruity between the articulating surfaces of the distal femur and proximal tibia to allow for load bearing and shock absorption.[3,18] The meniscus also plays a role in the distribution of nutrients to the articular cartilage.[21] The material properties of the various regions of the meniscus are determined largely by the composition and microstructure of the tissue. These properties are highly anisotropic and different in compression and tension. They also vary considerably, depending on the depth and circumferential location of the force in the tissue.[31] Approximately 50% of the compressive load is transmitted through the menisci in extension, and the load can increase to as much as 85% as the knee goes to flexion.[32] In humans, the medial meniscus bears about 50% of load whereas the lateral meniscus bears as much as 70%.[33] Under normal physiologic loading, the meniscus experiences large tensile and shear stresses as well as compressive stress. Under loaded conditions the meniscus is not only subjected to a vertical force to the inner portions of the menisci, there is also a radially directed force applied to the concave surface pushing the meniscus outward.[3,34] The strong peripheral attachments of the meniscus counteract this circumferentially directed force and the associated tensile hoop stress in the collagen fibers of the meniscal substrate (Figure 36.4).

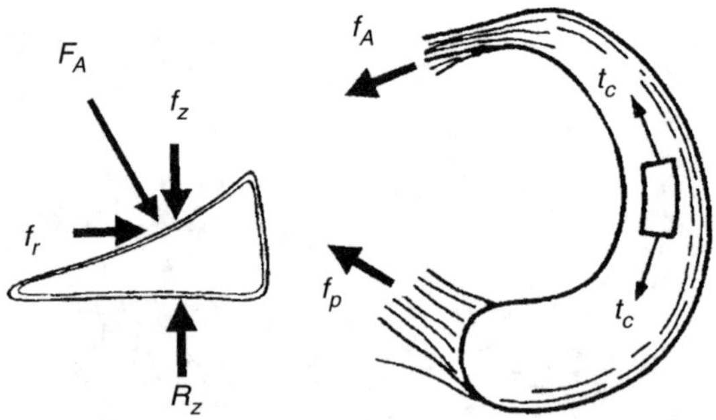

FIGURE 36.4 Schematic of the meniscus showing a joint load (F_A) transmitted normally to the proximal meniscal surface and the associated reaction force at the tibial plateau (R_2). An outward-directed force resultant (f_r) is balanced by forces generated in the anterior and posterior horns (f_a and f_p). Together these forces generate tensile hoop stress tractions (t_c) and axial and radial components of stress in the meniscus during loading. (From Setton et al., Biomechanical factors in tissue engineered meniscal repair, *Clin. Orthop.*, 367(Suppl.), S256, 1999. With permission.)

One can appreciate the extreme forces the meniscus is subject to during sporting activities, like alpine skiing, where severe torque can be placed on the knee joint. The extreme internal stresses applied to the meniscus when the limb is forced into abnormal positions can cause lesions of various types. These forces should be taken into consideration when attempting to engineer meniscal tissue.

IV. MENISCAL CELLS

The cellular population of the meniscus is heterogeneous, consisting of chondrocytes and fibrochondrocytes. The cells in the superficial layers of the tissue are fusiform, whereas cells in the deeper zones are more polygonal (Figure 36.2). Cell morphology also changes from the innermost avascular region, where the cells are nearly indistinguishable from articular chondrocytes, to the outer vascular regions, where the cells are more fibroblast-like (Figure 36.3). Although the cells share many morphologic similarities to articular chondrocytes, they predominantly produce Type I collagen. Thus, the cells are most often referred to as fibrochondrocytes.[35]

Meniscal cells can be isolated by enzymatic digestion of the tissue. The cells have been shown to retain their phenotypic expression over time under defined culture conditions.[35–39] They are also capable of expressing their differentiated phenotype when cultured suspended in a three-dimensional matrix.[36] Meniscal fibrochondrocytes from several species have been seeded onto scaffolds in attempts to generate meniscal tissue.[40] Histological evidence suggests that the meniscal cells were capable of generating fibrocartilagenous tissue resembling meniscus. It is not clear if the radial and circumferential collagen fibers, which contribute to the biomechanical properties of the meniscus, were produced in their normal spatial orientation. It is unlikely, however, that autologous meniscal fibrochondrocytes will be used to repair or engineer one's own menisci because there are no expendable donor sites. It is conceivable that allogeneic meniscal cells could be used, and this option will be discussed below.

V. CELL SOURCES FOR TISSUE ENGINEERING

The primary obstacle for engineering cartilage of any type is acquiring the appropriate numbers of chondrogenic cells for generating the tissue. The seeding density of cells capable of chondrogenesis in or onto a polymer carrier is a critical ingredient for successfully engineering cartilage. Although there is precedent for harvesting noninvolved joint cartilage to regenerate cartilage on the articular surface,[41,42] few would risk morbidity to normal meniscus to obtain cells for engineering new meniscal tissue. It is evident that the type of cells, the number of cells, and the physical environment into which they are placed are critical elements for successfully engineering meniscal cartilage tissue.

Suspending chondrocytes in a three-dimensional matrix, similar to their natural environment, can permit the cells to retain their native phenotype and produce their extracellular components. By restoring the three-dimensional spherical shape of the chondrocyte, its characteristic phenotype can be re-expressed.[43] Chondrocytes regain their spherical shape after being placed in biodegradable hydrogel polymers and allowed to begin producing their characteristic matrix macromolecules.[44–46] Using polymers that undergo a controllable bulk erosion process *in vivo*, the polymer can be made to resorb at a rate proportional to the rate at which cartilaginous extracellular matrix is being deposited into the intercellular spaces. When properly orchestrated, the result is the generation of cartilage with its characteristic microarchitecture, guided by the chondrocyte's intrinsic programming and facilitated by the polymer's engineering.[47] The influence of scaffolds on tissue formation will be covered in more detail below.

Chondrocytes from other areas of the body could be used to generate meniscal tissue or used to induce meniscal repair. For example, we have shown that articular chondrocytes are capable of forming new cartilage and forming bonds between cartilage discs and meniscal lesions *in vivo*.[48,49]

Extending these findings to a large animal model, we have shown that articular chondrocytes seeded onto a devitalized cartilage matrix are capable of inducing a healing process in a bucket-handle lesion of the meniscus in swine.[50] Other alternative cartilage sources for obtaining cells include the cartilage portion of ribs, which are expendable, or possibly small biopsies of ear cartilage. Chondrocytes from these sources have been shown to generate new cartilage matrix and are capable of forming bonds between cartilage discs.[51] These alternative cell sources could be used to engineer reparative meniscal tissue.

The use of other noncartilage cells with chondrogenic potential may permit the generation of cartilage given the correct polymer and appropriate conditions. For example, Mizuno et al have demonstrated that human dermal fibroblasts, seeded onto collagen Type I sponges containing demineralized bone matrix and grown in culture, can be stimulated to produce a cartilage-like tissue including collagen Type II.[52] Lorenz and colleagues have recently described the chondrogenic potential of cells derived from lipoaspirates when the appropriate culture conditions were presented *in vitro*.[53] Caplan and colleagues have published extensively on the use of bone marrow-derived, mesenchymal stem cells (MSC) to generate various musculoskeletal tissues, including cartilage, when these cells are given the appropriate signals.[54]

Multiple experiments using human, chicken, dog, and rabbit MSCs have shown that under controlled *in vitro* conditions these cells can differentiate into bone, fat, tendon, muscle and cartilage-like tissues.[55–63] However, conditions for differentiating the cells are somewhat species dependent. Chondrogenesis has been achieved by culturing MSC with chondrogenic media containing transforming growth factor beta (TGFβ), dexamethazone, insulin-like growth factor (IGF), basic fibroblast growth factor (bFGF), and other growth factors. Chondrogenic differentiation was evidenced by demonstration of cartilage extracellular matrix formation (Safranin-O staining and inmumohistochemistry using antibodies to Type II and X collagen) and expression of genes characteristic of chondrocytes (Type II and X collagen mRNA).[60,61] Unlike the *in vitro* differentiation conditions for osteogenesis and adipogenesis, chondrogenesis was obtained when the MSC were grown as a pelleted cell micromass, but not in monolayer culture.[61] This finding suggests that chondrogenesis by MSCs requires a three-dimensional environment much like that of their native microenvironment. These results demonstrate that cells from different sources can be used to engineer cartilage, possibly eliminating the need to collect chondrocytes directly from cartilage that could, in essence, cause unwarranted morbidity to fragile tissues. However, few *in vivo* studies on cartilage generation and repair using MSCs have been reported, and none have reached clinical application.

One option that has not been exploited is using chondrocytes from allogeneic or xenogeneic sources for engineering cartilage. Whereas the use of allogeneic and xenogeneic cartilage en block has not enjoyed favorable results clinically, the use of isolated chondrocytes from these sources may perform satisfactorily under the appropriate conditions.[64–69] The central issue involving the use of allogeneic or xenogeneic chondrocytes is their potential for eliciting an immune response once the extracellular matrix is removed and MHC antigens are exposed.[70–73] Therefore, means to stifle the immune response are necessary if allogeneic or xenogeneic chondrocytes are used to engineer cartilage. It may be possible to develop scaffolds that will permit the formation of new extracellular matrix while simultaneously preventing immune rejection of the isolated chondrocytes.

VI. SCAFFOLDS

The other critical element for engineering meniscal cartilage is finding or developing suitable scaffold materials that permit or accelerate the formation of new extracellular matrix. Using polymers, both natural and synthetic, that undergo controllable bulk erosion or resorption can be favorable for engineering cartilage tissues *in vitro* or *in vivo*. For example, polymers that degrade at a rate proportional to which cartilaginous extracellular matrix is being deposited into the intercellular spaces could be employed to generate meniscal tissue *in situ*. Several scaffolds, both natural and synthetic, have been tested in animal models for regenerating meniscus tissue. Whereas many favorable polymers are open lattice structures with large pores into which cartilage matrix is permitted to form, new synthetic hydrogels are also good candidate scaffolds for generating meniscal tissue. There has been more experience using articular chondrocytes to generate cartilage, but many of the same principles apply for engineering meniscal tissue.

A. Synthetic Polymers

Many early investigations for engineering cartilage from synthetic polymers focused on the use of polyesters of poly(α-hydroxy esters).[74] Biodegradable polyesters such as poly(L-lactic acid) (PLLA), poly(glycolic) acid (PGA), and the copolymer poly(DL-lactic-co-glycolic acid) (PLGA) possessed many desired properties to support cell transplantation. These polymers can be formed into open lattice structures with high porosity that allows the free exchange of nutrients and waste products. The lattice structure can be adjusted and provides a high surface area to volume ratio to allow for matrix production in the open interstices, and both PGA and PLLA have been used successfully to generate cartilage *in vitro* and *in vivo*. These polyesters degrade by hydrolysis into nontoxic metabolites and can be tailored to degrade at defined rates.

Polymer constructs made from PGA have been evaluated extensively by several investigators.[75–80] Constructs assessed *in vitro* using phase contrast microscopy demonstrate that chondrocytes adhere in multiple layers to the branching PGA polymer fibers and retain the rounded morphologic appearance typical of chondrocytes. Electron micrographs show the rounded chondrocytes attached to the polymer fibers after *in vitro* culture (Figure 36.5). Attachment of chondrocytes is enhanced if the polymer fibers are coated with PLLA, which is more hydrophilic. Freed and Vunjak-Novakovic have successfully grown cartilage *in vitro* using PGA polymer.[78] Crucial to matrix production *in vitro*, however, is the placement of the constructs in a dynamic environment.[79,80]

PGA polymer constructs seeded with chondrocytes have been implanted into various animal models by numerous investigators. Most studies involve implantation of cell/polymer constructs into subcutaneous pockets in nude mice. Kim et al. have demonstrated that implants of PGA and chondrocytes designed into specific shapes can produce cartilage matrix and retain these shapes during *in vivo* incubation in nude mice (Figure 36.6).[75] Using bovine meniscal fibroblasts and PGA scaffolds, Ibarra et al. described the generation of meniscal-like tissue when constructs were placed subcutaneously on the dorsum of nude mice. Like their predecessors, they reported that the constructs were able to retain their original shape and produced a matrix histologically similar to meniscal tissue (Figure 36.7).[81] Subsequent studies with meniscal fibrochondrocytes from ovine, canine, and human sources have shown similar results in the nude mouse model.[40] These studies demonstrate that biocompatible polymer fibers seeded with chondrocytes and meniscal fibrochondrocytes will form new cartilage tissue *in vivo* in nude mice models.

Ibarra et al. also reported on a pilot study in sheep where they used autologous meniscal fibrochondrocytes and PGA polymer.[40] Cells were harvested by performing a meniscectomy on the left hindlimb, enzymatically digesting the tissue, and expanding the cells in monolayer culture. The cells were seeded onto PGA scaffolds for several days and then implanted into a subcutaneous position for 4 weeks. After this period of subcutaneous incubation, the constructs were transferred to the right hind limb to replace the medial meniscus. The authors report that the constructs produced a new tissue with fibroblasts and chondrocytes. The presence of collagen fibers was observed histologically and the cells produced proteoglycans. The number of animals was small, but the authors demonstrated proof of principle for the technique. No follow-up studies have been reported.

FIGURE 36.5 Electron micrographs of PGA polymer fibers (top) sowing the open lattice structure. The sheep chondrocytes attached to the polymer are rounded and producing extracellular matrix (bottom).

B. Biological Scaffolds

Collagen is the prevalent structural biomolecule in the extracellular matrix of cartilage, making it a logical choice for composing a tissue engineering scaffold. Collagen sponges have many desirable properties as a biological scaffold for cartilage including porosity, biodegradability, and biocompatibility. Multiple methods for preparing collagen and collagen–GAG scaffolds have been reported. Generally, collagen scaffolds are made from animal tissues such as Type I collagen from bovine tendon. It is also possible to chemically modify the biomechanical and biological properties of the collagen scaffolds to enhance certain characteristics that promote cartilage formation.[82] Open lattice collagen scaffolds, some of which also include glycosaminoglycans, have been synthesized and used for generating new cartilage matrix. Scaffolds made from a single collagen type or composites of two or more types have been employed.

In their simplest form, collagen scaffolds can be used without adding cells and implanted into meniscal defects.[83] These scaffolds permit migration of cells into the scaffold network and

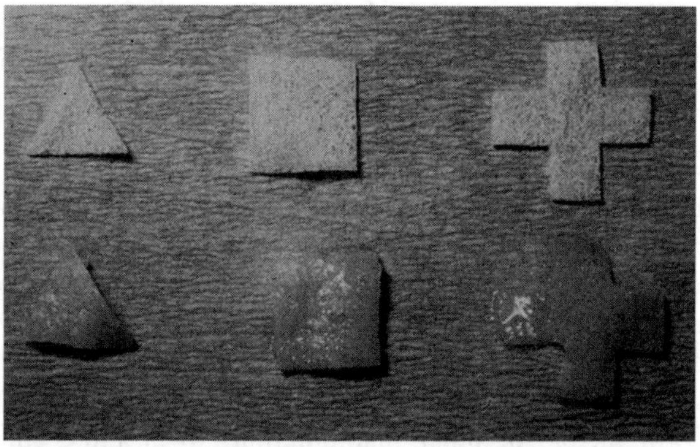

FIGURE 36.6 Chondrocytes seeded onto PGA polymer can be made into specific shapes determined by the geometry of the polymer. The top row is polymer cut into predesigned shapes. The bottom row shows the cartilage formation after *in vivo* incubation in nude mice. (From Kim et al., Cartilage engineered in predetermined shapes employing cell transplantation on synthetic biodegradable polymers, *Plast. Reconstr. Surg.*, 94, 233–237, 1994. With permission.)

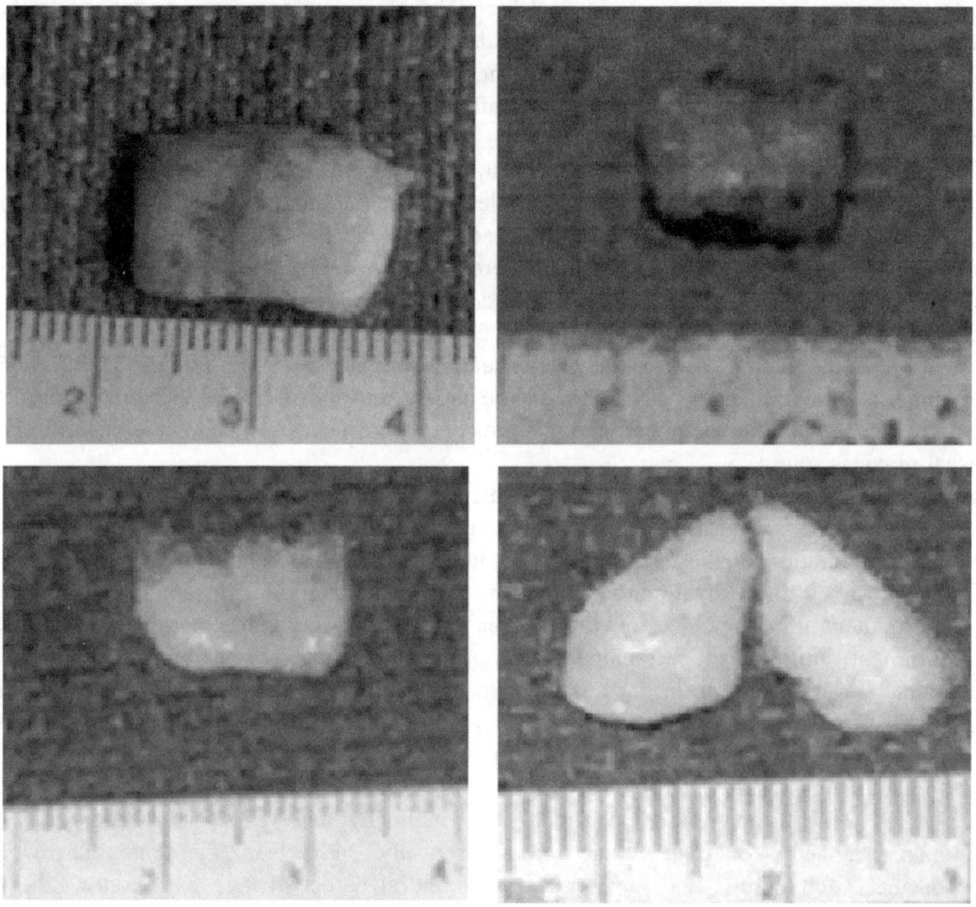

FIGURE 36.7 Examples of meniscal fibrochondrocytes from various species seeded onto PGA polymer and incubated *in vivo* in nude mice. From left to right: bovine, ovine, human, and canine. (From Ibarra et al., Tissue engineering meniscus: cells and matrix, *Orthop. Clin. North Am.*, 31(3), 413, 2000. With permission.)

formation of new meniscal cartilage matrix. Such scaffolds have been used with favorable results in a phase II clinical study to replace the inner portion of the meniscus that has been removed following trauma.[84] These scaffolds are not indicated for replacement of the entire meniscus, however.

Collagen scaffolds seem to show favorable results with regard to chondrocyte adherence and their ability to maintain a differentiated chondrocyte phenotype.[85,86] In comparison to other open lattice synthetic scaffolds, it appears that collagen sponges promote collagen production whereas synthetic polymers such as PGA promote proteoglycan synthesis.[87] Additionally, the porosity of the sponge and the surface characteristics of the meshwork are critical elements to consider for promoting cell growth and cartilage formation. Some characteristics, such as pore size, can be changed by varying the freeze-drying process or cross-linking.[86] Collagen sponges can be also be modified using growth factors or other manipulations to promote chondrocyte growth and cartilage matrix formation. For example, sponges can be impregnated with exogenous growth factors such as bFGF that promote matrix production.[88] Studies from Mizuno and Glowacki have shown that collagen sponges can also incorporate other components that induce differentiation of nonchondrocyte cells to differentiate and form cartilage. For instance, those studies demonstrated that the addition of demineralized bone powder to the collagen sponges would stimulate human dermal fibroblasts to synthesize an extracellular matrix resembling that of cartilaginous tissue.[52,89] Despite their appeal as a biological material for making a tissue engineering scaffold, collagen sponges have some disadvantages. These scaffolds can cause a foreign body reaction with a thin fibrous capsular tissue surrounding collagen sponge implants.[88] Such reactions may interfere with the integration of new cartilage formed in the scaffold with the surrounding native cartilage in the recipient site.

Our group has developed a scaffold of devitalized meniscal cartilage matrix that can be used as a scaffold for delivering cells into a bucket-handle lesion in the meniscus of swine. Once the cells are seeded onto the surface of the scaffold, the implant is secured into a lesion in the swine meniscus. Results using this implant were very favorable in a preliminary study using swine meniscus implanted on the dorsum of nude mice.[90] The results *in situ* in the swine knee were promising, despite incomplete healing.[50] The devitalized meniscal cartilage scaffold was integrated into the lesion with satisfactory healing on at least one margin (Figure 36.8). Interestingly, the devitalized matrix appeared to be undergoing a form of creeping substitution. This pilot study in a large animal demonstrated that the concept for using tissue engineered cartilage for meniscal repair is sound.

Hyaluronan is one of the major constituents of undifferentiated mesenchyme in the developing embryo, as well as in the cartilage extracellular matrix. Hyaluronan has been shown to support proliferation of mesenchymal progenitor cells and differentiation into chondrocytes. Additionally, in cartilage hyaluranon is believed to play a significant role in the physical microenvironment affecting chondrocyte function.[91–93] It is also possible that hyaluronan creates an environment during cartilage repair that recapitulates processes occurring during embryonic development. Hyaluronan can be formulated into many different chemical and physical entities that provide a favorable environment for cartilage generation, allowing both synthesis of matrix components and differentiation of the progenitor cells.[94,95] A recent study from Solchogá et al.[96] showed encouraging results in the *in vivo* treatment of an osteochondral defect in a rabbit knee model using a hyaluronan-based polymer. The main hypothesis of their study was that hyaluronan fragments could encourage migration of the MSC from the bone marrow to a chondrogenic differentiation and repair the osteochondral defect in the rabbit condyle. Results showed a higher reparative rate in the lesion repaired by the polymer sponges than those untreated or those treated with HYAFF-11®. Although hyaluronan-based polymers have some favorable attributes, there are no reports in the literature documenting the use of hyaluronan for engineering meniscal tissue.

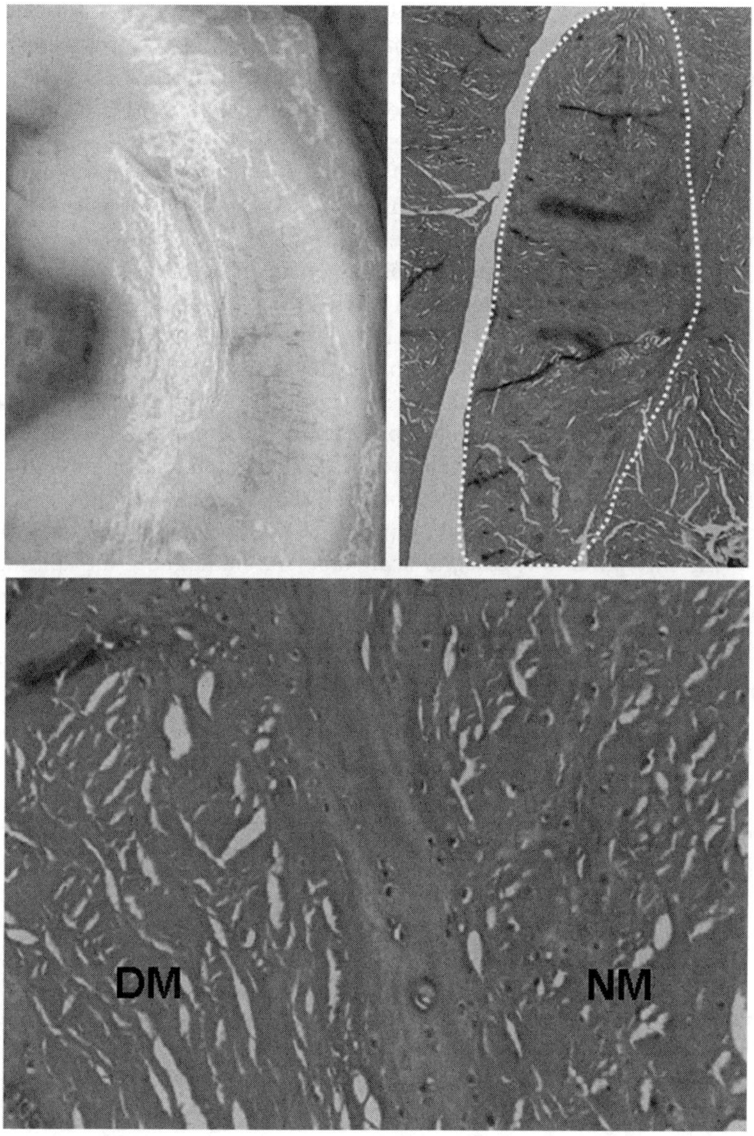

FIGURE 36.8 Meniscus from a swine with a bucket-handle lesion treated with a chondrocyte-seeded scaffold for 9 weeks. The meniscus appears to be closed on gross inspection (top left). Histologically, the scaffold (outlined with the white dotted line) is adhered to the meniscus on only one margin (H&E; ×25). A magnified view shows integration of the scaffold implant (DM) with the native matrix (NM) (H&E; ×200).

C. Hydrogel Scaffolds

The inability to deliver chondrocytes through minimally invasive techniques when using fibrous or open lattice-type polymers stimulated investigations into other types of polymer carriers such as hydrogels. Hydrogels are gelatinous colloids that, when maintained under controlled conditions, exhibit three-dimensional stability. They are produced by mixing a soluble polymer in water and adding a cross-linking agent to gel the mixture. As the polymer gels, there is usually sufficient opportunity to mold and shape the final three-dimensional configuration of the gel. Additionally, by existing in a liquid phase, these polymers have the potential for injectable delivery. The combined high water content and elasticity of polymer hydrogels leads to these materials having many tissue-like properties, making them ideal candidates for tissue engineering matrices. Hydrogels have been proven to be extremely effective in providing a hospitable, three-dimensional support matrix for the immobilization of cells. By suspending chondrocytes in a highly porous aqueous matrix, they can maintain their differentiated function and are capable of producing large quantities of extracellular matrix macromolecules.[46,97] Examples of hydrogels used to encapsulate chondrocytes include ionically cross-linked alginates[46,98–100] and chitosan,[101,102] hydrogen bonded block copolymers such as Pluronics,[103] and covalently cross-linked fibrin glue.[104]

One of the most basic hydrogel scaffolds is fibrin gel, which is formed through the interaction of thrombin and fibrinogen in a blood clot. Fibrin, which can be obtained from autologous blood products and is favorably biocompatible, can be formulated as an injectable vehicle with degradation controlled using agents that slow fibrinolysis, like aprotinin.[105,106] When fibrin forms a blood clot, platelets and blood cells are entrapped in a serum rich gel containing various cytokines and growth factors that promote migration and proliferation of fibroblasts. Arnoczky et al. reported on the use of an exogenous fibrin blood clot to successfully repair punch biopsy lesions in the avascular zone of canine menisci.[107] They concluded that the cells migrating from the adjacent meniscal tissue, surrounding synovium and joint capsule contributed to the repair. They described the tissue as passing through a fibrovascular scar stage and undergoing metaplasia into meniscal fibrocartilage. Similar work has been reported on goat meniscus where bone marrow cells were deliberately mixed into the exogenous blood clot.[108]

Subsequent studies in our laboratory have focused on using chondrocytes in fibrin gel polymer as a hydrogel scaffold for engineering cartilage.[105,109–111] Silverman et al. investigated the possibility of using a fibrin glue polymer to produce injectable tissue engineered cartilage but reported that a significant reduction in volume (>60%) occurred after implantation of swine articular chondrocytes into nude mice.[105] Balance between the absorption of polymer scaffold and the production of cartilaginous matrix is essential for controlling the volume of chondrocyte-fibrin glue constructs. Whereas many early studies used (juvenile bovine) articular chondrocytes, current research is focusing on auricular and costal chondrocytes and their ability to synthesize a matrix that maintains the implant volume in the subcutaneous environment.[112] Significant disadvantages of fibrin gel are inconsistency in polymerization, which is related to the fibrin content in human blood products, and the fragile nature of the gel. The latter limits the use of fibrin in meniscal repair where extreme mechanical forces can fracture the gel or squeeze it out of the defect.

Investigators have sought alternative synthetic gels in which the chemical parameters can be carefully controlled because biological hydrogels, like fibrin glue and alginate, are highly variable. Sims initially demonstrated the capacity for generating cartilage using a nonpolymerized form of poly(ethylene oxide) (PEO), a linear polyether with repeating molecular units of $(-CH_2CH_2O-)_n$.[113] PEO molecules can be cross-linked by adding methacrylate groups to PEO backbone and using a photo-sensitive initiator to form cross-links between molecules when activated with ultraviolet light. *In situ* polymerization can permit such hydrogels to be sculpted into desired shapes in defect sites.[114,115] For example, Elisseeff et al. demonstrated that chondrocyte–PEO constructs could be injected subcutaneously, molded to the desired shape, and then polymerized transdermally with ultraviolet light.[116,117] Other synthetic hydrogels have been used to generate cartilage as well. Ashiku and others have investigated the use of pluronics, a thermosensitive copolymer gel made from combinations of poly(ethylene oxide) and poly(propylene oxide), to encapsulate chondrocytes for injection to generate cartilage.[103,118] Poly(vinyl alcohol) (PVA) is another hydrogel with desirable chemistry that allows easy modification of the macromer backbone and has a long history in medical applications.[119–121] These polymers and gelation processes can be designed to supply mechanical and structural stability with desirable transport properties during the regeneration process, and to allow the formation of complex shapes with suitable adhesion to treat joint defects.[122–124] Rational design of synthetic materials could optimize the immunoprotective capacity of the photopolymerizeable gels through modifications of the network structure and chemistry.

The physical and mechanical behavior of a gel is highly dependent on the cross-linking density that controls the mesh size of the network, and the backbone chemistry, which controls the equilibrium water uptake. Alginate forms a stable gel by ionic cross-linking in the presence of divalent cations, such as calcium or barium.[100] Pluronics, a copolymer of poly(ethylene oxide) and poly(propylene oxide), forms a gel through physical cross-links by hydrophobic group association at higher temperatures of 37°C.[103] However, when considering the encapsulation of cells in hydrogel matrices for cartilage engineering, limitations exist with these current materials and encapsulation technologies. Specifically, limitations exist with respect to controlling the mechanics and uniformity of the gels, as well as spatial and temporal control during gel formation, which is especially important in injectable systems. For example, gels formed with Pluronics tend to lose their integrity under load because of their weak physical cross-links.[103,118] With alginate and fibrin glue, two components are mixed and reacted to form the gel, thus uniformity of the gel can be difficult to achieve. In alginate gelation, the outer layers of the gel are initially cross-linked, forcing the calcium ions to diffuse through the outer cross-linked layers in order to continue cross-linking the inside of the gel.[105] This diffusion limitation can result in unequal cross-linking throughout the gel.[100] Additionally, in each of these systems, it is difficult to maintain temporal control over the gelation process, since cross-linking begins as soon as the two components are mixed. In pluronics, gelation occurs once the polymer solution is injected inside the body as a result of the increase in temperature. Loss of temporal control can lead to spatial inhomogeneities as cells can settle before gelation occurs in the low viscosity solutions. In addition to temporal control of the gelation, these techniques also lack spatial control, which is often important in generating complex architectures that might be required *in vivo*.

Synthesis of scaffolds using polymers like poly(ethylene) glycol or polyvinyl alcohol could permit controlled cartilage matrix production. One possible means to obtain uniform cell distribution and homogeneous gel properties throughout the matrix during gelation is to employ photo-crosslinking with polymers like PEO.[116,117] PEO or PVA could be selected for the macromer backbone because of their long history in medical applications and desirable chemistry, which allow easy modification. These materials and techniques show that chondrocytes survive photoencapsulation, and permit the production of extracellular matrix.[116] It is possible to rationally design the materials to optimize the chondrogenic capacity of the photopolymerizeable gels through modifications to the network structure and chemistry. By using light to photoencapsulate the cells, gelation occurs on demand by simply shuttering the light source and only occurs in regions that are exposed to the light, thereby providing spatial control. As the polymer solidifies there is sufficient opportunity to mold and shape the final configuration of the hydrogel.

D. Inert Nonresorbable Materials

The engineering of cartilage to fill articular surface defects might easily be accomplished using a single polymer scaffold such as a hydrogel. The meniscus presents different challenges however, because of its inhomogeneities and its ligamentous attachments to the knee structures. As such, engineering meniscal tissue may require more than one degradable polymer and possibly even nondegradable elements in the scaffold. Investigators have demonstrated numerous techniques for improving the biological and biomechanical properties of tissue engineered cartilage. Some key strategies for improving these properties include techniques to: (1) improve the bioproperties of extracellular cartilaginous matrix, (2) provide internal support to tissue engineered cartilage, and (3) add external (pseudoperichondrium) support to tissue engineered cartilage. One objective is to improve the flexibility of a tissue engineered cartilage framework, particularly in cartilage tissues other than articulating joint cartilage. This section will discuss some possibilities for enhancing the neocartilage matrix properties by incorporating nonresorbable materials to meet the needs for reconstruction.

Cao et al. engineered cartilage in the shape of human auricles in nude mice using articular joint chondrocytes and a biodegradable internal PGA/PLLA scaffold to attain the desired shape of an ear.[125] Arevalo-Silva and colleagues[126] investigated the use of nonbiodegradable endoskeletal scaffolds made from the following materials: (1) high-density polyethylene, (2) soft acrylic, (3) polymethylmethacrylate, (4) extrapurified silastic, and (5) conventional silastic. They concluded that using a permanent biocompatible endoskeleton demonstrated success in limiting the inflammatory response to the scaffold, especially the high-density polyethylene, acrylic, and extrapurified silastic. Despite the success of using these materials, the investigators did not comment on the flexibility of the constructs or other biomechanical properties of the neocartilage.

Similarities in composition between auricular and meniscal cartilage, such as elastin, provide a basis for examining nonabsorbable or slowly absorbing materials in the production of meniscal tissue. However, extrusion and erosion could be limiting factors because of the motion and extreme forces placed on the meniscus. Expanded polytetrafluoroethylene (ePTFE) is a biocompatible material that has been used successfully in a multitude of biomedical and clinical applications.[127–129] One advantage of this material is its microporous structure that allows biointegration for soft tissue fixation, as well as overall mechanical integrity.[130] We have employed ePTFE to provide flexibility for making tissue engineered cartilage for craniofacial repair.[131] Extrapolating these results to meniscal tissue, ePTFE could be used as an endoskeleton to add internal mechanical integrity and flexibility to a meniscal implant.

VII. CONCLUSIONS

The geometric complexity of the meniscus, the composition of the tissue, and the forces on the meniscus present many obstacles for engineering a substitute. Finding suitable cell sources, providing a blood supply to the periphery, and developing new polymers that can withstand the rigor of the mechanical forces in the knee are key to successfully engineering meniscal tissue. At the present time, however, successful strategies are being developed using tissue engineering approaches to heal meniscal tears. Clinical application of these approaches appears to be in the very near future.

REFERENCES

1. Bland Sutton, J., *Ligaments, Their Nature and Morphology*, 2nd ed., HK Lewis, London, England, 1897.
2. Ahmed, A. M., and Burke, D. L., In vitro measurement of static pressure distribution in synovial joints: part I. Tibial surface of the knee, *J. Biomech. Eng.*, 105, 216–225, 1983.
3. King, D., The function of semilunar cartilages, *J. Bone Joint Surg.*, 18, 1069–1076, 1936.
4. Krause, W. R., Pope, H. M., Johnson, R. J., and Wilder, D. G., Mechanical changes in the knee after meniscectomy, *J. Bone Joint Surg.*, 58A, 599–604, 1976.
5. Levy, I. M., Torzilli, P. A., Gould, J. U. D., and Warren, R. F., The effect of lateral meniscectomy on motion of the knee, *J. Bone Joint Surg.*, 71A, 401–406, 1989.
6. Seedhom, B. B., and Hargreaves, D. J., Transmission of the load in the knee joint with special reference to the role of menisci: part II. Experimental results, discussion and conclusions, *Eng. Med.*, 8, 220–228, 1979.
7. Arnoczky, S. P., and Warren, R. F., Microvasculature of the human meniscus, *Am. J. Sports Med.*, 10, 90–95, 1982.
8. Arnoczky, S. P., and Warren, R. F., The microvasculature of the meniscus and its response to injury: an experimental study in the dog, *Am. J. Sports Med.*, 11, 131–141, 1983.

9. Allen, P. R., Denham, R. A., and Swan, A. V., Late degenerative changes after meniscectomy: factors affecting the knee after operation, *J. Bone Joint Surg.*, 66B, 666–671, 1984.
10. Appel, H., Late results after meniscectomy in the knee joint: a clinical and roentgenologic follow up investigation, *Acta Orthop. Scand.*, 133(Suppl.), 1–111, 1970.
11. Berjon, J. J., Munuera, L., and Calvo, M., Degenerative lesions in the articular cartilage after meniscectomy: preliminary experimental study in dogs, *J. Trauma*, 31, 342–350, 1991.
12. Ghosh, P., Sutherland, J., Bellenger, C., Read, R., and Darvodelsky, A., The influence of weight-bearing exercise on articular cartilage of meniscectomized joints: an experimental study in sheep, *Clin. Orthop.*, 252, 101–113, 1990.
13. Hoch, D. H., Grodzinsky, A. J., Koob, T. J., Albert, M. L., and Eyre, D. R., Early changes in material properties of rabbit articular cartilage after meniscectomy, *J. Orthop. Res.*, 1, 4–12, 1983.
14. Huckell, J. R., Is meniscectomy a benign procedure? A long term follow up study, *Can. J. Surg.*, 8, 254–260, 1965.
15. Jackson, J. P., Degenerative changes in the knee after meniscectomy, *Br. Med. J.*, 2, 525–527, 1968.
16. Lufti, A. M., Morphological changes in the articular cartilage after meniscectomy. An experimental study in the monkey, *J. Bone Joint Surg.*, 57B, 525–528, 1975.
17. Northmore-Ball, M. D., and Dandy, D. J., Long term results of arthroscopic partial meniscectomy, *Clin. Orthop.*, 167, 34–42, 1982.
18. Voloshin, A. S., and Wosk, J., Shock absorption of meniscectomized and painful knees: a comparative in vivo study, *J. Biomed. Eng.*, 5, 157–161, 1983.
19. Sommerlath, K. G., Results of meniscal repair and partial meniscectomy in stable knees, *Int. Orthop.*, 15, 347–350, 1991.
20. Shelbourne, K. D., and Carr, D. R., Meniscal repair compared with meniscectomy for bucket-handle medial meniscal tears in the anterior cruciate ligament-repaired knee, *Am. J. Sports Med.*, 31(5), 718–723, 2003.
21. Arnoczky, S. P., Adams, M. E., DeHaven, K., Eyre, D. R., and Mow, V. C., The meniscus, In *Injury and Repair of the Musculoskeletal Soft Tissues*, Woo, S. L.-Y., and Buckwalter, J., eds., Academy of Orthopaedic Surgeons, Park Ridge, IL, pp. 487–537, 1988.
22. O'Connor, B. L., and McConnaughey, J. S., The structure and innervation of cat knee menisci and their relation to a "sensory hypothesis" of meniscal function, *Am. J. Anat.*, 153, 431–442, 1978.
23. Wilson, A. S., Legg, P. G., and McNeur, J. C., Studies on the innervation of the medial meniscus in the human knee joint, *Anat. Rec.*, 165, 485–491, 1969.
24. Setton, L. A., Guilak, F., Hsu, E. W., and Vail, T. P., Biomechanical factors in tissue engineered meniscal repair, *Clin. Orthop.*, 367(Suppl.), S245–S272, 1999.
25. Renstrom, P., and Johnson, R. J., Anatomy and biomechanics of the menisci, *Clin. Sports Med.*, 9, 523–538, 1990.
26. Cooper, D. E., Arnoczky, S. P., and Warren, R. F., Arthoscopic meniscal repair, *Clin. Sports Med.*, 9, 589–607, 1990.
27. Cooper, D. E., Arnoczky, S. P., and Warren, R. F., Meniscal repair, *Clin. Sports Med.*, 9, 529–548, 1991.
28. Adams, M. E., and Hukins, D. W. L., The extracellular matrix of the meniscus, In *Knee Meniscus: Basic and Clinical Foundations*, Mow, V. C., Arnoczky, S. P., and Jackson, D. W., eds., Raven Press, New York, pp. 15–28, 1992.
29. Ghosh, P., and Taylor, T. K. F., The knee joint meniscus. A fibrocartilage of some distinction, *Clin. Orthop.*, 224, 52–63, 1987.
30. McDevitt, C. A., and Weber, R. J., The ultrastructure and biochemistry of meniscal cartilage, *Clin. Orthop.*, 252, 8–18, 1990.
31. Mow, V. C., Ratcliffe, A., Chern, K. Y., and Kelly, M., Structure and function relationships on the menisci in the knee, In *Knee Meniscus: Basic and Clinical Foundations*, Mow, V. C., Arnoczky, S. P., and Jackson, D. W., eds., Raven Press, New York, pp. 37–57, 1992.
32. Ahmed, A. M., and Burke, D. L., In vitro measurements of static pressure distribution in synovial joints: I. Tibial surface of the knee, *J. Biomech. Eng.*, 105, 216–225, 1983.
33. Walker, P. S., and Erkman, M. J., The role of the meniscus in force transmission across the knee, *Clin. Orthop.*, 109, 184–192, 1975.

34. Shrive, N. G., O'Connor, J. J., and Goodfellow, J. W., Load-bearing in the knee joint, *Clin. Orthop.*, 131, 279–287, 1978.
35. McDevitt, C. A., Miller, R. R., and Spindler, K. P., The cells and cell matrix interactions of the meniscus, In *Knee Meniscus: Basic and Clinical Foundations*, Mow, V. C., Arnoczky, S. P. and Jackson, D. W., eds., Raven Press, New York, pp. 29–36, 1992.
36. Webber, R. J., In vitro culture of meniscal tissue, *Clin. Orthop.*, 252, 114–120, 1990.
37. Webber, R. J., Harris, M. G., and Hough, A. J., Cell culture of rabbit meniscal fibrochondrocytes: proliferative and synthetic response to growth factors and ascorbate, *J. Orthop. Res.*, 3, 36–42, 1985.
38. Webber, R. J., Norby, D. P., Malemud, C. J., Goldberg, V. M., and Moskowitz, R. W., Characterization of newly synthesized proteoglycans from rabbit menisci in organ culture, *Biochem. J.*, 221, 875–884, 1984.
39. Webber, R. J., Zitaglio, T., and Hough, A. J., Serum-free culture of rabbit meniscal fibrochondrocytes: proliferative response, *J. Orthop. Res.*, 6, 13–23, 1988.
40. Ibarra, C., Koski, J. A., and Warren, R. F., Tissue engineering meniscus, *Orthop. Clin. North Am.*, 31, 411–418, 2000.
41. Brittberg, M., Lindahl, A., Nilsson, A., Ohlsson, C., Isaksson, O., and Peterson, L., Treatment of deep cartilage defects in the knee with autologous chondrocyte transplantation, *N. Engl. J. Med.*, 331(14), 889–895, 1994.
42. Brittberg, M., Nilsson, A., Lindahl, A., Ohlsson, C., and Peterson, L., Rabbit articular cartilage defects treated with autologous cultured chondrocytes, *Clin. Orthop.*, 326, 270–283, 1996.
43. Glowacki, J., Trepman, E., and Folkman, J., Cell shape and phenotypic expression in chondrocytes, *Proc. Soc. Exp. Biol. Med.*, 172, 93–98, 1983.
44. Benya, P. D., and Shaffer, J. D., Dedifferentiated chondrocytes re-express the differentiated phenotype when cultured in agarose cells, *Cell*, 30, 215–224, 1982.
45. Gibson, G. J., Schor, S. L., and Grant, M. E., Effects of matrix molecules on chondrocyte gene expression: synthesis of a low molecular weight collagen species by cells cultured within collagen gels, *J. Cell Biol.*, 93, 767–774, 1982.
46. Hauselmann, H. J., Aydelotte, M. B., Schumacher, B. L., Kuettner, K. E., Gitelis, S. H., and Thonar, E. J. A., Synthesis and turnover of proteoglycans by human and bovine adult articular chondrocytes cultured in alginate beads, *Matrix*, 12, 116–129, 1992.
47. Cima, L. G., Vacanti, J. P., Vacanti, C., Ingber, D., Mooney, D., and Langer, R., Tissue engineering by cell transplantation using biodegradable polymer substrates, *J. Biomech. Eng.*, 113, 143–151, 1991.
48. Peretti, G. M., Bonassar, L. J., Caruso, E. M., Randolph, M. A., Trahan, C. A., and Zaleske, D. J., Biomechanical analysis of a chondrocyte-based repair model of articular cartilage, *Tissue Eng.*, 5(4), 317–326, 1999.
49. Silverman, R. P., Bonassar, L. J., Passaretti, D., Randolph, M. A., and Yaremchuk, M. J., Adhesion of tissue engineered cartilage to native cartilage, *Plast. Reconstr. Surg.*, 105(4), 1393–1398, 2000.
50. Peretti, G. M., Gill, T. J., Xu, J. W., Randolph, M. A., Morse, K. R., and Zaleske, D. J., Cell-based therapy for meniscal repair: a large animal study, *Am. J. Sports Med.*, 32(1), 146–158, 2004.
51. Johnson, T. S., Xu, J. W., Zaporojan, V. V., Mesa, J. M., Weinand, C., Randolph, M. A., Bonassar, L. J., Winograd, J. M., and Yaremchuk, M. J., Integrative repair of cartilage with articular and non-articular chondrocytes, *Tissue Eng.*, 10(9–10), 1308–1315, 2004.
52. Mizuno, S., and Glowacki, J., Three-dimensional composite of demineralized bone powder and collagen for *in vitro* analysis of chondroinduction of human dermal fibroblasts, *Biomaterials*, 17(18), 1819–1825, 1996.
53. Zuk, P. A., Zhu, M., Mizuno, H., Huang, J., Futrell, J. W., Katz, A. J., Benhaim, P., Lorenz, H. P., and Hedrick, M. H., Multilineage cells from human adipose tissue: implications for cell-based therapies, *Tissue Eng.*, 7(2), 211–228, 2001.
54. Caplan, A. I., and Bruder, S. P., Mesenchymal stem cells: building blocks for molecular medicine in the 21st century. Review, *Trends Mol. Med.*, 7(6), 259–264, 2001.
55. Berry, L., Grant, M. E., McClure, J., and Rooney, P., Bone-marrow-derived chondrogenesis in vitro, *J. Cell Sci.*, 101(Pt 2), 333–342, 1992.

56. Pittenger, M. F., Mackay, A. M., Beck, S. C., Jaiswal, R. K., Douglas, R., Mosca, J. D., Moorman, M. A., Simonetti, D. W., Craig, S., and Marshak, D. R., Multilineage potential of adult human mesenchymal stem cells, *Science*, 284(5411), 143–147, 1999.
57. Prockop, D. J., Marrow stromal cells as stem cells for nonhematopoietic tissues, *Science*, 276(5309), 71–74, 1997.
58. Conget, P. A., and Minguell, J. J., Phenotypical and functional properties of human bone marrow mesenchymal progenitor cells, *J. Cell. Physiol.*, 181(1), 67–73, 1999.
59. Yoo, J. U., Barthel, T. S., Nishimura, K., Solchaga, L., Caplan, A. I., Goldberg, V. M., and Johnstone, B., The chondrogenic potential of human bone-marrow-derived mesenchymal progenitor cells, *J. Bone Joint Surg.*, 80A(12), 1745–1757, 1998.
60. Johnstone, B., Hering, T. M., Caplan, A. I., Goldberg, V. M., and Yoo, J. U., In vitro chondrogenesis of bone marrow-derived mesenchymal progenitor cells, *Exp. Cell Res.*, 238(1), 265–272, 1998.
61. Haynesworth, S. E., Goshima, J., Goldberg, V. M., and Caplan, A. I., Characterization of cells with osteogenic potential from human marrow, *Bone*, 13(1), 81–88, 1992.
62. Angele, P., Kujat, R., Nerlich, M., Yoo, J., Goldberg, V. M., and Johnstone, B., Engineering of osteochondral tissue with bone marrow mesenchymal progenitor cells in a derivatized hyaluronan–gelatin composite sponge, *Tissue Eng.*, 5(6), 545–554, 1999.
63. Gao, J., Dennis, J. E., Solchaga, L. A., Awadallah, A. S., Goldberg, V. M., and Caplan, A. I., Tissue-engineered fabrication of an osteochondral composite graft using rat bone marrow-derived mesenchymal stem cells, *Tissue Eng.*, 7(4), 363–371, 2001.
64. Sancho, B. V., and Molina, A. R., Use of septal cartilage homografts in rhinoplasty, *Aesthetic Plast. Surg.*, 24(5), 357–363, 2000.
65. Gibson, T., and Davis, W. B., The fate of preserved bovine cartilage grafts in man, *Br. J. Plast. Surg.*, 6, 4, 1953.
66. Elves, M. W., A study of the transplantation antigens on chondrocytes from articular cartilage, *J. Bone Joint Surg.*, 56B(1), 178–185, 1974.
67. Arnoczky, S. P., Warren, R. F., and McDevitt, C. A., Meniscal replacement using a cryopreserved allograft, *Clin. Orthop.*, 252, 121–128, 1990.
68. Jackson, D. W., McDevitt, C. A., Simon, T. M., Arnoczky, S. P., Atwell, E. A., and Silvino, N. J., Meniscal transplantation using fresh and cryopreserved allografts. An experimental study in goats, *Am. J. Sports Med.*, 20, 644–656, 1992.
69. Galili, U., LaTemple, D. C., Stone, K. R., and Walgenbach, A. W., Porcine and bovine cartilage transplants in Cynomologus monkey. II. Changes in anti-gal response during chronic rejection, *Transplantation*, 63, 646–651, 1997.
70. Tiku, M. L., Liu, S., Weaver, C. W., Teodorescu, M., and Skosey, J. L., Class II histocompatibility antigen-mediated immunologic function of normal articular chondrocytes, *J. Immunol.*, 135(5), 2923–2928, 1985.
71. Lance, E. M., Immunological reactivity towards chondrocytes in rat and man: relevance to autoimmune arthritis, *Immunol. Lett.*, 21, 63–73, 1989.
72. Mukerji, S., Randolph, M. A., Ma, P., and Lee, W. P. A., Immune reactions of autogenous and allogeneic cartilage cells and cartilage polymer constructs. *Presented at the American Association of Hand Surgeons*, Boca Raton, FL, 1997.
73. Moskalewski, S., Kawiak, J., and Rymaszewska, T., Local cellular response evoked by cartilage formed after auto- and allogeneic transplantation of isolated chondrocytes, *Transplantation*, 4, 572, 1966.
74. Langer, R., and Vacanti, J. P., Tissue engineering, *Science*, 260, 920–926, 1993.
75. Kim, W. S., Vacanti, J. P., Cima, L., Mooney, D., Upton, J., Puelacher, W. C., and Vacanti, C. A., Cartilage engineered in predetermined shapes employing cell transplantation on synthetic biodegradable polymers, *Plast. Reconstr. Surg.*, 94, 233–237, 1994.
76. Vacanti, C. A., Langer, R., Schloo, B., and Vacanti, J. P., Synthetic polymers seeded with chondrocytes provide a template for new cartilage formation, *Plast. Reconstr. Surg.*, 88, 753–759, 1991.
77. Cao, Y., Vacanti, J. P., Paige, K. T., Upton, J., and Vacanti, C. A., Transplantation of chondrocytes utilizing a polymer-cell construct to produce tissue-engineered cartilage in the shape of a human ear, *Plast. Reconst. Surg.*, 100, 297–302, 1997.

78. Freed, L. E., Hollander, A. P., Martin, I., Barry, J. R., Langer, R., and Vunjak-Novakovic, G., Chondrogenesis in a cell-polymer-bioreactor system, *Exp. Cell Res.*, 240, 58–65, 1998.
79. Vunjak-Novakovic, G., Freed, L. E., Biron, R. J., and Langer, R., Effects of mixing on the composition and morphology of tissue engineered cartilage, *J. Am. Inst. Chem. Eng.*, 42, 850–860, 1996.
80. Vunjak-Novakovic, G., Martin, I., Obradovic, B., Treppo, S., and Grodzinsky, A. J., Bioreactor cultivation conditions modulate the composition and mechanical properties of tissue-engineered cartilage, *J. Orthop. Res.*, 17, 130–138, 1999.
81. Ibarra, C., Jannetta, C., Vacanti, C. A., Cao, Y., Kim, T. H., Upton, J., and Vacanti, J. P., Tissue engineered meniscus: potential new alternative to allogeneic meniscus transplantation, *Transplant. Proc.*, 29, 986–988, 1999.
82. Speer, D. P., Chvapil, M., Volz, R. G., and Holmes, M. D., Enhancement of healing in osteochondral defects by collagen sponge implants, *Clin. Orthop.*, 144, 326–335, 1979.
83. Li, S.-T., Yuen, D., Li, P. C., Rodkey, W. G., and Stone, K. R., Collagen as a biomaterial: an application in knee meniscal fibrocartilage regeneration, *Mater. Res. Soc. Symp. Proc.*, 331, 25–32, 1994.
84. Rodkey, W. G., Steadman, J. R., and Li, S.-T., A clinical study of collagen meniscus implant to restore the injured meniscus, *Clin. Orthop.*, 367S, S281–S292, 1999.
85. Fuss, M., Ehlers, E. M., Russlies, M., Rohwedel, J., and Behrens, P., Characteristics of human chondrocytes, osteoblasts and fibroblasts seeded onto a type I/III collagen sponge under different culture conditions. A light, scanning and transmission electron microscopy study, *Anat. Anz.*, 182(4), 303–310, 2000.
86. Lee, C. R., Breinan, H. A., Nehrer, S., and Spector, M., Articular cartilage chondrocytes in type I and type II collagen-GAG matrices exhibit contractile behavior in vitro, *Tissue Eng.*, 6(5), 555–564, 2000.
87. Grande, D. A., Halberstadt, C., Naughton, G., Schwartz, R., and Manji, R., Evaluation of matrix scaffolds for tissue engineering of articular cartilage grafts, *J. Biomed. Mater. Res.*, 34(2), 211–220, 1997.
88. Fujisato, T., Sajiki, T., Liu, Q., and Ikada, Y., Effect of basic fibroblast growth factor on cartilage regeneration in chondrocyte-seeded collagen sponge scaffold, *Biomaterials*, 17(2), 155–162, 1996.
89. Mizuno, S., and Glowacki, J., Chondroinduction of human dermal fibroblasts by demineralized bone in three-dimensional culture, *J. Exp. Cell Res.*, 227(1), 89–97, 1996.
90. Peretti, G. M., Randolph, M. A., Caruso, E. M., Rossetti, F., and Zaleske, D. J., Bonding of cartilaginous matrices with cultured chondrocytes: an experimental model, *J. Orthop. Res.*, 16(1), 89–95, 1998.
91. Solchagá, L. A., Dennis, J. E., Goldberg, V. M., and Caplan, A. I., Hyaluronic acid-based polymers as cell carriers for tissue-engineered repair of bone and cartilage, *J. Orthop. Res.*, 17(2), 205–213, 1999.
92. Toole, B. P., Banerjee, S., Turner, R., Munaim, S., and Knudson, C., Hyaluronan–cell interactions in limb development, In *Developmental Patterning of the Vertebrate Limb*, Hinchliffe, J. R., Hurle, J. M. and Summerbell, D., eds., Plenum Press Inc., New York, pp. 215–223, 1997.
93. Toole, B. P., Hyaluronan in morphogenesis, *J. Intern. Med.*, 42, 35–40, 1997.
94. Brun, P., Abatangelo, G., Radice, M., Zacchi, V., Guidolin, D., Daga Gordini, D., and Cortivo, R., Chondrocyte aggregation and reorganization into three-dimensional scaffolds, *J. Biomed. Mater. Res.*, 46(3), 337–346, 1999.
95. Aigner, J., Tegeler, J., Hutzler, P., Campoccia, D., Pavesio, A., Hammer, C., Kastenbauer, E., and Naumann, A., Cartilage tissue engineering with novel nonwoven structured biomaterial based on hyaluronic acid benzyl ester, *J. Biomed. Mater. Res.*, 42(2), 172–181, 1998.
96. Solchagá, L. A., Yoo, J. U., Lundberg, M., Dennis, J. E., Huibregtse, B. A., Goldberg, V. M., and Caplan, A. I., Hyaluronan-based polymers in the treatment of osteochondral defects, *J. Orthop. Res.*, 18(5), 773–780, 2000.
97. Hauselmann, H. J., Fernandes, R. J., Mok, S. S., Schmid, T. M., Block, J. A., Aydelotte, M. B., Kuettner, K. E., and Thonar, E. J., Phenotypic stability of bovine articular chondrocytes after long-term culture in alginate beads, *J. Cell Sci.*, 107(Pt 1), 17–27, 1994.

98. Lanza, R. P., Kuhtreiber, W. M., Ecker, D., Staruk, J. E., and Chick, W. L., Xenotransplantation of porcine and bovine islets without immunosuppression using uncoated alginate microspheres, *Transplantation*, 59, 1377–1384, 1995.
99. Paige, K. T., Cima, L. G., Yaremchuk, M. J., Vacanti, J. P., and Vacanti, C. A., Injectable cartilage, *Plast. Reconstr. Surg.*, 96, 1390–1398, 1995.
100. Paige, K. T., Cima, L. G., Yaremchuk, M. J., Schloo, B. L., Vacanti, J. P., and Vacanti, C. A., De novo cartilage generation using calcium alginate-chondrocyte constructs, *Plast. Reconstr. Surg.*, 97(1), 168–178, 1996.
101. Sechriest, V. F., Miao, Y. J., Niyibizi, C., Westerhausen-Larsen, A., Matthew, H. W., Evans, C. H., Fu, F. H., and Suh, J. K., GAG-augmented polysaccharide hydrogel: a novel biocompatible and biodegradable material to support chondrogenesis, *J. Biomed. Mater. Res.*, 49(4), 534–541, 2000.
102. Lahiji, A., Sohrabi, A., Hungerford, D. S., and Frondoza, C. G., Chitosan supports the expression of extracellular matrix proteins in human osteoblasts and chondrocytes, *J. Biomed. Mater. Res.*, 51(4), 586–595, 2000.
103. Cao, Y., Rodriguez, A., Vacanti, M., Ibarra, C., Arevalo, C., and Vacanti, C. A., Comparative study of the use of poly(glycolic acid), calcium alginate and pluronics in the engineering of autologous porcine cartilage, *J. Biomater. Sci. Polym. Ed.*, 9(5), 475–487, 1998.
104. Itay, S., Abramovici, A., and Nevo, Z., Use of cultured embryonal chick epiphyseal chondrocytes as grafts for defects in chick articular cartilage, *Clin. Orthop.*, 220, 284–303, 1987.
105. Silverman, R. P., Passaretti, D., Huang, W., Randolph, M. A., and Yaremchuk, M. J., Injectable tissue-engineered cartilage using a fibrin glue polymer, *Plast. Reconstr. Surg.*, 103(7), 1809–1818, 1999.
106. Park, M. S., and Cha, C. I., Biochemical aspects of autologous fibrin glue derived from ammonium sulfate precipitation, *Laryngoscope*, 103(2), 193–196, 1993.
107. Arnoczky, S. P., Warren, R. F., and Spivak, J. M., Meniscal repair using an exogenous fibrin clot. An experimental study in dogs, *J. Bone Joint Surg.*, 70A, 1209–1217, 1988.
108. Port, J., Jackson, D. W., Lee, T. Q., and Simon, T. M., Meniscal repair supplemented with exogenous fibrin clot and autogenous cultured marrow cells in the goat model, *Am. J. Sports Med.*, 24, 547–555, 1996.
109. Sims, D. D., Butler, P. E., Cao, Y. L., Casanova, R., Randolph, M. A., Black, A., Vacanti, C. A., and Yaremchuk, M. J., Tissue engineered neocartilage using plasma derived polymer substrates and chondrocytes, *Plast. Reconstr. Surg.*, 101, 1580–1585, 1998.
110. Peretti, G. M., Randolph, M. A., Villa, M. T., Buragas, M. S., and Yaremchuk, M. J., Cell/based tissue engineered allogeneic implant for cartilage repair, *Tissue Eng.*, 6(5), 567–576, 2000.
111. Peretti, G. M., Randolph, M. A., Zaporojan, V., Bonassar, L. J., Xu, J. W., Fellers, J., and Yaremchuk, M. J., A biomechanical analysis of an engineered cell-scaffold implant for cartilage repair, *Ann. Plast. Surg.*, 46(5), 533–537, 2001.
112. Xu, J. W., Zaporojan, V., Randolph, M. A., Peretti, G. M., Roses, R. E., Roy, A., Bonassar, L. J., and Yaremchuk, M. J., Engineering cartilage using different chondrocyte sources, *Plast. Reconstr. Surg.*, 113(5), 1361–1371, 2004.
113. Sims, C. D., Butler, P. E., Casanova, R., Lee, B. T., Randolph, M. A., Lee, W. P., Vacanti, C. A., and Yaremchuk, M. J., Injectable cartilage using polyethylene oxide polymer substrates, *Plast. Reconstr. Surg.*, 98(5), 843–850, 1996.
114. Hill-West, J. L., Chowdhury, S. M., Slepian, M. J., and Hubbell, J. A., Inhibition of thrombosis and intimal thickening by in situ photopolymerization of thin hydrogel barriers, *Proc. Natl Acad. Sci.*, 91(13), 5967–5971, 1994.
115. Hill-West, J. L., Chowdhury, S. M., Sawhney, A. S., Pathak, C. P., Dunn, R. C., and Hubbell, J. A., Prevention of postoperative adhesions in the rat by in situ photopolymerization of bioresorbable hydrogel barriers, *Obstet. Gynecol.*, 83(1), 59–64, 1994.
116. Elisseeff, J., Anseth, K., Sims, D., McIntosh, W., Randolph, M., Yaremchuk, M., and Langer, R., Transdermal photopolymerization of poly(ethylene oxide)-based injectable hydrogels for tissue-engineered cartilage, *Plast. Reconstr. Surg.*, 104(4), 1014–1022, 1999.
117. Elisseeff, J., Anseth, K., Sims, D., McIntosh, W., Randolph, M., and Langer, R., Transdermal photopolymeriztion for minimally invasive implantation, *Proc. Natl Acad. Sci.*, 96, 3104–3107, 1999.

118. Ashiku, S., Randolph, M. A., Vacanti, C. A., Mathisen, D., and Yaremchuk, M. J., European Tissue Repair Society, Second ETRS Consensus Meeting, August 20–22, Frieburg, Germany, 1997.
119. Nuttelman, C. R., Mortisen, D. J., Henry, S. M., and Anseth, K. S., Attachment of fibronectin to poly(vinyl alcohol) hydrogels promotes NIH3T3 cell adhesion, proliferation, and migration, *J. Biomed. Mater. Res.*, 57(2), 217–223, 2001.
120. Poshusta, A. K., and Anseth, K. S., Photopolymerized biomaterials for application in the temporomandibular joint, *Cells Tissues Organs*, 169(3), 272–278, 2001.
121. Martens, P. J., Bryant, S. J., and Anseth, K. S., Tailoring the degradation of hydrogels formed from multivinyl poly(ethylene glycol) and poly(vinyl alcohol) macromers for cartilage tissue engineering, *Biomacromolecules*, 4(2), 283–292, 2003.
122. Bryant, S. J., and Anseth, K. S., The effects of scaffold thickness on tissue engineered cartilage in photocrosslinked poly(ethylene oxide) hydrogels, *Biomaterials*, 22(6), 619–626, 2001.
123. Bryant, S. J., and Anseth, K. S., Hydrogel properties influence ECM production by chondrocytes photoencapsulated in poly(ethylene glycol) hydrogels, *J. Biomed. Mater. Res.*, 59(1), 63–72, 2002.
124. Bryant, S. J., Nuttelman, C. R., and Anseth, K. S., The effects of crosslinking density on cartilage formation in photocrosslinkable hydrogels, *Biomed. Sci. Instrum.*, 35, 309–314, 1999.
125. Cao, Y., Vacanti, J. P., Paige, K. T., Upton, J., and Vacanti, C. A., Transplantation of chondrocytes utilizing a polymer-cell construct to produce tissue-engineered cartilage in the shape of a human ear, *Plast. Reconst. Surg.*, 100, 297–302, 1997.
126. Arevalo-Silva, C. A., Eavey, R. D., Cao, Y., Vacanti, M., Weng, Y., and Vacanti, C. A., Internal support of tissue-engineered cartilage, *Arch. Otolaryngol. Head Neck Surg.*, 126, 1448–1452, 2000.
127. Nishibe, T., Yasuda, K., Ohkashiwa, H., Watanabe, S., Okuda, Y., and Tanabe, T., High-porosity expanded polytetrafluoroethylene grafts for thoracic vena cava replacement with or without an omentum wrap, *Surg. Today*, 30, 631–635, 2000.
128. Stanec, S., and Stanec, Z., Ulnar nerve reconstruction with an expanded polytetrafluoroetylene conduit, *Br. J. Plast. Surg.*, 51, 637–639, 1998.
129. Maas, C. S., Gnepp, D. R., and Bumpous, J., Expanded polytetrafluoroethylene (Gore-Tex Soft-tissue patch) in facial augmentation, *Arch. Otolaryngol Head Neck Surg.*, 119, 1008–1014, 1993.
130. Catanese, J., Cooke, D., Maas, C., and Pruitt, L., Mechanical properties of medical grade expanded polytetrafluoroethylene: the effects of internodal distance, density, and displacement rate, *J. Biomed. Mater. Res.*, 48, 187–192, 1999.
131. Xu, J. W., Nazzal, J., Peretti, G. M., Kirchhoff, C. H., Randolph, M. A., and Yaremchuk, M. J., Tissue-engineered cartilage composite with expanded polytetrafluoroethylene membrane, *Ann. Plast. Surg.*, 46(5), 527–532, 2001.

37 Tissue Engineering for Insulin Replacement in Diabetes

Amy S. Lewis and Clark K. Colton

CONTENTS

I. INTRODUCTION

Type 1 diabetes is a disease that results from a person's impaired ability to produce insulin, a protein that regulates blood glucose concentration. Insulin is produced by β-cells in the Islets of Langerhans, which are aggregates of cells averaging about 150 μm in diameter and constituting about 1 to 2% of the pancreas volume. Type 1 diabetes is caused by auto-immune attack and destruction of the β-cells, which compose about 80% of the islets. Diabetes has a serious impact on the health care system. In 2002, there were an estimated 12.1 million people diagnosed (and about 50% more undiagnosed) with diabetes in the U.S. This number is expected to grow to 17.4 million by 2020.[1] Type 1 (insulin-dependent) diabetes patients represent 5 to 10% of the total number of diabetes cases.[2] The increased blood sugar level puts diabetic patients at a greater risk for various complications, including blindness, gangrene, stroke, ketoacidosis, and heart, kidney, nervous system, and periodontal disease. About 19% of all deaths in the U.S. in people aged 25 or older occurred among people with diabetes, and the risk of death among people with diabetes is about 2 times greater than that of people without diabetes.[2] The cost associated with diabetes in the U.S. was $132 billion dollars in 2002.[1]

II. EXOGENOUS INSULIN DELIVERY

Conventional treatment for type 1 diabetes involves one or two daily insulin injections and daily monitoring of blood glucose but without adjustments of insulin dosage. Insulin injections under these conditions are not optimal because blood glucose levels fluctuate incorrectly when there is no feedback control like that provided by β-cells in healthy patients.[3] In a recent clinical trial, intensive insulin therapy (three or more times daily by injection or external pump, adjusted according to results of self-monitoring of glucose at least four times per day) improved blood glucose control and substantially decreased occurrences of long term complications in comparison to conventional therapy.[4] The chief adverse event associated with intensive therapy was a two to threefold increase in severe hypoglycemia, a major concern because hypoglycemic unawareness can be life threatening and tends to occur when hypoglycemic episodes are more frequent.

Alternative methods of insulin delivery have been investigated to deliver insulin in a pain free manner, thereby increasing patient compliance and more effectively regulate blood glucose levels. Externally worn pumps that continuously deliver insulin subcutaneously are safe and effective with fewer hypoglycemic episodes, no increase in complications, and lower levels of glycosylated hemoglobin.[5] Methods are being developed to deliver insulin transdermally and through the pulmonary system, and to measure blood glucose without the use of a needle, for example, by permeabilizing the skin with an ultrasound, extracting interstitial fluid, and correlating the glucose concentration to blood glucose concentration.[6–8] Implantable glucose sensors would be the best way to measure blood glucose in real-time, but they have been limited by a need for frequent recalibration as a result of changes in transport properties of the tissue surrounding the sensor from fibrotic overgrowth.[9–13] Methods of administering insulin that do not provide feedback control do not usually produce normal glucose control and are prone to complications.

More advanced insulin delivery systems under study would provide feedback-controlled insulin delivery without patient intervention. One example is an implantable glucose sensor placed in a feedback control loop that controls the delivery of insulin from a reservoir with a pump.[12,13] Another example is implantation of a polymer that changes permeability in response to glucose. For example, glucose oxidase is immobilized in pH-responsive polyacrylic acid, which is grafted onto a porous support. When glucose concentration increases, its oxidation is accompanied by a decrease in pH that causes acrylic acid to assume a more compact conformation, thereby opening pores for insulin release.[14] Such systems capable of providing normoglycemia are possibly a long way off.

Islet transplantation promises normoglycemia. The islets themselves provide physiological feedback controlled insulin release and are capable of continuously producing insulin. In humans, this procedure involves implanting islets isolated from a cadaver donor into the recipient's liver via the portal vein. Islet transplantation would allow patients to be free of daily insulin injections and have tighter blood glucose control, thus eliminating many of the secondary complications of diabetes.[3] After several decades of frustration, a dramatic improvement in efficacy was reported in July of 2000 by a group from the University of Alberta, Edmonton, that renewed interest throughout the medical community in islet transplantation research.[15] Seven consecutive patients with type 1 diabetes were freed from the need for exogenous insulin with excellent metabolic control for 1 year or more. There were three new features in the work: (1) islet isolation from a human pancreas was followed by immediate (no more than 4 h) implantation in the recipient to prevent islet damage that may occur during culture, (2) patients received islets from two or more donors (until normoglycemia was achieved), and (3) patients were treated with a glucocorticoid-free immunosuppressive therapy, consisting of low-dose tacrolimus and daclizumab, that protected the islets from auto-immune and allo-immune reactivity.[15] In further clinical studies with the Edmonton protocol, 87% of patients are insulin free 1 year after implantation.[16]

The efficacy of islet cell transplantations has been demonstrated in humans, but obstacles remain for wide scale application.[17,18] This first obstacle is that there is a lack of available tissue for transplantation procedures. In the U.S. there are one million people with type 1 diabetes, and about 30,000 new cases per year. By contrast, there are currently about 5000 cadaver pancreases available. Even if all other problems were eliminated, islet transplantation would be limited to a small fraction of those who could benefit from it. Expansion to a much larger patient population requires identification and development of new sources of islets or glucose responsive insulin-secreting tissue. The second issue is that successful islet transplantation requires permanent use of multiple immunosuppressive agents that may have serious side effects and are a substantial financial burden.

Protecting the islets from immune attack with a semipermeable membrane barrier is the basic notion behind a wide variety of tissue-engineered constructs that have been investigated. In the remainder of the review, we discuss areas that are important for tissue engineering in support of islet transplantation. These include routes to new sources of glucose-responsive insulin-secreting tissue, the immune response to implanted tissue and how it is affected by a semipermeable barrier, the nature of tissue-engineered devices and encapsulants for protecting islets, recent developments with microencapsulating materials, the nature and amelioration of oxygen supply limitations that are exacerbated by the presence of immunobarrier materials, and noninvasive assessment of islets *in vitro* and *in vivo*. Several of these areas have been covered in prior reviews,[19–22] and emphasis will be placed on recent research.

III. TISSUE SOURCES

Ample supplies of islets are required to make islet transplantation a widespread treatment for patients with type 1 diabetes. Current clinical trials in humans are limited to use of allogeneic human tissue for which efficacious immunosuppressive drugs have been developed.

A. Xenogeneic Islets

Xenotransplantations, implantation of cells of a species different from the recipient, has been investigated in various studies.[23,24] Rat and porcine islets implanted in mice provide normoglycemia for extended periods with immunobarrier systems.[25,26] Porcine islets are an attractive source for use in humans because their availability is much greater than that of human donors, and diabetes was previously treated in humans with insulin from pigs, thereby demonstrating that porcine insulin is capable of controlling glucose levels in humans without adverse effects. The drawbacks of using porcine islets are the lack of immunosuppressive drugs for xenotransplantation and the need for assurance that no disease will be transferred from pigs to humans.[17,27,28] In order to provide immunoprotection, studies to date have made use of encapsulated porcine islets, in some cases together with immunosuppressive drugs.[24,29] Recipients of porcine islets have included rats, dogs, and monkeys.[24,29] Genetically engineered pigs have been created to reduce the immune reaction to the transplant after transplantation.[30] Tissue from these transgenic animals eliminates hyperacute rejection associated with naturally occurring IgM. It is unclear whether elimination of this phenomenon is a significant benefit for islet tissue.

Adult porcine islets and neonatal porcine islet cell clusters (NPCCs) have been considered for use in humans. Adult porcine islets are difficult to isolate and, as a result, the islet yield from the isolation is low. NPCCs are easier to isolate and can be expanded and matured in culture.[24,31] When encapsulated in alginate, NPCCs can reverse diabetes in nude mice and can survive when cultured in human serum and exposed to complement.[32]

B. Stem Cells

Effort is currently underway to develop an alternative cell source from either human adult, mouse embryonic, or human embryonic stem cells. Recent reviews[33–35] provide greater detail than what is presented in the following paragraphs. Adult cells contained in islets do not replicate *in vitro*, but new islets are generated throughout a person's life, thereby suggesting the existence of adult stem cell or islet progenitor cells in the pancreas.[33] Several different cells in the pancreas have been identified as possible pancreatic stem or progenitor cells — ductal tissue, oval cells, or intraislet stem cells.[33] Human ductal tissue from the pancreas differentiates into immature islets *in vitro*, but the rate of replication is too slow to make a sizable impact in providing adequate tissue for islet transplantation.[36] Murine pancreatic ductal tissue has also been expanded *in vitro* and implanted into mice that remained insulin free until sacrificed at 55 days.[37,38] The exact markers of islet progenitors or stem cells are under debate, but three possible markers have drawn the greatest interest. $PDX1^+$ cells are progenitors for all pancreatic tissue.[39] Nestin-positive and $NGN3^+$ cells are believed to be islet progenitors.[33,39] In contrast, nestin-positive cells were recently shown to be part of the microvasculature adjacent to β-cells in the islet, but they did not differentiate into β-cells, indicating that pancreatic endocrine cells arise independently of nestin-positive precursors.[40] Adult stem/progenitor cells are an attractive cell source for islet transplantation because they may allow for autologous transplantation and decrease the patient's immune response to the implant. Adult stem cells have the disadvantage of having a slower rate of replication in comparison to embryonic stem cells.[34]

Mouse embryonic stem cells can differentiate into insulin secreting cells. Early studies aimed at differentiating insulin producing cells from mouse embryonic stem cells selected only for cells that produced insulin.[41] In 2001, islet-like structures were produced from mouse embryonic stem cells, but with very low insulin levels as compared to adult β-cells.[42] Attempts to replicate this work failed and resulted in the belief that the previous results were caused by insulin uptake from the media.[43] Contemporaneous and subsequent work by other groups has proved that insulin producing cells from mouse embryonic stem cells can indeed be generated.[44–46] Their results were verified by showing colocalization of other proteins involved in insulin production, such as C-peptide in the insulin-positive cells, and by using insulin free media. The cells that are produced have lower insulin levels than adult β-cells, but when transplanted into streptozotocin diabetic mice they normalized blood glucose levels for the length of the study (14 or 21 days).[44,45] The research with mouse embryonic stem cells will be useful to learn about the maturation process and to apply these techniques to human embryonic stem cells.

Human embryonic stem cells are being studied for islet transplantation, but the level of knowledge about the differentiation process lags behind that of mouse embryonic stem cells. Human embryonic stem cells spontaneously differentiate into insulin producing cells, but they have not been shown to be glucose responsive.[47] Human embryonic stem cells are a promising source of islets for transplantations but are limited by ethical, moral, and legal issues associated with embryonic stem cell use.

C. Cell Lines

A glucose-responsive insulin-producing cell line would be advantageous because it could be cultured indefinitely, expanded in culture, and genetically modified without the need of a donor and complex isolation procedures. To be feasible for transplantation, a cell line must meet various constraints. The cell line must release insulin in response to glucose with the same sensitivity as β-cells. The need to be able to arrest cell growth after implantation, so that the transplant does not form a tumor or cause hypoglycemia, motivated development of the βTC-tet cell line (from βTC3, a mouse insulinoma cell line) that undergoes growth arrest *in vitro* and *in vivo* in the presence of tetracycline.[48] The cells remain viable, producing and secreting insulin, when implanted in mice.[48] Growth arrest of the cell line leads to a fourfold increase in insulin-content and improves its ability to regulate glucose levels, although the cell line only produces one third the amount of insulin of normal β-cells.[49,50] The INS-1 cell line is a rat insulinoma cell line that secretes insulin in response to glucose levels in a physiological range and is used to study the biochemical pathway for glucose stimulated insulin secretion or as a cell source for islet transplantation.[51] Transfection of the INS-1 cells with a human insulin gene can allow for selection of a cell line that has enhanced and stable glucose responsiveness.[51] The INS-1 cell line can develop a resistance to the cytokines that destroy β-cells in type 1 diabetes, IL-1β and INFγ, by culturing the cells in the presence of these two cytokines.[52] An insulin-secreting human tumor cell line has not yet been reported that does not loose its differentiated functions, such as insulin secretion, at a rate greater that that of mouse cell lines.[53]

IV. IMMUNE DESTRUCTION OF ISLETS

We consider here the immune response to transplanted islets, including pathways followed in auto-immune, allogeneic, and xenogeneic situations, as well as the response to an immunobarrier material. Type 1 diabetes is an auto-immune disease, and islet transplantation will trigger an immune attack by the same mechanisms that caused the patient to become diabetic in the first place. Additionally, if the transplant is an allograft or a xenograft, the immune system will recognize it as nonself and mount an additional response. The reaction to the xenograft is generally more intense because the immune system will also recognize the transplant as being from another species.[54] The final component that can elicit an immune response is an immunobarrier material surrounding the islets. Successful islet transplantations require that the immune system be blocked from attacking the transplant or responding deleteriously to the implanted material. The immune system can be blocked with drugs administered systemically, or by encapsulation of the islets in a semipermeable material that will allow for glucose, insulin, and oxygen transport while blocking the transport of large immune reactive molecules. Other approaches, such as inducing immune tolerance, immunomodulation and local immunosuppression are also under study.

A. Auto-Immune Reactions

The auto-immune destruction of β-cells in type 1 diabetes is carried out by macrophages and involves both $CD4^+$ and $CD8^+$ T-cells. Results with animal models indicate that β-cell death is manifested through apoptosis.[55] In the NOD mouse model of auto-immune diabetes, cytotoxic $CD8^+$ T-cells induce apoptosis in the target cell by the secretion of perforin and the action of membrane bound Fas ligand, whereas $CD4^+$ mediated β-cell destruction is dependent on the secreted cytokines IFNγ and TNFα.[56] Macrophages release free radicals, nitric oxide (NO) and cytokines, such as IL-1β and TNFα that induce apoptosis in β-cells. IL-1β induces apoptosis in β-cells by upregulating inducible nitric oxide synthetase (iNOS) mRNA expression, leading to destruction by NO as well as by NO-independent pathways.[55–57] Activated macrophages can destroy microencapsulated rat islets by producing NO.[58] The challenge in preventing auto-immune destruction of transplanted β-cells can therefore be addressed at different points:

1. Prevent T-cells and macrophages from being activated with immunosuppressive drugs.
2. Prevent cytokines from reaching the islets. For example, the presence in tissue of receptors for IL-1 and TNFα is protective in preventing destruction of cells by auto-immune mechanisms.[59,60] Enclosure of the islets within an immunobarrier membrane poses a difficult challenge because IL-1 and TNF-α have molecular weights (about 20,000) that fall between insulin (6800), albumin (67,000) and transferrin (about 100,000) that are needed by cells.[21]
3. Prevent the buildup of NO in the vicinity of the transplanted tissue, perhaps through agents that can remove nitric oxide from solution.

B. Allograft Reactions

Allograft rejection is also carried out by macrophages and T-cells, both of which can be activated either through the direct or indirect pathway.[61] In the direct pathway, T-cells directly bind to MHC molecules on the surface of the implanted cells and are activated. The indirect pathway is activated by T-cell recognition of peptides shed from the allogeneic cells and bound to self MHC as being nonself. Encapsulation of islets should prevent activation of the direct pathway because T-cells are prevented from contacting islets. The indirect pathway will still be active because the small molecules that are shed can diffuse through the encapsulating material. The role of the indirect pathway in allograft rejection is relatively minor, at least in acute rejection, although its role in chronic rejection is unclear.[61] Current research is aimed at developing new methods of immunosuppression[62] or understanding the processes involved in tolerance induction.[63] The reaction of both macrophages and T-cells, to an allograft, needs to be prevented. It has been shown that allotransplantation of islets with testicular Sertoli cells or genetically engineered cells that produce the Fas ligand can protect islets.[64–66]

C. Xenograft Reactions

T-cell activation against xenografts is by the indirect pathway because the MHC molecules are so different that the direct pathway is minimal.[61] The resulting immune response to xenografts includes both a humoral component (involving IgG and IgM antibodies and complement) and a cellular component (involving T-cells and macrophages and their secreted cytokines, free radicals, and NO).[54,67,68] Immune reactions to xenografts produce a florid cellular response that is much more severe than the reaction to empty capsules. Because no immunosuppressive agents exist for xenotransplantations, significant effort has been devoted to investigation of immunobarrier approaches. The implications of a predominant indirect pathway with humoral and cellular responses on the transport characteristics of immunobarriers has been discussed in detail.[19–21] Briefly, in addition to preventing contact with immune cells, it is essential to prevent attack by the humoral immune system, which is mediated by complement. Complement plays a major role in the rejection of xenografts.[69] The complement reaction is initiated either by the binding of the complement component C1q to an IgM or IgG molecule, or by the alternate pathway which involves C3.[20] This results in the initiation of a cascade of events that result in lysis of cells. A more difficult challenge is the antigen release from the islet tissue. A portion of every protein in the cell is displayed on its surface in the form of peptides 8 to 22 amino acids in length.[61] Some or all of these peptides are lost into the surrounding extracellular fluid and presumably picked up by antigen presenting cells (dendritic cells and macrophages) that initiate the immune response by T-cells.[61] There is no way for a size-selective immunobarrier to prevent passage of these shed antigens while simultaneously permitting passage of insulin. If the antigen load is sufficiently high, a cellular immune response can be elicited outside the capsule that can lead to starvation and oxygen deprivation of the islet within, and to release of cytokines, free radicals, and NO. One approach to the problem is by local immunosuppression or immunoisolation to inhibit T-cell activation. The complement reaction can be stopped by preventing the passage of C1q, IgM, and C3 to islets through a membrane with a maximum effective pore diameter of 30 nm,[19] through materials in the membrane that denature the complement proteins, or by adding molecules that inhibit a step in the formation of the membrane attack complex on cells of the encapsulated tissue.[69,70]

D. Inflammatory Reactions

Nonspecific inflammatory events occur at the transplant site immediately after implantation, mediated by macrophage secretion of cytokines, free radicals, and NO, that can contribute to islet cell damage. Reduction of these effects has been demonstrated following macrophage depletion.[71] The immune system initiates a similar nonspecific inflammatory response, the so-called foreign body response, to materials implanted into the body, which ultimately results in fibrotic overgrowth of the encapsulated material that inhibits transport to the islets and may also contribute to islet destruction. Reactions to biomaterials implanted without cells are common and much research has been done to find biomaterials that minimize this reaction. The degree of the reaction can sometimes be reduced when pure materials are used.[72]

V. ENGINEERING APPROACHES TO IMMUNOPROTECTION OF ISLETS

Various approaches have been investigated to impose a semipermeable barrier to immune system components between islets and the blood stream or surrounding tissue. These approaches can be usefully categorized into three approaches.[19,21]

1. Intravascular devices are shunts between an artery and a vein. Blood flows through the lumen of a cylindrical membrane with islets surrounding the outer surface. This type of device is beneficial because islets come into close proximity with arterial blood, which has high oxygen content, thereby somewhat ameliorating the problem of oxygen supply limitations. This approach is the most successful to date in terms of long-term xenotransplantations (porcine islets transplanted into the dog) which have been successful for more than a year.[21] Human clinical trials in the planning stages were stopped by the FDA several years ago because of a mechanical failure of the cannula leading from the device, and the studies never resumed for economic reasons.
2. Extravascular devices contain many islets in a device that is implanted into a cavity of the body. This type of device may be limited by the amount of oxygen that can be provided to islets if diffusion distances for oxygen are large, thus preventing adequate oxygen transfer to maintain islets without vascularization of the device. This problem can be exacerbated by the fibrotic tissue that can grow around implanted devices[73] but developments in improved interfacing materials can improve vascularization, as described below.
3. Islet microcapsules are spheres of a biomaterial that usually contain one or sometimes more islets and have a diameter as small as 200 to 300 μm (in the case of a conformal coating) to as large as 1 to 2 mm (sometimes called macrobeads). The most common material used is alginate. A large number of these small capsules containing islets are usually implanted into the peritoneal cavity. This approach has been found effective in rodents even when there is only a minimal acute insulin secretory response to glucose stimulation.[74]

The entire field is sometimes referred to simply as encapsulation. In what follows we will review recent progress with extravascular macroencapsulation devices and microencapsulation.

A. Extravascular Macroencapsulation Devices

Extravascular islet macroencapsulation devices contain many islets in an implantable device that is placed in a cavity of the body. Macroencapsulation devices can be advantageous because they are mechanically stable and easier to remove than microcapsules. Some configurations may be limited by greater transport barriers than that of microcapsules because of greater diffusion distances.[75,76] In this section we review examples of two different types, hollow fiber devices and planar diffusion chambers.

Hollow fiber devices are thin cylindrical semipermeable membranes in the lumen of which islets are encapsulated. In one example, canine islet xenografts in alginate were encapsulated in the lumen of permselective acrylic ultrafiltration-type tubular membranes. The membranes were 2 to 3 cm long with about 2 mm internal diameter and 80 μm wall thickness. Hollow fiber capsules containing about 50,000 to 100,000 islets/kg were implanted in the peritoneal cavity of spontaneously diabetic BB/Wor rats and produced normoglycemia for periods from 1 to 8 months. Upon explant, the external surfaces of the explants were free of fibrosis and host cell adherence, although some were broken, leading to xenograft rejection in the interior. This study demonstrated that xenografts could be protected within a permselective membrane without immunosuppression.[77] Histological examination of the alginate core of such devices has revealed that only islets near the membrane luminal surface are viable. Those near the center are not.[21]

Polysulphone capillary polymers that were chemically modified with hydroxyl-methyl groups provided improved glucose-induced insulin secretion compared to unmodified hollow fibers.[78] However, it is unclear whether this observation simply reflected the high protein binding affinity of unmodified polysulphone.

We consider two examples of a planar diffusion device: the Theracyte System developed by Baxter Healthcare Corporation and the islet sheet of The Islet Sheet Medical Company. The Theracyte device uses a laminated composite membrane composed of three different layers: a microporous 5 μm expanded polytetrafluoroethylene (PTFE) vascularizing membrane (about 15 μm thick) which interfaces with host tissue, a cell-impermeable 0.45 μm hydrophilic PTFE Biopore membrane (about 25 to 30 μm thick), and a highly porous external support layer.[21] Allograft[79] and xenograft[80] transplantations have been carried out in auto-immune diabetic NOD mice using the Theracyte device. In both studies, blood glucose normalization persisted well in excess of 100 days. This finding with xenografts is in conflict with previous observations[81] and with another study[82] using the Theracyte in which xenogeneic cells transplanted in normal mice were rejected but survived for more than 35 days if the mice were initially given a single dose of anti-CD4 antibody that prevents activation of $CD4^+$ T-cells via the indirect antigen presentation pathway.[83] In transplantation experiments with titanium ring devices using the same membranes, fibrotic tissue was observed covering the devices after explantation but there were blood vessels adjacent to the membrane within the fibrotic tissue.[25,84] Preimplantation of the device 3 months before loading with islets improves the survival of the encapsulated graft and reduces fibroblast growth in the device.[85–87] A method for quantitating the actual volume of islets in these devices has been described.[88] The Theracyte has also been used in a variety of other cell therapy applications.[89–93]

In the islet sheet, islets are placed in a reinforcing polymer mesh that is sandwiched between two acellular alginate layers.[94] The islet sheet was designed to minimize transplant volume and maintain the highest level of islet viability during its formation.[94] Following allotransplantation in a dog, fasting euglycemia was maintained until elective explantation at 84 days. Histological examination showed that the device folded on itself and elicited a strong fibrotic reaction, limiting transport to islets.[94]

Other materials have been investigated as potential membranes for macroencapsulation devices. Corona treatment of AN69 membranes, a copolymer of acrylonitrile and sodium methallyl sulphonate originally designed for kidney dialysis, removed layers of glycerol and improved insulin permeability while leaving the glucose permeability unchanged. When implanted intraperitoneally for 1 year, the corona treated membrane remained structurally intact and free of cellular adhesions indicating biocompatibility.[95] Micromachined silicon membranes allow for tight control over pore size. Membranes with 18 nm pores greatly reduced the effective diffusion coefficient of IgG while maintaining adequate glucose and insulin transport.[96] Tricontinuous membranes of hydrophilic poly-(ethylene glycol) and lipophilic polyisobutylene segments cross-linked by oxyphilic poly(pentamethylcyclopentasiloxane) combine several different material properties, characteristic of each component, that are advantageous for islet encapsulation.[97] The material is a hydrogel, has high oxygen solubility, and gets its strength from covalent cross linking and the lipophilic component.[97]

In a recent critical analysis[76] of the problems associated with the development of extravascular macroencapsulation devices, several unique features are suggested in the context of an immunobarrier pouch containing islets within a gel. One notable feature is use of a thermally reversible synthetic hydrogel made of N-isopropylacrylamide based copolymer that facilitates recharging with fresh islets whenever necessary. This would be a valuable feature in order to replace the islets with minimally invasive surgery.

B. Microencapsulation

Among the characteristics of the ideal microcapsule for islets are the following: (1) noncytotoxic to encapsulated cells, (2) biocompatible in transplantation location, (3) adequate permeability for oxygen, glucose and insulin, (4) impermeable to antibodies and complement proteins, (5) chemically and mechanically stable, (6) high purity materials, and (7) minimal shrinking or swelling after transplantation.[98,99] Various materials have been studied as potential encapsulating materials. The most common is alginate, derived from seaweed and considered safe for human use, which is used alone or in combination with other materials such as poly-L-lysine (PLL), chitosan, cellulose sulfate, aminopropyl silicate, polytheylene glycol, or by itself.[100–102] Alternatives to alginate, such as polylactide-*co*-glycoloide, poly(ethylene glycol) diacrylate, and hydroxyl-ethyl methacrylate–methyl methacrylate (HEMA–MMA), have also been studied.[103–105] Reviews are available that compare different types of microencapsulation materials, describe techniques for forming microcapsules; Refs. 98 and 106 discuss the current capabilities of microcapsules, and examine what issues still limit their effectiveness.[18,107]

1. Alginate and Poly-L-lysine

Alginate and poly-L-lysine microcapsules were first shown to provide immunoprotection in 1980,[108] and over 20 years later it is still the most widely used system. Alginate has remained the most common material used for microencapsulation because the encapsulation procedure is simple, and purified alginate is biocompatibile. Alginate is a polysaccharide made up of blocks of mannuronic (M) and guluronic (G) acid. Alginate beads are formed by dropwise addition of alginate into a solution of a bivalent cation, typically calcium chloride or barium chloride. The cation cross-links the monomers of alginate and rapidly forms a gel bead. Beads cross-linked with barium ions have a greater mechanical stability than those cross-linked with calcium ions.[109]

Droplets of alginate are formed by an air droplet generator or an electronic droplet generator. The electronic droplet generator is able to create beads of a smaller size, which is advantageous as it will reduce transplant volume.[110] Electronic droplet generation reduces the bead size because, in the presence of an applied voltage, the electrical forces and gravitational forces reduce the critical volume for drop detachment.[111] Smaller bead sizes are also produced by emulsification and internal gelation procedures.[112] Addition of nongelling sodium ions to the Ca^{2+} gellation solution produces beads that have a homogeneous concentration of alginate, whereas use of Ba^{2+} or Sr^{2+} for gellation leads to inhomogeneous beads that are more stable and less permeable.[109,113] The positively charged polycation PLL is used to coat the beads through brief incubation in a low concentration solution; the resulting polyelectrolyte complex forms a semipermeable layer that serves to enhance the molecular size selectivity of the bead. As incubation time and the molecular weight of PLL increase, so does the mechanical stability of the capsule[114] and the diffusion barrier to cytokine permeation.[115] The last step in microcapsule formation is to coat the bead outer surface with alginate by placing the PLL coated bead in a dilute alginate solution. This step hides PLL from the immune system because PLL is not biocompatible and initiates a fibrotic response.

Alginate is a biocompatible material, but virtually all materials induce some foreign body response by the immune system. The foreign body response inhibits transport to the islets and causes an increase in central islet necrosis.[116] Purification of alginate and smoother capsules greatly improve its biocompatibility, and, as a result, increases the survival of encapsulated islets.[72,116–118] Studies on the biocompatibility of alginate itself have shown that the M or MG blocks are potent stimulators of IL-1 and TNF production, while G blocks only stimulate a minimal secretion of these cytokines by macrophages.[119] In studies of purified alginate PLL microcapsules *in vivo*, the responses against intermediate-G alginate capsules are always less severe than against high-G alginate capsules.[120] Furthermore, explanted high-G alginate PLL capsules contain 20% more PLL but there are fewer molecular interactions between the alginate and PLL in high-G capsules, making it conceivable that some PLL diffuses out of the capsules after implantation and induces the inflammatory response.[121] The findings from these and related studies suggest that using highly purified alginate with a G content of 40 to 45% reduces the biological response after implantation.[122] Ninety percent of empty capsules, with intermediate-G content, implanted in the peritoneal cavity remain free of fibrotic overgrowth after 24 months, as compared to unpurified alginate capsules which are completely overgrown within a month.[122] Others find that the use of an epimerized alginate coating on alginate–PLL–alginate capsules can improve the biocompatibility of the capsules but does not completely eliminate the effects that PLL has on the biocompatibility of the capsules.[123]

Many studies have examined the ability of microcapsules to protect against a variety of immune responses. Increasing the incubation time in PLL, from 5 to 10 min, does not interfere with insulin diffusion from encapsulated islets but provides more protection against cytokines.[115] Tissue responses to implanted empty capsules are the strongest during the first week following implantation. The numbers of granulocytes, macrophages, and basophils in the vicinity of the capsule are highest initially and decrease over the course of the first week of implantation.[124] Overgrowth of a small fraction of the implanted capsules containing allogeneic islets can have detrimental effects on the function of encapsulated islets by activated macrophages. *In vitro*, only islet-containing capsules, but not empty capsules, cause macrophage activation. Therefore, it is the islets themselves and not the biomaterial that causes macrophage activation.[125] Prolonged survival of xenotransplantations of neonatal porcine islets into mice has been achieved with microencapsulation and a brief treatment of the recipient with costiumulatory blockade mediated with CTLA4-Ig. This agent blocks the costimulation signal of $CD4^+$ T-cells, thereby preventing activation of host effector cells.[126]

A common factor in many of the immune response pathways is the production of NO by immune cells, particularly macrophages. An *in vitro* study[58] has demonstrated that NO secreted by activated macrophages is able to cause lysis of islets despite being encapsulated in alginate. Inhibition of NO formation using enzyme inhibitors and scavenging of NO by hemoglobin from coencapsulated autologous red blood cells both protect encapsulated islets from destruction.

Alginate PLL microcapsules have been successfully transplanted in animals without immunosuppression. For example, islet allotransplantation in streptozotocin induced diabetic rats produced reversal of diabetes, but with impaired insulin secretion.[74] Islet isografts and allografts have survived for similar periods of time when implanted in intermediate-G purified alginate beads, indicating that the capsules provided adequate immunoprotection, although 10% of the capsules were overgrown after 200 days.[122] In the most dramatic demonstration to date, xenotransplantation of porcine islets, microencapsulated in Alginate PLL beads, into spontaneously diabetic monkeys, resulted in recipients that were insulin free for up to 2 years.[29] Unfortunately, there has been no reported replication of these results by others.

2. Alginate Alone

Alginate has been used alone to encapsulate islets in order to maximize biocompatibility by eliminating the use of PLL or other polycations that produce a surface immunobarrier. Highly purified barium alginate microcapsules protected islets against allorejection and autoimmunity in mice for up to 400 days.[127] Cytokines could still permeate the gel, but islet survival was not impaired because of the lack of a cellular response to the implant.[127] Barium alginate provided immune protection for porcine neonatal pancreatic cell clusters in streptozotocin induced diabetic mice against xenorejection for over 20 weeks.[24,128] The neonatal pancreatic cell clusters were able to survive and mature within the microcapsule. Beneficial effects of low-temperature long-term culture as an immunoaltering protocol have been demonstrated on initial and long-term function of allograft islet transplantation. The time of graft survival is similar to microencapsulated syngeniec transplants, which is longer than that of microencapsulated allogenic transplants.[129]

3. Other Alginate Systems

Alginate has been combined with other materials to improve its mechanical stability and biocompatibility as compared to that of alginate and PLL microcapsules. The polycation chitosan has been used instead of PLL to form a more thermodynamically stable outer permselective membrane.[100,101] Changing the number of chitosan layers that alternate with alginate can alter the membrane molecular weight cutoff.[100] Alginate beads with an aminopropyl-silicate membrane provide a 60 kDa molecular weight cutoff.[102] Improved mechanical stability has been obtained with beads made from a mixture of interpenetrating alginate and polyethylene glycol. The alginate is gelled first by ion cross-linking, and polyethylene glycol is then photopolymerized to form a more stable covalent bond.[130,131]

A comprehensive systematic study has been carried out to develop improved microcapsules for islet transplantation. A total of 75 polyanions (to form the inner core) and polycations (to form the outer membrane), leading to over a thousand binary polyelectrolyte combinations, were evaluated on the basis of cytotoxicity and capsule mechanical strength, shape, surface smoothness, stability, and swelling or shrinking. The use of anion and cation blends was most desirable because components that affected strength and permeability properties could be uncoupled and capsules could be created with the permeability and strength properties desired. The best alternative to alginate/PLL was a multicomponent system that employed alginate and cellulose sulfate as the anion blend and poly(methylene-*co*-guanidine) and $CaCl_2$ (together with NaCl) as the cation blend.[132–134] The multicomponent capsules with rat islets transplanted into streptozotocin-induced diabetic mice and NOD mice provided normalization of blood glucose levels for a median of 110 and 50 days, respectively.[135] Fibrosis was not observed in the streptozotocin transplants but was in the NOD mice transplants, indicating that the reaction in NOD mice was derived from an immune or auto-immune response.[135]

4. Other Microencapsulation Materials

Alternatives to alginate have been under study in order to improve on some of the limitations of alginate: instability for long periods of time, batch to batch variation, and elicitation of a foreign body reaction. Three examples are polylactide-*co*-glycolide (PLGA), HEMA–MMA, and poly(ethylene glycol) (PEG) diacrylate.[104,105] Islets are microencapsulated in PLGA through interfacial precipitation.[105] Streptozotocin-induced diabetic rats that were transplanted with PLGA microencapsulated islets had reduced blood glucose levels, compared to the untreated controls, but were still hyperglycemic.[105] Conformal coating of islets by HEMA–MMA produces capsules that have a very thin wall (10 to 25 μm) that greatly reduces the size of the capsule.[104,136] The conformal coated islets are placed within agarose slabs to prevent aggregation, prevent cellular attachment, and protect islets from stresses within the peritoneal cavity but the agarose is an additional transport barrier.[136] Conformal coatings of photopolymerized PEG diacrylate are formed with acceptable losses in porcine islet viability and can correct streptozotocin diabetes in rats.[103] Chitosan–polyvinyl pyrrolidone hydrogels have recently been suggested for islet microencapsulation. The material does not activate endothelial cells, does not promote fibroblast adhesion, and is noncytotoxic *in vitro* to islets.[137,138]

5. *In Vivo* Performance of Microencapsulated Islets

Table 37.1 summarizes the successful experiments that have cured animals of diabetes for varying periods of time. The entries are relatively complete from 1996 on; a few selected pre 1996 studies are shown. There have been very few studies with nonrodent models of diabetes. All but one of the studies have used alginate as a component of the microcapsule. Only two studies used canine[141] or monkey recipients.[29] Only two human studies were done, and both employed systemic immunosuppression and exogenous insulin, with long term graft functioning that resulted in better glycemic control than before transplantation.[143,144,146] When purified alginate was used, isografts and allografts remained functional for more than 200 days when less than 13,000 islets/kg were transplanted and for more than a year when 38,000 islets/kg were transplanted. The time of allograft functioning was similar for streptozotocin and spontaneous diabetic recipients, indicating that the capsule adequately protected against auto-immune and alloimmune reactions. The studies that performed xenotransplantations usually used more islets (40,000 to 50,000 islets/kg) and resulted in shorter times of graft functioning. Further improvements are needed with xenografts, and it is unlikely that changes in permeability properties can achieve what is needed if the external cellular response is caused by the binding of shed antigen to antigen presenting cells. Some approaches for further study include local or short-term systemic immunosuppression (e.g., use of CTLA4-Ig[126] to inhibit costimulation), immunomodulation to inhibit activation of T-cells and macrophages, and means for rendering inactive the harmful agents secreted by these cells.

C. Mass Transfer

1. Mass Transfer Limitations

Islets contained within the pancreas are very well vascularized with blood at arterial pO_2. When islets are removed from the pancreas and implanted elsewhere, they loose their blood supply and must rely on diffusion for nutrient delivery and oxygen transport to the tissues. In the case of naked islet transplantations, over time the islets are revascularized and regain their blood supply. Even then, transplanted islets may be exposed to a lower pO_2 than in the pancreas.[147] Islets contained within devices never become revascularized and must rely on diffusion through the device for nutrient supply, insulin removal, and waste removal. Fibrotic tissue formation around an implant increases the diffusion distance of nutrients to islets and consumes nutrients, which worsens the mass transfer problem and decreases the survival of transplanted islets.[148]

Hypoxic conditions can have deleterious effects on islet function and viability. Hypoxic conditions *in vitro* result in normal basal insulin secretion rates, but lower insulin secretion rates in response to glucose stimulation.[149–153] Viability of encapsulated islets *in vitro* is reduced as a result of necrosis associated with depletion of ATP and to some extent apoptosis caused by hypoxia.[18,153] Most importantly, hypoxia is associated with increased expression of inducible iNOS mRNA in the islets, which likely leads to increased secretion of NO by the islets themselves and to increased expression of monocyte chemoattractant protein mRNA, suggesting that islets contribute to their own graft failure by attracting cytokine-producing macrophages.

Theoretical studies have been done to investigate the relation between device design characteristics and transport limitations. Devices can be of many shapes, such as spheres, cylinders and slabs. Calculations show that a spherical geometry will have the greatest nonhypoxic depth.[154] Analysis also shows that fibrotic overgrowth of the implant greatly reduces oxygen transport and starving the enclosed islet, and that vascularization of the membrane increases the viable thickness of tissues.[73]

Insulin secretion kinetics are also influenced by mass transfer limitations. Finite element modeling has demonstrated that device geometry should be optimized and oxygen transport enhanced in order to maximize insulin secretion rate.[155] Successful islet transplantations also require an ideal transplantation site. Microcapsules are typically implanted within the

TABLE 37.1
Transplantation of Islets into the Peritoneal Cavity of Diabetic Animals

Microcapsule Materials	Donor Species	Recipient Species	Diabetes Type	Graft Type	Immuno Suppression	Islet Load #/kg	Function Days	Comments	Reference
Ca Alg/PLL/PEE	Rat	Rat	STZ	Iso		11,500	20		108
Ca Alg/PLL	Wistar rats	BALB/c mice	STZ	Xeno		50,000	144		139
Ca Alg/PLO	Wistar rats	BALB/c mice	STZ	Xeno		50,000	93		
Ca Alg/PLL	Wistar rats	BB rats	Spontaneous	Allo		17,408	190		140
Ca Alg/PLL	Dog	Dog	Spontaneous	Allo	Yes	17,500	172	Cyclosporine A (CsA)	141
Ca Alg/PLL	Pig	Monkey	Spontaneous	Xeno		22,000	803	Multiple transplants	29
Ca Alg/PLL	AO/G rats	AO/G rats	STZ	Iso		12,623	133	Purified alginate	142
Ca Alg/PLL	Lewis rats	AO/G rats	STZ	Allo		12,623	140		
Alginate/PLL	Neonatal pig	Nude mouse	STZ	Xeno		80,000	>294		32
Alg/Cellulose Sulfate/	SD rats	C57/Bl6 mice	STZ	Xeno		40,000	300		135
$NaCl/CaCl_2$/poly (methylene-*co*-guanidine)	SD rats	NOD mice	Spontaneous	Xeno		40,000	180		
Poly(Ethylene Glycol)	Pig	Athymic mice	STZ	Xeno		325,000	110		103
Ba Alg	Wistar rats	Lewis rats	STZ	Allo		11,667	21		129
Ba Alg	Lewis rats	Lewis rats	STZ	Iso		11,667	>105		
Ba Alg	Wistar rats	Lewis rats	STZ	Allo		11,667	>105	Low temperature culture	
Ca Alg/PLL	Human	Human	Spontaneous	Allo	Yes	15,000	1740	Exongenous insulin	143
Ca Alg/PLL	Neonatal pig	Human	Spontaneous	Xeno	Yes	15,000	>420	Exongenous insulin	144
Ca Alg/PLO	SD rats	CD-1 mice	STZ	Xeno		40,000	>100	With sertoli cells	66
Ca Alg/PLO	SD rats	CD-1 mice	STZ	Xeno		40,000	>100		
Ba Alg	B6AF1 mice	NOD mice	Spontaneous	Allo		38,000	>385		127
Ba Alg	NOD mice	NOD mice	Spontaneous	Iso		38,000	>343		
Ba Alg	B6AF1 mice	BALB/c	STZ	Allo		38,000	>385		
Ca Alg/PLL	SD rats	B6AF1 mice	STZ	Xeno		40,000	>63		145
Ca Alg/PLL	Neonatal pig	NOD mice	Spontaneous	Xeno	Yes	480,000	230	Treated with CTLA4-Ig	126
Ca Alg/PLL	Neonatal pig	NOD mice	Spontaneous	Xeno		480,000	25		
Ca Alg/PLL	Lewis rats	AO rats	STZ	Allo		12,623	>200	Purified alginate	122
Ca Alg/PLL	AO rats	AO rats	STZ	Iso		12,623	>200	Purified alginate	
Ba Alg	Neonatal pig	B6AF1 mice	STZ	Xeno		400,000	>140		24

peritoneal cavity. Comparison of insulin infused into the portal vein and the peritoneal cavity showed that insulin delivery from the peritoneal cavity is delayed and at lower levels, indicating that transplanted islets need to be in close contact with the vasculature to see proper insulin response kinetics.[142,156] Encapsulated rat islets implanted into the peritoneal cavity of mice showed a delayed rise of plasma C-peptide levels but efficient glucose clearance, indicating that the kinetics of insulin release from microencapsulated islets in the peritoneal cavity may be adequate for proper glucose control.[145] The implantation site also needs to expose islets to high oxygen partial pressures in order to deliver adequate amounts of oxygen, and this can also be provided by close proximity to the vasculature.

Successful outcomes for islet transplantations will require prolonged survival of islets after implantation. If the islets can be adequately protected from the immune system, they will still have to be properly nourished by devices that have been engineered to provide the necessary amount of oxygen and nutrients.

2. Methods to Improve Mass Transfer

To maintain islet viability and function within devices and microcapsules, methods are being developed to enhance mass transfer, especially that of oxygen. Methods include use of vascularizing membranes, *in situ* oxygen generation, use of thinner encapsulation membranes, and enhancing oxygen carrying capacity in encapsulated materials.

Transport will be increased to encapsulated islets if the vasculature is brought as close to the protective membrane as possible. The use of a microporous material that induces neovascularization at the material tissue interface[157–159] is the basis upon which the Theracyte planar diffusion chambers are built, since a planar device without close vascular structures would fail at islet encapsulation when implanted subcutaneously, as is the case for the Theracyte.[73] Membranes of this type derive their vascularizing capabilities from the geometric details of the membrane, which allows penetration by macrophages that maintain their rounded morphology, and not its chemical structure. A vascularizing membrane can be used in combination with an immunoprotective membrane, which has smaller pores to prevent inflammatory cell penetration (or even to prevent entry by soluble immune proteins), to create a device that can enhance mass transfer to the implanted islets, such as the Baxter Theracyte System. The vascularizing membrane used in the Theracyte is an expanded PTFE structure. An expanded PTFE structure has also been reported as a prevascularized solid support in the peritoneal space. The blood vessels begin to form within 5 days of implantation,[159] and this process can be enhanced when VEGF is infused in the device.[160] The addition of VEGF infusion through the membrane increases the delivery of insulin into the circulation, indicating an enhancement of mass transfer.[160]

Transport of insulin, glucose, and oxygen can generally be enhanced if the diffusion distance through the encapsulating material is decreased. A drawback of reducing the diffusion distance is that it may enhance the flux of shed antigen and thereby increase the susceptibility of the islets to an immune reaction. The electric droplet generation of alginate beads allows for smaller beads to be produced, which minimizes the transplant volume and diffusion distance.[110] The HEMA–MMA conformal coating of islets is an example of a minimal volume capsule that greatly decreases the diffusion distance.[104,136] Minimal volume alginate microcapsules can be prepared by a two-phase aqueous emulsification procedure in which the diameter of the microcapsule is equal to, or slightly greater than, the islet itself.[161] The minimal volume alginate microcapsule had limited immunoprotective capabilities due to the small diffusion distances; it could provide protection against allograft but not xenograft rejection.[161]

The supply of oxygen to islets can be increased if oxygen is generated near the transplanted islets. *In situ* oxygen generation can be carried out in a device that electrolytically decomposes water.[162] The device consisted of a planar diffusion chamber with an immunobarrier layer, vascularizing layer, and support layer on one side, combined with an electrolyzer on the other side.[162] When βTC3 cells are cultured under nitrogen a viable layer of cells survives, the thickness of which increases when the oxygen flux to the cells is increased.[22,162] Similarly, insulin secretion remains normal during anoxic culture when oxygen is provided by the electrolyzer.[163]

Oxygen carriers in the encapsulated material or islet culture medium have been investigated in order to minimize oxygen supply limitations, increase the partial pressure of oxygen around the islets, and thereby improve islet function and viability. Perfluorocarbon emulsions and hemoglobin, both of which have been under development as blood substitutes, have been used in the presence of islets *in vivo* and *in vitro*. Perfluorocarbon emulsions[164,165] are in phase three trials as blood substitutes. Perfluorocarbons have also been used to preserve organs prior to transplant, including the pancreas, and are more effective than blood allowing for longer preservation times.[166–168] A perfluorocarbon emulsion added to islet culture medium improves islet function and viability.[169] Perfluorocarbons have greatly increased (up to 25-fold) oxygen solubility, and therefore oxygen permeability, as compared to water. If used in a microcapsule, the effective permeability to oxygen can be increased substantially. Silicone oils, which have about 20-fold higher oxygen solubility, have been microencapsulated in membranes to enhance oxygen transport for bioreactor oxygenation.[170]

Hemoglobin is the oxygen carrier in red blood cells. When hemoglobin is contained within a liquid membrane, oxyhemoglobin facilitates the transport of oxygen by being a diffusive carrier between the low and high oxygen side of the membrane.[171] When immobilized, hemoglobin cannot facilitate transport of oxygen.[172,173] Nonetheless, encapsulating hemoglobin with islets has improved islet viability and function as compared to encapsulation without hemoglobin.[174,175] The basis for the improvement in not clear;[173] it may result from the ability of hemoglobin to scavenge NO,[174] which itself is produced by islets as a consequence of hypoxia.[153]

D. NMR Spectroscopy and Imaging

Successful islet transplantations require that the tissue at the time of implantation is healthy and viable. NMR spectroscopy is a useful tool to evaluate islet viability and identify damaged tissue because multiple metabolites can be identified and quantified from one spectrum.[176] By implanting alginate beads that contained liquid perfluorocarbons or perfluorocarbon emulsions ^{19}F-NMR can be used to image the alginate beads and detect the oxygen partial pressure in different implantation sites in rats.[177,178]

After implantation, it is desirable to monitor the viability of the transplanted islets. The function of a graft is usually monitored by following the blood glucose level of the recipient. NMR has been used to visualize islets or cells *in vitro* and determine the location of alginate microcapsults *in vivo*.[179] MRI, using a targeted contrast agent, has been used to detect apoptosis at an early stage in tumors with high resolution and could also be used to monitor islet viability.[180] Improved methods to monitor the functionality of an islet transplant will allow for better care of patients after the transplantation procedure.

VI. FUTURE DIRECTIONS

The goal remains to maintain viable islets after transplantation without immunosuppressive therapy, which will require the use of a material to encapsulate islets and protect them from the immune system. Many materials have been investigated, but to date no material can block all components of the immune system that may be attacking the transplanted islets and still facilitate adequate transport of oxygen and insulin. Work is needed to develop even better materials for islet encapsulation that do not initiate a foreign body reaction and protect the islets from immune attack. It is possible that no sole encapsulation material will ever be able to fully protect islets. Other means, such as locally delivered immunosuppressive agents and scavengers of toxic molecules, may be needed to provide the extra protection that islets require. It may be necessary to transfect islets with antiapoptotic genes to protect them when exposed to inflammatory stresses and metabolic limitations, enhancing their ability to survive these adverse conditions.[181]

Islet viability and function will not be maintained, even if the effects of the immune system have been blocked, unless sufficient oxygen supply is available. Methods to enhance oxygen transport will need to be employed, such as vascularizing membranes, *in situ* oxygen generation, minimizing diffusion distances, or increasing oxygen solubility in the encapsulating material to maintain islet viability and function.

Islets are continually generated throughout the lifetime of a person, but have not been shown to expand *in vitro* or after transplantation. Therefore, any successful islet transplantation has a limited lifetime for which it will be effective, as the islets will eventually die off. Therefore, methods to promote islet regeneration will need to be discovered or devices used for islet transplantation will need to be easily refillable or replaceable. This will enable transplanted islets to continue to provide the necessary insulin to a patient for the rest of their life.

REFERENCES

1. American Diabetes Association, Economic costs of diabetes in the US in 2002, *Diabetes Care*, 26(3), 917–932, 2003.
2. Centers for Disease Control. *National Diabetes Fact Sheet: General Information and National Estimates on Diabetes in The United States, 2000*, US Department of Health and Human Services, Centers for Disease Control and Prevention, Atlanta, GA, 2002.
3. Sutherland, D. E. R., Is immunosuppression justified for nonuremic diabetic patients to keep them insulin independent? (the argument for), *Transplant. Proc.*, 34, 1927–1928, 2002.
4. The Diabetes Control and Complications Trial Research Group, The effect of intensive treatment of diabetes on the development and progression of long-term complications in insulin-dependent diabetes mellitus, *N. Engl. J. Med.*, 329(14), 977–986, 1993.
5. Plotnick, L. P., Clark, L. M., Brancati, F. L., and Erlinger, T., Safety and effectiveness of insulin pump therapy in children and adolescents with type 1 diabetes, *Diabetes Care*, 26(4), 1142–1146, 2003.
6. Kost, J., Mitragotri, S., Gabbay, R. A., Pishko, M., and Langer, R., Transdermal monitoring of glucose and other analytes using ultrasound, *Nat. Med.*, 6(3), 347–350, 2000.
7. Mitragori, S., Coleman, M., Kost, J., and Langer, R., Analysis of ultrasonically extracted interstitial fluid as a predictor of blood glucose levels, *J. Appl. Physiol.*, 89, 961–966, 2000.
8. Cefalu, W. T., Evaluation of alternative strategies for optimizing glycemia: progress to date, *Am. J. Med.*, 113(6A), 23S–35S, 2002.
9. Gough, D., Armour, J., and Baker, D., Advances and prospects in glucose assay technologies, *Diabetologia*, 40, S102–S107, 1997.
10. Gerritsen, M., Jansen, J., and Lutterman, J., Performance of subcutaneously implanted sensors for continuous monitoring, *Netherlands J. Med.*, 54, 167–179, 1999.
11. Koschchinsky, T., and Heinemann, L., Sensors for glucose monitoring: technical and clinical aspects, *Diabetes Metab. Res. Rev.*, 17, 113–123, 2001.
12. Abel, P. U., and von Woedtke, T., Biosensors for in vivo glucose measurement: can we cross the experimental stage, *Biosens. Bioelectron.*, 17, 1059–1070, 2002.
13. Frost, M. C., and Meyerhoff, M. E., Implantable chemical sensors for real-time clinical monitoring: progress and challenges, *Curr. Opin. Chem. Biol.*, 6, 633–641, 2002.
14. Galaev, I. Y., and Mattiasson, B., "Smart" polymers and what they could do in biotechnology and medicine, *Trends Biotechnol.*, 17, 335–340, 1999.
15. Shapiro, A. M. J., Lakey, J. R. T., Ryan, E. A., Korbutt, G. S., Toth, E., Warnock, G. L., Kneteman, N. M., and Rajotte, R. V., Islet transplantation in seven patients with type 1 diabetes mellitus using a glucocorticoid-free immunosuppressive regimen, *N. Engl. J. Med.*, 343(4), 230–238, 2000.
16. Burridge, P. W., Shapiro, A. M. J., Ryan, E. A., and Lakey, J. R. T., Future trends in clinical islet transplantation, *Transplant. Proc.*, 34, 3347–3348, 2002.
17. Weir, G. C., and Bonner-Weir, S., Scientific and political impediments to successful islet transplantation, *Diabetes*, 46(8), 1247–1256, 1997.

18. De Vos, P., Van Straaten, J. F. M., Nieuwenhuizen, A. G., de Groot, M., Ploeg, R. J., De Haan, B. J., and Van Schilfgaarde, R., Why do microencapsulated islet grafts fail in the absence of fibrotic overgrowth?, *Diabetes*, 48(7), 1381–1388, 1999.
19. Colton, C. K., and Avgoustiniatos, E. S., Bioengineering in development of the hybrid artificial pancreas, *Trans. ASME*, 113, 152–170, 1991.
20. Colton, C. K., Engineering issues in islet immunoisolation, In *Pancreatic Islet Transplantation Volume III: Immunoisolation of Pancreatic Islets*, Lanza, R. P., and Chick, W. L., eds., R.G. Landes Company, Austin, pp. 13–20, 1994.
21. Colton, C. K., Implantable biohybrid artificial organs, *Cell Transplant.*, 4(4), 415–436, 1995.
22. Avgoustiniatos, E. S., Wu, H., and Colton, C. K., Engineering challenges in immunoisolation device development, In *Principles of Tissue Engineering*, Lanza, R. P., Langer, R., and Vacanti, J., eds., Academic Press, San Diego, CA, pp. 331–350, 2000.
23. O'Connell, P., Commentaries: pancreatic islet xenotransplantation, *Xenotransplantation*, 9, 367–375, 2002.
24. Omer, A., Duvivier-Kali, V. F., Trivedi, N., Wilmot, K., Bonner-Weir, S., and Weir, G. C., Survival and maturation of microencapsulated porcine neonatal pancreatic cell clusters transplanted into immunocompetent diabetic mice, *Diabetes*, 52, 69–75, 2003.
25. Tatarkiewicz, K., Hollister-Lock, J., Quickel, R. R., Colton, C. K., Bonner-Weir, S., and Weir, G. C., Reversal of hyperglycemia in mice after subcutaneous transplantation of macroencapsulated islets, *Transplantation*, 67(5), 665–671, 1999.
26. Hsu, B. R.-S., Hsueh, C., Fu, S.-H., Hsu, S., and Juang, J.-H., A removable tubing for implanting islet graft and studying immunosuppression, *Transplant. Proc.*, 34, 1462, 2002.
27. Colton, C. K., The engineering of xenogeneic islet transplantation by immunoisolation, *Diabetes, Nutrition Metab.*, 5(Suppl. 1), 145–149, 1992.
28. Bach, F. H., and Fineberg, H. V., Call for moratorium on xenotransplants, *Nature*, 391, 326, 1998.
29. Sun, Y., Ma, X., Zhou, D., Vacek, I., and Sun, A. M., Normalization of diabetes in spontaneously diabetic cynomologous monkeys by xenografts of microencapsulated porcine islets without immunosuppression, *J. Clin. Invest.*, 98(6), 1417–1422, 1996.
30. Boyce, N., Down on the organ farm, *US News World Rep.*, 134(21), 47, 2003.
31. Binette, T. M., Dufour, J. M., and Korbutt, G. S., *In vitro* maturation of neonatal porcine islets: a novel method for the study of islet development and xenotransplantation, *Ann. NY Acad. Sci.*, 944, 47–61, 2001.
32. Korbutt, G. S., Elliott, J. F., Ao, Z., Flashner, M., Warnock, G. L., and Rajotte, R. V., Microencapsulation of neonatal porcine islets: long-term reversal of diabetes in nude mice and in vitro protection from human complement mediated cytolysis, *Transplant. Proc.*, 29, 2128, 1997.
33. Bonner-Weir, S., and Sharma, A., Pancreatic stem cells, *J. Pathol.*, 197, 519–526, 2002.
34. Kaczorowki, D. J., Patterson, E. S., Jastromb, W. E., and Shamblott, M. J., Glucose-responsive insulin-producing cells from stem cells, *Diabetes/Metab. Res. Rev.*, 18, 442–450, 2002.
35. Scharfmann, R., Alternative sources of beta cells for cell therapy of diabetes, *Eur. J. Clin. Invest.*, 33, 595–600, 2003.
36. Bonner-Weir, S., Taneja, M., Weir, G. C., Tatarkiewicz, K., Song, K.-H., Sharma, A., and O'Neil, J. J., *In vitro* cultivation of human islets from expanded ductal tissue, *Proc. Natl Acad. Sci.*, 97(14), 7999–8004, 2000.
37. Ramiya, V. K., Maraist, M., Arfors, K. E., Schatz, D. A., Peck, A. B., and Cornelius, J. G., Reversal of insulin-dependent diabetes using islets generated *in vitro* from pancreatic stem cells, *Nat. Med.*, 6(3), 278–282, 2000.
38. Peck, A. B., Cornelius, J. G., Chaudhari, M., Shatz, D., and Ramiya, V. K., Use of *in vitro*-generated, stem cell-derived islets to cure type 1 diabetes: how close are we?, *Ann. NY Acad. Sci.*, 958, 59–68, 2002.
39. Gu, G., Dubauskaite, J., and Melton, D. A., Direct evidence for the pancreatic lineage: NGN3 + cells are islet progenitors and are distinct from duct progenitors, *Development*, 129, 2447–2457, 2002.
40. Treutelaar, M. K., Skidmore, J. M., Dias-Leme, C. L., Hara, M., Zhang, L., Simeone, D., Martin, D. M., and Burant, C. F., Nestin-lineage cells contribute to the microvasculature but not endocrine cells of the islet, *Diabetes*, 52, 2503–2512, 2003.

41. Soria, B., Roche, E., Berna, G., Leon-Quinto, T., Reig, J. A., and Martin, F., Insulin-secreting cells derived from embryonic stem cells normalize glycemia in streptozotocin-induced diabetic mice, *Diabetes*, 49(2), 157–162, 2000.
42. Lumelsky, N., Blondel, O., Laeng, P., Velasco, I., Ravin, R., and McKay, R., Differentiation of embryonic stem cells to insulin-secreting structures similar to pancreatic islets, *Science*, 292, 1389–1394, 2001.
43. Rajagopal, J., Anderson, W. J., Kume, S., Martinez, O. I., and Melton, D. A., Insulin staining of ES cell progeny from insulin uptake, *Science*, 299, 363, 2003.
44. Hori, Y., Rulifson, I. C., Tsai, B. C., Heit, J. J., Cahoy, J. D., and Kim, S. K., Growth inhibitors promote differentiation of insulin-producing tissue from embryonic stem cells, *Proc. Natl Acad. Sci.*, 99(25), 16105–16110, 2002.
45. Blyszczuk, P., Czyz, J., Kania, G., Wagner, M., Roll, U., St-Onge, L., and Wobus, A. M., Expression of Pax4 in embryonic stem cells promotes differentiation of nestin-positive progenitor and insulin-producing cells, *Proc. Natl Acad. Sci.*, 100(3), 998–1003, 2003.
46. Kahan, B. W., Jacobson, L. M., Hullett, D. A., Ochoada, J. M., Oberley, T. D., Lang, K. M., and Odorico, J. S., Pancreatic precursors and differentiated islet cell types from murine embryonic stem cells: an in vitro model to study islet differentiation, *Diabetes*, 52, 2016–2024, 2003.
47. Assady, S., Maor, G., Amit, M., Itskovitz-Eldor, J., Skorecki, K. L., and Tzukerman, M., Insulin production by human embryonic stem cells, *Diabetes*, 50(8), 1691–1697, 2001.
48. Efrat, S., Fusco-DeMane, D., Lemberg, H., Al Emran, O., and Wang, X., Conditional transformation of a pancreatic β-cell line derived from transgenic mice expressing a tetracycline-regulated oncogene, *Proc. Natl Acad. Sci.*, 92, 3576–3580, 1995.
49. Fleischer, N., Chen, C., Manju, S., Leiser, M., Rossetti, L., Pralong, W., and Efrat, S., Functional analysis of a conditionally transformed pancreatic beta-cell line, *Diabetes*, 47(9), 1419–1425, 1998.
50. Efrat, S., Genetically engineered pancreatic β-cell lines for cell therapy of diabetes, *Ann. NY Acad. Sci.*, 875, 286–293, 1999.
51. Hohmeier, H. E., Mulder, H., Chen, G., Henkel-Rieger, R., Prentki, M., and Newgard, C. B., Isolation of INS-1-derived cell lines with robust ATP-sensitive K^+ channel-dependent and independent glucose-stimulated insulin secretion, *Diabetes*, 49, 424–430, 2000.
52. Chen, G., Hohmeier, H. E., Gasa, R., Tran, V. V., and Newgard, C. B., Selection of insulinoma cell lines with resistance to interleukin-1β and γ-interferon-induced cytotoxicity, *Diabetes*, 49, 562–570, 2000.
53. Efrat, S., Cell replacement therapy for type 1 diabetes, *TRENDS Mol. Med.*, 8(7), 334–339, 2002.
54. Rokstad, A. M., Kulseng, B., Strand, B. L., Skjark-Braek, G., and Espevik, T., Transplantation of alginate microcapsules with proliferating cells in mice: capsular overgrowth and survival of encapsulated cells of mice and human origin, *Ann. NY Acad. Sci.*, 944, 216–225, 2001.
55. Eizirik, D. L., and Mandrup-Poulsen, T., A choice of death — the signal-transduction of immune-mediated beta-cell apoptosis, *Diabetologia*, 44, 2115–2133, 2001.
56. Thomas, H. E., and Kay, T. W. H., Beta cell destruction in the development of autoimmune diabetes in the non-obese diabetic (NOD) mouse, *Diabetes/Metab. Res. Rev.*, 16, 251–261, 2000.
57. Kaneto, H., Fujii, J., Seo, H. G., Suzuki, K., Matsuoka, T.-A., Nakamura, M., Tatsumi, H., Yamasaki, Y., Kamada, T., and Taniguchi, N., Apoptotic cell death triggered by nitric oxide in pancreatic beta-cells, *Diabetes*, 44(7), 733–738, 1995.
58. Wiegand, F., Kroncke, K.-D., and Kolb-Bachofen, V., Macrophage-generated nitric oxide as cytotoxic factor in destruction of alginate-encapsulated islets, *Transplantation*, 56(5), 1206–1212, 1993.
59. Hunger, R. E., Muller, S., Laissue, J. A., Hess, M. W., Carnaud, C., Garcia, I., and Mueller, C., Inhibition of submandibular and lacrimal gland infiltration in nonobese diabetic mice by transgenic expression of soluble TNF-receptor p55, *J. Clin. Invest.*, 98(4), 954–961, 1996.
60. Welsh, N., Bendtzen, K., and Welsh, M., Expression of an insulin/interleukin-1 receptor antagonist hybrid gene in insulin-producing cell lines (HIT-T15 and NIT-1) confers resistance against interleukin-1-induced nitric oxide production, *J. Clin. Invest.*, 95(4), 1717–1722, 1995.
61. Gray, D. W., An overview of the immune system with specific reference to membrane encapsulation and islet transplantation, *Ann. NY Acad. Sci.*, 944, 226–239, 2001.
62. Waegell, W., Babineau, M., Hart, M., Dixon, K., McRae, B., Wallace, C., Leach, M., Ratnofsky, S., Belanger, A., Hirst, G., Rossini, A., Appel, M., Mordes, J., Greiner, D., and Banerjee, S., A420983,

a novel, small molecule inhibitor of LCK prevents allograft rejection, *Transplant. Proc.*, 34, 1411–1417, 2002.
63. Sanchez-Fueyo, A., Weber, M., Domenig, C., Strom, T. B., and Zheng, X. X., Tracking the immunoregulatory mechanisms active during allograft tolerance, *J. Immunol.*, 168, 2274–2281, 2002.
64. Lau, H. T., Yu, M., Fontana, A., and Stoeckert, C. J., Prevention of islet allograft rejection with engineering myoblasts expressing FasL in mice, *Science*, 273, 109–112, 1996.
65. Korbutt, G. S., Elliott, J. F., and Rajotte, R. V., Cotransplantation of allogeneic islets with allogeneic testicular cell aggregates allows long-term graft survival without systemic immunosuppression, *Diabetes*, 46(2), 317–322, 1997.
66. Calafiore, R., Luca, G., Calvitti, M., Neri, L. M., Basta, G., Capitani, S., Becchetti, E., and Brunetti, P., Cellular support systems for alginate microcapsules containing islets, as composite bioartificial pancreas, *Ann. NY Acad. Sci.*, 944, 240–252, 2001.
67. Siebers, U., Horcher, A., Brandhorst, H., Brandhorst, D., Hering, B., Federlin, K., Bretzel, R. G., and Zekorn, T., Analysis of the cellular reaction towards microencapsulated xenogeneic islets after intrapertioneal transplantation, *J. Mol. Med.*, 77, 215–218, 1999.
68. Weber, C. J., Safley, S., Hagler, M., and Kapp, J., Evaluation of graft-host response for various tissue sources and animal models, *Ann. NY Acad. Sci.*, 875, 233–254, 1999.
69. Iwata, H., Murakami, Y., and Ikada, Y., Control of complement activities for immunoisolation, *Ann. NY Acad. Sci.*, 875, 7–23, 1999.
70. White, D. J. G., and Yannoutsos, N., Production of pigs transgenic for human DAF to overcome complement-mediated hyperacute xenograft rejection in man, *Res. Immunol.*, 147, 88–94, 1996.
71. Bottino, R., Fernandez, L. A., Ricordi, C., Lehmann, R., Tsan, M.-F., Oliver, R., and Inverardi, L., Transplantation of allogeneic islets of Langerhans in the rat liver: effects of macrophage depletion on graft survival and microenvironment activation, *Diabetes*, 47(3), 316–323, 1998.
72. De Vos, P., De Haan, B. J., Wolters, G. H. J., Strubbe, J. H., and Van Schilfgaarde, R., Improved biocompatibility but limited graft survival after purification of alginate for microencapsulation of pancreatic islets, *Diabetologia*, 40, 262–270, 1997.
73. Avgoustiniatos, E. S., and Colton, C. K., Effect of external oxygen mass transfer resistance on viability of immunoisolated tissue, *Ann. NY Acad. Sci.*, 831, 145–167, 1997.
74. Trivedi, N., Keegan, M., Steil, G. M., Hollister-Lock, J., Hasenkamp, W. M., Colton, C. K., Bonner-Weir, S., and Weir, G. C., Islets in alginate macrobeads reverse diabetes despite minimal acute insulin secretory responses, *Transplantation*, 71(2), 203–211, 2001.
75. Li, R. H., Materials for immunoisolated cell transplantation, *Adv. Drug Deliv. Rev.*, 33, 87–109, 1998.
76. Hou, Q. P., and Bae, Y. H., Biohybrid artificial pancreas based on macrocapsule device, *Adv. Drug Deliv. Rev.*, 35, 271–287, 1999.
77. Lanza, R. P., Borland, K. M., Staruk, J. E., Appel, M. C., Solomon, B. A., and Chick, W. L., Transplantation of encapsulated canine islets into spontaneously diabetic BB/Wor rats without immunosuppression, *Endocrinology*, 131(2), 637–642, 1992.
78. Lembert, N., Petersen, P., Wesche, J., Enderle, A., Doser, M., Planck, H., Becker, H. D., and Ammon, H. P. T., *In vitro* test of new biomaterials for the development of a bioartificial pancreas, *Ann. NY Acad. Sci.*, 944, 271–276, 2001.
79. Loudovaris, T., Jacobs, S., Young, S., Maryanov, D., Brauker, J., and Johnson, R. C., Correction of diabetic nod mice with insulinomas implanted within Baxter immunoisolated devices, *J. Mol. Med.*, 77, 219–222, 1999.
80. Yang, Z., Chen, M., Fialkow, L. B., Ellett, J. D., Wu, R., and Nadler, J. L., Survival of pancreatic islet xenografts in NOD mice with the Theracyte device, *Transplant. Proc.*, 34, 3349–3350, 2002.
81. Brauker, J., Martinson, L., Young, S., and Johnson, R. C., Local inflammatory response around diffusion chambers containing xenografts. Nonspecific destruction of tissues and decreased local vascularization, *Transplantation*, 61(12), 1671–1677, 1996.
82. McKenzie, A. W., Georgiou, H. M., Zhan, Y., Brady, J. L., and Lew, A. M., Protection of xenografts by a combination of immunoisolation and a single dose of anti-CD4 antibody, *Cell Transplant.*, 10, 183–193, 2001.
83. Loudovaris, T., Mandel, T., and Charlton, B., CD4 + T cell mediated destruction of xenografts within cell-impermeable membranes in the absence of CD8 + T cells and B cells, *Transplantation*, 61(12), 1678–1684, 1996.

84. Suzuki, K., Bonner-Weir, S., Trivedi, N., Yoon, K.-H., Hollister-Lock, J., Colton, C. K., and Weir, G. C., Function and survival of macroencapsulated syngeneic islets transplanted into streptozocin-diabetic mice, *Transplantation*, 66(1), 21–28, 1998.
85. Rafael, E., Wernerson, A., Arner, P., Wu, G. S., and Tibell, A., In vivo evaluation of glucose permeability of an immunoisolation device intended for islet transplantation: a novel application of the microdialysis technique, *Cell Transplant.*, 8, 317–326, 1999.
86. Rafael, E., Gazelius, B., Wu, G. S., and Tibell, A., Longitudinal studies on the microcirculation around Theracyte™ immunoisolated device, using the laser Doppler technique, *Cell Transplant.*, 9, 107–113, 2000.
87. Rafael, E., Wu, G. S., Hultenby, K., Tibell, A., and Wernerson, A., Improved survival of macroencapsulated islets of Langerhans by preimplantation of the immunoisolating device: a morphometric study, *Cell Transplant.*, 12, 407–412, 2003.
88. Suzuki, K., Bonner-Weir, S., Hollister-Lock, J., Colton, C. K., and Weir, G. C., Number and volume of islets transplanted in immunobarrier devices, *Cell Transplant.*, 7(1), 47–52, 1998.
89. Carr-Brendel, V., Geller, R., Thomas, T., Boggs, D., Young, S., Crudele, J., Martinson, L., Maryanov, D., Johnson, R., and Brauker, J., Transplantation of cells in an immunoisolation device for gene therapy, *Methods Mol. Biol.*, 63, 373–387, 1997.
90. Geller, R., Loudovaris, T., Neuenfeldt, S., Johnson, R., and Brauker, J., Use of an immunoisolation device for cell transplantation and tumor immunotherapy, *Ann. NY Acad. Sci.*, 831, 438–451, 1997.
91. Geller, R., Neuenfeldt, S., Levon, S., Maryanov, D., Thomas, T., and Brauker, J., Immunoisolation of tumor cells: generation of antitumor immunity through indirect presentation of antigen, *J. Immunother.*, 20(2), 131–137, 1997.
92. Brauker, J., Frost, G., Dwarki, V., Nijjar, T., Chin, R., Carr-Brendel, V., Jasunas, C., Hodgett, D., Stone, W., Cohen, L., and Johnson, R., Sustained expression of high levels of human factor IX from human cells implanted within an immunoisolation device into athymic rodents, *Hum. Gene Ther.*, 9(6), 879–888, 1998.
93. Tibell, A., Rafael, E., Wennberg, L., Nordenstrom, J., Bergstrom, M., Geller, R. L., Loudovaris, T., Johnson, R. C., Brauker, J. H., Neuenfeldt, S., and Wernerson, A., Survival of macroencapsulated allogeneic parathyroid tissue one year after transplantation in nonimmunosuppressed humans, *Cell Transplant.*, 10, 591–599, 2001.
94. Storrs, R., Dorian, R., King, S. R., Lakey, J., and Rilo, H., Preclinical development of the islet sheet, *Ann. NY Acad. Sci.*, 944, 252–266, 2001.
95. Kessler, L., Legeay, G., West, R., Belcourt, A., and Pinget, M., Physicochemical and biological studies of corona-treated artificial membranes used for pancreatic islets encapsulation: mechanism of diffusion and interface modification, *J. Biomed. Mater. Res.*, 34, 235–245, 1997.
96. Desai, T. A., Hansford, D., and Ferrari, M., Characterization of micromachined silicon membranes for immunoisolation and bioseparation applications, *J. Membr. Sci.*, 159, 221–231, 1999.
97. Kurian, P., and Kennedy, J. P., Novel tricontinuous hydrophilic–lipophilic–oxyphilic membranes: synthesis and characterization, *J. Polym. Sci.: Part A: Polym. Chem.*, 40, 1209–1217, 2002.
98. Renken, A., and Hunkeler, D., Microencapsulation: a review of polymers and technologies with a focus on bioartificial organs, *Polimery*, 43(9), 530–539, 1998.
99. Orive, G., Hernandez, R. M., Gascon, A. R., Calafiore, R., Chang, T. M. S., De Vos, P., Hortelano, G., Hunkeler, D., Lacik, I., Shapiro, A. M. J., and Pedraz, J. L., Cell encapsulation: promise and progress, *Nat. Med.*, 9(1), 104–107, 2003.
100. Sakai, S., Ono, T., Ijima, H., and Kawakami, K., Control of molecular weight cut-off for immunoisolation by multilayering glycol chitosan–alginate polyion complex on alginate-based microcapsules, *J. Microencapsulation*, 17(6), 691–699, 2000.
101. Bartkowiak, A., Optimal conditions of transplantable binary polyelectrolyte microcapsules, *Ann. NY Acad. Sci.*, 944, 120–134, 2001.
102. Sakai, S., Ono, T., Ijima, H., and Kawakami, K., Newly developed aminopropyl-silicate immunoisolation membrane for a microcapsule-shaped bioartificial pancreas, *Ann. NY Acad. Sci.*, 944, 277–283, 2001.
103. Cruise, G. M., Hegre, O. D., Lamberti, F. V., Hager, S. R., Hill, R., Scharp, D. S., and Hubbell, J. A., In vitro and in vivo performance of porcine islets encapsulated in interfacially photopolymerized poly(ethylene glycol) diacrylate membranes, *Cell Transplant.*, 8, 293–306, 1999.

104. May, M. H., and Sefton, M. V., Conformal coating of small particles and cell aggregates at a liquid–liquid interface, *Ann. NY Acad. Sci.*, 875, 126–134, 1999.
105. Abalovich, A., Jatimliansky, C., Digex, E., Arias, M., Altamirano, A., Amorena, C., Martinez, B., and Nacucchio, M., Pancreatic islets microencapsulation with polylactide-*co*-glycolide, *Transplant. Proc.*, 33, 1977–1979, 2001.
106. Hunkeler, D., Polymers for bioartificial organs, *Trends Polym. Sci.*, 5(9), 286–293, 1997.
107. De Vos, P., Hamel, A. F., and Tatarkiewicz, K., Considerations for successful transplantation of encapsulated pancreatic islets, *Diabetologia*, 45, 159–173, 2002.
108. Lim, F., and Sun, A. M., Microencapsulated islets as bioartificial endocrine pancreas, *Science*, 210(21), 908–910, 1980.
109. Strand, B. L., Morch, Y. A., Espevik, T., and Skjak-Braek, G., Visualization of alginate–poly-L-lysine–alginate microcapsules by confocal laser scanning microscopy, *Biotechnol. Bioeng.*, 82(4), 386–394, 2003.
110. Strand, B. L., Gaserod, O., Kulseng, B., Espevik, T., and Skjak-Braek, G., Alginate–polylysine–alginate microcapsules: effect of size reduction on capsule properties, *J. Microencapsulation*, 19(5), 615–630, 2002.
111. Goosen, M. A., Physico-chemical and mass transfer considerations in microencapsulation, *Ann. NY Acad. Sci.*, 875, 84–104, 1999.
112. Poncelet, D., Production of alginate beads by emulsification/internal gelation, *Ann. NY Acad. Sci.*, 944, 74–82, 2001.
113. Thu, B., Bruheim, P., Espevik, T., Smidsrod, O., Soon-Shiong, P., and Skjak-Braek, G., Alginate polycation microcapsules II. Some functional properties, *Biomaterials*, 17, 1069–1079, 1996.
114. Thu, B., Bruheim, P., Espevik, T., Smidsrod, O., Soon-Shiong, P., and Skjak-Braek, G., Alginate polycation microcapsules I. Interaction between alginate and polycation, *Biomaterials*, 17, 1031–1040, 1996.
115. De Groot, M., Keizer, P. P. M., De Haan, B. J., Schuurs, T. A., Leuvenink, H. G. D., Van Schilfgaarde, R., and De Vos, P., Microcapsules and their ability to protect against cytokin-mediated dysfunction, *Transplant. Proc.*, 33, 1711–1712, 2001.
116. Van Schilfgaarde, R., and De Vos, P., Factors influencing the properties and performance of microcapsules for immunoprotection of pancreatic islets, *J. Mol. Med.*, 77, 199–205, 1999.
117. Wandrey, C., and Sainz Vidal, D., Purification of polymeric biomaterials, *Ann. NY Acad. Sci.*, 944, 187–197, 2001.
118. Zimmermann, U., Thurmer, F., Jork, A., Weber, M., Mimietz, S., Hillgartner, M., Brunnenmeier, F., Zimmermann, H., Westphal, I., Fuhr, G., Noth, U., Haase, A., Steinert, A., and Hendrich, C., A novel class of amitogenic alginate microcapsules for long-term immunoisolated transplantation, *Ann. NY Acad. Sci.*, 944, 199–215, 2001.
119. Soon-Shiong, P., Otterlie, M., Skjak-Braek, G., Smidsrod, O., Heintz, R., Lanza, R. P., and Espevik, T., An immunologic basis for the fibrotic reaction to implanted microcapsules, *Transplant. Proc.*, 23(1), 758–759, 1991.
120. De Vos, P., Hoogmoed, C. G., and Busscher, H. J., Chemistry and biocompatibility of alginate–PLL capsules for immunoprotection of mammalian cells, *J. Biomed. Mater. Res.*, 60, 252–259, 2002.
121. Van Hoogmoed, C. G., Busscher, H. J., and De Vos, P., Fourier transform infrared spectroscopy studies of alginate–PLL capsules with varying compositions, *J. Biomed. Mater. Res.*, 67A, 172–178, 2003.
122. De Vos, P., van Hoogmoed, C. G., van Zanten, J., Netter, S., Strubbe, J. H., and Busscher, H. J., Long-term biocompatibility, chemistry, and function of microencapsulated pancreatic islets, *Biomaterials*, 24, 305–312, 2003.
123. King, A., Strand, B., Rokstad, A.-M., Kulseng, B., Andersson, A., Skjak-Braek, G., and Sandler, S., Improvement of the biocompatibility of alginate/poly-L-lysine/alginate microcapsules by the use of epimerized alginate as a coating, *J. Biomed. Mater. Res.*, 64A, 533–539, 2003.
124. De Vos, P., van Hoogmoed, C. G., De Haan, B. J., and Busscher, H. J., Tissue responses against immunoisolating alginate–PLL capsules in the immediate posttransplant period, *J. Biomed. Mater. Res.*, 62, 430–437, 2002.
125. De Vos, P., Smedema, I., Van Goor, H., Moes, H., Van Zanten, J., Netters, S., De Leij, L. F. M., De Haan, A., and De Haan, B. J., Association between macrophage activation and function of micro-encapsulated rat islets, *Diabetologia*, 46, 666–673, 2002.

126. Safley, S. A., Kapp, J. A., and Weber, C. J., Proliferative and cytokine responses in CTLA4-Ig-treated diabetic NOD mice transplanted with microencapsulated neonatal porcine ICCs, *Cell Transplant.*, 11, 695–705, 2002.
127. Duvivier-Kali, V. F., Omer, A., Parent, R. J., O'Neil, J. J., and Weir, G. C., Complete protection of islets against allorejection and autoimmunity by a simple barium-alginate membrane, *Diabetes*, 50, 1698–1705, 2001.
128. Omer, A., Keegan, M., Czismadia, E., De Vos, P., Van Rooijen, N., Bonner-Weir, S., and Weir, G. C., Macrophage depletion improves survival of porcine neonatal pancreatic cell clusters contained in alginate macrocapsules transplanted into rats, *Xenotransplantation*, 10, 240–251, 2003.
129. Zekorn, T. D. C., Horcher, A., Siebers, U., Federlin, K., and Bretzel, R. G., Synergistic effect of microencapsulation and immunoalteration on islet allograft survival in bioartificial pancreas, *J. Mol. Med.*, 77, 193–198, 1999.
130. Pathak, C. P., Sawhney, A. S., and Hubbell, J. A., Rapid photopolymerization of immunoprotective gels in contact with cells and tissues, *J. Am. Chem. Soc.*, 114(21), 8311–8312, 1992.
131. Desai, N. P., Sojomihardjo, A., Yao, Z., Ron, N., and Soon-Shiong, P., Interpenetrating polymer networks of alginate and polyethylene glycol for encapsulation of islets of Langerhans, *J. Microencapsulation*, 17(6), 677–690, 2000.
132. Prokop, A., Hunkeler, D., DiMari, S., Haralson, M. A., and Wang, T. G., Water soluble polymers for immunoisolation I: complex coacervation and cytotoxicity, *Adv. Polym. Sci.*, 136, 1–51, 1998.
133. Prokop, A., Hunkeler, D., Powers, A. C., Whitesell, R. R., and Wang, T. G., Water soluble polymers for immunoisolation II: evaluation of multicomponent microencapsulation systems, *Adv. Polym. Sci.*, 136, 53–73, 1998.
134. Bartkowiak, A., Canaple, L., Ceausoglu, I., Nurdin, N., Renken, A., Rindisbacher, L., Wandrey, C., Desvergne, B., and Hunkeler, D., New multicomponent capsules for immunoisolation, *Ann. NY Acad. Sci.*, 875, 135–145, 1999.
135. Wang, T., Lacik, I., Brissova, M., Anilkumar, A. V., Prokop, A., Hunkeler, D., Green, R., Shahrokhi, K., and Powers, A. C., An encapsulation system for the immunoisolation of pancreatic islets, *Nat. Biotechnol.*, 15, 358–362, 1997.
136. Sefton, M. V., May, M. H., Lahooti, S., and Babensee, J. E., Making microencapsulation work: conformal coating, immobilization gels and in vivo performance, *J. Control. Release*, 65, 173–186, 2000.
137. Risbud, M., Hardikar, A., and Bhonde, R., Chitosan–polyvinyl pyrrolidone hydrogels as candidate for islet immunoisolation: in vitro biocompatibility evaluation, *Cell Transplant.*, 9, 25–31, 2000.
138. Risbud, M. V., Bhonde, M. R., and Bhonde, R. R., Effects of chitosan–polyvinyl pyrrolidone hydrogel on proliferation and cytokine expression of endothelial cells: implications in islet immunoisolation, *J. Biomed. Mater. Res.*, 57, 300–305, 2001.
139. O'Shea, G. M., and Sun, A. M., Encapsulation of rat islets of Langerhans prolongs xenograft survival in diabetic mice, *Diabetes*, 35, 943–946, 1986.
140. Fan, M.-Y., Lum, Z.-P., Fu, X.-W., Levesque, L., Tai, I. T., and Sun, A. M., Reversal of diabetes in BB rats by transplantation of encapsulated pancreatic islets, *Diabetes*, 39, 519–522, 1990.
141. Soon-Shiong, P., Feldman, E., Nelson, R., Komtebedde, J., Smidsrod, O., Skjak-Braek, G., Espevik, T., Heintz, R., and Lee, M., Successful reversal of spontaneous diabetes in dogs by intraperitoneal microencapsulated islets, *Transplantation*, 54(5), 769–774, 1992.
142. De Vos, P., Vegter, D., Strubbe, J. H., De Haan, B. J., and Van Schilfgaarde, R., Impaired glucose tolerance in recipients of an intraperitoneally implanted microencapsulated islet allograft is caused by the slow diffusion of insulin through the peritoneal membrane, *Transplant. Proc.*, 29, 756–757, 1997.
143. Soon-Shiong, P., Treatment of type I diabetes using encapsulated islets, *Adv. Drug Deliv. Rev.*, 35, 259–270, 1999.
144. Elliott, R. B., Escobar, L., Garkavenko, O., Croxson, M. C., Schroeder, B. A., McGregor, M., Ferguson, G., Beckman, N., and Ferguson, S., No evidence of infection with porcine endogenous retrovirus in recipients of encapsulated porcine islet xenografts, *Cell Transplant.*, 9, 895–901, 2000.
145. Tatarkiewicz, K., Garcia, M., Omer, A., Van Schilfgaarde, R., and De Vos, P., C-peptide responses after meal challenge in mice transplanted with microencapsulated rat islets, *Diabetologia*, 44, 646–653, 2001.

146. Soon-Shiong, P., Heintz, R. E., Merideth, N., Yao, Q. X., Yao, Z., Zheng, T., Murphy, M., Moloney, M. K., Schmehl, M., Harris, M., Mendez, R., Mendez, R., and Sandford, P. A., Insulin independence in a type 1 diabetic patient after encapsulated islet transplantation, *Lancet*, 343, 950–951, 1994.
147. Carlsson, P.-O., Liss, P., Andersson, A., and Jansson, L., Measurements of oxygen tension in native and transplanted rat pancreatic islets, *Diabetes*, 47(7), 1027–1032, 1998.
148. Schrezenmeir, J., Gero, L., Laue, C., Kirchgessner, J., Muller, A., Huls, A., Passmann, R., Hahn, H. J., Kunz, L., Mueller-Klieser, W., and Altman, J. J., The role of oxygen supply in islet transplantation, *Transplant. Proc.*, 24(6), 2925–2929, 1992.
149. Dionne, K. E., Colton, C. K., and Yarmush, M. L., Effect of oxygen on isolated pancreatic tissue, *Trans. Am. Soc. Artif. Intern. Organs*, 35(3), 739–741, 1989.
150. Ohta, M., Nelson, D., Nelson, J., Meglasson, M. D., and Erecinska, M., Oxygen and temperature dependence of stimulated insulin secretion in isolated rat islets of Langerhans, *J. Biol. Chem.*, 265(29), 17525–17532, 1990.
151. Dionne, K. E., Colton, C. K., and Yarmush, M. L., Effect of hypoxia on insulin secretion by isolated rat and canine islets of Langerhans, *Diabetes*, 42, 12–21, 1993.
152. Schrezenmeir, J., Gero, L., Solhdju, M., Kirchgessner, J., Laue, C., Beyer, J., Stier, H., and Muller-Klieser, W., Relation between secretory function and oxygen supply in isolated islet organs, *Transplant. Proc.*, 26(2), 809–813, 1994.
153. De Groot, M., Schuurs, T. A., Keizer, P. P. M., Fekken, S., Leuvenink, H. G. D., and Van Schilfgaarde, R., Response of encapsulated rat pancreatic islets to hypoxia, *Cell Transplant.*, 12, 867–875, 2003.
154. Avgoustiniatos, E. S., and Colton, C. K., Design considerations in immunoisolation, In *Principles of Tissue Engineering*, Lanza, R., Langer, R., and Chick, W., eds., R.G. Landes Company, Austin, pp. 333–346, 1997.
155. Dulong, J.-L., Legallais, C., Darquy, S., and Reach, G., A novel model of solute transport in a hollow-fiber bioartificial pancreas based on a finite element method, *Biotechnol. Bioeng.*, 78(5), 576–582, 2002.
156. De Vos, P., Vegter, D., De Haan, B. J., Strubbe, J. H., Bruggink, J. E., and Van Schilfgaarde, R., Kinetics of intraperitoneally infused insulin in rats: functional implications for the bioartificial pancreas, *Diabetes*, 45(8), 1102–1107, 1996.
157. Brauker, J., Martinson, L., Hill, R., Young, S., Carr-Brendel, V., and Johnson, R. C., Neovascularization of immunoisolation membranes: the effect of membrane architecture and encapsulated tissue, *Transplant. Proc.*, 24(6), 2924, 1992.
158. Brauker, J., Carr-Brendel, V., Martinson, L., Crudele, J., Johnston, W., and Johnson, R. C., Neovascularization of synthetic membranes directed by membrane microarchitecture, *J. Biomed. Mater. Res.*, 29(12), 1517–1524, 1995.
159. Padera, R. F., and Colton, C. K., Time course of membrane microarchitecture-driven neovascularization, *Biomaterials*, 17, 277–284, 1996.
160. Trivedi, N., Steil, G. M., Colton, C. K., Bonner-Weir, S., and Weir, G. C., Improved vascularization of planar membrane diffusion devices following continuous infusion of vascular endothelial growth factor, *Cell Transplant.*, 9, 115–124, 2000.
161. Calafiore, R., Basta, G., Luca, G., Boselli, C., Bufalari, A., Bufalari, A., Cassarani, M. P., Giustozzi, G. M., and Brunetti, P., Transplantation of pancreatic islets contained in minimal volume microcapsules in diabetic high mammalians, *Ann. NY Acad. Sci.*, 875, 219–232, 1999.
162. Wu, H., Avgoustiniatos, E. S., Swette, L., Bonner-Weir, S., Weir, G. C., and Colton, C. K., *In situ* electrochemical oxygen generation with an immunoisolated device, *Ann. NY Acad. Sci.*, 875, 105–125, 1999.
163. Wu, H., The Effect of Oxygen on Cultured Islets. Ph.D. thesis, MIT, Cambridge, MA, in progress.
164. Riess, J. G., Fluorocarbon-based in vivo oxygen transport and delivery systems, *Vox Sang*, 61, 225–239, 1991.
165. Riess, J. G., Fluorocarbon emulsions-designing an efficient shuttle service for the respiratory gases — the so-called "blood substitutes", In *Fluorine Chemistry at the Millenium*, Banks, R. E., ed., Elsevier, Amsterdam, pp. 385–431, 2000.

166. Brasile, L., Clarke, J., Green, E., and Haisch, C., The feasibility of organ preservation at warmer temperatures, *Transplant. Proc.*, 128(1), 349–351, 1996.
167. Voiglio, E. J., Zarif, L., Gorry, F. C., Krafft, M.-P., Margonari, J., Martin, X., Riess, J., and Dubernard, J. M., Aerobic preservation of organs using a new perflubron/lecithin emulsion stabilized by molecular dowels, *J. Surg. Res.*, 63, 439–446, 1996.
168. Matsumoto, S., and Kuroda, Y., Perfluorocarbon for organ preservation before transplantation, *Transplantation*, 74(12), 1804–1809, 2002.
169. Zekorn, T., Siebers, U., Bretzel, R. G., Heller, S., Meder, U., Ruttkay, H., Zimmermann, U., and Federlin, K., Impact of the perfluorochemical FC43 on function of isolated islets, *Horm. Metab. Res.*, 23, 302–303, 1991.
170. Poncelet, D., Leung, R., Centomo, L., and Neufeld, R. J., Microencapsulation of silicone oils within polyacrylamide–polyethylene membranes as oxygen carriers for bioreactor oxygenation, *J. Chem. Technol. Biotechnol.*, 57, 253–263, 1993.
171. Nishide, H., Chen, X. S., and Tsuchida, E., Facilitated oxygen transport with modified and encapsulated hemoglobins across non-flowing solution membrane, *Artif. Cells Blood Substit. Immobil. Biotechnol.*, 25, 335–346, 1997.
172. Colton, C. K., Stroeve, P., and Zahka, J. G., Mechanism of oxygen transport augmentation by hemoglobin, *J. Appl. Physiol.*, 35(2), 307–309, 1973.
173. Schrezenmeir, J., Hyder, A., Vreden, M., Laue, C., and Mueller-Klieser, W., Oxygen profile of microencapsulated islets: effect of immobilised hemoglobin in the alginate matrix, *Transplant. Proc.*, 33, 3511–3516, 2001.
174. Chae, S. Y., Kim, S. W., and Bae, Y. H., Effect of cross-linked hemoglobin on functionality and viability of microencapsulated pancreatic islets, *Tissue Eng.*, 8(3), 379–394, 2002.
175. Schrezenmeir, J., Velten, F., and Beyer, J., Immobilized hemoglobin improves islet function and viability in the bioartificial in vitro and in vivo, *Transplant. Proc.*, 26(2), 792–800, 1994.
176. Papas, K. K., Colton, C. K., Gounarides, J. S., Roos, E. S., Jarema, M. A. C., Shapiro, M. J., Cheng, L. L., Cline, G. W., Shulman, G. I., Wu, H., Bonner-Weir, S., and Weir, G. C., NMR spectroscopy in beta cell engineering and islet transplantation, *Ann. NY Acad. Sci.*, 944, 96–119, 2001.
177. Noth, U., Grohn, P., Jork, A., Zimmermann, U., Hasse, A., and Lutz, J., 19F-MRI in vivo determination of the partial oxygen pressure in perfluorocarbon-loaded alginate capsules implanted into the peritoneal cavity and different tissues, *Magn. Reson. Med.*, 42, 1039–1047, 1999.
178. Zimmermann, U., Noth, U., Grohn, P., Jork, A., Ulrichs, K., Lutz, J., and Haase, A., Non-invasive evaluation of the location, the functional integrity, and the oxygen supply of implants: ^{19}F nuclear magnetic resonance imaging of perfluorocarbon-loaded Ba^{2+}-alginate beads, *Artif. Cells Blood Substit. Immobil. Biotechnol.*, 28(2), 129–146, 2000.
179. Constantinidis, I., Rober, J., Rober Long, J., Weber, C., Safley, S., and Sambanis, A., Non-invasive monitoring of a bioartificial pancreas *in vitro* and *in vivo*, *Ann. NY Acad. Sci.*, 944, 83–95, 2001.
180. Zhao, M., Beauregard, D. A., Loizou, L., Davletov, B., and Brindle, K. M., Non-invasive detection of apoptosis using magnetic resonance imaging and a targeted contrast agent, *Nat. Med.*, 7(11), 1241–1244, 2001.
181. Thorens, B., Dupraz, P., and Cottet, S., Engineering tolerance into transplanted beta cell lines, *Ann. NY Acad. Sci.*, 944, 267–270, 2001.

38 Three-Dimensional Tissue Fabrication: Application in Hepatic Tissue Engineering

Valerie Liu Tsang and Sangeeta N. Bhatia

CONTENTS

I. INTRODUCTION TO HEPATIC TISSUE ENGINEERING

Liver disease is the direct cause of death for over 25,000 Americans annually and is the underlying cause of death for another 15,000, with estimated direct health care costs ranging from $60 billion to $100 billion in the U.S. alone. Unlike other chronic diseases such as cancer and heart disease that have experienced declining rates over the past two decades, the rates of liver failure are increasing due to the spread of hepatitis C.[1] Currently, the only available medical treatment for liver failure is full or partial liver transplantation. While the demand for liver donors far outweighs the available supply, the rapidly increasing incidence of liver disease continues to aggravate the shortage. As a result, researchers are working to develop alternative cell-based strategies for treating hepatic failure. These therapies include cell transplantation, in which hepatocytes are injected directly into the body,[2] and dialysis-like bioartificial liver assist devices for patients while they wait for a transplant.[3] However, owing to inefficient engraftment of transplanted cells and loss of function of hepatocytes within the extracorporeal devices, these approaches present only a short-term solution to hepatic failure, and do not allow a life for patients outside the hospital.

Over the past two decades, tissue engineering as a solution for organ replacement and restoration of tissue function has become a growing target of research in the scientific and medical community. While the fabrication of a whole replacement organ remains a lofty goal, replacement organs such as the bladder and the skin have already emerged through the combination of live cells and artificial polymer scaffolds. An organ such as the liver, with its many physiological functions (protein synthesis, detoxification, bile production, metabolism of lipids, proteins, and carbohydrates), intricate microarchitecture, and large nutrient requirements is far more complex than the skin and bladder. Furthermore, issues of cell sourcing and the need to incorporate multiple cell types pose unique challenges in engineering liver tissue. With regard to cell sourcing, three coupled limitations have plagued the field: (1) the enormous number of hepatocytes estimated necessary to replace liver function ($\sim$ 10 billion cells), (2) the limited replication of hepatocytes *in vitro*, and (3) the relative instability of the isolated hepatocyte phenotype.[4] In the long-term, fundamental advances in liver biology that yield new cell sources (oval cells, hepatoblasts, adult or embryonic stem cells, reversibly immortalized hepatocytes, or successful proliferation of mature hepatocytes), and insight into the role of micro-environmental cues in stabilizing the hepatocyte phenotype will therefore have enormous impact on the field. In addition to hepatocyte sourcing, multiple cell types such as those found in the native liver are required for functions such as bile transport and a nonthrombogenic endothelium. Therefore, technologies that would enable the placement of nonparenchymal cells as well as hepatocytes in an organized three-dimensional arrangement may provide a means of creating both biliary and vascular networks in an engineered hepatic tissue. Clearly, engineering hepatic tissue is a complex undertaking and will likely take many years to accomplish. Nonetheless, progress is being made toward developing a functional implantable liver by initially focusing on implantation of the parenchymal work-horse of the liver, the mature hepatocyte.

Various types of hepatocyte implantation explored include attachment to microcarriers,[5] encapsulation of hepatocyte aggregates,[6,7] and biodegradable polymer scaffolds.[8,9] Transplantation of hepatocytes (preaggregated or attached to protein-coated microcarriers) has been shown to provide hepatic functions initially, but their potential for stabilized efficacy has not been demonstrated.[10] The ability of the liver to process blood and produce bile results from the interaction of three dendritic networks — two for blood flow and one for bile transport. Thus, the function of the liver emerges from a complex three-dimensional architecture. Takezawa et al. have attempted to reconstruct the liver architecture by perfusing the liver in multiple steps to remove blood, deconstructing the existing collagen scaffold, and then filling it with collagen gel.[11] However, large diffusion distances for oxygen and nutrients may limit this method from being efficacious. Others have incorporated collagen or growth factors into scaffolds seeded with hepatocytes to promote engraftment and angiogenesis.[12,13] For example, Vacanti and coworkers have addressed the issue of poor engraftment by coating poly-L-lactic acid discs with the angiogenic basic fibroblast growth factor. Implantation of the constructs into the mesentery of syngeneic animals for two weeks resulted in significantly greater microvascular density and hepatocyte engraftment as compared to the control group.[13] Their group has also cultured hepatocytes on scaffolds in flow perfusion systems and cultured monolayers of cells on a network of branching planar channels to increase oxygen and nutrient diffusion.[14–16] Despite significant progress in the field, hepatocytes transplanted into rats on biodegradable polymer matrices have been found to be inferior to liver grafts of equivalent liver mass at compensating for metabolic deficiencies.[17] Thus, methods of building tissue engineered liver constructs with intricate architectures that provide adequate transport of oxygen and nutrients, as well as an environment that promotes hepatocyte viability and liver-specific function, are still lacking. Recent developments in scaffold technology may help move towards the engineering of functional liver tissue.

II. OVERVIEW OF THREE-DIMENSIONAL TISSUE FABRICATION

In recent years, CAD-based technologies borrowed from the manufacturing sector have greatly improved tissue engineering techniques, as witnessed by the development from porous polymer scaffolds with unpredictable void space connectivity[18] to constructs with complex architectures containing fabricated micro and macro scale features. Three-dimensional fabrication techniques such as these may provide solutions for some of the major obstacles faced by tissue engineers, such as spatial positioning of multiple cell types, incorporation of vasculature, and opportunities for production at a large scale. Complex tissues such as the liver have well-defined three-dimensional architectures that involve not only the placement of multiple cell types at a microscale level, but also intricate vascular organization at a macroscale level. The application of fabrication technologies may allow tissue engineers to more closely mimic the *in vivo* micro and macroscale environments in the production of engineered tissues, resulting in better tissue function. The inclusion of vascular channels allows the production of larger tissue constructs, eliminating the problem of diffusion faced by traditional scaffolds. Finally, the ability to fabricate tissues with controlled spatial cell orientations may be of great use to explore structure–function relationships in model tissues. In two-dimensional model systems, the patterning of microscale features has indeed led to insights on the effect of cell–cell and cell–matrix interactions on hepatocyte and endothelial cell fate.[19] Therefore, three-dimensional fabrication techniques may also prove useful for examining structure/function relationships in tissues such as the liver. In this chapter, we examine several fabrication methods that are used to build three-dimensional tissue engineering scaffolds, and extend our discussion to the potential application of these techniques for tissue engineering of the liver.

III. 3D SCAFFOLD FABRICATION METHODS

Various CAD-based manufacturing techniques have been adapted to fabricate three-dimensional polymer scaffolds for tissue engineering applications, providing computerized and accurate control over the design and production of internal scaffold architectures. The scaffold fabrication techniques generally fall into four categories based on their modes of assembly: heat-based fabrication, light-based fabrication, adhesive-based fabrication, and scaffold molding. Comprehensive reviews of solid freeform fabrication techniques for tissue engineering are available elsewhere[20–22]; here we present a summary of these techniques and their potential applications for liver tissue engineering (Figure 38.1).

A. Heat-Based Scaffold Fabrication

One rapid prototyping technique that has been adapted for scaffold fabrication involves the heating of materials above their glass transition temperature, resulting in the fusion of polymer layers. An example of this technique is sheet lamination, in which shapes that are laser-cut from polymer sheets are fused together by applying heat and pressure.[22] Prototypes fabricated using this method have a very dense material composition with little void space, and therefore, may not be practical for construction of highly cellular tissues such as the liver.

Scaffolds that are more complex in architecture can also be created by laminating membranes of higher porosity and smaller features. Borenstein et al. created biodegradable membranes containing small (20 μm) grooves by casting thin films of poly(D,L-lactic-*co*-gycolic) acid (PLGA) onto microfabricated silicon wafers (Figure 38.2(c)).[23] Channels were then formed by laminating the patterned PLGA membranes to each other. In a similar manner, Bhatia and coworkers fabricated porous tissue engineering scaffolds using soft lithography (Figure 38.2(b)).[24] A mold consisting of polydimethylsiloxane (PDMS) elastomer is cast from a microfabricated silicon master,[25] into which PLGA solution is cast and baked. The resulting PLGA layer, which may incorporate micropores by particulate leaching, contains microstructures (20–30 μm) that replicate those on the silicon master. These porous PLGA membranes can then be laminated together sequentially to form a three-dimensional scaffold. The micropores that are created increase the surface area of the scaffold and may enable the building of highly cellular tissues.

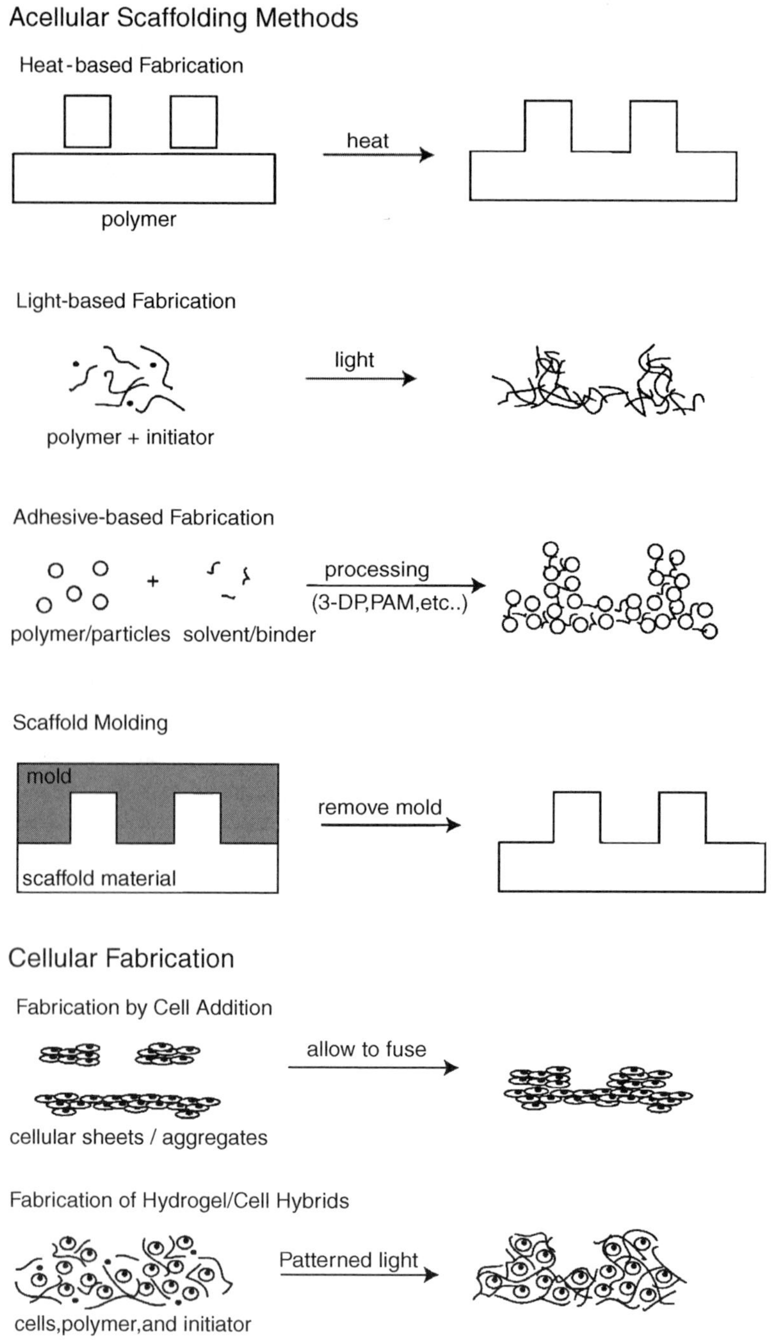

FIGURE 38.1 3D scaffold fabrication methods. Acellular scaffolds can be fabricated using a variety of techniques, including heat (FDM), light (SLA), adhesives (3DP) and molding. Cells themselves can also be incorporated in the fabrication process by cellular addition or by photopatterning of hydrogels.

Selective laser sintering (SLS) is a method that uses heat to fuse polymers into the desired shapes. A laser beam raises the local surface temperature of the powder bed to fuse particles, forming patterned structures within each layer. The resolution of SLS is limited by the laser beam diameter, which is typically in the range of 400 μm.[22] Unfused powders commonly result within

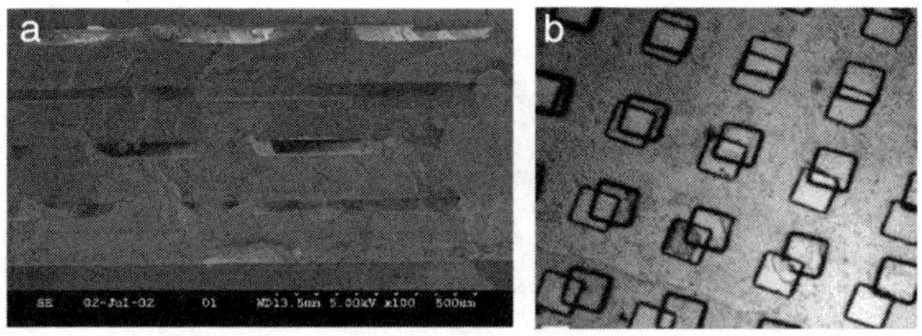

FIGURE 38.2 Heat-based fabrication: lamination. Membranes of the biodegradable polymer PLGA are cast from (a) silicon with layers of PLGA fused together to form microfluidic channels for vascular tissue engineering (photo courtesy of Jeff Borenstein and Kevin King, Draper Laboratory), or (b) PDMS, molds and then laminated to create 3D scaffolds. (From Vozzi, G., Flaim, C., Ahluwalia, A., and Bhatia, S., Fabrication of PLGA scaffolds using soft lithography and microsyringe deposition, *Biomaterials*, 24, 2533–2540, 2003. With permission.)

the structures and add to the porosity of the scaffold material. Lee and Barlow have applied SLS for the fabrication of polymer-coated calcium phosphate scaffolds, and have reported bone tissue ingrowth over several weeks in canine models.[26] In addition to ceramic/polymer blends, other improvements to the process are being developed, including biopolymer applications.[21]

Fused deposition modeling (FDM) is a computerized manufacturing technology that uses heat energy to create three-dimensional prototypes layer by layer. Each layer of the three-dimensional scaffold is produced by extrusion of molten plastics or ceramics through a nozzle and fusion to previous layers. Using FDM, Hutmacher and coworkers have fabricated bioresorbable poly(ε-caprolactone) (PCL) scaffolds with feature sizes of approximately 250 to 700 μm.[27] They have also seeded the scaffolds with primary human fibroblasts and have demonstrated cell proliferation and extracellular matrix production.[20] Other groups have also explored the application of FDM of bioceramic or polymer materials in scaffold fabrication (Figure 38.2(a)).[21] FDM offers an automated method of mass producing three-dimensional scaffolds for tissue engineering, but is limited in that the height of the pores created is predetermined by the size of the polymer filament extruded through the nozzle. In addition, the scaffolding materials in FDM have low porosity in comparison to other heat-based methods such as SLS.

The materials used in heat-based fabrication methods must be able to withstand elevated temperatures while retaining their desired properties (degradation, biocompatibility, etc.), limiting the choice of materials primarily to synthetic polymers. Some natural biomaterials are also compatible with heat-based free-form fabrication because of their phase transition properties. Three-dimensional plotting technology was used by Mulhaupt and coworkers to deposit heated agar and gelatin solutions (90°C) into a cooled plotting medium (10 to 15°C), resulting in three-dimensional hydrogel scaffolds (Figure 38.3).[28] The scaffolds were then sterilized in ethanol, coated with fibrin, and seeded with either a human osteosarcoma cell line or mouse fibroblasts to demonstrate cell adhesion to the surfaces.

Scaffolds created using heat-based fabrication techniques tend to have strong mechanical properties but low porosities, as a result of the material fusion that occurs at high temperatures. As a result, methods such as FDM and SLS may be better suited for tissues that bear mechanical loads such as bone, rather than highly cellular tissues such as the liver. The three-dimensional plotted hydrogel scaffolds possess mechanical properties that are similar to soft tissues, although the technology, like the other acellular fabrication methods described, does not provide a means for placing multiple cell types.

B. Light-Based Scaffold Fabrication

Light energy can also be used to build three-dimensional structures. Photopolymerization involves the application of light to initiate a chain reaction, resulting in the solidification of a liquid polymer

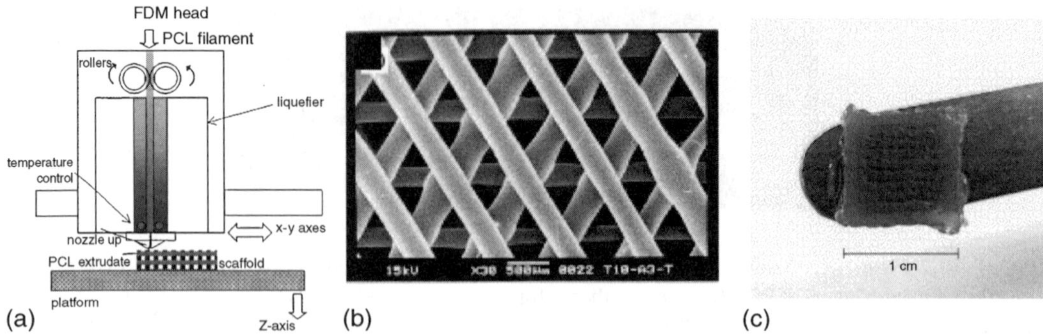

FIGURE 38.3 Heat-based fabrication: three-dimensional material deposition. (a, b) Fused deposition molding. Molten biomaterials are extruded through a nozzle to build three-dimensional scaffolds layer by layer. (From Zein, I., Hutmacher, D. W., Tan, K. C., and Teoh, S. H., Fused deposition modeling of novel scaffold architectures for tissue engineering applications, *Biomaterials,* 23, 1169–1185, 2002. With permission.) (c) Three-dimensional plotting. Heated liquid agar solidifies into a three-dimensional hydrogel scaffold when deposited into a cooled medium. (From Landers, R., Hubner, U., Schmelzeisen, R., and Mulhaupt R., Rapid prototyping of scaffolds derived from thermoreversible hydrogels and tailored for applications in tissue engineering, *Biomaterials,* 23, 4437–4447, 2002. With permission.)

solution. Stereolithography is a three-dimensional fabrication method using photopolymerization that can be used for construction of tissue engineering scaffolds. Spatial photopolymerization of the material is achieved by directing the position of a laser beam or by exposing certain areas of an entire layer through a mask with specific regions that are transparent to the light. The stage supporting the polymerized structure is then lowered and the process repeated to form additional layers. Mikos and coworkers reported the application of stereolithography to fabricate biodegradable three-dimensional polymer scaffolds using diethyl fumarate, poly(propylene fumarate), and the photoinitiator bisacylphosphine oxide (Figure 38.4).[29] Structures created using stereolithography typically contain features on the order of 250 μm, but certain systems using small-spot lasers have been shown to produce features as small as 70 μm.[22] Because stereolithography has traditionally been used to create models for manufacturing, the prototypes tend to be very rigid and nonporous, and may not be suited for hepatic tissue engineering, which may need greater surface areas for cell attachment.

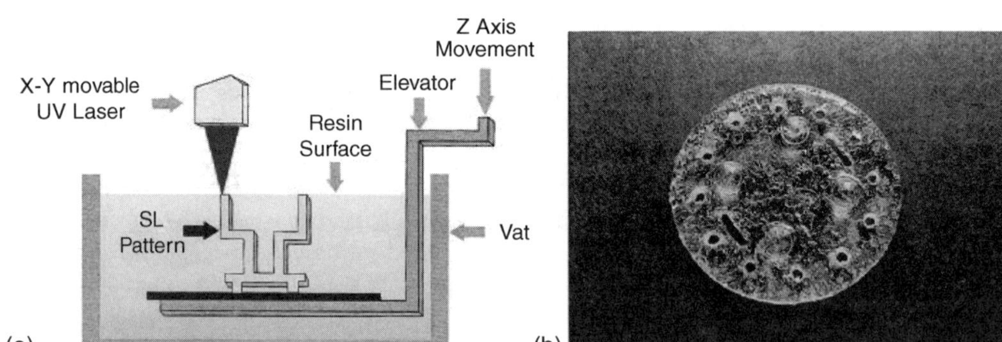

FIGURE 38.4 Light-based fabrication: stereolithography. (a) UV light is used to cross-link the material in specific regions of a layer. The elevator is then lowered to reveal a new layer of polymer, and the process is repeated to create the desired shape. (b) A prototype scaffold designed using SLA. (From Cooke, M. N., Fisher, J. P., Dean, D., Rimnac, C., and Mikos, A. G., Use of stereolithography to manufacture critical-sized 3D biodegradable scaffolds for bone ingrowth, *J. Biomed. Mater. Res.,* 64B, 65–69, 2003. With permission.)

In addition to the formation of rigid polymer constructs using stereolithography, light can also be used to pattern hydrogel polymer scaffolds that can swell in aqueous solution postprocessing. These cross-linked networks of hydrophilic polymers are becoming an increasingly popular material for tissue engineering due to their high water content and "tissue-like" mechanical properties. Ober and coworkers demonstrated the photolithographic patterning of 2-hydroxyethyl methacrylate, which was dried and later rehydrated before the addition of cells.[30] This method has resulted in the creation of single layer structures, although it could potentially be adapted for multilayer fabrication. However, some feature resolution may be lost during the dehydration and rehydration process. The photopatterning of hydrated hydrogels will be discussed later in this chapter.

C. Adhesive-Based Scaffold Fabrication

As an alternative to using heat or light energy, fabrication techniques have also been developed that use adhesives or binders for building scaffolds. Adhesive-based fabrication techniques do not require the use of a photo-initiator or elevated temperatures, and as a result can utilize some materials that cannot be used in light or heat-based fabrication. An example of adhesive-based fabrication is three-dimensional printing (3DP), in which an ink jet printer is used to deposit a binder solution onto a biomaterial powder bed to create structures (200 to 500 μm) one layer at a time (Figure 38.5).[31] Griffith and coworkers fabricated porous PLGA scaffolds by combining 3DP and particulate leaching, and demonstrated attachment rat hepatocyte and nonparenchymal cell cocultures.[32] Zeltinger et al. further explored this work by creating three-dimensional printed scaffolds with varying pore sizes and examining attachment, proliferation, and matrix deposition of various cell types.[33]

Another scaffold fabrication method that is based on layer by layer deposition is the pressure-assisted microsyringe (PAM). Unlike 3DP, in which the binder is printed onto a bed of powder, PAM involves the deposition of polymer dissolved in solvent through a syringe fitted with a 10 to 20 μm glass capillary needle.[24] The size of the polymer stream deposited can be varied by controlling the syringe pressure, solution viscosity, syringe tip diameter, and motor speed. This method is similar to FDM, but uses the solvent as binder rather than requiring heat and can produce structures with greater resolution. However, because of the small dimensions of the syringe, this method currently cannot be combined with particulate leaching to create micropores.

(a) (b)

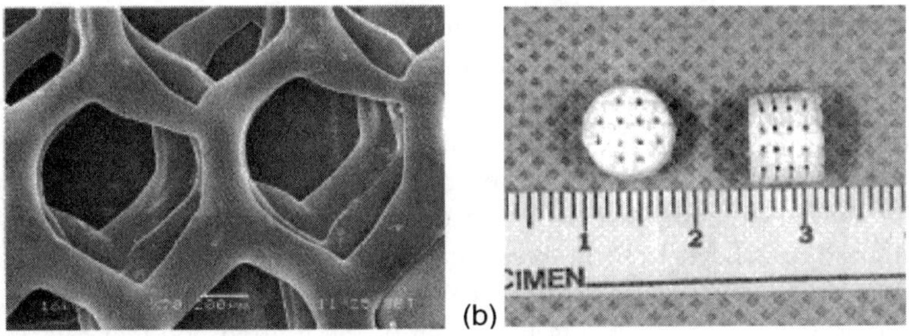

FIGURE 38.5 Adhesive-based fabrication. (a) Pressure assisted microsyringe. A polymer solution dissolved in solvent is deposited onto a substrate through a syringe. The solvent acts as a binder as polymer is added layer by layer (From Vozzi, G., Flaim, C., Ahluwalia, A., and Bhatia, S., Fabrication of PLGA scaffolds using soft lithography and microsyringe deposition, *Biomaterials,* 24, 2533-2540, 2003. With permission.) (b) 3D printing. Ink jet technology is used to print a binder solution onto a bed of polymer powder. An additional layer of powder is then deposited, and the process is repeated to form three-dimensional scaffolds. (From Park, A., Wu, B., and Griffith, L.G., Integration of surface modification and 3D fabrication techniques to prepare patterned poly(L-lactide) substrates allowing regionally selective cell adhesion. *J. Biomater. Sci. Polym. Ed.*, 9, 89–110, 1998. With permission.)

D. Molded Scaffold Fabrication

Many of the methods described above are limited in the repertoire of materials that could be used to fabricate the scaffolds. One way to take advantage of free-form fabrication technologies while broadening the scope of scaffolding materials is to use the three-dimensional prototypes as molds to indirectly fabricate scaffolds of materials that may not be compatible with the fabrication processes. For example, Orton and coworkers created a negative epoxy mold of the desired scaffold design using stereolithography, which was then filled with a hydroxyapatite/acrylate suspension.[34] Upon curing with heat, the mold and binder were incinerated in a furnace, leaving a three-dimensional hypoxyapatite scaffold. They also implanted the scaffolds into minipig mandibles for over two months and reported the ingrowth of bone.[35] While these rigid scaffolds may be useful for bone tissue engineering, they may not be ideal for building hepatic tissues.

Three-dimensional printing can also be used to create molds for casting polymer scaffolds. Wu and coworkers used this technique to cast PLGA for creating intestinal implants with three-dimensional protrusions to simulate villi. Using this technique, scaffolds were produced (Figure 38.6(a)) that mimic the large cell surface area found in the intestine. Ink jet printing technology can not only be used to print binders onto powder beds (3DP), but can also be used to deposit wax or other materials that have low melting points to fabricate molds (Solidscape, Inc.). These molds can then be used to cast scaffolding materials, such as hypoxyapatite, poly(L)lactide, and polyglycolide, which can be combined with particulate leaching to create micropores, as shown by Hollister and coworkers.[36,37] This process is milder for the scaffold than a mold produced by stereolithography, because the wax molds are removed by melting and washing with solvents rather than by incineration in a furnace. Ink jet printed molding technology was also used by Sachlos et al. to cast the extracellular matrix component collagen (Figure 38.6(b)).[38] Patterned collagen scaffolds with features of 200 μm were created after dissolving the mold in ethanol. The use of extracellular matrix as a building material may be an advantage due to its cellular adhesive properties; however, it cannot be tuned to designate regions of adhesivity and nonadhesivity, and is not specific to the adhesion of particular cell types, which may be problematic upon implantation. Nonetheless, this approach presents a valuable tool for *in vitro* studies of three-dimensional hepatocyte–matrix interactions.

IV. FABRICATION OF CELLULAR STRUCTURES

Tissue engineering typically involves the combination of functional cells with a biomaterial substrate, as described above. A problem that may arise from seeding cells on acellular scaffolds is that insufficient and heterogeneous engraftment that can occur. For other cell types, this can be partially addressed by allowing cell proliferation, but because hepatocytes typically do not replicate *in vitro* this is not a viable option. To address this issue, several groups are attempting to build tissues by directly constructing layers of live cells. Yamoto and coworkers have proposed the layering of cultured cell sheets to form three-dimensional tissues.[39] They allowed cardiomyocytes to proliferate on temperature-responsive poly(*N*-iso-propylacrylamide) culture surfaces, and then lowered the temperature by 5°C,causing hydration of the grafted polymer release of the sheet of cells. Multiple sheets of cardiomyocytes can then be layered to create an *in vitro* myocardial tissue construct. However, complex three-dimensional patterning is not possible using this method. Since hepatocytes do not typically proliferate in culture, the production of confluent hepatocyte sheets is unlikely; therefore, this method may not be of great utility for tissue engineering of the liver.

Cell sheets cultured *in vitro* have also been used for the construction of blood vessels. Auger and coworkers formed blood vessels by "rolling" sheets of cells and culturing them with pulsatile flow.[40] Multiple layers of the blood vessels were created by wrapping smooth muscle cell sheets around a tubular support and seeding endothelial cells within the lumen. Culture under pulsatile flow resulted in a tissue engineered blood vessel similar to an actual blood vessel in both

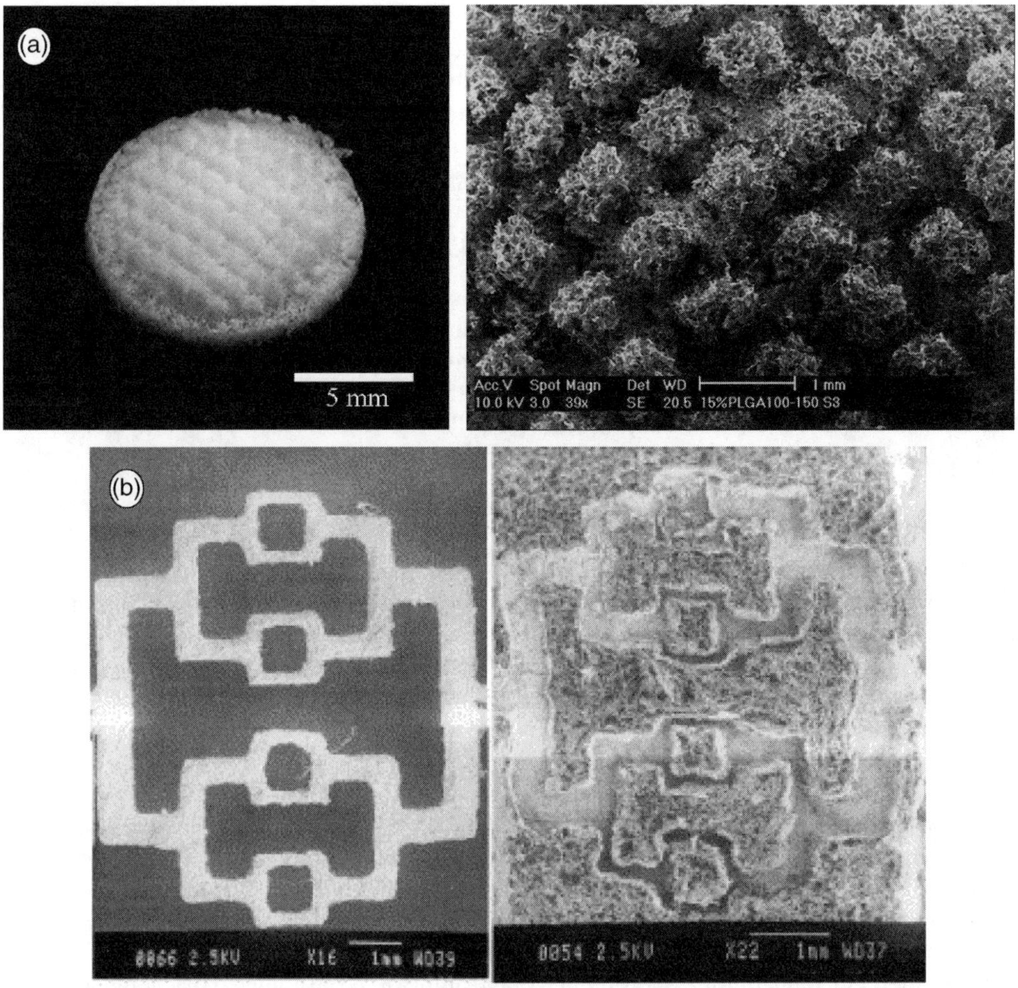

FIGURE 38.6 Molded scaffolds. (a) PLGA was cast from a mold that had been fabricated using three-dimensional printing to create intestinal implants (Photo courtesy of Ben Wu, UCLA Department of Bioengineering). (b) The extracellular matrix compound collagen was cast onto a negative mold that was printed using ink jet technology. The mold was then dissolved away with ethanol, leaving a patterned collagen scaffold. (From Sachlos, E., Reis, N., Ainsley, C., Derby, B., and Czernuszka, J.T., Novel collagen scaffolds with predefined internal morphology made by solid freeform fabrication, *Biomaterials*, 24, 1487–1497, 2003. With permission.)

mechanical strength and vascular-specific cellular markers. Like the cell sheet layering method, the sheet rolling method does not allow for patterning of arbitrary three-dimensional tissue structures, and would be difficult to adapt for tissue engineering of the liver.

Odde and coworkers have developed techniques of guiding cells onto a substrate using a laser. This cell "writing" method involves the plotting of a stream of cells using optical trapping forces to guide the cells.[41] While this technique allows for specific placement of individual cells, scaling up to produce tissues containing large quantities of cells may be limited by the serial nature of this technique. Others have developed a three-dimensional plotting mechanism to directly "print" living cells layer by layer with the hope that three-dimensional structures will be formed by spontaneous cell fusion.[42–44] Mironov et al. printed cell aggregates and embryonic heart mesenchymal fragments, observing fusion into tube-like structures when they were placed in collagen or thermosensitive gel.[43] If successful, this type of technology could result in the printing of three-dimensional tissue constructs containing precisely placed cells. Future studies in the emerging field of cellular self-assembly will help to elucidate which tissues can be assembled using a fusion-based three-dimensional printing approach. In order for this to approach to be successfully applied to hepatic tissue engineering, developments in liver biology and morphogenesis will be necessary.

V. FABRICATION OF HYBRID (CELL/SCAFFOLD) CONSTRUCTS

Acellular scaffolds generally possess superb mechanical integrity but may be difficult to populate with cells. Cellular constructs on the other hand provide excellent tissue density but may be mechanically unstable. Thus, hydrogel polymers have become increasingly popular for tissue engineering because of their potential to provide both structural support and high tissue density while maintaining an *in vivo*-like environment for cells.[45] These water-swollen polymer networks can be formed in mild conditions, and in some cases in the presence of cells.

A. Cell-Laden Hydrogels

Hydrogels of both biological (fibrin, collagen, etc.) and synthetic (PEG, PVA, etc.) composition have been explored to entrap cells. Hubbell and coworkers have functionalized fibrin gels by incorporating genetically engineered bioactive sites for cell adhesion and proteolytic remodeling.[46–48] Desai and coworkers have applied microfluidic molding methods to deposit patterned collagen gel structures containing cells.[49] While this method would be useful for the fabrication of some model tissues, it may be difficult to generalize due to the constraints of the microfluidic network on a flat surface.

Synthetic hydrogels have been of increasing interest because they can be tailored for specific applications (adhesive regions, biodegradable linkages, and so on). Many of the cross-linking reactions require photo-initiation of the polymer solutions to entrap the cells, allowing a homogenous distribution of cells at relatively high density throughout a hydrogel network. In addition, the transport properties of the hydrogels can be customized per application by adjusting polymer chain lengths and density. Poly(ethylene glycol) (PEG)-based hydrogels have been explored by many because of their biocompatibility, hydrophilicity, and ability to be customized by chemically adding biological molecules.[50] They have been used to immobilize various cell types including chondrocytes,[51,52] vascular smooth muscle cells,[53] osteoblasts,[54] and fibroblasts[55,56] that can attach, grow, and produce matrix. PEG-based hydrogels have also been customized by incorporating adhesion domains of extracellular matrix proteins to promote cell adhesion, growth factors to modulate cell function, and degradable linkages.[53,56–62] Clearly, hydrogel photo-crosslinking for tissue engineering is a rapidly growing field because of its tissue-like physical properties and its chemical flexibility; features which may also be useful for hepatic tissue engineering.

B. Three-Dimensional Photopatterning of Cell-Laden Hydrogels

Typically, the shape of a photo-crosslinked hydrogel is defined by the shape of its reaction chamber. For example, the disc-shaped structures depicted in Figure 38.7 were cross-linked in a cylindrical vial. One property of photosensitive hydrogel systems that, until recently, had not been exploited for tissue engineering is the potential to localize polymer cross-linking by directing the light to specific regions to produce patterned hydrogel features containing living cells. In other fields (nonmedical systems), photolithographic patterning has been applied to pattern hydrogel microstructures,[63] valves within flow systems,[64] and single-layer cell-laden microstructures on silicon.[65]

Based on the foundations of photolithography, spatial patterning in hydrogel tissue engineering may enable the construction of three-dimensional tissues. We have recently combined

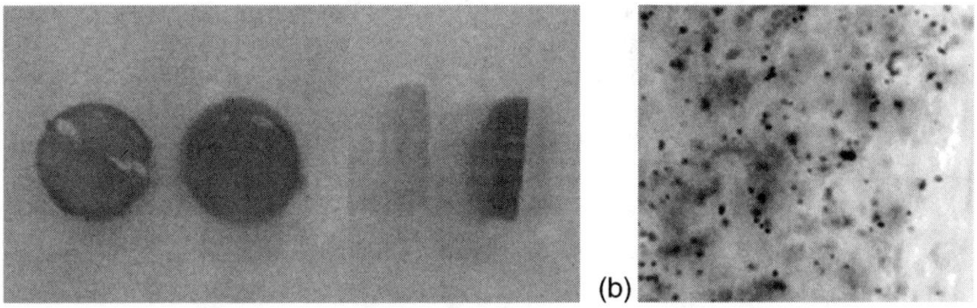

FIGURE 38.7 PEG-based hydrogels containing cells. (a) PEG-based hydrogels are cross-linked to form the shape of the container (dye added for clarity). (b) Living cells are suspended within the cross-linked hydrogel (MTT stain for viability). (Photos courtesy of Jennifer Elisseeff, Johns Hopkins University.)

photolithographic techniques with existing PEG-based cell encapsulation methods to build structural features within a three-dimensional cell/hydrogel network (Figure 38.8).[66] Using this method, live mammalian cells are localized within a controlled hydrogel architecture. Cells suspended in a polymer solution are photo-immobilized in multiple cellular domains. The uncross-linked polymer and cells are then rinsed away and the process can be repeated within the same layer or in additional layers. This allows different cell types to be introduced in each step, as represented in Figure 38.8 with cells that have been labeled fluorescently (red, green, and blue). By increasing the height of the photo-crosslinking chamber in between steps, additional layers can be added to create a three-dimensional cellular hydrogel tissue construct. Thus far, hydrogel features as small as 50 μm containing cells have been achieved. Furthermore, structures up to three layers have been fabricated using this method (Figure 38.8(b)). In complementary experiments, we have also developed a tool to specify cellular location within the prepolymer solution (as opposed to random dispersal) using electromagnetic fields, resulting in entrapment of cells in the designated configuration upon photo-crosslinking (data not shown). In conjunction with hydrogel technologies being explored by other groups (bioactive materials, incorporation of adhesion peptides and growth factors, and biodegradable linkages), photopatterning of hydrogels for tissue engineering may lead to the development of tissue constructs that can be customized spatially, physically, and chemically to improve upon existing polymer scaffold systems. The flexibility of these hydrogel systems shows great promise for hepatic tissue engineering because it would allow researchers to address the structural, multicellular, and biochemical complexity of the liver.

VI. SUMMARY AND FUTURE DIRECTIONS

Tissues in the body are diverse in their mechanical properties, cellular and extracellular matrix composition, and physiological functions. The vast array of tissue engineering scaffold assembly methods that have been developed reflects the diversity of both the technological creativity and goals of the field. The liver is one of the most physiologically and structurally complex organs in the body; therefore development of functional hepatic replacements is likely to benefit from novel technologies that enable arbitrary three-dimensional tissue fabrication. In this chapter, we have reviewed several fabrication methods that have been developed or adapted from manufacturing for tissue engineering applications with an eye towards hepatic tissue engineering in particular. Ultimately, the utility of each fabrication technique for other tissues will be a function of tissue-specific design criteria such as mechanical strength, chemical composition, degradation, cellular organization, and nutrient requirements. In moving towards three-dimensional tissue fabrication, greater insight into the structure/function relationships of cells in complex micro-environments will be invaluable. Therefore, we expect that three-dimensional fabrication tools may also prove useful

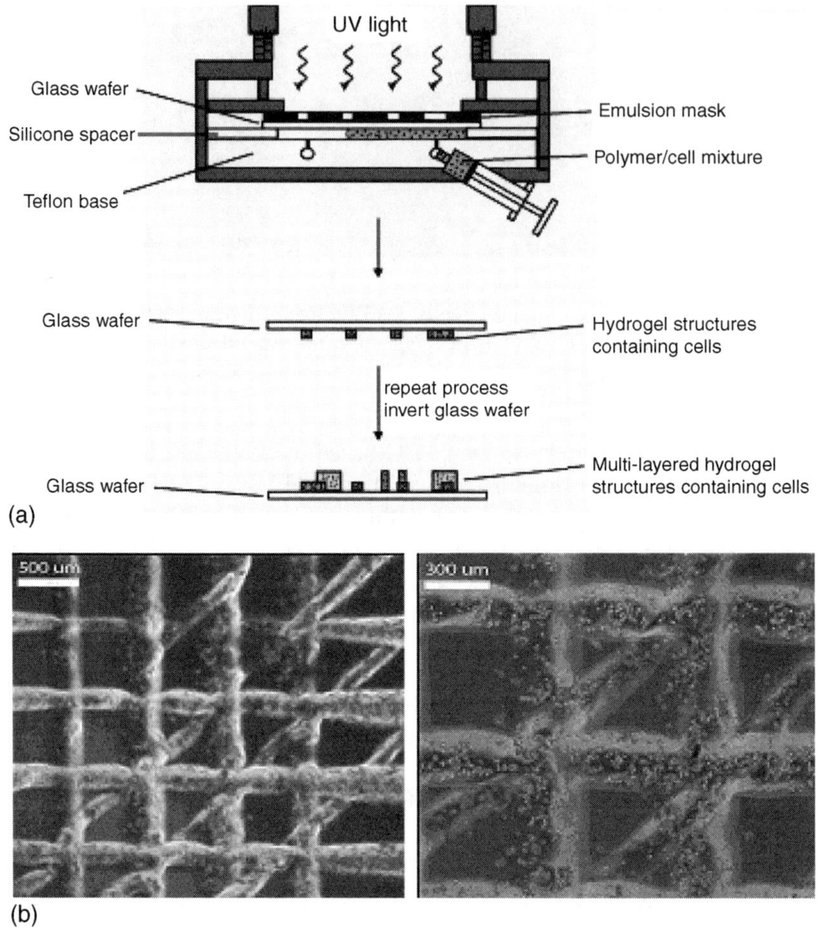

FIGURE 38.8 3D photopatterning of hydrogels. (a) Photopatterning method. Polymer solution and cells are introduced into a chamber. The unit is exposed to 365 nm light through an emulsion mask, causing cross-linking of the polymer in the exposed areas and trapping the cells within these regions. The uncross-linked polymer solution and cells are then washed away, and the process is repeated with thicker spacers and a new mask to create thee-dimensional cellular hydrogel structures. Each layer may contain the same type of polymer/cell mixture, or can be composed of different polymer properties or cell types. (b) Three layered hydrogel structure containing cells. (From Liu V.A., and Bhatia, S.N., Three-dimensional photopatterning of hydrogels containing living cells, *Biomed. Microdevices*, 4, 257–266, 2002. With permission.)

in defining the repertoire of physical and biochemical cues that are required for functional tissue engineering. Finally, it is worth noting that, in embryogenesis, nature teaches us how three-dimensional tissues adopt form and function. Moving forward, successful engineering of complex three-dimensional tissues is likely to result from advances in cell and developmental biology as well as new fabrication technologies and biomaterials.

ACKNOWLEDGMENTS

We would like to thank The Whitaker Foundation (V.L.), American Association of University Women (V.L.), NIH NIDDK, NSF CAREER, The David and Lucile Packard Foundation, and NASA for their generous support.

REFERENCES

1. I CellECT Bio. Focus on Liver Disease. http://www.cellectbio.com/html/liverdis.html. 2002.
2. Strom, S. C., Fisher, R. A., Thompson, M. T., Sanyal, A. J., Cole, P. E., Ham, J. M., and Posner, M. P., Hepatocyte transplantation as a bridge to orthotopic liver transplantation in terminal liver failure, *Transplantation*, 63, 559–569, 1997.
3. Allen, J. W., Hassanein, T., and Bhatia, S. N., Advances in bioartificial liver devices, *Hepatology*, 34, 447–455, 2001.
4. Dunn, J. C., Tompkins, R. G., and Yarmush, M. L., Long-term in vitro function of adult hepatocytes in a collagen sandwich configuration, *Biotechnol. Prog.*, 7, 237–245, 1991.
5. Demetriou, A. A., Whiting, J. F., Feldman, D., Levenson, S. M., Chowdhury, N. R., Moscioni, A. D., Kram, M., and Chowdhury, J. R., Replacement of liver function in rats by transplantation of microcarrier-attached hepatocytes, *Science*, 233, 1190–1192, 1986.
6. Dixit, V., Darvasi, R., Arthur, M., Brezina, M., Lewin, K., and Gitnick, G., Restoration of liver function in Gunn rats without immunosuppression using transplanted microencapsulated hepatocytes, *Hepatology*, 12, 1342–1349, 1990.
7. Hamazaki, K., Doi, Y., and Koide, N., Microencapsulated multicellular spheroid of rat hepatocytes transplanted intraperitoneally after 90% hepatectomy, *Hepatogastroenterology*, 49, 1514–1516, 2002.
8. Mooney, D. J., Sano, K., Kaufmann, P. M., Majahod, K., Schloo, B., Vacanti, J. P., and Langer, R., Long-term engraftment of hepatocytes transplanted on biodegradable polymer sponges, *J. Biomed. Mater. Res.*, 37, 413–420, 1997.
9. Vacanti, J. P., Morse, M. A., Saltzman, W. M., Domb, A. J., Perez-Atayde, A., and Langer, R., Selective cell transplantation using bioabsorbable artificial polymers as matrices, *J. Pediatr. Surg.*, 23, 3–9, 1988.
10. Davis, M. W., and Vacanti, J. P., Toward development of an implantable tissue engineered liver, *Biomaterials*, 17, 365–372, 1996.
11. Takezawa, T., Inoue, M., Aoki, S., Sekiguchi, M., Wada, K., Anazawa, H., and Hanai, N., Concept for organ engineering: a reconstruction method of rat liver for in vitro culture, *Tissue Eng.*, 6, 641–650, 2000.
12. Hasirci, V., Berthiaume, F., Bondre, S. P., Gresser, J. D., Trantolo, D. J., Toner, M., and Wise, D. L., Expression of liver-specific functions by rat hepatocytes seeded in treated poly(lactic-*co*-glycolic) acid biodegradable foams, *Tissue Eng.*, 7, 385–394, 2001.
13. Lee, H., Cusick, R. A., Browne, F., Ho, T., Ma, P. X., Utsunomiya, H., Langer, R., and Vacanti, J. P., Local delivery of basic fibroblast growth factor increases both angiogenesis and engraftment of hepatocytes in tissue-engineered polymer devices, *Transplantation*, 73, 1589–1593, 2002.
14. Kim, S. S., Sundback, C. A., Kaihara, S., Benvenuto, M. S., Kim, B. S., Mooney, D. J., and Vacanti, J. P., Dynamic seeding and in vitro culture of hepatocytes in a flow perfusion system, *Tissue Eng.*, 6, 39–44, 2000.
15. Kaihara, S., Kim, S., Kim, B. S., Mooney, D. J., Tanaka, K., and Vacanti, J. P., Survival and function of rat hepatocytes cocultured with nonparenchymal cells or sinusoidal endothelial cells on biodegradable polymers under flow conditions, *J. Pediatr. Surg.*, 35, 1287–1290, 2000.
16. Kaihara, S., Borenstein, J., Koka, R., Lalan, S., Ochoa, E. R., Ravens, M., Pien, H., Cunningham, B., and Vacanti, J. P., Silicon micromachining to tissue engineer branched vascular channels for liver fabrication, *Tissue Eng.*, 6, 105–117, 2000.
17. Uyama, S., Kaufmann, P. M., Kneser, U., Fiegel, H. C., Pollok, J. M., Kluth, D., Vacanti, J. P., and Rogiers, X., Hepatocyte transplantation using biodegradable matrices in ascorbic acid-deficient rats: comparison with heterotopically transplanted liver grafts, *Transplantation*, 71, 1226–1231, 2001.
18. Langer, R., and Vacanti, J. P., Tissue engineering, *Science*, 260, 920–926, 1993.
19. Bhatia, S. N., and Chen, C. S., Tissue engineering at the micro-scale, *Biomed. Microdevices*, 2, 131–144, 1999.
20. Hutmacher, D. W., Scaffold design and fabrication technologies for engineering tissues — state of the art and future perspectives, *J. Biomater. Sci. Polym. Ed.*, 12, 107–124, 2001.

21. Leong, K. F., Cheah, C. M., and Chua, C. K., Solid freeform fabrication of three-dimensional scaffolds for engineering replacement tissues and organs, *Biomaterials*, 24, 2363–2378, 2003.
22. Yang, S., Leong, K. F., Du, Z., and Chua, C. K., The design of scaffolds for use in tissue engineering. Part II. Rapid prototyping techniques, *Tissue Eng.*, 8, 1–11, 2002.
23. Borenstein, J. T., Terai, H., King, K. R., Weinberg, E. J., Kaazempur-Mofrad, M. R., and Vacanti, J. P., Microfabrication technology for vascularized tissue engineering, *Biomed. Microdevices*, 4, 167–175, 2002.
24. Vozzi, G., Flaim, C., Ahluwalia, A., and Bhatia, S., Fabrication of PLGA scaffolds using soft lithography and microsyringe deposition, *Biomaterials*, 24, 2533–2540, 2003.
25. Chen, C. S., Mrksich, M., Huang, S., Whitesides, G. M., and Ingber, D. E., Geometric control of cell life and death, *Science*, 276, 1425–1428, 1997.
26. Lee, G., Barlow, J., Fox, W., and Aufdermorste, T., *Biocompatibility of SLS-Formed Calcium Phosphate Implants*, *Solid Freeform Fabrication Symposium Proceedings*, 1996, pp. 15–22.
27. Zein, I., Hutmacher, D. W., Tan, K. C., and Teoh, S. H., Fused deposition modeling of novel scaffold architectures for tissue engineering applications, *Biomaterials*, 23, 1169–1185, 2002.
28. Landers, R., Hubner, U., Schmelzeisen, R., and Mulhaupt, R., Rapid prototyping of scaffolds derived from thermoreversible hydrogels and tailored for applications in tissue engineering, *Biomaterials*, 23, 4437–4447, 2002.
29. Cooke, M. N., Fisher, J. P., Dean, D., Rimnac, C., and Mikos, A. G., Use of stereolithography to manufacture critical-sized 3D biodegradable scaffolds for bone ingrowth, *J. Biomed. Mater. Res.*, 64B, 65–69, 2003.
30. Yu, T., Chiellini, F., Schmaljohann, D., Solaro, R., and Ober, C. K., Microfabrication of hydrogels as polymer scaffolds for tissue engineering applications, *Polymer Prepr.*, 41, 1699–1700, 2000.
31. Park, A., Wu, B., and Griffith, L. G., Integration of surface modification and 3D fabrication techniques to prepare patterned poly(L-lactide) substrates allowing regionally selective cell adhesion, *J. Biomater. Sci. Polym. Ed.*, 9, 89–110, 1998.
32. Kim, S. S., Utsunomiya, H., Koski, J. A., Wu, B. M., Cima, M. J., Sohn, J., Mukai, K., Griffith, L. G., and Vacanti, J. P., Survival and function of hepatocytes on a novel three-dimensional synthetic biodegradable polymer scaffold with an intrinsic network of channels, *Ann. Surg.*, 228, 8–13, 1998.
33. Zeltinger, J., Sherwood, J. K., Graham, D. A., Mueller, R., and Griffith, L. G., Effect of pore size and void fraction on cellular adhesion, proliferation, and matrix deposition, *Tissue Eng.*, 7, 557–572, 2001.
34. Chu, T. M., Hollister, S. J., Halloran, J. W., Feinberg, S. E., and Orton, D. G., Manufacturing and characterization of 3-d hydroxyapatite bone tissue engineering scaffolds, *Ann. NY Acad. Sci.*, 961, 114–117, 2002.
35. Chu, T. M., Orton, D. G., Hollister, S. J., Feinberg, S. E., and Halloran, J. W., Mechanical and in vivo performance of hydroxyapatite implants with controlled architectures, *Biomaterials*, 23, 1283–1293, 2002.
36. Taboas, J. M., Maddox, R. D., Krebsbach, P. H., and Hollister, S. J., Indirect solid free form fabrication of local and global porous, biomimetic and composite 3D polymer-ceramic scaffolds, *Biomaterials*, 24, 181–194, 2003.
37. Hollister, S. J., Maddox, R. D., and Taboas, J. M., Optimal design and fabrication of scaffolds to mimic tissue properties and satisfy biological constraints, *Biomaterials*, 23, 4095–4103, 2002.
38. Sachlos, E., Reis, N., Ainsley, C., Derby, B., and Czernuszka, J. T., Novel collagen scaffolds with predefined internal morphology made by solid freeform fabrication, *Biomaterials*, 24, 1487–1497, 2003.
39. Shimizu, T., Yamato, M., Kikuchi, A., and Okano, T., Cell sheet engineering for myocardial tissue reconstruction, *Biomaterials*, 24, 2309–2316, 2003.
40. L'Heureux, N., Paquet, S., Labbe, R., Germain, L., and Auger, F. A., A completely biological tissue-engineered human blood vessel, *FASEB J.*, 12, 47–56, 1998.
41. Odde, D. J., and Renn, M. J., Laser-guided direct writing of living cells, *Biotechnol. Bioeng.*, 67, 312–318, 2000.
42. EnvisionTec. www.envisiontec.de.

43. Mironov, V., Boland, T., Trusk, T., Forgacs, G., and Markwald, R. R., Organ printing: computer-aided jet-based 3D tissue engineering, *Trends Biotechnol.*, 21, 157–161, 2003.
44. I Sciperio. www.sciperio.com.
45. Nguyen, K. T., and West, J. L., Photopolymerizable hydrogels for tissue engineering applications, *Biomaterials*, 23, 4307–4314, 2002.
46. Halstenberg, S., Panitch, A., Rizzi, S., Hall, H., and Hubbell, J. A., Biologically engineered protein-graft-poly(ethylene glycol) hydrogels: a cell adhesive and plasmin-degradable biosynthetic material for tissue repair, *Biomacromolecules*, 3, 710–723, 2002.
47. Lutolf, M. P., Lauer-Fields, J. L., Schmoekel, H. G., Metters, A. T., Weber, F. E., Fields, G. B., and Hubbell, J. A., Synthetic matrix metalloproteinase-sensitive hydrogels for the conduction of tissue regeneration: engineering cell-invasion characteristics, *Proc. Natl Acad. Sci. USA*, 100, 5413–5418, 2003.
48. Sakiyama, S. E., Schense, J. C., and Hubbell, J. A., Incorporation of heparin-binding peptides into fibrin gels enhances neurite extension: an example of designer matrices in tissue engineering, *FASEB J.*, 13, 2214–2224, 1999.
49. Tan, W., and Desai, T. A., Microfluidic patterning of cells in extracellular matrix biopolymers: effects of channel size, cell type, and matrix composition on pattern integrity, *Tissue Eng.*, 9, 255–267, 2003.
50. Peppas, N. A., Bures, P., Leobandung, W., and Ichikawa, H., Hydrogels in pharmaceutical formulations, *Eur. J. Pharm. Biopharm.*, 50, 27–46, 2000.
51. Elisseeff, J., McIntosh, W., Anseth, K., Riley, S., Ragan, P., and Langer, R., Photoencapsulation of chondrocytes in poly(ethylene oxide)-based semi-interpenetrating networks, *J. Biomed. Mater. Res.*, 51, 164–171, 2000.
52. Bryant, S. J., and Anseth, K. S., Hydrogel properties influence ECM production by chondrocytes photoencapsulated in poly(ethylene glycol) hydrogels, *J. Biomed. Mater. Res.*, 59, 63–72, 2002.
53. Mann, B. K., Gobin, A. S., Tsai, A. T., Schmedlen, R. H., and West, J. L., Smooth muscle cell growth in photopolymerized hydrogels with cell adhesive and proteolytically degradable domains: synthetic ECM analogs for tissue engineering, *Biomaterials*, 22, 3045–3051, 2001.
54. Behravesh, E., Zygourakis, K., and Mikos, A. G., Adhesion and migration of marrow-derived osteoblasts on injectable in situ crosslinkable poly(propylene fumarate-*co*-ethylene glycol)-based hydrogels with a covalently linked RGDS peptide, *J. Biomed. Mater. Res.*, 65A, 260–270, 2003.
55. Gobin, A. S., and West, J. L., Cell migration through defined, synthetic ECM analogs, *FASEB J.*, 16, 751–753, 2002.
56. Hern, D. L., and Hubbell, J. A., Incorporation of adhesion peptides into nonadhesive hydrogels useful for tissue resurfacing, *J. Biomed. Mater. Res.*, 39, 266–276, 1998.
57. Kao, W. J., and Hubbell, J. A., Murine macrophage behavior on peptide-grafted polyethyleneglycol-containing networks, *Biotechnol. Bioeng.*, 59, 2–9, 1998.
58. Koo, L. Y., Irvine, D. J., Mayes, A. M., Lauffenburger, D. A., and Griffith, L. G., Co-regulation of cell adhesion by nanoscale RGD organization and mechanical stimulus, *J. Cell Sci.*, 115, 1423–1433, 2002.
59. Alsberg, E., Anderson, K. W., Albeiruti, A., Rowley, J. A., and Mooney, D. J., Engineering growing tissues, *Proc. Natl Acad. Sci. USA*, 99, 12025–12030, 2002.
60. Schmedlen, R. H., Masters, K. S., and West, J. L., Photocrosslinkable polyvinyl alcohol hydrogels that can be modified with cell adhesion peptides for use in tissue engineering, *Biomaterials*, 23, 4325–4332, 2002.
61. Sawhney, A. S., Pathak, C. P., van Rensburg, J. J., Dunn, R. C., and Hubbell, J. A., Optimization of photopolymerized bioerodible hydrogel properties for adhesion prevention, *J. Biomed. Mater. Res.*, 28, 831–838, 1994.
62. Nuttelman, C. R., Henry, S. M., and Anseth, K. S., Synthesis and characterization of photocrosslinkable, degradable poly(vinyl alcohol)-based tissue engineering scaffolds, *Biomaterials*, 23, 3617–3626, 2002.
63. Revzin, A., Russell, R. J., Yadavalli, V. K., Koh, W. G., Deister, C., Hile, D. D., Mellott, M. B., and Pishko, M. V., Fabrication of poly(ethylene glycol) hydrogel microstructures using photolithography, *Langmuir*, 17, 5440–5447, 2001.

64. Beebe, D. J., Moore, J. S., Bauer, J. M., Qing, Y., Liu, R. H., Devadoss, C., and Byung-Ho, J., Functional hydrogel structures for autonomous flow control inside microfluidic channels, *Nature (UK)*, 404, 588–590, 2000.
65. Koh, W. G., Revzin, A., and Pishko, M. V., Poly(ethylene glycol) hydrogel microstructures encapsulating living cells, *Langmuir*, 18, 2459–2462, 2002.
66. Liu, V. A., and Bhatia, S. N., Three-dimensional photopatterning of hydrogels containing living cells, *Biomed. Microdevices*, 4, 257–266, 2002.

Index

C

D

I

J

K

L

R